Lecture Notes in Computer Science 11923

More information about this series at http://www.springer.com/series/7410

Steven D. Galbraith · Shiho Moriai (Eds.)

Advances in Cryptology – ASIACRYPT 2019

25th International Conference on the Theory
and Application of Cryptology and Information Security
Kobe, Japan, December 8–12, 2019
Proceedings, Part III

Springer

Editors
Steven D. Galbraith
University of Auckland
Auckland, New Zealand

Shiho Moriai
Security Fundamentals Lab
NICT
Tokyo, Japan

ISSN 0302-9743 ISSN 1611-3349 (electronic)
Lecture Notes in Computer Science
ISBN 978-3-030-34617-1 ISBN 978-3-030-34618-8 (eBook)
https://doi.org/10.1007/978-3-030-34618-8

LNCS Sublibrary: SL4 – Security and Cryptology

This Springer imprint is published by the registered company Springer Nature Switzerland AG
The registered company address is: Gewerbestrasse 11, 6330 Cham, Switzerland

Preface

ASIACRYPT 2019, the 25th Annual International Conference on Theory and Application of Cryptology and Information Security, was held in Kobe, Japan, during December 8–12, 2019.

The conference focused on all technical aspects of cryptology, and was sponsored by the International Association for Cryptologic Research (IACR).

We received a total of 307 submissions from all over the world. This was a significantly higher number of submissions than recent Asiacrypt conferences, which necessitated a larger Program Committee (PC) than we had originally planned. We thank the seven additional PC members who accepted our invitation at extremely short notice. They are Gorjan Alagic, Giorgia Azzurra Marson, Zhenzhen Bao, Olivier Blazy, Romain Gay, Takanori Isobe, and Daniel Masny.

The PC selected 71 papers for publication in the proceedings of the conference. The two program chairs were supported by a PC consisting of 55 leading experts in aspects of cryptology. Each submission was reviewed by at least three Program Committee members (or their sub-reviewers) and five PC members were assigned to submissions co-authored by PC members. The strong conflict of interest rules imposed by the IACR ensure that papers are not handled by PC members with a close working relationship with authors. There were approximately 380 external reviewers, whose input was critical to the selection of papers.

The review process was conducted using double-blind peer review. The conference operated a two-round review system with a rebuttal phase. After the reviews and first-round discussions the PC selected 193 submissions to proceed to the second round. The authors of those 193 papers were then invited to provide a short rebuttal in response to the referee reports. The second round involved extensive discussions by the PC members. Indeed, the total number of text items in the online discussion (including reviews, rebuttals, questions to authors, and PC member comments) exceeded 3,000.

The three volumes of the conference proceedings contain the revised versions of the 71 papers that were selected, together with 1 invited paper. The final revised versions of papers were not reviewed again and the authors are responsible for their contents.

The program of Asiacrypt 2019 featured excellent invited talks by Krzysztof Pietrzak and Elaine Shi. The conference also featured a rump session which contained short presentations on the latest research results of the field.

The PC selected the work "Wave: A New Family of Trapdoor One-Way Preimage Sampleable Functions Based on Codes" by Thomas Debris-Alazard, Nicolas Sendrier, and Jean-Pierre Tillich for the Best Paper Award of Asiacrypt 2019. Two more papers were solicited to submit a full version to the *Journal of Cryptology*. They are "An LLL Algorithm for Module Lattices" by Changmin Lee, Alice Pellet-Mary, Damien Stehlé, and Alexandre Wallet, and "Numerical Method for Comparison on Homomorphically Encrypted Numbers" by Jung Hee Cheon, Dongwoo Kim, Duhyeong Kim, Hun Hee Lee, and Keewoo Lee.

The Program Chairs are delighted to recognize the outstanding work by Mark Zhandry and Shweta Agrawal, by awarding them jointly the Best PC Member Award.

Many people have contributed to the success of Asiacrypt 2019. We would like to thank the authors for submitting their research results to the conference. We are very grateful to the PC members and external reviewers for contributing their knowledge and expertise, and for the tremendous amount of work that was done with reading papers and contributing to the discussions.

We are greatly indebted to Mitsuru Matsui, the general chair, for his efforts and overall organization.

We thank Mehdi Tibouchi for expertly organizing and chairing the rump session.

We are extremely grateful to Lukas Zobernig for checking all the latex files and for assembling the files for submission to Springer.

Finally we thank Shai Halevi and the IACR for setting up and maintaining the Web Submission and Review software, used by IACR conferences for the paper submission and review process. We also thank Alfred Hofmann, Anna Kramer, Ingrid Haas, Anja Sebold, Xavier Mathew, and their colleagues at Springer for handling the publication of these conference proceedings.

December 2019

Steven Galbraith
Shiho Moriai

ASIACRYPT 2019

The 25th Annual International Conference on Theory and Application of Cryptology and Information Security

Sponsored by the International Association for Cryptologic Research (IACR)

Kobe, Japan, December 8–12, 2019

General Chair

Mitsuru Matsui	Mitsubishi Electric Corporation, Japan

Program Co-chairs

Steven Galbraith	University of Auckland, New Zealand
Shiho Moriai	NICT, Japan

Program Committee

Shweta Agrawal	IIT Madras, India
Gorjan Alagic	University of Maryland, USA
Shi Bai	Florida Atlantic University, USA
Zhenzhen Bao	Nanyang Technological University, Singapore
Paulo S. L. M. Barreto	UW Tacoma, USA
Lejla Batina	Radboud University, The Netherlands
Sonia Belaïd	CryptoExperts, France
Olivier Blazy	University of Limoges, France
Colin Boyd	NTNU, Norway
Xavier Boyen	Queensland University of Technology, Australia
Nishanth Chandran	Microsoft Research, India
Melissa Chase	Microsoft Research, USA
Yilei Chen	Visa Research, USA
Chen-Mou Cheng	Osaka University, Japan
Nils Fleischhacker	Ruhr-University Bochum, Germany
Jun Furukawa	NEC Israel Research Center, Israel
David Galindo	University of Birmingham and Fetch AI, UK
Romain Gay	UC Berkeley, USA
Jian Guo	Nanyang Technological University, Singapore
Seokhie Hong	Korea University, South Korea
Andreas Hülsing	Eindhoven University of Technology, The Netherlands
Takanori Isobe	University of Hyogo, Japan
David Jao	University of Waterloo and evolutionQ, Inc., Canada

Jérémy Jean	ANSSI, France
Elena Kirshanova	ENS Lyon, France
Virginie Lallemand	CNRS, France
Jooyoung Lee	KAIST, South Korea
Helger Lipmaa	Simula UiB, Norway
Feng-Hao Liu	Florida Atlantic University, USA
Atul Luykx	Swirlds Inc., USA
Hemanta K. Maji	Purdue, USA
Giorgia Azzurra Marson	NEC Laboratories Europe, Germany
Daniel Masny	Visa Research, USA
Takahiro Matsuda	AIST, Japan
Brice Minaud	Inria and ENS, France
David Naccache	ENS, France
Kartik Nayak	Duke University and VMware Research, USA
Khoa Nguyen	Nanyang Technological University, Singapore
Svetla Nikova	KU Leuven, Belgium
Carles Padró	UPC, Spain
Jiaxin Pan	NTNU, Norway, and KIT, Germany
Arpita Patra	Indian Institute of Science, India
Thomas Peters	UCL, Belgium
Duong Hieu Phan	University of Limoges, France
Raphael C.-W. Phan	Monash University, Malaysia
Carla Ràfols	Universitat Pompeu Fabra, Spain
Ling Ren	VMware Research and University of Illinois, Urbana-Champaign, USA
Yu Sasaki	NTT laboratories, Japan
Junji Shikata	Yokohama National University, Japan
Ron Steinfeld	Monash University, Australia
Qiang Tang	New Jersey Institute of Technology, USA
Mehdi Tibouchi	NTT Laboratories, Japan
Hoeteck Wee	CNRS and ENS, France
Mark Zhandry	Princeton University, USA
Fangguo Zhang	Sun Yat-sen University, China

Local Organizing Committee

General Chair

Mitsuru Matsui	Mitsubishi Electric Corporation, Japan

Honorary Advisor

Tsutomu Matsumoto	Yokohama National University, Japan

External Reviewers

Masayuki Abe
Parhat Abla
Victor Arribas Abril
Divesh Aggarwal
Martin Albrecht
Bar Alon
Prabhanjan Ananth
Elena Andreeva
Yoshinori Aono
Daniel Apon
Toshinori Araki
Seiko Arita
Tomer Ashur
Nuttapong Attrapadung
Man Ho Allen Au
Benedikt Auerbach
Saikrishna Badrinarayanan
Vivek Bagaria
Josep Balasch
Gustavo Banegas
Laasya Bangalore
Subhadeep Banik
Achiya Bar-On
Manuel Barbosa
James Bartusek
Carsten Baum
Arthur Beckers
Rouzbeh Behnia
Francesco Berti
Alexandre Berzati
Ward Beullens
Shivam Bhasin
Nina Bindel
Nicolas Bordes
Jannis Bossert
Katharina Boudgoust
Christina Boura
Florian Bourse
Zvika Brakerski
Anne Broadbent
Olivier Bronchain
Leon Groot Bruinderink
Megha Byali
Eleonora Cagli
Ignacio Cascudo
Pyrros Chaidos
Avik Chakraborti
Donghoon Chang
Hao Chen
Jie Chen
Long Chen
Ming-Shing Chen
Qian Chen
Jung Hee Cheon
Céline Chevalier
Ilaria Chillotti
Wonhee Cho
Wonseok Choi
Wutichai Chongchitmate
Jérémy Chotard
Arka Rai Choudhuri
Sherman Chow
Michele Ciampi
Michael Clear
Thomas De Cnudde
Benoît Cogliati
Sandro Coretti-Drayton
Edouard Cuvelier
Jan Czajkowski
Dana Dachman-Soled
Joan Daemen
Nilanjan Datta
Gareth T. Davies
Patrick Derbez
Apporva Deshpande
Siemen Dhooghe
Christoph Dobraunig
Rafael Dowsley
Yfke Dulek
Avijit Dutta
Sébastien Duval
Keita Emura
Thomas Espitau
Xiong Fan
Antonio Faonio
Oriol Farràs
Sebastian Faust
Prastudy Fauzi
Hanwen Feng
Samuele Ferracin
Dario Fiore
Georg Fuchsbauer
Thomas Fuhr
Eiichiro Fujisaki
Philippe Gaborit
Tatiana Galibus
Chaya Ganesh
Daniel Gardham
Luke Garratt
Pierrick Gaudry
Nicholas Genise
Esha Ghosh
Satrajit Ghosh
Kristian Gjøsteen
Aarushi Goel
Huijing Gong
Junqing Gong
Alonso González
Dahmun Goudarzi
Rishabh Goyal
Jiaxin Guan
Aurore Guillevic
Chun Guo
Kaiwen Guo
Qian Guo
Mohammad Hajiabadi
Carmit Hazay
Jingnan He
Brett Hemenway
Nadia Heninger
Javier Herranz
Shoichi Hirose
Harunaga Hiwatari
Viet Tung Hoang
Justin Holmgren
Akinori Hosoyamada
Kexin Hu
Senyang Huang

Yan Huang
Phi Hun
Aaron Hutchinson
Chloé Hébant
Kathrin Hövelmanns
Ilia Iliashenko
Mitsugu Iwamoto
Tetsu Iwata
Zahra Jafargholi
Christian Janson
Ashwin Jha
Dingding Jia
Sunghyun Jin
Charanjit S. Jutla
Mustafa Kairallah
Saqib A. Kakvi
Marc Kaplan
Emrah Karagoz
Ghassan Karame
Shuichi Katsumata
Craig Kenney
Mojtaba Khalili
Dakshita Khurana
Duhyeong Kim
Hyoseung Kim
Sam Kim
Seongkwang Kim
Taechan Kim
Agnes Kiss
Fuyuki Kitagawa
Michael Klooβ
François Koeune
Lisa Kohl
Stefan Kölbl
Yashvanth Kondi
Toomas Krips
Veronika Kuchta
Nishant Kumar
Noboru Kunihiro
Po-Chun Kuo
Kaoru Kurosawa
Ben Kuykendall
Albert Kwon
Qiqi Lai
Baptiste Lambin
Roman Langrehr
Jason LeGrow
ByeongHak Lee
Changmin Lee
Keewoo Lee
Kwangsu Lee
Youngkyung Lee
Dominik Leichtle
Christopher Leonardi
Tancrède Lepoint
Gaëtan Leurent
Itamar Levi
Baiyu Li
Yanan Li
Zhe Li
Xiao Liang
Benoît Libert
Fuchun Lin
Rachel Lin
Wei-Kai Lin
Eik List
Fukang Liu
Guozhen Liu
Meicheng Liu
Qipeng Liu
Shengli Liu
Zhen Liu
Alex Lombardi
Julian Loss
Jiqiang Lu
Xianhui Lu
Yuan Lu
Lin Lyu
Fermi Ma
Gilles Macario-Rat
Urmila Mahadev
Monosij Maitra
Christian Majenz
Nikolaos Makriyannis
Giulio Malavolta
Sogol Mazaheri
Bart Mennink
Peihan Miao
Shaun Miller
Kazuhiko Minematsu
Takaaki Mizuki
Amir Moradi
Kirill Morozov
Fabrice Mouhartem
Pratyay Mukherjee
Pierrick Méaux
Yusuke Naito
Mridul Nandi
Peter Naty
María Naya-Plasencia
Anca Niculescu
Ventzi Nikov
Takashi Nishide
Ryo Nishimaki
Anca Nitulescu
Ariel Nof
Sai Lakshmi Bhavana Obbattu
Kazuma Ohara
Emmanuela Orsini
Elena Pagnin
Wenlun Pan
Omer Paneth
Bo Pang
Lorenz Panny
Jacques Patarin
Sikhar Patranabis
Alice Pellet-Mary
Chun-Yo Peng
Geovandro Pereira
Olivier Pereira
Léo Perrin
Naty Peter
Cécile Pierrot
Jeroen Pijnenburg
Federico Pintore
Bertram Poettering
David Pointcheval
Yuriy Polyakov
Eamonn Postlethwaite
Emmanuel Prouff
Pille Pullonen
Daniel Puzzuoli
Chen Qian
Tian Qiu
Willy Quach
Håvard Raddum
Ananth Raghunathan

Somindu Ramanna
Kim Ramchen
Shahram Rasoolzadeh
Mayank Rathee
Divya Ravi
Joost Renes
Angela Robinson
Thomas Roche
Miruna Rosca
Mélissa Rossi
Mike Rosulek
Yann Rotella
Arnab Roy
Luis Ruiz-Lopez
Ajith Suresh
Markku-Juhani O. Saarinen
Yusuke Sakai
Kazuo Sakiyama
Amin Sakzad
Louis Salvail
Simona Samardjiska
Pratik Sarkar
Christian Schaffner
John Schanck
Berry Schoenmakers
Peter Scholl
André Schrottenloher
Jacob Schuldt
Sven Schäge
Sruthi Sekar
Srinath Setty
Yannick Seurin
Barak Shani
Yaobin Shen
Sina Shiehian
Kazumasa Shinagawa
Janno Siim
Javier Silva
Mark Simkin
Boris Skoric
Maciej Skórski
Yongsoo Song
Pratik Soni
Claudio Soriente
Florian Speelman
Akshayaram Srinivasan
François-Xavier Standaert
Douglas Stebila
Damien Stehlé
Patrick Struck
Valentin Suder
Bing Sun
Shifeng Sun
Siwei Sun
Jaechul Sung
Daisuke Suzuki
Katsuyuki Takashima
Benjamin Hong Meng Tan
Stefano Tessaro
Adrian Thillard
Yan Bo Ti
Jean-Pierre Tillich
Radu Ţiţiu
Yosuke Todo
Junichi Tomida
Viet Cuong Trinh
Rotem Tsabary
Hikaru Tsuchida
Yi Tu
Nirvan Tyagi
Bogdan Ursu
Damien Vergnaud
Jorge Luis Villar
Srinivas Vivek
Christine van Vredendaal
Satyanarayana Vusirikala
Sameer Wagh
Hendrik Waldner
Alexandre Wallet
Michael Walter
Han Wang
Haoyang Wang
Junwei Wang
Mingyuan Wang
Ping Wang
Yuyu Wang
Zhedong Wang
Yohei Watanabe
Gaven Watson
Weiqiang Wen
Yunhua Wen
Benjamin Wesolowski
Keita Xagawa
Zejun Xiang
Hanshen Xiao
Shota Yamada
Takashi Yamakawa
Kyosuke Yamashita
Avishay Yanai
Guomin Yang
Kan Yasuda
Masaya Yasuda
Aaram Yun
Alexandros Zacharakis
Michal Zajac
Bin Zhang
Cong Zhang
En Zhang
Huang Zhang
Xiao Zhang
Zheng Zhang
Chang-An Zhao
Raymond K. Zhao
Yongjun Zhao
Yuanyuan Zhou
Jiamin Zhu
Yihong Zhu
Lukas Zobernig

Contents – Part III

Symmetric Cryptography (2)

Functional Encryption

Public Key Encryption (2)

Zero Knowledge

Signatures

Approximate Trapdoors for Lattices and Smaller Hash-and-Sign Signatures

Yilei Chen[1(✉)], Nicholas Genise[2], and Pratyay Mukherjee[1]

[1] Visa Research, Palo Alto, USA
{yilchen,pratmukh}@visa.com
[2] University of California, San Diego, USA
ngenise@eng.ucsd.edu

Abstract. We study a relaxed notion of lattice trapdoor called *approximate trapdoor*, which is defined to be able to invert Ajtai's one-way function approximately instead of exactly. The primary motivation of our study is to improve the efficiency of the cryptosystems built from lattice trapdoors, including the hash-and-sign signatures.

Our main contribution is to construct an approximate trapdoor by modifying the gadget trapdoor proposed by Micciancio and Peikert [Eurocrypt 2012]. In particular, we show how to use the approximate gadget trapdoor to sample short preimages from a distribution that is simulatable without knowing the trapdoor. The analysis of the distribution uses a theorem (implicitly used in past works) regarding linear transformations of discrete Gaussians on lattices.

Our approximate gadget trapdoor can be used together with the existing optimization techniques to improve the concrete performance of the hash-and-sign signature in the random oracle model under (Ring-)LWE and (Ring-)SIS assumptions. Our implementation shows that the sizes of the public-key & signature can be reduced by half from those in schemes built from exact trapdoors.

1 Introduction

In the past two decades, lattice-based cryptography has emerged as one of the most active areas of research. It has enabled both advanced cryptographic capabilities, such as fully homomorphic encryption [29]; and practical post-quantum secure public-key encryptions and signatures, as observed in the ongoing NIST post-quantum cryptography (PQC) standardization procedure [4]. A large fraction of the lattice-based cryptosystems uses *lattice trapdoors*. Those cryptosystems include basic primitives like public-key encryption and signature schemes [31,33,38,39], as well as advanced primitives such as identity-based encryption [1,19,31], attribute-based encryption [34], and graded encodings [30].

In this work, we focus on the trapdoor for the lattice-based one-way function defined by Ajtai [2], and its application in digital signatures [31]. Given a wide, random matrix $\mathbf{A}$, and a target vector $\mathbf{y}$, the *inhomogeneous short integer solution* (ISIS) problem asks to find a short vector $\mathbf{x}$ as a preimage of $\mathbf{y}$, i.e.

S. D. Galbraith and S. Moriai (Eds.): ASIACRYPT 2019, LNCS 11923, pp. 3–32, 2019.
https://doi.org/10.1007/978-3-030-34618-8_1

$$\mathbf{A} \cdot \mathbf{x} = \mathbf{y} \pmod{q}.$$

Without a trapdoor for the matrix **A**, finding a short preimage is proven to be as hard as solving certain lattice problems in the worst case [2]. A trapdoor for the matrix **A**, on the other hand, allows its owner to efficiently produce a short preimage. An explicit construction of the trapdoor for Ajtai's function was first given in [3] and later simplified by [9,42].

Towards the proper use of lattice trapdoors in cryptography, what really gives the trapdoor a punch is the work of Gentry, Peikert and Vaikuntanathan [31]. They show how to sample a short preimage from a distribution that is simulatable without knowing the trapdoor, instead of a distribution which may leak information about the trapdoor (as observed by the attacks [32,46] on the initial attempts of building lattice-based signatures [33,38]). Such a preimage sampling algorithm allows [31] to securely build a hash-and-sign signature as follows. Let the matrix **A** be the public verification key, the trapdoor of **A** be the secret signing key. To sign a message m, first hash it to a vector $\mathbf{y}$, then use the trapdoor to sample a short preimage $\mathbf{x}$ as the signature. The secret signing key is guaranteed to be hidden from the signatures, since the signatures are simulatable without using the trapdoor.

Despite its elegant design, the hash-and-sign signature based on Ajtai's function suffers from practical inefficiency due to its large key size and signature size. Indeed, all the three lattice-based signature candidates that enter the second round of NIST PQC standardization [4] are built from two alternative approaches—Falcon [27] is based on the hash-and-sign paradigm over NTRU lattices; Dilithium [26] and qTESLA [8] are based on the rejection sampling approach [11,40]. The suggested parameters for the three candidates lead to competitive performance measures. For example, for 128-bit security, the sizes of the public keys & signatures for all the three candidates are below 5 kB & 4 kB (respectively). By contrast, for the hash-and-sign signature based on Ajtai's function, the sizes of the public keys & signatures are more than 35 kB & 25 kB according to the implementation benchmarks of [13,14,36].

1.1 Summary of Our Contributions

In this paper we develop new techniques to bring down the sizes of the public keys & signatures of the hash-and-sign signature based on Ajtai's one-way function. We define a relaxed notion of lattice trapdoor called *approximate trapdoor*, which can be used to solve the ISIS problem *approximately* instead of exactly. With a relaxation of the correctness requirement, it is possible to generate smaller public matrices, trapdoors, and preimages for Ajtai's function, which translate to smaller public-keys, secret-keys, and signatures for the hash-and-sign signature scheme.

Our main technical contribution is to show that the gadget trapdoor proposed by Micciancio and Peikert [42] can be modified to an approximate trapdoor. In particular, we show how to use the approximate gadget trapdoor to sample preimages from a distribution that is simulatable without knowing the trapdoor. The analysis of the distribution uses a theorem (implicitly used in past works) regarding linear transformations of discrete Gaussians on lattices.

Our approximate gadget trapdoor can be used together with all existing optimization techniques, such as using the Hermite normal form and using a bigger base in the gadget, to improve the concrete performance of the hash-and-sign signature in the random oracle model under RingLWE and RingSIS assumptions. Our proof-of-concept implementation shows that the sizes of the public-key & signature can be reduced to 5 kB & 4.45 kB for an estimation of 88-bit security, and 11.25 kB & 9.38 kB for an estimation of 184-bit security. Those are much closer to the sizes of the signatures based on the rejection sampling approach [8,11,26,40]. More details of the parameters are given in Sects. 1.3 and 5.2.

1.2 Technical Overview

Given a public matrix $\mathbf{A} \in \mathbb{Z}_q^{n\times m}$ where $m = O(n \log q)$, and a target $\mathbf{y}$, we call a vector $\mathbf{x} \in \mathbb{Z}^m$ an *approximate short preimage* of $\mathbf{y}$ if

$$\mathbf{A} \cdot \mathbf{x} = \mathbf{y} + \mathbf{z} \pmod q$$

for some $\mathbf{z} \in \mathbb{Z}^n$, and both $\mathbf{x}$ and $\mathbf{z}$ are short. An *approximate trapdoor* for $\mathbf{A}$ is defined to be a string that allows its owner to efficiently find an approximate short preimage given a target $\mathbf{y}$.

Of course, to make sense of the word "trapdoor", we first need to argue that solving the approximate version of ISIS is hard without the trapdoor. Under proper settings of parameters, we show the approximate ISIS problem is as hard as the standard ISIS problem, or no easier than LWE. The reductions extensively use the Hermite normal form (HNF) and are pretty straightforward.

The approximate ISIS problem and the approximate trapdoor are natural generalizations of their exact variants. Indeed, both notions have been used in the literature, at least on an informal level. For example, the approximate ISIS problem was used in the work of Bai et al. [12] to improve the combinatorial algorithms of the exact ISIS problem.

It is well-known that an exact trapdoor of a public matrix in the HNF, say a trapdoor for $\mathbf{A} = [\mathbf{I}_n \mid \mathbf{A}']$, can be used as an approximate trapdoor for $\mathbf{A}'$. Such a method was often used in the implementation of signatures to decrease the sizes of the public key and the signature by a dimension of n. Our goal is thus to further reduce the sizes compared to the HNF approach, while preserving the quality of the trapdoor, i.e. at least not increasing the norm of the preimage.

Approximate gadget trapdoor. Our main contribution is to show that the gadget trapdoor (G-trapdoor) proposed by Micciancio and Peikert [42] can be modified to an approximate trapdoor, in a way that further reduces the sizes of the public matrix, the trapdoor, and the preimage.

Recall the core of the G-trapdoor is a specific "gadget" matrix of base b,

$$\mathbf{G} := \mathbf{I}_n \otimes \mathbf{g}^t := \mathbf{I}_n \otimes (1, b, ..., b^{k-1}) \in \mathbb{Z}^{n\times(nk)},$$

where $k := \lceil \log_b q \rceil$. The base b is typically chosen to be 2 for simplicity, or a larger value in practical implementations.

Micciancio and Peikert [42] show how to generate a random matrix $\mathbf{A}$ together with a matrix $\mathbf{D}$ of small norm such that $\mathbf{A} \cdot \mathbf{D} = \mathbf{G} \pmod q$. In particular, $\mathbf{A}$ is designed to be

$$\mathbf{A} = [\bar{\mathbf{A}} | \mathbf{G} - \bar{\mathbf{A}}\mathbf{R}],$$

where $\mathbf{R}$ is a matrix with small entries and is the actual trapdoor. The matrix $\mathbf{D}$ is then equal to $\begin{bmatrix} \mathbf{R} \\ \mathbf{I}_{nk} \end{bmatrix}$. Since the kernel of the $\mathbf{G}$ matrix has a public short basis, one can first solve the ISIS problem under the public matrix $\mathbf{G}$, then use $\mathbf{D}$ to solve the ISIS problem under the public matrix $\mathbf{A}$.

We observe that if we drop a few (say l) entries corresponding to the small powers of b from the gadget matrix $\mathbf{G}$, i.e. let the following $\mathbf{F}$ matrix be a modified gadget matrix

$$\mathbf{F} := \mathbf{I}_n \otimes \mathbf{f}^t := \mathbf{I}_n \otimes (b^l, ..., b^{k-1}) \in \mathbb{Z}^{n \times n(k-l)},$$

then we are still able to solve the ISIS problem w.r.t. the public matrix $\mathbf{F}$ up to a b^l-approximation of the solution (i.e., the norm of the error vector is proportional to b^l). Replacing $\mathbf{G}$ by $\mathbf{F}$ in $\mathbf{A}$ gives

$$\mathbf{A} = [\bar{\mathbf{A}} | \mathbf{F} - \bar{\mathbf{A}}\mathbf{R}]. \tag{1}$$

Then the dimensions of the trapdoor $\mathbf{R}$ and the public matrix $\mathbf{A}$ can be reduced.

Sampling from a distribution that is simulatable without knowing the trapdoor. Given a public matrix $\mathbf{A}$ together with its approximate G-trapdoor $\mathbf{R}$, finding an arbitrary approximate short preimage of a given target $\mathbf{u}$ is quite straightforward, but sampling the preimage from a distribution that is simulatable without knowing the trapdoor turns out to be non-trivial. As mentioned earlier, the ability to sample from such a distribution is fundamental to most of the trapdoor applications including digital signatures.

We provide an algorithm that samples an approximate short preimage from a distribution that is simulatable without knowing the trapdoor. The algorithm itself is a fairly simple generalization of the perturbation-based discrete Gaussian sampler from [42], but the analyses of the preimage distribution from [42] are not easy to generalize. Our analyses of the preimage distribution and the approximation error distribution extensively use a linear transformation theorem on lattice distributions (cf. Lemma 4, or Theorem 1, implicitly used in [15,25,42,43]).

The details of the analyses are quite technical. Here let us mention the difference in the way of obtaining the main result of ours compared to the ones from [31,42]. The approach taken by [31,42] is to first spell out the distributions of the preimages for *all* the target images $\mathbf{u} \in \mathbb{Z}_q^n$, then show the distributions are simulatable for *uniformly random* target images. For the approximate preimage sampling, we are only able to simulate the distributions of the preimages and the errors for *uniformly random* targets, without being able to spell out the meaningful distributions for *all* the targets an intermediate step. Still, simulating the preimages of uniform targets suffices for the application of digital signatures.

To briefly explain the reason behind the difference, let us point out that the methods we have tried to analyze the preimage distribution for *all* the target images require significant increases in the smoothing parameters of the lattice intersections required in the linear transformation theorem (Theorem 1). In other words, the norm of the resulting preimage increases significantly rendering the result meaningless.

1.3 Improvement in the Efficiency Compared to the Exact Trapdoor

We now explain the efficiency gain of using our approximate trapdoor compared to the exact trapdoor and the other existing optimization techniques, with a focus on the signature application. Our goal is to set the parameters to achieve the following "win-win-win" scenario:

1. Save on the size of the preimage (i.e., the signature).
2. Save on the size for the public matrix $\mathbf{A}$.
3. Retain, or even gain, concrete security, which is related to the discrete Gaussian width of the preimage and the norm of the error term.

Parameters	Exact G-trapdoor	Approximate G-trapdoor
m	$n(2+k)$	$n(2+(k-l))$
σ	$\sqrt{b^2+1}\cdot\omega(\sqrt{\log n})$	$\sqrt{b^2+1}\cdot\omega(\sqrt{\log n})$
s	$C\cdot\tau\cdot(\sqrt{m}+2\sqrt{n})\cdot\sigma$	$C\cdot\tau\cdot(\sqrt{m}+2\sqrt{n})\cdot\sigma$
ν	0	$b^l\cdot\sigma$

Fig. 1. A brief comparison of the parameters. The parameters in the table are derived under a fixed lattice dimension n, a fixed modulus $q \geq \sqrt{n}$, and a fixed base b. Let $k = \lceil \log_b q \rceil$. Let l denote the number of entries removed from $\mathbf{g}$ $(1 \leq l < k)$. Then we list m as the dimension of the public matrix and the preimage; σ as the width of the gadget preimage distribution; s as the width of the final preimage distribution (where $C > 0$ is a universal constant); τ as the width, or subgaussian parameter, of the distribution of the entries in the trapdoor matrix $\mathbf{R}$; ν as the length bound of the error for each entry in the image.

Let us start with an understanding of the dependency of the savings on the variable l, i.e, the number of entries dropped from the gadget $\mathbf{g}$. In Fig. 1 we provide a comparison of the parameters between the exact G-trapdoor of [42] and the approximate G-trapdoor samplers in this paper. In both cases the public matrices are instantiated in the pseudorandom mode. For the approximate trapdoor, the dimension of the trapdoor decreases from nk to $n(k-l)$. The dimension m of the public matrix and the preimage decreases. The width s of the preimage distribution also decreases slightly following the decreasing of m. However, the norm of the error factor in the image grows with l. So in the concrete instantiation of the hash-and-sign signature discussed later, we need to coordinate the value of l with the norms of the preimage and the error, which will determine the cost of the attacks together.

Our algorithm inherits the $O(\log q)$-space, $O(n \log q)$-time G-preimage sample subroutine from [28,42]. So the saving of space and time in the sampling of the perturbation is proportional to the saving in the dimension m.

Concrete parameters for the signatures. We give a proof-of-concept implementation of the hash-and-sign signature based on our approximate trapdoor. The security is analyzed in the random oracle model, assuming the hardness of RingLWE for the pseudorandomness of the public key and RingSIS for the unforgeability of the signature. Here we provide a short summary and leave more details in Sect. 5.2.

Let us first remark that different implementation results of the hash-and-sign signatures [13,14,36] possibly use different ways of measuring sizes and security, and not all the details behind the parameters are recoverable from these papers. So we also implementation the exact trapdoor as a reference. For an estimation of 88-bit security, our reference implementation for the exact trapdoor under the modulus $q \approx 2^{24}$ and base $b = 2$ matches the parameters reported in [13].

We also use smaller moduli and bigger bases to reduce the size and increase the security level. The parameters in Fig. 2 suggest that for the 3 choices of q and b, using the approximate gadget trapdoor by setting $l = \lceil(\log_b q)/2\rceil$ saves about half of the sizes in the public key and signatures comparing to using the exact trapdoor, with even a slight increase in the expected cost for the attacking algorithms. Let us mention that some schemes in the literature (like [23]) use an extremely large base of size $b \approx \sqrt{q}$ (the resulting gadget is $\mathbf{g} = [1, \sqrt{q}]$). However, for the small moduli like 2^{16} or 2^{18}, such large bases lead to Gaussian widths larger than the moduli. So we only use moderately large bases.

Params	Exact	Approx	Approx	Exact	Approx	Approx	Exact	Approx	Approx
n	512	512	512	512	512	512	512	512	512
$\lceil \log_2 q \rceil$	24	24	24	16	16	16	16	16	16
b	2	2	2	2	2	2	4	4	4
l	0	12	15	0	7	9	0	2	4
τ	40	40	40	2.6	2.6	2.6	2.6	2.6	2.6
s	38317.0	29615.3	26726.3	2170.7	1756.3	1618.2	3114.2	2833.3	2505.6
m	13312	7168	5632	9216	5632	4608	5120	4096	3072
$\Vert\mathbf{x}\Vert_2$	4441737.7	2521387.0	2035008.5	211100.9	133305.5	109339.1	223740.1	183004.9	138145.7
$\Vert\mathbf{z}\Vert_2$	0	374014.0	2118987.6	0	11897.9	46428.4	0	1402.3	19807.1
PK	37.50	19.50	15.00	17.00	10.00	8.00	9.00	7.00	5.00
Sig	25.68	13.53	10.51	13.16	7.83	6.30	7.62	5.94	4.45
LWE	100.0	100.0	100.0	104.7	104.7	104.7	104.7	104.7	104.7
AISIS	80.2	85.8	81.1	83.7	89.0	88.1	82.8	85.5	87.8

Fig. 2. Summary of the concrete parameters. The size of PK and Sig are measured in kB. $\Vert\mathbf{x}\Vert_2$, $\Vert\mathbf{z}\Vert_2$ are the upper-bounds of the norms of the preimage and the error term. LWE and AISIS refer to the estimations of security levels for the pseudorandomness of the PK and finding a short approximate preimage.

Our implementation shows that the sizes of the public-key & signature can be reduced to 5 kB & 4.45 kB for an estimation of 88-bit security, and 11.25 kB & 9.38 kB for an estimation of 184-bit security. Those are much closer to the sizes of the signatures based on the rejection sampling approach [8, 11, 26, 40]. As a reference, the sizes of the public-key & signature for qTESLA [8] are 4.03 kB & 3.05 kB for an estimation of 128-bit security, and 8.03 kB & 6.03 kB for an estimation of 192-bit security. The sizes for Dilithium [26] are even smaller. Let us remark that our implementation has not adapted possible further optimizations used in Dilithium [26] and qTESLA [8]. So it is reasonable to expect we have more room to improve after adding making further optimizations. The parameters for Falcon [27] are the smallest due to the use of NTRU lattices, so they are rather incomparable with the ones based on RingLWE. As a side note, we do not know how to construct approximate trapdoors for NTRU lattices, and we leave it as an interesting question to investigate in future.

Using approximate trapdoors in the advanced lattice cryptosystems. Finally, let us briefly mention the possible applications of the approximate trapdoors in the cryptosystems built from the dual-Regev approach [1, 19, 31, 34] and the GGH15 approach [17, 18, 21, 30, 35, 52].

To use approximate trapdoors in the schemes based on the dual-Regev approach, we need to sample the LWE secret term with a small norm instead of from the uniform distribution to maintain the correctness of the schemes. For many of these schemes, the security analyses require the extensions of the Bonsai techniques in the approximate setting. We leave the extensions to future works.

For the schemes based on the GGH15-approach, the correctness of the schemes holds without any changes. The security also holds, except for the schemes in [21] which requires the extension of the Bonsai techniques. Let us remark that the saving in the dimension m is of significant importance to the applications built on the GGH15 graded encoding scheme (implemented in [20, 37]). In those applications, the modulus q is proportional to m^d (where $d \in \mathbb{N}$ is the number of "levels" of the graded encodings; larger d supports richer functionalities). So reducing the dimension m would dramatically reduce the overall parameter.

Organizations. The rest of the paper is organized as follows. Section 2 provides the necessary background of lattices. Section 3 provides the definition and the hardness reductions of the approximate ISIS problem. Section 4 presents the approximate gadget trapdoors. Section 5 provides an instantiation of the hash-and-sign signature scheme under the approximate trapdoor, with concrete parameters.

2 Preliminaries

Notations and terminology. In cryptography, the security parameter (denoted as λ) is a variable that is used to parameterize the computational complexity of the cryptographic algorithm or protocol, and the adversary's probability of breaking security. An algorithm is "efficient" if it runs in (probabilistic) polynomial time over λ.

When a variable v is drawn uniformly random from the set S we denote as $v \leftarrow U(S)$. We use $\approx_s$ and $\approx_c$ as the abbreviations for statistically close and computationally indistinguishable. For two distributions D_1, D_2 over the same support $\mathcal{X}$, we denote $D_1 \overset{\varepsilon}{\approx} D_2$ to denote that each $x \in \mathcal{X}$ has $D_1(x) \in [1 \pm \varepsilon]D_2(x)$ and $D_2(x) \in [1 \pm \varepsilon]D_1(x)$.

Let $\mathbb{R}, \mathbb{Z}, \mathbb{N}$ be the set of real numbers, integers and positive integers. Denote $\mathbb{Z}/q\mathbb{Z}$ by $\mathbb{Z}_q$. For $n \in \mathbb{N}$, $[n] := \{1, ..., n\}$. A vector in $\mathbb{R}^n$ (represented in column form by default) is written as a bold lower-case letter, e.g. $\mathbf{v}$. For a vector $\mathbf{v}$, the ith component of $\mathbf{v}$ will be denoted by v_i. For an integer base $b > 1$, we call a positive integer's "b-ary" decomposition the vector $(q_0, q_1, \ldots, q_{k-1}) \in \{0, \ldots, b-1\}^k$ where $k := \lceil \log_b q \rceil$, and $q = \sum q_i b^i$.

A matrix is written as a bold capital letter, e.g. $\mathbf{A}$. The ith column vector of $\mathbf{A}$ is denoted $\mathbf{a}_i$. The length of a vector is the ℓ_p-norm $\|\mathbf{v}\|_p := (\sum v_i^p)^{1/p}$, or the infinity norm given by its largest entry $\|\mathbf{v}\|_\infty := \max_i\{|v_i|\}$. The length of a matrix is the norm of its longest column: $\|\mathbf{A}\|_p := \max_i \|\mathbf{a}_i\|_p$. By default we use ℓ_2-norm unless explicitly mentioned. When a vector or matrix is called "small" or "short", we refer to its norm but not its dimension, unless explicitly mentioned. The thresholds of "small" or "short" will be precisely parameterized in the article when necessary.

2.1 Linear Algebra

Let $\{\mathbf{e}_i\}_{i=1}^n$ be the canonical basis for $\mathbb{R}^n$, with entries $\delta(j,k)$ where $\delta(j,k) = 1$ when $j = k$ and 0 otherwise. For any set $S \subseteq \mathbb{R}^n$, its span (denoted as span(S)) is the smallest subspace of $\mathbb{R}^n$ containing S. For a matrix, $\mathbf{M} \in \mathbb{R}^{n \times m}$, its span is the span of its column vectors, written as span$(\mathbf{M})$. We write matrix transpose as $\mathbf{M}^t$. Let $\tilde{\mathbf{B}}$ denote the Gram-Schmidt orthogonalization of $\mathbf{B}$. The GSO of an ordered basis $\mathbf{B} = [\mathbf{b}_1, \ldots, \mathbf{b}_k]$ is assumed to be from left to right, $\tilde{\mathbf{b}}_1 = \mathbf{b}_1$, unless stated otherwise.

Recall $\mathbf{M}$'s singular value decomposition (SVD), i.e. $\mathbf{M} = \mathbf{VDW} \in \mathbb{R}^{n \times m}$ where $\mathbf{V} \in \mathbb{R}^{n \times n}$ along with $\mathbf{W} \in \mathbb{R}^{m \times m}$ are unitary, and $\mathbf{D} \in \mathbb{R}^{n \times m}$ is a triangular matrix containing $\mathbf{M}$'s singular values. Further, let $q = \min\{n, m\}$ and $\mathbf{D}_q = \text{diag}(s_1, \ldots, s_q)$ be the diagonal matrix containing $\mathbf{M}$'s singular values $s_i = s_i(\mathbf{M})$. Throughout the paper, we are concerned with random, subgaussian [51] matrices $\mathbf{M}$ with $\{s_1 \geq \ldots \geq s_q > 0\}$. Then, $\mathbf{D} = \mathbf{D}_q$ when $n = m$, $\mathbf{D} = [\mathbf{D}_q\ \mathbf{0}]$ when $m > n$, and $\mathbf{D} = \begin{bmatrix} \mathbf{D}_q \\ \mathbf{0} \end{bmatrix}$ in the case $m < n$.

A symmetric matrix $\Sigma \in \mathbb{R}^{n \times n}$ is *positive semi-definite* if for all $\mathbf{x} \in \mathbb{R}^n$, we have $\mathbf{x}^t \Sigma \mathbf{x} \geq 0$. It is *positive definite*, $\Sigma > 0$, if it is positive semi-definite and $\mathbf{x}^t \Sigma \mathbf{x} = 0$ implies $\mathbf{x} = \mathbf{0}$. We say $\Sigma_1 > \Sigma_2$ $(\geq)$ if $\Sigma_1 - \Sigma_2$ is positive-(semi)definite. This forms a partial ordering on the set of positive semi-definite matrices, and we denote $\Sigma \geq \alpha \mathbf{I}$ often as $\Sigma \geq \alpha$ for constants $\alpha \in \mathbb{R}^+$. For any positive semi-definite matrix Σ, we write $\sqrt{\Sigma}$ to be any full rank matrix $\mathbf{T}$ such that $\Sigma = \mathbf{TT}^t$. We say $\mathbf{T}$ is a *square root* of Σ. For two positive semi-definite matrices, Σ_1 and Σ_2, we denote the positive semi-definite matrix formed by their block diagonal concatenation as $\Sigma_1 \oplus \Sigma_2$. Let $\mathbf{M}^*$ denote Hermitian transpose. The *(Moore-Penrose) pseudoinverse* for matrix $\mathbf{M}$ with SVD $\mathbf{M} = \mathbf{VDW}$ is $\mathbf{M}^+ = \mathbf{WD}^+\mathbf{V}^*$ where $\mathbf{D}^+$ is given by transposing $\mathbf{D}$ and inverting $\mathbf{M}$'s nonzero singular values. For example, $\mathbf{T} = s\mathbf{I}$ and $\mathbf{T}^+ = s^{-1}\mathbf{I}$ for a covariance $\Sigma = s^2\mathbf{I}$. (An analogous $\mathbf{T}^+ = \mathbf{T}^{-1}$ is given for the non-spherical, full-rank case $\Sigma > 0$ using Σ's diagonalization.)

2.2 Lattices Background

An n-dimensional lattice Λ of rank $k \leq n$ is a discrete additive subgroup of $\mathbb{R}^n$. Given k linearly independent basis vectors $\mathbf{B} = \{\mathbf{b}_1, ..., \mathbf{b}_k \in \mathbb{R}^n\}$, the lattice generated by $\mathbf{B}$ is

$$\Lambda(\mathbf{B}) = \Lambda(\mathbf{b}_1, ..., \mathbf{b}_k) = \{\sum_{i=1}^{k} x_i \cdot \mathbf{b}_i, x_i \in \mathbb{Z}\}.$$

Given $n, m \in \mathbb{N}$ and a modulus $q \geq 2$, we often use q-ary lattices and their cosets, denoted as

$$\text{for } \mathbf{A} \in \mathbb{Z}_q^{n\times m}, \text{ denote } \Lambda^\perp(\mathbf{A}) \text{ or } \Lambda_q^\perp(\mathbf{A}) \text{ as } \{\mathbf{x} \in \mathbb{Z}^m : \mathbf{A} \cdot \mathbf{x} = \mathbf{0} \pmod q\};$$

$$\text{for } \mathbf{A} \in \mathbb{Z}_q^{n\times m}, \mathbf{w} \in \mathbb{Z}_q^n, \quad \text{denote } \Lambda_{\mathbf{w}}^\perp(\mathbf{A}) \text{ as } \{\mathbf{x} \in \mathbb{Z}^m : \mathbf{A} \cdot \mathbf{x} = \mathbf{w} \pmod q\}.$$

Gaussians on lattices. For any $s > 0$ define the Gaussian function on $\mathbb{R}^n$ with parameter s:

$$\forall \mathbf{x} \in \mathbb{R}^n,\ \rho_s(\mathbf{x}) = e^{-\pi\|\mathbf{x}\|^2/s^2}.$$

For any $\mathbf{c} \in \mathbb{R}^n$, real $s > 0$, and n-dimensional lattice Λ, define the discrete Gaussian distribution $D_{\Lambda+\mathbf{c},s}$ as:

$$\forall \mathbf{x} \in \Lambda + \mathbf{c},\ D_{\Lambda+\mathbf{c},s}(\mathbf{x}) = \frac{\rho_s(\mathbf{x})}{\rho_s(\Lambda + \mathbf{c})}.$$

The subscripts s and $\mathbf{c}$ are taken to be 1 and $\mathbf{0}$ (respectively) when omitted.

For any positive semidefinite $\Sigma = \mathbf{T} \cdot \mathbf{T}^t$, define the non-spherical Gaussian function as

$$\forall \mathbf{x} \in \text{span}(\mathbf{T}) = \text{span}(\Sigma),\ \rho_{\mathbf{T}}(\mathbf{x}) = e^{-\pi\mathbf{x}^t\Sigma^+\mathbf{x}},$$

and $\rho_{\mathbf{T}}(\mathbf{x}) = 0$ for all $\mathbf{x} \notin \text{span}(\Sigma)$. Note that $\rho_{\mathbf{T}}(\cdot)$ only depends on Σ but not the specific choice of the $\mathbf{T}$, so we may write $\rho_{\mathbf{T}}(\cdot)$ as $\rho_{\sqrt{\Sigma}}(\cdot)$.

For any $\mathbf{c} \in \mathbb{R}^n$, any positive semidefinite Σ, and n-dimensional lattice Λ such that $(\Lambda + \mathbf{c}) \cap \text{span}(\Sigma)$ is non-empty, define the discrete Gaussian distribution $D_{\Lambda+\mathbf{c},\sqrt{\Sigma}}$ as:

$$\forall \mathbf{x} \in \Lambda + \mathbf{c},\ D_{\Lambda+\mathbf{c},\sqrt{\Sigma}}(\mathbf{x}) = \frac{\rho_{\sqrt{\Sigma}}(\mathbf{x})}{\rho_{\sqrt{\Sigma}}(\Lambda + \mathbf{c})}.$$

Smoothing parameter. We recall the definition of smoothing parameter and some useful facts.

Definition 1 (Smoothing parameter [44]**).** *For any lattice Λ and positive real $\epsilon > 0$, the smoothing parameter $\eta_\epsilon(\Lambda)$ is the smallest real $s > 0$ such that $\rho_{1/s}(\Lambda^* \setminus \{\mathbf{0}\}) \leq \epsilon$.*

Notice that for two lattices of the same rank $\Lambda_1 \subseteq \Lambda_2$, the denser lattice always has the smaller smoothing parameter, i.e. $\eta_\epsilon(\Lambda_2) \leq \eta_\epsilon(\Lambda_1)$.

We will need a generalization of the smoothing parameter to the non-spherical Gaussian.

Definition 2. *For a positive semi-definite $\Sigma = \mathbf{T}\mathbf{T}^t$, an $\epsilon > 0$, and a lattice Λ with span(Λ) $\subseteq$ span(Σ), we say $\eta_\epsilon(\Lambda) \leq \sqrt{\Sigma}$ if $\eta_\epsilon(\mathbf{T}^+\Lambda) \leq 1$.*

When the covariance matrix $\Sigma > 0$ and the lattice Λ are full-rank, $\sqrt{\Sigma} \geq \eta_\varepsilon(\Lambda)$ is equivalent to the minimum eigenvalue of Σ, $\lambda_{min}(\Sigma)$, being at least $\eta_\varepsilon^2(\Lambda)$.

Lemma 1 ([44])**.** *For any n-dimensional lattice Λ of rank k, and any real $\epsilon > 0$,*

$$\eta_\epsilon(\Lambda) \leq \lambda_k(\Lambda) \cdot \sqrt{\log(2k(1+1/\epsilon))/\pi}.$$

Lemma 2 ([44])**.** *Let Λ be a lattice, $\mathbf{c} \in$ span(Λ). For any $\Sigma \geq 0$, if $\sqrt{\Sigma} \geq \eta_\epsilon(\Lambda)$ for some $\epsilon > 0$, then*

$$\rho_{\sqrt{\Sigma}}(\Lambda + \mathbf{c}) \in \left[\frac{1-\epsilon}{1+\epsilon}, 1\right] \cdot \rho_{\sqrt{\Sigma}}(\Lambda)$$

The following is a generalization of [31, Corollary 2.8] for non-spherical Gaussian.

Corollary 1 (Smooth over the cosets). *Let Λ, Λ' be n-dimensional lattices s.t. $\Lambda' \subseteq \Lambda$. Then for any $\epsilon > 0$, $\sqrt{\Sigma} \geq \eta_\epsilon(\Lambda')$, and $\mathbf{c} \in$ span(Λ), we have*

$$\Delta(D_{\Lambda+\mathbf{c},\sqrt{\Sigma}} \bmod \Lambda',\ U(\Lambda \bmod \Lambda')) < 2\epsilon$$

Lemma 3 ([44,49])**.** *Let $\mathbf{B}$ be a basis of an n-dimensional lattice Λ, and let $s \geq \|\tilde{\mathbf{B}}\| \cdot \omega(\log n)$, then $\Pr_{\mathbf{x} \leftarrow D_{\Lambda,s}}[\|\mathbf{x}\| \geq s \cdot \sqrt{n} \vee \mathbf{x} = \mathbf{0}] \leq \mathsf{negl}(n)$.*

Linear Transformations of Discrete Gaussians. We will use the following general theorem, implicitly used in [15,42,43], regarding the linear transformation, $\mathbf{T}$, of a discrete Gaussian. It states that as long as the original discrete Gaussian over a lattice Λ is smooth enough in the lattice intersect the kernel of $\mathbf{T}$ ($\Lambda \cap \ker(\mathbf{T})$), then the distribution transformed by $\mathbf{T}$ is statistically close to another discrete Gaussian.

Theorem 1 ([41])**.** *For any positive definite Σ, vector $\mathbf{c}$, lattice coset $A := \Lambda + \mathbf{a} \subset \mathbf{c} +$ span(Σ), and linear transformation $\mathbf{T}$, if the lattice $\Lambda_\mathbf{T} = \Lambda \cap \ker(\mathbf{T})$ satisfies span($\Lambda_\mathbf{T}$) = ker($\mathbf{T}$) and $\eta_\epsilon(\Lambda_\mathbf{T}) \leq \sqrt{\Sigma}$, then*

$$\mathbf{T}(D_{A,\mathbf{c},\sqrt{\Sigma}}) \overset{\bar{\epsilon}}{\approx} D_{\mathbf{T}A,\mathbf{T}\mathbf{c},\mathbf{T}\sqrt{\Sigma}}$$

where $\bar{\epsilon} = 2\epsilon/(1-\epsilon)$.

We remark that if $\mathbf{T}$ is injective (i.e. ker($\mathbf{T}$) is trivial), then $\mathbf{T}(D_{A,\mathbf{c},\sqrt{\Sigma}}) = D_{\mathbf{T}A,\mathbf{Tc},\mathbf{T}\sqrt{\Sigma}}$.

Let us also remark that at the time of writing this article, the following lemma (which is a special case of Theorem 1) has already been proven in [25]. This lemma is suitable for all of our proofs using a non-injective linear transformation of a discrete gaussian.

In what follows, the max-log distance between two distributions with the same support S is $\Delta_{ML}(\mathcal{X},\mathcal{Y}) = \max_{s\in S}|\log \mathcal{X}(s) - \log \mathcal{Y}(s)|$ [45].

Lemma 4 (Lemma 3, [25]). *Let* $\mathbf{T} \in \mathbb{Z}^{n\times m}$ *such that* $\mathbf{T}\mathbb{Z}^m = \mathbb{Z}^n$ *and* $\Lambda^{\perp}(\mathbf{T}) = \{\mathbf{x} \in \mathbb{Z}^m : \mathbf{T}\mathbf{x} = \mathbf{0} \in \mathbb{Z}^n\}$. *Let* $\Sigma = \mathbf{T}\mathbf{T}^t$. *For* $\epsilon \in (0,1/2)$, $\hat{\epsilon} = \epsilon + O(\epsilon^2)$, $r \geq \eta_\epsilon(\Lambda^{\perp}(\mathbf{T}))$, *the max-log distance between* $\mathbf{T}\cdot D_{\mathbb{Z}^m,r}$ *and* $D_{\mathbb{Z}^n,r\sqrt{\Sigma}}$ *is at most* $4\hat{\epsilon}$.

2.3 Gadgets, or G-Lattices

Let $\mathbf{G} = \mathbf{I}_n \otimes \mathbf{g}^t \in \mathbb{Z}_q^{n\times nk}$ with $\mathbf{g}^t = (1, b, \ldots, b^{k-1})$, $k = \lceil \log_b q \rceil$. $\mathbf{G}$ is commonly referred to the gadget matrix. The gadget matrix's q-ary lattice, $\Lambda_q^{\perp}(\mathbf{G})$, is the direct sum of n copies of the lattice $\Lambda_q^{\perp}(\mathbf{g}^t)$. Further, $\Lambda_q^{\perp}(\mathbf{g}^t)$ has a simple basis,

$$\mathbf{B}_q = \begin{bmatrix} b & & & q_0 \\ -1 & \ddots & & \vdots \\ & \ddots & b & q_{k-2} \\ & & -1 & q_{k-1} \end{bmatrix}$$

where $(q_0, \ldots, q_{k-1}) \in \{0, 1, \ldots, b-1\}^k$ is the b-ary decomposition of the modulus, q. When $q = b^k$, we cheat by having $q_0 = q_1 = \ldots = q_{k-2} = 0$ and $q_{k-1} = b$. Either way, the integer cosets of $\Lambda_q^{\perp}(\mathbf{g}^t)$ can be viewed as the syndromes of $\mathbf{g}^t$ as a check matrix, in the terminology of coding theory. These cosets are expressed as $\Lambda_u^{\perp}(\mathbf{g}^t) = \{\mathbf{x} \in \mathbb{Z}^k : \mathbf{g}^t\mathbf{x} = u \mod q\} = \Lambda_q^{\perp}(\mathbf{g}^t) + \mathbf{u}$ where $\mathbf{u}$ can be any coset representative. A simple coset representative of $\Lambda_u^{\perp}(\mathbf{g}^t)$ is the b-ary decomposition of u. The integer cosets of $\Lambda_q^{\perp}(\mathbf{G})$ are expressed through the direct-sum construction, $\Lambda_{\mathbf{u}}^{\perp}(\mathbf{G}) = \Lambda_{u_1}^{\perp}(\mathbf{g}^t) \oplus \ldots \oplus \Lambda_{u_n}^{\perp}(\mathbf{g}^t)$ where $\mathbf{u} = (u_1, \ldots, u_n) \in \mathbb{Z}_q^n$. We call $\mathbf{G}$ a gadget matrix since the following problems, SIS and LWE, are easily solved on the matrix $\mathbf{G}$ [42].

2.4 SIS, LWE, and the Trapdoor

We first recall the short integer solution (SIS) problem.

Definition 3 (SIS [2]). *For any* $n, m, q \in \mathbb{Z}$ *and* $\beta \in \mathbb{R}$, *define the short integer solution problem* $\mathsf{SIS}_{n,m,q,\beta}$ *as follows: Given* $\mathbf{A} \in \mathbb{Z}_q^{n\times m}$, *find a non-zero vector* $\mathbf{x} \in \mathbb{Z}^m$ *such that* $\|\mathbf{x}\| \leq \beta$, *and*

$$\mathbf{A}\mathbf{x} = \mathbf{0} \bmod q.$$

Definition 4 (ISIS). *For any $n, m, q \in \mathbb{Z}$ and $\beta \in \mathbb{R}$, define the inhomogeneous short integer solution problem $\mathsf{ISIS}_{n,m,q,\beta}$ as follows: Given $\mathbf{A} \in \mathbb{Z}_q^{n\times m}$, $\mathbf{y} \in \mathbb{Z}_q^n$, find $\mathbf{x} \in \mathbb{Z}^m$ such that $\|\mathbf{x}\| \leq \beta$, and*

$$\mathbf{A}\mathbf{x} = \mathbf{y} \bmod q.$$

Lemma 5 (Hardness of (I)SIS based on the lattice problems in the worst case [2,31,44]). *For any $m = \mathsf{poly}(n)$, any $\beta > 0$, and any sufficiently large $q \geq \beta \cdot \mathsf{poly}(n)$, solving $\mathsf{SIS}_{n,m,q,\beta}$ or $\mathsf{ISIS}_{n,m,q,\beta}$ (where $\mathbf{y}$ is sampled uniformly from $\mathbb{Z}_q^n$) with non-negligible probability is as hard as solving GapSVP_γ and SIVP_γ on arbitrary n-dimensional lattices with overwhelming probability, for some approximation factor $\gamma = \beta \cdot \mathsf{poly}(n)$.*

All the (I)SIS problems and their variants admit the Hermite normal form (HNF), where the public matrix $\mathbf{A}$ is of the form $[\mathbf{I}_n \mid \mathbf{A}']$ where $\mathbf{A}' \in \mathbb{Z}_q^{n\times(m-n)}$. The HNF variant of (I)SIS is as hard as the standard (I)SIS. This can be seen by rewriting $\mathbf{A} \in \mathbb{Z}_q^{n\times m}$ as $\mathbf{A} =: [\mathbf{A}_1 \mid \mathbf{A}_2] = \mathbf{A}_1 \cdot [\mathbf{I}_n \mid \mathbf{A}_1^{-1} \cdot \mathbf{A}_2]$ (we always work with n, q such that $\mathbf{A}_1 \leftarrow U(\mathbb{Z}_q^{n\times n})$ is invertible with non-negligible probability).

Learning with errors. We recall the decisional learning with errors (LWE) problem.

Definition 5 (Decisional learning with errors [50]). *For $n, m \in \mathbb{N}$ and modulus $q \geq 2$, distributions for secret vectors, public matrices, and error vectors $\theta, \pi, \chi \subseteq \mathbb{Z}_q$. An LWE sample is obtained from sampling $\mathbf{s} \leftarrow \theta^n$, $\mathbf{A} \leftarrow \pi^{n\times m}$, $\mathbf{e} \leftarrow \chi^m$, and outputting $(\mathbf{A}, \mathbf{y}^t := \mathbf{s}^t\mathbf{A} + \mathbf{e}^t \mod q)$.*

We say that an algorithm solves $\mathsf{LWE}_{n,m,q,\theta,\pi,\chi}$ if it distinguishes the LWE sample from a random sample distributed as $\pi^{n\times m} \times U(\mathbb{Z}_q^m)$ with probability greater than 1/2 plus non-negligible.

Lemma 6 (Hardness of LWE based on the lattice problems in the worst case [16,47,48,50]). *Given $n \in \mathbb{N}$, for any $m = \mathsf{poly}(n)$, $q \leq 2^{\mathsf{poly}(n)}$. Let $\theta = \pi = U(\mathbb{Z}_q)$, $\chi = D_{\mathbb{Z},s}$ where $s \geq 2\sqrt{n}$. If there exists an efficient (possibly quantum) algorithm that breaks $\mathsf{LWE}_{n,m,q,\theta,\pi,\chi}$, then there exists an efficient (possibly quantum) algorithm for solving GapSVP_γ and SIVP_γ on arbitrary n-dimensional lattices with overwhelming probability, for some approximation factor $\gamma = \tilde{O}(nq/s)$.*

The next lemma shows that LWE with the secret sampled from the error distribution is as hard as the standard LWE.

Lemma 7 ([10,16]). *For n, m, q, s chosen as was in Lemma 6, $\mathsf{LWE}_{n,m',q,D_{\mathbb{Z},s},U(\mathbb{Z}_q),D_{\mathbb{Z},s}}$ is as hard as $\mathsf{LWE}_{n,m,q,U(\mathbb{Z}_q),U(\mathbb{Z}_q),D_{\mathbb{Z},s}}$ for $m' \leq m - (16n + 4\log\log q)$.*

Trapdoor. A *trapdoor* for a public matrix $\mathbf{A} \in \mathbb{Z}_q^{n\times m}$ is a string that allows its owner to efficiently solve both the (I)SIS and LWE problems w.r.t. $\mathbf{A}$.

3 The Approximate Trapdoor for Ajtai's Function

Given a matrix $\mathbf{A} \in \mathbb{Z}_q^{n\times m}$, define an *approximate trapdoor* of $\mathbf{A}$ as anything that allows us to efficiently solve the approximate version of the ISIS problem w.r.t. $\mathbf{A}$. We first define the approximate ISIS problem.

Definition 6 (Approximate ISIS). *For any $n, m, q \in \mathbb{N}$ and $\alpha, \beta \in \mathbb{R}$, define the approximate inhomogeneous short integer solution problem* $\mathsf{Approx.ISIS}_{n,m,q,\alpha,\beta}$ *as follows: Given $\mathbf{A} \in \mathbb{Z}_q^{n\times m}$, $\mathbf{y} \in \mathbb{Z}_q^n$, find a vector $\mathbf{x} \in \mathbb{Z}^m$ such that $\|\mathbf{x}\| \leq \beta$, and there is a vector $\mathbf{z} \in \mathbb{Z}^n$ satisfying*

$$\|\mathbf{z}\| \leq \alpha \quad \text{and} \quad \mathbf{A}\mathbf{x} = \mathbf{y} + \mathbf{z} \pmod q.$$

Let us remark that the approximate ISIS is only non-trivial when the bounds α, β are relatively small compared to the modulus q. Also, our definition chooses to allow the zero vector to be a valid solution, which means when $\|\mathbf{y}\| \leq \alpha$, the zero vector is trivially a solution. Such a choice in the definition does not cause a problem in the application, since the interesting case in the application is to handle all the $\mathbf{y} \in \mathbb{Z}_q^n$, or $\mathbf{y}$ sampled uniformly random from $\mathbb{Z}_q^n$.

Definition 7 (Approximate trapdoor). *A string τ is called an (α, β)-approximate trapdoor for a matrix $\mathbf{A} \in \mathbb{Z}_q^{n\times m}$ if there is a probabilistic polynomial time algorithm (in n, m, $\log q$) that given τ, $\mathbf{A}$ and any $\mathbf{y} \in \mathbb{Z}_q^n$, outputs a non-zero vector $\mathbf{x} \in \mathbb{Z}^m$ such that $\|\mathbf{x}\| \leq \beta$, and there is a vector $\mathbf{z} \in \mathbb{Z}^n$ satisfying*

$$\|\mathbf{z}\| \leq \alpha \quad \text{and} \quad \mathbf{A}\mathbf{x} = \mathbf{y} + \mathbf{z} \pmod q.$$

3.1 Hardness of the Approximate ISIS Problem

To make sense of the approximate trapdoor, we argue that for those who do not have the trapdoor, the approximate ISIS problem is a candidate one-way function under proper settings of parameters.

First, we observe a rather obvious reduction that bases the hardness of solving approximate ISIS (given an arbitrary target) on the hardness of decisional LWE with low-norm secret (e.g. when the secret is sampled from the error distribution). In the theorem statement below, when the norm symbol is applied on a distribution D, i.e. $\|D\|$, it denotes the lowest value $v \in \mathbb{R}^+$ such that $\Pr_{d\leftarrow D}[\|d\| < v] > 1 - \mathsf{negl}(\lambda)$.

Theorem 2. *For $n, m, q \in \mathbb{Z}$, $\alpha, \beta \in \mathbb{R}^+$, θ, χ be distributions over $\mathbb{Z}$ such that $q > 4(\|\theta\| \cdot (\alpha + 1) + \|\theta^n\| \cdot \alpha \cdot \sqrt{n} + \|\chi^m\| \cdot \beta \cdot \sqrt{m})$. Then $\mathsf{LWE}_{n,m,q,\theta,U(\mathbb{Z}_q),\chi} \leq_p \mathsf{Approx.ISIS}_{n,m,q,\alpha,\beta}$.*

Proof. Suppose there is a polynomial time adversary A that breaks $\mathsf{Approx.ISIS}_{n,m,q,\alpha,\beta}$, we build a polynomial time adversary B that breaks decisional LWE.

Let $r = \lfloor \alpha \rceil + 1$. Given an LWE challenge $(\mathbf{A}, \mathbf{w}) \in \mathbb{Z}_q^{n\times m} \times \mathbb{Z}_q^m$, where $\mathbf{w}$ is either an LWE sample or sampled uniformly from $\mathbb{Z}_q^m$. B picks a vector

$\mathbf{y} := (r, 0, ..., 0)^t \in \mathbb{Z}_q^n$, sends $\mathbf{A}$ and $\mathbf{y}$ to the adversary A as an approximate ISIS challenge. A replies with $\mathbf{x} \in \mathbb{Z}^m$ such that $\|\mathbf{x}\| \le \beta$, and there is a vector $\mathbf{z} \in \mathbb{Z}^n$ satisfying

$$\|\mathbf{z}\| \le \alpha \quad \text{and} \quad \mathbf{A}\mathbf{x} = \mathbf{y} + \mathbf{z} \pmod q.$$

Note that $\mathbf{x} \neq \mathbf{0}$ since $\|\mathbf{y}\| > \alpha$.

B then computes $v := \langle \mathbf{w}, \mathbf{x} \rangle$. If $\mathbf{w}^t = \mathbf{s}^t\mathbf{A} + \mathbf{e}^t$ for $\mathbf{s} \leftarrow \theta^n$, $\mathbf{e} \leftarrow \chi^m$, then

$$v = (\mathbf{s}^t\mathbf{A} + \mathbf{e}^t)\mathbf{x} = \mathbf{s}^t(\mathbf{y} + \mathbf{z}) + \mathbf{e}^t\mathbf{x} \Rightarrow$$
$$\|v\| \le \|\theta\| \cdot r + \|\theta^n\| \cdot \alpha \cdot \sqrt{n} + \|\chi^m\| \cdot \beta \cdot \sqrt{m} < q/4.$$

Otherwise v distributes uniformly random over $\mathbb{Z}_q$. So B can compare v with the threshold value and wins the decisional LWE challenge with probability 1/2 plus non-negligible.

Alternatively, we can also prove that the approximate ISIS problem is as hard as the standard ISIS. The reductions go through the HNFs of the ISIS and the approximate ISIS problems. All the reductions in the following theorem works for uniformly random target vectors.

Theorem 3. $\mathsf{ISIS}_{n,n+m,q,\beta} \ge_p \mathsf{Approx.ISIS}_{n,m,q,\alpha+\beta,\beta}$; $\mathsf{ISIS}_{n,n+m,q,\alpha+\beta} \le_p \mathsf{Approx.ISIS}_{n,m,q,\alpha,\beta}$.

Proof. We will show $\mathsf{ISIS} = \mathsf{HNF.ISIS} = \mathsf{HNF.Approx.ISIS} = \mathsf{Approx.ISIS}$ under proper settings of parameters.

Recall that $\mathsf{ISIS}_{n,m,q,\beta} = \mathsf{HNF.ISIS}_{n,m,q,\beta}$ as explained in the preliminary. Also, $\mathsf{HNF.ISIS}_{n,m,q,\beta} \ge_p \mathsf{HNF.Approx.ISIS}_{n,m,q,\alpha,\beta}$ for any $\alpha \ge 0$ by definition. It remains to show the rest of the connections.

Lemma 8. $\mathsf{HNF.ISIS}_{n,m,q,\alpha+\beta} \le_p \mathsf{HNF.Approx.ISIS}_{n,m,q,\alpha,\beta}$.

Proof. Suppose there is a polynomial time algorithm A that solves $\mathsf{HNF.Approx.ISIS}_{n,m,q,\alpha,\beta}$, we build a polynomial time algorithm B that solve $\mathsf{HNF.ISIS}_{n,m,q,\alpha+\beta}$. Given an $\mathsf{HNF.ISIS}$ instance $[\mathbf{I}_n \mid \mathbf{A}] \in \mathbb{Z}_q^{n\times m}$, $\mathbf{y}$, B passes the same instance to A, gets back a vector $\mathbf{x}$ such that

$$[\mathbf{I}_n \mid \mathbf{A}] \cdot \mathbf{x} = \mathbf{y} + \mathbf{z} \pmod q.$$

where $\|\mathbf{x}\| \le \beta$, $\|\mathbf{z}\| \le \alpha$. Now write $\mathbf{x} =: [\mathbf{x}_1^t \mid \mathbf{x}_2^t]^t$ where $\mathbf{x}_1 \in \mathbb{Z}^n$, $\mathbf{x}_2 \in \mathbb{Z}^m$. Then $\mathbf{x}' := [(\mathbf{x}_1 - \mathbf{z})^t \mid \mathbf{x}_2^t]^t$ satisfies

$$[\mathbf{I}_n \mid \mathbf{A}] \cdot \mathbf{x}' = \mathbf{y} \pmod q,$$

and $\|\mathbf{x}'\| \le \alpha + \beta$. So $\mathbf{x}'$ is a valid solution to $\mathsf{HNF.ISIS}$.

Lemma 9. $\mathsf{HNF.Approx.ISIS}_{n,n+m,q,\alpha,\beta} \le_p \mathsf{Approx.ISIS}_{n,m,q,\alpha,\beta}$.

Proof. Suppose there is a polynomial time algorithm A that solves $\mathsf{Approx.ISIS}_{n,m,q,\alpha,\beta}$, we build a polynomial time algorithm B that solves $\mathsf{HNF.Approx.ISIS}_{n,n+m,q,\alpha,\beta}$. Given $[\mathbf{I}_n \mid \mathbf{A}] \in \mathbb{Z}_q^{n\times(n+m)}$, $\mathbf{y} \in \mathbb{Z}_q^n$ as an $\mathsf{HNF.Approx.ISIS}$

instance, B passes $\mathbf{A} \in \mathbb{Z}_q^{n\times m}$, $\mathbf{y}$ to A, gets back a short vector $\mathbf{x} \in \mathbb{Z}^m$. Then $[\mathbf{0}_n^t \mid \mathbf{x}^t]^t$ is a valid solution to the HNF.Approx.ISIS instance.

Lemma 10. $\mathsf{HNF.Approx.ISIS}_{n,n+m,q,\alpha,\beta} \geq_p \mathsf{Approx.ISIS}_{n,m,q,\alpha+\beta,\beta}$.

Proof. Suppose there is a polynomial time algorithm A that solves HNF. $\mathsf{Approx.ISIS}_{n,n+m,q,\alpha,\beta}$, we build a polynomial time algorithm B that solves $\mathsf{Approx.ISIS}_{n,m,q,\alpha+\beta,\beta}$. Given an Approx.ISIS instance $\mathbf{A} \in \mathbb{Z}_q^{n\times m}$, $\mathbf{y} \in \mathbb{Z}^n$, B passes $[\mathbf{I}_n \mid \mathbf{A}] \in \mathbb{Z}_q^{n\times(n+m)}$, $\mathbf{y}$ as an HNF.Approx.ISIS instance to A, gets back an answer $\mathbf{x} \in \mathbb{Z}^{m+n}$ such that

$$[\mathbf{I}_n \mid \mathbf{A}] \cdot \mathbf{x} = \mathbf{y} + \mathbf{z} \pmod q, \tag{2}$$

where $\|\mathbf{x}\| \leq \beta$, $\|\mathbf{z}\| \leq \alpha$.

Now write $\mathbf{x} =: [\mathbf{x}_1^t \mid \mathbf{x}_2^t]^t$ where $\mathbf{x}_1 \in \mathbb{Z}^n$, $\mathbf{x}_2 \in \mathbb{Z}^m$. Rewriting Eq. (2) gives

$$\mathbf{A} \cdot \mathbf{x}_2 = \mathbf{y} + \mathbf{z} - \mathbf{x}_1 \pmod q,$$

so $\mathbf{x}_2$ is a valid solution to $\mathsf{Approx.ISIS}_{n,m,q,\alpha+\beta,\beta}$.

Theorem 3 then follows the lemmas above.

The following statement immediately follows the proof of Lemma 10.

Corollary 2. *An (α, β)-approximate trapdoor for $[\mathbf{I} \mid \mathbf{A}]$ is an $(\alpha + \beta, \beta)$-approximate trapdoor for $\mathbf{A}$.*

4 Approximate Gadget Trapdoor

We present an instantiation of an approximate trapdoor based on the gadget-based trapdoor generation and preimage sampling algorithms of Micciancio and Peikert [42] (without the tag matrices). In short, we show how to generate a pseudorandom $\mathbf{A} \in \mathbb{Z}_q^{n\times m}$ along with an approximate trapdoor $\mathbf{R}$ with small integer entries.

In the rest of this section, we first recall the exact G-trapdoor from [42], then present the approximate trapdoor generation algorithm and the approximate preimage sampling algorithm. Finally we show that the preimage and the error distributions for uniformly random targets are simulatable.

4.1 Recall the G-Trapdoor from [42]

Let $b \geq 2$ be the base for the G-lattice. Let q be the modulus, $k = \lceil \log_b q \rceil$. b is typically chosen to be 2 for simplicity, but often a higher base b is used for efficiency trade-offs in lattice-based schemes.

Recall the gadget-lattice trapdoor technique from [42]: the public matrix is

$$\mathbf{A} = [\bar{\mathbf{A}} | \mathbf{G} - \bar{\mathbf{A}}\mathbf{R}]$$

where $\mathbf{G}$ is the commonly used gadget matrix, $\mathbf{G} := \mathbf{I}_n \otimes \mathbf{g}_k^t$, $\mathbf{g}_k^t := (1, b, \ldots, b^{k-1})$, and $\mathbf{R}$ is a secret, trapdoor matrix with small, random entries. $\mathbf{A}$ is either statistically close to uniformly random or pseudorandom, depending on the structure of $\bar{\mathbf{A}}$ and the choice of χ (in the pseudorandom case $\chi \subseteq \mathbb{Z}$ is chosen to be a distribution such that $\mathsf{LWE}_{n,n,q,\chi,U(\mathbb{Z}_q),\chi}$ is hard). In this paper we focus on the pseudorandom case since the resulting public matrix $\mathbf{A}$ and preimage have smaller dimensions.

In order to sample a short element in $\Lambda_{\mathbf{u}}^{\perp}(\mathbf{A})$, we use the trapdoor to map short coset representatives of $\Lambda_q^{\perp}(\mathbf{G})$ to short coset representatives of $\Lambda_q^{\perp}(\mathbf{A})$ by the relation

$$\mathbf{A}\begin{bmatrix}\mathbf{R}\\ \mathbf{I}\end{bmatrix} = \mathbf{G}.$$

Using the trapdoor as a linear transformation alone leaks information about the trapdoor. Therefore, we perturb the sample to statistically hide the trapdoor. Let Σ_p be a positive definite matrix defined as $\Sigma_p := s^2\mathbf{I} - \sigma^2 \begin{bmatrix}\mathbf{R}\mathbf{R}^t & \mathbf{R}^t\\ \mathbf{R} & \mathbf{I}\end{bmatrix}$ where σ is at least $\eta_\varepsilon(\Lambda_q^{\perp}(\mathbf{G}))$. The perturbation can be computed offline as $\mathbf{p} \leftarrow D_{\mathbb{Z}^m,\sqrt{\Sigma_p}}$. We then sample a G-lattice vector in a coset dependent on $\mathbf{p}$ as $\mathbf{z} \leftarrow D_{\Lambda_{\mathbf{v}}^{\perp}(\mathbf{G}),\sigma}$ and $\mathbf{v} = \mathbf{u} - \mathbf{A}\mathbf{p} \in \mathbb{Z}_q^n$. Finally, the preimage is set to be

$$\mathbf{y} := \mathbf{p} + \begin{bmatrix}\mathbf{R}\\ \mathbf{I}\end{bmatrix}\mathbf{z}.$$

4.2 The Algorithms of the Approximate G-Trapdoor

As mentioned in the introduction, the main idea of obtaining an approximate trapdoor is to adapt the algorithms from [42] with a gadget matrix without the lower-order entries. Let $0 < l < k$ be the number of lower-order entries dropped from the gadget vector $\mathbf{g} \in \mathbb{Z}_q^k$. Define the resulting approximate gadget vector as $\mathbf{f} := (b^l, b^{l+1}, ..., b^{k-1})^t \in \mathbb{Z}_q^{(k-l)}$. Let $w = n(k-l)$ be the number of columns of the approximate gadget $\mathbf{F} := \mathbf{I}_n \otimes \mathbf{f}^t \in \mathbb{Z}^{n \times w}$. Then the number of columns of $\mathbf{A}$ will be $m := 2n + w$.

Once we replace the gadget matrix $\mathbf{G}$ with its truncated version, $\mathbf{F}$, our approximate trapdoor generation and approximate preimage sampling algorithms match the original gadget-based algorithms. The generation and preimage algorithms are given as Algorithms 2 and 3, respectively. Algorithm 1 represents our approximate $\mathbf{F}$-sampling algorithm. It simply runs the G-lattice preimage sampling algorithm and drops the first l entries from the preimage. The covariance of the perturbation in Algorithm 3 is chosen as

$$\Sigma_p := s^2\mathbf{I}_m - \sigma^2 \begin{bmatrix}\mathbf{R}\mathbf{R}^t & \mathbf{R}\\ \mathbf{R}^t & \mathbf{I}\end{bmatrix}.$$

Algorithm 1: GSAMP.CUT(v, σ)

Input: $v \in \mathbb{Z}_q$, $\sigma \in \mathbb{R}^+$
Output: $\mathbf{z} \in \mathbb{Z}^{k-l}$

Sample $\mathbf{x} \in \mathbb{Z}^k$ from $D_{\Lambda_v^\perp(\mathbf{g}^t),\sigma}$
Let $\mathbf{z}$ be the last $k-l$ entries of $\mathbf{x}$
return $\mathbf{z}$.

Algorithm 2: APPROX.TRAPGEN$_\chi$

Input: Security parameter λ
Output: matrix-approximate trapdoor pair $(\mathbf{A}, \mathbf{R})$.

Sample a uniformly random $\hat{\mathbf{A}} \leftarrow U(\mathbb{Z}_q^{n \times n})$.
Let $\bar{\mathbf{A}} := [\mathbf{I}_n, \hat{\mathbf{A}}]$.
Sample the approximate trapdoor $\mathbf{R} \leftarrow \chi^{2n \times w}$.
Form $\mathbf{A} := [\bar{\mathbf{A}} | \mathbf{F} - \bar{\mathbf{A}}\mathbf{R}] \in \mathbb{Z}_q^{n \times m}$.
return $(\mathbf{A}, \mathbf{R})$.

Algorithm 3: APPROX.SAMPLEPRE.

Input: $(\mathbf{A}, \mathbf{R}, \mathbf{u}, s)$ as in Thm. 4.
Output: An approximate preimage of $\mathbf{u}$ for $\mathbf{A}$, $\mathbf{y} \in \mathbb{Z}^m$.

Sample a perturbation $\mathbf{p} \leftarrow D_{\mathbb{Z}^m, \sqrt{\Sigma_p}}$.
Form $\mathbf{v} = \mathbf{u} - \mathbf{A}\mathbf{p} \in \mathbb{Z}_q^n$.
Sample the approximate gadget preimage $\mathbf{z} \in \mathbb{Z}^{n(k-l)}$ as $\mathbf{z} \leftarrow$ GSAMP.CUT$(\mathbf{v}, \sigma)$.
Form $\mathbf{y} := \mathbf{p} + \begin{bmatrix} \mathbf{R} \\ \mathbf{I} \end{bmatrix} \mathbf{z} \in \mathbb{Z}^m$.
return $\mathbf{y}$.

Fig. 3. Pseudocode for the approximate trapdoor sampling algorithm in Subsect. 4.3. We abuse notation and let GSAMP.CUT$(\mathbf{v}, \sigma)$ denote n independent calls to Algorithm 1 on each entries of $\mathbf{v} \in \mathbb{Z}_q^n$, and then concatenate the output vectors. The distribution $\chi \subseteq \mathbb{Z}$ is chosen so that $\mathsf{LWE}_{n,n,q,\chi,U(\mathbb{Z}_q),\chi}$ is hard.

The results of this section are summarized in the following theorem.

Theorem 4. *There exists probabilistic, polynomial time algorithms* APPROX.TRAPGEN$(\cdot)$ *and* APPROX.SAMPLEPRE$(\cdot,\cdot,\cdot,\cdot)$ *satisfying the following.*

1. APPROX.TRAPGEN(n) *takes as input a security parameter n and returns a matrix-approximate trapdoor pair* $(\mathbf{A}, \mathbf{R}) \in \mathbb{Z}_q^{n \times m} \times \mathbb{Z}^{2n \times n(k-l)}$.
2. *Let* $\mathbf{A}$ *be generated with an approximate trapdoor as above and let* APPROX.$\mathbf{A}^{-1}(\cdot)$ *denote the approximate preimage sampling algorithm,* APPROX.SAMPLEPRE$(\mathbf{A}, \mathbf{R}, s, \cdot)$. *The following two distributions are statistically indistinguishable:*

$$\{(\mathbf{A}, \mathbf{y}, \mathbf{u}, \mathbf{e}) : \quad \mathbf{u} \leftarrow U(\mathbb{Z}_q^n), \quad \mathbf{y} \leftarrow \text{APPROX}.\mathbf{A}^{-1}(\mathbf{u}), \quad \mathbf{e} = \mathbf{u} - \mathbf{A}\mathbf{y} \mod q\}$$

and

$$\{(\mathbf{A}, \mathbf{y}, \mathbf{u}, \mathbf{e}) : \mathbf{y} \leftarrow D_{\mathbb{Z}^m, s}, \mathbf{e} \leftarrow D_{\mathbb{Z}^n, \sigma\sqrt{(b^{2l}-1)/(b^2-1)}} \mod q, \mathbf{u} = \mathbf{A}\mathbf{y} + \mathbf{e} \mod q\}$$

for any $\sigma \geq \sqrt{b^2+1} \cdot \omega(\sqrt{\log n})$ *and* $s \gtrsim \sqrt{b^2+1}\frac{s_1^2(\mathbf{R})}{s_{2n}(\mathbf{R})}\eta_\epsilon(\mathbb{Z}^{nk})$[1]. *Furthermore, in the second distribution,* $\mathbf{A}$ *is computationally indistinguishable from random assuming* $\mathsf{LWE}_{n,n,q,\chi,U(\mathbb{Z}_q),\chi}$.

4.3 Simulate the Preimage and Error Distributions

This subsection is dedicated to proving Theorem 4. For the convenience of explanation, in this subsection we redefine the gadget $\mathbf{G}$ by permuting the columns so that the columns of smaller entries are all on the left, i.e.

$$\mathbf{G} := [\mathbf{M}|\mathbf{F}] := [\mathbf{I}_n \otimes (1, b, \ldots, b^{l-1})|\mathbf{F}]$$

Let $\mathbf{x} = (\mathbf{x}_1, \mathbf{x}_2) \in \mathbb{Z}^{nl} \times \mathbb{Z}^{n(k-l)}$ denote the short preimage of $\mathbf{v} := \mathbf{u} - \mathbf{Ap} \pmod q$ under the full gadget matrix $\mathbf{G}$, i.e. $\mathbf{Gx} = \mathbf{v} \pmod q$.

The first attempt of proving Theorem 4 is to first show that the joint distribution of $(\mathbf{p}, \mathbf{x})$ produced in Algorithm 3 is statistically close to $D_{\Lambda_{\mathbf{u}}^{\perp}[\mathbf{A},\mathbf{G}],\sqrt{\Sigma_p \oplus \sigma^2 \mathbf{I}_{nk}}}$ for *any* $\mathbf{u} \in \mathbb{Z}_q^n$, then apply the linear transformation theorem on $(\mathbf{p}, \mathbf{x})$ to obtain the distributions of the preimage $\mathbf{y}$ and the error term $\mathbf{e}$. However, applying the linear transformation theorem directly on the lattice coset $\Lambda_{\mathbf{u}}^{\perp}[\mathbf{A}, \mathbf{G}]$ leads to a technical problem. That is, the intermediate lattice intersections $\Lambda_{\mathbf{T}}$ required in Theorem 1 have large smoothing parameters, which means even if we go through that route, the Gaussian width of the resulting preimage would blow up significantly.

Instead, we work only with a uniformly random target $\mathbf{u}$ instead of an arbitrary target, and directly construct the simulation algorithm. We show that if the simulation algorithm produces $(\mathbf{p}, \mathbf{x}) \leftarrow D_{\mathbb{Z}^{m+nk},\sqrt{\Sigma_p \oplus \sigma^2 \mathbf{I}_{nk}}}$, then it is able to simulate the distributions of $\mathbf{y}$ and $\mathbf{e}$ correctly without using the trapdoor. Now the support of $(\mathbf{p}, \mathbf{x})$ is the integer lattice $\mathbb{Z}^{m+nk}$. Working with the integer lattice is important for two reasons. First, it allows us to treat $\mathbf{x}_1$ and $\mathbf{x}_2$ as statistically independent samples; and second, it gives us short vectors in the kernels summoned when using Lemma 4 or Theorem 1.

Formally, let $\varepsilon = \mathrm{negl}(\lambda) > 0$. We first prove three lemmas.

Lemma 11. *For any* $\sigma \geq \eta_\varepsilon(\Lambda^{\perp}(\mathbf{G}))$, *the following two distributions are statistically close.*

1. *First sample* $\mathbf{v} \leftarrow U(\mathbb{Z}_q^n)$, *then sample* $\mathbf{x} \leftarrow D_{\Lambda_{\mathbf{v}}^{\perp}(\mathbf{G}),\sigma}$, *output* $(\mathbf{x}, \mathbf{v})$;
2. *First sample* $\mathbf{x} \leftarrow D_{\mathbb{Z}^{nk},\sigma}$, *then compute* $\mathbf{v} = \mathbf{Gx} \pmod q$, *output* $(\mathbf{x}, \mathbf{v})$.

Proof. The proof follows directly from $\det(\Lambda_q^{\perp}(\mathbf{G})) = q^n$ and Corollary 1. Alternatively, one can use two applications of the fact $\rho_r(\Gamma + \mathbf{c}) \in (1 \pm \varepsilon)\sigma^n/\det(\Gamma)$ for any $r \geq \eta_\varepsilon(\Gamma)$. The latter yields $\Pr\{\text{Process returns } \mathbf{x}\} \in \left(\frac{1-\varepsilon}{1+\varepsilon}, \frac{1+\varepsilon}{1-\varepsilon}\right) \cdot D_{\mathbb{Z}^{nk},\sigma}(\mathbf{x})$.

[1] We remark that the ratio $\frac{s_1(\mathbf{R})}{s_{2n}(\mathbf{R})}$ is a small constant for commonly-used subgaussian distributions for $\mathbf{R}$'s entries [51].

Lemma 12. *The following random processes are statistically close for any* $\sigma \geq \sqrt{b^2+1} \cdot \omega(\sqrt{\log n}) \geq \eta_\varepsilon(\mathbf{g}^t)$*: sample* $\mathbf{x}_1 \leftarrow D_{\mathbb{Z}^l,\sigma}$ *and return* $e = [1, b, \ldots, b^{l-1}]\mathbf{x}_1$*; or, return* $e \leftarrow D_{\mathbb{Z},\sigma\sqrt{(b^{2l}-1)/(b^2-1)}}$.

Proof. We use Lemma 4 or Theorem 1 where $[1, b, \ldots, b^{l-1}]$ is the linear transformation. Notice that the kernel of $[1, b, \ldots, b^{l-1}]$ is the linear span of $[\mathbf{b}_1, \ldots, \mathbf{b}_{l-1}]$ where

$$\mathbf{b}_1 = (b, -1, 0, \ldots, 0), \mathbf{b}_2 = (0, b, -1, 0, \ldots, 0), \ldots, \mathbf{b}_{l-1} = (0, \ldots, 0, b, -1) \in \mathbb{Z}^l.$$

The support of $\mathbf{x}_1$, $\mathbb{Z}^l$, contains the $(l-1)$-dimensional lattice, $\Gamma = \mathbb{Z}^l \cap \mathrm{Ker}([1, b, \ldots, b^{l-1}])$, spanned by $[\mathbf{b}_1, \ldots, \mathbf{b}_{l-1}]$. Further, $\sigma \geq \eta_\epsilon(\mathbf{g}^t)$ implies σ is larger than the smoothing parameter of Γ since $\|\mathbf{b}_i\| \leq \sqrt{b^2+1}$ for $i = 1, \ldots, l-1$. Finally by routine calculation on the Gaussian width (and support), we have $e = [1, b, \ldots, b^{l-1}]\mathbf{x}_1 \approx_s D_{\mathbb{Z},\sigma\sqrt{(b^{2l}-1)/(b^2-1)}}$.

Let $\mathbf{R}' := \begin{bmatrix} \mathbf{R} \\ \mathbf{I}_{n(k-l)} \end{bmatrix}$. Next, we analyze the distribution given by the linear transformation representing the convolution step:

$$\mathbf{y} = \mathbf{p} + \mathbf{R}'\mathbf{x}_2 = [\mathbf{I}_m|\mathbf{R}'] \begin{pmatrix} \mathbf{p} \\ \mathbf{x}_2 \end{pmatrix}$$

for $(\mathbf{p}, \mathbf{x}_2) \leftarrow D_{\mathbb{Z}^{m+n(k-l)}, \sqrt{\Sigma_p \oplus \sigma^2 \mathbf{I}_{n(k-l)}}}$. Let $\mathbf{L} := [\mathbf{I}_m|\mathbf{R}']$ in Lemma 13 and its proof below.

Lemma 13. *For* $\sqrt{\Sigma_p \oplus \sigma^2 \mathbf{I}_{n(k-l)}} \geq \eta_\varepsilon\left(\Lambda\begin{pmatrix} \mathbf{R}' \\ -\mathbf{I}_{n(k-l)} \end{pmatrix}\right)$, $\mathbf{L}D_{\mathbb{Z}^{m+n(k-l)}, \sqrt{\Sigma_p \oplus \sigma^2 \mathbf{I}_{n(k-l)}}}$ *is statistically close to* $D_{\mathbb{Z}^m,s}$*. Further,* $\sqrt{\Sigma_p \oplus \sigma^2 \mathbf{I}_{n(k-l)}} \geq \eta_\varepsilon\left(\Lambda\begin{pmatrix} \mathbf{R}' \\ -\mathbf{I}_{n(k-l)} \end{pmatrix}\right)$ *is satisfied when* $s \gtrsim \sqrt{b^2+1}\frac{s_1^2(\mathbf{R})}{s_{2n}(\mathbf{R})}\eta_\epsilon(\mathbb{Z}^{nk})$.

Proof. The range and covariance are immediate. Next, we use Theorem 1. The kernel of $\mathbf{L}$ is given by all vectors $(\mathbf{a}, \mathbf{b})$ where $\mathbf{b} \in \mathbb{R}^{n(k-l)}$ and $\mathbf{a} = -\mathbf{R}'\mathbf{b}$. The integer lattice $\mathbb{Z}^{m+n(k-l)}$ contains all such integer vectors so $\Lambda_{\mathbf{L}} := \mathbb{Z}^{m+n(k-l)} \cap \ker(\mathbf{L})$ spans $\mathbf{L}$'s kernel. So $\begin{pmatrix} \mathbf{R}' \\ -\mathbf{I}_{n(k-l)} \end{pmatrix}$ is a basis of $\Lambda_{\mathbf{L}}$. Given that $\sqrt{\Sigma_p \oplus \sigma^2 \mathbf{I}_{n(k-l)}} \geq \eta_\varepsilon\left(\Lambda\begin{pmatrix} \mathbf{R}' \\ -\mathbf{I}_{n(k-l)} \end{pmatrix}\right)$, the lemma follows Theorem 1. Lastly, the implication that $\sqrt{\Sigma_p \oplus \sigma^2 \mathbf{I}_{n(k-l)}} \geq \eta_\varepsilon\left(\Lambda\begin{pmatrix} \mathbf{R}' \\ -\mathbf{I}_{n(k-l)} \end{pmatrix}\right)$ whenever $s \gtrsim \sqrt{b^2+1}\frac{s_1^2(\mathbf{R})}{s_{2n}(\mathbf{R})}\eta_\epsilon(\mathbb{Z}^{nk})$ is proved in Appendix A.

We are now ready to prove Theorem 4.

Proof. (of Theorem 4) The proof's overview is given via the following. Let

- $\mathbf{p} \leftarrow D_{\mathbb{Z}^m, \sqrt{\Sigma_p}}$ be a perturbation,
- $\mathbf{u} \in \mathbb{Z}_q^n$ be the input target coset,
- $\mathbf{v} = \mathbf{u} - \mathbf{Ap} \in \mathbb{Z}_q^n$ be the G-lattice coset,
- $\mathbf{x} = (\mathbf{x}_1, \mathbf{x}_2) \leftarrow D_{\mathbb{Z}^{nk}, \sigma}$ (G-lattice randomized over uniform coset $\mathbf{v}$ and $\sigma \geq \eta_\epsilon(\mathbf{g}^t)$, Lemma 11)
- $\mathbf{e} \leftarrow D_{\mathbb{Z}^n, \sigma\sqrt{(b^{2l}-1)/(b^2-1)}}$ be the concatenation of the errors, e, in Lemma 12,
- and $\mathbf{y} \leftarrow D_{\mathbb{Z}^m, s}$ as in Lemma 13.

The proof is best summarized via the sequence of hybrids below:

$$\begin{aligned}
\mathbf{u} &= \mathbf{v} + \mathbf{Ap} \\
&\approx_s \mathbf{Gx} + \mathbf{Ap} \\
&= \mathbf{Mx}_1 + \mathbf{Fx}_2 + \mathbf{Ap} \\
&\approx_s \mathbf{e} + \mathbf{Fx}_2 + \mathbf{Ap} \\
&= \mathbf{e} + \mathbf{AR}'\mathbf{x}_2 + \mathbf{Ap} \\
&= \mathbf{e} + \mathbf{AL}\begin{pmatrix}\mathbf{p} \\ \mathbf{x}_2\end{pmatrix} \\
&\approx_s \mathbf{e} + \mathbf{Ay}.
\end{aligned}$$

The first $\approx_s$ is through swapping the order of sampling $\mathbf{u}$ and $\mathbf{v}$ uniformly at random, then using the fact that $\sigma \geq \eta_\epsilon(\mathbf{G})$ (Lemma 11). The next $\approx_s$ is given by Lemma 12. Finally, the last $\approx_s$ is given by concatenating $(\mathbf{p}, \mathbf{x}_2) \leftarrow D_{\mathbb{Z}^{m+n(k-l)}, \sqrt{\Sigma_p \oplus \sigma^2 \mathbf{I}_{n(k-l)}}}$ and using Lemma 13.

We remark that the key in the equivalences above is that we can separate $\mathbf{x}$ into two statistically independent samples, $\mathbf{x}_1$ and $\mathbf{x}_2$, concatenate $\mathbf{p}$ and $\mathbf{x}_2$, then perform two instances of Theorem 1 (Lemma 4) on the statistically independent samples $\mathbf{L}(\mathbf{p}, \mathbf{x}_2)$ and $\mathbf{Mx}_1$. The statistical independence of $\mathbf{x}_1$ and $\mathbf{x}_2$ is due to the orthogonality of $\mathbb{Z}^{nk}$ and the same cannot be said if $\mathbf{x} \sim D_{\Lambda_{\mathbf{v}}^{\perp}(\mathbf{G}), \sigma}$ for a fixed $\mathbf{v}$ (via a fixed $\mathbf{u}$). This difference highlights why we must argue security for a uniformly random input coset $\mathbf{u}$ (and $\mathbf{v}$).

Real distribution: The real distribution of $\{(\mathbf{A}, \mathbf{y}, \mathbf{u}, \mathbf{e})\}$ is:
$\mathbf{A}$, $\mathbf{u} \leftarrow U(\mathbb{Z}_q^n)$, $\mathbf{p} \leftarrow D_{\mathbb{Z}^m, \sqrt{\Sigma_p}}$, $\mathbf{v} := \mathbf{u} - \mathbf{Ap}$, $\mathbf{x} = (\mathbf{x}_1, \mathbf{x}_2) \leftarrow D_{\Lambda_{\mathbf{v}}^{\perp}(\mathbf{G}), \sigma}$, $\mathbf{e} = \mathbf{Mx}_1$, and $\mathbf{y} = \mathbf{L}(\mathbf{p}, \mathbf{x}_2)$.

Hybrid 1: Here we swap the order of sampling $\mathbf{u}$ and $\mathbf{v}$. Let $\mathbf{v} \leftarrow U(\mathbb{Z}_q^n)$, $\mathbf{p} \leftarrow D_{\mathbb{Z}^m, \sqrt{\Sigma_p}}$, $\mathbf{u} = \mathbf{v} + \mathbf{Ap}$. We keep $\mathbf{x}, \mathbf{e}$, and $\mathbf{y}$ unchanged: $\mathbf{x} = (\mathbf{x}_1, \mathbf{x}_2) \leftarrow D_{\Lambda_{\mathbf{v}}^{\perp}(\mathbf{G}), \sigma}$, $\mathbf{e} = \mathbf{Mx}_1$, and $\mathbf{y} = \mathbf{L}(\mathbf{p}, \mathbf{x}_2)$. Then, the real distribution and Hybrid 1 are the same.

Hybrid 2: Instead of sampling a uniform $\mathbf{v} \in \mathbb{Z}_q^n$ and a G-lattice sample $\mathbf{x} = (\mathbf{x}_1, \mathbf{x}_2) \leftarrow D_{\Lambda_{\mathbf{v}}^{\perp}(\mathbf{G}), \sigma}$, we sample $\mathbf{x} \leftarrow D_{\mathbb{Z}^{nk}, \sigma}$ and let $\mathbf{v} = \mathbf{Gx} \in \mathbb{Z}_q^n$. The rest remains the same:

$\mathbf{A}$, $\mathbf{x} \leftarrow D_{\mathbb{Z}^{nk},\sigma}$, $\mathbf{v} = \mathbf{G}\mathbf{x}$, $\mathbf{p} \leftarrow D_{\mathbb{Z}^m,\sqrt{\Sigma_p}}$, $\mathbf{u} = \mathbf{v} + \mathbf{A}\mathbf{p}$, $\mathbf{e} = \mathbf{M}\mathbf{x}_1$, and $\mathbf{y} = \mathbf{L}(\mathbf{p}, \mathbf{x}_2)$. Lemma 11 implies Hybrid 1 and Hybrid 2 are statistically close.

Hybrid 3: We combine $\mathbf{p}, \mathbf{x}_2$ into the joint distribution $(\mathbf{p}, \mathbf{x}_2) \leftarrow D_{\mathbb{Z}^{m+n(k-l)},\sqrt{\Sigma_p \oplus \sigma^2 \mathbf{I}}}$:
$\mathbf{A}$, $(\mathbf{p}, \mathbf{x}_2) \leftarrow D_{\mathbb{Z}^{m+n(k-l)},\sqrt{\Sigma_p \oplus \sigma^2 \mathbf{I}}}$, $\mathbf{e} = \mathbf{M}\mathbf{x}_1$, $\mathbf{y} = \mathbf{L}(\mathbf{p}, \mathbf{x}_2)$, $\mathbf{v} = \mathbf{G}\mathbf{x}$, and $\mathbf{u} = \mathbf{v} + \mathbf{A}\mathbf{p}$.

Hybrid 4: Here we apply the linear transformation theorem on $\mathbf{L}$ and $\mathbf{M}$.
$\mathbf{A}$, $\mathbf{e} \leftarrow D_{\mathbb{Z}^{nl},\sigma\sqrt{(b^{2l}-1)/(b^2-1)}}$, $\mathbf{y} \leftarrow D_{\mathbb{Z}^m,s}$, $\mathbf{v} = \mathbf{A}\mathbf{y} + \mathbf{e}$.
Lemmas 12 and 13 imply Hybrids 3 and 4 are statistically close.

Final distribution: Sample $\mathbf{A} \leftarrow U(\mathbb{Z}_q^{n\times m})$ and keep the rest of the vectors from the same distribution as Hybrid 4 (notice that the trapdoor $\mathbf{R}$ of $\mathbf{A}$ is not used to sample $\mathbf{p}$, $\mathbf{x}$, $\mathbf{e}$ and $\mathbf{y}$). The final distribution is computationally indistinguishable from Hybrid 4 assuming $\mathsf{LWE}_{n,n,q,\chi,U(\mathbb{Z}_q),\chi}$.

5 Hash-and-Sign Signature Instantiated with the Approximate Trapdoor

We spell out the details of the hash-and-sign signature scheme from [31] instantiated with the approximate G-trapdoor instead of an exact trapdoor.

Recall the parameters from the last section. We set $k = \lceil \log_b q \rceil$, set l to be the number of entries dropped from the G-trapdoor such that $1 \leq l < k$ and $m = n(2 + (k - l))$. Let $\sigma, s \in \mathbb{R}^+$ be the discrete Gaussian widths of the distributions over the cosets of $\Lambda_q^\perp(\mathbf{G})$ and $\Lambda_q^\perp(\mathbf{A})$ respectively. Let χ be the distribution of the entries of the trapdoor $\mathbf{R}$ chosen so that $\mathsf{LWE}_{n,n,q,\chi,U(\mathbb{Z}_q),\chi}$ is hard.

Construction 5. *Given an approximate trapdoor sampler from Theorem 4, a hash function $H = \{H_\lambda : \{0,1\}^* \to R_\lambda\}$ modeled as a random oracle, we build a signature scheme as follows.*

- $\mathsf{Gen}(1^\lambda)$: *The key-generation algorithm samples $\mathbf{A} \in \mathbb{Z}_q^{n\times m}$ together with its (α, β)-approximate trapdoor $\mathbf{R}$ from* Approx.TrapGen(1^λ). *Let the range R_λ of H be $\mathbb{Z}_q^n$. It outputs $\mathbf{A}$ as the verification key, keeps $\mathbf{R}$ as the secret signing key.*
- $\mathsf{Sig}(\mathbf{R}, m)$: *The signing algorithm checks if the message-signature pair $(m, \mathbf{x}_m)$ has been produced before. If so, it outputs $\mathbf{x}_m$ as the signature of m; if not, computes $\mathbf{u} = H(m)$, and samples an approximate preimage $\mathbf{x}_m \leftarrow$* Approx.SamplePre$(\mathbf{A}, \mathbf{R}, \mathbf{u}, s)$. *It outputs $\mathbf{x}_m$ as the signature and stores $(m, \mathbf{x}_m)$ in the list.*
- $\mathsf{Ver}(\mathbf{A}, m, \mathbf{x})$: *The verification algorithm checks if $\|\mathbf{x}\| \leq \beta$ and $\|\mathbf{A} \cdot \mathbf{x} - H(m)\| \leq \alpha$. If so, it outputs accept; otherwise, it outputs reject.*

5.1 Security Analysis

In the security analysis we use the following property on the distributions produced by APPROX.SAMPLEPRE proven in Theorem 4. That is, the preimage and error term for a random target can be simulated from distributions denoted by D_{pre} and D_{err}. Both of them are independent of the public key $\mathbf{A}$ and the secret key $\mathbf{R}$.

To prove that the signature satisfies the strong EU-CMA security, we need an additional "near-collision-resistance" property for Ajtai's function, which can be based on the standard SIS assumption. Let us remark that without this property, we can still prove the signature scheme satisfies static security based on the hardness of the approximate ISIS problem, which is tighter by a factor of two according to Theorem 3.

Lemma 14 (The near-collision-resistance of Ajtai's function). *For any $n, m, q \in \mathbb{N}$ and $\alpha, \beta \in \mathbb{R}$. If there is an efficient adversary A that given $\mathbf{A} \leftarrow U(\mathbb{Z}_q^{n\times m})$, finds $\mathbf{x}_1 \neq \mathbf{x}_2 \in \mathbb{Z}^m$ such that*

$$\|\mathbf{x}_1\| \leq \beta \quad \text{and} \quad \|\mathbf{x}_2\| \leq \beta \quad \text{and} \quad \|\mathbf{A}\mathbf{x}_1 - \mathbf{A}\mathbf{x}_2 \pmod q\| \leq 2\alpha$$

Then there is an efficient adversary B that solves $\mathsf{SIS}_{n,n+m,q,2(\alpha+\beta)}$.

Proof Suppose B gets an $\mathsf{HNF.SIS}_{n,n+m,q,2(\alpha+\beta)}$ challenge (which is as hard as $\mathsf{SIS}_{n,n+m,q,2(\alpha+\beta)}$) with the public matrix $[\mathbf{I}_n \mid \mathbf{A}]$, B sends $\mathbf{A}$ to A, gets back $\mathbf{x}_1 \neq \mathbf{x}_2 \in \mathbb{Z}^m$ such that

$$\|\mathbf{x}_1\| \leq \beta \quad \text{and} \quad \|\mathbf{x}_2\| \leq \beta \quad \text{and} \quad \|\mathbf{y} := \mathbf{A}\mathbf{x}_1 - \mathbf{A}\mathbf{x}_2 \pmod q\| \leq 2\alpha$$

B then sets $\mathbf{z} := [-\mathbf{y}^t \mid (\mathbf{x}_1 - \mathbf{x}_2)^t]^t$ as the solution. $\mathbf{z}$ is then non-zero and satisfies $\|\mathbf{z}\| \leq 2(\alpha+\beta)$ and $[\mathbf{I}_n \mid \mathbf{A}]\mathbf{z} = \mathbf{0} \pmod q$.

Theorem 6. *Construction 5 is strongly existentially unforgeable under a chosen-message attack in the random oracle model assuming the hardness of $\mathsf{SIS}_{n,n+m,q,2(\alpha+\beta)}$ and $\mathsf{LWE}_{n,n,q,\chi,U(\mathbb{Z}_q),\chi}$.*

Proof. Suppose there is a polynomial time adversary A that breaks the strong EU-CMA of the signature scheme, we construct a polynomial time adversary B that breaks the near-collision-resistance of Ajtai's function, which is as hard as $\mathsf{SIS}_{n,n+m,q,2(\alpha+\beta)}$ due to Lemma 14.

To start, B sends Ajtai's function $\mathbf{A}$ to A as the public key for the signature scheme. Once A makes a random oracle query w.r.t. a message m, B samples $\mathbf{x} \leftarrow D_{\mathrm{pre}}$, computes $\mathbf{u} := \mathbf{A}\mathbf{x} + D_{\mathrm{err}} \pmod q$ as the random oracle response on m. B then replies $\mathbf{u}$ to A and stores $(m, \mathbf{u})$ in the random oracle storage, $(m, \mathbf{x})$ in the message-signature pair storage. Once A makes a signing query on the message m (wlog assume m has been queried to the random oracle before, since if not B can query it now), B finds $(m, \mathbf{x})$ in the storage and reply $\mathbf{x}$ as the signature. The signatures and the hash outputs produced by B are indistinguishable from the real ones due to the properties of the distributions D_{pre} and D_{err}, and the assumption that a real public key is indistinguishable from random under $\mathsf{LWE}_{n,n,q,\chi,U(\mathbb{Z}_q),\chi}$.

Without loss of generality, assume that before A tries to forge a signature on m^*, A has queried H on m^*. Denote the pair that B prepares and stores in the random oracle storage as $(m^*, \mathbf{u}^*)$, and the pair in the signature storage as $(m^*, \mathbf{x}^*)$. Finally A outputs $\mathbf{x}$ as the forged signature on m^*. So we have $\|\mathbf{A}(\mathbf{x} - \mathbf{x}^*) \pmod q\| \leq 2\alpha$. It remains to prove that $\mathbf{x} \neq \mathbf{x}^*$ so as to use them as a near-collision-pair. If m^* has been queried to the signing oracle before, then $\mathbf{x} \neq \mathbf{x}^*$ by the definition of a successful forgery; if m^* has not been queried to the signing oracle before, then $\mathbf{x}^*$ is with high min-entropy by the settings of the parameter, so $\mathbf{x} \neq \mathbf{x}^*$ with overwhelming probability.

5.2 Concrete Parameters

We provide a proof-of-concept implementation of the signature. Experiments are performed over several groups of parameters using different dimensions n, moduli q, bases b, targeting different security level (mainly around 80 to 90-bit and 170 to 185-bit security). In each group of parameters, we use fixed n, q, b, and compare the use of exact trapdoor (under our reference implementation) versus approximate trapdoor. In Figs. 4 and 5 we list 6 groups of parameters.

Params	Exact	Approx	Approx	Exact	Approx	Approx	Exact	Approx	Approx
n	512	512	512	512	512	512	512	512	512
$\lceil \log_2 q \rceil$	24	24	24	20	20	20	16	16	16
b	2	2	2	2	2	2	2	2	2
l	0	12	15	0	10	12	0	7	9
τ	40	40	40	10	10	10	2.6	2.6	2.6
s	38317.0	29615.3	26726.3	8946.4	6919.8	6416.4	2170.7	1756.3	1618.2
m	13312	7168	5632	11264	6144	5120	9216	5632	4608
$\|\mathbf{x}\|_2$	4441737.7	2521387.0	2035008.5	956758.1	545470.5	464022.0	211100.9	133305.5	109339.1
$\|\mathbf{x}\|_\infty$	184653	111909	94559	38507	25275	24762	8848	6853	6334
$\|\mathbf{z}\|_2$	0	374014.0	2118987.6	0	94916.6	343682.9	0	11897.9	46428.4
$\|\mathbf{z}\|_\infty$	0	46895	346439	0	13265	52789	0	1439	7213
PK	37.50	19.50	15.00	26.25	13.75	11.25	17.00	10.00	8.00
Sig	25.68	13.53	10.51	18.87	10.01	8.29	13.16	7.83	6.30
LWE	100.0	100.0	100.0	102.8	102.8	102.8	104.7	104.7	104.7
AISIS	80.2	85.8	81.1	82.0	87.5	84.3	83.7	89.0	88.1
δ	1.00685	1.00643	1.00678	1.00670	1.00631	1.00653	1.00658	1.00621	1.00628
k	174	193	177	180	199	188	186	204	201

Fig. 4. Summary of the concrete parameters, with base $b = 2$, aiming at around 80 to 90-bit security. The sizes of PK and Sig are measured in kB. τ is the Gaussian width of the secret matrix $\mathbf{R}$. s is the Gaussian width of the preimage. "LWE" refers to the security level of the pseudorandomness of the PK. "AISIS" refers to the security level of breaking approximate ISIS. δ and k are the variables used in the AISIS security estimation.

Methods for security estimation. Let us first explain how we make the security estimations. The concrete security estimation of lattice-based cryptographic primitive is a highly active research area and more sophisticated methods are proposed recently. Here we use relatively simple methods to estimate the

Params	Exact	Approx	Approx	Exact	Approx	Approx	Exact	Approx	Approx
n	512	512	512	1024	1024	1024	1024	1024	1024
$\lceil \log_2 q \rceil$	16	16	16	18	18	18	18	18	18
b	4	4	4	8	8	8	4	4	4
l	0	2	4	0	2	3	0	4	5
τ	2.6	2.6	2.6	2.8	2.8	2.8	2.8	2.8	2.8
s	3114.2	2833.3	2505.6	8861.1	7824.8	7227.9	5118.8	4297.8	4015.5
m	5120	4096	3072	8192	6144	5120	11264	7168	6144
$\|\mathbf{x}\|_2$	223740.1	183004.9	138145.7	805772.9	604711.5	516446.3	552713.4	369981.2	311153.9
$\|\mathbf{x}\|_\infty$	13320	11868	8948	35348	28823	30435	19274	18283	14927
$\|\mathbf{z}\|_2$	0	1402.3	19807.1	0	7316.5	54379.8	0	29958.0	115616.4
$\|\mathbf{z}\|_\infty$	0	174	2448	0	905	6680	0	3025	12070
PK	9.00	7.00	5.00	15.75	11.25	9.00	22.50	13.50	11.25
Sig	7.62	5.94	4.45	13.70	10.14	8.36	18.74	11.09	9.38
LWE	104.7	104.7	104.7	192.7	192.7	192.7	192.7	192.7	192.7
AISIS	82.8	85.5	87.8	165.3	172.9	174.9	175.8	185.7	183.7
δ	1.00664	1.00645	1.00629	1.0036	1.00347	1.00343	1.00342	1.00326	1.00329
k	183	192	200	462	488	495	498	532	525

Fig. 5. Summary of the concrete parameters, with base $b \geq 4$, aiming at around 80 to 90-bit and 170 to 184-bit security.

pseudorandomness of the public-key (henceforth "LWE security"), and the hardness of breaking approximate ISIS (henceforth "AISIS security"). Let us remark that our estimations may not reflect the state-of-art, but at least provide a fair comparison of the parameters for the exact trapdoor versus the approximate trapdoor.

LWE security depends on the choices of q, n, and the Gaussian width τ of the trapdoor $\mathbf{R}$. The estimation of LWE security was done with the online LWE bit security estimator with BKZ as the reduction model[2] [5].

For the approximate ISIS problem, the only direct cryptanalysis result we are aware of is the work of Bai et al. [12], but it is not clearly applicable to the parameters we are interested. Instead we estimate AISIS through $\mathsf{ISIS}_{n,m,q,\alpha+\beta}$ following the reduction in Lemma 8, where α and β are the upper-bounds of l_2 norm of the error $\mathbf{z}$ and preimage $\mathbf{x}$. We estimate the security level of $\mathsf{ISIS}_{n,m,q,\alpha+\beta}$ based on how many operations BKZ would take to find a vector in the lattice $\Lambda_q^\perp(\mathbf{A})$ of length $\alpha+\beta$. Further, we can throw away columns in $\mathbf{A}$. We choose to only use $2n$ columns of $\mathbf{A}$ as done in [14], denoted $\mathbf{A}_{2n}$, since Minkowski's theorem[3] tells us $\Lambda_q^\perp(\mathbf{A}_{2n})$ has a short enough vector. Following [5,7], we use sieving as the SVP oracle with time complexity $2^{.292k+16.4}$ in the block size, k. BKZ is expected to return a vector of length $\delta^{2n}\det^{1/2n}$ for a lattice of dimension $2n$. Hence, we found the smallest block size k achieving the needed δ corresponding to forging a signature, $\frac{\alpha+\beta}{\sqrt{q}} = \delta^{2n}$. Finally, we used the heuristic $\delta \approx (\frac{k}{2\pi e}(\pi k)^{1/k})^{1/2(k-1)}$ to determine the relation between k and δ, and we set the total time complexity of BKZ with block-size k, dimension $2n$ as

[2] https://bitbucket.org/malb/lwe-estimator.

[3] For any lattice $\mathbf{L}$, $\lambda_1 \leq \sqrt{r}\det(\mathbf{L})^{1/r}$ where r is the rank of the lattice.

$8 \cdot 2n \cdot \text{time}(SVP) = 8 \cdot 2n \cdot 2^{.292k+16.4}$ [7,22]. Here we use the "magic eight tour number" for BKZ to keep consistency with the LWE online estimator. We have not incorporated the more recent developments in [24] and [6] in the security estimation.

The comparison. For an estimation of 80-bit[4] security, our reference implementation for the exact trapdoor under the modulus $q \approx 2^{24}$ and base $b = 2$ matches the parameters reported in [13] (the parameters in the other implementation [14,36] are possibly measured in different ways). We also use smaller moduli and bigger bases to reduce the size and increase the security level. The parameters in Figs. 4 and 5 suggest that for all the choices of q and b, using the approximate gadget trapdoor by setting $l = \lceil (\log_b q)/2 \rceil$ saves about half of the sizes in the public key and signatures comparing to using the exact trapdoor, with even a slight increase in the security estimation.

Our implementation shows that the sizes of the public-key & signature can be reduced to 5 kB & 4.45 kB for an estimation of 88-bit security, and 11.25 kB & 9.38 kB for an estimation of 184-bit security. Those are still larger than, but much closer to the sizes for the signatures based on the rejection sampling approach [8,11,26,40]. As a reference, the sizes of the public-key & signature for qTESLA [8] are 4.03 kB & 3.05 kB for an estimation of 128-bit security, and 8.03 kB & 6.03 kB for an estimation of 192-bit security.

Acknowledgments. We are grateful to Daniele Micciancio for valuable advice and his generous sharing of ideas on the subject of this work. We would also like to thank Léo Ducas, Steven Galbraith, Thomas Prest, Yang Yu, Chuang Gao, Eamonn Postlethwaite, Chris Peikert, and the anonymous reviewers for their helpful suggestions and comments.

A The Smoothing Parameter of $\Lambda_{\mathbf{L}}$

Recall the notations that $\mathbf{R}' = \begin{bmatrix} \mathbf{R} \\ \mathbf{I}_{n(k-l)} \end{bmatrix} \in \mathbb{Z}^{m \times (n(k-l))}$, $\Sigma_p := s^2 \mathbf{I}_m - \mathbf{R}'(\mathbf{R}')^t$. Here we derive the conditions of s so that $\sqrt{\Sigma_p \oplus \sigma^2 \mathbf{I}_{n(k-l)}} \geq \eta_\epsilon(\Lambda_{\mathbf{L}})$ holds, where $\Lambda_{\mathbf{L}}$ is the lattice generated by

$$\mathbf{B} := \begin{bmatrix} -\mathbf{R}' \\ \mathbf{I}_{n(k-l)} \end{bmatrix}.$$

We do this in three steps: first we write out the dual basis of $\mathbf{B}$, then we reduce $\sqrt{\Sigma_p \oplus \sigma^2 \mathbf{I}_{n(k-l)}} \geq \eta_\epsilon(\Lambda_{\mathbf{L}})$ to a statement about the smoothing parameter of $\mathbb{Z}^{n(k-l)}$, and finally we find when $\sqrt{\Sigma_p \oplus \sigma^2 \mathbf{I}_{n(k-l)}} \geq \eta_\epsilon(\Lambda_{\mathbf{L}})$ as a function of s.

Dual basis, $\mathbf{B}^*$: Let $\Sigma = \Sigma_p \oplus \sigma^2 \mathbf{I}_{n(k-l)}$. By definition, we need $\rho(\sqrt{\Sigma}^t \Lambda_{\mathbf{L}}^*) \leq 1 + \epsilon$. In general, the dual basis Λ^* is generated by the dual basis $\mathbf{B}(\mathbf{B}^t\mathbf{B})^{-1}$. In the case of $\Lambda_{\mathbf{L}}$, we can write the dual basis as

[4] When one applies our security estimate methods to Table 1 of [13], one gets 82-bit security under the $\lambda = 97$, $n = 512$, $q = 2^{24}$ column.

$$\mathbf{B}^* := \begin{bmatrix} -\mathbf{R}' \\ \mathbf{I}_{n(k-l)} \end{bmatrix} \left[\mathbf{R}^t\mathbf{R} + 2\mathbf{I}\right]^{-1}.$$

Reducing to $\eta_\epsilon(\mathbb{Z}^{n(k-l)})$ **:** Next, the gaussian sum $\rho(\sqrt{\Sigma}^t \Lambda_{\mathbf{L}}^*)$ is equal to

$$\sum_{\mathbf{x}\in\mathbb{Z}^{n(k-l)}} \exp(-\pi\mathbf{x}^t(\mathbf{B}^*)^t\Sigma\mathbf{B}^*\mathbf{x}).$$

This reduces to showing $\sqrt{(\mathbf{B}^*)^t\Sigma\mathbf{B}^*} \geq \eta_\epsilon(\mathbb{Z}^{n(k-l)})$.

Now we write out the matrix product $(\mathbf{B}^*)^t\Sigma\mathbf{B}^*$,

$$\begin{aligned}(\mathbf{B}^*)^t\Sigma\mathbf{B}^* &= \left[\mathbf{R}^t\mathbf{R} + 2\mathbf{I}\right]^{-t} \left[-(\mathbf{R}')^t \ \mathbf{I}\right] \begin{bmatrix} \Sigma_p & \mathbf{0} \\ \mathbf{0} & \sigma^2\mathbf{I} \end{bmatrix} \begin{bmatrix} -\mathbf{R}' \\ \mathbf{I} \end{bmatrix} \left[\mathbf{R}^t\mathbf{R} + 2\mathbf{I}\right]^{-1} \\ &= \left[\mathbf{R}^t\mathbf{R} + 2\mathbf{I}\right]^{-t} \left[(\mathbf{R}')^t\Sigma_p\mathbf{R}' + \sigma^2\mathbf{I}\right] \left[\mathbf{R}^t\mathbf{R} + 2\mathbf{I}\right]^{-1}.\end{aligned}$$

Before we continue, we consider the structure of the middle matrix:

$$\begin{aligned}\Sigma_s := (\mathbf{R}')^t\Sigma_p\mathbf{R}' &= \left[\mathbf{R}^t \ \mathbf{I}\right] \left(s^2\mathbf{I} - \sigma^2 \begin{bmatrix} \mathbf{R} \\ \mathbf{I} \end{bmatrix} \left[\mathbf{R}^t \ \mathbf{I}\right]\right) \begin{bmatrix} \mathbf{R} \\ \mathbf{I} \end{bmatrix} \\ &= \left[\mathbf{R}^t\mathbf{R} + \mathbf{I}\right] \left(s^2\mathbf{I} - \sigma^2 \left[\mathbf{R}^t\mathbf{R} + \mathbf{I}\right]\right).\end{aligned}$$

Derive the condition for s**:** Now we will derive the condition for s so that

$$\left[\mathbf{R}^t\mathbf{R} + 2\mathbf{I}\right]^{-t} \left[\Sigma_s + \sigma^2\mathbf{I}\right] \left[\mathbf{R}^t\mathbf{R} + 2\mathbf{I}\right]^{-1} \geq \eta_\epsilon^2(\mathbb{Z}^{n(k-l)}).$$

Claim. All invertible matrices of the form $(\mathbf{R}^t\mathbf{R}+\alpha\mathbf{I})^i$ for $i \in \mathbb{Z}, \alpha \in \mathbb{R}$ commute.

Proof. Let $\mathbf{QSV}^t$ be $\mathbf{R}$'s singular value decomposition. Now, $\mathbf{R}^t\mathbf{R} + \alpha\mathbf{I} = \mathbf{VDV}^t + \mathbf{V}(\alpha\mathbf{I})\mathbf{V}^t$ where $\mathbf{D} = \mathbf{S}^t\mathbf{S} = \text{diag}(s_i^2(\mathbf{R}))$ since $\mathbf{V}, \mathbf{Q}$ are orthogonal. Equivalently, we have $\mathbf{R}^t\mathbf{R} + \alpha\mathbf{I} = \mathbf{VD}_\alpha\mathbf{V}^t$ where $\mathbf{D}_\alpha = \text{diag}(s_i^2(\mathbf{R}) + \alpha) = \mathbf{S}^t\mathbf{S} + \alpha\mathbf{I}_{2n}$. By induction, we have $(\mathbf{R}^t\mathbf{R}+\alpha\mathbf{I})^i = \mathbf{VD}_\alpha^i\mathbf{V}^t$, $i \in \mathbb{Z}$. Finally, $\mathbf{D}_\alpha^i$ is a diagonal matrix so $\mathbf{D}_\alpha^i$ and $\mathbf{D}_{\alpha'}^j$ commute for all α, α' since diagonal matrices commute. The result follows from the orthogonality of $\mathbf{V}$ ($\mathbf{V}^t\mathbf{V} = \mathbf{I}$).

Claim A allows us to lower-bound the smallest eigenvalue of

$$\begin{aligned}(\mathbf{B}^*)^t\Sigma\mathbf{B}^* &= \left[\mathbf{R}^t\mathbf{R} + 2\mathbf{I}\right]^{-2} \left(\left[\mathbf{R}^t\mathbf{R} + \mathbf{I}\right] \left[s^2\mathbf{I} - \sigma^2 \left[\mathbf{R}^t\mathbf{R} + \mathbf{I}\right]\right] + \sigma^2\mathbf{I}\right) \\ &= \left[\mathbf{R}^t\mathbf{R} + 2\mathbf{I}\right]^{-2} \left(s^2[\mathbf{R}^t\mathbf{R} + \mathbf{I}] - \sigma^2[2\mathbf{R}^t\mathbf{R} + (\mathbf{R}^t\mathbf{R})^2]\right).\end{aligned}$$

Viewing these matrices as their diagonal matrices of eigenvalues, we see $(\mathbf{B}^*)^t\Sigma\mathbf{B}^*$'s least eigenvalue is lower-bounded by

$$\lambda_{lb}(s, \mathbf{R}) := \frac{s^2(s_{2n}^2(\mathbf{R}) + 1) - \sigma^2(s_1^4(\mathbf{R}) + 2s_1^2(\mathbf{R}))}{(s_1^2(\mathbf{R}) + 2)^2}.$$

Next, we assume $\sigma = \sqrt{b^2 + 1}\eta_\epsilon(\mathbb{Z}^{nk}) \geq \eta_\epsilon(\Lambda_q^\perp(\mathbf{G}))$ and solve for s using $\lambda_{lb}(s, \mathbf{R}) \geq \eta_\epsilon^2(\mathbb{Z}^{n(k-l)})$,

$$s^2 \geq \frac{s_1^2(\mathbf{R}) + 1}{s_{2n}^2(\mathbf{R}) + 1}\eta_\epsilon^2(\mathbb{Z}^{n(k-l)}) + \frac{(b^2 + 1)(s_1^4(\mathbf{R}) + 2s_1^2(\mathbf{R}))}{s_{2n}^2(\mathbf{R}) + 1}\eta_\epsilon^2(\mathbb{Z}^{nk}).$$

This is

$$s \gtrsim \sqrt{b^2+1}\frac{s_1^2(\mathbf{R})}{s_{2n}(\mathbf{R})}\eta_\epsilon(\mathbb{Z}^{nk}).$$

We remark that the ratio $\frac{s_1(\mathbf{R})}{s_{2n}(\mathbf{R})}$ is a constant for commonly-used subgaussian distributions for **R**'s entries [51].

References

1. Agrawal, S., Boneh, D., Boyen, X.: Efficient lattice (H)IBE in the standard model. In: Gilbert, H. (ed.) EUROCRYPT 2010. LNCS, vol. 6110, pp. 553–572. Springer, Heidelberg (2010). https://doi.org/10.1007/978-3-642-13190-5_28
2. Ajtai, M.: Generating hard instances of lattice problems (extended abstract). In: STOC, pp. 99–108 (1996)
3. Ajtai, M.: Generating hard instances of the short basis problem. In: Wiedermann, J., van Emde Boas, P., Nielsen, M. (eds.) ICALP 1999. LNCS, vol. 1644, pp. 1–9. Springer, Heidelberg (1999). https://doi.org/10.1007/3-540-48523-6_1
4. Alagic, G., et al.: Status report on the first round of the NIST post-quantum cryptography standardization process. US Department of Commerce, National Institute of Standards and Technology (2019)
5. Albrecht, M.R., et al.: Estimate all the LWE, NTRU schemes!. In: Catalano, D., De Prisco, R. (eds.) SCN 2018. LNCS, vol. 11035, pp. 351–367. Springer, Cham (2018). https://doi.org/10.1007/978-3-319-98113-0_19
6. Albrecht, M.R., Ducas, L., Herold, G., Kirshanova, E., Postlethwaite, E.W., Stevens, M.: The general sieve kernel and new records in lattice reduction. In: Ishai, Y., Rijmen, V. (eds.) EUROCRYPT 2019. LNCS, vol. 11477, pp. 717–746. Springer, Cham (2019). https://doi.org/10.1007/978-3-030-17656-3_25
7. Albrecht, M.R., Player, R., Scott, S.: On the concrete hardness of learning with errors. J. Math. Cryptol. **9**(3), 169–203 (2015)
8. Alkim, E., Barreto, P.S.L.M., Bindel, N., Longa, P., Ricardini, J.E.: The lattice-based digital signature scheme qTESLA. IACR Cryptology ePrint Archive 2019, p. 85 (2019)
9. Alwen, J., Peikert, C.: Generating shorter bases for hard random lattices. Theory Comput. Syst. **48**(3), 535–553 (2011)
10. Applebaum, B., Cash, D., Peikert, C., Sahai, A.: Fast cryptographic primitives and circular-secure encryption based on hard learning problems. In: Halevi, S. (ed.) CRYPTO 2009. LNCS, vol. 5677, pp. 595–618. Springer, Heidelberg (2009). https://doi.org/10.1007/978-3-642-03356-8_35
11. Bai, S., Galbraith, S.D.: An improved compression technique for signatures based on learning with errors. In: Benaloh, J. (ed.) CT-RSA 2014. LNCS, vol. 8366, pp. 28–47. Springer, Cham (2014). https://doi.org/10.1007/978-3-319-04852-9_2
12. Bai, S., Galbraith, S.D., Li, L., Sheffield, D.: Improved combinatorial algorithms for the inhomogeneous short integer solution problem. J. Cryptol. **32**(1), 35–83 (2019)
13. El Bansarkhani, R., Buchmann, J.: Improvement and efficient implementation of a lattice-based signature scheme. In: Lange, T., Lauter, K., Lisoněk, P. (eds.) SAC 2013. LNCS, vol. 8282, pp. 48–67. Springer, Heidelberg (2014). https://doi.org/10.1007/978-3-662-43414-7_3

14. Bert, P., Fouque, P.-A., Roux-Langlois, A., Sabt, M.: Practical implementation of ring-SIS/LWE based signature and IBE. In: Lange, T., Steinwandt, R. (eds.) PQCrypto 2018. LNCS, vol. 10786, pp. 271–291. Springer, Cham (2018). https://doi.org/10.1007/978-3-319-79063-3_13
15. Bourse, F., Del Pino, R., Minelli, M., Wee, H.: FHE circuit privacy almost for free. In: Robshaw, M., Katz, J. (eds.) CRYPTO 2016. LNCS, vol. 9815, pp. 62–89. Springer, Heidelberg (2016). https://doi.org/10.1007/978-3-662-53008-5_3
16. Brakerski, Z., Langlois, A., Peikert, C., Regev, O., Stehlé, D.: Classical hardness of learning with errors. In: Proceedings of the Forty-Fifth Annual ACM Symposium on Theory of Computing, pp. 575–584. ACM (2013)
17. Brakerski, Z., Vaikuntanathan, V., Wee, H., Wichs, D.: Obfuscating conjunctions under entropic ring LWE. In: ITCS, pp. 147–156. ACM (2016)
18. Canetti, R., Chen, Y.: Constraint-hiding constrained PRFs for NC^1 from LWE. In: Coron, J.-S., Nielsen, J.B. (eds.) EUROCRYPT 2017. LNCS, vol. 10210, pp. 446–476. Springer, Cham (2017). https://doi.org/10.1007/978-3-319-56620-7_16
19. Cash, D., Hofheinz, D., Kiltz, E., Peikert, C.: Bonsai trees, or how to delegate a lattice basis. J. Cryptol. **25**(4), 601–639 (2012)
20. Chen, C., Genise, N., Micciancio, D., Polyakov, Y., Rohloff, K.: Implementing token-based obfuscation under (ring) LWE. IACR Cryptology ePrint Archive 2018, p. 1222 (2018)
21. Chen, Y., Vaikuntanathan, V., Wee, H.: GGH15 beyond permutation branching programs: proofs, attacks, and candidates. In: Shacham, H., Boldyreva, A. (eds.) CRYPTO 2018. LNCS, vol. 10992, pp. 577–607. Springer, Cham (2018). https://doi.org/10.1007/978-3-319-96881-0_20
22. Chen, Y.: Réduction de réseau et sécurité concréte du chiffrement complétement homomorphe. PhD thesis, Paris 7 (2013)
23. del Pino, R., Lyubashevsky, V., Seiler, G.: Lattice-based group signatures and zero-knowledge proofs of automorphism stability. In: Proceedings of the 2018 ACM SIGSAC Conference on Computer and Communications Security, CCS 2018, Toronto, ON, Canada, 15–19 October 2018, pp. 574–591 (2018)
24. Ducas, L.: Shortest vector from lattice sieving: a few dimensions for free. In: Nielsen, J.B., Rijmen, V. (eds.) EUROCRYPT 2018. LNCS, vol. 10820, pp. 125–145. Springer, Cham (2018). https://doi.org/10.1007/978-3-319-78381-9_5
25. Ducas, L., Galbraith, S., Prest, T., Yang, Y.: Integral matrix gram root and lattice Gaussian sampling without floats. IACR Cryptology ePrint Archive 2019, p. 320 (2019)
26. Ducas, L., et al.: CRYSTALS-Dilithium: a lattice-based digital signature scheme. IACR Trans. Cryptogr. Hardw. Embed. Syst. **2018**(1), 238–268 (2018)
27. Fouque, P.-A., et al.: Falcon: fast-fourier lattice-based compact signatures over NTRU (2018)
28. Genise, N., Micciancio, D.: Faster Gaussian sampling for trapdoor lattices with arbitrary modulus. In: Nielsen, J.B., Rijmen, V. (eds.) EUROCRYPT 2018. LNCS, vol. 10820, pp. 174–203. Springer, Cham (2018). https://doi.org/10.1007/978-3-319-78381-9_7
29. Gentry, C.: Fully homomorphic encryption using ideal lattices. In: STOC, pp. 169–178 (2009)
30. Gentry, C., Gorbunov, S., Halevi, S.: Graph-induced multilinear maps from lattices. In: Dodis, Y., Nielsen, J.B. (eds.) TCC 2015. LNCS, vol. 9015, pp. 498–527. Springer, Heidelberg (2015). https://doi.org/10.1007/978-3-662-46497-7_20
31. Gentry, C., Peikert, C., Vaikuntanathan, V.: Trapdoors for hard lattices and new cryptographic constructions. In: STOC, pp. 197–206 (2008)

32. Gentry, C., Szydlo, M.: Cryptanalysis of the revised NTRU signature scheme. In: Knudsen, L.R. (ed.) EUROCRYPT 2002. LNCS, vol. 2332, pp. 299–320. Springer, Heidelberg (2002). https://doi.org/10.1007/3-540-46035-7_20
33. Goldreich, O., Goldwasser, S., Halevi, S.: Public-key cryptosystems from lattice reduction problems. In: Kaliski, B.S. (ed.) CRYPTO 1997. LNCS, vol. 1294, pp. 112–131. Springer, Heidelberg (1997). https://doi.org/10.1007/BFb0052231
34. Gorbunov, S., Vaikuntanathan, V., Wee, H.: Attribute-based encryption for circuits. In: STOC, pp. 545–554. ACM (2013)
35. Goyal, R., Koppula, V., Waters, B.: Lockable obfuscation. In: FOCS, pp. 612–621. IEEE Computer Society (2017)
36. Gür, K.D., Polyakov, Y., Rohloff, K., Ryan, G.W., Savas, E.: Implementation and evaluation of improved gaussian sampling for lattice trapdoors. In: Proceedings of the 6th Workshop on Encrypted Computing and Applied Homomorphic Cryptography, pp. 61–71. ACM (2018)
37. Halevi, S., Halevi, T., Shoup, V., Stephens-Davidowitz, N.: Implementing BP-obfuscation using graph-induced encoding. In: ACM Conference on Computer and Communications Security, pp. 783–798. ACM (2017)
38. Hoffstein, J., Howgrave-Graham, N., Pipher, J., Silverman, J.H., Whyte, W.: NTRUSign: digital signatures using the NTRU lattice. In: Joye, M. (ed.) CT-RSA 2003. LNCS, vol. 2612, pp. 122–140. Springer, Heidelberg (2003). https://doi.org/10.1007/3-540-36563-X_9
39. Hoffstein, J., Pipher, J., Silverman, J.H.: NTRU: a ring-based public key cryptosystem. In: Buhler, J.P. (ed.) ANTS 1998. LNCS, vol. 1423, pp. 267–288. Springer, Heidelberg (1998). https://doi.org/10.1007/BFb0054868
40. Lyubashevsky, V.: Lattice signatures without trapdoors. In: Pointcheval, D., Johansson, T. (eds.) EUROCRYPT 2012. LNCS, vol. 7237, pp. 738–755. Springer, Heidelberg (2012). https://doi.org/10.1007/978-3-642-29011-4_43
41. Micciancio, D.: Personal communication (2018)
42. Micciancio, D., Peikert, C.: Trapdoors for lattices: simpler, tighter, faster, smaller. In: Pointcheval, D., Johansson, T. (eds.) EUROCRYPT 2012. LNCS, vol. 7237, pp. 700–718. Springer, Heidelberg (2012). https://doi.org/10.1007/978-3-642-29011-4_41
43. Micciancio, D., Peikert, C.: Hardness of SIS and LWE with small parameters. In: Canetti, R., Garay, J.A. (eds.) CRYPTO 2013. LNCS, vol. 8042, pp. 21–39. Springer, Heidelberg (2013). https://doi.org/10.1007/978-3-642-40041-4_2
44. Micciancio, D., Regev, O.: Worst-case to average-case reductions based on Gaussian measure. SIAM J. Comput. **37**(1), 267–302 (2007)
45. Micciancio, D., Walter, M.: On the bit security of cryptographic primitives. In: Nielsen, J.B., Rijmen, V. (eds.) EUROCRYPT 2018. LNCS, vol. 10820, pp. 3–28. Springer, Cham (2018). https://doi.org/10.1007/978-3-319-78381-9_1
46. Nguyen, P.Q., Regev, O.: Learning a parallelepiped: cryptanalysis of GGH and NTRU signatures. In: Vaudenay, S. (ed.) EUROCRYPT 2006. LNCS, vol. 4004, pp. 271–288. Springer, Heidelberg (2006). https://doi.org/10.1007/11761679_17
47. Peikert, C.: Public-key cryptosystems from the worst-case shortest vector problem: extended abstract. In: Proceedings of the 41st Annual ACM Symposium on Theory of Computing, STOC 2009, Bethesda, MD, USA, 31 May - 2 June 2009, pp. 333–342 (2009)
48. Peikert, C., Regev, O., Stephens-Davidowitz, N.: Pseudorandomness of ring-LWE for any ring and modulus. In: STOC, pp. 461–473. ACM (2017)

49. Peikert, C., Rosen, A.: Efficient collision-resistant hashing from worst-case assumptions on cyclic lattices. In: Halevi, S., Rabin, T. (eds.) TCC 2006. LNCS, vol. 3876, pp. 145–166. Springer, Heidelberg (2006). https://doi.org/10.1007/11681878_8
50. Regev, O.: On lattices, learning with errors, random linear codes, and cryptography. J. ACM **56**(6), 34 (2009)
51. Vershynin, R.: Introduction to the non-asymptotic analysis of random matrices. In: Compressed Sensing, pp. 210–268. Cambridge University Press (2012)
52. Wichs, D., Zirdelis, G.: Obfuscating compute-and-compare programs under LWE. In: FOCS, pp. 600–611. IEEE Computer Society (2017)

Decisional Second-Preimage Resistance: When Does SPR Imply PRE?

Daniel J. Bernstein[1,2](✉) and Andreas Hülsing[3](✉)

[1] Department of Computer Science, University of Illinois at Chicago, Chicago, IL 60607-7045, USA

[2] Horst Görtz Institute for IT Security, Ruhr University Bochum, Bochum, Germany
djb@cr.yp.to

[3] Department of Mathematics and Computer Science, Technische Universiteit Eindhoven, P.O. Box 513, 5600 MB Eindhoven, The Netherlands
andreas@huelsing.net

Abstract. There is a well-known gap between second-preimage resistance and preimage resistance for length-preserving hash functions. This paper introduces a simple concept that fills this gap. One consequence of this concept is that tight reductions can remove interactivity for multi-target length-preserving preimage problems, such as the problems that appear in analyzing hash-based signature systems. Previous reduction techniques applied to only a negligible fraction of all length-preserving hash functions, presumably excluding all off-the-shelf hash functions.

Keywords: Cryptographic hash functions · Preimage resistance · Second-preimage resistance · Provable security · Tight reductions · Multi-target attacks · Hash-based signatures

1 Introduction

Define $S : \{0,1\}^{256} \rightarrow \{0,1\}^{256}$ as the SHA-256 hash function restricted to 256-bit inputs. Does second-preimage resistance for S imply preimage resistance for S?

The classic Rogaway–Shrimpton paper "Cryptographic hash-function basics" [15] shows that second-preimage resistance tightly implies preimage resistance for an efficient hash function that maps fixed-length inputs to *much shorter* outputs. The idea of the proof is that one can find a second preimage of a

Author list in alphabetical order; see https://www.ams.org/profession/leaders/culture/CultureStatement04.pdf. This work was supported by the U.S. National Science Foundation under grant 1314919, by the Cisco University Research Program, and by DFG Cluster of Excellence 2092 "CASA: Cyber Security in the Age of Large-Scale Adversaries". "Any opinions, findings, and conclusions or recommendations expressed in this material are those of the author(s) and do not necessarily reflect the views of the National Science Foundation" (or other funding agencies). Permanent ID of this document: 36ecc3ad6d0fbbe65ce36226c2e3eb875351f326. Date: 2019.09.12.

S. D. Galbraith and S. Moriai (Eds.): ASIACRYPT 2019, LNCS 11923, pp. 33–62, 2019.
https://doi.org/10.1007/978-3-030-34618-8_2

random input x with high probability by finding a preimage of the hash of x. But this probability depends on the difference in lengths, and the proof breaks down for length-preserving hash functions such as S.

The same paper also argues that second-preimage resistance cannot imply preimage resistance for length-preserving hash functions. The argument, in a nutshell, is that the identity function from $\{0,1\}^{256}$ to $\{0,1\}^{256}$ provides unconditional second-preimage resistance—second preimages do not exist—even though preimages are trivial to find.

A counterargument is that this identity-function example says nothing about real hash functions such as S. The identity-function example shows that there cannot be a theorem that for *all* length-preserving hash functions proves preimage resistance from second-preimage resistance; but this is only the beginning of the analysis. The example does not rule out the possibility that second-preimage resistance, together with a mild additional assumption, implies preimage resistance.

1.1 Contributions of This Paper

We show that preimage resistance (PRE) follows tightly from the conjunction of second-preimage resistance (SPR) and decisional second-preimage resistance (DSPR). **Decisional second-preimage resistance** is a simple concept that we have not found in the literature: it means that the attacker has negligible advantage in deciding, given a random input x, whether x has a second preimage.

There is a subtlety in the definition of advantage here. For almost all length-preserving hash functions, always guessing that x *does* have a second preimage succeeds with probability approximately 63%. (See Sect. 3.) We define DSPR advantage as an increase in probability compared to this trivial attack.

We provide three forms of evidence that DSPR is a reasonable assumption. First, we show that DSPR holds for random functions even against quantum adversaries that get quantum access to a function. Specifically, a q-query quantum adversary has DSPR advantage at most $32q^2/2^n$ against an oracle for a uniform random hash function from $\{0,1\}^n$ to $\{0,1\}^n$. In [9] the same bound was shown for PRE and SPR together with matching attacks demonstrating the bounds are tight. This means that DSPR is at least as hard to break as PRE or SPR for uniform random hash functions from $\{0,1\}^n$ to $\{0,1\}^n$.

Second, the subtlety mentioned above means that DSPR, when generalized in the most natural way to m-bit-to-n-bit hash functions, becomes unconditionally provable when m is much larger than n. This gives a new proof of PRE from SPR, factoring the original proof by Rogaway and Shrimpton into two steps: first, prove DSPR when m is much larger than n; second, prove PRE from SPR and DSPR.

Third, we have considered ways to attack DSPR for real hash functions such as S, and have found nothing better than taking the time necessary to *reliably* compute preimages. A curious feature of DSPR is that there is no obvious way for a fast attack to achieve *any* advantage. A fast attack that occasionally finds a preimage of $H(x)$ will occasionally find a second preimage, but the baseline is already guessing that x has a second preimage; to do better than the baseline, one needs to have enough evidence to be reasonably confident that x does *not* have a second preimage. Formally, there exists a fast attack (in the non-uniform model) that achieves a nonzero advantage (by returning 0 if the input matches some no-second-preimage values built into the attack, and returning 1 otherwise), but we do not have a fast way to recognize this attack. See Sect. 2.3.

1.1.1 Multi-target Attacks. We see DSPR as showing how little needs to be assumed beyond SPR to obtain PRE. However, skeptics might object that SPR and DSPR are still two separate assumptions for cryptanalysts to study, that DSPR has received less study than PRE, and that DSPR could be easier to break than PRE, even assuming SPR. Why is assuming both SPR and DSPR, and deducing PRE, better than assuming both SPR and PRE, and ignoring DSPR? We give the following answer.

Consider the following simple interactive game T-openPRE. The attacker is given T targets $H(1, x_1), \ldots, H(T, x_T)$, where $x_1, \ldots, x_T$ are chosen independently and uniformly at random. The attacker is also given access to an "opening" oracle that, given i, returns x_i. The attacker's goal is to output (i, x') where $H(i, x') = H(i, x_i)$ and i was not an oracle query. Games of this type appear in, e.g., analyzing the security of hash-based signatures: legitimate signatures reveal preimages of some hash outputs, and attackers try to find preimages of other hash outputs.

One can try to use an attack against this game to break PRE as follows. Take the PRE challenge, insert it at a random position into a list of $T - 1$ randomly generated targets, and run the attack. Abort if there is an oracle query for the position of the PRE challenge; there is no difficulty answering oracle queries for other positions. The problem here is that a successful attack could query as many as $T - 1$ out of T positions, and then the PRE attack succeeds with probability only $1/T$. What happens if T is large and one wants a tight proof?

If T-openPRE were modified to use targets $H(x_i)$ instead of $H(i, x_i)$ then the attacker could try many guesses for x', checking each $H(x')$ against all of the targets. This generic attack is T times more likely to succeed than a generic attack against PRE using the same number of guesses. However, the inclusion of the prefix i (as in [9]) seems to force attackers to focus on single targets, and opens up the possibility of a security proof that does not quantitatively degrade with T.

One might try to tightly prove security of T-openPRE assuming security of a simpler non-interactive game T-PRE in which the opening oracle is removed: the attacker's goal is simply to find some (i, x') with $H(i, x') = H(i, x_i)$, given T targets $H(1, x_1), \ldots, H(T, x_T)$. This game T-PRE is simple enough that cryptanalysts can reasonably be asked to study it (and have already studied it without the i prefixes). However, the difficulty of answering the oracle queries in T-openPRE seems to be an insurmountable obstacle to a proof of this type.

We show that the security of T-openPRE follows tightly from the conjunction of two simple non-interactive assumptions, T-SPR and T-DSPR. This shows an important advantage of introducing DSPR, allowing a reduction to remove the interactivity of T-openPRE.

The advantage of SPR (and T-SPR) over PRE (and T-PRE) in answering oracle queries inside reductions was already pointed out in [9]. The remaining issue, the reason that merely assuming T-SPR is not enough, is that there might be an attack breaking PRE (and T-PRE and T-openPRE) only for hash outputs that have unique preimages. Such an attack would never break SPR.

To address this issue, [9] assumes that each hash-function output has at least two preimages. This is a restrictive assumption: it is not satisfied by most length-preserving functions, and presumably it is not satisfied by (e.g.) SHA-256 for 256-bit inputs. Building a hash function that can be reasonably conjectured to satisfy the assumption is not hard—for example, apply SHA-256, truncate the result to 248 bits (see Theorem 11), and apply SHA-256 again to obtain a random-looking 256-bit string—but the intermediate truncation here produces a noticeably smaller security level, and having to do twice as many SHA-256 computations is not attractive.

We instead observe that an attack of this type must somehow be able to recognize hash outputs with unique preimages, and, consequently, must be able to recognize hash inputs without second preimages, breaking DSPR. Instead of assuming that there are always two preimages, we make the weaker assumption that breaking DSPR is difficult. This assumption is reasonable for a much wider range of hash functions.

1.1.2 The Strength of SPR.

There are some hash functions H where SPR is easy to break, or at least seems easier to break than PRE (and T-PRE and T-openPRE):

- Define $H(x) = 4^x \bmod p$, where p is prime, 4 has order $(p-1)/2$ modulo p, and x is in the range $\{0, 1, \ldots, p-2\}$. Breaking PRE is then solving the discrete-logarithm problem, which seems difficult when p is large, but breaking SPR is a simple matter of adding $(p-1)/2$ modulo $p-1$. (Quantum computers break PRE in this example, but are not known to break PRE for analogous examples based on isogenies.)
- Define $H : \{0,1\}^{2^k n} \to \{0,1\}^n$ by Merkle–Damgård iteration of an n-bit compression function. Then, under reasonable assumptions, breaking SPR for H takes only 2^{n-k} simple operations. See [10]. See also [1] for attacks covering somewhat more general iterated hash functions.

In the first example, proving PRE from SPR+DSPR is useless. In the second example, proving PRE from SPR+DSPR is unsatisfactory, since it seems to underestimate the quantitative security of PRE. This type of underestimate raises the same difficulties as a loose proof: users have to choose larger and slower parameters for the proof to guarantee the desired level of security, or have to take the risk of the "nightmare scenario" that there is a faster attack.

Fortunately, modern "wide-pipe" hash functions and "sponge" hash functions such as SHA-3 are designed to eliminate the internal collisions exploited in attacks such as [10]. Furthermore, input lengths are restricted in applications to hash-based signatures, and this restriction seems to strengthen SPR even for older hash functions such as SHA-256. The bottom line is that one can easily select hash functions for which SPR and DSPR (and T-SPR and T-DSPR) seem to be as difficult to break as PRE, such as SHA3-256 and SHA-256 restricted to 256-bit inputs.

1.2 Organization of the Paper

In Sect. 2 we define DSPR and show how it can be used to relate SPR and PRE. A consequence of our definition is that a function does not provide DSPR if noticeably more than half the domain elements have no colliding value. In Sect. 3 we show that the overwhelming majority of length-preserving hash functions have the property that more than half of the domain elements have a colliding value. In Sect. 4 we extend the analysis to keyed hash functions. We show in Sect. 5 that DSPR is hard in the quantum-accessible-random-oracle model (QROM). We define T-DSPR in Sect. 6. We show in Sect. 7 how to use T-DSPR to eliminate the interactivity of T-openPRE. We close our work with a discussion of the implications for hash-based signatures in Sect. 8.

2 Decisional Second-Preimage Resistance

In this section we give a formal definition of decisional second-preimage resistance (DSPR) for cryptographic hash functions. We start by defining some notation and recalling some standard notions for completeness before we move on to the actual definition.

2.1 Notation

Fix nonempty finite sets $\mathcal{X}$ and $\mathcal{Y}$ of finite-length bit strings. In this paper, a *hash function* means a function from $\mathcal{X}$ to $\mathcal{Y}$.

As shorthands we write $M = |\mathcal{X}|$; $N = |\mathcal{Y}|$; $m = \log_2 M$; and $n = \log_2 N$. The *compressing* case is that $M > N$, i.e., $|\mathcal{X}| > |\mathcal{Y}|$; the *expanding* case is that $M < N$, i.e., $|\mathcal{X}| < |\mathcal{Y}|$; the *length-preserving* case is that $M = N$, i.e., $|\mathcal{X}| = |\mathcal{Y}|$.

We focus on bit strings so that it is clear what it means for elements of $\mathcal{X}$ or $\mathcal{Y}$ to be algorithm inputs or outputs. Inputs and outputs are required to be bit strings in the most common formal definitions of algorithms. These bit strings are often encodings of more abstract objects, and one could generalize all the definitions in this paper to work with more abstract concepts of algorithms.

2.2 Definitions

We now give several definitions of security concepts for a hash function H. We have not found decisional second-preimage resistance (DSPR) in the literature. We also define a second-preimage-exists predicate (SPexists) and the second-preimage-exists probability (SPprob) as tools to help understand DSPR. The definitions of preimage resistance (PRE) and second-preimage resistance (SPR) are standard but we repeat them here for completeness.

Definition 1 (PRE). *The success probability of an algorithm $\mathcal{A}$ against the preimage resistance of a hash function* H *is*

$$\mathrm{Succ}_{\mathrm{H}}^{\mathrm{PRE}}(\mathcal{A}) \stackrel{def}{=} \Pr\left[x \leftarrow_R \mathcal{X}; x' \leftarrow \mathcal{A}(\mathrm{H}(x)) : \mathrm{H}(x) = \mathrm{H}(x')\right].$$

Definition 2 (SPR). *The success probability of an algorithm $\mathcal{A}$ against the second-preimage resistance of a hash function* H *is*

$$\mathrm{Succ}_{\mathrm{H}}^{\mathrm{SPR}}(\mathcal{A}) \stackrel{def}{=} \Pr\left[x \leftarrow_R \mathcal{X}; x' \leftarrow \mathcal{A}(x) : \mathrm{H}(x) = \mathrm{H}(x') \wedge x \neq x'\right].$$

Definition 3 (SPexists). *The second-preimage-exists predicate* SPexists(H) *for a hash function* H *is the function $P : \mathcal{X} \to \{0,1\}$ defined as follows:*

$$P(x) \stackrel{def}{=} \begin{cases} 1 & \text{if } |\mathrm{H}^{-1}(\mathrm{H}(x))| \geq 2 \\ 0 & \text{otherwise.} \end{cases}$$

If $P(x) = 0$ then x has no second preimages under H: any $x' \neq x$ has $\mathrm{H}(x') \neq \mathrm{H}(x)$. The only possible successes of an SPR attack are for inputs x where $P(x) = 1$.

Definition 4 (SPprob). *The second-preimage-exists probability* SPprob(H) *for a hash function* H *is* $\Pr\left[x \leftarrow_R \mathcal{X} : P(x) = 1\right]$, *where* $P = \mathrm{SPexists}(\mathrm{H})$.

In other words, $p = \mathrm{SPprob}(\mathrm{H})$ is the maximum of $\mathrm{Succ}_{\mathrm{H}}^{\mathrm{SPR}}(\mathcal{A})$ over all algorithms $\mathcal{A}$, without any limits on the cost of $\mathcal{A}$. Later we will see that almost all length-preserving hash functions H have $p > 1/2$. More precisely, $p \approx 1 - e^{-1} \approx 0.63$. For comparison, $p = 0$ for an injective function H, such as the n-bit-to-n-bit identity function; and $p = 1$ for a function where every output has multiple preimages.

Definition 5 (DSPR). *Let $\mathcal{A}$ be an algorithm that always outputs* 0 *or* 1*. The advantage of $\mathcal{A}$ against the decisional second-preimage resistance of a hash function* H *is*

$$\mathrm{Adv}_{\mathrm{H}}^{\mathrm{DSPR}}(\mathcal{A}) \stackrel{def}{=} \max\left\{0, \Pr\left[x \leftarrow_R \mathcal{X}; b \leftarrow \mathcal{A}(x) : P(x) = b\right] - p\right\}$$

where $P = \mathrm{SPexists}(\mathrm{H})$ *and* $p = \mathrm{SPprob}(\mathrm{H})$.

2.3 Examples of DSPR Advantages

Here are some examples of computing DSPR advantages. As above, write $P = \mathrm{SPexists}(\mathrm{H})$ and $p = \mathrm{SPprob}(\mathrm{H})$.

If $\mathcal{A}(x) = 1$ for all x, then $\Pr\left[x \leftarrow_R \mathcal{X}; b \leftarrow \mathcal{A}(x) : P(x) = b\right] = p$ by definition, so $\mathrm{Adv}_{\mathrm{H}}^{\mathrm{DSPR}}(\mathcal{A}) = 0$.

If $\mathcal{A}(x) = 0$ for all x, then $\mathrm{Adv}_{\mathrm{H}}^{\mathrm{DSPR}}(\mathcal{A}) = \max\{0, 1-2p\}$. In particular, $\mathrm{Adv}_{\mathrm{H}}^{\mathrm{DSPR}}(\mathcal{A}) = 0$ if $p \geq 1/2$, while $\mathrm{Adv}_{\mathrm{H}}^{\mathrm{DSPR}}(\mathcal{A}) = 1$ for an injective function H.

More generally, say $\mathcal{A}(x)$ flips a biased coin and returns the result, where the probability of 1 is c, independently of x. Then $\mathcal{A}(x) = P(x)$ with probability $cp + (1-c)(1-p)$, which is between $\min\{1-p, p\}$ and $\max\{1-p, p\}$, so again $\mathrm{Adv}_{\mathrm{H}}^{\mathrm{DSPR}}(\mathcal{A}) = 0$ if $p \geq 1/2$.

As a more expensive example, say $\mathcal{A}(x)$ searches through all $x' \in \mathcal{X}$ to see whether x' is a second preimage for x, and returns 1 if any second preimage is found, otherwise 0. Then $\mathcal{A}(x) = P(x)$ with probability 1, so $\mathrm{Adv}_{\mathrm{H}}^{\mathrm{DSPR}}(\mathcal{A}) = 1-p$. This is the maximum possible DSPR advantage.

More generally, say $\mathcal{A}(x)$ runs a second-preimage attack $\mathcal{B}$ against H, and returns 1 if $\mathcal{B}$ is successful (i.e., the output x' from $\mathcal{B}$ satisfies $x' \neq x$ and $\mathrm{H}(x') = \mathrm{H}(x)$), otherwise 0. By definition $\mathcal{A}(x) = 1$ with probability $\mathrm{Succ}_{\mathrm{H}}^{\mathrm{SPR}}(\mathcal{B})$, and if $\mathcal{A}(x) = 1$ then also $P(x) = 1$, so $\mathcal{A}(x) = 1 = P(x)$ with probability $\mathrm{Succ}_{\mathrm{H}}^{\mathrm{SPR}}(\mathcal{B})$. Also $P(x) = 0$ with probability $1-p$ and if $P(x) = 0$ also $\mathcal{A}(x) = 0$ as there simply does not exist any second-preimage for $\mathcal{B}$ to find. Hence, $\mathcal{A}(x) = 0 = P(x)$ with probability $1-p$. Overall $\mathcal{A}(x) = P(x)$ with probability $1-p+\mathrm{Succ}_{\mathrm{H}}^{\mathrm{SPR}}(\mathcal{B})$, so

$$\mathrm{Adv}_{\mathrm{H}}^{\mathrm{DSPR}}(\mathcal{A}) = \max\{0, 1-2p+\mathrm{Succ}_{\mathrm{H}}^{\mathrm{SPR}}(\mathcal{B})\}.$$

This advantage is 0 whenever $0 \leq \mathrm{Succ}_{\mathrm{H}}^{\mathrm{SPR}}(\mathcal{B}) \leq 2p-1$: even if $\mathcal{B}$ breaks second-preimage resistance with probability as high as $2p-1$ (which is approximately 26% for almost all length-preserving H), $\mathcal{A}$ breaks DSPR with advantage 0. If $\mathcal{B}$ breaks second-preimage resistance with probability p, the maximum possible, then $\mathrm{Adv}_{\mathrm{H}}^{\mathrm{DSPR}}(\mathcal{A}) = 1-p$, the maximum possible advantage.

As a final example, say $x_1 \in \mathcal{X}$ has no second preimage, and say $\mathcal{A}(x)$ returns 0 if $x = x_1$, otherwise 1. Then $\mathcal{A}(x) = P(x)$ with probability $p + 1/2^m$, so $\mathrm{Adv}_{\mathrm{H}}^{\mathrm{DSPR}}(\mathcal{A}) = 1/2^m$. This example shows that an efficient algorithm can achieve a (very small) nonzero DSPR advantage. We can efficiently generate an algorithm $\mathcal{A}$ of this type with probability $1-p$ by choosing $x_1 \in \mathcal{X}$ at random (in the normal case that $\mathcal{X} = \{0,1\}^m$), but for typical hash functions H we do not have an efficient way to recognize whether $\mathcal{A}$ is in fact of this type, i.e., whether x_1 in fact has no second preimage: recognizing this is exactly the problem of breaking DSPR!

2.4 Why DSPR Advantage Is Defined This Way

Many security definitions require the attacker to distinguish two possibilities, each of which naturally occurs with probability 1/2. Any sort of blind guess is correct with probability 1/2. Define a as the probability of a correct output minus 1/2; a value of a noticeably larger than 0 means that the algorithm is noticeably more likely than a blind guess to be correct.

If an algorithm is noticeably *less* likely than a blind guess to be correct then one can do better by (1) replacing it with a blind guess or (2) inverting its output. The first option replaces a with $\max\{0, a\}$; the second option replaces a with $|a|$; both options have the virtue of eliminating negative values of a. Advantage is most commonly defined as $|a|$, or alternatively as $2|a|$, the distance between the probability of a correct output and the probability of an incorrect output. These formulas are simpler than $\max\{0, a\}$.

For DSPR, the two possibilities are not naturally balanced. A second preimage exists with probability p, and almost all length-preserving (or compressing) hash functions have $p > 1/2$. Guessing 1 is correct with probability p; guessing 0 is correct with probability $1-p$; random guesses can trivially achieve any desired intermediate probability. What is interesting—and what is naturally considered in our proofs—is an algorithm $\mathcal{A}$ that guesses correctly with probability larger than p. We thus define the advantage as $\max\{0, \text{Succ}(\mathcal{A}) - p\}$, where $\text{Succ}(\mathcal{A})$ is the probability of $\mathcal{A}$ generating a correct output.

An algorithm $\mathcal{A}$ that guesses correctly with probability smaller than $1-p$ is also useful. We could define advantage as $\max\{0, \text{Succ}(\mathcal{A}) - p, (1 - \text{Succ}(\mathcal{A})) - p\}$ to take this into account, rather than leaving it to the attack developer to invert the output. However, this formula is more complicated than $\max\{0, \text{Succ}(\mathcal{A}) - p\}$.

If $p < 1/2$ then, with our definitions, guessing 0 has advantage $1 - 2p > 0$. In particular, if $p = 0$ then guessing 0 has advantage 1: our definitions state that injective functions are trivially vulnerable to DSPR attacks. It might seem intuitive to define DSPR advantage as beating the best blind guess, i.e., as probability minus $\max\{p, 1-p\}$ rather than probability minus p. This, however, would break the proof that SPR $\wedge$ DSPR implies PRE: the identity function would have both SPR and DSPR but not PRE. We could add an assumption that $p \geq 1/2$, but the approach we have taken is simpler.

2.5 DSPR Plus SPR Implies PRE

We now present the main application of DSPR in the simplest case: We show that a second-preimage-resistant and decisional-second-preimage-resistant hash function is preimage resistant.

We first define the two reductions we use, SPfromP and DSPfromP, and then give a theorem statement analyzing success probabilities. The algorithm $\text{SPfromP}(\text{H}, \mathcal{A})$ is the standard algorithm that tries to break SPR using an algorithm $\mathcal{A}$ that tries to break PRE. The algorithm $\text{DSPfromP}(\text{H}, \mathcal{A})$ is a variant that tries to break DSPR. Each algorithm uses one computation of H, one call to $\mathcal{A}$, and (for DSPfromP) one string comparison, so each algorithm has essentially the same cost as $\mathcal{A}$ if H is efficient.

Definition 6 (SPfromP). *Let* H *be a hash function. Let* $\mathcal{A}$ *be an algorithm. Then* $\text{SPfromP}(\text{H}, \mathcal{A})$ *is the algorithm that, given* $x \in \mathcal{X}$*, outputs* $\mathcal{A}(\text{H}(x))$.

Definition 7 (DSPfromP). *Let* H *be a hash function. Let* $\mathcal{A}$ *be an algorithm. Then* $\text{DSPfromP}(\text{H}, \mathcal{A})$ *is the algorithm that, given* $x \in \mathcal{X}$*, outputs* $[x \neq \mathcal{A}(\text{H}(x))]$.

This output is 0 if $\mathcal{A}(\text{H}(x))$ returns the preimage x that was already known for $\text{H}(x)$, and 1 otherwise. Note that the 0 case provides some reason to believe that there is only one preimage. If there are $i > 1$ preimages then x, which is not known to $\mathcal{A}$ except via $\mathrm{H}(x)$, is information-theoretically hidden in a set of size i, so $\mathcal{A}$ cannot return x with probability larger than $1/i$.

Theorem 8 (DSPR $\wedge$ SPR $\Rightarrow$ PRE). *Let* H *be a hash function. Let* $\mathcal{A}$ *be an algorithm. Then*

$$\mathrm{Succ}_{\mathrm{H}}^{\mathrm{PRE}}(\mathcal{A}) \leq \mathrm{Adv}_{\mathrm{H}}^{\mathrm{DSPR}}(\mathcal{B}) + 3 \cdot \mathrm{Succ}_{\mathrm{H}}^{\mathrm{SPR}}(\mathcal{C})$$

where $\mathcal{B} = \mathrm{DSPfromP}(\mathrm{H}, \mathcal{A})$ *and* $\mathcal{C} = \mathrm{SPfromP}(\mathrm{H}, \mathcal{A})$.

Proof. This is a special case of Theorem 25 below, modulo a change of syntax. To apply Theorem 25 we set $\mathcal{K}$ to be $\{()\}$, where $()$ is the empty string. The change of syntax views a keyed hash function with an empty key as an unkeyed hash function. □

3 The Second-Preimage-Exists Probability

This section mathematically analyzes SPprob(H), the probability that a uniform random input to H has a second preimage. The DSPR advantage of any attacker is information-theoretically bounded by $1 - \mathrm{SPprob}(\mathrm{H})$.

3.1 Simple Cases

In retrospect, the heart of the Rogaway–Shrimpton SPR-PRE reduction [15, Theorem 7] is the observation that SPprob(H) is very close to 1 for all highly compressing hash functions H. See Theorem 9. We show that SPprob(H) is actually *equal* to 1 for almost all hash functions H that compress more than a few bits; see Theorem 11.

Theorem 9 (lower bound on SPprob in the compressing case). *If* H *is a hash function and* $M > N$ *then* $\mathrm{SPprob}(\mathrm{H}) \geq 1 - (N-1)/M$.

The maximum possible DSPR advantage in this case is $(N-1)/M$. For example, if $M > 1$ and $N = 1$ then $\mathrm{SPprob}(\mathrm{H}) = 1$ and the DSPR advantage is always 0. As another example, a 320-bit-to-256-bit hash function H has $\mathrm{SPprob}(\mathrm{H}) \geq 1 - (2^{256} - 1)/2^{320}$, and the DSPR advantage is at most $(2^{256} - 1)/2^{320} < 1/2^{64}$.

Proof. Define I as the set of elements of $\mathcal{X}$ that have no second preimages; i.e., the set of $x \in \mathcal{X}$ such that $|\mathrm{H}^{-1}(\mathrm{H}(x))| = 1$.

The image set $\mathrm{H}(I) \subseteq \mathcal{Y}$ has size $|I|$, so $|I| \leq |\mathcal{Y}| = N < M = |\mathcal{X}|$. The complement $\mathcal{X} - I$ is thus nonempty, so the image set $\mathrm{H}(\mathcal{X} - I)$ is also nonempty. This image set cannot overlap $\mathrm{H}(I)$: if $\mathrm{H}(x') = \mathrm{H}(x)$ with $x' \in \mathcal{X} - I$ and $x \in I$ then x', x are distinct elements of $\mathrm{H}^{-1}(\mathrm{H}(x))$, but $|\mathrm{H}^{-1}(\mathrm{H}(x))| = 1$ by definition of I. Hence $|I| \leq N - 1$.

By definition SPprob(H) is the probability that $|\mathrm{H}^{-1}(\mathrm{H}(x))| \geq 2$ where x is a uniform random element of $\mathcal{X}$, i.e., the probability that x is not in I. This is at least $1 - (N-1)/M$. □

Theorem 10 (average of SPprob). *The average of* SPprob(H) *over all hash functions* H *is* $1 - (1 - 1/N)^{M-1}$.

For example, the average is $1 - (1 - 1/2^{256})^{2^{256}-1} \approx 1 - 1/e \approx 0.63212$ if $M = 2^{256}$ and $N = 2^{256}$; see also Theorem 12. The average converges rapidly to 1 as N/M drops: for example, the average is approximately $1 - 2^{-369.33}$ if $M = 2^{256}$ and $N = 2^{248}$, and is approximately $1 - 2^{-94548}$ if $M = 2^{256}$ and $N = 2^{240}$, while the lower bounds from Theorem 9 are approximately $1 - 2^{-16}$ and approximately $1 - 2^{-32}$ respectively.

The average converges to 0 as N/M increases. The average crosses below 1/2, making DSPR trivially breakable for the average function, as N/M increases past about $1/\log 2 \approx 1.4427$.

Proof. For each $x \in \mathcal{X}$, there are exactly $N(N-1)^{M-1}$ hash functions H for which x has no second preimages. Indeed, there are N choices of $\mathrm{H}(x)$, and then for each $i \in \mathcal{X} - \{x\}$ there are $N - 1$ choices of $\mathrm{H}(i) \in \mathcal{Y} - \{\mathrm{H}(x)\}$.

Hence there are exactly $M(N^M - N(N-1)^{M-1})$ pairs (H, x) where x has a second preimage under H; i.e., the total of SPprob(H) over all N^M hash functions H is $N^M - N(N-1)^{M-1}$; i.e., the average of SPprob(H) over all hash functions H is $1 - N(N-1)^{M-1}/N^M = 1 - (1 - 1/N)^{M-1}$. □

Theorem 11 (how often SPprob is 1). *If* H *is a uniform random hash function then* SPprob(H) = 1 *with probability at least* $1 - M(1 - 1/N)^{M-1}$.

This is content-free in the length-preserving case but becomes more useful as N/M drops. For example, if $M = 2^{256}$ and $N = 2^{248}$, then the chance of SPprob(H) < 1 is at most $2^{256}(1 - 1/2^{248})^{2^{256}-1} \approx 2^{-113.33}$. Hence almost all 256-bit-to-248-bit hash functions have second preimages for all inputs, and therefore have perfect DSPR (DSPR advantage 0) against all attacks.

Proof. Write q for the probability that SPprob(H) = 1. Then SPprob(H) $\leq 1 - 1/M$ with probability $1 - q$. The point here is that SPprob(H) is a probability over M inputs, and is thus a multiple of $1/M$.

The average of SPprob(H) is at most $q + (1-q)(1-1/M) = 1 - (1-q)/M$. By Theorem 10, this average is exactly $1 - (1-1/N)^{M-1}$. Hence $1 - (1-1/N)^{M-1} \leq 1 - (1-q)/M$; i.e., $q \geq 1 - M(1 - 1/N)^{M-1}$. □

Theorem 12 (average of SPprob vs. 1 − 1/e in the length-preserving case). *If* $M = N > 1$ *then the average* a *of* SPprob(H) *over all hash functions* H *has* $1 - (1/e)N/(N-1) < a < 1 - 1/e$.

The big picture is that almost all length-preserving hash functions H have SPprob(H) close to $1 - 1/e$. This theorem states part of the picture: the average of SPprob(H) is extremely close to $1 - 1/e$ if N is large. Subsequent theorems fill in the rest of the picture.

Proof. The point is that $(N/(N-1))^{N-1} < e < (N/(N-1))^N$ for $N \geq 2$. See, e.g., [4]. In other words, $e(N-1)/N < (N/(N-1))^{N-1} < e$. Invert to see that $1/e < (1-1/N)^{N-1} < (1/e)N/(N-1)$. Finally, the average a of SPprob(H) is $1-(1-1/N)^{N-1}$ by Theorem 10. □

3.2 How SPprob Varies

This subsection analyzes the distribution of SPprob(H) as H varies. Theorem 14 amounts to an algorithm that computes the probability of each possible value of SPprob(H) in time polynomial in $M+N$. Theorem 16, used in Sect. 3.3, gives a simple upper bound on each term in the probability.

Theorem 13. *Let a, b be nonnegative integers. Define $c(a, b)$ as the coefficient of x^b in the power series $b!(e^x-1-x)^a/a!$. Then $a!c(a, b)$ is the number of functions from $\{1, \ldots, b\}$ to $\{1, \ldots, a\}$ for which each of $\{1, \ldots, a\}$ has at least two preimages.*

This is a standard example of "generatingfunctionology". See, e.g., [16, sequence A000478, "E.g.f."] for $a = 3$ and [16, sequence A058844, "E.g.f."] for $a = 4$.

Note that $c(a, b) = 0$ for $b < 2a$, and that $c(0, b) = 0$ for $b > 0$.

Proof. Choose integers $i_1, \ldots, i_a \geq 2$ with $i_1 + \cdots + i_a = b$, and consider any function f built as follows. Let π be a permutation of $\{1, \ldots, b\}$. Define $f(\pi(1)) = f(\pi(2)) = \ldots = f(\pi(i_1)) = 1$; note that 1 has $i_1 \geq 2$ preimages. Define $f(\pi(i_1 + 1)) = f(\pi(i_1+2)) = \ldots = f(\pi(i_1+i_2)) = 2$; note that 2 has $i_2 \geq 2$ preimages. Et cetera.

There are exactly $b!$ choices of π, producing exactly $b!/i_1! \cdots i_a!$ choices of f. This covers all functions f for which 1 has exactly i_1 preimages, 2 has exactly i_2 preimages, etc.

The total number of functions being counted is thus the sum of $b!/i_1! \cdots i_a!$ over all $i_1, \ldots, i_a \geq 2$ with $i_1 + \cdots + i_a = b$.

For comparison, the power series $e^x - 1 - x$ is $\sum_{i \geq 2} x^i/i!$, so

$$(e^x - 1 - x)^a = \sum_{i_1, \ldots, i_a \geq 2} x^{i_1 + \cdots + i_a}/i_1! \cdots i_a!.$$

The coefficient of x^b is the sum of $1/i_1! \cdots i_a!$ over all $i_1, \ldots, i_a \geq 2$ with $i_1 + \cdots + i_a = b$. By definition $a!c(a, b)/b!$ is this coefficient, so $a!c(a, b)$ is the sum of $b!/i_1! \cdots i_a!$ over all $i_1, \ldots, i_a \geq 2$ with $i_1 + \cdots + i_a = b$. □

Theorem 14 (exact distribution of SPprob). *There are exactly*

$$\binom{M}{j} \sum_{j \leq k \leq N} c(k-j, M-j) \frac{N!}{(N-k)!}$$

hash functions H *with* $\text{SPprob(H)} = 1 - j/M$.

Fig. 1. Cumulative distribution of SPprob(H) for $M = N = 1$; $M = N = 2$; $M = N = 4$; $M = N = 8$; $M = N = 16$; $M = N = 32$; $M = N = 64$. The probabilities that $\text{SPprob(H)} \le 0.5$ are, respectively, 1; 0.5; 0.65625; $\approx$0.417366; $\approx$0.233331; $\approx$0.100313; and $\approx$0.023805. As $N \to \infty$ with $M = N$, the distribution converges to a vertical line at $1 - 1/e$.

The summand is 0 if $k > (M + j)/2$, i.e., if $M - j < 2(k - j)$, since then $c(k - j, M - j) = 0$. The summand is also 0 if $k = j$ and $M > j$, since then $c(0, M - j) = 0$.

In particular, if $j > N$ then $\text{SPprob(H)} = 1 - j/M$ with probability 0; and if $j = N < M$ then $\text{SPprob(H)} = 1 - j/M$ with probability 0. This calculation shows that Theorem 14 includes Theorem 9.

The distribution of $M - j$ here, for a uniform random hash function H, is equal to the distribution of "K_1" in [3, formula (2.21)], but the formulas are different. The sum in [3, formula (2.21)] is an alternating sum with cancellation between large terms. The sum in Theorem 14 is a sum of nonnegative terms; this is important for our asymptotic analysis.

Figure 1 shows the cumulative distribution of SPprob(H) when $M = N \in \{1, 2, 4, 8, 16, 32, 64\}$. Each graph ranges from 0 through 1 horizontally, and from 0 through 1 vertically. At horizontal position p, the (maximum) vertical position is the probability that $\text{SPprob(H)} \le p$. We computed these probabilities using Theorem 14.

Proof. We count the hash functions that (1) have exactly $k \ge j$ outputs and (2) have exactly j inputs with no second preimages.

Choose the j inputs. There are $\binom{M}{j}$ ways to do this.

Choose a partition of the N outputs into

- j outputs that will be used (without second preimages) by the j inputs;
- $k - j$ outputs that will be used (with second preimages) by the other $M - j$ inputs; and
- $N - k$ outputs that will not be used.

There are $N!/j!(k - j)!(N - k)!$ ways to do this.

Choose an injective function from the j inputs to the j outputs. There are $j!$ ways to do this.

Choose a function from the other $M - j$ inputs to the other $k - j$ outputs for which each of these $k - j$ outputs has at least two preimages. By Theorem 13, there are $(k - j)!c(k - j, M - j)$ ways to do this.

This produces a hash function that, as desired, has exactly k outputs and has exactly j inputs with no second preimages. Each such function is produced exactly once. Hence there are $\binom{M}{j}c(k - j, M - j)N!/(N - k)!$ such functions.

Finally, sum over k to see that there are

$$\binom{M}{j} \sum_{j \le k \le N} c(k-j, M-j) \frac{N!}{(N-k)!}$$

hash functions H that have exactly j inputs with no second preimages, i.e., hash functions H that have $\mathrm{SPprob}(\mathrm{H}) = 1 - j/M$. □

Theorem 15. *Let a, b be positive integers. Let ζ be a positive real number. Assume that $b/a = \zeta + \zeta^2/(e^\zeta - 1 - \zeta)$. Then $c(a, b) \le (e^\zeta - 1 - \zeta)^a \zeta^{-b} b!/a!$.*

Our proof applies [5, Proposition VIII.7], which is an example of the "saddle-point method" in analytic combinatorics. With more work one can use the saddle-point method to improve bounds by a polynomial factor, but our main concern here is exponential factors.

Proof. Define $B(z) = \sum_{i \ge 2} z^{i-2}/i! = 1/2 + z/6 + z^2/24 + \cdots$. Note that $z^2 B(z) = e^z - 1 - z$, and that $zB'(z) = \sum_{i \ge 3} (i-2) z^{i-2}/i! = (z-2)B(z) + 1$. Also define $A(z) = 1$; $R = \infty$; $T = \infty$; $N = b - 2a$; $n = a$; and $\lambda = b/a - 2$.

Check the hypotheses of [5, Proposition VIII.7]: A and B are analytic functions of the complex variable z, with all coefficients nonnegative; $B(0) = 1/2 \ne 0$; the coefficient of z in B is nonzero; the radius of convergence of B is ∞; the radius of convergence of A is also ∞; the limit of $xB'(x)/B(x)$ as $x \to \infty$ is ∞; λ is a positive real number; $N = \lambda n$; and $\zeta B'(\zeta)/B(\zeta) = \zeta - 2 + 1/B(\zeta) = b/a - 2 = \lambda$.

Now [5, Proposition VIII.7] states that the coefficient of z^N in $A(z)B(z)^n$ is at most $A(\zeta)B(\zeta)^n \zeta^{-N}$; i.e., the coefficient of z^{b-2a} in $((e^z - 1 - z)/z^2)^a$ is at most $B(\zeta)^a \zeta^{2a-b}$; i.e., the coefficient of z^b in $(e^z - 1 - z)^a$ is at most $B(\zeta)^a \zeta^{2a-b}$. Hence $c(a, b) \le B(\zeta)^a \zeta^{2a-b} b!/a! = (e^\zeta - 1 - \zeta)^a \zeta^{-b} b!/a!$. □

Theorem 16 (exponential convergence of SPprob). *Let j be an integer with $0 < j < M$. Let k be an integer with $j < k < N$. Define $\mu = M/N$, $\alpha = j/N$, and $\kappa = k/N$. Let ζ be a positive real number. Assume that $(\mu - \alpha)/(\kappa - \alpha) = \zeta + \zeta^2/(e^\zeta - 1 - \zeta)$. Then*

$$\binom{M}{j} c(k-j, M-j) \frac{N!}{(N-k)!} \le \frac{M! N! e^N \tau^N}{N^N}$$

where $\tau = (e^\zeta - 1 - \zeta)^{\kappa - \alpha} / \zeta^{\mu - \alpha} \alpha^\alpha (\kappa - \alpha)^{\kappa - \alpha} (1 - \kappa)^{1 - \kappa}$.

The proof combines Theorem 15 with the weak Stirling bound $N! \ge (N/e)^N$. See [14] for a proof that $(N/e)^N \sqrt{2\pi N} e^{1/(12N+1)} \le N! \le (N/e)^N \sqrt{2\pi N} e^{1/12N}$.

Proof. Define $a = k - j$ and $b = M - j$. Then a and b are positive integers, and $b/a = (\mu - \alpha)/(\kappa - \alpha) = \zeta + \zeta^2/(e^\zeta - 1 - \zeta)$, so

$$c(k-j, M-j) = c(a, b) \le \frac{(e^\zeta - 1 - \zeta)^a b!}{\zeta^b a!}$$

by Theorem 15, so

$$\binom{M}{j} c(k-j, M-j)\frac{N!}{(N-k)!} \le \frac{M!N!(e^\zeta - 1 - \zeta)^a}{j!\zeta^b a!(N-k)!}$$
$$\le \frac{M!N!(e^\zeta - 1 - \zeta)^a}{(j/e)^j \zeta^b (a/e)^a ((N-k)/e)^{N-k}}$$

by the weak Stirling bound. Now substitute $j = \alpha N$, $k = \kappa N$, $a = (\kappa - \alpha)N$, and $b = (\mu - \alpha)N$:

$$\binom{M}{j} c(k-j, M-j)\frac{N!}{(N-k)!}$$
$$\le \frac{M!N!(e^\zeta - 1 - \zeta)^{(\kappa-\alpha)N}}{(\alpha N/e)^{\alpha N} \zeta^{(\mu-\alpha)N} ((\kappa-\alpha)N/e)^{(\kappa-\alpha)N} ((N-\kappa N)/e)^{N-\kappa N}}$$
$$= \frac{M!N!(e^\zeta - 1 - \zeta)^{(\kappa-\alpha)N}}{(N/e)^N \alpha^{\alpha N} \zeta^{(\mu-\alpha)N} (\kappa-\alpha)^{(\kappa-\alpha)N} (1-\kappa)^{N-\kappa N}} = \frac{M!N!\tau^N}{(N/e)^N}$$

as claimed. □

3.3 Maximization

This subsection formalizes and proves our claim that SPprob(H) is close to $1 - 1/e$ for almost all length-preserving hash functions H: as N increases (with $M = N$), the distributions plotted in Fig. 1 converge to a vertical line.

The basic idea here is that τ in Theorem 16 is noticeably below e when j/N is noticeably below or above $1/e$. One can quickly see this by numerically plotting τ as a function of α and ζ: note that any choice of α and ζ (along with $\mu = 1$) determines $\kappa = \alpha + (\mu - \alpha)/(\zeta + \zeta^2/(e^\zeta - 1 - \zeta))$ and thus determines τ. The plot suggests that $\zeta = 1$ maximizes τ for each α, and that moving α towards $1/e$ from either side increases τ up to its maximum value e. One could use interval arithmetic to show, e.g., that $\tau/e < 0.998$ for $j/N > 0.4$, but the required number of subintervals would rapidly grow as j/N approaches $1/e$. Our proof also handles some corner cases that are not visible in the plot.

Theorem 17. *Let $\mu, \alpha, \kappa, \zeta$ be positive real numbers with $\alpha < \mu$; $\alpha < \kappa < 1$; and $(\mu-\alpha)/(\kappa-\alpha) = \zeta + \zeta^2/(e^\zeta - 1 - \zeta)$. First, there is a unique positive real number Z such that $Z(e^Z - 1)/(e^Z - Z) = (\mu - \alpha)/(1 - \alpha)$. Second, there is a unique real number K such that $\alpha < K < 1$ and $(\mu - \alpha)/(K - \alpha) = Z + Z^2/(e^Z - 1 - Z)$. Third,*

$$\frac{(e^\zeta - 1 - \zeta)^{\kappa-\alpha}}{\zeta^{\mu-\alpha}\alpha^\alpha(\kappa-\alpha)^{\kappa-\alpha}(1-\kappa)^{1-\kappa}} \le \frac{(e^Z - 1 - Z)^{K-\alpha}}{Z^{\mu-\alpha}\alpha^\alpha(K-\alpha)^{K-\alpha}(1-K)^{1-K}}.$$

Fourth, if $\mu = 1$ then

$$\frac{(e^Z - 1 - Z)^{K-\alpha}}{Z^{\mu-\alpha}\alpha^\alpha(K-\alpha)^{K-\alpha}(1-K)^{1-K}} = \frac{(e-1)^{1-\alpha}}{\alpha^\alpha(1-\alpha)^{1-\alpha}}.$$

Proof. See full version of this paper online. □

Theorem 18. *Let α, κ, ζ, A be positive real numbers. Assume that $\alpha < \kappa < 1$; that $(1-\alpha)/(\kappa-\alpha) = \zeta + \zeta^2/(e^\zeta - 1 - \zeta)$; and that $1/e \le A \le \alpha$ or $\alpha \le A \le 1/e$. Then*

$$\frac{(e^\zeta - 1 - \zeta)^{\kappa-\alpha}}{\zeta^{1-\alpha}\alpha^\alpha(\kappa-\alpha)^{\kappa-\alpha}(1-\kappa)^{1-\kappa}} \le \frac{(e-1)^{1-A}}{A^A(1-A)^{1-A}}.$$

Proof. See full version of this paper online. □

Theorem 19. *Assume that $M = N$. Let A be a real number with $0 < A < 1$. Let* H *be a uniform random hash function. If $A > 1/e$, define E as the event that* SPprob(H) $\le 1 - A$. *If $A \le 1/e$, define E as the event that* SPprob(H) $\ge 1 - A$. *Then E occurs with probability at most $(T/e)^N 2\pi N^2(N+1)e^{1/6N}$ where*

$$T = \max\{1 + \sqrt{2}, (e-1)^{1-A}/A^A(1-A)^{1-A}\}.$$

Any $A \ne 1/e$ has $T/e < 1$, and then the important factor in the probability for large N is $(T/e)^N$. For example, if $A = 0.4$ then $T/e < 0.99780899$, so $(T/e)^N$ is below $1/2^{2^{247}}$ for $N = 2^{256}$. As another example, if $A = 0.37$ then $T/e < 0.99999034$, so $(T/e)^N$ is below $1/2^{2^{239}}$ for $N = 2^{256}$.

Proof. See full version of this paper online. □

4 DSPR for Keyed Hash Functions

In this section we lift the discussion to the setting of keyed hash functions. We model keyed hash functions as functions $\mathrm{H} : \mathcal{K} \times \mathcal{X} \to \mathcal{Y}$ that take a dedicated key as additional input argument. One might also view a keyed hash function as a family of hash functions where elements of the family H are obtained by fixing the first input argument which we call the function key. We write $\mathrm{H}_k \stackrel{\text{def}}{=} \mathrm{H}(k, \cdot)$ for the function that is obtained from H by fixing the first input as $k \in \mathcal{K}$.

We assume that $\mathcal{K}$, like $\mathcal{X}$ and $\mathcal{Y}$, is a nonempty finite set of finite-length bit strings. We define the *compressing*, *expanding*, and *length-preserving* cases as the cases $|\mathcal{X}| > |\mathcal{Y}|$, $|\mathcal{X}| < |\mathcal{Y}|$, and $|\mathcal{X}| = |\mathcal{Y}|$ respectively, ignoring the size of $\mathcal{K}$.

We recall the definitions of preimage and second-preimage resistance for keyed hash functions for completeness:

Definition 20 (PRE for keyed hash functions). *The success probability of adversary $\mathcal{A}$ against the preimage resistance of a keyed hash function* H *is*

$$\mathrm{Succ}_{\mathrm{H}}^{\mathrm{PRE}}(\mathcal{A}) \stackrel{def}{=} \Pr\left[x \leftarrow_R \mathcal{X}; k \leftarrow_R \mathcal{K}; x' \leftarrow \mathcal{A}(\mathrm{H}_k(x), k) : \mathrm{H}_k(x) = \mathrm{H}_k(x')\right].$$

Definition 21 (SPR for keyed hash functions). *The success probability of adversary $\mathcal{A}$ against the second-preimage resistance of a keyed hash function* H *is*

$$\mathrm{Succ}_{\mathrm{H}}^{\mathrm{SPR}}(\mathcal{A}) \stackrel{def}{=} \Pr\left[x \leftarrow_R \mathcal{X}; k \leftarrow_R \mathcal{K}; x' \leftarrow \mathcal{A}(x, k) : \mathrm{H}_k(x) = \mathrm{H}_k(x') \wedge x \ne x'\right].$$

Our definition of DSPR for a keyed hash function H relies on the second-preimage-exists predicate SPexists and the second-preimage-exists probability SPprob for the functions H_k. If H is chosen uniformly at random then, for large N and any reasonable size of $\mathcal{K}$, it is very likely that *all* of the functions H_k have $\text{SPprob}(H_k)$ close to $1 - 1/e$; see Theorem 19.

Definition 22 (DSPR for keyed hash functions). *Let $\mathcal{A}$ be an algorithm that always outputs* 0 *or* 1. *The advantage of $\mathcal{A}$ against the decisional second-preimage resistance of a keyed hash function* H *is*

$$\text{Adv}_{\mathcal{H}}^{\text{DSPR}}(\mathcal{A}) \stackrel{def}{=} \max\{0, \Pr[x \leftarrow_R \mathcal{X}, k \leftarrow_R \mathcal{K}, b \leftarrow \mathcal{A}(x,k) : P_k(x) = b] - p\}$$

where $P_k = \text{SPexists}(H_k)$ and p is the average of $\text{SPprob}(H_k)$ *over all k.*

As an example, consider the keyed hash function H with $\mathcal{X} = \mathcal{Y} = \{0,1\}^{256}$, $\mathcal{K} = \{0,1\}$, $H_0(x) = x$, and $H_1(x) = (x_1, x_2, \ldots, x_{255}, 0)$ where the x_i denote the bits of x. Then $P_k(x) = k$, $\text{SPprob}(H_k) = k$, and $p = 1/2$. A trivial adversary that outputs k has success probability 1 and thus DSPR advantage 1/2, the maximum possible DSPR advantage: this function does not have decisional second-preimage resistance.

It might seem natural to define SPprob(H) as the average mentioned in the theorem. However, we will see later in the multi-target context that p is naturally replaced by a more complicated quantity influenced by the algorithm.

4.1 DSPR Plus SPR Implies PRE

Before we show that DSPR is hard in the QROM (see Sect. 5), we give a generalization of Theorem 8 for keyed hash functions. This theorem states that second-preimage and decisional second-preimage resistance together imply preimage resistance.

As in Theorem 8, we first define the two reductions we use, and then give a theorem statement analyzing success probabilities. The special case that $\mathcal{K} = \{()\}$, where () means the empty string, is the same as Theorem 8, modulo syntactic replacements such as replacing the pair $((), x)$ with x.

Definition 23 (SPfromP for keyed hash functions). *Let* H *be a keyed hash function. Let $\mathcal{A}$ be an algorithm. Then* $\text{SPfromP}(H, \mathcal{A})$ *is the algorithm that, given $(k, x) \in \mathcal{K} \times \mathcal{X}$, outputs $\mathcal{A}(H_k(x), k)$.*

Definition 24 (DSPfromP for keyed hash functions). *Let* H *be a keyed hash function. Let $\mathcal{A}$ be an algorithm. Then* $\text{DSPfromP}(H, \mathcal{A})$ *is the algorithm that, given $(k, x) \in \mathcal{K} \times \mathcal{X}$, outputs $[x \neq \mathcal{A}(H_k(x), k)]$.*

Theorem 25 (DSPR $\wedge$ SPR $\Rightarrow$ PRE for keyed hash functions). *Let* H *be a keyed hash function. Let $\mathcal{A}$ be an algorithm. Then*

$$\text{Succ}_{H}^{\text{PRE}}(\mathcal{A}) \leq \text{Adv}_{H}^{\text{DSPR}}(\mathcal{B}) + 3 \cdot \text{Succ}_{H}^{\text{SPR}}(\mathcal{C})$$

where $\mathcal{B} = \text{DSPfromP}(H, \mathcal{A})$ and $\mathcal{C} = \text{SPfromP}(H, \mathcal{A})$.

Proof. To analyze the success probabilities, we split the universe of possible events into mutually exclusive events across two dimensions: the number of preimages of $\mathrm{H}_k(x)$, and whether $\mathcal{A}$ succeeds or fails in finding a preimage. Specifically, define

$$S_i \stackrel{\text{def}}{=} \left[\left|\mathrm{H}_k^{-1}(\mathrm{H}_k(x))\right| = i \wedge \mathrm{H}_k(\mathcal{A}(\mathrm{H}_k(x),k)) = \mathrm{H}_k(x)\right]$$

as the event that there are exactly i preimages and that $\mathcal{A}$ succeeds, and define

$$F_i \stackrel{\text{def}}{=} \left[\left|\mathrm{H}_k^{-1}(\mathrm{H}_k(x))\right| = i \wedge \mathrm{H}_k(\mathcal{A}(\mathrm{H}_k(x),k)) \neq \mathrm{H}_k(x)\right]$$

as the event that there are exactly i preimages and that $\mathcal{A}$ fails.

Note that there are only finitely many i for which the events S_i and F_i can occur, namely $i \in \{1, 2, \dots, M\}$. All sums below are thus finite sums.

Define s_i and f_i as the probabilities of S_i and F_i respectively. The probability space here includes the random choices of x and k, and any random choices made inside $\mathcal{A}$. The conditional probabilities mentioned below are conditional probabilities given S_i.

PRE success probability. By definition, $\mathrm{Succ}_{\mathrm{H}}^{\mathrm{PRE}}(\mathcal{A})$ is the probability of the event that $\mathrm{H}_k(x) = \mathrm{H}_k(\mathcal{A}(\mathrm{H}_k(x),k))$. This event is the union of S_i, so $\mathrm{Succ}_{\mathrm{H}}^{\mathrm{PRE}}(\mathcal{A}) = \sum_i s_i$.

DSPR success probability. Define $P_k = \mathrm{SPexists}(\mathrm{H}_k)$. For the $i = 1$ cases, we have $P_k(x) = 0$ by definition of SPexists, so $\mathcal{B}$ is correct if and only if $\mathcal{A}$ succeeds. For the $i > 1$ cases, we have $P_k(x) = 1$, so $\mathcal{B}$ is correct as long as $\mathcal{A}$ does not output x. There are two disjoint ways for this to occur:

- $\mathcal{A}$ succeeds (case S_i). Then $\mathcal{A}$ outputs x with conditional probability exactly $\frac{1}{i}$, since x is information-theoretically hidden in a set of size i; so there is conditional probability exactly $\frac{i-1}{i}$ that $\mathcal{A}$ does not output x.
- $\mathcal{A}$ fails (case F_i). Then $\mathcal{A}$ does not output x.

Together we get

$$\Pr[\mathcal{B}(x,k) = P_k(x)] = s_1 + \sum_{i>1} \frac{i-1}{i} s_i + \sum_{i>1} f_i.$$

DSPR advantage. By definition $\mathrm{Adv}_{\mathrm{H}}^{\mathrm{DSPR}}(\mathcal{B}) = \max\{0, \Pr[\mathcal{B}(x,k) = P_k(x)] - p\}$ where p is the average of $\mathrm{SPprob}(\mathrm{H}_k)$ over all k.

By definition $\mathrm{SPprob}(\mathrm{H}_k)$ is the probability over all choices of x that x has a second preimage under H_k. Hence p is the same probability over all choices of x and k; i.e., $p = \sum_{i>1} s_i + \sum_{i>1} f_i$. Now subtract:

$$\begin{aligned}
\mathrm{Adv}_{\mathrm{H}}^{\mathrm{DSPR}}(\mathcal{B}) &= \max\{0, \Pr[\mathcal{B}(x,k) = P_k(x)] - p\} \\
&\geq \Pr[\mathcal{B}(x,k) = P_k(x)] - p \\
&= s_1 + \sum_{i>1} \frac{i-1}{i} s_i + \sum_{i>1} f_i - \sum_{i>1} s_i - \sum_{i>1} f_i \\
&= s_1 - \sum_{i>1} \frac{1}{i} s_i.
\end{aligned}$$

SPR success probability. For the $i = 1$ cases, $\mathcal{C}$ never succeeds. For the $i > 1$ cases, $\mathcal{C}$ succeeds if and only if $\mathcal{A}$ succeeds and returns a value different from x. This happens with conditional probability $\frac{i-1}{i}$ for the same reason as above. Hence

$$\mathrm{Succ}_{\mathrm{H}}^{\mathrm{SPR}}(\mathcal{C}) = \sum_{i>1} \frac{i-1}{i} s_i.$$

Combining the probabilities. We have

$$\begin{aligned}\mathrm{Adv}_{\mathrm{H}}^{\mathrm{DSPR}}(\mathcal{B}) + 3 \cdot \mathrm{Succ}_{\mathrm{H}}^{\mathrm{SPR}}(\mathcal{C}) &\geq s_1 - \sum_{i>1} \frac{1}{i} s_i + 3 \sum_{i>1} \frac{i-1}{i} s_i \\ &= s_1 + \sum_{i>1} \frac{3i-4}{i} s_i \\ &\geq s_1 + \sum_{i>1} s_i = \mathrm{Succ}_{\mathrm{H}}^{\mathrm{PRE}}(\mathcal{A})\end{aligned}$$

as claimed.

The formal structure of the proof is concluded at this point, but we close with some informal comments on how to interpret this proof. What happens is the following. The cases where the plain reduction from SPR ($\mathcal{C}$ in the above) fails are the S_1 cases, i.e., $\mathcal{A}$ succeeds when there is only one preimage. If the probability that they occur (s_1) gets close to $\mathcal{A}$'s total success probability, the success probability of $\mathcal{C}$ goes towards zero. However, s_1 translates almost directly to the DSPR advantage of $\mathcal{B}$. This is also intuitively what we want. For a brute-force attack, one would expect s_1 to be less than a $1 - p$ fraction of $\mathcal{A}$'s success probability. If it is higher, this allows to distinguish. On the extreme: If $s_1 = s$, then $\mathcal{B}$'s DSPR advantage is exactly $\mathcal{A}$'s success probability and the reduction is tight. If $s_1 = 0$, $\mathcal{B}$ has no advantage over guessing, but $\mathcal{C}$ wins with at least half the success probability of $\mathcal{A}$ (in this case our generic $1/3$ bound can be tightened). As mentioned above, in general one would expect s_1 to be a recognizable fraction of s but clearly smaller than s. In these cases, both reductions succeed. □

5 DSPR is Hard in the QROM

So far we have highlighted relations between DSPR and other hash function properties. However, all this is useful only if DSPR is a hard problem for the hash functions we are interested in. In the following we show that DSPR is hard for a quantum adversary as long as the hash function behaves like a random function. We do this presenting a lower bound on the quantum query complexity for DSPR.

To make previous results reusable, we first need a result that relates the success probability of an adversary in a biased distinguishing game like the DSPR game to its success probability in the balanced version of the game.

Theorem 26. *Let B_λ denote the Bernoulli distribution that assigns probability λ to 1, $\mathcal{X}_b$ for $b \in \{0,1\}$ a non-empty set,*

$$\mathrm{Succ}_\lambda(\mathcal{A}) \stackrel{def}{=} \Pr[b \leftarrow_R B_\lambda; x \leftarrow_R \mathcal{X}_b; g \leftarrow \mathcal{A}(x) : g = b],$$

and

$$\mathrm{Adv}_\lambda(\mathcal{A}) \stackrel{def}{=} \max\{0, \mathrm{Succ}_\lambda(\mathcal{A}) - \lambda\}$$

Then for $p \geq 1/2$ we have

$$\mathrm{Adv}_p(\mathcal{A}) \leq p\left|\Pr[x \leftarrow_R \mathcal{X}_1 : 1 \leftarrow \mathcal{A}(x)] - \Pr[x \leftarrow_R \mathcal{X}_0 : 1 \leftarrow \mathcal{A}(x)]\right|.$$

More specifically

$$\mathrm{Succ}_{\frac{1}{2}}(\mathcal{A}) \geq \frac{1}{2p}\mathrm{Succ}_p(\mathcal{A}), \quad \mathrm{Adv}_{\frac{1}{2}}(\mathcal{A}) \geq \frac{1}{2p}\mathrm{Adv}_p(\mathcal{A}),$$

and

$$\frac{1}{2}\left|\Pr[x \leftarrow_R \mathcal{X}_1 : 1 \leftarrow \mathcal{A}(x)] - \Pr[x \leftarrow_R \mathcal{X}_0 : 1 \leftarrow \mathcal{A}(x)]\right| \geq \mathrm{Adv}_{\frac{1}{2}}(\mathcal{A})$$

Proof. Let $s_0 = \Pr[b=0 \wedge g=0] = \Pr[b=g \mid b=0]\Pr[b=0]$ and $s'_0 = \Pr[b=g \mid b=0]$. Define $s_1 = \Pr[b=1 \wedge g=1] = \Pr[b=g \mid b=1]\Pr[b=1]$ and $s'_1 = \Pr[b=g \mid b=1]$, accordingly. Then

$$\begin{aligned}\frac{1}{2p}\mathrm{Succ}_p(\mathcal{A}) &= \frac{1}{2p}((1-p)s'_0 + ps'_1)\\ &= \frac{1-p}{p}\cdot\frac{1}{2}s'_0 + \frac{1}{2}s'_1 \leq \frac{1}{2}s'_0 + \frac{1}{2}s'_1 = \mathrm{Succ}_{\frac{1}{2}}(\mathcal{A}),\end{aligned}$$

where we used $p \geq 1/2$. Now, for a zero advantage in the biased game the second sub-claim is trivially true. For a non-zero advantage $\mathrm{Adv}_p(\mathcal{A})$ we get

$$\begin{aligned}\mathrm{Adv}_p(\mathcal{A}) &= \max\{0, \mathrm{Succ}_p(\mathcal{A}) - p\}\\ \mathrm{Adv}_p(\mathcal{A}) + p &= \mathrm{Succ}_p(\mathcal{A})\\ \frac{1}{2p}(\mathrm{Adv}_p(\mathcal{A}) + p) &\leq \mathrm{Succ}_{\frac{1}{2}}(\mathcal{A})\\ \frac{1}{2p}\mathrm{Adv}_p(\mathcal{A}) + \frac{1}{2} &\leq \mathrm{Succ}_{\frac{1}{2}}(\mathcal{A})\\ \frac{1}{2p}\mathrm{Adv}_p(\mathcal{A}) &\leq \mathrm{Succ}_{\frac{1}{2}}(\mathcal{A}) - \frac{1}{2} = \mathrm{Adv}_{\frac{1}{2}}(\mathcal{A}).\end{aligned}$$

The last sub-claim follows from

$$\begin{aligned}\mathrm{Adv}_{\frac{1}{2}}(\mathcal{A}) &= \max\left\{0, \mathrm{Succ}_{\frac{1}{2}}(\mathcal{A}) - \frac{1}{2}\right\} \leq \left|\mathrm{Succ}_{\frac{1}{2}}(\mathcal{A}) - \frac{1}{2}\right|\\ &= \left|\frac{1}{2}(\Pr[x \leftarrow_R \mathcal{X}_1 : 1 \leftarrow \mathcal{A}(x)] + \Pr[x \leftarrow_R \mathcal{X}_0 : 0 \leftarrow \mathcal{A}(x)]) - \frac{1}{2}\right|\\ &= \left|\frac{1}{2}(\Pr[x \leftarrow_R \mathcal{X}_1 : 1 \leftarrow \mathcal{A}(x)] + 1 - \Pr[x \leftarrow_R \mathcal{X}_0 : 1 \leftarrow \mathcal{A}(x)]) - \frac{1}{2}\right|\\ &= \frac{1}{2}\left|\Pr[x \leftarrow_R \mathcal{X}_1 : 1 \leftarrow \mathcal{A}(x)] - \Pr[x \leftarrow_R \mathcal{X}_0 : 1 \leftarrow \mathcal{A}(x)]\right|\end{aligned}$$

The main statement follows from plugging the last two sub-claims together. □

Our approach to show that DSPR is hard is giving a reduction from an average-case distinguishing problem that was used in the full version of [9]. The problem makes use of the following distribution D_λ over boolean functions.

Definition 27 [9]. *Let* $\mathcal{F} \stackrel{def}{=} \{f : \{0,1\}^m \to \{0,1\}\}$ *be the collection of all boolean functions on* $\{0,1\}^m$. *Let* $\lambda \in [0,1]$ *and* $\varepsilon > 0$. *Define a family of distributions* D_λ *on* $\mathcal{F}$ *such that* $f \leftarrow_R D_\lambda$ *satisfies*

$$f : x \mapsto \begin{cases} 1 & \textit{with prob. } \lambda, \\ 0 & \textit{with prob. } 1-\lambda \end{cases}$$

for any $x \in \{0,1\}^m$.

In [9] the following bound on the distinguishing advantage of any q-query quantum adversary was shown.

Theorem 28 [9]. *Let* D_λ *be defined as in Definition 27, and* $\mathcal{A}$ *be any quantum algorithm making at most* q *quantum queries to its oracle. Then*

$$\mathrm{Adv}_{D_0,D_\lambda}(\mathcal{A}) \stackrel{def}{=} \left| \Pr_{f \leftarrow D_0}[\mathcal{A}^f(\cdot) = 1] - \Pr_{f \leftarrow D_\lambda}[\mathcal{A}^f(\cdot) = 1] \right| \le 8\lambda q^2.$$

We still have to briefly discuss how DSPR is defined in the (quantum-accessible) random oracle model. Instead of giving a description of the hash function H as implicitly done in Definition 5, we provide $\mathcal{A}$ with an oracle $\mathcal{O}$ that implements a function $\mathrm{F} : \mathcal{X} \to \mathcal{Y}$. As for most other notions that can be defined for unkeyed hash functions, DSPR in the (Q)ROM becomes the same for keyed and non-keyed hash functions. For keyed functions, instead of giving a description of the keyed hash function H and a key k to the adversary $\mathcal{A}$, we provide $\mathcal{A}$ with an oracle that implements a function $\mathrm{F} : \mathcal{X} \to \mathcal{Y}$ which now models H for a fixed key k. Hence, the following result applies to both cases. This can be seen as the key space might contain just a single key.

Now we got all tooling we need to show that DSPR is a hard problem.

Theorem 29. *Let* $n \in \mathbb{N}$, $N = 2^n$, $\mathrm{H} : \mathcal{K} \times \{0,1\}^n \to \{0,1\}^n$ *as defined above be a random, length-preserving keyed hash function. Any quantum adversary* $\mathcal{A}$ *that solves DSPR making* q *quantum queries to* H *can be used to construct a quantum adversary* $\mathcal{B}$ *that makes* $2q$ *queries to its oracle and distinguishes* D_0 *from* $D_{1/N}$ *with success probability*

$$\mathrm{Adv}_{D_0,D_{1/N}}(\mathcal{B}) \ge \mathrm{Adv}_{\mathrm{H}}^{\mathrm{DSPR}}(\mathcal{A}).$$

Proof. By construction. The algorithm $\mathcal{B}$ generates an DSPR instance as in Fig. 2 and runs $\mathcal{A}$ on it. It outputs whatever $\mathcal{A}$ outputs. To answer an H query $\mathcal{B}$ needs two f queries as it also has to uncompute the result of the f query after

Given: Oracle access to $f : \mathcal{X} \to \{0,1\}$.

1. Sample $x' \leftarrow \mathcal{X}$ and $y' \leftarrow \mathcal{Y}$ independently and uniformly at random.
2. Let $g : \mathcal{X} \to \mathcal{Y} \backslash \{y'\}$ be a random function. We construct $\mathrm{H} : \mathcal{X} \to \mathcal{Y}$ as follows: for any $x \in \mathcal{X}$

$$x \mapsto \begin{cases} y' & \text{if } x = x' \\ y' & \text{if } x \neq x' \wedge f(x) = 1 \\ g(x) & \text{otherwise.} \end{cases}$$

Output: DSPR instance (H, x'). Namely an adversary is given x' and oracle access to H, and the goal is to decide if x' has a second preimage under H.

Fig. 2. Reducing distinguishing D_0 from $D_{1/N}$ to DSPR.

it was used. The random function g can be efficiently simulated using $2q$-wise independent hash functions as discussed in [9].

Now, if $f \leftarrow_R D_0$, (H, x') is a random DSPR challenge from the set of all DSPR challenges with $P_{\mathrm{H}}(x') = 0$ (slightly abusing notation as we do not know a key for our random function). Similarly, if $f \leftarrow_R D_{1/N}$, (H, x') is a random DSPR challenge from the set of all DSPR challenges.

$$\begin{aligned}
\mathrm{Adv}_{D_0, D_\lambda}(\mathcal{B}) &= \left| \Pr_{f \leftarrow D_0}[\mathcal{B}^f(\cdot) = 1] - \Pr_{f \leftarrow D_\lambda}[\mathcal{B}^f(\cdot) = 1] \right| \\
&= \left| \Pr_{f \leftarrow D_0}[\mathcal{A}^{\mathrm{H}}(x') = 1] - \Pr_{f \leftarrow D_\lambda}[\mathcal{A}^{\mathrm{H}}(x') = 1] \right| \\
&= \big| \Pr[\mathcal{A}^{\mathrm{H}}(x') = 1 \mid P_{\mathrm{H}}(x') = 0] - \big(p \cdot \Pr[\mathcal{A}^{\mathrm{H}}(x') = 1 \mid P_{\mathrm{H}}(x') = 1] \\
&\quad + (1-p) \cdot \Pr[\mathcal{A}^{\mathrm{H}}(x') = 1 \mid P_{\mathrm{H}}(x') = 0]\big) \big| \\
&= p \cdot \big| \Pr[\mathcal{A}^{\mathrm{H}}(x') = 1 \mid P_{\mathrm{H}}(x') = 1] - \Pr[\mathcal{A}^{\mathrm{H}}(x') = 1 \mid P_{\mathrm{H}}(x') = 0] \big| \\
&\geq \mathrm{Adv}_{\mathrm{H}}^{\mathrm{DSPR}}(\mathcal{A}),
\end{aligned}$$

where the last inequality follows from Theorem 26.

□

Theorem 30. *Let $n \in \mathbb{N}$, $N = 2^n$, $\mathrm{H} : \mathcal{K} \times \{0,1\}^n \to \{0,1\}^n$ as defined above be a random, length-preserving keyed hash function. Any quantum adversary $\mathcal{A}$ that makes no more than q quantum queries to its oracle can only solve the decisional second-preimage problem with advantage*

$$\mathrm{Adv}_{\mathrm{H}}^{\mathrm{DSPR}}(\mathcal{A}) \leq 32q^2/N.$$

Proof. Use Theorem 29 to construct an adversary $\mathcal{B}$ that makes $2q$ queries and that has advantage at least $\mathrm{Adv}_{\mathrm{H}}^{\mathrm{DSPR}}(\mathcal{A})$ of distinguishing D_0 from $D_{1/N}$. This advantage is at most $8(1/N)(2q)^2 = 32q^2/N$ by Theorem 28. □

6 DSPR for Multiple Targets

Multi-target security considers an adversary that is given T independent targets and is asked to solve a problem for one out of the T targets. This section defines T-DSPR, a multi-target version of DSPR.

We draw attention to an unusual feature of this definition: the advantage of an adversary $\mathcal{A}$ is defined as the improvement from p to q, where p and q are two probabilities that can *both* be influenced by $\mathcal{A}$. The second probability q is $\mathcal{A}$'s chance of correctly predicting whether the input selected by $\mathcal{A}$ has a second preimage. The first probability p is the chance that the input selected by $\mathcal{A}$ does have a second preimage.

This deviates from the usual view of advantage as how much $\mathcal{A}$ improves upon success probability compared to some trivial baseline attack. What we are doing, for multi-target attacks, is asking how much $\mathcal{A}$ improves upon success probability compared to the baseline attack *against the same target that* $\mathcal{A}$ *selected.* In most of the contexts considered in the literature, the success probability of the baseline attack is independent of the target, so this matches the usual view. DSPR is different, because the success probability of the baseline attack depends on the target.

One can object that this allows the baseline attack to be affected (positively or negatively) by $\mathcal{A}$'s competence in target selection. We give two responses to this objection. First, our definition enables a proof (Theorem 33) that T-DSPR is at most T times easier to break than DSPR. Second, our definition enables an interactive multi-target generalization (Theorem 38) of our proof that DSPR and SPR together imply PRE.

Definition 31 (T-DSPR). *Let T be a positive integer. Let $\mathcal{A}$ be an algorithm with output in $\{1,\ldots,T\} \times \{0,1\}$. The advantage of $\mathcal{A}$ against the T-target decisional second-preimage resistance of a keyed hash function* H *is*

$$\mathrm{Adv}_{\mathrm{H}}^{T\text{-DSPR}}(\mathcal{A}) \overset{def}{=} \max\{0, q-p\}$$

where

$$\begin{aligned} q = \Pr\big[&(x_1,k_1,\ldots,x_T,k_T) \leftarrow_R (\mathcal{X}\times\mathcal{K})^T;\\ &(j,b) \leftarrow \mathcal{A}(x_1,k_1,\ldots,x_T,k_T) : P_{k_j}(x_j) = b\big];\\ p = \Pr\big[&(x_1,k_1,\ldots,x_T,k_T) \leftarrow_R (\mathcal{X}\times\mathcal{K})^T;\\ &(j,b) \leftarrow \mathcal{A}(x_1,k_1,\ldots,x_T,k_T) : P_{k_j}(x_j) = 1\big]; \end{aligned}$$

and $P_{k_j} = \mathrm{SPexists}(\mathrm{H}_{k_j})$.

The only difference between the formulas for q and p is that q compares $P_{k_j}(x_j)$ to b while p compares it to 1. If $T > 1$ then an algorithm might be able to influence p up or down, compared to any particular $\mathrm{SPprob}(\mathrm{H}_{k_i})$, through the choice of j. Obtaining a significant T-DSPR advantage then means obtaining q significantly larger than p, i.e., making a prediction of $P_{k_j}(x_j)$ significantly better than always predicting that it is 1.

As an extreme case, consider the following slow algorithm. Compute each $P_{k_j}(x_j)$ by brute force; choose j where $P_{k_j}(x_j) = 0$ if such a j exists, else $j = 1$; and output $P_{k_j}(x_j)$. This algorithm has $q = 1$ and thus T-DSPR advantage $1-p$. The probability p for this algorithm is the probability that *all* of $x_1, \ldots, x_T$ have second preimages. For most length-preserving functions, this probability is approximately $(1 - 1/e)^T$, which rapidly converges to 0 as T increases, so the T-DSPR advantage rapidly converges to 1.

Definition 32. *Let $\mathcal{A}$ be an algorithm, and let T be a positive integer. Then* $\mathrm{Plant}_T(A)$ *is the following algorithm:*

- *Input* $(x, k) \in \mathcal{X} \times \mathcal{K}$.
- *Generate* $i \leftarrow_R \{1, \ldots, T\}$.
- *Generate* $(x_1, k_1, \ldots, x_T, k_T) \leftarrow_R (\mathcal{X} \times \mathcal{K})^T$.
- *Overwrite* $(x_i, k_i) \leftarrow (x, k)$.
- *Compute* $(j, b) \leftarrow \mathcal{A}(x_1, k_1, \ldots, x_T, k_T)$.
- *Output* b *if* $j = i$, *or* 1 *if* $j \neq i$.

This uses the standard technique of planting a single-target challenge at a random position in a multi-target challenge. With probability $1/T$, the multi-target attack chooses the challenge position; in the other cases, this reduction outputs 1. The point of Theorem 33 is that this reduction interacts nicely with the subtraction of probabilities in the DSPR and T-DSPR definitions.

The cost of $\mathrm{Plant}_T(A)$ is the cost of generating a random number i between 1 and T, generating $T - 1$ elements of $\mathcal{X} \times \mathcal{K}$, running $\mathcal{A}$, and comparing j to i. The algorithm has essentially the same cost as $\mathcal{A}$ if $\mathcal{X}$ and $\mathcal{K}$ can be efficiently sampled.

Theorem 33 (T-loose implication DSPR $\Rightarrow$ T-DSPR). *Let* H *be a keyed hash function. Let T be a positive integer. Let $\mathcal{A}$ be an algorithm with output in* $\{1, \ldots, T\} \times \{0, 1\}$. *Then*

$$\mathrm{Adv}_{\mathrm{H}}^{T\text{-DSPR}}(\mathcal{A}) = T \cdot \mathrm{Adv}_{\mathrm{H}}^{\mathrm{DSPR}}(\mathcal{B})$$

where $\mathcal{B} = \mathrm{Plant}_T(\mathcal{A})$.

Proof. By definition $\mathrm{Adv}_{\mathrm{H}}^{T\text{-DSPR}}(\mathcal{A})$ runs $\mathcal{A}$ with T independent uniform random targets $(x_1, k_1, \ldots, x_T, k_T)$. Write (j, b) for the output of $\mathcal{A}(x_1, k_1, \ldots, x_T, k_T)$. Then $\mathrm{Adv}_{\mathrm{H}}^{T\text{-DSPR}}(\mathcal{A}) = \max\{0, q - p\}$, where q is the probability that $P_{k_j}(x_j) = b$, and p is the probability that $P_{k_j}(x_j) = 1$.

To analyze q and p, we split the universe of possible events into four mutually exclusive events:

$$E_{00} \stackrel{\text{def}}{=} [b = 0 \wedge P_{k_j}(x_j) = 0];$$

$$E_{01} \stackrel{\text{def}}{=} [b = 0 \wedge P_{k_j}(x_j) = 1];$$

$$E_{10} \stackrel{\text{def}}{=} [b = 1 \wedge P_{k_j}(x_j) = 0];$$

$$E_{11} \stackrel{\text{def}}{=} [b = 1 \wedge P_{k_j}(x_j) = 1].$$

Then $q = \Pr E_{00} + \Pr E_{11}$ and $p = \Pr E_{01} + \Pr E_{11}$, so $q - p = \Pr E_{00} - \Pr E_{01}$.

For comparison, $\mathrm{Adv}_{\mathrm{H}}^{\mathrm{DSPR}}(\mathcal{B})$ runs $\mathcal{B}$, which in turn runs $\mathcal{A}$ with T independent uniform random targets $(x_1, k_1, \ldots, x_T, k_T)$. One of these targets (x_i, k_i) is the uniform random target (x, k) provided to $\mathcal{B}$ as a challenge; $\mathcal{B}$ randomly selects i and the remaining targets. The output b' of $\mathcal{B}(x, k)$ is b if $j = i$, and 1 if $j \neq i$.

The choice of i is not visible to $\mathcal{A}$, so the event that $i = j$ has probability $1/T$. Furthermore, this event is independent of $E_{00}, E_{01}, E_{10}, E_{11}$: i.e., $i = j$ has conditional probability $1/T$ given E_{00}, conditional probability $1/T$ given E_{01}, etc.

Write q' for the chance that $P_k(x) = b'$, and p' for the chance that $P_k(x) = 1$. Then $\mathrm{Adv}_{\mathrm{H}}^{\mathrm{DSPR}}(\mathcal{B}) = \max\{0, q' - p'\}$. To analyze q' and p', we split into mutually exclusive events as follows:

- E_{00} occurs and $i = j$. This has probability $(\Pr E_{00})/T$. Then $(x_j, k_j) = (x_i, k_i) = (x, k)$ so $P_k(x) = P_{k_j}(x_j) = 0 = b = b'$. This contributes to q' and not to p'.
- E_{01} occurs and $i = j$. This has probability $(\Pr E_{01})/T$. Then $(x_j, k_j) = (x, k)$ so $P_k(x) = 1$, while $b' = b = 0$. This contributes to p' and not to q'.
- All other cases: $b' = 1$ (since $b' = 0$ can happen only if $b = 0$ and $i = j$). We further split this into two cases:
  - $P_k(x) = 1$. This contributes to q' and to p'.
  - $P_k(x) = 0$. This contributes to neither q' nor p'.

To summarize, $q' - p' = (\Pr E_{00})/T - (\Pr E_{01})/T = (q - p)/T$. Hence

$$\max\{0, q - p\} = \max\{0, T(q' - p')\} = T \max\{0, q' - p'\};$$

i.e., $\mathrm{Adv}_{\mathrm{H}}^{T\text{-DSPR}}(\mathcal{A}) = T \cdot \mathrm{Adv}_{\mathrm{H}}^{\mathrm{DSPR}}(\mathcal{B})$. □

7 Removing Interactivity

The real importance of DSPR for security proofs is that it allows interactive versions of preimage resistance to be replaced by non-interactive assumptions without penalty. Interactive versions of preimage resistance naturally arise in, e.g., the context of hash-based signatures; see Sect. 8.

The example discussed in this section is the T-openPRE notion already informally introduced in Sect. 1.1.1. We first review T-SPR, a multi-target version of second-preimage resistance. Then we formally define the interactive notion T-openPRE and show that its security tightly relates to T-SPR and T-DSPR.

T-SPR is what is called multi-function, multi-target second-preimage resistance in [9]. It was shown in [9] that a generic attack against T-SPR has the same complexity as a generic attack against SPR.

Definition 34 [9] **(*T*-SPR).** *The success probability of an algorithm $\mathcal{A}$ against the T-target second-preimage resistance of a keyed hash function* H *is*

$$\begin{aligned}\mathrm{Succ}_{\mathrm{H}}^{T\text{-SPR}}(\mathcal{A}) \stackrel{def}{=} \Pr\big[&(x_1, k_1, \ldots, x_T, k_T) \leftarrow_R (\mathcal{X} \times \mathcal{K})^T;\\ &(j, x) \leftarrow \mathcal{A}(x_1, k_1, \ldots, x_T, k_T) :\\ &\mathrm{H}_{k_j}(x) = \mathrm{H}_{k_j}(x_j) \wedge x \neq x_j\big].\end{aligned}$$

T-openPRE is essentially what would be T-PRE (which we did not define) but with the additional tweak that the adversary gets access to an opening oracle. The adversary is allowed to query the oracle for the preimages of all but one of the targets and has to output a preimage for the remaining one.

Definition 35 (T-openPRE). *Let* H *be a keyed hash function. The success probability of an algorithm $\mathcal{A}$ against the T-target opening-preimage resistance of* H *is defined as*

$$\begin{aligned}\mathrm{Succ}_{\mathrm{H}}^{T\text{-openPRE}}(\mathcal{A}) \stackrel{def}{=} \mathsf{Pr}\big[&(x_1, k_1, \ldots, x_T, k_T) \leftarrow_R (\mathcal{X} \times \mathcal{K})^T; \\ &(j, x') \leftarrow \mathcal{A}^{\mathsf{Open}}(\mathrm{H}_{k_1}(x_1), k_1, \ldots, \mathrm{H}_{k_T}(x_T), k_T) : \\ &\mathrm{H}_{k_j}(x') = \mathrm{H}_{k_j}(x_j) \wedge j \textit{ was no query of } \mathcal{A}\big]\end{aligned}$$

where $\mathsf{Open}(i) = x_i$.

Now, it is of course possible to reduce PRE to T-openPRE. However, such a reduction has to guess the index j for which $\mathcal{A}$ will output a preimage (and hence does not make a query) correctly. Otherwise, if the reduction embeds its challenge image in any of the other positions, it cannot answer $\mathcal{A}$'s query for that index. As $\mathcal{A}$ does not lose anything by querying all indices but j, we can assume that it actually does so. Hence, such a reduction from PRE must incur a loss in tightness of a factor T. For some applications discussed below, T can reach the order of $\sqrt[4]{N}$. This implies a quarter loss in the security level.

Theorem 38 shows that T-openPRE is tightly related to the non-interactive assumptions T-DSPR and T-SPR: if H is T-target decisional-second-preimage resistant and T-target second-preimage resistant then it is T-target opening-preimage-resistant. As before, we first define the reductions and then state a theorem regarding probabilities.

Definition 36 (T-target SPfromP). *Let* H *be a keyed hash function. Let $\mathcal{A}$ be an algorithm using an oracle. Let T be a positive integer. Then* $\mathrm{SPfromP}_T(\mathrm{H}, \mathcal{A})$ *is the following algorithm:*

- *Input* $(x_1, k_1, \ldots, x_T, k_T) \in (\mathcal{X} \times \mathcal{K})^T$.
- *Output* $\mathcal{A}^{\mathsf{Open}}(\mathrm{H}_{k_1}(x_1), k_1, \ldots, \mathrm{H}_{k_T}(x_T), k_T)$, *where* $\mathsf{Open}(i) = x_i$.

This generalizes the standard SPfromP reduction: it handles multiple targets in the obvious way, and it easily answers oracle queries with no failures since it knows all the x_i inputs. The algorithm $\mathrm{SPfromP}_T(\mathrm{H}, \mathcal{A})$ uses T calls to H (which can be deferred until their outputs are used) and one call to $\mathcal{A}$.

Definition 37 (T-target DSPfromP). *Let* H *be a keyed hash function. Let $\mathcal{A}$ be an algorithm. Then* $\mathrm{DSPfromP}_T(\mathrm{H}, \mathcal{A})$ *is the following algorithm:*

- *Input* $(x_1, k_1, \ldots, x_T, k_T) \in (\mathcal{X} \times \mathcal{K})^T$.
- *Compute* $(j, x') \leftarrow \mathcal{A}^{\mathsf{Open}}(\mathrm{H}_{k_1}(x_1), k_1, \ldots, \mathrm{H}_{k_T}(x_T), k_T)$, *where* $\mathsf{Open}(i) = x_i$.
- *Compute* $b \leftarrow ((x' \neq x_j) \vee j \textit{ was a query of } \mathcal{A})$.
- *Output* (j, b).

This is an analogous adaptation of our DSPfromP reduction to the interactive multi-target context. Again oracle queries are trivial to answer. Note that the case that $\mathcal{A}$ cheats, returning an index j that it used for an Open query, is a failure case for $\mathcal{A}$ by definition; the algorithm $\text{DSPfromP}_T(\text{H}, \mathcal{A})$ outputs 1 in this case, exactly as if $\mathcal{A}$ had failed to find a preimage. In other words, this algorithm returns 0 whenever $\mathcal{A}$ returns a solution that contains the preimage that was already known by the reduction (but *not* given to $\mathcal{A}$ via Open), and 1 otherwise.

Theorem 38 (T-DSPR $\wedge$ T-SPR $\Rightarrow$ T-openPRE). *Let* H *be a keyed hash function. Let T be a positive integer. Let $\mathcal{A}$ be an algorithm. Then*

$$\text{Succ}_{\text{H}}^{T\text{-openPRE}}(\mathcal{A}) \leq \text{Adv}_{\text{H}}^{T\text{-DSPR}}(\mathcal{B}) + 3 \cdot \text{Succ}_{\text{H}}^{T\text{-SPR}}(\mathcal{C})$$

where $\mathcal{B} = \text{DSPfromP}_T(\text{H}, \mathcal{A})$ *and* $\mathcal{C} = \text{SPfromP}_T(\text{H}, \mathcal{A})$.

The core proof idea is the following. As noted above, the reductions attacking T-SPR and T-DSPR can perfectly answer all of $\mathcal{A}$'s oracle queries as they know preimages. However, for the index for which $\mathcal{A}$ outputs a preimage (without cheating), it did not learn the preimage known to the reduction. Hence, from there on we can apply a similar argument as in the proof of Theorem 25. We include a complete proof below to aid in verification.

Proof. Write (j, x') for the output of $\mathcal{A}^{\mathsf{Open}}(\text{H}_{k_1}(x_1), k_1, \ldots, \text{H}_{k_T}(x_T), k_T)$. As in the proof of Theorem 25, we split the universe of possible events into mutually exclusive events across two dimensions: the number of preimages of $\text{H}_{k_j}(x_j)$, and whether $\mathcal{A}$ succeeds or fails in finding a preimage. Specifically, define

$$S_i \stackrel{\text{def}}{=} \left[\left|\text{H}_{k_j}^{-1}(\text{H}_{k_j}(x_j))\right| = i \wedge \text{H}_{k_j}(x') = \text{H}_{k_j}(x_j) \wedge j \text{ was no query of } \mathcal{A}\right],$$

as the event that there are exactly i preimages and that $\mathcal{A}$ succeeds, and define

$$F_i \stackrel{\text{def}}{=} \left[\left|\text{H}_{k_j}^{-1}(\text{H}_{k_j}(x_j))\right| = i \wedge \left(\text{H}_{k_j}(x') \neq \text{H}_{k_j}(x_j) \vee j \text{ was a query of } \mathcal{A}\right)\right]$$

as the event that there are exactly i preimages and that $\mathcal{A}$ fails. Note that there are only finitely many i for which the events S_i and F_i can occur.

Define s_i and f_i as the probabilities of S_i and F_i respectively. The probability space here includes the random choices of $(x_1, k_1, \ldots, x_T, k_T)$, and any random choices made inside $\mathcal{A}$.

***T*-openPRE success probability.** By definition, $\text{Succ}_{\text{H}}^{T\text{-openPRE}}(\mathcal{A})$ is the probability that x' is a non-cheating preimage of $\text{H}_{k_j}(x_j)$; i.e., that $\text{H}_{k_j}(x') = \text{H}_{k_j}(x_j)$ and j was not a query to the oracle. This event is the union of the events S_i, so $\text{Succ}_{\text{H}}^{T\text{-openPRE}}(\mathcal{A}) = \sum_i s_i$.

***T*-DSPR success probability.** By definition $\mathcal{B}$ outputs the pair (j, b), where $b = ((x' \neq x_j) \vee j \text{ was a query of } \mathcal{A})$.

Define $P_{k_j} = \mathrm{SPexists}(\mathrm{H}_{k_j})$, and define q as in the definition of $\mathrm{Adv}_{\mathrm{H}}^{T\text{-DSPR}}(\mathcal{B})$. Then q is the probability that $\mathcal{B}$ is correct, i.e., that $b = P_{k_j}(x_j)$. There are four cases:

- If the event S_1 occurs, then there is exactly 1 preimage of $\mathrm{H}_{k_j}(x_j)$, so $P_{k_j}(x_j) = 0$ by definition of SPexists. Also, $\mathcal{A}$ succeeds: i.e., j was not a query, and x' is a preimage of $\mathrm{H}_{k_j}(x_j)$, forcing $x' = x_j$. Hence $b = 0 = P_{k_j}(x_j)$.
- If the event F_1 occurs, then again $P_{k_j}(x_j) = 0$, but now $\mathcal{A}$ fails: i.e., j was a query, or x' is not a preimage of $\mathrm{H}_{k_j}(x_j)$. Either way $b = 1 \neq P_{k_j}(x_j)$. (We could skip this case in the proof, since we need only a lower bound on q rather than an exact formula for q.)
- If the event S_i occurs for $i > 1$, then $P_{k_j}(x_j) = 1$ and $\mathcal{A}$ succeeds. Hence j was not a query, and x' is a preimage of $\mathrm{H}_{k_j}(x_j)$, so $x' = x_j$ with conditional probability exactly $\frac{1}{i}$. Hence $b = 1 = P_{k_j}(x_j)$ with conditional probability exactly $\frac{i-1}{i}$.
- If the event F_i occurs for $i > 1$, then $P_{k_j}(x_j) = 1$ and $\mathcal{A}$ fails. Failure means that x' is not a preimage, so in particular $x' \neq x_j$, or that j was a query. Either way $b = 1 = P_{k_j}(x_j)$.

To summarize, $q = s_1 + \sum_{i>1} \frac{i-1}{i} s_i + \sum_{i>1} f_i$.

***T*-DSPR advantage.** Define p as in the definition of $\mathrm{Adv}_{\mathrm{H}}^{T\text{-DSPR}}(\mathcal{B})$. Then $\mathrm{Adv}_{\mathrm{H}}^{T\text{-DSPR}}(\mathcal{B}) = \max\{0, q - p\}$.

The analysis of p is the same as the analysis of q above, except that we compare $P_{k_j}(x_j)$ to 1 instead of comparing it to b. We have $1 = P_{k_j}(x_j)$ exactly for the events S_i and F_i with $i > 1$. Hence $p = \sum_{i>1} s_i + \sum_{i>1} f_i$. Subtract to see that

$$\mathrm{Adv}_{\mathrm{H}}^{T\text{-DSPR}}(\mathcal{B}) = \max\{0, q - p\} \geq q - p = s_1 - \sum_{i>1} \frac{1}{i} s_i.$$

***T*-SPR success probability.** By definition $\mathcal{C}$ outputs (j, x'). The T-SPR success probability $\mathrm{Succ}_{\mathrm{H}}^{T\text{-SPR}}(\mathcal{C})$ is the probability that x' is a second preimage of x_j under H_{k_j}, i.e., that $\mathrm{H}_{k_j}(x') = \mathrm{H}_{k_j}(x_j)$ while $x' \neq x_j$.

It is possible for $\mathcal{C}$ to succeed while $\mathcal{A}$ fails: perhaps $\mathcal{A}$ learns $x_j = \mathsf{Open}(j)$ and then computes a second preimage for x_j, which does not qualify as an T-openPRE success for $\mathcal{A}$ but does qualify as a T-SPR success for $\mathcal{C}$. We ignore these cases, so we obtain only a lower bound on $\mathrm{Succ}_{\mathrm{H}}^{T\text{-SPR}}(\mathcal{C})$; this is adequate for the proof.

Assume that event S_i occurs with $i > 1$. Then x' is a preimage of $\mathrm{H}_{k_j}(x_j)$. Furthermore, $\mathcal{A}$ did not query j, so x_j is not known to $\mathcal{A}$ except via $\mathrm{H}_{k_j}(x_j)$. There are i preimages, so $x' = x_j$ with conditional probability exactly $\frac{1}{i}$. Hence $\mathcal{C}$ succeeds with conditional probability $\frac{i-1}{i}$.

To summarize, $\mathrm{Succ}_{\mathrm{H}}^{T\text{-SPR}}(\mathcal{C}) \geq \sum_{i>1} \frac{i-1}{i} s_i$.

Combining the probabilities. We conclude as in the proof of Theorem 25:

$$\begin{aligned}\mathrm{Adv}_{\mathrm{H}}^{T\text{-DSPR}}(\mathcal{B}) + 3 \cdot \mathrm{Succ}_{\mathrm{H}}^{T\text{-SPR}}(\mathcal{C}) &\geq s_1 - \sum_{i>1} \frac{1}{i} s_i + 3 \sum_{i>1} \frac{i-1}{i} s_i \\ &= s_1 + \sum_{i>1} \frac{3i-4}{i} s_i \\ &\geq s_1 + \sum_{i>1} s_i = \mathrm{Succ}_{\mathrm{H}}^{T\text{-openPRE}}(\mathcal{A}).\end{aligned}$$

□

8 Applications to Hash-Based Signatures

The interactive notion of T-openPRE with a huge number of targets naturally arises in the context of hash-based signatures. This was already observed and extensively discussed in [9]. One conclusion of the discussion there is to use keyed hash functions with new (pseudo)random keys for each hash-function call made in a hash-based signature scheme.

When applying this idea to Lamport one-time signatures (L-OTS) [11], the standard security notion for OTS of existential unforgeability under one chosen message attacks (EU-CMA) becomes T-openPRE where $\mathcal{A}$ is allowed to make $T/2$ queries. Using L-OTS in a many-time signature scheme such as the Merkle Signature Scheme [13] and variants like [2,8,12] can easily amplify the difference in tightness between a reduction that uses (T-)PRE and a reduction from T-SPR and T-DSPR to 2^{70}.

Indeed, the general idea of using T-SPR instead of (T-)PRE in security reductions for hash-based signatures already occurs in [9]. However, there the authors make use of the assumption that for the used hash function every input has a colliding value for all keys, i.e., SPprob(H) $= 1$ in our notation. This is unlikely to hold for common length-preserving keyed hash functions as Sect. 3 shows SPprob(H) $\approx 1 - 1/e$ for random H. However, as shown above, it is also not necessary to require SPprob(H) $= 1$. Instead, it suffices to require (T-)DSPR.

For modern hash-based signatures like XMSS [7] L-OTS is replaced by variants [6] of the Winternitz OTS (W-OTS) [13]. For W-OTS the notion of EU-CMA security does not directly translate to T-openPRE. Indeed, the security reduction gets far more involved as W-OTS uses hash chains. However, as shown in [9] one can replace (T-)PRE in this context by T-SPR and the assumption that SPprob(H) $= 1$. Along the lines of the above approach we can then replace the assumption that SPprob(H) $= 1$ by T-DSPR.

References

1. Andreeva, E., Bouillaguet, C., Dunkelman, O., Fouque, P.-A., Hoch, J.J., Kelsey, J., Shamir, A., Zimmer, S.: New second-preimage attacks on hash functions. J. Cryptol. **29**(4), 657–696 (2016). https://www.di.ens.fr/~fouque/pub/joc11.pdf

2. Buchmann, J., Dahmen, E., Klintsevich, E., Okeya, K., Vuillaume, C.: Merkle signatures with virtually unlimited signature capacity. In: Katz, J., Yung, M. (eds.) ACNS 2007. LNCS, vol. 4521, pp. 31–45. Springer, Heidelberg (2007). https://doi.org/10.1007/978-3-540-72738-5_3
3. Charalambides, C.A.: Distributions of random partitions and their applications. Methodol. Comput. Appl. Probab. **9**(2), 163–193 (2007)
4. Dörrie, H.: 100 Great Problems of Elementary Mathematics. Courier Corporation (2013)
5. Flajolet, P., Sedgewick, R.: Analytic Combinatorics. Cambridge University Press, Cambridge (2009). http://ac.cs.princeton.edu/home/AC.pdf
6. Hülsing, A.: W-OTS+ – shorter signatures for hash-based signature schemes. In: Youssef, A., Nitaj, A., Hassanien, A.E. (eds.) AFRICACRYPT 2013. LNCS, vol. 7918, pp. 173–188. Springer, Heidelberg (2013). https://doi.org/10.1007/978-3-642-38553-7_10. https://eprint.iacr.org/2017/965
7. Hülsing, A., Butin, D., Gazdag, S.-L., Rijneveld, J., Mohaisen, A.: XMSS: eXtended Merkle Signature Scheme. RFC 8391, May 2018. https://rfc-editor.org/rfc/rfc8391.txt
8. Hülsing, A., Rausch, L., Buchmann, J.: Optimal parameters for $XMSS^{MT}$. In: Cuzzocrea, A., Kittl, C., Simos, D.E., Weippl, E., Xu, L. (eds.) CD-ARES 2013. LNCS, vol. 8128, pp. 194–208. Springer, Heidelberg (2013). https://doi.org/10.1007/978-3-642-40588-4_14. https://eprint.iacr.org/2017/966
9. Hülsing, A., Rijneveld, J., Song, F.: Mitigating multi-target attacks in hash-based signatures. In: Cheng, C.-M., Chung, K.-M., Persiano, G., Yang, B.-Y. (eds.) PKC 2016. LNCS, vol. 9614, pp. 387–416. Springer, Heidelberg (2016). https://doi.org/10.1007/978-3-662-49384-7_15. https://eprint.iacr.org/2015/1256
10. Kelsey, J., Schneier, B.: Second preimages on n-bit hash functions for much less than 2^n work. In: Cramer, R. (ed.) EUROCRYPT 2005. LNCS, vol. 3494, pp. 474–490. Springer, Heidelberg (2005). https://doi.org/10.1007/11426639_28. https://eprint.iacr.org/2004/304.pdf
11. Lamport, L.: Constructing digital signatures from a one way function. Technical report SRI-CSL-98, SRI International Computer Science Laboratory (1979). https://lamport.azurewebsites.net/pubs/dig-sig.pdf
12. Malkin, T., Micciancio, D., Miner, S.: Efficient generic forward-secure signatures with an unbounded number of time periods. In: Knudsen, L.R. (ed.) EUROCRYPT 2002. LNCS, vol. 2332, pp. 400–417. Springer, Heidelberg (2002). https://doi.org/10.1007/3-540-46035-7_27. https://cseweb.ucsd.edu/~daniele/papers/MMM.html
13. Merkle, R.C.: A certified digital signature. In: Brassard, G. (ed.) CRYPTO 1989. LNCS, vol. 435, pp. 218–238. Springer, New York (1990). https://doi.org/10.1007/0-387-34805-0_21. https://merkle.com/papers/Certified1979.pdf
14. Robbins, H.: A remark on Stirling's formula. Am. Math. Mon. **62**(1), 26–29 (1955)
15. Rogaway, P., Shrimpton, T.: Cryptographic hash-function basics: definitions, implications, and separations for preimage resistance, second-preimage resistance, and collision resistance. In: Roy, B., Meier, W. (eds.) FSE 2004. LNCS, vol. 3017, pp. 371–388. Springer, Heidelberg (2004). https://doi.org/10.1007/978-3-540-25937-4_24. https://eprint.iacr.org/2004/035
16. Sloane, N.J.A.: The on-line encyclopedia of integer sequences (2019). https://oeis.org

A Some Single-Variable Functions

This appendix proves features of some functions used in the proofs of theorems in Sect. 3. The proofs in this appendix are split into small lemmas to support verification, and proofs of the lemmas appear in the full version online. The notation $\mathbf{R}_{>0}$ means the set of positive real numbers.

Lemma 39. *If* $x \neq 0$ *then* $e^x > 1 + x$.

Lemma 40. *Any* $x \in \mathbf{R}$ *has* $e^x - 2x \geq 2 - 2\log 2 > 0$.

Lemma 41. *If* $x > 0$ *then* $e^x - 1 + x - x^2 > 0$.

Lemma 42. *Define* $\varphi_1(x) = x(e^x - 1)/(e^x - x)$. *Then* φ_1 *is increasing, and maps* $\mathbf{R}_{>0}$ *bijectively to* $\mathbf{R}_{>0}$.

Lemma 43. *If* $x \neq 0$ *then* $e^x + e^{-x} > 2$.

Lemma 44. *If* $x > 0$ *then* $e^x - e^{-x} - 2x > 0$.

Lemma 45. *If* $x > 0$ *then* $e^x + e^{-x} - 2 - x^2 > 0$.

Lemma 46. *Define* $\varphi_2(x) = x(e^x - 1)/(e^x - 1 - x)$ *for* $x > 0$. *Then* φ_2 *is increasing, and maps* $\mathbf{R}_{>0}$ *bijectively to* $\mathbf{R}_{>2}$.

Lemma 47. *The ratio* $(e-1)^{1-x}/x^x(1-x)^{1-x}$ *for* $0 < x < 1$ *increases for* $0 < x < 1/e$, *has maximum value* e *at* $x = 1/e$, *and decreases for* $1/e < x < 1$.

Lemma 48. *The maximum value of* $1/(2x-1)^{2x-1}(1-x)^{2(1-x)}2^{1-x}$ *for* $1/2 < x < 1$ *is* $1 + \sqrt{2}$.

Lemma 49. *Define* $\varphi_5(x) = xe^x - e^x + 1$. *Then* φ_5 *decreases for* $x < 0$, *has minimum value* 0 *at* $x = 0$, *and increases for* $x > 0$.

Lemma 50. *Let* x *be a positive real number. Define* $y = e^x - 1 - x$ *and* $z = 1/(x + x^2/y)$; *then* $0 < z < 1/2$. *Define* $\gamma = y^z/xz^z(1-z)^{1-z}$; *then* $\gamma \leq e - 1$.

Structure-Preserving Signatures on Equivalence Classes from Standard Assumptions

Mojtaba Khalili[1(✉)], Daniel Slamanig[2], and Mohammad Dakhilalian[1]

[1] Isfahan University of Technology, Isfahan, Iran
{m.khalili,mdalian}@ec.iut.ac.ir
[2] AIT Austrian Institute of Technology, Vienna, Austria
daniel.slamanig@ait.ac.at

Abstract. Structure-preserving signatures on equivalence classes (SPS-EQ) introduced at ASIACRYPT 2014 are a variant of SPS where a message is considered as a projective equivalence class, and a new representative of the same class can be obtained by multiplying a vector by a scalar. Given a message and corresponding signature, anyone can produce an updated and randomized signature on an arbitrary representative from the same equivalence class. SPS-EQ have proven to be a very versatile building block for many cryptographic applications.

In this paper, we present the first EUF-CMA secure SPS-EQ scheme under standard assumptions. So far only constructions in the generic group model are known. One recent candidate under standard assumptions are the weakly secure equivalence class signatures by Fuchsbauer and Gay (PKC'18), a variant of SPS-EQ satisfying only a weaker unforgeability and adaption notion. Fuchsbauer and Gay show that this weaker unforgeability notion is sufficient for many known applications of SPS-EQ. Unfortunately, the weaker adaption notion is only proper for a semi-honest (passive) model and as we show in this paper, makes their scheme unusable in the current models for almost all of their advertised applications of SPS-EQ from the literature.

We then present a new EUF-CMA secure SPS-EQ scheme with a tight security reduction under the SXDH assumption providing the notion of perfect adaption (under malicious keys). To achieve the strongest notion of perfect adaption under malicious keys, we require a common reference string (CRS), which seems inherent for constructions under standard assumptions. However, for most known applications of SPS-EQ we do not require a trusted CRS (as the CRS can be generated by the signer during key generation). Technically, our construction is inspired by a recent work of Gay et al. (EUROCRYPT'18), who construct a tightly secure message authentication code and translate it to an SPS scheme adapting techniques due to Bellare and Goldwasser (CRYPTO'89).

M. Khalili—Work partly done while visiting Universitat Pompeu Fabra, Barcelona, Spain.

S. D. Galbraith and S. Moriai (Eds.): ASIACRYPT 2019, LNCS 11923, pp. 63–93, 2019.
https://doi.org/10.1007/978-3-030-34618-8_3

1 Introduction

Structure-preserving signatures (SPS) [4] are signatures where the messages, public keys and signatures only consists of elements of groups equipped with an efficient bilinear map, and the verification algorithm just consists of group membership checks and evaluation of pairing product equations (PPEs). SPS schemes [2,4–8,43–45,60,63,64] are compatible with efficient pairing-based NIZK proofs [50], and are a useful building-block for many cryptographic applications, such as blind signatures [4,39], group signatures [4,68], traceable signatures [3], group encryption [23], homomorphic signatures [66], delegatable anonymous credentials [34], compact verifiable shuffles [24], network coding [10], oblivious transfer [48], tightly secure encryption [56] and anonymous e-cash [17]. SPS schemes come in various different flavors such as being able to sign elements in either one or both source groups of the bilinear group or requiring certain conditions for messages (e.g., messages need to be Diffie-Hellman tuples [33,45]). They come with different provable security guarantees, ranging from ones that are directly analyzed in the generic group model (GGM) to ones that can be constructed from standard assumptions such as SXDH or SXDLin (typically within the Matrix-Diffie-Hellman assumption framework [31]) and under different qualities of the reduction (from very loose to tight reductions). A desirable goal is to construct schemes with tight security reductions from standard assumptions which are at the same time highly efficient. Some SPS schemes are also randomizable (e.g., [4,6]), meaning that a signature can be randomized to another unlinkable valid signature on the same message.

Structure-preserving signatures on equivalence classes (SPS-EQ) [38,40,52] are a variant of SPS where anyone can randomize not only signatures, but a message-signature pair publicly, i.e., in addition to randomizing the signature also the message can be randomized. They have proven to be useful in many applications such as attribute-based anonymous credentials [29,40,52], delegatable anonymous credentials [27], self-blindable certificates [11], blind signatures [37,39], group signatures [11,12,26,30], sanitizable signatures [22], verifiably encrypted signatures [51], access control encryption [36] or proving the correctness of a shuffle in mix-nets (i.e., for anonymous communication or electronic voting) [59]. In many of these applications, the idea of randomizing signatures and messages offers the same functionality as when using SPS schemes combined with a NIZK proof, but without the need for any NIZK. Consequently, this allows for the design of more efficient constructions.

More concretely, in an SPS-EQ scheme, given a signature on an equivalence class defined over the message space, anyone can update the signature to another representative of the same class. Defined on $(\mathbb{G}^*)^\ell$ (where $\mathbb{G}$ is of prime order p), this equivalence relation $\sim_\mathcal{R}$ is as follows ($\ell > 1$):

$$\mathbf{M} \in (\mathbb{G}^*)^\ell \sim_\mathcal{R} \mathbf{N} \in (\mathbb{G}^*)^\ell \Leftrightarrow \exists \mu \in \mathbb{Z}_p^* : \mathbf{M} = \mu\mathbf{N}$$

An SPS-EQ scheme signs an equivalence class $[\mathbf{M}]_\mathcal{R}$ for $\mathbf{M} \in (\mathbb{G}_i^*)^\ell$ by signing a representative $\mathbf{M}$ of $[\mathbf{M}]_\mathcal{R}$. It then allows for switching to other representatives of $[\mathbf{M}]_\mathcal{R}$ and updating the signature without access to the secret key. Two important properties of SPS-EQ are *unforgeability* (EUF-CMA security) defined on equivalence classes and *perfect adaption* (potentially even under malicious signing keys), where the latter requires that updated signatures (output by the algorithm ChgRep) are distributed identically to new signatures on the respective representative (if signatures or even if signing keys are computed maliciously). Latter together with the DDH assumption on the message space then yields a notion of unlinkability, i.e., that original signatures and those output by ChgRep cannot be linked. As it turns out, coming up with constructions that achieve *both notions simultaneously* is a challenging task.

We note that, as observed in [39], every SPS-EQ yields a (randomizable) SPS scheme by appending some fixed group element to the message vector before signing and which is checked on verification, to allow only one single representative of each class. Recently, the concept of SPS-EQ has even been further extended to consider also equivalence classes on the public keys, denoted as signatures with flexible public key [11] and equivalence classes on messages and public keys simultaneously, denoted as mercurial signatures [27]. This further extends the scope of applications.

Prior Approaches to Construct SPS-EQ. The first instantiation of SPS-EQ in [52] was secure only against random message attacks, and later Fuchsbauer et al. [38,40] presented a revised scheme that achieves EUF-CMA security in the generic group model (GGM). In [39], Fuchsbauer et al. present another EUF-CMA secure scheme under a q-type assumption, which by construction does not provide the perfect adaption notion and thus is not interesting for existing applications of SPS-EQ. Recently, Fuchsbauer and Gay [35], presented a version of SPS-EQ (called equivalence class signatures or EQS) which can be proven secure under standard assumptions, i.e., in the Matrix-Diffie-Hellman assumption framework [31]. In order to prove their scheme secure, they have introduced a weakened unforgeability notion called existential unforgeability under chosen open message attacks (EUF-CoMA), in which the adversary does not send group element vectors to the signing oracle but vectors of $\mathbb{Z}_p^*$ elements. Moreover, in contrast to the original definition of SPS-EQ in [52] and the scheme of Fuchsbauer et al. [38,40], which allows to randomize a given signature (change the representative) an arbitrary number of times, the scheme of Fuchsbauer and Gay [35] distinguishes two types of signatures. The first type comes from the signing algorithm and when randomized yields a signature of the second type, which cannot be randomized any further. As argued by Fuchsbauer and Gay in [35], for most of the known applications of SPS-EQ the combination of EUF-CoMA notion and the one-time randomizability is sufficient. Actually, as argued in [35], it is sufficient for all applications in the literature, except for the one to round-optimal blind signatures from SPS-EQ [39].

The construction of Fuchsbauer and Gay in [35] does also rely on a weakened notion of adaption (weaker than the original one from [39] in that it only considers honestly generated keys and honestly computed signatures). We will show that even though their weaker unforgeability notion is sufficient for applications, the weaker adaption notion makes the scheme suitable only for restricted applications, i.e., access control encryption (ACE) or attribute-based credentials (ABCs) with an honest credential issuer. Moreover, the application to verifiably encrypted signatures in [51] requires another notion called perfect composition, which [35] seem to assume implicitly. Unfortunately, their scheme does not satisfy this notion. Consequently, for the interesting schemes providing the perfect adaption notion from [39], the current state of affairs is that there is only the EUF-CMA secure scheme from [38,40] secure in the GGM.

Tight Security for SPS-EQ Schemes. Tight security allows to choose cryptographic parameters of a scheme in a way that is supported by a security proof, without the need to sacrifice efficiency by compensating the security loss of a reduction with larger parameters. Latter can be significant if the reduction is very loose. In case of SPS, quite some progress has been made in recent years on constructing tightly-secure SPS [7,8,43,55,60], though the state-of-the-art tightly-secure schemes under standard assumptions are still less efficient than for instance schemes proven secure in the generic group model (GGM). While tight security is quite well studied within SPS (and other primitives such as encryption [41,54,55], signatures [25,46,54,55], identity-based encryption [25,57,58], key exchange [13,46,53], or zero-knowledge proofs [41,55]), there are no such results for SPS-EQ schemes so far.

1.1 Our Contributions

Our contributions in this paper can be summarized as follows:

Analysis of FG18: Firstly, we revisit the concrete approach to construct EUF-CoMA secure EQS from Fuchsbauer and Gay in [35], representing the only known candidate towards perfectly adapting SPS-EQ under standard assumptions so far. Thereby, we identify various problems with the applications of the scheme presented in [35]. We stress that we do not present attacks on the scheme itself (which is secure in their model), but show that their adaption notion is too weak for most applications claimed in [35] (apart from access control encryption (ACE) [36]). Briefly summarizing, we first show that their scheme cannot be used for the application to attribute-based credentials (ABCs) [38,40]. We demonstrate an attack based on a trapdoor in the signing key that invalidates the anonymity proof for ABCs. Secondly, we show an attack that demonstrates that the scheme in [35] cannot be used even for applications that assume honest generation of signing keys and in particular for ABCs under honest-keys [52] and dynamic group signatures [30]. We stress that due to this too weak adaption notion concrete instantiations presented in follow up works by Backes et al. [11,12], that rely on the FG18 scheme from [35], are invalidated and need to be reconsidered. Our results allow to repair their now broken claims in part.[1] Thirdly, we show that the FG18 scheme does not satisfy another notion called perfect composition [51], invalidating the use of their scheme for application to verifiably encrypted signatures as discussed in [35]. Consequently, this means that contrary to their claim, the EQS framework and scheme in [35] can only

[1] For the group signatures in [12] it will only work with our construction when relying on a CRS, or by using the construction secure in the GGM in [38].

be used for the construction of access control encryption (ACE) in [36] and for all other applications no instantiations under standard assumptions remain. We stress that one could relax the security models of the applications to make [35] usable again, but such models where signatures and keys are assumed to be generated honestly, i.e., that only guarantee semi-honest (passive) security, limits the practical applications. For example, one could consider ABCs with anonymity against honest credential issuers and use the EQS from [35].

SPS-EQ from Standard Assumptions and Applications: As our main contribution, we provide the first construction of SPS-EQ under standard assumptions and in particular the Matrix-Diffie-Hellman assumption framework. We therefore have to revise the model of SPS-EQ in some aspects: (1) we introduce tags, where the signing algorithm outputs a signature and a tag, randomization (i.e., ChgRep) requires a signature and a tag, whereas for verification only the signature is required; signatures that have been randomized using a tag can not further be randomized, i.e., only a single randomization is possible. This definition is comparable to the one in [35], apart that FG18 does not use tags. We stress that as demonstrated in [35], this restriction does not affect existing applications of SPS-EQ. (2) we require that signers generate their signing keys with respect to a common reference string (CRS) for achieving the perfect adaption notion in the malicious setting (prior works on SPS-EQ did not consider having a CRS). We will show that this does not impact the applications discussed in [35] with the exception of anonymous credentials in the malicious key model, as the security models in all other applications assume honest generation of the signing keys and thus every signer can produce its own CRS as part of the signing key. As we, however, cannot avoid a CRS in the malicious key setting, we are not able to instantiate round-optimal blind signatures in the standard model from SPS-EQ [39] under standard assumptions, which [35] could not achieve either. On the positive side, however, it allows us to obtain the most efficient round-optimal blind signatures in the CRS model from standard assumptions.

On the Use of a CRS. Although our scheme does not require a CRS for nearly all of the applications of SPS-EQ, avoiding a CRS in the malicious setting would be good. The use of a CRS in general seems to be debatable, as it needs to be generated by some trusted third party that is hard to find in the real world. Within recent years, we have seen a number of deployed real-world applications that require a CRS when using zk-SNARKS (e.g., Zcash[2] being probably the most prominent one) and which have used multi-party computation ceremonies to construct the CRS in a way that no entity provably knows the trapdoor. A number of such ceremonies has been run in real-world[3] and various works discuss approaches to achieve it [16,20,21]. In the light of this, we do not consider it unrealistic to generate a CRS for the use within practical applications of SPS-EQ that require security under malicious keys, especially since the CRS does not depend on the message length ℓ and so a single CRS can be used for all types

[2] https://z.cash/.

[3] See e.g., https://z.cash/blog/the-design-of-the-ceremony/ or https://www.zfnd.org/blog/conclusion-of-powers-of-tau/.

of SPS-EQ keys for different applications. Furthermore, it seems interesting to investigate the application of recent approaches towards subversion resistant (QA)-NIZK [1,14] or updatable CRS [49,69], though this typically comes at the cost of rather strong knowledge assumptions. Clearly, ultimately it would be good to find SPS-EQ in the malicious key model without a CRS, which we leave as a challenging open problem.

1.2 Outline of Our Construction

Fuchsbauer and Gay [35] modify an affine MAC of Blazy et al. [18] to obtain a linear structure-preserving MAC. Then, they make the scheme publicly verifiable using a known technique from Kiltz and Wee [65] already used previously in context of SPS [64]. Unfortunately, the structure-preserving MAC has an inherent problem in the security game, where both messages and Matrix Decision Diffie-Hellman (MDDH) challenges belong to the same source group of the bilinear group. This forces them to use the weaker EUF-CoMA instead of EUF-CMA security. Consequently, as we are interested in EUF-CMA security, we need to look for a different framework when trying to construct EUF-CMA secure SPS-EQ schemes.

Therefore, we borrow a central idea from the recent work of Gay et al. [43]. In particular, they use a specific OR-proof [71] to then construct tightly secure structure-preserving MACs based on the key encapsulation mechanism of Gay et al. in [42]. More precisely, they make use of adaptive partitioning [54] to randomize all tags in their MAC. Their work is based on the observation (core lemma in [43]) that for all $[\mathbf{t}]_1 = [\mathbf{A_0}]_1\mathbf{r}$ with $\mathbf{r} \xleftarrow{R} \mathbb{Z}_p^k$ chosen freshly for each instance, fixed matrices $\mathbf{A_0}, \mathbf{A_1} \xleftarrow{R} \mathcal{D}_{2k,k}$, and a NIZK proof π for $\mathbf{t} \in \text{span}(\mathbf{A_0}) \cup \text{span}(\mathbf{A_1})$, the following values

$$\mathbf{k_0}^\top[\mathbf{t}]_1 \ , \qquad (\mathbf{k_0}^\top + \mathbf{s}^\top)[\mathbf{t}]_1 \tag{1}$$

are indistinguishable under the MDDH assumption, where $\mathbf{k_0} \leftarrow \mathbb{Z}_p^{2k}$ is a key, and $\mathbf{s} \in \mathbb{Z}_p^{2k}$ is a fresh random value for each instance. Actually, they show that $[\mathbf{k_0}^\top\mathbf{t}]_1$ is pseudorandom.

In this paper, we are going to present an approach to obtain malleability for this pseudorandom function, which we use as one part of our signature, and the NIZK proof as another part. Therefore, we first add a tag (to allow a homomorphism on the pseudorandom part) to our signature, such that everyone who knows it can re-randomize the pseudorandom part. Second, we revise the NIZK proof and give a proof for well-formedness of both the pseudorandom part and the tag, such that it can be re-randomized and that we finally get a fresh signature, including fresh pseudorandom part and a proof for it. More precisely, we first show that for all $[\mathbf{t}]_1 = [\mathbf{A_0}]_1\mathbf{r}_1$ and $[\mathbf{w}]_1 = [\mathbf{A_0}]_1\mathbf{r}_2$ for $\mathbf{r}_1, \mathbf{r}_2 \xleftarrow{R} \mathbb{Z}_p^k$ chosen freshly for each instance, and a NIZK proof π for $\mathbf{t}, \mathbf{w} \in \text{span}(\mathbf{A_0}) \cup \text{span}(\mathbf{A_1})$ (to be discussed later), the following tuples are indistinguishable under the MDDH assumption

$$(\mathbf{k_0}^\top[\mathbf{t}]_1, \mathbf{k_0}^\top[\mathbf{w}]_1) \ , \qquad ((\mathbf{k_0}^\top + \mathbf{s}^\top)[\mathbf{t}]_1, \mathbf{k_0}^\top[\mathbf{w}]_1). \tag{2}$$

We then use this MAC (for $k = 1$)[4] to construct an SPS-EQ scheme on a message $[\mathbf{m}]_1 \in (\mathbb{G}_1^*)^\ell$. Our signature has a basic form like $\sigma = \mathbf{k_0}^\top [\mathbf{t}]_1 + \mathbf{k}^\top [\mathbf{m}]_1$, with a tag $\tau = \mathbf{k}_0^\top [\mathbf{w}]_1$ (which is only required for randomization), where $\mathbf{k_0} \xleftarrow{R} \mathbb{Z}_p^2$ and $\mathbf{k} \xleftarrow{R} \mathbb{Z}_p^\ell$. We can use (2) to add some randomness to the signature as $\sigma = \mathbf{k_0}^\top [\mathbf{t}]_1 + \mathbf{k}^\top [\mathbf{m}]_1 + \zeta$ for $\zeta \xleftarrow{R} \mathbb{Z}_p$. At a high level, by adding randomness to each signature, we can make every signature independent of each other. So, we completely hide the values $\mathbf{k}$, and an adversary has negligible chance to compute a valid forgery. On the other hand, everyone can obtain a fresh tag, using previous tag τ, and add it to the signature to obtain a fresh pseudorandom part. From a high level perspective, we have a basic MAC which is additively homomorphic and our signatures and tags are two instances of it, one on message $[\mathbf{m}]_1$ and another one on message zero. This allows deriving a signature on $\mu[\mathbf{m}]_1$ for $\mu \xleftarrow{R} \mathbb{Z}_p^*$, i.e., to adapt the signature part to representative $\mu[\mathbf{m}]_1$, using a multiplication of the signature part with μ and then add it to the fresh tag. Note that, in our scheme we do not need to have access to the tag τ in the verification algorithm, but it is required for randomizing messages and signatures (changing representatives in the language of SPS-EQ). We note that in the EUF-CMA game, we model it in a way that on a signature query the challenger returns both the signature and the tag, while the adversary only needs to output a signature without the tag as its forgery attempt.

Now, we will discuss how to randomize the NIZK proof. At this point, there is an obvious problem with the OR-proof used in [43] and we need to revise their approach such that the proof is randomizable (proofs can be re-randomized to look like fresh proofs) and malleable (statements for given proofs can be updated), where latter is required to switch between representatives of a class. In particular, to obtain these properties we change a part of the OR-proof and replace it with a QA-NIZK. In the NIZK proof of [43], we have a permanent CRS including $[\mathbf{D}]_2 \in \mathbb{G}_2^2$ and $[\mathbf{z}]_2 \in \mathbb{G}_2^2$, where $\mathbf{z} \notin \mathrm{span}(\mathbf{D})$ be parameters of the system. On the other hand, their scheme has an updatable CRS including $[\mathbf{z}_0]_2$ and $[\mathbf{z}_1]_2$. Now, given the permanent CRS, the complements of the parts of the updatable CRS are computed in each instance. The idea is that exactly these CRS generate a sound system (i.e., one of the parts of the updatable CRS is outside the span of $[\mathbf{D}]_2$) and in the other case we have a simulatable system (i.e., both parts of the updatable CRS are in the span of $[\mathbf{D}]_2$). As the public parameter $[\mathbf{z}]_2$ is not in the span of $[\mathbf{D}]_2$, we can obtain soundness by letting $[\mathbf{z}_0]_2 = [\mathbf{D}]_2 v$ and $[\mathbf{z}_1]_2 = [\mathbf{z}]_2 - [\mathbf{z}_0]_2$, for $v \xleftarrow{R} \mathbb{Z}_p$, where the sum of them is equal to the value $[\mathbf{z}]_2$, i.e., $[\mathbf{z}_0]_2 + [\mathbf{z}_1]_2 = [\mathbf{z}]_2$. So, it proves that at least one of $[\mathbf{z}_0]_2$ and $[\mathbf{z}_1]_2$ has a part in the span($\mathbf{z}$). The fact that this sum of the updatable CRS is a fixed value is of course not good to enable the randomization of the updatable CRS. To circumvent this state of affairs and obtain malleability, we need to compute a NIZK proof π for $\mathbf{t}, \mathbf{w} \in \mathrm{span}(\mathbf{A_0}) \cup \mathrm{span}(\mathbf{A}_1)$ with the shared updatable CRS, for $\mathbf{t}$ and $\mathbf{w}$, and adapt other proof parts, while we

[4] We note that we can only instantiate our construction for $k = 1$, i.e., under the SXDH assumption, and leave the construction of SPS-EQ under the more general Matrix Decision Diffie-Hellman assumption as an interesting open problem.

Table 1. Comparison of SPS-EQ and EQS Schemes when signing vectors of length ℓ and Q is the number of queries to the signing oracle. **A** means adaption. ✓✓ means perfect adaption under honest and malicious keys; ✓ means perfect adaption under honest keys and under malicious keys in the honest parameters model (i.e., using a CRS); $\approx$ means adaption under honest keys and honest signatures.

Scheme	|Signature|	|PK|	Model	Ass.	Loss	A
[38]	$2\|\mathbb{G}_1\| + 1\|\mathbb{G}_2\|$	$\ell\|\mathbb{G}_2\|$	EUF-CMA (strong)	GGM	–	✓✓
[35]	$(4\ell+2)\|\mathbb{G}_1\| + 4\|\mathbb{G}_2\|$	$(4\ell+2)\|\mathbb{G}_2\|$	EUF-CoMA (weak)	$\mathcal{D}_{4,2}$-MDDH, $\mathcal{D}_1$-KerMDH	$\mathcal{O}(Q)$	$\approx$
Section 5	$8\|\mathbb{G}_1\| + 9\|\mathbb{G}_2\|$	$3\ell\|\mathbb{G}_2\|$	EUF-CMA (strong)	SXDH	$\mathcal{O}(\log Q)$	✓

remain sound. Our approach is to set $[\mathbf{z}_0]_2 = [\mathbf{D}]_2 v$ and $[\mathbf{z}_1]_2 = [\mathbf{z}]_2 v$, and give a proof using a one-time homomorphic QA-NIZK due to Jutla and Roy [62] that $\mathbf{z}_0 + \mathbf{z}_1$ is in the linear subspace of $\mathbf{D} + \mathbf{z}$. This means that at least one of $[\mathbf{z}_0]_2$ and $[\mathbf{z}_1]_2$ has a part in span($\mathbf{z}$). Fortunately, after this change other parts of the proof adapt properly, and only moving to using a QA-NIZK comes at the cost of having computationally soundness instead of perfect soundness.[5]

For realizing the change representative algorithm ChgRep, our Prove algorithm of the OR-proof computes two proofs with shared randomness and QA-NIZK (where the second proof is part of the tag), which allows to randomize the first proof and update its word. This yields to have randomized signatures output by ChgRep to be distributed identical to a fresh signature for the new representative, i.e., we obtain perfect adaption. As explained above, we use a NIZK OR-proof and a QA-NIZK proof in the construction of the SPS-EQ. In order to guarantee perfect adaption even in front of a signer that generates the keys in a potentially malicious way (i.e., remembers a trapdoor), we need to have a CRS for these proof systems.[6] Consequently, the perfect adaption of our SPS-EQ is guaranteed in the common parameter model where the parameters include a common reference string. However, we stress again that for most applications the CRS generation can simply be part of the key generation and no trusted setup is required.

Comparison with Other Schemes. In the following Table 1 we provide a comparison of previous SPS-EQ schemes with the one proposed in this paper. We only consider schemes satisfying some reasonable adaption notion, i.e., we

[5] Thus, we will formally have a NIZK argument, but in the text we will usually not make a distinction between NIZK proofs and arguments.

[6] Even if all involved proof systems provide zero-knowledge definitions in the style of composable zero-knowledge [50], i.e., even if the adversary knows the trapdoor and still simulated and honestly computed proofs cannot be distinguished, we still have the problem of maliciously generated proofs and thus we cannot avoid a CRS.

exclude the one under q-type assumptions in [39]. We note that while for [38] original and randomized signatures are identical, for [35] and our scheme presented in this paper we only consider sizes of randomized signatures, i.e., those output by ChgRep and signatures without the tag respectively. For [35] we consider a concrete setting where $\mathcal{U}_{4,2}$-MDDH reduces to the SXDLin assumption [2], i.e., assuming DLin in $\mathbb{G}_1$ and $\mathbb{G}_2$, and $\mathcal{D}_1$-KerMDH in $\mathbb{G}_2$ reduces to the DDH assumption in $\mathbb{G}_2$. For our scheme $k = 1$ and thus we have the $\mathcal{L}_1$-MDDH assumption in $\mathbb{G}_1$ and the $\mathcal{L}_1$-KerMDH assumption in $\mathbb{G}_2$. Latter representing the 1-KerLin assumption which by Lemma 1 is implied by DDH. Consequently, our scheme is secure under SXDH, i.e., assuming DDH in $\mathbb{G}_1$ and $\mathbb{G}_2$.

2 Preliminaries

Notation. Let GGen be a probabilistic polynomial time (PPT) algorithm that on input 1^λ returns a description $\mathcal{G} = (\mathbb{G}, p, P)$ of an additive cyclic group $\mathbb{G}$ of order p for a λ-bit prime p, whose generator is P. We use implicit representation of group elements as introduced in [31]. For $a \in \mathbb{Z}_p$, define $[a] = aP \in \mathbb{G}$ as the implicit representation of a in $\mathbb{G}$. We will always use this implicit notation of elements in $\mathbb{G}$, i.e., we let $[a] \in \mathbb{G}$ be an element in $\mathbb{G}$, and note that from $[a] \in \mathbb{G}$ it is generally hard to compute the value a (discrete logarithm problem in $\mathbb{G}$).

Let BGGen be a PPT algorithm that returns a description $\mathsf{BG} = (\mathbb{G}_1, \mathbb{G}_2, \mathbb{G}_T, p, P_1, P_2, e)$ of an asymmetric bilinear group where $\mathbb{G}_1, \mathbb{G}_2, \mathbb{G}_T$ are cyclic groups of order p , P_1 and P_2 are generators of $\mathbb{G}_1$ and $\mathbb{G}_2$, respectively, and $e : \mathbb{G}_1 \times \mathbb{G}_2 \rightarrow \mathbb{G}_T$ is an efficiently computable (non-degenerate) bilinear map and for $s \in \{1, 2, T\}$ and $a \in \mathbb{Z}_p$, analogous to above, we write $[a]_s = aP_s \in \mathbb{G}_s$ as the implicit representation of a in $\mathbb{G}_s$. For two matrices (vectors) $\mathbf{A}, \mathbf{B}$ define $e([\mathbf{A}]_1, [\mathbf{B}]_2) := [\mathbf{AB}]_T \in \mathbb{G}_T$. With $\overline{\mathbf{B}}$ we denote the upper square matrix of $\mathbf{B}$. Let $r \xleftarrow{R} \mathcal{S}$ denotes sampling r from set $\mathcal{S}$ uniformly at random. We denote by λ the security parameter, and by ϵ any negligible function of λ.

Assumptions. We recall the definition of the Matrix Decision Diffie-Hellman assumption [31] and a natural computational analogue of it, called the Kernel-Diffie-Hellman assumption [70].

Definition 1 (Matrix Distribution). *Let $k \in \mathbb{N}$. We call $\mathcal{D}_k$ a matrix distribution if it outputs matrices in $\mathbb{Z}_q^{(k+1)\times k}$ of full rank k in polynomial time.*

Definition 2 ($\mathcal{D}_k$-Matrix Decision Diffie-Hellman Assumption). *Let $\mathcal{D}_k$ be a matrix distribution. We say that the $\mathcal{D}_k$-Matrix Diffie-Hellman ($\mathcal{D}_k$-MDDH) Assumption holds relative to BGGen in group $\mathbb{G}_s$ if for all PPT adversaries $\mathcal{A}$, we have:*

$$\mathbf{Adv}^{\mathsf{MDDH}}_{\mathcal{D}_k,\mathbb{G}_s}(\mathcal{A}) := \big|\Pr\big[\mathcal{A}(\mathsf{BG}, [\mathbf{A}]_s, [\mathbf{Aw}]_s) = 1\big] - \Pr\big[\mathcal{A}(\mathsf{BG}, [\mathbf{A}]_s, [\mathbf{u}]_s) = 1\big]\big| \leq \epsilon(\lambda)$$

where the probability is taken over $\mathsf{BG} \leftarrow \mathsf{BGGen}(1^\lambda), \mathbf{A} \leftarrow \mathcal{D}_k, \mathbf{w} \leftarrow \mathbb{Z}_q^k, \mathbf{u} \leftarrow \mathbb{Z}_q^{k+1}$

Definition 3 (Kernel Matrix Diffie-Hellman Assumption). *Let $\mathcal{D}_k$ be a matrix distribution and $s \in \{1,2\}$. We say that the $\mathcal{D}_k$-Kernel Diffie-Hellman Assumption ($\mathcal{D}_k$-KerMDH) holds relative to* BGGen *in group $\mathbb{G}_s$ if for all PPT adversaries $\mathcal{A}$,*

$$\mathbf{Adv}^{\mathsf{KerMDH}}_{\mathcal{D}_k,\mathbb{G}_s}(\mathcal{A}) = \ \mathsf{Pr}\Big[[\mathbf{c}]_{3-s} \leftarrow \mathcal{A}(\mathsf{BG},[\mathbf{A}]_s) : \mathbf{c}^\top\mathbf{A} = \mathbf{0} \wedge \mathbf{c} \neq \mathbf{0}\Big] \leq \epsilon(\lambda)$$

where $\mathbf{A} \xleftarrow{R} \mathcal{D}_k$.

Lemma 1 ($\mathcal{D}_k$-MDDH $\Longrightarrow$ $\mathcal{D}_k$-KerMDH [70]). *Let $k \in \mathbb{N}$ and let $\mathcal{D}_k$ be a matrix distribution. For any PPT adversary $\mathcal{A}$, there exists a PPT adversary $\mathcal{B}$ such that $\mathbf{Adv}^{\mathsf{KerMDH}}_{\mathcal{D}_k,\mathbb{G}_s}(\mathcal{A}) \leq \mathbf{Adv}^{\mathsf{MDDH}}_{\mathcal{D}_k,\mathbb{G}_s}(\mathcal{B})$.*

2.1 Structure-Preserving Signatures on Equivalence Classes

In this section, we recall the definition and the security model of SPS-EQ scheme, as introduced in [52]. We note that in order to cover a broader range of potential constructions, we rename the algorithm BGGen that generates the bilinear group BG to ParGen generating public parameters par, i.e., now the parameters par can potentially include additional values such as a common reference string. Moreover, our construction is tag-based where the tag output by Sign is just used as input to ChgRep, where no new tag is output, and required for randomization (for normal SPS-EQ, every occurrence of the tag τ is just ignored).

Definition 4 (SPS-EQ). *A SPS-EQ scheme is tuple of PPT algorithms:*

- $\mathsf{ParGen}(1^\lambda)$. *On security parameter λ and returns* par *including an asymmetric bilinear group* BG. par *is implicitly used as input by all of the algorithms.*
- $\mathsf{KeyGen}(\mathsf{par},\ell)$*: This algorithm takes* pp *and vector length $\ell > 1$ as input and outputs a key pair* $(\mathsf{sk},\mathsf{pk})$.
- $\mathsf{Sign}([\mathbf{m}]_i,\mathsf{sk})$*: This algorithm given a representative $[\mathbf{m}]_i \in (\mathbb{G}_i^*)^\ell$ for class $[\mathbf{m}]_\mathcal{R}$ and a secret key* sk *outputs a signature $\sigma' = (\sigma,\tau)$ (potentially including a tag τ).*
- $\mathsf{ChgRep}([\mathbf{m}]_i,(\sigma,\tau),\mu,\mathsf{pk})$*: This algorithm on input a representative $[\mathbf{m}]_i \in (\mathbb{G}_i^*)^\ell$ and signature σ (and potentially a tag τ), a scalar μ and* pk *as public key, computes an updated signature σ' on new representative $[\mathbf{m}']_i = [\mu\mathbf{m}]_i$ and returns $([\mathbf{m}']_i,\sigma')$.*
- $\mathsf{Verify}([\mathbf{m}]_i,(\sigma,\tau),\mathsf{pk})$*: This verification algorithm when given a representative $[\mathbf{m}]_i$, a signature σ (potentially including a tag τ) and public key* pk*, outputs 1 if it accepts and 0 otherwise.*
- $\mathsf{VKey}(\mathsf{sk},\mathsf{pk})$*: This algorithm on input key pair* $(\mathsf{sk},\mathsf{pk})$ *outputs 1 if secret key and public key are consistent and 0 otherwise.*

We recall correctness, EUF-CMA security and the notion of perfect adaption (latter being a stronger notion than the original class-hiding notion which we omit here).

Definition 5 (Correctness). *An* SPS-EQ *over* $(\mathbb{G}_i^*)^\ell$ *correct if for any* $\lambda \in N$, *any* $\ell > 1$, *any* $\mathsf{par} \leftarrow \mathsf{ParGen}(1^\lambda)$, *any pair* $(\mathsf{sk}, \mathsf{pk}) \leftarrow \mathsf{KeyGen}(\mathsf{par}, \ell)$, *any message* $[\mathbf{m}]_i \in (\mathbb{G}_i^*)^\ell$ *and any* $\mu \in \mathbb{Z}_p$ *the following holds:*

$$\begin{aligned}&\mathsf{VKey}(\mathsf{sk}, \mathsf{pk}) = 1, \ \ \textit{and}\\ &\Pr[\mathsf{Verify}([\mathbf{m}]_i, \mathsf{Sign}([\mathbf{m}]_i, \mathsf{sk}), \mathsf{pk}) = 1] = 1, \ \ \textit{and}\\ &\Pr[\mathsf{Verify}(\mathsf{ChgRep}([\mathbf{m}]_i, \mathsf{Sign}([\mathbf{m}]_i, \mathsf{sk}), \mu, \mathsf{pk}), \mathsf{pk}) = 1] = 1.\end{aligned}$$

Definition 6 (EU-CMA). *An* SPS-EQ *over* $(\mathbb{G}_i^*)^\ell$ *is* existentially unforgeable under adaptively chosen-message attacks, *if for all* $\ell > 1$ *and PPT adversaries* $\mathcal{A}$ *with access to a signing oracle* $\mathcal{O}^{\mathsf{Sign}}$, *there is a negligible function* $\epsilon(\cdot)$*:*

$$\Pr\left[\begin{array}{l}\mathsf{par} \leftarrow \mathsf{ParGen}(1^\lambda),\\ (\mathsf{sk}, \mathsf{pk}) \leftarrow \mathsf{KeyGen}(\mathsf{par}, \ell),\\ ([\mathbf{m}]_i^*, \sigma^*) \leftarrow \mathcal{A}^{\mathcal{O}^{\mathsf{Sign}(\mathsf{sk},\cdot)}}(\mathsf{pk})\end{array} : \begin{array}{r}[\mathbf{m}^*]_\mathcal{R} \neq [\mathbf{m}]_\mathcal{R} \ \ \forall [\mathbf{m}]_i \in Q^{\mathsf{Sign}} \ \ \wedge\\ \mathsf{Verify}([\mathbf{m}]_i^*, \sigma^*, \mathsf{pk}) = 1\end{array}\right] \leq \epsilon(\lambda),$$

where $Q^{\mathsf{Sign}_\mathcal{R}}$ *is the set of queries that* $\mathcal{A}$ *has issued to the signing oracle* $\mathcal{O}^{\mathsf{Sign}}$. *Note that in the tag-based case this oracle returns* (σ_i, τ_i).

Perfect adaption introduced in [39] by Fuchsbauer et al. requires signatures output by ChgRep are distributed like fresh signatures on the new representative. We present both variants here, as we will require them later. We do not yet adapt them to the tag-based variant of SPS-EQ (this is done afterwards). Note that in the following variant signatures are only required to verify (so may be maliciously computed) while we only consider keys need to satisfy VKey.

Definition 7 (Perfect adaption of signatures). *An* SPS-EQ *over* $(\mathbb{G}_i^*)^\ell$ *perfectly adapts signatures if for all tuples* $(\mathsf{sk}, \mathsf{pk}, [\mathbf{m}]_i, \sigma, \mu)$ *with:*

$$\mathsf{VKey}(\mathsf{sk}, \mathsf{pk}) = 1 \qquad \mathsf{Verify}([\mathbf{m}]_i, \sigma, \mathsf{pk}) = 1 \qquad [\mathbf{m}]_i \in (\mathbb{G}_i^*)^\ell \qquad \mu \in \mathbb{Z}_p^*$$

we have that $\mathsf{ChgRep}([\mathbf{m}]_i, \sigma, \mu, \mathsf{pk})$ *and* $([\mu \cdot \mathbf{m}]_i, \mathsf{Sign}([\mu \cdot \mathbf{m}]_i, \mathsf{sk}))$ *are identically distributed.*

In the subsequent definition, the strongest adaption notion, one in addition to potentially maliciously generated signatures one also considers maliciously generated keys (i.e., does not require that VKey needs to hold).

Definition 8 (Perfect adaption of signatures under malicious keys). *An* SPS-EQ *over* $(\mathbb{G}_i^*)^\ell$ *perfectly adapts signatures under malicious keys if for all tuples* $(\mathsf{pk}, [\mathbf{m}]_i, \sigma, \mu)$ *with:*

$$[\mathbf{m}]_i \in (\mathbb{G}_i^*)^\ell \qquad \mathsf{Verify}([\mathbf{m}]_i, \sigma, \mathsf{pk}) = 1 \qquad \mu \in \mathbb{Z}_p^*$$

we have that ChgRep *outputs* $([\mu \cdot \mathbf{m}]_i, \sigma')$ *such that* σ' *is a random element in the space of signatures, conditioned on* $\mathsf{Verify}([\mu \cdot \mathbf{m}]_i, \sigma', \mathsf{pk}) = 1$.

Perfect Adaption in Context of a CRS and for Tag-Based SPS-EQ. If par contains a CRS (as in the case of our construction), we need to consider this in the adaption notion. For Definition 7 we just replace $(\mathsf{sk}, \mathsf{pk}, [\mathbf{m}]_i, \sigma, \mu)$ with $(\mathsf{par}, \mathsf{sk}, \mathsf{pk}, [\mathbf{m}]_i, \sigma, \mu)$ where $\mathsf{par} \leftarrow \mathsf{ParGen}(1^\lambda)$ is honestly generated. We introduce it subsequently, for completeness.

Definition 9 (Perfect adaption in the honest parameter model). *An SPS-EQ scheme* (ParGen, Sign, ChgRep, Verify, VKey) *perfectly adapts signatures if for all* (par, sk, pk, $[\mathbf{m}]_i, \sigma, \tau, \mu$) *with*

$$\mathsf{VKey}(\mathsf{sk}, \mathsf{pk}) = 1 \quad \mathsf{Verify}([\mathbf{m}]_i, (\sigma, \tau), \mathsf{pk}) = 1 \quad [\mathbf{m}]_i \in (\mathbb{G}_i^*)^\ell \quad \mu \in \mathbb{Z}_p^*$$
$$\mathsf{par} \leftarrow \mathsf{ParGen}(1^\lambda)$$

the following are identically distributed:

$$(\sigma, \mathsf{ChgRep}([\mathbf{m}]_i, \sigma, \tau, \mu, \mathsf{pk})) \text{ and}$$

$$((\sigma', \cdot) \leftarrow \mathsf{Sign}(\mathsf{sk}, [\mathbf{m}]_i), \mathsf{ChgRep}([\mathbf{m}]_i, \mathsf{Sign}(\mathsf{sk}, [\mu \cdot \mathbf{m}]_i), 1, \mathsf{pk}))$$

Definition 8 does not change and also considers a potentially malicious generation of the parameters which may include a CRS (which is not satisfied by our construction). Moreover, we introduce an intermediate notion, where keys may be generated maliciously, but par is generated honestly. We formally define it in the following for completeness (this is satisfied by our construction).

Definition 10 (Perfect adaption of signatures under malicious keys in the honest parameters model). *An* SPS-EQ *over* $(\mathbb{G}_i^*)^\ell$ *perfectly adapts signatures under malicious keys in the honest parameter model if for all tuples* (par, pk, $[\mathbf{m}]_i, \sigma, \tau, \mu$) *with:*

$$[\mathbf{m}]_i \in (\mathbb{G}_i^*)^\ell \quad \mathsf{Verify}([\mathbf{m}]_i, (\sigma, \tau), \mathsf{pk}) = 1 \quad \mu \in \mathbb{Z}_p^* \quad \mathsf{par} \leftarrow \mathsf{ParGen}(1^\lambda)$$

we have that ChgRep *outputs* $([\mu \cdot \mathbf{m}]_i, \sigma')$ *such that* σ' *is a random element in the space of signatures, conditioned on* $\mathsf{Verify}([\mu \cdot \mathbf{m}]_i, \sigma', \mathsf{pk}) = 1$.

2.2 Non-Interactive Zero-Knowledge Proofs

Let $\mathcal{R}_\mathcal{L}$ be an efficiently computable relation of pairs (x, w) of words and witnesses. Let $\mathcal{L}$ be the language defined as $\mathcal{L} = \{x | \exists w : \mathcal{R}_\mathcal{L}(x, w) = 1\}$. We recall the definition of a NIZK proof system [19] for a relation $\mathcal{R}_\mathcal{L}$, where we use the formalization in [43] (based on [50]) for the sake of consistency. We note that we focus on NIZK argument systems, where soundness only holds for computationally bounded adversaries.

- $\mathsf{PGen}(1^\lambda, \mathsf{par})$: On input a security parameter λ and parameters par outputs a common reference string crs.
- $\mathsf{PTGen}(1^\lambda, \mathsf{par})$: On input a security parameter λ and parameters par outputs a common reference string crs and a trapdoor td.
- $\mathsf{PPro}(\mathsf{crs}, x, w)$: On input a common reference string crs, a statement x, and a witness w such that $\mathcal{R}_\mathcal{L}(x, w) = 1$, returns a proof Ω.
- $\mathsf{PVer}(\mathsf{crs}, x, \Omega)$: On input a reference string crs and a proof Ω, Returns accept if Ω is valid and reject otherwise.
- $\mathsf{PSim}(\mathsf{crs}, \mathsf{td}, x)$: On input common reference string crs, and the trapdoor td and word x and outputs a simulated proof Ω.

A NIZK argument system needs to satisfy the following properties:

- **Perfect Completeness:** For all possible public parameters par, all $\lambda \in \mathbb{N}$, all words $x \in \mathcal{L}$, and all witnesses w such that $\mathcal{R}_{\mathcal{L}}(x, w) = 1$, we have
$$\Pr\left[\begin{array}{l}\mathsf{crs} \leftarrow \mathsf{PGen}(1^{\kappa}, \mathsf{par}),\\ \Omega \leftarrow \mathsf{PPro}(\mathsf{crs}, x, w)\end{array} : \mathsf{PVer}(\mathsf{crs}, x, \Omega) = 1\right] = 1.$$
- **Computational Soundness:** For all PPT adversaries $\mathcal{A}$ and for all words $x \notin \mathcal{L}$ we have:
$$\Pr\left[\begin{array}{l}\mathsf{crs} \leftarrow \mathsf{PGen}(1^{\kappa}, \mathsf{par}),\\ \Omega \leftarrow \mathcal{A}(\mathsf{crs}, x)\end{array} : \mathsf{PVer}(\mathsf{crs}, x, \Omega) = 0\right] \approx 1.$$
- **Composable Zero-Knowledge:** For all PPT adversaries $\mathcal{A}$, we have
$$\Pr\left[\mathsf{crs} \leftarrow \mathsf{PGen}(1^{\lambda}, \mathsf{par}) : \mathcal{A}(1^{\lambda}, \mathsf{crs}) = 1\right] \approx$$
$$\Pr\left[(\mathsf{crs}, \mathsf{td}) \leftarrow \mathsf{PTGen}(1^{\lambda}, \mathsf{par}) : \mathcal{A}(1^{\lambda}, \mathsf{crs}) = 1\right].$$
Furthermore, for all for all $x \in \mathcal{L}$ with witness w such that $\mathcal{R}_{\mathcal{L}}(x, w) = 1$, the following are identically distributed:
$$\mathsf{PPro}(\mathsf{crs}, x, w) \text{ and } \mathsf{PSim}(\mathsf{crs}, \mathsf{td}, x)$$
where $(\mathsf{crs}, \mathsf{td}) \leftarrow \mathsf{PTGen}(1^{\lambda}, \mathsf{par})$. Note that the composable zero knowledge requires indistinguishability even for adversaries that get access to $(\mathsf{crs}, \mathsf{trap})$.

Quasi-Adaptive NIZK Proofs. Quasi-Adaptive NIZK (QA-NIZK) proofs [8,28,47,61,62,65,67] are NIZK proofs where the generation of the common reference string (CRS), for a class of languages $\mathcal{L}_{\rho}$, parametrized by ρ, is allowed to depend on the language parameter ρ. Moreover the common CRS includes a fixed part par, generated by an algorithm pargen. Here, we recall the definitions QA-NIZK proofs, as presented in [65].

Definition 11 (QA-NIZK). A non-interactive proof system $(\mathsf{pargen}, \mathsf{crsgen}, \mathsf{prove}, \mathsf{verify}, \mathsf{sim})$ is said to be a QA-NIZK proof system for an ensemble of distributions $\{\mathcal{D}_{\mathsf{par}}\}$ on collection of witness-relations $\mathcal{R} = \{\mathcal{R}_{\rho}\}$ with associated language parameter ρ if the following holds (cf. [65]):

Perfect Completeness: For all λ, all par output by $\mathsf{pargen}(1^{\lambda})$, all ρ output by $\mathcal{D}_{\mathsf{par}}$, all (x, y) with $\mathcal{R}_{\rho}(x, y) = 1$, we have
$$\Pr\left[\begin{array}{l}(\mathsf{crs}, \mathsf{trap}) \leftarrow \mathsf{crsgen}(\mathsf{par}, \rho),\\ \pi \leftarrow \mathsf{prove}(\mathsf{crs}, x, w)\end{array} : \mathsf{verify}(\mathsf{crs}, x, \pi) = 1\right] = 1$$

Computational Adaptive Soundness: For all PPT adversaries $\mathcal{A}$,
$$\Pr\left[\begin{array}{l}\rho \leftarrow \mathcal{D}_{\mathsf{par}}, \mathsf{par} \leftarrow \mathsf{pargen}(1^{\lambda}),\\ \mathsf{crs} \leftarrow \mathsf{crsgen}(\mathsf{par}, \rho),\\ (x, \pi) \leftarrow \mathcal{A}_1(\mathsf{crs}, \mathsf{par}, \rho)\end{array} : \begin{array}{c}\mathsf{verify}(\mathsf{crs}, x, \pi) = 1 \wedge\\ x \notin \mathcal{L}_{\rho}\end{array}\right] \leq \epsilon(\lambda)$$

Perfect Zero-Knowledge: For all λ, all par output by $\mathsf{pargen}(1^\lambda)$, all ρ output by $\mathcal{D}_{\mathsf{par}}$, all $(\mathsf{crs}, \mathsf{trap})$ output by $\mathsf{crsgen}(\mathsf{par}, \rho)$, all (x, y) with $\mathcal{R}_\rho(x, y) = 1$, the distributions

$$\mathsf{prove}(\mathsf{crs}, x, w) \text{ and } \mathsf{sim}(\mathsf{crs}, \mathsf{td}, x)$$

are identical. Note that the formalization of perfect zero-knowledge is similar to that of composable zero knowledge in [50] and requires indistinguishability even for adversaries that get access to $(\mathsf{crs}, \mathsf{trap})$.

2.3 Malleable Proof Systems

Let $\mathcal{R}_\mathcal{L}$ be the witness relation associated to language $\mathcal{L}$, then a controlled malleable proof system [24] is accompanied by a family of efficiently computable n-ary transformations $T = (T_x, T_w)$ such that for any n-tuple $\{(x_1, w_1), \ldots, (x_n, w_n)\} \in \mathcal{R}_\mathcal{L}^n$ it holds that $(T_x(x_1, \ldots, x_n), T_w(w_1, \ldots, w_n)) \in \mathcal{R}_\mathcal{L}$ (the family of admissible transformations is denoted by $\mathcal{T}$). Intuitively, such a proof system allows when given valid proofs $\{\Omega_i\}_{i\in[n]}$ for words $\{x_i\}_{i\in[n]}$ with associated witnesses $\{w_i\}_{i\in[n]}$ to publicly compute a valid proof Ω for word $x := T_x(x_1, \ldots, x_n)$ corresponding to witness $w := T_w(w_1, \ldots, w_n)$ using an additional algorithm denoted as ZKEval. More formally, the additional algorithms is defined as follows:

- $\mathsf{ZKEval}(\mathsf{crs}, T, (x_i, \Omega_i)_{i\in[n]})$: takes as input common reference string crs, a transformation $T \in \mathcal{T}$, words $x_1, \ldots x_n$ and corresponding proofs $\Omega_1, \ldots, \Omega_2$, and outputs a new word $x' := T_x(x_1, \ldots, x_n)$ and proof Ω'.

It is desirable that proofs computed by applying ZKEval are indistinguishable from freshly computed proofs for the resulting word $x' := T_x(x_1, \ldots, x_n)$ and corresponding witness $w' := T_w(w_1, \ldots, w_n)$ (this property is called (strong) derivation privacy). We recall the weaker notion of derivation privacy below.

Definition 12 (Derivation Privacy [24]**).** *A NIZK proof system* $\{\mathsf{PGen}, \mathsf{PTGen}, \mathsf{PPro}, \mathsf{PVer}, \mathsf{PSim}, \mathsf{ZKEval}\}$ *being malleable with respect to a set of transformations* $\mathcal{T}$ *defined on some relation* $\mathcal{R}$ *is derivation private, if for all PPT adversaries* $\mathcal{A}$,

$$\Pr\left[\begin{array}{l} \mathsf{crs} \leftarrow \mathsf{PGen}(1^\kappa), b \xleftarrow{R} \{0,1\}, \\ (\mathsf{st}, ((x_i, w_i), \Omega_i)_{i\in[q]}, T) \leftarrow \mathcal{A}(\mathsf{crs}), \\ \mathsf{Return} \perp \mathsf{if}\ (T \notin \mathcal{T}\ \vee\ \exists i \in [q] : (\mathsf{PVer}(\mathsf{crs}, x_i, \Omega_i) = 0\ \vee \\ (x_i, w_i) \notin \mathcal{R}), \\ \mathsf{Else\ if}\ b = 0 : \Omega \leftarrow \mathsf{PPro}(\mathsf{crs}, T_x((x_i)_{i\in[q]}), T_w((w_i)_{i\in[q]}), \\ \mathsf{Else\ if}\ b = 1 : \Omega \leftarrow \mathsf{ZKEval}(\mathsf{crs}, T, (x_i, \pi_i)_{i\in[q]}), \\ b^* \leftarrow \mathcal{A}(\mathsf{st}, \Omega) \end{array} : b = b^*\right] \le \epsilon(\lambda)$$

3 Revisiting the FG18 Model and Applications

In this section we recall the construction in [35] (denoted FG18 henceforth) and point out some issues regarding their signature adaption notion and the implicitly assumed notion of perfect composition from [51] for concrete applications. We again stress that FG18 scheme is secure in FG18 model (honestly signature and key generation or semi-honest), but we are going to show its problems in the stronger model, which is current acceptable model. In order to make it more convenient for the reader we adapt the notion used in [35] to the original SPS-EQ notion (but keep their name EQS).

First, we recall that their scheme has a one-time randomizability property and therefore FG18 need to modify the perfect adaption notion from [39] (Definition 7 in Sect. 2.1) to exclude trivial distinguishers, i.e., they always consider the pairs of original and adapted signatures in their distributions. We recall their version in Definition 13. The most important difference[7] is that while the original notion in Definition 7 considers maliciously generated signatures, the definition in [35] is restricted to *honestly generated* signatures.

Definition 13 (Signature Adaption [35]**).** *An EQS scheme* $(\mathsf{ParGen}, \mathsf{Sign}, \mathsf{ChgRep}, \mathsf{Verify}, \mathsf{VKey})$ *perfectly adapts signatures if for all* $(\mathsf{sk}, \mathsf{pk}, [\mathbf{m}]_i, \mu)$ *with*

$$\mathsf{VKey}(\mathsf{sk}, \mathsf{pk}) = 1 \qquad [\mathbf{m}]_i \in (\mathbb{G}_i^*)^\ell \qquad \mu \in \mathbb{Z}_p^*$$

the following are identically distributed:

$$(\rho := \mathsf{Sign}(\mathsf{sk}, [\mathbf{m}]_i), \mathsf{ChgRep}(\mathsf{pk}, \rho, \mu)) \text{ and}$$

$$(\rho := \mathsf{Sign}(\mathsf{sk}, [\mathbf{m}]_i), \mathsf{ChgRep}(\mathsf{pk}, \mathsf{Sign}(\mathsf{sk}, [\mu \cdot \mathbf{m}]_i), 1))$$

In Fig. 1 we recall the FG18 scheme and then proceed to discuss problems of Definition 13 and their scheme in context of applications.

3.1 Problem With Key Verification and the Need for a CRS

Fuchsbauer and Gay require for signature adaption that the respective EQS scheme provides a VKey algorithm that checks consistency of keys sk and pk. When looking at their keys $\mathsf{pk} := ([\mathbf{B}]_2, \{[\mathbf{K}_i\mathbf{B}]_2\}_{i\in[\ell]})$ and $\mathsf{sk} := (\mathbf{A}, \{\mathbf{K}_i\}_{i\in\ell})$, a potential VKey algorithm can check the consistency of pk with the part of the secret key $\{\mathbf{K}_i\}_{i\in\ell}$. They did not specify the VKey algorithm, but any reasonable VKey would check if sk contains the trapdoor $\mathbf{B}$, as honest keys would not contain it. Now an interesting aspect is that this does not per se present a problem in their definition, as they do not consider perfect adaption under malicious keys (in the vein of Definition 8; cf. Sect. 2.1). However, the existence of the potential trapdoor $\mathbf{B}$ and no means to proving the absence of it represents a problem with

[7] One syntactical difference is that for EQS they do not input the message $[\mathbf{m}]_i$ in their ChgRep algorithm, but this does not matter for our discussion.

$\mathsf{Setup}(\mathcal{PG})$:

$\mathbf{A} \xleftarrow{R} \mathcal{D}_{2k,k}, \mathbf{B} \xleftarrow{R} \mathcal{D}_{k'}$
for $i \in [\ell]$ **do**
$\quad \mathbf{K}_i \xleftarrow{R} \mathbb{Z}_p^{2k \times (k'+1)}$
endfor
$\mathsf{pk} := ([\mathbf{B}]_2, \{[\mathbf{K}_i\mathbf{B}]_2\}_{i\in[\ell]})$
$\mathsf{sk} := (\mathbf{A}, \{\mathbf{K}_i\}_{i\in[\ell]})$
return (pk,sk)

$\mathsf{Sign}(\mathsf{sk}, [\mathbf{m}]_1 \in (\mathbb{G}_1^{\ell})^*)$:

$\mathbf{U} \xleftarrow{R} \mathsf{GL}_k, \mathbf{S} = \mathbf{AU}$
for $i \in [\ell]$ **do**
$\quad [\mathbf{S}_i]_1 = [m_i]_1\mathbf{S}$
endfor
$$[\mathbf{S}_{\ell+1}]_1 = \sum_{i=1}^{\ell} [m_i]_1 \mathbf{K}_i^{\top} \mathbf{S}$$
return $\rho = (\{[\mathbf{S}_i]_1\}_{i\in[\ell+1]}, [\mathbf{S}]_2)$

$\mathsf{ChgRep}(\mathsf{pk}, \rho = (\{[\mathbf{S}_i]_1\}_{i\in[\ell+1]}, [\mathbf{S}]_2), \mu)$:

$\mathbf{r} \xleftarrow{R} (\mathbb{Z}_p^k)^*, [\mathbf{s}]_2 = [\mathbf{S}]_2\mathbf{r}$
for $i \in [\ell+1]$ **do**
$\quad [\mathbf{s}_i]_1 = \mu[\mathbf{S}_i]_1\mathbf{r}$
endfor
return $\sigma = (\{[\mathbf{s}_i]_1\}_{i\in[\ell+1]}, [\mathbf{s}]_2)$

$\mathsf{Ver}(\mathsf{pk}, [\mathbf{m}]_1, \sigma = (\{[\mathbf{s}_i]_1\}_{i\in[\ell+1]}, [\mathbf{s}]_2)$:

if $[\mathbf{s}]_2 \neq [\mathbf{0}]_2$
$\quad$ **and** $\forall i \in [\ell] : [\mathbf{s}_i]_1 \cdot [1]_2 = [m_i]_1 \cdot [\mathbf{s}]_2$
$$\text{and } \sum_{i=1}^{\ell} [\mathbf{s}_i^{\top}]_1 \cdot [\mathbf{K}_i\mathbf{B}]_2 = [\mathbf{s}_{\ell+1}^{\top}]_1 \cdot [\mathbf{B}]_2$$
$\quad$ **return** 1
else return 0

Fig. 1. EQS Scheme from [35].

the application of the FG18 scheme to attribute-based credentials (ABCs) (cf. Section 5 in [35]).

In the ABC construction from [40], the issuer generates an SPS-EQ key pair and in the Issue protocol, the issuer needs to provide a ZKPoK that $\mathsf{VKey}(\mathsf{sk}, \mathsf{pk}) = 1$. Note that for FG18 no realization of this ZKPoK can prove the absence of $\mathbf{B}$ (as the issuer could simply pretend to not knowing it and the ZKPoK cannot cover this) and a malicious issuer may remember $\mathbf{B}$. Now in the anonymity proof of the ABC scheme (Theorem 8 in [40]), the reduction can extract the signing key sk from the ZKPoK and in the transition from $\mathbf{Game}_1$ to $\mathbf{Game}_2$, for all calls to the oracle $\mathcal{O}_{LoR}$ the computation of ChgRep is replaced with Sign of the SPS-EQ, i.e., instead of adapting existing signatures fresh signatures are computed. Now, this is argued under their signature adaption notion. However, without additional means, by the strategy we discuss below (i.e., a way to construct malicious signatures that verify), an adversary can detect with overwhelming probability that the simulation deviates from the original anonymity game and thus this proof breaks down when instantiated with EQS in [35]. The reason is, that their adaption notion in Definition 13 is too weak to be useful to constructing ABCs following the approach in [40].

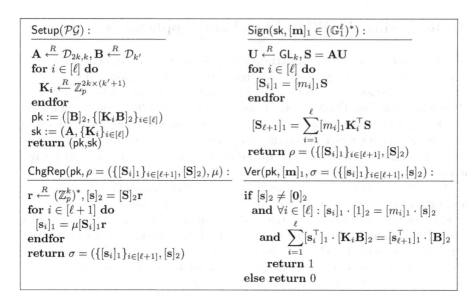

Attack Strategy. Let us assume that the adversary who generates the key-pair $\mathsf{pk} = ([\mathbf{B}]_2, \{[\mathbf{K}_i\mathbf{B}]_2\}_{i\in[\ell]})$ and $\mathsf{sk} = (\mathbf{A}, \{\mathbf{K}_i\}_{i\in[\ell]})$ remembers the trapdoor $\mathbf{B}$. For simplicity we set $k = 2$ and $k' = 1$ in Scheme 1 and so we have $\mathbf{B} = \begin{pmatrix} b_1 \\ b_2 \end{pmatrix}$. Let us for the sake of exposition assume that the signer (credential issuer)

wants to track a specific instance of signing (issuing) and generates all signatures honestly, except for the one instance (lets say Alice's credential). Latter signature is computed differently by the issuer, but in a way that it is indistinguishable for verifiers, i.e., it still verifies correctly. Actually, instead of computing $\mathbf{S}_{\ell+1} = \begin{pmatrix} S_1 & S_2 \\ S_3 & S_4 \end{pmatrix}$ as dictated by the Sign algorithm (cf. Fig. 1), he uses $\mathbf{S}_{\ell+1}$ (as in Sign) but also his trapdoor $\mathbf{B}$ to compute $\mathbf{S}'_{\ell+1} = \begin{pmatrix} S_1 - b_2 & S_2 + b_2 \\ S_3 + b_1 & S_4 - b_1 \end{pmatrix}$. Then, he includes $\mathbf{S}'_{\ell+1}$ instead of $\mathbf{S}_{\ell+1}$ in the first part of the signature ρ. Note that we have $\mathbf{S}^\top_{\ell+1}\mathbf{B} = \mathbf{S}'^\top_{\ell+1}\mathbf{B}$, and for a verifier this alternative signature computation is not noticeable. When Alice wants to randomize ρ (i.e., run ChgRep in Fig. 1), she chooses $\mathbf{r} \xleftarrow{R} \mathbb{Z}_p^2$ and obtains $\mathbf{s}'_{\ell+1} = \mu\mathbf{S}'_{\ell+1}\mathbf{r} = \mu \begin{pmatrix} (S_1 - b_2)r_1 + (S_2 + b_2)r_2 \\ (S_3 + b_1)r_1 + (S_4 - b_1)r_2 \end{pmatrix}$. Note that the signer knows $\mathbf{K}_i$, and so he can check for any given randomized signature the following:

$$\sum_{i=1}^{\ell} [\mathbf{s}_i^\top]_1 \mathbf{K}_i = [\mathbf{s}_{\ell+1}^\top]_1 \tag{3}$$

which does not use pairing evaluations and thus does not eliminate $\mathbf{B}$. Now it is easy to see that all randomized signatures including the randomized signature issued for Alice pass the original verification using Ver. However, the randomized signature of Alice has an additional part (i.e., $\mathbf{B}$) and so Eq. (3) cannot be satisfied. So, the signer can easily distinguish the signature issued to Alice from all other honestly computed signatures.

Trying to Fix the Problem. A modification of the FG18 scheme to prevent this attack would be to put $[\mathbf{B}]_2$ in a common reference string (CRS) used by all signers when generating their keys so that no signer knows $\mathbf{B}$. As we show subsequently, however, the adaption notion in Definition 13 used for FG18 still remains too weak for ABCs and group signatures.

3.2 Distinguishing Signatures

Now, we show how a malicious signer can distinguish signatures even if keys are generated honestly. In the case of dynamic group signatures (GS) in [30] (or ABCs under honest keys), the adversary in the anonymity game is allowed to compute signatures on its own and we will show how this enables the adversary to track signatures, which breaks the anonymity proof. We stress that this attack works independently of whether there is a trapdoor in the secret key, as the GS in [30] rely on the BSZ model [15] and thus assume honest key generation (mitigating the attack in Sect. 3.1 by construction).

Attack Strategy. First we show how a signer who remembers $\mathbf{S}$ during running Sign can obtain the value of $[\mathbf{r}]_2$, which was used as a randomizer for the signature during ChgRep, and then how he can use it to distinguish two signatures. Again, let us set $k = 2$ and $k' = 1$. So, we have $\mathbf{S} = \begin{pmatrix} S_1 & S_2 \\ S_3 & S_4 \\ S_5 & S_6 \\ S_7 & S_8 \end{pmatrix}$, and when ChgRep multiplies $[\mathbf{S}]_2$ on $\mathbf{r} = \begin{pmatrix} r_1 \\ r_2 \end{pmatrix}$, we receive $[\mathbf{s}]_2 = \begin{bmatrix} s_1 \\ s_2 \\ s_3 \\ s_4 \end{bmatrix}_2 = \begin{bmatrix} r_1S_1 + r_2S_2 \\ r_1S_3 + r_2S_4 \\ r_1S_5 + r_2S_6 \\ r_1S_7 + r_2S_8 \end{bmatrix}_2$. Taking $[\mathbf{s}]_2$ and $\mathbf{S}$, we compute $[\frac{s_1}{S_1}]_2 - [\frac{s_2}{S_3}]$, and then multiply it to $(\frac{S_2}{S_1} - \frac{S_4}{S_3})^{-1}$ to obtain $[r_2]_2$. Now, we also can recover $[r_1]_2$ and so we obtain $[\mathbf{r}]_2$.

Now, let the signer generate two signatures, say for Alice and Bob, where he later wants to link the received randomized signature to one of them.

The signer picks $\mathbf{S} = \begin{pmatrix} S_1 & S_2 \\ S_3 & S_4 \\ S_5 & S_6 \\ S_7 & S_8 \end{pmatrix}$ for Alice, and picks different S_5', S_6', S_7', S_8', and sets $\mathbf{S}' = \begin{pmatrix} S_1 & S_2 \\ S_3 & S_4 \\ S_5' & S_6' \\ S_7' & S_8' \end{pmatrix}$ for Bob in their respective signatures. When the signer receives $[\mathbf{s}]_2$, a candidate for a signature obtained from ChgRep, based on the approach discussed above he obtains $[\mathbf{r}]_2$. Now he checks whether $[s_3]_2 = [r_1S_5 + r_2S_6]_2$ holds, in which case the randomized signature is related to Alice. On the other hand, if $[s_3]_2 = [r_1S_5' + r_2S_6']_2$ holds, then the randomized signature is related to Bob.

3.3 No Perfect Composition

Subsequently, in Definition 14 we recall the perfect composition notion from [51] required to construct VES from SPS-EQ. This notion intuitively requires that ChgRep executed with random coins fixed to 1 updates only the parts of the given signature that are affected by updating the representative from $[\mathbf{m}]_i$ to $\mu[\mathbf{m}]_i$ and not changing the randomness ω previously used by Sign.

Definition 14 (Perfect Composition [51]**).** *An SPS-EQ scheme* (ParGen, $\mathsf{Sign}, \mathsf{ChgRep}, \mathsf{Verify}, \mathsf{VKey}$) *allows perfect composition if for all random tapes* ω *and tuples* $(\mathsf{sk}, \mathsf{pk}, [\mathbf{m}]_i, \sigma, \mu)$*:*

$$\mathsf{VKey}(\mathsf{sk}, \mathsf{pk}) = 1 \qquad \sigma \leftarrow \mathsf{Sign}([\mathbf{m}]_i, \mathsf{sk}; \omega) \qquad [\mathbf{m}]_i \in (\mathbb{G}_i^*)^\ell \qquad \mu \in \mathbb{Z}_p^*$$

it holds that $(\mu[\mathbf{m}]_i, \mathsf{Sign}(\mu[\mathbf{m}]_i, \mathsf{sk}; \omega)) = \mathsf{ChgRep}([\mathbf{m}]_i, \sigma, \mu, \mathsf{pk}; 1)$.

Since this notion does not require any assumption on the distribution of original and adapted signatures, the issues discussed so far do not yield to any problem. However, it is quite easy to see that this notion is not satisfied by the FG18 scheme and this is actually an inherent problem for EQS (SPS-EQ) schemes where signatures output by Sign and ChgRep have different forms. To illustrate this for the FG18 scheme (cf. Fig. 1), signatures resulting from Sign contain a matrix $[\mathbf{S}]_2$, whereas signatures output by ChgRep contain the vector $[\mathbf{s}]_2 := [\mathbf{S}]_2\mathbf{r}$ (where in context of Definition 14, $\mathbf{r}$ represents the all all-ones vector).

4 Our OR-Proof and Core Lemma

Subsequently, we present the concrete instantiation of our malleable OR-proof that we use for our SPS-EQ scheme. Firstly, PPro computes as a proof two copies Ω_1 and Ω_2 of an OR-proof for statements $[x_1]_1$ and $[x_2]_1$, which use the same randomness v and share a QA-NIZK proof π (denoted by Ω). Consequently, instead of ending up with two independent proofs, we end up with a single proof $\Omega = (\Omega_1 = ([\mathbf{C}_{1,i}]_2, [\mathbf{\Pi}_{1,i}]_1), \Omega_2 = ([\mathbf{C}_{2,i}]_2, [\mathbf{\Pi}_{2,i}]_1), [\mathbf{z}_i]_2, \pi)$ for $i = 0, 1$ where both proofs share $[\mathbf{z}_i]_2$ and π. We also have PVer and PSim which take two statements and proofs with shared randomness and QA-NIZK denoted by π as input. Our ZKEval is restricted to any two words $[\mathbf{x}_1]_1$ and $[\mathbf{x}_2]_1$ corresponding to witnesses r_1 and r_2 where the associated proofs Ω_1 and Ω_2 have been computed using the same randomness v and thus have shared $[\mathbf{z}_i]_2$ and π. The output of ZKEval is a proof $\Omega' = (\Omega'_1, [\mathbf{z}'_i]_2, \pi')$ for word $[\mathbf{x}'_1]_1$ corresponding to witness $r' = r_1 + \psi r_2$ with $\psi \xleftarrow{R} \mathbb{Z}_p$ chosen by ZKEval (i.e., ψ indexes a concrete transformation in the family $\mathcal{T}$). Finally, we also provide a verification algorithm (PRVer) that verifies a single OR-proof (as we use it in the SPS-EQ).

Our OR-Proof. Now, we present our malleable proof for OR language $\mathcal{L}^{\vee}_{\mathbf{A}_0,\mathbf{A}_1}$ based upon the one in [43]. We recall their NIZK proof as well as the QA-NIZK used by us in our NIZK proof in the full version. The language is

$$\mathcal{L}^{\vee}_{\mathbf{A_0},\mathbf{A_1}} = \{[\mathbf{x}]_1 \in \mathbb{G}_1^{2k} | \exists \mathbf{r} \in \mathbb{Z}_p^k : [\mathbf{x}]_1 = [\mathbf{A_0}]_1 \cdot \mathbf{r} \vee [\mathbf{x}]_1 = [\mathbf{A_1}]_1 \cdot \mathbf{r}\}$$

and par $:= (\mathsf{BG}, [\mathbf{A_0}]_1, [\mathbf{A_1}]_1)$ with $\mathsf{BG} \leftarrow \mathsf{BGGen}(1^\lambda)$ and $\mathbf{A_0}, \mathbf{A_1} \xleftarrow{R} \mathcal{D}_{2k,k}$ for $k \in \mathbb{N}$. We henceforth denote our proof by PS and set $k = 1$ and consider the class of admissible transformations $\mathcal{T} := \{(T_x^\psi, T_w^\psi)\}_{\psi \in \mathbb{Z}_p^*}$ and $T_x^\psi([\mathbf{x}_1]_1, [\mathbf{x}_2]_1) := [\mathbf{x}_1]_1 + \psi[\mathbf{x}_2]_1$ and $T_w^\psi(r_1, r_2) := r_1 + \psi r_2$. Observe that the output of ZKEval is a proof with new randomness $v' = \alpha v$, $s'_0 = \alpha s_{1,0} + \alpha\psi s_{2,0} + \beta_0$ and $s'_1 = \alpha s_{1,1} + \alpha\psi s_{2,1} + \beta_1$ as well as new witness $r' = r_1 + \psi r_2$.

Below, we show that the protocol in Fig. 2 is indeed a NIZK argument.

Theorem 1. *The protocol in Fig. 2 is a malleable non-interactive zero-knowledge argument for the language* $\mathcal{L}^{\vee}_{\mathbf{A}_0,\mathbf{A}_1}$ *with respect to allowable transformations* $\mathcal{T}$.

Proof. We need to prove three properties, perfect completeness, composable zero-knowledge, computational soundness and derivation privacy.

Completeness: This is easy to verify.

Zero-Knowledge: The challenger sends an MDDH challenge $([\mathbf{D}]_2, [\mathbf{z}]_2)$ to the adversary $\mathcal{B}$. Then $\mathcal{B}$ picks $\mathbf{A}_0, \mathbf{A}_1 \xleftarrow{R} \mathcal{D}_{2,1}$, $\mathbf{A} \xleftarrow{R} \mathcal{D}_1$, $\mathbf{K} \xleftarrow{R} \mathbb{Z}_p^{2\times 1}$ and computes $[\mathbf{P}]_2 = [\mathbf{z}^\top + \mathbf{D}^\top]_2\mathbf{K}$ and $\mathbf{C} = \mathbf{K}\overline{\mathbf{A}}$.

PGen(par, 1^λ) :

$\mathbf{D}, \mathbf{A} \xleftarrow{R} \mathcal{D}_1, \mathbf{z} \xleftarrow{R} \mathbb{Z}_p^2 \setminus \text{span}(\mathbf{D})$
$\mathbf{K} \xleftarrow{R} \mathbb{Z}_p^{2\times 1}$
$\mathbf{M} := \mathbf{D} + \mathbf{z}$
$\mathbf{P} := \mathbf{M}^\top \mathbf{K}$
$\mathbf{C} := \mathbf{K}\overline{\mathbf{A}}$
$\mathsf{crs} = (\mathsf{par}, [\mathbf{D}]_2, [\mathbf{z}]_2, [\mathbf{P}]_2, [\overline{\mathbf{A}}]_1, [\mathbf{C}]_1)$
return crs

PPro(crs, $[\mathbf{x}_1]_1, r_1, [\mathbf{x}_2]_1, r_2$) :

Let $b \in \{0,1\}, j \in \{1,2\}$ **s.t.** $[\mathbf{x}_j]_1 = [\mathbf{A_b}]_1 r_j$
$v \xleftarrow{R} \mathbb{Z}_p$
$[\mathbf{z}_{1-b}]_2 := v[\mathbf{D}]_2$
$[\mathbf{z}_b]_2 := v[\mathbf{z}]_2$
$\pi := v[\mathbf{P}]_2$
$s_{1,0}, s_{1,1}, s_{2,0}, s_{2,1} \xleftarrow{R} \mathbb{Z}_p$
$[\mathbf{C}_{1,b}]_2 := s_{1,b}[\mathbf{D}]_2^\top + r_1[\mathbf{z}_b]_2$
$[\mathbf{\Pi}_{1,b}]_1 := [\mathbf{A}_b]_1^\top s_{1,b}$
$[\mathbf{C}_{1,1-b}]_2 := s_{1,1-b}[\mathbf{D}]_2^\top$
$[\mathbf{\Pi}_{1,1-b}]_1 := [\mathbf{A}_{1-b}]_1 \cdot s_{1,1-b} - [\mathbf{x}_1]_1 v$
$[\mathbf{C}_{2,b}]_2 := s_{2,b}[\mathbf{D}]_2^\top + r_2[\mathbf{z}_b]_2$
$[\mathbf{\Pi}_{2,b}]_1 := [\mathbf{A}_b]_1^\top s_{2,b}$
$[\mathbf{C}_{2,1-b}]_2 := s_{2,1-b}[\mathbf{D}]_2^\top$
$[\mathbf{\Pi}_{2,1-b}]_1 := [\mathbf{A}_{1-b}]_1 \cdot s_{2,1-b} - [\mathbf{x}_2]_1 v$
$\Omega := ([\mathbf{C}_{j,i}]_2, [\mathbf{\Pi}_{j,i}]_1, [\mathbf{z}_i]_2, \pi)_{j\in\{1,2\}, i\in\{0,1\}}$
return Ω

PVer(crs, $[\mathbf{x}_1]_1, [\mathbf{x}_2]_1, \Omega$) :

if $e([\overline{\mathbf{A}}]_1, \pi) = e([\mathbf{C}]_1, [\mathbf{z}_1]_2 + [\mathbf{z}_0]_2)$
  and for all $i \in \{0,1\}, j \in \{1,2\}$ **it holds**
    $e([\mathbf{A}_i]_1, [\mathbf{C}_{j,i}]_2)$
      $e([\Pi_{j,i}]_1, [\mathbf{D}]_2^\top) + e([\mathbf{x}_j]_1, [\mathbf{z}_i]_2^\top)$
    return 1
else return 0

PRVer(crs, $[\mathbf{x}'_1]_1, \Omega'_1$) :

if $e([\overline{\mathbf{A}}]_1, \pi') = e([\mathbf{C}]_1, [\mathbf{z}_1]_2 + [\mathbf{z}_0]_2)$
  and for all $i \in \{0,1\}$ **it holds**
    $e([\mathbf{A}_i]_1, [\mathbf{C}'_i]_2) =$
      $e([\mathbf{\Pi}'_i]_1, [\mathbf{D}]_2^\top) + e([\mathbf{x}'_1]_1, [\mathbf{z}'_i]_2^\top)$
    return 1
else return 0

PTGen(par, 1^λ) :

$\mathbf{D}, \mathbf{A} \xleftarrow{R} \mathcal{D}_1, u \xleftarrow{R} \mathbb{Z}_p$
$\mathbf{K} \xleftarrow{R} \mathbb{Z}_p^{2\times 1}$
$\mathbf{z} := \mathbf{D}u$
$\mathbf{M} := \mathbf{D} + \mathbf{z}$
$\mathbf{P} := \mathbf{M}^\top \mathbf{K}$
$\mathbf{C} := \mathbf{K}\overline{\mathbf{A}}$
$\mathsf{crs} := (\mathsf{par}, [\mathbf{D}]_2, [\mathbf{z}]_2, [\mathbf{P}]_2, [\overline{\mathbf{A}}]_1, [\mathbf{C}]_1)$
$\mathsf{trap} := (u, \mathbf{K})$
return (crs, trap)

PSim(crs, trap, $[\mathbf{x}_1]_1, [\mathbf{x}_2]_1$) :

$v \xleftarrow{R} \mathbb{Z}_p$
$[\mathbf{z}_0]_2 := v[\mathbf{D}]_2$
$[\mathbf{z}_1]_2 := v[\mathbf{z}]_2$
$\pi := v[\mathbf{P}]_2$
$s_{1,0}, s_{1,1}, s_{2,0}, s_{2,1} \xleftarrow{R} \mathbb{Z}_p$
$[\mathbf{C}_{1,0}]_2 := s_{1,0}[\mathbf{D}]_2^\top$
$[\mathbf{\Pi}_{1,0}]_1 := [\mathbf{A}_0]_1^\top s_{1,0} - [\mathbf{x}_1]_1 v$
$[\mathbf{C}_{1,1}]_2 := s_{1,1}[\mathbf{D}]_2^\top$
$[\mathbf{\Pi}_{1,1}]_1 := [\mathbf{A}_1]_1 \cdot s_{1,1} - [\mathbf{x}_1]_1(vu)$
$[\mathbf{C}_{2,0}]_2 := s_{2,0}[\mathbf{D}]_2^\top$
$[\mathbf{\Pi}_{2,0}]_1 := [\mathbf{A}_0]_1^\top s_{2,0} - [\mathbf{x}_2]_1 v$
$[\mathbf{C}_{2,1}]_2 := s_{2,1}[\mathbf{D}]_2^\top$
$[\mathbf{\Pi}_{2,1}]_1 := [\mathbf{A}_1]_1 \cdot s_{2,1} - [\mathbf{x}_2]_1(vu)$
$\Omega := ([\mathbf{C}_{j,i}]_2, [\mathbf{\Pi}_{j,i}]_1, [\mathbf{z}_i]_2, \pi)_{j\in\{1,2\}, i\in\{0,1\}}$
return Ω

ZKEval(crs, $[\mathbf{x}_1]_1, [\mathbf{x}_2]_1, \Omega$) :

Parse $\Omega = (\Omega_1, \Omega_2, [\mathbf{z}_i]_2, \pi)$
if PVer(crs, $[\mathbf{x}_1]_1, [\mathbf{x}_2]_1, \Omega) = 0$
  return $\perp$
else $\psi, \alpha, \beta_0, \beta_1 \xleftarrow{R} \mathbb{Z}_p^*$
  and for all $b \in \{0,1\}$
    $[\mathbf{z}'_b]_2 := \alpha[\mathbf{z}_b]_2$
    $[\mathbf{C}'_b]_2 := \alpha[\mathbf{C}_{1,b}]_2 + \alpha\psi[\mathbf{C}_{2,b}]_2 + \beta_b[\mathbf{D}]_2$
    $[\mathbf{\Pi}'_b]_1 := \alpha[\mathbf{\Pi}_{1,b}]_1 + \alpha\psi[\mathbf{\Pi}_{2,b}]_1 + \beta_b[\mathbf{A}_b]_1$
  $\pi' := \alpha\pi$
  $\Omega' := (\Omega'_1, [\mathbf{z}'_i]_2, \pi')$
return Ω'

Fig. 2. Malleable NIZK argument for language $\mathcal{L}^{\vee}_{\mathbf{A_0},\mathbf{A_1}}$

Then $\mathcal{B}$ sends $([\mathbf{A}_0]_1, [\mathbf{A}_1]_1, [\mathbf{z}]_2, [\mathbf{D}]_2, [\mathbf{P}]_2, [\overline{\mathbf{A}}]_1, [\mathbf{C}]_1)$ to $\mathcal{A}$ as crs. When $\mathcal{B}$ receives a real MDDH tuple, where $[\mathbf{z}]_2 = [\mathbf{D}u]_2$ for some $u \in \mathbb{Z}_p$, $\mathcal{B}$ simulates crs as PTGen. In the other case, where $[\mathbf{z}]_2 \xleftarrow{R} \mathbb{G}_2^2$, using the fact that the uniform distribution over $\mathbb{Z}_p^2$ and the uniform distribution over $\mathbb{Z}_p^2 \backslash \text{span}(\mathbf{D})$ are $1/p$-statistically close distributions, since $\mathbf{D}$ is of rank 1, we can conclude that $\mathcal{B}$ simulates the crs as output by PGen, within a $1/p$ statistical distance. Now, note that PPro and PSim compute the vectors $[\mathbf{z}_0]_2$ and $[\mathbf{z}_1]_2$ in the exact same way, i.e., for all $b \in \{0,1\}$, $\mathbf{z}_b := \mathbf{D}v_b$ where v_0, v_1 are uniformly random over $\mathbb{Z}_p$ subject to $v_1 = v_0 u$ (recall $\mathbf{z} := \mathbf{D}u$).

Also for case $j = 1$, on input $[\mathbf{x}_1]_1 := [\mathbf{A}_b r_1]_1$, for some $b \in \{0,1\}$, $\mathsf{PPro}(\mathsf{crs}, [\mathbf{x}_1]_1, [\mathbf{x}_2]_1, r_1, r_2)$ computes $[\mathbf{C}_{1,1-b}]_2$ and $[\Pi_{1,1-b}]_1$ exactly as PSim, that is: $[C_{1,1-b}]_2 = s_{1,1-b}[\mathbf{D}]_2$ and $[\Pi_{1,1-b}]_1 = [\mathbf{A}_{1-b}]_1 s_{1,1-b} - [\mathbf{x}_1]_1 v_{1-b}$. The algorithm PPro additionally computes $[\mathbf{C}_{1,b}]_2 = s_{1,b}[\mathbf{D}]_2 + r_1[\mathbf{z}]_2$ and $[\Pi_{1,b}]_1 = [\mathbf{A}_b]_1 s_{1,b}$, with $s_{1,b} \xleftarrow{R} \mathbb{Z}_p$. Since the following are identically distributed:

$$s_{1,b} \quad \text{and} \quad s_{1,b} - r_1 v_b$$

for $s_{1,b} \xleftarrow{R} \mathbb{Z}_p$, we can re-write the commitment and proof computed by PPro as $[\mathbf{C}_{1,b}]_2 = s_{1,b}[\mathbf{D}]_2 - r_1 v_b[\mathbf{D}]_2 + r_1[\mathbf{z}_b]_2 = [s_{1,b}\mathbf{D}]_2$ and $[\Pi_{1,b}]_1 = [\mathbf{A}_b]_1 s_{1,b} - [\mathbf{A}_b r_1 v_b]_2 = [\mathbf{A}_b s_{1,b}]_1 - [\mathbf{x}_1 \mathbf{v}_b]_2$, which is exactly as the output of PSim.

For case $j = 2$ the argumentation is analogous.

Computational Soundness: Based on the computational soundness of the QA-NIZK proofs [65], we have $\mathbf{z}_0 + \mathbf{z}_1 \notin span(\mathbf{D})$. So, there is a $b \in \{0,1\}$ such that $\mathbf{z}_b \notin span(\mathbf{D})$. This implies that there exists a $\mathbf{d}^\perp \in \mathbb{Z}_p^2$ such that $\mathbf{D}^\top \mathbf{d}^\perp = 0$, and $\mathbf{z}_b^\top \mathbf{d}^\perp = 1$. Furthermore, as the row vectors of $\mathbf{D}$ together with $\mathbf{z}_b$ form a basis of $\mathbb{Z}_p^2$, we can write $[\mathbf{C}_{j,b}]_2 := [s_{j,b}\mathbf{D} + r_j \mathbf{z}_b]_2$ for some $s_{j,b}, r_j \xleftarrow{R} \mathbb{Z}_p$. Multiplying the verification equation by $\mathbf{d}$ thus yields $[\mathbf{A}_b r_j]_1 = [\mathbf{x}_j]_1$, which proves a successful forgery outside $\mathcal{L}^\vee_{\mathbf{A}_0,\mathbf{A}_1}$ impossible.

Derivation Privacy: As can be seen, the algorithm ZKEval outputs a proof with new independent randomness. So, the algorithm ZKEval and the algorithm PPro, when only compute a single proof, have identical distribution, i.e., we have perfect derivation privacy. More precisely, under the CRS $([\mathbf{A}_0]_1, [\mathbf{A}_1]_1, [\mathbf{z}]_2, [\mathbf{D}]_2, [\mathbf{P}]_2)$, a proof $\Omega' = (\Omega'_1, [\mathbf{z}'_i]_2, \pi')$ for word $[\mathbf{x}'_1]_1$ corresponding to witness r' has form $[\mathbf{z}'_{1-b}]_2 = v'[\mathbf{D}]_2$, $[\mathbf{z}'_b]_2 = v'[\mathbf{z}]_2$ and $\pi = v'[\mathbf{P}]_2$, and $[\mathbf{C}'_b]_2 = s'_b[\mathbf{D}]_2^\top + r'[\mathbf{z}'_b]_2$, $[\mathbf{\Pi}'_b]_1 = [\mathbf{A}_b]_1^\top s'_b$, $[\mathbf{C}'_{1-b}]_2 = s'_{1-b}[\mathbf{D}]_2^\top$ and $[\mathbf{\Pi}'_{1-b}]_1 = [\mathbf{A}_{1-b}]_1 \cdot s'_{1-b} - [\mathbf{x}'_1]_1 v'$ for new independent randomness r', v', s'_b, s'_{1-b} and so is a random element in the space of all proofs. Concluding, the proof output by ZKEval is distributed identically to a fresh proof output by PPro. □

4.1 Our Core Lemma

We now give a new core lemma, which we denote by $\text{Exp}_\beta^{\text{core}}$. Note that we set $k = 1$, as it is sufficient for our construction of SPS-EQ. Consider following experiments (for two cases $\beta = 0$ and $\beta = 1$), where $\mathbf{F} : \mathbb{Z}_p \to \mathbb{Z}_p^2$ is a random function computed on the fly:

$\mathsf{Exp}^{\mathsf{core}}_{\beta}(\lambda), \beta \in \{0,1\}:$	$\mathsf{TAGO}():$
$\mathsf{ctr} := 0$	$\mathsf{ctr} := \mathsf{ctr} + 1$
$\mathsf{BG} \leftarrow \mathsf{BGGen}(1^\lambda)$	$r_1, r_2 \xleftarrow{R} \mathbb{Z}_p$
$\mathbf{A}_0, \mathbf{A}_1 \xleftarrow{R} \mathcal{D}_1$	$[\mathbf{t}]_1 := [\mathbf{A}_0]_1 r_1, [\mathbf{w}]_1 := [\mathbf{A}_0]_1 r_2$
$\mathsf{par} := (\mathsf{BG}, [\mathbf{A}_0]_1, [\mathbf{A}_1]_1)$	$\Omega := (\Omega_1, \Omega_2, [\mathbf{z}_0]_2, [\mathbf{z}_1]_2, \pi) \leftarrow \mathsf{PPro}(\mathsf{crs}, [\mathbf{t}]_1, r_1, [\mathbf{w}]_1, r_2)$
$\mathsf{crs} \leftarrow \mathsf{PGen}(\mathsf{par}, 1^\lambda)$	$[u']_1 := (\mathbf{k}_0 + \beta \cdot \mathbf{F}(\mathsf{ctr}))^\top [\mathbf{t}]_1, [u'']_1 := (\mathbf{k}_0 + \beta \cdot \mathbf{k}_1)^\top [\mathbf{w}]_1$
$\mathbf{k}_0, \mathbf{k}_1 \xleftarrow{R} \mathbb{Z}_p^2$	$\mathsf{Tag} := ([\mathbf{t}]_1, [\mathbf{w}]_1, \Omega = (\Omega_1, \Omega_2, [\mathbf{z}_0]_2, [\mathbf{z}_1]_2, \pi), [u']_1, [u'']_1)$
$\mathsf{pp} := (\mathsf{BG}, [\mathbf{A}_0]_1, \mathsf{crs})$	**return** Tag
$\mathsf{tag} \leftarrow \mathcal{A}^{\mathsf{TAGO}()}(\mathsf{pp})$	$\mathsf{VERO}(\mathsf{tag}):$
return $\mathsf{VERO}(\mathsf{tag})$	Parse $\mathsf{tag} = ([\mathbf{t}]_1, \Omega_1, [\mathbf{z}_0]_2, [\mathbf{z}_1]_2, \pi, [u']_1)$
	if $1 \leftarrow \mathsf{PVer}(\mathsf{crs}, [\mathbf{t}]_1, (\Omega_1, [\mathbf{z}_0]_2, [\mathbf{z}_1]_2, \pi))$
	and $\exists \mathsf{ctr}' \leq \mathsf{ctr} : [u']_1 = (\mathbf{k}_0 + \beta \cdot \mathbf{F}(\mathsf{ctr}'))^\top [\mathbf{t}]_1$
	return 1
	else return 0

Lemma 2 (Core lemma). *If the* $\mathcal{D}_1$-MDDH *(DDH) assumption holds in* $\mathbb{G}_1$ *and the tuple of algorithms* (PGen, PTGen, PPro, PVer) *is a non-interactive zero-knowledge proof system for* $\mathcal{L}^{\vee}_{\mathbf{A}_0,\mathbf{A}_1}$, *then going from experiment* Exp_0^{core} *to* Exp_1^{core} *can (up to negligible terms) only increase the winning chance of an adversary. More precisely, for every adversary* $\mathcal{A}$, *there exist adversaries* $\mathcal{B}$, $\mathcal{B}_1$ *and* $\mathcal{B}_2$ *such that*

$$\mathbf{Adv}_0^{core}(\mathcal{A}) - \mathbf{Adv}_1^{core}(\mathcal{A}) \leq \Delta_{\mathcal{A}}^{core},$$

where

$$\Delta_{\mathcal{A}}^{core} = (2 + 2\lceil \log Q \rceil)\mathbf{Adv}_{\mathsf{PS}}^{\mathsf{zk}}(\mathcal{B}) + (8\lceil \log Q \rceil + 4)\mathbf{Adv}_{\mathcal{D}_1, \mathbb{G}_s}^{\mathsf{MDDH}}(\mathcal{B}_1)$$
$$2\lceil \log Q \rceil \mathbf{Adv}_{\mathsf{PS}}^{\mathsf{snd}}(\mathcal{B}_2) + \lceil \log Q \rceil \Delta_{\mathcal{D}_1} + \frac{(8\lceil \log Q \rceil + 4)}{p-1} + \frac{(\lceil \log Q \rceil)Q}{p}$$

and the term $\Delta_{\mathcal{D}_1}$ *is statistically small.*

Due to the lack of space and the similarity of the proof to the approach in [43] we present the full proof in the full version.

5 Our SPS-EQ Scheme

In Fig. 3 we present our SPS-EQ scheme in the common parameter model under simple assumptions. We set $k = 1$ as we need randomizability and note that our scheme is based on the malleable OR-proof presented in Sect. 4. Observe that in ChgRep the new randomness is $v' = \alpha v$, $s_0' = \alpha\mu s_{1,0} + \alpha\psi s_{2,0} + \beta_0$ and $s_1' = \alpha\mu s_{1,1} + \alpha\psi s_{2,1} + \beta_1$ and the new witness is $r' = \mu r_1 + \psi r_2$.

$\mathsf{ParGen}(1^\lambda)$:

$\mathsf{BG} \leftarrow \mathsf{BGGen}(1^\kappa)$
$\mathbf{A}_0, \mathbf{A}_1 \xleftarrow{R} \mathcal{D}_1$
$\mathsf{crs} \leftarrow \mathsf{PGen}((\mathsf{BG}, [\mathbf{A}_0]_1, [\mathbf{A}_1]_1), 1^\lambda)$
$\mathsf{par} := (\mathsf{BG}, [\mathbf{A}_0]_1, [\mathbf{A}_1]_1, \mathsf{crs})$
return par

$\mathsf{Sign}([\mathbf{m}]_1, \mathsf{sk})$:

$r_1, r_2 \xleftarrow{R} \mathbb{Z}_p$
$[\mathbf{t}]_1 := [\mathbf{A}_0]_1 r_1$
$[\mathbf{w}]_1 := [\mathbf{A}_0]_1 r_2$
$\Omega \leftarrow \mathsf{PPro}(\mathsf{crs}, [\mathbf{t}]_1, r_1, [\mathbf{w}]_1, r_2)$
Parse $\Omega = (\Omega_1, \Omega_2, [\mathbf{z}_0]_2, [\mathbf{z}_1]_2, \pi)$
$\mathbf{u}_1 := \mathbf{K}_0^\top [\mathbf{t}]_1 + \mathbf{K}^\top [\mathbf{m}]_1$
$\mathbf{u}_2 := \mathbf{K}_0^\top [\mathbf{w}]_1$
$\sigma := ([\mathbf{u}_1]_1, \Omega_1, [\mathbf{z}_0]_2, [\mathbf{z}_1]_2, \pi, [\mathbf{t}]_1)$
$\tau := ([\mathbf{u}_2]_1, \Omega_2, [\mathbf{w}]_1)$
return (σ, τ)

$\mathsf{Verify}([\mathbf{m}]_1, (\sigma, \tau), \mathsf{pk})$:

Parse $\sigma = ([\mathbf{u}_1]_1, \Omega_1, [\mathbf{z}_0]_2, [\mathbf{z}_1]_2, \pi, [\mathbf{t}]_1)$
Parse $\tau \in \{([\mathbf{u}_2]_1, \Omega_2, [\mathbf{w}]_1) \cup \perp\}$
1: **if** $1 = \mathsf{PVer}(\mathsf{crs}, [\mathbf{t}]_1, (\Omega_1, [\mathbf{z}_0]_2, [\mathbf{z}_1]_2, \pi))$
2:   **if** $e([\mathbf{u}_1]_1^\top, [\mathbf{A}]_2) =$
    $e([\mathbf{t}]_1^\top, [\mathbf{K}_0\mathbf{A}]_2) + e([\mathbf{m}]_1^\top, [\mathbf{K}\mathbf{A}]_2)$
    if $\tau \neq \perp$
3:     **if** $1 \leftarrow \mathsf{PVer}(\mathsf{crs}, [\mathbf{w}]_1, (\Omega_2, [\mathbf{z}_0]_2, [\mathbf{z}_1]_2, \pi))$
4:     **if** $e([\mathbf{u}_2]_1^\top, [\mathbf{A}]_2) = e([\mathbf{w}]_1^\top, [\mathbf{K}_0\mathbf{A}]_2)$
      return 1
    return 1
else return 0

$\mathsf{KeyGen}(\mathsf{par}, \ell)$:

$\mathbf{A} \xleftarrow{R} \mathcal{D}_1$
$\mathbf{K}_0 \xleftarrow{R} \mathbb{Z}_p^{2\times 2}$
$\mathbf{K} \xleftarrow{R} \mathbb{Z}_p^{\ell\times 2}$
$\mathsf{sk} := (\mathbf{K}_0, \mathbf{K})$
$\mathsf{pk} := ([\mathbf{A}]_2, [\mathbf{K}_0\mathbf{A}]_2, [\mathbf{K}\mathbf{A}]_2)$
return $(\mathsf{pk}, \mathsf{sk})$

$\mathsf{ChgRep}([\mathbf{m}]_1, \sigma, \tau, \mu, \mathsf{pk})$:

Parse $\sigma = ([\mathbf{u}_1]_1, \Omega_1, [\mathbf{z}_0]_2, [\mathbf{z}_1]_2, \pi, [\mathbf{t}]_1)$
Parse $\tau = ([\mathbf{u}_2]_1, \Omega_2, [\mathbf{w}]_1)$
$\Omega := (\Omega_1, \Omega_2, [\mathbf{z}_0]_2, [\mathbf{z}_1]_2, \pi)$
if $1 \neq \mathsf{PVer}(\mathsf{crs}, [\mathbf{t}]_1, [\mathbf{w}]_1, \Omega)$
  or $e([\mathbf{u}_2]_1^\top, [\mathbf{A}]_2) \neq e([\mathbf{w}]_1^\top, [\mathbf{K}_0\mathbf{A}]_2)$
  or $e([\mathbf{u}_1]_1^\top, [\mathbf{A}]_2) \neq$
    $e([\mathbf{t}]_1^\top, [\mathbf{K}_0\mathbf{A}]_2) + e([\mathbf{m}]_1^\top, [\mathbf{K}\mathbf{A}]_2)$
    return $\perp$
else $\psi, \alpha, \beta_0, \beta_1 \xleftarrow{R} \mathbb{Z}_p^*$
$[\mathbf{u}_1]_1' := \mu[\mathbf{u}_1]_1 + \psi[\mathbf{u}_2]_1$
$[\mathbf{t}']_1 := \mu[\mathbf{t}]_1 + \psi[\mathbf{w}]_1 = [\mathbf{A}_0]_1(\mu r_1 + \psi r_2)$
for all $b \in \{0, 1\}$
  $[\mathbf{z}_b']_2 := \alpha[\mathbf{z}_b]_2$
  $[\mathbf{C}_b']_2 := \alpha\mu[\mathbf{C}_{1,b}]_2 + \alpha\psi[\mathbf{C}_{2,b}]_2 + \beta_b[\mathbf{D}]_2$
  $[\mathbf{\Pi}_b']_1 := \alpha\mu[\mathbf{\Pi}_{1,b}]_1 + \alpha\psi[\mathbf{\Pi}_{2,b}]_1 + \beta_b[\mathbf{A}_b]_1$
$\pi' := \alpha\pi$
$\Omega' := (\Omega_1', [\mathbf{z}_i']_2, \pi')$
$\sigma' := ([\mathbf{u}_1']_1, \Omega', [\mathbf{t}']_1)$
return $(\mu[\mathbf{m}]_1, \sigma')$

Fig. 3. Our SPS-EQ scheme.

Theorem 2. *If* KerMDH *and* MDDH *assumptions holds, our SPS scheme is unforgeable.*

Proof. We prove the claim by using a sequence of Games and we denote the advantage of the adversary in the j-th game as $\mathbf{Adv}_j$.

Game 0: This game is the original game and we have:

$$\mathbf{Adv}_0 = \mathbf{Adv}_{\mathsf{SPS\text{-}EQ}}^{\mathsf{EUF\text{-}CMA}}(\mathcal{A})$$

Game 1: In this game, in Verify, we replace the verification in line (2:) with the following equation:

$$[\mathbf{u}_1^*]_1 = \mathbf{K_0}^\top [\mathbf{t}^*]_1 + \mathbf{K}^\top [\mathbf{m}^*]_1$$

For any signature $\sigma = ([\mathbf{u}_1^*]_1, \Omega_1^*, [\mathbf{z}_0^*]_2, [\mathbf{z}_1^*]_2, \pi^*, [\mathbf{t}^*]_1)$ that passes the original verification but not verification of Game 1 the value

$$[\mathbf{u}_1^*]_1 - \mathbf{K_0}^\top[\mathbf{t}^*]_1 - \mathbf{K}^\top[\mathbf{m}^*]_1$$

is a non-zero vector in the kernel of $\mathbf{A}$. Thus if $\mathcal{A}$ outputs such a signature, we can construct an adversary $\mathcal{B}$ that breaks the $\mathcal{D}_1$-KerMDH assumption in $\mathbb{G}_2$. To do this we proceed as follows: The adversary $\mathcal{B}$ receives $(\mathsf{BG}, [\mathbf{A}]_2)$, samples all other parameters and simulates Game 1 for $\mathcal{A}$. When $\mathcal{B}$ receives the forgery from $\mathcal{A}$ as tuple $\sigma = ([\mathbf{u}_1^*]_1, \Omega_1^*, [\mathbf{z}_0^*]_2, [\mathbf{z}_1^*]_2, \pi^*, [\mathbf{t}^*]_1)$ for message $[\mathbf{m}^*]_1$, he passes following values to its own challenger:

$$[\mathbf{u}_1^*]_1 - \mathbf{K_0}^\top[\mathbf{t}^*]_1 - \mathbf{K}^\top[\mathbf{m}^*]_1$$

We have:

$$|\mathbf{Adv}_1 - \mathbf{Adv}_0| \leqslant \mathbf{Adv}^{\mathsf{KerMDH}}_{\mathcal{D}_1,\mathbb{G}_2}(\mathcal{B})$$

Game 2: In this game, we set $\mathbf{K}_0 = \mathbf{K}_0 + \mathbf{k}_0(\mathbf{a}^\perp)^\top$ (in key generation we can pick $\mathbf{k}_0 \in \mathbb{Z}_p^2$ and $\mathbf{K}_0 \in \mathbb{Z}_p^{2\times 2}$ and set $\mathbf{K}_0$; we have $\mathbf{a}^\perp\mathbf{A} = 0$). We compute $[\mathbf{u}_1]_1 = \mathbf{K}_0^\top[\mathbf{t}]_1 + \mathbf{K}^\top[\mathbf{m}]_1 + \mathbf{a}^\perp(\mathbf{k}_0)^\top[\mathbf{t}]_1$ and $[\mathbf{u}_2]_1 = \mathbf{K}_0^\top[\mathbf{w}]_1 + \mathbf{a}^\perp(\mathbf{k}_0)^\top[\mathbf{w}]_1$. There is no difference to the previous game since both are distributed identically. So, we have:

$$\mathbf{Adv}_2 = \mathbf{Adv}_1$$

Game 3: In this game, we add the part of $\mathbf{F}(\mathsf{ctr})$ for $\mathsf{ctr} = \mathsf{ctr} + 1$, where $\mathbf{F}$ is a random function, and obtain $[\mathbf{u}_1]_1 = \mathbf{K}_0^\top[\mathbf{t}]_1 + \mathbf{K}^\top[\mathbf{m}]_1 + \mathbf{a}^\perp(\mathbf{k}_0 + \mathbf{F}(\mathsf{ctr}))^\top[\mathbf{t}]_1$ and $[\mathbf{u}_2]_1 = \mathbf{K}_0^\top[\mathbf{w}]_1 + \mathbf{a}^\perp(\mathbf{k}_0 + \mathbf{k}')^\top[\mathbf{w}]_1$. In the verification we have:

$$1 \leftarrow \mathsf{PVer}(\mathsf{crs}, [\mathbf{t}]_1, (\Omega_1, [\mathbf{z}_0]_2, [\mathbf{z}_1]_2, \pi)) \text{ and}$$
$$\exists \mathsf{ctr}' \leq \mathsf{ctr} :$$
$$[\mathbf{u}_1]_1 = \mathbf{K}_0^\top[\mathbf{t}]_1 + \mathbf{a}^\perp(\mathbf{k}_0 + \mathbf{F}(\mathsf{ctr}'))^\top + \mathbf{K}^\top[\mathbf{m}]_1$$

Let $\mathcal{A}$ be an adversary that distinguishes between Game 3 and Game 2. We can construct an adversary $\mathcal{B}_1$ that breaks the core lemma. $\mathcal{B}_1$ receives $\mathsf{par} = (\mathsf{BG}, [\mathbf{A}_0]_1, \mathsf{crs})$ from $\mathrm{Exp}^{\mathrm{core}}_{\beta,\mathcal{B}_1}$. $\mathcal{B}_1$ picks $\mathbf{A} \xleftarrow{R} \mathcal{D}_k$, $\boldsymbol{a}^\perp \in \mathrm{orth}(\mathbf{A})$, $\mathbf{K}_0 \xleftarrow{R} \mathbb{Z}_p^{2\times 2}$, $\mathbf{K} \xleftarrow{R} \mathbb{Z}_p^{2\times \ell}$, and sends public key $\mathsf{pk} = ([\mathbf{A}_0]_1, [\mathbf{A}]_2, [\mathbf{K}_0\mathbf{A}]_2, [\mathbf{K}\mathbf{A}]_2)$ to $\mathcal{A}$. $\mathcal{B}_1$ uses the oracle $\mathsf{TAGO}()$ to construct the signing algorithm. This oracle takes no input and returns $\mathsf{tag} = ([\mathbf{t}]_1, [\mathbf{w}]_1, \Omega = (\Omega_1, \Omega_2, [\mathbf{z}_0]_2, [\mathbf{z}_1]_2, \pi), [u']_1, [u'']_1)$. Then $\mathcal{B}_1$ computes $[\mathbf{u}_1]_1 = \mathbf{K}_0^\top[\mathbf{t}]_1 + \mathbf{a}^\perp[u']_1 + \mathbf{K}^\top[\mathbf{m}]_1$, $[\mathbf{u}_2]_1 = \mathbf{K}_0^\top[\mathbf{w}]_1 + \mathbf{a}^\perp[u'']_1$, and sends the signature $\sigma = ([\mathbf{u}_1]_1, [\mathbf{z}_0]_2, [\mathbf{z}_1]_2, \pi, [\mathbf{t}]_1)$ and tag $\tau = ([\mathbf{u}_2]_1, \Omega_2, [\mathbf{w}]_1,)$ to $\mathcal{A}$. When the adversary $\mathcal{A}$ sends his forgery $([\mathbf{m}^*]_1, \sigma^*) = (\mathbf{u}_1^*, [\mathbf{t}^*]_1, \Omega_1^*, [\mathbf{z}_0^*]_2, [\mathbf{z}_1^*]_2, \pi^*)$, $\mathcal{B}_1$ returns 0 if $[\mathbf{u}_1]_1 = 0$; otherwise he checks whether there exists $[u'^*]_1$ such that $[\mathbf{u}_1^*]_1 - \mathbf{K}_0^\top[\mathbf{t}^*]_1 - \mathbf{K}^\top[\mathbf{m}^*]_1 = \mathbf{a}^\perp[u'^*]_1$. If it does not hold, then it returns 0 to $\mathcal{A}$, otherwise $\mathcal{B}_1$ computes $[u'^*]_1$, and calls the verification oracle $\mathsf{VERO}()$ on the tag $\mathsf{tag}^* = ([\mathbf{t}^*]_1, \Omega_1^*, [\mathbf{z}_0^*]_2, [\mathbf{z}_1^*]_2, \pi^*, [u'^*]_1)$ and returns the answer to $\mathcal{A}$. Using the core lemma, we have:

$$\mathbf{Adv}_2 - \mathbf{Adv}_3 \leqslant \mathbf{Adv}^{\mathrm{core}}_{\mathsf{BG}}(\mathcal{B}_1)$$

Game 4: In this game, we pick r_1, r_2 from $\mathbb{Z}_p^*$ instead of $\mathbb{Z}_p$. The difference of advantage between Game 3 and Game 4 is bounded by the statistical distance between the two distributions of r_1, r_2. So, under Q adversarial queries, we have:

$$|\mathbf{Adv}_4 - \mathbf{Adv}_3| \leqslant \frac{Q}{p}$$

Game 5: In this game, we pick $\tilde{\mathsf{ctr}} \xleftarrow{R} [1, Q]$, and we add a condition $\mathsf{ctr}' = \tilde{\mathsf{ctr}}$ to verification. Actually, now we have this conditions:

$$1 \leftarrow \mathsf{PVer}(\mathsf{pk}, [\mathbf{t}]_1, (\Omega_1, [\mathbf{z}_0]_2, [\mathbf{z}_1]_2, \pi)) \quad \text{and}$$
$$\exists \mathsf{ctr}' \leq \mathsf{ctr} : \mathsf{ctr}' = \tilde{\mathsf{ctr}} \quad \text{and}$$
$$[\mathbf{u}_1]_1 = \mathbf{K}_0^\top [\mathbf{t}]_1 + \mathbf{a}^\perp (\mathbf{k}_0 + \mathbf{F}(\mathsf{ctr}'))^\top + \mathbf{K}^\top [\mathbf{m}]_1$$

Since the view of the adversary is independent of $\tilde{\mathsf{ctr}}$, we have

$$\mathbf{Adv}_5 = \frac{\mathbf{Adv}_4}{Q}$$

Game 6: In this game, we can replace $\mathbf{K}$ by $\mathbf{K} + \mathbf{v}(\mathbf{a}^\perp)^\top$ for $\mathbf{v} \xleftarrow{R} \mathbb{Z}_p^\ell$. Also, we replace $\{\mathbf{F}(i) : i \in [1, Q], i \neq \tilde{\mathsf{ctr}}\}$ by $\{\mathbf{F}(i) + \mathbf{w}_i : i \in [1, Q], i \neq \tilde{\mathsf{ctr}}\}$, for $\mathbf{w}_i \xleftarrow{R} \mathbb{Z}_p^{2k}$ and $i \neq \hat{\mathsf{ctr}}$. So, in each i-th query, where $i \neq \hat{\mathsf{ctr}}$, we compute

$$[\mathbf{u}_1]_1 = \mathbf{K}_0^\top [\mathbf{t}]_1 + (\mathbf{K}^\top + \mathbf{a}^\perp \mathbf{v}^\top)[\mathbf{m}_i]_1 + \mathbf{a}^\perp (\mathbf{k}_0 + \mathbf{F}(i) + \mathbf{w}_i)^\top [\mathbf{t}]_1$$

Also, for $\tilde{\mathsf{ctr}}$-th query for the message $[\mathbf{m}_{\tilde{\mathsf{ctr}}}]_1$, we compute

$$[\mathbf{u}_1]_1 = \mathbf{K}_0^\top [\mathbf{t}]_1 + (\mathbf{K}^\top + \mathbf{a}^\perp \mathbf{v}^\top)[\mathbf{m}_{\tilde{\mathsf{ctr}}}]_1 + \mathbf{a}^\perp (\mathbf{k}_0 + \mathbf{F}(\tilde{\mathsf{ctr}}) + \mathbf{w}_i)^\top [\mathbf{t}]_1$$

So, $\mathcal{A}$ must compute the following:

$$[\mathbf{u}_1^*]_1 = \mathbf{K}_0^\top [\mathbf{t}^*]_1 + (\mathbf{K}^\top + \mathbf{a}^\perp \mathbf{v}^\top)[\mathbf{m}^*]_1 + \mathbf{a}^\perp (\mathbf{k}_0 + \mathbf{F}(\tilde{\mathsf{ctr}}) + \mathbf{w}_i)^\top [\mathbf{t}^*]_1$$

Since $\mathbf{m}^* \neq [\mathbf{m}_{\tilde{\mathsf{ctr}}}]_{\mathcal{R}}$ (in different classes) by definition of the security game, we can argue $\mathbf{v}^\top \mathbf{m}^*$ and $\mathbf{v}^\top \mathbf{m}_{\tilde{\mathsf{ctr}}}$ are two independent values, uniformly random over $\mathbb{G}_1$. So, $\mathcal{A}$ only can guess it with probability of $\frac{1}{p}$. So, we have

$$\mathbf{Adv}_{\mathsf{SPS\text{-}EQ}}^{\mathsf{EUF\text{-}CMA}}(\mathcal{A}) \leqslant \mathbf{Adv}_{\mathsf{BG}}^{\mathsf{KerMDH}}(\mathcal{B}) + \mathbf{Adv}_{\mathsf{BG}}^{core}(\mathcal{B}_1) + \frac{2Q}{p}.$$

Theorem 3. *Our scheme satisfies perfect adaption under malicious keys in the honest parameters model, i.e., Definition 10.*

Proof. For any message $[\mathbf{m}]_1$, and pk which is generated according to the CRS $([\mathbf{A}]_2, [\mathbf{A}_0]_1, [\mathbf{A}_1]_1, [\mathbf{z}]_2, [\mathbf{D}]_2, [\mathbf{P}]_2)$, a signature $\sigma = ([\mathbf{u}_1]_1, \Omega, [\mathbf{t}]_1,)$ satisfying the verification algorithm must be of the form $\sigma = (\mathbf{K}_0^\top [\mathbf{A}_0]_1 r + \mathbf{K}^\top [\mathbf{m}]_1, v[\mathbf{z}]_2, v[\mathbf{D}]_2, v[\mathbf{P}]_2, s_0[\mathbf{D}^\top] + rv[\mathbf{z}]_2, s_1[\mathbf{D}^\top]_2, [\mathbf{A}_0]_1 s_0, [\mathbf{A}_1]_1 s_1 - [\mathbf{A}_0]_1 rv,$ $[\mathbf{A}_0]_1 r)$. A signature output by ChgRep has the form $\sigma' = (\mathbf{K}_0^\top [\mathbf{A}_0]_1 r' + \mathbf{K}^\top [\mathbf{m}]_1,$ $v'[\mathbf{z}]_2, v'[\mathbf{D}]_2, v'[\mathbf{P}]_2, s_0'[\mathbf{D}^\top] + r'v'[\mathbf{z}]_2, s_1'[\mathbf{D}^\top]_2, [\mathbf{A}_0]_1 s_0', [\mathbf{A}_1]_1 s_1 - [\mathbf{A}_0]_1 r'v',$ $[\mathbf{A}_0]_1 r')$ for new independent randomness r', v', s_0', s_1' and so is a random element in the space of all signatures. Actually, the signature output by ChgRep is distributed identically to a fresh signature on message $[\mathbf{m}]_1$ output by Sign. □

6 Applications

As already discussed in [35], there are no known applications of SPS-EQ where signatures that have been randomized need to be randomized again by an entity that does not know the original signature. Consequently, and as shown in [35], tag-based schemes as the one introduced in this paper can be used within all the known applications without restrictions. Now let us summarize and clarify how our SPS-EQ scheme can be used in existing applications of SPS-EQ.

Using our scheme we can instantiate the group signatures in [30] and [11] as well as access control encryption (ACE) in [36]. As already mentioned earlier, both models assume honest key generation and so we can merge ParGen and KeyGen of the SPS-EQ scheme and do not need a trusted party to generate the CRS, i.e., it can be done by the signer during key generation.

Also we can instantiate attribute-based credentials [38,40,52] in the honest key model or under malicious keys (for latter requiring a CRS), but not in the malicious key model without a CRS. Due to an argumentation following a reasoning related to the one in Sect. 3.3, our scheme cannot be used to instantiate the verifiable encrypted signatures from [51].

Round-Optimal Blind Signatures in the CRS Model. What remains to be discussed is the application to round-optimal blind signatures as introduced in [37,39]. As already mentioned, as our SPS-EQ scheme does not provide the strongest notion of perfect adaption under malicious keys, we are only able to construct round-optimal blind signatures in the CRS model. In contrast to existing schemes in the CRS model relying on non-standard and non-static q-type assumptions such as [9,33] which require around 30 group elements in the signature, the most recent scheme under standard assumptions, i.e., SXDH, by Abe et al. [8] requires $(42, 40)$ elements in $\mathbb{G}_1$ and $\mathbb{G}_2$ respectively. In contrast to other existing schemes which follow the framework of Fischlin [32], we can take our SPS-EQ scheme to instantiate the framework in [39]. We note that when we are in the CRS model, we can move the commitment parameters Q and $\hat{Q}$ from [39] in the CRS, and thus obtain a round optimal blind signature scheme under SXDH. This is the same assumption as used by Abe et al. in [8], but our signature sizes are only $(10, 9)$ elements in $\mathbb{G}_1$ and $\mathbb{G}_2$ respectively, improving over [8] by about a factor of 4 and even beating constructions proven secure under q-type assumptions.

Acknowledgments. We are grateful to the anonymous reviewers from ASIACRYPT 2019 and Romain Gay for their careful reading of the paper, their valuable feedback and suggestions to improve the presentation. We also thanks Carla Ràfols and Alonso González for their comments on earlier versions of this work. This work was supported by the EU's Horizon 2020 ECSEL Joint Undertaking project SECREDAS under grant agreement n°783119 and by the Austrian Science Fund (FWF) and netidee SCIENCE project PROFET (grant agreement P31621-N38).

References

1. Abdolmaleki, B., Lipmaa, H., Siim, J., Zając, M.: On QA-NIZK in the BPK model. Cryptology ePrint Archive, Report 2018/877 (2018)
2. Abe, M., Chase, M., David, B., Kohlweiss, M., Nishimaki, R., Ohkubo, M.: Constant-size structure-preserving signatures: generic constructions and simple assumptions. In: Wang, X., Sako, K. (eds.) ASIACRYPT 2012. LNCS, vol. 7658, pp. 4–24. Springer, Heidelberg (2012). https://doi.org/10.1007/978-3-642-34961-4_3
3. Abe, M., Chow, S.S., Haralambiev, K., Ohkubo, M.: Double-trapdoor anonymous tags for traceable signatures. In: Lopez, J., Tsudik, G. (eds.) ACNS 2011. LNCS, vol. 6715, pp. 183–200. Springer, Heidelberg (2011). https://doi.org/10.1007/978-3-642-21554-4_11
4. Abe, M., Fuchsbauer, G., Groth, J., Haralambiev, K., Ohkubo, M.: Structure-preserving signatures and commitments to group elements. In: Rabin, T. (ed.) CRYPTO 2010. LNCS, vol. 6223, pp. 209–236. Springer, Heidelberg (2010). https://doi.org/10.1007/978-3-642-14623-7_12
5. Abe, M., Groth, J., Haralambiev, K., Ohkubo, M.: Optimal structure-preserving signatures in asymmetric bilinear groups. In: Rogaway, P. (ed.) CRYPTO 2011. LNCS, vol. 6841, pp. 649–666. Springer, Heidelberg (2011). https://doi.org/10.1007/978-3-642-22792-9_37
6. Abe, M., Groth, J., Ohkubo, M., Tibouchi, M.: Unified, minimal and selectively randomizable structure-preserving signatures. In: Lindell, Y. (ed.) TCC 2014. LNCS, vol. 8349, pp. 688–712. Springer, Heidelberg (2014). https://doi.org/10.1007/978-3-642-54242-8_29
7. Abe, M., Hofheinz, D., Nishimaki, R., Ohkubo, M., Pan, J.: Compact structure-preserving signatures with almost tight security. In: Katz, J., Shacham, H. (eds.) CRYPTO 2017, Part II. LNCS, vol. 10402, pp. 548–580. Springer, Cham (2017). https://doi.org/10.1007/978-3-319-63715-0_19
8. Abe, M., Jutla, C.S., Ohkubo, M., Roy, A.: Improved (almost) tightly-secure simulation-sound QA-NIZK with applications. In: Peyrin, T., Galbraith, S. (eds.) ASIACRYPT 2018, Part I. LNCS, vol. 11272, pp. 627–656. Springer, Cham (2018). https://doi.org/10.1007/978-3-030-03326-2_21
9. Abe, M., Ohkubo, M.: A framework for universally composable non-committing blind signatures. In: Matsui, M. (ed.) ASIACRYPT 2009. LNCS, vol. 5912, pp. 435–450. Springer, Heidelberg (2009). https://doi.org/10.1007/978-3-642-10366-7_26
10. Attrapadung, N., Libert, B., Peters, T.: Computing on authenticated data: new privacy definitions and constructions. In: Wang, X., Sako, K. (eds.) ASIACRYPT 2012. LNCS, vol. 7658, pp. 367–385. Springer, Heidelberg (2012). https://doi.org/10.1007/978-3-642-34961-4_23
11. Backes, M., Hanzlik, L., Kluczniak, K., Schneider, J.: Signatures with flexible public key: introducing equivalence classes for public keys. In: Peyrin, T., Galbraith, S. (eds.) ASIACRYPT 2018, Part II. LNCS, vol. 11273, pp. 405–434. Springer, Cham (2018). https://doi.org/10.1007/978-3-030-03329-3_14
12. Backes, M., Hanzlik, L., Schneider, J.: Membership privacy for fully dynamic group signatures. Cryptology ePrint Archive, Report 2018/641 (2018)
13. Bader, C., Hofheinz, D., Jager, T., Kiltz, E., Li, Y.: Tightly-secure authenticated key exchange. In: Dodis, Y., Nielsen, J.B. (eds.) TCC 2015, Part I. LNCS, vol. 9014, pp. 629–658. Springer, Heidelberg (2015). https://doi.org/10.1007/978-3-662-46494-6_26

14. Bellare, M., Fuchsbauer, G., Scafuro, A.: NIZKs with an untrusted CRS: security in the face of parameter subversion. In: Cheon, J.H., Takagi, T. (eds.) ASIACRYPT 2016, Part II. LNCS, vol. 10032, pp. 777–804. Springer, Heidelberg (2016). https://doi.org/10.1007/978-3-662-53890-6_26
15. Bellare, M., Shi, H., Zhang, C.: Foundations of group signatures: the case of dynamic groups. In: Menezes, A. (ed.) CT-RSA 2005. LNCS, vol. 3376, pp. 136–153. Springer, Heidelberg (2005). https://doi.org/10.1007/978-3-540-30574-3_11
16. Ben-Sasson, E., Chiesa, A., Green, M., Tromer, E., Virza, M.: Secure sampling of public parameters for succinct zero knowledge proofs. In: 2015 IEEE Symposium on Security and Privacy, pp. 287–304. IEEE Computer Society Press (2015)
17. Blazy, O., Canard, S., Fuchsbauer, G., Gouget, A., Sibert, H., Traoré, J.: Achieving optimal anonymity in transferable e-cash with a judge. In: Nitaj, A., Pointcheval, D. (eds.) AFRICACRYPT 2011. LNCS, vol. 6737, pp. 206–223. Springer, Heidelberg (2011). https://doi.org/10.1007/978-3-642-21969-6_13
18. Blazy, O., Kiltz, E., Pan, J.: (Hierarchical) identity-based encryption from affine message authentication. In: Garay, J.A., Gennaro, R. (eds.) CRYPTO 2014, Part I. LNCS, vol. 8616, pp. 408–425. Springer, Heidelberg (2014). https://doi.org/10.1007/978-3-662-44371-2_23
19. Blum, M., Feldman, P., Micali, S.: Non-interactive zero-knowledge and its applications (extended abstract). In: 20th ACM STOC, pp. 103–112. ACM Press (1988)
20. Bowe, S., Gabizon, A., Green, M.D.: A multi-party protocol for constructing the public parameters of the pinocchio zk-SNARK. In: Zohar, A., et al. (eds.) FC 2018. LNCS, vol. 10958, pp. 64–77. Springer, Heidelberg (2019). https://doi.org/10.1007/978-3-662-58820-8_5
21. Bowe, S., Gabizon, A., Miers, I.: Scalable multi-party computation for zk-SNARK parameters in the random beacon model. IACR Cryptology ePrint Archive 2017, 1050 (2017)
22. Bultel, X., Lafourcade, P., Lai, R., Malavolta, G., Schröder, D., Thyagarajan, S.: Efficient invisible and unlinkable sanitizable signatures. In: PKC 2019 (2019, to appear)
23. Cathalo, J., Libert, B., Yung, M.: Group encryption: non-interactive realization in the standard model. In: Matsui, M. (ed.) ASIACRYPT 2009. LNCS, vol. 5912, pp. 179–196. Springer, Heidelberg (2009). https://doi.org/10.1007/978-3-642-10366-7_11
24. Chase, M., Kohlweiss, M., Lysyanskaya, A., Meiklejohn, S.: Malleable proof systems and applications. In: Pointcheval, D., Johansson, T. (eds.) EUROCRYPT 2012. LNCS, vol. 7237, pp. 281–300. Springer, Heidelberg (2012). https://doi.org/10.1007/978-3-642-29011-4_18
25. Chen, J., Wee, H.: Fully, (almost) tightly secure IBE and dual system groups. In: Canetti, R., Garay, J.A. (eds.) CRYPTO 2013, Part II. LNCS, vol. 8043, pp. 435–460. Springer, Heidelberg (2013). https://doi.org/10.1007/978-3-642-40084-1_25
26. Clarisse, R., Sanders, O.: Short group signature in the standard model. IACR Cryptology ePrint Archive 2018, 1115 (2018)
27. Crites, E.C., Lysyanskaya, A.: Delegatable anonymous credentials from mercurial signatures. In: Matsui, M. (ed.) CT-RSA 2019. LNCS, vol. 11405, pp. 535–555. Springer, Cham (2019). https://doi.org/10.1007/978-3-030-12612-4_27
28. Daza, V., González, A., Pindado, Z., Ràfols, C., Silva, J.: Shorter quadratic QA-NIZK proofs. In: Lin, D., Sako, K. (eds.) PKC 2019, Part I. LNCS, vol. 11442, pp. 314–343. Springer, Cham (2019). https://doi.org/10.1007/978-3-030-17253-4_11

29. Derler, D., Hanser, C., Slamanig, D.: A new approach to efficient revocable attribute-based anonymous credentials. In: Groth, J. (ed.) IMACC 2015. LNCS, vol. 9496, pp. 57–74. Springer, Cham (2015). https://doi.org/10.1007/978-3-319-27239-9_4
30. Derler, D., Slamanig, D.: Highly-efficient fully-anonymous dynamic group signatures. In: Kim, J., Ahn, G.J., Kim, S., Kim, Y., López, J., Kim, T. (eds.) ASIACCS 18, pp. 551–565. ACM Press (2018)
31. Escala, A., Herold, G., Kiltz, E., Rafols, C., Villar, J.: An algebraic framework for Diffie-Hellman assumptions. J. Cryptol. **30**(1), 242–288 (2017)
32. Fischlin, M.: Round-optimal composable blind signatures in the common reference string model. In: Dwork, C. (ed.) CRYPTO 2006. LNCS, vol. 4117, pp. 60–77. Springer, Heidelberg (2006). https://doi.org/10.1007/11818175_4
33. Fuchsbauer, G.: Automorphic signatures in bilinear groups and an application to round-optimal blind signatures. Cryptology ePrint Archive, Report 2009/320 (2009)
34. Fuchsbauer, G.: Commuting Signatures and verifiable encryption. In: Paterson, K.G. (ed.) EUROCRYPT 2011. LNCS, vol. 6632, pp. 224–245. Springer, Heidelberg (2011). https://doi.org/10.1007/978-3-642-20465-4_14
35. Fuchsbauer, G., Gay, R.: Weakly secure equivalence-class signatures from standard assumptions. In: Abdalla, M., Dahab, R. (eds.) PKC 2018, Part II. LNCS, vol. 10770, pp. 153–183. Springer, Cham (2018). https://doi.org/10.1007/978-3-319-76581-5_6
36. Fuchsbauer, G., Gay, R., Kowalczyk, L., Orlandi, C.: Access control encryption for equality, comparison, and more. In: Fehr, S. (ed.) PKC 2017, Part II. LNCS, vol. 10175, pp. 88–118. Springer, Heidelberg (2017). https://doi.org/10.1007/978-3-662-54388-7_4
37. Fuchsbauer, G., Hanser, C., Kamath, C., Slamanig, D.: Practical round-optimal blind signatures in the standard model from weaker assumptions. In: Zikas, V., De Prisco, R. (eds.) SCN 2016. LNCS, vol. 9841, pp. 391–408. Springer, Cham (2016). https://doi.org/10.1007/978-3-319-44618-9_21
38. Fuchsbauer, G., Hanser, C., Slamanig, D.: Structure-preserving signatures on equivalence classes and constant-size anonymous credentials. Cryptology ePrint Archive, Report 2014/944 (2014)
39. Fuchsbauer, G., Hanser, C., Slamanig, D.: Practical round-optimal blind signatures in the standard model. In: Gennaro, R., Robshaw, M. (eds.) CRYPTO 2015, Part II. LNCS, vol. 9216, pp. 233–253. Springer, Heidelberg (2015). https://doi.org/10.1007/978-3-662-48000-7_12
40. Fuchsbauer, G., Hanser, C., Slamanig, D.: Structure-preserving signatures on equivalence classes and constant-size anonymous credentials. J. Cryptol. **32**(2), 498–546 (2019)
41. Gay, R., Hofheinz, D., Kiltz, E., Wee, H.: Tightly CCA-secure encryption without pairings. In: Fischlin, M., Coron, J.-S. (eds.) EUROCRYPT 2016, Part I. LNCS, vol. 9665, pp. 1–27. Springer, Heidelberg (2016). https://doi.org/10.1007/978-3-662-49890-3_1
42. Gay, R., Hofheinz, D., Kohl, L.: Kurosawa-Desmedt meets tight security. In: Katz, J., Shacham, H. (eds.) CRYPTO 2017, Part III. LNCS, vol. 10403, pp. 133–160. Springer, Cham (2017). https://doi.org/10.1007/978-3-319-63697-9_5
43. Gay, R., Hofheinz, D., Kohl, L., Pan, J.: More efficient (almost) tightly secure structure-preserving signatures. In: Nielsen, J.B., Rijmen, V. (eds.) EUROCRYPT 2018, Part II. LNCS, vol. 10821, pp. 230–258. Springer, Cham (2018). https://doi.org/10.1007/978-3-319-78375-8_8

44. Ghadafi, E.: Short structure-preserving signatures. In: Sako, K. (ed.) CT-RSA 2016. LNCS, vol. 9610, pp. 305–321. Springer, Cham (2016). https://doi.org/10.1007/978-3-319-29485-8_18
45. Ghadafi, E.: More efficient structure-preserving signatures - or: bypassing the type-III lower bounds. In: Foley, S.N., Gollmann, D., Snekkenes, E. (eds.) ESORICS 2017, Part II. LNCS, vol. 10493, pp. 43–61. Springer, Cham (2017). https://doi.org/10.1007/978-3-319-66399-9_3
46. Gjøsteen, K., Jager, T.: Practical and tightly-secure digital signatures and authenticated key exchange. In: Shacham, H., Boldyreva, A. (eds.) CRYPTO 2018, Part II. LNCS, vol. 10992, pp. 95–125. Springer, Cham (2018). https://doi.org/10.1007/978-3-319-96881-0_4
47. González, A., Hevia, A., Ràfols, C.: QA-NIZK arguments in asymmetric groups: new tools and new constructions. In: Iwata, T., Cheon, J.H. (eds.) ASIACRYPT 2015, Part I. LNCS, vol. 9452, pp. 605–629. Springer, Heidelberg (2015). https://doi.org/10.1007/978-3-662-48797-6_25
48. Green, M., Hohenberger, S.: Universally composable adaptive oblivious transfer. In: Pieprzyk, J. (ed.) ASIACRYPT 2008. LNCS, vol. 5350, pp. 179–197. Springer, Heidelberg (2008). https://doi.org/10.1007/978-3-540-89255-7_12
49. Groth, J., Kohlweiss, M., Maller, M., Meiklejohn, S., Miers, I.: Updatable and universal common reference strings with applications to zk-SNARKs. In: Shacham, H., Boldyreva, A. (eds.) CRYPTO 2018, Part III. LNCS, vol. 10993, pp. 698–728. Springer, Cham (2018). https://doi.org/10.1007/978-3-319-96878-0_24
50. Groth, J., Sahai, A.: Efficient non-interactive proof systems for bilinear groups. In: Smart, N. (ed.) EUROCRYPT 2008. LNCS, vol. 4965, pp. 415–432. Springer, Heidelberg (2008). https://doi.org/10.1007/978-3-540-78967-3_24
51. Hanser, C., Rabkin, M., Schröder, D.: Verifiably encrypted signatures: security revisited and a new construction. In: Pernul, G., Ryan, P.Y.A., Weippl, E. (eds.) ESORICS 2015, Part I. LNCS, vol. 9326, pp. 146–164. Springer, Cham (2015). https://doi.org/10.1007/978-3-319-24174-6_8
52. Hanser, C., Slamanig, D.: Structure-preserving signatures on equivalence classes and their application to anonymous credentials. In: Sarkar, P., Iwata, T. (eds.) ASIACRYPT 2014, Part I. LNCS, vol. 8873, pp. 491–511. Springer, Heidelberg (2014). https://doi.org/10.1007/978-3-662-45611-8_26
53. Hesse, J., Hofheinz, D., Kohl, L.: On tightly secure non-interactive key exchange. In: Shacham, H., Boldyreva, A. (eds.) CRYPTO 2018, Part II. LNCS, vol. 10992, pp. 65–94. Springer, Cham (2018). https://doi.org/10.1007/978-3-319-96881-0_3
54. Hofheinz, D.: Adaptive partitioning. In: Coron, J.-S., Nielsen, J.B. (eds.) EUROCRYPT 2017, Part III. LNCS, vol. 10212, pp. 489–518. Springer, Cham (2017). https://doi.org/10.1007/978-3-319-56617-7_17
55. Hofheinz, D., Jager, T.: Tightly secure signatures and public-key encryption. In: Safavi-Naini, R., Canetti, R. (eds.) CRYPTO 2012. LNCS, vol. 7417, pp. 590–607. Springer, Heidelberg (2012). https://doi.org/10.1007/978-3-642-32009-5_35
56. Hofheinz, D., Jager, T.: Tightly secure signatures and public-key encryption. Designs Codes Cryptogr. **80**(1), 29–61 (2016)
57. Hofheinz, D., Jia, D., Pan, J.: Identity-based encryption tightly secure under chosen-ciphertext attacks. In: Peyrin, T., Galbraith, S. (eds.) ASIACRYPT 2018, Part II. LNCS, vol. 11273, pp. 190–220. Springer, Cham (2018). https://doi.org/10.1007/978-3-030-03329-3_7

58. Hofheinz, D., Koch, J., Striecks, C.: Identity-based encryption with (almost) tight security in the multi-instance, multi-ciphertext setting. In: Katz, J. (ed.) PKC 2015. LNCS, vol. 9020, pp. 799–822. Springer, Heidelberg (2015). https://doi.org/10.1007/978-3-662-46447-2_36
59. Hébant, C., Phan, D.H., Pointcheval, D.: Linearly-homomorphic signatures and scalable mix-nets. Cryptology ePrint Archive, Report 2019/547 (2019)
60. Jutla, C.S., Ohkubo, M., Roy, A.: Improved (almost) tightly-secure structure-preserving signatures. In: Abdalla, M., Dahab, R. (eds.) PKC 2018, Part II. LNCS, vol. 10770, pp. 123–152. Springer, Cham (2018). https://doi.org/10.1007/978-3-319-76581-5_5
61. Jutla, C.S., Roy, A.: Shorter quasi-adaptive NIZK proofs for linear subspaces. In: Sako, K., Sarkar, P. (eds.) ASIACRYPT 2013, Part I. LNCS, vol. 8269, pp. 1–20. Springer, Heidelberg (2013). https://doi.org/10.1007/978-3-642-42033-7_1
62. Jutla, C.S., Roy, A.: Switching lemma for bilinear tests and constant-size NIZK proofs for linear subspaces. In: Garay, J.A., Gennaro, R. (eds.) CRYPTO 2014, Part II. LNCS, vol. 8617, pp. 295–312. Springer, Heidelberg (2014). https://doi.org/10.1007/978-3-662-44381-1_17
63. Jutla, C.S., Roy, A.: Improved structure preserving signatures under standard bilinear assumptions. In: Fehr, S. (ed.) PKC 2017, Part II. LNCS, vol. 10175, pp. 183–209. Springer, Heidelberg (2017). https://doi.org/10.1007/978-3-662-54388-7_7
64. Kiltz, E., Pan, J., Wee, H.: Structure-preserving signatures from standard assumptions, revisited. In: Gennaro, R., Robshaw, M. (eds.) CRYPTO 2015, Part II. LNCS, vol. 9216, pp. 275–295. Springer, Heidelberg (2015). https://doi.org/10.1007/978-3-662-48000-7_14
65. Kiltz, E., Wee, H.: Quasi-adaptive NIZK for linear subspaces revisited. In: Oswald, E., Fischlin, M. (eds.) EUROCRYPT 2015, Part II. LNCS, vol. 9057, pp. 101–128. Springer, Heidelberg (2015). https://doi.org/10.1007/978-3-662-46803-6_4
66. Libert, B., Peters, T., Joye, M., Yung, M.: Linearly homomorphic structure-preserving signatures and their applications. In: Canetti, R., Garay, J.A. (eds.) CRYPTO 2013, Part II. LNCS, vol. 8043, pp. 289–307. Springer, Heidelberg (2013). https://doi.org/10.1007/978-3-642-40084-1_17
67. Libert, B., Peters, T., Joye, M., Yung, M.: Non-malleability from malleability: simulation-sound quasi-adaptive NIZK proofs and CCA2-secure encryption from homomorphic signatures. In: Nguyen, P.Q., Oswald, E. (eds.) EUROCRYPT 2014. LNCS, vol. 8441, pp. 514–532. Springer, Heidelberg (2014). https://doi.org/10.1007/978-3-642-55220-5_29
68. Libert, B., Peters, T., Yung, M.: Short group signatures via structure-preserving signatures: standard model security from simple assumptions. In: Gennaro, R., Robshaw, M. (eds.) CRYPTO 2015, Part II. LNCS, vol. 9216, pp. 296–316. Springer, Heidelberg (2015). https://doi.org/10.1007/978-3-662-48000-7_15
69. Lipmaa, H.: Key-and-argument-updatable QA-NIZKS. Cryptology ePrint Archive, Report 2019/333 (2019)
70. Morillo, P., Ràfols, C., Villar, J.L.: The kernel matrix Diffie-Hellman assumption. In: Cheon, J.H., Takagi, T. (eds.) ASIACRYPT 2016, Part I. LNCS, vol. 10031, pp. 729–758. Springer, Heidelberg (2016). https://doi.org/10.1007/978-3-662-53887-6_27
71. Ràfols, C.: Stretching groth-sahai: NIZK proofs of partial satisfiability. In: Dodis, Y., Nielsen, J.B. (eds.) TCC 2015, Part II. LNCS, vol. 9015, pp. 247–276. Springer, Heidelberg (2015). https://doi.org/10.1007/978-3-662-46497-7_10

Public Key Encryption (1)

Simple and Efficient KDM-CCA Secure Public Key Encryption

Fuyuki Kitagawa[1(✉)], Takahiro Matsuda[2], and Keisuke Tanaka[3]

[1] NTT Secure Platform Laboratories, Tokyo, Japan
fuyuki.kitagawa.yh@hco.ntt.co.jp
[2] National Institute of Advanced Industrial Science and Technology (AIST), Tokyo, Japan
t-matsuda@aist.go.jp
[3] Tokyo Institute of Technology, Tokyo, Japan
keisuke@is.titech.ac.jp

Abstract. We propose two efficient public key encryption (PKE) schemes satisfying key dependent message security against chosen ciphertext attacks (KDM-CCA security). The first one is KDM-CCA secure with respect to affine functions. The other one is KDM-CCA secure with respect to polynomial functions. Both of our schemes are based on the KDM-CPA secure PKE schemes proposed by Malkin, Teranishi, and Yung (EUROCRYPT 2011). Although our schemes satisfy KDM-CCA security, their efficiency overheads compared to Malkin et al.'s schemes are very small. Thus, efficiency of our schemes is drastically improved compared to the existing KDM-CCA secure schemes.

We achieve our results by extending the construction technique by Kitagawa and Tanaka (ASIACRYPT 2018). Our schemes are obtained via semi-generic constructions using an IND-CCA secure PKE scheme as a building block. We prove the KDM-CCA security of our schemes based on the decisional composite residuosity (DCR) assumption and the IND-CCA security of the building block PKE scheme.

Moreover, our security proofs are *tight* if the IND-CCA security of the building block PKE scheme is tightly reduced to its underlying computational assumption. By instantiating our schemes using existing tightly IND-CCA secure PKE schemes, we obtain the first tightly KDM-CCA secure PKE schemes whose ciphertext consists only of a constant number of group elements.

Keywords: Key dependent message security · Chosen ciphertext security

1 Introduction

1.1 Background

Key dependent message (KDM) security, introduced by Black, Rogaway, and Shrimpton [3], guarantees confidentiality of communication even if an adversary

S. D. Galbraith and S. Moriai (Eds.): ASIACRYPT 2019, LNCS 11923, pp. 97–127, 2019.
https://doi.org/10.1007/978-3-030-34618-8_4

can get a ciphertext of secret keys. KDM security is defined with respect to a function family $\mathcal{F}$. Informally, a public key encryption (PKE) scheme is said to be $\mathcal{F}$-KDM secure if confidentiality of messages is protected even when an adversary can see a ciphertext of $f(\mathsf{sk}_1, \cdots, \mathsf{sk}_\ell)$ under the k-th public key for any $f \in \mathcal{F}$ and $k \in \{1, \cdots, \ell\}$, where ℓ denotes the number of keys. KDM security is useful for many practical applications including anonymous credential systems [7] and hard disk encryption systems (e.g., BitLocker [4]).

In this paper, we focus on constructing *efficient* PKE schemes that satisfy KDM security against chosen ciphertext attacks, namely *KDM-CCA* security, in the standard model. As pointed out by Camenisch, Chandran, and Shoup [6] who proposed the first KDM-CCA secure PKE scheme, KDM-CCA security is well motivated since it resolves key wrapping problems that arise in many practical applications. Moreover, in some applications of KDM secure schemes such as anonymous credential systems, we should consider active adversaries and need KDM-CCA security.

The first attempt to construct an efficient KDM secure PKE scheme was made by Applebaum, Cash, Peikert, and Sahai [1]. They proposed a PKE scheme that is KDM-CPA secure with respect to affine functions ($\mathcal{F}_{\mathsf{aff}}$-KDM-CPA secure) under a lattice assumption. Their scheme is as efficient as IND-CPA secure schemes based on essentially the same assumption.

Malkin, Teranishi, and Yung [22] later proposed a more efficient KDM-CPA secure PKE scheme under the decisional composite residuosity (DCR) assumption [9,24]. Moreover, their scheme is KDM-CPA secure with respect to polynomial functions ($\mathcal{F}_{\mathsf{poly}}$-KDM-CPA secure), which is much richer than affine functions. A ciphertext of their scheme contains $d+1$ group elements, where d is the maximum degree of polynomial functions with respect to which their scheme is KDM-CPA secure. As a special case of $d = 1$, their scheme is an $\mathcal{F}_{\mathsf{aff}}$-KDM-CPA secure PKE scheme whose ciphertext consists of only two group elements.

Due to these works, we now have efficient KDM-CPA secure PKE schemes. As we can see, the above $\mathcal{F}_{\mathsf{aff}}$-KDM-CPA secure schemes are as efficient as PKE schemes that are IND-CPA secure under the same assumptions. However, the situation is somewhat unsatisfactory when considering KDM-CCA secure PKE.

Camenisch et al. [6] proposed the first KDM-CCA secure PKE scheme based on the Naor-Yung paradigm [23]. They showed that for any function class $\mathcal{F}$, an $\mathcal{F}$-KDM-CPA secure PKE scheme can be transformed into an $\mathcal{F}$-KDM-CCA secure one assuming a non-interactive zero knowledge (NIZK) proof system. They also showed a concrete instantiation based on the decisional Diffie-Hellman (DDH) assumption on bilinear groups. A ciphertext of their scheme contains $O(\lambda)$ group elements, where λ is the security parameter. Subsequently, Hofheinz [12] showed a more efficient KDM-CCA secure PKE scheme. His scheme is circular-CCA secure, relying on both the DCR and DDH assumptions, and decisional linear (DLIN) assumption on bilinear groups. A ciphertext of his scheme contains more than 50 group elements. Recently, Libert and Qian [20] improved the construction of Hofheinz based on the 3-party DDH (D3DH) assumption on bilinear groups, and shortened the ciphertext size by about 20 group elements.

The first KDM-CCA secure PKE scheme using neither NIZK proofs nor bilinear maps was proposed by Lu, Li, and Jia [21]. They claimed their scheme is $\mathcal{F}_{\mathsf{aff}}$-KDM-CCA secure based on both the DCR and DDH assumptions. However, a flaw in their security proof was later pointed out by Han, Liu, and Lyu [11]. Han et al. also showed a new $\mathcal{F}_{\mathsf{aff}}$-KDM-CCA secure scheme based on Lu et al.'s construction methodology, and furthermore constructed a $\mathcal{F}_{\mathsf{poly}}$-KDM-CCA secure PKE scheme. Their schemes rely on both the DCR and DDH assumptions. A ciphertext of their $\mathcal{F}_{\mathsf{aff}}$-KDM-CCA secure scheme contains around 20 group elements. A ciphertext of their $\mathcal{F}_{\mathsf{poly}}$-KDM-CCA secure scheme contains $O(d^9)$ group elements, where d is the maximum degree of polynomial functions.

Recently, Kitagawa and Tanaka [18] showed a new framework for constructing KDM-CCA secure schemes, and they constructed an $\mathcal{F}_{\mathsf{aff}}$-KDM-CCA secure PKE scheme based solely on the DDH assumption (without bilinear maps). However, their scheme is somewhat inefficient and its ciphertext consists of $O(\lambda)$ group elements.

The currently most efficient KDM-CCA secure PKE scheme is that of Han et al. Their schemes are much efficient compared to other KDM-CCA secure schemes. However, there are still a large overhead compared to efficient KDM-CPA secure schemes. Especially, its overhead compared to Malkin et al.'s scheme is large even though Han et al.'s schemes are based on both the DDH and DCR assumptions while Malkin et al.'s scheme is based only on the DCR assumption.

In order to use a KDM-CCA secure PKE scheme in practical applications, we need a more efficient scheme.

1.2 Our Results

We propose two efficient KDM-CCA secure PKE schemes. The first one is $\mathcal{F}_{\mathsf{aff}}$-KDM-CCA secure, and the other one is $\mathcal{F}_{\mathsf{poly}}$-KDM-CCA secure. Both of our schemes are based on the KDM-CPA secure scheme proposed by Malkin et al. [22]. Although our schemes satisfy KDM-CCA security, its efficiency overheads compared to Malkin et al.'s schemes are very small. Thus, efficiency of our schemes is drastically improved compared to the previous KDM-CCA secure schemes.

We achieve our results by extending the construction technique by Kitagawa and Tanaka [18]. Our schemes are obtained via semi-generic constructions using an IND-CCA secure PKE scheme as a building block. By instantiating the underlying IND-CCA secure PKE scheme with the factoring-based scheme by Hofheinz and Kiltz [16] (and with some optimization techniques), we obtain KDM-CCA secure PKE schemes (with respect to affine functions and with respect to polynomials) such that the overhead of the ciphertext size of our schemes compared to Malkin et al.'s KDM-CPA secure scheme can be less than a single DCR-group element. (See Figs. 1 and 2.)

Moreover, our security proofs are *tight* if the IND-CCA security of the building block PKE scheme is tightly reduced to its underlying computational assumption. By instantiating our schemes using existing tightly IND-CCA secure PKE schemes [10,13], we obtain the first tightly KDM-CCA secure PKE schemes whose ciphertext consists only of a constant number of group elements. To the best of our knowledge, prior to our work, the only way to construct a tightly KDM-CCA secure PKE scheme is to instantiate the construction proposed by Camenisch et al. [6] using a tightly secure NIZK proof system such as the one proposed by Hofheinz and Jager [14]. A ciphertext of such schemes consists of $O(\lambda)$ group elements, where λ is the security parameter.

For a comparison of efficiency between our schemes and existing schemes, see Figs. 1 and 2. In the figures, for reference, we include [22] on which our schemes are based but which is not KDM-CCA secure. In the figures, we also show concrete instantiations of our constructions. The details of these instantiations are explained in Sect. 7.

We note that the plaintext space of the schemes listed in Figs. 1 and 2 except for our schemes and Malkin et al.'s [22], is smaller than the secret key space, and some modifications are needed for encrypting a whole secret key, which will result in a larger ciphertext size in the resulting PKE schemes. On the other hand, our and Malkin et al.'s schemes can encrypt a whole secret key without any modification by setting $s \geq 3$. (We provide a more detailed explanation on the plaintext space of our scheme in Sect. 5.1.)

Organization. In Sect. 2, we give a technical overview behind our proposed PKE schemes. In Sect. 3, we review definitions of cryptographic primitives and assumptions. In Sect. 4, we introduce a new primitive that we call symmetric key encapsulation mechanism (SKEM) and provide concrete instantiations. In Sect. 5, we present our KDM-CCA secure PKE scheme with respect to affine functions, and in Sect. 6, we present our KDM-CCA secure PKE scheme with respect to polynomials. Finally, in Sect. 7, we give instantiation examples of KDM-CCA secure PKE schemes.

2 Technical Overview

We provide an overview of our construction. Our starting point is the construction of KDM-CPA secure PKE proposed by Malkin et al. [22]. Their scheme is highly efficient, but only KDM-CPA secure. Our basic idea is to construct KDM-CCA secure PKE by adopting a construction technique used in the recent work by Kitagawa and Tanaka [18] into Malkin et al.'s scheme. However, since a simple combination of them does not work, we introduce a new primitive that ties them together. We first review Malkin et al.'s scheme. Below, we explain the overview by focusing on constructing a PKE scheme that is $\mathcal{F}_{\mathsf{aff}}$-KDM-CCA secure. The actual Malkin et al.'s scheme is $\mathcal{F}_{\mathsf{poly}}$-KDM-CPA secure, and we can construct a $\mathcal{F}_{\mathsf{poly}}$-KDM-CCA secure scheme analogously.

2.1 KDM-CPA Secure Scheme by Malkin et al.

Malkin et al.'s scheme is secure under the DCR assumption and all procedures of their scheme are performed on $\mathbb{Z}_{N^s}^*$, where $N = PQ$ is an RSA modulus with safe primes P and Q of the same length, and $s \geq 2$ is an integer. Below, let $n = \frac{\phi(N)}{4}$.

Scheme	Assumption	Ciphertext size	Tight?
[23] (not CCA)	DCR	$2\lvert\mathbb{Z}_{N^s}\rvert$	
[7] with [15, § 4]	DLIN	$O(\lambda)\lvert\mathbb{G}_{\mathsf{bi}}\rvert$	✓
[13] (Circular)	DCR+DDH$^{(\dagger)}$ & DLIN	$6\lvert\mathbb{Z}_{N^3}\rvert + 50\lvert\mathbb{G}_{\mathsf{bi}}\rvert + \mathsf{OH}_{\mathsf{ch\&sig}}$	
[21] (Circular)	DCR+DDH$^{(\dagger)}$ & D3DH	$6\lvert\mathbb{Z}_{N^3}\rvert + 31\lvert\mathbb{G}_{\mathsf{bi}}\rvert + \mathsf{OH}_{\mathsf{ch\&sig}}$	
[12]	DCR+DDH$^{(\ddagger)}$	$9\lvert\mathbb{Z}_{N^s}\rvert + 9\lvert\mathbb{Z}_{N^2}\rvert + 2\lvert\mathbb{Z}_{\bar{N}}\rvert + \lvert\mathbb{Z}_N\rvert + \mathsf{OH}_{\mathsf{ae}}$	
[19]	DDH	$O(\lambda)\lvert\mathbb{G}_{\mathsf{ddh}}\rvert$	
Ours (§ 5)	DCR & CCAPKE	$2\lvert\mathbb{Z}_{N^s}\rvert + \lvert\pi_{\mathsf{phf}}\rvert + \mathsf{OH}_{\mathsf{cca}}$	
with [17]+CRHF	DCR	$2\lvert\mathbb{Z}_{N^s}\rvert + 2\lvert\mathbb{Z}_{N'}\rvert + \mathsf{len}_{\mathsf{crhf}}$	
with [14]	DCR	$3\lvert\mathbb{Z}_{N^s}\rvert + 28\lvert\mathbb{Z}_{N'^2}\rvert + \mathsf{OH}_{\mathsf{ae}}$	✓
with [11]	DCR & DDH	$3\lvert\mathbb{Z}_{N^s}\rvert + 3\lvert\mathbb{G}_{\mathsf{ddh}}\rvert + \mathsf{OH}_{\mathsf{ae}}$	✓

Fig. 1. Comparison of KDM-CCA secure PKE schemes with respect to affine functions. The last three rows are instantiation examples of our scheme. In the "Ciphertext size" column, we use the following notations: N and N' are RSA moduli, and $s \geq 2$ is the exponent of N in the DCR setting; $\bar{N} = 2N + 1$; For a group G, $|G|$ denotes the size of an element in G; $\mathbb{G}_{\mathsf{bi}}$ denotes a group equipped with a bilinear map, and $\mathbb{G}_{\mathsf{ddh}}$ denotes a DDH-hard group (without bilinear maps); $|\pi_{\mathsf{phf}}|$ denotes the output size of the underlying projective hash function; $\mathsf{OH}_{\mathsf{cca}}$ (resp. $\mathsf{OH}_{\mathsf{ae}}$) denotes the ciphertext overhead of the underlying IND-CCA secure PKE (resp. authenticated encryption) scheme; $\mathsf{OH}_{\mathsf{ch\&sig}}$ denotes an overhead caused by the underlying chameleon hash function and one-time signature scheme; $\mathsf{len}_{\mathsf{crhf}}$ denotes the output size of a collision resistant hash function; For λ-bit security, $\mathsf{OH}_{\mathsf{ae}} = \lambda$, $\mathsf{len}_{\mathsf{crhf}} = 2\lambda$, and $\mathsf{OH}_{\mathsf{ch\&sig}}$ can be smaller than $|\mathbb{Z}_N|$. $^{(\dagger)}$ DDH in the order-$\frac{\phi(N)}{4}$ subgroup of $\mathbb{Z}^*_{N^3}$. $^{(\ddagger)}$ DDH in $\mathbb{QR}_{\bar{N}} := \{a^2 \bmod \bar{N} | a \in \mathbb{Z}^*_{\bar{N}}\}$.

Scheme	Assumption	Ciphertext size	Tight?
[23] (not CCA)	DCR	$(d+1)\lvert\mathbb{Z}_{N^s}\rvert$	
[12]	DCR+DDH$^{(\ddagger)}$	$(8d^9+1)\lvert\mathbb{Z}_{N^s}\rvert + 9\lvert\mathbb{Z}_{N^2}\rvert + 2\lvert\mathbb{Z}_{\bar{N}}\rvert + \lvert\mathbb{Z}_N\rvert + \mathsf{OH}_{\mathsf{ae}}$	
Ours (§ 6)	DCR & CCAPKE	$(d+1)\lvert\mathbb{Z}_{N^s}\rvert + \lvert\pi_{\mathsf{phf}}\rvert + \mathsf{OH}_{\mathsf{cca}}$	
with [17]+CRHF	DCR	$(d+1)\lvert\mathbb{Z}_{N^s}\rvert + 2\lvert\mathbb{Z}_{N'}\rvert + \mathsf{len}_{\mathsf{crhf}}$	
with [14]	DCR	$(2d+1)\lvert\mathbb{Z}_{N^s}\rvert + 28\lvert\mathbb{Z}_{N'^2}\rvert + \mathsf{OH}_{\mathsf{ae}}$	✓
with [11]	DCR & DDH	$(2d+1)\lvert\mathbb{Z}_{N^s}\rvert + 3\lvert\mathbb{G}_{\mathsf{ddh}}\rvert + \mathsf{OH}_{\mathsf{ae}}$	✓

Fig. 2. Comparison of KDM-CCA secure PKE schemes with respect to degree-d polynomial functions. We use the same notation as in Fig. 1.

We can decompose $\mathbb{Z}^*_{N^s}$ as the internal direct product $G_{N^{s-1}} \otimes \langle -1 \rangle \otimes G_n \otimes G_2$, where $\langle -1 \rangle$ is the subgroup of $\mathbb{Z}^*_{N^s}$ generated by $-1 \bmod N^s$, and $G_{N^{s-1}}$, G_n, and G_2 are cyclic groups of order N^{s-1}, n, and 2, respectively. Note that $T := 1 + N \in \mathbb{Z}^*_{N^s}$ has order N^{s-1} and it generates $G_{N^{s-1}}$. Moreover, we can efficiently compute discrete logarithms on $G_{N^{s-1}}$. In addition, we can generate a random generator of G_n.[1]

We can describe Malkin et al.'s scheme by using generators T and g of $G_{N^{s-1}}$ and G_n, respectively, and for simplicity we consider the single user setting for now. Below, all computations are done mod N^s unless stated otherwise, and

[1] This is done by generating $\mu \xleftarrow{r} \mathbb{Z}^*_{N^s}$ and setting $g := \mu^{2N^{s-1}} \bmod N^s$. Then, g is a generator of G_n with overwhelming probability.

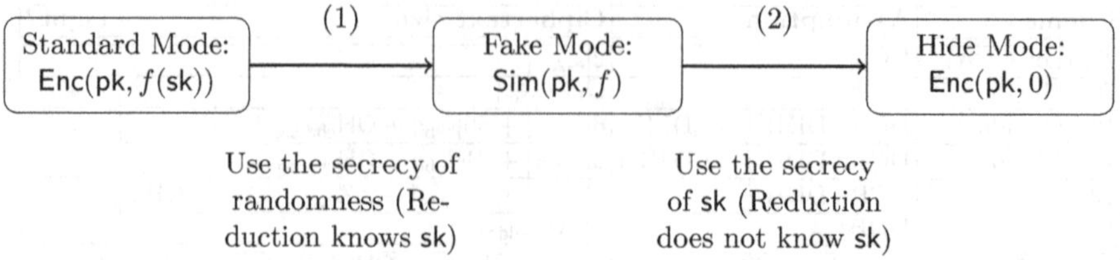

Fig. 3. The triple mode proof. "XX Mode: YY" indicates that in XX Mode, the challenger returns YY as the answer to a KDM query from an adversary.

we omit to write $\bmod N^s$. When generating a key pair, we sample[2] a secret key as $x \xleftarrow{r} \mathbb{Z}_n$ and compute a public key as $h = g^x$. When encrypting a message $m \in \mathbb{Z}_{N^{s-1}}$, we first sample $r \xleftarrow{r} \mathbb{Z}_n$ and set a ciphertext as $(g^r, T^m \cdot h^r)$. If we have the secret key x, we can decrypt the ciphertext by computing the discrete logarithm of $(T^m \cdot h^r) \cdot (g^r)^{-x} = T^m$.

Triple Mode Proof Framework. We say that a PKE scheme is KDM secure if an encryption of $f(\mathsf{sk})$ is indistinguishable from that of some constant message such as 0, where sk is a secret key and f is a function. Malkin et al. showed the $\mathcal{F}_{\mathsf{aff}}$-KDM-CPA security of their scheme based on the DCR assumption via the proof strategy that they call the *triple mode proof*.

In the triple mode proof framework, we prove KDM security using three main hybrid games. We let f be a function queried by an adversary as a KDM query. In the first hybrid called Standard Mode, the challenger returns an encryption of $f(\mathsf{sk})$. In the second hybrid called Fake Mode, the challenger returns a simulated ciphertext from f and the public key corresponding to sk. In the final hybrid called Hide Mode, the challenger returns an encryption of 0. See Fig. 3.

If we can prove that the behavior of the adversary does not change between Standard Mode and Hide Mode, we see that the scheme is KDM secure. However, it is difficult to prove it directly by relying on the secrecy of the secret key. This is because a reduction algorithm needs the secret key to simulate answers to KDM queries in Standard Mode. Then, we consider the intermediate hybrid, Fake Mode, and we try to prove the indistinguishability between Standard Mode and Fake Mode based on the secrecy of encryption randomness. We call this part Step (1). If we can do that, by showing the indistinguishability between Fake Mode and Hide Mode based on the secrecy of the secret key, we can complete the proof. We call this part Step (2). Note that a reduction for Step (2) does not need the secret key to simulate answers to KDM queries.

Using this framework, we can prove the KDM-CPA security of Malkin et al.'s scheme as follows. Let $f(x) = ax + b \bmod N^{s-1}$ be an affine function queried by an adversary, where $a, b \in \mathbb{Z}_{N^{s-1}}$. In Standard Mode, the adversary is given $(g^r, T^{ax+b} \cdot h^r)$. In Fake Mode, the adversary is given $(T^{-a} \cdot g^r, T^b \cdot h^r)$. We can prove the indistinguishability of these two hybrids using the indistinguishability

[2] In the actual scheme, we sample a secret key from $[\frac{N-1}{4}]$. We ignore this issue in this overview.

of g^r and $T^{-a} \cdot g^r$. Namely, we use the DCR assumption and the secrecy of encryption randomness r in this step. Then, in Hide Mode, the adversary is given (g^r, h^r) that is an encryption of 0. We can prove the indistinguishability between Fake Mode and Hide Mode based on the interactive vector (IV) lemma [5] that is in turn based on the DCR assumption. The IV lemma says that for every constant $c_1, c_2 \in \mathbb{Z}_{N^{s-1}}$, $(T^{c_1} \cdot g^r, T^{c_2} \cdot h^r)$ is indistinguishable from (g^r, h^r) if in addition to r, x satisfying $h = g^x$ is hidden from the view of an adversary. This completes the proof of Malkin et al.'s scheme.

2.2 Problem When Proving KDM-CCA Security

Malkin et al.'s scheme is malleable thus is not KDM-CCA secure. In terms of the proof, Step (2) of the triple mode proof does not go through when considering KDM-CCA security. In Step (2), a reduction does not know the secret key and thus the reduction cannot simulate answers to decryption queries correctly.

On the other hand, we see that Step (1) of the triple mode proof goes through also when proving KDM-CCA security since a reduction algorithm knows the secret key in this step. Thus, to construct a KDM-CCA secure scheme based on Malkin et al.'s scheme, all we need is a mechanism that enables us to complete Step (2) of the triple mode proof.

2.3 The Technique by Kitagawa and Tanaka

To solve the above problem, we adopt the technique used by Kitagawa and Tanaka [18]. They constructed a KDM-CCA secure PKE scheme Π_{kdm} by combining projective hash functions PHF and PHF' and an IND-CCA secure PKE scheme Π_{cca}. Their construction is a double layered construction. Namely, when encrypting a message by their scheme, we first encrypt the message by the inner scheme constructed from PHF and PHF', and then encrypt the ciphertext again by Π_{cca}. The inner scheme is the same as the IND-CCA secure PKE scheme based on projective hash functions proposed by Cramer and Shoup [8] except that PHF used to mask a message is required to be *homomorphic* and on the other hand PHF' is required to be only universal (not 2-universal).

The security proof for this scheme can be captured by the triple mode proof framework. We first perform Step (1) of the triple mode proof based on the homomorphism of PHF and the hardness of a subset membership problem on the group behind projective hash functions. Then, we perform Step (2) of the triple mode proof using the IND-CCA security of Π_{cca}. In this step, a reduction algorithm can simulate answers to decryption queries. This is because the reduction algorithm can generate secret keys for PHF and PHF' by itself and access to the decryption oracle for Π_{cca}. When proving the CCA security of a PKE scheme based on projective hash functions, at some step in the proof, we need to estimate the probability that an adversary makes an "illegal" decryption query. In the proof of the scheme by Kitagawa and Tanaka, this estimation can be done in Hide Mode of the triple mode proof. Due to this, the underlying PHF' needs to be only universal.

If the secret key csk of Π_{cca} is included as a part of the secret key of Π_{kdm}, to complete the proof, we need to change the security game so that csk is not needed to simulate answers to KDM queries in Step (1). It seems difficult unless we require an additional property for secret keys of Π_{cca} such as homomorphism. Instead, Kitagawa and Tanaka designed their scheme so that csk is included in the public key of Π_{kdm} after encrypting it by PHF. Then, by eliminating this encrypted csk from an adversary's view by using the security of PHF before Step (2) of the triple mode proof, the entire proof goes through. Note that, similarly to the proof for the construction by Cramer and Shoup [8], a reduction algorithm attacking the security of PHF can simulate answers to decryption queries due to the fact that the security property of PHF is statistical and an adversary for Π_{kdm} is required to make a proof that the query is "legal" using PHF'.

2.4 Adopting the Technique by Kitagawa and Tanaka

We now consider adopting the technique by Kitagawa and Tanaka into Malkin et al.'s scheme. Namely, we add a projective hash function for proving that an inner layer ciphertext of Malkin et al.'s scheme is well-formed, and also add an IND-CCA secure PKE scheme Π_{cca} as the outer layer. In order to prove the KDM-CCA security of this construction, we need to make the secret key csk of Π_{cca} as part of the public key of the resulting scheme after encrypting it somehow. Moreover, we have to eliminate this encrypted csk before Step (2) of the triple mode proof. However, this is not straightforward.

One naive way to do this is encrypting csk by the inner scheme based on the DCR assumption, but this idea does not work. Since the security of the inner scheme is computational unlike a projective hash function, a reduction algorithm attacking the inner scheme cannot simulate answers to decryption queries. One might think the problem is solved by modifying the scheme so that the security property of the inner scheme becomes statistical as a projective hash function, but this modification causes another problem. In order to do this, similarly to the DCR-based projective hash function by Cramer and Shoup [8], a secret key of the inner scheme needs to be sampled from a space whose size is as large as the order of $G_{N^{s-1}} \otimes G_n$ (that is, $N^{s-1} \cdot n$). However, the message space of this scheme is $\mathbb{Z}_{N^{s-1}}$, and thus we cannot encrypt such a large secret key by this scheme. The problem is more complicated when considering KDM-CCA security in the multi-user setting. Therefore, we need another solution to hide the secret key csk of Π_{cca}.

2.5 Solution: Symmetric Key Encapsulation Mechanism (SKEM)

To solve the above problem, we introduce a new primitive we call symmetric key encapsulation mechanism (SKEM). It is a key encapsulation mechanism in which we can use the same key for both the encapsulation algorithm Encap and decapsulation algorithm Decap. Moreover, it satisfies the following properties.

Encap can take an arbitrary integer $x \in \mathbb{Z}$ as an input secret key, but its computation is done by $x \bmod z$, where z is an integer determined in the setup. Then, for correctness, we require $\mathsf{Decap}(x \bmod z, \mathsf{ct}) = \mathsf{K}$, where $(\mathsf{ct}, \mathsf{K}) \leftarrow \mathsf{Encap}(x)$. Moreover, for security, the pseudorandomness of the session-time key K is required to hold as long as $x \bmod z$ is hidden from an adversary even if any other information of x is revealed.

Using SKEM ($\mathsf{Encap}, \mathsf{Decap}$) in addition to an IND-CCA secure PKE scheme Π_{cca} and a projective hash function PHF, we can construct a KDM-CCA secure PKE scheme based on Malkin et al.'s scheme as follows. When generating a key pair, we first sample $x \xleftarrow{r} [n \cdot z]$ and compute $h \leftarrow g^x$, where z is an integer that is co-prime to n and satisfies $n \cdot z \leq N^{s-1}$. Then, we generate a key pair $(\mathsf{ppk}, \mathsf{psk})$ of PHF and $(\mathsf{cpk}, \mathsf{csk})$ of Π_{cca}, and $(\mathsf{ct}, \mathsf{K}) \leftarrow \mathsf{Encap}(x)$, and encrypt psk and csk to $\mathsf{ct}_{\mathsf{sk}}$ using the one-time key K. The resulting secret key is just x and public key is h, psk, cpk, and $(\mathsf{ct}, \mathsf{ct}_{\mathsf{sk}})$.[3] When encrypting a message m, we encrypt it in the same way as the Malkin et al.'s scheme and prove that those ciphertext components are included in G_n by using PHF. Then, we encrypt them by Π_{cca}. When decrypting the ciphertext, we first retrieve csk and psk from $(\mathsf{ct}, \mathsf{ct}_{\mathsf{sk}})$ and x using Decap, and decrypt the ciphertext using x, psk, and csk.

We can prove the $\mathcal{F}_{\mathsf{aff}}$-KDM-CCA security of this scheme basically based on the triple mode proof framework. By doing the same process as Step (1) of the triple mode proof for Malkin et al.'s scheme, we can change the security game so that we can simulate answers to KDM queries using only $x \bmod n$. Moreover, due to the use of the projective hash function PHF, we can change the security game so that we can reply to decryption queries using only $x \bmod n$. Therefore, at this point, we do not need $x \bmod z$ to simulate the security game, and thus we can use the security of the SKEM. We now delete csk and psk from $\mathsf{ct}_{\mathsf{sk}}$ using the security of the SKEM. Then, by using the security of Π_{cca}, we can accomplish Step (2) of the triple mode proof. Note that, similarly to the proof by Kitagawa and Tanaka [18], we estimate the probability that an adversary makes an "illegal" decryption query after Step (2) using the security of PHF.

2.6 Extension to the Multi-user Setting Using RKA Secure SKEM

The above overview of the proof considers KDM-CCA security in the single user setting. We can extend it to the multi-user setting. When considering KDM-CCA security in the multi-user setting, we modify the scheme so that we sample a secret key x from $[n \cdot z \cdot 2^{\xi}]$ such that $n \cdot z \cdot 2^{\xi} \leq N^{s-1}$. In the security proof, we sample a single x from $[n \cdot z]$ and generate the secret key x_i of the i-th user by sampling $\Delta_i \xleftarrow{r} [n \cdot z \cdot 2^{\xi}]$ and setting $x_i = x + \Delta_i$, where the addition is done over $\mathbb{Z}$. In this case, an affine function f of $x_1 \ldots, x_\ell$ is also an affine function of only x whose coefficients are determined by those of f and $\Delta_1, \ldots, \Delta_\ell$. Moreover, the statistical distance between a secret key generated in this way and that generated honestly is at most $2^{-\xi}$. Then, we can proceed the security proof in the same way as above, except for the part using the security of the SKEM.

---

[3] In the actual construction, we derive key pairs $(\mathsf{csk}, \mathsf{cpk})$ and $(\mathsf{ppk}, \mathsf{psk})$ using K as a random coin. This modification reduces the size of a public key.

The secret key x_i of the i-th user is now generated as $x + \Delta_i$ by using a single source x. Thus, each user's one-time key K_i used to hide the user's $(\mathsf{psk}, \mathsf{csk})$ is derived from a single source x and a "shift" value Δ_i. Standard security notations do not capture such a situation.

To address this problem, we require a *security property against related key attacks (RKA security)* for SKEM. However, a very weak form of RKA security is sufficient to complete the proof. We show that such an RKA secure SKEM can be constructed based only on the DCR assumption. Therefore, we can prove the KDM-CCA security in the multi-user setting of our scheme based only on the DCR assumption and the IND-CCA security of the underlying PKE scheme.

2.7 Differences in Usage of RKA Secure Primitive with Han et al.

We note that the previous most efficient KDM-CCA secure PKE schemes of Han et al. [11] (and the scheme of Lu et al. [21] on which the constructions of [11] are based), also use a "symmetric key" primitive that is "RKA secure". Specifically, Han et al. use a primitive called *authenticated encryption with auxiliary-input* (AIAE, for short), for which they define confidentiality and integrity properties both under some appropriate forms of affine-RKA. Here, we highlight the differences between our proposed schemes and the schemes by Han et al. regarding the usage of a symmetric primitive with RKA security.

In our schemes, an RKA secure SKEM is used to derive the secret keys $(\mathsf{psk}, \mathsf{csk})$ of the underlying projective hash function and IND-CCA secure PKE scheme, and an SKEM ciphertext is put as part of a public key of the resulting scheme. In a modified security game considered in our security proofs, a KDM-CCA adversary sees multiple SKEM ciphertexts $\{\mathsf{ct}_i\}$ (contained in the public keys initially given to the adversary), where each ct_i is computed by using $x + \Delta_i \bmod z$ as a secret key, where $\Delta_i \in [n \cdot z \cdot 2^{\xi}]$ is chosen uniformly at random. Consequently, an SKEM used as a building block in our proposed schemes needs to be secure only against "passive" addition-RKA, in which the shift values $\{\Delta_i\}$ are chosen randomly by the challenger (rather than by an RKA adversary). Such an SKEM is easy to construct, and we will show several simple and efficient instantiations based on the DCR assumption, the DDH assumption, and hash functions with some appropriate form of "correlation-robustness" [2,17].

On the contrary, in the Han et al.'s schemes, an AIAE ciphertext is directly contained as part of a ciphertext of the resulting scheme, and thus AIAE ciphertexts are exposed to a CCA. This is a main reason of the necessity of the integrity property for AIAE. Furthermore, in a modified security game considered in the security proofs of their schemes, a KDM-CCA adversary is able to observe multiple AIAE ciphertexts that are computed under secret keys that are derived via (some restricted from of) an affine function of a single (four-dimensional) vector of elements in $\mathbb{Z}_N$ through affine/poly-KDM queries, and thus their AIAE scheme needs to be secure under standard "active" affine-RKA (where key derivation functions are chosen by an RKA adversary, rather than the challenger). Han et al.'s instantiation of AIAE is essentially the Kurosawa-Desmedt encryption scheme [19] used as a symmetric encryption scheme, which is why they require the DDH assumption in addition to the DCR assumption.

2.8 Tightness of Our Construction

Our construction can be tightly instantiated by using a tightly IND-CCA secure PKE scheme as a building block. In our security proof, we can accomplish Step (1) of the triple mode proof by applying the DCR assumption only once via the IV lemma [5]. In Step (2), we need only a single application of the IND-CCA security of the outer scheme by requiring IND-CCA security in the multi-challenge multi-user setting. Thus, if the underlying IND-CCA secure scheme satisfies tight security in the setting, this step is also tight. In the estimation of the probability of "illegal" decryption queries, we only use a statistical property, and thus we do not lose any factor to the underlying assumption. The remaining part of our proof is eliminating secret keys of projective hash function and IND-CCA secure PKE encrypted by SKEM from an adversary's view. To make the entire proof tight, we have to accomplish this step tightly.

To achieve this, we show the RKA security of our SKEM can be tightly reduced to the underlying assumptions. Especially, in the proof of the DCR based construction, we show this using the IV lemma that is different from that we use in Step (1) of the triple mode proof. Namely, in this work, we use two flavors of the IV lemmas to make the security proof for the DCR-based instantiation tight.

To the best of our knowledge, prior to our work, the only way to construct tightly KDM-CCA secure PKE is instantiating the construction proposed by Camenisch et al. [6] using a tightly secure NIZK proof system such as that proposed by Hofheinz and Jager [14]. Schemes instantiated in such a way are not so practical and a ciphertext of them consists of $O(\lambda)$ group elements, where λ is the security parameter. We observe that the DDH-based construction of Kitagawa and Tanaka [18] can be tightly instantiated by using a tightly IND-CCA secure PKE scheme as a building block, though they did not state that explicitly. However, its ciphertext also consists of $O(\lambda)$ group elements. Thus, our schemes are the first tightly KDM-CCA secure PKE scheme whose ciphertext consists of a constant number of group elements.

3 Preliminaries

Here, we review basic notations, cryptographic primitives, and assumptions.

Notations. In this paper, $x \xleftarrow{\mathsf{r}} X$ denotes choosing an element from a finite set X uniformly at random, and $y \leftarrow \mathsf{A}(x)$ denotes assigning to y the output of an algorithm A on an input x. For an integer $\ell > 0$, $[\ell]$ denote the set of integers $\{1, \ldots, \ell\}$. For a function f, $\mathsf{Sup}(f)$ denotes the support of f. For a finite set S, $|S|$ denotes its cardinality, and U_S denotes the uniform distribution over S.

λ denotes a security parameter. PPT stands for probabilistic polynomial time. A function $f(\lambda)$ is a negligible function if $f(\lambda)$ tends to 0 faster than $\frac{1}{\lambda^c}$ for every constant $c > 0$. We write $f(\lambda) = \mathsf{negl}(\lambda)$ to denote $f(\lambda)$ being a negligible function.

Let X and Y be distributions over a set S. The *min-entropy* of X, denoted by $\mathbf{H}_\infty(X)$, is defined by $\mathbf{H}_\infty(X) := -\log_2 \max_{z \in S} \Pr[X = z]$. The *statistical distance* between X and Y, denoted by $\mathbf{SD}(X, Y)$, is defined by $\mathbf{SD}(X, Y) := \frac{1}{2} \sum_{z \in S} |\Pr[X = z] - \Pr[Y = z]|$. X and Y are said to be ϵ-close if $\mathbf{SD}(X, Y) \leq \epsilon$.

3.1 Assumptions

We review the algebraic structure and assumptions used in this paper.

Let $N = PQ$ be an RSA modulus with len-bit safe primes $P = 2p + 1$ and $Q = 2q + 1$ where p and q are also primes. Let $n = pq$. Throughout the paper, we assume $\mathsf{len} \geq \lambda$, and we will frequently use the fact that $\mathbf{SD}(\mathsf{U}_{[n]}, \mathsf{U}_{[\frac{N-1}{4}]}) = \frac{P+Q-2}{N-1} = O(2^{-\mathsf{len}})$.

Let $s \geq 2$ be an integer and $T := 1 + N$. We can decompose $\mathbb{Z}^*_{N^s}$ as the internal direct product $G_{N^{s-1}} \otimes \langle -1 \rangle \otimes G_n \otimes G_2$, where $\langle -1 \rangle$ is the subgroup of $\mathbb{Z}^*_{N^s}$ generated by $-1 \bmod N^s$, and $G_{N^{s-1}}$, G_n, and G_2 are cyclic groups of order N^{s-1}, n, and 2, respectively. Note that $T = 1 + N \in \mathbb{Z}^*_{N^s}$ has order N^{s-1} and it generates $G_{N^{s-1}}$. In addition, we can generate a random generator of G_n by generating $\mu \xleftarrow{r} \mathbb{Z}^*_{N^s}$ and setting $g := \mu^{2N^{s-1}} \bmod N^s$. Then, g is a generator of G_n with overwhelming probability. We also note that the discrete logarithm (base T) is easy to compute in $G_{N^{s-1}}$.

Let $\mathbb{QR}_{N^s} := \{x^2 | x \in \mathbb{Z}^*_{N^s}\}$. Then, we have $\mathbb{QR}_{N^s} = G_{N^{s-1}} \otimes G_n$. We denote $\langle -1 \rangle \otimes \mathbb{QR}_{N^s}$ by $\mathbb{J}_{N^s}$. We can efficiently check the membership of $\mathbb{J}_{N^s}$ by computing the Jacobi symbol with respect to N, without P and Q.

Let GGen be an algorithm, which we call the DCR group generator, that given 1^λ and an integer $s \geq 2$, outputs $\mathsf{param} = (N, P, Q, T, g)$, where N, P, Q, and T are defined as above, and g is a random generator of G_n.

We adopt the definition of the DCR assumption [9,24] used by Hofheinz [12].

Definition 1 (DCR assumption). *We say that the DCR assumption holds with respect to* GGen *if for any integer* $s \geq 2$ *and PPT adversary* $\mathcal{A}$*, we have* $\mathsf{Adv}^{\mathsf{dcr}}_{s,\mathcal{A}}(\lambda) = |\Pr[\mathcal{A}(N, g, g^r \bmod N^s) = 1] - \Pr[\mathcal{A}(N, g, T \cdot g^r \bmod N^s) = 1]| = \mathsf{negl}(\lambda)$*, where* $(N, P, Q, T, g) \leftarrow \mathsf{GGen}(1^\lambda, s)$ *and* $r \xleftarrow{r} [n]$*.*

We recall the *interactive vector game* [5].

Definition 2 (Interactive vector game). *Let* $s \geq 2$ *be an integer and* ℓ *be a polynomial of* λ*. We define the following* $\mathsf{IV}_{s,\ell}$ *game between a challenger and an adversary* $\mathcal{A}$*.*

1. *The challenger chooses a challenge bit* $b \xleftarrow{r} \{0, 1\}$ *and generates* $(N, P, Q, T, g) \leftarrow \mathsf{GGen}(1^\lambda, s)$*. If* $\ell = 1$*, the challenger sends* N *and* $g_1 := g$ *to* $\mathcal{A}$*. Otherwise, the challenger generates* $\alpha_i \xleftarrow{r} [\frac{N-1}{4}]$ *and computes* $g_i \leftarrow g^{\alpha_i} \bmod N^s$ *for every* $i \in [\ell]$*, and sends* N*,* g*, and* $g_1, \ldots, g_\ell$ *to* $\mathcal{A}$*.*
2. $\mathcal{A}$ *can adaptively make sample queries.*
   Sample queries $\mathcal{A}$ *sends* $(a_1, \ldots, a_\ell) \in \mathbb{Z}^\ell_{N^{s-1}}$ *to the challenger. The challenger generates* $r \xleftarrow{r} [\frac{N-1}{4}]$ *and computes* $e_i \leftarrow T^{b \cdot a_i} \cdot g_i^r \bmod N^s$ *for every* $i \in [\ell]$*. The challenger then returns* $(e_1, \ldots, e_\ell)$ *to* $\mathcal{A}$*.*

3. $\mathcal{A}$ outputs $b' \in \{0,1\}$.

We say that $\mathsf{IV}_{s,\ell}$ *is hard if for any PPT adversary* $\mathcal{A}$*, we have* $\mathsf{Adv}^{\mathsf{IV}}_{s,\ell,\mathcal{A}}(\lambda) = 2 \cdot \left|\Pr[b = b'] - \frac{1}{2}\right| = \mathsf{negl}(\lambda)$.

For any s and ℓ, $\mathsf{IV}_{s,\ell}$ is hard under the DCR assumption [5,22]. We show the following lemmas related to $\mathsf{IV}_{s,\ell}$ that are useful to prove the tight security of our constructions. The proofs of the lemmas are given in the full version.

Lemma 1. *Let $s \geq 2$ be an integer. Let $\mathcal{A}$ be a PPT adversary that plays the* $\mathsf{IV}_{s,1}$ *game and makes at most* q_{iv} *queries. Then, there exists a PPT adversary* $\mathcal{B}$ *satisfying* $\mathsf{Adv}^{\mathsf{iv}}_{s,1,\mathcal{A}}(\lambda) \leq 2 \cdot \mathsf{Adv}^{\mathsf{dcr}}_{s,\mathcal{B}}(\lambda) + \frac{O(q_{\mathsf{iv}})}{2^{\mathsf{len}}}$.

Lemma 2. *Let $s \geq 2$ be an integer. Let ℓ be a polynomial of λ. Let $\mathcal{A}$ be a PPT adversary that plays the* $\mathsf{IV}_{s,\ell}$ *game and makes exactly one sample query. Then, there exists a PPT adversary* $\mathcal{B}$ *satisfying* $\mathsf{Adv}^{\mathsf{iv}}_{s,\ell,\mathcal{A}}(\lambda) \leq 2 \cdot \mathsf{Adv}^{\mathsf{dcr}}_{s,\mathcal{B}}(\lambda) + \frac{O(\ell)}{2^{\mathsf{len}}}$.

3.2 Projective Hash Function

We review the notion of *projective hash functions* (PHF) introduced by Cramer and Shoup [8] (which is also called *hash proof systems* in the literature). In this work, we will use PHFs defined with respect to the DCR group generator GGen.

Definition 3 (Projective hash function family). *A PHF family* PHF *with respect to* GGen *consists of a tuple* $(\mathsf{Setup}, \Pi_{\mathsf{yes}}, \Pi_{\mathsf{no}}, \mathcal{SK}, \mathcal{PK}, \mathcal{K}, \Lambda, \mu, \mathsf{Pub})$ *with the following properties:*

- Setup *is a PPT algorithm that takes* $\mathsf{param} = (N, P, Q, T, g)$ *output by* $\mathsf{GGen}(1^\lambda, s)$ *(for some $s \geq 2$) as input, and outputs a public parameter* pp *that parameterizes the remaining components of* PHF. *(In the following, we always make the existence of* pp *implicit and suppress it from the notation).*
- Π_{yes}*,* Π_{no}*,* $\mathcal{SK}$*,* $\mathcal{PK}$*, and* $\mathcal{K}$ *are sets parameterized by* pp *(and also by* param*).* Π_{yes} *and* Π_{no} *form an NP-language,*[4] *where for all* $c \in \Pi_{\mathsf{yes}}$*, there exists a witness r with which one can efficiently check the fact of* $c \in \Pi_{\mathsf{yes}}$*. An element in* Π_{yes} *(resp.* Π_{no}*) is called an yes (resp. no) instance.*
  Furthermore, it is required that given pp*, one can efficiently sample a uniformly random element from* $\mathcal{SK}$*.*
- Λ *is an efficiently computable (deterministic) hash function that takes a secret key* $\mathsf{sk} \in \mathcal{SK}$ *and an yes or no instance* $c \in \Pi_{\mathsf{yes}} \cup \Pi_{\mathsf{no}}$ *as input, and outputs a hash value* $\pi \in \mathcal{K}$*.*
- μ *is an efficiently computable (deterministic) projection map that takes a secret key* $\mathsf{sk} \in \mathcal{SK}$ *as input, and outputs a public key* $\mathsf{pk} \in \mathcal{PK}$*.*
- Pub *is an efficiently computable algorithm that takes a public key* $\mathsf{pk} \in \mathcal{PK}$*, an yes instance* $c \in \Pi_{\mathsf{yes}}$*, and a witness r that* $c \in \Pi_{\mathsf{yes}}$ *as input, and outputs a hash value* $\pi \in \mathcal{K}$*.*

[4] Strictly speaking, since Π_{yes} and Π_{no} may not cover the entire input space of the function $\Lambda_{\mathsf{sk}}(\cdot)$ introduced below, they form an NP-promise problem.

– *Projective property: For all* $\mathsf{sk} \in \mathcal{SK}$*, the action of* $\Lambda_{\mathsf{sk}}(\cdot)$ *for yes instances* $c \in \Pi_{\mathsf{yes}}$ *is completely determined by* $\mathsf{pk} = \mu(\mathsf{sk})$*. Furthermore, for all* $c \in \Pi_{\mathsf{yes}}$ *and a corresponding witness* r*, it holds that* $\Lambda_{\mathsf{sk}}(c) = \mathsf{Pub}(\mu(\mathsf{sk}), c, r)$*.*

We next introduce the universal property for a PHF family. In this paper, we consider the statistical and computational variants. Our definition of the computational universal property is based on the "computational universal2" property for a hash proof system introduced by Hofheinz and Kiltz [15]. We adapt their definition to the "universal1" case, and also relax the notion so that we only require that guessing a hash value for a no instance is hard, rather than requiring that a hash value of a no instance is pseudorandom.

Definition 4 (Statistical/computational universal). *Let* $s \geq 2$, GGen *be the DCR group generator, and* $\mathsf{PHF} = (\mathsf{Setup}, \Pi_{\mathsf{yes}}, \Pi_{\mathsf{no}}, \mathcal{SK}, \mathcal{PK}, \mathcal{K}, \Lambda, \mu, \mathsf{Pub})$ *be a PHF family with respect to* GGen*. We say that* PHF *is*

– ϵ-universal *if for any* param *output by* $\mathsf{GGen}(1^\lambda, s)$*, any* pp *output by* $\mathsf{Setup}(\mathsf{param})$*, any* $\mathsf{pk} \in \mathcal{PK}$*, any* $c \in \Pi_{\mathsf{no}}$*, and any* $\pi \in \mathcal{K}$*, we have*

$$\Pr_{\mathsf{sk} \leftarrow \mathcal{SK}} \left[\Lambda_{\mathsf{sk}}(c) = \pi \middle| \mu(\mathsf{sk}) = \mathsf{pk}\right] \leq \epsilon. \tag{1}$$

Furthermore, we simply say that PHF *is* universal *if it is* ϵ*-universal for some negligible function* $\epsilon = \epsilon(\lambda)$*.*

– computationally universal *if for any PPT adversary* $\mathcal{A}$*, the advantage* $\mathsf{Adv}^{\mathsf{cu}}_{\mathsf{PHF},\mathcal{A}}(\lambda)$ *in the following game played by* $\mathcal{A}$ *and a challenger is negligible in* λ*:*
  1. *First, the challenger executes* $\mathsf{param} = (N, P, Q, T, g) \leftarrow \mathsf{GGen}(1^\lambda, s)$ *and* $\mathsf{pp} \leftarrow \mathsf{Setup}(\mathsf{param})$*. The challenger then chooses* $\mathsf{sk} \xleftarrow{\mathsf{r}} \mathcal{SK}$*, and computes* $\mathsf{pk} \leftarrow \mu(\mathsf{sk})$*. Then, the challenger sends* $(N, T, g, \mathsf{pp}, \mathsf{pk})$ *to* $\mathcal{A}$*.*
  2. $\mathcal{A}$ *can adaptively make evaluation queries.*
     Evaluation queries $\mathcal{A}$ *sends an yes or no instance* $c \in \Pi_{\mathsf{yes}} \cup \Pi_{\mathsf{no}}$ *to the challenger. If* $c \in \Pi_{\mathsf{yes}}$*, the challenger returns* $\pi \leftarrow \Lambda_{\mathsf{sk}}(c)$ *to* $\mathcal{A}$*. Otherwise (i.e.* $c \in \Pi_{\mathsf{no}}$*), the challenger returns* $\perp$ *to* $\mathcal{A}$*.*
  3. $\mathcal{A}$ *outputs a pair* $(c^*, \pi^*) \in \Pi_{\mathsf{no}} \times \mathcal{K}$*. The advantage of* $\mathcal{A}$ *is defined by* $\mathsf{Adv}^{\mathsf{cu}}_{\mathsf{PHF},\mathcal{A}}(\lambda) := \Pr[\Lambda_{\mathsf{sk}}(c^*) = \pi^*]$*.*

Remark 1 (Statistical implies computational). It is not hard to see that the (statistical) universal property implies the computational one (even against computationally unbounded adversaries). To see this, recall that the projective property ensures that the action of $\Lambda_{\mathsf{sk}}(\cdot)$ for yes instances is determined by pk. Thus, the evaluation results $\Lambda_{\mathsf{sk}}(c)$ for yes instances $c \in \Pi_{\mathsf{yes}}$ do not reveal the information of sk beyond the fact that $\mathsf{pk} = \mu(\mathsf{sk})$. Also, evaluation queries with no instances $c \in \Pi_{\mathsf{no}}$ are answered with $\perp$. These imply that throughout the game, the information of sk does not leak to an adversary beyond what is already leaked from pk. Thus, at the point of outputting (c^*, π^*), sk is uniformly distributed over the subset $\mathcal{SK}|_{\mathsf{pk}} := \{\mathsf{sk}' \in \mathcal{SK} | \mu(\mathsf{sk}') = \mathsf{pk}\}$ from an adversary's viewpoint, which is exactly the distribution of sk in the probability defining the universal property. Hence, if a PHF family is ϵ-universal, the probability that $\Lambda_{\mathsf{sk}}(c^*) = \pi^*$ occurs is upper bounded by ϵ.

3.3 Public Key Encryption

A public key encryption (PKE) scheme PKE is a four tuple $(\mathsf{Setup}, \mathsf{KG}, \mathsf{Enc}, \mathsf{Dec})$ of PPT algorithms. Let $\mathcal{M}$ be the message space of PKE. The setup algorithm Setup, given a security parameter 1^λ, outputs a public parameter pp. The key generation algorithm KG, given a public parameter pp, outputs a public key pk and a secret key sk. The encryption algorithm Enc, given a public key pk and message $m \in \mathcal{M}$, outputs a ciphertext CT. The decryption algorithm Dec, given a public key pk, a secret key sk, and a ciphertext CT, outputs a message $\tilde{m} \in \{\bot\} \cup \mathcal{M}$. As correctness, we require $\mathsf{Dec}(\mathsf{pk}, \mathsf{sk}, \mathsf{Enc}(\mathsf{pk}, m)) = m$ for every $m \in \mathcal{M}$, $\mathsf{pp} \leftarrow \mathsf{Setup}(1^\lambda)$, and $(\mathsf{pk}, \mathsf{sk}) \leftarrow \mathsf{KG}(\mathsf{pp})$.

Next, we define key dependent message security against chosen ciphertext attacks (KDM-CCA security) for PKE.

Definition 5 (KDM-CCA security). *Let* PKE *be a PKE scheme,* $\mathcal{F}$ *function family, and* ℓ *the number of keys. We define the* $\mathcal{F}$*-KDM-CCA game between a challenger and an adversary* $\mathcal{A}$ *as follows. Let* $\mathcal{SK}$ *and* $\mathcal{M}$ *be the secret key space and message space of* PKE*, respectively.*

1. *The challenger chooses a challenge bit* $b \xleftarrow{r} \{0,1\}$ *and generates* $\mathsf{pp} \leftarrow \mathsf{Setup}(1^\lambda)$ *and* ℓ *key pairs* $(\mathsf{pk}_k, \mathsf{sk}_k) \leftarrow \mathsf{KG}(\mathsf{pp})\ (k \in [\ell])$*. The challenger sets* $\mathbf{sk} := (\mathsf{sk}_1, \ldots, \mathsf{sk}_\ell)$ *and sends* $(\mathsf{pk}_1, \ldots, \mathsf{pk}_\ell)$ *to* $\mathcal{A}$*. Finally, the challenger prepares a list* L_{kdm} *which is initially empty.*
2. $\mathcal{A}$ *may adaptively make the following queries polynomially many times.*
   KDM queries $\mathcal{A}$ *sends* $(j, f^0, f^1) \in [\ell] \times \mathcal{F} \times \mathcal{F}$ *to the challenger. We require that* f^0 *and* f^1 *be functions such that* $f : \mathcal{SK}^\ell \to \mathcal{M}$*. The challenger returns* $\mathsf{CT} \leftarrow \mathsf{Enc}\left(\mathsf{pk}_j, f^b(\mathbf{sk})\right)$ *to* $\mathcal{A}$*. Finally, the challenger adds* (j, CT) *to* L_{kdm}*.*
   Decryption queries $\mathcal{A}$ *sends* (j, CT) *to the challenger. If* $(j, \mathsf{CT}) \in L_{\mathsf{kdm}}$*, the challenger returns* $\bot$ *to* $\mathcal{A}$*. Otherwise, the challenger returns* $m \leftarrow \mathsf{Dec}\left(\mathsf{pk}_j, \mathsf{sk}_j, \mathsf{CT}\right)$ *to* $\mathcal{A}$*.*
3. $\mathcal{A}$ *outputs* $b' \in \{0,1\}$*.*

We say that PKE *is* $\mathcal{F}$*-KDM-CCA secure if for any polynomial* $\ell = \ell(\lambda)$ *and PPT adversary* $\mathcal{A}$*, we have* $\mathsf{Adv}^{\mathsf{kdmcca}}_{\mathsf{PKE},\mathcal{F},\ell,\mathcal{A}}(\lambda) = 2 \cdot \left|\Pr[b = b'] - \frac{1}{2}\right| = \mathsf{negl}(\lambda)$*.*

The above definition is slightly different from the standard definition where an adversary is required to distinguish encryptions of $f(\mathsf{sk}_1, \ldots, \mathsf{sk}_\ell)$ from encryptions of some fixed message. However, the two definitions are equivalent if the function class $\mathcal{F}$ contains a constant function, and this is the case for affine functions and polynomials treated in this paper.

The definition of IND-CCA security (in the multi-user/challenge setting) is recovered by restricting the functions used in KDM queries in the KDM-CCA game to constant functions, and thus we omit the description of the security game for it. We denote an adversary $\mathcal{A}$'s IND-CCA advantage by $\mathsf{Adv}^{\mathsf{indcca}}_{\mathsf{PKE},\ell,\mathcal{A}}(\lambda)$.

4 Symmetric KEM and Passive RKA Security

In our proposed PKE schemes, we will use a secret key variant of a key encapsulation mechanism (KEM) satisfying a weak form of RKA security with respect to addition, as one of the main building blocks. Since several instantiations for this building block from various assumptions are possible, in this section we formalize it as a stand-alone primitive called *symmetric KEM (SKEM)*, together with its RKA security in the form we use in the security proofs of the proposed PKE schemes.

4.1 Definition

We first give the formal syntax and functional requirements of an SKEM, and then give some remarks.

Definition 6 (Symmetric key encapsulation mechanism). *An SKEM* SKEM *is a three tuple* $(\mathsf{Setup}, \mathsf{Encap}, \mathsf{Decap})$ *of PPT algorithms.*

- *The setup algorithm* Setup*, given a security parameter* 1^λ*, outputs a public parameter* pp *and a pair of natural numbers* $(z, \widetilde{z})$*, where* z *represents the size of the secret key space, and the secret key space is* $[z]$*, and* $\widetilde{z}$ *is an approximation of* z*. We assume that* $\widetilde{z}$ *(but not necessarily* z*) can be efficiently derived from* pp*. We also assume that* pp *specifies the session-key space* $\mathcal{K}$*.*
- *The encapsulation algorithm* Encap*, given a public parameter* pp *and a secret key* $\mathsf{sk} \in \mathbb{Z}$*, outputs a ciphertext* ct *and a session-key* $\mathsf{K} \in \mathcal{K}$*.*
- *The decapsulation algorithm* Decap*, given a public parameter* pp*, a secret key* $\mathsf{sk} \in \mathbb{Z}$*, and a ciphertext* ct*, outputs a session-key* $\mathsf{K} \in \mathcal{K}$*.*

As the functional (syntactical) requirements, we require the following three properties to hold for all $(\mathsf{pp}, z, \widetilde{z}) \leftarrow \mathsf{Setup}(1^\lambda)$*:*

1. *(Approximate samplability of secret keys:)* $\mathbf{SD}(\mathsf{U}_{[z]}, \mathsf{U}_{[\widetilde{z}]})) \le O(2^{-\lambda})$ *holds.*
2. *(Correctness of decapsulation:)* $\mathsf{Decap}(\mathsf{pp}, \mathsf{sk} \bmod z, \mathsf{ct}) = \mathsf{K}$ *holds for every* $\mathsf{sk} \in \mathbb{Z}$ *and* $(\mathsf{ct}, \mathsf{K}) \leftarrow \mathsf{Encap}(\mathsf{pp}, \mathsf{sk})$*.*
3. *(Implicit modular-reduction in encapsulation:)* $\mathsf{Encap}(\mathsf{pp}, \mathsf{sk}; r) = \mathsf{Encap}(\mathsf{pp}, \mathsf{sk} \bmod z; r)$ *holds for every* $\mathsf{sk} \in \mathbb{Z}$ *and randomness* r *for* Encap*.*

Remark 2 (On the syntax and functional requirements).

- As mentioned above, when $(\mathsf{pp}, z, \widetilde{z})$ is output by $\mathsf{Setup}(1^\lambda)$, the secret key space under pp is $[z]$. For security reasons, however, in some constructions, the exact order z cannot be made public even for an entity executing Encap and Decap. (In particular, this is the case in our concrete instantiation from the DCR assumption, in which we set $z = \frac{\phi(N)}{4}$ and $\widetilde{z} = \frac{N-1}{4}$). Hence, we instead require its approximation $\widetilde{z}$ to be public via pp.
- We allow Encap and Decap to take any integer $\mathsf{sk} \in \mathbb{Z}$ (rather than $\mathsf{sk} \in [z]$ or $\mathsf{sk} \in [\widetilde{z}]$) as a secret key, but their "correctness guarantees" expressed by the second and third items of the functional requirements, are with respect to the modular-reduced value $\mathsf{sk} \bmod z$. Such flexible interface is convenient when an SKEM is used as a building block in the proposed PKE schemes.

- The third item in the functional requirements ensures that a ciphertext/session-key pair $(\mathsf{ct}, \mathsf{K})$ generated by using $\mathsf{sk} \in \mathbb{Z}$ does not leak the information of sk beyond $\mathsf{sk} \bmod z$. This property plays an important role in the security proofs of our proposed PKE schemes.
- Note that an SKEM can satisfy our syntactical and functional requirements even if its ciphertext is empty. (Say, Encap and Decap output some deterministic function of pp and $\mathsf{sk} \bmod \widetilde{z}$).

In the following, we give the formalization of passive RKA security. It is essentially the definition of the same name defined for symmetric encryption by Applebaum, Harnik, and Ishai [2], with the slight difference that we allow an adversary to specify the upper bound B of the interval from which key-shifting values $\{\Delta_k\}$ are chosen randomly by the challenger.

Definition 7 (Passive RKA security). *Let* $\mathsf{SKEM} = (\mathsf{Setup}, \mathsf{Encap}, \mathsf{Decap})$ *be an SKEM, and let* ℓ *be a natural number. Consider the following game between a challenger and an adversary* $\mathcal{A}$*:*

1. *First, the challenger chooses a challenge bit* $b \xleftarrow{r} \{0,1\}$ *and generates* $(\mathsf{pp}, z, \widetilde{z}) \leftarrow \mathsf{Setup}(1^\lambda)$. *Then, the challenger sends* $\widetilde{z}$ *to* $\mathcal{A}$.
2. $\mathcal{A}$ *sends an integer* $B \geq \widetilde{z}$ *specifying the upper bound of the interval from which key-shifting values* $\{\Delta_k\}_{k\in[\ell]}$ *are chosen, to the challenger.*
3. *The challenger samples* $\mathsf{sk} \xleftarrow{r} [z]$ *and* $\Delta_k \xleftarrow{r} [B]$ *for every* $k \in [\ell]$. *Then, the challenger computes* $(\mathsf{ct}_k, \mathsf{K}_k^1) \leftarrow \mathsf{Encap}(\mathsf{pp}, \mathsf{sk} + \Delta_k)$[5] *and also samples* $\mathsf{K}_k^0 \leftarrow \mathcal{K}$ *for every* $k \in [\ell]$. *Finally, the challenger sends* pp, $(\Delta_k)_{k\in[\ell]}$, *and* $(\mathsf{ct}_k, \mathsf{K}_k^b)_{k\in[\ell]}$ *to* $\mathcal{A}$.
4. $\mathcal{A}$ *outputs* $b' \in \{0,1\}$.

We say that SKEM *is* passively RKA secure, *if for any polynomial* $\ell = \ell(\lambda)$ *and PPT adversary* $\mathcal{A}$, *we have* $\mathsf{Adv}^{\mathsf{rka}}_{\mathsf{SKEM},\ell,\mathcal{A}}(\lambda) = 2 \cdot \left|\Pr[b = b'] - \frac{1}{2}\right| = \mathsf{negl}(\lambda)$.

Remark 3 (Stretching a session-key with a pseudorandom generator). From the definition, it is easy to see that a session-key of an SKEM can be stretched by using a pseudorandom generator (PRG) while preserving its passive RKA security. More specifically, let $\mathsf{SKEM} = (\mathsf{Setup}, \mathsf{Encap}, \mathsf{Decap})$ be an SKEM with session-key space $\mathcal{K}$, and let $\mathsf{PRG} : \mathcal{K} \to \mathcal{K}'$ be a PRG such that $|\mathcal{K}| < |\mathcal{K}'|$. Let $\mathsf{SKEM}' = (\mathsf{Setup}, \mathsf{Encap}', \mathsf{Decap}')$ be the SKEM with session-key space $\mathcal{K}'$ that is obtained by naturally composing SKEM with PRG, namely, $\mathsf{Encap}'(\mathsf{pp}, \mathsf{sk})$ runs $(\mathsf{ct}, \mathsf{K}) \leftarrow \mathsf{Encap}(\mathsf{pp}, \mathsf{sk})$ and outputs $(\mathsf{ct}, \mathsf{PRG}(\mathsf{K}))$, and $\mathsf{Decap}'(\mathsf{pp}, \mathsf{sk}, \mathsf{ct}) := \mathsf{PRG}(\mathsf{Decap}(\mathsf{pp}, \mathsf{sk}, \mathsf{ct}))$. Then, if SKEM is passively RKA secure and PRG is a secure PRG, then SKEM' is also passively RKA secure. Moreover, if the passive RKA security of SKEM is tightly reduced to some assumption and the multi-instance version of the security of PRG is also tightly reduced to the same assumption, then so is the passive RKA security of SKEM'. (Since the proof is straightforward, we omit a formal proof of this simple fact). Note that we can easily construct tightly secure PRG based on the DDH or DCR assumption.

[5] The addition $\mathsf{sk} + \Delta_k$ is done over $\mathbb{Z}$.

Setup(1^λ) :	Encap(pp, sk $\in \mathbb{Z}$) :	Decap(pp, sk $\in \mathbb{Z}$, ct) :
$(N', P', Q', T', g') \leftarrow \mathsf{GGen}(1^\lambda, s)$ $H \xleftarrow{r} \mathcal{H}$ $\mathsf{pp} \leftarrow (N', T', g', H)$ Return $(\mathsf{pp}, z := \frac{\phi(N')}{4}, \widetilde{z} := \frac{N'-1}{4})$.	$(N', T', g', H) \leftarrow \mathsf{pp}$ $\alpha \xleftarrow{r} [\frac{N'-1}{4}]$ $\mathsf{ct} \leftarrow g'^{\alpha} \bmod N'^s$ $\mathsf{K} \leftarrow H(\mathsf{ct}^{\mathsf{sk}} \bmod N'^s)$ Return (ct, K).	$(N', T', g', H) \leftarrow \mathsf{pp}$ $\mathsf{K} \leftarrow H(\mathsf{ct}^{\mathsf{sk}} \bmod N'^s)$ Return K.

Fig. 4. The DCR-based instantiation of an SKEM.

4.2 Concrete Instantiations

Our definition of passive RKA security for an SKEM is sufficiently weak so that simple and efficient constructions are possible from the DCR or DDH assumption, which are essentially the symmetric-key version of the ElGamal KEM. We can also realize it from a hash function satisfying an appropriate form of "correlation robustness" [2,17]. We only give a concrete instantiation based on the DCR assumption here. The other instantiations are given in the full version.

Let $s \geq 2$, GGen be the DCR group generator, and $\mathcal{H} = \{H : \{0,1\}^{2s \cdot \mathsf{len}} \rightarrow \mathcal{K}\}$ be a universal hash family. Then, we can construct an SKEM SKEM = (Setup, Encap, Decap) whose session-key space is $\mathcal{K}$, as described in Fig. 4.[6]

It is obvious to see that SKEM satisfies the three functional requirements of SKEM. Specifically, let $(\mathsf{pp}, z, \widetilde{z})$ be output by Setup. Then, we have $\mathbf{SD}\left(\mathsf{U}_{[z]}, \mathsf{U}_{[\widetilde{z}]}\right) = \mathbf{SD}(\mathsf{U}_{\left[\frac{\phi(N')}{4}\right]}, \mathsf{U}_{\left[\frac{N'-1}{4}\right]}) = O(2^{-\mathsf{len}}) \leq O(2^{-\lambda})$. The other two properties of the functional requirements are also satisfied due to the fact that in Encap and Decap, a secret key is treated only in the exponent of elements in $G_{n'}$ (where $n' = (P'-1)(Q'-1)/4$, and $G_{n'}$ is the subgroup of $Z^*_{N'^s}$ of order n').

The passive RKA security of SKEM is guaranteed by the following lemma, which is proved via Lemma 2 and the leftover hash lemma. We provide the formal proof in the full version.

Lemma 3. *If the DCR assumption holds with respect to* GGen, *and* $\epsilon_{\mathsf{LHL}} := \frac{1}{2} \cdot \sqrt{2^{-(s-1)\cdot(2\mathsf{len}-1)} \cdot |\mathcal{K}|} = \mathsf{negl}(\lambda)$, *then* SKEM *is passively RKA secure.*

Specifically, for any polynomial $\ell = \ell(\lambda)$ *and PPT adversary* $\mathcal{A}$ *that attacks the passive RKA security of* SKEM, *there exists a PPT adversary* $\mathcal{B}$ *such that* $\mathsf{Adv}^{\mathsf{rka}}_{\mathsf{SKEM},\ell,\mathcal{A}}(\lambda) \leq 2 \cdot \mathsf{Adv}^{\mathsf{dcr}}_{s,\mathcal{B}}(\lambda) + \ell \cdot \left(\epsilon_{\mathsf{LHL}} + O(2^{-\mathsf{len}})\right)$.

5 KDM-CCA Secure PKE with Respect to Affine Functions

In this section, we show a PKE scheme that is KDM-CCA secure with respect to affine functions based on the DCR assumption.

---

[6] Since the RSA modulus used in the SKEM has to be generated independently of that in the main constructions presented in Sects. 5 and 6, here we use characters with a prime (e.g. N') for values in param.

$\mathsf{Setup}_{\mathsf{aff}}(1^\lambda)$: $\mathsf{param} = (N, P, Q, T, g) \leftarrow \mathsf{GGen}(1^\lambda, s)$ $\mathsf{pp}_{\mathsf{phf}} \leftarrow \mathsf{Setup}_{\mathsf{phf}}(\mathsf{param})$ $(\mathsf{pp}_{\mathsf{skem}}, z, \widetilde{z}) \leftarrow \mathsf{Setup}_{\mathsf{skem}}(1^\lambda)$ $\mathsf{pp}_{\mathsf{cca}} \leftarrow \mathsf{Setup}_{\mathsf{cca}}(1^\lambda)$ $\mathsf{pp}_{\mathsf{aff}} \leftarrow (N, T, g, \mathsf{pp}_{\mathsf{phf}}, \mathsf{pp}_{\mathsf{skem}}, \mathsf{pp}_{\mathsf{cca}})$ Return $\mathsf{pp}_{\mathsf{aff}}$.	$\mathsf{KG}_{\mathsf{aff}}(\mathsf{pp}_{\mathsf{aff}})$: $(N, T, g, \mathsf{pp}_{\mathsf{phf}}, \mathsf{pp}_{\mathsf{skem}}, \mathsf{pp}_{\mathsf{cca}}) \leftarrow \mathsf{pp}_{\mathsf{aff}}$ $x \xleftarrow{\mathsf{r}} [\frac{N-1}{4} \cdot \widetilde{z} \cdot 2^\xi]$ $(\mathsf{ct}, \mathsf{K}) \leftarrow \mathsf{Encap}(\mathsf{pp}_{\mathsf{skem}}, x)$ Parse K as $(r^{\mathsf{KG}}, \mathsf{psk}) \in \mathcal{R}^{\mathsf{KG}} \times \mathcal{SK}$. $h \leftarrow g^{2x} \bmod N^s$ $\mathsf{ppk} \leftarrow \mu(\mathsf{psk})$ $(\mathsf{cpk}, \mathsf{csk}) \leftarrow \mathsf{KG}_{\mathsf{cca}}(\mathsf{pp}_{\mathsf{cca}}; r^{\mathsf{KG}})$ Return $\mathsf{PK} := (h, \mathsf{ct}, \mathsf{ppk}, \mathsf{cpk})$ and $\mathsf{SK} := x$.
$\mathsf{Enc}_{\mathsf{aff}}(\mathsf{PK}, m \in \mathbb{Z}_{N^{s-1}})$: $(h, \mathsf{ct}, \mathsf{ppk}, \mathsf{cpk}) \leftarrow \mathsf{PK}$ $r \xleftarrow{\mathsf{r}} [\frac{N-1}{4}]$ $u \leftarrow g^r \bmod N^s$ $v \leftarrow T^m \cdot h^r \bmod N^s$ $\pi \leftarrow \mathsf{Pub}(\mathsf{ppk}, u^2 \bmod N^s, 2r)$ $\mathsf{CT} \leftarrow \mathsf{Enc}_{\mathsf{cca}}(\mathsf{cpk}, (u, v, \pi))$ Return CT.	$\mathsf{Dec}_{\mathsf{aff}}(\mathsf{PK}, \mathsf{SK}, \mathsf{CT})$: $(h, \mathsf{ct}, \mathsf{ppk}, \mathsf{cpk}) \leftarrow \mathsf{PK}$; $x \leftarrow \mathsf{SK}$ $\mathsf{K} \leftarrow \mathsf{Decap}(\mathsf{pp}_{\mathsf{skem}}, x, \mathsf{ct})$ Parse K as $(r^{\mathsf{KG}}, \mathsf{psk}) \in \mathcal{R}^{\mathsf{KG}} \times \mathcal{SK}$. $(\mathsf{cpk}, \mathsf{csk}) \leftarrow \mathsf{KG}_{\mathsf{cca}}(\mathsf{pp}_{\mathsf{cca}}; r^{\mathsf{KG}})$ $(u, v, \pi) \leftarrow \mathsf{Dec}_{\mathsf{cca}}(\mathsf{cpk}, \mathsf{csk}, \mathsf{CT})$ If $(u, v) \notin \mathbb{J}_{N^s}^2$ then return $\perp$. If $\pi \neq \Lambda_{\mathsf{psk}}(u^2 \bmod N^s)$ then return $\perp$. Return $m \leftarrow \log_T(v \cdot u^{-2x} \bmod N^s)$.

Fig. 5. The proposed KDM-CCA secure PKE scheme Π_{aff} with respect to affine functions. (The public parameter $\mathsf{pp}_{\mathsf{aff}}$ is omitted from the inputs to $\mathsf{Enc}_{\mathsf{aff}}$ and $\mathsf{Dec}_{\mathsf{aff}}$).

We first specify the *DCR language* with respect to which the underlying PHF family used in our proposed scheme is considered. Then, we give our proposed PKE scheme in Sect. 5.1. We also give two instantiations for the underlying PHF family, the first one in Sect. 5.2 and the second one in Sect. 5.3.

DCR Language. Let $s \geq 2$, GGen be the DCR group generator, and $\mathsf{param} = (N, P, Q, T, g) \leftarrow \mathsf{GGen}(1^\lambda, s)$. The set of yes instances Π_{yes} is the subgroup G_n of $\mathbb{J}_{N^s}$, and the set of no instances Π_{no} is $G_{N^{s-1}} \otimes G_n \setminus G_n$. Note that we can represent any yes instance $c \in G_n$ as $c = g^r \bmod N^s$, where $r \in \mathbb{Z}$. Thus, such r works as a witness for $c \in \Pi_{\mathsf{yes}}$.

5.1 Proposed PKE Scheme

Let $s \geq 2$, and GGen be the DCR group generator. Let $\Pi_{\mathsf{cca}} = (\mathsf{Setup}_{\mathsf{cca}}, \mathsf{KG}_{\mathsf{cca}}, \mathsf{Enc}_{\mathsf{cca}}, \mathsf{Dec}_{\mathsf{cca}})$ be a PKE scheme such that the randomness space of $\mathsf{KG}_{\mathsf{cca}}$ is $\mathcal{R}^{\mathsf{KG}}$. Let $\mathsf{PHF} = (\mathsf{Setup}_{\mathsf{phf}}, \Pi_{\mathsf{yes}}, \Pi_{\mathsf{no}}, \mathcal{SK}, \mathcal{PK}, \mathcal{K}, \Lambda, \mu, \mathsf{Pub})$ be a PHF family with respect to GGen for the DCR language (defined as above). Let $\mathsf{SKEM} = (\mathsf{Setup}_{\mathsf{skem}}, \mathsf{Encap}, \mathsf{Decap})$ be an SKEM whose session key space is $\mathcal{R}^{\mathsf{KG}} \times \mathcal{SK}$.[7] Finally, let $\xi = \xi(\lambda)$ be any polynomial such that $2^{-\xi} = \mathsf{negl}(\lambda)$. Using these building blocks, our proposed PKE scheme $\Pi_{\mathsf{aff}} = (\mathsf{Setup}_{\mathsf{aff}}, \mathsf{KG}_{\mathsf{aff}}, \mathsf{Enc}_{\mathsf{aff}}, \mathsf{Dec}_{\mathsf{aff}})$ is constructed as described in Fig. 5. The plaintext space of Π_{aff} is $\mathbb{Z}_{N^{s-1}}$, where N is the modulus generated in $\mathsf{Setup}_{\mathsf{aff}}$.

[7] Strictly speaking, the concrete format of $\mathcal{SK}$ could be dependent on a public parameter $\mathsf{pp}_{\mathsf{phf}}$ of PHF. However, as noted in Remark 3, the session-key space of an SKEM can be flexibly adjusted by using a pseudorandom generator. Hence, for simplicity we assume that such an adjustment of the spaces is applied.

The correctness of Π_{aff} follows from that of SKEM and Π_{cca}, and the projective property of PHF.

We note that although our scheme has correctness and can be proved secure for any $s \geq 2$, the plaintext space of our scheme is $\mathbb{Z}_{N^{s-1}}$, and thus if $s = 2$, then the plaintext space $\mathbb{Z}_N$ becomes smaller than the secret key space $[\frac{N-1}{4} \cdot \widetilde{z} \cdot 2^{\xi}]$, in which case KDM security for affine functions does not even capture circular security. (Malkin et al.'s scheme [22] has exactly the same issue.) If $\widetilde{z} \cdot 2^{\xi}$ is smaller than N, then the secret key space can be contained in $\mathbb{Z}_{N^2}$, in which case $s \geq 3$ is sufficient in practice.[8]

We also note that even if the building block SKEM SKEM and/or PKE scheme Π_{cca} are instantiated also from the DCR assumption (or any other factoring-related assumption), the DCR groups formed by (N, T, g) in $\mathsf{pp}_{\mathsf{aff}}$ should not be shared with those used in SKEM and/or Π_{cca}. This is because in our security proof, the reduction algorithms for SKEM and Π_{cca} will use the information of P and Q behind N. (See our security proof below.) We also remark that in our construction, N has to be generated by a trusted party, or by users jointly via some secure computation protocol, so that no user knows its factorization. (The same applies to our DCR-based SKEM.) This is the same setting as in the previous DCR-based (KDM-)CCA secure PKE schemes [11,13,22].

Before proving the KDM-CCA security of Π_{aff}, we also note the difference between the "inner scheme" of Π_{aff} and Malkin et al.'s scheme [22]. Although these schemes are essentially the same, there is a subtle difference. Specifically, when generating h contained in PK of Π_{aff}, we generate it as $h \leftarrow g^{2x} \bmod N^s$ while it is generated as $h \leftarrow g^x \bmod N^s$ in Malkin et al.'s scheme. Moreover, such additional squarings are performed on u in the decryption procedure of our scheme. By these additional squarings, if it is guaranteed that an element u appearing in the decryption procedure belongs to $\mathbb{J}_{N^s} = G_{N^{s-1}} \otimes \langle -1 \rangle \otimes G_n$, it can be converted to an element in $G_{N^{s-1}} \otimes G_n$. Thus, we can consider a PHF family on $G_{N^{s-1}} \otimes G_n$ rather than $G_{N^{s-1}} \otimes \langle -1 \rangle \otimes G_n$, and as a result, we need not worry about a case that an adversary for Π_{aff} may learn $x \bmod 2$ through decryption queries. This helps us to simplify the security proof. Note that we cannot explicitly require that group elements contained in a ciphertext be elements in $G_{N^{s-1}} \otimes G_n$ since it is not known how to efficiently check the membership in $G_{N^{s-1}} \otimes G_n$ without the factorization of N, while we can efficiently check the membership in $\mathbb{J}_{N^s}$ using only N.

KDM-CCA Security. Let ℓ be the number of keys in the security game. We will show that Π_{aff} is KDM-CCA secure with respect to the function family $\mathcal{F}_{\mathsf{aff}}$ consisting of functions described as

$$f(x_1, \ldots, x_\ell) = \sum_{k \in [\ell]} a_k x_k + a_0 \bmod N^{s-1},$$

where $a_0, \ldots, a_\ell \in \mathbb{Z}_{N^{s-1}}$. Formally, we prove the following theorem.

[8] Actually, if $s = 3$ and our DCR-based instantiation in Sect. 4.2 is used as the underlying SKEM, then the RSA modulus N generated at the setup of our PKE construction has to be ξ-bit larger than the RSA modulus generated at the setup of SKEM to satisfy $[\frac{N-1}{4} \cdot \widetilde{z} \cdot 2^{\xi}] \subset \mathbb{Z}_{N^2}$. We do not need this special treatment if $s \geq 4$.

Theorem 1. *Assume that the DCR assumption holds with respect to* GGen, SKEM *is passively RKA secure,* PHF *is computationally universal, and* Π_{cca} *is IND-CCA secure. Then,* Π_{aff} *is* $\mathcal{F}_{\mathsf{aff}}$*-KDM-CCA secure.*

Specifically, for any polynomial $\ell = \ell(\lambda)$ *and PPT adversary* $\mathcal{A}$ *that attacks the* $\mathcal{F}_{\mathsf{aff}}$*-KDM-CCA security of* Π_{aff} *and makes* $q_{\mathsf{kdm}} = q_{\mathsf{kdm}}(\lambda)$ *KDM queries and* $q_{\mathsf{dec}} = q_{\mathsf{dec}}(\lambda)$ *decryption queries, there exist PPT adversaries* $\mathcal{B}_{\mathsf{dcr}}$, $\mathcal{B}_{\mathsf{rka}}$, $\mathcal{B}'_{\mathsf{rka}}$, $\mathcal{B}_{\mathsf{cca}}$, $\mathcal{B}'_{\mathsf{cca}}$, *and* $\mathcal{B}_{\mathsf{cu}}$ *such that*

$$\begin{aligned}\mathsf{Adv}^{\mathsf{kdmcca}}_{\Pi_{\mathsf{aff}},\mathcal{F}_{\mathsf{aff}},\ell,\mathcal{A}}(\lambda) \le 2 \cdot \Big(2 \cdot \mathsf{Adv}^{\mathsf{dcr}}_{s,\mathcal{B}_{\mathsf{dcr}}}(\lambda) + \mathsf{Adv}^{\mathsf{rka}}_{\mathsf{SKEM},\ell,\mathcal{B}_{\mathsf{rka}}}(\lambda) + \mathsf{Adv}^{\mathsf{rka}}_{\mathsf{SKEM},\ell,\mathcal{B}'_{\mathsf{rka}}}(\lambda) \\ + \mathsf{Adv}^{\mathsf{indcca}}_{\Pi_{\mathsf{cca}},\ell,\mathcal{B}_{\mathsf{cca}}}(\lambda) + \mathsf{Adv}^{\mathsf{indcca}}_{\Pi_{\mathsf{cca}},\ell,\mathcal{B}'_{\mathsf{cca}}}(\lambda) + \ell \cdot (q_{\mathsf{dec}} \cdot \mathsf{Adv}^{\mathsf{cu}}_{\mathsf{PHF},\mathcal{B}_{\mathsf{cu}}}(\lambda) + 2^{-\xi})\Big) \\ + O(q_{\mathsf{kdm}} \cdot 2^{-\mathsf{len}}) + O(2^{-\lambda}). \qquad (2)\end{aligned}$$

Remark 4 (Tightness of the reduction). Note that our reductions to the DCR assumption and the security of the building blocks are tight, except for the reduction to the computational universal property of the underlying PHF family PHF, which has the factor $\ell \cdot q_{\mathsf{dec}}$. However, if PHF satisfies the *statistical* universal property, the term $\mathsf{Adv}^{\mathsf{cu}}_{\mathsf{PHF},\mathcal{B}_{\mathsf{cu}}}(\lambda)$ can be replaced with a negligible function that is independent of a computational assumption, and thus our reduction becomes fully tight. Hence, if we use an SKEM and an IND-CCA PKE scheme with a tight security reduction to the DCR assumption (or another assumption A), the overall reduction to the DCR(& A) assumption becomes fully tight as well.

Proof of Theorem 1. We proceed the proof via a sequence of games argument using 8 games (Game 0 to Game 7). For every $t \in \{0, \dots, 7\}$, let $\mathtt{SUC}_t$ be the event that $\mathcal{A}$ succeeds in guessing the challenge bit b in Game t. Our goal is to upper bound every term appearing in $\mathsf{Adv}^{\mathsf{kdmcca}}_{\Pi_{\mathsf{aff}},\mathcal{F}_{\mathsf{aff}},\ell,\mathcal{A}}(\lambda) = 2 \cdot \left|\Pr[\mathtt{SUC}_0] - \frac{1}{2}\right| \le 2 \cdot \sum_{t \in \{0,\dots,6\}} \left|\Pr[\mathtt{SUC}_t] - \Pr[\mathtt{SUC}_{t+1}]\right| + 2 \cdot \left|\Pr[\mathtt{SUC}_7] - \frac{1}{2}\right|$.

Game 0: This is the original $\mathcal{F}_{\mathsf{aff}}$-KDM-CCA game regarding Π_{aff}.

Game 1: Same as Game 0, except for how KDM queries are replied. When $\mathcal{A}$ makes a KDM query $\left(j, \left(a_0^0, \dots, a_\ell^0\right), \left(a_0^1, \dots, a_\ell^1\right)\right)$, the challenger generates v and π respectively by $v \leftarrow T^m \cdot u^{2x_j} \bmod N^s$ and $\pi \leftarrow \Lambda_{\mathsf{psk}_j}\left(u^2 \bmod N^s\right)$, instead of $v \leftarrow T^m \cdot h_j^r \bmod N^s$ and $\pi \leftarrow \mathsf{Pub}\left(\mathsf{ppk}_j, u^2 \bmod N^s, 2r\right)$, where $r \xleftarrow{\mathsf{r}} \left[\frac{N-1}{4}\right]$ and $u = g^r \bmod N^s$.

v is generated identically in both games. Moreover, by the projective property of PHF, $\Lambda_{\mathsf{psk}_j}\left(u^2 \bmod N^s\right) = \mathsf{Pub}\left(\mathsf{ppk}_j, u^2 \bmod N^s, 2r\right)$ holds, and thus π is also generated identically in both games. Hence, we have $|\Pr[\mathtt{SUC}_0] - \Pr[\mathtt{SUC}_1]| = 0$.

Game 2: Same as Game 1, except for how the challenger generates $\{x_k\}_{k \in [\ell]}$. The challenger first generates $x \xleftarrow{\mathsf{r}} \left[\frac{N-1}{4} \cdot \widetilde{z}\right]$. Then, for every $k \in [\ell]$, the challenger generates $\Delta_k \xleftarrow{\mathsf{r}} \left[\frac{N-1}{4} \cdot \widetilde{z} \cdot 2^\xi\right]$ and computes $x_k \leftarrow x + \Delta_k$, where the addition is done over $\mathbb{Z}$.

$|\Pr[\mathtt{SUC}_1] - \Pr[\mathtt{SUC}_2]| \leq \ell \cdot 2^{-\xi}$ holds since the distribution of x_k in Game 2 and that in Game 1 are $2^{-\xi}$-close for every $k \in [\ell]$.

Next, we will change the game so that we can respond to KDM queries made by $\mathcal{A}$ using only $x \bmod n = x \bmod \frac{\phi(N)}{4}$. To this end, we make some preparation. Observe that in Game 2, the answer to a KDM query $\left(j, \left(a_0^0, \ldots, a_\ell^0\right), \left(a_0^1, \ldots, a_\ell^1\right)\right)$ is $\mathsf{Enc_{cca}}\left(\mathsf{cpk}_j, (u, v, \pi)\right)$, where

$$u = g^r \bmod N^s, v = T^{\sum_{k\in[\ell]} a_k^b x_k + a_0^b} \cdot u^{2x_j} \bmod N^s, \pi = \Lambda_{\mathsf{psk}_j}\left(u^2 \bmod N^s\right),$$

and $r \xleftarrow{\mathsf{r}} \left[\frac{N-1}{4}\right]$. We also have

$$\sum_{k\in[\ell]} a_k^b x_k + a_0^b = \sum_{k\in[\ell]} a_k^b \left(x + \Delta_k\right) + a_0^b = \left(\sum_{k\in[\ell]} a_k^b\right) x + \sum_{k\in[\ell]} a_k^b \Delta_k + a_0^b,$$

where the addition is done over $\mathbb{Z}$. Thus, by defining

$$A^b = \sum_{k\in[\ell]} a_k^b \quad \text{and} \quad B^b = \sum_{k\in[\ell]} a_k^b \Delta_k + a_0^b, \tag{3}$$

we have $v = T^{A^b x + B^b} \cdot u^{2x_j} \bmod N^s = T^{A^b x + B^b} \cdot \left(g^r\right)^{2x_j} \bmod N^s$. Note that A^b and B^b are computed only from $\left(a_0^b, \cdots, a_\ell^b\right)$ and $\{\Delta_k\}_{k\in[\ell]}$.

Game 3: Same as Game 2, except that for a KDM query $\left(j, \left(a_0^0, \ldots, a_\ell^0\right), \left(a_0^1, \ldots, a_\ell^1\right)\right)$ made by $\mathcal{A}$, the challenger responds as follows. (The difference from Game 2 is only in Step 3).

1. Compute A^b and B^b as in Eq. 3.
2. Generate $r \xleftarrow{\mathsf{r}} \left[\frac{N-1}{4}\right]$.
3. Compute $u \leftarrow T^{-\frac{A^b}{2}} \cdot g^r \bmod N^s$.
4. Compute $v \leftarrow T^{A^b x + B^b} \cdot u^{2x_j} \bmod N^s$.
5. Compute $\pi \leftarrow \Lambda_{\mathsf{psk}_j}\left(u^2 \bmod N^s\right)$.
6. Return $\mathsf{CT} \leftarrow \mathsf{Enc_{cca}}\left(\mathsf{cpk}_j, (u, v, \pi)\right)$ and add (j, CT) to L_{kdm}.

Under the hardness of $\mathsf{IV}_{s,1}$, the distributions of $g^r \bmod N^s$ and $T^{-\frac{A^b}{2}} \cdot g^r \bmod N^s$ are computationally indistinguishable. More specifically, there exists a PPT adversary $\mathcal{B}_{\mathsf{iv}}$ that makes q_{kdm} sample queries in the $\mathsf{IV}_{s,1}$ game and satisfies $|\Pr[\mathtt{SUC}_2] - \Pr[\mathtt{SUC}_3]| = \mathsf{Adv}^{\mathsf{iv}}_{s,1,\mathcal{B}_{\mathsf{iv}}}(\lambda)$. Due to Lemma 1, this means that there exists another PPT adversary $\mathcal{B}_{\mathsf{dcr}}$ such that $|\Pr[\mathtt{SUC}_2] - \Pr[\mathtt{SUC}_3]| \leq 2 \cdot \mathsf{Adv}^{\mathsf{dcr}}_{s,\mathcal{B}_{\mathsf{dcr}}}(\lambda) + O(q_{\mathsf{kdm}} \cdot 2^{-\mathsf{len}})$.

In Game 3, the answer to a KDM query $\left(j, \left(a_0^0, \ldots, a_\ell^0\right), \left(a_0^1, \ldots, a_\ell^1\right)\right)$ is $\mathsf{Enc_{cca}}\left(\mathsf{cpk}_j, (u, v, \pi)\right)$, where

$$u = T^{-\frac{A^b}{2}} \cdot g^r \bmod N^s,$$
$$v = T^{A^b x + B^b} \cdot u^{2x_j} \bmod N^s = T^{B^b - A^b \Delta_j} \cdot g^{2r(x \bmod n)} \cdot g^{2r\Delta_j} \bmod N^s,$$
$$\pi = \Lambda_{\mathsf{psk}_j}\left(u^2 \bmod N^s\right),$$

$r \xleftarrow{r} \left[\frac{N-1}{4}\right]$, and A^b and B^b are computed as in Eq. 3. Thus, we can reply to a KDM query made by $\mathcal{A}$ using only $x \bmod n = x \bmod \frac{\phi(N)}{4}$.

We next change how decryption queries made by $\mathcal{A}$ are replied.

Game 4: Same as Game 3, except for how the challenger responds to decryption queries made by $\mathcal{A}$. For a decryption query (j, CT) made by $\mathcal{A}$, the challenger returns $\perp$ to $\mathcal{A}$ if $(j, \mathsf{CT}) \in L_{\mathsf{kdm}}$, and otherwise responds as follows. (The difference from Game 3 is adding Step 2 to the procedure).

1. Compute $(u, v, \pi) \leftarrow \mathsf{Dec_{cca}}\left(\mathsf{cpk}_j, \mathsf{csk}_j, \mathsf{CT}\right)$. If $(u, v) \notin \mathbb{J}_{N^s}^2$, return $\perp$. Otherwise, compute as follows.
2. If $u \notin \langle -1 \rangle \otimes G_n$, return $\perp$. Otherwise, compute as follows.
3. Return $\perp$ if $\pi \neq \Lambda_{\mathsf{psk}_j}\left(u^2 \bmod N^s\right)$ and $m \leftarrow \log_T\left(v \cdot u^{-2x_j} \bmod N^s\right)$ otherwise.

We define the following event in Game $i \in \{4, 5, 6, 7\}$.

$\mathtt{BDQ}_i$: $\mathcal{A}$ makes a decryption query $(j, \mathsf{CT}) \notin L_{\mathsf{kdm}}$ which satisfies the following conditions, where $(u, v, \pi) \leftarrow \mathsf{Dec_{cca}}\left(\mathsf{cpk}_j, \mathsf{csk}_j, \mathsf{CT}\right)$.

- $(u, v) \in \mathbb{J}_{N^s}^2$.
- $u \notin \langle -1 \rangle \otimes G_n$. Note that $\mathbb{J}_{N^s} = \langle -1 \rangle \otimes G_{N^{s-1}} \otimes G_n$.
- $\pi = \Lambda_{\mathsf{psk}_j}(u^2 \bmod N^s)$.

We call such a decryption query a "**bad decryption query**".

Games 3 and 4 are identical unless $\mathcal{A}$ makes a bad decryption query in each game. Therefore, we have $|\Pr[\mathtt{SUC}_3] - \Pr[\mathtt{SUC}_4]| \leq \Pr[\mathtt{BDQ}_4]$. Combining this with the triangle inequality, we will also bound the terms in $|\Pr[\mathtt{SUC}_3] - \Pr[\mathtt{SUC}_4]| \leq \sum_{t \in \{4,5,6\}} |\Pr[\mathtt{BDQ}_t] - \Pr[\mathtt{BDQ}_{t+1}]| + \Pr[\mathtt{BDQ}_7]$.

We let (j, CT) be a decryption query made by $\mathcal{A}$. We also let $(u, v, \pi) \leftarrow \mathsf{Dec_{cca}}\left(\mathsf{cpk}_j, \mathsf{csk}_j, \mathsf{CT}\right)$. If the query is not a bad decryption query and $u \in \mathbb{J}_{N^s}$, then $(u^2 \bmod N^s) \in G_n$. Thus,

$$u^{2x_j} \bmod N^s = (u^2)^{x + \Delta_j} \bmod N^s = (u^2 \bmod N^s)^{(x \bmod n)} \cdot u^{2\Delta_j} \bmod N^s.$$

Thus, if the query is not a bad decryption query, the answer to it can be computed by using only $x \bmod n$.

Furthermore, recall that due to the "implicit modular-reduction in encapsulation" property of SKEM, for every $k \in [\ell]$, the SKEM-ciphertext/session-key pair $(\mathsf{ct}_k, \mathsf{K}_k)$ computed for generating the k-th public key PK_k at the initial phase, can be generated by using only $x_k \bmod z = x + \Delta_k \bmod z$.

Hence, due to the change in Game 4, now we have done the preparation for "decomposing" x into its "mod n"-component and its "mod z"-component.

Game 5: Same as Game 4, except that the challenger generates $\widehat{x} \xleftarrow{r} [n]$ and $\bar{x} \xleftarrow{r} [z]$ and then uses them for $x \bmod n$ and $x \bmod z$, respectively.

Note that when $x \xleftarrow{r} [\frac{N-1}{4} \cdot \widetilde{z}]$, the statistical distance between $(x \bmod n, x \bmod z)$ and $(\widehat{x} \bmod n, \bar{x} \bmod z)$ is bounded by $\mathbf{SD}(\mathsf{U}_{[\frac{N-1}{4}\cdot\widetilde{z}]}, \mathsf{U}_{[n\cdot z]})$, because if $x \xleftarrow{r} [n \cdot z]$, then the distribution of $(x \bmod n, x \bmod z)$ and that of $(\widehat{x} \bmod n, \bar{x} \bmod z)$ are identical due to the Chinese remainder theorem.[9] Note also that $\mathbf{SD}(\mathsf{U}_{[\frac{N-1}{4}\cdot\widetilde{z}]}, \mathsf{U}_{[n\cdot z]}) \leq \mathbf{SD}(\mathsf{U}_{[\frac{N-1}{4}]}, \mathsf{U}_{[n]}) + \mathbf{SD}(\mathsf{U}_{[\widetilde{z}]}, \mathsf{U}_{[z]})$. Here, the former statistical distance is $\frac{P+Q-2}{N-1} = O(2^{-\mathsf{len}}) \leq O(2^{-\lambda})$, and the latter statistical distance is bounded by $O(2^{-\lambda})$ due to the "approximate samplability of a secret key" property of SKEM. Hence, we have $|\Pr[\mathtt{SUC}_4] - \Pr[\mathtt{SUC}_5]| \leq O(2^{-\lambda})$ and $|\Pr[\mathtt{BDQ}_4] - \Pr[\mathtt{BDQ}_5]| \leq O(2^{-\lambda})$.

Game 6: Same as Game 5, except that for every $k \in [\ell]$, the challenger generates $\mathsf{K}_k \xleftarrow{r} \mathcal{R}^{\mathsf{KG}} \times \mathcal{SK}$ from which $r_k^{\mathsf{KG}} \in \mathcal{R}^{\mathsf{KG}}$ and $\mathsf{psk}_k \in \mathcal{SK}$ are generated, instead of using K_k associated with ct_k.

By the passive RKA security of SKEM, the view of $\mathcal{A}$ in Game 6 is indistinguishable from that of Game 5. Namely, there exist PPT adversaries $\mathcal{B}_{\mathsf{rka}}$ and $\mathcal{B}'_{\mathsf{rka}}$ that attack the passive RKA security of SKEM so that $|\Pr[\mathtt{SUC}_5] - \Pr[\mathtt{SUC}_6]| = \mathsf{Adv}^{\mathsf{rka}}_{\mathsf{SKEM},\ell,\mathcal{B}_{\mathsf{rka}}}(\lambda)$ and $|\Pr[\mathtt{BDQ}_5] - \Pr[\mathtt{BDQ}_6]| = \mathsf{Adv}^{\mathsf{rka}}_{\mathsf{SKEM},\ell,\mathcal{B}'_{\mathsf{rka}}}(\lambda)$ hold, respectively. We provide the descriptions of them in the full version.

Game 7: Same as Game 6, except that the challenger responds to KDM queries (j, CT) made by $\mathcal{A}$ with $\mathsf{CT} \leftarrow \mathsf{Enc}_{\mathsf{cca}}(\mathsf{cpk}_j, (0, 0, 0))$.

We can consider straightforward reductions to the security of the underlying PKE scheme Π_{cca} for bounding $|\Pr[\mathtt{SUC}_6] - \Pr[\mathtt{SUC}_7]|$ and $|\Pr[\mathtt{BDQ}_6] - \Pr[\mathtt{BDQ}_7]|$. Note that the reduction algorithms can check whether $\mathcal{A}$ makes a bad decryption query or not by using decryption queries for Π_{cca}, and $\phi(N)$ and $\{\mathsf{psk}_k\}_{k\in[\ell]}$ that could be generated by the reductions themselves. Thus, there exist PPT adversaries $\mathcal{B}_{\mathsf{cca}}$ and $\mathcal{B}'_{\mathsf{cca}}$ such that $|\Pr[\mathtt{SUC}_6] - \Pr[\mathtt{SUC}_7]| = \mathsf{Adv}^{\mathsf{indcca}}_{\Pi_{\mathsf{cca}},\ell,\mathcal{B}_{\mathsf{cca}}}(\lambda)$ and $|\Pr[\mathtt{BDQ}_6] - \Pr[\mathtt{BDQ}_7]| = \mathsf{Adv}^{\mathsf{indcca}}_{\Pi_{\mathsf{cca}},\ell,\mathcal{B}'_{\mathsf{cca}}}(\lambda)$.

In Game 7, the challenge bit b is information-theoretically hidden from the view of $\mathcal{A}$. Thus, we have $\left|\Pr[\mathtt{SUC}_7] - \frac{1}{2}\right| = 0$.

Finally, $\Pr[\mathtt{BDQ}_7]$ is bounded by the computational universal property of PHF. More specifically, there exists a PPT adversary $\mathcal{B}_{\mathsf{cu}}$ such that $\Pr[\mathtt{BDQ}_7] \leq \ell \cdot q_{\mathsf{dec}} \cdot \mathsf{Adv}^{\mathsf{cu}}_{\mathsf{PHF},\mathcal{B}_{\mathsf{cu}}}(\lambda) + O(2^{-\mathsf{len}})$. We provide the description of $\mathcal{B}_{\mathsf{cu}}$ in the full version.

From the above arguments, we conclude that there exist PPT adversaries $\mathcal{B}_{\mathsf{dcr}}, \mathcal{B}_{\mathsf{rka}}, \mathcal{B}'_{\mathsf{rka}}, \mathcal{B}_{\mathsf{cca}}, \mathcal{B}'_{\mathsf{cca}}$, and $\mathcal{B}_{\mathsf{cu}}$ satisfying Eq. 2. □ (**Theorem** 1)

5.2 Basic Construction of Projective Hash Function

For the PHF family for the DCR language used in our construction Π_{aff}, we provide two instantiations: the basic construction $\mathsf{PHF}_{\mathsf{aff}}$ that achieves the statistical

[9] Here, we are implicitly assuming that $n = pq$ and z are relatively prime. This occurs with overwhelming probability due to the DCR assumption. We thus ignore the case of n and z are not relatively prime in the proof for simplicity.

universal property in this subsection, and its "space-efficient" variant $\mathsf{PHF}_{\mathsf{aff}}^{\mathsf{hash}}$ that achieves only the computational universal property in the next subsection.

Let $s \geq 2$, and GGen be the DCR group generator. The basic construction $\mathsf{PHF}_{\mathsf{aff}} = (\mathsf{Setup}, \Pi_{\mathsf{yes}}, \Pi_{\mathsf{no}}, \mathcal{SK}, \mathcal{PK}, \mathcal{K}, \Lambda, \mu, \mathsf{Pub})$ is as follows. (The construction here is basically the universal PHF family for the DCR setting by Cramer and Shoup [8], extended for general $s \geq 2$). Recall that $\Pi_{\mathsf{yes}} = G_n$ and $\Pi_{\mathsf{no}} = G_{N^{s-1}} \otimes G_n \setminus G_n$ for the DCR language. Given param output from $\mathsf{GGen}(1^\lambda, s)$, Setup outputs a public parameter pp that concretely specifies $(\mathcal{SK}, \mathcal{PK}, \mathcal{K}, \Lambda, \mu, \mathsf{Pub})$ defined as follows. We define $\mathcal{SK} := \left[N^{s-1} \cdot \frac{N-1}{4}\right]$, $\mathcal{PK} := G_n$, and $\mathcal{K} := G_{N^{s-1}} \otimes G_n$. For every $\mathsf{sk} \in \left[N^{s-1} \cdot \frac{N-1}{4}\right]$ and $c \in G_{N^{s-1}} \otimes G_n$, we also define μ and Λ as $\mu(\mathsf{sk}) := g^{\mathsf{sk}} \bmod N^s$ and $\Lambda_{\mathsf{sk}}(c) := c^{\mathsf{sk}} \bmod N^s$.

Projective Property. Let $\mathsf{sk} \in \left[N^{s-1} \cdot \frac{N-1}{4}\right]$, $\mathsf{pk} = g^{\mathsf{sk}} \bmod N^s$, and $c = g^r \bmod N^s$, where $r \in \mathbb{Z}$ is regarded as a witness for $c \in G_n$. We define the public evaluation algorithm Pub as $\mathsf{Pub}(\mathsf{pk}, c, r) := \mathsf{pk}^r \bmod N^s$. We see that $\mathsf{pk}^r \equiv \left(g^{\mathsf{sk}}\right)^r \equiv (g^r)^{\mathsf{sk}} \equiv \Lambda_{\mathsf{sk}}(c) \bmod N^s$, and thus $\mathsf{PHF}_{\mathsf{aff}}$ satisfies the projective property.

Universal Property. We can prove that $\mathsf{PHF}_{\mathsf{aff}}$ satisfies the statistical universal property. The proof is almost the same as that for the statistical universal property of the DCR-based projective hash function by Cramer and Shoup [8]. We provide the formal proof in the full version.

5.3 Space-Efficient Construction of Projective Hash Function

The second instantiation is a "space-efficient" variant of the first construction. Specifically, it is obtained from $\mathsf{PHF}_{\mathsf{aff}}$ by "compressing" the output of the function Λ in $\mathsf{PHF}_{\mathsf{aff}}$ with a collision resistant hash function.

More formally, let $\mathcal{H} = \left\{H : \{0,1\}^* \rightarrow \{0,1\}^{\mathsf{len}_{\mathsf{crhf}}}\right\}$ be a collision resistant hash family. Then, consider the "compressed"-version of the PHF family $\mathsf{PHF}_{\mathsf{aff}}^{\mathsf{hash}} = (\mathsf{Setup}', \Pi_{\mathsf{yes}}, \Pi_{\mathsf{no}}, \mathcal{SK}, \mathcal{PK}, \mathcal{K}' := \{0,1\}^{\mathsf{len}_{\mathsf{crhf}}}, \Lambda', \mu, \mathsf{Pub}')$, in which Setup' picks $H \xleftarrow{r} \mathcal{H}$ in addition to generating $\mathsf{pp} \leftarrow \mathsf{Setup}$, Λ' is defined simply by composing Λ and H by $\Lambda'_{\mathsf{sk}}(\cdot) := H(\Lambda_{\mathsf{sk}}(\cdot))$, Pub' is defined similarly by composing Pub and H, and the remaining components are unchanged from $\mathsf{PHF}_{\mathsf{aff}}$. $\mathsf{PHF}_{\mathsf{aff}}^{\mathsf{hash}}$ preserves the projective property of $\mathsf{PHF}_{\mathsf{aff}}$ and it is possible to show that the "compressed" construction $\mathsf{PHF}_{\mathsf{aff}}^{\mathsf{hash}}$ satisfies the computational universal property.

This "compressing technique" is applicable to not only the specific instantiation $\mathsf{PHF}_{\mathsf{aff}}$, but also more general PHF families PHF, so that if the underlying PHF is (statistically) universal and satisfies some additional natural properties (that are satisfied by our instantiation in Sect. 5.2) and $\mathcal{H}$ is collision resistant, then the resulting "compressed" version $\mathsf{PHF}^{\mathsf{hash}}$ is computationally universal. In the full version, we formally show the additional natural properties, and the formal statement for the compressing technique as well as its proof.

The obvious merit of using $\mathsf{PHF}^{\mathsf{hash}}_{\mathsf{aff}}$ instead of $\mathsf{PHF}_{\mathsf{aff}}$ is its smaller output size. The disadvantage is that unfortunately, the computational universal property of $\mathsf{PHF}^{\mathsf{hash}}_{\mathsf{aff}}$ is only loosely reduced to the collision resistance of $\mathcal{H}$. Specifically, the advantage of a computational universal adversary is bounded only by the square root of the advantage of the collision resistance adversary (reduction algorithm). For the details, see the full version.

6 KDM-CCA Secure PKE with Respect to Polynomials

In this section, we show a PKE scheme that is KDM-CCA secure with respect to polynomials based on the DCR assumption. More specifically, our scheme is KDM-CCA secure with respect to modular arithmetic circuits (MAC) defined by Malkin et al. [22].

Our scheme is based on the *cascaded ElGamal encryption* scheme used by Malkin et al., and uses a PHF family for a language that is associated with it, which we call the *cascaded ElGamal language*. Furthermore, for considering a PHF family for this language, we need to make a small extension to the syntax of the functions μ, and thus we also introduce it here as well.

After introducing the cascaded ElGamal language as well as the extension to a PHF family below, we will show our proposed PKE scheme, and explain the instantiations of the underlying PHF family.

Augmenting the Syntax of PHFs. For our construction in this section, we use a PHF family whose syntax is slightly extended from Definition 3. Specifically, we introduce an auxiliary key $\mathsf{ak} \in \mathcal{AK}$ that is used as part of a public parameter pp output by Setup, where $\mathcal{AK}$ itself could also be parameterized by param output by GGen. Then, we allow this ak to (1) affect the structure of the witnesses for Π_{yes}, and (2) be taken as input by the projection map μ so that it takes $\mathsf{ak} \in \mathcal{AK}$ and $\mathsf{sk} \in \mathcal{SK}$ as input. We simply refer to a PHF family with such augmentation as an augmented PHF family.

For an augmented PHF family, we have to slightly adapt the definition of the statistical/computational universal property from Definition 4. Specifically,

- for the definition of the ϵ-universal property, in addition to param, pp, $\mathsf{pk} \in \mathcal{PK}$, $c \in \Pi_{\mathsf{no}}$, and $\pi \in \mathcal{K}$, we also take the universal quantifier for all $\mathsf{ak} \in \mathcal{AK}$ for considering the probability in Eq. 1.
- for the definition of the computational universal property, we change the initial phase (Step 1) of the game to allow an adversary to choose $\mathsf{ak} \in \mathcal{AK}$ in the following way:
  1. First, the challenger executes $\mathsf{param} = (N, P, Q, T, g) \leftarrow \mathsf{GGen}(1^\lambda, s)$, and sends (N, T, g) to $\mathcal{A}$. $\mathcal{A}$ sends $\mathsf{ak} \in \mathcal{AK}$ to the challenger. The challenger then executes $\mathsf{pp} \leftarrow \mathsf{Setup}(\mathsf{param})$, chooses $\mathsf{sk} \xleftarrow{\mathsf{r}} \mathcal{SK}$, and computes $\mathsf{pk} \leftarrow \mu(\mathsf{ak}, \mathsf{sk})$. Then, the challenger sends $(\mathsf{pp}, \mathsf{pk})$ to $\mathcal{A}$.

  The remaining description of the game and the definition of the adversary's advantage are unchanged.

We note that the implication of the statistical universal property to the computational one, is also true for an augmented PHF family.

$\mathsf{Setup}_{\mathsf{poly}}(1^\lambda)$: $\mathsf{param} = (N, P, Q, T, g) \leftarrow \mathsf{GGen}(1^\lambda, s)$ $\mathsf{pp}_{\mathsf{phf}} \leftarrow \mathsf{Setup}_{\mathsf{phf}}(\mathsf{param})$ $(\mathsf{pp}_{\mathsf{skem}}, z, \widetilde{z}) \leftarrow \mathsf{Setup}_{\mathsf{skem}}(1^\lambda)$ $\mathsf{pp}_{\mathsf{cca}} \leftarrow \mathsf{Setup}_{\mathsf{cca}}(1^\lambda)$ $\mathsf{pp}_{\mathsf{poly}} \leftarrow (N, T, g, \mathsf{pp}_{\mathsf{phf}}, \mathsf{pp}_{\mathsf{skem}}, \mathsf{pp}_{\mathsf{cca}})$ Return $\mathsf{pp}_{\mathsf{poly}}$.	$\mathsf{KG}_{\mathsf{poly}}(\mathsf{pp}_{\mathsf{poly}})$: $(N, T, g, \mathsf{pp}_{\mathsf{phf}}, \mathsf{pp}_{\mathsf{skem}}, \mathsf{pp}_{\mathsf{cca}}) \leftarrow \mathsf{pp}_{\mathsf{poly}}$ $x \xleftarrow{r} [\frac{N-1}{4} \cdot \widetilde{z} \cdot 2^{\xi}]$ $(\mathsf{ct}, \mathsf{K}) \leftarrow \mathsf{Encap}(\mathsf{pp}_{\mathsf{skem}}, x)$ Parse K as $(r^{\mathsf{KG}}, \mathsf{psk}) \in \mathcal{R}^{\mathsf{KG}} \times \mathcal{SK}$. $h \leftarrow g^{2x} \bmod N^s$ $\mathsf{ppk} \leftarrow \mu(h, \mathsf{psk})$ //h is used as an aux. key $(\mathsf{cpk}, \mathsf{csk}) \leftarrow \mathsf{KG}_{\mathsf{cca}}(\mathsf{pp}_{\mathsf{cca}}; r^{\mathsf{KG}})$ Return $\mathsf{PK} := (h, \mathsf{ct}, \mathsf{ppk}, \mathsf{cpk})$ and $\mathsf{SK} := x$.
$\mathsf{Enc}_{\mathsf{poly}}(\mathsf{PK}, m \in \mathbb{Z}_{N^s})$: $(h, \mathsf{ct}, \mathsf{ppk}, \mathsf{cpk}) \leftarrow \mathsf{PK}$ $\forall i \in [d]$: $r_i \xleftarrow{r} [\frac{N-1}{4}]$; $y_i \leftarrow g^{r_i} \bmod N^s$ $u_d \leftarrow y_d$ $\forall i \in [d-1]$: $u_i \leftarrow y_i \cdot h^{r_{i+1}} \bmod N^s$ $r \leftarrow (2r_1, \ldots, 2r_d)$ $u \leftarrow (u_1^2 \bmod N^s, \ldots, u_d^2 \bmod N^s)$ $v \leftarrow T^m \cdot h^{r_1} \bmod N^s$ $\pi \leftarrow \mathsf{Pub}(\mathsf{ppk}, u, r)$ $\mathsf{CT} \leftarrow \mathsf{Enc}_{\mathsf{cca}}(\mathsf{cpk}, (\{u_i\}_{i\in[d]}, v, \pi))$ Return CT.	$\mathsf{Dec}_{\mathsf{poly}}(\mathsf{PK}, \mathsf{SK}, \mathsf{CT})$: $(h, \mathsf{ct}, \mathsf{ppk}, \mathsf{cpk}) \leftarrow \mathsf{PK}$; $x \leftarrow \mathsf{SK}$ $\mathsf{K} \leftarrow \mathsf{Decap}(\mathsf{pp}_{\mathsf{skem}}, x, \mathsf{ct})$ Parse K as $(r^{\mathsf{KG}}, \mathsf{psk}) \in \mathcal{R}^{\mathsf{KG}} \times \mathcal{SK}$. $(\mathsf{cpk}, \mathsf{csk}) \leftarrow \mathsf{KG}_{\mathsf{cca}}(\mathsf{pp}_{\mathsf{cca}}; r^{\mathsf{KG}})$ $(\{u_i\}_{i\in[d]}, v, \pi) \leftarrow \mathsf{Dec}_{\mathsf{cca}}(\mathsf{cpk}, \mathsf{csk}, \mathsf{CT})$ If $(\{u\}_{i\in[d]}, v) \notin \mathbb{J}_{N^s}^{d+1}$ then return $\perp$. $u \leftarrow (u_1^2 \bmod N^s, \ldots, u_d^2 \bmod N^s)$ If $\pi \neq \Lambda_{\mathsf{psk}}(u)$ then return $\perp$. $y_d \leftarrow u_d$ $\forall i \in [d-1]$: $y_i \leftarrow u_i \cdot (y_{i+1})^{-2x} \bmod N^s$ Return $m \leftarrow \log_T(v \cdot y_1^{-2x} \bmod N^s)$.

Fig. 6. The proposed KDM-CCA secure PKE scheme Π_{poly} with respect to polynomials. (The public parameter $\mathsf{pp}_{\mathsf{poly}}$ is omitted from the inputs to $\mathsf{Enc}_{\mathsf{poly}}$ and $\mathsf{Dec}_{\mathsf{poly}}$).

Cascaded ElGamal Language. Let $s \geq 2$, GGen be the DCR group generator, and $\mathsf{param} = (N, P, Q, T, g) \leftarrow \mathsf{GGen}(1^\lambda, s)$. Let $d = d(\lambda)$ be a polynomial. Let the auxiliary key space $\mathcal{AK}$ be defined as G_n, and let $\mathsf{ak} \in \mathcal{AK}$ (which will be a public key of the underlying cascaded ElGamal encryption scheme in our concrete instantiations of PHFs). The set of yes instances Π_{yes} is G_n^d, and the set of no instances is $(G_{N^{s-1}} \otimes G_n)^d \setminus G_n^d$. Any yes instance $c \in G_n^d$ can be expressed in the form $c = (c_1, \ldots, c_d)$ such that $c_d = g^{r_d} \bmod N^s$ and $c_i = g^{r_i} \cdot \mathsf{ak}^{r_{i+1}} \bmod N^s$ for every $i \in [d-1]$, where $r = (r_1, \ldots, r_d) \in \mathbb{Z}^d$. Thus, such r works as a witness for $c \in \Pi_{\mathsf{yes}}$ under $\mathsf{ak} \in \mathcal{AK}$.

The Proposed PKE Scheme. Let $s \geq 2$, and GGen be the DCR group generator. Let $d = d(\lambda)$ be a polynomial. Let $\Pi_{\mathsf{cca}} = (\mathsf{Setup}_{\mathsf{cca}}, \mathsf{KG}_{\mathsf{cca}}, \mathsf{Enc}_{\mathsf{cca}}, \mathsf{Dec}_{\mathsf{cca}})$ be a PKE scheme such that the randomness space of $\mathsf{KG}_{\mathsf{cca}}$ is $\mathcal{R}^{\mathsf{KG}}$. Let $\mathsf{PHF} = (\mathsf{Setup}_{\mathsf{phf}}, \Pi_{\mathsf{yes}}, \Pi_{\mathsf{no}}, \mathcal{SK}, \mathcal{PK}, \mathcal{K}, \mu, \Lambda, \mathsf{Pub})$ be an augmented PHF family with respect to GGen for the cascaded ElGamal language (defined as above). Let $\mathsf{SKEM} = (\mathsf{Setup}_{\mathsf{skem}}, \mathsf{Encap}, \mathsf{Decap})$ be an SKEM whose session-key space is $\mathcal{R}^{\mathsf{KG}} \times \mathcal{SK}$.[10] Finally, let $\xi = \xi(\lambda)$ be any polynomial such that $2^{-\xi} = \mathsf{negl}(\lambda)$. Our proposed PKE scheme $\Pi_{\mathsf{poly}} = (\mathsf{Setup}_{\mathsf{poly}}, \mathsf{KG}_{\mathsf{poly}}, \mathsf{Enc}_{\mathsf{poly}}, \mathsf{Dec}_{\mathsf{poly}})$ is constructed as described in Fig. 6. The plaintext space of Π_{poly} is $\mathbb{Z}_{N^{s-1}}$, where N is the RSA modulus generated in $\mathsf{Setup}_{\mathsf{poly}}$.

For the scheme Π_{poly}, the same remarks as those for Π_{aff} apply. Namely, the correctness and the security proof work for any $s \geq 2$, while to capture circular

[10] The same format adjustment as in Π_{aff} can be applied. See the footnote in Sect. 5.1.

security, we should use $s \geq 3$. Furthermore, if we use a statistically universal PHF family, the KDM-CCA security of Π_{poly} is tightly reduced to the DCR assumption and the security properties of the building blocks Π_{cca} and SKEM.

Π_{poly} is KDM-CCA secure with respect to the class of circuits $\mathcal{MAC}_d$, consisting of circuits satisfying the following conditions.

- Inputs are variables and constants of $\mathbb{Z}_{N^{s-1}}$.
- Gates are $+$, $-$, or $\cdot$ over $\mathbb{Z}_{N^{s-1}}$ and the number of gates is polynomial in λ.
- Each circuit in $\mathcal{MAC}_d$ computes a polynomial whose degree is at most d. For a circuit $C \in \mathcal{MAC}_d$, we denote the polynomial computing C by f_C.

The formal statement for the security of Π_{poly} is as follows. Its proof goes similarly to that of Theorem 1, and we provide it in the full version.

Theorem 2. *Assume that the DCR assumption holds with respect to* GGen, SKEM *is passively RKA secure,* PHF *is computationally universal, and* Π_{cca} *is IND-CCA secure. Then,* Π_{poly} *is* $\mathcal{MAC}_d$*-KDM-CCA secure.*

Specifically, for any polynomial $\ell = \ell(\lambda)$ *and PPT adversary* $\mathcal{A}$ *that attacks the* $\mathcal{MAC}_d$*-KDM-CCA security of* Π_{poly} *and makes* $q_{\mathsf{kdm}} = q_{\mathsf{kdm}}(\lambda)$ *KDM queries and* $q_{\mathsf{dec}} = q_{\mathsf{dec}}(\lambda)$ *decryption queries, there exist PPT adversaries* $\mathcal{B}_{\mathsf{dcr}}$, $\mathcal{B}_{\mathsf{rka}}$, $\mathcal{B}'_{\mathsf{rka}}$, $\mathcal{B}_{\mathsf{cca}}$, $\mathcal{B}'_{\mathsf{cca}}$, *and* $\mathcal{B}_{\mathsf{cu}}$ *such that*

$$\begin{aligned}\mathsf{Adv}^{\mathsf{kdmcca}}_{\Pi_{\mathsf{poly}},\mathcal{MAC}_d,\ell,\mathcal{A}}(\lambda) \leq 2 \cdot \Big(2 \cdot \mathsf{Adv}^{\mathsf{dcr}}_{s,\mathcal{B}_{\mathsf{dcr}}}(\lambda) + \mathsf{Adv}^{\mathsf{rka}}_{\mathsf{SKEM},\ell,\mathcal{B}_{\mathsf{rka}}}(\lambda) + \mathsf{Adv}^{\mathsf{rka}}_{\mathsf{SKEM},\ell,\mathcal{B}'_{\mathsf{rka}}}(\lambda) \\ + \mathsf{Adv}^{\mathsf{indcca}}_{\Pi_{\mathsf{cca}},\ell,\mathcal{B}_{\mathsf{cca}}}(\lambda) + \mathsf{Adv}^{\mathsf{indcca}}_{\Pi_{\mathsf{cca}},\ell,\mathcal{B}'_{\mathsf{cca}}}(\lambda) + \ell \cdot (q_{\mathsf{dec}} \cdot \mathsf{Adv}^{\mathsf{cu}}_{\mathsf{PHF},\mathcal{B}_{\mathsf{cu}}}(\lambda) + 2^{-\xi})\Big) \\ + O(d \cdot q_{\mathsf{kdm}} \cdot 2^{-\mathsf{len}}) + O(2^{-\lambda}).\end{aligned}$$

Instantiations of PHF Families. We propose two instantiations of an augmented PHF family used in Π_{poly}: The basic construction and its space-efficient variant, which are constructed similarly to those provided in Sects. 5.2 and 5.3, respectively. We provide the details in the full version.

The basic construction $\mathsf{PHF}_{\mathsf{poly}}$ is a simple extension of $\mathsf{PHF}_{\mathsf{aff}}$, so that they become identical in case $d = 1$. The output size of the function Λ in $\mathsf{PHF}_{\mathsf{poly}}$ consists of d elements of $\mathbb{Z}_{N^s}$, and its statistical universal property is shown very similarly to that for $\mathsf{PHF}_{\mathsf{aff}}$. The space-efficient construction $\mathsf{PHF}^{\mathsf{hash}}_{\mathsf{poly}}$ is the combination of $\mathsf{PHF}_{\mathsf{poly}}$ and a collision resistant hash function, and is identical to $\mathsf{PHF}^{\mathsf{hash}}_{\mathsf{aff}}$ in case $d = 1$. Although it is only computationally universal, the remarkable advantage of $\mathsf{PHF}^{\mathsf{hash}}_{\mathsf{poly}}$ is that its output size is independent of d.

7 Instantiations

We give some instantiation examples of $\mathcal{F}_{\mathsf{aff}}$-KDM-CCA secure PKE schemes and $\mathcal{F}_{\mathsf{poly}}$-KDM-CCA secure PKE schemes from our proposed schemes Π_{aff} in Sect. 5 and Π_{poly} in Sect. 6. These instantiations are summarized in Figs. 1 and 2 in Sect. 1.2. In all of the following instantiations, the plaintext space of the resulting schemes is $\mathbb{Z}_{N^{s-1}}$, where N is the RSA modulus generated in the setup algorithm and $s \geq 3$, and we assume that the underlying SKEM is instantiated with the one presented in Sect. 4.2.

The first instantiations are obtained by instantiating the underlying PHF family with the "space-efficient" PHF families ($\mathsf{PHF}_{\mathsf{aff}}^{\mathsf{hash}}$ for Π_{aff} and $\mathsf{PHF}_{\mathsf{poly}}^{\mathsf{hash}}$ for Π_{poly}), and the underlying IND-CCA secure PKE scheme with the scheme based on the factoring assumption proposed by Hofheinz and Kiltz [16]. The KDM-CCA security of the resulting PKE schemes is not tightly reduced to the DCR assumption, but a ciphertext of the $\mathcal{F}_{\mathsf{aff}}$-KDM-CCA secure scheme consists of only two elements of $\mathbb{Z}_{N^s}$, two elements of $\mathbb{Z}_{N'}$ (caused by the Hofheinz-Kiltz scheme), and a hash value output by a collision-resistant hash function, where N' is the RSA modulus generated in the Hofheinz-Kiltz scheme. Note that if $s \geq 3$, the size of two elements of $\mathbb{Z}_{N'}$ plus the size of a hash value is typically (much) smaller than one element of $\mathbb{Z}_{N^s}$! Furthermore, the improvement on the ciphertext size of $\mathcal{F}_{\mathsf{poly}}$-KDM-CCA secure scheme from the previous works is much more drastic. For KDM security with respect to degree-d polynomials, a ciphertext of our instantiation consists of $(d+1)$ elements of $\mathbb{Z}_{N^s}$, two elements of $\mathbb{Z}_{N'}$, and a hash value, and its size overhead compared to Malkin et al.'s scheme [22] is independent of d. In contrast, the ciphertext size of the previous best construction of Han et al. [11] is $O(d^9)$ elements of $\mathbb{Z}_{N^s}$ and more (and in addition its security relies on both the DCR and DDH assumptions).

The second instantiations are PKE schemes obtained by instantiating the underlying PHF family with the "basic" PHF families ($\mathsf{PHF}_{\mathsf{aff}}$ for Π_{aff} and $\mathsf{PHF}_{\mathsf{poly}}$ for Π_{poly}), and the underlying IND-CCA secure PKE scheme with the scheme proposed by Hofheinz [13]. Hofheinz' scheme is tightly IND-CCA secure under the DCR assumption, and its ciphertext overhead is 28 group elements plus the ciphertext overhead caused by authenticated encryption. The advantage of the second instantiations is that we obtain the first tightly $\mathcal{F}_{\mathsf{aff}}$-KDM-CCA secure PKE scheme and a tightly $\mathcal{F}_{\mathsf{poly}}$-KDM-CCA PKE scheme based solely on the DCR assumption. The disadvantage is the relatively large ciphertext size.

The third instantiations are obtained by replacing the underlying PKE scheme in the second ones with the PKE scheme proposed by Gay, Hofheinz, and Kohl [10]. Gay et al.'s scheme is tightly IND-CCA secure under the DDH assumption, and its ciphertext overhead is just three group elements of a DDH-hard group plus the ciphertext overhead caused by authenticated encryption. By the third instantiations, relying on both the DCR and DDH assumptions, we obtain a tightly $\mathcal{F}_{\mathsf{aff}}$-KDM-CCA secure PKE scheme whose ciphertext consists of essentially only three elements of $\mathbb{Z}_{N^s}$ and three elements of the DDH-hard group. We also obtain a tightly $\mathcal{F}_{\mathsf{poly}}$-KDM-CCA secure PKE scheme with much smaller ciphertexts than our second instantiation achieving the same security.

Acknowledgement. A part of this work was supported by NTT Secure Platform Laboratories, JST OPERA JPMJOP1612, JST CREST JPMJCR19F6 and JPMJCR14D6, and JSPS KAKENHI JP16H01705 and JP17H01695.

References

1. Applebaum, B., Cash, D., Peikert, C., Sahai, A.: Fast cryptographic primitives and circular-secure encryption based on hard learning problems. In: Halevi, S. (ed.) CRYPTO 2009. LNCS, vol. 5677, pp. 595–618. Springer, Heidelberg (2009). https://doi.org/10.1007/978-3-642-03356-8_35
2. Applebaum, B., Harnik, D., Ishai, Y.: Semantic security under related-key attacks and applications. In: ICS 2011, pp. 45–60 (2011)
3. Black, J., Rogaway, P., Shrimpton, T.: Encryption-scheme security in the presence of key-dependent messages. In: Nyberg, K., Heys, H. (eds.) SAC 2002. LNCS, vol. 2595, pp. 62–75. Springer, Heidelberg (2003). https://doi.org/10.1007/3-540-36492-7_6
4. Boneh, D., Halevi, S., Hamburg, M., Ostrovsky, R.: Circular-secure encryption from decision Diffie-Hellman. In: Wagner, D. (ed.) CRYPTO 2008. LNCS, vol. 5157, pp. 108–125. Springer, Heidelberg (2008). https://doi.org/10.1007/978-3-540-85174-5_7
5. Brakerski, Z., Goldwasser, S.: Circular and leakage resilient public-key encryption under subgroup indistinguishability (or: Quadratic residuosity strikes back). In: Rabin, T. (ed.) CRYPTO 2010. LNCS, vol. 6223, pp. 1–20. Springer, Heidelberg (2010). https://doi.org/10.1007/978-3-642-14623-7_1
6. Camenisch, J., Chandran, N., Shoup, V.: A public key encryption scheme secure against key dependent chosen plaintext and adaptive chosen ciphertext attacks. In: Joux, A. (ed.) EUROCRYPT 2009. LNCS, vol. 5479, pp. 351–368. Springer, Heidelberg (2009). https://doi.org/10.1007/978-3-642-01001-9_20
7. Camenisch, J., Lysyanskaya, A.: An efficient system for non-transferable anonymous credentials with optional anonymity revocation. In: Pfitzmann, B. (ed.) EUROCRYPT 2001. LNCS, vol. 2045, pp. 93–118. Springer, Heidelberg (2001). https://doi.org/10.1007/3-540-44987-6_7
8. Cramer, R., Shoup, V.: Universal hash proofs and a paradigm for adaptive chosen ciphertext secure public-key encryption. In: Knudsen, L.R. (ed.) EUROCRYPT 2002. LNCS, vol. 2332, pp. 45–64. Springer, Heidelberg (2002). https://doi.org/10.1007/3-540-46035-7_4
9. Damgård, I., Jurik, M.: A generalisation, a simplication and some applications of paillier's probabilistic public-key system. In: Kim, K. (ed.) PKC 2001. LNCS, vol. 1992, pp. 119–136. Springer, Heidelberg (2001). https://doi.org/10.1007/3-540-44586-2_9
10. Gay, R., Hofheinz, D., Kohl, L.: Kurosawa-Desmedt meets tight security. In: Katz, J., Shacham, H. (eds.) CRYPTO 2017. LNCS, vol. 10403, pp. 133–160. Springer, Cham (2017). https://doi.org/10.1007/978-3-319-63697-9_5
11. Han, S., Liu, S., Lyu, L.: Efficient KDM-CCA secure public-key encryption for polynomial functions. In: Cheon, J.H., Takagi, T. (eds.) ASIACRYPT 2016. LNCS, vol. 10032, pp. 307–338. Springer, Heidelberg (2016). https://doi.org/10.1007/978-3-662-53890-6_11
12. Hofheinz, D.: Circular chosen-ciphertext security with compact ciphertexts. In: Johansson, T., Nguyen, P.Q. (eds.) EUROCRYPT 2013. LNCS, vol. 7881, pp. 520–536. Springer, Heidelberg (2013). https://doi.org/10.1007/978-3-642-38348-9_31
13. Hofheinz, D.: Adaptive partitioning. In: Coron, J.-S., Nielsen, J.B. (eds.) EUROCRYPT 2017. LNCS, vol. 10212, pp. 489–518. Springer, Cham (2017). https://doi.org/10.1007/978-3-319-56617-7_17

14. Hofheinz, D., Jager, T.: Tightly secure signatures and public-key encryption. In: Safavi-Naini, R., Canetti, R. (eds.) CRYPTO 2012. LNCS, vol. 7417, pp. 590–607. Springer, Heidelberg (2012). https://doi.org/10.1007/978-3-642-32009-5_35
15. Hofheinz, D., Kiltz, E.: Secure hybrid encryption from weakened key encapsulation. In: Menezes, A. (ed.) CRYPTO 2007. LNCS, vol. 4622, pp. 553–571. Springer, Heidelberg (2007). https://doi.org/10.1007/978-3-540-74143-5_31
16. Hofheinz, D., Kiltz, E.: Practical chosen ciphertext secure encryption from factoring. In: Joux, A. (ed.) EUROCRYPT 2009. LNCS, vol. 5479, pp. 313–332. Springer, Heidelberg (2009). https://doi.org/10.1007/978-3-642-01001-9_18
17. Ishai, Y., Kilian, J., Nissim, K., Petrank, E.: Extending oblivious transfers efficiently. In: Boneh, D. (ed.) CRYPTO 2003. LNCS, vol. 2729, pp. 145–161. Springer, Heidelberg (2003). https://doi.org/10.1007/978-3-540-45146-4_9
18. Kitagawa, F., Tanaka, K.: A framework for achieving KDM-CCA secure public-key encryption. In: Peyrin, T., Galbraith, S. (eds.) ASIACRYPT 2018. LNCS, vol. 11273, pp. 127–157. Springer, Cham (2018). https://doi.org/10.1007/978-3-030-03329-3_5
19. Kurosawa, K., Desmedt, Y.: A new paradigm of hybrid encryption scheme. In: Franklin, M. (ed.) CRYPTO 2004. LNCS, vol. 3152, pp. 426–442. Springer, Heidelberg (2004). https://doi.org/10.1007/978-3-540-28628-8_26
20. Libert, B., Qian, C.: Lossy algebraic filters with short tags. In: Lin, D., Sako, K. (eds.) PKC 2019. LNCS, vol. 11442, pp. 34–65. Springer, Cham (2019). https://doi.org/10.1007/978-3-030-17253-4_2
21. Lu, X., Li, B., Jia, D.: KDM-CCA security from RKA secure authenticated encryption. In: Oswald, E., Fischlin, M. (eds.) EUROCRYPT 2015. LNCS, vol. 9056, pp. 559–583. Springer, Heidelberg (2015). https://doi.org/10.1007/978-3-662-46800-5_22
22. Malkin, T., Teranishi, I., Yung, M.: Efficient circuit-size independent public key encryption with KDM security. In: Paterson, K.G. (ed.) EUROCRYPT 2011. LNCS, vol. 6632, pp. 507–526. Springer, Heidelberg (2011). https://doi.org/10.1007/978-3-642-20465-4_28
23. Naor, M., Yung, M.: Public-key cryptosystems provably secure against chosen ciphertext attacks. In: 22nd ACM STOC 1990, pp. 427–437 (1990)
24. Paillier, P.: Public-key cryptosystems based on composite degree residuosity classes. In: Stern, J. (ed.) EUROCRYPT 1999. LNCS, vol. 1592, pp. 223–238. Springer, Heidelberg (1999). https://doi.org/10.1007/3-540-48910-X_16

Non-Committing Encryption with Quasi-Optimal Ciphertext-Rate Based on the DDH Problem

Yusuke Yoshida[1(✉)], Fuyuki Kitagawa[2], and Keisuke Tanaka[1]

[1] Tokyo Institute of Technology, Tokyo, Japan
yoshida.y.aw@m.titech.ac.jp, keisuke@is.titech.ac.jp
[2] NTT Secure Platform Laboratories, Tokyo, Japan
fuyuki.kitagawa.yh@hco.ntt.co.jp

Abstract. Non-committing encryption (NCE) was introduced by Canetti et al. (STOC '96). Informally, an encryption scheme is non-committing if it can generate a dummy ciphertext that is indistinguishable from a real one. The dummy ciphertext can be opened to any message later by producing a secret key and an encryption random coin which "explain" the ciphertext as an encryption of the message. Canetti et al. showed that NCE is a central tool to achieve multi-party computation protocols secure in the adaptive setting. An important measure of the efficiently of NCE is the ciphertext rate, that is the ciphertext length divided by the message length, and previous works studying NCE have focused on constructing NCE schemes with better ciphertext rates.

We propose an NCE scheme satisfying the ciphertext rate $\mathcal{O}(\log \lambda)$ based on the decisional Diffie-Hellman (DDH) problem, where λ is the security parameter. The proposed construction achieves the best ciphertext rate among existing constructions proposed in the plain model, that is, the model without using common reference strings. Previously to our work, an NCE scheme with the best ciphertext rate based on the DDH problem was the one proposed by Choi et al. (ASIACRYPT '09) that has ciphertext rate $\mathcal{O}(\lambda)$. Our construction of NCE is similar in spirit to that of the recent construction of the trapdoor function proposed by Garg and Hajiabadi (CRYPTO '18).

Keywords: Non-committing encryption · Decisional Diffie-Hellman problem · Chameleon encryption

1 Introduction

1.1 Background

Secure multi-party computation (MPC) allows a set of parties to compute a function of their inputs while maintaining the privacy of each party's input. Depending on when corrupted parties are determined, two types of adversarial settings called static and adaptive have been considered for MPC. In the static

S. D. Galbraith and S. Moriai (Eds.): ASIACRYPT 2019, LNCS 11923, pp. 128–158, 2019.
https://doi.org/10.1007/978-3-030-34618-8_5

setting, an adversary is required to declare which parties it corrupts before the protocol starts. On the other hand, in the adaptive setting, an adversary can choose which parties to corrupt on the fly, and thus the corruption pattern can depend on the messages exchanged during the protocol. Security guarantee in the adaptive setting is more desirable than that in the static setting since the former naturally captures adversarial behaviors in the real world while the latter is somewhat artificial.

In this work, we study *non-committing encryption (NCE)* which is introduced by Canetti, Feige, Goldreich, and Naor [4] and known as a central tool to achieve MPC protocols secure in the adaptive setting. NCE is an encryption scheme that has a special property called non-committing property. Informally, an encryption scheme is said to be non-committing if it can generate a dummy ciphertext that is indistinguishable from real ones, but can later be opened to any message by producing a secret key and an encryption random coin that "explain" the ciphertext as an encryption of the message. Cannetti et al. [4] showed how to create adaptively secure MPC protocols by instantiating the private channels in a statically secure MPC protocol with NCE.

Previous Constructions of NCE and their Ciphertext Rate. The ability to open a dummy ciphertext to any message is generally achieved at the price of efficiency. This is in contrast to ordinary public-key encryption for which we can easily obtain schemes the size of whose ciphertext is $n + \mathsf{poly}(\lambda)$ by using hybrid encryption methodology, where n is the length of an encrypted message and λ is the security parameter. The first NCE scheme proposed by Canetti et al. [4] only needs the optimal number of rounds (that is, two rounds), but it has ciphertexts of $O(\lambda^2)$-bits for every bit of an encrypted message. In other words, the ciphertext rate of their scheme is $O(\lambda^2)$, which is far from that of ordinary public-key encryption schemes. Subsequent works have focused on building NCE schemes with better efficiency.

Beaver [1] proposed a three-round NCE scheme with the ciphertext rate $\mathcal{O}(\lambda)$ based on the decisional Diffie-Hellman (DDH) problem. Damgård and Nielsen [8] generalized Beaver's scheme and achieved a three-round NCE scheme with ciphertext rate $O(\lambda)$ based on a primitive called simulatable PKE which in turn can be based on concrete problems such as the DDH, computational Diffie-Hellman (CDH), and learning with errors (LWE) problems. Choi, Dachman-Soled, Malkin, and Wee [7] further improved these results and constructed a two-round NCE scheme with ciphertext rate $\mathcal{O}(\lambda)$ based on a weaker variant of simulatable PKE called trapdoor simulatable PKE which can be constructed the factoring problem.

The first NCE scheme achieving a sub-linear ciphertext rate was proposed by Hemenway, Ostrovsky, and Rosen [20]. Their scheme needs only two rounds and achieves the ciphertext rate $\mathcal{O}(\log n)$ based on the ϕ-hiding problem which is related to (and generally believed to be easier than) the RSA problem, where n is the length of messages. Subsequently, Hemenway, Ostrovsky, Richelson, and Rosen [19] proposed a two-round NCE scheme with the ciphertext rate $\mathsf{poly}(\log \lambda)$ based on the LWE problem. Canetti, Poburinnaya, and Raykova [5]

Table 1. Comparison of existing NCE schemes. The security parameter is denoted by λ, and the message length n. Common-domain TDP can be instantiated based on the CDH and RSA problems. Simulatable and trapdoor simulatable PKE can be instantiated based on many computational problems realizing ordinary PKE. (*) This scheme uses common reference strings.

	Rounds	Ciphertext rate	Assumption
Canetti et al. [4]	2	$\mathcal{O}(\lambda^2)$	Common-domain TDP
Beaver [1]	3	$\mathcal{O}(\lambda)$	DDH
Damgård and Nielsen [8]	3	$\mathcal{O}(\lambda)$	Simulatable PKE
Choi et al. [7]	2	$\mathcal{O}(\lambda)$	Trapdoor simulatable PKE
Hemenway et al. [19]	2	$\mathsf{poly}(\log \lambda)$	LWE, Ring-LWE
Hemenway et al. [20]	2	$\mathcal{O}(\log n)$	Φ-hiding
Canetti et al. [5](*)	2	$1 + o(1)$	Indistinguishability obfuscation
This work	2	$\mathcal{O}(\log \lambda)$	DDH

showed that by using indistinguishability obfuscation, an NCE scheme with the asymptotically optimal ciphertext rate (that is, $1 + o(1)$) can be constructed. Their scheme needs only two rounds but was proposed in the common reference string model.

Despite the many previous efforts, as far as we know, we have only a single NCE scheme satisfying a sub-linear ciphertext rate based on widely and classically used problems, that is, the scheme proposed by Hemenway et al. [19] based on the LWE problem. Since NCE is an important cryptographic tool in constructing MPC protocols secure in the adaptive setting, it is desirable to have more constructions of NCE satisfying a better ciphertext rate.

1.2 Our Contribution

We propose an NCE scheme satisfying the ciphertext rate $\mathcal{O}(\log \lambda)$ based on the DDH problem. The proposed construction achieves the best ciphertext rate among existing constructions proposed in the plain model, that is, the model without using common reference strings. The proposed construction needs only two rounds, which is the optimal number of rounds for NCE. Previously to our work, an NCE scheme with the best ciphertext rate based on the DDH problem was the one proposed by Choi et al. [7] that satisfies the ciphertext rate $\mathcal{O}(\lambda)$. We summarize previous results on NCE and our result in Table 1.

We first show an NCE scheme that we call basic construction, which satisfies the ciphertext rate $\mathsf{poly}(\log \lambda)$. Then, we give our full construction satisfying the ciphertext rate $\mathcal{O}(\log \lambda)$ by extending the basic construction using error-correcting codes. Especially, in the full construction, we use a linear-rate error-correcting code which can correct errors of weight up to a certain constant proportion of the codeword length.

Our construction of NCE utilizes a variant of *chameleon encryption*. Chameleon encryption was originally introduced by Döttling and Garg [10] as an intermediate tool for constructing an identity-based encryption scheme based on the CDH problem. Roughly speaking, chameleon encryption is public-key encryption in which we can use a hash value of a chameleon hash function and its pre-image as a public key and a secret key, respectively. We show a variant of chameleon encryption satisfying *oblivious samplability* can be used to construct an NCE scheme with a sub-linear ciphertext rate. Informally, oblivious samplability of chameleon encryption requires that a scheme can generate a dummy hash key obliviously to the corresponding trapdoor, and sample a dummy ciphertext that is indistinguishable from a real one, without using any randomness except the dummy ciphertext itself.

Need for the DDH Assumption. A key and a ciphertext of the CDH based chameleon encryption proposed by Döttling and Garg [10] together form multiple Diffie-Hellman tuples. Thus, it seems difficult to sample them obliviously unless we prove that the knowledge of exponent assumption [2,18] is false. In order to solve this issue, we rely on the DDH assumption instead of the CDH assumption. Under the DDH assumption, a hash key and a ciphertext of our chameleon encryption are indistinguishable from independent random group elements, and thus we can perform oblivious sampling of them by sampling random group elements directly from the underlying group.

Public Key Size. As noted above, we first give the basic construction satisfying the ciphertext rate $\mathsf{poly}(\log \lambda)$, and then extend it to the full construction satisfying the ciphertext rate $\mathcal{O}(\log \lambda)$. In addition to satisfying only the ciphertext rate $\mathsf{poly}(\log \lambda)$, the basic construction also has a drawback that its public key size depends on the length of a message quadratically.

A public key of the basic construction contains ciphertexts of our obliviously samplable chameleon encryption. The size of those ciphertexts is quadratic in the length of an input to the associated chameleon hash function similarly to the construction by Döttling and Garg [10]. Since the input length of the chameleon hash function is linear in the message length of the basic construction, the public key size of the basic construction depends on the message length quadratically.

Fortunately, we can remove this quadratic dependence by a simple block-wise encryption technique. Thus, in the full construction, we utilize such a block-wise encryption technique in addition to the error-correcting code. By doing so, we reduce not only the ciphertext rate to $\mathcal{O}(\log \lambda)$, but also the public key size to linear in the length of a message as in the previous constructions of NCE.

Relation with Trapdoor Function by Garg and Hajiabadi [14]. There has been a line of remarkable results shown by using variants of chameleon encryption, starting from the one by Cho, Döttling, Garg, Gupta, Miao, and Polychroniadou [6]. This includes results on identity-based encryption [3,9–11], secure MPC [6,16], adaptive garbling schemes [15,17], and so on. Garg and Hajiabadi [14] showed how to realize trapdoor function (TDF) based on the CDH problem using a variant of chameleon encryption called one-way function with encryption.[1]

[1] Their technique is further extended by Garg, Gay, and Hajiabadi [13] and Döttling, Garg, Ishai, Malavolta, Mour, and Ostrovsky [12].

Our construction of NCE can be seen as an extension of that of TDF by Garg and Hajiabadi. Our formulation of chameleon encryption is based on that of one-way function with encryption. Concretely, we define chameleon encryption so that it has *recyclability* introduced by Garg and Hajiabadi as a key property in their work.

1.3 Paper Organization

Hereafter, in Sect. 2, we first review the definition of NCE. Then, in Sect. 3, we provide high-level ideas behind our construction of NCE. In Sect. 4, we formally define and construct obliviously samplable chameleon encryption. In Sect. 5, using obliviously samplable chameleon encryption, we construct an NCE scheme that we call the basic construction satisfying the ciphertext rate $\mathsf{poly}(\log \lambda)$. Finally, in Sect. 6, we improve the basic construction and provide the full construction that achieves the ciphertext rate $\mathcal{O}(\log \lambda)$.

2 Preliminaries

Let PPT denote probabilistic polynomial time. In this paper, λ always denotes the security parameter. For a finite set X, we denote the uniform sampling of x from X by $x \xleftarrow{\$} X$. $y \leftarrow \mathsf{A}(x;r)$ denotes that given an input x, a PPT algorithm A runs with internal randomness r, and outputs y. A function f is said to be negligible if $f(\lambda) = 2^{-\omega(\lambda)}$, and we write $f(\lambda) = \mathsf{negl}(\lambda)$ to denote that f is negligible. Let $\mathsf{Ham}(\mathsf{x})$ denotes the Hamming weight of $\mathsf{x} \in \{0,1\}^n$. $\mathbb{E}[X]$ denotes expected value of X. $[n]$ denotes $\{1,\ldots,n\}$.

Lemma 1 (Chernoff bound). *For a binomial random variable X. If $\mathbb{E}[X] \leq \mu$, then for all $\delta > 0$, $\Pr[X \geq (1+\delta)\mu)] \leq e^{-\frac{\delta^2}{2+\delta}\mu}$ holds.*

We provide the definition of the DDH assumption and its variants used in the proof of Theorem 1. We first introduce the leftover hash lemma.

Lemma 2 (Leftover hash lemma). *Let X and Y are sets. Let $\mathcal{H} := \{\mathsf{H} : X \to Y\}$ be a universal hash family. Then, the distributions $(\mathsf{H}, \mathsf{H}(x))$ and (H, y) are $\sqrt{\frac{|Y|}{4|X|}}$-close, where $\mathsf{H} \xleftarrow{\$} \mathcal{H}$, $x \xleftarrow{\$} X$, and $y \xleftarrow{\$} Y$.*

We review some computational assumptions. Below, we let $\mathbb{G}$ be a cyclic group of order p with a generator g. We also define the function $\mathsf{dh}(\cdot,\cdot)$ as $\mathsf{dh}(g^a, g^b) := g^{ab}$ for every $a, b \in \mathbb{Z}_p$. We start with the decisional Diffie-Hellman (DDH) assumption.

Definition 1 (Decisional Diffie-Hellman Assumption). *We say that the DDH assumption holds if for any PPT adversary $\mathcal{A}$,*

$$|\Pr[\mathcal{A}(g_1, g_2, \mathsf{dh}(g_1, g_2)) = 1] - \Pr[\mathcal{A}(g_1, g_2, g_3) = 1]| = \mathsf{negl}(\lambda)$$

holds, where $g_1, g_2, g_3 \xleftarrow{\$} \mathbb{G}$.

We introduce a lemma that is useful for the proof of oblivious samplability of our chameleon encryption. We can prove this lemma by using the self reducibility of the DDH problem.

Lemma 3. *Let n be a polynomial of λ. Let $g_{i,b} \xleftarrow{\$} \mathbb{G}$ for every $i \in [n]$ and $b \in \{0,1\}$. We set $M := (g_{i,b})_{i\in[n],b\in\{0,1\}} \in \mathbb{G}^{2\times n}$.*

Then, if the DDH assumption holds, for any PPT adversary $\mathcal{A}$, we have

$$|\Pr[\mathcal{A}(M, M^\rho) = 1] - \Pr[\mathcal{A}(M, R) = 1]| = \mathsf{negl}(\lambda),$$

where $M^\rho = (g_{i,b}^\rho)_{i\in[n],b\in\{0,1\}} \in \mathbb{G}^{2\times n}$ and $R \leftarrow \mathbb{G}^{2\times n}$.

We next define the hashed DDH assumption which is a variant of the DDH assumption.

Definition 2 (Hashed DDH Assumption). *Let $\mathcal{H} = \{\mathsf{H}_\mathbb{G} : \mathbb{G} \rightarrow \{0,1\}^\ell\}$ be a family of hash functions. We say that the hashed DDH assumption holds with respect to $\mathcal{H}$ if for any PPT adversary $\mathcal{A}$,*

$$|\Pr[\mathcal{A}(\mathsf{H}_\mathbb{G}, g_1, g_2, \mathbf{e}) = 1] - \Pr[\mathcal{A}(\mathsf{H}_\mathbb{G}, g_1, g_2, \mathbf{e}') = 1]| = \mathsf{negl}(\lambda)$$

holds, where $\mathsf{H}_\mathbb{G} \xleftarrow{\$} \mathcal{H}, g_1, g_2, \xleftarrow{\$} \mathbb{G}$, $\mathbf{e} = \mathsf{H}_\mathbb{G}(\mathsf{dh}(g_1, g_2))$, and $\mathbf{e}' \xleftarrow{\$} \{0,1\}^\ell$.

In this work, we use the hashed DDH assumption with respect to a hash family $\mathcal{H}$ whose output length ℓ is small enough such as $\ell = \mathsf{poly}(\log \lambda)$ or $\mathcal{O}(\log \lambda)$. In this case, by using a family of universal hash functions $\mathcal{H}$, we can reduce the hardness of the hashed DDH problem to that of the DDH problem by relying on the leftover hash lemma. Formally, we have the following lemma.

Lemma 4. *Let $\mathcal{H} = \{\mathsf{H}_\mathbb{G} : \mathbb{G} \rightarrow \{0,1\}^\ell\}$ be a family of universal hash functions, where $\ell = \mathsf{poly}(\log \lambda)$. Then, if the DDH assumption holds, the hashed DDH assumption with respect to $\mathcal{H}$ also holds by the leftover hash lemma.*

Non-Committing Encryption. A non-committing encryption (NCE) scheme is a public-key encryption scheme that has efficient simulator algorithms $(\mathsf{Sim}_1, \mathsf{Sim}_2)$ satisfying the following properties. The simulator Sim_1 can generate a simulated public key pk and a simulated ciphertext CT. Later Sim_2 can explain the ciphertext CT as encryption of any plaintext. Concretely, given a plaintext m, Sim_2 can output a pair of random coins for key generation r^{Gen} and encryption r^{Enc}, as if pk was generated by the key generation algorithm with the random coin r^{Gen}, and CT is encryption of m with the random coin r^{Enc}.

Some previous works proposed NCE schemes that are three-round protocols. In this work, we focus on NCE that needs only two rounds, which is also called non-committing public-key encryption, and we use the term NCE to indicate it unless stated otherwise. Below, we introduce the definition of NCE according to Hemenway et al. [19].

Definition 3 (Non-Committing Encryption). *A non-committing encryption scheme* NCE *consists of the following PPT algorithms* $(\mathsf{Gen}, \mathsf{Enc}, \mathsf{Dec}, \mathsf{Sim}_1, \mathsf{Sim}_2)$.

- $\mathsf{Gen}\left(1^\lambda; r^{\mathsf{Gen}}\right)$: *Given the security parameter* 1^λ, *using a random coin* r^{Gen}, *it outputs a public key pk and a secret key sk.*
- $\mathsf{Enc}\left(pk, m; r^{\mathsf{Enc}}\right)$: *Given a public key pk and a plaintext* $m \in \{0,1\}^\mu$, *using a random coin* r^{Enc}, *it outputs a ciphertext* CT.
- $\mathsf{Dec}\left(sk, CT\right)$: *Given a secret key sk and a ciphertext* CT, *it outputs* m *or* $\bot$.
- $\mathsf{Sim}_1\left(1^\lambda\right)$: *Given the security parameter* 1^λ, *it outputs a simulated public key pk, a simulated ciphertext* CT, *and an internal state st.*
- $\mathsf{Sim}_2\left(m, st\right)$: *Given a plaintext* m *and a state st, it outputs random coins for key generation* r^{Gen} *and encryption* r^{Enc}.

We require NCE *to satisfy the following correctness and security.*

Correctness. NCE *is called* γ*-correct, if for any plaintext* m,

$$\Pr[(pk, sk) \leftarrow \mathsf{Gen}\left(1^\lambda; r^{\mathsf{Gen}}\right), CT \leftarrow \mathsf{Enc}\left(pk, m; r^{\mathsf{Enc}}\right), \\ m' = \mathsf{Dec}\left(sk, CT\right); m = m'] \geq \gamma.$$

When $\gamma = 1 - \mathsf{negl}\left(\lambda\right)$, *we call it correct. Note that* γ *cannot be equal to* 1 *in the plain model (i.e., the model without using common reference strings).*

Security. *For any stateful PPT adversary* $\mathcal{A}$, *we define two experiments as follows.*

$\mathsf{Exp}^{Real}_{\mathtt{NCE},\mathcal{A}}$	$\mathsf{Exp}^{Ideal}_{\mathtt{NCE},\mathcal{A}}$
$(pk, sk) \leftarrow \mathsf{Gen}\left(1^\lambda; r^{\mathsf{Gen}}\right)$	$(pk, CT, st) \leftarrow \mathsf{Sim}_1\left(1^\lambda\right)$
$m \leftarrow \mathcal{A}\left(pk\right)$	$m \leftarrow \mathcal{A}\left(pk\right)$
$CT \leftarrow \mathsf{Enc}\left(pk, m; r^{\mathsf{Enc}}\right)$	$\left(r^{\mathsf{Gen}}, r^{\mathsf{Enc}}\right) \leftarrow \mathsf{Sim}_2\left(m, st\right)$
$\mathsf{out} \leftarrow \mathcal{A}\left(CT, r^{\mathsf{Gen}}, r^{\mathsf{Enc}}\right)$	$\mathsf{out} \leftarrow \mathcal{A}\left(CT, r^{\mathsf{Gen}}, r^{\mathsf{Enc}}\right)$

We say that NCE *is secure if*

$$\mathsf{Adv}_{\mathtt{NCE},\mathcal{A}}\left(\lambda\right) := \left|\Pr\left[\mathsf{out} = 1 \textit{ in } \mathsf{Exp}^{Real}_{\mathtt{NCE},\mathcal{A}}\right] - \Pr\left[\mathsf{out} = 1 \textit{ in } \mathsf{Exp}^{Ideal}_{\mathtt{NCE},\mathcal{A}}\right]\right| = \mathsf{negl}\left(\lambda\right)$$

holds for every PPT adversary $\mathcal{A}$.

3 Ideas of Our Construction

In this section, we provide high-level ideas behind our construction of NCE.

As a starting point, we review the three-round NCE protocol proposed by Beaver [1], which contains a fundamental idea to build NCE from the DDH problem. Next, we show how to extend it and construct a two-round NCE scheme whose ciphertext rate is $\mathcal{O}\left(\lambda\right)$. Then, we show how to reduce the ciphertext rate to $\mathcal{O}\left(\log \lambda\right)$, and obtain our main result. Finally, we state that our resulting construction can be described by using a variant of chameleon encryption, and it can be seen as an extension of trapdoor function proposed by Garg and Hajiabadi [14].

Sender		Receiver
Input: $m \in \{0,1\}$		
$\mathsf{z} \xleftarrow{\$} \{0,1\}$		$\mathsf{x} \xleftarrow{\$} \{0,1\}$
$\rho_\mathsf{z} \xleftarrow{\$} \mathbb{Z}_p$		$\alpha_\mathsf{x} \xleftarrow{\$} \mathbb{Z}_p$
$A_\mathsf{z} = g^{\rho_\mathsf{z}}$	(A_0, A_1)	$B_\mathsf{x} = g^{\alpha_\mathsf{x}}$
$A_{1-\mathsf{z}} \xleftarrow{\$} \mathbb{G}$	$\longrightarrow$	$B_{1-\mathsf{x}} \xleftarrow{\$} \mathbb{G}$
	$(\mathbf{e}_0, \mathbf{e}_1), (B_0, B_1)$	$\mathbf{e}_\mathsf{x} = {A_\mathsf{x}}^{\alpha_\mathsf{x}}$
	$\longleftarrow$	$\mathbf{e}_{1-\mathsf{x}} \xleftarrow{\$} \mathbb{G}$
if ${B_\mathsf{z}}^{\rho_\mathsf{z}} = \mathbf{e}_\mathsf{z}$	w	
$w := \mathsf{z} \oplus m$	$\longrightarrow$	if $w \neq \perp$
else $w := \perp$		Output: $m = w \oplus \mathsf{x}$

Fig. 1. The description of Beaver's protocol [1].

3.1 Starting Point: Beaver's Protocol

Beaver's NCE protocol essentiality executes two Diffie-Hellman key exchange protocols in parallel. This protocol can send a 1-bit message. The ciphertext rate is $\mathcal{O}(\lambda)$. We describe the protocol below and in Fig. 1.

Step1. Let $\mathbb{G}$ be a group of order p with a generator g. The sender picks a random bit $\mathsf{z} \xleftarrow{\$} \{0,1\}$ and an exponent $\rho_\mathsf{z} \xleftarrow{\$} \mathbb{Z}_p$, and then sets $A_\mathsf{z} = g^{\rho_\mathsf{z}}$. The sender also generates a random group element $A_{1-\mathsf{z}} \xleftarrow{\$} \mathbb{G}$ *obliviously*, i.e., without knowing the discrete log of $A_{1-\mathsf{z}}$. The sender sends (A_0, A_1) to the receiver and stores the secret $sk = (\mathsf{z}, \rho_\mathsf{z})$. The random coin used in this step is $(\mathsf{z}, \rho_\mathsf{z}, A_{1-\mathsf{z}})$.

Step2. The receiver picks a random bit $\mathsf{x} \xleftarrow{\$} \{0,1\}$ and an exponent $\alpha_\mathsf{x} \xleftarrow{\$} \mathbb{Z}_p$, and then sets $B_\mathsf{x} = g^{\alpha_\mathsf{x}}$. The receiver also obliviously generates $B_{1-\mathsf{x}} \xleftarrow{\$} \mathbb{G}$. Moreover, the receiver computes $\mathbf{e}_\mathsf{x} = {A_\mathsf{x}}^{\alpha_\mathsf{x}}$ and obliviously samples $\mathbf{e}_{1-\mathsf{x}} \xleftarrow{\$} \mathbb{G}$. The receiver sends $((B_0, B_1), (\mathbf{e}_0, \mathbf{e}_1))$ to the sender. The random coin used in this step is $(\mathsf{x}, \alpha_\mathsf{x}, B_{1-\mathsf{x}}, \mathbf{e}_{1-\mathsf{x}})$.

Step3. The sender checks whether $\mathsf{x} = \mathsf{z}$ holds or not, by checking if ${B_\mathsf{z}}^{\rho_\mathsf{z}} = \mathbf{e}_\mathsf{z}$ holds. With overwhelming probability, this equation holds if and only if $\mathsf{x} = \mathsf{z}$. If $\mathsf{x} = \mathsf{z}$, the sender sends $w := \mathsf{z} \oplus m$, and otherwise quits the protocol.

Step4. The receiver recovers the message by $w \oplus \mathsf{x}$.

We next describe the simulator for this protocol.

Simulator. The simulator simulates a transcript $(A_0, A_1), ((B_0, B_1), (\mathbf{e}_0, \mathbf{e}_1))$, and w as follows. It generates $\rho_0, \rho_1, \alpha_0, \alpha_1 \xleftarrow{\$} \mathbb{Z}_p$ and sets

$$((A_0, A_1), (B_0, B_1), (\mathbf{e}_0, \mathbf{e}_1)) = ((g^{\rho_0}, g^{\rho_1}), (g^{\alpha_0}, g^{\alpha_1}), (g^{\rho_0 \alpha_0}, g^{\rho_1 \alpha_1})).$$

The simulator also generates $w \xleftarrow{\$} \{0,1\}$.

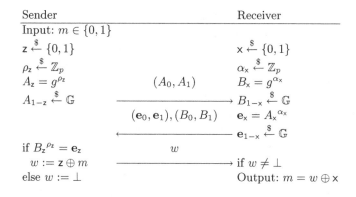

The simulator can later open this transcript to both messages 0 and 1. In other words, for both messages, the simulator can generate consistent sender and receiver random coins. For example, when opening it to $m = 0$, the simulator sets $\mathsf{x} = \mathsf{z} = w$, and outputs (w, ρ_w, A_{1-w}) and $(w, \alpha_w, B_{1-w}, \mathbf{e}_{1-w})$ as the sender's and receiver's opened random coins, respectively.

Security. Under the DDH assumption on $\mathbb{G}$, we can prove that any PPT adversary $\mathcal{A}$ cannot distinguish the pair of transcript and opened random coins generated in the real protocol from that generated by the simulator. The only difference of them is that $\mathbf{e}_{1-\mathsf{x}}$ is generated as a random group element in the real protocol, but it is generated as ${A_{1-\mathsf{x}}}^{\alpha_{1-\mathsf{x}}} = g^{\rho_{1-\mathsf{x}}\alpha_{1-\mathsf{x}}}$ in the simulation. When the real protocol proceeds to Step. 4, we have $\mathsf{x} = \mathsf{z}$ with overwhelming probability. Then, the random coins used by the sender and receiver (and thus given to $\mathcal{A}$) does not contain exponents of $A_{1-\mathsf{x}}$ and $B_{1-\mathsf{x}}$, that is, $\rho_{1-\mathsf{x}}$ and $\alpha_{1-\mathsf{x}}$. Thus, under the DDH assumption, $\mathcal{A}$ cannot distinguish randomly generated $\mathbf{e}_{1-\mathsf{x}} \stackrel{\$}{\leftarrow} \mathbb{G}$ from ${A_{1-\mathsf{x}}}^{\alpha_{1-\mathsf{x}}} = g^{\rho_{1-\mathsf{x}}\alpha_{1-\mathsf{x}}}$. Thus, this protocol is a secure NCE protocol.

This protocol succeeds in transmitting a message only when $\mathsf{z} = \mathsf{x}$, and otherwise it fails. Note that even when $\mathsf{z} \neq \mathsf{x}$, the protocol can transmit a message because in Step. 3, the sender knows the receiver's secret x. However, in that case, we cannot construct a successful simulator. In order to argue the security based on the DDH assumption, we have to ensure that either one pair of exponents (ρ_0, α_0) or (ρ_1, α_1) is not known to the adversary, but when $\mathsf{z} \neq \mathsf{x}$, we cannot ensure this.

Next, we show how to extend this protocol into a (two-round) NCE scheme and obtain an NCE scheme with the ciphertext rate $\mathcal{O}(\lambda)$.

3.2 Extension to Two-Round NCE Scheme

As a first attempt, we consider an NCE scheme $\mathtt{NCE}^1_{\mathtt{lin}}$ that is a natural extension of Beaver's three-round NCE protocol. Intuitively, $\mathtt{NCE}^1_{\mathtt{lin}}$ is Beaver's protocol in which the role of the sender and receiver is reversed, and the sender sends a message even when z and x are different. Specifically, the receiver generates the public key $pk = (A_0, A_1)$ and secret key $(\mathsf{z}, \rho_\mathsf{z})$, and the sender generates the ciphertext $CT = ((B_0, B_1), (\mathbf{e}_0, \mathbf{e}_1), w)$, where $(A_0, A_1), (B_0, B_1), (\mathbf{e}_0, \mathbf{e}_1)$, and $w := \mathsf{x} \oplus m$ are generated in the same way as those in Beaver's protocol. When decrypting the CT, the receiver first recovers the value of x by checking whether $B_\mathsf{z}^{\rho_\mathsf{z}} = \mathbf{e}_\mathsf{z}$ holds or not, and then computes $w \oplus \mathsf{x}$.

Of course, $\mathtt{NCE}^1_{\mathtt{lin}}$ is not a secure NCE scheme in the sense that we cannot construct a successful simulator when $\mathsf{z} \neq \mathsf{x}$ for a similar reason stated above. However, we can fix this problem and construct a secure NCE scheme by running multiple instances of $\mathtt{NCE}^1_{\mathtt{lin}}$.

In $\mathtt{NCE}^1_{\mathtt{lin}}$, if z coincides with x, we can construct a simulator similarly to Beaver's protocol, which happens with probability $\frac{1}{2}$. Thus, if we run multiple instances of it, we can construct simulators successfully for some fraction of them.

Receiver		Sender
		Input: $m \in \{0,1\}^\mu$
$\mathsf{z} \xleftarrow{\$} \{0,1\}^n$		$\mathsf{x} \xleftarrow{\$} \{0,1\}^n$
$\forall i \in [n], \rho_i \xleftarrow{\$} \mathbb{Z}_p$		$\forall i \in [n], \alpha_i \xleftarrow{\$} \mathbb{Z}_p$
$A_{i,\mathsf{z}_i} = g^{\rho_i}$		$B_{i,\mathsf{x}_i} = g^{\alpha_i}$
$A_{i,1-\mathsf{z}_i} \xleftarrow{\$} \mathbb{G}$	$\begin{pmatrix} A_{1,0},\ldots,A_{n,0} \\ A_{1,1},\ldots,A_{n,1} \end{pmatrix}$ $\longrightarrow$	$B_{i,1-\mathsf{x}_i} \xleftarrow{\$} \mathbb{G}$
		$\mathbf{e}_{i,\mathsf{x}_i} = A_{i,\mathsf{x}_i}{}^{\alpha_i}$
		$\mathbf{e}_{i,1-\mathsf{x}_i} \xleftarrow{\$} \mathbb{G}$
if $B_{i,\mathsf{z}_i}{}^{\rho_i} = \mathbf{e}_{i,\mathsf{z}_i}$	$\begin{pmatrix} B_{1,0},\ldots,B_{n,0} \\ B_{1,1},\ldots,B_{n,1} \end{pmatrix}, \begin{pmatrix} \mathbf{e}_{1,0},\ldots,\mathbf{e}_{n,0} \\ \mathbf{e}_{1,1},\ldots,\mathbf{e}_{n,1} \end{pmatrix}, w$	$w = H(\mathsf{x}) \oplus m$
$\mathsf{x}_i := \mathsf{z}_i$	$\longleftarrow$	
else $\mathsf{x}_i := 1 - \mathsf{z}_i$		
Output: $m = w \oplus H(\mathsf{x})$		

Fig. 2. The description of $\mathtt{NCE}_{\mathtt{lin}}$.

Based on this observation, we construct an NCE scheme $\mathtt{NCE}_{\mathtt{lin}}$ as follows. We also describe $\mathtt{NCE}_{\mathtt{lin}}$ in Fig. 2.

Let the length of messages be μ and $n = \mathcal{O}(\mu)$. We later specify the concrete relation of μ and n. The receiver first generates $\mathsf{z}_1 \cdots \mathsf{z}_n = \mathsf{z} \xleftarrow{\$} \{0,1\}^n$. Then, for every $i \in [n]$, the receiver generates a pubic key of $\mathtt{NCE}^1_{\mathtt{lin}}$, $(A_{i,0}, A_{i,1})$ in which the single bit randomness is z_i. We let the exponent of A_{i,z_i} be ρ_i, that is, $A_{i,\mathsf{z}_i} = g^{\rho_i}$. The receiver sends these n public keys of $\mathtt{NCE}^1_{\mathtt{lin}}$ as the public key of $\mathtt{NCE}_{\mathtt{lin}}$ to the sender. The secret key is $(\mathsf{z}, \rho_1, \ldots, \rho_n)$.

When encrypting a message m, the sender first generates $\mathsf{x}_1 \cdots \mathsf{x}_n = \mathsf{x} \xleftarrow{\$} \{0,1\}^n$. Then, for every $i \in [n]$, the sender generates $((B_{i,0}, B_{i,1}), (\mathbf{e}_{i,0}, \mathbf{e}_{i,1}))$ in the same way as $\mathtt{NCE}^1_{\mathtt{lin}}$ (and thus Beaver's protocol) "encapsulates" x_i by using the i-th public key $(A_{i,0}, A_{i,1})$. We call it i-th encapsulation. Finally, the sender generates $w = m \oplus H(x)$, where H is a hash function explained later in more detail.

The resulting ciphertext is

$$\left(\begin{pmatrix} B_{1,0},\ldots,B_{n,0} \\ B_{1,1},\ldots,B_{n,1} \end{pmatrix}, \begin{pmatrix} \mathbf{e}_{1,0},\ldots,\mathbf{e}_{n,0} \\ \mathbf{e}_{1,1},\ldots,\mathbf{e}_{n,1} \end{pmatrix}, w\right).$$

Decryption is done by recovering each x_i in the same way as $\mathtt{NCE}^1_{\mathtt{lin}}$ and computing $w \oplus H(\mathsf{x})$.

The simulator for this scheme runs as follows. It first generates $\mathsf{z}_1 \cdots \mathsf{z}_n = \mathsf{z} \xleftarrow{\$} \{0,1\}^n$ and $\mathsf{x}_1 \cdots \mathsf{x}_n = \mathsf{x} \xleftarrow{\$} \{0,1\}^n$. Then, for every index $i \in [n]$ such that $\mathsf{z}_i = \mathsf{x}_i$, it simulates the i-th public key and encapsulation in the same way as the simulator for $\mathtt{NCE}^1_{\mathtt{lin}}$ (and thus Beaver's protocol). For every index $i \in [n]$ such that $\mathsf{z}_i \neq \mathsf{x}_i$, it simply generates i-th public key and encapsulation in the same way as $\mathtt{NCE}_{\mathtt{lin}}$ does in the real execution. Finally, it generates $w \xleftarrow{\$} \{0,1\}^\mu$.

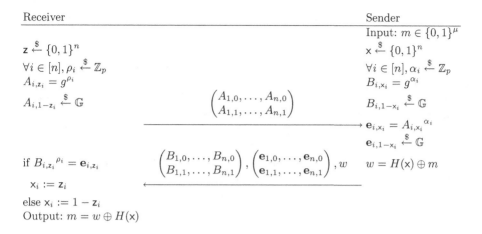

Although the ciphertext generated by the simulator is not "fully non-committing" about x, it loses the information of bits of x such that $\mathsf{x}_i = \mathsf{z}_i$. Thus, if we can program the output value of the hash function H freely by programming only those bits of x, the simulator can later open the ciphertext to any message, and we see that $\mathtt{NCE_{lin}}$ is a secure NCE scheme.

To realize this idea, we first set $n = 8\mu$ in order to ensure that the simulated ciphertext loses the information of at least μ-bits of x with overwhelming probability. This is guaranteed by the Chernoff bound. Moreover, as the hash function H, we use a matrix $R \in \{0,1\}^{\mu \times n}$, such that randomly picked μ out of n column vectors of length μ are linearly independent. The ciphertext rate of $\mathtt{NCE_{lin}}$ is $\mathcal{O}(\lambda)$, that is already the same as the best rate based on the DDH problem achieved by the construction of Choi et al. [7].

3.3 Reduce the Ciphertext Rate

Finally, we show how to achieve the ciphertext rate $\mathcal{O}(\log \lambda)$ by compressing the ciphertext of $\mathtt{NCE_{lin}}$. This is done by two steps. In the first step, we reduce the size of the first part of a ciphertext of $\mathtt{NCE_{lin}}$, that is, $\{B_{i,b}\}_{i\in[n],b\in\{0,1\}}$. By this step, we compress it into just a single group element. Then, in the second step, we reduce the size of the second part of a ciphertext of $\mathtt{NCE_{lin}}$, that is, $\{\mathbf{e}_{i,b}\}_{i\in[n],b\in\{0,1\}}$. In this step, we compress each $\mathbf{e}_{i,b}$ into a $\mathcal{O}(\log \lambda)$-bit string. By applying these two steps, we can achieve the ciphertext rate $\mathcal{O}(\log \lambda)$.

The second step is done by replacing each group element $\mathbf{e}_{i,b}$ with a hash value of it. In $\mathtt{NCE_{lin}}$, they are used to recover the value of x_i by checking $B_{i,\mathsf{z}_i}^{\rho_i} = \mathbf{e}_{i,\mathsf{z}_i}$. We can successfully perform this recovery process with overwhelming probability even if $\mathbf{e}_{i,b}$ is hashed to a $\mathsf{poly}(\log \lambda)$-bit string. Furthermore, with the help of an *error-correcting code*, we can reduce the length of the hash value to $\mathcal{O}(\log \lambda)$-bit.

In the remaining part, we explain how to perform the first step.

Compressing a Matrix of Group Elements into a Single Group Element. We realize that we do not need all of the elements $\{B_{i,b}\}_{i\in[n],b\in\{0,1\}}$ to decrypt the ciphertext. Although the receiver gets both $B_{i,0}$ and $B_{i,1}$ for every $i \in [n]$, the receiver uses only B_{i,z_i}. Recall that the receiver recovers the value of x_i by checking whether $B_{i,\mathsf{z}_i}^{\rho_i} = \mathbf{e}_{i,\mathsf{z}_i}$ holds. This recovery of x_i can be done even if the sender sends only B_{i,x_i}, and not $B_{i,1-\mathsf{x}_i}$.

This is because, similarly to the equation $B_{i,\mathsf{z}_i}^{\rho_i} = \mathbf{e}_{i,\mathsf{z}_i}$, with overwhelming probability, the equation $B_{i,\mathsf{x}_i}^{\rho_i} = \mathbf{e}_{i,\mathsf{z}_i}$ holds if and only if $\mathsf{z}_i = \mathsf{x}_i$. For this reason, we can compress the first part of the ciphertext on $\mathtt{NCE_{lin}}$ into $(B_{1,\mathsf{x}_1}, \ldots, B_{n,\mathsf{x}_n})$.

We further compress $(B_{1,\mathsf{x}_1}, \ldots, B_{n,\mathsf{x}_n})$ into a single group element generated by multiplying them, that is, $\mathsf{y} = \prod_{j\in[n]} B_{j,\mathsf{x}_j}$. In order to do so, we modify the scheme so that the receiver can recover x_i for every $i \in [n]$ using y instead of B_{i,x_i}. Concretely, for every $i \in [n]$, the sender computes $\mathbf{e}_{i,\mathsf{x}_i}$ as

$$\mathbf{e}_{i,\mathsf{x}_i} = \prod_{j\in[n]} A_{i,\mathsf{x}_i}^{\alpha_j},$$

where α_j is the exponent of B_{j,x_j} for every $j \in [n]$ generated by the sender. The sender still generates $\mathbf{e}_{i,1-\mathsf{x}_i}$ as a random group element for every $i \in [n]$. In this case, with overwhelming probability, the receiver can recover x_i by checking whether $\mathbf{e}_{i,\mathsf{z}_i} = \mathsf{y}^{\rho_i}$ holds.

However, unfortunately, it seems difficult to prove the security of this construction. In order to delete the information of x_i for indices $i \in [n]$ such that $\mathsf{z}_i = \mathsf{x}_i$ as in the proof of $\mathtt{NCE_{lin}}$, we have to change the distribution of $\mathbf{e}_{i,1-\mathsf{x}_i}$ from a random group element to $\prod_{j\in[n]} A_{i,1-\mathsf{x}_i}^{\alpha_j}$ so that $\mathbf{e}_{i,0}$ and $\mathbf{e}_{i,1}$ are symmetrically generated. However, we cannot make this change by relying on the DDH assumption since all α_j are given to the adversary as a part of the sender random coin. Thus, in order to solve this problem, we further modify the scheme and construct an NCE scheme $\mathtt{NCE}$ as follows.

The Resulting NCE Scheme $\mathtt{NCE}$. In $\mathtt{NCE}$, the receiver first generates $\mathsf{z} \xleftarrow{\$} \{0,1\}^n$ and $\{A_{i,b}\}_{i\in[n],b\in\{0,1\}}$ in the same way as $\mathtt{NCE_{lin}}$. Moreover, instead of the sender, the receiver *obliviously* generates $B_{i,b} = g^{\alpha_{i,b}}$ for every $i \in [n]$ and $b \in \{0,1\}$, and adds them into the public key. Moreover, for every $i \in [n]$, the receiver adds

$$\{B_{j,b}^{\rho_i} = A_{i,\mathsf{z}_i}^{\alpha_{j,b}}\}_{j\in[n],b\in\{0,1\} \text{ s.t. } (j,b)\neq(i,1-\mathsf{z}_i)}$$

to the public key. In order to avoid the leakage of the information of z from the public key, for every $i \in [n]$, we have to add

$$\{A_{i,1-\mathsf{z}_i}^{\alpha_{j,b}}\}_{j\in[n],b\in\{0,1\} \text{ s.t. } (j,b)\neq(i,\mathsf{z}_i)}$$

to the public key. However, the receiver cannot do it since the receiver generates $A_{i,1-\mathsf{z}_i}$ obliviously. Thus, instead, the receiver adds the same number of random group elements into the public key. At the beginning of the security proof, we can replace them with $\{A_{i,1-\mathsf{z}_i}^{\alpha_{j,b}}\}_{j\in[n],b\in\{0,1\} \text{ s.t. } (j,b)\neq(i,\mathsf{z}_i)}$ by relying on the DDH assumption, and eliminate the information of z from the public key. For simplicity, below, we suppose that the public key includes $\{A_{i,1-\mathsf{z}_i}^{\alpha_{j,b}}\}_{j\in[n],b\in\{0,1\} \text{ s.t. } (j,b)\neq(i,\mathsf{z}_i)}$ instead of random group elements.

When encrypting a message m by $\mathtt{NCE}$, the sender first generates $\mathsf{x} \xleftarrow{\$} \{0,1\}^n$ and computes $\mathsf{y} = \prod_{j\in[n]} B_{j,\mathsf{x}_j}$. Then, for every $i \in [n]$, the sender computes $\mathbf{e}_{i,\mathsf{x}_i}$ as

$$\mathbf{e}_{i,\mathsf{x}_i} = \prod_{j\in[n]} A_{i,\mathsf{x}_i}^{\alpha_{j,\mathsf{x}_j}} = \mathsf{y}^{\rho_i}$$

just multiplying $A_{i,\mathsf{x}_i}^{\alpha_{1,\mathsf{x}_1}}, \ldots, A_{i,\mathsf{x}_i}^{\alpha_{n,\mathsf{x}_n}}$ included in the pubic key. Recall that $A_{i,\mathsf{x}_i} = g^{\rho_i}$. Note that $A_{i,\mathsf{z}_i}^{\alpha_{i,1-\mathsf{z}_i}}$ is not included in the public key, but we do not need it to compute $\mathbf{e}_{i,\mathsf{x}_i}$. The sender generates $\mathbf{e}_{i,\mathsf{x}_i}$ as a random group element for every $i \in [n]$ as before. The resulting ciphertext is

$$\left(\mathsf{y}, \begin{pmatrix} \mathbf{e}_{1,0}, \ldots, \mathbf{e}_{n,0} \\ \mathbf{e}_{1,1}, \ldots, \mathbf{e}_{n,1} \end{pmatrix}, R\mathsf{x} \oplus m\right).$$

The receiver can recover x_i by checking whether $\mathbf{e}_{i,\mathsf{z}_i} = \mathsf{y}^{\rho_i}$ holds, and decrypt the ciphertext.

By defining the simulator appropriately, the security proof of NCE proceeds in a similar way to that of $\mathtt{NCE_{lin}}$. In NCE, for indices $i \in [n]$ such that $\mathsf{z}_i = \mathsf{x}_i$, we can eliminate the information of x_i. We can change $\mathbf{e}_{i,1-\mathsf{x}_i}$ from a random group element to $\prod_{j\in[n]} A_{i,1-\mathsf{x}_i}^{\alpha_{j,\mathsf{x}_j}}$ by relying on the fact that $A_{i,1-\mathsf{x}_i}^{\alpha_{i,\mathsf{x}_i}}$ is indistinguishable from a random group element by the DDH assumption. By this change, $\mathbf{e}_{i,0}$ and $\mathbf{e}_{i,1}$ become symmetric and the ciphertext loses the information of x_i. Then, the remaining part of the proof goes through in a similar way as that of $\mathtt{NCE_{lin}}$ except the following point. In NCE, the first component of the ciphertext, that is, $\mathsf{y} = \prod_{j\in[n]} B_{j,\mathsf{x}_j}$ has the information of x. In order to deal with the issue, in our real construction, we replace y with $g^{\mathsf{r}} \prod_{j\in[n]} B_{j,\mathsf{x}_j}$, where $\mathsf{r} \xleftarrow{\$} \mathbb{Z}_p$. Then, y no longer leaks any information of x. Moreover, after y is fixed, for any $\mathsf{x}' \in \{0,1\}^n$, we can efficiently find r' such that $\mathsf{y} = g^{\mathsf{r}'} \prod_{j\in[n]} B_{j,\mathsf{x}'_j}$. This is important to ensure that the simulator of NCE runs in polynomial time.

3.4 Abstraction by Chameleon Encryption

We can describe NCE by using obliviously samplable chameleon encryption. If we consider $\{B_{i,b}\}_{i\in[n],b\in\{0,1\}}$ as a hash key k of chameleon hash function, the first element of the ciphertext $g^{\mathsf{r}} \prod_{j\in[n]} B_{j,\mathsf{x}_j}$ can be seen as the output of the hash $\mathsf{H}(\mathsf{k}, \mathsf{x}; \mathsf{r})$. Moreover, group elements contained in the public key are considered as ciphertexts of an chameleon encryption scheme. Oblivious samplability of chameleon encryption makes it possible to deal with the above stated issue of sampling random group elements instead of $\{A_{i,1-\mathsf{z}_i}^{\alpha_{j,b}}\}_{j\in[n],b\in\{0,1\}\ s.t.\ (j,b)\neq(i,\mathsf{z}_i)}$ for every $i \in [n]$.

Relation with Trapdoor Function of Garg and Hajiabadi. We finally remark that the construction of NCE can be seen as an extension of that of trapdoor function (TDF) proposed by Garg and Hajiabadi [14].

If we do not add the random mask g^{r} to $\mathsf{y} = \prod_{j\in[n]} B_{j,\mathsf{x}_j}$, the key encapsulation part of a ciphertext of NCE, that is,

$$\left(\mathsf{y}, \begin{pmatrix} \mathbf{e}_{1,0}, \ldots, \mathbf{e}_{n,0} \\ \mathbf{e}_{1,1}, \ldots, \mathbf{e}_{n,1} \end{pmatrix}\right)$$

is the same as an output of the TDF constructed by Garg and Hajiabadi. The major difference between our NCE scheme and their TDF is the secret key. A secret key of their TDF contains all discrete logs of $\{A_{i,b}\}_{i\in[n],b\in\{0,1\}}$, that is, $\{\rho_{i,b}\}_{i\in[n],b\in\{0,1\}}$. On the other hand, a secret key of our NCE scheme contains half of them corresponding to the bit representation of z, that is, $\{\rho_{i,\mathsf{z}_i}\}_{i\in[n]}$. Garg and Hajiabadi already stated that their TDF can be inverted with $\{\rho_{i,\mathsf{z}_i}\}_{i\in[n]}$ for any $\mathsf{z} \in \{0,1\}^n$, and use this fact in the security proof of a chosen ciphertext security of a public-key encryption scheme based on their TDF. By explicitly using this technique in the construction, we achieve non-committing property.

We observe that construction techniques for TDF seem to be useful for achieving NCE. Encryption schemes that can recover an encryption random coin with a message in the decryption process, such as those based on TDFs, is said to be randomness recoverable. For randomness recoverable schemes, receiver non-committing property is sufficient to achieve full (that is, both sender and receiver) non-committing property. This is because an encryption random coin can be recovered from a ciphertext by using a key generation random coin.

4 Obliviously Samplable Chameleon Encryption

Chameleon encryption was originally introduced by Döttling and Garg [10]. In this work, we introduce a variant of chameleon encryption satisfying *oblivious samplability.*

4.1 Definiton

We start with the definition of the chameleon hash function.

Definition 4 (Chameleon Hash Function). *A chameleon hash function consists of the following PPT algorithms* $(\mathsf{K}, \mathsf{H}, \mathsf{H}^{-1})$. *Below, we let the input space and randomness space of* H *be* $\{0,1\}^n$ *and* $\mathcal{R}_{\mathsf{H}}$, *respectively, where* $n = O(\lambda)$.

- $\mathsf{K}(1^\lambda)$: *Given the security parameter* 1^λ, *it outputs a hash key* k *and a trapdoor* t.
- $\mathsf{H}(\mathsf{k}, \mathsf{x}; \mathsf{r})$: *Given a hash key* k *and input* $\mathsf{x} \in \{0,1\}^n$, *using randomness* $\mathsf{r} \in \mathcal{R}_{\mathsf{H}}$, *it outputs a hash value* y.
- $\mathsf{H}^{-1}(\mathsf{t}, (\mathsf{x}, \mathsf{r}), \mathsf{x}')$: *Given a trapdoor* t, *an input to the hash* x, *randomness for the hash* r *and another input to the hash* x', *it outputs randomness* r'.

A chameleon hash function is required to satisfy the following trapdoor collision property.[2]

Trapdoor Collision. *For all* $\mathsf{x}, \mathsf{x}' \in \{0,1\}^n$ *and hash randomness* $\mathsf{r} \in \mathcal{R}_{\mathsf{H}}$, $\mathsf{H}(\mathsf{k}, \mathsf{x}; \mathsf{r}) = \mathsf{H}(\mathsf{k}, \mathsf{x}'; \mathsf{r}')$ *holds, where* $(\mathsf{k}, \mathsf{t}) \leftarrow \mathsf{K}(1^\lambda)$, $\mathsf{r}' \leftarrow \mathsf{H}^{-1}(\mathsf{t}, (\mathsf{x}, \mathsf{r}), \mathsf{x}')$. *Moreover, if* r *is sampled uniformly at random, then so is* r'.

Next, we define the chameleon encryption.

Definition 5 (Chameleon Encryption). *Chameleon encryption (CE) consists of a chameleon hash function* $(\mathsf{K}, \mathsf{H}, \mathsf{H}^{-1})$ *and the following PPT algorithms* $(\mathsf{E}_1, \mathsf{E}_2, \mathsf{D})$. *Below, we let the input space and randomness space of* H *are* $\{0,1\}^n$ *and* $\mathcal{R}_{\mathsf{H}}$, *respectively, where* $n = O(\lambda)$. *We also let the randomness space of* E_1 *and* E_2 *be* $\mathcal{R}_{\mathsf{E}}$. *Moreover, we let the output space of* E_2 *be* $\{0,1\}^\ell$, *where* ℓ *be a polynomial of* λ.

[2] Usually, a chameleon hash function is required to be collision resistant, but we omit it since it is implied by the security of chameleon encryption defined later.

- $\mathsf{E}_1(\mathsf{k}, (i, b); \rho)$: *Given a hash key* k *and index* $i \in [n]$ *and* $b \in \{0,1\}$, *using a random coin* $\rho \in \mathcal{R}_\mathsf{E}$, *it outputs a ciphertext* ct.
- $\mathsf{E}_2(\mathsf{k}, \mathsf{y}; \rho)$: *Given a hash key* k *and a hash value* y, *using a random coin* $\rho \in \mathcal{R}_\mathsf{E}$, *it outputs* $\mathbf{e} \in \{0,1\}^\ell$.
- $\mathsf{D}(\mathsf{k}, (\mathsf{x}, \mathsf{r}), \mathsf{ct})$: *Given a hash key* k, *a pre-image of the hash* (x, r) *and a ciphertext* ct, *it outputs* $\mathbf{e} \in \{0,1\}^\ell$ *or* $\perp$.

Chameleon encryption must satisfy the following correctness and security.

Correctness. *For all* k *output by* $\mathsf{K}(1^\lambda)$, $i \in [n]$, $\mathsf{x} \in \{0,1\}^n$, $\mathsf{r} \in \mathcal{R}_\mathsf{H}$, *and* $\rho \in \mathcal{R}_\mathsf{E}$, $\mathsf{E}_2(\mathsf{k}, \mathsf{y}; \rho) = \mathsf{D}(\mathsf{k}, (\mathsf{x}, \mathsf{r}), \mathsf{ct})$ *holds, where* $\mathsf{y} \leftarrow \mathsf{H}(\mathsf{k}, \mathsf{x}; \mathsf{r})$ *and* $\mathsf{ct} \leftarrow \mathsf{E}_1(\mathsf{k}, (i, \mathsf{x}_i); \rho)$.

Security. *For any stateful PPT adversary* $\mathcal{A}$, *we define the following experiments.*

$\mathsf{Exp}^0_{\mathsf{CE},\mathcal{A}}$	$\mathsf{Exp}^1_{\mathsf{CE},\mathcal{A}}$
$(\mathsf{x}, \mathsf{r}, i) \leftarrow \mathcal{A}(1^\lambda)$	$(\mathsf{x}, \mathsf{r}, i) \leftarrow \mathcal{A}(1^\lambda)$
$(\mathsf{k}, \mathsf{t}) \leftarrow \mathsf{K}(1^\lambda)$	$(\mathsf{k}, \mathsf{t}) \leftarrow \mathsf{K}(1^\lambda)$
$\mathsf{ct} \leftarrow \mathsf{E}_1(\mathsf{k}, (i, 1 - \mathsf{x}_i); \rho)$	$\mathsf{ct} \leftarrow \mathsf{E}_1(\mathsf{k}, (i, 1 - \mathsf{x}_i); \rho)$
$\mathbf{e} \leftarrow \mathsf{E}_2(\mathsf{k}, \mathsf{H}(\mathsf{k}, \mathsf{x}; \mathsf{r}); \rho)$	$\mathbf{e} \xleftarrow{\$} \{0,1\}^\ell$
$\mathsf{out} \leftarrow \mathcal{A}(\mathsf{k}, \mathsf{ct}, \mathbf{e})$	$\mathsf{out} \leftarrow \mathcal{A}(\mathsf{k}, \mathsf{ct}, \mathbf{e})$

We say CE *is secure if*

$$\begin{aligned}\mathsf{Adv}_{\mathsf{CE},\mathcal{A}}(\lambda) := \Big|&\Pr\left[\mathsf{out} = 1 \text{ in } \mathsf{Exp}^0_{\mathsf{CE},\mathcal{A}}\right] \\ &- \Pr\left[\mathsf{out} = 1 \text{ in } \mathsf{Exp}^1_{\mathsf{CE},\mathcal{A}}\right]\Big| = \mathsf{negl}(\lambda)\end{aligned}$$

holds for every PPT adversary $\mathcal{A}$.

Remark 1 (On the recyclability). The above definition of chameleon encryption is slightly different from that of Döttling and Garg [10] since we define it so that it satisfies a property called recyclability introduced by Garg and Hajiabadi [14] when defining a primitive called one-way function with encryption that is similar to chameleon encryption.

More specifically, in our definition, there are two encryption algorithms E_1 and E_2. E_1 outputs only a key encapsulation part and E_2 outputs only a session key part. In the original definition by Döttling and Garg, there is a single encryption algorithm that outputs the key encapsulation part and a message masked by the session key part at once. Importantly, an output of E_1 does not depend on a hash value y. This makes possible to relate a single output of E_1 with multiple hash values. (In other words, a single output of E_1 can be recycled for multiple hash values.) We need this property in the construction of NCE and thus adopt the above definition.

We then introduce our main tool, that is, obliviously samplable chameleon encryption (obliviously samplable CE).

Definition 6 (Obliviously Samplable Chameleon Encryption). *Let* $\mathsf{CE} = (\mathsf{K}, \mathsf{H}, \mathsf{H}^{-1}, \mathsf{E}_1, \mathsf{E}_2, \mathsf{D})$ *be a chameleon encryption scheme. We define two associated PPT algorithms* $\widehat{\mathsf{K}}$ *and* $\widehat{\mathsf{E}_1}$ *as follows.*

- $\widehat{\mathsf{K}}(1^\lambda)$: *Given the security parameter* 1^λ, *it outputs only a hash key* $\widehat{\mathsf{k}}$ *without using any randomness other than* $\widehat{\mathsf{k}}$ *itself.*
- $\widehat{\mathsf{E}_1}\left(\widehat{\mathsf{k}}, (i, b)\right)$: *Given a hash key* $\widehat{\mathsf{k}}$ *and index* $i \in [n]$ *and* $b \in \{0, 1\}$, *it outputs a ciphertext* $\widehat{\mathsf{ct}}$ *without using any randomness except* $\widehat{\mathsf{ct}}$ *itself.*

For any PPT adversary $\mathcal{A}$, *we also define the following experiments.*

$\mathsf{Exp}^{os\text{-}0}_{\mathsf{CE},\mathcal{A}}$	$\mathsf{Exp}^{os\text{-}1}_{\mathsf{CE},\mathcal{A}}$
$(\mathsf{k}, \mathsf{t}) \leftarrow \mathsf{K}(1^\lambda)$	$\widehat{\mathsf{k}} \leftarrow \widehat{\mathsf{K}}(1^\lambda)$
$\mathsf{out} \leftarrow \mathcal{A}^{O(\cdot,\cdot)}(\mathsf{k})$	$\mathsf{out} \leftarrow \mathcal{A}^{\widehat{O}(\cdot,\cdot)}\left(\widehat{\mathsf{k}}\right)$

The oracles $O(\cdot, \cdot)$ *and* $\widehat{O}(\cdot, \cdot)$ *are defined as follows.*

- $O(i, b)$: *Given an index* $i \in [n]$ *and* $b \in \{0, 1\}$, *it returns* $\mathsf{ct} \leftarrow \mathsf{E}_1(\mathsf{k}, (i, b); \rho)$ *using uniformly random* ρ.
- $\widehat{O}(i, b)$: *Given an index* $i \in [n]$ *and* $b \in \{0, 1\}$, *it returns* $\widehat{\mathsf{ct}} \leftarrow \widehat{\mathsf{E}_1}\left(\widehat{\mathsf{k}}, (i, b)\right)$.

We say that CE *is obliviously samplable if*

$$\mathsf{Adv}^{os}_{\mathsf{CE},\mathcal{A}}(\lambda) := \left|\Pr\left[\mathsf{out} = 1 \textit{ in } \mathsf{Exp}^{os\text{-}0}_{\mathsf{CE},\mathcal{A}}\right] - \Pr\left[\mathsf{out} = 1 \textit{ in } \mathsf{Exp}^{os\text{-}1}_{\mathsf{CE},\mathcal{A}}\right]\right| = \mathsf{negl}(\lambda)$$

holds for every PPT adversary $\mathcal{A}$.

We define another correctness of obliviously samplable CE necessary to assure the correctness of our NCE.

Definition 7 (Correctness under Obliviously Sampled Keys). *An obliviously samplable CE* $(\mathsf{CE}, \widehat{\mathsf{K}}, \widehat{\mathsf{E}_1})$ *is correct under obliviously sampled keys if for all* $\widehat{\mathsf{k}}$ *output by* $\widehat{\mathsf{K}}$, $i \in [n], \mathsf{x} \in \{0, 1\}^n, \mathsf{r} \in \mathcal{R}_\mathsf{H}$, *and* $\rho \in \mathcal{R}_\mathsf{E}$, $\mathsf{E}_2\left(\widehat{\mathsf{k}}, (i, b); \rho\right) = \mathsf{D}\left(\widehat{\mathsf{k}}, (\mathsf{x}, \mathsf{r}), \mathsf{ct}\right)$ *holds, where* $\mathsf{y} \leftarrow \mathsf{H}\left(\widehat{\mathsf{k}}, \mathsf{x}; \mathsf{r}\right)$ *and* $\mathsf{ct} \leftarrow \mathsf{E}_1\left(\widehat{\mathsf{k}}, (i, \mathsf{x}_i); \rho\right)$.

4.2 Construction

We construct an obliviously samplable CE $\mathsf{CE} = \left(\mathsf{K}, \mathsf{H}, \mathsf{H}^{-1}, \mathsf{E}_1, \mathsf{E}_2, \mathsf{D}, \widehat{\mathsf{K}}, \widehat{\mathsf{E}_1}\right)$ based on the hardness of the DDH problem.

Let $\mathbb{G}$ be a cyclic group of order p with a generator g. In the construction, we use a universal hash family $\mathcal{H} = \{\mathsf{H}_\mathbb{G} : \mathbb{G} \to \{0, 1\}^\ell\}$. Below, let $\mathsf{H}_\mathbb{G}$ be a hash function sampled from $\mathcal{H}$ uniformly at random, and it is given to all the algorithms implicitly.

$\mathsf{K}(1^\lambda)$:

- For all $i \in [n], b \in \{0,1\}$, sample $\alpha_{i,b} \xleftarrow{\$} \mathbb{Z}_p$ and set $g_{i,b} := g^{\alpha_{i,b}}$.
- Output

$$\mathsf{k} := \left(g, \begin{pmatrix} g_{1,0}, \ldots, g_{n,0} \\ g_{1,1}, \ldots, g_{n,1} \end{pmatrix}\right) \text{ and } \mathsf{t} := \begin{pmatrix} \alpha_{1,0}, \ldots, \alpha_{n,0} \\ \alpha_{1,1}, \ldots, \alpha_{n,1} \end{pmatrix}. \quad (1)$$

$\mathsf{H}(\mathsf{k}, \mathsf{x}; \mathsf{r})$:

- Sample $\mathsf{r} \xleftarrow{\$} \mathcal{R}_{\mathsf{H}} = \mathbb{Z}_p$ and output $\mathsf{y} = g^{\mathsf{r}} \prod_{i\in[n]} g_{i,\mathsf{x}_i}$.

$\mathsf{H}^{-1}(\mathsf{t}, (\mathsf{x}, \mathsf{r}), \mathsf{x}')$:

- Parse t as in Eq. 1.
- Output $\mathsf{r}' := \mathsf{r} + \sum_{i\in[n]} (\alpha_{i,\mathsf{x}_i} - \alpha_{i,\mathsf{x}'_i})$.

$\mathsf{E}_1(\mathsf{k}, (i, b); \rho)$:

- Parse k as in Eq. 1.
- Sample $\rho \xleftarrow{\$} \mathcal{R}_{\mathsf{E}} = \mathbb{Z}_p$ and compute $c := g^\rho$.
- Compute $c_{i,b} := (g_{i,b})^\rho$ and $c_{i,1-b} := \bot$.
- For all $j \in [n]$ such that $j \neq i$, compute $c_{j,0} := (g_{j,0})^\rho$ and $c_{j,1} := (g_{j,1})^\rho$
- Output

$$\mathsf{ct} := \left(c, \begin{pmatrix} c_{1,0}, \ldots, c_{n,0} \\ c_{1,1}, \ldots, c_{n,1} \end{pmatrix}\right). \quad (2)$$

$\mathsf{E}_2(\mathsf{k}, \mathsf{y}; \rho)$:

- Output $\mathbf{e} \leftarrow \mathsf{H}_{\mathbb{G}}(\mathsf{y}^\rho)$.

$\mathsf{D}(\mathsf{k}, (\mathsf{x}, \mathsf{r}), \mathsf{ct})$:

- Parse ct as in Eq. 2.
- Output $\mathbf{e} \leftarrow \mathsf{H}_{\mathbb{G}}\left(c^{\mathsf{r}} \prod_{i\in[n]} c_{i,\mathsf{x}_i}\right)$.

$\widehat{\mathsf{K}}(1^\lambda)$:

- For all $i \in [n]$ and $b \in \{0,1\}$, sample $g_{i,b} \xleftarrow{\$} \mathbb{G}$.
- Output $\widehat{\mathsf{k}} := \left(g, \begin{pmatrix} g_{1,0}, \ldots, g_{n,0} \\ g_{1,1}, \ldots, g_{n,1} \end{pmatrix}\right)$.

$\widehat{\mathsf{E}_1}\left(\widehat{\mathsf{k}}, (i, b)\right)$:

- Set $\widehat{c}_{i,1-b} := \bot$, and sample $\widehat{c} \xleftarrow{\$} \mathbb{G}$ and $\widehat{c}_{i,b} \xleftarrow{\$} \mathbb{G}$.
- For all $j \in [n]$ such that $j \neq i$, sample $\widehat{c}_{j,0} \xleftarrow{\$} \mathbb{G}$ and $\widehat{c}_{j,1} \xleftarrow{\$} \mathbb{G}$.
- Output $\widehat{\mathsf{ct}} := \left(\widehat{c}, \begin{pmatrix} \widehat{c}_{1,0}, \ldots, \widehat{c}_{n,0} \\ \widehat{c}_{1,1}, \ldots, \widehat{c}_{n,1} \end{pmatrix}\right)$.

Theorem 1. CE *is an obliviously samplable CE scheme assuming the hardness of the DDH problem.*

The trapdoor collision property, correctness, and correctness under obliviously sampled keys of CE directly follow from the construction of CE. Below, we first prove the security of CE under the hashed DDH assumption with respect to $\mathcal{H}$. We then prove the oblivious samplability of CE under the DDH assumption.

Security. Let $\mathcal{A}$ be an adversary against the security of CE. We construct a reduction algorithm $\mathcal{A}'$ which solves the hashed DDH problem using $\mathcal{A}$.

Given $(\mathsf{H}_{\mathbb{G}}, g_1, g_2, \mathbf{e})$, $\mathcal{A}'$ first runs $(\mathsf{x}, \mathsf{r}, i) \leftarrow \mathcal{A}(1^\lambda)$, and generates k as follows. For all $(j, b) \in [n] \times \{0,1\}$ such that $(j, b) \neq (i, \mathsf{x}_i)$, $\mathcal{A}'$ samples $\alpha_{j,b} \xleftarrow{\$} \mathbb{Z}_p$ and sets $g_{j,b} := g^{\alpha_{j,b}}$, $g_{i,\mathsf{x}_i} := g_1 / \left(g^{\mathsf{r}} \prod_{j \neq i} g_{j,\mathsf{x}_j}\right)$ and

$$\mathsf{k} := \left(g, \begin{pmatrix} g_{1,0}, \ldots, g_{n,0} \\ g_{1,1}, \ldots, g_{n,1} \end{pmatrix}\right).$$

Next, $\mathcal{A}'$ generates ct as follows. $\mathcal{A}'$ first sets $c := g_2$ and $c_{i,\mathsf{x}_i} := \bot$. Then for all $(j, b) \in [n] \times \{0,1\}$ such that $(j, b) \neq (i, \mathsf{x}_i)$, $\mathcal{A}'$ sets $c_{j,b} := g_2{}^{\alpha_{j,b}}$. $\mathcal{A}'$ sets the ciphertext to

$$\mathsf{ct} := \left(c, \begin{pmatrix} c_{1,0}, \ldots, c_{n,0} \\ c_{1,1}, \ldots, c_{n,1} \end{pmatrix}\right).$$

Finally, $\mathcal{A}'$ outputs what $\mathcal{A}(\mathsf{k}, \mathsf{ct}, \mathbf{e})$ does.

k and ct generated by $\mathcal{A}'$ distribute identically to those output by $\mathsf{K}(1^\lambda)$ and $\mathsf{E}_1(\mathsf{k}, (i, 1 - \mathsf{x}_i); \rho)$, respectively. $\mathcal{A}'$ perfectly simulates $\mathsf{Exp}^0_{\mathsf{CE},\mathcal{A}}$ to $\mathcal{A}$ if $\mathbf{e} = \mathsf{H}_{\mathbb{G}}(\mathsf{dh}(g_1, g_2))$ because we have

$$\mathsf{E}_2(\mathsf{k}, \mathsf{y}; \rho) = \mathsf{H}_{\mathbb{G}}\left(\mathsf{dh}\left(g^{\mathsf{r}} \prod_{i \in [n]} g_{i,\mathsf{x}_i}, c\right)\right) = \mathsf{H}_{\mathbb{G}}(\mathsf{dh}(g_1, g_2)) = \mathbf{e}.$$

On the other hand, if $\mathbf{e} \xleftarrow{\$} \{0,1\}^\ell$, $\mathcal{A}'$ perfectly simulates $\mathsf{Exp}^1_{\mathsf{CE},\mathcal{A}}$ to the adversary. Thus, it holds that $\mathsf{Adv}_{\mathsf{CE},\mathcal{A}}(\lambda) = \mathsf{negl}(\lambda)$ under the hash DDH assumption with respect to $\mathcal{H}$.

This completes the security proof of CE.

Oblivious Samplability. Let $\mathcal{A}$ be an PPT adversary that attacks oblivious samplability of CE and makes q queries to its oracle. We prove that the probability that $\mathcal{A}$ outputs 1 in $\mathsf{Exp}^{\text{os-0}}_{\mathsf{CE},\mathcal{A}}$ is negligibly close to that in $\mathsf{Exp}^{\text{os-1}}_{\mathsf{CE},\mathcal{A}}$. The detailed description of these experiments is as follows.

$\mathsf{Exp}^{\text{os-0}}_{\mathsf{CE},\mathcal{A}}$: $\mathcal{A}$ is given a hash key k output by K and can access to the oracle $O(i, b) = \mathsf{E}_1(\mathsf{k}, (i, b); \rho)$, where $i \in [n]$, $b \in \{0,1\}$, and $\rho \leftarrow \mathbb{Z}_p$. Concretely, $O(i, b)$ behaves as follows.

- Sample ρ uniformly from $\mathbb{Z}_p$, and let $c := g^\rho$. For all $j \neq i$, let $c_{j,0} := (g_{j,0})^\rho$ and $c_{j,1} := (g_{j,1})^\rho$, and let $c_{i,b} := (g_{i,b})^\rho$ and $c_{i,1-b} := \bot$. Return $\mathsf{ct} := \left(c, \begin{pmatrix} c_{1,0}, \ldots, c_{n,0} \\ c_{1,1}, \ldots, c_{n,1} \end{pmatrix}\right)$.

$\mathsf{Exp}^{\text{os-1}}_{CE,\mathcal{A}}$: $\mathcal{A}$ is given a hash key $\widehat{\mathsf{k}}$ output by $\widehat{\mathsf{K}}$ and can access to the oracle $\widehat{O}(i, b) = \widehat{\mathsf{E}_1}\left(\widehat{\mathsf{k}}, (i, b)\right)$, where $i \in [n]$ and $b \in \{0,1\}$. Concretely, $\widehat{O}(i, b)$ behaves as follows.

- Let $\widehat{c}_{i,1-b} := \perp$, and sample $\widehat{c}, \widehat{c}_{i,b}$, and $\widehat{c}_{j,0}$ and $\widehat{c}_{j,1}$ for all $j \neq i$ uniformly from $\mathbb{G}$. Return $\widehat{\mathsf{ct}} := \left(\widehat{c}, \begin{pmatrix} \widehat{c}_{1,0}, \ldots, \widehat{c}_{n,0} \\ \widehat{c}_{1,1}, \ldots, \widehat{c}_{n,1} \end{pmatrix}\right)$.

We define $\mathsf{Exp}\ j$ for every $j \in \{0, \ldots, q\}$ that are intermediate experiments between $\mathsf{Exp}^{\text{os-0}}_{\mathsf{CE},\mathcal{A}}$ and $\mathsf{Exp}^{\text{os-1}}_{\mathsf{CE},\mathcal{A}}$ as follows. Below, for two experiments $\mathsf{Exp}\ X$ and $\mathsf{Exp}\ Y$, we write $\mathsf{Exp}\ X \approx \mathsf{Exp}\ Y$ to denote that the probability that $\mathcal{A}$ outputs 1 in $\mathsf{Exp}\ X$ is negligibly close to that in $\mathsf{Exp}\ Y$.

$\mathsf{Exp}\ j$**:** This experiment is exactly the same as $\mathsf{Exp}^{\text{os-0}}_{\mathsf{CE},\mathcal{A}}$ except how queries made by $\mathcal{A}$ are answered. For the j'-th query $(i, b) \in [n] \times \{0, 1\}$ made by $\mathcal{A}$, the experiment returns $\mathsf{E}_1(\mathsf{k}, (i, b); \rho)$ if $j < j'$, and $\widehat{\mathsf{E}_1}(\mathsf{k}, (i, b))$ otherwise.

We see that $\mathsf{Exp}\ 0$ and $\mathsf{Exp}\ q$ are exactly the same experiment as $\mathsf{Exp}^{\text{os-0}}_{\mathsf{CE},\mathcal{A}}$ and $\mathsf{Exp}^{\text{os-1}}_{\mathsf{CE},\mathcal{A}}$, respectively. Note that $\mathcal{A}$ is given k output by $\mathsf{K}(1^\lambda)$ and can access to the oracle $\widehat{\mathsf{E}_1}(\mathsf{k}, (i, b))$ in $\mathsf{Exp}\ q$, but on the other hand, $\mathcal{A}$ is given $\widehat{\mathsf{k}}$ output by $\widehat{\mathsf{K}}(1^\lambda)$ and can access to the oracle $\widehat{\mathsf{E}_1}\left(\widehat{\mathsf{k}}, (i, b)\right)$ in $\mathsf{Exp}^{\text{os-1}}_{\mathsf{CE},\mathcal{A}}$. However, this is not a problem since k output by $\mathsf{K}(1^\lambda)$ and $\widehat{\mathsf{k}}$ output by $\widehat{\mathsf{K}}(1^\lambda)$ distribute identically in our construction. For every $j \in [q]$, $\mathsf{Exp}\ j-1 \approx \mathsf{Exp}\ j$ directly follows from Lemma 3. Therefore, we have $\mathsf{Exp}^{\text{os-0}}_{\mathsf{CE},\mathcal{A}} \approx \mathsf{Exp}^{\text{os-1}}_{\mathsf{CE},\mathcal{A}}$ under the DDH assumption. From the above arguments, $\mathtt{CE}$ satisfies oblivious samplability under the DDH assumption.

This completes the proof of Theorem 1.

5 Basic Construction of Proposed NCE

In this section, we present our NCE scheme with ciphertext rate $\mathsf{poly}(\log \lambda)$ from an obliviously samplable CE. We call this construction basic construction. In Sect. 6, improving the basic construction, we describe our full construction of NCE which achieves ciphertext rate $\mathcal{O}(\log \lambda)$.

5.1 Construction

We use three parameters μ, n, and ℓ, all of which are polynomials of λ and concretely determined later.

Let $\mathtt{CE} = \left(\mathsf{K}, \mathsf{H}, \mathsf{H}^{-1}, \mathsf{E}_1, \mathsf{E}_2, \mathsf{D}, \widehat{\mathsf{K}}, \widehat{\mathsf{E}_1}\right)$ be an obliviously samplable CE scheme. We let the input length of H be n and let the output length of E_2 (and thus D) be ℓ. We also let the randomness spaces of H and E_1 be $\mathcal{R}_\mathsf{H}$ and $\mathcal{R}_\mathsf{E}$, respectively. Below, using $\mathtt{CE}$, we construct an NCE scheme $\mathtt{NCE} = (\mathsf{Gen}, \mathsf{Enc}, \mathsf{Dec}, \mathsf{Sim}_1, \mathsf{Sim}_2)$ whose message space is $\{0, 1\}^\mu$.

In the construction, we use a matrix $R \in \{0, 1\}^{\mu \times n}$, such that randomly picked μ out of n column vectors of length μ are linearly independent. A random matrix satisfies such property except for negligible probability [21].

We first describe $(\mathsf{Gen}, \mathsf{Enc}, \mathsf{Dec})$ and show the correctness of $\mathtt{NCE}$ below. We also describe a protocol when using $\mathtt{NCE}$ in Fig. 3.

Receiver		Sender
$\mathsf{z} \xleftarrow{\$} \{0,1\}^n, \mathsf{k} \leftarrow \widehat{\mathsf{K}}(1^\lambda)$		Input: $m \in \{0,1\}^\mu$
$\forall i \in [n], \rho_i \xleftarrow{\$} \mathcal{R}_{\mathsf{E}}$.		
$\mathsf{ct}_{i,\mathsf{z}_i} = \mathsf{E}_1(\mathsf{k},(i,\mathsf{z}_i);\rho_i)$		$\mathsf{x} \xleftarrow{\$} \{0,1\}^n$
$\mathsf{ct}_{i,1-\mathsf{z}_i} \leftarrow \widehat{\mathsf{E}_1}(\mathsf{k},(i,1-\mathsf{z}_i))$		$\mathsf{r} \xleftarrow{\$} \mathcal{R}_{\mathsf{H}}$.
	$\left(\mathsf{k}, \begin{pmatrix} \mathsf{ct}_{1,0},\dots,\mathsf{ct}_{n,0} \\ \mathsf{ct}_{1,1},\dots,\mathsf{ct}_{n,1} \end{pmatrix}\right)$ $\longrightarrow$	$\mathsf{y} \leftarrow \mathsf{H}(\mathsf{k},\mathsf{x};\mathsf{r})$
		$\forall i \in [n]$,
		$\mathbf{e}_{i,\mathsf{x}_i} = \mathsf{D}(\mathsf{k},(\mathsf{x},\mathsf{r}),\mathsf{ct}_{i,\mathsf{x}_i})$
		$\mathbf{e}_{i,1-\mathsf{x}_i} \xleftarrow{\$} \{0,1\}^\ell$
if $\mathbf{e}_{i,\mathsf{z}_i} = \mathsf{E}_2(\mathsf{k},\mathsf{y};\rho_i)$	$\left(\mathsf{y}, \begin{pmatrix} \mathbf{e}_{1,0},\dots,\mathbf{e}_{n,0} \\ \mathbf{e}_{1,1},\dots,\mathbf{e}_{n,1} \end{pmatrix}, w\right)$	$w = R\mathsf{x} \oplus m$
$\mathsf{x}_i := \mathsf{z}_i$	$\longleftarrow$	
else $\mathsf{x}_i := 1-\mathsf{z}_i$		
Output: $m = w \oplus R\mathsf{x}$		

Fig. 3. The description of NCE.

$\mathsf{Gen}\left(1^\lambda; r^{\mathsf{Gen}}\right)$:

- Sample $\mathsf{k} \leftarrow \widehat{\mathsf{K}}\left(1^\lambda\right)$ and $\mathsf{z} \xleftarrow{\$} \{0,1\}^n$.
- For all $i \in [n]$, sample $\rho_i \xleftarrow{\$} \mathcal{R}_{\mathsf{E}}$.
- For all $i \in [n]$ and $b \in \{0,1\}$, compute

$$\mathsf{ct}_{i,b} \leftarrow \begin{cases} \mathsf{E}_1(\mathsf{k},(i,b);\rho_i) & (b = \mathsf{z}_i) \\ \widehat{\mathsf{E}_1}(\mathsf{k},(i,b)) & (b \neq \mathsf{z}_i) \end{cases}.$$

- Output

$$pk := \left(\mathsf{k}, \begin{pmatrix} \mathsf{ct}_{1,0},\dots,\mathsf{ct}_{n,0} \\ \mathsf{ct}_{1,1},\dots,\mathsf{ct}_{n,1} \end{pmatrix}\right) \quad \text{and} \quad sk := (\mathsf{z}, (\rho_1,\dots,\rho_n)). \qquad (3)$$

The random coin r^{Gen} used in Gen is $\left(\mathsf{k}, \mathsf{z}, \{\rho_i\}_{i\in[n]}, \{\mathsf{ct}_{i,1-\mathsf{z}_i}\}_{i\in[n]}\right)$.

$\mathsf{Enc}\left(pk, m; r^{\mathsf{Enc}}\right)$:

- Sample $\mathsf{x} \xleftarrow{\$} \{0,1\}^n$ and $\mathsf{r} \xleftarrow{\$} \mathcal{R}_{\mathsf{H}}$.
- Compute $\mathsf{y} \leftarrow \mathsf{H}(\mathsf{k},\mathsf{x};\mathsf{r})$.
- For all $i \in [n]$ and $b \in \{0,1\}$, compute

$$\mathbf{e}_{i,b} \leftarrow \begin{cases} \mathsf{D}(\mathsf{k},(\mathsf{x},\mathsf{r}),\mathsf{ct}_{i,b}) & (b = \mathsf{x}_i) \\ \{0,1\}^\ell & (b \neq \mathsf{x}_i) \end{cases}.$$

- Compute $w \leftarrow R\mathsf{x} \oplus m$.

- Output

$$CT := \left(\mathsf{y}, \begin{pmatrix} \mathbf{e}_{1,0}, \ldots, \mathbf{e}_{n,0} \\ \mathbf{e}_{1,1}, \ldots, \mathbf{e}_{n,1} \end{pmatrix}, w\right). \tag{4}$$

The random coin r^{Enc} used in Enc is $\left(\mathsf{x}, \mathsf{r}, \{\mathbf{e}_{i,1-\mathsf{x}_i}\}_{i\in[n]}\right)$.

$\mathsf{Dec}(sk, CT)$:
- Parse sk and CT as the Eqs. 3 and 4, respectively.
- For all $i \in [n]$, set

$$\mathsf{x}_i := \begin{cases} \mathsf{z}_i & (\mathbf{e}_{i,\mathsf{z}_i} = \mathsf{E}_2(\mathsf{k}, \mathsf{y}; \rho_i)) \\ 1 - \mathsf{z}_i & (\text{otherwise}) \end{cases}.$$

- Output $m := R\mathsf{x} \oplus w$.

By setting $\ell = \mathsf{poly}(\log \lambda)$, NCE is correct. Formally, we have the following theorem.

Theorem 2. *Let $\ell = \mathsf{poly}(\log \lambda)$. If* CE *is correct under obliviously sampled keys, then* NCE *is correct.*

Proof. Due to the correctness under obliviously sampled keys of CE, the recovery of x_i fails only when $\mathsf{z}_i \neq \mathsf{x}_i$ happens and $\mathbf{e}_{i,1-\mathsf{x}_i} \xleftarrow{\$} \{0,1\}^\ell$ coincides with $\mathsf{E}_2(\mathsf{k}, \mathsf{y}; \rho_i)$. Thus, the probability of decryption failure is bounded by

$$\begin{aligned} &\Pr[m \neq \mathsf{Dec}(sk, CT)] \\ \leq &\Pr\left[\exists i \in [n], \mathbf{e}_{i,1-\mathsf{x}_i} \xleftarrow{\$} \{0,1\}^\ell; \mathbf{e}_{i,1-\mathsf{x}_i} = \mathsf{E}_2(\mathsf{k}, \mathsf{y}; \rho_i)\right] \leq \frac{n}{2^\ell}. \end{aligned}$$

Note that at the last step, we used the union bound. Since $n = \mathcal{O}(\lambda)$, the probability is negligible by setting $\ell = \mathsf{poly}(\log \lambda)$. Therefore NCE is correct.

Intuition for the Simulators and Security Proof. The description of the simulators $(\mathsf{Sim}_1, \mathsf{Sim}_2)$ of NCE is somewhat complex. Thus, we give an overview of the security proof for NCE before describing them. We think this will help readers understand the construction of simulators.

In the proof, we start from the real experiment $\mathsf{Exp}^{\mathrm{Real}}_{\mathtt{NCE},\mathcal{A}}$, where $\mathcal{A}$ is an PPT adversary attacking the security of NCE. We then change the experiment step by step so that, in the final experiment, we can generate the ciphertext CT given to $\mathcal{A}$ without the message m chosen by $\mathcal{A}$, which can later be opened to any message. The simulators $(\mathsf{Sim}_1, \mathsf{Sim}_2)$ are defined so that they simulate the final experiment.

In $\mathsf{Exp}^{\mathrm{Real}}_{\mathtt{NCE},\mathcal{A}}$, CT is of the form

$$CT := \left(\mathsf{y}, \begin{pmatrix} \mathbf{e}_{1,0}, \ldots, \mathbf{e}_{n,0} \\ \mathbf{e}_{1,1}, \ldots, \mathbf{e}_{n,1} \end{pmatrix}, R\mathsf{x} \oplus m\right).$$

Informally, $\left(\mathsf{y}, \begin{pmatrix} \mathbf{e}_{1,0}, \ldots, \mathbf{e}_{n,0} \\ \mathbf{e}_{1,1}, \ldots, \mathbf{e}_{n,1} \end{pmatrix}\right)$ encapsulates $\mathsf{x} \in \{0,1\}^n$, and $R\mathsf{x} \oplus m$ is a one-time encryption of $m \in \{0,1\}^\mu$ by x. If we can eliminate the information of x from the encapsulation part, CT becomes statistically independent of m. Thus, if we can do that, the security proof is almost complete since in that case, CT can be simulated without m and later be opened to any message. While we cannot eliminate the entire information of x from the encapsulation part, we can eliminate the information of μ out of n bits of x from the encapsulation part, and it is enough to make CT statistically independent of m. Below, we briefly explain how to do it.

We first change $\begin{pmatrix} \mathsf{ct}_{1,0}, \ldots, \mathsf{ct}_{n,0} \\ \mathsf{ct}_{1,1}, \ldots, \mathsf{ct}_{n,1} \end{pmatrix}$ contained in pk so that every $\mathsf{ct}_{i,b}$ is generated as $\mathsf{ct}_{i,b} \leftarrow \mathsf{E}_1(\mathsf{k}, (i,b); \rho_{i,b})$, and set $\rho_i := \rho_{i,\mathsf{z}_i}$, where $\mathsf{z} \in \{0,1\}^n$ is a random string generated in Gen. We can make this change by the oblivious samplability of CE.

Next, by using the security of CE, we try to change the experiment so that for every $i \in [n]$, $\mathbf{e}_{i,0}$ and $\mathbf{e}_{i,1}$ contained in CT are symmetrically generated in order to eliminate the information of x_i from the encapsulation part. Concretely, for every $i \in [n]$, we try to change $\mathbf{e}_{i,1-\mathsf{x}_i}$ from a random string to

$$\mathbf{e}_{i,b} \leftarrow \mathsf{D}(\mathsf{k}, (\mathsf{x}, \mathsf{r}), \mathsf{ct}_{i,1-\mathsf{x}_i}) = \mathsf{E}_2(\mathsf{k}, \mathsf{y}; \rho_{i,1-\mathsf{x}_i}).$$

Unfortunately, we cannot change the distribution of every $\mathbf{e}_{i,1-\mathsf{x}_i}$ because some of $\rho_{i,1-\mathsf{x}_i}$ is given to $\mathcal{A}$ as a part of r^{Gen}. Concretely, for $i \in [n]$ such that $\mathsf{z}_i \neq \mathsf{x}_i$, $\rho_i = \rho_{i,\mathsf{z}_i} = \rho_{i,1-\mathsf{x}_i}$ is given to $\mathcal{A}$ and we cannot change the distribution of $\mathbf{e}_{i,1-\mathsf{x}_i}$. On the other hand, for $i \in [n]$ such that $\mathsf{z}_i = \mathsf{x}_i$, we can change the distribution of $\mathbf{e}_{i,1-\mathsf{x}_i}$.

In order to make clear which index $i \in [n]$ we can change the distribution of $\mathbf{e}_{i,1-\mathsf{x}_i}$, in the proof, we replace z with $\mathsf{z}' = \mathsf{x} \oplus \mathsf{z}$. Then, we can say that for $i \in [n]$ such that $\mathsf{z}_i = 0$, we can change the distribution of $\mathbf{e}_{i,1-\mathsf{x}_i}$. Since z is chosen uniformly at random, due to the Chernoff bound, we can ensure that the number of such indices is greater than μ with overwhelming probability by setting n and μ appropriately. Namely, we can eliminate the information of μ out of n bits of x from CT. At this point, CT becomes statistically independent of m, and we almost complete the security proof. Note that y itself does not have any information of x. To make this fact clear, in the proof, we add another step using the trapdoor collision property of CE after using the security of CE.

To complete the proof formally, we have to ensure that CT can later be opened to any message *efficiently (i.e., in polynomial time)*. This is possible by using a matrix $R \in \{0,1\}^{\mu \times n}$, such that randomly picked μ out of n column vectors of length μ are linearly independent. For more details, see the formal security proof in Sect. 5.2.

We now show the simulators $(\mathsf{Sim}_1, \mathsf{Sim}_2)$.

$\mathsf{Sim}_1(1^\lambda)$:
- Sample $(\mathsf{k}, \mathsf{t}) \leftarrow \mathsf{K}(1^\lambda)$.

- For all $i \in [n]$ and $b \in \{0,1\}$, sample $\rho_{i,b} \xleftarrow{\$} \mathcal{R}_{\mathsf{E}}$ and compute $\mathsf{ct}_{i,b} \leftarrow \mathsf{E}_1(\mathsf{k},(i,b);\rho_{i,b})$.
- Sample $\mathsf{z} \xleftarrow{\$} \{0,1\}^n, \mathsf{x} \xleftarrow{\$} \{0,1\}^n$ [3], and $\mathsf{r} \xleftarrow{\$} \mathcal{R}_{\mathsf{H}}$.
- Compute $\mathsf{y} \leftarrow \mathsf{H}(\mathsf{k},0^n;\mathsf{r})$ and sample $w \xleftarrow{\$} \{0,1\}^{\mu}$.
- For all $i \in [n]$ and $b \in \{0,1\}$, compute

$$\mathbf{e}_{i,b} \leftarrow \begin{cases} \mathsf{E}_2(\mathsf{k},\mathsf{y};\rho_{i,b}) & (b = \mathsf{x}_i \vee \mathsf{z}_i = 0) \\ \{0,1\}^{\ell} & (b \neq \mathsf{x}_i \wedge \mathsf{z}_i = 1) \end{cases}.$$

- Output

$$pk := \left(\mathsf{k}, \begin{pmatrix} \mathsf{ct}_{1,0},\dots,\mathsf{ct}_{n,0} \\ \mathsf{ct}_{1,1},\dots,\mathsf{ct}_{n,1} \end{pmatrix}\right),\quad CT := \left(\mathsf{y}, \begin{pmatrix} \mathbf{e}_{1,0},\dots,\mathbf{e}_{n,0} \\ \mathbf{e}_{1,1},\dots,\mathbf{e}_{n,1} \end{pmatrix}, w\right),$$

and $st := (\mathsf{t},\mathsf{z},\mathsf{x},\mathsf{r})$.

$\mathsf{Sim}_2(m, st)$:
- Sample x' at random from $\{0,1\}^n$ under the condition that $R\mathsf{x}' = m \oplus w$ and $\mathsf{x}_i = \mathsf{x}'_i$ hold for every $i \in [n]$ such that $\mathsf{z}_i = 1$.
- Compute $\mathsf{r}' \leftarrow \mathsf{H}^{-1}(\mathsf{t},(0^n,\mathsf{r}),\mathsf{x}')$ and $\mathsf{z}' := \mathsf{z} \oplus \mathsf{x}'$.
- Output

$$r^{\mathsf{Gen}} := \left(\mathsf{k},\mathsf{z}',\{\rho_{i,\mathsf{z}'_i}\}_{i\in[n]},\{\mathsf{ct}_{i,1-\mathsf{z}'_i}\}_{i\in[n]}\right) \text{ and } r^{\mathsf{Enc}} := \left(\mathsf{x}',\mathsf{r}',\{\mathbf{e}_{i,1-\mathsf{x}'_i}\}_{i\in[n]}\right).$$

5.2 Security Proof

In this section, we prove the security of NCE. Formally, we prove the following theorem.

Theorem 3. *Let $\mu = \mathcal{O}(\lambda)$ and $n = 8\mu$. If* CE *is an obliviously samplable CE, then* NCE *is secure.*

Proof. Let $\mathcal{A}$ is a PPT adversary attacking the security of NCE. We define a sequence of experiments Exp 0, ..., Exp 6. Below, for two experiments Exp X and Exp Y, we write Exp $X \approx$ Exp Y (resp. Exp $X \equiv$ Exp Y) to denote that the probability that $\mathcal{A}$ outputs 1 in Exp X is negligibly close to (resp. the same as) that in Exp Y.

Exp 0: This experiment is exactly the same as $\mathsf{Exp}^{\mathrm{Real}}_{\mathtt{NCE},\mathcal{A}}$. The detailed description is as follows.

1. The experiment first samples $\mathsf{k} \leftarrow \widehat{\mathsf{K}}(1^{\lambda})$ and $\mathsf{z} \xleftarrow{\$} \{0,1\}^n$. Then, for all $i \in [n]$, it samples $\rho_i \xleftarrow{\$} \mathcal{R}_{\mathsf{E}}$. Next, for all $i \in [n]$ and $b \in \{0,1\}$, it computes

$$\mathsf{ct}_{i,b} \leftarrow \begin{cases} \mathsf{E}_1(\mathsf{k},(i,b);\rho_i) & (b = \mathsf{z}_i) \\ \widehat{\mathsf{E}_1}(\mathsf{k},(i,b)) & (b \neq \mathsf{z}_i) \end{cases}.$$

[3] Sim_1 and Sim_2 do not use x_i for i such that $\mathsf{z}_i = 0$, but for simplicity, we generate whole x.

It sets

$$pk := \left(\mathsf{k}, \begin{pmatrix} \mathsf{ct}_{1,0}, \ldots, \mathsf{ct}_{n,0} \\ \mathsf{ct}_{1,1}, \ldots, \mathsf{ct}_{n,1} \end{pmatrix}\right) \quad \text{and} \quad r^{\mathsf{Gen}} := \left(\mathsf{k}, \mathsf{z}, \{\rho_i\}_{i\in[n]}, \{\mathsf{ct}_{i,1-\mathsf{z}_i}\}_{i\in[n]}\right).$$

Finally, it runs $m \leftarrow \mathcal{A}(pk)$. Note that r^{Gen} is used in the next step.

2. The experiment samples $\mathsf{x} \xleftarrow{\$} \{0,1\}^n$ and $\mathsf{r} \xleftarrow{\$} \mathcal{R}_{\mathsf{H}}$. It then computes $\mathsf{y} \leftarrow \mathsf{H}(\mathsf{k}, \mathsf{x}; \mathsf{r})$. For all $i \in [n]$ and $b \in \{0,1\}$, it also computes

$$\mathbf{e}_{i,b} \leftarrow \begin{cases} \mathsf{D}(\mathsf{k}, (\mathsf{x}, \mathsf{r}), \mathsf{ct}_{i,b}) & (b = \mathsf{x}_i) \\ \{0,1\}^{\ell} & (b \neq \mathsf{x}_i) \end{cases}.$$

It sets

$$CT := \left(\mathsf{y}, \begin{pmatrix} \mathbf{e}_{1,0}, \ldots, \mathbf{e}_{n,0} \\ \mathbf{e}_{1,1}, \ldots, \mathbf{e}_{n,1} \end{pmatrix}, R\mathsf{x} \oplus m\right) \quad \text{and} \quad r^{\mathsf{Enc}} = \left(\mathsf{x}, \mathsf{r}, \{\mathbf{e}_{i,1-\mathsf{x}_i}\}_{i\in[n]}\right).$$

Finally, it outputs $\mathsf{out} \leftarrow \mathcal{A}\left(CT, r^{\mathsf{Gen}}, r^{\mathsf{Enc}}\right)$.

Exp 1: This experiment is the same as Exp 0 except the followings. First, pk is generated together with a trapdoor of the chameleon hash function t as $(\mathsf{k}, \mathsf{t}) \leftarrow \mathsf{K}(1^\lambda)$ instead of $\mathsf{k} \leftarrow \widehat{\mathsf{K}}(1^\lambda)$. Moreover, all ciphertexts of chameleon encryption $\mathsf{ct}_{i,b}$ are computed by E_1, instead of $\widehat{\mathsf{E}_1}$. Specifically, for every $i \in [n]$ and $b \in \{0,1\}$, the experiment samples $\rho_{i,b} \xleftarrow{\$} \mathcal{R}_{\mathsf{E}}$ and compute $\mathsf{ct}_{i,b} \leftarrow \mathsf{E}_1(\mathsf{k}, (i,b); \rho_{i,b})$. Also, it sets $r^{\mathsf{Gen}} = (\mathsf{k}, \mathsf{z}, \{\rho_{i,\mathsf{z}_i}\}_{i\in[n]}, \{\mathsf{ct}_{i,1-\mathsf{z}_i}\}_{i\in[n]})$.

Lemma 5. *Assuming the oblivious samplability of* CE, Exp 0 $\approx$ Exp 1 *holds.*

Proof. Using $\mathcal{A}$, we construct a reduction algorithm $\mathcal{A}'^{O^*(\cdot,\cdot)}$ that attacks the oblivious samplability of CE and makes n oracle queries.

1. On receiving a hash key k^*, $\mathcal{A}'$ generates $\rho_i \xleftarrow{\$} \mathcal{R}_{\mathsf{E}}$ for every $i \in [n]$ and sets the public key as $pk = \left(\mathsf{k}^*, \begin{pmatrix} \mathsf{ct}_{1,0}, \ldots, \mathsf{ct}_{n,0} \\ \mathsf{ct}_{1,1}, \ldots, \mathsf{ct}_{n,1} \end{pmatrix}\right)$, where

$$\mathsf{ct}_{i,b} \leftarrow \begin{cases} \mathsf{E}_1(\mathsf{k}^*, (i,b); \rho_i) & (b = \mathsf{z}_i) \\ O^*(i,b) & (b \neq \mathsf{z}_i) \end{cases}.$$

   $\mathcal{A}'^{O^*(\cdot,\cdot)}$ also sets $r^{\mathsf{Gen}} = \left(\mathsf{k}, \mathsf{z}, \{\rho_i\}_{i\in[n]}, \{\mathsf{ct}_{i,1-\mathsf{z}_i}\}_{i\in[n]}\right)$. Then, $\mathcal{A}'^{O^*(\cdot,\cdot)}$ runs $\mathcal{A}(pk)$ and obtains m.
2. $\mathcal{A}'^{O^*(\cdot,\cdot)}$ simulates the step 2. of Exp 0 and Exp 1, and outputs what $\mathcal{A}$ does. Note that the step 2. of Exp 0 is exactly the same as that of Exp 1.

When playing $\mathsf{Exp}^{\text{os-0}}_{\mathsf{CE},\mathcal{A}}$ and $\mathsf{Exp}^{\text{os-1}}_{\mathsf{CE},\mathcal{A}}$, $\mathcal{A}'$ perfectly simulates Exp 0 and Exp 1 for $\mathcal{A}$, respectively. By the oblivious samplability of CE,

$$|\Pr[\mathsf{out} = 1 \text{ in Exp } 0] - \Pr[\mathsf{out} = 1 \text{ in Exp } 1]| = \mathsf{Adv}^{\mathsf{os}}_{\mathsf{CE},\mathcal{A}'}(\lambda) = \mathsf{negl}(\lambda)$$

holds. This proves Exp 0 $\approx$ Exp 1.

Exp 2: This experiment is the same as Exp 1, except that we replace z contained in r^{Gen} by $\mathsf{z}' := \mathsf{z} \oplus \mathsf{x}$.

Because z distributes uniformly at random, so does z'. Therefore, the distribution of the inputs to $\mathcal{A}$ does not change between Exp 1 and Exp 2, and thus Exp 1 $\equiv$ Exp 2 holds.

Exp 3: The essential difference from Exp 2 in this experiment is that when $\mathsf{z}_i = 0, \mathbf{e}_{i,1-\mathsf{x}_i}$ is computed by $\mathsf{E}_2(\mathsf{k}, \mathsf{y}; \rho_{i,1-\mathsf{x}_i})$ instead of uniformly sampled from $\{0,1\}^\ell$.
Additionally, each $\mathbf{e}_{i,\mathsf{x}_i}$ is replaced to $\mathsf{E}_2(\mathsf{k}, \mathsf{y}; \rho_{i,\mathsf{x}_i})$ from $\mathsf{D}(\mathsf{k}, (\mathsf{x}, \mathsf{r}), \mathsf{ct}_{i,\mathsf{x}_i})$, though this does not change the distribution due to the correctness of CE. After all, for every $i \in [n]$ and $b \in \{0,1\}$, the experiment computes

$$\mathbf{e}_{i,b} \leftarrow \begin{cases} \mathsf{E}_2(\mathsf{k}, \mathsf{y}; \rho_{i,b}) & (b = \mathsf{x}_i \vee \mathsf{z}_i = 0) \\ \{0,1\}^\ell & (b \neq \mathsf{x}_i \wedge \mathsf{z}_i = 1) \end{cases}.$$

Lemma 6. *If* CE *is correct and secure,* Exp 2 $\approx$ Exp 3 *holds.*

Proof. This proof is done by hybrid arguments. We define Exp 2_j for every $j \in \{0, \dots, n\}$ that are intermediate experiments between Exp 2 and Exp 3 as follows.

Exp 2_j: This experiment is exactly the same as Exp 2 except how $\mathbf{e}_{i,b}$ is generated for every $i \in [n]$. For $j < i \leq n, \mathbf{e}_{i,b}$ is generated as in Exp 2. For $1 \leq i \leq j, \mathbf{e}_{i,b}$ is generated as in Exp 3.

Exp 2_0 is equal to Exp 2, and Exp 2_n is equal to Exp 3. In the following, we show Exp $2_{j-1} \approx$ Exp 2_j for all $j \in [n]$.

In the case of $\mathsf{z}_j = 1$, except negligible probability, $\mathbf{e}_{j,\mathsf{x}_j}$ distributes identically in Exp 2_{j-1} and Exp 2_j because $\mathsf{E}_2(\mathsf{k}, \mathsf{y}; \rho_{j,\mathsf{x}_j}) = \mathsf{D}(\mathsf{k}, (\mathsf{x}, \mathsf{r}), \mathsf{ct}_{j,\mathsf{x}_j})$ holds with overwhelming probability due to the correctness of CE. Moreover, $\mathbf{e}_{j,1-\mathsf{x}_j}$ is generated in the same way in both experiments. Thus Exp $2_{j-1} \approx$ Exp 2_j holds.

In the case of $\mathsf{z}_j = 0$, we show Exp $2_{j-1} \approx$ Exp 2_j by constructing a reduction algorithm $\mathcal{A}'$ that uses $\mathcal{A}$ and attacks the security of CE. The description of $\mathcal{A}'$ is as follows.

1. $\mathcal{A}'$ samples $\mathsf{x} \xleftarrow{\$} \{0,1\}^n$ and $\mathsf{r} \xleftarrow{\$} \mathcal{R}_{\mathsf{H}}$, outputs $(\mathsf{x}, \mathsf{r}, j)$, and receives $(\mathsf{k}^*, \mathsf{ct}^*, \mathbf{e}^*)$. Then, $\mathcal{A}'$ generates pk as follows. $\mathcal{A}'$ first samples $\mathsf{z} \xleftarrow{\$} \{0,1\}^n$ and sets $\mathsf{z}' = \mathsf{x} \oplus \mathsf{z}$. For every $(i,b) \in [n] \times \{0,1\}$ such that $(i,b) \neq (j, 1-\mathsf{x}_j)$, $\mathcal{A}'$ samples $\rho_{i,b} \xleftarrow{\$} \mathcal{R}_{\mathsf{E}}$ and computes $\mathsf{ct}_{i,b} \leftarrow \mathsf{E}_1(\mathsf{k}, (i,b); \rho_{i,b})$. $\mathcal{A}'$ sets $\mathsf{ct}_{j,1-\mathsf{x}_j} := \mathsf{ct}^*$,
$$pk := \left(\mathsf{k}^*, \begin{pmatrix} \mathsf{ct}_{1,0}, \dots, \mathsf{ct}_{n,0} \\ \mathsf{ct}_{1,1}, \dots, \mathsf{ct}_{n,1} \end{pmatrix}\right) \text{ and } r^{\mathsf{Gen}} = \left(\mathsf{k}^*, \mathsf{z}', \{\rho_{i,\mathsf{z}'_i}\}_{i\in[n]}, \{\mathsf{ct}_{i,1-\mathsf{z}'_i}\}_{i\in[n]}\right).$$
Finally, $\mathcal{A}'$ runs $m \leftarrow \mathcal{A}(pk)$. Note that $\rho_{j,\mathsf{z}'_i} = \rho_{j,\mathsf{x}_j \oplus \mathsf{z}_j} = \rho_{j,\mathsf{x}_j}$ since we consider the case of $\mathsf{z}_j = 0$, and thus $\mathcal{A}'$ generates ρ_{i,z'_i} by itself for every $i \in [n]$.

2. $\mathcal{A}'$ computes $\mathsf{y} \leftarrow \mathsf{H}(\mathsf{k}^*, \mathsf{x}; \mathsf{r})$. For $j < i \leq n$, $\mathcal{A}'$ computes $\mathbf{e}_{i,b}$ as in Exp 2, and for $1 \leq i < j$, it does as in Exp 3. For $i = j$, $\mathcal{A}'$ computes $\mathbf{e}_{j,\mathsf{x}_j} \leftarrow \mathsf{E}_2\left(\mathsf{k}, \mathsf{y}; \rho_{j,\mathsf{x}_j}\right)$ and sets $\mathbf{e}_{j,1-\mathsf{x}_j} := \mathbf{e}^*$. Finally, $\mathcal{A}'$ sets

$$CT := \left(\mathsf{y}, \begin{pmatrix} \mathbf{e}_{1,0}, \ldots, \mathbf{e}_{n,0} \\ \mathbf{e}_{1,1}, \ldots, \mathbf{e}_{n,1} \end{pmatrix}, R\mathsf{x} \oplus m\right) \quad \text{and} \quad r^{\mathsf{Enc}} = \left(\mathsf{x}, \mathsf{r}, \{\mathbf{e}_{i,1-\mathsf{x}_i}\}_{i \in [n]}\right),$$

and outputs $\mathsf{out} \leftarrow \mathcal{A}\left(CT, r^{\mathsf{Gen}}, r^{\mathsf{Enc}}\right)$.

When playing $\mathsf{Exp}^1_{\mathsf{CE},\mathcal{A}'}$, $\mathcal{A}'$ simulates Exp 2_{j-1} for $\mathcal{A}$. Also, when playing $\mathsf{Exp}^0_{\mathsf{CE},\mathcal{A}'}$, $\mathcal{A}'$ simulates Exp 2_j for $\mathcal{A}$. By the security of CE,

$$|\Pr[\mathsf{out} = 1 \text{ in } \mathsf{Exp}\ 2_{j-1}] - \Pr[\mathsf{out} = 1 \text{ in } \mathsf{Exp}\ 2_j]| = \mathsf{Adv}_{\mathsf{CE},\mathcal{A}'}(\lambda) = \mathsf{negl}(\lambda)$$

holds. From the above, we have

$$\begin{aligned} &|\Pr[\mathsf{out} = 1 \text{ in } \mathsf{Exp}\ 2] - \Pr[\mathsf{out} = 1 \text{ in } \mathsf{Exp}\ 3]| \\ \leq &\sum_{j \in [n]} |\Pr[\mathsf{out} = 1 \text{ in } \mathsf{Exp}\ 2_{j-1}] - \Pr[\mathsf{out} = 1 \text{ in } \mathsf{Exp}\ 2_j]| = \mathsf{negl}(\lambda). \end{aligned}$$

We can conclude Exp 2 $\approx$ Exp 3.

Exp 4: This experiment is the same as Exp 3 except how y and r are computed. In this experiment, y is computed as $\mathsf{y} \leftarrow \mathsf{H}(\mathsf{k}, 0^n; \mathsf{r})$. Moreover, the randomness r contained in r^{Enc} is replaced with $\mathsf{r}' \leftarrow \mathsf{H}^{-1}(\mathsf{t}, (0^n, \mathsf{r}), \mathsf{x})$.

Due to the trapdoor collision property of CE, the view of $\mathcal{A}$ does not change between Exp 3 and Exp 4. Thus, Exp 3 $\equiv$ Exp 4 holds.

Exp 5: This experiment is the same as Exp 4, except that $R\mathsf{x}$ is replaced with $w \xleftarrow{\$} \{0,1\}^\mu$. Moreover, the experiment computes r' as $\mathsf{r}' \leftarrow \mathsf{H}^{-1}(\mathsf{t}, (0^n, \mathsf{r}), \mathsf{x}')$, and replaces x in r^{Enc} with x', where x' is a uniformly random string sampled from $\{0,1\}^n$ under the following two conditions:
- $R\mathsf{x}' = w$ holds.
- $\mathsf{x}'_i = \mathsf{x}_i$ holds for every $i \in [n]$ such that $\mathsf{z}_i = 1$.

Before showing Exp 4 $\approx$ Exp 5, we review a basic lemma on inversion sampling.

Lemma 7. *For a function $f : \mathcal{X} \to \mathcal{Y}$, we define two distributions $\mathcal{D}_1$ and $\mathcal{D}_2$ as $\mathcal{D}_1 = \left\{(x, y) \mid x \xleftarrow{\$} \mathcal{X}, y = f(x)\right\}$ and $\mathcal{D}_2 = \left\{(x', y) \mid x \xleftarrow{\$} \mathcal{X}, y = f(x), x' \xleftarrow{\$} f^{-1}(y)\right\}$, where $f^{-1}(y)$ denotes the set of pre-images of y. Then, $\mathcal{D}_1$ and $\mathcal{D}_2$ are identical.*

Furthermore, we define a distribution $\mathcal{D}_3$ as $\mathcal{D}_3 = \left\{(x', y) \mid y \xleftarrow{\$} \mathcal{Y}, x' \xleftarrow{\$} f^{-1}(y)\right\}$. If f has a property that $f(x)$ distributes uniformly at random over $\mathcal{Y}$ if the input x distributes uniformly at random over $\mathcal{X}$, $\mathcal{D}_1$ and $\mathcal{D}_3$ are identical.

Lemma 8. $\mathsf{Exp}\ 4 \approx \mathsf{Exp}\ 5$ *holds.*

Proof. According to the Chernoff bound on z,

$$\Pr\left[\mathsf{Ham}(\mathsf{z}) \geq (1+\delta)\frac{n}{2}\right] \leq e^{-\frac{\delta^2}{2+\delta}\frac{n}{2}}$$

holds for any $\delta > 0$. By taking $\delta = 1 - \frac{2\mu}{n}$, we have

$$\Pr[\mathsf{Ham}(\mathsf{z}) \geq n - \mu] \leq 2^{-\lambda} = \mathsf{negl}(\lambda).$$

Below, we show that $(\mathsf{x}, R\mathsf{x})$ in $\mathsf{Exp}\ 4$ has the same distribution as (x', w) in $\mathsf{Exp}\ 5$ in the case of $\mathsf{Ham}(\mathsf{z}) < n - \mu$, and complete the proof of this lemma.

We first introduce some notations. For an integer ordered set $\mathcal{I} \subset [n]$, we define $R_{\mathcal{I}}$ as the restriction of R to $\mathcal{I}$, that is $R_{\mathcal{I}} = (\mathbf{r}_1|\cdots|\mathbf{r}_{|\mathcal{I}|})$, where $R = (\mathbf{r}_1|\cdots|\mathbf{r}_n)$. We define $\mathsf{x}_{\mathcal{I}}$ in a similar way.

Fix any z which satisfies $\mathsf{Ham}(\mathsf{z}) < n - \mu$ and set $\mathcal{I} = \{i_k \in [n] \mid \mathsf{z}_{i_k} = 0\}$. Because $|\mathcal{I}| \geq \mu$, $R_{\mathcal{I}}$ is full rank due to the choice of R. Hence, $R_{\mathcal{I}} \cdot u$ is uniformly random over $\{0,1\}^{\mu}$ if u is uniformly random over $\{0,1\}^{|\mathcal{I}|}$.

Then, from Lemma 7 when setting $\mathcal{X} := \{0,1\}^{|\mathcal{I}|}, \mathcal{Y} := \{0,1\}^{\mu}$, and $f(u) = R_{\mathcal{I}} \cdot u$, the distribution of $(\mathsf{x}_{\mathcal{I}}, R_{\mathcal{I}} \cdot \mathsf{x}_{\mathcal{I}})$ and (u, w) are the same, where $\mathsf{x} \stackrel{\$}{\leftarrow} \{0,1\}^n, u \stackrel{\$}{\leftarrow} f^{-1}(w) = \{u' \in \{0,1\}^{|\mathcal{I}|} \mid R_{\mathcal{I}} \cdot u' = w\}$, and $w \stackrel{\$}{\leftarrow} \{0,1\}^{\mu}$. Moreover, we have $R\mathsf{x} = R_{\mathcal{I}} \cdot \mathsf{x}_{\mathcal{I}} \oplus R_{[n]\setminus\mathcal{I}} \cdot \mathsf{x}_{[n]\setminus\mathcal{I}}$. Since x' sampled in $\mathsf{Exp}\ 5$ is a bit string generated by replacing i_k-th bit of x with k-th bit of u for every $k \in [|\mathcal{I}|]$, we see that $(\mathsf{x}, R\mathsf{x})$ has the same distribution as $(\mathsf{x}', w \oplus R_{[n]\setminus\mathcal{I}} \cdot \mathsf{x}_{[n]\setminus\mathcal{I}})$. $(\mathsf{x}', w \oplus R_{[n]\setminus\mathcal{I}} \cdot \mathsf{x}_{[n]\setminus\mathcal{I}})$ also has the same distribution as (x', w) because w is sampled uniformly at random, and thus $(\mathsf{x}, R\mathsf{x})$ has the same distribution as (x', w). This completes the proof of Lemma 8.

Note that we can sample the above u in polynomial time, by computing a particular solution $v \in \{0,1\}^{|\mathcal{I}|}$ of $R_{\mathcal{I}} \cdot v = w$, and add a vector sampled uniformly at random from the kernel of $R_{\mathcal{I}}$.

$\mathsf{Exp}\ 6$: This experiment is the same as $\mathsf{Exp}\ 6$ except that w is replaced with $w \oplus m$. By this change, CT is of the form

$$CT := \left(\mathsf{y}, \begin{pmatrix} \mathbf{e}_{1,0}, \ldots, \mathbf{e}_{n,0} \\ \mathbf{e}_{1,1}, \ldots, \mathbf{e}_{n,1} \end{pmatrix}, w\right).$$

Moreover, x' contained in r^{Enc} is sampled so that $R\mathsf{x}' = m \oplus w$ holds.

Since w is uniformly at random, so is $w \oplus m$. Thus, $\mathsf{Exp}\ 5 \equiv \mathsf{Exp}\ 6$ holds.

We see that $\mathsf{Exp}\ 6$ is the same as $\mathsf{Exp}^{\mathrm{Ideal}}_{\mathsf{NCE},\mathcal{A}}$. Put all the above arguments together, we have

$$\mathsf{Adv}_{\mathsf{NCE},\mathcal{A}}(\lambda) \leq |\Pr[\mathsf{out} = 1 \text{ in } \mathsf{Exp}\ 0] - \Pr[\mathsf{out} = 1 \text{ in } \mathsf{Exp}\ 6]| = \mathsf{negl}(\lambda).$$

Hence $\mathtt{NCE}$ is secure. This completes the proof of Theorem 3.

5.3 Ciphertext Rate

Finally, we evaluate the ciphertext rate of NCE. From Theorem 2, in order to make NCE correct, it is sufficient to set $\ell = \mathsf{poly}(\log \lambda)$. Moreover, from Theorem 3, in order to make NCE secure, it is sufficient to set $\mu = \mathcal{O}(\lambda)$ and $n = 8\mu$. In this setting, the ciphertext length of NCE is $|CT| = \lambda + 2n\ell + \mu$. Note that we assume a group element of $\mathbb{G}$ is described as a λ-bit string. Then, the ciphertext rate of NCE is evaluated as

$$\frac{|CT|}{\mu} = \frac{\lambda + 2n\ell + \mu}{\mu} = \mathcal{O}(\ell) = \mathsf{poly}(\log \lambda).$$

6 Full Construction of Proposed NCE

In the basic construction, we construct an NCE scheme with correctness $\gamma = 1 - \mathsf{negl}(\lambda)$, by setting $\ell = \mathsf{poly}(\log \lambda)$ which is the output length of E_2 (and thus D) of the underlying CE. Of course, if we set ℓ to $\mathcal{O}(\log \lambda)$, we can make the ciphertext rate of the resulting NCE scheme $\mathcal{O}(\log \lambda)$. However, this modification also affects the correctness of the resulting NCE scheme. γ is no longer $= 1 - \mathsf{negl}(\lambda)$, and is at most $1 - 1/\mathsf{poly}(\lambda)$.

Fortunately, we can amplify the correctness of the scheme to $1 - \mathsf{negl}(\lambda)$ from enough large constant without changing the ciphertext rate. For that purpose, we use a constant-rate error-correcting code which can correct errors up to some constant fraction. Concretely, we modify the scheme as follows. In the encryption, we first encode the plaintext by the error-correcting code and parse it into N blocks of length μ. Then, we encrypt each block by the γ-correct NCE scheme for a constant γ using different public keys. The decryption is done naturally, i.e., decrypt each ciphertext, concatenate them, and decode it. The ciphertext rate is still $\mathcal{O}(\log \lambda)$ because the rate of error-correcting code is constant.

This block-wise encryption technique not only amplifies the correctness but also reduces the public key size. In the basic construction, the size of a public key depends on the length of a message quadratically. However, by applying the block-wise encryption technique, it becomes linear in the length of a message.

The description of the full construction is as follows. Let ECC $=$ $(\mathsf{Encode}, \mathsf{Decode})$ be a constant-rate error-correcting code which can correct errors up to ϵ-fraction of the codeword where $\epsilon > 0$ is some constant.

Specifically, given a message $m \in \{0,1\}^{\mu M}$, Encode outputs a codeword $\overrightarrow{CW} \in \{0,1\}^{\mu N}$. If $\mathsf{Ham}\left(\overrightarrow{CW} - \overrightarrow{CW'}\right) \leq \epsilon\mu N$, $\mathsf{Decode}\left(\overrightarrow{CW'}\right) = m$. The rate of ECC is some constant N/M.

Let NCE $= (\mathsf{Gen}, \mathsf{Enc}, \mathsf{Dec}, \mathsf{Sim}_1, \mathsf{Sim}_2)$ be an NCE scheme whose message space is $\{0,1\}^{\mu}$, ciphertext rate is $\mathcal{O}(\log \lambda)$, and correctness is $\gamma = 1 - \frac{\epsilon}{2}$. We construct $\overrightarrow{\mathtt{NCE}} = (\overrightarrow{\mathsf{Gen}}, \overrightarrow{\mathsf{Enc}}, \overrightarrow{\mathsf{Dec}}, \overrightarrow{\mathsf{Sim}_1}, \overrightarrow{\mathsf{Sim}_2})$ as follows. The message space of $\overrightarrow{\mathtt{NCE}}$ is $\{0,1\}^{\mu M}$.

$\overrightarrow{\mathsf{Gen}}\left(1^\lambda; \overrightarrow{r^{\mathsf{Gen}}}\right)$:

- Parse the given random coin to $\overrightarrow{r^{\mathsf{Gen}}} = \left(r_1^{\mathsf{Gen}}, \ldots, r_N^{\mathsf{Gen}}\right)$.
- For all $i \in [N]$, generate key pairs $(pk_i, sk_i) \leftarrow \mathsf{Gen}\left(1^\lambda; r_i^{\mathsf{Gen}}\right)$.
- Output $\overrightarrow{pk} := (pk_1, \ldots, pk_N)$ and $\overrightarrow{sk} := (sk_1, \ldots, sk_N)$.

$\overrightarrow{\mathsf{Enc}}\left(\overrightarrow{pk}, m; \overrightarrow{r^{\mathsf{Enc}}}\right)$:

- Parse $\overrightarrow{r^{\mathsf{Enc}}} = \left(r_1^{\mathsf{Enc}}, \ldots, r_N^{\mathsf{Enc}}\right)$.
- Compute $\overrightarrow{CW} \leftarrow \mathsf{Encode}(m)$ and parse $\overrightarrow{CW} = (CW_1, \ldots, CW_N)$.
- For all $i \in [N]$, compute $CT_i \leftarrow \mathsf{Enc}\left(pk_i, CW_i; r_i^{\mathsf{Enc}}\right)$.
- Output $\overrightarrow{CT} := (CT_1, \ldots, CT_N)$.

$\overrightarrow{\mathsf{Dec}}\left(\overrightarrow{sk}, \overrightarrow{CT}\right)$:

- For all $i \in [N]$, Compute $CW_i' \leftarrow \mathsf{Dec}(sk_i, CT_i)$.
- Concatenate them as $\overrightarrow{CW'} := (CW_1', \ldots, CW_N')$.
- Output $m \leftarrow \mathsf{Decode}\left(\overrightarrow{CW'}\right)$.

$\overrightarrow{\mathsf{Sim}_1}\left(1^\lambda\right)$:

- For all $i \in [N]$, compute $(pk_i, CT_i, st_i) \leftarrow \mathsf{Sim}_1\left(1^\lambda\right)$,
- Output $\overrightarrow{pk} := (pk_1, \ldots, pk_N)$, $\overrightarrow{CT} := (CT_1, \ldots, CT_N)$, and $\overrightarrow{st} := (st_1, \ldots, st_N)$.

$\overrightarrow{\mathsf{Sim}_2}\left(m, \overrightarrow{st}\right)$:

- Compute $\overrightarrow{CW} \leftarrow \mathsf{Encode}(m)$ and parse $(CW_1, \ldots, CW_N) \leftarrow \overrightarrow{CW}$.
- For all $i \in [N]$, compute $\left(r_i^{\mathsf{Gen}}, r_i^{\mathsf{Enc}}\right) \leftarrow \mathsf{Sim}_2(CW_i, st_i)$.
- Output $\overrightarrow{r^{\mathsf{Gen}}} := \left(r_1^{\mathsf{Gen}}, \ldots, r_N^{\mathsf{Gen}}\right)$ and $\overrightarrow{r^{\mathsf{Enc}}} := \left(r_1^{\mathsf{Enc}}, \ldots, r_N^{\mathsf{Enc}}\right)$.

Correctness. We can prove the correctness of $\overrightarrow{\mathtt{NCE}}$ by the Chernoff bound. Formally, we have the following theorem. See the full version for the proof.

Theorem 4. *Let* ECC *be an constant-rate error-correcting code which can correct errors up to ϵ-fraction of a codeword. Let* NCE *be a γ-correct NCE scheme, where $\gamma = 1 - \frac{\epsilon}{2}$. If the number of parsed codeword $N \geq \mathsf{poly}(\log \lambda)$, the above $\overrightarrow{\mathtt{NCE}}$ is correct.*

Security. For the security of $\overrightarrow{\mathtt{NCE}}$, we have the following theorem. Since we can prove it via a straightforward hybrid argument, we omit it.

Theorem 5. *If* NCE *is an secure NCE scheme, then $\overrightarrow{\mathtt{NCE}}$ is also secure.*

Ciphertext Rate. Since rate of the error-correcting code N/M is constant, the ciphertext rate of $\overrightarrow{\mathtt{NCE}}$ is $\frac{N|CT|}{\mu M} = \mathcal{O}(\ell) = \mathcal{O}(\log \lambda)$.

Acknowledgements. A part of this work was supported by NTT Secure Platform Laboratories, JST OPERA JPMJOP1612, JST CREST JPMJCR14D6, JSPS KAKENHI JP16H01705, JP17H01695, JP19J22363.

References

1. Beaver, D.: Plug and play encryption. In: Kaliski, B.S. (ed.) CRYPTO 1997. LNCS, vol. 1294, pp. 75–89. Springer, Heidelberg (1997). https://doi.org/10.1007/BFb0052228
2. Bellare, M., Palacio, A.: The knowledge-of-exponent assumptions and 3-round zero-knowledge protocols. In: Franklin, M. (ed.) CRYPTO 2004. LNCS, vol. 3152, pp. 273–289. Springer, Heidelberg (2004). https://doi.org/10.1007/978-3-540-28628-8_17
3. Brakerski, Z., Lombardi, A., Segev, G., Vaikuntanathan, V.: Anonymous IBE, leakage resilience and circular security from new assumptions. In: Nielsen, J.B., Rijmen, V. (eds.) EUROCRYPT 2018, Part I. LNCS, vol. 10820, pp. 535–564. Springer, Cham (2018). https://doi.org/10.1007/978-3-319-78381-9_20
4. Canetti, R., Feige, U., Goldreich, O., Naor, M.: Adaptively secure multi-party computation. In: 28th ACM STOC, pp. 639–648 (1996)
5. Canetti, R., Poburinnaya, O., Raykova, M.: Optimal-rate non-committing encryption. In: Takagi, T., Peyrin, T. (eds.) ASIACRYPT 2017, Part III. LNCS, vol. 10626, pp. 212–241. Springer, Cham (2017). https://doi.org/10.1007/978-3-319-70700-6_8
6. Cho, C., Döttling, N., Garg, S., Gupta, D., Miao, P., Polychroniadou, A.: Laconic oblivious transfer and its applications. In: Katz, J., Shacham, H. (eds.) CRYPTO 2017, Part II. LNCS, vol. 10402, pp. 33–65. Springer, Cham (2017). https://doi.org/10.1007/978-3-319-63715-0_2
7. Choi, S.G., Dachman-Soled, D., Malkin, T., Wee, H.: Improved non-committing encryption with applications to adaptively secure protocols. In: Matsui, M. (ed.) ASIACRYPT 2009. LNCS, vol. 5912, pp. 287–302. Springer, Heidelberg (2009). https://doi.org/10.1007/978-3-642-10366-7_17
8. Damgård, I., Nielsen, J.B.: Improved non-committing encryption schemes based on a general complexity assumption. In: Bellare, M. (ed.) CRYPTO 2000. LNCS, vol. 1880, pp. 432–450. Springer, Heidelberg (2000). https://doi.org/10.1007/3-540-44598-6_27
9. Döttling, N., Garg, S.: From selective IBE to full IBE and selective HIBE. In: Kalai, Y., Reyzin, L. (eds.) TCC 2017, Part I. LNCS, vol. 10677, pp. 372–408. Springer, Cham (2017). https://doi.org/10.1007/978-3-319-70500-2_13
10. Döttling, N., Garg, S.: Identity-based encryption from the Diffie-Hellman assumption. In: Katz, J., Shacham, H. (eds.) CRYPTO 2017, Part I. LNCS, vol. 10401, pp. 537–569. Springer, Cham (2017). https://doi.org/10.1007/978-3-319-63688-7_18
11. Döttling, N., Garg, S., Hajiabadi, M., Masny, D.: New constructions of identity-based and key-dependent message secure encryption schemes. In: Abdalla, M., Dahab, R. (eds.) PKC 2018, Part I. LNCS, vol. 10769, pp. 3–31. Springer, Cham (2018). https://doi.org/10.1007/978-3-319-76578-5_1
12. Döttling, N., Garg, S., Ishai, Y., Malavolta, G., Mour, T., Ostrovsky, R.: Trapdoor hash functions and their applications. In: Boldyreva, A., Micciancio, D. (eds.) CRYPTO 2019, Part III. LNCS, vol. 11694, pp. 3–32. Springer, Cham (2019). https://doi.org/10.1007/978-3-030-26954-8_1
13. Garg, S., Gay, R., Hajiabadi, M.: New techniques for efficient trapdoor functions and applications. In: Ishai, Y., Rijmen, V. (eds.) EUROCRYPT 2019, Part III. LNCS, vol. 11478, pp. 33–63. Springer, Cham (2019). https://doi.org/10.1007/978-3-030-17659-4_2

14. Garg, S., Hajiabadi, M.: Trapdoor functions from the computational Diffie-Hellman assumption. In: Shacham, H., Boldyreva, A. (eds.) CRYPTO 2018, Part II. LNCS, vol. 10992, pp. 362–391. Springer, Cham (2018). https://doi.org/10.1007/978-3-319-96881-0_13
15. Garg, S., Ostrovsky, R., Srinivasan, A.: Adaptive garbled RAM from laconic oblivious transfer. In: Shacham, H., Boldyreva, A. (eds.) CRYPTO 2018, Part III. LNCS, vol. 10993, pp. 515–544. Springer, Cham (2018). https://doi.org/10.1007/978-3-319-96878-0_18
16. Garg, S., Srinivasan, A.: Adaptively secure garbling with near optimal online complexity. In: Nielsen, J.B., Rijmen, V. (eds.) EUROCRYPT 2018, Part II. LNCS, vol. 10821, pp. 535–565. Springer, Cham (2018). https://doi.org/10.1007/978-3-319-78375-8_18
17. Garg, S., Srinivasan, A.: Two-round multiparty secure computation from minimal assumptions. In: Nielsen, J.B., Rijmen, V. (eds.) EUROCRYPT 2018, Part II. LNCS, vol. 10821, pp. 468–499. Springer, Cham (2018). https://doi.org/10.1007/978-3-319-78375-8_16
18. Hada, S., Tanaka, T.: On the existence of 3-round zero-knowledge protocols. In: Krawczyk, H. (ed.) CRYPTO 1998. LNCS, vol. 1462, pp. 408–423. Springer, Heidelberg (1998). https://doi.org/10.1007/BFb0055744
19. Hemenway, B., Ostrovsky, R., Richelson, S., Rosen, A.: Adaptive security with quasi-optimal rate. In: Kushilevitz, E., Malkin, T. (eds.) TCC 2016, Part I. LNCS, vol. 9562, pp. 525–541. Springer, Heidelberg (2016). https://doi.org/10.1007/978-3-662-49096-9_22
20. Hemenway, B., Ostrovsky, R., Rosen, A.: Non-committing encryption from ϕ-hiding. In: Dodis, Y., Nielsen, J.B. (eds.) TCC 2015, Part I. LNCS, vol. 9014, pp. 591–608. Springer, Heidelberg (2015). https://doi.org/10.1007/978-3-662-46494-6_24
21. Tao, T., Vu, V.: On the singularity probability of random Bernoulli matrices. J. Am. Math. Soc. **20**(3), 603–628 (2007)

Structure-Preserving and Re-randomizable RCCA-Secure Public Key Encryption and Its Applications

Antonio Faonio[1(✉)], Dario Fiore[1], Javier Herranz[2], and Carla Ràfols[3]

[1] IMDEA Software Institute, Madrid, Spain
antonio.faonio@imdea.org
[2] Cybercat and Universitat Politècnica de Catalunya, Barcelona, Spain
[3] Cybercat and Universitat Pompeu Fabra, Barcelona, Spain

Abstract. Re-randomizable RCCA-secure public key encryption (Rand-RCCA PKE) schemes reconcile the property of re-randomizability of the ciphertexts with the need of security against chosen-ciphertexts attacks. In this paper we give a new construction of a Rand-RCCA PKE scheme that is perfectly re-randomizable. Our construction is structure-preserving, can be instantiated over Type-3 pairing groups, and achieves better computation and communication efficiency than the state of the art perfectly re-randomizable schemes (e.g., Prabhakaran and Rosulek, CRYPTO'07). Next, we revive the Rand-RCCA notion showing new applications where our Rand-RCCA PKE scheme plays a fundamental part: (1) We show how to turn our scheme into a *publicly-verifiable* Rand-RCCA scheme; (2) We construct a malleable NIZK with a (variant of) simulation soundness that allows for re-randomizability; (3) We propose a new UC-secure Verifiable Mix-Net protocol that is secure in the common reference string model. Thanks to the structure-preserving property, all these applications are efficient. Notably, our Mix-Net protocol is the most efficient universally verifiable Mix-Net (without random oracle) where the CRS is an uniformly random string of size independent of the number of senders. The property is of the essence when such protocols are used in large scale.

1 Introduction

Security against chosen ciphertext attacks (CCA) is considered by many the gold standard for public key encryption (PKE). Since the seminal paper of Micali,

First and second authors are supported by the Spanish Government through the projects Datamantium (ref. RTC-2016-4930-7), SCUM (RTI2018-102043-B-I00), and ERC2018-092822, and by the Madrid Regional Government under project BLOQUES (ref. S2018/TCS-4339).
The work of the third author is partially supported by Spanish Government through project MTM2016-77213-R.
The fourth author was supported by a Marie Curie "UPF Fellows" Postdoctoral Grant and by Project RTI2018-102112-B-I00 (AEI/FEDER,UE).

S. D. Galbraith and S. Moriai (Eds.): ASIACRYPT 2019, LNCS 11923, pp. 159–190, 2019.
https://doi.org/10.1007/978-3-030-34618-8_6

Rackoff and Sloan [30], the research community has spent a great effort on this fundamental topic by both interconnecting different security notions and producing a large body of efficient public encryption schemes.

Challenging the overwhelming agreement that CCA security is **the** right notion of security for PKE, a paper of Canetti, Krawczyk and Nielsen [6] showed that for many use cases a weaker security notion than CCA security is already sufficient. More in details, the paper introduced the notion of Replayable CCA (RCCA) and showed that the notion is sufficient to realize a variant of the public key encryption functionality in the universal composability (UC) model of Canetti [3] where only replay attacks, namely attacks in which the data could be maliciously repeated, can be mounted by the adversary.

In a nutshell, the main fundamental difference between RCCA security and CCA security is that, in a RCCA secure scheme (which is not CCA secure) an adversary is able to maul the challenge ciphertext to obtain new decryptable ciphertexts, the only limitation is that the adversary still cannot break the integrity of the underlying plaintext. To explain this with an example, in a RCCA secure PKE scheme an adversary might append an extra 0 at the end of the ciphertext and still be able to obtain a valid decryption of the mauled ciphertext (to the same plaintext), on the other hand, for a CCA secure PKE, this attack should by definition result into an invalid decryption.

Later, Groth [21] showed that the capability to maul a ciphertext to obtain a new ciphertext which decrypts to the same plaintext should be seen as a feature and not a weakness. In his paper, he introduced the notion of re-randomizable RCCA (Rand-RCCA) PKE, namely a RCCA-secure PKE which comes with an algorithm that re-randomizes the ciphertexts in a way that cannot be linked.

PKE schemes that are both re-randomizable and RCCA-secure have been shown to have several applications, such as: anonymous and secure message transmissions (see Prabhakaran and Rosulek [34]), Mix-Nets (see Faonio and Fiore [14], and Pereira and Rivest [32]), Controlled Functional Encryption (see Naveed *et al.* [31]), and one-round message-transmission protocols with reverse firewalls (see Dodis, Mironov, and Stephens-Davidowitz [11]).

When it comes to constructing these objects, if we look at the literature it is striking to observe that there are extremely efficient constructions of schemes that are only RCCA-secure but not re-randomizable (e.g., Cramer-Shoup [8] or Phan-Pointcheval [33]), or are re-randomizable but only CPA-secure (e.g., ElGamal [12]). In contrast, when the two properties are considered in conjunction, a considerable gap in the efficiency of the schemes seems to arise. More in concrete, the most efficient Rand-RCCA scheme in the standard model of [34] has ciphertexts of 20 groups elements,[1] while, for example, the celebrated Cramer-Shoup PKE [8] has ciphertexts of only 4 groups elements.

In the following paragraphs we state the main contributions of our work.

[1] A recent work of Faonio and Fiore [14] takes this down to 11 group elements at the price of achieving a strictly weaker notion of re-randomizability, in the random oracle model.

Rand-RCCA PKE. Our first contribution is a new structure-preserving[2] Rand-RCCA PKE scheme which significantly narrows the efficiency gap described above. The scheme is secure under the Matrix Diffie-Hellman Assumption (MDDH) in bilinear groups, and for its strongest instantiation, namely, under the Symmetric External Diffie-Hellman Assumption (SXDH), has ciphertexts of 6 groups elements (3 elements in $\mathbb{G}_1$, 2 elements in $\mathbb{G}_2$ and 1 element in $\mathbb{G}_T$).

From a practical perspective, the advantage of a re-randomizable PKE over a standard (non-re-randomizable) PKE strikes when the re-randomizable PKE scheme is part of a larger protocol. To this end, we notice that the structure-preserving property is indeed vital as it allows for modularity and easy integration, which are basic principles for protocol design. However, we can substantiate further our assertion by giving three applications where structure-preserving Rand-RCCA PKE schemes are essential.

Publicly-Verifiable Rand-RCCA PKE. Our first application is a publicly-verifiable (pv) Rand-RCCA PKE scheme. A PKE scheme is publicly verifiable when the validity of a ciphertext can be checked without the secret key. This property is for example convenient in the setting of threshold decryption with CCA security [4,36], as the task, roughly speaking, reduces to first publicly check the validity of the ciphertext and then CPA-threshold-decrypt it. Very roughly speaking, we can obtain our pv-Rand-RCCA PKE scheme by appending a Groth-Sahai (GS) NIZK proof [23] of the validity of the ciphertext. We notice that the ciphertext of our Rand-PKE scheme contains[3] an element in $\mathbb{G}_T$. The verification equation does not admit a GS NIZK proof, but only NIWI. We overcome this problem by constructing an additional commitment type for elements in $\mathbb{G}_T$. This gives us a *new* general technique that extends the class of pairing product equations which admit GS NIZK proofs, enlarging therefore the notion of structure preserving. The latter is a contribution of independent interest which might have applications in the field of structure-preserving cryptography in general.

Controlled-Malleable NIZKs. Our second application is a general framework for true-simulation extractable (tSE) and re-randomizable (more generally, controlled-malleable) NIZK systems. The notion of tSE-NIZK was introduced by Dodis *et al.* [10] and found a long series of applications (see for example [9,16,18]). Briefly, the notion assures soundness of the NIZK proofs even when the adversary gets to see simulated NIZK proofs for *true* statements of its choice. In comparison with simulation-extractable (SE) NIZKs (see [22,35]), tSE-NIZKs are considerably more efficient and keep many of the benefits which motivated the introduction of SE-NIZKs[4]. However, if one would like a *controlled malleable*

[2] A scheme is structure preserving if all its public materials, such as messages, public keys, etc. are group elements and the correctness can be verified via pairing-product equations.

[3] In the lingo of structure-preserving cryptography, the scheme is not *strongly* structure preserving.

[4] As an example, tSE-NIZKs are sufficient for the CCA2-secure Naor-Yung PKE of Sahai [35], simulation-sound (SS) NIZKs were introduced in the same paper with exactly this application in mind.

tSE-NIZK, the only available scheme is an SE-NIZK obtained through the general result of Chase *et al.* [7], which is not very efficient. As main result, we scale down the framework of Chase *et al.* to true-simulation extractability, and by using our new Rand-RCCA PKE we construct a new re-randomizable tSE-NIZK scheme. Compared to [7], our scheme can handle a more restricted class of relations and transformations,[5] but our proofs are significantly more efficient. For example, for simple re-randomizable NIZK proofs our tSE NIZKs have an overhead of the order of *tens* more pairing operations for verification, opposed to an overhead of the order of *hundreds* more pairing operations for verification of the simulation-extractable with controlled malleability NIZK systems of [7]. The overhead is computed as the difference with the adaptive sound Groth-Sahai NIZK proof for the same statement.

Mix-Net. Our third application is a universally verifiable and UC-secure Mix-Net based on our pv-Rand-RCCA PKE scheme. Recently, Faonio and Fiore [14] gave a new paradigm to obtain UC-secure verifiable Mix-Net protocols based on Rand-RCCA PKE scheme. Their construction makes use of a non-publicly verifiable Rand-RCCA PKE scheme and obtains a weaker notion of security called *optimistic* (*àla* Golle *et al.* [20]). More in details, the mixing paradigm of [14] is conceptually simple: a mixer receives a list of Rand-RCCA ciphertexts and outputs a randomly permuted list of re-randomized ciphertexts together with a simple NIZK proof that they informally dub "loose shuffling". Such "loose shuffling" proof guarantees that if all the ciphertexts correctly decrypt then the output list is a shuffle of the input one. Hence, in their scheme, cheating can be caught at decryption time, that is after the last mixer returned its list. The problem is that, cheating might be caught too late, thus, their scheme is only optimistic secure. Namely, the scheme is an universal verifiable mix-net optimized to quickly produce a correct output when all the mixers run the protocol correctly. If instead one or more mixers cheat, then no privacy is guaranteed but one can "back up" to a different, slow, mix-net execution.

In this paper, we show that by leveraging the public verifiability of the Rand-RCCA PKE scheme we can obtain a simple design for Mix-Net protocols. In fact, since it is possible to publicly check that a mixer did not invalidate any ciphertext, the proof of loose shuffling turns out to be, indeed, a proof of shuffle.

Interestingly, our use of publicly verifiable ciphertexts come with additional benefits. As mentioned in the paragraph above, our pv-RCCA-PKE scheme can support threshold decryption very easily, and more efficiently than Faonio and Fiore [14]. Finally, our protocol can be fully instantiated in the standard model, whereas the one in [14] rely on non-programmable random oracles.

[5] Yet, our framework is powerful enough for the application of controlled-malleable CCA security of Chase *et al.* Interestingly, we can obtain another pv-Rand-RCCA PKE through their paradigm, although less efficient than our construction. We believe that analyzing what other kinds of CM-CCA notions are supported by our scheme is interesting future work.

Most notably, our protocol is the *first efficient universally verifiable Mix-Net in the common random string model*, namely where the common reference string is a (small) uniformly random string. In fact, a popular approach to achieve a universally verifiable Mix-Net is to use a NIZK proof of shuffle. However, the most efficient protocols for this task either rely on random oracles to become non-interactive (such as the protocol of Bayer and Groth [1] or Verificatum [39]), or need a structured common reference string (as is the case for the most efficient state-of-the-art NIZK proof of shuffle of Fauzi *et al.* [17]). Furthermore, the common reference string of [17] has size that depends on the number of senders (which in practical scenarios can be huge), whereas our common reference string is made by a number of group elements that is linear in the number of mixers.

Our Mix-Net protocol is proved secure based only on general properties of the pv-Rand-RCCA PKE scheme, and can be instantiated with other schemes in literature (for example with the schemes in [7,29]).

Controlled-Malleable Smooth Projective Hash Functions. At the core of our Rand-RCCA PKE scheme is a new technique that can be seen as a re-randomizable version of smooth projective hash functions (SPHFs) [8]. Given the pervasive use of SPHFs in cryptographic constructions, we believe that our technique may find more applications in the realm of re-randomizable cryptographic primitives. For this reason, we formalize our technique as a primitive called *controlled-malleable SPHF*. Briefly, we define it as an SPHF with tags that allows to re-randomize both instances and tags (inside appropriate spaces), and for which soundness (i.e., smoothness) holds even if the adversary can see a hash value for an invalid instance. We elaborate on this notion in the full version of this paper [15].

Comparison with Related Work. If we consider the state of the art of Rand-RCCA PKE schemes, the most relevant works are the work of Groth, which introduced the notion of Rand-RCCA PKE scheme [21], the aforementioned scheme of Prabhakaran and Rosulek [34], the Rand-RCCA PKE scheme of Chase *et al.* derived from their malleable NIZK systems [7], and two recent works of Libert, Peters and Qian [29] and of Faonio and Fiore [14]. In Table 1 we offer a comparison, in terms of security and functionality properties, of our schemes of Sect. 3 ($\mathcal{PKE}_1$) and Sect. 4 ($\mathcal{PKE}_2$) against previous schemes.

From a technical point of view, the scheme of [34] and our scheme $\mathcal{PKE}_1$, although both based on the Cramer-Shoup paradigm, have little in common. The main differences are: (1) a different design to handle the tags (see next section); (2) a different approach for the re-randomization of the ciphertext. In particular, the Rand-PKE scheme of [34] uses the double-strand technique of Golle *et al.* [19] to re-randomize the ciphertext, while our re-randomization technique, as far as we know, is novel. Furthermore, the scheme of [34] works in two special groups, $\hat{\mathbb{G}}$ and $\tilde{\mathbb{G}}$ that are the subgroups of quadratic residues of $\mathbb{Z}^*_{2q+1}$ and $\mathbb{Z}^*_{4q+3}$ respectively, for a prime q such that $(q, 2q+1, 4q+3)$ is a sequence of primes (a Cunningham Chain of the first kind of length 3).

In Table 2 we compare the efficiency of our new schemes (in the most efficient instantiation with $k = 1$) with the most efficient ones among the Rand-RCCA

Table 1. Comparison of the properties of a selection of Rand-RCCA-secure PKE schemes. For group setting, – means any group where the assumption holds; Cunn. refers to a pair of groups whose prime orders form a Cunningham chain (see [34]); Bil. stands for bilinear groups. For model, GGM refers to generic group and NPRO refers to non-programmable random oracle. * the structure-preserving property of the two schemes in this paper is not strict, since ciphertexts contain some elements in $\mathbb{G}_T$.

PKE	Group Setting	Assumption	Model	Struc. Pres.	Pub. Ver.	Re-Rand
[21] Groth	–	DDH	GGM			perfect
[34] PR07	Cunn.	DDH	std			perfect
[7,29] CKLM12, LPQ17	Bilin.	SXDH	std	✓	✓	perfect
[14] FF18	–	DDH	NPRO			weak
$\mathcal{PKE}_1$	Bilin.	$\mathcal{D}_k$-MDDH	std	✓*		perfect
$\mathcal{PKE}_2$	Bilin.	$\mathcal{D}_k$-MDDH	std	✓*	✓	perfect

schemes: the ones in [34] and [14] for the case of secret verifiability, and the scheme in [29] for publicly verifiable Rand-RCCA encryption.

Among the schemes with private verifiability, the most efficient one is that in [14], but its re-randomizability property is weak and the security is in the random oracle model. Among the other two, our scheme $\mathcal{PKE}_1$ is more efficient than that in [34], because the special groups $\tilde{\mathbb{G}}$ required in [34] are large, at least 3072 bits for a security level of 128 bits. Turning to comparing with publicly verifiable schemes, the computational costs for the scheme in [29], in the table, are roughly approximate, because not all the exact computations in the algorithms of the scheme (involving Groth-Sahai proofs) are explicitly described. The size of the ciphertexts reported in [29] is $34|\mathbb{G}_1| + 18|\mathbb{G}_2|$. After personal communication with the authors, we realized that this number is not correct; the correct one is $42|\mathbb{G}_1| + 20|\mathbb{G}_2|$. Our scheme $\mathcal{PKE}_2$ is the most efficient Rand-RCCA scheme with public verifiability up to date: ciphertext size is comparable to that in [29] whereas the computational costs are significantly lower. Even for ciphertext size, ours is comparable to [29] only due to the size of the 4 $\mathbb{G}_T$ elements in our scheme. Besides that, our ciphertexts have many fewer group elements, which is conceptually simpler and, we believe, leaves hope for further improvements. For the two publicly verifiable schemes, the number of pairings required for decryption can be decreased, at the cost of increasing the number of exponentiations, by applying the batching techniques in [24]. The resulting number would be 22P for $\mathcal{PKE}_2$ and something between 40P and 50 P for the scheme in [29].

Technical Overview. We recall that the main technical contributions of this paper are: (1) a new technique for Rand-RCCA PKE scheme (which we also formalize in terms of SPHFs), (2) a new general technique that extends significantly the class of pairing product equations which admits GS NIZK proofs, and (3) a new technique for standard-model UC-secure verifiable Mix-Nets. For space reason, in this technical overview we concentrate on (1).

Table 2. Efficiency comparison among the best Rand-RCCA-secure PKE schemes; only the last two rows include schemes with public verifiability. For our schemes we consider $k = 1$, so based on SXDH assumption. We use $\tilde{\mathbb{G}}$ for the special groups used in [34], $\mathbb{G}$ for standard DDH groups as considered in [14], and then groups in asymmetric bilinear pairings $e : \mathbb{G}_1 \times \mathbb{G}_2 \rightarrow \mathbb{G}_T$ as considered both in [29] and in this work. Similarly, we denote as $E, \tilde{E}, E_1, E_2, E_T$ the cost of an exponentiation in groups $\mathbb{G}, \tilde{\mathbb{G}}, \mathbb{G}_1, \mathbb{G}_2, \mathbb{G}_T$, respectively. Finally, P denotes the cost of computing a bilinear pairing.

PKE	Enc ≈ Rand	Dec	$\lvert \mathsf{C} \rvert$	$\lvert \mathsf{pk} \rvert$
PR07	$22\ \tilde{E}$	$32\ \tilde{E}$	$20\tilde{\mathbb{G}}$	$11\tilde{\mathbb{G}}$
FF18	$16\ E$	$18\ E$	$11\mathbb{G}$	$11\mathbb{G}$
$\mathcal{PKE}_1$	$4E_1{+}5E_2{+}2E_T{+}5\mathsf{P}$	$8E_1{+}4E_2{+}4\mathsf{P}$	$3\mathbb{G}_1{+}2\mathbb{G}_2{+}\mathbb{G}_T$	$7\mathbb{G}_1{+}7\mathbb{G}_2{+}2\mathbb{G}_T$
LPQ17	$79E_1{+}64E_2$	$1E_1{+}142\mathsf{P}$	$42\mathbb{G}_1{+}20\mathbb{G}_2$	$11\mathbb{G}_1{+}16\mathbb{G}_2$
$\mathcal{PKE}_2$	$35E_1{+}31E_2{+}6E_T{+}5\mathsf{P}$	$2E_1{+}46\mathsf{P}$	$12\mathbb{G}_1{+}11\mathbb{G}_2{+}4\mathbb{G}_T$	$8\mathbb{G}_1{+}8\mathbb{G}_2$

A common technique of many CCA-secure PKE schemes in the standard model consists in explicitly labeling each ciphertext produced by the encryption algorithm with a unique tag. Some notable examples of CCA-secure PKE schemes that use tags are the Cramer-Shoup PKE [8], the tag-based PKE of Kiltz [27], and IBE-to-CCA transform of Canetti, Halevi and Katz [5].

Unfortunately, unique tags are not a viable option when designing a re-randomizable PKE scheme. In fact, a ciphertext and its re-randomization would share the same tag, and so they could be trivially linked by an attacker. The main consequence is that many well-known techniques in CCA security cannot be easily exported in the context of Rand-RCCA security. A remarkable exception is the work on Rand-RCCA PKE of Prabhakaran and Rosulek [34]. In this work, the authors managed to reconcile tags and re-randomizability with an ingenious technique: the tag for a new ciphertext is computed as a re-randomizable encoding of the plaintext itself, the tag is then encrypted and attached to the rest of the ciphertext. The decryptor first decrypts the tag and then uses it to check the validity of the payload ciphertext. More in details, the PKE scheme follows the Cramer-Shoup paradigm, therefore their tag (more accurately, a part of their tag) is a $\mathbb{Z}_q$ element (for a properly chosen q). Unfortunately, the restriction on the type of the tags implies that the scheme can be instantiated only in special groups $\mathbb{G}$ of prime order q where the DDH assumption simultaneously holds for both $\mathbb{Z}_q$ and $\mathbb{G}$. Conclusively, the main drawback is a quite large ciphertext size.

We use bilinear-pairing cryptography to overcome the problem of the tags in $\mathbb{Z}_q$. Our starting point is the structure-preserving CCA-PKE of Camenisch *et al.* [2]. Briefly, their PKE scheme is based on the Cramer-Shoup paradigm, with the main twist of performing the validity check in $\mathbb{G}_T$. This trick allows to move the tags from $\mathbb{Z}_q$ to the source group. We give a brief description of the ideas underlying our PKE scheme. We use the implicit notation of Escala *et al.* [13], that uses additive notation for groups and where elements in $\mathbb{G}_i$, are denoted as $[a]_i := a\mathcal{P}_i$ where $\mathcal{P}_i$ is a generator for $\mathbb{G}_i$. The PKE scheme of [2] uses Type-1 pairing groups (where $\mathbb{G}_1 = \mathbb{G}_2$) which are less efficient and secure than Type-3 pairing groups (where no efficient isomorphism from $\mathbb{G}_2$ to $\mathbb{G}_1$ is known to exist). As a first step, we convert their scheme to Type-3 pairing groups; however, for simplicity, in this overview we present the Type-1 version.

Following the blue print of Cramer and Shoup, a ciphertext of the PKE scheme of Camenisch *et al.* consists of three elements: a vector $[\mathbf{c}]_1 \in \mathbb{G}_1^3$ which we call the *instance* (for the DLIN problem described by a matrix $[\mathbf{D}]_1 \in \mathbb{G}_1^{3\times 2}$), an element $[p]_1$ which we call the *payload*, and an element $[\pi]_T$ which we call the *hash*. Together, the instance and the payload form the *tag*, that we denote as $[\mathbf{x}]_1 = [(\mathbf{c}^\top, p)^\top]_1$. The hash is, briefly speaking, a tag-based designated-verifier zero-knowledge proof of the randomness of $[\mathbf{c}]_1$ (namely, that $[\mathbf{c}]_1 = [\mathbf{D}]_1 \cdot \mathbf{r}$). The main difference is that in Cramer-Shoup PKE the tag is computed as a collision-resistant hash of $[\mathbf{x}]_1$, while in our scheme the is the value $[\mathbf{x}]_1$ itself. More in details, the public key material consists of $[\mathbf{D}^*]_1 = [(\mathbf{D}^\top, (\mathbf{a}^\top \mathbf{D})^\top)^\top]_1$, $[\mathbf{f}^\top \mathbf{D}]_T$, and $[\mathbf{F}^\top \mathbf{D}]_1$, where $\mathbf{a}, \mathbf{f} \in \mathbb{Z}_q^3$ and $\mathbf{F} \in \mathbb{Z}_q^{3\times 4}$ are uniformly random, and the encryption algorithm on message $[m]_1$ computes the tag as $[\mathbf{x}]_1 = [\mathbf{D}^*]_1 \cdot \mathbf{r} + [(\mathbf{0}^\top, m)^\top]_1$, and the proof of consistency as $([\mathbf{f}^\top \mathbf{D}]_T + [(\mathbf{F}^\top \mathbf{D})^\top \cdot \mathbf{x}]_T) \cdot \mathbf{r}$, where the addend $[(\mathbf{F}^\top \mathbf{D})^\top \cdot \mathbf{x}]_T$ can be efficiently computed using the pairing. Using the terminology of SPHFs, the hash of the instance $[\mathbf{c}]_1$ and tag $[\mathbf{x}]_1$ is produced using the projective hash algorithm which takes as input the witness $\mathbf{r}$ for $[\mathbf{c}]_1 \in span([\mathbf{D}])$, the tag $[\mathbf{x}]_1$ and the projection key $([\mathbf{f}^\top \mathbf{D}]_T, [\mathbf{F}^\top \mathbf{D}]_1)$. The decryption procedure can re-compute the hash as $e(\mathbf{f}^\top [\mathbf{c}]_1, [1]_1) + e([\mathbf{x}]_1, \mathbf{F}^\top [\mathbf{c}]_1)$, without the knowledge of the witness $\mathbf{r}$ but only using the hash key $(\mathbf{f}, \mathbf{F})$.

To validly re-randomize a ciphertext, the goal would be to compute, using only public information, a new ciphertext where the tag is of the form $[\mathbf{x}'] = [\mathbf{D}^*](\mathbf{r} + \hat{\mathbf{r}}) + [(\mathbf{0}^\top, m)^\top]_1$ (and therefore the instance is of the form $[\mathbf{c}'] = [\mathbf{D}](\mathbf{r} + \hat{\mathbf{r}})$) and the hash is of the form $([\mathbf{f}^\top \mathbf{D}]_T + [(\mathbf{F}^\top \mathbf{D})^\top \mathbf{x}']_T)(\mathbf{r} + \hat{\mathbf{r}})$. However, computing such a re-randomization of the hash is actually infeasible since the scheme is CCA secure.

To overcome this problem, our idea is to reveal enough information about the secret key so as to allow re-randomizability while keeping the scheme secure. To this end, our first observation is to rewrite the equation defining the re-randomized hash considering what we know about $\mathbf{x}'$. Specifically, we use the fact that $(\mathbf{F}^\top \mathbf{D})^\top \mathbf{x}' = (\mathbf{F}^\top \mathbf{D})^\top (\mathbf{x} + \mathbf{D}^* \hat{\mathbf{r}}) = (\mathbf{F}^\top \mathbf{D})^\top \mathbf{x} + (\mathbf{F}^\top \mathbf{D})^\top\ \mathbf{D}^* \hat{\mathbf{r}}$. So the re-randomized hash can be decomposed in three addends as:

$$[\mathbf{f}^\top \mathbf{D} + (\mathbf{F}^\top \mathbf{D})^\top \mathbf{x}]_T (\mathbf{r} + \hat{\mathbf{r}}) \ + \ [(\mathbf{F}^\top \mathbf{D})^\top (\mathbf{D}^* \hat{\mathbf{r}})]_T \hat{\mathbf{r}} \ + \ [(\mathbf{F}^\top \mathbf{D})^\top (\mathbf{D}^* \hat{\mathbf{r}})]_T \mathbf{r}$$

Notice that the first and the second addends can be easily computed knowing the randomizer $\hat{\mathbf{r}}$, the hash $[\pi]_T$ and thanks to the pairing function. So only the third addend is missing.

The second key observation is that we can include the value $[\mathbf{F}\mathbf{D}^*]_1$ in the public key. It is easy to check that, due to the bilinearity of the pairing function, we can compute the missing part as a function of tag $\mathbf{x}$, the randomizer $\hat{\mathbf{r}}$ and this extra piece of information. The third addend can be rewritten as:

$$[(\mathbf{F}^\top \mathbf{D})^\top (\mathbf{D}^* \hat{\mathbf{r}})]_T \mathbf{r} = [\mathbf{D}^\top \mathbf{F} \mathbf{D}^* \hat{\mathbf{r}}]_T \mathbf{r} = [(\mathbf{r}^\top \mathbf{D}^\top)(\mathbf{F}\mathbf{D}^*)\hat{\mathbf{r}}]_T = [\mathbf{x}^\top (\mathbf{F}\mathbf{D}^* \hat{\mathbf{r}})]_T$$

(The last equation can be computed using the pairing $e([\mathbf{x}]_1, [\mathbf{FD}^*]\hat{\mathbf{r}})$). However, at first look, it is not clear why the scheme should still be secure. To understand it, let us strip away all the computational pieces of the scheme, keeping only the information-theoretic core. In a nutshell, the (one-time simulation) soundness property of the hash boils down to the fact that the function $f(\mathbf{x}) = \mathbf{f} + \mathbf{F} \cdot \mathbf{x}$ is pair-wise independent, meaning that, with knowledge of $f(\mathbf{x})$ one cannot predict $f(\mathbf{x}')$ for $\mathbf{x} \neq \mathbf{x}'$ better than guessing it. However, once we publish the value $\mathbf{FD}^*$ we lose this property. Indeed, given $f(\mathbf{x})$ and $\mathbf{FD}^*$, now we can easily compute the function f over all the points in the affine space $\{\mathbf{x}' \mid \mathbf{x}' = \mathbf{x} + \mathbf{D}^*\mathbf{r}, \mathbf{r} \in \mathbb{Z}_q^2\}$. On one hand, this is good as it allows us to re-randomize. On the other hand, we should prove that one cannot do more than this honest manipulation. Our main technical lemma shows that for any $\mathbf{x}'$ outside this affine space we still have pair-wise independence, i.e., the value $f(\mathbf{x}')$ is unpredictable.

2 Preliminaries and Definitions

A function is negligible in λ if it vanishes faster than the inverse of any polynomial in λ, we write $f(\lambda) \in \mathtt{negl}(\lambda)$ when f is negligible in λ. An asymmetric bilinear group is a tuple $\mathcal{G}$ is a tuple $(q, \mathbb{G}_1, \mathbb{G}_2, \mathbb{G}_T, e, \mathcal{P}_1, \mathcal{P}_2)$, where $\mathbb{G}_1, \mathbb{G}_2$ and $\mathbb{G}_T$ are groups of prime order q, the elements $\mathcal{P}_1, \mathcal{P}_2$ are generators of $\mathbb{G}_1, \mathbb{G}_2$ respectively, $e : \mathbb{G}_1 \times \mathbb{G}_2 \to \mathbb{G}_T$ is an efficiently computable, non-degenerate bilinear map, and there is no efficiently computable isomorphism between $\mathbb{G}_1$ and $\mathbb{G}_2$. Let GGen be some probabilistic polynomial time algorithm which on input 1^λ, where λ is the security parameter returns a description of an asymmetric bilinear group $\mathcal{G}$. Elements in $\mathbb{G}_i$, are denoted in implicit notation as $[a]_i := a\mathcal{P}_i$, where $i \in \{1, 2, T\}$ and $\mathcal{P}_T := e(\mathcal{P}_1, \mathcal{P}_2)$. Every element in $\mathbb{G}_i$ can be written as $[a]_i$ for some $a \in \mathbb{Z}_q$, but note that given $[a]_i$, $a \in \mathbb{Z}_q$ is in general hard to compute (discrete logarithm problem). Given $a, b \in \mathbb{Z}_q$ we distinguish between $[ab]_i$, namely the group element whose discrete logarithm base $\mathcal{P}_i$ is ab, and $[a]_i \cdot b$, namely the execution of the multiplication of $[a]_i$ and b, and $[a]_1 \cdot [b]_2 = [a \cdot b]_T$, namely the execution of a pairing between $[a]_1$ and $[b]_2$. Vectors and matrices are denoted in boldface. We extend the pairing operation to vectors and matrices as $e([\mathbf{A}]_1, [\mathbf{B}]_2) = [\mathbf{A}^\top \cdot \mathbf{B}]_T$. $span(\mathbf{A})$ denotes the linear span of the columns of $\mathbf{A}$.

Let ℓ, k be positive integers. We call $\mathcal{D}_{\ell,k}$ a matrix distribution if it outputs (in PPT time, with overwhelming probability) matrices in $\mathbb{Z}_q^{\ell \times k}$. We define $\mathcal{D}_k := \mathcal{D}_{k+1,k}$. Our results will be proven secure under the following decisional assumption in $\mathbb{G}_\gamma$, for some $\gamma \in \{1, 2\}$.

Definition 1 (Matrix Decisional Diffie-Hellman Assumption in $\mathbb{G}_\gamma$, [13]). *The $\mathcal{D}_{\ell,k}$-MDDH assumption holds if for all non-uniform PPT adversaries* A,

$$|\Pr[\mathsf{A}(\mathcal{G}, [\mathbf{A}]_\gamma, [\mathbf{Aw}]_\gamma) = 1] - \Pr[\mathsf{A}(\mathcal{G}, [\mathbf{A}]_\gamma, [\mathbf{z}]_\gamma) = 1]| \in \mathtt{negl}(\lambda),$$

where the probability is taken over $\mathcal{G} = (q, \mathbb{G}_1, \mathbb{G}_2, \mathbb{G}_T, e, \mathcal{P}_1, \mathcal{P}_2) \leftarrow \mathsf{GGen}(1^\lambda)$, $\mathbf{A} \leftarrow \mathcal{D}_{\ell,k}, \mathbf{w} \leftarrow \mathbb{Z}_q^k, [\mathbf{z}]_\gamma \leftarrow \mathbb{G}_\gamma^\ell$ *and the coin tosses of adversary* A.

Experiment $\mathbf{Exp}^{\mathsf{RCCA}}_{\mathsf{A},\mathcal{PKE}}(\lambda)$:
$\mathtt{prm} \leftarrow \mathsf{Setup}(1^\lambda), b^* \leftarrow_\$ \{0,1\}$
$(\mathsf{pk},\mathsf{sk}) \leftarrow \mathsf{KGen}(\mathtt{prm})$
$(\mathtt{M}_0,\mathtt{M}_1) \leftarrow \mathsf{A}^{\mathsf{Dec}(\mathsf{sk},\cdot)}(\mathsf{pk})$
$\mathtt{C} \leftarrow \mathsf{Enc}(\mathsf{pk},\mathtt{M}_{b^*})$
$b' \leftarrow \mathsf{A}^{\mathsf{Dec}^\diamond(\mathsf{sk},\cdot)}(\mathsf{pk},\mathtt{C})$
return $(b' = b^*)$

Oracle $\mathsf{Dec}^\diamond(\mathsf{sk},\cdot)$:
Upon input $\mathtt{C}$;
$\mathtt{M}' \leftarrow \mathsf{Dec}(\mathsf{sk},\mathtt{C})$;
if $\mathtt{M}' \in \{\mathtt{M}_0,\mathtt{M}_1\}$ then output $\diamond$
else output $\mathtt{M}'$

Fig. 1. The RCCA Security Experiment.

2.1 Re-randomizable RCCA PKE

A re-randomizable PKE (Rand-PKE) scheme $\mathcal{PKE}$ is a tuple of five algorithms: (I) $\mathsf{Setup}(1^\lambda)$ upon input the security parameter λ produces public parameters $\mathtt{prm}$, which include the description of the message and ciphertext space $\mathcal{M},\mathcal{C}$. (II) $\mathsf{KGen}(\mathtt{prm})$ upon input the parameters $\mathtt{prm}$, outputs a key pair $(\mathsf{pk},\mathsf{sk})$; (III) $\mathsf{Enc}(\mathsf{pk},\mathtt{M})$ upon inputs a public key pk and a message $\mathtt{M} \in \mathcal{M}$, outputs a ciphertext $\mathtt{C} \in \mathcal{C}$; (IV) $\mathsf{Dec}(\mathsf{pk},\mathsf{sk},\mathtt{C})$ upon input the secret key sk and a ciphertext $\mathtt{C}$, outputs a message $\mathtt{M} \in \mathcal{M}$ or an error symbol $\perp$; (V) $\mathsf{Rand}(\mathsf{pk},\mathtt{C})$ upon inputs a public key pk and a ciphertext $\mathtt{C}$, outputs another ciphertext $\mathtt{C}'$.

The RCCA security notion is formalized with a security experiment similar to the CCA security one except that in RCCA the decryption oracle (called the guarded decryption oracle) can be queried on any ciphertext and, when decryption leads to one of the challenge messages $\mathtt{M}_0,\mathtt{M}_1$, it answers with a special symbol $\diamond$ (meaning "same").

Definition 2 (Replayable CCA Security, [6]**).** *Consider the experiment* $\mathbf{Exp}^{\mathsf{RCCA}}$ *in Fig. 1, with parameters* λ*, an adversary* A*, and a PKE scheme* $\mathcal{PKE}$*. We say that* $\mathcal{PKE}$ *is* indistinguishable secure under replayable chosen-ciphertext attacks *(RCCA-secure) for any PPT adversary* A*:*

$$\mathbf{Adv}^{\mathsf{RCCA}}_{\mathsf{A},\mathcal{PKE}}(\lambda) := \left|\Pr\left[\mathbf{Exp}^{\mathsf{RCCA}}_{\mathsf{A},\mathcal{PKE}}(\lambda) = 1\right] - \frac{1}{2}\right| \in \mathtt{negl}(\lambda).$$

We formally define perfect re-randomizability in the full version of this paper [15]. Here we give a simplified description of the notion. The notion of perfect re-randomizability consists of three conditions: (i) the re-randomization of a valid ciphertext and a fresh ciphertext (for the same message) are equivalently distributed; (ii) the re-randomization procedure maintains correctness, meaning the randomized ciphertext and the original decrypt to the same value, in particular, invalid ciphertexts keep being invalid; (iii) it is hard to find a valid ciphertext that is not in the support of the encryption scheme. The last condition, coupled with the first one, implies that for any (possibly malicious) ciphertext that decrypts correctly the distribution of the re-randomized ciphertext and a fresh ciphertext are statistically close. This stronger property is particularly useful in applications, like our Mix-Net of Sect. 6, where we need to re-randomize adversarially chosen ciphertexts.

$\mathbf{Exp}^{\mathtt{der\text{-}priv}}_{\mathsf{A},\mathcal{NIZK}}$:
$\mathtt{prm}_G \leftarrow_\$ \mathsf{Setup}_G(1^\lambda)$; $b^* \leftarrow_\$ \{0,1\}$;
$(\mathsf{crs}, tp_e, tp_s) \leftarrow \mathsf{Init}(\mathtt{prm}_G)$;
$(x, w, \pi, T) \leftarrow \mathsf{A}(\mathsf{crs}, tp_s)$; Assert $\mathsf{V}(\mathsf{crs}, x, \pi) = 1$;
If $b^* = 0$ then $\pi' \leftarrow_\$ \mathsf{P}(\mathsf{crs}, T_x(x), T_w(w))$;
else $\pi' \leftarrow_\$ \mathsf{ZKEval}(\mathsf{crs}, \pi, T)$;
$b \leftarrow \mathsf{A}(\pi')$;
Output $b = b^*$.

Fig. 2. The security experiments for the derivation privacy.

Definition 3 (Public Verifiability). *$\mathcal{PKE}$ = (Setup, KGen, Enc, Dec, Rand) is a public key scheme with publicly verifiable ciphertexts if there is a deterministic algorithm* Ver *which, on input* (pk, C) *outputs an error symbol* $\perp$ *whenever* $\mathsf{Dec}(\mathsf{pk}, \mathsf{sk}, \mathsf{C}) = \perp$, *else it outputs valid.*

2.2 Malleable NIZKs

Recall that a non-interactive zero-knowledge proof system (NIZK) is a tuple (Init, P, V) of PPT algorithms. Briefly, the algorithm Init upon input group parameters outputs a common reference string and, possibly, trapdoor information (we will consider algorithms that outputs a trapdoor tp_e for extraction and a trapdoor tp_s for simulation). We use the definitional framework of Chase *et al.* [7] for malleable proof systems. For simplicity of the exposition we consider only the unary case for transformations (see the aforementioned paper for more details). Let $T = (T_x, T_r)$ be a pair of efficiently computable functions, that we refer as a *transformation.*

Definition 4 (Admissible transformations, [7]). *An efficient relation $\mathcal{R}$ is* closed *under a transformation $T = (T_x, T_w)$ if for any $(x, w) \in \mathcal{R}$ the pair $(T_x(x), T_w(w)) \in \mathcal{R}$. If $\mathcal{R}$ is closed under T then we say that T is an* admissible *for $\mathcal{R}$. Let $\mathcal{T}$ be a set of transformations, if for every $T \in \mathcal{T}$, T is admissible for $\mathcal{R}$, then $\mathcal{T}$ is* allowable *set of transformations.*

Definition 5 (Malleable NIZK, [7]). *Let $\mathcal{NIZK}$ = (Init, P, V) be a NIZK for a relation $\mathcal{R}$. Let $\mathcal{T}$ be an allowable set of transformations for $\mathcal{R}$. The proof system is* malleable with respect to $\mathcal{T}$ *if there exists an PPT algorithm* ZKEval *that on input $(\mathsf{crs}, T, (x, \pi))$, where $T \in \mathcal{T}$ and $\mathsf{V}(\mathsf{crs}, x, \pi) = 1$ outputs a valid proof π' for the statement $x' = T_x(x)$.*

We would like the property that two NIZK proofs where one is derived from the other cannot be linked. This is formalized with the notion of *derivation privacy.*

Definition 6. *Let $\mathcal{NIZK}$ = (Init, P, V, ZKEval) be a malleable NIZK argument for a relation $\mathcal{R}$ and an allowable set of transformations $\mathcal{T}$. We say that $\mathcal{NIZK}$ is* derivation private *if for any PPT adversary* A *we have that*

$$\mathbf{Adv}^{\mathtt{der\text{-}priv}}_{\mathsf{A},\mathcal{NIZK}}(\lambda) := \left|\Pr\left[\mathbf{Exp}^{\mathtt{der\text{-}priv}}_{\mathsf{A},\mathcal{NIZK}}(1^\lambda) = 1\right] - \tfrac{1}{2}\right| \in \mathtt{negl}(\lambda)$$

$\underline{\mathsf{Setup}(1^\lambda)}$:
$\mathcal{G} \leftarrow_\$ \mathsf{GGen}(1^\lambda)$ where
$\mathcal{G} = (q, \mathbb{G}_1, \mathbb{G}_2, \mathbb{G}_T, e, \mathcal{P}_1, \mathcal{P}_2)$;
$\mathcal{M} = \mathbb{G}_1$;
$\mathcal{C} = \mathbb{G}_1^{k+2} \times \mathbb{G}_2^{k+1} \times \mathbb{G}_T$;
Output $\mathtt{prm} = (\mathcal{G}, \mathcal{M}, \mathcal{C})$.

$\underline{\mathsf{KGen}(\mathtt{prm})}$:
Sample $\mathbf{D}, \mathbf{E} \leftarrow_\$ \mathcal{D}_k$;
Sample $\mathbf{a}, \mathbf{f}, \mathbf{g} \leftarrow_\$ \mathbb{Z}_q^{k+1}$;
$\mathbf{F} \leftarrow_\$ \mathbb{Z}_q^{k+1 \times k+1}$ and $\mathbf{G} \leftarrow_\$ \mathbb{Z}_q^{k+1 \times k+2}$;
Set $\mathbf{D}^* = (\mathbf{D}^\top, (\mathbf{a}^\top \mathbf{D})^\top)^\top$;
Set $\mathsf{sk} = (\mathbf{a}, \mathbf{f}, \mathbf{g}, \mathbf{F}, \mathbf{G})$ and
Set $\mathsf{pk} =$
$([\mathbf{D}]_1, [\mathbf{E}]_2, [\mathbf{a}^\top \mathbf{D}]_1,$
$[\mathbf{f}^\top \mathbf{D}]_T, [\mathbf{F}^\top \mathbf{D}]_1,\ [\mathbf{g}^\top \mathbf{E}]_T, [\mathbf{G}^\top \mathbf{E}]_2,$
$[\mathbf{G}\mathbf{D}^*]_1, [\mathbf{F}\mathbf{E}]_2)$;
Output $(\mathsf{pk}, \mathsf{sk})$.

$\underline{\mathsf{Enc}(\mathsf{pk}, [\mathtt{M}]_1)}$:
Sample $\mathbf{r}, \mathbf{s} \leftarrow_\$ \mathbb{Z}_q^k$;
$[\mathbf{u}]_1 \leftarrow [\mathbf{D}]_1 \cdot \mathbf{r}$, $[p]_1 \leftarrow [\mathbf{a}^\top \mathbf{D}]_1 \cdot \mathbf{r} + [\mathtt{M}]_1$;
$[\mathbf{x}]_1 \leftarrow ([\mathbf{u}^\top]_1, [p]_1)^\top$;
$[\mathbf{v}]_2 \leftarrow [\mathbf{E}]_2 \cdot \mathbf{s}$;
$[\pi_1]_T = [\mathbf{f}^\top \mathbf{D}]_T \cdot \mathbf{r} + e([\mathbf{F}^\top \mathbf{D}]_1 \cdot \mathbf{r}, [\mathbf{v}]_2)$;
$[\pi_2]_T = [\mathbf{g}^\top \mathbf{E}]_T \cdot \mathbf{s} + e([\mathbf{x}]_1, [\mathbf{G}^\top \mathbf{E}]_2 \cdot \mathbf{s})$;
Set $\pi = \pi_1 + \pi_2$;
Output $\mathsf{C} = ([\mathbf{x}]_1, [\mathbf{v}]_2, [\pi]_T)$;

$\underline{\mathsf{Dec}(\mathsf{sk}, \mathsf{C})}$:
Parse $\mathsf{C} = ([\mathbf{x}]_1, [\mathbf{v}]_2, \pi)$;
parse $[\mathbf{x}^\top]_1 = ([\mathbf{u}^\top]_1, [p]_1)$;
set $[\mathtt{M}]_1 \leftarrow [p]_1 - [\mathbf{a}^\top \mathbf{u}]_1$;
set $[\pi_1]_T \leftarrow [(\mathbf{f} + \mathbf{F}\mathbf{v})^\top \mathbf{u}]_T$;
set $[\pi_2]_T \leftarrow [(\mathbf{g} + \mathbf{G}\mathbf{x})^\top \mathbf{v}]_T$;
If $\pi \neq \pi_1 + \pi_2$ then output $\perp$
else output $[\mathtt{M}]_1$.

$\underline{\mathsf{Rand}(\mathsf{pk}, \mathsf{C})}$:
Parse $\mathsf{C} = ([\mathbf{x}]_1, [\mathbf{v}]_2, [\pi]_T)$, $[\mathbf{x}^\top]_1 = ([\mathbf{u}^\top]_1, [p]_1)$;
Sample $\hat{\mathbf{r}}, \hat{\mathbf{s}} \leftarrow_\$ \mathbb{Z}_q^k$
$[\hat{\mathbf{x}}]_1 \leftarrow [\mathbf{x}]_1 + [\mathbf{D}^*]_1 \cdot \hat{\mathbf{r}}$;
$[\hat{\mathbf{v}}]_2 \leftarrow [\mathbf{v}]_2 + [\mathbf{E}]_2 \cdot \hat{\mathbf{s}}$;
$[\hat{\pi}_1]_T = [\mathbf{f}^\top \mathbf{D}]_T \cdot \hat{\mathbf{r}} + e([\mathbf{F}^\top \mathbf{D}]_1 \cdot \hat{\mathbf{r}}, [\hat{\mathbf{v}}]_2) + e([\mathbf{u}]_1, [\mathbf{F}\mathbf{E}]_2 \cdot \hat{\mathbf{s}})$;
$[\hat{\pi}_2]_T = [\mathbf{g}^\top \mathbf{E}]_T \cdot \hat{\mathbf{s}} + e([\hat{\mathbf{x}}]_1, [\mathbf{G}^\top \mathbf{E}]_2 \cdot \hat{\mathbf{s}}) + e([\mathbf{G}\mathbf{D}^*]_1 \cdot \hat{\mathbf{r}}, [\mathbf{v}]_2)$;
Output the ciphertext $\hat{\mathsf{C}} = ([\hat{\mathbf{x}}]_1, [\hat{\mathbf{v}}]_2, [\hat{\pi}]_T)$, with $[\hat{\pi}]_T \leftarrow [\pi]_T + [\hat{\pi}_1]_T + [\hat{\pi}_2]_T$.

Fig. 3. Our Rand-RCCA encryption scheme $\mathcal{PKE}_1$ based on the $\mathcal{D}_k$-MDDH assumption for $k \in \mathbb{N}^*$.

where $\mathbf{Exp}^{\mathtt{der\text{-}priv}}$ *is the game described in Fig. 2. Moreover we say that* $\mathcal{NIZK}$ *is* perfectly derivation private *(resp.* statistically derivation private*) when for any (possibly unbounded) adversary the advantage above is 0 (resp. negligible).*

Finally, we assume that an adversary cannot find a verifying proof for a valid statement which is not in the support of the proof generated by the proving algorithm. We notice that this property is true for both GS proof systems and for quasi-adaptive proof system of Kiltz and Wee [28]. In particular, for GS proofs, for any commitment to the witness, the prover generates a proof that is uniformly distributed over the set of all the possible valid proofs. On the other hand, the proofs of Kiltz and Wee are unique, therefore the condition is trivially true.

3 Our Rand-RCCA PKE Scheme

We present our scheme in Fig. 3. We refer to the introduction for an informal exposition of our techniques. We notice that the check in the decryption procedure can be efficiently computed using the pairing function and the knowledge of $\mathbf{f}, \mathbf{F}, \mathbf{g}, \mathbf{G}$. In the next paragraphs we first show correctness of the scheme, secondly, we give an information-theoretic lemma which is the basic core of the security of our PKE scheme, then we proceed with the RCCA-security of the scheme.

Correctness of Decryption. For correctness of decryption, it is easy to see that for a honestly generated ciphertext $([\mathbf{x}]_1, [\mathbf{v}]_2, [\pi]_T) \leftarrow_{\$} \mathsf{Enc}(\mathsf{pk}, [\mathtt{M}]_1)$, the first line of decryption $[p]_1 - [\mathbf{a}^\top \mathbf{u}]_1$ yields $[\mathtt{M}]_1$. Hence, we are left with showing that the test $[\pi]_T = [(\mathbf{f} + \mathbf{F}\mathbf{v})^\top \mathbf{u}]_T + [(\mathbf{g} + \mathbf{G}\mathbf{x})^\top \mathbf{v}]_T$ is satisfied:

$$\begin{aligned}\pi = \pi_1 + \pi_2 &= (\mathbf{f}^\top \mathbf{D})\mathbf{r} + (\mathbf{F}^\top \mathbf{D}\mathbf{r})^\top \mathbf{v} + (\mathbf{g}^\top \mathbf{E})\mathbf{s} + \mathbf{x}^\top (\mathbf{G}^\top \mathbf{E})\mathbf{s} \\ &= (\mathbf{f} + \mathbf{F}\mathbf{v})^\top \mathbf{u} + (\mathbf{g} + \mathbf{G}\mathbf{x})^\top \mathbf{v} \end{aligned} \tag{1}$$

Before analyzing the perfect re-randomizability and RCCA security of the scheme we state and prove a powerful information-theoretic lemma. Very informally speaking, the lemma proves that the smooth projective hash proof system at the core of our scheme remains sound even if the adversary gets to see a proof for an instance of its choice. As we want to allow for re-randomization, we relax the notion of soundness by requiring that the instance forged by the adversary does not lie in the set of possible re-randomizations of its query.

Lemma 1. *Let k be a positive integer. For any matrices $\mathbf{D} \in \mathbb{Z}_q^{k+1 \times k}, \mathbf{E} \in \mathbb{Z}_q^{k+1 \times k}$ and any (possibly unbounded) adversary A:*

$$\Pr\left[\begin{array}{c} \mathbf{u} \notin span(\mathbf{D}) \\ (\mathbf{v} - \mathbf{v}^*) \notin span(\mathbf{E}) \\ z = (\mathbf{f} + \mathbf{F}\mathbf{v})^\top \mathbf{u} \end{array} \middle| \begin{array}{l} \mathbf{f} \leftarrow_{\$} \mathbb{Z}_q^{k+1}, \mathbf{F} \leftarrow_{\$} \mathbb{Z}_q^{k+1 \times k+1}; \\ (z, \mathbf{u}, \mathbf{v}) \leftarrow_{\$} \mathsf{A}^{\mathcal{O}(\cdot)}(\mathbf{D}, \mathbf{E}, \mathbf{D}^\top \mathbf{f}, \mathbf{D}^\top \mathbf{F}, \mathbf{F}\mathbf{E}) \end{array}\right] \leq 1/q,$$

where the adversary outputs a single query $\mathbf{v}^*$ to $\mathcal{O}(\cdot)$ which returns $\mathbf{f} + \mathbf{F} \cdot \mathbf{v}^*$.

Proof. Let $\mathbf{K} = (\mathbf{f}, \mathbf{F}) \in \mathbb{Z}_q^{k+1 \times k+2}$. We can rewrite the information that the adversary sees about $\mathbf{f}, \mathbf{F}$ in matrix form:

$$(\mathbf{D}, \mathbf{E}, \mathbf{D}^\top \mathbf{f}, \mathbf{D}^\top \mathbf{F}, \mathbf{F}\mathbf{E}, \mathbf{f} + \mathbf{F} \cdot \mathbf{v}^*) = \left(\mathbf{D}, \mathbf{E}, \mathbf{D}^\top \mathbf{K}, \mathbf{K}\begin{pmatrix}\mathbf{0} \\ \mathbf{E}\end{pmatrix}, \mathbf{K}\begin{pmatrix}1 \\ \mathbf{v}^*\end{pmatrix}\right).$$

We now have to argue that $z = \mathbf{u}^\top \mathbf{K}\begin{pmatrix}1 \\ \mathbf{v}\end{pmatrix}$ is independent of the adversary's view when $\mathbf{u} \notin span(\mathbf{D})$ and $(\mathbf{v} - \mathbf{v}^*) \notin span(\mathbf{E})$. Without loss of generality we assume the matrices $\mathbf{D}, \mathbf{E}$ to be full rank. Otherwise this means there is a redundancy in the information provided to the adversary and this clearly does not give him more chances of being successful. Define the following matrices:

$$\tilde{\mathbf{D}} = (\mathbf{D}, \mathbf{u}) \in \mathbb{Z}_q^{k+1 \times k+1}, \qquad \tilde{\mathbf{E}} = \begin{pmatrix}\mathbf{0}, & 1, & 1 \\ \mathbf{E}, & \mathbf{v}^*, & \mathbf{v}\end{pmatrix} \in \mathbb{Z}_q^{k+2 \times k+2}.$$

By the condition that $\mathbf{u} \notin span(\mathbf{D})$ and $(\mathbf{v} - \mathbf{v}^*) \notin span(\mathbf{E})$, $\tilde{\mathbf{D}}$ and $\tilde{\mathbf{E}}$ are invertible matrices.

Let us consider the matrix $\mathbf{Z} = \tilde{\mathbf{D}}^\top \mathbf{K} \tilde{\mathbf{E}} \in \mathbb{Z}_q^{k+1 \times k+2}$ and the information that the adversary has on this matrix. Note that for $z_{k+1,k+2}$, namely the term in last row and last column of $\mathbf{Z}$, the following holds:

$$z_{k+1,k+2} = \mathbf{u}^\top \mathbf{K} \begin{pmatrix} 1 \\ \mathbf{v} \end{pmatrix} = z.$$

Since the view of the adversary contains invertible matrix $\tilde{\mathbf{E}}$, knowledge of $\mathbf{D}^\top \mathbf{K}$ (in the view of the adversary) is equivalent to knowledge of $\mathbf{D}^\top \mathbf{K} \tilde{\mathbf{E}}$, which are the first k rows of $\mathbf{Z}$.

Similarly, let $\hat{\mathbf{E}}$ be the first $k+1$ columns of $\tilde{\mathbf{E}}$, since $\tilde{\mathbf{D}}$ is invertible and is known by the adversary, knowledge of $\mathbf{K}\hat{\mathbf{E}}$ (in the view of the adversary) is equivalent to knowledge of $\tilde{\mathbf{D}}^\top \mathbf{K} \hat{\mathbf{E}}$, the first $k+1$ columns of $\mathbf{Z}$. Therefore, the view of the adversary includes all the matrix $\mathbf{Z}$ except for $z_{k+1 \times k+2}$.

On the other hand, since $\tilde{\mathbf{D}}$ and $\tilde{\mathbf{E}}$ are invertible matrices, if we see $\mathbf{Z} = \tilde{\mathbf{D}}^\top \mathbf{K} \tilde{\mathbf{E}} \in \mathbb{Z}_q^{k+1 \times k+2}$ as a system of equations with unknown $\mathbf{K}$, there exists a unique solution $\mathbf{K}$ for any choice of $\mathbf{Z}$, namely, $\mathbf{K} = (\tilde{\mathbf{D}}^\top)^{-1} \mathbf{Z} \tilde{\mathbf{E}}^{-1}$.

Therefore, from the point of view of the adversary, every value of $z_{k+1 \times k+2} \in \mathbb{Z}_q$ is equally likely, since $\mathbf{K} \leftarrow_{\$} \mathbb{Z}_q^{k+1 \times k+2}$ is sampled uniformly at random. This concludes the proof.

Security. For space reason we prove perfect re-randomizability in the full version of this paper [15]. We prove that the security of the scheme reduces to the $\mathcal{D}_k$-MDDH assumption. Below we state the main theorem:

Theorem 1. *For any matrix distribution $\mathcal{D}_k$ such that the $\mathcal{D}_k$-MDDH assumption holds for the groups $\mathbb{G}_1$ and $\mathbb{G}_2$ generated by* GGen*, the Rand-PKE scheme $\mathcal{PKE}_1$ described above is RCCA-secure.*

Proof. We start by describing a sequence of hybrid games. For readability purposes, we underline the main differences between each consecutive hybrid. In hybrids $\mathbf{H}_0$ and from $\mathbf{H}_3$ until $\mathbf{H}_7$ we progressively change the way the decryption procedure works. In the description of the games, the changes correspond to the underlined formulae. We summarize the main changes in Fig. 4.

Hybrid $\mathbf{H}_0$. This hybrid experiment is equivalent to the RCCA experiment described in Fig. 1 but the oracle $\mathsf{Dec}^\diamond$ is instantiated with a slightly different decryption procedure. Decryption proceeds exactly as in the description of the PKE scheme, except that, before setting each variable $\mathtt{M}, \pi_1, \pi_2$ it additionally checks if the variable was not set already. For future reference, we label these commands as the decryption rule (*).

Notice that, in this hybrid, this change is merely syntactical, as at each invocation of the decryption procedure all the three variables are unset. The hybrid $\mathbf{H}_0$ is equivalent to the experiment $\mathbf{Exp}^{\mathsf{RCCA}}_{\mathsf{A},\mathcal{PKE}}(\lambda)$ of Fig. 1.

Hybrid $\mathbf{H}_1$. The hybrid $\mathbf{H}_1$ is the same as $\mathbf{H}_0$ but it computes the challenge ciphertext $\mathtt{C}^* = ([\mathbf{x}^*]_1, [\mathbf{v}^*]_2, [\pi^*]_T)$ by using the secret key. Let $\mathbf{x}^*$ be $((\mathbf{u}^*)^\top, p^*)^\top$ and $\pi^* = \pi_1^* + \pi_2^*$.

$$[\mathbf{u}^*]_1 \leftarrow [\mathbf{D}]_1 \cdot \mathbf{r}^*, \quad \underline{[p^*]_1 \leftarrow \mathbf{a}^\top \cdot [\mathbf{u}^*]_1 + [\mathtt{M}_{b^*}]_1} \text{ where } \mathbf{r}^* \leftarrow_\$ \mathbb{Z}_q^k$$
$$[\mathbf{v}^*]_2 \leftarrow [\mathbf{E}]_2 \cdot \mathbf{s}^* \text{ where } \mathbf{s}^* \leftarrow_\$ \mathbb{Z}_q^k$$
$$[\pi_1^*]_T \leftarrow \underline{e([\mathbf{u}^*]_1, [\mathbf{f}]_2 + \mathbf{F} \cdot [\mathbf{v}^*]_2)}, \quad [\pi_2^*]_T \leftarrow \underline{e([\mathbf{g}]_1 + \mathbf{G} \cdot [\mathbf{x}^*]_1, [\mathbf{v}^*]_2)}.$$

Notice that $[\pi_1^*]_T$ and $[\pi_2^*]_T$ can be efficiently computed using the secret key and the pairing function. The only differences introduced are in the way we compute $[p^*]_1$ and $[\pi^*]_T$. However, notice that such differences are only syntactical, as, by the correctness of the scheme, we compute exactly the same values the hybrid $\mathbf{H}_0$ would compute.

Hybrid $\mathbf{H}_2$. The hybrid $\mathbf{H}_2$ is the same as $\mathbf{H}_1$ but the challenger, upon challenge messages $[\mathtt{M}_0]_1, [\mathtt{M}_1]_1 \in \mathbb{G}_1$, computes the challenge ciphertext $\mathtt{C}^* = ([\mathbf{x}^*]_1, [\mathbf{v}^*]_2, [\pi^*]_T)$ where $\mathbf{x}^*$ is $((\mathbf{u}^*)^\top, p^*)^\top$ by sampling :

$$\underline{\mathbf{u}^* \leftarrow_\$ \mathbb{Z}_q^{k+1} \setminus span(\mathbf{D})} \qquad \underline{\mathbf{v}^* \leftarrow_\$ \mathbb{Z}_q^{k+1} \setminus span(\mathbf{E})}.$$

The hybrids $\mathbf{H}_1$ and $\mathbf{H}_2$ are computationally indistinguishable. This follows by applying the $\mathcal{D}_k$-MDDH Assumption on $[\mathbf{D}, \mathbf{u}^*]_1$ in $\mathbb{G}_1$ and $[\mathbf{E}, \mathbf{v}^*]_2$ in $\mathbb{G}_2$, respectively, and then a standard statistical argument to show that sampling $\mathbf{u}^*$ uniformly at random in $\mathbb{Z}_q^{k+1}$ is statistically close to sampling it at random in $\mathbb{Z}_q^{k+1} \setminus span(\mathbf{D})$. The reduction is straightforward and is omitted.

From now on, we prove that each pair of consecutive hybrids is statistically close. In particular, this means that the hybrids (and in principle also the adversary) are allowed to run in unbounded time.

Hybrid $\mathbf{H}_3$. The hybrid $\mathbf{H}_3$ is the same as $\mathbf{H}_2$ but adds the following decryption rules that upon input a ciphertext $([\mathbf{u}]_1, [p]_1, [\mathbf{v}]_2, [\pi]_T)$:

(i) If $\mathbf{u} = \mathbf{D}\mathbf{r}$ for some $\mathbf{r} \in \mathbb{Z}_q^k$, then compute

$$\underline{[\pi_1]_T \leftarrow [(\mathbf{f}^\top \mathbf{D} + \mathbf{v}^\top \mathbf{F}^\top \mathbf{D})]_T \cdot \mathbf{r}} \qquad \underline{[\mathtt{M}]_1 \leftarrow [p]_1 - [\mathbf{a}^\top \mathbf{D}]_1 \cdot \mathbf{r}}$$

(ii) If $\mathbf{v} = \mathbf{E}\mathbf{s}$ for some $\mathbf{s} \in \mathbb{Z}_q^k$, letting $\mathbf{x} = (\mathbf{u}^\top, p)^\top$, then compute:

$$\underline{[\pi_2]_T \leftarrow [(\mathbf{g}^\top \mathbf{E} + \mathbf{x}^\top \mathbf{G}^\top \mathbf{E})]_T \cdot \mathbf{s}}$$

Specifically, in the first rule the decryption of $\mathtt{M}$ and π_1 are computed using the public key components $[\mathbf{a}^\top \mathbf{D}]_1, [\mathbf{f}^\top \mathbf{D}]_T$ and $[\mathbf{F}^\top \mathbf{D}]_1$ instead of the secret key components $\mathbf{a}, \mathbf{f}, \mathbf{F}$ for all the ciphertexts with $\mathbf{u} \in span(\mathbf{D})$. Recall that this strategy is not efficient, but it is possible because the simulator does not need to run in polynomial time (since we want to argue the games are statistically close). If $\mathbf{v} = \mathbf{E}\mathbf{s}$, then by the second rule, the hybrid computes the proof π_2 using only the components $[\mathbf{g}^\top \mathbf{E}]_T$ and $[\mathbf{G}^\top \mathbf{E}]_2$ of the public key.

We notice that, again by correctness of the PKE scheme, the computation of π_1, π_2 and $\mathtt{M}$ in the hybrids $\mathbf{H}_3$ and $\mathbf{H}_2$ is equivalent. In particular, let π_1' be the proof as computed in $\mathbf{H}_2$, then $[\pi_1']_T = [(\mathbf{f} + \mathbf{F}\mathbf{v})^\top \mathbf{u}]_T = [(\mathbf{f} + \mathbf{F}\mathbf{v})^\top \mathbf{D}\mathbf{r}]_T = [(\mathbf{f}^\top \mathbf{D} + \mathbf{v}^\top \mathbf{F}^\top \mathbf{D})]_T \cdot \mathbf{r} = [\pi_1]_T$. (An equivalent derivation holds for π_2 and $\mathtt{M}$.) The difference is then only syntactical.

Procedure $\mathsf{Dec}^*(\mathsf{sk}, \mathsf{C})$: Parse $\mathsf{C} = ([\mathbf{x}]_1, [\mathbf{v}]_2, [\pi]_T)$ and $[\mathbf{x}^\top]_1 = ([\mathbf{u}^\top]_1, [p]_1)$ (i) If $\mathbf{u} \in span(\mathbf{D})$, let $\mathbf{u} = \mathbf{D}\mathbf{r}$ then $[\mathsf{M}]_1 \leftarrow [p - \mathbf{a}^\top \mathbf{D}\mathbf{r}]_1$; $[\pi_1]_T \leftarrow [(\mathbf{f}^\top \mathbf{D} + \mathbf{v}^\top \mathbf{F}^\top \mathbf{D})\mathbf{r}]_T$; (ii) If $\mathbf{v} \in span(\mathbf{E})$, let $\mathbf{v} = \mathbf{E}\mathbf{s}$ then $[\pi_2]_T \leftarrow [(\mathbf{g}_0^\top \mathbf{E} + \mathbf{x}^\top \mathbf{G}^\top \mathbf{E})\mathbf{s}]_T$; (iii) If $\mathbf{u} \notin span(\mathbf{D})$ and $(\mathbf{v} - \mathbf{v}^* \notin span(\mathbf{E})$ or $\mathbf{v}^*$ unset) then output $\perp$. (iv) If $\mathbf{v} \notin span(\mathbf{E})$ and $(\mathbf{x} - \mathbf{x}^* \notin span(\mathbf{D}^*)$ or $\mathbf{u}^*$ unset) then output $\perp$. (v) If $\mathbf{x} - \mathbf{x}^* \in span(\mathbf{D}^*)$ and $\mathbf{v} - \mathbf{v}^* \in span(\mathbf{E})$ then $\mathsf{M} \leftarrow \diamond$; $[\pi_1]_T \leftarrow [\pi^*]_T + [(\mathbf{f}^\top \mathbf{D} + \tilde{\mathbf{v}}^\top \mathbf{F}^\top \mathbf{D})\tilde{\mathbf{x}}]_T$ $[\pi_2]_T \leftarrow [(\mathbf{g}_0^\top \mathbf{E} + \tilde{\mathbf{x}}^\top \mathbf{G}^\top \mathbf{E})\tilde{\mathbf{x}}]_T$ (*) If $[\mathsf{M}]_1$ is unset set $[\mathsf{M}]_1 \leftarrow [p]_1 - \mathbf{a}^\top [\mathbf{u}]$; (*) If $[\pi_1]_T$ is unset set $[\pi_1]_T \leftarrow [(\mathbf{f} + \mathbf{F}\mathbf{v})^\top \mathbf{u}]_T$; (*) If $[\pi_2]_T$ is unset set $[\pi_2]_T \leftarrow [(\mathbf{g}_0 + \mathbf{G}\mathbf{x})^\top \mathbf{v}]_T$; If $[\pi]_T = [\pi_1]_T + [\pi_2]_T$ output M else $\perp$.

Fig. 4. The decryption procedure in the hybrids experiment. The decryption procedure of the hybrid $\mathbf{H}_0$ executes only the rules (*) and the last decryption check. The decryption procedure of the hybrid $\mathbf{H}_3$ additionally executes (i) and (ii). The decryption procedure of the hybrid $\mathbf{H}_4$ additionally executes (iii). The decryption procedure of the hybrid $\mathbf{H}_5$ additionally executes (iv). The decryption procedure of the hybrid $\mathbf{H}_6$ additionally executes (V). The decryption procedure of the hybrid $\mathbf{H}_7$ stops to execute the rules (*).

Hybrid $\mathbf{H}_4$. The hybrid $\mathbf{H}_4$ is the same as $\mathbf{H}_3$ but adds the following decryption rule, on input a ciphertext $\mathsf{C} = ([\mathbf{u}]_1, [p]_1, [\mathbf{v}]_2, [\pi]_T)$:

(iii) If $\mathbf{u} \notin span(\mathbf{D})$ and $(\mathbf{v} - \mathbf{v}^* \notin span(\mathbf{E})$ or $\mathbf{v}^*$ is unset) then output $\perp$.

Recall that the challenge ciphertext is $\mathsf{C}^* = ([\mathbf{u}^*]_1, [p^*]_1, [\mathbf{v}^*]_2, [\pi]_T)$. Notice that we check either if $\mathbf{v} - \mathbf{v}^* \notin span(\mathbf{E})$ or $\mathbf{v}^*$ is unset. We do so to handle simultaneously the decryption queries before and after the challenge ciphertext is computed. In particular, before the challenge ciphertext is computed the decryption rule simply checks if $\mathbf{u} \notin span(\mathbf{D})$ (as in the classical Cramer-Shoup proof strategy).

We show in Lemma 3 that $\mathbf{H}_4$ is statistically close to $\mathbf{H}_3$. Here we continue describing the hybrid games.

Hybrid $\mathbf{H}_5$. The hybrid $\mathbf{H}_5$ is the same as $\mathbf{H}_4$ but adds the following decryption rule, on input a ciphertext $\mathsf{C} = ([\mathbf{x}]_1, [\mathbf{v}]_2, [\pi]_T)$:

(iv) If $\mathbf{v} \notin span(\mathbf{E})$ and $(\mathbf{x} - \mathbf{x}^* \notin span(\mathbf{D}^*)$ or $\mathbf{x}^*$ is unset) then output $\perp$.

We show that $\mathbf{H}_5$ is statistically close to $\mathbf{H}_4$ in the full version of this paper [15]. The proof of the lemma is almost identical to the proof of Lemma 3.

Hybrid $\mathbf{H}_6$. The hybrid $\mathbf{H}_6$ is the same as $\mathbf{H}_5$ but adds the following decryption rule, on input a ciphertext $\mathsf{C} = ([\mathbf{x}]_1, [\mathbf{v}]_2, [\pi]_T)$:

(v) If $\mathbf{x} - \mathbf{x}^* \in span(\mathbf{D}^*)$ and $\mathbf{v} - \mathbf{v}^* \in span(\mathbf{E})$ then let $\tilde{\mathbf{r}}, \tilde{\mathbf{s}}$ be such that $\mathbf{x} - \mathbf{x}^* = \tilde{\mathbf{x}} = \mathbf{D}\tilde{\mathbf{r}}$ and $\mathbf{v} - \mathbf{v}^* = \tilde{\mathbf{v}} = \mathbf{E}\tilde{\mathbf{s}}$, and compute $[\pi_1]_T, [\pi_2]_T$ as follows:

$$[\pi_1]_T \leftarrow \underline{[\pi^*]_T + [(\mathbf{f}^\top \mathbf{D} + \tilde{\mathbf{v}}^\top \mathbf{F}^\top \mathbf{D})\tilde{\mathbf{x}}]_T}, \quad [\pi_2]_T \leftarrow \underline{[(\mathbf{g}\mathbf{E} + \tilde{\mathbf{x}}^\top \mathbf{G}^\top \mathbf{E})\tilde{\mathbf{v}}]_T},$$

This hybrid is equivalent to $\mathbf{H}_5$. The conditions of the decryption rule (v) imply that, if the proof π is correct, then the ciphertext C is a re-randomization of C^*.

Hybrid $\mathbf{H}_7$. The hybrid $\mathbf{H}_7$ is the same as $\mathbf{H}_6$ but its decryption procedure does not execute the rules (*) introduced in the hybrid $\mathbf{H}_0$.

In Lemma 4 we show that $\mathbf{H}_7$ and $\mathbf{H}_6$ are identically distributed, while in the following we prove that the challenge bit b^* is perfectly hidden.

Lemma 2. $\Pr[\mathbf{H}_7 = 1] = \frac{1}{2}$.

Proof. We notice that in $\mathbf{H}_7$ the decryption procedure does not use the secret key $\mathbf{a}$ to perform the decryption; this can be easily confirmed by inspection of the decryption procedure in Fig. 4. Notice also that given the value $\mathbf{a}^\top\mathbf{D}$ the random variable $\mathbf{a}^\top \cdot \mathbf{u}^*$ is uniformly distributed. Thus, both the challenge ciphertext C^* and the answers of the decryption oracle are independent of the bit b^*.

Lemma 3. *The hybrids $\mathbf{H}_4$ and $\mathbf{H}_3$ are statistically close.*

Proof. We prove the statement with a hybrid argument over the number of decryption queries of the adversary. Let the hybrid $\mathbf{H}_{3,i}$ be the experiment that answers the first i-th oracle queries as in $\mathbf{H}_4$ (namely, considering the decryption rule (iii)) and answers the remaining queries as in $\mathbf{H}_3$. Let Q_D be the number of decryption queries performed by the adversary A. It is easy to check that $\mathbf{H}_{3,0} \equiv \mathbf{H}_3$ and $\mathbf{H}_{3,Q_D} \equiv \mathbf{H}_4$.

On the other hand $\mathbf{H}_{3,i}$ and $\mathbf{H}_{3,i+1}$ differ when the $(i+1)$-th ciphertext $\mathsf{C} = (([\mathbf{u}]_1, [p]_1), [\mathbf{v}]_2, [\pi]_T)$ is such that "$\mathbf{u} \notin span(\mathbf{D})$ and $((\mathbf{v} - \mathbf{v}^*) \notin span(\mathbf{E})$ or $\mathbf{v}^*$ is unset)", but the decryption oracle (as it would be computed in $\mathbf{H}_3$) outputs a value different from $\perp$. In particular, the latter implies that the proof $[\pi]_T$ verifies correctly. Let $\mathtt{Sound_i}$ be such event. To conclude the proof of the lemma we prove that $\Pr[\mathtt{Sound}_i] \leq 1/q$. Then a standard union bound gives us that the statistical distance between $\mathbf{H}_4$ and $\mathbf{H}_3$ is at most Q_D/q, which is negligible.

We reduce an adversary A that causes event $\mathtt{Sound}_i$ to occur into an adversary A' for the game of Lemma 1. Namely, we define an adversary A' for the experiment in the lemma which internally simulates the experiment $\mathbf{H}_{3,i+1}$ running with the adversary A.

<u>Adversary $\mathsf{A}'(\mathbf{D}, \mathbf{E}, \mathbf{f}^\top\mathbf{D}, \mathbf{F}^\top\mathbf{D}, \mathbf{F}\mathbf{E})$</u> with oracle access to $\mathcal{O}$:

1. Sample $\mathbf{a} \leftarrow_\$ \mathbb{Z}_q^{k+1}, \mathbf{g} \leftarrow_\$ \mathbb{Z}_q^{k+1}, \mathbf{G} \leftarrow_\$ \mathbb{Z}_q^{k+1 \times k+2}$.

2. Set the public key as:

$$\mathsf{pk} = \begin{pmatrix} [\mathbf{D}]_1, [\mathbf{E}]_2, [\mathbf{a}^\top \mathbf{D}]_1, [\mathbf{f}^\top \mathbf{D}]_T, [\mathbf{F}^\top \mathbf{D}]_1, \\ [\mathbf{g}^\top \mathbf{E}]_T, [\mathbf{G}^\top \mathbf{E}]_2, [\mathbf{G}\mathbf{D}^*]_1, [\mathbf{F}\mathbf{E}]_2 \end{pmatrix}$$

as described by the key generation algorithm and set the secret key $\mathsf{sk} = (\mathbf{a}, \cdot, \mathbf{g}, \cdot, \mathbf{G})$.

3. Run the adversary A with input the public key pk. Answer the j-th decryption oracle query with ciphertext $\mathtt{C} = ([\mathbf{u}]_1, [p]_1, [\mathbf{v}]_2, [\pi]_T)$ as follows:
   (a) If $j \le i$ and $\mathbf{u} \in \mathit{span}(\mathbf{D})$ compute, let $\mathbf{u} = \mathbf{D}\mathbf{r}$:

$$[\mathtt{M}]_1 \leftarrow [p - \mathbf{a}^\top \mathbf{D} \cdot \mathbf{r}]_1, \qquad [\pi_1]_T \leftarrow [(\mathbf{f}^\top \mathbf{D} + \mathbf{v}^\top \cdot \mathbf{F}^\top \mathbf{D})]_T \cdot \mathbf{r},$$
$$[\pi_2]_T \leftarrow [(\mathbf{g} + \mathbf{G} \cdot \mathbf{x})^\top \cdot \mathbf{v}]_T$$

   If $\pi = \pi_1 + \pi_2$ then answer with $[\mathtt{M}]_1$, else answer $\perp$;
   (b) If $\mathbf{u} \notin \mathit{span}(\mathbf{D})$ answer $\perp$;
   (c) If $j = i + 1$ then stop and return $(\pi - (\mathbf{g} + \mathbf{G}\mathbf{x})^\top \mathbf{v}, \mathbf{u}, \mathbf{v})$.

4. Eventually, A outputs $[\mathtt{M}_0]_1, [\mathtt{M}_1]_1$. Sample $\mathbf{v}^* \leftarrow_\$ \mathbb{Z}_q^{k+1} \setminus \mathit{span}(\mathbf{E})$, and sample $\mathbf{u}^* \leftarrow_\$ \mathbb{Z}_q^{k+1} \setminus \mathit{span}(\mathbf{D})$, query the oracle $\mathcal{O}$ with the element $\mathbf{v}^*$ and receive $\Pi = \mathbf{f} + \mathbf{F} \cdot \mathbf{v}^*$. Set $p^* = \mathbf{a}^\top \mathbf{u}^* + \mathtt{M}_{b^*}$ and $\mathbf{x}^* = ((\mathbf{u}^*)^\top, p^*)^\top$, and:

$$[\pi^*]_T \leftarrow [\Pi^\top \cdot \mathbf{u}^* + (\mathbf{g} + \mathbf{G}\mathbf{x}^*)^\top \mathbf{v}]_T \tag{2}$$

   and send to the adversary the challenge ciphertext $\mathtt{C}^* = ([\mathbf{c}^*]_1, [p^*]_1, [\mathbf{v}]_2, [\pi^*]_T)$.

5. Answer the j-th decryption oracle query with ciphertext $\mathtt{C} = ([\mathbf{u}]_1, [p]_1, [\mathbf{v}]_2, [\pi]_T)$ as follows:
   (a) If $j \le i$ and $\mathbf{u} \in \mathit{span}(\mathbf{D})$ execute the same as in step 3a.
   (b) If $j \le i$ and $\mathbf{u} \notin \mathit{span}(\mathbf{D})$ do as follows:
      i. if $(\mathbf{v}^* - \mathbf{v}) \in \mathit{span}(\mathbf{E})$ let $\mathbf{v} = \mathbf{v}^* + \mathbf{E}\gamma$, compute

$$[\pi_1]_T \leftarrow [(\Pi + \mathbf{F}\mathbf{E}\gamma)^\top)\mathbf{u}]_T, \qquad [\pi_2]_T \leftarrow [(\mathbf{g}^\top + \mathbf{G}\mathbf{x})^\top \mathbf{v}]_T$$

      if $\pi = \pi_1 + \pi_2$ then answer $[p - \mathbf{a}^\top \cdot \mathbf{u}]_1$ else answer $\perp$.
      ii. if $(\mathbf{v}^* - \mathbf{v}) \notin \mathit{span}(\mathbf{E})$ then output $\perp$.
   (c) If $j = i + 1$ then stop and return $(\pi - (\mathbf{g} + \mathbf{G}\mathbf{x})^\top \mathbf{v}, \mathbf{u}, \mathbf{v})$.

We show that the adversary perfectly simulates the hybrid $\mathbf{H}_{3,i}$ up to the i-th decryption query. By inspection, it is easy to check that up to step 3, the simulation is perfect[6].

More interestingly, at step 4 the adversary A′ uses its oracle to compute $\Pi = \mathbf{f} + \mathbf{F}\mathbf{v}^*$. Thanks to this information the adversary can compute the challenge ciphertext exactly as the hybrid experiment would do as shown in Eq. 2. After this step, the adversary A′ can easily answer the decryption queries whenever $j \le i$ and $\mathbf{u} \in \mathit{span}(\mathbf{D})$ or $\mathbf{u} \notin \mathit{span}(\mathbf{D})$ and $(\mathbf{v}^* - \mathbf{v}) \notin \mathit{span}(\mathbf{E})$. We show

[6] The adversary computes π_2 in step 3 as the original decryption procedure would do, but by the modification in $\mathbf{H}_1$ we are assured that this is equivalent.

that the answers for the decryption queries where $j \leq i$, $\mathbf{u} \notin span(\mathbf{D})$ and $(\mathbf{v}^* - \mathbf{v}) \in span(\mathbf{E})$ are distributed exactly as in the hybrid experiment, in fact:

$$(\Pi + \mathbf{FE}\gamma)^\top \mathbf{u} = \mathbf{f}^\top \mathbf{u} + (\mathbf{Fv}^*)^\top \mathbf{u} + (\mathbf{FE}\gamma)^\top \mathbf{u} = \mathbf{f}^\top \mathbf{u} + (\mathbf{F}(\mathbf{v}^* + \mathbf{E}\gamma))^\top \mathbf{u} = (\mathbf{f} + \mathbf{Fv})^\top \mathbf{u}.$$

Finally, by definition of $\mathtt{Sound}_i$, the adversary A at the $(j+1)$-th query outputs a ciphertext that would correctly decrypt in the hybrid experiment and where $\mathbf{u} \notin span(\mathbf{D})$ and $(\mathbf{v}^* - \mathbf{v}) \notin span(\mathbf{E})$ with probability $\Pr[\mathtt{Sound}_i]$. Since the ciphertext correctly decrypts, it means that $\pi = (\mathbf{f} + \mathbf{Fv})^\top \mathbf{u} + (\mathbf{g} + \mathbf{Gx})^\top \mathbf{v}$, therefore the output of A' is a valid guess for the experiment of Lemma 1. However, the adversary A' can win with probability at most $1/q$, and thus the lemma follows.

Lemma 4. *The hybrids* $\mathbf{H}_6$ *and* $\mathbf{H}_7$ *are identically distributed.*

Proof. We prove this lemma by showing that in $\mathbf{H}_6$ the decryption procedure never executes the lines with rules (*). To do this, for any ciphertext queried to the decryption oracle we partition over all possible cases and show that the decryption procedure used for the oracle queries either sets the values $\mathtt{M}, \pi_1, \pi_2$ (and thus the rules (*) are not executed) or it stops before reaching those rules as it outputs $\perp$ or $\diamond$. Let $\mathtt{C} = ([\mathbf{x}]_1, [\mathbf{v}]_2, [\pi]_T)$ be the ciphertext queried to the oracle, where $[\mathbf{x}^\top]_1 = ([\mathbf{u}^\top]_1, [p]_1)$. We consider all the possible alternatives:

- $\mathbf{u} \in span(\mathbf{D})$: notice that in this case, by the rule (i), $\mathtt{M}$ and π_1 are set;
  - $\mathbf{v} \in span(\mathbf{E})$: notice that in this case, by rule (ii), π_2 is also set. Therefore, since in this branch $\mathtt{M}, \pi_1, \pi_2$ are set, the rules (*) are not executed.
  - $\mathbf{v} \notin span(\mathbf{E})$: in this case we enter rule (iv) and thus decryption stops and outputs $\perp$. To see why this rule is entered, notice that either $\mathbf{u}^*$ is unset, or, if it is set, then $\mathbf{u}^* \notin span(\mathbf{D})$, and so $\mathbf{x} - \mathbf{x}^* \notin span(\mathbf{D}^*)$.
- $\mathbf{u} \notin span(\mathbf{D})$, in this case the output could be either $\diamond$ or $\perp$, more in details:
  - $\mathbf{v}^*$ is unset: by rule (iii) decryption stops and outputs $\perp$.
  - $\mathbf{v}^*$ is set and $(\mathbf{v} - \mathbf{v}^*) \notin span(\mathbf{E})$: by rule (iii) decryption outputs $\perp$.
  - $\mathbf{v}^*$ is set and $(\mathbf{v} - \mathbf{v}^*) \in span(\mathbf{E})$:
    - $(\mathbf{x} - \mathbf{x}^*) \notin span(\mathbf{D}^*)$: notice that since $\mathbf{v}^* \notin span(\mathbf{E})$ then it must be that $\mathbf{v} \notin span(\mathbf{E})$. Hence, rule (iv) is entered and decryption outputs $\perp$.
    - $(\mathbf{x} - \mathbf{x}^*) \in span(\mathbf{D}^*)$: rule (v) is entered, decryption outputs $\diamond$, so $\mathtt{M}, \pi_1, \pi_2$ are set, and thus the rules (*) are not executed.

4 Our Publicly-Verifiable Rand-RCCA PKE

Here we show that our RCCA scheme from the previous section can be turned into a publicly verifiable one. Very informally, the idea is to append a malleable proof (essentially a GS proof) that $[\pi]_T$ is well formed. The decryption procedure of the publicly verifiable scheme can simply check the validity of the proof and then CPA-decrypt the ciphertext $[\mathbf{x}]_1$. Let $\mathcal{PKE}_1 = (\mathsf{KGen}_1, \mathsf{Enc}_1, \mathsf{Dec}_1, \mathsf{Rand}_1)$

$\underline{\mathsf{KGen}_2(\mathtt{prm})}$:
$(\mathsf{pk}', \mathsf{sk}') \leftarrow_\$ \mathsf{KGen}'(\mathtt{prm})$, $\mathsf{crs} \leftarrow \mathsf{Init}(\mathtt{prm})$;
Parse $\mathsf{sk}' = (\mathbf{a}, \mathbf{f}, \mathbf{F}, \mathbf{g}, \mathbf{G})$;
Set $\mathsf{sk} = (\mathbf{a}, \mathsf{crs})$, $\mathsf{pk} = (\mathsf{pk}', \mathsf{crs})$;
Output $(\mathsf{pk}, \mathsf{sk})$.

$\underline{\mathsf{Rand}_2(\mathsf{pk}, \mathsf{C})}$:
Parse $\mathsf{C} = ([\mathbf{x}]_1, [\mathbf{v}]_2, \Pi)$,
$T \leftarrow_\$ \mathcal{T}$, (with associated $\hat{\mathbf{r}}, \hat{\mathbf{s}} \in \mathbb{Z}_q^k$)
$\hat{\mathbf{x}} = \mathbf{x} + \mathbf{D}^* \cdot \hat{\mathbf{r}}$;
$\hat{\mathbf{v}} = \mathbf{v} + \mathbf{E} \cdot \hat{\mathbf{s}}$;
$\hat{\Pi} = \mathsf{ZKEval}(\mathsf{crs}, T, ([\mathbf{x}]_1, [\mathbf{v}]_2), \Pi)$;
Output $([\hat{\mathbf{x}}]_1, [\hat{\mathbf{v}}]_2, \hat{\Pi})$.

$\underline{\mathsf{Enc}_2(\mathsf{pk}, [\mathtt{M}]_1)}$:
$\mathbf{r}, \mathbf{s} \leftarrow_\$ \mathbb{Z}_q^k$;
$([\mathbf{x}]_1, [\mathbf{v}]_2, [\pi]_T) \leftarrow \mathsf{Enc}'(\mathsf{pk}, [\mathtt{M}]_1; \mathbf{r}, \mathbf{s})$;
$\Pi \leftarrow_\$ \mathsf{P}(\mathsf{crs}, ([\mathbf{x}]_1, [\mathbf{v}]_2), ([\pi]_T, \mathbf{r}, \mathbf{s}))$;
Output $\mathsf{C} = ([\mathbf{x}]_1, [\mathbf{v}]_2, \Pi)$.

$\underline{\mathsf{Dec}_2(\mathsf{sk}, \mathsf{C})}$:
Parse $\mathsf{C} = ([\mathbf{x}]_1, [\mathbf{v}]_2, \Pi)$;
if $\mathsf{V}(\mathsf{crs}, ([\mathbf{x}]_1, [\mathbf{v}]_2), \Pi) = 1$
output $(-\mathbf{a}^\top, 1) \cdot [\mathbf{x}]_1$;
else output $\perp$.

$\underline{\mathsf{Ver}(\mathsf{pk}, \mathsf{C})}$:
Parse $\mathsf{C} = ([\mathbf{x}]_1, [\mathbf{v}]_2, \Pi)$;
Output $\mathsf{V}(\mathsf{crs}, ([\mathbf{x}]_1, [\mathbf{v}]_2), \Pi)$.

Fig. 5. Our publicly-verifiable re-randomizable RCCA encryption scheme $\mathcal{PKE}_2$. The NIZK is for the relation $\mathcal{R}_{\mathcal{PKE}_1}$ and transformation $\mathcal{T}_{\mathcal{PKE}_1}$.

be the scheme of Sect. 3 and let $\mathcal{NIZK} = (\mathsf{Init}, \mathsf{P}, \mathsf{V}, \mathsf{ZKEval})$ be a malleable NIZK system for membership in the relation defined below:

$$\mathcal{R}_{\mathcal{PKE}_1} = \left\{([\mathbf{x}]_1, [\mathbf{v}]_2), ([\pi]_T, \mathbf{r}, \mathbf{s}) : [\pi]_T = [(\mathbf{f} + \mathbf{F}\mathbf{v})^\top \mathbf{u} + (\mathbf{g} + \mathbf{G}\mathbf{x})^\top \mathbf{v}]_T\right\},$$

and with allowable set of transformations:

$$\mathcal{T}_{\mathcal{PKE}_1} = \left\{ T : \exists \hat{\mathbf{r}}, \hat{\mathbf{s}} \in \mathbb{Z}_q^k : \begin{array}{c} T_x([\mathbf{x}]_1, [\mathbf{v}]_2) = ([\hat{\mathbf{x}}]_1, [\hat{\mathbf{v}}]_2) \\ T_w([\pi]_T, \mathbf{r}, \mathbf{s}) = ([\hat{\pi}]_T, \mathbf{r} + \hat{\mathbf{r}}, \mathbf{s} + \hat{\mathbf{s}}) \\ ([\hat{\mathbf{x}}]_1, [\hat{\mathbf{v}}]_2, [\hat{\pi}]_T) = \mathsf{Rand}_1(\mathsf{pk}, ([\mathbf{x}]_1, [\mathbf{v}]_2, [\pi]_T); \hat{\mathbf{r}}, \hat{\mathbf{s}}) \end{array} \right\}.$$

We write $T \leftarrow_\$ \mathcal{T}_{\mathcal{PKE}_1}$ for the operation that samples the uniquely defined $\hat{\mathbf{r}}, \hat{\mathbf{s}}$ associated to the transformation T. The pv-Rand-PKE scheme $\mathcal{PKE}_2 = (\mathsf{Init}, \mathsf{KGen}_2, \mathsf{Enc}_2, \mathsf{Dec}_2, \mathsf{Rand}_2, \mathsf{Ver})$ is described in Fig. 5. We defer the proof of the following theorem in the full version of this paper [15].

Theorem 2. *If the $\mathcal{NIZK}$ is adaptive sound and perfect derivation private then the pv-Rand-PKE scheme $\mathcal{PKE}_2$ described in Fig. 5 is publicly verifiable, perfect re-randomizable and RCCA-secure.*

Malleable NIZK. The equations we would like to prove do not admit Groth-Sahai NIZK proofs [23], but only NIWI. We overcome this problem by developing a new technique that extends the class of pairing product equations which admit GS NIZK proofs. This technique is *per se* a result of independent interest.

More in detail, we produce an additional commitment to $[\pi]_T$, using a new commitment type defined over $\mathbb{G}_T$ with good bilinear properties. This allows us to construct a NIZK proof that the ciphertext is valid with perfect completeness and soundness and composable zero-knowledge. The latter notion refers to the fact that if the common reference string is defined in a "witness indistinguishable mode", the proof system is perfect zero-knowledge. By replacing $[\pi_T]$ in

$\mathbf{Exp}^{\mathtt{tse\text{-}cm}}_{\mathsf{A},\mathsf{Ext},\mathcal{NIZK}}$:
$\mathtt{prm}_G \leftarrow_\$ \mathsf{Setup}_G(1^\lambda)$; Set $\mathcal{Q}_w \leftarrow \emptyset$;
$(\mathsf{crs}, tp_e, tp_s) \leftarrow \mathsf{Init}(\mathtt{prm}_G)$;
$(x, \pi) \leftarrow \mathsf{A}(\mathsf{crs}, \mathcal{R})^{\mathcal{SIM}()}$; $z \leftarrow \mathsf{Ext}(tp_e, x, \pi, \mathcal{R})$;
Output 1 if $\mathsf{V}(\mathsf{crs}, x, \pi) = 1$ and either:
(a) $z \neq \circ$ and $\forall w$ s.t. $z = f(w)$ we have $(x, w) \notin \mathcal{R}$ or
(b) $z = \circ$ and $\forall x' \in \mathcal{Q}_x, \forall T \in \mathcal{T}$ we have $T_x(x) \neq x$.

$\mathcal{SIM}(x, w)$:
if $(x, w) \in \mathcal{R}$ then
$\pi \leftarrow \mathsf{Sim}(tp_s, x)$;
$\mathcal{Q}_x \leftarrow \mathcal{Q}_x \cup \{x\}$;

Fig. 6. The security experiments for the NIZK argument system.

the ciphertext by its commitment, in the witness indistinguishable mode we can simulate a proof of validity of the ciphertext by setting $\pi = 0$ and in an undetectable manner. The proof will be correctly distributed because of the perfect zero-knowledge property in these modes.

All the details on how to compute the proof are given in the full version of this paper [15]. Beyond GS Proofs, it also makes use of the QANIZK proof of membership in linear spaces [25,26,28]. The size of the ciphertexts for the SXDH instantiation of the publicly verifiable scheme is $12|\mathbb{G}_1| + 11|\mathbb{G}_2| + 4|\mathbb{G}_T|$. The number of pairings for verification is 32 for the GS proof and 14 for the argument of linear spaces, which can be reduced to $8+14$ by batch verifying the GS equation using the techniques of [24].

5 Malleable and True-Simulation Extractable NIZK

In this section we show an application of our Rand-RCCA scheme to build a malleable and true-simulation extractable NIZK.

True-Simulation Extractability. We recall the notion of true-simulation f-extractability (f-tSE-NIZK, for short) of Dodis *et al.* [10]. The notion is a weakening of the concept of simulation extractability where the extractor can compute a function of the witness and the adversary sees simulated proofs only for true statements. Here, we give a variation of the notion that allows for re-randomizability (and malleability). Consider the experiment described in Fig. 6, the main difference respect to the notion of [10], is that the winning condition (b) allows the extractor to give up and output a special symbol $\circ$. The restriction is that the extractor can safely do this without losing the game only when the proof π produced by the adversary is derived from a simulated proof.

Definition 7. *Let f be an efficiently computable function, let $\mathcal{NIZK} = (\mathsf{Init}, \mathsf{P}, \mathsf{V})$ be a NIZK argument for a relation $\mathcal{R}$, and consider the experiment $\mathbf{Exp}^{\mathtt{tse\text{-}cm}}$ described in Fig. 6. We say that $\mathcal{NIZK}$ is* true-simulation controlled-malleable *f-extractable (f-tSE-cm) iff there exists a PPT algorithm* Ext *such that for all PPT* A *we have that*

$$\mathbf{Adv}^{\mathtt{tse-cm}}_{\mathsf{A},\mathsf{Ext},\mathcal{NIZK}}(\lambda) := \Pr\left[\mathbf{Exp}^{\mathtt{tse\text{-}cm}}_{\mathsf{A},\mathsf{Ext},\mathcal{NIZK}}(1^\lambda) = 1\right] \in \mathtt{negl}(\lambda).$$

Construction. The construction follows the blueprint of Dodis *et al.* [10] with the twist that we use a Rand RCCA-PKE scheme instead of a CCA-PKE scheme. Our compiler works for a special class of tuples, consisting of a function f, an NP relation $\mathcal{R}$ and a transformation $\mathcal{T}$, that we define below:

Definition 8. *A tuple $(f, \mathcal{R}, \mathcal{T})$, where f is efficiently computable, $\mathcal{R}$ is an NP-relation and $\mathcal{T}$ is an admissible transformation for $\mathcal{R}$, is* suitable *if:*

1. *there exists an efficiently computable decision procedure g such that for any (x, w) the function $g(x, f(w)) = 1$ if and only if $(x, w) \in \mathcal{R}$;*
2. *For any $T \in \mathcal{T}$ and any $(x, w) \in \mathcal{R}$ the transformation of the witness is invariant respect to the function f, namely $f(w) = f(T_w(w))$.*

The restrictions above still allow for many interesting malleabilities. For example, the condition (2) clearly applies to re-randomizable NIZKs, as in this case $T_w(\cdot)$ is the identity function. Condition (1) holds in all those cases where the relation $\mathcal{R}$ can be sampled together with a trapdoor information that allows to compute w from x. The condition (1) applies also to the NIZKs of [10]. More importantly, the conjunction of (1) and (2) allows to efficiently check the condition (b) of the security experiment, which makes the tSE-cm NIZK primitive easier to use.

Let $\mathcal{PKE} = (\mathsf{KGen}, \mathsf{Enc}, \mathsf{Dec}, \mathsf{Rand})$ be a Rand-RCCA PKE scheme, we additionally assume there exists an integer $\ell \in \mathbb{N}$ such that the random coins of both the encryption procedure and the re-randomization procedure are in $\mathbb{Z}_q^\ell$ and that, for any $\mathsf{pk}, \mathtt{M}$, given $\mathsf{Rand}(\mathsf{pk}, \mathsf{Enc}(\mathsf{pk}, \mathtt{M}; \rho_0); \rho_1) = \mathsf{Enc}(\mathsf{pk}, \mathtt{M}; \rho_0 + \rho_1)$ where $\rho_0, \rho_1 \in \mathbb{Z}_q^\ell$. Notice that the schemes in Sects. 3 and 4 have this property. Let $\mathcal{R}$ be a NP relation and $\mathcal{T}$ be a set of allowable transformations for the relation $\mathcal{R}$. Let $\mathcal{NIZK}' = (\mathsf{Init}', \mathsf{P}', \mathsf{V}', \mathsf{ZKEval}')$ be a malleable NIZK argument for $\mathcal{R}'$ with the allowable set of transformations $\mathcal{T}'$ as described below:

$$\mathcal{R}' = \{((\mathsf{pk}, c, x), (w, \rho)) : \ (x, w) \in \mathcal{R} \ \wedge \ c = \mathsf{Enc}(\mathsf{pk}, f(w); \rho)\}$$

$$\mathcal{T}' = \left\{ T' : \ \exists \hat{\rho}, T : \begin{array}{c} T'_x(\mathsf{pk}, c, x) = (\mathsf{pk}, \mathsf{Rand}(\mathsf{pk}, c; \hat{\rho}), T_x(x)), \\ T'_w(w, \rho) = (T_w(w), \rho + \hat{\rho}), \ \ T \in \mathcal{T} \end{array} \right\}$$

We also assume that any transformation $T' \in \mathcal{T}'$ can be efficiently parsed as a tuple $(\hat{\rho}, T)$ and viceversa. We define a malleable NIZK argument $\mathcal{NIZK} = (\mathsf{Init}, \mathsf{P}, \mathsf{V}, \mathsf{ZKEval})$ for the relation $\mathcal{R}$ with allowable set of transformations $\mathcal{T}$ in Fig. 7. Notice that the co-domain of the function f for which we can prove f-tSE soundness is the message space of the underlying Rand-RCCA PKE scheme. We remark that, although our scheme is presented with a message space $\mathcal{M} = \mathbb{G}_1$, we could easily extend our construction to encrypt vectors in $\mathbb{G}_1^{\ell_0} \times \mathbb{G}_2^{\ell_1}$.

Theorem 3. *For any suitable $(f, \mathcal{R}, \mathcal{T})$ the proof system $\mathcal{NIZK}$ is a malleable NIZK for $\mathcal{R}$ with allowable transformations $\mathcal{T}$, and if $\mathcal{NIZK}'$ is perfectly (resp. statistically) derivation private (Def. 6) and $\mathcal{PKE}$ is perfectly re-randomizable then $\mathcal{NIZK}$ is perfectly (resp. statistically) derivation private.*

Theorem 4. *For any suitable $(f, \mathcal{R}, \mathcal{T})$ the proof system $\mathcal{NIZK}$ described above is true-simulation controlled-malleable f-extractable.*

$\underline{\mathsf{Init}(\mathtt{prm})}$:	$\underline{\mathsf{V}(\mathsf{crs}, x, \pi)}$:
$(\mathsf{crs}', tp'_s) \leftarrow \mathsf{Init}'(\mathtt{prm})$;	Output $\mathsf{V}'(\mathsf{crs}', (\mathsf{pk}, \mathtt{C}, x), \pi')$
$(\mathsf{pk}, \mathsf{sk}) \leftarrow \mathsf{KGen}(\mathtt{prm})$;	
$\mathsf{crs} \leftarrow (\mathsf{crs}', \mathsf{pk}), tp_e \leftarrow \mathsf{sk}, tp_s \leftarrow (\mathsf{pk}, tp'_s)$	$\underline{\mathsf{ZKEval}(\mathsf{crs}, T, (x, \pi))}$:
Output $(\mathsf{crs}, tp_e, tp_s)$.	Let $\pi = (\mathtt{C}, \pi'), \rho \leftarrow_\$ \mathbb{Z}_q^\ell$;
$\underline{\mathsf{P}(\mathsf{crs}, x, w)}$:	Let $T' = (\rho, T)$;
$\mathtt{C} \leftarrow \mathsf{Enc}(\mathsf{pk}, f(w); r)$;	$\hat{\mathtt{C}} \leftarrow \mathsf{Rand}(\mathsf{pk}, \mathtt{C}; \rho)$;
$\pi' \leftarrow \mathsf{P}'(\mathsf{crs}', (\mathsf{pk}, \mathtt{C}, x), (w, r))$;	$\hat{\pi}' \leftarrow_\$ \mathsf{ZKEval}'(\mathsf{crs}', T', (x, \pi'))$;
Output $\pi = (\mathtt{C}, \pi')$.	Output $(\hat{\mathtt{C}}, \hat{\pi}')$.

Fig. 7. Our f-tSE-cm $\mathcal{NIZK}$ compiler.

The proofs of Theorems 3 and 4 are in the full version of this paper [15]. We give an intuition for the proof of Theorem 4, which proceeds with a two-steps hybrid argument. We start with the true-simulation extractability experiment, we can switch to an experiment where each simulated proof for $\mathcal{NIZK}$ contains an encryption of the $f(w)$. This step can be easily argued using the RCCA security of the scheme. In particular, the guarded decryption oracle and the suitability of $(f, \mathcal{R}, \mathcal{T})$ are necessary to check the winning condition of the tSE experiment. In the second step, we switch to valid proofs for $\mathcal{NIZK}'$, instead of simulated proofs, the indistinguishability follows trivially by the zero-knowledge of $\mathcal{NIZK}'$. At this point we are in an experiment where the proofs provided by the $\mathcal{SIM}$ are not simulated, so the standard adaptive soundness of $\mathcal{NIZK}'$ is sufficient to bound the winning probability of the adversary.

Instantiation. For any suitable $(f, \mathcal{R}, \mathcal{T})$ where the co-domain of f is $\mathbb{G}_1$, we can instantiate the tSE-cm NIZK scheme with the pv-Rand-RCCA Scheme $\mathcal{PKE}_2$. The public verifiability enables for a simpler malleable NIZK proof for the associated $\mathcal{R}'$. In fact, we can subdivide the proof in: (1) a malleable GS proof Π_1 for $\mathcal{R}$ with transformations $\mathcal{T}$, in particular Π_1 contains GS commitments $[\mathbf{c}_w]_1$ of the witness; (2) a malleable GS proof Π_2 to prove that commitments $[\mathbf{c}_w]_1$ and $[\mathbf{c}_{w'}]_1$ open to w, w' an $w' = f(w)$; (3) a malleable proof Π_3 to prove $w' = (-\mathbf{a}^T, 1) \cdot [\mathbf{x}]$, in particular, from the linearity of GS commitments the relation for the last proof is a linear subspace relationship. The verification checks the proofs Π_1, Π_2, Π_3 and verifies the validity of the ciphertext $\mathtt{C}$.

For the case where f is the identity function, namely, re-randomizable NIZK, the proof Π_2 is trivial as we can set $[\mathbf{c}_w]_1 = [\mathbf{c}_{w'}]_1$. The overhead in proof size between a adaptive sound re-randomizable GS proof for $\mathcal{R}$ based on SXDH and an tSE-cm NIZK based on SXDH is equal to $13|\mathbb{G}_1| + 11|\mathbb{G}_2| + 4|\mathbb{G}_T|$.

6 An UC-Secure Mix-Net

In this section we propose an application of pv-Rand-PKE schemes with RCCA security to Mix-Net protocols. Our starting point is a recent work of Faonio and Fiore [14] who build an UC-secure Optimistic Mix-Net using a new paradigm that relies on a specific re-randomizable and RCCA-secure PKE scheme. Here we extend the main idea of [14] and use the power of public verifiability in order to obtain a full fledged Mix-Net protocol (not only optimistic secure).

The Universal Composability Model. We review some basic notions of the Universal Composability model and the extension to *auditable protocols* of Faonio and Fiore. In a nutshell, a protocol Π UC-realizes an ideal functionality $\mathcal{F}$ with setup assumption $\mathcal{G}$ if there exists a PPT simulator S such that no PPT environment $\mathcal{Z}$ can distinguish an execution of the protocols Π which can interact with the setup assumption $\mathcal{G}$ from a joint execution of the simulator S with the ideal functionality $\mathcal{F}$. The environment $\mathcal{Z}$ provides the inputs to all the parties of the protocols, decides which party to corrupt (we consider static corruption, where the environment decides the corrupted parties before the protocol starts), and schedules the order of the messages in the networks. When specifying an ideal functionality, we use the "delayed outputs" terminology of Canetti [3]. Namely, when a functionality $\mathcal{F}$ *sends a public delayed output* M *to party* $\mathcal{P}_{P_i}$ we mean that M is first sent to the simulator and then forwarded to $\mathcal{P}_{P_i}$ only after acknowledgement by the simulator. Faonio and Fiore consider a variation of the UC model where, roughly speaking, a bulletin board functionality **BB** acts as global setup assumption. More in details, the bulletin board is present in both the ideal world and the real world, so that the simulator does not have any advantage over the real-world adversary and all the parties of the protocol can register their message on the board. An *auditable protocol* is a tuple (Π, Audit) where Π is a protocol and Audit is a PPT algorithm. The model additionally includes an external off-line party, the auditor. The auditor is an incorruptible party which, whenever is called on an input y', runs the audit algorithm Audit on this input and the transcript written in the bulletin boards and forwards its output to the environment. In the ideal world, the auditor always replies according to the output of the ideal functionality, for example, if the ideal functionality has output y and the auditor is called on input y', the auditor replies with `valid` if and only if $y = y'$.

Defining Mix-Net Protocols. Our protocol UC-realizes the ideal functionality $\mathcal{F}_{\mathsf{Mix}}$ described in Fig. 8 with setup assumptions: the ideal functionality $\mathcal{F}_{\mathsf{TDec}}$ for threshold decryption of our PKE scheme and the ideal functionality for a common-reference string $\mathcal{F}_{\mathsf{CRS}}$ (and the bulletin board of the auditable framework of Faonio and Fiore). The functionality $\mathcal{F}_{\mathsf{Mix}}$ (similarly to [14]) is slightly weaker than the one considered by Wikström in [37,38]. The difference is that the corrupted senders can replace their inputs, however, they loose this ability when the first honest mixer sends its message `mix`. On the other hand, in the ideal functionality of Wikström, the senders can cast their messages only during the inputs submission phase.

Building Blocks. The main building blocks of our mix-net construction are:

(i) An *linear* pv-Rand-RCCA PKE scheme $\mathcal{PKE}$. We say that a pv-Rand-RCCA PKE scheme is *linear* if there exist a group $\mathbb{G}$ (for example $\mathbb{G} = \mathbb{G}_1$)

Functionality $\mathcal{F}_{\mathsf{Mix}}$:

The functionality has n sender parties $\mathcal{P}_{S_i}$ and m mixer parties $\mathcal{P}_{M_i}$:

Input: On message $(\texttt{input}, \texttt{M}_i)$ from $\mathcal{P}_{S_i}$ (or the adversary if $\mathcal{P}_{S_i}$ is corrupted) register the index i in the list of the senders and register the entry $(i, \texttt{M}_i)$ in the database of the inputs. Notify the adversary that the sender $\mathcal{P}_{S_i}$ has sent its input.

Mix: On message $\texttt{mix}$ from $\mathcal{P}_{M_i}$ (or the adversary if $\mathcal{P}_{M_i}$ is corrupted), register the index i in the list of the mixers and notify the adversary.

Delivery: If all the senders are in the list of the senders and at least one honest mixer is in the list of the mixers send a public delayed output $\mathcal{O} \leftarrow \mathsf{Sort}(\langle \texttt{M}_j \rangle_{j\in[n]})$ to all the mixers.

Fig. 8. Ideal Functionality for Mixing.

and parameters $\ell, \ell', \ell'' \in \mathbb{N}$ such that (1) every key pair $(\mathsf{pk}, \mathsf{sk})$ we can parse $\mathsf{pk} = ([\mathbf{P}], \hat{\mathsf{pk}})$ and $\mathsf{sk} = (\mathbf{S}, \hat{\mathsf{sk}})$, where $[\mathbf{P}] \in \mathbb{G}^{\ell \times \ell''}$ and $\mathbf{S} \in \mathbb{Z}_q^{\ell' \times \ell}$, (2) any ciphertext $\mathsf{C} \in \mathcal{C}$ can be parsed as $([\mathbf{y}], \hat{\mathsf{C}})$ where $[\mathbf{y}] \in \mathbb{G}^{\ell}$, (3) for any ciphertext C such that $\mathsf{Ver}(\mathsf{pk}, \mathsf{C}) = 1$ the decryption procedure is linear, i.e., we have $\mathsf{Dec}(\mathsf{sk}, \mathsf{C}) = \mathbf{S} \cdot [\mathbf{y}]$ (4) let $\mathsf{C}' = \mathsf{Rand}(\mathsf{pk}, \mathsf{C}; \mathbf{r}, r)$ where $\mathsf{C}' = ([\mathbf{y}'], \hat{\mathsf{C}}')$ be a re-randomization of $\mathsf{C} = ([\mathbf{y}], \hat{\mathsf{C}})$ and $\mathbf{r} \in \mathbb{Z}_q^{\ell''}$ then $([\mathbf{y}] - [\mathbf{y}']) = [\mathbf{P}]\mathbf{r}$. We notice that both the scheme $\mathcal{PKE}_2$ in Sect. 4 and the pv-Rand-RCCA PKE scheme of [7,29] are linear. Indeed, our abstraction is made to include the three schemes under the same template.

(ii) An All-but-One label-based NIZK. An ABO label-based $\mathcal{NIZK}_{\mathsf{sd}} = (\mathsf{Init}_{\mathsf{sd}}, \mathsf{P}_{\mathsf{sd}}, \mathsf{V}_{\mathsf{sd}})$ for knowledge of the plaintext of the linear PKE. More in details a ABO label-based $\mathcal{NIZK}$ is a NIZK system with labels where there exists an algorithm $\mathsf{ABOInit}(\texttt{prm}, \tau)$ which creates a common reference string crs together with a trapdoor tp_s such that for any label $\tau' \neq \tau$ the trapdoor allows for zero-knowledge while for τ the proof system is adaptive sound. A ABO label-based $\mathcal{NIZK}$ in the random-string model can be easily obtained from GS NIZK proof system.

(iii) An adaptive sound NIZK. $\mathcal{NIZK}_{\mathtt{mx}} = (\mathsf{Init}_{\mathtt{mx}}, \mathsf{P}_{\mathtt{mx}}, \mathsf{V}_{\mathtt{mx}})$ for proving membership in the relation $\mathcal{R}_{\mathtt{mx}} = \{([\mathbf{P}], [\mathbf{y}]) : [\mathbf{y}] \in \mathit{span}([\mathbf{P}])\}$. We recall that GS proof system is in the random-string model.

(iv) An ideal functionality $\mathcal{F}_{\mathsf{TDec}}$ for threshold decryption of the pv-Rand-RCCA PKE $\mathcal{PKE}$ scheme. More in details, $\mathcal{F}_{\mathsf{TDec}}$ takes as parameters the definition of the PKE scheme and group parameters $\texttt{prm}$ for the key generation. The functionality initializes a fresh key pair and accepts input of the form $(\mathtt{dec}, \mathsf{C})$ from the mixers: when a mixer sends a message of this kind, we say that the mixer *asks for the decryption of* C. When all the mixers have sent a message of the form $(\mathtt{dec}, \mathsf{C})$ the functionality sends a public delayed output $\mathsf{Dec}(\mathsf{sk}, \mathsf{C})$: in this case we say that the mixers *agreed on the decryption of* C. In the full version of this paper [15] we show a protocol for the functionality $\mathcal{F}_{\mathsf{TDec}}$ in the $\mathcal{F}_{\mathsf{CRS}}$-hybrid world.

(v) An ideal functionality for the common reference string of the above NIZKs. The functionality initializes m different CRS $\{\mathsf{crs}^i_{\mathtt{mx}}\}_{i=1,\dots,m}$, one for each mixer,[7] for $\mathcal{NIZK}_{\mathtt{mx}}$ and a CRS $\mathsf{crs}_{\mathtt{sd}}$ for $\mathcal{NIZK}_{\mathtt{sd}}$. We stress that all the CRSs can be sampled as uniformly random strings in the real protocol.

Also we recall that our auditable protocol uses a Bulletin Board functionality. We do not mention it as a "building block" because every auditable protocol, as defined by [14], necessarily needs a bulletin board as setup assumption.

Our Mix-Net Protocol. Following the design rationale of Faonio and Fiore, given two lists of ciphertexts $\mathcal{L} = \langle \mathtt{C}_1, \dots, \mathtt{C}_n \rangle$ and $\mathcal{L}' = \langle \mathtt{C}'_1, \dots, \mathtt{C}'_n \rangle$, we define the *checksum* of these lists as the output of the following procedure:

Procedure **CkSum**($\mathcal{L}, \mathcal{L}'$):

1. For all $j \in [n]$ parse $\mathtt{C}_j = ([\mathbf{y}_j], \hat{\mathtt{C}}_j)$ and $\mathtt{C}'_j = ([\mathbf{y}'_j], \hat{\mathtt{C}}'_j)$;
2. Output $\sum_j [\mathbf{y}_j] - [\mathbf{y}'_j]$.

We describe our mix-net protocol Π between n sender parties $\mathcal{P}_{S_i}$ and m mixer parties $\mathcal{P}_{M_i}$ and with resources the ideal functionalities $\mathcal{F}_{\mathsf{TDec}}$ and $\mathcal{F}_{\mathsf{CRS}}$:

Inputs Submission. Every sender $\mathcal{P}_{S_j}$, with $j \in [n]$, encrypts its message $\mathtt{M}_j$ by computing $\mathtt{C}_j \leftarrow \mathsf{Enc}(\mathsf{pk}, \mathtt{M}_j; r)$, and creates a NIZK proof of knowledge $\pi^{\mathtt{sd}}_j \leftarrow \mathsf{P}_{\mathtt{sd}}(\mathsf{crs}_{\mathtt{sd}}, j, (\mathsf{pk}, \mathtt{C}), (\mathtt{M}_j, r))$ (the label for the proof is j). The party $\mathcal{P}_{S_j}$ posts $(\mathtt{C}_j, \pi^{\mathtt{sd}}_j)$ on the bulletin board.

Mix. Once all the senders are done with the previous phase, let $\mathcal{L}_0 = \langle \mathtt{C}_{0,j} \rangle_{j \in [n]}$ be the list of ciphertexts they posted on the bulletin board. To simplify the exposition of the result, we assume that all the NIZK proofs $\{\pi^{\mathtt{sd}}_j\}_{j \in [n]}$ and all the ciphertexts in $\mathcal{L}_0$ verify.

For $i = 1$ to m, the mixer $\mathcal{P}_{M_i}$ waits for the $\mathcal{P}_{M_{i-1}}$ to complete and does:

1. Sample a permutation $\tau_i \leftarrow\$ \mathcal{S}_n$;
2. Read from the BB the message $(\mathcal{L}_{i-1}, \pi^{\mathtt{mx}}_{i-1})$ posted by $\mathcal{P}_{M_{i-1}}$ (or read $\mathcal{L}_0$ if this is the first mixer), and parse $\mathcal{L}_{i-1} = \langle \mathtt{C}_{i-1,j} \rangle_{j \in [n]}$;
3. Build the list $\mathcal{L}_i \leftarrow \langle \mathtt{C}_{i,j} \rangle_{j \in [n]}$ of shuffled and re-randomized ciphertexts by sampling randomness $\mathbf{r}_j, r_j$ and computing $\mathtt{C}_{i,\tau_i(j)} \leftarrow \mathsf{Rand}$ $(\mathsf{pk}, \mathtt{C}_{i-1,j}; \mathbf{r}_j, r_j)$.
4. Compute a NIZK proof $\pi^{\mathtt{mx}}_i \leftarrow\$ \mathsf{P}_{\mathtt{mx}}(\mathsf{crs}^i_{\mathtt{mx}}, ([\mathbf{P}], \mathbf{CkSum}(\mathcal{L}_{i-1}, \mathcal{L}_i)), \sum_j \mathbf{r}_j)$,
5. Post in the BB the tuple $(\mathcal{L}_i, \pi^{\mathtt{mx}}_i)$.

Verification. Once all mixers are done, every mixer $\mathcal{P}_{M_i}$ executes:

1. Read the messages $(\mathcal{L}_i, \pi^{\mathtt{mx}}_i)$ posted by every mixer on the BB, as well as the messages $(\mathtt{C}_{0,j}, \pi^{\mathtt{sd}}_j)$ posted by the senders;
2. For all $i \in [m]$ and for all $j \in [n]$ check that $\mathsf{Ver}(\mathsf{pk}, \mathtt{C}_{i,j}) = 1$;

[7] We could modify our protocol to let the mixers share the same CRS, at the price of requiring $\mathcal{NIZK}_{\mathtt{mx}}$ be simulation sound. Since in most applications the number of mixers is small, we go for the simpler option of one crs per mixer.

3. For all $i \in [m]$, check $\mathsf{V}_{\mathtt{mx}}(\mathsf{crs}^i_{\mathtt{mx}}, ([\mathbf{P}], \mathbf{CkSum}(\mathcal{L}_{i-1}, \mathcal{L}_i)), \pi^{\mathtt{mx}}_i) = 1$;
4. If one of the checks does not verify abort and write `invalid` in the BB.

Decrypt. All the mixers $\mathcal{P}_{M_i}$ execute the following in parallel (using the ideal functionality $\mathcal{F}_{\mathsf{TDec}}$ to compute decryptions):

1. let $\mathcal{L}_m = \langle \mathtt{C}^*_j \rangle_{j \in [n]}$ be the list of ciphertexts returned by the last mixer. For $j = 1$ to n, ask $\mathcal{F}_{\mathsf{TDec}}$ for the decryption of $\mathtt{C}^*_j$. Once all the mixers agreed on the decryption, receive $\mathtt{M}_j \leftarrow \mathsf{Dec}(\mathsf{sk}, \mathtt{C}^*_j)$ from the functionality;
2. Post $\mathsf{Sort}(\langle \mathtt{M}_j \rangle_{j \in [n]})$ on the BB.

Audit Message. The mixers $\mathcal{P}_{M_i}$ post the message `valid` on the BB.

Algorithm Audit: the algorithm reads from the BB and computes the verification step of the protocol above (notice that this only relies on public information).

Theorem 5. *The auditable protocol* (Π, Audit) *described above* UC-realizes $\mathcal{F}_{\mathsf{Mix}}$ *with setup assumptions* $\mathcal{F}_{\mathsf{TDec}}$ *and* $\mathcal{F}_{\mathsf{CRS}}$.

Proof (Sketch). We prove the theorem via a sequence of hybrid experiments. In the last experiment we define a simulator and highlight its interaction with the ideal functionality.

In the proof, we let h^* be the index of the first honest mixer. Also, we consider two sets Ψ_{in} and Ψ_{hide}, both consisting of tuples $(X, Y) \in \mathbb{G}_1^2$. For Ψ_{in} (resp. Ψ_{hide}) we define a corresponding *map* $\psi_{\mathsf{in}} : \mathbb{G}_1 \to \mathbb{G}_1$ (resp. ψ_{hide}) such that $\psi_{\mathsf{in}}(X)$ (resp. $\psi_{\mathsf{hide}}(X)$) is equal to Y if $(X, Y) \in \Psi_{\mathsf{in}}$ (resp. $(X, Y) \in \Psi_{\mathsf{hide}}$), otherwise X. We assume that all the NIZK proofs verify and that all the ciphertexts verify (as otherwise the protocol would abort without producing any output).

For space reason, in this proof sketch, we group together the hybrid experiments according to their function in the overall strategy.

Hybrids $\mathbf{H}_1$ to $\mathbf{H}_3$: In the first step we program the CRSs of both the NIZKs so that we can simulate the proof of the h^*-th mixer and of all the senders but one corrupted sender (whose index is hidden to the adversary by the CRS indistinguishability). For this step we can use the zero-knowledge property of the NIZKs. In the second and third step we use perfect-rerandomizability and RCCA security to introduce a change in the output of the h^*-th mixer. Specifically, the mixer $\mathcal{P}_{M_{h^*}}$ outputs ciphertexts which are fresh encryptions of random and independent messages $\mathtt{H}_1, \ldots, \mathtt{H}_n$. Moreover, we populate the set Ψ_{hide} with the pairs $(\mathtt{M}_{h^*-1,j}, \mathtt{H}_j)_{j \in [n]}$ to associate $\mathtt{H}_j$ with $\mathtt{M}_{h^*-1,j} \leftarrow \mathsf{Dec}(\mathsf{sk}, \mathsf{C}_{h^*-1,j})$, and then we simulate the ideal functionality $\mathcal{F}_{\mathsf{TDec}}$ to output $\Psi_{\mathsf{hide}}(\mathtt{M})$ instead of $\mathtt{M}$. This way the modification is not visible by looking at the decrypted ciphertexts.

Hybrid $\mathbf{H}_4$: Let $\mathcal{V}_m$ (resp. $\mathcal{V}_{h^*}$) be the decryption of the list of ciphertexts output by the last mixer $\mathcal{P}_{M_m}$ (resp. by the first honest mixer $\mathcal{P}_{M_{h^*}}$). The hybrid $\mathbf{H}_4$ aborts if $\mathcal{V}_m \neq \mathcal{V}_{h^*}$. Using the perfect adaptive soundness of $\mathcal{NIZK}_{\mathtt{mx}}$ and the RCCA security and the public-verifiability of our PKE, we can show that this abort can happen only with negligible probability. We adapt the security argument of Faonio and Fiore [14] to our pv-Rand-PKE and our NIZK proof of "checksum". The idea is that the proofs of checksum $\pi^{\mathtt{mx}}_{h^*+1}, \ldots, \pi^{\mathtt{mx}}_{n}$ establish a linear relationship between the plaintexts encrypted in the list of ciphertexts output by $\mathcal{P}_{M_{h^*}}$ and the plaintexts in the list of ciphertext output by $\mathcal{P}_{M_m}$. The reduction to RCCA security can install a challenge ciphertext in the first list and then learn information about the underlying plaintext by decrypting the second list. The idea is that the condition $\mathcal{V}_m \neq \mathcal{V}_{h^*}$ guarantees that the RCCA decryption oracle would not answer $\diamond$ on ciphertexts from the second list, and the linear relationship guaranteed by the proofs allows to extract the information on the challenge ciphertext.

Hybrid $\mathbf{H}_5$: Simulate the ideal functionality $\mathcal{F}_{\mathsf{TDec}}$ in different way. Whenever the mixers agree on the decryption of a ciphertext $\mathtt{C} \in \mathcal{L}_m$, simulate the functionality $\mathcal{F}_{\mathsf{TDec}}$ by outputting a message chosen uniformly at random (without re-introduction) from the list $\mathcal{V}_{h^*-1}$. Notice, we don't need to compile the list Ψ_{hide} anymore as the mixers would only agree to decrypt ciphertexts from the last list $\mathcal{L}_m$ and $\mathcal{V}_m = \mathcal{V}_{h^*} = \Psi_{\mathsf{hide}}(\mathcal{V}_{h^*-1})$.

We can prove that $\mathbf{H}_5$ and $\mathbf{H}_4$ are identically distributed. In fact in $\mathbf{H}_4$, after the first honest mixer outputs $\mathcal{L}_{h^*}$, an unbounded environment $\mathcal{Z}$ knows that in Ψ_{hide} the element $\mathtt{H}_j$ for $j \in [n]$ is mapped to some other value in $\mathcal{V}_{h^*-1}$ but, from its view, it cannot know to which value. Such information is revealed only during decryption time. In other words, we could sample the permutation τ_{h^*} (uniformly at random) at decryption time.

It is easy to check that, at this point of the hybrid argument, the list of ciphertexts received by the first honest mixers is (a permutation of) the output of the protocol. Moreover, the ordering of the ciphertexts in the former list and in the latter list are uncorrelated. With the next hybrids we make sure that the inputs of the honest senders are not discarded along the way from the first mixer to first honest mixer.

Hybrids $\mathbf{H}_6$ to $\mathbf{H}_7$: Notice that at this point the output of the mix-net is already distributed uniformly over the set of all the possible permutations of the inputs. However, the input messages of the honest senders are still (at least information theoretically) in the view of the adversary, as the honest senders still encrypt their inputs. In the next hybrids we switch into a hybrid experiment where all the honest senders encrypt dummy messages from a set $\mathcal{M}_H$, that we call the set of honest simulated messages. To do so we first program the map ψ_{in} to map the simulated messages to the (real) honest ones, and we simulate the functionality $\mathcal{F}_{\mathsf{TDec}}$ to pick messages $\mathtt{M}$ chosen uniformly at random (without re-introduction) from the list $\mathcal{V}_{h^*-1}$ and return $\psi_{\mathsf{in}}(\mathtt{M})$ instead of $\mathtt{M}$. Then in the second step we switch and encrypt the simulated messages, relying on RCCA security.

Hybrid $\mathbf{H}_8$ to $\mathbf{H}_9$: In the last two hybrids we make sure that (1) the malicious senders do not copy the ciphertexts of the honest senders, for this step we rely on the ABO soundness of the $\mathcal{NIZK}_{\mathsf{sd}}$ proof system, and (2) the malicious mixers do not duplicate or remove the messages of the honest senders, this argument is almost the same as in the step $\mathbf{H}_4$.

We can proceed to present the simulator S. For space reason, here we describe the most important parts.

Extraction of the Inputs: Let $\mathcal{L}_{h^*-1}$ be the list produced by the malicious mixer $\mathcal{P}_{M_{h^*-1}}$. For any j, the simulator S decrypts $\hat{\mathtt{M}}_j \leftarrow \mathsf{Dec}(\mathsf{sk}, \mathtt{C}_{h^*-1,j})$ and if $\hat{\mathtt{M}}_j \notin \mathcal{M}_H$ then it submits it as input to the ideal functionality $\mathcal{F}_{\mathsf{Mix}}$.

Decryption Phase: The simulator S receives from the ideal functionality $\mathcal{F}_{\mathsf{Mix}}$ the sorted output $\langle \mathtt{M}^o_1, \ldots, \mathtt{M}^o_n \rangle$. Whenever the mixers agree on the decryption of a ciphertext, it simulates the ideal functionality $\mathcal{F}_{\mathsf{TDec}}$ by outputting a message from the sorted output randomly chosen (without reinsertion).

We notice that the hybrid compiles the map ψ_{in} by setting a correspondence between the inputs of the honest senders and the simulated ones, and, during the decryption phase, uses the map ψ_{in} to revert this correspondence. On the other hand, the simulator does not explicitly set the map, as it does not know the inputs of the honest senders (which are sent directly to the functionality). However, at inputs submission phase the simulator picks a simulated input for any honest sender, and at decryption phase it picks a message from the ordered list in output, which contains the inputs of the honest senders. By doing so, the simulator is implicitly defining the map ψ_{in}. The second difference is that the simulator picks the outputs from the list $\langle \mathtt{M}^o_1, \ldots, \mathtt{M}^o_n \rangle$ while the hybrid $\mathbf{H}_9$ uses the list $\psi_{\mathsf{in}}(\mathcal{V}_{h^*-1})$. However, recall that the simulator extracts the corrupted inputs from the same list $\mathcal{V}_{h^*-1}$, and that, by the change introduced in $\mathbf{H}_9$, we are assured that all the inputs of the honest senders will be in the list $\psi_{\mathsf{in}}(\mathcal{V}_{h^*-1})$.

References

1. Bayer, S., Groth, J.: Efficient zero-knowledge argument for correctness of a shuffle. In: Pointcheval, D., Johansson, T. (eds.) EUROCRYPT 2012. LNCS, vol. 7237, pp. 263–280. Springer, Heidelberg (2012). https://doi.org/10.1007/978-3-642-29011-4_17
2. Camenisch, J., Haralambiev, K., Kohlweiss, M., Lapon, J., Naessens, V.: Structure preserving CCA secure encryption and applications. In: Lee, D.H., Wang, X. (eds.) ASIACRYPT 2011. LNCS, vol. 7073, pp. 89–106. Springer, Heidelberg (2011). https://doi.org/10.1007/978-3-642-25385-0_5
3. Canetti, R.: Universally composable security: a new paradigm for cryptographic protocols. In: 42nd FOCS, pp. 136–145. IEEE Computer Society Press (2001)
4. Canetti, R., Goldwasser, S.: An efficient *threshold* public key cryptosystem secure against adaptive chosen ciphertext attack (Extended Abstract). In: Stern, J. (ed.) EUROCRYPT 1999. LNCS, vol. 1592, pp. 90–106. Springer, Heidelberg (1999). https://doi.org/10.1007/3-540-48910-X_7
5. Canetti, R., Halevi, S., Katz, J.: Chosen-ciphertext security from identity-based encryption. In: Cachin, C., Camenisch, J.L. (eds.) EUROCRYPT 2004. LNCS, vol. 3027, pp. 207–222. Springer, Heidelberg (2004). https://doi.org/10.1007/978-3-540-24676-3_13

6. Canetti, R., Krawczyk, H., Nielsen, J.B.: Relaxing chosen-ciphertext security. In: Boneh, D. (ed.) CRYPTO 2003. LNCS, vol. 2729, pp. 565–582. Springer, Heidelberg (2003). https://doi.org/10.1007/978-3-540-45146-4_33
7. Chase, M., Kohlweiss, M., Lysyanskaya, A., Meiklejohn, S.: Malleable proof systems and applications. In: Pointcheval, D., Johansson, T. (eds.) EUROCRYPT 2012. LNCS, vol. 7237, pp. 281–300. Springer, Heidelberg (2012). https://doi.org/10.1007/978-3-642-29011-4_18
8. Cramer, R., Shoup, V.: Universal hash proofs and a paradigm for adaptive chosen ciphertext secure public-key encryption. In: Knudsen, L.R. (ed.) EUROCRYPT 2002. LNCS, vol. 2332, pp. 45–64. Springer, Heidelberg (2002). https://doi.org/10.1007/3-540-46035-7_4
9. Damgård, I., Faust, S., Mukherjee, P., Venturi, D.: Bounded tamper resilience: how to go beyond the algebraic barrier. In: Sako, K., Sarkar, P. (eds.) ASIACRYPT 2013, Part II. LNCS, vol. 8270, pp. 140–160. Springer, Heidelberg (2013). https://doi.org/10.1007/978-3-642-42045-0_8
10. Dodis, Y., Haralambiev, K., López-Alt, A., Wichs, D.: Efficient public-key cryptography in the presence of key leakage. In: Abe, M. (ed.) ASIACRYPT 2010. LNCS, vol. 6477, pp. 613–631. Springer, Heidelberg (2010). https://doi.org/10.1007/978-3-642-17373-8_35
11. Dodis, Y., Mironov, I., Stephens-Davidowitz, N.: Message transmission with reverse firewalls—secure communication on corrupted machines. In: Robshaw, M., Katz, J. (eds.) CRYPTO 2016, Part I. LNCS, vol. 9814, pp. 341–372. Springer, Heidelberg (2016). https://doi.org/10.1007/978-3-662-53018-4_13
12. ElGamal, T.: A public key cryptosystem and a signature scheme based on discrete logarithms. In: Blakley, G.R., Chaum, D. (eds.) CRYPTO 1984. LNCS, vol. 196, pp. 10–18. Springer, Heidelberg (1985). https://doi.org/10.1007/3-540-39568-7_2
13. Escala, A., Herold, G., Kiltz, E., Ràfols, C., Villar, J.: An algebraic framework for Diffie-Hellman assumptions. In: Canetti, R., Garay, J.A. (eds.) CRYPTO 2013, Part II. LNCS, vol. 8043, pp. 129–147. Springer, Heidelberg (2013). https://doi.org/10.1007/978-3-642-40084-1_8
14. Faonio, A., Fiore, D.: Optimistic mixing, revisited. Cryptology ePrint Archive, Report 2018/864 (2018). https://eprint.iacr.org/2018/864
15. Faonio, A., Fiore, D., Herranz, J., Ràfols, C.: Structure-preserving and rerandomizable RCCA-secure public key encryption and its applications. Cryptology ePrint Archive, Report 2019/955 (2019). https://eprint.iacr.org/2019/955
16. Faonio, A., Venturi, D.: Efficient public-key cryptography with bounded leakage and tamper resilience. In: Cheon, J.H., Takagi, T. (eds.) ASIACRYPT 2016, Part I. LNCS, vol. 10031, pp. 877–907. Springer, Heidelberg (2016). https://doi.org/10.1007/978-3-662-53887-6_32
17. Fauzi, P., Lipmaa, H., Siim, J., Zając, M.: An efficient pairing-based shuffle argument. In: Takagi, T., Peyrin, T. (eds.) ASIACRYPT 2017, Part II. LNCS, vol. 10625, pp. 97–127. Springer, Cham (2017). https://doi.org/10.1007/978-3-319-70697-9_4
18. Garg, S., Jain, A., Sahai, A.: Leakage-resilient zero knowledge. In: Rogaway, P. (ed.) CRYPTO 2011. LNCS, vol. 6841, pp. 297–315. Springer, Heidelberg (2011). https://doi.org/10.1007/978-3-642-22792-9_17
19. Golle, P., Jakobsson, M., Juels, A., Syverson, P.: Universal re-encryption for mixnets. In: Okamoto, T. (ed.) CT-RSA 2004. LNCS, vol. 2964, pp. 163–178. Springer, Heidelberg (2004). https://doi.org/10.1007/978-3-540-24660-2_14

20. Golle, P., Zhong, S., Boneh, D., Jakobsson, M., Juels, A.: Optimistic mixing for exit-polls. In: Zheng, Y. (ed.) ASIACRYPT 2002. LNCS, vol. 2501, pp. 451–465. Springer, Heidelberg (2002). https://doi.org/10.1007/3-540-36178-2_28
21. Groth, J.: Rerandomizable and replayable adaptive chosen ciphertext attack secure cryptosystems. In: Naor, M. (ed.) TCC 2004. LNCS, vol. 2951, pp. 152–170. Springer, Heidelberg (2004). https://doi.org/10.1007/978-3-540-24638-1_9
22. Groth, J.: Simulation-sound NIZK proofs for a practical language and constant size group signatures. In: Lai, X., Chen, K. (eds.) ASIACRYPT 2006. LNCS, vol. 4284, pp. 444–459. Springer, Heidelberg (2006). https://doi.org/10.1007/11935230_29
23. Groth, J., Sahai, A.: Efficient non-interactive proof systems for bilinear groups. In: Smart, N. (ed.) EUROCRYPT 2008. LNCS, vol. 4965, pp. 415–432. Springer, Heidelberg (2008). https://doi.org/10.1007/978-3-540-78967-3_24
24. Herold, G., Hoffmann, M., Klooß, M., Ràfols, C., Rupp, A.: New techniques for structural batch verification in bilinear groups with applications to Groth-Sahai proofs. In: ACM CCS 17, pp. 1547–1564. ACM Press (2017)
25. Jutla, C.S., Roy, A.: Shorter quasi-adaptive NIZK proofs for linear subspaces. In: Sako, K., Sarkar, P. (eds.) ASIACRYPT 2013, Part I. LNCS, vol. 8269, pp. 1–20. Springer, Heidelberg (2013). https://doi.org/10.1007/978-3-642-42033-7_1
26. Jutla, C.S., Roy, A.: Switching lemma for bilinear tests and constant-size NIZK proofs for linear subspaces. In: Garay, J.A., Gennaro, R. (eds.) CRYPTO 2014, Part II. LNCS, vol. 8617, pp. 295–312. Springer, Heidelberg (2014). https://doi.org/10.1007/978-3-662-44381-1_17
27. Kiltz, E.: Chosen-ciphertext security from tag-based encryption. In: Halevi, S., Rabin, T. (eds.) TCC 2006. LNCS, vol. 3876, pp. 581–600. Springer, Heidelberg (2006). https://doi.org/10.1007/11681878_30
28. Kiltz, E., Wee, H.: Quasi-adaptive NIZK for linear subspaces revisited. In: Oswald, E., Fischlin, M. (eds.) EUROCRYPT 2015, Part II. LNCS, vol. 9057, pp. 101–128. Springer, Heidelberg (2015). https://doi.org/10.1007/978-3-662-46803-6_4
29. Libert, B., Peters, T., Qian, C.: Structure-preserving chosen-ciphertext security with shorter verifiable ciphertexts. In: Fehr, S. (ed.) PKC 2017, Part I. LNCS, vol. 10174, pp. 247–276. Springer, Heidelberg (2017). https://doi.org/10.1007/978-3-662-54365-8_11
30. Micali, S., Rackoff, C., Sloan, B.: The notion of security for probabilistic cryptosystems (Extended Abstract). In: Odlyzko, A.M. (ed.) CRYPTO 1986. LNCS, vol. 263, pp. 381–392. Springer, Heidelberg (1987). https://doi.org/10.1007/3-540-47721-7_27
31. Naveed, M., et al.: Controlled functional encryption. In: ACM CCS 14, pp. 1280–1291. ACM Press (2014)
32. Pereira, O., Rivest, R.L.: Marked mix-nets. In: Brenner, M., et al. (eds.) FC 2017. LNCS, vol. 10323, pp. 353–369. Springer, Cham (2017). https://doi.org/10.1007/978-3-319-70278-0_22
33. Phan, D.H., Pointcheval, D.: OAEP 3-round: a generic and secure asymmetric encryption padding. In: Lee, P.J. (ed.) ASIACRYPT 2004. LNCS, vol. 3329, pp. 63–77. Springer, Heidelberg (2004). https://doi.org/10.1007/978-3-540-30539-2_5
34. Prabhakaran, M., Rosulek, M.: Rerandomizable RCCA encryption. In: Menezes, A. (ed.) CRYPTO 2007. LNCS, vol. 4622, pp. 517–534. Springer, Heidelberg (2007). https://doi.org/10.1007/978-3-540-74143-5_29
35. Sahai, A.: Non-malleable non-interactive zero knowledge and adaptive chosen-ciphertext security. In: 40th FOCS, pp. 543–553. IEEE Computer Society Press (1999)

36. Shoup, V., Gennaro, R.: Securing threshold cryptosystems against chosen ciphertext attack. In: Nyberg, K. (ed.) EUROCRYPT 1998. LNCS, vol. 1403, pp. 1–16. Springer, Heidelberg (1998). https://doi.org/10.1007/BFb0054113
37. Wikström, D.: A universally composable mix-net. In: Naor, M. (ed.) TCC 2004. LNCS, vol. 2951, pp. 317–335. Springer, Heidelberg (2004). https://doi.org/10.1007/978-3-540-24638-1_18
38. Wikström, D.: A sender verifiable mix-net and a new proof of a shuffle. In: Roy, B. (ed.) ASIACRYPT 2005. LNCS, vol. 3788, pp. 273–292. Springer, Heidelberg (2005). https://doi.org/10.1007/11593447_15
39. Wikström, D.: Verificatum (2010). https://www.verificatum.com

iUC: Flexible Universal Composability Made Simple

Jan Camenisch[1], Stephan Krenn[2], Ralf Küsters[3(✉)], and Daniel Rausch[3]

[1] Dfinity, Zurich, Switzerland
jan@dfinity.org
[2] AIT Austrian Institute of Technology GmbH, Vienna, Austria
stephan.krenn@ait.ac.at
[3] University of Stuttgart, Stuttgart, Germany
{ralf.kuesters,daniel.rausch}@sec.uni-stuttgart.de

Abstract. Proving the security of complex protocols is a crucial and very challenging task. A widely used approach for reasoning about such protocols in a modular way is universal composability. A perfect model for universal composability should provide a sound basis for formal proofs and be very flexible in order to allow for modeling a multitude of different protocols. It should also be easy to use, including useful design conventions for repetitive modeling aspects, such as corruption, parties, sessions, and subroutine relationships, such that protocol designers can focus on the core logic of their protocols.

While many models for universal composability exist, including the UC, GNUC, and IITM models, none of them has achieved this ideal goal yet. As a result, protocols cannot be modeled faithfully and/or using these models is a burden rather than a help, often even leading to under-specified protocols and formally incorrect proofs.

Given this dire state of affairs, the goal of this work is to provide a framework for universal composability which combines soundness, flexibility, and usability in an unmatched way. Developing such a security framework is a very difficult and delicate task, as the long history of frameworks for universal composability shows.

We build our framework, called iUC, on top of the IITM model, which already provides soundness and flexibility while lacking sufficient usability. At the core of iUC is a single simple template for specifying essentially arbitrary protocols in a convenient, formally precise, and flexible way. We illustrate the main features of our framework with example functionalities and realizations.

Keywords: Universal Composability · Foundations

This work was in part funded by the European Commission through grant agreements n°s 321310 (PERCY) and 644962 (PRISMACLOUD), and by the *Deutsche Forschungsgemeinschaft* (DFG) through Grant KU 1434/9-1. We would like to thank Robert Enderlein for helpful discussions.

S. D. Galbraith and S. Moriai (Eds.): ASIACRYPT 2019, LNCS 11923, pp. 191–221, 2019.
https://doi.org/10.1007/978-3-030-34618-8_7

1 Introduction

Universal composability [4,25] is an important concept for reasoning about the security of protocols in a modular way. It has found wide spread use, not only for the modular design and analysis of cryptographic protocols, but also in other areas, for example for modeling and analyzing OpenStack [16], network time protocols [11], OAuth v2.0 [14], the integrity of file systems [8], as well as privacy in email ecosystems [13].

The idea of universal composability is that one first defines an *ideal protocol* (or ideal functionality) $\mathcal{F}$ that specifies the intended behavior of a target protocol/system, abstracting away implementation details. For a concrete realization (real protocol) $\mathcal{P}$, one then proves that "$\mathcal{P}$ behaves just like $\mathcal{F}$" in arbitrary contexts. Therefore, it is ensured that the real protocol enjoys the security and functional properties specified by $\mathcal{F}$.

Several models for universal composability have been proposed in the literature [4,5,7,9,10,15,18,23–25]. Ideally, a framework for universal composability should support a protocol designer in easily creating full, precise, and detailed specifications of various applications and in various adversary models, instead of being an additional obstacle. In particular, such frameworks should satisfy at least the following requirements:

Soundness: This includes the soundness of the framework itself and the general theorems, such as composition theorems, proven in it.

Flexibility: The framework must be flexible enough to allow for the precise design and analysis of a wide range of protocols and applications as well as security models, e.g., in terms of corruption, setup assumptions, etc.

Usability: It should be easy to precisely and fully formalize protocols; this is also an important prerequisite for carrying out formally/mathematically correct proofs. There should exist (easy to use) modeling conventions that allow a protocol designer to focus on the core logic of protocols instead of having to deal with technical details of the framework or repeatedly taking care of recurrent issues, such as modeling standard corruption behavior.

Unfortunately, despite the wide spread use of the universal composability approach, existing models and frameworks are still unsatisfying in these respects as none combines all of these requirements simultaneously (we discuss this in more detail below). Thus, the goal of this paper is to provide a universal composability framework that is *sound*, *flexible*, and *easy to use*, and hence constitutes a solid framework for designing and analyzing essentially any protocol and application in a modular, universally composable, and sound way. Developing such a security framework is a difficult and very delicate task that takes multiple years if not decades as the history on models for universal composability shows. Indeed, this paper is the result of many years of iterations, refinements, and discussions.

Contributions: To achieve the above described goal, we here propose a new universal composability framework called iUC ("IITM based Universal Composability"). This framework builds on top of the IITM model with its extension to so-called responsive environments [1]. The IITM model was originally proposed in [18], with a full and revised version – containing a simpler and more general runtime notion – presented in [22].

The IITM model already meets our goals of *soundness* and *flexibility*. That is, the IITM model offers a very general and at the same time simple runtime notion so that protocol designers do not have to care much about runtime issues, making sound proofs easier to carry out. Also, protocols are defined in a very general way, i.e., they are essentially just arbitrary sets of Interactive Turing Machines (ITMs), which may be connected in some way. In addition, the model offers a general addressing mechanism for machine instances. This gives great flexibility as arbitrary protocols can be specified; all theorems, such as composition theorems, are proven for this very general class of protocols. Unfortunately, this generality hampers *usability*. The model does not provide design conventions, for example, to deal with party IDs, sessions, subroutine relationships, shared state, or (different forms of) corruption; all of this is left to the protocol designer to manually specify for every design and analysis task, distracting from modeling the actual core logic of a protocol.

In essence, iUC is an instantiation of the IITM model that provides a convenient and powerful framework for specifying protocols. In particular, iUC greatly improves upon *usability* of the IITM model by adding missing conventions for many of the above mentioned repetitive aspects of modeling a protocol, while also abstracting from some of the (few) technical aspects of the underlying model; see below for the comparison of iUC with other frameworks.

At the core of iUC is *one* convenient template that supports protocol designers in specifying arbitrary types of protocols in a precise, intuitive, and compact way. This is made possible by new concepts, including the concept of entities as well as public and private roles. The template comes with a clear and intuitive syntax which further facilitates specifications and allows others to quickly pick up protocol specifications and use them as subroutines in their higher-level protocols.

A key difficulty in designing iUC was to preserve the *flexibility* of the original IITM model in expressing (and composing) arbitrary protocols while still improving *usability* by fixing modeling conventions for certain repetitive aspects. We solve this tension between flexibility and usability by, on the one hand, allowing for a high degree of customization and, on the other hand, by providing sensible defaults for repetitive and standard specifications. Indeed, as further explained and discussed in Sect. 3 and also illustrated by our case study (cf. Sect. 4), iUC preserves flexibility and supports a wide range of protocol types, protocol features, and composition operations, such as: ideal and global functionalities with arbitrary protocol structures, i.e., rather than being just monolithic machines, they may, for example, contain subroutines; protocols with joint-state and/or global state; shared state between multiple protocol sessions (without resorting to joint-state realizations); subroutines that are partially globally available while other parts are only locally available; realizing global functionalities with other protocols (including joint-state realizations that combine multiple global functionalities); different types of addressing mechanisms via globally unique and/or locally chosen session IDs; global functionalities that can be changed to be local when used as a subroutine; many different highly customizable corruption types (including incorruptability, static corruption, dynamic corruption, corruption only under certain conditions, automatic corruption upon subroutine corruptions); a corruption model that is fully compatible with joint-state realizations; arbitrary protocol structures that are not necessarily hierarchical trees and which allow for, e.g., multiple highest-level protocols that are accessible to the environment.

Importantly, all of the above is supported by just *a single template* and *two* composition theorems (one for parallel composition of multiple protocols and one for unbounded self composition of the same protocol). This makes iUC quite user friendly as protocol designers can leverage the full flexibility with just the basic framework; there are no extensions or special cases required to support a wide range of protocol types.

We emphasize that we do not claim specifications done in iUC to be shorter than the informal descriptions commonly found in the universal composability literature. A full, non-ambiguous specification cannot compete with such informal descriptions in terms of brevity, as these descriptions are often underspecified and ignore details, including model specific details and the precise corruption behavior. iUC is rather meant as a *powerful and sound tool for protocol designers that desire to specify protocols fully, without sweeping or having to sweep anything under the rug, and at the same time without being overburdened with modeling details and technical artifacts.* Such specifications are crucial for being able to understand, reuse, and compose results and to carry out sound proofs.

Related Work: The currently most relevant universal composability models are the UC model [4] (see [3] for the latest version), the GNUC model [15], the IITM model [18] (see [22] for the full and revised version), and the CC model [23]. The former three models are closely related in that they are based on polynomial runtime machines that can be instantiated during a run. In contrast, the CC model follows a more abstract approach that does not fix a machine model or runtime notion, and is thus not directly comparable to the other models (including iUC). Indeed, it is still an open research question if and how typical UC-style specifications, proofs, and arguments can be modeled in the CC model. In what follows, we therefore relate iUC with the UC and GNUC models; as already explained and further detailed in the rest of the paper, iUC is an instantiation of the IITM model.

While both the UC and GNUC models also enjoy the benefits of established protocol modeling conventions, those are, however, less flexible and less expressive than iUC. Let us give several concrete examples: conventions in UC and GNUC are built around the assumption of having globally unique SIDs that are shared between all participants of a protocol session, and thus locally managed SIDs cannot directly be expressed (cf. Sects. 3, 4, and 4.3 for details including a discussion of local SIDs). Both models also assume protocols to have disjoint sessions and thus their conventions do not support expressing protocols that directly share state between sessions, such as signature keys (while both models support joint-state realizations to somewhat remedy this drawback, those realizations have to modify the protocols at hand, which is not always desirable; cf. Sect. 4.3). Furthermore, in both models there is only a single highest-level protocol machine with potentially multiple instances, whereas iUC supports arbitrarily many highest-level protocol machines. This is very useful as it, for example, allows for seamlessly modeling global state without needing any extensions or modifications to our framework or protocol template (as illustrated in Sect. 4). In the case of GNUC, there are also several additional restrictions imposed on protocols, such as a hierarchical tree structure where all subroutines have a single uniquely defined caller (unless they are globally available also to the environment) and a fixed top-down corruption mechanism; none of which is required in iUC.

There are also some major differences between UC/GNUC and iUC on a technical level which further affect overall usability as well as expressiveness. Firstly, both UC and GNUC had to introduce various extensions of the basic computational model to support new types of protocols and composition, including new syntax and new composition theorems for joint-state, global state, and realizations of global functionalities [5,7,12,15]. This not only forces protocol designers to learn new protocol syntax and conventions for different types of composition, but also indicates a lack of flexibility in supporting new types of composition (say, for example, a joint-state realization that combines several separate global functionalities, cf. Sect. 4.3). In contrast, both composition theorems in iUC as well as our single template for protocols seamlessly support all of those types of protocols and composition, including some not considered in the literature so far (cf. Sect. 4.3). Secondly, there are several technical aspects in the UC model a protocol designer has to take care of in order to perform sound proofs: a runtime notion that allows for exhaustion of machines, even ideal functionalities, and that forces protocols to manually send runtime tokens between individual machine instances; a directory machine where protocols have to register all instances when they are created; "subroutine respecting" protocols that keep sessions disjoint. Technical requirements of the GNUC model mainly consist of several restrictions imposed on protocol structures (as mentioned above) which in particular keep protocol sessions disjoint. Unlike UC, the runtime notion of GNUC supports modeling protocols that cannot be exhausted, however, GNUC introduces additional flow-bounds to limit the number of bits sent between certain machines. In contrast, as also illustrated by our case study, iUC does not require directory machines, iUC's notion for protocols with disjoint sessions is completely optional and can be avoided entirely, and iUC's runtime notion allows for modeling protocols without exhaustion, without manual runtime transfers, and without requiring flow bounds (exhaustion and runtime transfers can of course be modeled as special cases, if desired).

The difference in flexibility and expressiveness of iUC compared to UC and GNUC is further explained in Sect. 3 and illustrated by our case study in Sect. 4, where we model a real world key exchange protocol exactly as it would be deployed in practice. This case study is not directly supported by the UC and GNUC models (as further discussed in Sect. 4.3). A second illustrative example is given in the full version of this paper [2], where we show that iUC can capture the SUC model [10] as a mere special case. The SUC model was proposed as a simpler version of the UC model specifically designed for secure multi party computation (MPC), but has to break out of (some technical aspects of) the UC model.

Structure of This Paper: We describe the iUC framework in Sect. 2, with a discussion of the main concepts and features in Sect. 3. A case study further illustrates and highlights some features of iUC in Sect. 4. We conclude in Sect. 5. Full details are given in our full version [2].

2 The iUC Framework

In this section, we present the iUC framework which is built on top of the IITM model. As explained in Sect. 1, the main shortcoming of the IITM model is a lack of usability due to missing conventions for protocol specifications. Thus, protocol designers have to manually define many repetitive modeling related aspects such as a corruption model, connections between machines, specifying the desired machine instances (e.g., does an instance model a single party, a protocol session consisting of multiple parties, a globally available resource), the application specific addressing of individual instances, etc. The iUC framework solves this shortcoming by adding convenient and powerful conventions for protocol specifications to the IITM model. A key difficulty in crafting these conventions is preserving the flexibility of the original IITM model in terms of expressing a multitude of various protocols in natural ways, while at the same time not overburdening a protocol designer with too many details. We solve this tension by providing *a single template* for specifying arbitrary types of protocols, including real, ideal, joint-state, global state protocols, which needed several sets of conventions and syntax in other frameworks, and sometimes even new theorems. Our template includes many optional parts with sensible defaults such that a protocol designer has to define only those parts relevant to her specific protocol. As the iUC framework is an instantiation of the IITM model, all composition theorems and properties of the IITM model carry over.

The following description of the iUC framework is kept independently of the IITM model, i.e., one can understand and use the iUC framework without knowing the IITM model. More details of the underlying IITM model are available in the full version [2]. Here we explain the IITM model not explicitly, but rather explain relevant parts as part of the description of the iUC framework. We start with some preliminaries in Sect. 2.1, mainly describing the general computational model, before we explain the general structure of protocols in iUC in Sect. 2.2, with corruption explained in Sect. 2.3. We then present our protocol template in Sect. 2.4. In Sect. 2.5, we explain how protocol specifications can be composed in iUC to create new, more complex protocol specification. Finally, in Sect. 2.6, we present the realization relation and the composition theorem of iUC. As mentioned, concrete examples are given in our case study (cf. Sect. 4). We provide a precise mapping from iUC protocols to the underlying IITM model in the full version, which is crucial to verify that our framework indeed is an instantiation of the IITM model, and hence, inherits soundness and all theorems of the IITM model. We note, however, that it is not necessary to read this technical mapping to be able to use our framework. The abstraction level provided by iUC is entirely sufficient to understand and use this framework.

2.1 Preliminaries

Just as the IITM model, the iUC framework uses interactive Turing machines as its underlying computational model. Such interactive Turing machines can be connected to each other to be able to exchange messages. A set of machines $\mathcal{Q} = \{M_1, \ldots, M_k\}$ is called a *system*. In a run of $\mathcal{Q}$, there can be one or more instances (copies) of each machine in $\mathcal{Q}$. One instance can send messages to another instance. At any point in a run, only a single instance is active, namely, the one to receive the last message; all other instances wait for input. The active instance becomes inactive once it has sent a message; then the instance that receives the message becomes active instead and can perform arbitrary computations. The first machine to run is the so-called *master*. The master is also triggered if the last active machine did not output a message. In iUC, the environment (see next) will take the role of the master. Jumping ahead, in the iUC framework a special user-specified **CheckID** algorithm is used to determine which instance of a machine receives a message and whether a new instance is to be created (cf. Sect. 2.4).

To define the universal composability security experiment (cf. Fig. 1 and Sect. 2.5), one distinguishes between three types of systems: protocols, environments, and adversaries. Intuitively, the security experiment in any universal composability model compares a protocol $\mathcal{P}$ with another protocol $\mathcal{F}$, where $\mathcal{F}$ is typically an ideal specification of some task, called *ideal protocol* or *ideal functionality*. The idea is that if one cannot distinguish $\mathcal{P}$ from $\mathcal{F}$, then $\mathcal{P}$ must be "as good as" $\mathcal{F}$. More specifically, the protocol $\mathcal{P}$ is considered secure (written $\mathcal{P} \leq \mathcal{F}$) if for all adversaries $\mathcal{A}$ controlling the network of $\mathcal{P}$ there exists an (ideal) adversary $\mathcal{S}$, called *simulator*, controlling the network of $\mathcal{F}$ such that $\{\mathcal{A}, \mathcal{P}\}$ and $\{\mathcal{S}, \mathcal{F}\}$ are indistinguishable for all environments $\mathcal{E}$. Indistinguishability means that the probability of the environment outputting 1 in runs of the system $\{\mathcal{E}, \mathcal{A}, \mathcal{P}\}$ is negligibly close to the probability of outputting 1 in runs of the system $\{\mathcal{E}, \mathcal{S}, \mathcal{F}\}$ (written $\{\mathcal{E}, \mathcal{A}, \mathcal{P}\} \equiv \{\mathcal{E}, \mathcal{S}, \mathcal{F}\}$).

In the security experiment, systems are connected as follows (cf. arrows in Fig. 1): Every (machine in a) protocol has an I/O interface that is used to connect to other protocol machines, higher-level protocols, or an environment, which, in turn, can simulate higher-level protocols. Every (machine in a) protocol also has a network interface to connect to a network adversary. We sometimes let the environment subsume the network adversary. That is, the environment performs both roles: on the left-hand side of Fig. 1, instead of having the systems $\mathcal{E}$ and $\mathcal{A}$ we can have an environment $\mathcal{E}'$ that connects to both the I/O interface and the network interface of $\mathcal{P}$.

The iUC framework includes support for so-called responsive environments and responsive adversaries introduced in [1]. Such environments/adversaries can

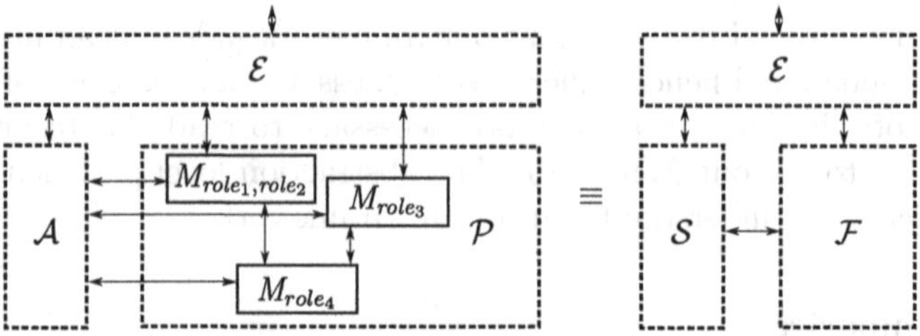

Fig. 1. The setup for the universal composability experiment ($\mathcal{P} \leq \mathcal{F}$) and internal structure of protocols. Here $\mathcal{E}$ is an environment, $\mathcal{A}$ and $\mathcal{S}$ are adversaries, and $\mathcal{P}$ and $\mathcal{F}$ are protocols. Arrows between systems denote connections/interfaces that allow for exchanging messages. The boxes M_i in $\mathcal{P}$ are different machines modeling various tasks in the protocol. Note that the machines in $\mathcal{P}$ and the way they are connected is just an example; other protocols can have a different internal structure.

be forced to answer certain messages on the network interface of the protocol immediately, without interfering with the protocol in between. These messages are called *restricting messages*. This mechanism is very handy to, e.g., exchange meta information such as the corruption state of a protocol participant or obtain cryptographic keys from the adversary; see our full version [2] and [1] for a more detailed discussion.

We require environments to be *universally bounded*, i.e., there is a fixed polynomial in the security parameter (and possibly external input) that upper bounds the runtime of an environment no matter to which protocol and adversary it is connected to. A system $\mathcal{Q}$ is called *environmentally bounded* if for every (universally bounded) environment $\mathcal{E}$ there is a polynomial that bounds the runtime of the system $\mathcal{Q}$ connected to $\mathcal{E}$ (except for potentially a negligible probability). This will mostly be required for protocols; note that natural protocols used in practice are typically environmentally bounded, including all protocols that run in polynomial time in their inputs received so far and the security parameter. This is the same runtime notion used in the IITM model. Compared to other models, this notion is very general and particularly simple (see [22] for a discussion).

We define $\mathsf{Env}(\mathcal{Q})$ to be the set of all universally bounded (responsive) environments that connect to a system $\mathcal{Q}$ via network and I/O interfaces. We further define $\mathsf{Adv}(\mathcal{P})$ to be the set of (responsive) adversaries that connect to the network interface of a protocol $\mathcal{P}$ such that the combined system $\{\mathcal{A}, \mathcal{P}\}$ is environmentally bounded.

2.2 Structure of Protocols

A protocol $\mathcal{P}$ in our framework is specified via a system of machines $\{M_1, \ldots, M_l\}$. Each machine M_i implements one or more roles of the protocol, where a role describes a piece of code that performs a specific task. For example, a (real) protocol $\mathcal{P}_{\text{sig}}$ for digital signatures might contain a `signer` role for signing

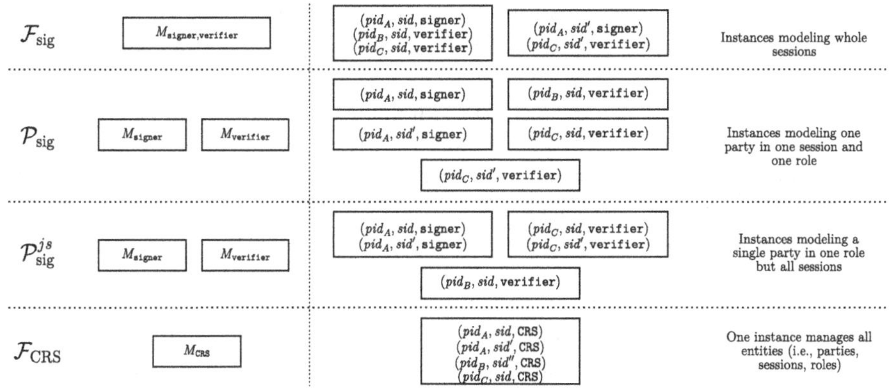

Fig. 2. Examples of static and dynamic structures of various protocol types. $\mathcal{F}_{\text{sig}}$ is an ideal protocol, $\mathcal{P}_{\text{sig}}$ a real protocol, $\mathcal{P}^{js}_{\text{sig}}$ a so-called joint-state realization, and $\mathcal{F}_{\text{CRS}}$ a global state protocol. On the left-hand side: static structures, i.e., (specifications of) machines/protocols. On the right-hand side: possible dynamic structures (i.e., several machine instances managing various entities).

messages and a `verifier` role for verifying signatures. In a run of a protocol, there can be several instances of every machine, interacting with each other (and the environment) via I/O interfaces and interacting with the adversary (and possibly the environment) via network interfaces. An instance of a machine M_i manages one or more so-called *entities*. An entity is identified by a tuple $(pid, sid, role)$ and describes a specific party with party ID (PID) pid running in a session with session ID (SID) sid and executing some code defined by the role $role$ where this role has to be (one of) the role(s) of M_i according to the specification of M_i. Entities can send messages to and receive messages from other entities and the adversary using the I/O and network interfaces of their respective machine instances. In the following, we explain each of these parts in more detail, including roles and entities; we also provide examples of the static and dynamic structure of various protocols in Fig. 2.

Roles: As already mentioned, a role is a piece of code that performs a specific task in a protocol $\mathcal{P}$. Every role in $\mathcal{P}$ is implemented by a single unique machine M_i, but one machine can implement more than one role. This is useful for sharing state between several roles: for example, consider an ideal functionality $\mathcal{F}_{\text{sig}}$ for digital signatures consisting of a `signer` and a `verifier` role. Such an ideal protocol usually stores all messages signed by the `signer` role in some global set that the `verifier` role can then use to prevent forgery. To share such a set between roles, both roles must run on the same (instance of a) machine, i.e., $\mathcal{F}_{\text{sig}}$ generally consists of a single machine $M_{\texttt{signer,verifier}}$ implementing both roles. In contrast, the real protocol $\mathcal{P}_{\text{sig}}$ uses two machines $M_{\texttt{signer}}$ and $M_{\texttt{verifier}}$ as those roles do not and cannot directly share state in a real implementation (cf. left-hand side of Fig. 2). Machines provide an I/O interface and a network interface for every role that they implement. The I/O interfaces of two roles of two different machines can be connected. This means that, in a run of a system, two entities (managed by two instances of machines) with connected roles can then directly send and receive messages to/from each other; in contrast, entities of unconnected roles cannot directly send and receive messages to/from each other. Jumping ahead, in a protocol specification (see below) it is specified for each machine in that protocol to which other roles (subroutines) a machine connects to (see, e.g., also Fig. 3a where the arrows denote connected roles/machines). The network interface of every role is connected to the adversary (or simulator), allowing for sending and receiving messages to and from the adversary. For addressing purposes, we assume that each role in $\mathcal{P}$ has a unique name. Thus, role names can be used for communicating with a specific piece of code, i.e., sending and receiving a message to/from the correct machine.

Public and Private Roles: We, in addition, introduce the concept of public and private roles, which, as we will explain, is a very powerful tool. Every role of a protocol $\mathcal{P}$ is either *private* or *public*. Intuitively, a private role can be called/used only internally by other roles of $\mathcal{P}$ whereas a public role can be called/used by any protocol and the environment. Thus, private roles provide their functionality only internally within $\mathcal{P}$, whereas public roles provide their functionality also to other protocols and the environment. More precisely, a private role connects via its I/O interface only to (some of the) other roles in $\mathcal{P}$ such that only those roles can send messages to and receive messages from a private role; a public role additionally provides its I/O interface for arbitrary other protocols and the environment such that they can also send messages to and receive messages from a public role. We illustrate the concept of public and private roles by an example below.

Using Other Protocols as Subroutines: Protocols can be combined to construct new, more complex protocols. Intuitively, two protocols $\mathcal{P}$ and $\mathcal{R}$ can be combined if they connect to each other only via (the I/O interfaces of) their public roles. (We give a formal definition of connectable protocols in Sect. 2.5.) The new combined protocol $\mathcal{Q}$ consists of all roles of $\mathcal{P}$ and $\mathcal{R}$, where private roles remain private while public roles can be either public or private in $\mathcal{Q}$; this is up to the protocol designer to decide. To keep role names unique within $\mathcal{Q}$, even if the same role name was used in both $\mathcal{P}$ and $\mathcal{R}$, we (implicitly) assume that role names are prefixed with the name of their original protocol. We will often also explicitly write down this prefix in the protocol specification for better readability (cf. Sect. 2.4).

Examples Illustrating the Above Concepts: Figure 3a, which is further explained in our case study (cf. Sect. 4), illustrates the structure of the protocols we use to model a real key exchange protocol. This protocol as a whole forms a protocol in the above sense and at the same time consists of three separate (sub-) protocols: The highest-level protocol $\mathcal{P}_{\text{KE}}$ has two public roles `initiator` and `responder` executing the actual key exchange and one private role `setup` that

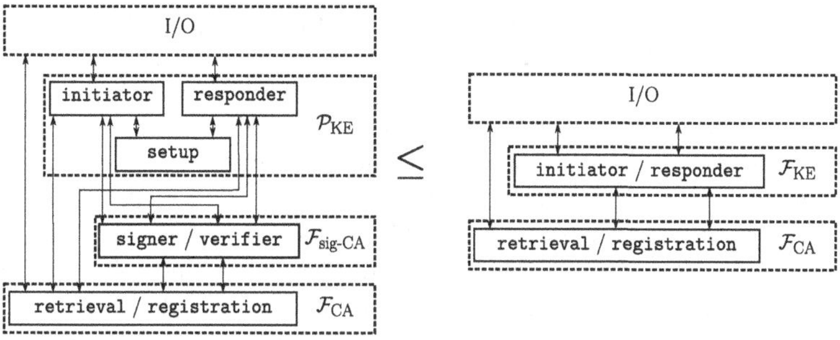

(a) Real world (b) Ideal world

Fig. 3. The static structures of the ideal key exchange functionality $\mathcal{F}_{\mathrm{KE}}$ (right side) and its realization $\mathcal{P}_{\mathrm{KE}}$ (left side), including their subroutines, in our case study. Arrows denote direct connections of I/O interfaces; network connections are omitted for simplicity. Solid boxes (labeled with one or two role names) denote individual machines, dotted boxes denote (sub-)protocols that are specified by one instance of our template each (cf. Sect. 2.4).

generates some global system parameters. The protocol $\mathcal{P}_{\mathrm{KE}}$ uses two other protocols as subroutines, namely the ideal functionality $\mathcal{F}_{\text{sig-CA}}$ for digital signatures with roles `signer` and `verifier`, for signing and verifying messages, and an ideal functionality $\mathcal{F}_{\mathrm{CA}}$ for certificate authorities with roles `registration` and `retrieval`, for registering and retrieving public keys (public key infrastructure). Now, in the context of the combined key exchange protocol, the `registration` role of $\mathcal{F}_{\mathrm{CA}}$ is private as it should be used by $\mathcal{F}_{\text{sig-CA}}$ only; if everyone could register keys, then it would not be possible to give any security guarantees in the key exchange. The `retrieval` role of $\mathcal{F}_{\mathrm{CA}}$ remains public, modeling that public keys are generally considered to be known to everyone, so not only $\mathcal{P}_{\mathrm{KE}}$ but also the environment (and possibly other protocols later using $\mathcal{P}_{\mathrm{KE}}$) should be able to access those keys. This models so-called global state. Similarly to role `registration`, the `signer` role of $\mathcal{F}_{\text{sig-CA}}$ is private too. For simplicity of presentation, we made the `verifier` role private, although it could be made public. Note that this does not affect the security statement: the environment knows the public verification algorithm and can obtain all verification keys from $\mathcal{F}_{\mathrm{CA}}$, i.e., the environment can locally compute the results of the verification algorithm. Altogether, with the concept of public and private roles, we can easily decide whether we want to model global state or make parts of a machine globally available while others remain local subroutines. We can even change globally available roles to be only locally available in the context of a new combined protocol.

As it is important to specify which roles of a (potentially combined) protocol are public and which ones are private, we introduce a simple notation for this. We write $(role_1, \ldots, role_n \mid role_{n+1}, \ldots, role_m)$ to denote a protocol $\mathcal{P}$ with public roles $role_1, \ldots, role_n$ and private roles $role_{n+1}, \ldots, role_m$. If there are no private roles, we just write $(role_1, \ldots, role_n)$, i.e., we omit "|". Using this notation, the example key exchange protocol from Fig. 3a can be written as (**initiator**, **responder**, **retrieval** | **setup**, **signer**, **verifier**, **registration**).

Entities and Instances: As mentioned before, in a run of a protocol there can be several instances of every protocol machine, and every instance of a protocol machine can manage one or more, what we call, *entities*. Recall that an entity is identified by a tuple $(pid, sid, role)$, which represents party *pid* running in a session with SID *sid* and executing some code defined by the role *role*. As also mentioned, such an entity can be managed by an instance of a machine only if this machine implements *role*. We note that *sid* does not necessarily identify a protocol session in a classical sense. The general purpose is to identify multiple instantiations of the role *role* executed by party *pid*. In particular, entities with different SIDs may very well interact with each other, if so desired, unlike in many other frameworks.

The novel concept of entities allows for easily customizing the interpretation of a machine instance by managing appropriate sets of entities. An important property of entities managed by the same instance is that they have access to the same internal state, i.e., they can share state; entities managed by different instances cannot access each others internal state directly. This property is usually the main factor for deciding which entities should be managed in the same instance. With this concept of entities, we obtain a *single* definitional framework for modeling various types of protocols and protocol components in a uniform way, as illustrated by the examples in Fig. 2, explained next.

One instance of an ideal protocol in the literature, such as a signature functionality $\mathcal{F}_{\text{sig}}$, often models a single session of a protocol. In particular, such an instance contains all entities for all parties and all roles of one session. Figure 2 shows two instances of the machine $M_{\texttt{signer,verifier}}$, managing sessions *sid* and *sid'*, respectively. In contrast, instances of real protocols in the literature, such as the realization $\mathcal{P}_{\text{sig}}$ of $\mathcal{F}_{\text{sig}}$, often model a single party in a single session of a single role, i.e., every instance manages just a single unique entity, as also illustrated in Fig. 2. If, instead, we want to model one global common reference string (CRS), for example, we have one instance of a machine $M_{\texttt{CRS}}$ which manages all entities, for all sessions, parties, and roles. To give another example, the literature also considers so-called joint-state realizations [7,20] where a party re-uses some state, such as a cryptographic key, in multiple sessions. An instance of such a joint-state realization thus contains entities for a single party in one role and in all sessions. Figure 2 shows an example joint-state realization $\mathcal{P}^{js}_{\text{sig}}$ of $\mathcal{F}_{\text{sig}}$ where a party uses the same signing key in all sessions. As illustrated by these examples, instances model different things depending on the entities they manage.

Exchanging Messages: Entities can send and receive messages using the I/O and network interfaces belonging to their respective roles. When an entity sends a message it has to specify the receiver, which is either the adversary in the case of the network interface or some other entity (with a role that has a connected I/O interface) in the case of the I/O interface. If a message is sent to another entity $(pid_{rcv}, sid_{rcv}, role_{rcv})$, then the message is sent to the machine M implementing

$role_{rcv}$; a special user-defined **CheckID** algorithm (see Sect. 2.4) is then used to determine the instance of M that manages $(pid_{rcv}, sid_{rcv}, role_{rcv})$ and should hence receive the message. When an entity $(pid_{rcv}, sid_{rcv}, role_{rcv})$ receives a message on the I/O interface, i.e., from another entity $(pid_{snd}, sid_{snd}, role_{snd})$, then the receiver learns pid_{snd}, sid_{snd}[1] and either the actual role name $role_{snd}$ (if the sender is a known subroutine of the receiver, cf. Sect. 2.4) or an arbitrary but fixed number i (from an arbitrary but fixed range of natural numbers) denoting a specific I/O connection to some (unknown) sender role (if the sender is an unknown higher-level protocol or the environment[2]). The latter models that a receiver/subroutine does not necessarily know the exact machine code of a caller in some arbitrary higher-level protocol, but the receiver can at least address the caller in a consistent way for sending a response. If a message is received from the network interface, then the receiving entity learns only that it was sent from the adversary.

We note that we do not restrict which entities can communicate with each other as long as their roles are connected via their I/O interfaces, i.e., entities need not share the same SID or PID to communicate via an I/O connection. This, for example, facilitates modeling entities in different sessions using the same resource, as illustrated in our case study. It, for example, also allows us to model the global functionality $\mathcal{F}_{\text{CRS}}$ from Fig. 2 in the following natural way: $\mathcal{F}_{\text{CRS}}$ could manage only a single (dummy) entity $(\epsilon, \epsilon, \texttt{CRS})$ in one machine instance, which can be accessed by all entities of higher-level protocols.

2.3 Modeling Corruption

We now explain on an abstract level how our framework models corruption of entities. In Sect. 2.4, we then explain in detail how particular aspects of the corruption model are specified and implemented. Our framework supports five different modes of corruption: *incorruptible, static corruption, dynamic corruption with/without secure erasures,* and *custom corruption.* Incorruptible protocols do not allow the adversary to corrupt any entities; this can, e.g., be used to model setup assumptions such as common reference strings which should not be controllable by an adversary. Static corruption allows adversaries to corrupt entities when they are first created, but not later on, whereas dynamic corruption allows for corruption at arbitrary points in time. In the case of dynamic corruption, one can additionally choose whether by default only the current internal state (known as dynamic corruption *with secure erasures*) or also a history of the entire state, including all messages and internal random coins (known as dynamic corruption *without secure erasures*) is given to the adversary upon corruption. Finally, custom corruption is a special case that allows a protocol designer to disable corruption handling of our framework and instead define her own corruption model while still taking advantage of our template and the defaults that we provide; we will ignore this custom case in the following description.

[1] The environment can claim arbitrary PIDs and SIDs as sender.

[2] The environment can choose the number that it claims as a sender as long as it does not collide with a number used by another (higher-level) role in the protocol.

To corrupt an entity (*pid*, *sid*, *role*) in a run, the adversary can send the special message corrupt on the network interface to that entity. Note that, depending on the corruption model, such a request might automatically be rejected (e.g., because the entity is part of an incorruptible protocol). In addition to this automatic check, protocol designers are also able to specify an algorithm **AllowCorruption**, which can be used to specify arbitrary other conditions that must be met for a corrupt request to be accepted. For example, one could require that all subroutines must be corrupted before a corruption request is accepted (whether or not subroutines are corrupted can be determined using `CorruptionStatus?` requests, see later), modeling that an adversary must corrupt the entire protocol stack running on some computer instead of just individual programs, which is often easier to analyze (but yields a less fine grained security result). One could also prevent corruption during a protected/trusted "setup" phase of the protocol, and allow corruption only afterwards.

If a corrupt request for some entity (*pid*, *sid*, *role*) passes all checks and is accepted, then the state of the entity is leaked to the adversary (which can be customized by specifying an algorithm **LeakedData**) and the entity is considered *explicitly corrupted* for the rest of the protocol run. The adversary gains full control over explicitly corrupted entities: messages arriving on the I/O interface of (*pid*, *sid*, *role*) are forwarded on the network interface to the adversary, while the adversary can tell (*pid*, *sid*, *role*) (via its network interface) to send messages to arbitrary other entities on behalf of the corrupted entity (as long as both entities have connected I/O interfaces). The protocol designer can control which messages the adversary can send in the name of a corrupted instance by specifying an algorithm **AllowAdvMessage**. This can be used, e.g., to prevent the adversary from accessing uncorrupted instances or from communicating with other (disjoint) sessions, as detailed in Sect. 2.4.

In addition to the corruption mechanism described above, entities that are activated for the first time also determine their initial corruption status by actively asking the adversary whether he wants to corrupt them. More precisely, once an entity (*pid*, *sid*, *role*) has finished its initialization (see Sect. 2.4), it asks the adversary via a *restricting message*[3] whether he wants to corrupt (*pid*, *sid*, *role*) before performing any other computations. The answer of the adversary is processed as discussed before, i.e., the entity decides whether to accept or reject a corruption request. This gives the adversary the power to corrupt new entities right from the start, if he desires; note that in the case of static corruption, this is also the last point in time where an adversary can explicitly corrupt (*pid*, *sid*, *role*).

---

[3] Recall from Sect. 2.1 that by sending a restricting message, the adversary is forced to answer, and hence, decide upon corruption right away, before he can interact in any other way with the protocol, preventing artificial interference with the protocol run. This is a very typical use of restricting messages, which very much simplifies corruption modeling (see also [1]).

For modeling purposes, we allow other entities and the environment to obtain the current corruption status of an entity $(pid, sid, role)$.[4] This is done by sending a special `CorruptionStatus`? request on the I/O interface of $(pid, sid, role)$. If $(pid, sid, role)$ has been explicitly corrupted by the adversary, the entity returns `true` immediately. Otherwise, the entity is free to decide whether `true` or `false` is returned, i.e., whether it considers itself corrupted nevertheless (this is specified by the protocol designer via an algorithm **DetermineCorrStatus**). For example, a higher level protocol might consider itself corrupted if at least one of its subroutines is (explicitly or implicitly) corrupted, which models that no security guarantees can be given if certain subroutines are controlled by the adversary. To figure out whether subroutines are corrupted, a higher level protocol can send `CorruptionStatus`? requests to subroutines itself. We call an entity that was not explicitly corrupted but still returns `true` *implicitly corrupted.* We note that the responses to `CorruptionStatus`? request are guaranteed to be consistent in the sense that if an entity returns `true` once, it will always return `true`. Also, according to the defaults of our framework, `CorruptionStatus`? request are answered immediately (without intervention of the adversary) and processing these requests does not change state. These are important features which allow for a smooth handling of corruption.

2.4 Specifying Protocols

We now present our template for fully specifying a protocol $\mathcal{Q}$, including its uncorrupted behavior, its corruption model, and its connections to other protocols. As mentioned previously, the template is sufficiently general to capture many different types of protocols (real, ideal, hybrid, joint-state, global, ...) and includes several optional parts with reasonable defaults. Thus, our template combines freedom with ease of specification.

The template is given in Fig. 4. Some parts are self-explanatory; the other parts are described in more detail in the following. The first section of the template specifies properties of the whole protocol that apply to all machines.

Participating Roles: This list of sets of roles specifies which roles are (jointly) implemented by a machine. To give an example, the list "$\{role_1, role_2\}$, $role_3$, $\{role_4, role_5, role_6\}$" specifies a protocol $\mathcal{Q}$ consisting of three machines $M_{role_1,role_2}$, M_{role_3}, and $M_{role_4,role_5,role_6}$, where $M_{role_1,role_2}$ implements $role_1$ and $role_2$, and so on.

Corruption Model: This fixes one of the default corruption models supported by iUC, as explained in Sect. 2.3: *incorruptible*, *static*, *dynamic with erasures*, and

[4] This operation is purely for modeling purposes and does of course not exist in reality. It is crucial for obtaining a reasonable realization relation: The environment needs a way to check that the simulator in the ideal world corrupts exactly those entities that are corrupted in the real world, i.e., the simulation should be perfect also with respect to the corruption states. If we did not provide such a mechanism, the simulator could simply corrupt all entities in the ideal world which generally allows for a trivial simulation of arbitrary protocols.

Setup for the protocol $\mathcal{Q} = \{M_1, \ldots, M_n\}$:

Participating roles: list of all n sets of roles participating in this protocol. Each set corresponds to one machine M_i.
Corruption model: incorruptible, static, dynamic with/without erasures, custom.
Protocol parameters*: e.g., externally provided algorithms parametrizing a machine.

Implementation of M_i for each set of roles:

Implemented role(s): the set of roles that is implemented by this machine.
Subroutines*: a list of all (other) roles that this machine uses as subroutines.
Internal state*: state variables used to store data across different invocations.
CheckID*: algorithm for deciding whether this machine is responsible for an entity $(pid, sid, role)$.
Corruption behavior*: description of **DetermineCorrStatus**, **AllowCorruption**, **LeakedData**, and/or **AllowAdvMessage** algorithms.
Initialization*: this block is executed only the first time an instance of the machine accepts a message; useful to, e.g., assign initial values that are globally used for all entities managed by this instance.
EntityInitialization*: this block is executed only the first time that some message for a (new) entity is received; useful to, e.g., assign initial values that are specific for single entities.
MessagePreprocessing*: this algorithm is executed every time a message for an uncorrupted entity is received.
Main: specification of the actual behavior of an uncorrupted entity.

Fig. 4. Template for specifying protocols. Blocks labeled with an asterisk (*) are optional. Note that the template does not specify public and private roles as those change depending on how several protocols (each defined via a copy of this template) are connected.

dynamic without erasures. Moreover, if the corruption model is set to *custom*, the protocol designer has to manually define his own corruption model and process corruption related messages, such as `CorruptionStatus?`, using the algorithms **MessagePreprocessing** and/or **Main** (see below), providing full flexibility.

Apart from the protocol setup, one has to specify each procotol machine M_i, and hence, the behavior of each set of roles listed in the protocol setup.

Subroutines: Here the protocol designer lists all roles that M_i uses as subroutines. These roles may be part of this or potentially other protocols, but may not include roles that are implemented by M_i. The I/O interface of (all roles of) the machine M_i will then be connected to the I/O interfaces of those roles, allowing M_i to access and send messages to those subroutines.[5] We note that (subroutine) roles are uniquely specified by their name since we assume globally unique names for each role. We also note that subroutines are specified on the level of roles, instead of the level of whole protocols, as this yields more flexibility and a more fine grained subroutine relationship, and hence, access structure.

If roles of some other protocol $\mathcal{R}$ are used, then protocol authors should prefix the roles with the protocol name to improve readability, e.g., "$\mathcal{R}$: `roleInR`" to

[5] We emphasize that we do not put any restrictions on the graph that the subroutine relationships of machines of several protocols form. For example, it is entirely possible to have machines in two different protocols that specify each other as subroutines.

denote a connection to the role `roleInR` in the protocol $\mathcal{R}$. This is mandatory if the same role name is used in several protocols to avoid ambiguity. If a machine is supposed to connect to all roles of some protocol $\mathcal{R}$, then, as a short-hand notation, one can list the name $\mathcal{R}$ of the protocol instead.

Internal State: State variables declared here (henceforth denoted by sans-serif fonts, e.g., a, b) preserve their values across different activations of an instance of M_i.

In addition to these user-specified state variables, every machine has some additional framework-specific state variables that are set and changed automatically according to our conventions. Most of these variables are for internal bookkeeping and need not be accessed by protocol designers. Those that might be useful in certain algorithms are mentioned and explained further below (we provide a complete list of all framework specific variables in the full version).

CheckID: As mentioned before, instances of machines in our framework manage (potentially several) entities $(pid_i, sid_i, role_i)$. The algorithm **CheckID** allows an instance of a machine to decide which of those entities are accepted and thus managed by that instance, and which are not. Furthermore, it allows for imposing a certain structure on pid_i and sid_i; for example, SIDs might only be accepted if they encode certain session parameters, e.g., $sid_i = (parameter_1, parameter_2, sid'_i)$.

More precisely, the algorithm **CheckID**$(pid, sid, role)$ is a *deterministic algorithm* that computes on the input $(pid, sid, role)$, the internal state of the machine instance, and the security parameter. It runs in *polynomial time* in the length of the current input, the internal state, and the security parameter and outputs `accept` or `reject`.

Whenever one (entity in one) instance of a machine, the adversary, or the environment sends a message m to some entity $(pid, sid, role)$ (via the entity's I/O interface or network interface), the following happens: m is delivered to the first instance of the machine, say M, that implements *role*, where instances of a machine are ordered by the time of their creation. That instance then runs **CheckID**$(pid, sid, role)$ to determine whether it manages $(pid, sid, role)$, and hence, whether the message m (that was sent to $(pid, sid, role)$) should be accepted. If **CheckID** accepts the entity, then the instance gets to process the message m; otherwise, it resets itself to the state before running **CheckID** and the message is given to the next instance of M (according to the order of instances mentioned before) which then runs **CheckID**$(pid, sid, role)$, and so on. If no instance accepts, or no instance exists yet, then a new one is created that also runs **CheckID**$(pid, sid, role)$. If that final instance accepts, it also gets to process m; otherwise, the new instance is deleted, the message m is dropped, and the environment is triggered (with a default trigger message).

We require that **CheckID** behaves consistently, i.e., it never accepts an entity that has previously been rejected, and it never rejects an entity that has previously been accepted; this ensures that there are no two instances that manage the same entity. For this purpose, we provide access to a convenient framework specific list $\mathsf{acceptedEntities}$ that contains all entities that have been accepted so far (in the order in which they were first accepted). We note that **CheckID** cannot change the (internal) state of an instance; all changes caused by running **CheckID** are dropped after outputting a decision, i.e., the state of an instance is set back to the state before running **CheckID**.

If **CheckID** is not specified, its default behavior is as follows: Given input $(pid, sid, role)$, if the machine instance in which **CheckID** is running has not accepted an entity yet, it outputs `accept`. If it has already accepted an entity $(pid', sid', role')$, then it outputs `accept` iff $pid = pid'$ and $sid = sid'$. Otherwise, it outputs `reject`. Thus, by default, a machine instance accepts, and hence, manages, not more than one entity per role for the roles the machine implements.

Corruption Behavior: This element of the template allows for customization of corruption related behavior of machines by specifying one or more of the optional algorithms **DetermineCorrStatus**, **AllowCorruption**, **LeakedData**, and **AllowAdvMessage**, as explained and motivated in Sect. 2.3, with the formal definition of these algorithms, including their default behavior if not specified, given in the full version. A protocol designer can access two useful framework specific variables for defining these algorithms: transcript, which, informally, contains a transcript of all messages sent and received by the current machine instance, and CorruptionSet, which contains all explicitly corrupted entities that are managed by the current machine instance. As these algorithms are part of our corruption conventions, they are used only if **Corruption Model** is not set to *custom*.

Initialization, **EntityInitialization**, **MessagePreprocessing**, **Main:** These algorithms specify the actual behavior of a machine for uncorrupted entities.

The **Initialization** algorithm is run exactly once per machine instance (*not per entity* in that instance) and is mainly supposed to be used for initializing the internal state of that instance. For example, one can generate global parameters or cryptographic key material in this algorithm.

The **EntityInitialization**$(pid, sid, role)$ algorithm is similar to **Initialization** but is run once for each entity $(pid, sid, role)$ instead of once for each machine instance. More precisely, it runs directly after a potential execution of **Initialization** if **EntityInitialization** has not been run for the current entity $(pid, sid, role)$ yet. This is particularly useful if a machine instance manages several entities, where not all of them might be known from the beginning.

After the algorithms **Initialization** and, for the current entity, the algorithm **EntityInitialization** have finished, the current entity determines its initial corruption status (if not done yet) and processes a corrupt request from the network/adversary, if any. Note that this allows for using the initialization algorithms to setup some internal state that can be used by the entity to determine its corruption status.

Finally, after all of the previous steps, if the current entity has not been explicitly corrupted,[6] the algorithms **MessagePreprocessing** and **Main** are run. The **MessagePreprocessing** algorithm is executed first. If it does not

[6] As mentioned in Sect. 2.3, if an entity is explicitly corrupted, it instead acts as a forwarder for messages to and from the adversary.

end the current activation, **Main** is executed directly afterwards. While we do not fix how authors have to use these algorithms, one would typically use **MessagePreprocessing** to prepare the input m for the **Main** algorithm, e.g., by dropping malformed messages or extracting some key information from m. The algorithm **Main** should contain the core logic of the protocol.

If any of the optional algorithms are not specified, then they are simply skipped during computation. We provide a convenient syntax for specifying these algorithms in the full version; see our case study in Sect. 4 for examples.

This concludes the description of our template. As already mentioned, in the full version of this paper we give a formal mapping of this template to protocols in the sense of the IITM model, which provides a precise semantics for the templates and also allows us to carry over all definitions, such as realization relations, and theorems, such as composition theorems, of the IITM model to iUC (see Sect. 2.6).

2.5 Composing Protocol Specifications

Protocols in our framework can be composed to obtain more complex protocols. More precisely, two protocols $\mathcal{Q}$ and $\mathcal{Q}'$ that are specified using our template are called *connectable* if they connect via their public roles only. That is, if a machine in $\mathcal{Q}$ specifies a subroutine role of $\mathcal{Q}'$, then this subroutine role has to be public in $\mathcal{Q}'$, and vice versa.

Two connectable protocols can be composed to obtain a new protocol $\mathcal{R}$ containing all roles of $\mathcal{Q}$ and $\mathcal{Q}'$ such that the public roles of $\mathcal{R}$ are a subset of the public roles of $\mathcal{Q}$ and $\mathcal{Q}'$. Which potentially public roles of $\mathcal{R}$ are actually declared to be public in $\mathcal{R}$ is up the protocol designer and depends on the type of protocol that is to be modeled (see Sect. 2.2 and our case study in Sect. 4). In any case, the notation from Sect. 2.2 of the form $(role_1^{pub} \ldots role_i^{pub} \mid role_1^{priv} \ldots role_j^{priv})$ should be used for this purpose.

For pairwise connectable protocols $\mathcal{Q}_1, \ldots, \mathcal{Q}_n$ we define $\mathsf{Comb}(\mathcal{Q}_1, \ldots, \mathcal{Q}_\mathsf{n})$ to be the (finite) set of all protocols $\mathcal{R}$ that can be obtained by connecting $\mathcal{Q}_1, \ldots, \mathcal{Q}_n$. Note that all protocols $\mathcal{R}$ in this set differ only by their sets of public roles. We define two shorthand notations for easily specifying the most common types of combined protocols: by $(\mathcal{Q}_1, \ldots, \mathcal{Q}_i \mid \mathcal{Q}_{i+1}, \ldots, \mathcal{Q}_n)$ we denote the protocol $\mathcal{R} \in \mathsf{Comb}(\mathcal{Q}_1, \ldots, \mathcal{Q}_\mathsf{n})$, where the public roles of $\mathcal{Q}_1, \ldots, \mathcal{Q}_i$ remain public in $\mathcal{R}$ and all other roles are private. This notation can be mixed with the notation from Sect. 2.2 in the natural way by replacing a protocol $\mathcal{Q}_j$ with its roles, some of which might be public while others might be private in $\mathcal{R}$. Furthermore, by $\mathcal{Q}_1 \,||\, \mathcal{Q}_2$ we denote the protocol $\mathcal{R} \in \mathsf{Comb}(\mathcal{Q}_1, \mathcal{Q}_2)$ where exactly those public roles of $\mathcal{Q}_1$ and $\mathcal{Q}_2$ remain public that are not used as a subroutine by any machine in $\mathcal{Q}_1$ or $\mathcal{Q}_2$.

We call a protocol $\mathcal{Q}$ *complete* if every subroutine *role* used by a machine in $\mathcal{Q}$ is also part of $\mathcal{Q}$. In other words, $\mathcal{Q}$ fully specifies the behavior of all subroutines. Since security analysis makes sense only for a fully specified protocol, we will (implicitly) consider this to be the default in the following.

2.6 Realization Relation and Composition Theorems

In the following, we define the universal composability experiment and state the main composition theorem of iUC. Since iUC is an instantiation of the IITM model, as shown by our mapping mentioned in Sect. 2.4, both the experiment and theorem are directly carried over from the IITM model and hence do not need to be re-proven.

Definition 1 (Realization relation in iUC). *Let $\mathcal{P}$ and $\mathcal{F}$ be two environmentally bounded complete protocols with identical sets of public roles. The protocol $\mathcal{P}$* realizes *$\mathcal{F}$ (denoted by $\mathcal{P} \leq \mathcal{F}$) iff there exists a simulator (system) $\mathcal{S} \in \mathsf{Adv}(\mathcal{F})$ such that for all $\mathcal{E} \in \mathsf{Env}(\mathcal{P})$ it holds true that $\{\mathcal{E}, \mathcal{P}\} \equiv \{\mathcal{E}, \mathcal{S}, \mathcal{F}\}$.*[7]

Note that $\mathcal{E}$ in $\{\mathcal{E}, \mathcal{P}\}$ connects to the I/O interfaces of public roles as well as the network interfaces of all roles of $\mathcal{P}$. In contrast, $\mathcal{E}$ in the system $\{\mathcal{E}, \mathcal{S}, \mathcal{F}\}$ connects to the I/O interfaces of public roles of $\mathcal{F}$ and the network interface of $\mathcal{S}$. The simulator $\mathcal{S}$ connects to $\mathcal{E}$ (simulating the network interface of $\mathcal{P}$) and the network interface of $\mathcal{F}$; see also Fig. 1, where here we consider the case that $\mathcal{E}$ subsumes the adversary $\mathcal{A}$. (As shown in [1], whether or not the adversary $\mathcal{A}$ is considered does not change the realization relation. The resulting notions are equivalent.)

Now, the main composition theorem of iUC, which is a corollary of the composition of the IITM model, is as follows:

Corollary 1 (Concurrent composition in iUC). *Let $\mathcal{P}$ and $\mathcal{F}$ be two protocols such that $\mathcal{P} \leq \mathcal{F}$. Let $\mathcal{Q}$ be another protocol such that $\mathcal{Q}$ and $\mathcal{F}$ are connectable. Let $\mathcal{R} \in \mathsf{Comb}(\mathcal{Q}, \mathcal{P})$ and let $\mathcal{I} \in \mathsf{Comb}(\mathcal{Q}, \mathcal{F})$ such that $\mathcal{R}$ and $\mathcal{I}$ have the same sets of public roles. If $\mathcal{R}$ is environmentally bounded and complete, then $\mathcal{R} \leq \mathcal{I}$.*

Just as in the IITM model, we emphasize that this corollary also covers the special cases of protocols with joint-state and global state. Furthermore, a second composition theorem for secure composition of an unbounded number of sessions of a protocol is also available, again a corollary of a more general theorem in the IITM model (see the full version [2]).

3 Concepts and Discussion

Recall from the introduction that a main goal of iUC is to provide a flexible yet easy to use framework for universally composable protocol analysis and design. In this section, we briefly summarize and highlight some of the core concepts that allow us to retain the flexibility and expressiveness of the original IITM model while adding the usability with a handy set of conventions. We then highlight a selection of features that are supported by iUC due to the concepts iUC

[7] Intuitively, the role names are used to determine which parts of $\mathcal{F}$ are realized by which parts of $\mathcal{P}$, hence they must have the same sets of public roles.

uses and that are not supported by other (conventions of) models, including the prominent UC and GNUC models. Our case study in Sect. 4 further illustrates the expressiveness of iUC. An extended discussion of concepts and features is available in the full version [2]. Some of the most crucial concepts of iUC, discussed next, are the separation of entities and machine instances, public and private roles, a model independent interpretation of SIDs, support for responsive environments as well as a general addressing mechanism, which enables some of these concepts.

Separation of Entities and Machine Instances: Traditionally, universal composability models do not distinguish between a machine instance and its interpretation. Instead, they specify that, e.g., a *real protocol instance* always represents a single party in a single session running a specific piece of code. Sometimes even composition theorems depend on this view. This has the major downside that, if the interpretation of a machine instance needs to be changed, then existing models, conventions, and composition theorems are no longer applicable and have to be redefined (and, in the case of theorems, reproven). For example, a typical *joint state protocol instance* [7,20] manages a single party in *all sessions* and one role. Thus, in the case of the UC and GNUC models, the models had to be extended and reproven, including conventions and composition theorems. This is in contrast to iUC, which introduces the concept of *entities*. A protocol designer can freely define the interpretation of a machine instance by specifying the set of entities managed by that instance; the resulting protocol is still supported by our single template and the main composition theorem. This is a crucial feature that allows for the unified handling of real, ideal, joint-state, and (in combination with the next concept) also global state protocols.

We emphasize that this generality is made possible by the highly customizable addressing mechanism (**CheckID** in the template) used in iUC, which in turn is based on the very general addressing mechansim of the IITM model.

Public and Private Roles: Similar to the previous point, traditionally global state is defined by adding a special new global functionality with its own sets of conventions and proving specific global state composition theorems. However, whether or not state is global is essentially just a matter of access to that state. Our framework captures this property via the natural concept of *public roles*, which provides a straightforward way to make parts of a protocol accessible to the environment and other protocols. Thus, there is actually no difference between protocols with and without global state in terms of conventions or composition theorems in our framework.

A Model Independent Interpretation of SIDs: In most other models, such as UC and GNUC, SIDs play a crucial role in the composition theorems. Composition theorems in these frameworks require protocols to either have disjoint sessions, where a session is defined via the SID, or at least behave as if they had disjoint sessions (in the case of joint-state composition theorems). This has two major implications: Firstly, one cannot directly model a protocol where different sessions share the same state and influence each other. This, however, is often the case for real world protocols that were not built with session separation in mind. For example, many protocols such as our case study (cf. Sect. 4) use the same signing key in multiple sessions, but do not include a session specific SID in the signature (as would be required for a joint-state realization). Secondly, sessions in ideal functionalities can consist only of parties sharing the same SID, which models so-called *global SIDs* or *pre-shared SIDs* [21]. That is, participants of a protocol session must share the same SID. This is in contrast to so-called *local SIDs* often used in practice, where participants with different SIDs can be part of the same protocol session (cf. 4.3). Because our main composition theorem is independent of (the interpretation of) SIDs, and in particular does not require state separation, we can also capture shared state and local SIDs in our framework.

Just as for the concept of entities and instances, this flexibility is made possible by the general addressing mechanism of iUC (and its underlying IITM model).

Support for Responsive Environments: Recall that responsive environments [1] allow for sending special messages on the network interface, called restricting messages, that have to be answered immediately by the adversary and environment. This is a very handy mechanism that allows protocols to exchange modeling related meta information with the adversary without disrupting the protocol run. For example, entities in our framework request their initial corruption status via a restricting message. Hence, the adversary has to provide the corruption status right away and the protocol run can continue as expected. Without responsive environments, one would have to deal with undesired behavior such as delayed responses, missing responses, as well as state changes and unexpected activations of (other parts of) the protocol before the response is provided. In the case of messages that exist only for modeling purposes, this adversarial behavior just complicates the protocol design and analysis without relating to any meaningful attack in reality, often leading to formally wrong security proofs and protocol specifications that cannot be re-used in practice. See our full version and [1] for more information.

Selected Features of iUC. The iUC framework uses and combines the above concepts to support a wide range of protocols and composition types, some of which have not even been considered in the literature so far, using just *a single template* and *one main composition theorem.* We list some important examples:

(i) Protocols with *local SIDs* and *global SIDs*, arbitrary forms of *shared state* including state that is shared across multiple protocol sessions, as well as *global state.* Our case study in Sect. 4 is an example of a protocol that uses and combines all of these protocol features, with a detailed explanation and discussion provided in Sect. 4.3.
(ii) Ideal protocols that are structured into several subcomponents, unlike the monolithic ideal functionalities considered in other (conventions of) models. Parts of such structured ideal protocols can also be defined to be global, allowing for easily mixing traditional ideal protocols with global state.

Again, this is also illustrated in our case study in Sect. 4. We also note that in iUC there is no need to consider so-called dummy machines in ideal protocols, which are often required in other models that do not allow for addressing the same machine instance with different IDs (entities).

(iii) The general composition theorem, which in particular is agnostic to the specific protocols at hand, allows for combining and mixing classical composition of protocols with disjoint session, composition of joint-state protocols, composition of protocols with global state, and composition of protocols with arbitrarily shared state. One can also, e.g., realize a global functionality with another protocol (this required an additional composition theorem for the UC model [12] and is not yet supported by GNUC, whereas in iUC this is just another trivial special case of protocol composition). iUC even supports new types of compositions that have not been considered in the literature so far, such as joint-state realizations of two separate independent protocols (in contrast to traditional joint-state realizations of multiple independent sessions of the same protocol; cf. Sect. 4.3).

Besides our case study in Sect. 4, the flexibility and usability of iUC is also illustrated by another example in the full version, where we discuss that the iUC framework can capture the SUC model [10] as a mere special case. As already mentioned in the introdcution, the SUC model has been specifically designed for secure multi party computation (MPC) as a simpler version of the UC model, though it has to break out of (some technical aspects of) the UC model.

4 Case Study

In this section, we illustrate the usage of iUC by means of a concrete example, demonstrating usability, flexibility, and soundness of our framework. More specifically, we model and analyze a key exchange protocol of the ISO/IEC 9798-3 standard [17], an authenticated version of the Diffie-Hellman key exchange protocol, depicted in Fig. 5. While this protocol has already been analyzed previously in universal composability models (e.g., in [6,19]), these analyses were either for modified versions of the protocol (as the protocol could not be modeled precisely as deployed in practice) or had to manually define many recurrent modeling related aspects (such as a general corruption model and an interpretation of machine instances), which is not only cumbersome but also hides the core logic of the protocol.

We have chosen this relatively simple protocol for our case study as it allows for showing how protocols can be modeled in iUC and highlighting several core features of the framework without having to spend much time on first explaining the logic of the protocol.

More specifically, our case study illustrates that our framework manages to combine *soundness* and *usability*: the specifications of the ISO protocol given in the figures below are formally complete, no details are swept under the rug, unlike the informal descriptions commonly encountered in the literature on universal composability. This allows for a precise understanding of the protocol, enabling

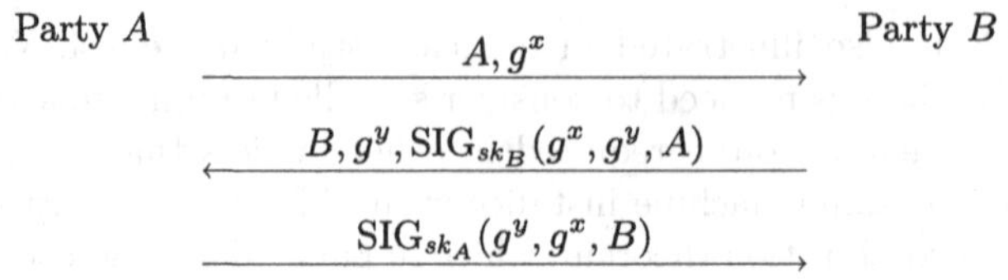

Fig. 5. ISO 9798-3 key exchange protocol for mutual authentication. A and B are the names of two parties that, at the end of the protocol, share a session key g^{xy}.

formally sound proofs and re-using the protocol in higher-level protocols. At the same time, specifications of the ISO protocol are not overburdened by recurrent modeling related aspects as they make use of convenient defaults provided by the iUC framework. All parts of the ISO protocol are specified using *a single* template with one set of syntax rules, including real, ideal, and global state (sub-)protocols, allowing for a uniform treatment.

This case study also shows the *flexibility* of our framework: entites are grouped in different ways into machine instances to model different types of protocols and setup assumptions; we are able to share state across several sessions; we make use of the highly adjustable corruption model to precisely capture the desired corruption behavior of each (sub-)protocol; we are able to model both global state and locally chosen SIDs in a very natural way (we discuss some of these aspects, including locally chosen SIDs, in detail in Sect. 4.3).

We start by giving a high-level overview of how we model this ISO key exchange protocol in Sect. 4.1, then state our security result in Sect. 4.2, and finally discuss some of the features of our modeling in Sect. 4.3.

4.1 Overview of Our Modeling

We model the ISO protocol in a modular way using several smaller protocols. The static structure of all protocols, including their I/O connections for direct communication, is shown in Fig. 3, which was partly explained already in Sect. 2.2. We provide a formal specification of $\mathcal{F}_{\mathrm{CA}}$ using our template and syntax in Fig. 6. The remaining protocols specifications are given in the full version due to space limitations. The syntax is mostly self-explanatory, except for $(\mathsf{pid}_{\mathsf{cur}}, \mathsf{sid}_{\mathsf{cur}}, \mathsf{role}_{\mathsf{cur}})$, which denotes the currently active entity (that was accepted by **CheckID**), $(\mathsf{pid}_{\mathsf{call}}, \mathsf{sid}_{\mathsf{call}}, \mathsf{role}_{\mathsf{call}})$, which denotes the entity that called the currently active entity on the I/O interface, and "_", which is a wildcard symbol. In the following, we give a high-level overview of each protocol.

The ISO key exchange (Fig. 5) is modeled as a real protocol $\mathcal{P}_{\mathrm{KE}}$ that uses two ideal functionalities as subroutines: an ideal functionality $\mathcal{F}_{\text{sig-CA}}$ for creating and verifying ideal digital signatures and an ideal functionality $\mathcal{F}_{\mathrm{CA}}$ modeling a certificate authority (CA) that is used to distribute public verification keys generated by $\mathcal{F}_{\text{sig-CA}}$. The real protocol $\mathcal{P}_{\mathrm{KE}}$, as already mentioned in Sect. 2.2, consists of three roles, `initiator`, `responder`, and `setup`. The `setup` role models secure generation and distribution of a system parameter, namely, a description of a cyclic group (G, n, g). As this parameter must be shared between all runs of a key exchange protocol, `setup` is implemented by a single machine which spawns a single instance that manages all entities and always outputs the same parameter. The roles `initiator` and `responder` implement parties A and B, respectively, from Fig. 5. Each role is implemented by a separate machine and every instance of those machines manages exactly one entity. Thus, these instances directly correspond to an actual implementation where each run of a key exchange protocol spawns a new program instance. We emphasize that two entities can perform a key exchange together even if they do not share the same SID, which models so-called local SIDs (cf. [21]) and is the expected behavior for many real-world protocols; we discuss this feature in more detail below.

During a run of $\mathcal{P}_{\mathrm{KE}}$, entities use the ideal signature functionality $\mathcal{F}_{\text{sig-CA}}$ to sign messages. The ideal functionality $\mathcal{F}_{\text{sig-CA}}$ consists of two roles, `signer` and `verifier`, that allow for the corresponding operations. Both roles are implemented by the same machine and instances of that machine manage entities that share the same SID. The SID *sid* of an entity is structured as a tuple (pid_{owner}, sid'), modeling a specific key pair of the party pid_{owner}. More specifically, in protocol $\mathcal{P}_{\mathrm{KE}}$, every party *pid* owns a single key pair, represented by SID (pid, ϵ)[8], and uses this single key pair to *sign messages throughout all sessions of the key exchange.* Again, this is precisely what is done in reality, where the same signing key is re-used several times. The behavior of $\mathcal{F}_{\text{sig-CA}}$ is closely related to the standard ideal signature functionalities found in the literature (such as [20]), except that public keys are additionally registered with $\mathcal{F}_{\mathrm{CA}}$ when being generated.

As also mentioned in Sect. 2.2, the ideal CA functionality $\mathcal{F}_{\mathrm{CA}}$ allows for storing and retrieving public keys. Both roles, `registration` and `retrieval`, are implemented by one machine and a single instance of that machine accepts all entities, as $\mathcal{F}_{\mathrm{CA}}$ has to output the same keys for all sessions and parties. Keys are stored for arbitrary pairs of PIDs and SIDs, where the SID allows for storing different keys for a single party. In our protocol, keys can only be registered by $\mathcal{F}_{\text{sig-CA}}$, and the SID is chosen in a matter that it always has the form (pid, ϵ), denoting the single public key of party *pid*. We emphasize again that arbitrary other protocols and the environment are able to retrieve public keys from $\mathcal{F}_{\mathrm{CA}}$, which models so-called global state.

In summary, the real protocol that we analyze is the combined protocol ($\mathcal{P}_{\mathrm{KE}}, \mathcal{F}_{\mathrm{CA}}$: `retrieval` | $\mathcal{F}_{\text{sig-CA}}, \mathcal{F}_{\mathrm{CA}}$: `registration`) (cf. left side of Fig. 3). We note that we analyze this protocol directly in a multi-session setting. That is, the environment is free to spawn arbitrarily many entities belonging to arbitrary parties and having arbitrary local SIDs and thus there can be multiple key exchanges running in parallel. Analyzing a single session of this key exchange in isolation is not possible due to the shared signing keys and the use of local

8 Since we need only a single key pair per party, we set *sid'* to be the fixed value ϵ, i.e., the empty string.

Description of the protocol $\mathcal{F}_{CA} = (\texttt{registration}, \texttt{retrieval})$:

Participating roles: $\{\texttt{registration}, \texttt{retrieval}\}$ **Corruption model:** incorruptible

Description of $M_{\texttt{registration},\texttt{retrieval}}$:

Implemented role(s): $\{\texttt{registration}, \texttt{retrieval}\}$

Internal state:

- $\mathsf{keys} : (\{0,1\}^*)^2 \to \{0,1\}^* \cup \{\bot\}$ — *Mapping from a tuple (PID,SID) to stored keys; initially $\bot$.*

CheckID$(pid, sid, role)$: Accept all entities. — *By this there is only a single machine instance that manages all entities.*

Main:

recv (Register, key) **from** I/O **to** $(\_, \_, \texttt{registration})$:
  if $\mathsf{keys}[\mathsf{pid}_{\mathsf{call}}, \mathsf{sid}_{\mathsf{call}}] \neq \bot$:
    reply (Register, failed).
  else:
    $\mathsf{keys}[\mathsf{pid}_{\mathsf{call}}, \mathsf{sid}_{\mathsf{call}}] = key$
    reply (Register, success).

Allows every higher level protocol that connects to the registration *role to register a key. The key is stored for the PID and SID of the caller of $\mathcal{F}_{CA}$.*

recv (Retrieve, (pid, sid)) **from** _ **to** $(\_, \_, \texttt{retrieval})$:
  reply (Retrieve, $\mathsf{keys}[pid, sid]$).

Everyone, including NET, *can retrieve keys registered by someone with PID pid and SID sid.*

Fig. 6. The ideal CA functionality $\mathcal{F}_{CA}$ models a public key infrastructure based on a trusted certificate authority.

SIDs, which, as mentioned, precisely models how this protocol would usually be deployed in practice.[9]

We model the security properties of a multi-session key exchange via an ideal key exchange functionality $\mathcal{F}_{KE}$. This functionality consists of two roles, `initiator` and `responder`, and uses $\mathcal{F}_{CA}$ as a subroutine, thus providing the same interfaces (including the public role `retrieval` of $\mathcal{F}_{CA}$) as $\mathcal{P}_{KE}$ in the real world. Both `initiator` and `responder` roles are implemented via a single machine, and one instance of this machine manages all entities. This is due to the fact that, at the start of a run, it is not yet clear which entities will interact with each other to form a "session" and perform a key exchange (recall that entities need not share the same SID to do so, i.e., they use locally chosen SIDs, see also Sect. 4.3). Thus, a single instance of $\mathcal{F}_{KE}$ must manage all entities such that it can internally group entities into appropriate sessions that then obtain the same session key. Formally, the adversary/simulator is allowed to decide which entities are grouped into a session, subject to certain restrictions that ensure the expected security guarantees of a key exchange, including authentication. If

[9] Note that this is true in *all* UC-like models that can express this setting: the assumption of disjoint sessions, which is necessary for performing a single session analysis, is simply not fulfilled by this protocol. This issue cannot even be circumvented by using a so-called joint-state realization for digital signatures, as such a realization not only requires global SIDs (cf. Sect. 4.3) but also changes the messages that are signed, thus creating a modified protocol with different security properties.

two honest entities finish a key exchange in the same session, then $\mathcal{F}_{\mathrm{KE}}$ ensures that they obtain an ideal session key that is unknown to the adversary. The adversary may also use $\mathcal{F}_{\mathrm{KE}}$ to register arbitrary keys in the subroutine $\mathcal{F}_{\mathrm{CA}}$, also for honest parties, i.e., no security guarantees for public keys in $\mathcal{F}_{\mathrm{CA}}$ are provided.

4.2 Security Result

For the above modeling, we obtain the following result, with a proof provided in the full version.

Theorem 1. *Let* $\mathsf{groupGen}(1^\eta)$ *be an algorithm that outputs descriptions* (G, n, g) *of cyclical groups (i.e.,* G *is a group of size* n *with generator* g*) such that* n *grows exponentially in* η *and the DDH assumption holds true. Then we have:*

$$(\mathcal{P}_{\mathrm{KE}}, \mathcal{F}_{\mathrm{CA}} : \mathtt{retrieval} \mid \mathcal{F}_{\text{sig-CA}}, \mathcal{F}_{\mathrm{CA}} : \mathtt{registration}) \\ \leq (\mathcal{F}_{\mathrm{KE}}, \mathcal{F}_{\mathrm{CA}} : \mathtt{retrieval} \mid \mathcal{F}_{\mathrm{CA}} : \mathtt{registration})\,.$$

Note that we can realize $\mathcal{F}_{\text{sig-CA}}$ via a generic implementation $\mathcal{P}_{\text{sig-CA}}$ of a digital signature scheme (we provide a formal definition of $\mathcal{P}_{\text{sig-CA}}$ in the full version):

Lemma 1. *If the digital signature scheme used in* $\mathcal{P}_{\text{sig-CA}}$ *is existentially unforgable under chosen message attacks (EUF-CMA-secure), then*

$$(\mathcal{P}_{\text{sig-CA}}, \mathcal{F}_{\mathrm{CA}} : \mathtt{retrieval} \mid \mathcal{F}_{\mathrm{CA}} : \mathtt{registration}) \\ \leq (\mathcal{F}_{\text{sig-CA}}, \mathcal{F}_{\mathrm{CA}} : \mathtt{retrieval} \mid \mathcal{F}_{\mathrm{CA}} : \mathtt{registration})\,.$$

Proof. Analogous to the proof in [20].

By Corollary 1, we can thus immediately replace the subroutine $\mathcal{F}_{\text{sig-CA}}$ of $\mathcal{P}_{\mathrm{KE}}$ with its realization $\mathcal{P}_{\text{sig-CA}}$ to obtain an actual implementation of Fig. 3 based on an ideal trusted CA:

Corollary 2. *If the conditions of Theorem 1 and Lemma 1 are fulfilled, then*

$$(\mathcal{P}_{\mathrm{KE}}, \mathcal{F}_{\mathrm{CA}} : \mathtt{retrieval} \mid \mathcal{P}_{\text{sig-CA}}, \mathcal{F}_{\mathrm{CA}} : \mathtt{registration}) \\ \leq (\mathcal{F}_{\mathrm{KE}}, \mathcal{F}_{\mathrm{CA}} : \mathtt{retrieval} \mid \mathcal{F}_{\mathrm{CA}} : \mathtt{registration})\,.$$

4.3 Discussion

In the following, we highlight some of the key details of our protocol specification where we are able to model reality very precisely and in a natural way, illustrating the *flexibility* of iUC, also compared to (conventions of) the UC and GNUC models.

Local SIDs: Many real-world protocols, including the key exchange in our case study, use so-called local session IDs in practice (cf. [21]). That is, the SID of an entity (*pid*, *sid*, *role*) models a value that is locally chosen and managed by each party *pid* and used only for locally addressing a specific instance of a protocol run of that party, but is not used as part of the actual protocol logic. In particular, multiple entities can form a "protocol session" even if they use different SIDs. This is in contrast to using so-called pre-established SIDs (or global SIDs), where entities in the same "protocol session" are assumed to already share some globally unique SID that was created prior to the actual protocol run, e.g., by adding an additional roundtrip to exchange nonces, or that is chosen by and then transmitted from one entity to the others during the protocol run. As illustrated by the protocols $\mathcal{P}_{\mathrm{KE}}$ (and $\mathcal{F}_{\mathrm{KE}}$) in our case study, iUC can easily model such local SIDs in a natural way. This is in contrast to several other UC-like models, including the UC and GNUC models, that are built around global SIDs and thus do not directly support local SIDs with their conventions. While it might be possible to find workarounds by ignoring conventions, e.g., by modeling all sessions of a protocol in a single machine instance M, i.e., essentially ignoring the model's intended SID mechanism and taking care of the addressing of different sessions with another layer of SIDs within M itself, this has two major drawbacks: Firstly, it decreases overall usability of the models as this workaround is not covered by existing conventions of these models. Secondly, existing composition theorems of UC and GNUC do not allow one to compose such a protocol with a higher-level protocol modeled in the "standard way" where different sessions use different SIDs.[10] We emphasize that the difference between local and global SIDs is not just a minor technicality or a cosmetic difference: as argued by Küsters et al. [21], there are natural protocols that are insecure when using locally chosen SIDs but become secure if a global SID for all participants in a session has already been established, i.e., security results for protocols with global SIDs do not necessarily carry over to actual implementations using local SIDs.

Shared State: In iUC, entities can easily and naturally share arbitrary state in various ways, even across multiple protocol sessions, if so desired. This is illustrated, e.g., by $\mathcal{P}_{\mathrm{KE}}$ in our case study, where every party uses just a single signature key pair across arbitrarily many key exchanges. This allows for a very flexible and precise modeling of protocols. In particular, for many real-world protocols this modeling is much more precise than so-called joint-state realizations that are often used to share state between sessions in UC-like models that assume disjoint sessions to be the default, such as the UC and GNUC models. Joint-state realizations have to modify protocols by, e.g., prefixing signed messages with some globally unique SID for every protocol session (which is not done by many real-world protocols, including our case study). Thus, even if the modified

[10] This is because such a higher level protocol would then access the same subroutine session throughout many different higher-level sessions, which violates session disjointness as required by both UC and GNUC.

protocol is proven to be secure, this does not imply security of the unmodified one. The UC and GNUC models do not directly support state sharing without resorting to joint-state realizations or global functionalities. While one might be able to come up with workarounds similar to what we described for local SIDs above, this comes with the same drawbacks in terms of usability and flexibility.

Global State: Our concept of public and private roles allows us to not only easily model global state but also to specify, in a convenient and flexible way, machines that are only partially global. This is illustrated by $\mathcal{F}_{\text{CA}}$ in our case study, which allows arbitrary other protocols to retrieve keys but limits key registration to one specific protocol to model that honest users will not register their signing keys for other contexts (which, in general, otherwise voids all security guarantees). This feature makes $\mathcal{F}_{\text{CA}}$ easier to use as a subroutine than the existing global functionality $\mathcal{G}_{bb}$ for certificate authorities by Canetti et al. [12], which does not support making parts of the functionality "private". Thus, everyone has full access to all operations of $\mathcal{G}_{bb}$, including key registration, allowing the environment to register keys in the name of (almost) arbitrary parties, even if they are supposed to be honest.

Note that our formulation of $\mathcal{F}_{\text{CA}}$ means that, if the ideal protocol ($\mathcal{F}_{\text{KE}}$, $\mathcal{F}_{\text{CA}}$: `retrieval` | $\mathcal{F}_{\text{CA}}$: `registration`) is used as a subroutine for a new hybrid protocol, then only $\mathcal{F}_{\text{KE}}$ but not the higher-level protocol can register keys in $\mathcal{F}_{\text{CA}}$. If desired, one can, however, also obtain a single global $\mathcal{F}_{\text{CA}}$ where both $\mathcal{F}_{\text{KE}}$ and the higher-level protocol can store keys in the following way: First analyze the whole hybrid protocol while using a second separate copy of $\mathcal{F}_{\text{CA}}$, say $\mathcal{F}'_{\text{CA}}$, where only the higher-level protocol can register keys. After proving this to be secure (which is simpler than directly using a global CA where multiple protocols register keys), one can replace both $\mathcal{F}_{\text{CA}}$ and $\mathcal{F}'_{\text{CA}}$ with a joint-state realization where keys are stored in and retrieved from the same $\mathcal{F}_{\text{CA}}$ subroutine along with a protocol dependent tag (we discuss this novel type of joint-state realization in detail in the full version). Of course, this approach can be iterated to support arbitrarily many protocols using the same $\mathcal{F}_{\text{CA}}$. This modeling reflects reality where keys are certified for certain contexts/purposes.

5 Conclusion

We have introduced the iUC framework for universal composability. As illustrated by our case study, iUC is highly *flexible* in that it supports a wide range of protocol types, protocol features, and composition operations. This flexibility is combined with greatly improved *usability* compared to the IITM model due to its protocol template that fixes recurring modeling related aspects while providing sensible defaults for optional parts. Adding usability while preserving flexibility is a difficult task that is made possible, among others, due to the concepts of roles and entities; these concepts allow for having just *a single template* and *two composition theorems* that are able to handle arbitrary types of protocols, including real, ideal, joint-state, and global ones, and combinations thereof.

The flexibility and usability provided by iUC also significantly facilitates the precise modeling of protocols, which is a prerequisite for carrying out formally complete and sound proofs. Our formal mapping from iUC to the IITM shows that iUC indeed is an instantiation of the IITM, and hence, immediately inherits all theorems, in particular, all composition theorems, of the IITM model. Since we formulate these theorems also in the iUC terminology, protocol designers can completely stay in the iUC realm when designing and analyzing protocols.

Altogether, the iUC framework is a well-founded framework for universal composability which combines soundness, flexibility, and usability in an unmatched way. As such, it is an important and convenient tool for the precise modular design and analysis of security protocols and applications.

References

1. Camenisch, J., Enderlein, R.R., Krenn, S., Küsters, R., Rausch, D.: Universal composition with responsive environments. In: Cheon, J.H., Takagi, T. (eds.) ASIACRYPT 2016, Part II. LNCS, vol. 10032, pp. 807–840. Springer, Heidelberg (2016). https://doi.org/10.1007/978-3-662-53890-6_27
2. Camenisch, Krenn, S., Küsters, R., Rausch, D.: iUC: flexible universal composability made simple (full version). Technical report 2019/1073, Cryptology ePrint Archive (2019). http://eprint.iacr.org/2019/1073
3. Canetti, R.: Universally composable security: a new paradigm for cryptographic protocols. Technical report 2000/067, Cryptology ePrint Archive (2000). http://eprint.iacr.org/2000/067 with new versions from December 2005, July 2013, December 2018
4. Canetti, R.: Universally composable security: a new paradigm for cryptographic protocols. In: FOCS 2001, pp. 136–145. IEEE Computer Society (2001)
5. Canetti, R., Dodis, Y., Pass, R., Walfish, S.: Universally composable security with global setup. In: Vadhan, S.P. (ed.) TCC 2007. LNCS, vol. 4392, pp. 61–85. Springer, Heidelberg (2007). https://doi.org/10.1007/978-3-540-70936-7_4
6. Canetti, R., Krawczyk, H.: Universally composable notions of key exchange and secure channels. In: Knudsen, L.R. (ed.) EUROCRYPT 2002. LNCS, vol. 2332, pp. 337–351. Springer, Heidelberg (2002). https://doi.org/10.1007/3-540-46035-7_22
7. Canetti, R., Rabin, T.: Universal composition with joint state. In: Boneh, D. (ed.) CRYPTO 2003. LNCS, vol. 2729, pp. 265–281. Springer, Heidelberg (2003). https://doi.org/10.1007/978-3-540-45146-4_16
8. Canetti, R., Chari, S., Halevi, S., Pfitzmann, B., Roy, A., Steiner, M., Venema, W.: Composable security analysis of OS services. In: Lopez, J., Tsudik, G. (eds.) ACNS 2011. LNCS, vol. 6715, pp. 431–448. Springer, Heidelberg (2011). https://doi.org/10.1007/978-3-642-21554-4_25
9. Canetti, R., et al.: Analyzing security protocols using time-bounded task-PIOAs. Discret. Event Dyn. Syst. **18**(1), 111–159 (2008)
10. Canetti, R., Cohen, A., Lindell, Y.: A simpler variant of universally composable security for standard multiparty computation. In: Gennaro, R., Robshaw, M. (eds.) CRYPTO 2015, Part II. LNCS, vol. 9216, pp. 3–22. Springer, Heidelberg (2015). https://doi.org/10.1007/978-3-662-48000-7_1
11. Canetti, R., Hogan, K., Malhotra, A., Varia, M.: A universally composable treatment of network time. In: CSF 2017, pp. 360–375. IEEE Computer Society (2017)

12. Canetti, R., Shahaf, D., Vald, M.: Universally composable authentication and key-exchange with global PKI. In: Cheng, C.-M., Chung, K.-M., Persiano, G., Yang, B.-Y. (eds.) PKC 2016, Part II. LNCS, vol. 9615, pp. 265–296. Springer, Heidelberg (2016). https://doi.org/10.1007/978-3-662-49387-8_11
13. Chaidos, P., Fourtounelli, O., Kiayias, A., Zacharias, T.: A universally composable framework for the privacy of email ecosystems. In: Peyrin, T., Galbraith, S. (eds.) ASIACRYPT 2018, Part III. LNCS, vol. 11274, pp. 191–221. Springer, Cham (2018). https://doi.org/10.1007/978-3-030-03332-3_8
14. Chari, S., Jutla, C.S., Roy, A.: Universally Composable Security Analysis of OAuth v2.0. IACR Cryptology ePrint Archive 2011/526 (2011)
15. Hofheinz, D., Shoup, V.: GNUC: a new universal composability framework. J. Cryptol. **28**(3), 423–508 (2015)
16. Hogan, K., et al.: On the Universally Composable Security of OpenStack. IACR Cryptology ePrint Archive 2018/602 (2018)
17. ISO/IEC IS 9798-3, Entity authentication mechanisms – Part 3: Entity authentication using assymetric techniques (1993)
18. Küsters, R.: Simulation-based security with inexhaustible interactive turing machines. In: CSFW 2006, pp. 309–320. IEEE Computer Society (2006). See [22] for a full and revised version
19. Küsters, R., Rausch, D.: A framework for universally composable Diffie-Hellman key exchange. In: S&P 2017, pp. 881–900. IEEE Computer Society (2017)
20. Küsters, R., Tuengerthal, M.: Joint state theorems for public-key encryption and digital signature functionalities with local computation. In: CSF 2008, pp. 270–284. IEEE Computer Society (2008). The full version is available at https://eprint.iacr.org/2008/006 and will appear in Journal of Cryptology
21. Küsters, R., Tuengerthal, M.: Composition theorems without pre-established session identifiers. In: CCS 2011, pp. 41–50. ACM (2011)
22. Küsters, R., Tuengerthal, M., Rausch, D.: The IITM model: a simple and expressive model for universal composability. Technical report 2013/025, Cryptology ePrint Archive (2013). http://eprint.iacr.org/2013/025. To appear in Journal of Cryptology
23. Maurer, U.: Constructive cryptography – a new paradigm for security definitions and proofs. In: Mödersheim, S., Palamidessi, C. (eds.) TOSCA 2011. LNCS, vol. 6993, pp. 33–56. Springer, Heidelberg (2012). https://doi.org/10.1007/978-3-642-27375-9_3
24. Maurer, U., Renner, R.: Abstract cryptography. In: Chazelle, B. (ed.) Innovations in Computer Science - ICS 2010. Proceedings, pp. 1–21. Tsinghua University Press (2011)
25. Pfitzmann, B., Waidner, M.: A model for asynchronous reactive systems and its application to secure message transmission. In: S&P 2001, pp. 184–201. IEEE Computer Society (2001)

Side Channels

Leakage Resilience of the Duplex Construction

Christoph Dobraunig(✉) and Bart Mennink

Digital Security Group, Radboud University, Nijmegen, The Netherlands
{cdobraunig,b.mennink}@cs.ru.nl

Abstract. Side-channel attacks, especially differential power analysis (DPA), pose a serious threat to cryptographic implementations deployed in a malicious environment. One way to counter side-channel attacks is to design cryptographic schemes to withstand them, an area that is covered amongst others by leakage resilient cryptography. So far, however, leakage resilient cryptography has predominantly focused on block cipher based designs, and insights in permutation based leakage resilient cryptography are scarce. In this work, we consider leakage resilience of the keyed duplex construction: we present a model for leakage resilient duplexing, derive a fine-grained bound on the security of the keyed duplex in said model, and map it to ideas of Taha and Schaumont (HOST 2014) and Dobraunig et al. (ToSC 2017) in order to use the duplex in a leakage resilient manner.

Keywords: Duplex · Sponge · Security proof · Leakage resilience

1 Introduction

With the selection of Keccak [9] as SHA-3 [20], cryptography based on public permutations has become more and more popular. This is especially caused by the fact that the sponge [7] and the duplex [8] constructions provide a huge flexibility by enabling various cryptographic tasks besides hashing, such as encryption, authenticated encryption, and message authentication, by just relying on a public permutation. Keyed versions of the sponge and duplex constructions have been analyzed in a series of papers [1,8,10,12,15,21,26,30,31], however, this analysis has been done in a black-box scenario, not considering the leakage of information that occurs in applications where side-channel attacks are feasible.

Ever since the threat of side-channel attacks has become evident to the public [27,28], finding suitable protection mechanisms against this attack vector has become of increasing importance. One can identify two different ways to protect against side-channel attacks. The first one deals with hardening the implementation of cryptographic schemes by means of countermeasures like hiding [14] or masking [11,13,22,32,33]. The other one aims at developing dedicated schemes that provide easier protection against side-channel attacks in the first place,

S. D. Galbraith and S. Moriai (Eds.): ASIACRYPT 2019, LNCS 11923, pp. 225–255, 2019.
https://doi.org/10.1007/978-3-030-34618-8_8

like fresh re-keying [29] or leakage resilient cryptography [18]. With respect to the sponge and duplex constructions, there exist proposals of Taha and Schaumont [38] and ISAP [16] that introduce dedicated algorithms that are claimed to provide protection against side-channel attacks.

Unfortunately, a closer look at the field of leakage resilient symmetric cryptography [6,17,19,34–36,41] reveals that the focus lies on constructions that can be instantiated with block ciphers. Hence, results regarding the leakage resilience of the keyed sponge, or more generally the keyed duplex construction that solely rely on unkeyed cryptographic permutations as building block are scarce. This particularly means that proposals such as those of [16,38] lack formal support regarding their leakage resilience.

1.1 Our Contribution

The contributions of this paper are manifold.

First, in Sect. 3, we describe a security model for leakage resilient duplexing. To do so, we start from the "ideal equivalent" of the keyed duplex of Daemen et al. [15], called an ideal extendable input function (IXIF), and present an adjusted version AIXIF. AIXIF is semantically equivalent to the IXIF if there is no leakage, but it allows to properly model leakage resilience of the keyed duplex. The model of leakage resilience of the duplex is now conceptually simple: as we argue in detail in Sect. 3.4, we consider a scheme leakage resilient if no attacker can distinguish a keyed duplex *that leaks for every query* from the random AIXIF. Here, we focus on non-adaptive leakage, where the leakage function is fixed in advance, akin to [17,19,35,37,41]. At this point our approach seems to be different from the typical models: the typical approach is to give a distinguisher access to a leaky version and a leak-free version of the cryptographic construction, and it has to distinguish the latter from a random function. The reason that we adopted a different model is that the duplex is just used as building block for encryption, authenticated encryption, or other types of functionalities. To prove that the use of a leakage resilient duplex gives rise to a leakage resilient construction with one of above-mentioned functionalities, the typical approach to give a distinguisher access to a leaky version and a leak-free version of the cryptographic construction has to be used again, as we will show later.

Second, in Sect. 5, we perform an in-depth and fine-grained analysis of the keyed duplex in the newly developed model. We take inspiration from Daemen et al. [15], who presented a detailed analysis of the keyed duplex in the black-box scenario, but the proof is not quite the same. To the contrary, due to various obstacles, it is not possible to argue similar to Daemen et al., nor to reduce the leakage resilience of a keyed duplex to its black-box security. Instead, we adopt ideas from the analysis of the NORX authenticated encryption scheme of Jovanovic et al. [26], and reason about the security of the keyed duplex in a sequential manner. One of the difficulties then is to determine the amount of min-entropy of a state in the duplex construction, given that the distinguisher may learn leakage from a duplex construction at different points in time. On the way, in Sect. 4 we give a detailed and accessible rationale of how leakage resilience proofs are performed in general and in our case.

Third, in Sect. 6, we interpret our results on the leakage resilience of the keyed duplex in the context of the proposals of Taha and Schaumont [38] and ISAP [16]. In a nutshell, these proposals can be seen to consist of a sequential evaluation of two duplex constructions: one that "gains entropy" by absorbing a nonce with small portions at a time, and one that "maintains entropy" in the sense that after the nonce is absorbed any state that will be visited by the duplex has high entropy and will be visited only once. We will then have a closer look at one use case of such a keyed duplex, nonce-based stream encryption, in Sect. 7. We build this scheme using aforementioned ideas, and prove that it is leakage resilient in the conventional security model. The proof is hybrid and reduces security of the stream cipher to that of the underlying duplex.

1.2 Related Work

Guo et al. [23] independently considered leakage resilience of duplex based modes. Their work is more specifically targeted to authenticated encryption (rather than to the duplex as building block). A second difference is that it considers a more generous leakage assumption. We consider a bounded leakage model, that upper bounds the amount of information that an attacker learns by λ, whereas Guo et al. assume hard-to-invert leakages. As such, Guo et al. [23] follow a different approach that is complementary to ours, and that might likewise be relevant in many different use cases.

1.3 Notation

For $b \in \mathbb{N}$, the set of b-bit strings is denoted $\{0,1\}^b$ and the set of arbitrarily length strings is denoted $\{0,1\}^*$. We define by $\mathrm{func}(b)$ the set of all functions $\mathsf{f} : \{0,1\}^b \to \{0,1\}^b$ and by $\mathrm{perm}(b)$ the set of all permutations $\mathsf{p} : \{0,1\}^b \to \{0,1\}^b$. By $X \leftarrow Y$ we denote the assignment of the value Y to X, and by $X \xleftarrow{\$} \mathcal{X}$ we denote the uniformly random drawing of an element X from a finite set $\mathcal{X}$. For $X \in \{0,1\}^b$ and for $c \in \mathbb{N}$ with $c \leq b$, we denote by $\mathrm{left}_c(X)$ the c leftmost bits of X and by $\mathrm{right}_c(X)$ the c rightmost bits of X. We denote by $\mathsf{rot}_c(X)$ the right-rotation of X by c bits.

A random variable S has *min-entropy* at least h, denoted $H_\infty(S) \geq h$, if $\max_{s \in S} \mathbf{Pr}\left(S = s\right) \leq 2^{-h}$. The conditional min-entropy is straightforward to define: the probability term gets expanded by the condition.

2 Keyed Duplex Construction

Let $b, c, r, k, u, \alpha \in \mathbb{N}$, with $c + r = b$, $k \leq b$, and $\alpha \leq b - k$. We describe the keyed duplex construction KD in Algorithm 1. The keyed duplex construction gets as input a key array $\boldsymbol{K} = (K[1], \ldots, K[u]) \in (\{0,1\}^k)^u$ consisting of u keys, and it is instantiated using a b-bit permutation $\mathsf{p} \in \mathrm{perm}(b)$. The construction internally maintains a b-bit state S, and has two interfaces: $\mathsf{KD.init}$ and $\mathsf{KD.duplex}$.

The initialization interface gets as input a key index $\delta \in [1, u]$ and an initialization vector $IV \in \mathcal{IV} \subseteq \{0,1\}^{b-k}$, and initializes the state with the δ-th

Algorithm 1. Keyed duplex construction $\mathsf{KD}[\mathsf{p}]_{\boldsymbol{K}}$

Interface: KD.init
Input: $(\delta, IV) \in [1, u] \times \mathcal{IV}$
Output: $\varnothing$
  $S \leftarrow \mathsf{rot}_\alpha(\boldsymbol{K}[\delta] \parallel IV)$
  $S \leftarrow \mathsf{p}(S)$
  return $\varnothing$

Interface: KD.duplex
Input: $(\mathit{flag}, P) \in \{\mathit{true}, \mathit{false}\} \times \{0,1\}^b$
Output: $Z \in \{0,1\}^r$
  $Z \leftarrow \mathrm{left}_r(S)$
  $S \leftarrow S \oplus [\mathit{flag}] \cdot (Z \| 0^{b-r}) \oplus P$ ▷ if *flag*, overwrite outer part
  $S \leftarrow \mathsf{p}(S)$
  return Z

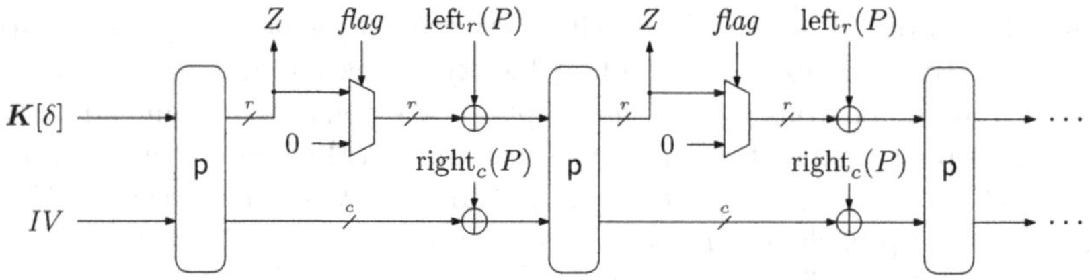

Fig. 1. The duplexing interface of KD.

key and the initialization vector IV as $S \leftarrow \mathsf{rot}_\alpha(\boldsymbol{K}[\delta] \parallel IV)$, followed by an evaluation of the underlying permutation p on the state S. It outputs nothing. Note that the constant α simply determines the bit positions where to place the key. We will see different examples of the value α in Sect. 6.

The duplexing interface gets as input a flag $\mathit{flag} \in \{\mathit{true}, \mathit{false}\}$ and a new data block $P \in \{0,1\}^b$. The interface outputs an r-bit block $Z \in \{0,1\}^r$ off the internal state S, transforms the state using the new data block P, and finally evaluates the underlying permutation p on the state. The flag *flag* describes how absorption is done on the r leftmost bits of the state that are squeezed: those r bits are either overwritten (if $\mathit{flag} = \mathit{true}$) or XORed with r bits of the input block P (if $\mathit{flag} = \mathit{false}$). See also Fig. 1, where the duplex is depicted for key offset $\alpha = 0$.

This description is another rephasing of how the duplex construction can be viewed compared to the original description used by Bertoni et al. [8], but also differs from the rephased description of Daemen et al. [15]. Compared to Daemen et al. the call of the underlying permutation is done at the end of the duplexing call instead of the beginning. This way of describing the duplex eases the proof in the leakage resilient setting, while at the same time empowers a leakage-aware attacker to adaptively react to the leakage of the permutation before providing new inputs. However, it still reflects the usage of the duplex in the same way as the description of Daemen et al. [15]. In particular, Daemen et al. also already considered multi-user security by default, and likewise had two different types of duplexing calls (for *flag* $\in \{true, false\}$) to allow implementation of SpongeWrap and variants using the duplex construction. Indeed, whereas SpongeWrap encryption can be performed using $\mathsf{KD.duplex}(\mathit{false}, \cdot)$, the decryption function must be performed using evaluations of $\mathsf{KD.duplex}(\mathit{true}, \cdot)$.

3 Security Model

In this section, we will describe our leakage resilience security model for the keyed duplex. We consider sampling of keys in Sect. 3.1. We settle the basic notation of distinguishers in Sect. 3.2. For reference, the black-box duplex security model of Daemen et al. [15] is treated in Sect. 3.3. We lift the model to leakage resilience in Sect. 3.4.

3.1 Sampling of Keys

The duplex construction of Sect. 2 is based on an array of u k-bit keys. These keys may be generated uniformly at random, as $\boldsymbol{K} \xleftarrow{\mathcal{D}_K} (\{0,1\}^k)^u$. In our analysis of leakage resilience, however, we will require the scheme to be still secure if the keys are not uniformly random but as long as they have sufficient min-entropy. Henceforth, we will adopt the approach of Daemen et al. [15] to consider keys sampled using a distribution $\mathcal{D}_{\boldsymbol{K}}$, that distributes the key independently[1] and with sufficient min-entropy, i.e., for which

$$H_\infty(\mathcal{D}_{\boldsymbol{K}}) = \min_{\delta \in [1,u]} H_\infty(\boldsymbol{K}[\delta])$$

is sufficiently high. Note that if $\mathcal{D}_{\boldsymbol{K}}$ is the random distribution, $H_\infty(\mathcal{D}_{\boldsymbol{K}}) = k$.

3.2 Distinguishers

A distinguisher D is an algorithm that is given access to one or more oracles O, denoted D^{O}, and that outputs a bit $b \in \{0,1\}$ after interaction with O. If O and P are oracles, we denote by $\Delta_{\mathsf{D}}(\mathsf{O}\ ;\ \mathsf{P})$ the advantage of a distinguisher D in distinguishing O from P. In our work, we will only be concerned with information-theoretic distinguishers: these have unbounded computational power, and their success probabilities are solely measured by the number of queries made to the oracles.

3.3 Black-Box Security

Daemen et al. [15] described the ideal extendable input function (IXIF) as ideal equivalent for the keyed duplex. We will also consider this function, modulo syntactical changes based on the changes we made on the keyed duplex in Sect. 2. The function is described in Algorithm 2.

[1] In Daemen et al. [15], the keys need not be mutually independent, but omitting this conditions will give various tricky corner cases in the analysis of leakage resilience.

Algorithm 2. Ideal extendable input function $\mathsf{IXIF}[\mathsf{ro}]_{\boldsymbol{K}}$

Interface: $\mathsf{IXIF.init}$
Input: $(\delta, IV) \in [1,u] \times \mathcal{IV}$
Output: $\varnothing$
  $path \leftarrow \mathsf{encode}[\delta] \parallel IV$
  return $\varnothing$

Interface: $\mathsf{IXIF.duplex}$
Input: $(flag, P) \in \{true, false\} \times \{0,1\}^b$
Output: $Z \in \{0,1\}^r$
  $Z \leftarrow \mathsf{ro}(path, r)$
  $path \leftarrow path \parallel ([flag] \cdot (Z \| 0^{b-r}) \oplus P)$ ▷ if *flag*, overwrite outer part
  return Z

The IXIF has the same interface as the keyed duplex, but instead of being based on a key array $\boldsymbol{K} \in (\{0,1\}^k)^u$ and being built on primitive $\mathsf{p} \in \text{perm}(b)$, it is built on a random oracle $\mathsf{ro} : \{0,1\}^* \times \mathbb{N} \to \{0,1\}^\infty$, that is defined as follows. Let $\mathsf{ro}_\infty : \{0,1\}^* \to \{0,1\}^\infty$ be a random oracle in the sense of Bellare and Rogaway [3]. For $P \in \{0,1\}^*$, $\mathsf{ro}(P, r)$ outputs the first r bits of $\mathsf{ro}(P)$. The IXIF maintains a path *path*, in which it unambiguously stores all data input by the user. It is initialized by $\mathsf{encode}[\delta] \parallel IV$ for some suitable injective encoding function $\mathsf{encode} : [1,u] \to \{0,1\}^k$, and upon each duplexing call, the new message block is appended to the path. Duplexing output is generated by evaluating the random oracle on *path*.

Let $b, c, r, k, u, \alpha \in \mathbb{N}$, with $c + r = b$, $k \leq b$, and $\alpha \leq b - k$. Let $\mathsf{p} \xleftarrow{\$} \text{perm}(b)$ be a random transformation, ro be a random oracle, and $\boldsymbol{K} \xleftarrow{\mathcal{D}_K} (\{0,1\}^k)^u$ a random array of keys. In the black-box security model, one considers a distinguisher that has access to either $(\mathsf{KD}[\mathsf{p}]_{\boldsymbol{K}}, \mathsf{p}^{\pm})$ in the real world or $(\mathsf{IXIF}[\mathsf{ro}], \mathsf{p}^{\pm})$ in the ideal world, where "$\pm$" stands for the fact that the distinguisher has bi-directional query access:

$$\mathbf{Adv}^{\mathsf{bb}}_{\mathsf{KD}}(\mathsf{D}) = \Delta_{\mathsf{D}}\left(\mathsf{KD}[\mathsf{p}]_{\boldsymbol{K}}, \mathsf{p}^{\pm} \; ; \; \mathsf{IXIF}[\mathsf{ro}], \mathsf{p}^{\pm}\right). \tag{1}$$

This is the model explicitly considered by Daemen et al. [15].

3.4 Leakage Resilience

We consider non-adaptive leakage resilience of the keyed duplex construction. Non-adaptive leakage has been considered before in [17,19,35,37,41], among others, and we will use the description of $\mathcal{L}$-resilience of Dodis and Pietrzak [17]. These models, however, consider the underlying primitive to be a block cipher or weak PRF, whereas in our setting it is a public permutation. In addition, the duplex has its characteristic property that it allows variable length input *and* variable length output. A final, and technically more delicate difference (as becomes clear below), is that the duplex consists of two oracles init and duplex, which the distinguisher may call interchangeably at its own discretion.

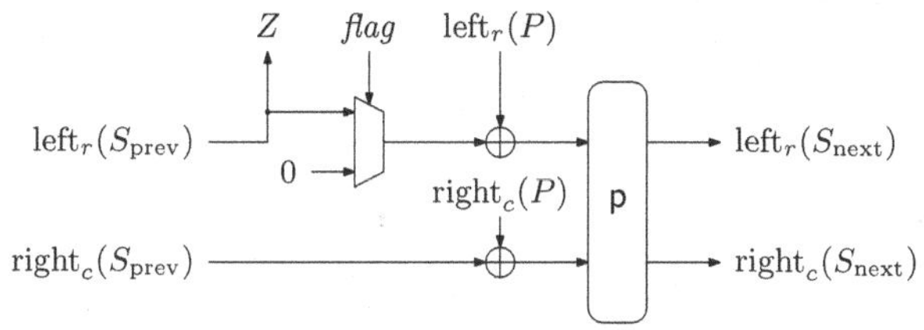

Fig. 2. An evaluation of KD.duplex, with its previous state S_{prev} and next state S_{next} are indicated. Intuitively, leakage occurs on both states, and the leakage function L returns λ bits of leakage.

We will assume that only values leak information that take part in the current computation, i.e., only leakage can occur from information that is used in calls to init and duplex. Note that this general way to describe what information can leak does not put restrictions on how this leakage occurs. For instance, this model covers even very strong attackers that can directly probe a limited amount of bits in a circuit, or that can get some limited amount of information about all values that are used in the current computation.

Recall from (1) that in the black-box model, one compares $(\mathsf{KD}[\mathsf{p}]_{\boldsymbol{K}}, \mathsf{p}^{\pm})$ with $(\mathsf{IXIF}[\mathsf{ro}], \mathsf{p}^{\pm})$, where $\mathsf{p} \xleftarrow{\$} \text{perm}(b)$ and ro is a random oracle. In order to prove leakage resilience of the construction, we have to demonstrate that "leakage does not help". For the real keyed duplex $\mathsf{KD}[\mathsf{p}]_{\boldsymbol{K}}$, modeling this is as simple as giving the distinguisher the leakage value $\ell \leftarrow \mathsf{L}(S_{\text{prev}}, \mathit{flag}, P, S_{\text{next}})$, where $\mathsf{L} : \{0,1\}^b \times \{\mathit{true}, \mathit{false}\} \times \{0,1\}^b \times \{0,1\}^b \rightarrow \{0,1\}^\lambda$ is the leakage function, S_{prev} the state before the call, and S_{next} the state after the call. See also Fig. 2.

For the ideal world $\mathsf{IXIF}[\mathsf{ro}]$, there is no such thing as a state, and simply generating random leakage allows for a trivial win for the distinguisher, as leaked bits may happen to coincide with the actual squeezed bits. For example, if L is defined as $\mathsf{L}(S_{\text{prev}}, \mathit{flag}, P, S_{\text{next}}) = \text{left}_\lambda(S_{\text{next}})$, in the real world, any leakage ℓ satisfies $\ell = \text{left}_\lambda(Z)$, whereas in the ideal world this equation holds with probability around $1/2^\lambda$, only. We resolve this by making a minor tweak to the duplexing interface of IXIF: the oracle maintains a dummy state S, and instead of $Z \leftarrow \mathsf{ro}(\mathit{path}, r)$, it gets Z from this dummy state $Z \leftarrow \text{left}_r(S)$ and updates the dummy state constantly by doing $S \leftarrow \mathsf{ro}(\mathit{path}, b)$. The dummy state is initialized as in the normal duplex (Algorithm 1). The resulting adjusted IXIF (AIXIF) is given in Algorithm 3.

It is important to note that the change from IXIF to AIXIF is purely administrative, in that for any distinguisher D,

$$\Delta_{\mathsf{D}}\left(\mathsf{IXIF}[\mathsf{ro}]\ ;\ \mathsf{AIXIF}[\mathsf{ro}]_{\boldsymbol{K}}\right) = 0.$$

The reason is that (i) an initialized state $S = \mathsf{rot}_\alpha(\boldsymbol{K}[\delta] \parallel IV)$ is never used for outputting data to the distinguisher, and (ii) later versions of the dummy state are always updated with b bits of ro-output of which only r bits are squeezed a

Algorithm 3. Adjusted ideal extendable input function $\mathsf{AIXIF}[\mathsf{ro}]_{\boldsymbol{K}}$

Interface: $\mathsf{AIXIF.init}$
Input: $(\delta, IV) \in [1,u] \times \mathcal{IV}$
Output: $\varnothing$
  $path \leftarrow \mathsf{encode}[\delta] \parallel IV$
  $S \leftarrow \mathsf{rot}_\alpha(\boldsymbol{K}[\delta] \parallel IV)$
  $S \leftarrow \mathsf{ro}(path, b)$
  return $\varnothing$

Interface: $\mathsf{AIXIF.duplex}$
Input: $(\mathit{flag}, P) \in \{\mathit{true}, \mathit{false}\} \times \{0,1\}^b$
Output: $Z \in \{0,1\}^r$
  $Z \leftarrow \mathrm{left}_r(S)$
  $path \leftarrow path \parallel ([\mathit{flag}] \cdot (Z \| 0^{b-r}) \oplus P)$ ▷ if *flag*, overwrite outer part
  $S \leftarrow \mathsf{ro}(path, b)$
  return Z

single time. Therefore, the original black-box security model could just as well be defined based on AIXIF. The good thing of AIXIF, now, is that it allows to easily formalize security in the leakage resilience setting where each construction call leaks.

Let $b, c, r, k, u, \alpha, \lambda \in \mathbb{N}$, with $c+r = b$, $k \leq b$, $\alpha \leq b-k$, and $\lambda \leq 2b$. Let $\mathsf{p} \xleftarrow{\$} \mathrm{perm}(b)$ be a random permutation, ro be a random oracle, and $\boldsymbol{K} \xleftarrow{\mathcal{D}_K} (\{0,1\}^k)^u$ a random array of keys. Let $\mathcal{L} = \{\mathsf{L} : \{0,1\}^b \times \{\mathit{true}, \mathit{false}\} \times \{0,1\}^b \times \{0,1\}^b \to \{0,1\}^\lambda\}$ be a class of leakage functions, and for any leakage function $\mathsf{L} \in \mathcal{L}$, define by $\mathsf{KD}[\mathsf{p}]^{\mathsf{L}}_{\boldsymbol{K}}$ (resp., $\mathsf{AIXIF}[\mathsf{ro}]^{\mathsf{L}}_{\boldsymbol{K}}$) the keyed duplex (resp., adjusted ideal extendable input function) that for each construction call leaks $\mathsf{L}(S_{\text{prev}}, \mathit{flag}, P, S_{\text{next}})$, where S_{prev} is the state before the call and S_{next} the state after the call. In the leakage resilience security model, one considers a distinguisher that has access to either $(\mathsf{KD}[\mathsf{p}]^{\mathsf{L}}_{\boldsymbol{K}}, \mathsf{p}^{\pm})$ in the real world, and $(\mathsf{AIXIF}[\mathsf{ro}]^{\mathsf{L}}_{\boldsymbol{K}}, \mathsf{p}^{\pm})$ in the ideal world, maximized over all possible leakage functions $\mathsf{L} \in \mathcal{L}$:

$$\mathbf{Adv}^{\mathcal{L}\text{-naLR}}_{\mathsf{KD}}(\mathsf{D}) = \max_{\mathsf{L} \in \mathcal{L}} \Delta_{\mathsf{D}} \left(\mathsf{KD}[\mathsf{p}]^{\mathsf{L}}_{\boldsymbol{K}}, \mathsf{p}^{\pm} \ ; \ \mathsf{AIXIF}[\mathsf{ro}]^{\mathsf{L}}_{\boldsymbol{K}}, \mathsf{p}^{\pm} \right). \tag{2}$$

Note that we indeed consider non-adaptive leakage resilience, as we maximize over all possible leakage functions L. Note furthermore that we do not consider future computation: the keyed duplex construction is based on the random permutation p and the set of allowed leakage functions is independent of p; the functions simply operate on the state right before and right after the transformation that leaks.

Remark 1. It is important to observe that, in our model, *any* duplex call leaks. In this way, our model seems to be conceptually different to the established models of, e.g., [17,19,35,37,41]. At a high level, in these models, the distinguisher has access to a leak-free version of the construction, which it has to distinguish from random, and a leaky version of the construction, which it may use to gather information. The intuition is that, whatever the distinguisher may learn from leakage, any new evaluation of the construction still looks random. In comparison, in our model of (2), we simply assume that the construction *always leaks*: the real construction KD.duplex leaks actual data of the state, whereas AIXIF.duplex leaks random data. This can be tolerated in our model as, typically, the KD.duplex will be used as *building block* for constructions that enable functionalities like, e.g., encryption. When we realize leakage resilient encryption with the help of the keyed duplex in Sect. 7, we consider the established model where the distinguisher has access to a leaky and a leak-free version of the construction, and the latter has to be distinguished from random.

4 Proof Rationale

In this section, we outline the rationale of proving leakage resilience of the keyed duplex. The section is extensive, but should give a high-level overview of how the security analysis is performed. First, in Sect. 4.1, we detail how typically leakage resilience of sequential constructions is proven. Then, in Sect. 4.2, we explain to what degree these approaches apply to permutation based cryptography. In Sect. 4.3, we consider the keyed duplex construction in more detail, and explain at a high level how the security proof is performed and how it relies on existing research on the keyed duplex construction in the black-box model. The discussion will form a stepping stone to the formal analysis of the keyed duplex in Sect. 5 and of the application of the result in Sect. 6.

4.1 Proving Leakage Resilience

The rationale of leakage resilience security proofs is not straightforward, and the main cause of this is the delicate selection of entropy measure for a leaky state. First off, it is important to know that starting from the seminal work of Dziembowski and Pietrzak [18], almost all leakage resilient PRGs and PRFs in literature [5,6,17,19,34,35,40,41] are sequential: they maintain a state, and use a cryptographic primitive to evolve the state in a sequential manner and to output a random stream. The cryptographic primitive is, in most of these cases, a block cipher modeled as a weak PRF $\mathsf{F} : \{0,1\}^k \times \{0,1\}^m \to \{0,1\}^n$.

A measure to identify the amount of randomness of a value is the min-entropy. Informally, a value S has min-entropy $H_\infty(S) \geq h$ if the success probability of guessing S is at most $1/2^h$. Unfortunately, the min-entropy is not fully suited to deal with leakage in above-mentioned sequential constructions: each round, certain information of a state leaks, and the min-entropy will only *decrease* with the leakage over time. Dziembowski and Pietrzak [18] observed that one does not strictly need the min-entropy of the state to be high enough: all that is needed is that the state is *computationally indistinguishable from a state with sufficiently high min-entropy*, in the eye of the computationally bounded distinguisher. This is formalized by the HILL-pseudoentropy [24] (or formally the conditional HILL-pseudoentropy [25], taking into account leakage data). The security proofs of above constructions now all exist of an iterative execution of the following steps:

(1) If the input to the wPRF F has sufficiently high min-entropy, then with high probability the output is an n-bit pseudorandom value S;
(2) If λ bits of the n-bit pseudorandom state S are leaked, then with high probability the state has HILL-pseudoentropy at least $n - 2\lambda$;
(3) By definition of the HILL-pseudoentropy, the state is computationally indistinguishable from a state with min-entropy at least $n - 2\lambda$;
(4) The resulting state will be (part of the) input to next round's wPRF.

A formalization of the first three steps can be found in [35, Lemma 2], [35, Lemma 6], and [35, Definition 3]. We note that the original introduction of leakage resilient cryptography of Dziembowski and Pietrzak [18] did not consider a weak PRF but a (stronger) PRG.

It is clear that an iterative execution of above steps allows to prove security of a sequential wPRF-based construction, provided that the state after step (4) has enough min-entropy to make the application of step (1) in next round go through. The iterative execution allows to prove security of the construction, with a security loss quantified by a sum of the individual losses in steps (1)–(3) for each of the rounds. More importantly, the security proof stands under the assumption that the block cipher is a weak PRF, or can be used to construct a weak PRF (see also Standaert et al. [37]). At this point, it requires cryptanalysts to investigate the weak PRF security of actual block ciphers.

4.2 Towards Permutation-Based Constructions

The focus in our work is on constructions based on cryptographic permutations. In the black-box model, both the keyed sponge [1,10,12,21,26,30,31] and the keyed duplex [8,15,30] have received thorough investigation.

The security analyses are different from black-box analyses of block cipher based constructions: whereas for the latter one argues security under the assumption that the block cipher is a (strong) pseudorandom permutation, in the former one assumes that the permutation is perfect and considers a distinguisher that is computationally unbounded and whose complexity is only measured by the online complexity (the amount of construction queries) and the offline complexity (the amount of primitive queries).

The approach is well-established, and in our analysis of the leakage resilience of the duplex, we adopt the approach. This gives two significant advantages in the analysis. First off, we consider computationally unbounded adversaries, and there is no need to make the HILL-detour. In other words, we can *directly* argue that an n-bit pseudorandom state S has min-entropy at least $n - \lambda$ after λ bits are leaked. Second, there is no issue with repeated min-entropy degradation: the state is transformed through a perfectly random permutation that outputs a random value (bar repetition) for each new input. We remark that concurrent work [23] also builds upon the random permutation model (and, in addition, an ideal tweakable block cipher).

These two advantages clearly simplify the rationale and simplicity of the leakage resilience security analysis of the duplex, yet do not make the security analysis a trivial extension of earlier leakage resilience analyses: in the new setting, the amount of entropy of a state is not only dependent on the leakage, but *also* on the primitive queries that the distinguisher makes, recalling that the distinguisher has *direct access* to the primitive. Indeed, this is not the case in ordinary wPRF-based security proofs.

There is another complication in the analysis of our construction: the distinguisher can re-initialize the state and start over. This is in line with the particular application of the duplex: authenticated encryption, where different authenticated encryptions may start from the same state and even have identical first permutation calls. Even if we had the possibility to argue that the duplex primitive is a weak PRF, repeated or mutually related states would invalidate step (1) of above reasoning, as the query history would skew the distribution of the weak PRF. In detail, step (1) requires the inputs to be close-to-random, a condition that appears to be more delicate than one would expect (cf., [41]), and that is false for repeated states in the duplex.

In a nutshell, one can say that the main overlap in our leakage resilience analysis compared with earlier approaches [5,6,17,19,34,35,40,41] is that we use the min-entropy to express the amount of randomness that is left after leakage, and we argue security based on the assumption that all state values in a keyed duplex have enough entropy.

4.3 Proving Security of Duplex Construction

Our proof uses many ideas from the solid black-box research already performed on keyed sponges and duplexes [1,8,10,12,15,21,26,30,31]. However, not all techniques from this line of research are suited in the leakage resilience setting. Most importantly, a notable technique [1,12,15,30] is to view the keyed sponge/duplex as a mode based on an Even-Mansour construction on top of the permutation $\mathsf{p} \in \mathrm{perm}(b)$. The trick is to XOR two copies of a dummy key with the inner part in-between every two evaluations of the permutation p. The change is purely syntactical, and a distinguisher cannot note the difference. However, in the leakage resilience setting, the distinguisher may have chosen the leakage function L so as to leak part of the state that is keyed, and XORing dummy keys turns out to become tricky. In particular, adoption of the approach to the leakage resilience setting would require us to be able to "split" leakages into input leakages and output leakages, but this is not always possible, depending on the leakage function L.

Instead, the proof resembles much of the approach of Jovanovic et al. [26], who performed a direct security proof of the NORX nonce-based authenticated encryption scheme that also applied to other CAESAR candidates. At a high level, the proof of Jovanovic et al. consists of observing that the output states are always uniformly random (bar repetition, as a permutation is evaluated), as long as no bad event occurs. A bad event, in turn, occurs if there are two construction queries with colliding states or if there is a construction query and a primitive query with colliding states. The absence of collisions is dealt with in the first phase by replacing the random permutation by a function that samples values from random at the cost of an RP-to-RF switch.

In our leakage resilience proofs, we follow the same approach. We also start by replacing the random permutation by a function f, that samples values from random and provides two-sided oracle access. Then, as long as the state of the keyed duplex has enough entropy, the result after applying f is random and also has enough entropy. Clearly, the entropy of the state reduces with the amount of leakage that occurs on the state, and consequently, bad events happen with a slightly larger probability as before. This also shows that estimating (formally, lower bounding) the amount of min-entropy of the states in the keyed duplex construction is important for deriving a tight security bound.

Focus on the keyed duplex (KD) of Algorithm 1, based on a function $\mathsf{f} \xleftarrow{\$} \mathrm{func}(b)$, and consider a duplex state $S_{\mathrm{prev}} \in \{0,1\}^b$. Assume that the interface $\mathsf{KD.duplex}$ is evaluated on this state for R different inputs,

$$\{(\mathit{flag}_i, P_i)\}_{i=1}^R .$$

As the previous state S_{prev} is the direct output of a call to a function f that samples b-bit values from random, S_{prev} is a value with min-entropy b minus the leakage occurred on this function call. Clearly, the R evaluations of the duplex in question are made for the same state S_{prev}, and hence, in total they reduce the entropy of S_{prev} further by at most $R \cdot \lambda$ bits due to the next function call. In addition, by regular squeezing, the distinguisher learns r bits of the state. In total, S_{prev} has conditional min-entropy at least

$$b - r - (R+1)\lambda .$$

If this entropy is sufficiently high, we get R new states S_{next} with min-entropy b minus the leakage occurred from one function call. The main lesson learned from this: a state that could be duplexed for different message blocks *should have small-rate absorption* (as this bounds R), and a unique state can be used for larger rates *even up to full-state absorption.*

5 Leakage Resilience of Keyed Duplex Construction

We will prove non-adaptive leakage resilience of the keyed duplex construction based on a cryptographic permutation $\mathsf{p} \xleftarrow{\$} \mathrm{perm}(b)$ in the model of Sect. 3.4 (see (2)). Although the generic construction and the model are based on the work of Daemen et al. [15], the security proof approach differs, as explained in Sect. 4.3. We quantify distinguishers in Sect. 5.1. The main security result is stated in Sect. 5.2, and an interpretation of it is given in Sect. 5.3. The proof is given in Sect. 5.4.

5.1 Distinguisher's Resources

We consider an information-theoretic distinguisher D that has access to either the real world $(\mathsf{KD}[\mathsf{p}]^{\mathsf{L}}_{\boldsymbol{K}}, \mathsf{p}^{\pm})$ or the ideal world $(\mathsf{AIXIF}[\mathsf{ro}]^{\mathsf{L}}_{\boldsymbol{K}}, \mathsf{p}^{\pm})$, where p is some permutation and L some leakage function. Two basic measures to quantify the distinguisher's resources are its online complexity M and offline complexity N:

- M: the number of distinct construction calls, either initialization or duplexing calls;
- N: the number of distinct primitive queries.

For each construction call, we define a path *path* that "registers" the data that got absorbed in the duplex up to the point that the cryptographic primitive (p in the real world and ro in the ideal world) is evaluated. For an initialization call $(\delta, IV) \mapsto \varnothing$, the associated path is defined as $path = \mathsf{encode}[\delta] \parallel IV$. For each duplexing call $(flag, P) \mapsto Z$, the value $[flag] \cdot (Z \| 0^{b-r}) \oplus M$ is appended to the path of the previous construction query. Not surprisingly, the definition matches the actual definition of *path* in the $\mathsf{AIXIF}[\mathsf{ro}]_{\boldsymbol{K}}$ construction of Algorithm 3, but defining the same thing for the real world will allow us to better reason about the security of the keyed duplex. Note that the value *path* contains no information that is secret to the distinguisher. In order to reason about duplexing calls, we will also define a *subpath* of a *path*, which is the path leading to the particular duplexing call. In other words, for a path *path*, it *subpath* is simply *path* with the last b bits removed.

In order to derive a detailed and versatile security bound, that in particular well-specifies how leakage influences the bound, we further parameterize the distinguisher as follows. For initialization calls:

- q: the number of initialization calls;
- q_{IV}: the maximum number of initialization calls for a single IV;
- q_δ: the maximum number of initialization calls for a single δ.

For duplexing calls:

- Ω: the number of duplexing queries with $flag = true$;
- L: the number of duplexing calls with repeated subpath, i.e., M minus the number of distinct subpaths;
- R: the maximum number of duplexing calls for a single non-empty *subpath*.

Note that these parameters can all be described as a function of the duplexing calls and the related *path*'s, and the distinguisher can compute these values based on the queries it made so far. The parametrization of the distinguisher is roughly as that of Daemen et al. [15], but we have added parameter R: it maximizes the number of occurrences of a path *subpath* for different inputs $(flag, P)$. The parameter will be used to determine, factually upper bound, the amount of leakage that the distinguisher learns on a state *after* the duplexing call. Indeed, if a certain path *subpath* occurs R times, this means that these R duplexing calls have the same input-state, and any evaluation of p in one of these duplexing calls leaks information about that state. In total, this results in a maximum amount of $R+1$ leakages. The parameter R is related to parameter L, but it is not quite the same. The parameters Ω and L are, as in [15], used to upper bound the number of duplexing calls for which the distinguisher may have *set* the r leftmost bits of the input to the permutation in the duplexing call to a certain value of its choice. This brings us to the last parameter:

- ν_{fix}: the maximum number of duplexing calls for which the adversary has set the outer part to a single value $\text{left}_r(T)$.

Note that $\nu_{\mathsf{fix}} \leq L + \Omega$, but it may be much smaller in specific use cases of the duplex, for example, if overwrites only happen for unique values.

5.2 Main Result

We will use a notion from Daemen et al. [15], namely that of the multicollision limit function.

Definition 1 (multicollision limit function). *Let $M, c, r \in \mathbb{N}$. Consider the experiment of throwing M balls uniformly at random in 2^r bins, and let μ be the maximum number of balls in a single bin. We define the multicollision limit function $\nu_{r,c}^M$ as the smallest natural number x that satisfies*

$$\mathbf{Pr}\left(\mu > x\right) \leq \frac{x}{2^c}.$$

We derive the following result on the keyed duplex under leakage.

Theorem 1. *Let $b, c, r, k, u, \alpha, \lambda \in \mathbb{N}$, with $c + r = b$, $k \leq b$, $\alpha \leq b - k$, and $\lambda \leq 2b$. Let $\mathsf{p} \xleftarrow{\$} \text{perm}(b)$ be a random permutation, and $\boldsymbol{K} \xleftarrow{\mathcal{D}_K} (\{0,1\}^k)^u$ a random array of keys. Let $\mathcal{L} = \{\mathsf{L} : \{0,1\}^b \times \{0,1\}^b \to \{0,1\}^\lambda\}$ be a class of leakage functions. For any distinguisher D quantified as in Sect. 5.1,*

$$\begin{aligned}
&\mathbf{Adv}_{\mathsf{KD}}^{\mathcal{L}\text{-naLR}}(\mathsf{D}) \\
&\quad\leq \frac{\nu_{\mathsf{fix}} N}{2^{c-(R+1)\lambda}} + \frac{2\nu_{r,c}^M N}{2^{c-(R+1)\lambda}} + \frac{2\nu_{r,c}^M}{2^c} + \frac{\nu_{r,c}^M (L+\Omega) + \frac{\nu_{\mathsf{fix}}-1}{2}(L+\Omega)}{2^{c-R\lambda}} \\
&\quad+ \frac{\binom{M-L-q}{2} + (M-L-q)(L+\Omega)}{2^{b-\lambda}} + \frac{\binom{M+N}{2} + \binom{N}{2}}{2^b} \\
&\quad+ \frac{q(M-q)}{2^{H_\infty(\mathcal{D}_K)+\min\{c,\max\{b-\alpha,c\}-k\}-(R+q_\delta)\lambda}} + \frac{q_{IV} N}{2^{H_\infty(\mathcal{D}_K)-q_\delta\lambda}} + \frac{\binom{u}{2}}{2^{H_\infty(\mathcal{D}_K)}}.
\end{aligned}$$

In addition, except with probability at most the same bound, the final output states have min-entropy at least $b - \lambda$.

The proof is given in Sect. 5.4; we first give an interpretation of the bound in Sect. 5.3.

5.3 Interpretation

By rephasing the duplex and by going over the duplex in a sequential manner (as [26]), and by only absorbing isolated concepts from Daemen et al. [15] (the quantification and the multicollision limit function), the proof is intuitively simpler to follow than the black-box variant. This is in part due to the fact that we start the proof with a transformation reminiscent of the RP-to-RF switch. This simplifies the proof at various aspects (for example, at the application of the multicollision limit function) but is not for free, as it induces an extra term of around $\binom{M+N}{2}/2^b$.

The proof is still fairly general, in part due to the presence of the term $\nu_{r,c}^M$. A naive bounding akin to the derivation of Jovanovic et al. [26] would give a bound

$$\nu_{r,c}^M \leq \max\left\{r, \left(\frac{2eM2^c}{2^r}\right)^{1/2}\right\},$$

but the bound is loose, in particular for small r. Daemen et al. [15] gave a more detailed analysis of the term, including two lemmas upper bounding it. Omitting details, one can think of the multicollision limit function to behave as follows [15]:

$$\nu_{r,c}^M \lesssim \begin{cases} b/\log_2\left(\frac{2^r}{M}\right), & \text{for } M \lesssim 2^r, \\ b \cdot \frac{M}{2^r}, & \text{for } M \gtrsim 2^r. \end{cases}$$

Beyond this multicollision term, the bound of Theorem 1 is complicated due to the multivariate quantification of the distinguisher's resources, and most importantly the terms L and Ω. In Sect. 6, we will consider how the duplex can be used to create leakage resilient cryptographic schemes, and see how the bound simplifies drastically for specific use cases.

5.4 Proof of Theorem 1

Let $\mathsf{L} \in \mathcal{L}$ be any leakage function. Consider any information-theoretic distinguisher D. Our goal is to bound

$$\Delta_{\mathsf{D}}\left(\mathsf{KD}[\mathsf{p}]_{K}^{\mathsf{L}}, \mathsf{p}^{\pm} \; ; \; \mathsf{AIXIF}[\mathsf{ro}]_{K}^{\mathsf{L}}, \mathsf{p}^{\pm}\right). \tag{3}$$

The first step is to replace p with a function $\mathsf{f} : \{0,1\}^b \to \{0,1\}^b$ that has the same interface as p. The function f maintains an initially empty list $\mathcal{F}$ of input/output tuples (X, Y). For a new query $\mathsf{f}(X)$ with $(X, \cdot) \notin \mathcal{F}$, it generates $Y \xleftarrow{\$} \{0,1\}^b$ and returns this value. For a new query $\mathsf{f}^{-1}(Y)$ with $(\cdot, Y) \notin \mathcal{F}$, it generates $X \xleftarrow{\$} \{0,1\}^b$ and returns this value. In both cases, the primitive adds (X, Y) to $\mathcal{F}$, and it aborts if this addition yields a collision in X or in Y. Clearly, as long as f does not abort, the function is perfectly indistinguishable from p, so we get:

$$\Delta_{\mathsf{D}}\left(\mathsf{KD}[\mathsf{p}]_{\boldsymbol{K}}^{\mathsf{L}}, \mathsf{p}^{\pm}\ ;\ \mathsf{KD}[\mathsf{f}]_{\boldsymbol{K}}^{\mathsf{L}}, \mathsf{f}^{\pm}\right) \leq \frac{\binom{M+N}{2}}{2^b},$$
$$\Delta_{\mathsf{D}}\left(\mathsf{AIXIF}[\mathsf{ro}]_{\boldsymbol{K}}^{\mathsf{L}}, \mathsf{p}^{\pm}\ ;\ \mathsf{AIXIF}[\mathsf{ro}]_{\boldsymbol{K}}^{\mathsf{L}}, \mathsf{f}^{\pm}\right) \leq \frac{\binom{N}{2}}{2^b},$$

as in the former there are $M + N$ evaluations of p and in the latter there are N. Note that this is a purely probabilistic case, and the switch does not involve/concern any leakage. From (3) we get

$$\Delta_{\mathsf{D}}\left(\mathsf{KD}[\mathsf{p}]_{\boldsymbol{K}}^{\mathsf{L}}, \mathsf{p}^{\pm}\ ;\ \mathsf{AIXIF}[\mathsf{ro}]_{\boldsymbol{K}}^{\mathsf{L}}, \mathsf{p}^{\pm}\right) \leq \Delta_{\mathsf{D}}\left(\mathsf{KD}[\mathsf{f}]_{\boldsymbol{K}}^{\mathsf{L}}, \mathsf{f}^{\pm}\ ;\ \mathsf{AIXIF}[\mathsf{ro}]_{\boldsymbol{K}}^{\mathsf{L}}, \mathsf{f}^{\pm}\right) + \frac{\binom{M+N}{2} + \binom{N}{2}}{2^b}. \tag{4}$$

We proceed with the remaining distance of (4).

The distinguisher makes M construction calls, each of which is either an initialization call $(\delta_i, IV_i) \mapsto (\varnothing, \ell_i)$ or a duplexing call $(\mathit{flag}_i, P_i) \mapsto (Z_i, \ell_i)$, where ℓ_i is the λ bits of leakages obtained in this i-th construction call. In addition, associated to each call is a path path_i as described in Sect. 5.1. Noting that for an initialization call, δ_i and IV_i are implicit in $\mathit{path}_i = \mathsf{encode}[\delta_i] \parallel IV_i$, we can unify the description as follows. For any initialization call, we define $(\mathit{flag}_i, P_i, Z_i) := (0, 0^b, 0^r)$; all M construction calls – either initialization or duplex – can be summarized in a transcript

$$\mathcal{Q}_c := ((\mathit{path}_i, \mathit{flag}_i, P_i, Z_i, \ell_i))_{i=1}^{M}. \tag{5}$$

For each construction call, we define a triplet of states (S_i, T_i, U_i). The state S_i is the previous or incoming state. For initialization queries it is defined as $\mathsf{rot}_\alpha(\boldsymbol{K}[\delta_i] \parallel IV_i)$. The state U_i is the next or outgoing state. These are properly defined for both the real and ideal world. The state T_i is an intermediate state, which is defined as $T_i := S_i \oplus [\mathit{flag}_i] \cdot (Z_i \| 0^{b-r}) \oplus P_i$. Note that the intermediate state is only meaningful for the real world, but the value we add to it is known to the adversary. Without loss of generality, each leakage satisfies $\ell_i = \mathsf{L}(T_i, U_i)$.

Furthermore, the distinguisher makes N primitive calls that are summarized in a transcript

$$\mathcal{Q}_p := ((X_j, Y_j))_{j=1}^{N}. \tag{6}$$

We define the following two collisions events, one that captures collisions between two construction calls and one that captures collisions between a construction call and a primitive call:

$$\mathsf{col}_{\mathsf{cc}}\ :\ \exists\ i, i' \text{ such that } \mathit{path}_i \neq \mathit{path}_{i'} \wedge T_i = T_{i'}, \tag{7}$$
$$\mathsf{col}_{\mathsf{cp}}\ :\ \exists\ i, j \text{ such that } T_i = X_j \vee U_i = Y_j. \tag{8}$$

We write $\mathsf{col} = \mathsf{col}_{\mathsf{cc}} \vee \mathsf{col}_{\mathsf{cp}}$. The bad events are comparable with those of Daemen et al. [15], but they are not the same. One notable difference: Daemen et al. consider (in our terminology) $\mathsf{col}_{\mathsf{cc}}$ for *both input and output collisions.* We do not need to do so, thanks to the RP-RF switch made before.

In Lemma 1 below, we will prove that $(\mathsf{KD}[\mathsf{f}]^{\mathsf{L}}_{\boldsymbol{K}}, \mathsf{f}^{\pm})$ and $(\mathsf{AIXIF}[\mathsf{ro}]^{\mathsf{L}}_{\boldsymbol{K}}, \mathsf{f}^{\pm})$ are identical until col is triggered in the real world. Lemma 2 subsequently derives an upper bound on the event that col is triggered in the real world. These two results, together with (4) above, complete the proof of Theorem 1. Note that from the result of Lemma 1, we can particularly conclude that the final states of the keyed duplex, i.e., all states before re-initializations, have min-entropy $b - \lambda$.

Lemma 1. *As long as* $\mathsf{D}^{\mathsf{KD}[\mathsf{f}]^{\mathsf{L}}_{\boldsymbol{K}}, \mathsf{f}^{\pm}}$ *does not set* col, *the worlds* $(\mathsf{KD}[\mathsf{f}]^{\mathsf{L}}_{\boldsymbol{K}}, \mathsf{f}^{\pm})$ *and* $(\mathsf{AIXIF}[\mathsf{ro}]^{\mathsf{L}}_{\boldsymbol{K}}, \mathsf{f}^{\pm})$ *are identical, or formally,*

$$\Delta_{\mathsf{D}}\left(\mathsf{KD}[\mathsf{f}]^{\mathsf{L}}_{\boldsymbol{K}}, \mathsf{f}^{\pm} \; ; \; \mathsf{AIXIF}[\mathsf{ro}]^{\mathsf{L}}_{\boldsymbol{K}}, \mathsf{f}^{\pm}\right) \leq \mathbf{Pr}\left(\mathsf{D}^{\mathsf{KD}[\mathsf{f}]^{\mathsf{L}}_{\boldsymbol{K}}, \mathsf{f}^{\pm}} \text{ sets col}\right). \tag{9}$$

Proof. By the fundamental lemma of game playing [4], it suffices to prove that, as long as the real world $(\mathsf{KD}[\mathsf{f}]^{\mathsf{L}}_{\boldsymbol{K}}, \mathsf{f}^{\pm})$ does not set col, the real and ideal world are indistinguishable.

Clearly, in the ideal world $(\mathsf{AIXIF}[\mathsf{ro}]^{\mathsf{L}}_{\boldsymbol{K}}, \mathsf{f}^{\pm})$, the construction oracle is independent of the primitive oracle $\mathsf{f}^{\pm}$. Also in the real world, the construction oracle $\mathsf{KD}[\mathsf{f}]^{\mathsf{L}}_{\boldsymbol{K}}$ is independent of $\mathsf{f}^{\pm}$, by exclusion of duplex-primitive collisions $\mathsf{col}_{\mathsf{cp}}$ and as each new query to $\mathsf{f}^{\pm}$ is replied with a uniformly generated value. Therefore, we can drop the primitive oracle, and focus on proving that $\mathsf{KD}[\mathsf{f}]^{\mathsf{L}}_{\boldsymbol{K}}$ is indistinguishable from $\mathsf{AIXIF}[\mathsf{ro}]^{\mathsf{L}}_{\boldsymbol{K}}$ under the assumption that $\neg\mathsf{col}_{\mathsf{cc}}$ holds.

We will not only consider the output values (Z_i, ℓ_i), but we will rather prove a stronger result, namely that output states are identically distributed in both worlds. Note that in the real world, the output state is computed as $U_i \leftarrow \mathsf{f}(T_i)$, whereas in the ideal world, it is computed as $U_i \leftarrow \mathsf{ro}(path_i, b)$. Consider the i-th construction call. Clearly, $path_i \neq path_{i'}$, as otherwise the query would be a repeated call. By $\neg\mathsf{col}_{\mathsf{cc}}$, also $T_i \neq T_{i'}$ for all $i' < i$. This means that in both worlds, U_i is a uniformly randomly generated value from $\{0,1\}^b$. □

Lemma 2. *The probability that* $\mathsf{D}^{\mathsf{KD}[\mathsf{f}]^{\mathsf{L}}_{\boldsymbol{K}}, \mathsf{f}^{\pm}}$ *sets* col *satisfies:*

$$\begin{aligned}
\mathbf{Pr}&\left(\mathsf{D}^{\mathsf{KD}[\mathsf{f}]^{\mathsf{L}}_{\boldsymbol{K}}, \mathsf{f}^{\pm}} \text{ sets col}\right) \\
&\leq \frac{\nu_{\mathsf{fix}} N}{2^{c-(R+1)\lambda}} + \frac{2\nu^M_{r,c} N}{2^{c-(R+1)\lambda}} + \frac{2\nu^M_{r,c}}{2^c} + \frac{\nu^M_{r,c}(L+\Omega) + \frac{\nu_{\mathsf{fix}}-1}{2}(L+\Omega)}{2^{c-R\lambda}} \\
&\quad + \frac{\binom{M-L-q}{2} + (M-L-q)(L+\Omega)}{2^{b-\lambda}} \\
&\quad + \frac{q(M-q)}{2^{H_\infty(\mathcal{D}_K)+\min\{c,\max\{b-\alpha,c\}-k\}-(R+q_\delta)\lambda}} + \frac{q_{IV} N}{2^{H_\infty(\mathcal{D}_K)-q_\delta\lambda}} + \frac{\binom{u}{2}}{2^{H_\infty(\mathcal{D}_K)}}.
\end{aligned}$$

Proof. Consider any distinguisher D that has query access to $(\mathsf{KD}[\mathsf{f}]^{\mathsf{L}}_{\boldsymbol{K}}, \mathsf{f}^{\pm})$, and is bound to the parameters $(M, N, q, q_{IV}, q_\delta, \Omega, L, R, \nu_{\mathsf{fix}})$ listed in Sect. 5.1. Our goal is to bound

$$\mathbf{Pr}(\mathsf{col}) := \mathbf{Pr}\left(\mathsf{D}^{\mathsf{KD}[\mathsf{f}]^{\mathsf{L}}_{\boldsymbol{K}}, \mathsf{f}^{\pm}} \text{ sets col}\right). \tag{10}$$

Additional Notation. One can consider duplexing-calls to occur in a tree fashion, as long as $\mathsf{col}_{\mathsf{cc}}$ never happens. To proper reasoning about the probability that col is set, we will have to define parents, siblings, and children of a duplex call. Consider any construction query $(\mathit{path}_i, \mathit{flag}_i, P_i, Z_i, \ell_i)$.

The parent of this construction query, $\mathrm{parent}(i) \in \{\bot, 1, \ldots, i-1\}$, is defined as follows: if i corresponds to an initialization call, so if $|\mathit{path}_i| = b$, then $\mathrm{parent}(i) = \bot$; otherwise, $\mathrm{parent}(i)$ is the index of the unique duplexing call that satisfies

$$\mathit{path}_i = \mathit{path}_{\mathrm{parent}(i)} \parallel ([\mathit{flag}_{\mathrm{parent}(i)}] \cdot (Z_{\mathrm{parent}(i)} \| 0^{b-r}) \oplus P_{\mathrm{parent}(i)}). \tag{11}$$

If the i-th query is not an initialization call, its siblings $\mathrm{sibling}(i) \subseteq \{1, \ldots, i\}$ are the set of queries *up to the i-th one* (later siblings have yet to be born) with the same parent:

$$\mathrm{sibling}(i) = \left\{ l \in \{1, \ldots, i\} \mid \mathit{path}_{\mathrm{parent}(l)} = \mathit{path}_{\mathrm{parent}(i)} \right\}. \tag{12}$$

Note that we have $|\mathrm{sibling}(i)| \leq R$ for any $i \in \{1, \ldots, M\}$. The children of the i-th query are the set of all queries that have i as parent:

$$\mathrm{child}(i) = \{l \in \{i+1, \ldots, M\} \mid \mathrm{parent}(l) = i\}. \tag{13}$$

We define the type type_i of a construction query $(\mathit{path}_i, \mathit{flag}_i, P_i, Z_i, \ell_i)$:

$$\mathit{type}_i = \begin{cases} \mathit{init}\,, \text{ if } |\mathit{path}_i| = b\,, \\ \mathit{full}\,, \text{ if } |\mathit{path}_i| > b \wedge (|\mathrm{sibling}(i)| = 1 \wedge \mathit{flag}_i = \mathit{false})\,, \\ \mathit{fix}\,, \text{ if } |\mathit{path}_i| > b \wedge (|\mathrm{sibling}(i)| > 1 \vee \mathit{flag}_i = \mathit{true})\,. \end{cases} \tag{14}$$

Note that we have q queries of type *init*. Type *full* corresponds to duplex calls of which the input state S_i is a random value from $\{0,1\}^b$ of which the adversary may have learned the outer r bits, but it had no possibility to *set* the outer part to a certain value of its choice. By definition, there are at most $M-L-q$ queries of type *full*. Finally, type *fix* corresponds to duplex calls of which distinguisher might have set the outer part to a certain value of its choice; this happens if the preceding duplex call had siblings, or if the adversary has turned $\mathit{flag}_i = \mathit{true}$, i.e., enabled the overwrite functionality in the duplex. There are at most $L + \Omega$ queries of type *fix*.

Analyzing Bad Events. We define three additional collision events. The first two correspond to multicollisions among the construction queries exceeding an threshold $\nu := \nu_{r,c}^{M}$, and the third one corresponds to plain key collisions in the key array $\boldsymbol{K}$:

$$\begin{aligned} \mathsf{mc}_{\mathsf{in}} &: \exists \text{ distinct } i_1, \ldots, i_{\nu+1} \text{ with } \mathit{type}_{i_j} = \mathit{full} \text{ such that} \\ &\qquad \mathrm{left}_r(T_{i_1}) = \cdots = \mathrm{left}_r(T_{i_{\nu+1}})\,, && (15) \\ \mathsf{mc}_{\mathsf{out}} &: \exists \text{ distinct } i_1, \ldots, i_{\nu+1} \text{ such that } \mathrm{left}_r(U_{i_1}) = \cdots = \mathrm{left}_r(U_{i_{\nu+1}})\,, && (16) \\ \mathsf{key} &: \exists \text{ distinct } \delta, \delta' \text{ such that } \boldsymbol{K}[\delta] = \boldsymbol{K}[\delta']. && (17) \end{aligned}$$

We define $\mathsf{mc} = \mathsf{mc}_{\mathsf{in}} \vee \mathsf{mc}_{\mathsf{out}}$. By basic probability theory,

$$\mathbf{Pr}(\mathsf{col}) = \mathbf{Pr}(\mathsf{col}_{\mathsf{cc}} \vee \mathsf{col}_{\mathsf{cp}}) \leq \mathbf{Pr}(\mathsf{col}_{\mathsf{cc}} \vee \mathsf{col}_{\mathsf{cp}} \mid \neg(\mathsf{mc} \vee \mathsf{key})) + \mathbf{Pr}(\mathsf{mc} \vee \mathsf{key}) .$$

Note that key is an event independent of the number of queries, whereas $\mathsf{col}_{\mathsf{cc}}$, $\mathsf{col}_{\mathsf{cp}}$, and mc are. The distinguisher can make $M+N$ queries, which it makes in a certain order. For $l \in \{1, \ldots, M+N\}$, denote by $\mathsf{col}_{\mathsf{cc}}(l)$, $\mathsf{col}_{\mathsf{cp}}(l)$, and $\mathsf{mc}(l)$ the event that the l-th query sets the respective event. For brevity of notation, write $\mathsf{col}(l) = \mathsf{col}_{\mathsf{cc}}(l) \vee \mathsf{col}_{\mathsf{cp}}(l)$. By basic probability theory,

$$\mathbf{Pr}(\mathsf{col}) \leq \sum_{l=1}^{M+N} \mathbf{Pr}(\mathsf{col}_{\mathsf{cc}}(l) \mid \neg\mathsf{col}(1 \ldots l-1) \wedge \neg\mathsf{mc}(1 \ldots l) \wedge \neg\mathsf{key}) \tag{18a}$$

$$+ \sum_{l=1}^{M+N} \mathbf{Pr}(\mathsf{col}_{\mathsf{cp}}(l) \mid \neg\mathsf{col}(1 \ldots l-1) \wedge \neg\mathsf{mc}(1 \ldots l) \wedge \neg\mathsf{key}) \tag{18b}$$

$$+ \mathbf{Pr}(\mathsf{mc}) \tag{18c}$$

$$+ \mathbf{Pr}(\mathsf{key}) . \tag{18d}$$

Based on this, we will proceed as follows. We will consider any query made by the distinguisher and consider the probability that *this query* sets either of the events $\mathsf{col}_{\mathsf{cc}}$, $\mathsf{col}_{\mathsf{cp}}$, and mc under the assumption that no earlier query set the event. Note that $\mathsf{col}_{\mathsf{cc}}$ and mc may only be set by a construction query; $\mathsf{col}_{\mathsf{cp}}$ may be set by a construction or a primitive query.

Probability of $\mathsf{col}_{\mathsf{cc}}$ of Eq. (18a). The event can only be set in duplex queries. Consider any two $i \neq i'$, and assume that at the point that the latest of the two queries is made, the events col, mc, and key are still false. We will make a distinction depending on the type of queries of i and i'.

- $type_i = type_{i'} = init$. Note that $T_i = \mathsf{rot}_\alpha(\boldsymbol{K}[\delta_i] \parallel IV_i)$, where δ_i and IV_i can be deduced from $path_i$, and $T_{i'} = \mathsf{rot}_\alpha(\boldsymbol{K}[\delta_{i'}] \parallel IV_{i'})$, where $\delta_{i'}$ and $IV_{i'}$ can be deduced from $path_{i'}$. As $path_i \neq path_i'$, a collision $T_i = T_{i'}$ implies that necessarily $\delta_i \neq \delta_{i'}$ and $\boldsymbol{K}[\delta_i] = \boldsymbol{K}[\delta_{i'}]$. This is impossible under the assumption that $\neg\mathsf{key}$ holds;
- $type_i = init$ and $type_{i'} \neq init$. Note that $T_i = \mathsf{rot}_\alpha(\boldsymbol{K}[\delta_i] \parallel IV_i)$, where δ_i and IV_i can be deduced from $path_i$. Also, $T_{i'} = U_{\text{parent}(i')} \oplus [flag_{i'}] \cdot (Z_{i'} \| 0^{b-r}) \oplus P_{i'}$.
  - $i < i'$. The conditional min-entropy of bits $\alpha \ldots \alpha + k$ of T_i is at least $H_\infty(\mathcal{D}_{\boldsymbol{K}}) - q_\delta\lambda$ and the conditional min-entropy of $\text{right}_c(T_{i'})$ is at least $c - |\text{sibling}(i')|\lambda$. The value T_i hits $T_{i'}$ with probability at most $1/2^{H_\infty(\mathcal{D}_{\boldsymbol{K}})+\min\{c,\max\{b-\alpha,c\}-k\}-(|\text{sibling}(i')|+q_\delta)\lambda}$;
  - $i' < i$. The conditional min-entropy of bits $\alpha \ldots \alpha + k$ of T_i is at least $H_\infty(\mathcal{D}_{\boldsymbol{K}}) - (q_\delta - 1)\lambda$ and the conditional min-entropy of $\text{right}_c(T_{i'})$ is at least $c - (|\text{sibling}(i')| + 1)\lambda$. The value T_i hits $T_{i'}$ with probability at most $1/2^{H_\infty(\mathcal{D}_{\boldsymbol{K}})+\min\{c,\max\{b-\alpha,c\}-k\}-(|\text{sibling}(i')|+q_\delta)\lambda}$.

  Note that $|\text{sibling}(i')| \leq R$. There are at most q queries i with $type_i = init$, and at most $M - q$ with $type_{i'} \neq init$. By the union bound, $\mathsf{col}_{\mathsf{cc}}$ is set in this case with probability at most $q(M-q)/2^{H_\infty(\mathcal{D}_{\boldsymbol{K}})+\min\{c,\max\{b-\alpha,c\}-k\}-(R+q_\delta)\lambda}$;

- *type*$_i \neq$ *init* and *type*$_{i'} \neq$ *init*. We will argue based on the randomness generated in any query l, which generates a random output state $U_l \xleftarrow{\$} \{0,1\}^b$. The probability bound will follow through a union bound, as any query i with *type*$_i \neq$ *init* is the child any such query.
  - Consider any $i \in \text{child}(l)$ with *type*$_i$ = *full*. So far, the distinguisher learned λ bits of leakage on state S_i in query l. Thus, T_i has conditional min-entropy at least $b - \lambda$. It hits any other $T_{i'}$ with probability at most $1/2^{b-\lambda}$. There are at most $M-L-q$ queries i, i' with *type*$_i$ = *type*$_{i'}$ = *full*, and furthermore, there are at most $L + \Omega$ queries i' with *type*$_{i'}$ = *fix*. By the union bound, omitting duplicate counting:
  $$\frac{\binom{M-L-q}{2} + (M-L-q)(L+\Omega)}{2^{b-\lambda}};$$
  - Consider any $i \in \text{child}(l)$ with *type*$_i$ = *fix*. So far, the distinguisher learned λ bits of leakage on state S_i in query l, and $(|\text{sibling}(i)|-1)\lambda$ bits of leakage on state S_i from its sibling queries. Thus, T_i has conditional min-entropy at least $c - |\text{sibling}(l)|\lambda \geq c - R\lambda$. It hits any other $T_{i'}$ with probability at most $1/2^{c-R\lambda}$.
  There are at most $L + \Omega$ queries i with *type*$_i$ = *fix*. By $\neg\mathsf{mc}_{\mathsf{in}}$, there are at most ν out of at most $M - L - q$ queries i' with *type*$_{i'}$ = *full* whose outer part equals $\text{left}_r(T_l)$. There are at most $\nu_{\mathsf{fix}} - 1$ queries i' with *type*$_{i'}$ = *fix* whose outer part equals $\text{left}_r(T_l)$. By the union bound, omitting duplicate counting:
  $$\frac{\nu(L+\Omega) + \frac{\nu_{\mathsf{fix}}-1}{2}(L+\Omega)}{2^{c-R\lambda}}.$$

  $\mathsf{col}_{\mathsf{cc}}$ is set in this case with probability the sum of above two bounds.

By the union bound,

$$(18a) \leq \frac{q(M-q)}{2^{H_\infty(\mathcal{D}_K)+\min\{c,\max\{b-\alpha,c\}-k\}-(R+q_\delta)\lambda}} + \frac{\binom{M-L-q}{2} + (M-L-q)(L+\Omega)}{2^{b-\lambda}} + \frac{\nu(L+\Omega) + \frac{\nu_{\mathsf{fix}}-1}{2}(L+\Omega)}{2^{c-R\lambda}}. \quad (19)$$

Probability of $\mathsf{col}_{\mathsf{cp}}$ of Eq. (18b). The event can be set in duplex and in primitive queries. Consider any duplex query i or any primitive query j, and assume that at the point of querying, the events col, mc, and key are still false. Note that the bad event consists of two parts, namely input collisions $T_i = X_j$ and output collisions $U_i = Y_j$. For both cases, we will make a distinction depending on the type of query of i.

- Event $T_i = X_j$.
  - *type*$_i$ = *init*. Note that $T_i = \mathsf{rot}_\alpha(\boldsymbol{K}[\delta_i] \parallel IV_i)$, where δ_i and IV_i can be deduced from *path*$_i$. For fixed primitive query, regardless of whether it is in forward or inverse direction, there are at most q_{IV} possible duplexing calls

with matching rightmost $b-k$ bits, i.e., for which $IV_i = \text{right}_{b-k}(X_j)$. In addition, the conditional min-entropy of $\boldsymbol{K}[\delta_i]$ is at least $H_\infty(\mathcal{D}_{\boldsymbol{K}}) - q_\delta\lambda$, and a collision $T_i = X_j$ happens with probability at most $1/2^{H_\infty(\mathcal{D}_{\boldsymbol{K}})-q_\delta\lambda}$. Summing over all queries, $\mathsf{col}_{\mathsf{cp}}$ is set in this case with probability at most $q_{IV}N/2^{H_\infty(\mathcal{D}_{\boldsymbol{K}})-q_\delta\lambda}$;

- *type*$_i$ = *full*. As query i is of the type *full*, its preceding duplexing call parent(i) generated $U_{\text{parent}(i)} = S_i$ uniformly at random from $\{0,1\}^b$. However, the distinguisher has learned $\text{left}_r(T_i)$, where $T_i = S_i \oplus [\mathit{flag}_i] \cdot (Z_i \| 0^{b-r}) \oplus P_i$, and it may have learned leakage on the other part. For fixed primitive query, regardless of whether it is in forward or inverse direction, by $\neg\mathsf{mc}_{\mathsf{in}}$ there are at most ν possible duplexing calls with matching leftmost r bits, i.e., for which $\text{left}_r(T_i) = \text{left}_r(X_j)$. In addition, the conditional min-entropy of $\text{right}_c(T_i)$ is at least $c - (R+1)\lambda$, and a collision $T_i = X_j$ happens with probability at most $1/2^{c-(R+1)\lambda}$. Summing over all queries, $\mathsf{col}_{\mathsf{cp}}$ is set in this case with probability at most $\nu N/2^{c-(R+1)\lambda}$;
- *type*$_i$ = *fix*. As query i is of the type *fix*, the earliest sibling of its preceding duplex call min(sibling(parent(i))) generated $T_{\text{min(sibling(parent}(i)))}$ uniformly at random from $\{0,1\}^b$, but in duplexing call i the distinguisher might have set the outer part to a certain value of its choice, and the distinguisher may have learned leakage on the other part. For fixed primitive query, regardless of whether it is in forward or inverse direction, there are at most ν_{fix} possible duplexing calls with matching leftmost r bits, i.e., for which $\text{left}_r(T_i) = \text{left}_r(X_j)$. In addition, the conditional min-entropy of $\text{right}_c(T_i)$ is at least $c-(R+1)\lambda$, and a collision $T_i = X_j$ happens with probability at most $1/2^{c-(R+1)\lambda}$. Summing over all queries, $\mathsf{col}_{\mathsf{cp}}$ is set in this case with probability at most $\nu_{\mathsf{fix}}N/2^{c-(R+1)\lambda}$;

– Event $U_i = Y_j$. The duplex call generates U_i uniformly at random from $\{0,1\}^b$. However, the distinguisher may have learned $\text{left}_r(U_i)$ in any subsequent call in child(i), and it may have learned leakage on the other part. For fixed primitive query, regardless of whether it is in forward or inverse direction, by $\neg\mathsf{mc}_{\mathsf{out}}$ there are at most ν possible duplexing calls with matching leftmost r bits, i.e., for which $\text{left}_r(U_i) = \text{left}_r(Y_j)$. In addition, the conditional min-entropy of $\text{right}_c(U_i)$ is at least $c - (R+1)\lambda$, and a collision $U_i = Y_j$ happens with probability at most $1/2^{c-(R+1)\lambda}$. Summing over all queries, $\mathsf{col}_{\mathsf{cp}}$ is set in this case with probability at most $\nu N/2^{c-(R+1)\lambda}$;

By the union bound,

$$(18\text{b}) \leq \frac{q_{IV}N}{2^{H_\infty(\mathcal{D}_{\boldsymbol{K}})-q_\delta\lambda}} + \frac{2\nu N}{2^{c-(R+1)\lambda}} + \frac{\nu_{\mathsf{fix}}N}{2^{c-(R+1)\lambda}}. \tag{20}$$

Probability of mc of Eq. (18c)**.** For $\mathsf{mc}_{\mathsf{in}}$, note that the state values T_i are randomly generated using a random function f and $M-L-q$ drawings are made (we only consider queries of the type *full*). For $\mathsf{mc}_{\mathsf{out}}$, the state values U_i are randomly generated using a random function f and M drawings are made. The event $\mathsf{mc}_{\mathsf{in}}$ is thus identical to a balls-and-bins experiment with $M-L-q$

balls that are uniformly randomly thrown into 2^r bins, and the event is set if there is a bin with more than ν balls. The event $\mathsf{mc}_{\mathsf{out}}$ is the same experiment but with M balls. By definition of $\nu := \nu^M_{r,c}$ (see Definition 1), any of the two happens with probability at most

$$(18c) \le \frac{2\nu}{2^c} . \tag{21}$$

Probability of key of Eq. (18d). This is a simple birthday bound collision event for u randomly drawn k-bit values, as $\boldsymbol{K} = (K[1], \ldots, K[u]) \xleftarrow{\mathcal{D}_K} (\{0,1\}^k)^u$. As the keys are mutually independent, we obtain:

$$(18d) \le \frac{\binom{u}{2}}{2^{H_\infty(\mathcal{D}_K)}} . \tag{22}$$

Conclusion. The proof is completed by plugging the individual bounds (19), (20), (21), and (22) into main inequality (18). □

6 Limiting Leakage of Keyed Duplex Construction

As it can be seen in Theorem 5.2, the advantage that an attacker can gain from the leakage rises by an increase of either the maximum number of duplexing calls for a single *path* R, or the maximum number of different initialization calls q_δ for a single key. Taha and Schaumont [38] and the developers of ISAP [16] presented ways to limit R and q_δ. Their usage of the keyed duplex, generalized to our description of the keyed duplex, is shown in Fig. 3.

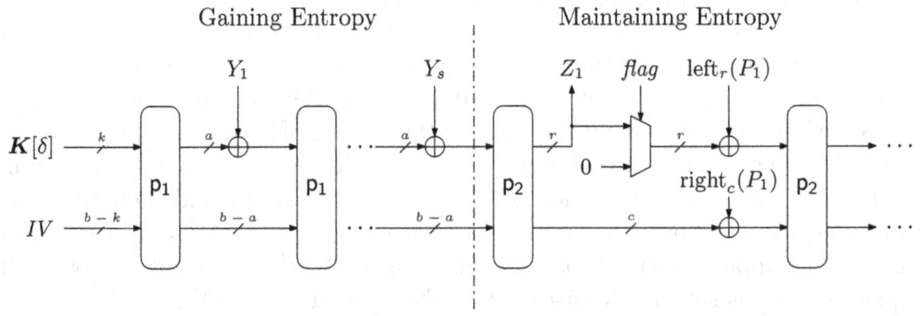

Fig. 3. The duplex as used by Taha and Schaumont [38] and ISAP [16].

The limit on q_δ is simply put by limiting the number of different IV's to a small number, typically to one or two different IV's. The role of the IV is then emulated by a value Y, which is typically a nonce in the case of encryption. Y is absorbed directly after the initialization in a-bit portions, where $a \le r$. Then, duplexing is performed the normal way, starting from the final state obtained after absorption of Y.

As becomes clear from Fig. 3, this approach splits the construction into two different keyed duplex constructions, KD_1 and KD_2, that use two different random permutations (p_1 and p_2) as well as different rate (a and r). The first part KD_1 is responsible for "gaining entropy", where the resulting output states are sufficiently random and mutually independent as long as no two values Y are the same. In the second part KD_2, entropy is "maintained" and used to perform cryptographic operations. In this separation, the last block Y_s is considered to be absorbed in KD_2.

The use of different permutations p_1 and p_2 may seem artificial, and to a certain extent it is: we will rely on mutual independence of the two permutations for easier composability. But also in practical scenarios different permutations for p_1 and p_2 would be used, yet, p_1 would often just be a permutation with a very small number of rounds and it could in a strict sense not be considered to be cryptographically strong.

In what follows, we will apply our general result of Theorem 1 to the construction of Fig. 3. For simplicity of reasoning, we will restrict our focus to the case of $a = 1$, where two different uniformly randomly generated keys are possible (so $u \leq 2$), and where two different IV's are possible (so $|\mathcal{IV}| \leq 2$). This matches the description of ISAP [16]. We will consider a distinguisher that makes Q evaluations, each consisting of a unique s-bit Y and an arbitrary amount of duplexing calls in the second part. The distinguisher makes N offline evaluations of p_1 and N offline evaluations of p_2. The remaining parameters of Sect. 5.1 will be bounded by the specific use case of the two duplexes in the construction of Fig. 3.

6.1 Gaining Entropy

The keyed duplex construction KD_1 matches the construction of Sect. 2 with capacity $c = b - 1$, rate $r = 1$, and key offset $\alpha = 0$. The number of initialization calls is at most $q \leq 4$, as there are at most two keys and two IV's. Likewise, $q_{IV}, q_\delta \leq 2$. The number of overwrites satisfies $\Omega = 0$. For the number of repeated paths, note that if a query Y is made, and Y' is an older query with the longest common prefix, then the new query will add one new repeated path, namely the one that ends at the absorption of the bit where Y and Y' differ. In other words, $L \leq Q$, and thus also $\nu_{\mathsf{fix}} \leq Q$. The total number of duplexing calls is at most $M \leq q + Q \cdot s$, noting that each query consists of an initialization and s duplexing calls. We adopt a non-tight $\nu_{1,b-1}^{M} \leq M$ for simplicity. Finally, as the absorbed bits Y_i can be considered as b-bit blocks P_i where $b - 1$ bits are zero-padded, we obtain that R, the maximum number of duplexing calls for a single non-empty *subpath*, is at most 2.

We obtain the following corollary from Theorem 1, where we have simplified the bound by gathering some fractions with leakage in the denominator. Here, we have also assumed that there is at least 1 bit of leakage, and at least 3 bits of input, and at least 2 queries.

Corollary 1. *Let $b, k, s, \lambda \in \mathbb{N}$, with $k \leq b$, $s \geq 3$, and $1 \leq \lambda \leq 2b$. Let $\mathsf{p}_1 \xleftarrow{\$} \text{perm}(b)$ be a random permutation, and $\boldsymbol{K} \xleftarrow{\$} (\{0,1\}^{k_1})^2$ a random array of keys. Let $\mathcal{L} = \{\mathsf{L} : \{0,1\}^b \times \{0,1\}^b \to \{0,1\}^\lambda\}$ be a class of leakage functions. For any distinguisher D making $Q \geq 2$ queries of length at most s bits, and making N primitive queries,*

$$\mathbf{Adv}_{\mathsf{KD}_1}^{\mathcal{L}\text{-naLR}}(\mathsf{D}) \leq \frac{4sQN + s^2Q^2}{2^{b-4\lambda}} + \frac{\binom{4+sQ+N}{2} + \binom{N}{2}}{2^b} + \frac{2N}{2^{k-2\lambda}} + \frac{1}{2^k}.$$

In addition, except with probability at most the same bound, all output states after absorption of the values Y have min-entropy at least $b - \lambda$.

6.2 Maintaining Entropy

For the keyed duplex construction KD_2, we consider Y_s to be not yet absorbed by KD_1, but instead, it forms the IV for KD_2. More detailed, KD_2 matches the construction of Sect. 2 with arbitrary c, r such that $c + r = b$, with $k = b - 1$, and key offset $\alpha = 1$ meaning that the key is in the bottom $b - 1$ bits. Note that, in fact, Y_s is XORed to the leftmost bit of the state, but for simplicity of reasoning, we simply consider it to *overwrite* it, making the key to KD_2 of size $b - 1$ bits. The number of initialization calls is Q, all of which may potentially be under different keys (so $u \leq Q$ and $q = Q$), one for every $Y \in \{0,1\}^s$ that goes through KD_1. The keys are not uniformly distributed, yet by Corollary 1 they are independent and all have min-entropy $b - 1 - \lambda$. The number of IV's is bounded by 2 (it corresponds to the single bit Y_s), so $q_\delta \leq 2$, but each IV may appear up to Q times, so $q_{IV} \leq q = Q$. The value R, the maximum number of duplexing calls for a single non-empty *subpath*, as it most the maximum number of repetitions of Y, so $R = 1$. There are no repeating paths, hence $L = 0$. As we make no a priori restriction on the choice of the *flag*'s, Ω is yet undetermined and $\nu_{\text{fix}} \leq \Omega$.

We obtain the following corollary from Theorem 1, where we have simplified the bound by gathering some fractions with leakage in the denominator. Here, we have also assumed that there is at least 1 bit of leakage.

Corollary 2. *Let $b, c, r, \lambda \in \mathbb{N}$, with $c+r = b$ and $1 \leq \lambda \leq 2b$. Let $\mathsf{p}_2 \xleftarrow{\$} \text{perm}(b)$ be a random permutation, and $\boldsymbol{K} \xleftarrow{\mathcal{D}_K} (\{0,1\}^b)^Q$ a random array of keys each with min-entropy at least $b - 1 - \lambda$. Let $\mathcal{L} = \{\mathsf{L} : \{0,1\}^b \times \{0,1\}^b \to \{0,1\}^\lambda\}$ be a class of leakage functions. For any distinguisher D making M construction queries, of which Q initialization calls, and N primitive queries,*

$$\mathbf{Adv}_{\mathsf{KD}_2}^{\mathcal{L}\text{-naLR}}(\mathsf{D}) \leq \frac{2\nu_{r,c}^M(N+1)}{2^{c-2\lambda}} + \frac{QN + 2M^2}{2^{b-4\lambda}} + \frac{\binom{M+N}{2} + \binom{N}{2}}{2^b} + \frac{(\nu_{r,c}^M + N + \Omega)\Omega}{2^{c-2\lambda}} + \frac{(M-Q)\Omega}{2^{b-\lambda}}.$$

The bound clearly reveals the impact of overwriting: if the distinguisher may make all its M duplexing calls with *flag* = *true*, the dominating term becomes $MN/2^{c-2\lambda}$.

7 Application to Encryption

We will put the results in practice, and show how Corollaries 1 and 2 guarantee leakage resilient nonce-based stream encryption in a modular manner. Let $b, c, r, k \in \mathbb{N}$ with $c + r = b$ and $k \leq b$. Consider the stream cipher encryption scheme $\mathcal{E}$ of Fig. 4, that gets as input a key K of k bits, a public nonce $\aleph$ of k bits, and an arbitrarily large plaintext P, and it outputs a ciphertext C. The ciphertext C is computed by adding $|P|$ bits of key stream generated by the duplex to P. The IV is a fixed constant.

7.1 Security of Stream Encryption

We consider security of $\mathcal{E}$ in the random permutation model. Let $\mathsf{p}_1, \mathsf{p}_2 \xleftarrow{\$} \mathrm{perm}(b)$ be two random permutations, and $K \xleftarrow{\$} \{0,1\}^k$. Let \$ be a function that for each $(\aleph, P)$ outputs a string of length $|P|$ bits (noting that a nonce should never be repeated). In the black-box security model, one would consider a distinguisher that has access to either $(\mathcal{E}[\mathsf{p}_1, \mathsf{p}_2]_K, \mathsf{p}_1^{\pm}, \mathsf{p}_2^{\pm})$ in the real world or $(\$, \mathsf{p}_1^{\pm}, \mathsf{p}_2^{\pm})$ in the ideal world, where again "±" stands for bi-directional query access:

$$\mathbf{Adv}_{\mathcal{E}}^{\text{bb-cpa}}(\mathsf{D}) = \Delta_{\mathsf{D}}\left(\mathcal{E}[\mathsf{p}_1, \mathsf{p}_2]_K, \mathsf{p}_1^{\pm}, \mathsf{p}_2^{\pm} \; ; \; \$, \mathsf{p}_1^{\pm}, \mathsf{p}_2^{\pm}\right).$$

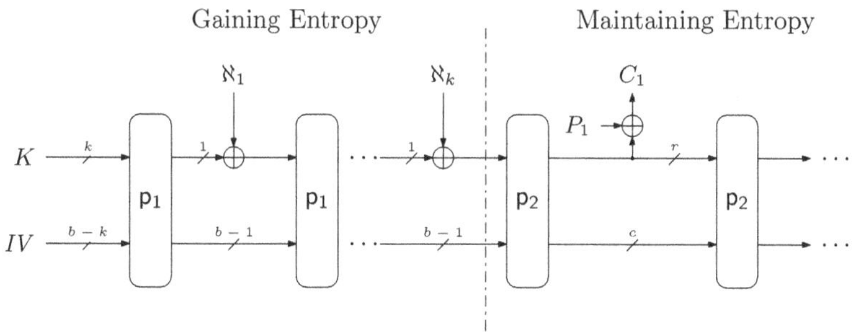

Fig. 4. Leakage-resilient stream encryption using the duplex.

In case of leakage resilience, we will stick to non-adaptive $\mathcal{L}$-resilience of Dodis and Pietrzak [17], as we did in Sect. 3.4. In the current case, however, we cannot simply consider *any* evaluation of the construction to leak, as this would allow for a trivial break of the scheme. Instead, we adopt the conventional approach of, e.g., [17,19,35,37,41], where the distinguisher has access to a leak-free version of the construction, which it has to distinguish from random, and a leaky version, which it may use to gather information. Formally, we obtain the following model, which follows Barwell et al. [2] with the difference that we consider security in the ideal permutation model. Let $\mathsf{p}_1, \mathsf{p}_2, K, \$$ be as above.

Let $\mathcal{L} = \{\mathsf{L} : \{0,1\}^b \times \{\mathit{true}, \mathit{false}\} \times \{0,1\}^b \times \{0,1\}^b \to \{0,1\}^\lambda\}$ be a class of leakage functions, and for any leakage function $\mathsf{L} \in \mathcal{L}$, define by $\mathcal{E}[\mathsf{p}_1, \mathsf{p}_2]_K^{\mathsf{L}}$ encryption such that for each call leaks $\mathsf{L}(S_{\text{prev}}, \mathit{flag}, P, S_{\text{next}})$, where S_{prev} is the state before the call and S_{next} the state after the call. In the leakage resilience security model, one considers a distinguisher that *in addition* to the oracles in the black-box model has access to $\mathcal{E}[\mathsf{p}_1, \mathsf{p}_2]_K^{\mathsf{L}}$:

$$\mathbf{Adv}_{\mathcal{E}}^{\mathcal{L}\text{-naLR-cpa}}(\mathsf{D}) = \max_{\mathsf{L}\in\mathcal{L}} \Delta_{\mathsf{D}}\left(\mathcal{E}[\mathsf{p}_1, \mathsf{p}_2]_K^{\mathsf{L}}, \mathcal{E}[\mathsf{p}_1, \mathsf{p}_2]_K, \mathsf{p}_1^{\pm}, \mathsf{p}_2^{\pm} ; \mathcal{E}[\mathsf{p}_1, \mathsf{p}_2]_K^{\mathsf{L}}, \$, \mathsf{p}_1^{\pm}, \mathsf{p}_2^{\pm}\right). \quad (23)$$

The distinguisher is not allowed to make an encryption query (to the leaky or leak-free oracle) under a repeated nonce.

7.2 Security of $\mathcal{E}$

We will demonstrate that the stream cipher encryption is leakage resilient, by relying on Corollaries 1 and 2.

Theorem 2. *Let $b, c, r, k, \lambda \in \mathbb{N}$, with $c + r = b$, $4 \le k \le b$, and $1 \le \lambda \le 2b$. Let $\mathsf{p}_1, \mathsf{p}_2 \xleftarrow{\$} \text{perm}(b)$ be two random permutations, and $K \xleftarrow{\$} \{0,1\}^k$ a random key. Let $\mathcal{L} = \{\mathsf{L} : \{0,1\}^b \times \{0,1\}^b \to \{0,1\}^\lambda\}$ be a class of leakage functions. For any distinguisher making $Q \ge 2$ queries with unique nonces, with a total amount of M plaintext blocks, N primitive queries to p_1 and N primitive queries to p_2,*

$$\begin{aligned}\mathbf{Adv}_{\mathcal{E}}^{\mathcal{L}\text{-naLR-cpa}}(\mathsf{D}) &\le \frac{(16k+2)QN + 4M^2 + 4k^2Q^2}{2^{b-4\lambda}} + \frac{4\binom{4+kQ+N}{2} + 2\binom{M+N}{2} + 6\binom{N}{2}}{2^b} \\ &+ \frac{4\nu_{r,c}^M(N+1)}{2^{c-2\lambda}} + \frac{8N}{2^{k-2\lambda}} + \frac{4}{2^k}.\end{aligned}$$

Proof. Let $\mathsf{KD}_1[\mathsf{p}_1]$ and $\mathsf{KD}_2[\mathsf{p}_2]$ be the two duplexes described in Sects. 6.1 and 6.2, with the difference that $\mathit{flag} = \mathit{false}$ and no data is absorbed for all calls to $\mathsf{KD}_2[\mathsf{p}_2]$. One can equivalently describe $\mathcal{E}[\mathsf{p}_1, \mathsf{p}_2]_K$ based on $\mathsf{KD}_1[\mathsf{p}_1]_K$ and $\mathsf{KD}_2[\mathsf{p}_2]_{\boldsymbol{K}^\star}$ as in Algorithm 4, where $\boldsymbol{K}^\star$ is defined as the output states of $\mathsf{KD}_1[\mathsf{p}_1]_K$ (we use the * to remind of this fact).

Let $\mathsf{L} \in \mathcal{L}$ be any leakage and D be any distinguisher. Our goal is to bound

$$\begin{aligned}&\Delta_{\mathsf{D}}\left(\mathcal{E}[\mathsf{p}_1, \mathsf{p}_2]_K^{\mathsf{L}}, \mathcal{E}[\mathsf{p}_1, \mathsf{p}_2]_K, \mathsf{p}_1^{\pm}, \mathsf{p}_2^{\pm} ; \mathcal{E}[\mathsf{p}_1, \mathsf{p}_2]_K^{\mathsf{L}}, \$, \mathsf{p}_1^{\pm}, \mathsf{p}_2^{\pm}\right) \\ &= \Delta_{\mathsf{D}}\left(\mathcal{E}[\mathsf{KD}_1[\mathsf{p}_1]_K^{\mathsf{L}}, \mathsf{KD}_2[\mathsf{p}_2]_{\boldsymbol{K}^\star}^{\mathsf{L}}], \mathcal{E}[\mathsf{KD}_1[\mathsf{p}_1]_K, \mathsf{KD}_2[\mathsf{p}_2]_{\boldsymbol{K}^\star}], \mathsf{p}_1^{\pm}, \mathsf{p}_2^{\pm} ;\right. \\ &\qquad\qquad \left.\mathcal{E}[\mathsf{KD}_1[\mathsf{p}_1]_K^{\mathsf{L}}, \mathsf{KD}_2[\mathsf{p}_2]_{\boldsymbol{K}^\star}^{\mathsf{L}}], \$, \mathsf{p}_1^{\pm}, \mathsf{p}_2^{\pm}\right). \quad (24)\end{aligned}$$

Let $\mathsf{AIXIF}_1[\mathsf{ro}_1]$ be an AIXIF with the same parameter setting as $\mathsf{KD}_1[\mathsf{p}_1]$, and similarly for $\mathsf{AIXIF}_2[\mathsf{ro}_2]$.

We recall from Corollary 1 that, except with probability at most the bound stated in that corollary, the final output states of $\mathsf{KD}_1[\mathsf{p}_1]$ have min-entropy at

Algorithm 4. Equivalent description of $\mathcal{E}[\mathsf{p}_1, \mathsf{p}_2]$

Interface: $\mathcal{E}[\mathsf{KD}_1[\mathsf{p}_1], \mathsf{KD}_2[\mathsf{p}_2]]$
Input: $(K, \aleph, P) \in \{0,1\}^k \times \{0,1\}^k \times \{0,1\}^*$
Output: $C \in \{0,1\}^{|P|}$

$\mathsf{KD}_1.\mathsf{init}(1, IV)$ ▷ only one key K, only one IV
$\aleph_1 \| \ldots \| \aleph_k \leftarrow \aleph$
for $i = 1, \ldots, k-1$ **do**
  $Z \leftarrow \mathsf{KD}_1.\mathsf{duplex}(\mathit{false}, \aleph_i \| 0^{n-1})$ ▷ discard output
$\boldsymbol{K}^\star[\mathsf{encode}(\aleph_1 \ldots \aleph_{k-1})] \leftarrow \mathrm{right}_{b-1}(S)$ ▷ store state of KD_1 in key array of KD_2
$\mathsf{KD}_2.\mathsf{init}(\mathsf{encode}(\aleph_1 \ldots \aleph_{k-1}), Z \oplus \aleph_k)$ ▷ KD_2 has key offset $\alpha = 1$
$Z \leftarrow \varnothing$
$\ell \leftarrow \lceil |P|/r \rceil$
for $i = 1, \ldots, \ell$ **do**
  $Z \leftarrow Z \,\|\, \mathsf{KD}_2.\mathsf{duplex}(\mathit{false}, 0^b)$
return $\mathrm{left}_{|P|}(P \oplus Z)$

least $b - \lambda$. This means that we can replace the generation of $\boldsymbol{K}^\star$ in Algorithm 4 by a dummy $\boldsymbol{K} \xleftarrow{\mathcal{D}_{\boldsymbol{K}}} (\{0,1\}^b)^Q$ consisting of keys with min-entropy $b - 1 - \lambda$ at negligible cost. Formally, denoting the resulting scheme by $\mathcal{E}^\star$, we have obtained:

$$\begin{aligned}
&\Delta_{\mathsf{D}}\big(\mathcal{E}[\mathsf{KD}_1[\mathsf{p}_1]_K^{\mathsf{L}}, \mathsf{KD}_2[\mathsf{p}_2]_{\boldsymbol{K}^\star}^{\mathsf{L}}]\,,\ \mathcal{E}[\mathsf{KD}_1[\mathsf{p}_1]_K, \mathsf{KD}_2[\mathsf{p}_2]_{\boldsymbol{K}^\star}]\,,\ \mathsf{p}_1^\pm\,,\ \mathsf{p}_2^\pm\ ; \\
&\qquad\qquad \mathcal{E}[\mathsf{KD}_1[\mathsf{p}_1]_K^{\mathsf{L}}, \mathsf{KD}_2[\mathsf{p}_2]_{\boldsymbol{K}^\star}^{\mathsf{L}}]\,,\ \$\,,\ \mathsf{p}_1^\pm\,,\ \mathsf{p}_2^\pm\big) \\
&\leq \Delta_{\mathsf{D}}\big(\mathcal{E}^\star[\mathsf{KD}_1[\mathsf{p}_1]_K^{\mathsf{L}}, \mathsf{KD}_2[\mathsf{p}_2]_{\boldsymbol{K}}^{\mathsf{L}}]\,,\ \mathcal{E}^\star[\mathsf{KD}_1[\mathsf{p}_1]_K, \mathsf{KD}_2[\mathsf{p}_2]_{\boldsymbol{K}}]\,,\ \mathsf{p}_1^\pm\,,\ \mathsf{p}_2^\pm\ ; \\
&\qquad\qquad \mathcal{E}^\star[\mathsf{KD}_1[\mathsf{p}_1]_K^{\mathsf{L}}, \mathsf{KD}_2[\mathsf{p}_2]_{\boldsymbol{K}}^{\mathsf{L}}]\,,\ \$\,,\ \mathsf{p}_1^\pm\,,\ \mathsf{p}_2^\pm\big) \\
&+ 2 \cdot \Delta_{\mathsf{D}'}\big(\mathsf{KD}_1[\mathsf{p}_1]_K^{\mathsf{L}}, \mathsf{p}_1^\pm\ ;\ \mathsf{AIXIF}_1[\mathsf{ro}_1]_K^{\mathsf{L}}, \mathsf{p}_1^\pm\big)\,,
\end{aligned} \tag{25}$$

where D' is some distinguisher making Q queries of length $k-1$ bits, and making N primitive queries. The factor 2 comes from the fact that we perform the change in both the real and ideal world.

For the remaining distance of (25), we can perform a hybrid argument:

$$\begin{aligned}
&\Delta_{\mathsf{D}}\big(\mathcal{E}^\star[\mathsf{KD}_1[\mathsf{p}_1]_K^{\mathsf{L}}, \mathsf{KD}_2[\mathsf{p}_2]_{\boldsymbol{K}}^{\mathsf{L}}]\,,\ \mathcal{E}^\star[\mathsf{KD}_1[\mathsf{p}_1]_K, \mathsf{KD}_2[\mathsf{p}_2]_{\boldsymbol{K}}]\,,\ \mathsf{p}_1^\pm\,,\ \mathsf{p}_2^\pm\ ; \\
&\qquad\qquad \mathcal{E}^\star[\mathsf{KD}_1[\mathsf{p}_1]_K^{\mathsf{L}}, \mathsf{KD}_2[\mathsf{p}_2]_{\boldsymbol{K}}^{\mathsf{L}}]\,,\ \$\,,\ \mathsf{p}_1^\pm\,,\ \mathsf{p}_2^\pm\big) \\
&\leq \Delta_{\mathsf{D}}\big(\mathcal{E}^\star[\mathsf{AIXIF}_1[\mathsf{ro}_1]_K^{\mathsf{L}}, \mathsf{KD}_2[\mathsf{p}_2]_{\boldsymbol{K}}^{\mathsf{L}}]\,,\ \mathcal{E}^\star[\mathsf{AIXIF}_1[\mathsf{ro}_1]_K, \mathsf{KD}_2[\mathsf{p}_2]_{\boldsymbol{K}}]\,,\ \mathsf{p}_2^\pm\ ; \\
&\qquad\qquad \mathcal{E}^\star[\mathsf{AIXIF}_1[\mathsf{ro}_1]_K^{\mathsf{L}}, \mathsf{KD}_2[\mathsf{p}_2]_{\boldsymbol{K}}^{\mathsf{L}}]\,,\ \$\,,\ \mathsf{p}_2^\pm\big) \\
&+ 2 \cdot \Delta_{\mathsf{D}'}\big(\mathsf{KD}_1[\mathsf{p}_1]_K^{\mathsf{L}}, \mathsf{p}_1^\pm\ ;\ \mathsf{AIXIF}_1[\mathsf{ro}_1]_K^{\mathsf{L}}, \mathsf{p}_1^\pm\big)\,,
\end{aligned} \tag{26}$$

where D' is some distinguisher making Q queries of length $k-1$ bits, and making N primitive queries. The distinguisher D' operates as follows: it generates a dummy key $\boldsymbol{K} \xleftarrow{\mathcal{D}_{\boldsymbol{K}}} (\{0,1\}^b)^Q$ and dummy permutation $\mathsf{p}_2 \xleftarrow{\$} \mathrm{perm}(b)$ on its own; for each query $(\aleph, P)$ that D makes, D' pads $\aleph_1 \| \ldots \| \aleph_k \leftarrow \aleph$ and evaluates its own oracle for $\aleph_1 \| \ldots \| \aleph_{k-1}$; it uses the output value Z, the last nonce bit $\aleph_k$, and its freshly generated $\boldsymbol{K}$ and p_2 to simulate the encryption of P, and it outputs the result. If D' is communicating with the real world $\mathsf{KD}_1[\mathsf{p}_1]_K$, this perfectly simulates $\mathcal{E}^\star[\mathsf{KD}_1[\mathsf{p}_1]_K, \mathsf{KD}_2[\mathsf{p}_2]_{\boldsymbol{K}}]$, and if it is communicating with the ideal world $\mathsf{AIXIF}_1[\mathsf{ro}_1]_K$, this perfectly simulates $\mathcal{E}^\star[\mathsf{AIXIF}_1[\mathsf{ro}_1]_K, \mathsf{KD}_2[\mathsf{p}_2]_{\boldsymbol{K}}]$.

The isolated distances on KD_1 in both (25) and (26) can be bounded directly by Corollary 1:

$$\Delta_{\mathsf{D}'}\left(\mathsf{KD}_1[\mathsf{p}_1]_K^{\mathsf{L}}, \mathsf{p}_1^{\pm} \;;\; \mathsf{AIXIF}_1[\mathsf{ro}_1]_K^{\mathsf{L}}, \mathsf{p}_1^{\pm}\right) \leq \frac{4kQN + k^2Q^2}{2^{b-4\lambda}} + \frac{\binom{4+kQ+N}{2} + \binom{N}{2}}{2^b} + \frac{2N}{2^{k-2\lambda}} + \frac{1}{2^k}. \tag{27}$$

We can proceed from the remaining distance in (26):

$$\begin{aligned}
&\Delta_{\mathsf{D}}\left(\mathcal{E}^\star[\mathsf{AIXIF}_1[\mathsf{ro}_1]_K^{\mathsf{L}}, \mathsf{KD}_2[\mathsf{p}_2]_{\boldsymbol{K}}^{\mathsf{L}}]\,,\, \mathcal{E}^\star[\mathsf{AIXIF}_1[\mathsf{ro}_1]_K, \mathsf{KD}_2[\mathsf{p}_2]_{\boldsymbol{K}}]\,,\, \mathsf{p}_2^{\pm} \;;\right.\\
&\qquad\qquad\qquad\qquad \left.\mathcal{E}^\star[\mathsf{AIXIF}_1[\mathsf{ro}_1]_K^{\mathsf{L}}, \mathsf{KD}_2[\mathsf{p}_2]_{\boldsymbol{K}}^{\mathsf{L}}]\,,\, \$\,,\, \mathsf{p}_2^{\pm}\right)\\
&\leq \Delta_{\mathsf{D}}\left(\mathcal{E}^\star[\mathsf{AIXIF}_1[\mathsf{ro}_1]_K^{\mathsf{L}}, \mathsf{AIXIF}_2[\mathsf{ro}_2]_{\boldsymbol{K}}^{\mathsf{L}}]\,,\, \mathcal{E}^\star[\mathsf{AIXIF}_1[\mathsf{ro}_1]_K, \mathsf{AIXIF}_2[\mathsf{ro}_2]_{\boldsymbol{K}}] \;;\right.\\
&\qquad\qquad\qquad\qquad \left.\mathcal{E}^\star[\mathsf{AIXIF}_1[\mathsf{ro}_1]_K^{\mathsf{L}}, \mathsf{AIXIF}_2[\mathsf{ro}_2]_{\boldsymbol{K}}^{\mathsf{L}}]\,,\, \$\right)\\
&+ 2 \cdot \Delta_{\mathsf{D}''}\left(\mathsf{KD}_2[\mathsf{p}_2]_{\boldsymbol{K}}^{\mathsf{L}}, \mathsf{p}_2^{\pm} \;;\; \mathsf{AIXIF}_2[\mathsf{ro}_2]_{\boldsymbol{K}}^{\mathsf{L}}, \mathsf{p}_2^{\pm}\right)\,,
\end{aligned} \tag{28}$$

where D'' is some distinguisher making M construction queries, of which Q initialization calls, and N primitive queries. Distinguisher D'' works symmetrically to distinguisher D' above, and its description is omitted.

The second distance of (28) can be bounded directly by Corollary 2 for $\Omega = 0$:

$$\Delta_{\mathsf{D}''}\left(\mathsf{KD}_2[\mathsf{p}_2]_{\boldsymbol{K}}^{\mathsf{L}}, \mathsf{p}_2^{\pm} \;;\; \mathsf{AIXIF}_2[\mathsf{ro}_2]_{\boldsymbol{K}}^{\mathsf{L}}, \mathsf{p}_2^{\pm}\right) \leq \frac{2\nu_{r,c}^M(N+1)}{2^{c-2\lambda}} + \frac{QN + 2M^2}{2^{b-4\lambda}} + \frac{\binom{M+N}{2} + \binom{N}{2}}{2^b}. \tag{29}$$

It remains to consider the first distance of (28). As the adversary may never query its oracle (leaky nor leak-free) for the same nonce, the leaky and leak-free oracles are mutually independent, and we obtain:

$$\begin{aligned}
&\Delta_{\mathsf{D}}\left(\mathcal{E}^\star[\mathsf{AIXIF}_1[\mathsf{ro}_1]_K^{\mathsf{L}}, \mathsf{AIXIF}_2[\mathsf{ro}_2]_{\boldsymbol{K}}^{\mathsf{L}}]\,,\, \mathcal{E}^\star[\mathsf{AIXIF}_1[\mathsf{ro}_1]_K, \mathsf{AIXIF}_2[\mathsf{ro}_2]_{\boldsymbol{K}}] \;;\right.\\
&\qquad\qquad\qquad\qquad \left.\mathcal{E}^\star[\mathsf{AIXIF}_1[\mathsf{ro}_1]_K^{\mathsf{L}}, \mathsf{AIXIF}_2[\mathsf{ro}_2]_{\boldsymbol{K}}^{\mathsf{L}}]\,,\, \$\right)\\
&= \Delta_{\mathsf{D}}\left(\mathcal{E}^\star[\mathsf{AIXIF}_1[\mathsf{ro}_1]_K, \mathsf{AIXIF}_2[\mathsf{ro}_2]_{\boldsymbol{K}}] \;;\; \$\right) = 0.
\end{aligned} \tag{30}$$

The proof is completed by combining (24)–(30). □

7.3 Towards Authentication

The stream cipher encryption construction considered in this section can be extended to cover authentication as well. One way of doing so is by absorbing the plaintext blocks P_i during streaming and outputting a tag at the end; another approach is by evaluating a MAC function (with a different key and IV, noting that Corollary 1 supports two keys and two IV's) after encryption has taken place. Note that in the first case, authenticated decryption would require to turn *flag* = *true* (see Sect. 2). In either case, one must take care of the fact that, upon decryption, nonces may get reused. In terms of the general picture of Fig. 3, this means that a same nonce can be "tried" for different blocks P_i, leading to repeating paths (hence $L > 0$) and to a higher leakage per evaluation of p_2 (hence $R > 1$). An authenticated encryption scheme that prohibits such "trial" of the same nonce with different inputs is ISAP [16].

Acknowledgments. We thank the ISAP team, the ESCADA team, and the authors of [23] for fruitful discussions. Christoph Dobraunig is supported by the Austrian Science Fund (FWF): J 4277-N38. Bart Mennink is supported by a postdoctoral fellowship from the Netherlands Organisation for Scientific Research (NWO) under Veni grant 016.Veni.173.017.

References

1. Andreeva, E., Daemen, J., Mennink, B., Van Assche, G.: Security of keyed sponge constructions using a modular proof approach. In: Leander, G. (ed.) FSE 2015. LNCS, vol. 9054, pp. 364–384. Springer, Heidelberg (2015). https://doi.org/10.1007/978-3-662-48116-5_18
2. Barwell, G., Martin, D.P., Oswald, E., Stam, M.: Authenticated encryption in the face of protocol and side channel leakage. In: Takagi, T., Peyrin, T. (eds.) ASIACRYPT 2017. LNCS, vol. 10624, pp. 693–723. Springer, Cham (2017). https://doi.org/10.1007/978-3-319-70694-8_24
3. Bellare, M., Rogaway, P.: Random oracles are practical: a paradigm for designing efficient protocols. In: Denning, D.E., Pyle, R., Ganesan, R., Sandhu, R.S., Ashby, V. (eds.) CCS 1993, pp. 62–73. ACM (1993)
4. Bellare, M., Rogaway, P.: The security of triple encryption and a framework for code-based game-playing proofs. In: Vaudenay, S. (ed.) EUROCRYPT 2006. LNCS, vol. 4004, pp. 409–426. Springer, Heidelberg (2006). https://doi.org/10.1007/11761679_25
5. Berti, F., Koeune, F., Pereira, O., Peters, T., Standaert, F.X.: Leakage-Resilient and Misuse-Resistant Authenticated Encryption. Cryptology ePrint Archive, Report 2016/996 (2016)
6. Berti, F., Pereira, O., Peters, T., Standaert, F.X.: On leakage-resilient authenticated encryption with decryption leakages. IACR Trans. Symmetric Cryptol. **2017**(3), 271–293 (2017)
7. Bertoni, G., Daemen, J., Peeters, M., Van Assche, G.: Sponge functions. In: Ecrypt Hash Workshop 2007, May 2007
8. Bertoni, G., Daemen, J., Peeters, M., Van Assche, G.: Duplexing the sponge: single-pass authenticated encryption and other applications. In: Miri, A., Vaudenay, S. (eds.) SAC 2011. LNCS, vol. 7118, pp. 320–337. Springer, Heidelberg (2012). https://doi.org/10.1007/978-3-642-28496-0_19
9. Bertoni, G., Daemen, J., Peeters, M., Van Assche, G.: The Keccak reference, January 2011
10. Bertoni, G., Daemen, J., Peeters, M., Van Assche, G.: On the security of the keyed sponge construction. In: Symmetric Key Encryption Workshop, February 2011

11. Bloem, R., Gross, H., Iusupov, R., Könighofer, B., Mangard, S., Winter, J.: Formal verification of masked hardware implementations in the presence of glitches. In: Nielsen, J.B., Rijmen, V. (eds.) EUROCRYPT 2018. LNCS, vol. 10821, pp. 321–353. Springer, Cham (2018). https://doi.org/10.1007/978-3-319-78375-8_11
12. Chang, D., Dworkin, M., Hong, S., Kelsey, J., Nandi, M.: A keyed sponge construction with pseudorandomness in the standard model. In: NIST SHA-3 Workshop, March 2012
13. Chari, S., Jutla, C.S., Rao, J.R., Rohatgi, P.: Towards sound approaches to counteract power-analysis attacks. In: Wiener [39], pp. 398–412
14. Clavier, C., Coron, J.-S., Dabbous, N.: Differential power analysis in the presence of hardware countermeasures. In: Koç, Ç.K., Paar, C. (eds.) CHES 2000. LNCS, vol. 1965, pp. 252–263. Springer, Heidelberg (2000). https://doi.org/10.1007/3-540-44499-8_20
15. Daemen, J., Mennink, B., Van Assche, G.: Full-state keyed duplex with built-in multi-user support. In: Takagi, T., Peyrin, T. (eds.) ASIACRYPT 2017. LNCS, vol. 10625, pp. 606–637. Springer, Cham (2017). https://doi.org/10.1007/978-3-319-70697-9_21
16. Dobraunig, C., Eichlseder, M., Mangard, S., Mendel, F., Unterluggauer, T.: ISAP - towards side-channel secure authenticated encryption. IACR Trans. Symmetric Cryptol. **2017**(1), 80–105 (2017)
17. Dodis, Y., Pietrzak, K.: Leakage-resilient pseudorandom functions and side-channel attacks on feistel networks. In: Rabin, T. (ed.) CRYPTO 2010. LNCS, vol. 6223, pp. 21–40. Springer, Heidelberg (2010). https://doi.org/10.1007/978-3-642-14623-7_2
18. Dziembowski, S., Pietrzak, K.: Leakage-resilient cryptography. In: FOCS 2008, pp. 293–302. IEEE Computer Society (2008)
19. Faust, S., Pietrzak, K., Schipper, J.: Practical leakage-resilient symmetric cryptography. In: Prouff, E., Schaumont, P. (eds.) CHES 2012. LNCS, vol. 7428, pp. 213–232. Springer, Heidelberg (2012). https://doi.org/10.1007/978-3-642-33027-8_13
20. FIPS 202: SHA-3 Standard: Permutation-Based Hash and Extendable-Output Functions, August 2015
21. Gaži, P., Pietrzak, K., Tessaro, S.: The exact PRF security of truncation: tight bounds for keyed sponges and truncated CBC. In: Gennaro, R., Robshaw, M. (eds.) CRYPTO 2015. LNCS, vol. 9215, pp. 368–387. Springer, Heidelberg (2015). https://doi.org/10.1007/978-3-662-47989-6_18
22. Goubin, L., Patarin, J.: DES and differential power analysis the "duplication" method. In: Koç, Ç.K., Paar, C. (eds.) CHES 1999. LNCS, vol. 1717, pp. 158–172. Springer, Heidelberg (1999). https://doi.org/10.1007/3-540-48059-5_15
23. Guo, C., Pereira, O., Peters, T., Standaert, F.X.: Towards Lightweight Side-Channel Security and the Leakage-Resilience of the Duplex Sponge. Cryptology ePrint Archive, Report 2019/193 (2019)
24. Håstad, J., Impagliazzo, R., Levin, L.A., Luby, M.: A pseudorandom generator from any one-way function. SIAM J. Comput. **28**(4), 1364–1396 (1999)
25. Hsiao, C.-Y., Lu, C.-J., Reyzin, L.: Conditional computational entropy, or toward separating pseudoentropy from compressibility. In: Naor, M. (ed.) EUROCRYPT 2007. LNCS, vol. 4515, pp. 169–186. Springer, Heidelberg (2007). https://doi.org/10.1007/978-3-540-72540-4_10
26. Jovanovic, P., Luykx, A., Mennink, B.: Beyond $2^{c/2}$ security in sponge-based authenticated encryption modes. In: Sarkar, P., Iwata, T. (eds.) ASIACRYPT

2014. LNCS, vol. 8873, pp. 85–104. Springer, Heidelberg (2014). https://doi.org/10.1007/978-3-662-45611-8_5
27. Kocher, P.C.: Timing attacks on implementations of Diffie-Hellman, RSA, DSS, and other systems. In: Koblitz, N. (ed.) CRYPTO 1996. LNCS, vol. 1109, pp. 104–113. Springer, Heidelberg (1996). https://doi.org/10.1007/3-540-68697-5_9
28. Kocher, P.C., Jaffe, J., Jun, B.: Differential power analysis. In: Wiener [39], pp. 388–397
29. Medwed, M., Standaert, F.-X., Großschädl, J., Regazzoni, F.: Fresh re-keying: security against side-channel and fault attacks for low-cost devices. In: Bernstein, D.J., Lange, T. (eds.) AFRICACRYPT 2010. LNCS, vol. 6055, pp. 279–296. Springer, Heidelberg (2010). https://doi.org/10.1007/978-3-642-12678-9_17
30. Mennink, B., Reyhanitabar, R., Vizár, D.: Security of full-state keyed sponge and duplex: applications to authenticated encryption. In: Iwata, T., Cheon, J.H. (eds.) ASIACRYPT 2015. LNCS, vol. 9453, pp. 465–489. Springer, Heidelberg (2015). https://doi.org/10.1007/978-3-662-48800-3_19
31. Naito, Y., Yasuda, K.: New bounds for keyed sponges with extendable output: independence between capacity and message length. In: Peyrin, T. (ed.) FSE 2016. LNCS, vol. 9783, pp. 3–22. Springer, Heidelberg (2016). https://doi.org/10.1007/978-3-662-52993-5_1
32. Nikova, S., Rechberger, C., Rijmen, V.: Threshold implementations against side-channel attacks and glitches. In: Ning, P., Qing, S., Li, N. (eds.) ICICS 2006. LNCS, vol. 4307, pp. 529–545. Springer, Heidelberg (2006). https://doi.org/10.1007/11935308_38
33. Nikova, S., Rijmen, V., Schläffer, M.: Secure hardware implementation of nonlinear functions in the presence of glitches. J. Cryptol. **24**(2), 292–321 (2011)
34. Pereira, O., Standaert, F.X., Vivek, S.: Leakage-resilient authentication and encryption from symmetric cryptographic primitives. In: Ray, I., Li, N., Kruegel, C. (eds.) CCS 2015, pp. 96–108. ACM (2015)
35. Pietrzak, K.: A leakage-resilient mode of operation. In: Joux, A. (ed.) EUROCRYPT 2009. LNCS, vol. 5479, pp. 462–482. Springer, Heidelberg (2009). https://doi.org/10.1007/978-3-642-01001-9_27
36. Standaert, F.-X., Pereira, O., Yu, Y.: Leakage-resilient symmetric cryptography under empirically verifiable assumptions. In: Canetti, R., Garay, J.A. (eds.) CRYPTO 2013. LNCS, vol. 8042, pp. 335–352. Springer, Heidelberg (2013). https://doi.org/10.1007/978-3-642-40041-4_19
37. Standaert, F.X., Pereira, O., Yu, Y., Quisquater, J.J., Yung, M., Oswald, E.: Leakage resilient cryptography in practice. In: Sadeghi, A.R., Naccache, D. (eds.) Towards Hardware-Intrinsic Security - Foundations and Practice. ISC, pp. 99–134. Springer, Heidelberg (2010). https://doi.org/10.1007/978-3-642-14452-3_5
38. Taha, M.M.I., Schaumont, P.: Side-channel countermeasure for SHA-3 at almost-zero area overhead. In: HOST 2014, pp. 93–96. IEEE Computer Society (2014)
39. Wiener, M.J. (ed.): CRYPTO 1999. LNCS, vol. 1666. Springer, Heidelberg (1999). https://doi.org/10.1007/3-540-48405-1
40. Yu, Y., Standaert, F.-X.: Practical leakage-resilient pseudorandom objects with minimum public randomness. In: Dawson, E. (ed.) CT-RSA 2013. LNCS, vol. 7779, pp. 223–238. Springer, Heidelberg (2013). https://doi.org/10.1007/978-3-642-36095-4_15
41. Yu, Y., Standaert, F.X., Pereira, O., Yung, M.: Practical leakage-resilient pseudorandom generators. In: Al-Shaer, E., Keromytis, A.D., Shmatikov, V. (eds.) CCS 2010, pp. 141–151. ACM (2010)

A Critical Analysis of ISO 17825 ('Testing Methods for the Mitigation of Non-invasive Attack Classes Against Cryptographic Modules')

Carolyn Whitnall[1] and Elisabeth Oswald[1,2](✉)

[1] University of Bristol, Bristol, UK
{carolyn.whitnall,elisabeth.oswald}@bristol.ac.uk
[2] University of Klagenfurt, Klagenfurt, Austria

Abstract. The ISO standardisation of 'Testing methods for the mitigation of non-invasive attack classes against cryptographic modules' (ISO/IEC 17825:2016) specifies the use of the Test Vector Leakage Assessment (TVLA) framework as the sole measure to assess whether or not an implementation of (symmetric) cryptography is vulnerable to differential side-channel attacks. It is the only publicly available standard of this kind, and the first side-channel assessment regime to exclusively rely on a TVLA instantiation.

TVLA essentially specifies statistical leakage detection tests with the aim of removing the burden of having to test against an ever increasing number of attack vectors. It offers the tantalising prospect of 'conformance testing': if a device passes TVLA, then, one is led to hope, the device would be secure against all (first-order) differential side-channel attacks.

In this paper we provide a statistical assessment of the specific instantiation of TVLA in this standard. This task leads us to inquire whether (or not) it is possible to assess the side-channel security of a device via leakage detection (TVLA) only. We find a number of grave issues in the standard and its adaptation of the original TVLA guidelines. We propose some innovations on existing methodologies and finish by giving recommendations for best practice and the responsible reporting of outcomes.

Keywords: Side-channel analysis · Leakage detection · Security certification · Statistical power analysis

1 Introduction

In the late 1990s, Kocher et al. [23] raised awareness of the fact that 'provably secure' cryptography is potentially vulnerable to attacks exploiting auxiliary information not accounted for in traditional security models (e.g. power consumption or other measureable characteristics of devices in operation). Since then, designers and certification bodies have been increasingly concerned with ensuring and evaluating the physical security of cryptographic implementations.

S. D. Galbraith and S. Moriai (Eds.): ASIACRYPT 2019, LNCS 11923, pp. 256–284, 2019.
https://doi.org/10.1007/978-3-030-34618-8_9

Adapting theoretical security models to incorporate the full range of realistic physical threats is difficult (likely infeasible) [36], so that it is typically considered necessary to subject actual products to experimental testing in a laboratory setting.

The approach taken by testing regimes within the context of Common Criteria (CC) or EMVCo evaluations is to test 'all' of the most effective known attacks developed in the side-channel literature to date (the JHAS group decides on the strategies to be considered). But the growing number of such attacks and the difficulty of determining *a priori* which are the most pertinent to a particular scenario (see e.g. [9,35]) makes this unsustainable. An alternative option could be to rely on *leakage detection* testing along the lines of the Test Vector Leakage Assessment (TVLA) framework first proposed by Cryptography Research, Inc. (now Rambus) [17].

Rather than aim at the successful extraction of sensitive information from side-channel measurements, as an attack-based evaluation would do, leakage detection simply seeks evidence (or convincing lack of evidence) of sensitive data dependencies in the measured traces. TVLA does this via a suite of Welch's t-tests targeting mean differences in carefully chosen partitions of trace measurements. For example, the fixed-versus-random test looks for a statistically significant difference between a trace set associated with a fixed plaintext input and another trace set associated with randomly varying inputs. Alternatively, the leakage associated with a specific intermediate value (such as an S-box output) can be targeted by comparing a trace set that has been partitioned into two according to the value of that bit or byte. Both the 'specific' and the 'non-specific' type tests are univariate and are performed on each point in a trace set separately in order to draw conclusions about the overall vulnerability of the implementation. So-called 'higher order' tests exist to target leakage, more complex in its functional form, that does not present via differences in the mean but can be found in higher order (joint) statistical moments; these typically entail pre-processing the traces before performing the same univariate point-wise test procedures [32].

TVLA is the most well-established and widely-adopted suite of leakage detection tests despite the lack of a comprehensive analysis of its performance. Significantly, the ISO standard ISO/IEC 17825:2016 ('Testing methods for the mitigation of non-invasive attack classes against cryptographic modules'; we will refer to it as ISO 17825) [20] specifies TVLA (in its full first-order form, as we describe in Sect. 2) as the sole required measure for testing against differential side-channel attacks on symmetric key cryptosystems[1]. ISO 17825 ties in with ISO 19790, which is the intended replacement/revision of FIPS 140-2[2] (the

[1] Other detection methodologies exist outside of the TVLA framework (including approaches based on mutual information [6,7,25], correlation [13] and the F-statistic [3] – all variants on statistical hypothesis tests, with differing degrees of formalism). These other tests and 'higher order' tests are not part of ISO 17825 and therefore outside the scope of this submission.

[2] https://csrc.nist.gov/Projects/cryptographic-module-validation-program/Standards.

main evaluation scheme in the US). ISO 19790 specifies the much broader goals of a security evaluation, and ISO 17825 focuses on susceptibility to non-invasive attacks for devices aiming for security level 3 or 4.

Within the cryptographic community, publicly available standards are a key mechanism to ensure the widespread adoption of good practice, and we would argue that the same should hold in the area of security evaluations. Yet this is sadly not the case: high-security evaluations according to (e.g.) CC, or EMVCo, do not release the list of threats that JHAS has agreed are relevant for evaluation. Thus ISO 17825 is the only publicly available standard that covers side channel evaluations. As such it is positioned to become *the* standard methodology for side-channel testing outside the existing smart card market (which is dominated by CC and EMVCo). Much is therefore at stake from a commercial as well as an academic perspective when we come to consider how good ISO 17825/TVLA is at the task for which it was designed (conformance testing in the context of side-channel leakage).

We begin this submission by considering the goal(s) of leakage detection in the context of external evaluations generally, followed by some background on TVLA in particular and the relevant ISO standards (see Sect. 2). We introduce statistical power analysis[3] in Sect. 3 and, in Sects. 4 and 5 use these tools to examine the false positive and false negative error rates implied by the standard recommendations, with appropriate consideration for the fact that multiple tests are performed as part of a single evaluation. We also introduce the notion of coverage, inspired by that of code coverage in software testing, and use this to comment on how thoroughly the recommendations take account of realistic threats. We explore some alternative approaches in Sect. 6 and conclude with some recommendations for best practice in Sect. 7. Our analysis is enabled by adapting a novel method for complex statistical power simulations by Porter [27], as well as deriving real-world effect sizes from some actual devices. Interested readers can find more details about statistical power analysis for leakage detection, including in relation to the subtly different goals of *in-house* evaluation, in our companion paper *A Cautionary Note Regarding the Usage of Leakage Detection Tests in Security Evaluation* [42].

2 Background: Leakage Detection in a Security Evaluation

Leakage detection is often carried out as part of an exercise to evaluate the security of a cryptographic device. It might be performed by an evaluation laboratory in order to provide security certification when the device goes on sale, or it might be an in-house effort during the development process to highlight and fix potential problems prior to formal external evaluation. We address both scenarios in [42], while here we focus on the context of external evaluations, where there are two potential end results aimed at by a detection test:

[3] 'Power,' as we will explain later in the paper, is a statistical concept and should not be confused with the 'P' of DPA which refers to power consumption.

Certifying vulnerability: Find a leak in **at least one** trace point. In such a case it is important to control the number of false positives (that is, concluding there is a leak where there isn't one).

Certifying security: Find **no leaks** having tested thoroughly. Here false negatives (failure to find leaks that are really there) become a concern.

As we will see, the statistical methods used for leakage detection cannot 'prove' that there is no effect, they can at best conclude that there is evidence of a leak or that there is no evidence of a leak. Hence it is especially important to design tests with **'statistical power'** in mind – that is, to make sure the sample size is large enough to detect a present effect of a certain size with reasonable probability (see Sect. 3). Then, in the event that no leak is discovered, these constructed features of the test form the basis of a reasoned interpretation. A further, considerable challenge implicit to this goal is the necessity to be convincingly exhaustive in the range of tests performed – that is, to target 'all possible' intermediates and all relevant higher-order combinations of points. (This suggests analogues with the idea of *coverage* in code testing, which we discuss in Sect. 5.1).

2.1 TVLA and its Adoption Within Standards

The TVLA framework was presented by researchers from Cryptography Research Inc. (now Rambus) at the 2011 Non-Invasive Attack Testing workshop organised by NIST [17]. It describes a series of statistical hypothesis tests to reject (or not) the null of 'no sensitive information leakage' against various alternative hypotheses designed to capture a large range of possible leakage forms and sources. In summary form (see the paper for full details) the procedure is follows:

- An acquisition of size n is taken as the device operates with a fixed key on a fixed plaintext chosen to induce certain values in one of the middle rounds. It is then divided into two disjoint sets FIXED1 and FIXED2, each of size $n/2$.
- An acquisition of size $2n$ is taken as the device operates with the same fixed key on random inputs. It is then divided into two disjoint sets RANDOM1 and RANDOM2, each of size n.
- Welch's t-tests [41] are performed, with an (implied, for large samples) significance level of $\alpha \approx 0.00001$, comparing the population means of:
  - The fixed-plaintext traces FIXED1 with the random-plaintext traces RANDOM1.
  - The RANDOM1 traces such that a target intermediate takes a certain value, versus the remainder of the RANDOM1 traces, for the following targets: each bit of the XOR between round R input and output; each bit of the Rth round SubBytes output; each bit of the round R output; each byte of the round R output (repeated for all possible values in a one-versus-all manner).

- The above is repeated identically for trace sets FIXED2, RANDOM2. The module is considered to fail the overall test if any *pair* of repeated individual tests both conclude that there is a statistically significant difference (in the same direction) at any trace index.

The TVLA specification provides no discussion of the statistical power of this procedure, nor does it explicitly discuss the chosen parameters, nor whether the multiple comparisons problem was accounted for in the design.

2.2 ISO Standards for Physical Security

ISO/IEC 19790 [21] specifies four increasingly rigorous security levels and the criteria for achieving them. Levels 3 and 4 require (among other things) that the modules mitigate successfully (to a specified degree) against non-invasive physical attacks including simple power analysis (SPA) and differential power analysis (DPA).

ISO/IEC 17825:2016 [20] specifies the tests that the modules must undergo and the different parameters (sample size, laboratory time, pass/fail criteria) for running the tests according to each security level.

Under this latter standard, the DPA resilience of symmetric key cryptosystems is essentially determined by performing the full suite of first-order TVLA tests as detailed above, with the following main differences:

- Fixed plaintexts are required to have the same special characteristics as the particular values specified by Goodwill et al., but the method of choosing suitable candidates is left up to the analyst.
- The specified risk of false positives (a.k.a. the significance level, typically denoted α) is 0.05, which is considerably higher than the level of 0.00001 implied by Goodwill et al.'s t-value threshold of 4.5.

Security levels 3 and 4 are separated by the resources available to perform the analysis, and the degree of data pre-processing, as per Table 1. These criteria seem to be directly inherited from FIPS 140-2, which originally was based on attacks (like CC and EMVCo evaluations).

Table 1. Configuration of the tests to attain security levels 3 and 4. (Note that the overall acquisition time includes tests not related to DPA vulnerability).

	Level 3	Level 4
Maximum acquisition time per test (hours)	6	24
Maximum overall acquisition time (hours)	72	288
Sample size	10,000	100,000
Synchronisation signal available	Yes	Yes
Noise reduction	Averaging (over 10)	Spectrum analysis
Static alignment attempted	No	Yes
Dynamic alignment attempted	No	Yes?

The standard leaves ambiguous whether the sample size specifications apply per acquisition or for both fixed and random trace sets combined; similarly whether they are intended per repetition or for both the first and the confirmatory analysis combined. We have assumed fixed and random are counted separately and the two repetitions are counted jointly, so that there are 10,000 or 100,000 each of the fixed input and random input traces, split across the two 'independent' evaluations.

The remaining questions of interest are then how well TVLA, when applied as specified in ISO 17825, succeeds in the goals of certifying vulnerability and/or certifying security – and whether or not (and how) the recommendations could be adapted to do so more effectively. To address these questions we first introduce statistical power analysis, which will give us the tools to analyse (and potentially improve) the theoretical properties of the tests.

3 Statistical Power Analysis for Leakage Detection Tests

It is *impossible to eliminate* errors in statistical hypothesis testing; the aim is rather to understand and minimise them. The decision to reject a null hypothesis when it is in fact true is called a Type I error, a.k.a. 'false positive' (e.g. finding leakage when in fact there is none). The acceptable rate of false positives is explicitly set by the analyst at a significance level α. A Type II error, a.k.a. 'false negative' is a failure to reject the null when it is in fact false (e.g. failing to find leakage when in reality there is some). The Type II error rate of an hypothesis test is denoted β and the **power** of the test is $1 - \beta$, that is, the probability of correctly rejecting a false null in favour of a true alternative. The two errors can be traded-off against one another, and mitigated (but not eliminated) by:

- Increasing the **sample size** N, intuitively resulting in more evidence from which to draw a conclusion.
- Increasing the minimum **effect size** of interest ζ, which in our case implies increasing the magnitude of leakage that one would be willing to dismiss as 'negligible'.
- Choosing a different statistical test that is more efficient with respect to the sample size.

For a given test (i.e. leaving aside the latter option) the techniques of **statistical power analysis** are concerned with the mutually determined relationship between α, $1 - \beta$, ζ and N. For the simple case of a t-test with equal sample sizes and population variances σ_1 and σ_2[4], the following formula can be derived (see Appendix A):

$$N = 2 \cdot \frac{(z_{\alpha/2} + z_{\beta})^2 \cdot ({\sigma_1}^2 + {\sigma_2}^2)}{\zeta^2} \tag{1}$$

[4] We consider these conditions to approximately hold in the case of most of the ISO standard tests, where the partitions are determined by uniformly distributed intermediates.

where $\zeta = \mu_1 - \mu_2$ is the true difference in means between the two populations (this relationship can be found in any standard statistics textbook). Note that Eq. (1) can be straightforwardly rearranged to alternatively compute any of the significance level, effect size or power in terms of the other three quantities.

3.1 Configuring Tests via an *A Priori* Power Analysis

Ideally, a power analysis is performed before a leakage evaluation takes place as an aid to experimental design; this is known as *a priori* power analysis and can help to ensure (e.g.) the collection of a large enough sample to detect data-dependencies of the expected magnitude with the desired probability of success [25]. Power analysis can be performed *after* data collection in order to make statements about the power to detect a particular effect size of interest, or the minimum effect size that the test would be able to detect with a certain power. This can be useful when it comes to responsibly interpreting the non-rejection of a null hypothesis. However, it is crucial that the effect sizes are chosen independently of the test, based on external criteria, as it has been shown that attempts to estimate 'true' effect sizes from the test data produce circular reasoning. In fact, there is a direct correspondence between the *p*-value and the power to detect the observed effect, so that 'post hoc power analysis' merely re-expresses the information contained already in the test outcome [18].

Also needed in order to perform statistical power analysis are the population standard deviations of the partitioned samples, which may or may not be the same. These are usually assumed to have been obtained from previous experiments and/or already-published results, which can be especially tricky when approaching a new target for evaluation.

3.2 Effect Size

This requirement for information *about* the data sample which cannot be estimated *from* the data sample is the main obstacle to statistical power analysis. The choice of effect sizes for the computations can be guided by previous experiments (e.g., in our case, leakage evaluation on a similar device with a similar measurement set up) or (ideally) by some rationale about the practical implications of a given magnitude (e.g. in terms of loss of security). Note that we always eventually need some rationale of this latter type: what is ultimately of interest is not just whether we are able to detect effects but whether the effects that we detect are of practical concern. With a large enough sample we will always be able to find 'arbitrarily small' differences; the question then remains, at what threshold do they *become* 'arbitrary'?

It is convenient (and bypasses some of the reliance on prior information) to express effect sizes in standardised form. Cohen's d is defined as the mean difference divided by the pooled standard deviation of two samples of (univariate) random variables A and B:

$$d = \frac{\overline{a} - \overline{b}}{\sqrt{\frac{(n_A-1)s_A^2+(n_B-1)s_B^2}{n_A+n_B-2}}}$$

where $\overline{a}$, $\overline{b}$ are the sample means, s_A^2, s_B^2 are the sample variances and n_A, n_B are the sample sizes. Notice that this is essentially a measure of signal-to-noise ratio (SNR), closely related to (and therefore tracking) the various notions that already appear in the side-channel literature. The formula for the sample size required for the t-test can be expressed in terms of the standardised effect size as follows:

$$N = 4 \cdot \frac{(z_{\alpha/2} + z_{\beta})^2}{d^2} \tag{2}$$

Cohen [8] proposed that effects of 0.2 or less should be considered 'small', effects around 0.5 are 'medium', and effects of 0.8 or more are 'large'. Sawilowsky [30] expanded the list to incorporate 'very small' effects of 0.01 or less, and 'very large' and 'huge' effects of over 1.2 or 2.0 respectively. The relative cheapness of sampling leakage traces (and subsequent large sample sizes) compared with studies in other fields (such as medicine, psychology and econometrics), as well as the high security stakes of side-channel analysis, make 'very small' effects of more interest than they typically are in other statistical applications.

Focusing on standardised effects helps to put the analysis on a like-for-like footing for all implementations, but it doesn't remove the need for specific knowledge about a device in order for meaningful interpretation.

3.3 The Impact of Multiple Testing

Statistical hypothesis testing is generally introduced under the implicit assumption that a single null/alternative pair is up for consideration. Unfortunately, controlling error rates becomes even more complicated when multiple tests are performed as part of the same experiment. Without appropriate modifications, test conclusions are no longer formally supported. This is because, if each test has (by design) a probability α of falsely rejecting the null hypothesis, then the probability of rejecting *at least one* true null hypothesis across all m tests (that is, the overall false positive rate as opposed to the per-test rate) might be as high as $\alpha_{\text{overall}} = 1 - (1 - \alpha_{\text{per-test}})^m$ if those tests are independent. (Otherwise, the rate will be lower but will depend on the form of the dependencies).

Multiplicity Corrections. In the statistics literature there are two main approaches to correcting for multiple tests: controlling the *family-wise error rate* (FWER) and controlling the *false discovery rate* (FDR). Both of these were discussed and evaluated in the context of leakage detection by Mather *et al.* [25].

FWER-based methods work by adjusting the per-test significance criteria in such a way that the *overall* rate of Type I errors is no greater than the desired α level. For example:

- Bonferroni correction [12]: per-test significance level obtained by dividing the desired overall significance level by the number of tests m, i.e. $\alpha_{\text{per-test}} = \frac{\alpha}{m}$. Controls the FWER for the 'worst case' scenario that the tests are independent, and is conservative otherwise.

- Šidák correction [40]: explicitly *assumes* independence, and that all null hypotheses are false, and sets $\alpha_{\text{per-test}} = 1 - (1 - \alpha)^{\frac{1}{m}}$. These assumptions potentially gain power but are unlikely to suit a leakage evaluation setting.
- Holm adjustment [19]: a 'step up' procedure; tests are ordered according to p-value (smallest to largest), and criteria set such that $\alpha_i = \frac{\alpha}{m-i+1}$ for the ith test.

It should be clear that any such downward adjustment to the per-test Type I error rates (i.e. in order to prevent concluding that there is a leak when there isn't) inevitably increases the rate of Type II errors (the probability of missing a leak which is present). Erring on the "safe side" with respect to the former criterion may not be at all "safe" in terms of the cost to the latter. The relative undesirability of the two error types depends heavily on the application and must be carefully considered.

FDR-based methods take a slightly different approach which is more relaxed with respect to Type I errors and subsequently less prone to Type II errors. Rather than minimise the probability of *any* false positives they instead seek to bound the proportion of total 'discoveries' (i.e. rejected nulls) which are false positives. The main FDR-controlling method, and the one that we will consider in the following, is the Benjamini–Hochberg procedure, which (like the Holm correction) operates in a 'step up' manner as follows:

1. For the ordered (small to large) p-values $p_{(1)}, \ldots, p_{(m)}$, find the largest k such that $p_{(k)} \leq \frac{k}{m}\alpha$.
2. Reject the null hypothesis for all tests $i = 1, \ldots, k$.

A recent proposal in the side-channel literature [11] takes an alternative third way, using methods developed for the purpose of performing a meta-analysis based on multiple independent studies: the decision to collectively reject or not reject a set of null hypotheses is based on the *distribution* of the p-values. (We do not analyse this method in the following due to its heavy reliance on the independence assumption).

In addition to the inevitable loss of power associated with all of the above adjustments, a substantial obstacle to their use is the difficulty of analysing (and controlling) the power, which is essential if we want to draw meaningful and comparable conclusions from test outcomes. In cases where a single per-test significance level $\alpha_{\text{per-test}}$ is derived (e.g. Bonferroni and Šidák), this can simply be substituted into the power analysis formulae to gain the per-test power. However, consensus is lacking when it comes to performing equivalent computations for FDR-controlling procedures (compare, e.g., [4,14,24,28,39]; in Sect. 6 we adopt an approach by Porter that operates by simulating test statistics but is constrained to fully specified test scenarios [27]). Moreover, depending on the over-arching goal of the analysis, per-test power may not even be the relevant quantity to consider, as we next discuss.

Different Notions of Power. Just as multiple tests raise the notion of an 'overall' Type I error rate which is not equal to the per-test error rate, so it is worth giving thought to the 'overall' Type II error and what precisely we mean by that. We have seen above that multiplicity corrections reduce the per-test power – the probability of detecting a true effect wherever one exists. Porter [27] describes this as 'individual' power, and contrasts it with the notion of 'r-minimal' power[5] – the probability of detecting at least r true effects. We propose that the 1-minimal power is the relevant notion in the context of certifying vulnerability/security, since a single detected leak is sufficient to fail a device.

The probability of detecting *all* true effects (as might be the goal of an in-house development-time evaluation) is known as the 'complete power'. The r-minimal power is naturally greater than or equal to this quantity. In particular, the 1-minimal power can actually be *higher* in a multiple testing scenario than in a single test – as long as the true number of false positives is greater than 1, each such test represents an additional opportunity to find an effect.

4 ISO 17825 for Certifying Vulnerability

In this section we examine how reliable ISO 17825 is for certifying vulnerability – demonstrating a sensitive dependency in the trace measurements. Since a single significant test outcome is sufficient to fail the device, it is crucial that the probability of a false positive be kept very low.

Under the standard, the per-test rate is controlled at $\alpha_{\text{per-test}} = 0.05$ (see [21, Subsect. 11.1]), and no adjustment is made for the fact that each test is performed against multiple (potentially thousands of) trace points. However, any discovered vulnerability is required to be confirmed by a second test on a separate, identically acquired dataset. In either one of the two sets of tests we would expect that (on average, under the assumption of independence) 5 in every hundred true null hypotheses will be falsely rejected, so that for long traces the overall probability of a false detection becomes almost one. The probability of *both* sets of tests producing a false positive is $(1-(1-\alpha_{\text{per-test}})^m)^2$; the probability of this happening such that the sign of both the effects is the same is $(1-(1-\alpha_{\text{per-test}})^m) \times (1-(1-\alpha_{\text{per-test}}/2)^m)$ (the product of an error of any direction in the first test and an error of fixed direction in the second; see the red lines in Fig. 1). However, the probability of observing two false positives (of the same sign) *in the same position* is $\alpha_{repeat} = 1 - \left(1 - \frac{\alpha^2_{\text{per-test}}}{2}\right)^m$, which grows much slower as m increases (see the yellow lines in Fig. 1). Still, under the standard-recommended significance criterion of $\alpha_{\text{per-test}} = 0.05$, the probability of at least one coinciding detection is over a half once the length of the trace reaches 600. By contrast, under the original TVLA recommendations (which imply $\alpha_{\text{per-test}} \approx 0.00001$), the probability of a coinciding detection is close to zero even for traces that are millions of points long. (Only once the number of points is on the order of 10^{10} do coinciding false detections become non-negligibly probable).

[5] Porter uses the terminology d-minimal; we use r instead of d to avoid confusion with Cohen's d.

The standard fails to provide adequate assurance that detected vulnerabilities are real unless leakage traces are extremely short. Either a stricter per-test significance criterion (combined with the repetition step) or an established method to control the FWER (see the purple lines in Fig. 1) would be preferable for this purpose.

The probability of a false detection under an FDR-controlling procedure depends on the density of true leaks within the trace and is less easy to state in advance in this way; note however that such methods do not *claim* to avoid false detections altogether, rather to ensure that they are few relative to the number of true effects identified. We provide some analysis in Sect. 6, essentially confirming that they are ill-suited to the goal of certifying vulnerability, where a single false positive is enough to fail a device altogether according to the standard.

The question of how best to handle multiple comparisons depends not just on the ability of each option to avoid false positives but on the power of each to detect true positives (i.e. their ability to avoid false negatives). We address this within the next section, as we turn our attention to the standard's capabilities when it comes to certifying security.

5 ISO 17825 for Certifying Security

We have argued so far that the discovery of a leak when the standard recommendations are followed does not reliably certify vulnerability, due to the high risk of a false positive. We now ask the complementary question: what, if anything, can be concluded if a leak is *not* detected? Can non-discovery be interpreted to 'certify security'?

This question is best separated into two: have all realistic vulnerabilities been tested for? and can we trust the conclusions of all the tests that were performed? The first of these is the simpler to answer.

5.1 Have All Realistic Vulnerabilities Been Tested For?

In code testing, the extent to which everything that *could* be tested *has* been tested is referred to as 'coverage' [26]. Typical metrics in this setting include

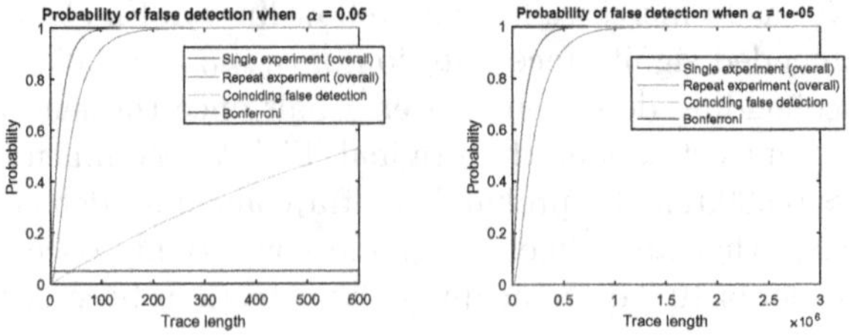

Fig. 1. Overall probability of a false positive as the length of the trace increases, for two different per-test significance levels.

code coverage (have all lines of code been touched by the test procedure?), function coverage (has each function been reached?), and branch coverage (have all branches been executed?) [1]. In a hardware setting one might alternatively (or additionally) test for toggle coverage (have all binary nodes in the circuit been switched?) [37]. These examples all assume white-box access to the source code; in black-box testing scenarios, coverage might alternatively be defined in functional terms.

We suggest that the concept of coverage is a useful one for thinking about the (in)adequacy of a side-channel evaluation. The types of questions we might consider include:

- Have all possible intermediates been tested?
- Have all possible leakage forms been taken into account? For example, some circuits might leak in function of the intermediate values; some in function of the *transitions between* certain intermediates; some in combination of both. Differences might present in distribution means or more subtly, such as in higher order moments (e.g. in the presence of countermeasures).
- Have all possible locations in the trace been tested (with each intermediate and leakage form in mind)? This includes all relevant *tuples* of trace points in the case where higher order leakage of protected intermediates is of concern.
- What proportion of the input space has been sampled? Some key/input combinations might be more 'leaky' than others; with a total possible input space of, e.g. (in the case of AES-128) $2^{128} \times 2^{128} = 2^{256}$ (key, plaintext) pairs, it is unavoidable that one can only test a tiny fraction, and we are typically obliged to rely on simplifying assumptions (e.g. 'Equal Images under different Subkeys (EIS)' [31]) in order to interpret outcomes as representative.
- Have all possible side-channels been tested?! With most of the literature typically focused on power (and sometimes EM radiation [16,29]) it is easy to forget that other potentially exploitable characteristics (timing [22], temperature [5], light [15,34] and sound [2,33] emissions) can also be observed.

It should be clear from the description in Sect. 2.2 that the coverage of ISO 17825 is quite limited. It considers first-order univariate leakages only, relies on one fixed key and (in the case of the fixed-versus-random tests) one fixed input to be representative of the entire sample space, and is confined to a small number of target values (although the fixed-versus-random tests do aim at non-specific leakages). Moreover, by relying solely on the t-test the evaluations are only able to discover differences that exhibit in the *means* of the partitioned populations – more general distributional differences (such as those produced by masking in parallel) will remain completely undetected.

5.2 How Reliably do the Performed Tests Find Leakage?

Formally, a statistical hypothesis test either rejects the null hypothesis in favour of the alternative, or it 'fails to reject' the null hypothesis. It does not 'prove' nor even 'accept' the null hypothesis. Moreover, it does this with a certain probability of error.

Whilst the Type I error rate α is provided by the standard (albeit chosen badly), the Type II error rate (denoted β) – i.e. concluding that there is no leak when there is – is opaque to the evaluator without further effort. If this rate is very high (equivalently, we say that the 'statistical power' $1-\beta$ is low) then the failure of the test to detect leakage really doesn't mean very much at all.

So, if a test fails to reject the null of 'no leakage' in the context of an evaluation, we must be able to say something about its power. The ability of a device to withstand a well-designed test which is known to be powerful indicates far more about its security than its ability to withstand an ad-hoc test which may or may not be suitable for purpose. In addition, the more the statistical properties of the applied methodologies are known and managed, the easier it becomes to compare evaluations across different targets and measurement set-ups, and to establish criteria for fairness. We therefore turn to the tools of statistical power analysis.

Recall from Sect. 3 that the power of a test depends on the sample size, the standardised effect size of interest (alternatively, the raw effect size and the variance of the data), and the significance criteria (the pre-chosen rate of Type I errors). The standard specifies sample sizes of 10,000 and 100,000 for each of the security levels 3 and 4 respectively, and an (unadjusted) per-test significance criteria of $\alpha_{\texttt{per-test}} = 0.05$. The actual effect size (if an effect exists) is necessarily unknown (if it was known the evaluator wouldn't need to test for its existence) and depends on the target implementation even if a perfect measurement set-up were available. But we *can* answer the following:

- What is the power of the tests (as specified) to detect the standardised effects as categorised by Cohen and Sawilowsky?
- What effect sizes can the tests (as specified) detect for a given power (for example, if the analyst wishes to balance the rates of the two types of error)?
- What effect sizes have been observed in practice, and would the current specifications need to be revised in order to detect these?

Power of a Single Test. The LHS of Table 2 shows that, of the standardised effects as categorised by Cohen and Sawilowsky, all but the 'very small' are detected with high probability under the sample size criteria defined by the standard. Meanwhile, level 3 and 4 criteria are both inadequate to detect standardised effects of 0.01. (Remember though that a single test essentially corresponds to a leakage trace of unrealistic length 1).

The RHS of the table shows the effect sizes that *are* detectable; for example, an analyst who wishes to control Type II errors at the same rate as Type I errors ($\beta = \alpha = 0.05$) is able to detect effects of size 0.072 under the level 3 criteria and 0.023 under the level 4 criteria. By comparison, the minimum detectable effect sizes for balanced error rates are more than doubled under the original TVLA significance criterion (which approximates to $\alpha = 0.00001$): 0.174 with a sample size of 10,000 and 0.055 with a sample size of 100,000. (See Table 6 in Appendix B).

Table 2. LHS: Power to detect Cohen's and Sawilowsky's standardised effects under the level 3 ($N = 10,000$) and level 4 ($N = 100,000$) criteria; RHS: Minimum effect sizes detectable for increasing power thresholds, under the level 3 ($N = 10,000$) and level 4 ($N = 100,000$) criteria.

Cohen's *d*	**Power** Level 3	**Power** Level 4
Very small (0.01)	0.072	0.352
Small (0.2)	1.000	1.000
Medium (0.5)	1.000	1.000
Large (0.8)	1.000	1.000
Very large (1.2)	1.000	1.000
Huge (2)	1.000	1.000

Power	**Cohen's** *d* Level 3	**Cohen's** *d* Level 4
0.75	0.053	0.017
0.80	0.056	0.018
0.90	0.065	0.021
0.95	0.072	0.023
0.99	0.086	0.027
0.99999	0.124	0.039

A natural next question is what size *are* the effects exhibited in actual trace acquisitions, and are the criteria laid out in the standard adequate to detect real-world vulnerabilities? We seek indicative answers via analysis of some example scenarios.

Observed Effect Sizes from Realistic Devices. It is not straightforward to 'simply' observe magnitudes in existing acquisitions; all estimated differences will be non-zero, and deciding which ones are 'meaningful' essentially corresponds to the task of detection itself. Choosing 'real' effects based on the outcomes of t-tests, and then using the magnitudes of those effects to make claims about 'detectable' effect sizes, amounts to circular reasoning, and depends on the choice of significance criteria. Fortunately the motivation behind leakage detection provides us with a natural, slightly more objective, criterion for identifying 'real' effects, via the outcomes of key recovery attacks. That is, if leakage detection is geared towards identifying (without having to perform attacks) points in the trace which are vulnerable to attack, then an effect size which is 'large enough' to be of interest is one that can be successfully exploited.

We take this approach, and perform distance-of-means attacks on all 128 bits of the first round SubBytes output for three AES acquisitions, taken on an ARM-M0 processor, an 8051 microcontroller and an RFID (i.e. custom ASIC) device. We also compute the sample effects for each of those bits, which enables us to report estimated effect sizes of interest.

To mitigate for false positives we (adapting from [38]) take measures to confirm the stability of an outcome before classifying a point as 'interesting': we repeat the attack on 99% of the full sample and retain only those points where the correct subkey is ranked first in both instances.

Figure 2 shows the raw (top) and standardised (bottom) observed effect sizes (i.e. mean differences associated with an S-box bit) of first round AES traces measured from an ARM-M0 processor, an 8051 microcontroller and an RFID (custom ASIC) device respectively. As expected, because of the different scales of the measurements (arising from different pre-processing, etc), the raw effects are not necessarily useful to compare. The ARM effects range up to about 0.8, while effects on the 8051 and the RFID implementation range up to 3 and 2 respectively. The standardised effects are much more comparable (≈ 0.6 and ≈ 1 for ARM and 8051 respectively; ≈ 0.4 for the RFID, although this is for the second rather than the first S-box as the latter is less 'leaky' in this instance).[6]

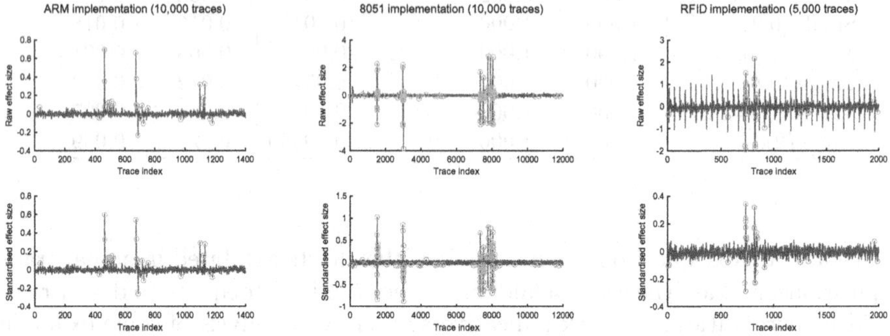

Fig. 2. Difference of means (top) and standardised equivalent (bottom) associated with the first bit of the first S-box of two software AES implementations and the first bit of the second S-box of one hardware implementation. Red circles denote points where a distance-of-means attack achieves stable key recovery. (Color figure online)

Table 3 summarises the standardised and raw effect sizes associated with distance-of-means key recoveries over *all* bits of all S-boxes. The smallest standardised effect detected is 0.0413 for the 8051 microcontroller; the ARM and RFID smallest effects are in a similar ballpark.

Table 3. Summary of effect magnitudes associated with stable distance-of-means key recovery attacks.

Implementation	Proportion	Standardised			Raw		
	interesting	Min	Max	Median	Min	Max	Median
ARM	0.0226	0.0444	0.9087	0.1155	0.0388	1.0265	0.1073
8051	0.0150	0.0413	1.4265	0.1670	0.0254	5.3808	0.1469
RFID	0.0049	0.0624	0.3935	0.0933	0.2272	3.4075	0.3836

[6] In a non-specific fixed-versus-random experiment (even more so in a fixed-versus-fixed one) the differences depend on more than a single bit so, depending on the value of a given intermediate under the fixed input, can potentially be several times larger (see e.g. [32]) – or they can be smaller (e.g. if the leakage of the fixed intermediate coincides with the average case, such as the (decimal) value 15 in an approximately Hamming weight leakage scenario). It is typically assumed in the non-specific case that, as the input propagates through the algorithm, at least some of the intermediates will correspond to large (efficiently detected) class differences [13].

Taking 0.04 as an indicative standardised effect size for actual trace measurements would lead us to conclude that the level 4 criterion is adequate if the full sample of size 100,000 is used in an individual (non-repeated) test, but that the level 3 criterion of 10,000 is not. Using the sample size formula we obtain that a minimum of 32,487 traces are needed to detect an effect of size 0.04 in a single test with balanced error rates $\alpha = \beta = 0.05$. (In reality, one type of error may be deemed more or less of a concern than the other; we state results for balanced rates merely by way of example).

However, data-intensive research has been carried out into the exploitable leakage of devices with far less 'neat' side-channel characteristics than the (comparatively) favourable scenarios exampled above. De Cnudde et al. [10], for example, perform successful attacks against masked hardware implementations with up to 500 million traces, implying both that *extremely* small effects exist and that researchers (and, presumably, some 'worst case' attackers) have the resources and determination to detect and exploit them. FIPS 140-2 (and thus ISO 19790) was conceived to be more economic than CC, but this comes at the cost of not being adequate for state of the art hardware implementations.

We would argue that effects of real world relevance should be extended to include a new category: 'tiny' effects of standardised size $d = 0.001$. An evaluation with $\alpha = 0.05$ and a sample of size of 10,000 or 100,000 (as per the levels 3 and 4 criteria respectively) would have power of just 0.028 or 0.036 respectively to detect such an effect. To achieve a power of 0.95 (that is, balanced error rates) would require a sample of size nearly 52,000,000. Clearly, leakage of this nature is beyond the scope of the ISO standard to detect, whilst still representing a demonstrably exploitable vulnerability.

Furthermore, in practice, of course, evaluators are *not* just checking for a single effect via a single test, but for a range of different effects all in a series of separate (possibly correlated) trace points. This adds considerably to the challenge of rigorous and convincing analysis, due to the problem of multiple comparisons discussed above – corrections for which inevitably impact on the power.

'Overall' Power in an Example Scenario. The per-test power can be computed via the formulae in Sect. 3, but the r-minimal and the complete power of a set of tests depends on the total number of tests and the ratio of true to false null hypotheses, as well as the covariance structure of the test statistics. This information is not available if an evaluation is set up according to ISO 17825 (it would need to be determined in preliminary experiments).

By way of illustrative analysis we consider the scenario described above in Sect. 5.2, where there appeared to be around 30 true leak points in a (truncated) first round AES software trace of length 1,400, and we make the simplifying assumption that the tests are independent (we will relax this in Sect. 6 and show that it makes little difference).

Table 4 shows the per-test and the 1-minimal power under the standard specifications to detect two different effect sizes: the empirically observed effect of standardised size 0.04, and the 'worst case adversary' inspired 'tiny' effect of 0.001. The level 3 sample size is just short of that required to achieve an overall (i.e. 1-minimal) power of $1 - \alpha$ to detect at least one effect of the observed size when the repetition is performed[7]; the level 4 sample size detects it with high probability (even at the stricter TVLA-recommended α-level, see Table 7 in Appendix B); however, to detect the 'tiny' effect would require 170 times as many measurements (1,700 more for $\alpha = 0.00001$). Thus, for this scenario at least (and under our simplifying assumptions) we conclude that the standard recommendations are adequate to certify security with respect to modest effect sizes.

Recall, though, that the standard recommendations are inadequate to certify vulnerability, as the overall false positive rates are considerably higher than should be tolerated by a procedure that fails a device based on a single rejected null hypothesis (see Sect. 4)–this is a prime example that error rates can be 'traded off'. The question is therefore whether any set of parameters or alternative method for multiplicity correction is able to make a better trade-off between the overall false negative and false positive rates.

Table 4. Average ('per-test') and 1-minimal ('overall') power to detect observed and 'tiny' effect sizes under the level 3 and 4 criteria, and the sample size required to achieve balanced errors for a significance criterion of $\alpha = 0.05$. (30 leak points in a trace set of length 1,400).

Effect	Repeat test?	Level 3		Level 4		Required sample size	
		Ave	1-min	Ave	1-min	Ave	1-min
0.04	No	0.516	1.000	1.000	1.000	32,487	1,055
0.04	Yes	0.086	0.932	0.988	1.000	76,615	10,647
0.001	No	0.028	0.574	0.036	0.665	51,978,840	1,687,843
0.001	Yes	0.001	0.022	0.001	0.031	122,584,748	17,034,581

6 Exploring Alternative Test Configurations

We wish to extend the analysis above to a wider range of adjustment methods in order to see if any emerge as being promising alternatives to the current recommendations. Porter suggests a way to approximate the different types of power by simulating large numbers of test statistics under a suitable alternative hypothesis, performing the multiplicity adjustments and simply counting the proportion of instances where 1, r, or all the false nulls are rejected (for the 1-, r-minimal and complete powers) as well as the total proportion of false nulls rejected (for the average individual power) [27]. An advantage of this approach

[7] We compute the per-test power under the repetition step as the square of the power to detect with half the sample, deriving from the assumption that the two iterations of the test are independent.

is that it also allows us to relax the independence assumptions underpinning the computations in Table 4 – but this introduces the considerable limitation that specific and detailed information about the particular leakage scenario is needed. In a real evaluation we do not typically have this; however, for the purposes of illustration we take the dataset analysed in Sect. 5.2 as an example scenario from which to construct a realistic set of null and alternative hypotheses, with the aim of showing how the different notions of power evolve as the sample size increases.

Suppose the t-statistics corresponding to a trace set of length 1,400 have the same correlation structure as the observed ARM traces, characterised by the covariance matrix Σ. The null hypothesis is that none of the points leak; the alternative is that there are 30 effects of standardised size 0.04, located as per the analysis presented in Fig. 2, where $\mathcal{T}$ denotes the set of indices of successful attacks. Under the null hypothesis, for a large enough trace set (which we need anyway to detect such a small effect) the joint distribution of the t-statistics under the alternative hypothesis can be approximated by a multivariate normal with mean $\mu = [\mu_1, \ldots, \mu_{1400}]$ such that $\mu_t = 0.04$ for all $t \in \mathcal{T}$ and $\mu_t = 0$ for all $t \notin \mathcal{T}$, and covariance matrix Σ. By drawing repeatedly from this distribution and noting which of the (individual) tests, with and without correction, reject the null hypothesis and which do not, we can estimate the power and the error rates for tests in this particular scenario.

We performed the analysis for two different significance levels ($\alpha_{ISO} = 0.05$ and $\alpha_{TVLA} = 0.00001$) and six different methods: no correction, Bonferroni, Šidák and Holm corrections to control the FWER, the Benjamini–Hochberg procedure to control the FDR, and the experiment repetition (for a given overall sample size) as per ISO and TVLA recommendations. Figure 3 shows, for $\alpha_{ISO} = 0.05$, what we consider to be the most relevant results, based on 5,000 random draws from the distribution under the alternative hypothesis. (In particular, the three FWER-controlling corrections perform near-identically, and so we only display a single representative). Figure 6 in Appendix B shows the corresponding results for $\alpha_{TVLA} = 0.00001$.

It is clear that the different approaches have substantially different characteristics in practice. The FWER-controlling procedures, represented by Bonferroni, successfully keep false positives down at only a small cost to the power relative to the repetition step. The FDR-controlling procedure, meanwhile, has better power than the repetition step but a comparable false positive rate as the sample size increases. At the lower α level implied by the TVLA criteria Bonferroni (as well as the BH procedure) actually has higher power than the repetition step, and all methods keep false positives low for the (short) trace length in question. Moreover, they all achieve high probability of detecting at least one of the 30 leaks within the level 4 sample size threshold.

We repeated the experiment assuming independence between the tests, and found that it made very little difference to either error rate. This is *not* to say that *taking the dependence structure into account in the tests themselves* would not improve the performance of the tests, but it does imply that (at least in

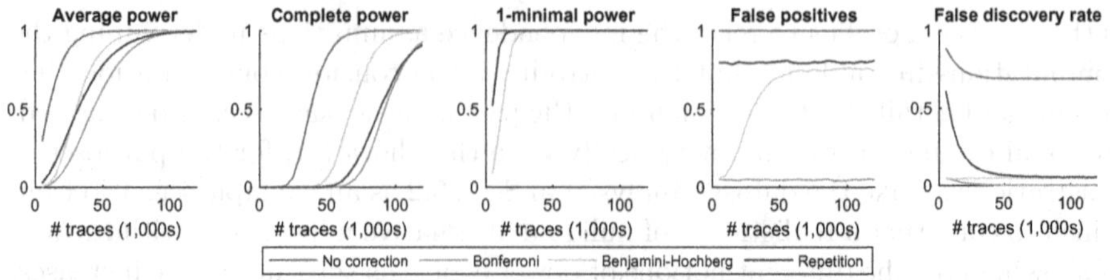

Fig. 3. Different types of power and error to detect 30 true effects of size 0.04 in a trace set of length 1,400, as sample size increases, for an overall significance level of $\alpha = 0.05$. (Based on 5,000 random draws from the multivariate test statistic distribution under the alternative hypothesis).

this instance) a power analysis which assumes independence need not give a misleading account of the capabilities of the chosen tests.

In this example scenario, then, the FWER controlling procedures (but not the FDR controlling one) appear favourable to the ISO standard confirmation requirement, holding all other parameters of the ISO standard fixed. However, we have not yet fully explored the impact of the *length* of the trace on their performance, and many real-world evaluations involve considerably more tests than the 1,400 we here consider. Porter's methodology does not readily scale – and, besides, requires specifying a covariance structure. Instead, then, given the similarity of our results under the independence assumption, we proceed on that simplifying basis and take advantage of the fact that the Bonferroni-corrected tests (by contrast with the BH procedure, which we have already been able to rule out) are relatively straightforward to examine analytically.

The obstacle remains, though, that *overall* notions of power – such as 1-minimal, which we have argued is the relevant quantity for our purposes – will always be highly dependent on the (*a priori* unknown) particulars of the evaluation scenario under consideration. In particular, if a longer trace implies more leakage points, then the increased opportunity to detect leakage might help to compensate for the stricter criteria enforced by the Bonferroni procedure (and similar). On the other hand, if the number of leakage points stays fixed as the trace length increases, there is no compensation for the loss of per-test power. We therefore consider a range of hypothetical scenarios: fixed leakage density of 1 in 1,000 and 1 in 100 as the trace length increases; fixed number of leaks at 1 and (as per our example scenario) 30 as the trace length increases. (In the latter, we suppose that the first 30 trace points are the vulnerable ones and all those subsequently added are random).

Figure 4 presents the FWER and the 1-minimal ('overall') power of the unadjusted, repeated and Bonferroni-corrected tests under the level 3 and level 4 sample size (10,000, 100,000) and significance level (0.05) criteria. It is clear that the relative effectiveness of the approaches is sensitive to the combinations of various parameters and scenario configurations.

Of the three methods only the Bonferroni succeeds in controlling the FWER at an acceptable level (recall that a device fails to meet the standard if a single point of leakage is discovered). Under the level 3 criteria it has lower power than the repetition in all leakage scenarios; however, at the level 4 sample size it is *more* powerful in the case that the density of leak points is fixed. In these fixed density cases the power of all the methods grows as the trace length increases; in the case that the *number* is fixed the unadjusted and repeated tests have a fixed overall probability of detection whilst the Bonferroni tests peak when there are no non-leaky points and then decrease at a speed which depends on the sample size. Note that, at level 4, the power to detect at least one of 30 leaks is still very close to 1 for traces of length up to 10 million; at level 3 it is close to zero from traces of 1 million or more.

At the TVLA significance level (see Fig. 7 in Appendix B) the FWER is (as we've already seen) still very low for both adjustment methods, even up to traces of length 10 million or more (not shown on the graph). The level 3 sample size is completely inadequate to detect effects of this size regardless of trace length. Interestingly, for the level 4 sample size the advantage displayed by the Bonferroni method has widened. We again see a decrease in power to detect a fixed number of leaks as the total length increases, however it should be pointed out that the power to detect one of at least 30 leaks is still above 0.999 for a trace of length 10 million (although it is lower than the power of the repetition step by this point).

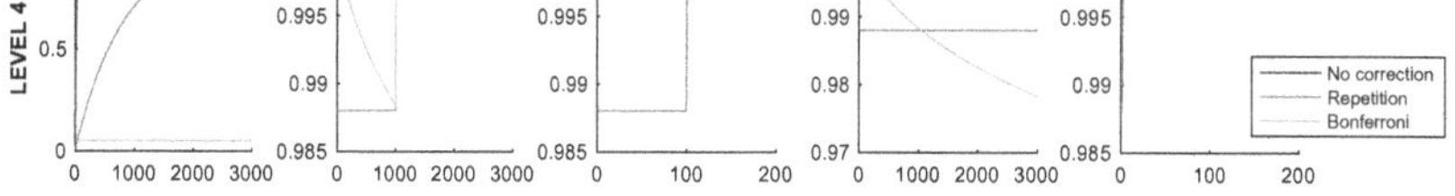

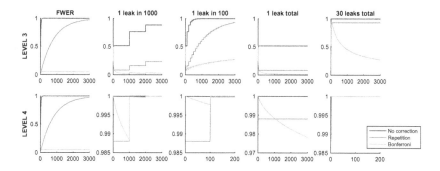

Fig. 4. FWER and 1-minimal ('overall') power of the tests to detect effects of the 'observed' size 0.04 for various leakage scenarios as the trace length increases, under the level 3 and level 4 standard criteria with a significance level of $\alpha = 0.05$. Note that some of the axes have been truncated in order to focus on the interesting regions of the graphs.

We remark that the level 4 standard criteria swapping the repetition step for the Bonferroni method seems an adequate measure to certify vulnerability and/or security for effect sizes of 0.04, even as the trace length increases. Swapping the significance level for the original TVLA recommendation of 0.00001 also achieves this, although we note that the Bonferroni adjustment is anyway more powerful than the repetition step in this instance. However, we already know from Table 4 that the level 4 sample size is too small to reliably detect 'tiny' effects (repeating the Fig. 4 analysis confirms this and reveals no new insights). A reasonable question to ask is then what methods/parameter choices would enable certification with respect to these types of (still realistic) vulnerabilities.

As should be clear by now, appropriate configuration necessarily depends on the type of leakage scenario that we envisage. For example, a typical software implementation might produce very long (e.g. 100,000-point) traces; in the case that it is unprotected (and especially for the non-specific fixed-versus-random tests) the number of leak points could be high, say, 1 in 100; in the presence of countermeasures and/or in the case of a specific test the number could be far lower, say, 10 total, or even just one (which it remains crucial to be able to find). By contrast, hardware implementations are faster and typically produce shorter (e.g. 1,000-point) traces, with any leakage concentrated at one or a few indices.

Table 5 shows suitable parameter choices for Bonferroni-adjusted tests in each of these settings. The large sample sizes (especially when we are concerned with finding very sparse leakage) are something of a reality check on the popular view that leakage detection is a 'more efficient' alternative to performing attacks: the advantages of the former are best understood in terms of its potential to find a *wider variety* of possible sensitive dependencies than an attack-based approach. Meanwhile, precisely because an adversary is targeting a specific vulnerability – with a tailored tool, using information (if available) about the form of the data dependency – we should always expect attacks to be more data efficient than detection tests. It follows (importantly) that we should never interpret the sample sizes required for leakage detection as quantitative markers of a device's resistance to attack. Reciprocally, attack-based configurations should not be used to inform the specifications of detection-based approaches: the influence of the (originally attack-based) FIPS 140-2 on the (detection-based) ISO 17825 likely explains why the level 3 and 4 sample sizes are as limitingly small as they are.

Table 5. Parameter combinations for reliably certifying vulnerability/security in different realistic leakage scenarios using the Bonferroni adjustment to control the false positive rate at an overall level α.

Scenario type	**Trace length**	**# leaks**	**ISO** $\alpha = 0.05$		**TVLA** $\alpha = 0.00001$	
			$d = 0.04$	$d = 0.001$	$d = 0.04$	$d = 0.001$
Software (generic leaks)	100,000	100	2.5×10^4	3.9×10^7	6.8×10^4	1.1×10^8
Software (specific leaks)	100,000	10	4.8×10^4	7.7×10^7	1.2×10^5	1.9×10^8
Software (protected)	100,000	1	1.1×10^5	1.8×10^8	2.9×10^5	4.6×10^8
Hardware (unprotected)	1,000	10	2.9×10^4	4.6×10^7	9.6×10^4	1.5×10^8
Hardware (protected)	1,000	1	8.1×10^4	1.3×10^8	2.5×10^5	4.0×10^8

Remark: At this point it is important to recall that in an actual evaluation the entire process has to be applied to several/many intermediate values as part of the specific detection tests. These further tests are synonymous with considering longer traces and an extended analysis would be possible given a specified number of tests.

7 Conclusions and Recommendations

TVLA was originally conceived as a structured set of leakage detection tests to overcome the issue of having to test against an ever increasing number of attack vectors (thus the concern was coverage of an evaluation rather than trace efficiency). An in-dept statistical analysis was never carried out yet these recommendations became the basis for leakage evaluations as specified in ISO 17825.

We have shown that **following the ISO 17825 recommendations to the letter would result in the failure of *all* target devices** (at security levels 3 and 4) with extremely high probability. This is because of the inflation of Type I errors (false positives) as the number of jointly performed statistical hypothesis tests increases.

The problem can be mitigated by **replacing the (somewhat ad hoc) test repetition step (inherited from TVLA) with an established statistical method** to control the overall error rate, such as the Bonferroni adjustment, and/or by replacing the threshold for significance with the stricter one originally implied by the TVLA standard. In the latter case, the repetition step is anyway shown to be less efficient than Bonferroni-style adjustments, so we recommend against adhering to that part of TVLA.

There are some ambiguities in ISO 17825 about how to interpret the acquisition criterion. Even opting for the most generous interpretation, the **level 3 sample size specification is shown to be inadequate to certify vulnerability/security against effects of the size and frequency that we observe in a range of typical 'easy to attack' implementations**. The level 4 specification *is* able to detect these with high probability, even with the stricter TVLA-based significance threshold provided the leakages are of sufficient density as the length of the trace increases. **However, neither are sufficient to detect the types of 'tiny' effects that have been shown to exist (and to be exploitable) by larger-scale academic studies.**

We therefore recommend the **necessity for larger acquisitions** than those specified by the standard. A difficulty here is that, although statistical power analysis provides tools to derive the appropriate sample sizes for a particular test scenario, it requires considerable *a priori* information about that scenario to do so (even more so in the case of multiple tests and their corresponding adjustment procedures). Whilst it is possible to broadly identify common expected features across classes of scenario, **a preferable approach would be to develop a two-stage evaluation procedure** combining an exploratory phase with a pared-down confirmatory analysis in which information about the covariance structure and likely location/nature of the leaks is used to inform the acquisition process and to chose a (reduced set) of carefully-formulated hypothesis tests to perform. We leave the precise details of such a strategy as an interesting avenue for further work.

However the standard procedures (or adaptations therefore) are applied it is **important that outcomes are presented responsibly**. An evaluator needs to decide – and to give a justification for – the false positive and false negative rates that are acceptable. For example, even if a multiplicity adjustment is used to successfully control the overall false positive rate at the level specified by the standard, this still implies that 5 in every 100 secure devices will fail the test at random. If this is considered too high, then a stricter significance criterion will need to be chosen, inevitably implying greater data complexity. Either way, **the error rates must be made transparent** – **as should the effect size** the test is able to detect, **the coverage limitations** that we identified in Sect. 5.1, and the fact that the sample size needed for a successful *attack* may be much smaller than that required for detection.

Acknowledgements. Our work has been funded by the European Commission through the H2020 project 731591 (acronym REASSURE). A fuller report on this aspect of the project can be found in *A Cautionary Note Regarding the Usage of Leakage Detection Tests in Security Evaluation* [42].

A Sample Size for the t-Test

We begin with a simple visual example that illustrates the concepts of α and β values and their relationship to the sample size.

Consider the following two-sided hypothesis test for the mean of a Gaussian-distributed variable $A \sim \mathcal{N}(\mu, \sigma)$, where μ and σ are the (unknown) parameters:

$$H_0 : \mu = \mu_0 \text{ vs. } H_{alt} : \mu \neq \mu_0. \tag{3}$$

Note that, in the leakage detection setting, where one typically wishes to test for a non-zero difference in means between *two* Gaussian distributions Y_1 and Y_2, this can be achieved by defining $A = Y_1 - Y_2$ and (via the properties of the Gaussian distribution) performing the above test with $\mu_0 = 0$.

Suppose the alternative hypothesis is true and that $\mu = \mu_{alt}$. This is called a 'specific alternative'[8], in recognition of the fact that it is not usually possible to compute power for *all* the alternatives when H_{alt} defines a set or range. In the leakage detection setting one typically chooses $\mu_{alt} > 0$ to be the smallest difference $|\mu_1 - \mu_2|$ that is considered of practical relevance; this is called the effect size. Without loss of generality, we suppose that $\mu_{alt} > \mu_0$.

Figure 5 illustrates the test procedure when the risk of a Type I error is set to α and the sample size is presumed large enough (typically $n > 30$) that the

[8] The overloading of terminology between 'specific alternatives' and 'specific' TVLA tests is unfortunate but unavoidable.

distributions of the test statistic under the null and alternative hypotheses can be approximated by Gaussian distributions. The red areas together sum to α; the blue area indicates the overlap of H_0 and H_{alt} and corresponds to β (the risk of a Type II error). The power of the test – that is, the probability of correctly rejecting the null hypothesis when the alternative in true – is then $1 - \beta$, as depicted by the shaded area.

There are essentially three ways to raise the power of the test. One is to increase the effect size of interest which, as should be clear from Fig. 5, serves to push the distributions apart, thereby diminishing the overlap between them. Another is to increase α – that is, to make a trade-off between Type II and Type I errors – or (if appropriate) to perform a one-sided test, either of which has the effect (in this case) of shifting the critical value to the left so that the shaded region becomes larger. (In the leakage detection case the one-sided test is unlikely to be suitable as differences in either direction are equally important and neither can be ruled out *a priori*). The third way to increase the power is to increase the sample size for the experiment. This reduces the standard error on the sample means, which again pushes the alternative distribution of the test statistic further away from null (note from Fig. 5 that it features in the denominator of the distance).

Suppose you have an effect size in mind – based either on observations made during similar previous experiments, or on a subjective value judgement about how large an effect needs to be before it is practically relevant (e.g. the level of leakage which is deemed intolerable) – and you want your test to have a given confidence level α and power $1-\beta$. The relationship between confidence, power, effect size and sample size can then be used to derive the minimum sample size necessary to achieve this.

The details of the argumentation that now follows are specific to a two-tailed t-test, but the general procedure can be adapted to any test for which the distribution of the test statistic is known under the null and alternative hypotheses.

For the sake of simplicity (i.e. to avoid calculating effectively irrelevant degrees of freedom) we will assume that our test will in any case require the acquisition of more than 30 observations, so that the Gaussian approximations for the test statistics hold as in Fig. 5. Without loss of generality we also assume that the difference of means is positive (otherwise the sets can be easily swapped). Finally, we assume that we seek to populate both sets with equal numbers $n = |Y|/2$ of observed traces.

Theorem 1. *Let Y_1 be a set of traces of size $N/2$ drawn via repeat sampling from a normal distribution $\mathcal{N}(\mu_1, \sigma_1^2)$ and Y_2 be a set of traces of size $N/2$ drawn via repeat sampling from a normal distribution $\mathcal{N}(\mu_2, \sigma_2^2)$. Then, in a two-tailed test for a difference between the sample means:*

$$H_0\colon \mu_1 = \mu_2 \text{ vs. } H_{alt}\colon \mu_1 \neq \mu_2, \tag{4}$$

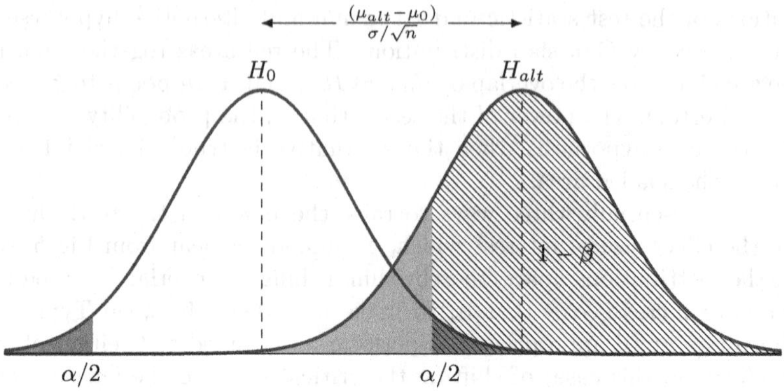

Fig. 5. Figure showing the Type I and II error probabilities, α and β as well as the effect size $\mu_{alt} - \mu_0$ for a specific alternative such that $\mu_{alt} > \mu_0$.

in order to achieve significance level α *and power* $1-\beta$*, the overall number of traces N needs to be chosen such that:*

$$N \geq 2 \cdot \frac{(z_{\alpha/2} + z_\beta)^2 \cdot ({\sigma_1}^2 + {\sigma_2}^2)}{(\mu_1 - \mu_2)^2}. \tag{5}$$

Note that Eq. 5 can be straightforwardly rearranged to alternatively compute any of the significance level, effect size or power in terms of the other three quantities.

B Results for Original TVLA-Recommended Threshold

Table 6. LHS: Power to achieve Cohen's and Sawilowsky's standardised effects under the TVLA significance criteria (which approximates to $\alpha = 0.00001$) and the standard level 3 ($N = 10,000$) and level 4 ($N = 100,000$) sample size criteria; RHS: Minimum effect sizes detectable for increasing power thresholds.

Cohen's d	**Power**	
	Level 3	**Level 4**
Very small (0.01)	0.000	0.002
Small (0.2)	1.000	1.000
Medium (0.5)	1.000	1.000
Large (0.8)	1.000	1.000
Very large (1.2)	1.000	1.000
Huge (2)	1.000	1.000

Power	**Cohen's d**	
	$N = 10,000$	$N = 100,000$
0.75	0.102	0.032
0.80	0.105	0.033
0.90	0.114	0.036
0.95	0.121	0.038
0.99	0.135	0.043
0.99999	0.174	0.055

Table 7. Average ('per-test') and 1-minimal ('overall') power to detect observed and 'tiny' effect sizes under the level 3 and 4 criteria, and the sample size required to achieve balanced errors for a significance criterion of $\alpha = 0.00001$. (30 leak points in a trace set of length 1,400).

Effect	Repeat test?	Level 3		Level 4		Required sample size	
		Ave	1-min	Ave	1-min	Ave	1-min
0.04	No	0.008	0.210	0.972	1.000	188,446	38,924
0.04	Yes	0.000	0.000	0.272	1.000	390,228	104,867
0.001	No	0.000	0.000	0.000	0.000	301,512,956	62,279,197
0.001	Yes	0.000	0.000	0.000	0.000	624,365,394	167,786,951

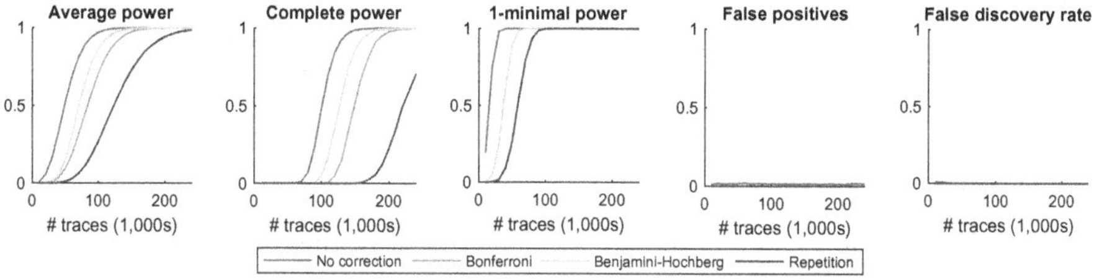

Fig. 6. Different types of power and error to detect 30 true effects of size 0.04 in a trace set of length 1,400, as sample size increases, for an overall significance level of $\alpha = 0.00001$. (Based on 5,000 random draws from the multivariate test statistic distribution under the alternative hypothesis).

Table 8. Different types of power and error to detect 30 true effects of size 0.04 in a trace set of length 1,400, under the level 3 and level 4 sample size criteria and with an overall significance level of $\alpha = 0.00001$. (Based on 5,000 random draws from the multivariate test statistic distribution under the alternative hypothesis).

Correction strategy	Level 3		Level 4	
	1-min power	FWER	1-min power	FWER
None	0.1912	0.0156	1.0000	0.0134
Bonferroni	0.0020	0.0000	1.0000	0.0000
Šidák	0.0020	0.0000	1.0000	0.0000
Holm	0.0020	0.0000	1.0000	0.0000
Benjamini-Hochberg	0.0020	0.0000	1.0000	0.0000
Repetition	0.0000	0.0000	0.9986	0.0000

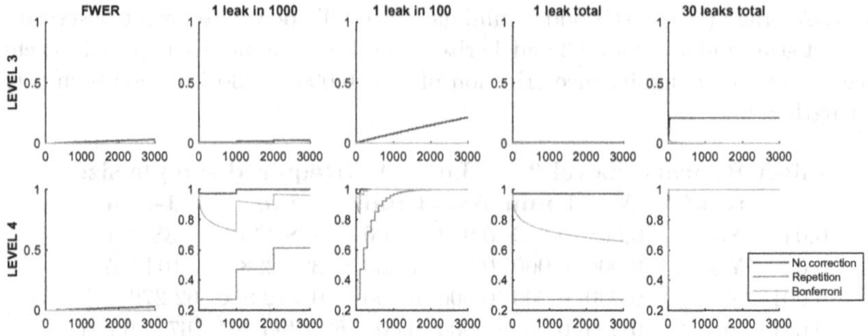

Fig. 7. FWER and 1-minimal ('overall') power of the tests to detect effects of the 'observed' size 0.04 for various leakage scenarios as the trace length increases, under the level 3 and level 4 standard criteria with an overall significance level of $\alpha = 0.00001$.

References

1. Ammann, P., Offutt, J.: Introduction to Software Testing, 1st edn. Cambridge University Press, New York (2008)
2. Asonov, D., Agrawal, R.: Keyboard acoustic emanations. In: IEEE Symposium on Security and Privacy, pp. 3–11. IEEE Computer Society (2004)
3. Bhasin, S., Danger, J.L., Guilley, S., Najm, Z.: Side-channel leakage and trace compression using normalized inter-class variance. In: Lee, R.B., Shi, W. (eds.) HASP 2014, Hardware and Architectural Support for Security and Privacy, pp. 7:1–7:9. ACM (2014)
4. Bi, R., Liu, P.: Sample size calculation while controlling false discovery rate for differential expression analysis with RNA-sequencing experiments. BMC Bioinform. **17**(1), 146 (2016)
5. Brouchier, J., Kean, T., Marsh, C., Naccache, D.: Temperature attacks. IEEE Secur. Priv. **7**(2), 79–82 (2009)
6. Chatzikokolakis, K., Chothia, T., Guha, A.: Statistical measurement of information leakage. In: Esparza, J., Majumdar, R. (eds.) TACAS 2010. LNCS, vol. 6015, pp. 390–404. Springer, Heidelberg (2010). https://doi.org/10.1007/978-3-642-12002-2_33
7. Chothia, T., Guha, A.: A statistical test for information leaks using continuous mutual information. In: CSF, pp. 177–190 (2011)
8. Cohen, J.: Statistical Power Analysis for the Behavioral Sciences. Routledge (1988)
9. Danger, J.-L., Duc, G., Guilley, S., Sauvage, L.: Education and open benchmarking on side-channel analysis with the DPA contests. In: NIST Non-Invasive Attack Testing Workshop (2011)
10. De Cnudde, T., Ender, M., Moradi, A.: Hardware masking, revisited. IACR Trans. Cryptogr. Hardw. Embed. Syst. **2018**(2), 123–148 (2018)
11. Ding, A.A., Zhang, L., Durvaux, F., Standaert, F.-X., Fei, Y.: Towards sound and optimal leakage detection procedure. In: Eisenbarth, T., Teglia, Y. (eds.) CARDIS 2017. LNCS, vol. 10728, pp. 105–122. Springer, Cham (2018). https://doi.org/10.1007/978-3-319-75208-2_7
12. Dunn, O.J.: Multiple comparisons among means. J. Am. Stat. Assoc. **56**(293), 52–64 (1961)

13. Durvaux, F., Standaert, F.-X.: From improved leakage detection to the detection of points of interests in leakage traces. In: Fischlin, M., Coron, J.-S. (eds.) EUROCRYPT 2016. LNCS, vol. 9665, pp. 240–262. Springer, Heidelberg (2016). https://doi.org/10.1007/978-3-662-49890-3_10
14. Efron, B.: Size, power and false discovery rates. Ann. Stat. **35**(4), 1351–1377 (2007)
15. Ferrigno, J., Hlaváč, M.: When AES blinks: introducing optical side channel. IET Inf. Secur. **2**(3), 94–98 (2008)
16. Gandolfi, K., Mourtel, C., Olivier, F.: Electromagnetic analysis: concrete results. In: Koç, Ç.K., Naccache, D., Paar, C. (eds.) CHES 2001. LNCS, vol. 2162, pp. 251–261. Springer, Heidelberg (2001). https://doi.org/10.1007/3-540-44709-1_21
17. Goodwill, G., Jun, B., Jaffe, J., Rohatgi, P.: A testing methodology for side-channel resistance validation. In: NIST Non-Invasive Attack Testing Workshop (2011)
18. Hoenig, J.M., Heisey, D.M.: The abuse of power. Am. Stat. **55**(1), 19–24 (2001)
19. Holm, S.: A simple sequentially rejective multiple test procedure. Scand. J. Stat. **6**, 65–70 (1979)
20. Information technology - Security techniques - Testing methods for the mitigation of non-invasive attack classes against cryptographic modules. Standard, International Organization for Standardization, Geneva, CH (2016)
21. Information technology - Security techniques - Security requirements for cryptographic modules. Standard, International Organization for Standardization, Geneva, CH (2012)
22. Kocher, P.C.: Timing attacks on implementations of Diffie-Hellman, RSA, DSS, and other systems. In: Koblitz, N. (ed.) CRYPTO 1996. LNCS, vol. 1109, pp. 104–113. Springer, Heidelberg (1996). https://doi.org/10.1007/3-540-68697-5_9
23. Kocher, P., Jaffe, J., Jun, B.: Differential power analysis. In: Wiener, M. (ed.) CRYPTO 1999. LNCS, vol. 1666, pp. 388–397. Springer, Heidelberg (1999). https://doi.org/10.1007/3-540-48405-1_25
24. Liu, P., Hwang, J.T.G.: Quick calculation for sample size while controlling false discovery rate with application to microarray analysis. Bioinformatics **23**(6), 739–746 (2007)
25. Mather, L., Oswald, E., Bandenburg, J., Wójcik, M.: Does my device leak information? An *a priori* statistical power analysis of leakage detection tests. In: Sako, K., Sarkar, P. (eds.) ASIACRYPT 2013. LNCS, vol. 8269, pp. 486–505. Springer, Heidelberg (2013). https://doi.org/10.1007/978-3-642-42033-7_25
26. Miller, J.C., Maloney, C.J.: Systematic mistake analysis of digital computer programs. Commun. ACM **6**(2), 58–63 (1963)
27. Porter, K.E.: Statistical power in evaluations that investigate effects on multiple outcomes: A guide for researchers. J. Res. Educ. Eff. **11**, 1–29 (2017)
28. Pounds, S., Cheng, C.: Sample size determination for the false discovery rate. Bioinformatics **21**(23), 4263–4271 (2005)
29. Quisquater, J.-J., Samyde, D.: ElectroMagnetic Analysis (EMA): Measures and counter-measures for smart cards. In: Attali, I., Jensen, T. (eds.) E-smart 2001. LNCS, vol. 2140, pp. 200–210. Springer, Heidelberg (2001). https://doi.org/10.1007/3-540-45418-7_17
30. Sawilowsky, S.S.: New effect size rules of thumb. J. Mod. Appl. Stat. Methods **8**(2), 597–599 (2009)
31. Schindler, W., Lemke, K., Paar, C.: A stochastic model for differential side channel cryptanalysis. In: Rao, J.R., Sunar, B. (eds.) CHES 2005. LNCS, vol. 3659, pp. 30–46. Springer, Heidelberg (2005). https://doi.org/10.1007/11545262_3

32. Schneider, T., Moradi, A.: Leakage assessment methodology. In: Güneysu, T., Handschuh, H. (eds.) CHES 2015. LNCS, vol. 9293, pp. 495–513. Springer, Heidelberg (2015). https://doi.org/10.1007/978-3-662-48324-4_25
33. Shamir, A., Tromer, E.: Acoustic cryptanalysis (website). http://theory.csail.mit.edu/~tromer/acoustic/. Accessed 9 Sept 2019
34. Skorobogatov, S.: Using optical emission analysis for estimating contribution to power analysis. In: Breveglieri, L., Koren, I., Naccache, D., Oswald, E., Seifert, J.-P. (eds.) Fault Diagnosis and Tolerance in Cryptography - FDTC 2009, pp. 111–119. IEEE Computer Society (2009)
35. Standaert, F.-X., Gierlichs, B., Verbauwhede, I.: Partition *vs.* comparison side-channel distinguishers: An empirical evaluation of statistical tests for univariate side-channel attacks against two unprotected CMOS devices. In: Lee, P.J., Cheon, J.H. (eds.) ICISC 2008. LNCS, vol. 5461, pp. 253–267. Springer, Heidelberg (2009). https://doi.org/10.1007/978-3-642-00730-9_16
36. Standaert, F.-X., Pereira, O., Yu, Y., Quisquater, J.-J., Yung, M., Oswald, E.: Leakage resilient cryptography in practice. In: Sadeghi, A.-R., Naccache, D. (eds.) Towards Hardware-Intrinsic Security: Foundations and Practice, pp. 99–134. Springer, Heidelberg (2010). https://doi.org/10.1007/978-3-642-14452-3_5
37. Tasiran, S., Keutzer, K.: Coverage metrics for functional validation of hardware designs. IEEE Des. Test **18**(4), 36–45 (2001)
38. Thillard, A., Prouff, E., Roche, T.: Success through confidence: Evaluating the effectiveness of a side-channel attack. In: Bertoni, G., Coron, J.-S. (eds.) CHES 2013. LNCS, vol. 8086, pp. 21–36. Springer, Heidelberg (2013). https://doi.org/10.1007/978-3-642-40349-1_2
39. Tong, T., Zhao, H.: Practical guidelines for assessing power and false discovery rate for fixed sample size in microarray experiments. Stat. Med. **27**, 1960–1972 (2008)
40. Šidák, Z.: Rectangular confidence regions for the means of multivariate normal distributions. J. Am. Stat. Assoc. **62**(318), 626–633 (1967)
41. Welch, B.L.: The generalization of "Student's" problem when several different population variances are involved. Biometrika **34**(1–2), 28–35 (1947)
42. Whitnall, C., Oswald, E.: A cautionary note regarding the usage of leakage detection tests in security evaluation. IACR Cryptology ePrint Archive, Report 2019/703 (2019). https://eprint.iacr.org/2019/703

Location, Location, Location: Revisiting Modeling and Exploitation for Location-Based Side Channel Leakages

Christos Andrikos[1(✉)], Lejla Batina[2], Lukasz Chmielewski[2,4], Liran Lerman[5], Vasilios Mavroudis[6], Kostas Papagiannopoulos[2,3], Guilherme Perin[4], Giorgos Rassias[1], and Alberto Sonnino[6]

[1] National Techical University of Athens, Athens, Greece
{candrikos,grassias}@cslab.ece.ntua.gr
[2] Radboud University, Nijmegen, The Netherlands
lejla@cs.ru.nl
[3] NXP Semiconductors, Hamburg, Germany
kostaspap88@gmail.com
[4] Riscure BV, Delft, The Netherlands
chmielewski@riscure.com, guilhermeperin7@gmail.com
[5] Thales Belgium, Herstal, Belgium
liran.lerman@be.thalesgroup.com
[6] University College London, London, England
v.mavroudis@cs.ucl.ac.uk, alberto.sonnino@ucl.ac.uk

Abstract. Near-field microprobes have the capability to isolate small regions of a chip surface and enable precise measurements with high spatial resolution. Being able to distinguish the activity of small regions has given rise to the location-based side-channel attacks, which exploit the spatial dependencies of cryptographic algorithms in order to recover the secret key. Given the fairly uncharted nature of such leakages, this work revisits the location side-channel to broaden our modeling and exploitation capabilities. Our contribution is threefold. First, we provide a simple spatial model that partially captures the effect of location-based leakages. We use the newly established model to simulate the leakage of different scenarios/countermeasures and follow an information-theoretic approach to evaluate the security level achieved in every case. Second, we perform the first successful location-based attack on the SRAM of a modern ARM Cortex-M4 chip, using standard techniques such as difference of means and multivariate template attacks. Third, we put forward neural networks as classifiers that exploit the location side-channel and showcase their effectiveness on ARM Cortex-M4, especially in the context of single-shot attacks and small memory regions. Template attacks and neural network classifiers are able to reach high spacial accuracy, distinguishing between 2 SRAM regions of 128 bytes each with 100% success rate and distinguishing even between 256 SRAM byte-regions with 32% success rate. Such improved exploitation capabilities revitalize the interest for location vulnerabilities on various implementations, ranging from RSA/ECC with large memory footprint, to lookup-table-based AES with smaller memory usage.

S. D. Galbraith and S. Moriai (Eds.): ASIACRYPT 2019, LNCS 11923, pp. 285–314, 2019.
https://doi.org/10.1007/978-3-030-34618-8_10

Keywords: Side-channel analysis · Location leakage · Microprobe · Template attack · Neural network · ARM Cortex-M

1 Introduction

Side-channel analysis (SCA) allows adversaries to recover sensitive information, by observing and analyzing the physical characteristics and emanations of a cryptographic implementation. Usually, physical observables such as the power consumption and electromagnetic (EM) emission of a device [13,24] are closely related to the data that is being accessed, stored or processed. Such *data-based leakage* compromises the device's security and may allow the adversary to infer the implemented cipher's secret key.

Location-based leakage is a less common form of side-channel leakage when compared to data-based leakages, yet it arises in many practical scenarios. This form of leakage stems from the fact that chip components such as registers, memory regions, storage units, as well as their respective addressing mechanisms (control logic, buses) exhibit leakage when accessed and such leakage is identifiable and data-independent. Thus, the power or EM side-channel potentially conveys information about the *location* of the accessed component, i.e. it can reveal the particular register or memory address that has been accessed, regardless of the data stored in it. If there exists any dependence between the secret key and the location of the activated component, then a side-channel adversary can exploit it to his advantage and recover the key.

1.1 Previous Research and Terminology

The work of Sugawara et al. [48] demonstrates the presence of location-based leakage in an ASIC. In particular, they show that the power consumption of the chip's SRAM conveys information about the memory address that is being accessed. They refer to this effect as "geometric" leakage since it relates to the memory layout. Similarly, Andrikos et al. [2] performed preliminary analyses using the EM-based location leakage exhibited at the SRAM of an ARM Cortex-M4. The work of Heyszl et al. [18] manages to recover the secret scalar by exploiting the spatial dependencies of the double-and-add-always algorithm for elliptic curve cryptography. The experiments were carried out on a decapsulated FPGA, using near-field microprobes that identify the accessed register. Schlösser et al. [40] use the photonic side-channel in order to recover the exact SRAM location that is accessed during the activation of an AES Sbox lookup table. This location information can assist in key recovery, thus even cases of photonic emission analysis can be classified as location-based leakage. Moreover, countermeasures such as RSM [31] rely on rotating lookup tables to mask the data. Location-based leakage can identify which lookup table is currently under use and potentially weaken masking.

For the sake of clarity, we distinguish between "location leakage" and "localized leakage". Location leakage arises when knowing the location of a component (register, memory region, etc.) is assisting towards key recovery. On the contrary, localized leakage arises when the adversary is able to focus on the leakage of a specific (usually small) region of the chip. For example, recovering the memory address accessed during an Sbox lookup implies location leakage. Being able to measure the leakage right on top of a processor's register file implies that the adversary is capturing localized leakage. Note that capturing localized leakage can be useful for data-based attacks as well as for location-based attacks. The works of Unterstein et al. [51], Immler et al. [20] and Specht et al. [43–45] acquire localized leakage via a microprobe in order to improve the signal-to-noise ratio of their data-dependent leakage. The work of Heyszl et al. [18] uses the same technique in order to improve the signal-to-noise ratio of their location-dependent leakage. The current work follows experimental techniques similar to Heyszl et al. (localized EM) to showcase a potent location-based attack on ARM Cortex-M4 devices.

Again, for the sake of clarity we distinguish between "location leakage" and "address leakage" [21]. In our work, address leakage implies the leakage of addressing mechanisms, e.g. the leakage of the control logic of a storage unit. Such leakage can even be observed far from the storage unit itself, e.g. at memory buses or at the CPU. Location leakage implies the leakage caused by such address leakage *and* the leakage of the unit itself, which is often observed near it. We refer to the latter as "spatial leakage", i.e. location leakage encapsulates both address-related and spatial effects. For example accessing a table in memory requires indexing and memory addressing in the CPU (address leakage). In addition, accessing causes the memory itself to be activated (spatial leakage). The adversary is usually able to observe both types of leakage and it is often hard to distinguish between them.

1.2 Contribution and Organization

This work presents the following results in the field on location-based leakage by expanding our modeling and exploitation capabilities.

1. We provide a simple model that captures the effect of spatial leakages. The model is motivated by experimental data observed in the SRAM of an ARM Cortex-M4.
2. Using the newly established model, we simulate the different theoretical scenarios that enhance or diminish spatial leakage. We investigate the security of every scenario using the perceived information (PI) metric.
3. We perform the first practical location-based attack on the SRAM of a modern ARM Cortex-M4, using difference-of-means, multivariate template attacks and neural networks.
4. We showcase attacks where it is possible to distinguish consecutive SRAM regions of 128 bytes each, with 100% success rate and to distinguish between 256 consecutive SRAM bytes with 32% success rate. We conclude that EM location-based leakages are potent enough to compromise the security of AES implementations that use SRAM lookup-tables.

Notation. Capital letters denote random variables and small case letters denote instances of random variables or constant values. Bold font denotes vectors. For instance, side-channel leakage variables are denoted by L and their instances by l; and likewise leakage vectors are denoted by $\mathbf{L}$ and their instances by $\mathbf{l}$. The notation $Unif(\{a,b\})$, $Bern(p)$, $Binomial(n,p)$ and finally $Norm(\mu,\sigma^2)$ denotes random variables with uniform, Bernoulli, binomial and normal probability distributions respectively. Parameter p denotes the probability of Bernoulli/binomial trials and μ, σ^2 denote the mean and variance of the normal distribution. The set $\{a,b\}$ denotes that the discrete uniform distribution can receive value a or b equiprobably. The notation $E[\cdot]$, $Var[\cdot]$ and $H[\cdot]$ describes the expected value, variance and entropy of a random variable. Finally, the notation $H_{p,q}[\cdot]$ shows the cross entropy of a random variable, between probability distributions p and q.

Organization. Section 2 describes the microprobe-based experimental setup on ARM Cortex-M4, shows a simple location analysis using difference-of-means, and motivates experimentally the spatial part of location leakage. Section 3 puts forward the spatial leakage model, describes several theoretical scenarios, and performs an evaluation using the perceived information metric. Section 4 demonstrates real-world template attacks on ARM Cortex-M4 for various cases and Sect. 5 demonstrates the attacks using neural networks on the same device. We conclude and discuss future directions in Sect. 6.

2 Experimental Setup and T-Test Analysis

This section describes a high-precision EM-based setup that is able to detect location leakage on the surface of an ARM Cortex-M4 (Sects. 2.1, 2.2). Using the setup, we obtain intuition about the location leakage that is caused by switching circuitry and is observable via EM emissions on the die surface (Sect. 2.3). Throughout the text, we concentrate on the following adversarial scenario. The device has implemented a key-dependent cipher operation that uses a lookup-table and the adversary aims to infer which part of the table is active, i.e. uncover the location information leading to key recovery.

2.1 Experimental Setup

The main goal of our experimental evaluation is to examine whether it is possible to detect the access to different SRAM regions in a modern ARM-based device. Rephrasing, we examine the device's susceptibility to location-based attacks during e.g. key-dependent memory lookups, similarly to AES LUT. Our measurement setup consists of a decapsulated Riscure Piñata device[1], on a modified board, fabricated with 90 nm technology. The decapsulated chip surface (roughly

[1] https://tinyurl.com/y9tmnklr.

$6\,mm^2 \approx 2.4\,mm \times 2.4\,mm$) is scanned using an ICR HH 100-27 Langer microprobe[2] with diameter of $100\,\mu m$ (approximately $0.03\,mm^2$). The scan is performed on a rectangular grid of dimension 300, using the Inspector tooling[3] and resulting in 300×300 measurement spots. The near-field probe is moved over the chip surface with the assistance of an XYZ-table with positioning accuracy of $50\,\mu m$. At every position of the scan grid, a single measurement is performed, using sampling rate of 1 Gsample/s and resulting in 170k samples. Due to the complex and non-homogeneous nature of a modern chip, several types of EM emissions are present on the surface, most of which are unrelated to the SRAM location. In this particular case study, the signals of interest were observed in amplitudes of roughly 70 mV, so we set the oscilloscope voltage range accordingly. In addition, several device peripherals (such as USB communication) have been disabled in order to reduce interference. The decapsulated surface where the scan is performed is visible in Fig. 1 and the approximate microprobe area is also overlaid on the figure (in red) for comparison.

Fig. 1. The chip surface of the device-under-test (ARM Cortex-M4) after removal of the plastic layer. The approximate area of the ICR HH 100-27 Langer microprobe is shown by the red circle ($0.03\,mm^2$). (Color figure online)

To effectively cause location-dependent leakage, we perform sequential accesses to a continuous region of 16 KBytes in the SRAM by loading data from all memory positions. The data at all accessed memory positions have been fixed to value zero prior to the experiment in order to remove any data-based leakage. The word size of this ARM architecture is 32 bits, i.e. we accessed 4096 words in memory. We opted to access the SRAM using ARM assembly instead of a high-level language in order to avoid compiler-induced optimizations that could alter the side-channel behavior.

[2] https://tinyurl.com/mcd3ntp.
[3] https://tinyurl.com/jlgfx95.

2.2 Difference-of-Means T-Test

The initial scan measurements were analyzed using a simple difference-of-means test. To demonstrate the presence of location-based leakage, we partitioned every trace (170k samples) into two classes. The first class contains SRAM accesses from the beginning of the memory until word no. 2047 and the second class contains SRAM accesses from word 2048 until word 4096. Each class corresponds to 8 KBytes of SRAM. For every grid position (x, y), we averaged the leakages samples of class 1 and class 2 producing $\bar{l}_{class1} = \frac{1}{85k}\sum_{j=1}^{85k} l_{x,y}^j$ and $\bar{l}_{class2} = \frac{1}{85k}\sum_{j=85k}^{170k} l_{x,y}^j$ respectively. Continuing, we computed the difference of means $\bar{l}_{class1} - \bar{l}_{class2}$ and we performed a Welch t-test with significance level of 0.1% in order to determine if location-based leakage is present. The results are visible in Fig. 2a, which is focusing on a specific part of the chip surface that exhibits high difference.

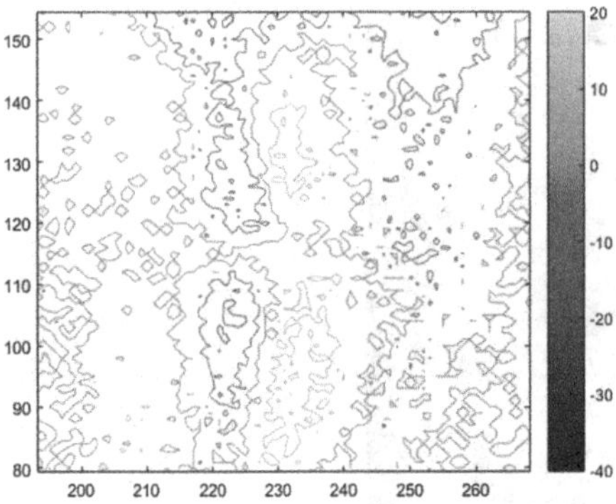

(a) Distinguishing two 8 KByte regions of the SRAM with difference-of-means. Yellow region indicates stronger leakage from class 1 while blue region from class 2.

(b) Chip surface of ARM Cortex-M4 after removal of the top metal layer. The red rectangular region corresponds to the difference-of-means plot of Figure 2a, i.e. it shows the location where the highest differences were observed.

Fig. 2. Spatial properties of chip leakage. (Color figure online)

2.3 Motivating the Location Leakage Model

In Fig. 2a we can observe that the spatial part of location leakage is indeed present in the ARM Cortex-M4 and it can even be detected through simple visual inspection if memory regions are large enough (8 KBytes). Repeating the same difference-of-means test for SRAM regions of 4 KBytes yields similar results, i.e. the regions remain visually distinct. In both cases, we observe that these location dependencies demonstrate strong *spatial characteristics*. That is, in Fig. 2a we see two regions at close proximity (yellow and blue) where the yellow region shows positive difference between class 1 and 2, while the blue region shows negative difference between class 1 and 2. To investigate this proximity, we performed additional chemical etching on the chip surface in order to remove the top metal layer (Fig. 2b).

The different regions (yellow, blue) shown in Fig. 2a are observed directly above the chip area enclosed by the red rectangle of Fig. 2b. Interestingly, after the removal of the top metal layer, we see that the red rectangular region contains large continuous chip components, possibly indicating that SRAM circuitry is present at this location. This hypothesis is corroborated by the following fact: when we perform difference-of-means test for 4 KByte regions, the yellow and blue regions shrink, indicating that the leakage area is *proportional* to the memory size that is being activated.

The approximate surface area of an SRAM component can be estimated as $a = \frac{m \cdot a_{bit}}{e}$, where m is the number of bits in the memory region, a_{bit} is the area of a single-bit memory cell and e is the array layout efficiency (usually around 70%) [54]. The value of a_{bit} ranges from $600\lambda^2$ to $1000\lambda^2$, where λ is equal to half the feature size, i.e. for the current device-under-test $\lambda = 0.5 * 90\,\text{nm}$, thus the area of a 32-bit word is between 55 and 92 μm^2. Likewise, an 8 KByte region of the ARM Cortex-M4 amounts to an area of approximately 0.12 until 0.19 mm^2, depending on the fabrication process. Notably, this area estimation is quite close to the area of the yellow or the blue region of Fig. 2a (approximately half of the red rectangle). Similar spatial characteristics have been observed by Heyszl et al. [18] in the context of FPGA registers.

Thus, experimental evidence that suggest that (A) proximity exists between leaky regions and (B) the area of leaky regions is approximately proportional to the memory size that we activate. Section 3 builds up on these observations and develops a simple model that describes *spatial leakage*, yet we first need to provide the following disclaimer.

Word of Caution. The activation of a memory region can indeed be inferred by observing spatial leakage, which according to experimental data is quite rich in location information. Still, this does not imply that spatial leakage is the *sole* source of location leakage. It is possible that location information is also revealed through address leakage on the CPU and the memory control logic or buses when they process SRAM addresses, or even by other effects such as imperfect routing [52]. Thus, modeling spatial leakage captures *part* of the available information and can be considered as the first step towards full modeling of location leakage.

3 A Spatial Model for Location Leakage

Unlike the well-established power and EM data leakage models [12,46], high-resolution EM-based location leakage remains less explored. The main reason is the semi-invasive nature of location attacks (often requiring chemical decapsulation), the time-consuming chip surface scanning and the lengthy measurement procedures involved. Still, we maintain that such attacks are increasingly relevant due to the fairly average cost (approx. 15k euros), along with the widespread protection against data leakages [33,39], which encourages attackers towards different exploitation strategies.

Hence, this section puts forward a theoretical model that describes the spatial part of location leakage on a chip surface. The model can be viewed as an extension of the standard data-based model to the spatial domain, encapsulating the complexity of surface-scanning experiments. The proposed simulation of Sect. 3.1, in conjunction with the analysis of Sect. 3.2 can significantly enhance the design and evaluation cycle of SCA-resistant devices. Our approach allows the countermeasure designer to gauge the amount of experimental work an adversary would need to breach the device using spatial leakage. Thus, the designer can fine-tune protection mechanisms, provide customized security and avoid lengthy design-evaluation cycles by capturing certain security hazards at an early stage. The time-consuming leakage certification on the physical device can be carried out at a later stage, once obvious defects have been fixed. Naturally, all simulation-driven models (including this work) have inherent limitations, i.e. they are incapable to describe all the underlying physical phenomena, as we shall see in Sect. 4. Still, avoiding core issues early on, can free up valuable time that evaluators can invest towards device-specific effects such as coupling [10] and leakage combination [35].

3.1 Model Definition and Assumptions

Experimental Parameters. We define a side-channel experiment ϵ as any valid instance of the random variable set $\mathcal{E} = \{S, O, G, \mathbf{A}, \mathbf{P}\}$. The experimental parameters are shown in Table 1. We designate the experiment's goal to be the acquisition of spatial leakage $\mathbf{L}$, i.e. obtain $(\mathbf{L}|\mathcal{E} = \epsilon)$ or $(\mathbf{L}|\epsilon)$ for short. Much like in Sect. 2, the experiment consists of a probe scan over the chip surface in order to distinguish between different components (or regions) and ultimately between different memory addresses, registers, etc. The parameter S denotes the *area of the chip surface* on which we perform measurements, e.g. s can be the whole chip die (6 mm^2) or any smaller surface. Parameter O denotes the *area of the measuring probe* that we use in our experiments, e.g. the area o of the ICR HH 100-27 microprobe is roughly 0.03 mm^2. Continuing, parameter G denotes the *measurement grid dimensions*, i.e. it specifies the resolution of a uniform rectangular array of antennas [53]. In Sect. 2 we opted for $g = 300$. Continuing,

Table 1. Parameters of simulated experiment

Parameter	Description	Unit
S	chip surface area	u^2
O	probe area	u^2
G	scan grid dimension	¡no unit¿
A	component areas	vector with 1D entries of u^2
P	component positions	vector with 2D entries of u

the vector parameters $\mathbf{A}, \mathbf{P}$ describe the n_c surface components that emit EM-based spatial leakage. The parameter $\mathbf{A} = [A_1, A_2, \ldots, A_{n_c}]$ describes the surface *area occupied by each component*, e.g. in Sect. 2.3 we estimated the area of an 32-bit word component to be at most 92 μm^2. The parameter $\mathbf{P} = [\mathbf{P}_1, \mathbf{P}_2, \ldots, \mathbf{P}_{n_c}]$ describes the *position of every component* on the chip surface, i.e. $\mathbf{P}_i$ is a 2-dimensional vector. For simplicity, we assume the geometry of the surface, probe and components to be square, yet we note that the model can be extended to different geometrical shapes in a straightforward manner. Moreover, we assume that the measuring probe can capture only emissions that are directly beneath it, i.e. it functions like an identity spatial filter with area o.

Control Parameter. Every device can use program code to activate different components of the chip surface, e.g. by accessing different SRAM words through load/store instructions. To describe this, we use an additional control parameter $\mathbf{C}$ that denotes which components (indexed $1, \ldots, n_c$) are accessed during a particular experiment ϵ. Analytically, $\mathbf{C} = [C_1, C_2, \ldots, C_{n_c}]$, where $C_i = 1$ if component i is active during the experiment and $C_i = 0$ if it is inactive; for instance the vector $\mathbf{c} = [0, 1, 0]$ implies that the surface has 3 components ($n_c = 3$) and only component no. 2 is currently active. Note also that in our model only one out of n_c components can be active at a given point in time, since we assume that the ordinary microcontrollers do not support concurrent memory access.[4]. Thus the parameter $\mathbf{c}$ uses the one-hot encoding and we define $\mathbf{v}^i$ as an n_c-dimensional vector where all entries are zero except for the ith entry. For instance, if $n_c = 3$, then $\mathbf{v}^3$ is equal to $[0, 0, 1]$ and it describes the program state where only component no. 3 is active. In general, we use the notation $(\mathbf{L}|\mathcal{E} = \epsilon, \mathbf{C} = \mathbf{v}^i)$ or equivalently $(\mathbf{L}|\epsilon, \mathbf{v}^i)$ to describe a side-channel experiment ϵ that captures the leakage when the ith component is active. For an attack to be successful, we need to distinguish between two (or more) different components using this spatial leakage. Formally, we need to be able to distinguish between $(\mathbf{L}|\epsilon, \mathbf{v}^i)$ and $(\mathbf{L}|\epsilon, \mathbf{v}^j)$, for $i \neq j$.

Representative Example. To elucidate the model, Fig. 3 presents an experiment ϵ with parameters $\{s, o, g, \mathbf{a}, \mathbf{p}\} = \{25, 3, 2, [0.8, 3], [[0.6, 1.5], [1.6, 4.1]]\}$, where all position parameters are in arbitrary units u and all area parameters are in square units u^2. The experiment targets two components ($n_c = 2$) and their position is $[0.6\ u, 1.5\ u]$ and $[1.6\ u, 4.1\ u]$ respectively. The surface area s, probe area o, and component areas a_1 and a_2 are respectively 25 u^2, 3 u^2, 0.8 u^2 and 3 u^2. The dimension g of the measurement grid is 2, resulting in a 2×2 scan and we capture a single measurement (trace) in every grid spot. We use the program code (control parameter) to activate components 1 and 2, generating $(\mathbf{L}|\epsilon, [1, 0])$ and $(\mathbf{L}|\epsilon, [1, 0])$ respectively. Note that in general $(\mathbf{L}|\epsilon, \mathbf{v}^i)$ results in leakage with g^2 dimensions, e.g. $(\mathbf{L}|\epsilon, [1, 0])$ is a 4-dimensional vector. We refer to the leakage measured at any specified position $[x, y]$ as $(L_{[x,y]}|\epsilon, \mathbf{v}^i)$ or simply $L_{[x,y]}$.

[4] Parallel word processing can be easily included.

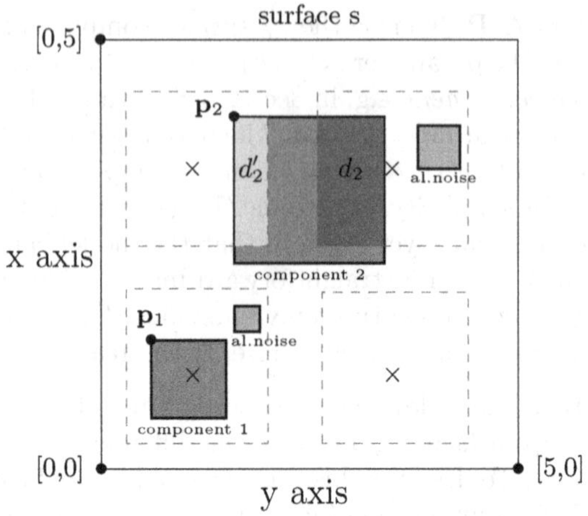

Fig. 3. Sample experiment ϵ. The $\times$ spots show the measurement points of the 2×2 scan grid. Dashed black-line rectangles enclosing these spots denote the measuring probe area o. Vectors $\mathbf{p}_1, \mathbf{p}_2$ show the position of two components ($n_c = 2$), whose areas (a_1, a_2) are enclosed by the solid black-line rectangles. The blue area d_2 shows the area of component 2 captured by the top-right measurement point and the yellow area d_2' shows the area of component 2 captured by the top-left measurement point. (Color figure online)

Independent Noise. In accordance with standard data-based leakage models, we assume that for given parameters $\epsilon, \mathbf{v}^i$, the leakage $L_{[x,y]}$ at any grid position $[x, y]$ consists of a deterministic part $l^{det}_{[x,y]}$, an algorithmic noise part N^{algo} and an electrical noise part N^{el}, thus: $L_{[x,y]} = l^{det}_{[x,y]} + N^{algo} + N^{el}$.

Deterministic Leakage. We assume that the deterministic part of the leakage $l^{det}_{[x,y]}$ at position $[x, y]$ is caused by the activation (switching behavior) of any component that is captured by the probe at this grid position. Based on the experimental observations of Sect. 2.3, we assume the deterministic leakage to be proportional to the area of the active component located underneath the probe surface, thus:

$$l^{det}_{[x,y]}|\mathbf{v}^i = \begin{cases} 0, & \text{if comp. } i \text{ is not captured at [x,y]} \\ d_i,\ 0 < d_i < a_i, & \text{if comp. } i \text{ is partially captured at [x,y]} \\ a_i, & \text{if comp. } i \text{ is fully captured at [x,y]} \end{cases}$$

For example, Fig. 3 shows that component 1 is fully captured by the probe on the bottom-left grid spot, thus $(l^{det}_{[down,left]}|\mathbf{v}^1) = a_1$. Since no other measurement position can capture component 1, it holds that $(l^{det}|\mathbf{v}^1) = 0$ for the other three grid positions. On the contrary, component 2 is partially captured in two grid positions. Thus, it holds that $(l^{det}_{[up,right]}|\mathbf{v}^2) = d_2$ (blue area), $(l^{det}_{[up,left]}|\mathbf{v}^2) = d'_2$ (yellow area) and zero elsewhere.

Electrical and Algorithmic Noise. We employ the common assumption that the electrical noise N^{el} follows a normal distribution with zero mean and variance σ^2_{el}, i.e. $N^{el} \sim Norm(0, \sigma^2_{el})$. The variance σ^2_{el} is related to the specific device-under-test and measurement apparatus that we use.

The algorithmic noise in our model is caused by components that, like the targeted components, leak underneath the probe on measurement spot of the scan grid. However, unlike our targeted components, they exhibit uniformly random switching activity (equiprobable 'on' and 'off' states) that is independent of the control parameter $\mathbf{c}$. If n_a such components, with area parameter $\mathbf{b} = [b_1, b_2, \ldots, b_{n_a}]$ are located under the probe, then we assume again their leakage to be proportional to the respective captured area. The leakage of these independent, noise-generating components is denoted by N^{algo}_i, $i = 1, \ldots, n_a$. Thus, N^{algo} constitutes of the following sum.

$$N^{algo} = \sum_{i=1}^{n_a} N^{algo}_i, \text{where } N^{algo}_i \sim Unif(\{0, b_i\})$$

The algorithmic noise is highly dependent on the device-under-test, i.e. we could potentially encounter cases where there is little or no random switching activity around the critical (targeted) components, or we may face tightly packed implementations that induce such noise in large quantities. For example, in Fig. 3 the top-left and bottom-right spots have no algorithmic noise, while the top-right and bottom-left spots contain randomly switching components (red rectangles) that induce noise. Note that the larger the probe area o, the more likely we are to capture leakage from such components. Appendix A elaborates on the form of algorithmic noise on tightly-packed surfaces.

3.2 Information-Theoretic Analysis

The proposed spatial leakage model of Sect. 3.1 is able to simulate the EM emission over a chip surface and provide us with side-channel observables. Due to the complexity of surface-scanning experiments, the model needs to take into account multiple parameters in ϵ (component area, grid size, noise level, etc.), all of which can directly impact our ability to distinguish between different regions.

In order to demonstrate and gauge the impact of the experimental parameters on the side-channel security level, this section introduces an information-theoretic apporach to analyze the following simple location-leakage scenario. Using the model of Sect. 3.1, we simulate the spatial leakage emitted by the ARM Cortex-M4 SRAM, while accessing a lookup-table (LUT) of 256 bytes. This LUT computation, emulates the spatial leakage of an AES LUT, while excluding any data-based leakages. The processor uses a 32-bit architecture, thus we represent the 256-byte lookup table with 64 words (4 bytes each) stored consecutively in SRAM. The LUT memory region is placed randomly[5] on a chip surface with $s = 0.6\,\text{mm}^2$. Then our model generates leakage stemming from 64 chip components ($n_c = 64$), where each one occupies surface area pertaining to 4 SRAM bytes. Using the simulated traceset, we perform template attacks [6] after PCA-based dimensionality reduction [3], in order to distinguish between different LUT regions (consisting of one or more words). Being able to infer which LUT/SRAM region was accessed can substantially reduce the number of AES key candidates. For instance, the adversary may template separately the leakage of all 64 words (high granularity) in order to recover the exact activated word and reduce the possible AES key candidates from 256 to 4. Alternatively, he can partition the LUT to two regions (words 0 until 31 and words 32 until 63), profile both regions (low granularity), in order to recover the activated 128-byte region and reduce the AES key candidates from 256 to 128.

Formally, at a certain point in time, the microcontroller is able to access only one out of 64 components (high granularity), thus the control variable $\mathbf{c} \in \mathcal{V} = \{\mathbf{v}^1, \mathbf{v}^2, \ldots, \mathbf{v}^{64}\}$ and the adversary can observe the leakage of word-sized regions $(\mathbf{L}|\mathbf{C} = \mathbf{v}^i)$, for $i = 1, 2, \ldots, 64$. Alternatively (low granularity), he can focus on $|\mathcal{R}|$ memory regions and partition the set $\mathcal{V}$ to sets $\mathcal{V}^1, \mathcal{V}^2, \ldots, \mathcal{V}^{|\mathcal{R}|}$, where usually $\mathcal{V}^r \subset \mathcal{V}$ and $\mathcal{V}^i \cap \mathcal{V}^j = \emptyset$, for $i \neq j$. We define random variable $R \in \mathcal{R} = \{1, 2, \ldots, k\}$ to denote the activated region and we represent the leakage of region r as $(\mathbf{L}|R = r) = (\mathbf{L}|\mathbf{c} \in \mathcal{V}^r)$. For example, in the high granularity scenario, the adversary observes and profiles $(\mathbf{L}|\mathbf{v}^1), (\mathbf{L}|\mathbf{v}^2), \ldots, (\mathbf{L}|\mathbf{v}^{64})$, while in the low granularity scenario he profiles two regions ($\mathcal{R} = \{1, 2\}$) with $\mathcal{V}^1 = \{\mathbf{v}^1, \mathbf{v}^2, \ldots, \mathbf{v}^{32}\}$ and $\mathcal{V}^2 = \{\mathbf{v}^{33}, \mathbf{v}^{34}, \ldots, \mathbf{v}^{64}\}$. Thus he can obtain $(\mathbf{L}|R = 1) = (\mathbf{L}|\mathbf{c} \in \mathcal{V}^1)$ and $(\mathbf{L}|R = 2) = (\mathbf{L}|\mathbf{c} \in \mathcal{V}^2)$.

Having completed the profiling of regions for a certain experiment ϵ, we quantify the leakage, using the perceived information metric (PI) [38] as follows.

$$PI(\mathbf{L}; R) = H[R] - H_{true,model}[\mathbf{L}|R] =$$

$$H[R] + \sum_{r \in \mathcal{R}} Pr[r] \cdot \int_{\mathbf{l} \in \mathcal{L}^{g^2}} Pr_{true}[\mathbf{l}|r] \cdot log_2 Pr_{model}[r|\mathbf{l}] \; d\mathbf{l}$$

$$Pr_{model}[r|\mathbf{l}] = \frac{Pr_{model}[\mathbf{l}|r]}{\sum_{r^* \in \mathcal{R}} Pr_{model}[\mathbf{l}|r^*]} \quad Pr_{true}[\mathbf{l}|r] = \frac{1}{n_{test}}, n_{test} \text{ size of test set}$$

PI can quantify the amount of information that leakage $\mathbf{L}$ conveys about the activated region R, taking into account the divergence between the real and estimated distributions. Computing PI requires the distribution $Pr_{model}[\mathbf{l}|r]$, i.e. the template that is estimated from the training dataset. In addition, it requires the true leakage distribution $Pr_{true}[\mathbf{l}|r]$, which is unknown and can only be sampled directly from the test dataset. We opt for this metric since it indicates when degraded (under-trained) leakage models are present, due to our choice of experimental parameters. Negative PI values indicate that the trained model

[5] Unless specified otherwise, we place every word directly next to each other, starting from a random position in the surface.

is incapable of distinguishing regions, while a positive value indicates a sound model that can lead to classification.

Using the proposed leakage simulation and the PI metric we evaluate several scenarios for the LUT case. Sections 3.2 until 3.2 showcase how different experimental parameters hinder or enhance leakage, offering several design options. To apply the theoretical model in an evaluation context we can simply set our current device SNR to the PI graphs.

Area and Number of Regions. The first simulation scenario examines the core attack question: using a certain experimental setup with parameters $\epsilon = \{s, o, g, \mathbf{a}, \mathbf{p}\}$, what is the smallest region size that I can distinguish reliably? Rephrasing, we assess how much location information can be extracted from the observed leakage by plotting the $PI(\mathbf{L}; R)$ metric against the electrical noise variance σ_{el}^2 for certain ϵ and $\mathbf{c}$ parameters. We simulate an adversary that distinguishes regions of a 256-byte LUT using the following three LUT partitions of increasing granularity. First, he partitions the 256-byte LUT to 2 regions of 128 bytes each (depicted by the solid line in Fig. 4). Second, he partitions the LUT to 8 regions of 32 bytes (dashed line) and third to 16 regions of 16 bytes (dotted line). For every partition we profile the regions' leakage $(\mathbf{L}|R = r) = (\mathbf{L}|\mathbf{c} \in \mathcal{V}^r)$ for $r = 1, 2, \ldots, |\mathcal{R}|$, where $|\mathcal{R}| = 2$ or 8 or 16 and subsequently tries to distinguish. Note that surface $s = 6\,\text{mm}^2$, probe size $o = 0.03\,\text{mm}^2$ (ICR HH 100-27), feature size 90 nm and $g = 100$, i.e. the scan resolution is 100×100. The component area $a = 92\,\mu\text{m}^2$ for all SRAM words and the words are placed adjacent to each other, starting from a random surface position; we denote this as $\mathbf{p} = random$. Along with parameters ϵ and $\mathbf{c}$, we need to include the measurement complexity in our simulation. Thus, we specify the amount of traces measured at every grid spot, resulting in an acquisition of $g^2 \cdot \#traces$. As expected, the experiments with higher region granularity yield more location information, as shown by the vertical gaps of the PI metric in Fig. 4. Still, we also observe that smaller regions are harder to distinguish, even for low noise levels. Partitioning to 8 or 16 regions could optimally yield 3 or 4 bits of information respectively, yet the dashed and dotted curves remain well below this limit. Thus, we note that the adversary may need to improve his experiment ϵ by measuring more traces, using smaller probes or increasing the grid dimension in order to extract the maximum information.

Measurement Grid Dimension. Any side-channel experiment involving surface scanning can be particularly time-consuming. Moving the microprobe between adjacent positions takes approximately 2 s, thus the 300×300 surface scan carried out in Sect. 2 takes almost 2 days to conduct. Using the spatial leakage simulation, we can specify the grid dimension g and find the minimum scan resolution required to distinguish between certain SRAM regions. Figure 5 demonstrates the information captured when conducting scans with resolutions 100×100, 40×40 and 20×20, taking approximately 6 h, 1 h and 15 min respectively. Across the three simulations we maintain constant data complexity of

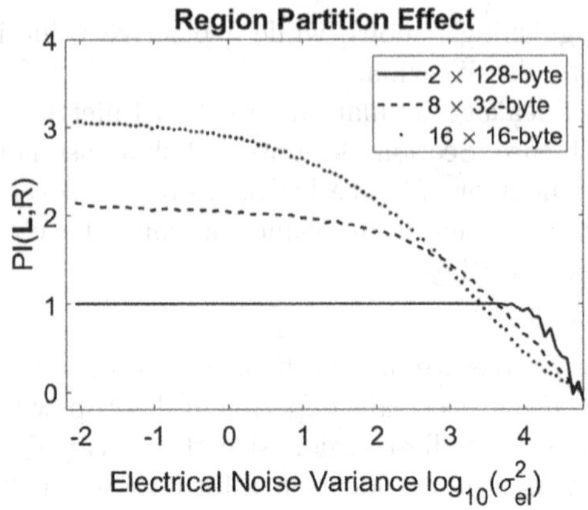

Fig. 4. Region partition of 256-byte LUT to 2, 8, 16 regions. Parameters $\epsilon = \{6\,\text{mm}^2, 0.03\,\text{mm}^2, 100, 92\,\mu\text{m}^2, \text{random}\}$, capturing 10 traces/spot 100k traces in total.

100k traces, distributed to grid spots accordingly (10, 62 and 250 traces per spot). Figure 5 shows information loss (vertical gap) as the grid dimension is decreasing, i.e. when trying to distinguish 4 regions only the 6 hour-experiment with 100×100 grid is able reach maximum information (2 bits). Notably, we see that for larger noise levels, small grid sizes with many traces per spot (dense measurements) are able to outperform larger grid sizes with less traces per spot (spread measurements).

Feature Size. A common issue encountered in the side-channel literature is the scaling of attacks and countermeasures as devices become more complicated and feature size decreases [22, 30, 32]. This section uses our simple leakage model to describe the effect of feature size on SCA. We simulate the location leakage of SRAM cells fabricated with 180 nm, 120 nm and 90 nm technologies, resulting in bit cell areas of approximately $8\,\mu\text{m}^2$, $3.5\,\mu\text{m}^2$ and $2\,\mu\text{m}^2$. The results are visible in Fig. 6. Naturally, smaller technology sizes can potentially limit the amount of available information, as they decrease the region's area and force the adversary towards more expensive tooling.

Algorithmic Noise. This section simulates the countermeasure of spatial algorithmic noise, when implemented on the ARM device. Analytically, we examine the case where the designer is able to place word-sized noise-generating components on the chip surface in order to "blur" the location leakage of a targeted region and hinder recovery. The simulation (Fig. 7) uses the analysis of Appendix A to approximate the algorithmic noise when the probe captures the leakage of 11 SRAM words, one of which is the target word (and reveals the critical region information) and the ten remaining words are randomly activated at the same time. Observing Fig. 7, we see the algorithmic-noise PI curve (dashed

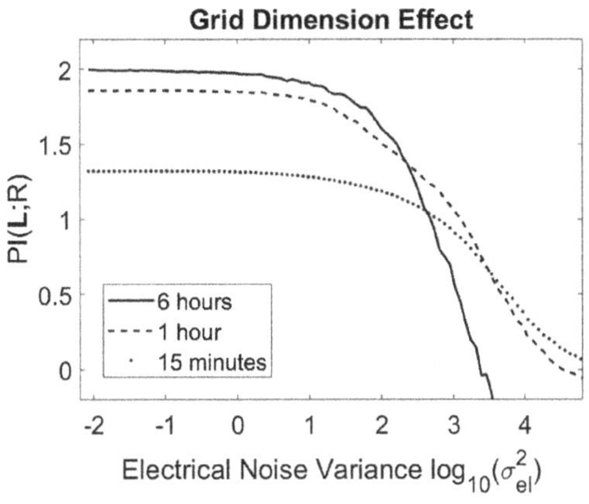

Fig. 5. Grid dimension $g = 100, 40$ and 20. Parameters $\epsilon = \{6\,\text{mm}^2,\ 0.03\,\text{mm}^2,\ \text{g},\ 92\,\mu\text{m}^2,\ \text{random}\}$, distinguishing 4 regions of 64 bytes each, 100k traces in total.

line) shifting to the left of the PI curve without algorithmic noise (solid line). Thus, much like data-based algorithmic noise [47], we see that randomly activating words functions indeed as an SCA countermeasure.

Region Proximity and Interleaving. Last, we simulate the countermeasure of region proximity and region interleaving on the ARM device, which was considered by He et al. [16] and Heyszl [17]. Analytically, we assume that the designer controls the place-and-route process and can place two memory regions on the chip surface using the following three configurations.

1. Distant placement: the distance between the two regions is roughly 1 *mm*.
2. Close placement: adjacent placement of the two regions.
3. Interleaved placement: the words of the two regions are interleaved together in a checkered fashion, i.e. word 0, 2, 4, ... of SRAM belongs to 1st region and word no. 1, 3, 5, ... belongs to 2nd region.

Figure 8 demonstrates the effect of different placement choices, confirming the basic intuition that higher proximity is essentially a countermeasure against location-based leakage. The vertical gap in PI between distant, close and interleaved placement shows that as components get closer, the attainable information decreases, forcing the adversary to increase the grid size or use a smaller probe.

4 Exploitation Using Template Attacks

Having established a theoretical model for spatial leakages, we move towards a practical scenario. In particular, we exploit the available *location leakages* in the ARM Cortex-M4 so as to infer the accessed memory position of an 256-byte, data-independent LUT. Note that in the real chip we cannot isolate spatial

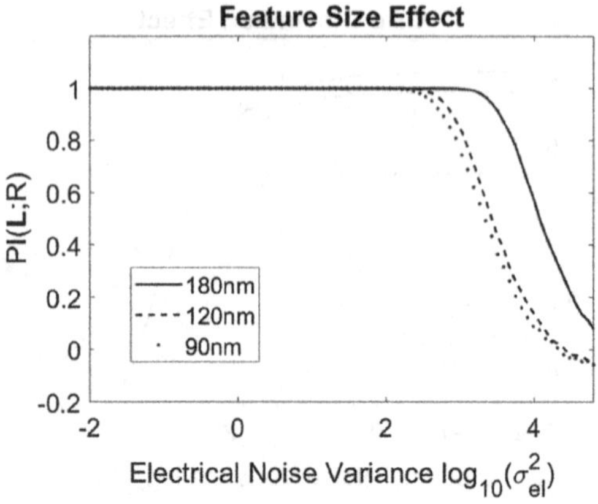

Fig. 6. Feature size of 180, 120, 90 nm, word area a = 368, 163, 92 μm^2. Parameters $\epsilon = \{6\,\text{mm}^2,\ 0.03\,\text{mm}^2,\ 40,\ \text{a},\ \text{random}\}$, for 2 regions of 128 bytes each, 250 measurements/spot, 400k traces in total.

from address leakage, i.e. we observe location leakage in its entirety. We use the template attack [6], i.e. we model the leakage using a multivariate normal distribution and attack trying to identify the key, or in our case region r of the SRAM.

The leakage vector $(\mathbf{L}|R = r)$ exhibits particularly large dimensionality and can generate a sizeable dataset, even for modest values of the grid dimension g. Thus, we employ dimensionality reduction techniques based on the correlation heuristic so as to detect points of interest (POIs) in the 300 × 300 grid and use a train-test ratio of 70–30. In addition, when performing template matching, we combine several time samples from the test set together (multi-sample/multi-shot attack), in order to reduce the noise and improve our detection capabilities[6]. We also opt for the improved template formulas by Choudary et al. [9] with pooled covariance matrix and numerical speedups. The goal of our template-based evaluation is not only to answer whether location exploitation is possible but also to gauge the effect of the experimental parameters ϵ on the exploitation process. Thus, similarly to Sects. 3.2, 3.2 and 3.2, we will investigate the effect of region partition, grid dimension and region placement in the real-world scenario. Unfortunately Sects. 3.2 and 3.2 would require control over the manufacturing process (i.e. chips of different feature size) or control over regions with algorithmic noise (i.e. parallel memory activation), thus they cannot be tested in our current context. Throughout this section we will engage in comparisons between the theoretical model of Sect. 2 and our real-world attack, i.e. we will put the

[6] Whether this constitutes an option depends on the situation. If any sort of randomization such as masking or re-keying is present in the device then the adversary is limited in the number of shots that he can combine.

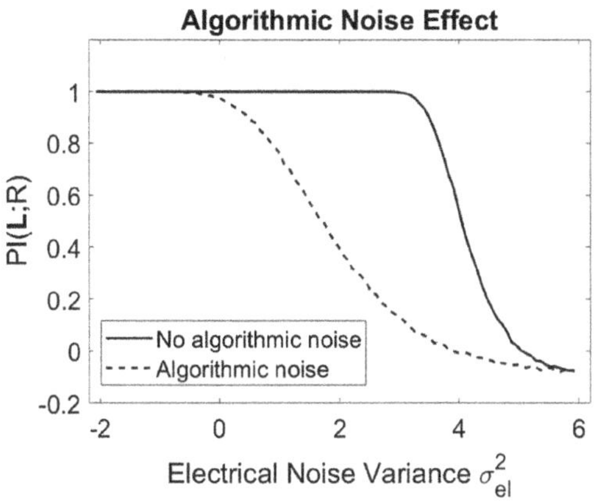

Fig. 7. Deploying 10 noise-generating words. Parameters $\epsilon = \{6\,\text{mm}^2,\ 0.03\,\text{mm}^2,\ 40,\ 92\,\mu\text{m}^2,\ \text{random}\}$, for 2 regions of 128 bytes each, 250 measurements/spot, 400k traces in total.

model's assumptions to test, discover its limitations and obtain more insight into the source of location leakage.

4.1 Area and Number of Regions

To observe the effect of partitioning, we gradually split the 256 bytes of the LUT into classes and built the corresponding template for each class. We perform a template attack on 2, 4, 8 and 16 partitions (with 128, 64, 32 and 16 bytes each respectively), i.e. we gauge the distinguishing capability of the adversary, as the number of components increases and their respective areas decrease. The results are visible in Fig. 9, which showcases how the number of grid positions (spatial POIs) and samples/shots per attack affects the success rate (SR). The adversary can achieve a success rate of 100% when distinguishing between 2 or 4 regions, assuming that he uses multiple samples in his attack. The success rate drops to 75% for 8 and 50% for 16 regions, an improvement compared to random guess SRs of 12.5% and 6.25% respectively. Although we are not able to reach successful byte-level classification, we can safely conclude that location-based attacks are definitely possible on small LUTs and they can reduce the security level of an LUT-based implementations, unless address randomization countermeasures are deployed. When performing *single-shot* attacks, the template strategy becomes less potent, achieving SR of 57%, 33%, 17% and 11% for 2, 4, 8 and 16 regions, i.e. only slightly better than a random guess. In order to compare the success rate of the real attack to the theoretical model, we compute the model's SR for current device SNR under the same data complexity[7]. The model's *single-shot* SR is 99%, 50%, 13% and 12% for 2, 4, 8 and 16 regions respectively. We

[7] The template attack uses the experimental data, while the theoretical SR uses simulated data of the same size and dimensionality.

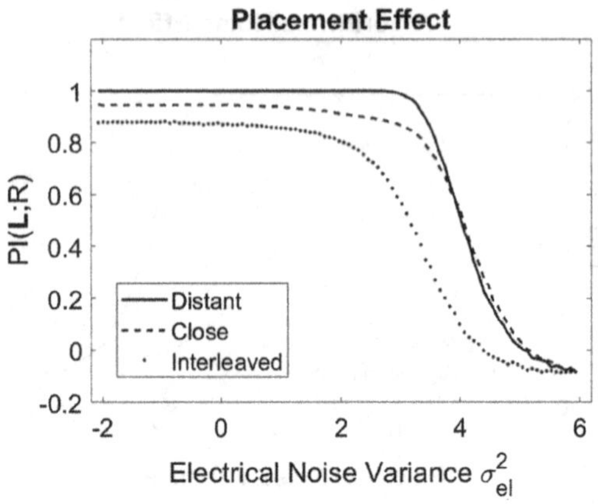

Fig. 8. Distant, close, interleaved placement. Parameters $\epsilon = \{6\,\text{mm}^2,\ 0.03\,\text{mm}^2,\ 20,\ 92\,\mu\text{m}^2,\ \text{random}\}$, distinguishing 2 regions of 128 bytes each and using 250 measurements/spot, 100k traces in total.

observe that the model follows the same trend, yet the device leakage exhibits divergences that indicate modeling imperfections.

4.2 Measurement Grid Dimension

Using the same approach, we evaluate the effect of grid dimension on the success rate of the template attack. We commence with the full 300 × 300 grid (2-day experiment) and subsequently scale down to 40×40 grid (1-hour) and 10×10 grid (2-minutes), as shown in Fig. 10. We observe that for small grid sizes such as 10× 10 the reduced dataset makes training harder, yet the multi-shot template attack is able to distinguish with SR equal to 100%. On the contrary, the theoretical model is unable classify correctly because the spatial POIs are often missed by such a coarse grid. To pinpoint this model limitation, we assess the spread of the POIs across the die surface and we visualize the best (according to correlation) grid positions in Fig. 11. Interestingly, we discover numerous surface positions that leak location information, while being far away from the SRAM circuitry itself. This finding is in accordance with Unterstein et al. [51] on FPGAs. The figure suggests that location leakage is a combination of SRAM spatial leakage (as in the model), address leakage in the control logic and potentially out-of-model effects.

4.3 Region Proximity and Interleaving

Finally, we evaluate the effect of region proximity and interleaving on the SR of templates. We examine close placement (adjacent SRAM regions), distant

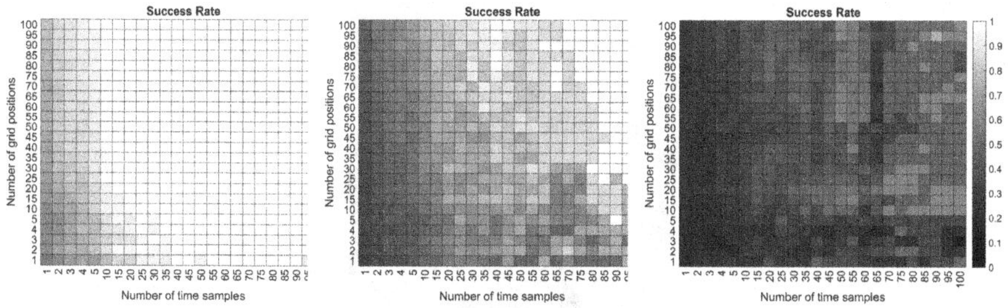

(a) 2 regions of 128 bytes each (b) 4 regions of 64 bytes each (c) 8 regions of 32 bytes each

Fig. 9. The success rate of the template classifier as we partition the LUT. Y-axis denotes the number of spatial POIs used in model, X-axis denotes the number of samples used in attack. Scale denotes SR where white is 100% and black is 0%.

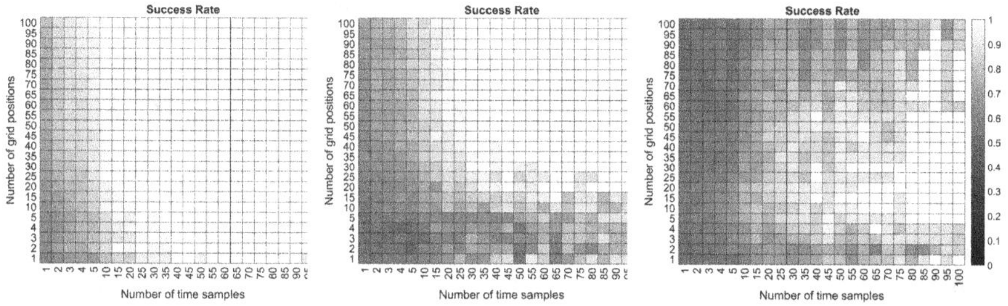

(a) 300 × 300 grid (b) 40 × 40 grid (c) 10 × 10 grid

Fig. 10. The success rate of the 2-region template classifier as we decrease the experiment's grid size.

placement (SRAM regions at a large distance[8]) and word-interleaved placement (checkered SRAM regions). The results are visible in Fig. 12. We observe that in all cases we reach multi-sample SR of 100%, in accordance with the theoretical model at the device SNR. However, attacking the word-interleaved LUT requires a bigger effort in modeling in terms of both grid POIs and samples per attack. Likewise, distinguishing between distant regions puts considerably less strain on the model. Thus, we conclude that distance and interleaving does indeed function like a countermeasure against location leakage, albeit it offers only mild protection in our ARM device.

[8] Without knowledge of the chip layout we cannot be fully certain about the distance between memory addresses. Here we assume that the low addresses of the SRAM are sufficiently distant from mid ones, which are approx. 8 KBytes away.

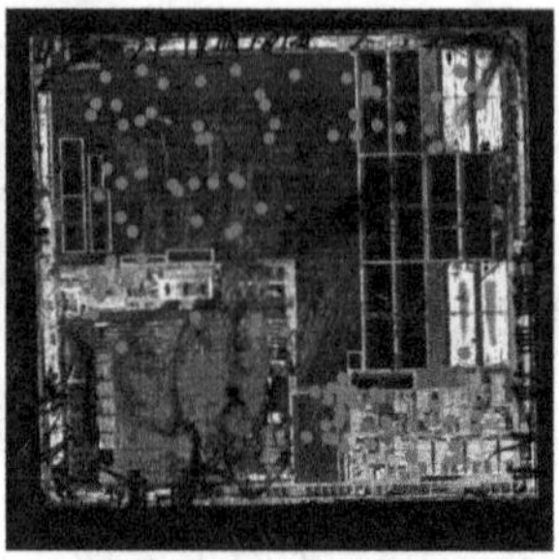

Fig. 11. Spread of spatial POIs on chip surface.

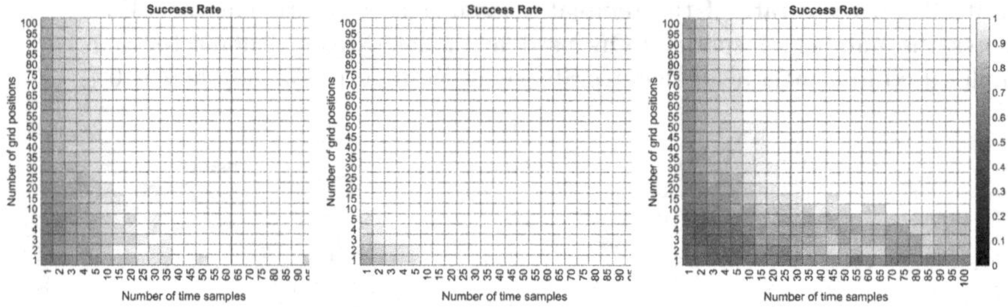

(a) close (b) distant (c) word-interleaved

Fig. 12. The success rate of the 2-region template-based classifier as we change the placement of regions.

5 Exploitation Using Neural Networks

Despite the fact that the multivariate normal leakage assumption is fairly realistic in the side channel context, applying distribution-agnostic techniques appears to be another rational approach [26]. Over the past few years, there has been a resurgence of interest in Deep Learning (DL) techniques, powered by the rapid hardware evolution and the need for rigorous SCA modeling [5,27–29,37,55].

In this section, we evaluate the performance of various DL methods, including convolutional neural networks (CNNs, Subsect. 5.1) and multi layer perceptrons (MLPs, Subsect. 5.2), in inferring the activated region of the 256-byte, data-independent LUT on the ARM Cortex-M4. First, motivated by reusable neural networks, we use trained CNNs with all grid positions (no POI selection), but the results are fairly unsuccessful. We suspect that the results are affected by noise coming from spatial points that contain no location information. This approach can be compared to using full pictures for DL training.

To solve the above problem, we apply a dimensionality reduction based spatial POI selection similarly to Sect. 4.1. We notice that even a simple CNN already provides good results in a limited number of epochs and there is no gain in using a complex CNN. Thus, we move our attention to simpler MLPs. Interestingly, this approach surpasses the template attacks in effectiveness, enabling stronger location-based attacks that use less attack samples and can distinguish between smaller regions. The above method can be compared to using only the most meaningful parts of pictures for DL training.

5.1 Convolutional Neural Network Analysis

Fully Pretrained CNNs. Before developing and customizing our own CNN model, we evaluate the performance of existing, state-of-the-art pre-trained networks. Pre-trained models are usually large networks that have been trained for several weeks over vast image datasets. As a result, their first layers tend to learn very good, generic discriminative features. Transfer Learning [34] is a set of techniques that, given such a pre-trained network, repurposes its last few layers for another similar (but not necessarily identical) task. Indeed, the objectives of our spacial identification task appear to be very close to those of standard image classification. Moreover, as outlined in Sect. 2, our data is formulated as 300×300 grid images, which makes them compatible with the input format of several computer vision classification networks. For this first attempt at CNN classification we use several state-of-the-art networks, namely Oxford VGG16 and VGG19 [42], Microsoft ResNet50 [15], Google InceptionV3 [50] and Google InceptionResNetV2 [49]. It should be noted that the input format of these networks is often RGB images, while our 300×300 heatmaps resemble single-channel, grayscale images. To address this and recreate the three color channels that the original networks were trained for, we experiment with two techniques; (1) we assemble triplets of randomly chosen heatmaps, and (2) we recreate the three color channels by replicating the heatmaps of the samples three times.

We apply the pretrained CNN classification on 2 closely placed SRAM regions of 128 bytes each. In accordance with the standard transfer learning methodology, during re-training we freeze the first few layers of the networks to preserve the generic features they represent. In each re-training cycle, we perform several thousand training-testing iterations. Despite all these and multiple hours of training, none of the aforementioned CNNs results into a retrained network with high classification success rate.

Custom Pretrained CNNs. As a result of the low success rate of fully pretrained networks, we choose to proceed with Xavier [14] weight initialization and training from scratch. We observe that, despite the transformation of the sequential problem (SRAM accesses over time) to a spatial one, our dataset is dissimilar to visual classification datasets. Rephrasing, the images that we have to cope with feature intricate characteristics having little resemblance with those of the datasets that the pretrained CNN versions have been trained on, such as the ImageNet dataset [11]. Moreover, due to the fully distribution-agnostic approach, any randomly initialized CNN may suffer the effect of vanishing or exploding gradients, a danger that Xavier initialization should eliminate. The framework that was used for training and evaluating our customized CNNs is

Keras [7] over TensorFlow [1] backend and the customized CNNs tested were VGG19 [42], InceptionV3 [50], ResNet50 [15], DenseNet121 [19] and Xception [8]. We also made use of the *scikit-learn* Python library [36] for the preprocessing of our data. The execution of this customized CNN training and testing was carried out in ARIS GRNET HPC (High-Performance Computing) infrastructure[9].

To gauge the effect of SRAM memory addressing on the training, all five CNNs are trained in two ways: one-batch training and multiple-batch training. During one-batch training we use location leakage from a single SRAM LUT, while for multiple-batch training we use four LUTs placed within a 16 KByte SRAM address range. The dataset is split into training, validation and test sets using a 70-20-10 ratio and is standardized by removing the median and scaling the data according to the quantile range. The networks are trained for 150 epochs of 32 images each, using the Adam optimizer [23] with default parameters. The results are visible in Fig. 13.

We observe that the single-shot success rate of the Xception network (green line) exceeds by far all others' at 84% and the SR improves in stability when using multiple-batch training. It is worth noting that some CNNs, especially VGG19, remain incapable of learning anything meaningful about the discrimination of the two 128-byte regions. Another troubling fact is the sudden drops of validation accuracy during training time for both best-performing networks, Xception and ResNet50, a phenomenon rather indicative of overfitting. In our efforts to squeeze the best possible performance without sacrificing training stability and generalization capacity, we investigated the tolerance of the best performing network against two additional preprocessing techniques, namely sample-wise standardization and feature-wise standardization. The test set success rate of the three alternative techniques is visible in Table 2. Comparing with Sect. 4, we observe that CNNs are capable of surpassing the single-shot accuracy of template attacks, reaching 88% and making the CNN-based attack particularly useful against randomization countermeasures that limit the number of samples we can combine. Moreover we observe that spreading the training phase over several SRAM addresses (multiple batch) can assist classification, showing that the knowledge learned in a certain address range may be applicable elsewhere in the SRAM.

Table 2. Success rate of Xception network for alternative preprocessing techniques.

Alternative Pipeline	Preprocessing step	Success Rate
Xception-V1	dataset-wise, robust to outliers standardization	84.47 %
Xception-V2	sample-wise standardization	88.636 %
Xception-V3	feature-wise standardization	84.848 %

[9] https://hpc.grnet.gr/en/.

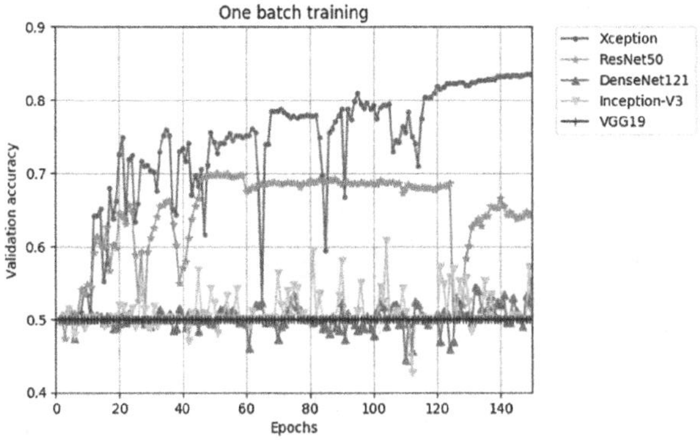

(a) Single-batch training.

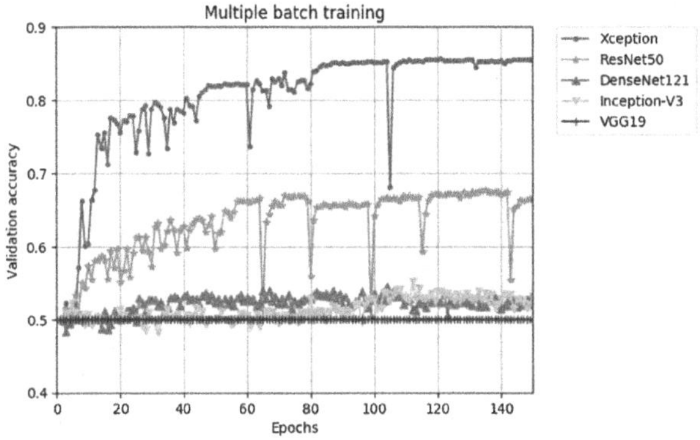

(b) Multiple-batch training.

Fig. 13. CNN validation accuracy for single/multiple-batch training.

5.2 Multi Layer Perceptron Network Analysis

We believe that the results from the previous section are affected by noise coming from spatial points that contain no location information. To eliminate this issue, we apply a dimensionality reduction techniques based on the correlation heuristic to detect the best spatial POI in the 300 × 300 grid, like in Sect. 4.1 Initial results using CNN show that even a simple CNN already provides good results in a limited number of epochs. Therefore there is no gain in using a complex CNN and we move our focus to simpler MLPs.

In [27], the authors presented how to use Multi Layer Perceptron (MLP) network to perform SCA on AES. In this section we present how to use MLP to recognize accesses to different addresses in the memory. Based on experiments, we discover that 5000 POI yielded the best network training.

Table 3. Hyper-Parameters for training and validation.

Epochs	30 – 80 (depends on the number of regions)
Mini-Batch	100
Learning Rate	0.003
Learning Rate Decay Rate	0.5%
Learning Rate Decay Interval	100 epochs
L1	0.001
L2 (weight decay)	0.001
Weight Initialization	RELU
Activation Output Layer	SOFTMAX
Loss Function	NEGATIVELOGLIKELIHOOD
Updater	NESTEROVS
1 Dense Layer:	
- Number of Neurons:	20
- Activation Dense Layer:	TANH

We define our MLP to contain a single dense layer and used the back-propagation with NESTEROVS updater, with momentum 0.9, during training. The weights are initialized at random and applied to a RELU activation. The MLP is also configured with L1 and L2 regularization in order to improve the generalization. The analysis presented in this section is performed using Deep Learning for Java[10] in conjunction with Riscure Inspector deep learning functionality. To observe the effectiveness of MLPs, we gradually partition the 256 bytes of the LUT into classes and built the corresponding MLP for each class. We perform an MLP analysis on 2, 4, 8, 16, 32, 64, 128, and 256 partitions (with 128, 64, 32, 16, 8, 4, 2, and 1 bytes each, respectively). The dataset is split into training, validation and test sets using a 40-30-30 ratio. Then we select best hyper-parameters for training and validation[11] of our MLP network using a trial and error method. The chosen parameters are listed in Table 3.

The validation accuracy for 2, 4, 8, 16, 32, 64, 128, and 256 partitions is visible in Fig. 14a. We have discovered that we achieve the best results for various numbers of epochs depending on the number of partitions. We have used 30 epochs for the 2 and 4 partitions, 40 epochs for the 16 and 32 partitions, 40 epochs for the 8 partitions, 70 for the 128 partitions, and 80 for the 64 and 256 partitions. Figure 14a indicates that the MLP network reaches high accuracy even for a large number of regions. To visualize the validation set success rate we present the validation final partitioning (for 16 partitions) in Appendix B. The greatest values are located on the diagonal and this indicates that the MLP learns correctly with high probability. The attack success rates for the test traces for 2, 4, 8, 16, 32, 64, 128, and 256 partitions are presented in Fig. 14b; the exact

[10] https://deeplearning4j.org/.

[11] The MLP parameters are chosen to maximize the attack success rate (which is equivalent to accuracy).

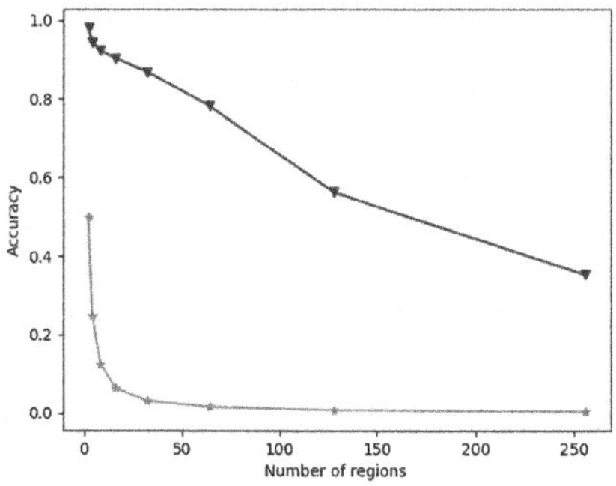

(a) Blue line denotes validation accuracy and red line denotes random guess success rate.

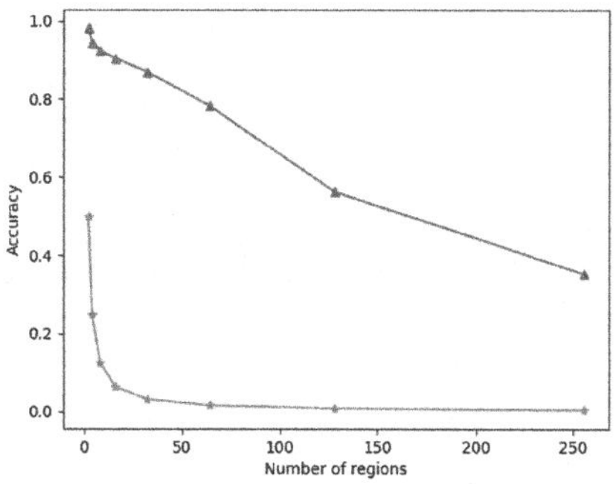

(b) Green line denotes attack success rate and red line denotes random guess success rate.

Fig. 14. Validation accuracy for training and success rate for testing in MLP. (Color figure online)

accuracy values are 96%, 91%, 90%, 88%, 83%, 75%, 57%, and 32%, respectively. As expected, these values are slightly lower for the attacking phase than the validation ones in the learning phase. We observe that even the SR for the 256 partitions, namely the 32% SR is significantly higher then a SR of a random guess: $1/256 = 4\%$.

Observing these results, we conclude that the MLP network can be substantially stronger than the template attacks when exploiting location leakage. It can achieve high SR using single-shot attacks, reaching 98% for 2 regions and 32% even when targeting single bytes in the SRAM. Notably, the MLP classification can strongly enhance the SR of a microprobe setup making it almost on par with the substantially more expensive photonic emission setup.

6 Conclusions and Future Directions

In this work, we have revisited the potent, yet often overlooked location-based leakage. We take the first steps towards theoretical modeling of such effects and we put forward a simple spatial model to capture them. Continuing, we demonstrate successful location-based attacks on a modern ARM Cortex-M4 using both standard template attacks, CNNs and MLPs. Throughout these attacks we assess the impact of various experimental parameters in order to elucidate the nature and exploitability of location-based leakage.

Regarding future work, we note that during the last years of side-channel research, the community has established a multitude of potent tools (ranging from Bayesian techniques to neural networks), all of which are particularly good at extracting the available leakage. Still, we remain far less capable of finding the exact cause behind it, especially in complex modern chips [4,35]. Thus, a natural extension to this work is to delve deeper into the electrical layer of a system-on-chip, try to identify the "culprits" behind location leakage and ultimately diminish the emitted information. In the same spirit, we should strive towards improved circuit modeling, similarly to the works of Šijačić et al. [41] and Kumar et al. [25], adapt them to the spatial model and use it in order to shorten the development-testing cycle of products. Finally, going back to algorithmic countermeasures, we can start designing masking schemes that account for data *and* location leakage in order to provide a fully-fledged security that encapsulates multiple side-channel vulnerabilities.

Acknowledgments. We would like to thank Riscure BV, Rafael Boix Carpi, Ilya Kizhvatov and Tin Soerdien for supporting the process of chip decapsulation and scan.

7 Appendix A

Algorithmic Noise in Tightly-Packed Surfaces. Since countermeasure designers opt often for algorithmic noise countermeasures, we investigate the statistical variance of N^{algo} for a tightly packed circuit that contains a large number of randomly switching components which try to hide the targeted component. We assume every noise-generating component to have area $b_i \approx d$, where d is the area of the targeted component Since we assume large n_a, both the noise-generating components as well as the targeted component are small w.r.t. the probe size, i.e. $d \ll o$. In a tightly packed circuit, the probe area o contains roughly $\frac{o}{d}$ randomly switching components, i.e. $n_a \approx \frac{o}{d}$. In this particular scenario, the following formula approximates N^{algo}.

$$N^{algo} = \sum_{i=1}^{n_a} N_i^{algo} = d \cdot \sum_{i=1}^{n_a} B_i = d \cdot A,$$

$$B_i \sim Bern(0.5)\ ,\ A \sim Binomial(n_a, 0.5)$$

$$\text{Thus, } N^{algo} \xrightarrow[\text{Theorem}]{\text{Central Limit}} Norm(\frac{d \cdot n_a}{2}, \frac{d \cdot n_a}{4})$$

Using the approximation of the Central Limit Theorem, we see that $Var[N^{algo}] = \frac{d \cdot n_a}{4} = \frac{o}{4}$. Thus, for the tightly-packed, small-component scenario we have established a direct link between the probe area o and the level of algorithmic noise, demonstrating how increasing the probe area induces extra noise.

8 Appendix B

Predicted versus actual values, visualizing the validation set success rate.

Predicted:	0	1	2	3	4	5	6	7	8	9	10	11	12	13	14	15
Actual:																
0	35	0	1	0	0	0	1	0	0	0	0	0	0	0	0	0
1	1	30	1	0	0	0	0	0	0	0	0	0	0	0	0	0
2	1	0	27	0	0	0	0	0	0	0	0	0	0	0	0	0
3	0	0	0	40	1	1	0	0	0	0	0	0	0	0	0	0
4	2	0	0	1	31	0	0	1	0	0	0	0	1	0	0	0
5	0	1	0	1	0	28	1	0	0	0	0	0	0	0	0	0
6	0	1	0	0	0	0	37	0	0	0	0	0	0	1	0	0
7	0	0	0	1	2	1	0	26	0	0	0	0	0	0	1	0
8	0	0	0	1	0	0	0	0	26	0	0	0	0	0	0	0
9	0	0	1	0	0	1	1	0	1	30	0	0	0	0	0	0
10	0	0	0	0	0	2	0	0	1	1	37	0	1	0	0	2
11	0	1	0	1	1	0	0	1	0	1	0	34	0	1	0	0
12	0	1	0	0	0	0	0	2	1	0	0	0	34	0	0	0
13	0	0	0	0	0	0	1	1	0	0	3	0	0	32	2	0
14	0	0	0	0	0	0	0	0	0	0	0	0	0	0	27	0
15	0	0	0	0	1	0	0	0	0	0	0	1	0	1	0	31

References

1. Abadi, M., et al.: TensorFlow: large-scale machine learning on heterogeneous systems (2015). Software available from tensorflow.org
2. Andrikos, C., Rassias, G., Lerman, L., Papagiannopoulos, K., Batina, L.: Location-based leakages: new directions in modeling and exploiting. In: 2017 International Conference on Embedded Computer Systems: Architectures, Modeling, and Simulation (SAMOS), pp. 246–252, July 2017
3. Archambeau, C., Peeters, E., Standaert, F.-X., Quisquater, J.-J.: Template attacks in principal subspaces. In: Goubin, L., Matsui, M. (eds.) CHES 2006. LNCS, vol. 4249, pp. 1–14. Springer, Heidelberg (2006). https://doi.org/10.1007/11894063_1
4. Balasch, J., Gierlichs, B., Grosso, V., Reparaz, O., Standaert, F.-X.: On the cost of lazy engineering for masked software implementations. In: Joye, M., Moradi, A. (eds.) CARDIS 2014. LNCS, vol. 8968, pp. 64–81. Springer, Cham (2015). https://doi.org/10.1007/978-3-319-16763-3_5
5. Cagli, E., Dumas, C., Prouff, E.: Convolutional neural networks with data augmentation against jitter-based countermeasures. In: Fischer, W., Homma, N. (eds.) CHES 2017. LNCS, vol. 10529, pp. 45–68. Springer, Cham (2017). https://doi.org/10.1007/978-3-319-66787-4_3
6. Chari, S., Rao, J.R., Rohatgi, P.: Template attacks. In: Kaliski, B.S., Koç, K., Paar, C. (eds.) CHES 2002. LNCS, vol. 2523, pp. 13–28. Springer, Heidelberg (2003). https://doi.org/10.1007/3-540-36400-5_3
7. Chollet, F., et al.: Keras (2015). https://keras.io
8. Chollet, F.: Xception: deep learning with depthwise separable convolutions. CoRR, abs/1610.02357 (2016)

9. Choudary, O., Kuhn, M.G.: Efficient template attacks. In: Francillon, A., Rohatgi, P. (eds.) CARDIS 2013. LNCS, vol. 8419, pp. 253–270. Springer, Cham (2014). https://doi.org/10.1007/978-3-319-08302-5_17
10. De Cnudde, T., Bilgin, B., Gierlichs, B., Nikov, V., Nikova, S., Rijmen, V.: Does coupling affect the security of masked implementations? In: Guilley, S. (ed.) COSADE 2017. LNCS, vol. 10348, pp. 1–18. Springer, Cham (2017). https://doi.org/10.1007/978-3-319-64647-3_1
11. Deng, J., Dong, W., Socher, R., Li, L.-J., Li, K., Fei-Fei, L.: ImageNet: a large-scale hierarchical image database. In: CVPR 2009 (2009)
12. Doget, J., Prouff, E., Rivain, M., Standaert, F.-X.: Univariate side channel attacks and leakage modeling. J. Cryptogr. Eng. **1**(2), 123–144 (2011)
13. Gandolfi, K., Mourtel, C., Olivier, F.: Electromagnetic analysis: concrete results. In: Koç, Ç.K., Naccache, D., Paar, C. (eds.) CHES 2001. LNCS, vol. 2162, pp. 251–261. Springer, Heidelberg (2001). https://doi.org/10.1007/3-540-44709-1_21
14. Glorot, X., Bengio, Y. Understanding the difficulty of training deep feedforward neural networks. In: JMLR W&CP: Proceedings of the Thirteenth International Conference on Artificial Intelligence and Statistics (AISTATS 2010), vol. 9, pp. 249–256, May 2010
15. He, K., Zhang, X., Ren, S., Sun, J.: Deep residual learning for image recognition. In: Proceedings of the IEEE Conference on Computer Vision and Pattern Recognition, pp. 770–778 (2016)
16. He, W., de la Torre, E., Riesgo, T.: An interleaved EPE-immune PA-DPL structure for resisting concentrated EM side channel attacks on FPGA implementation. In: Schindler, W., Huss, S.A. (eds.) COSADE 2012. LNCS, vol. 7275, pp. 39–53. Springer, Heidelberg (2012). https://doi.org/10.1007/978-3-642-29912-4_4
17. Heyszl, J.: Impact of Localized Electromagnetic Field Measurements on Implementations of Asymmetric Cryptography. https://mediatum.ub.tum.de/doc/1129375/1129375.pdf
18. Heyszl, J., Mangard, S., Heinz, B., Stumpf, F., Sigl, G.: Localized electromagnetic analysis of cryptographic implementations. In: Dunkelman, O. (ed.) CT-RSA 2012. LNCS, vol. 7178, pp. 231–244. Springer, Heidelberg (2012). https://doi.org/10.1007/978-3-642-27954-6_15
19. Huang, G., Liu, Z., Weinberger, K.Q.: Densely connected convolutional networks. CoRR, abs/1608.06993 (2016)
20. Immler, V., Specht, R., Unterstein, F.: Your rails cannot hide from localized EM: how dual-rail logic fails on FPGAs. In: Fischer, W., Homma, N. (eds.) CHES 2017. LNCS, vol. 10529, pp. 403–424. Springer, Cham (2017). https://doi.org/10.1007/978-3-319-66787-4_20
21. Itoh, K., Izu, T., Takenaka, M.: Address-bit differential power analysis of cryptographic schemes OK-ECDH and OK-ECDSA. In: Kaliski, B.S., Koç, K., Paar, C. (eds.) CHES 2002. LNCS, vol. 2523, pp. 129–143. Springer, Heidelberg (2003). https://doi.org/10.1007/3-540-36400-5_11
22. Kamel, D., Standaert, F.X., Flandre, D.: Scaling trends of the AES S-box low power consumption in 130 and 65 nm CMOS technology nodes. In: International Symposium on Circuits and Systems (ISCAS 2009), 24–17 May 2009, Taipei, Taiwan, pp. 1385–1388 (2009)
23. Kingma, D.P., Ba, J.: Adam: a method for stochastic optimization. CoRR, abs/1412.6980 (2014)
24. Kocher, P.C., Jaffe, J., Jun, B.: Differential power analysis. In: Wiener, M. (ed.) CRYPTO 1999. LNCS, vol. 1666, pp. 388–397. Springer, Heidelberg (1999). https://doi.org/10.1007/3-540-48405-1_25

25. Kumar, A., Scarborough, C., Yilmaz, A., Orshansky, M.: Efficient simulation of EM side-channel attack resilience. In 2017 IEEE/ACM International Conference on Computer-Aided Design (ICCAD), pp. 123–130, November 2017
26. Lerman, L., Poussier, R., Markowitch, O., Standaert, F.-X.: Template attacks versus machine learning revisited and the curse of dimensionality in side-channel analysis: extended version. J. Cryptogr. Eng. **8**(4), 301–313 (2018)
27. Maghrebi, H., Portigliatti, T., Prouff, E.: Breaking cryptographic implementations using deep learning techniques. In: Carlet, C., Hasan, M.A., Saraswat, V. (eds.) SPACE 2016. LNCS, vol. 10076, pp. 3–26. Springer, Cham (2016). https://doi.org/10.1007/978-3-319-49445-6_1
28. Martinasek, Z., Hajny, J., Malina, L.: Optimization of power analysis using neural network. In: Francillon, A., Rohatgi, P. (eds.) CARDIS 2013. LNCS, vol. 8419, pp. 94–107. Springer, Cham (2014). https://doi.org/10.1007/978-3-319-08302-5_7
29. Martinasek, Z., Zeman, V.: Innovative method of the power analysis. Radioengineering **22**(2), 586–594 (2013)
30. Maurine, P.: Securing SoCs in advanced technologies. https://cosade.telecom-paristech.fr/presentations/invited2.pdf
31. Nassar, M., Souissi, Y., Guilley, S., Danger, J.L.: RSM: a small and fast countermeasure for AES, secure against 1st and 2nd-order zero-offset SCAs. In: 2012 Design, Automation Test in Europe Conference Exhibition (DATE), pp. 1173–1178, March 2012
32. Nawaz, K., Kamel, D., Standaert, F.-X., Flandre, D.: Scaling trends for dual-rail logic styles against side-channel attacks: a case-study. In: Guilley, S. (ed.) COSADE 2017. LNCS, vol. 10348, pp. 19–33. Springer, Cham (2017). https://doi.org/10.1007/978-3-319-64647-3_2
33. Nikova, S., Rechberger, C., Rijmen, V.: Threshold implementations against side-channel attacks and glitches. In: Ning, P., Qing, S., Li, N. (eds.) ICICS 2006. LNCS, vol. 4307, pp. 529–545. Springer, Heidelberg (2006). https://doi.org/10.1007/11935308_38
34. Pan, S.J., Yang, Q., et al.: A survey on transfer learning. IEEE Trans. Knowl. Data Eng. **22**(10), 1345–1359 (2010)
35. Papagiannopoulos, K., Veshchikov, N.: Mind the gap: towards secure 1st-order masking in software. In: Guilley, S. (ed.) COSADE 2017. LNCS, vol. 10348, pp. 282–297. Springer, Cham (2017). https://doi.org/10.1007/978-3-319-64647-3_17
36. Pedregosa, F., et al.: Scikit-learn: machine learning in Python. J. Mach. Learn. Res. **12**, 2825–2830 (2011)
37. Prouff, E., Strullu, R., Benadjila, R., Cagli, E., Dumas, C.: Study of deep learning techniques for side-channel analysis and introduction to ASCAD database. IACR Cryptology ePrint Archive 2018, 53 (2018)
38. Renauld, M., Standaert, F.-X., Veyrat-Charvillon, N., Kamel, D., Flandre, D.: A formal study of power variability issues and side-channel attacks for nanoscale devices. In: Paterson, K.G. (ed.) EUROCRYPT 2011. LNCS, vol. 6632, pp. 109–128. Springer, Heidelberg (2011). https://doi.org/10.1007/978-3-642-20465-4_8
39. Rivain, M., Prouff, E.: Provably secure higher-order masking of AES. In: Mangard, S., Standaert, F.-X. (eds.) CHES 2010. LNCS, vol. 6225, pp. 413–427. Springer, Heidelberg (2010). https://doi.org/10.1007/978-3-642-15031-9_28
40. Schlösser, A., Nedospasov, D., Krämer, J., Orlic, S., Seifert, J.-P.: Simple photonic emission analysis of AES. In: Prouff, E., Schaumont, P. (eds.) CHES 2012. LNCS, vol. 7428, pp. 41–57. Springer, Heidelberg (2012). https://doi.org/10.1007/978-3-642-33027-8_3

41. Sijacic, D., Balasch, J., Yang, B., Ghosh, S., Verbauwhede, I.: Towards efficient and automated side channel evaluations at design time. In: Batina, L., Kühne, U., Mentens, N., (eds.) PROOFS 2018. 7th International Workshop on Security Proofs for Embedded Systems. Kalpa Publications in Computing, vol. 7, pp. 16–31. EasyChair (2018)
42. Simonyan, K., Zisserman, A.: Very deep convolutional networks for large-scale image recognition. arXiv preprint arXiv:1409.1556 (2014)
43. Specht, R., Immler, V., Unterstein, F., Heyszl, J., Sig, G.: Dividing the threshold: Multi-probe localized EM analysis on threshold implementations. In: 2018 IEEE International Symposium on Hardware Oriented Security and Trust (HOST), pp. 33–40, April 2018
44. Specht, R., Heyszl, J., Kleinsteuber, M., Sigl, G.: Improving non-profiled attacks on exponentiations based on clustering and extracting leakage from multi-channel high-resolution EM measurements. In: Mangard, S., Poschmann, A.Y. (eds.) COSADE 2014. LNCS, vol. 9064, pp. 3–19. Springer, Cham (2015). https://doi.org/10.1007/978-3-319-21476-4_1
45. Specht, R., Heyszl, J., Sigl, G.: Investigating measurement methods for high-resolution electromagnetic field side-channel analysis. In: 2014 International Symposium on Integrated Circuits (ISIC), Singapore, 10–12 December 2014, pp. 21–24 (2014)
46. Standaert, F.-X., Malkin, T.G., Yung, M.: A unified framework for the analysis of side-channel key recovery attacks. In: Joux, A. (ed.) EUROCRYPT 2009. LNCS, vol. 5479, pp. 443–461. Springer, Heidelberg (2009). https://doi.org/10.1007/978-3-642-01001-9_26
47. Standaert, F.-X., Peeters, E., Archambeau, C., Quisquater, J.-J.: Towards security limits in side-channel attacks. In: Goubin, L., Matsui, M. (eds.) CHES 2006. LNCS, vol. 4249, pp. 30–45. Springer, Heidelberg (2006). https://doi.org/10.1007/11894063_3
48. Sugawara, T., Suzuki, D., Saeki, M., Shiozaki, M., Fujino, T.: On measurable side-channel leaks inside ASIC design primitives. J. Cryptogr. Eng. **4**(1), 59–73 (2014)
49. Szegedy, C., Ioffe, S., Vanhoucke, V., Alemi, A.A.: Inception-v4, inception-ResNet and the impact of residual connections on learning. In: AAAI, vol. 4, p. 12 (2017)
50. Szegedy, C., Vanhoucke, V., Ioffe, S., Shlens, J., Wojna, Z.: Rethinking the inception architecture for computer vision. In: Proceedings of the IEEE Conference on Computer Vision and Pattern Recognition, pp. 2818–2826 (2016)
51. Unterstein, F., Heyszl, J., De Santis, F., Specht, R.: Dissecting leakage resilient PRFs with multivariate localized EM attacks - a practical security evaluation on FPGA. COSADE 2017/272 (2017)
52. Unterstein, F., Heyszl, J., De Santis, F., Specht, R., Sigl, G.: High-resolution EM attacks against leakage-resilient PRFs explained - and an improved construction. Cryptology ePrint Archive, Report 2018/055 (2018). https://eprint.iacr.org/2018/055
53. Van Trees, H.L.: Detection, Estimation, and Modulation Theory: Part IV: Optimum Array Processing. Wiley, Hoboken (2002)
54. Weste, N., Harris, D.: CMOS VLSI Design: A Circuits and Systems Perspective, 4th edn. Addison-Wesley Publishing Company, USA (2010)
55. Yang, S., Zhou, Y., Liu, J., Chen, D.: Back propagation neural network based leakage characterization for practical security analysis of cryptographic implementations. In: Kim, H. (ed.) ICISC 2011. LNCS, vol. 7259, pp. 169–185. Springer, Heidelberg (2012). https://doi.org/10.1007/978-3-642-31912-9_12

Simple Refreshing in the Noisy Leakage Model

Stefan Dziembowski[1], Sebastian Faust[2], and Karol Żebrowski[1(✉)]

[1] University of Warsaw, Warsaw, Poland
s.dziembowski@crypto.edu.pl, k.zebrowski@mimuw.edu.pl
[2] TU Darmstadt, Darmstadt, Germany

Abstract. Masking schemes are a prominent countermeasure against power analysis and work by concealing the values that are produced during the computation through randomness. The randomness is typically injected into the masked algorithm using a so-called *refreshing* scheme, which is placed after each masked operation, and hence is one of the main bottlenecks for designing efficient masking schemes. The main contribution of our work is to investigate the security of a very simple and efficient refreshing scheme and prove its security in the *noisy leakage model* (EUROCRYPT'13). Compared to earlier constructions our refreshing is significantly more efficient and uses only n random values and $<2n$ operations, where n is the security parameter. In addition we show how our refreshing can be used in more complex masked computation in the presence of noisy leakage. Our results are established using a new methodology for analyzing masking schemes in the noisy leakage model, which may be of independent interest.

1 Introduction

Over the last decade cryptographic research has made tremendous progress in developing solid foundations for cryptography in the presence of side-channel leakage (see, e.g., [19] for a recent overview). The common approach in this area – often referred to as "leakage resilient cryptography" – is to first extend the black-box model to incorporate side-channel leakage, and then to propose countermeasures that are provable secure within this model. The typical leakage model considered in the literature assumes an adversary that obtains some partial knowledge about the internal state of the device. For instance, the adversary may learn a few bits of the intermediate values that are produced by the device during its computation.

One of the countermeasures that significantly benefits from such a formal treatment are masking schemes (see, e.g., [6,8,9,11,18,21] and many more). Masking is a frequently used countermeasure against power analysis attacks, which de-correlates the internal computation of a device from the observable leakage (e.g., the power consumption). A core ingredient of any secure masking scheme is a *refreshing* algorithm. At a very high level (we will explain this in much more detail below) the refreshing algorithm introduces new randomness

S. D. Galbraith and S. Moriai (Eds.): ASIACRYPT 2019, LNCS 11923, pp. 315–344, 2019.
https://doi.org/10.1007/978-3-030-34618-8_11

into the masked computation, thereby preventing that an adversary can exploit correlations between different intermediate values of the computation. Since refreshing schemes are computationally expensive a large body of work has explored how to securely improve their efficiency. Unfortunately, one of the most simple and efficient (in terms of computation and randomness) refreshing schemes due to Rivain and Prouff [22] cannot be proven secure; even worse, it was shown in [5] that a simple – though impractical – attack breaks the scheme in the common threshold probing leakage model [18]. In this work we show – somewhat surprisingly – that the simple refreshing of Rivain and Prouff [22] is secure under noisy leakages [21]. Noisy leakages are considered generally to accurately model physical side-channel leakage, and hence our result implies that the simple refreshing can securely replace more complex and expensive schemes in practice.

1.1 Masking Schemes

Ingredients of a Masking Scheme. One of the most common countermeasures against power analysis attacks are masking schemes. Masking schemes work by randomizing the intermediate values produced during the computation of an algorithm through secret sharing. To this end each sensitive variable x is represented by an encoding $\mathrm{Enc}(x) := (x_1, \ldots, x_n)$ and the corresponding decoding function $\mathrm{Dec}(\cdot)$ recovers $x := \mathrm{Dec}(x_1, \ldots, x_n)$. A simple encoding function uses the additive encoding function, which works by sampling x_i uniformly at random from some finite field $\mathbb{F}$ subject to the constraint that $x := \sum_{i=1}^{n} x_i$. If $\mathbb{F}$ is the binary field, then such a masking scheme is typically called *Boolean masking*.

In addition to an encoding scheme, we need secure algorithms to compute with encoded elements. To this end, the algorithm's computation is typically modeled as an *arithmetic circuit* over a finite field $\mathbb{F}$. In such circuits the wires carry values from $\mathbb{F}$ and the gates perform operations from $\mathbb{F}$. At a high-level the circuit is made out of gates that represent the basic field operations (i.e., addition gate denoted "$\oplus$" and multiplication gate denoted "$\otimes$"). Moreover, it may consist of gates for inversion (i.e., outputting $-x$ on input x), and so-called randomness gates *RND* that take no input and produce an output that is distributed uniformly over $\mathbb{F}$. We often assign unique labels to the wires. Each label can be interpreted as a variable whose value is equal to the value that the corresponding wire carries.

Given a circuit built from these gates, a masking scheme then typically works by replacing each of the above operations by a "masked" version of the gate. For instance, in case of the aforementioned additive encoding scheme (Enc, Dec) the masked version of the $\oplus$ takes as input two encodings $\mathrm{Enc}(x)$ and $\mathrm{Enc}(y)$ and outputs an encoding $\mathrm{Enc}(z)$, where $\sum_i z_i := \sum_i x_i + \sum_i y_i$. Informally, the masked version of a gate is said to be secure if leakage emitted from the internal computation of the masked version of the gate does not reveal any sensitive information.

Refreshing Schemes. A key building block to securely compose multiple masked operations to a complex masked circuit is the *refreshing scheme*. The refreshing scheme takes as input an encoding $\overrightarrow{x}^j := (x_1^j, \ldots, x_n^j) = \mathrm{Enc}(x)$ and outputs a new encoding $(x_1^{j+1}, \ldots, x_n^{j+1}) = \overrightarrow{x}^{j+1}$ of x. By "new encoding" we mean that this procedure should inject new randomness into the encoding, in such a way that the leakage from the previous encodings should not accumulate. In other words: if we periodically refresh the encodings of x (which leads to a sequence of encodings: $\overrightarrow{x}^0 \mapsto \overrightarrow{x}^1 \mapsto \overrightarrow{x}^2 \mapsto \cdots$) then x should remain secret even if bounded partial information about each $\overrightarrow{x}^j$ leaks to the adversary. The operations of computing $\overrightarrow{x}^{j+1}$ from $\overrightarrow{x}^j$ is also called a *refreshing round*, and a circuit that consists of some number of such rounds (and not other operations) is called a *multi-round refreshing circuit.*

A common approach for securely refreshing additive encodings is to exploit the homomorphism of the underlying encoding with respect to addition:[1] one starts by designing an algorithm that samples $(b_1, \ldots, b_n)$ from the distribution $\mathrm{Enc}(0)$, and then, in order to refresh an encoding $(x_1^j, \ldots, x_n^j)$ one adds $(b_1, \ldots, b_n)$ to it. Therefore, the refreshed encoding is equal to $(x_1^j + b_1, \ldots, x_n^j + b_n)$. Observe that after $\mathrm{Enc}(0)$ is generated, the refreshing can be done without any further computation, by just adding b_i to every x_i^j. Of course in this approach the whole technical difficulty is to generate the encodings of 0 in a secure way (without relying on any assumptions on leakage-freeness of the encoding generation).

The most simple and efficient refreshing scheme originally introduced in [22] uses the "encoding of 0 approach" mentioned above and works as follows (see also Fig. 1 on page 8). In order to refresh $\overrightarrow{x}^j = (x_1^j, \ldots, x_n^j)$, we first sample $b_1^j, \ldots, b_{n-1}^j$ uniformly at random from $\mathbb{F}$ and set $b_n^j := -b_1^j - \ldots - b_{n-1}^j$. Then, we compute the fresh encoding of x as $(x_1^{j+1}, \ldots, x_n^{j+1}) := (x_1^j + b_1^j, \ldots, x_n^j + b_n^j)$. Notice that besides its simplicity the above refreshing enjoys additional beneficial properties including optimal randomness complexity (only $n-1$ random values are used) and minimal circuit size (only $2n-1$ field operations are required). Somewhat surprisingly this simple refreshing scheme turns out to be insecure in the security model of *threshold probing attacks* introduced in the seminal work of Ishai, Sahai and Wagner [18].

Insecurity of Simple Refreshing. The standard model to analyze the security of masking schemes is the *t-probing model* [18]. In the t-probing model the adversary can (adaptively) select up to t wires of the internal masked computation and learn the values carried on these wires during computation. While originally it was believed that the simple refreshing from above guarantees security for $t = n - 1$ [22], Coron et al. [10] showed that when it is combined with certain other masked operations (e.g., in a masked AES) the resulting construction can be broken using only $\leq t := n/2 + 1$ probes.

An even more devastating attack against this natural refreshing can be shown in the following setting. Consider a circuit that consists of a sequence of n refreshings of an encoding $\overrightarrow{x}^0$. This may naturally happen in a masked key schedule

[1] By this we mean that for every x and y we have $\mathrm{Dec}(\mathrm{Enc}(x) + \mathrm{Enc}(y)) = x + y$, where "+" on the left-hand-side denotes the vector addition.

of the AES algorithm, where the secret key is encoded and after each use for encrypting/decrypting is refreshed. If for each of these refreshings the adversary can learn 2 values, then a simple attack allows to recover the secret (we describe this attack in more detail below). The attack, however, is rather impossible to carry out in practice. In particular, it requires the adversary to learn for the n consecutive executions of the refreshing scheme specific (different) intermediate values.

The Noisy Leakage Model. The attack against the simple refreshing illustrates that in some sense the probing model is too strong. An alternative model is the so-called *noisy leakage model* of Prouff and Rivain [21]. In the noisy leakage model the leakage is not quantitatively bounded but instead it is assumed that the adversary obtains a "noisy distribution" of each value carried on a wire. The noisy leakage model is believed to model real-world physical leakage accurately, and hence is prominently used in practice to analyze the real-world security of physical devices [12].

In [11] it was shown that the noisy leakage model of [21] can be reduced to the *p-random probing model.* In the p-random probing model we assume that the value carried on each wire is revealed independently with probability p. Since in the p-random probing model the adversary looses control over the choice of wire that he learns, the attack against the simple refreshing ceases to work. This raises the question if the simple and most natural refreshing scheme is secure in the p-random probing model. The main contribution of this paper is to answer this question affirmatively.

1.2 Our Contribution

We provide a technical outline of our contributions in Sect. 2 and give in the following only a high-level summary of our results.

Simple Refreshing. Our main contribution is to analyze the security of the simple refreshing scheme from [22] in the noisy leakage model. In particular, we show that refreshing an encoding $(x_1, \ldots, x_n)$ is secure even if each wire in the refreshing circuit is revealed with constant probability p. Our result directly implies that refreshing an encoded secret k times (where k may be much larger than the security parameter n) remains secure under noisy leakages for constant noise parameter. Such consecutive use of refreshings naturally appears in many practical settings such as the key schedule of the AES mentioned above, or in general for refreshing the secret key between multiple runs of any cryptographic primitive. Since the simple refreshing is *optimal* in terms of circuit size and randomness complexity our result significantly improves the practicality of the masking countermeasure.

Concretely, the simple refreshing requires $n-1$ random values and uses $2n-1$ addition gates to securely refresh an encoding $(x_1, \ldots, x_n)$ in the random probing model (and hence implying security in the noisy leakage model of [21]). In contrast, the most widely used refreshing scheme from Ishai, Sahai and Wagner [18] requires $(n-1)^2/2$ randoms and $2n^2+n$ addition gates and has been proven secure only for $p \approx 1/n$, which is significantly worse than ours.[2] Recently, various works provide asymptotically improved refreshing algorithms. In particular, in [1,3] it was shown how to build a secure refreshing with circuit size $O(n)$, and randomness complexity $O(n)$ for a constant noise parameter p. While asymptotically these constructions are the same as for the simple refreshing analyzed in our work, from a concrete practical point of view these schemes are very inefficient as they are based on expander graphs.

New Techniques for Proving Security. At the technical level, our main contribution is to introduce a new technique for proving security in the random probing model. Our main observation is that probing security can be translated into a question of connectivity between nodes in certain graphs. As an example consider the circuit $\widehat{C}$ executing k times the simple refreshing. It can be represented as a grid G with $k+1$ rows and $n+1$ columns, where in each row we have $n+1$ nodes. The edges between the nodes represent intermediate values that are computed during the execution of the circuit. Leakage of a certain sub-set of wires then corresponds to a sub-graph of G, which we call *leakage diagram.*

We then show that if the "leftmost side" and the "rightmost side" of the grid are connected by a path in the leakage diagram, then this leads to an attack that allows to recover the encoded secret that is refreshed by the circuit $\widehat{C}$. On the other hand, and more importantly, if the two sides of the diagram are *not connected* by a path in the leakage diagram, then we show that the adversary does not learn any information about the encoded secret from the leakage. The above can be extended to arbitrary masked arithmetic circuits, in which the graphs representing the circuit are slightly more involved.

The above allows us to cast security against probing leakage as a question about connectivity of nodes within a graph. To show security in the p-random probing model we then need to bound the probability that the random subgraph of G representing the leakage contains a path that connects the two sides of G. The main challenge is that although in the p-random probing model each wire leaks independently with probability p, in our graph representation certain edges are more likely to be part of the leakage diagram. Even worse, the events of particular edges of G ending up in the leakage diagram are not independent. This significantly complicates our analysis. We believe that the techniques introduced in our paper are of independent interest and provide a novel tool set for analyzing security of masked computation in the random probing model.

Extension to Any Masked Computation. As our last contribution we show how to use the simple refreshing as part of a more complex masked computation. To this end, we study the security of the masking compiler provided by Ishai, Sahai and Wagner [18] when using the simple refreshing described above. Notably, we first show that the simple refreshing can be used to securely compose any

[2] We expect that also the refreshing from [18] is secure for some constant probability p, but we did not analyze its security.

affine masked operations. This result is important because it shows for the first time that the most natural and efficient way to carry out affine computation in the masked domain is secure against noisy leakages. Compared to the standard construction of [18] we save a factor of n in circuit and randomness complexity. Moreover, at the concrete level we make huge practical improvements when compared to the recent works of [1,3], which use expander graphs and algebraic geometric codes.

Finally, we show that the simple refreshing can also be securely composed with the masked multiplication of [18]. Since the masked multiplication of [18] itself is a composable refreshing [11], this result is maybe not so surprising. Nevertheless, it shows that combining from a complexity point of view optimal masked computation with the ISW masked multiplication results into general masked computation that is secure in the random probing model.[3]

1.3 Other Related Work

A large amount of work proves different formal security guarantees of masking schemes (see, e.g., [1,7,18,20,21] and many more), and we only discuss the most relevant work.

Noisy Leakage Model. As already mentioned most relevant for us is the so-called noisy leakage model introduced in the work of Prouff and Rivain [21] and further refined by Duc et al. [11]. In the later it was also shown that the p-random probing model is closely related to the noisy leakage model. Since both [11,21] require $p \approx 1/n$, one important goal of research is to improve the noise parameter p. There has recently been significant progress on this. In [1,3] it was shown how to securely compute in the random probing model for constant p. Further improvements are made in [2,16], where the later achieves security under a quasi-constant noise for a construction with complexity $O(n \log(n))$ avoiding heavy tools such as expander graphs and AG codes. Another line of work investigates relations between different noisy leakage models [14,17] and provides tight relations between them. A more practical view on noisy leakage – and in particular a quantitative study of its relation to real-world leakage – was given by Duc et al. [12].

Refreshing Schemes and Their Usage. Refreshing schemes have always been a core ingredient of masking schemes. Their randomness consumption is, however, often the bottleneck for an efficient masked implementation[4]. Hence, an important goal of research is to minimize the overheads resulting from the use of

[3] Recall that for the ISW scheme it is known that $p \approx 1/n$ as otherwise there is an attack against the masked multiplication. Thus, our result requires a similar bound on p in the general case. It is an interesting open question if we can combine the simple refreshing with masked multiplications that are secure for constant p, e.g., the schemes from [1,3].

[4] Notice that true randomness is hard to generate in practice, and producing securely pseudorandomness is costly as we need to run, e.g., an AES.

refreshing. There are two main directions to achieve this. First, we may improve the refreshing algorithm itself. In particular, in [1,3] it was shown how to build a secure refreshing with circuits size and randomness complexity $O(n)$ for a constant noise parameter p. While asymptotically optimal from a concrete practical point of view these schemes are very inefficient as they are based on expander graphs. A second direction to improve on the costs for refreshing is to reduce the number of times the refreshing algorithms are used. This approach was taken by several works [6,8,9] which develop tools for placing the refreshing algorithm in an efficiency optimizing way without compromising on security. It is an interesting question for future research to develop tools and methods that securely place the simple refreshing within a complex masked circuit.

2 Our Approach Informally

As a simple example of circuit to present our approach let us consider a circuit $\widehat{C}$ (in the following the "hat notation" will denote masked/transformed circuits) that is a k-round refreshing circuit. This circuit consist of k consecutive subcircuits that we call *refreshing gadgets* $\widehat{R}$, presented in Fig. 1. Note that in addition to the notation from Sect. 1.1 we also use terms c_i^j that denote the partial sums: $c_i^j = b_1^j + \cdots + b_i^j$ (for consistency define c_0^j and c_n^j to be always equal to 0). It is a simple fact that the adversary can learn the encoded secret for $k = n$ even if just 2 wires from each refreshing gadget leak to her (and no additional leakage is given), namely x_{j+1}^j and c_{j+1}^j. We recall this attack in the full version of the paper [15]. Similar attacks for different refreshing schemes have been shown in [5,13]. This attack strongly relies on the fact that the adversary can *choose* which wires she learns. As discussed in the introduction, in the weaker p-random probing model it is very unlikely that the adversary will be lucky enough to learn x_{j+1}^j and c_{j+1}^j in each round (unless p is close to 1). Of course, the fact that one particular attack does not work, does not immediately imply that the scheme is secure.

Relaxing the Leakage Model. As already mentioned in Sect. 1.2 our first main contribution is a formal proof that indeed this simple refreshing procedure is secure in the p-random probing model. Our starting point is the natural question: *can we characterize the leakages which allow the adversary to compute the secret?* We answer this question affirmatively by introducing the notion of *leakage diagrams*, which we explain below (for formal definitions see Sect. 4.4).

Leakage Diagrams. Essentially, the leakage diagrams are graphs that can be viewed as abstract representations of the leakage that occurred during the evaluation of a circuit. For a moment let us focus only on leakage diagrams that correspond to k-round refreshing circuits $\widehat{C}$. Let $x_1^0, \ldots, x_n^0$ be some initial encoding of the secret x. In this case the leakage diagram will be a subgraph of a $(n+1) \times (k+1)$ grid G with edges labeled x_i^j and c_i^j as on Fig. 2.

To illustrate how the leakage diagrams are constructed take as an example a 2-round refreshing circuit (with $n = 3$) that is depicted on Fig. 3a. Note that this picture omits the part of circuit that is responsible for generating the b_i^j's,

$(b_1^j, \ldots, b_{n-1}^j) \leftarrow \mathbb{F}^{n-1}$
$c_0^j := 0$
for $i = 1, \ldots, n-1$ **do**
  $c_i^j := c_{i-1}^j + b_i^j$
$b_n^j := -c_{n-1}^j$
for $i = 1, \ldots, n$ **do**
  $x_i^{j+1} := x_i^j + b_i^j$

(a) Pseudocode of the simple refreshing gadget $\widehat{R}$.

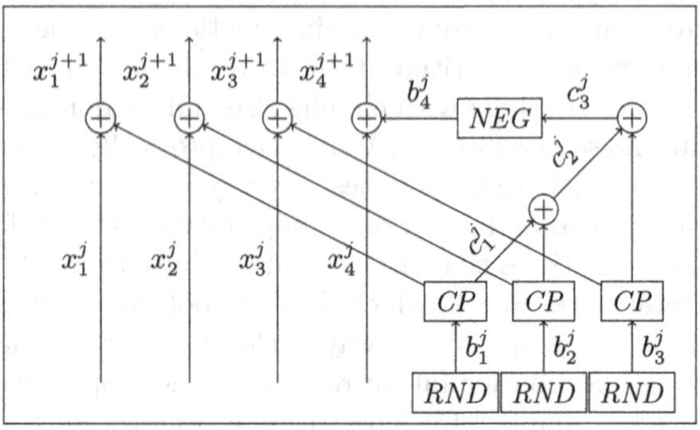

(b) Corresponding circuit (for $n = 4$).

Fig. 1. The refreshing gadget. The "j" superscript is added for the future reference (e.g. on Fig. 3a).

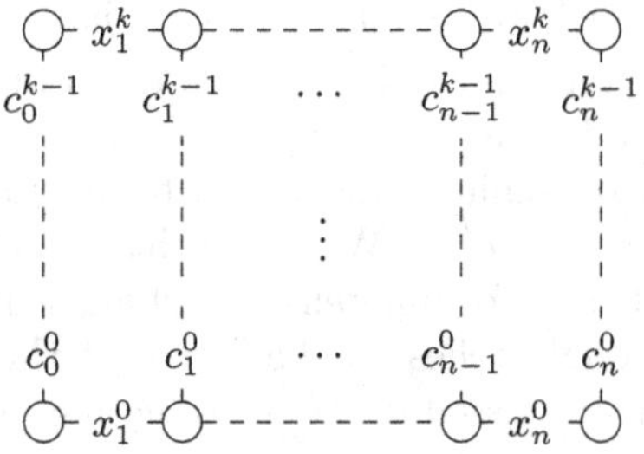

Fig. 2. Graph G corresponding to the k-round refreshing circuit. It has $k+1$ rows. In each jth row (for $j = 0, \ldots, k$) it has $n+1$ vertices connected with edges (there is an edge labeled with "x_i" between the ith and $(i+1)$st vertex). It also has an edge between every pair of ith vertices (for $i = 0, \ldots, n$) in the jth and $j+1$st row. This edge is labeled with "c_i^j".

and in particular the wires carrying the c_i^j values are missing on it. This is done in order to save space on the picture. Let L be the wires that leaked in the refreshing procedure. Suppose the leaking wires are x_3^0, x_1^1, x_2^3, and b_2^1, which is indicated by double color lines over the corresponding edges on Fig. 3a. We also have to remember about the c_i^j's that were omitted on the figure and can also leak. Recall that every c_i^j is equal to a sum $b_1^j + \cdots + b_i^j$. Hence, the leakage from c_i^j is indicated by a shaded colored region around $b_1^j, \ldots, b_i^j$. Let us assume that c_2^0, c_1^1, and c_2^1 are leaking, and therefore the shaded regions on Fig. 3a are placed over b_1^1, and the pairs $(b_1^0, b_2^0), (b_1^1, b_2^1)$.

The corresponding leakage diagram is a subgraph of the graph G from Fig. 2 with $k := 2$ and $n := 3$. The leakage diagram $S(L)$ has the same vertices as

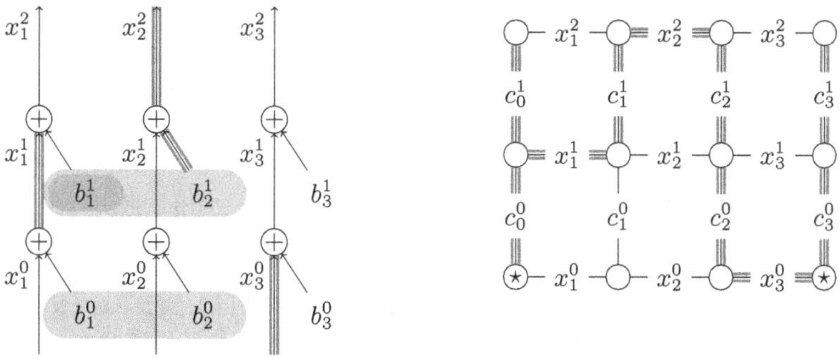

(a) A circuit with leaking wires marked with colored double lines. Additionally wires $c_2^0(= b_1^0 + b_2^0)$, $c_1^1(= b_1^1)$, and $c_2^1(= b_1^1 + b_2^1)$ leak, which is indicated by colored shaded areas around "b_1^0 b_2^0", "b_1^1", "b_1^1 b_2^1".

(b) The corresponding leakage diagram. We show how the adversary can compute the sum of edges x_1^0, x_2^0, and x_3^0. The leftmost and the rightmost vertices of the row containing these edges are marked with "$\circledast$".

Fig. 3. A leaking circuit and its corresponding leakage diagram.

G, but it has only a subset of its edges. Informally, the labels on the edges of $S(L)$ are variables that suffice to fully reconstruct the leakage from the circuit. More precisely: given these values one can compute the same leakage information that the adversary received. Going back to our example: the leakage diagram corresponding to the leakage presented on Fig. 3a is depicted on Fig. 3b, on which the members of $S(L)$ are marked with double colored lines. The set $S(L)$ is created according to the following rules. First, we add to $S(L)$ all the edges labeled x_i^j and c_i^j if the corresponding wires are in L. For this reason $S(L)$ on Fig. 3b contains $x_3^0, x_1^1, x_2^2, c_2^0, c_1^1$, and c_2^1. Handling leaking b_i^j's is slightly less natural, since graph G does not contain edges labeled with the b_i^j's. To deal with this, we make use of the fact that every b_i^j can be computed from c_i^j and c_{i-1}^j (as $b_i^j = c_i^j - c_{i-1}^j$). Hence, for every b_i^j from L we simply add c_i^j and c_{i-1}^j to $S(L)$. For this reason we add c_1^1 and c_2^1 to L (as b_2^1 is in L). This approach works, since, as mentioned above, the edges in $S(L)$ should suffice to fully reconstruct L. Note that in some sense we are "giving out too much" in the leakage diagram (as c_i^j and c_{i-1}^j cannot be uniquely determined from b_i^j). Fortunately, this "looseness" does not cost us much in terms of parameters, while at the same time it greatly simplifies our proofs. Finally, we add to $S(L)$ all the edges labeled with c_0^j and c_n^j (i.e.: the leftmost and the rightmost columns in G). We can do it since these edges are always equal to 0 and hence the adversary knows them "for free".

What the Adversary Can Learn from a Leakage Diagram. The ultimate goal of the adversary is to gain some information about the encoded secret. To achieve this it is enough that she learns the sum of all the x_i^j's from some row of the diagram. We now show how in case of leakage from Fig. 3 the adversary can compute $x_1^0+x_2^0+x_3^0$ from the values that belong to the $S(L)$ (i.e. those that are marked with double colored lines on Fig. 3b). Using the facts that $x_i^{j+1}=x_i^j+b_i^j$ and $c_{i+1}^j=c_i^j+b_{i+1}^j$ several times we have:

$$x_1^0+x_2^0+x_3^0=(x_1^1-b_1^0)+(x_2^1-b_2^0)+x_3^0=x_1^1+x_2^1-(b_1^0+b_2^0)+x_3^0=$$
$$x_1^1+(x_2^2-b_2^1)-c_2^0+x_3^0=x_1^1+x_2^2+c_1^1-c_2^1-c_2^0+x_3^0$$

where all the variables on the right hand side belong to $S(L)$. It is easy to see that the reason why the adversary is able to compute $x_1^0+x_2^0+x_3^0$ is that the leftmost and the rightmost nodes in the row containing edges labeled with variables were connected. These nodes are indicated with the "⊛" symbol on Fig. 3b.

Since the leftmost and the rightmost columns always belong to the leakage diagram, thus in general a similar computation is possible when these two columns are connected. Our first key observation is that if these columns are *not* connected, then the secret x remains secure. We state this fact below in the form of a following informal lemma.

Informal Lemma 1. *Consider a multi-round refreshing circuit. Let L be the set of leaking wires. Let E denote the event that the leftmost and the rightmost columns of S(L) are connected. If E did not occur then the adversary gains no information about the secret.*

This informal lemma is formalized as Claim 5 (in the full version of this paper [15]), where it is also stated in a more general form, covering the case of more complicated circuits (i.e. those that perform some operations in addition to refreshing). The rest of this section is organized as follows. In Sect. 2.1 we outline the main ideas behind the proof on Informal Lemma 1, in Sect. 2.2 we sketch the proof of the upper bound on the probability of E. This, together with the Informal Lemma 1 shows the security of our multi-round refreshing construction. Then in Sect. 2.3 we describe how these ideas can be generalized to arbitrary circuits. Besides of presenting the intuitions behind our formal proof, the goal of this part is also to introduce some more terminology that is useful later (e.g.: the "modification vectors"). In the sequel we use the following convention: if G is a labeled graph such that the labels on its edges are unique, then we sometimes say "edge λ" as a shortcut for "edge *labeled* with λ". The same convention applies to circuits and wires.

2.1 Proof Sketch of Informal Lemma 1

Here we present the main ideas behind the proof of Informal Lemma 1. Consider a k-round refreshing circuit $\widehat{C}$ that takes as input a secret shared over n wires. For two arbitrary field elements $x^0, x^1 \in \mathbb{F}$ consider experiments of applying $\widehat{C}$ to their random encodings. In the proof we consider a fixed set L of leaking wires in $\widehat{C}$. Assume that event E did not occur, i.e., the leftmost and the rightmost columns of the leakage diagram are disconnected. To prove Informal Lemma 1, it is enough to show that for the distributions of the values of wires in L are identical in both experiments (following the standard approach in cryptography this formally captures the fact that the adversary "gains no information about the secret"). We do it using a hybrid argument. Namely, we consider a sequence of experiments denoted $\mathrm{Exp}_A^0, \mathrm{Exp}_B^0, \mathrm{Exp}_C, \mathrm{Exp}_B^1$, and Exp_A^1 (see below), such that: (a) Exp_A^ℓ (for $\ell = 0, 1$) is equal to the original experiment in which x^ℓ is refreshed k times, and (b) the view of the adversary is identical for each pair of consecutive experiments on this list (and hence it is identical for all of them).

Exp_A^ℓ:

Sample $\overrightarrow{x}^{0,\ell} \leftarrow \mathrm{Enc}(x^\ell)$.
For $j = 0$ to $k-1$ do:
1. sample $\overrightarrow{b}^{j,\ell} \leftarrow \mathrm{Enc}(0)$,
2. let $\overrightarrow{c}^{j,\ell} := f(\overrightarrow{b}^{j,\ell})$,
3. let $\overrightarrow{x}^{j+1,\ell} := \overrightarrow{x}^{j,\ell} + \overrightarrow{b}^{j,\ell}$,

Exp_B^ℓ:	Exp_C:
Sample $\overrightarrow{x}^{0,\ell} \leftarrow \mathrm{Enc}(x^\ell)$.	Sample $\overrightarrow{x}^{0,1} \leftarrow \mathrm{Enc}(x^0) + (x^1 - x^0) \cdot \overrightarrow{m}^0$.
For $j = 0$ to $k-1$ do:	For $j = 0$ to $k-1$ do:
1. sample $\overrightarrow{x}^{j+1,\ell} \leftarrow \mathrm{Enc}(x^\ell)$,	1. sample $\overrightarrow{x}^{j+1,1} \leftarrow \mathrm{Enc}(x^0) + (x^1 - x^0) \cdot \overrightarrow{m}^{j+1}$,
2. let $\overrightarrow{b}^{j,\ell} := \overrightarrow{x}^{j+1,\ell} - \overrightarrow{x}^{j,\ell}$,	2. let $\overrightarrow{b}^{j,1} := \overrightarrow{x}^{j+1,1} - \overrightarrow{x}^{j,1}$,
3. let $\overrightarrow{c}^{j,\ell} := f(\overrightarrow{b}^{j,\ell})$.	3. let $\overrightarrow{c}^{j,1} := f(\overrightarrow{b}^{j,1})$.

Fig. 4. The sequence of experiments.

Extending the notation from the pseudocode given in Fig. 1a, we will add for future reference to the procedure that refreshes a secret x^ℓ (with $\ell \in \{0,1\}$) a superscript "ℓ" to all the labels, i.e., denote $\overrightarrow{x}^{j,\ell} := (x_1^{j,\ell}, \ldots, x_n^{j,\ell})$, $\overrightarrow{b}^{j,\ell} := (b_1^{j,\ell}, \ldots, b_n^{j,\ell})$ and $\overrightarrow{c}^{j,\ell} := (c_1^{j,\ell}, \ldots, c_n^{j,\ell})$. Note that all the operations in the refreshing circuit are linear, and in terms of linear algebra this experiment (repeated k times) can be described as Exp_A^ℓ on Fig. 4, where f is a linear function defined as $f(\overrightarrow{b}^{j,\ell}) = (b_1^{j,\ell}, b_1^{j,\ell} + b_2^{j,\ell}, \ldots, b_1^{j,\ell} + \cdots + b_n^{j,\ell})$. It is easy to see that the experiment Exp_B^ℓ depicted on Fig. 4 (where the $\overrightarrow{x}^{j,\ell}$'s are chosen *first*, and then $\overrightarrow{b}^{j,\ell}$ is computed as their difference) has the same distribution of the variables as Exp_A^ℓ. To finish the proof of the Informal Lemma 1 we need to construct an experiment Exp_C, such that the view of the adversary in experiments Exp_B^0, Exp_C and Exp_B^1 is identical. Our approach to this is as follows. Based on the leakage diagram $S(L)$ (and *independently* from the choice of the $x_i^{j,\ell}$'s) we construct carefully crafted vectors $\overrightarrow{m}^0, \ldots, \overrightarrow{m}^k \in \{-1, 0, 1\}^n$ that we call *basic modification vectors* such that for every j we have that $m_1^j + \cdots + m_n^j = 1$ (where $(m_1^j, \ldots, m_n^j) = \overrightarrow{m}^j$). These vectors have to satisfy also some other conditions (that we define in the full version). See Fig. 5 for an example. The modification of Exp_B^ℓ is denoted Exp_C and presented on Fig. 4.

Two claims (discussed extensively in the full version of this paper [15]) that allow the proof to go through are that (1) the joint distribution of the variables $\overrightarrow{x}^{j,1}$, $\overrightarrow{b}^{j,1}$, and $\overrightarrow{c}^{j,1}$ in Exp_C is the same as in Exp_B^1, and (2) the view of the adversary are distributed identically in Exp_C and in Exp_B^0. What remains is to show how the basic modification vectors are constructed. Let LS be the connected component of $S(L)$ that contains its leftmost column. By assumption, E did not occur so LS does *not* contain the rightmost column of $S(L)$. This makes it possible to construct the basic modification vectors with desired properties. For each j construct $\overrightarrow{m}^j = (m_1^j, \ldots, m_n^j)$ according to the following rules: (i) if the left node of the edge "x_i^j" *does* belong to LS and its right node *does not* belong to LS, then let m_i^j be equal to $+1$, (ii) if the left node of the edge "x_i^j" *does not* belong to LS and its right node *does belong* to LS, then let m_i^j be equal to -1, and (iii) let all the other m_i^j's be equal to 0. An example of how the basic modification vectors are constructed is presented on Fig. 5 (these vectors and their coordinates are marked there with numbers in boxes). As it turns out (see Lemma 6 in the full version of this paper [15] for a generalization of this statement) these rules guarantee that the requirement that "$m_1^j + \cdots + m_n^j = 1$", and all other necessary conditions, are satisfied.

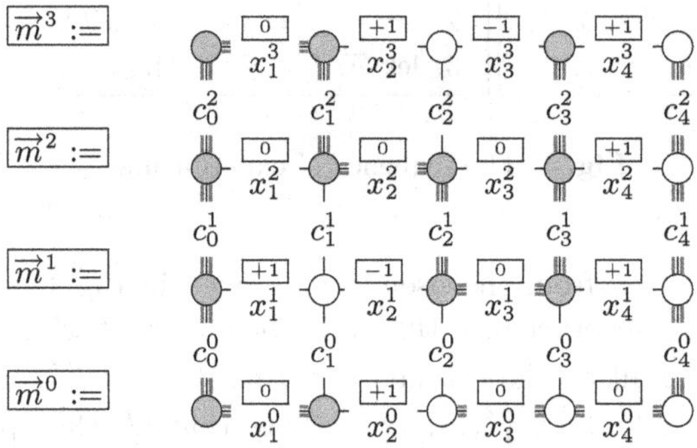

Fig. 5. The example of the leakage diagram with leakage indicated with double colored lines. The nodes of the connected component LS (containing the leftmost column) are indicated with gray color. The modification vectors $\overrightarrow{m}^j$ and their coordinates are placed in boxes (e.g.: $\overrightarrow{m}^0 := (0, +1, 0, 0)$).

2.2 Bounding the Probability of E

To show how we derive a bound on the probability of E we take a closer look at how, from the probabilistic point of view, the leakage diagram is constructed (see p. 9). By definition, it is a subgraph of a graph G from Fig. 2. Recall that in our experiment every wire of the circuit $\widehat{C}$ leaks independently at random with probability p. The leakage diagram $S(L)$ corresponding to leakage L is a random subgraph of G.

Let us now analyze the distribution of $S(L)$. It is easy to see that every edge "x_i^j" is added to $S(L)$ independently with probability p. Unfortunately, the situation is slightly more complicated when it comes to the c_i^j's. Recall that c_i^j's can be added to $S(L)$ for three reasons. The first (trivial) reason is that $i = 0$ or $i = n$. The second reason is that the wire "c_i^j" leaks in $\widehat{C}$ (i.e.: it belongs to L). The third reason is that the wire b_i^j or b_{i+1}^j leaks in $\widehat{C}$. Because of this, the events $\{$"c_i^j belongs to $S(L)$"$\}_{i,j}$ are *not* independent, and the probability of each of them may *not* equal to p.[5]

Let us look at the "non-trivial" edges in $S(L)$, i.e., the x_i^j's and the c_i^j's such that $i \in \{1, \ldots, n-1\}$. Let $\mathcal{U}$ be the variable equal to the set of non-trivial edges in $S(L)$. To make the analysis of the leakage diagram simpler it will be very useful to eliminate the dependencies between the "$c_i^j \in \mathcal{U}$" events. We do it by defining another random variable $\mathcal{Q}$ (that takes the same values as $\mathcal{U}$), and that has the following properties.

1. It is "more generous to the adversary", i.e., for every set $\mathcal{C}$ of the edges we have that
$$\Pr[\mathcal{C} \subset \mathcal{Q}] \geq \Pr[\mathcal{C} \subset \mathcal{U}] \tag{1}$$
(we will also say that the distribution of Q *covers the distribution of* $\mathcal{U}$, see Definition 1 on p. 18), and
2. The events $\{v \in \mathcal{Q}\}$ (where v is a non-trivial edge) are independent and have equal probability. Denote this probability q, and say that $\mathcal{Q}$ has a *standard distribution* (see Definition 2 on p. 18).

Now, consider an experiment $\text{Exp}_{\mathcal{Q}}$ of constructing a leakage diagram when the "$c_{i,j}$" and "$x_{i,j}$" edges are chosen according to $\mathcal{Q}$. More precisely: let the edges in the leakage diagram be sampled independently according to the following rules: the $\{c_0^j\}$'s and $\{c_n^j\}$'s are chosen with probability 1, and the remaining $\{c_i^j\}$'s are chosen with probability q. It is easy to see that, thanks to Eq. (1), the probability of E in $\text{Exp}_{\mathcal{Q}}$ is at least as high as in the probability in the original experiment. Hence, to give a bound on the probability of E it suffices to bound the probability of this probability in $\text{Exp}_{\mathcal{Q}}$. Thanks to the independence of the events $\{c_i^j \in \mathcal{Q}\}_{i,j} \cup \{x_i^j \in \mathcal{Q}\}_{i,j}$ bounding the probability of E in $\text{Exp}_{\mathcal{Q}}$ becomes a straightforward probability-theoretic exercise. For the details on how it is done see full version of this paper [15].

2.3 Generalizations to Arbitrary Circuits

As mentioned in Sect. 1.2, our final main contribution is a circuit compiler that uses the simple refreshing together with gadgets that perform the field operations. We follow the standard method of constructing compilers in a "gate-by-gate" fashion (see, e.g., [18], and the follow up work). A compiler takes as input

[5] For example: it is easy to see that if we know that $c_i^j \in S(L)$ then the event "c_{i+1}^j belongs to $S(L)$" becomes more likely (because leakage of b_{i+1}^j is more likely).

a circuit C (for simplicity assume it has no randomness gates) and produces as output a transformed circuit $\widehat{C}$ (that contains randomness gates RND). More concretely a wire carrying x in C gets transformed into a *bundle* of n wires carrying a random encoding of x. Every gate Γ in C is transformed into a "masked gate" $\widehat{\Gamma}$. For example, an addition gadget will have $2n$ inputs for n-share encodings of two values a and b, and n output wires that will carry some encoding of $a+b$. The masked input gates simply encode the secret (they have one input and n outputs). The masked output gates decode the secret (they have n inputs and one outputs). These two gadgets are assumed to be leak-free. They are also called: *input encoder* $\widehat{I}$ and *output decoder* $\widehat{O}$, respectively. For technical reasons, in our construction we insert the refreshing gadgets between the connected gadgets.

The main challenge in extending our ideas to such general circuits is that we need to take into account the leakage from wires of the individual gadgets, and represent them in the leakage diagram. We do it in such a way that unless an event E occurs, we are guaranteed that the adversary gained no information about the secret input. By the "event E" we mean a generalization of the event E (from the previous sections) to more complicated leakage diagrams. More concretely (see Sect. 4.1 for details) our approach is to represent each gadget $\widehat{\Gamma}$ in the graph G with a path $N_0^{\widehat{\Gamma}} - \cdots - N_n^{\widehat{\Gamma}}$ of length n and to "project" the leaking wires of the given gadget onto the edges of the path. Technically, this is done be defining, for every gadget $\widehat{\Gamma}$, a leakage *projection function* (see Sect. 4.3) that describes how a leakage from an internal wire is mapped on the path.

A projection function P, by definition, takes as argument a leaking wire w in a gadget $\widehat{\Gamma}$, and returns a subset of $[n]$ (usually of size 1 except for some wires in the multiplication gadget). We can refer to a projection of a set of wires in $\widehat{\Gamma}$ defined in a natural way as $P(\{w_1, \ldots, w_l\}) := P(w_1) \cup \ldots \cup P(w_l)$. One of the requirements that we impose on the function P is the following: every set of probes $\{w_1, \ldots, w_l\}$ (regardless of its size) from $\widehat{\Gamma}$ can be simulated knowing only input shares of indices in the projection $P(\{w_1, \ldots, w_l\})$ within each input bundle. Notice that it makes our definition of the gadget security similar in spirit to the existing definitions for the t-probing leakage model, like d-non-interference. One of the differences is that we care not only about the number of input shares that suffice to simulate the leakage, but also take into account their indices in a particular input bundle. Having a leakage projection function P defined for a gadget $\widehat{\Gamma}$, we will represent a leakage from that gadget in the leakage diagram as a subset of the edges from the path in G: $N_0^{\widehat{\Gamma}} - \cdots - N_n^{\widehat{\Gamma}}$. The positions (with the edge $N_0^{\widehat{\Gamma}} - N_1^{\widehat{\Gamma}}$ being the 1st one) of these edges in the path are taken from the set $P(\{w_1, \ldots, w_l\})$, when the wires $w_1, \ldots, w_l$ are leaking. This way we can "project" any given leakage from a gadget onto the path of length n in the leakage diagram.

As an example consider the addition gadget "$\widehat{\oplus}$" that computes an encoding $\overrightarrow{z}$ of $z = x + y$ as $\overrightarrow{z} := \overrightarrow{x} + \overrightarrow{y}$ (where $\overrightarrow{x}$ and $\overrightarrow{y}$ are encodings of x and y, respectively). The leakage projection function $P_{\widehat{\oplus}}$ for this gadget is defined as follows. Each input wire that is on ith position in the input bundle is projected onto the set $\{i\}$, i.e., $P_{\widehat{\oplus}}(x_i) = \{i\}$ and $P_{\widehat{\oplus}}(y_i) = \{i\}$. Moreover, projection of the output wires is defined similarly, namely $P_{\widehat{\oplus}}(z_i) = \{i\}$. It is easy to see that with such projection function the above mentioned simulation requirement is satisfied. For example, the leakage illustrated on Fig. 6a can be simulated knowing 3 input shares from each input bundle, namely x_2, x_4, x_5 and y_2, y_4, y_5. On the leakage diagram we represent this particular leakage from the addition gadget with 3 edges, as illustrated on Fig. 6b. Note that the addition gate is simple, and hence the projection function for it is rather straightforward. The projection function for a multiplication gadget is more involved (see Sect. 4.2).

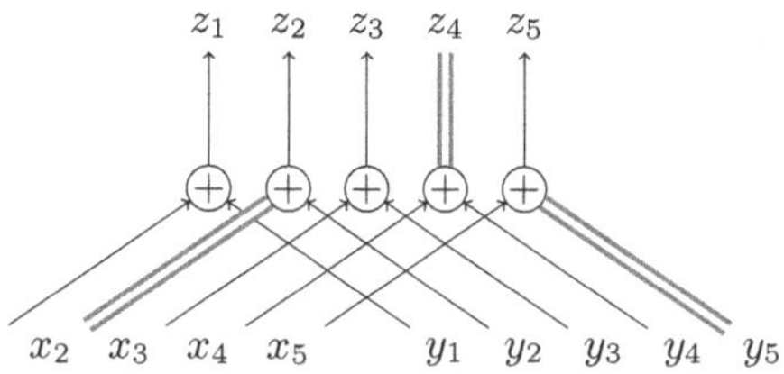

(a) An example of leakage from the addition gadget "$\widehat{\oplus}$" (marked with double colored lines).

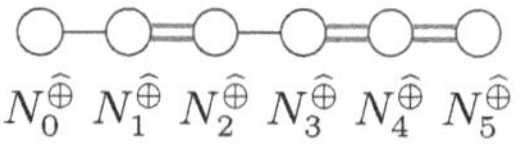

(b) The corresponding "projected" leakage in the leakage diagram (marked with double colored lines).

Fig. 6. Leakage from an addition gadget and the corresponding "projected" leakage. This is a valid projection, since it is enough to know x_2, x_4, x_5 and y_2, y_4, y_5 to simulate the leakage.

Having the projections of leakages for individual gadgets defined, we can generalize the idea of a leakage diagram $S(L)$ presented in previous sections from simple sequential k-round refreshing circuits to arbitrary private circuits built according to our construction. Recall that we insert a refreshing gadget between each pair of connected gadgets. The leakage from each individual gadget is projected onto a respective path in the leakage diagram, and the leakage from the remaining wires, i.e., wires used to generate encodings Enc(0) between two gadgets is "projected" onto the edges connecting the respective paths (analogue of the edges c_i^j's from previous sections). See Sect. 4.4 for the details. Overall, we obtain a graph that is similar to the leakage diagrams from the previous sections, but it is more general. In case of an example depicted on Fig. 7 the leakage from the gadget $\widehat{\Gamma}_1$ induces a projection set $\{3\}$. This fact is represented by including the edge $N_2^{\widehat{\Gamma}_1} - N_3^{\widehat{\Gamma}_1}$ into the leakage diagram.

A crucial property of such leakage diagrams is that the generalization of the Informal Lemma 1 still holds: the notion of the leftmost and the rightmost column are generalized to the leftmost and the rightmost *sides* (respectively). On Fig. 7 the leftmost side is a graph consisting of nodes $N_0^{\widehat{\Gamma}_1}, N_0^{\widehat{\Gamma}_2}, N_0^{\widehat{\Gamma}_3}, N_0^{\widehat{\Gamma}_4}$, and $N_0^{\widehat{\Gamma}_5}$, while the rightmost one consists of nodes $N_3^{\widehat{\Gamma}_1}, N_3^{\widehat{\Gamma}_2}, N_3^{\widehat{\Gamma}_3}, N_3^{\widehat{\Gamma}_4}$, and

$N_3^{\widehat{\Gamma}_5}$. We now define the event E as: "the leftmost and the rightmost sides are connected". For example E does *not* hold for the diagram on Fig. 7. To make it easier to verify this fact, we indicate (with gray color) the nodes connected with the leftmost side.

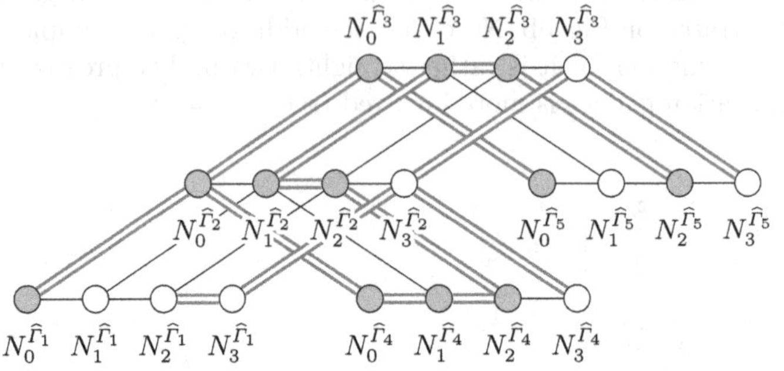

Fig. 7. An example of a leakage diagram for a transformed circuit $\widehat{C}$ with 5 gadgets. The nodes connected with the leftmost side are marked in gray.

When using the leakage projection functions we encounter the following problem that is similar to the "lack of independency problem" described in Sect. 2.2. Namely, it may happen that the events different edges become part of the projected set are *not* independent (this is, e.g., the case for the multiplication gadget in Sect. 4.2). We handle this problem in a similar way as before (see points 1 and 2 on page 13). That is: we define a "more generous" *leakage projection distribution* that (1) "covers" the original distribution, and (2) is "standard" (see the aforementioned points for the definition). Let q be the parameter denoting the probability in the standard distribution. This parameter, of course, depends on the probability p with which a wire leaks. A function that describes this dependence is called *projection probability function.* Every gadget in our construction comes with such a function. See Sect. 4.1 for a formalization of these notions.

Our construction is modular and works for different implementations of the addition and multiplication gadgets, assuming that they come with the leakage projection that satisfies certain conditions (see Theorem 1 on p. 26). We show (see Sect. 4.2) that the standard gadgets from the literature (including the ISW multiplication gadget [18]) satisfy this condition. Note that the construction and reasoning regarding the refreshing circuit presented in previous sections are special case of the construction and the security proof for the general arithmetic circuit. Indeed, we can treat each bundle between refreshing gadgets as an "identity gadget" (see Sect. 4.2).

Organization of the Rest of the Paper. In the next two sections we describe the technical details of the ideas outlined above. Section 3 consists of formal definitions, and Sect. 4 contains the details of our constructions. Due to the lack of space, some parts of these sections are moved to the full version of this paper [15].

3 Formal Definitions

We start by presenting formal definitions of some notions that were introduced informally in Sect. 2. Let us start with introducing some standard notation. In the sequel $[n]$ denotes the set $\{1, 2, \ldots, n\}$. We write $x \leftarrow \mathcal{X}$ when the element x is chosen uniformly at random from the finite set $\mathcal{X}$. A circuit C is *affine* if it does not use product gates. A *statistical distance* between two random variables X_0 and X_1 (distributed over some set $\mathcal{X}$) is defined as $\Delta(X_0; X_1) := 1/2 \cdot \sum_{x \in \mathcal{X}} |\Pr[X_0 = x] - \Pr[X_1 = x]|$. If $\Delta(X_0; X_1) \leq \epsilon$ then we say that X_0 and X_1 *are* ϵ*-close.* We assume a fixed security parameter n, i.e. every wire in C will be represented by a bundle of n wires in $\widehat{C}$.

Assumptions About the Circuit. For syntactic purposes we introduce special input encoding I gate and output decoding O gate used in original circuit C that simply implement the identity function, but will be transformed to $\widehat{I}$ and $\widehat{O}$ gadgets in $\widehat{C}$ (see Sect. 2.3). Gate I is required at every input wire of the circuit C that will be a subject to our compiler, and similarly O gate is required at every output gate of C. We call such circuits satisfying that requirement *complete.* However, in our proofs we consider also transformations of circuits that do not use gates I on the input wires and O on the output wires. Such circuits will be called *incomplete.* For incomplete circuit C we denote by its *completion* a circuit C with added gates I at every input and O at every output.

We also assume that the original C is deterministic, i.e. it has no randomness gates. This can be done without loss of generality, as the randomness can be provided to C as an additional input.

Partial Order of the Distributions over Subsets. We now provide formal definition of what it means that one probability distribution "covers" another one. The motivation and the intuition behind this concept were described in Sect. 2.2.

Definition 1. *Consider a fixed finite set A and its power set $P(A)$. Let D_1 and D_2 be some probability distributions over $P(A)$. We will say that distribution D_2* covers *distribution D_1 if it is possible to obtain D_2 from D_1 by a sequence of the following operations on a distribution D:*

> 1. *Pick two subsets satisfying $A_1 \subset A_2 \subset A$.*
> 2. *Pick a real value $0 < d < D(A_1)$.*
> 3. *Subtract d from $D(A_1)$ and add d to $D(A_2)$.*

It is clear from the definition above that the relation of covering is indeed a partial order on the probability distributions. We will write $D_2 \geq D_1$ to denote the coverage relationship of the distributions (when it is clear from the context over which power set these distributions are). As already mentioned in Sect. 2.2, one specific distribution over a power set $P(A)$ that we will consider is *a standard distribution* $D_p(A)$ where $0 < p < 1$.

Definition 2. *Let A be a finite set and let $0 < p < 1$. We define a random subset S of A as follows: any element of A belongs to S with probability p, independently. We call the distribution over $P(A)$ determined by the random subset S* a standard distribution $D_p(A)$.

For a random variable X with a domain $P(A)$ we denote by $\mathcal{D}(X)$ the probability distribution over $P(A)$ generated by that variable. For two random variables X, Y with the same domain $P(A)$ we will say that Y covers X if $\mathcal{D}(Y)$ covers $\mathcal{D}(X)$.

3.1 Security Definitions

In this section we present the formal definitions of soundness and privacy of a circuit transformation. Soundness is defined as follows.

Definition 3. *We say that transformation $\widehat{C}$ of k-input complete circuit C is* sound *if it preserves the functionality of C, that is*

$$\widehat{C}(\overrightarrow{x}) = C(\overrightarrow{x})$$

for every input $\overrightarrow{x}$ of length k. In case of incomplete circuit C, we say that its transformation is sound if transformation of its completion is sound.

To reason about privacy we consider the following experiment.

Definition 4. *For a fixed circuit C with k input wires, its input $\overrightarrow{x} = (x_1, \ldots, x_k)$ and probability p we define an experiment $Leak(C, \overrightarrow{x}, p)$ that outputs an adversarial view as follows:*

1. *Transformed circuit $\widehat{C}$ is fed with $(x_1, \ldots, x_k)$ resulting with some assignment of the wires of $\widehat{C}$.*
   In case when C is incomplete, the i-th input wire bundle of transformed circuit $\widehat{C}$ is fed with an encoding of respective input value x_i, chosen uniformly at random.
2. *Each wire of $\widehat{C}$ leaks independently with probability p. Note that in case of complete circuit C input and output wires do not leak, as part of $\widehat{I}$ and $\widehat{O}$ gadgets.*
3. Output: *(LW: set of leaking wires in $\widehat{C}$, A: values assigned to the leaking wires in LW during the circuit evaluation).*

We are now ready to define privacy of a circuit transformation.

Definition 5. *We say that transformation $\widehat{C}$ of circuit C is (p, ϵ)-private if leakage in experiment Leak$(C, \overrightarrow{x}, p)$ can be simulated up to ϵ statistical distance, for any input $\overrightarrow{x}$. More precisely, there exist a simulation algorithm that, not knowing input $\overrightarrow{x}$, outputs a random variable that is ϵ-close to the actual output of Leak$(C, \overrightarrow{x}, p)$.*

4 Technical Details of the Circuit Transformation

Let us now present the technical details of the ideas outlined in Sect. 2, i.e., our construction of the transformed circuit $\widehat{C}$ together with a proof of the privacy. For syntactic purposes we introduce a special single-input single-output refreshing gate R that acts as an identity function, similarly to I and O gates, but can be placed anywhere in the circuit C. The general transformation of the original circuit C consists of two phases. We start with the *preprocessing phase.* In this phase, if circuit C is incomplete then we add I gate to every input wire and O to every output wire. Moreover, we add refreshing gate R on every wire of C that connects any two gates, except for I and O (see Fig. 8). We call the resulting circuit C'. We then proceed to the *actual transformation phase* in which each wire in C' carrying value x is replaced with a bundle of n wires that carry an encoding of x. Each gate Γ in C' is replaced with a respective gadget subcircuit $\widehat{\Gamma}$ that operates on the encodings. Below we give a detailed description of the gadget subcircuits.

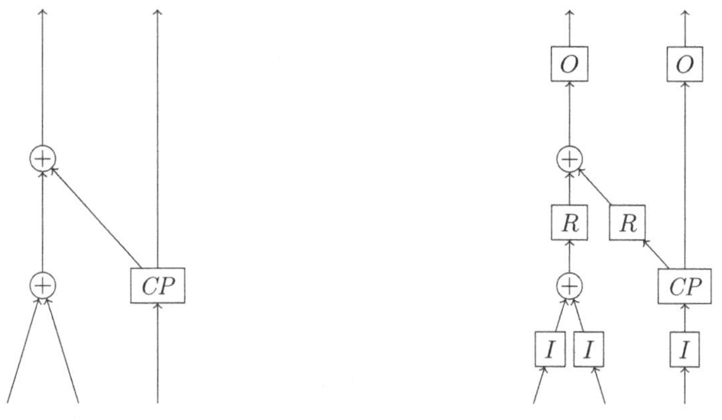

(a) An original circuit C (b) Circuit C' after preprocessing phase

Fig. 8. Example of the preprocessing phase of the transformation.

We say that two regular gadgets (not refreshing gadgets) $\widehat{\Gamma}_1$ and $\widehat{\Gamma}_2$ in $\widehat{C}$ are *connected* if there is a refreshing gadget between them. More precisely, if there is a refreshing gadget $\widehat{R}$ that takes as input the output bundle of $\widehat{\Gamma}_1$, and outputs the input bundle to $\widehat{\Gamma}_2$.

4.1 General Gadget Description

In this section we give a general definition of a gadget and the required properties. Every gadget used in our construction, except for the refreshing gadget $\widehat{R}$, satisfies the given definition.

Input and Output Wires of the Gadget. Let us consider a gate Γ in circuit C of $0 \leq i \leq 2$ inputs and $1 \leq o \leq 2$ outputs excluding the case $(i, o) = (2, 2)$, for example Γ might be a sum gate $\oplus$ or a product gate $\otimes$. A respective gadget $\widehat{\Gamma}$ will have i input wire bundles and o output wire bundles, that is $i \cdot n$ inputs and $o \cdot n$ outputs in total. We will denote with $IN_k^b(\widehat{\Gamma})$ the k-th wire of its b-th input bundle and with $OUT_k^b(\widehat{\Gamma})$ the k wire of its b-th output bundle. We denote with $IN_k(\widehat{\Gamma})$ all the input wires of index k in its input bundle. More precisely, $IN_k(\widehat{\Gamma}) := \{IN_k^b(\widehat{\Gamma})|1 \leq b \leq i\}$. Similarly, we define $OUT_k(\widehat{\Gamma})$ as $OUT_k(\widehat{\Gamma}) := \{OUT_k^b(\widehat{\Gamma})|1 \leq b \leq o\}$. Moreover, we use $IN^b(\widehat{\Gamma})$ to denote the b-th input bundle of $\widehat{\Gamma}$ and OUT^b to denote the b-th output bundle. That is, $IN^b(\widehat{\Gamma}) := \{IN_k^b(\widehat{\Gamma})|1 \leq k \leq n\}$, and $OUT^b(\widehat{\Gamma}) = \{OUT_k^b(\widehat{\Gamma})|1 \leq k \leq n\}$. Let $g : \mathbb{F}^i \to \mathbb{F}^o$ be the function computed by the gate Γ. The gadget $\widehat{\Gamma}$ should implement the same functionality as Γ. More precisely, if $g(x_1, \ldots, x_i) = (y_1, \ldots, y_o)$ then for *any* encoding $(\overrightarrow{x_1}, \ldots, \overrightarrow{x_i})$ of $(x_1, \ldots, x_i)$ fed to $\widehat{\Gamma}$ as input, it outputs *some* encoding $(\overrightarrow{y_1}, \ldots, \overrightarrow{y_o})$ of $(y_1, \ldots, y_o)$.

Leakage Projections. We now define the "leakage projections" already informally discussed in Sect. 2.3. Every gadget comes with *a leakage projection function* P that takes as input a leaking wire w in $\widehat{\Gamma}$ and outputs an associated subset $P(w)$ of $[n]$, usually an one-element subset. We can refer to *the projection set* of a subset W of wires in $\widehat{\Gamma}$ defined as $P(W) = \bigcup_{w \in W} P(w)$. We require the following properties of the projection P. Firstly, for any subset LG of leaking wires in $\widehat{\Gamma}$, it is enough to know the values carried by wires of the indices $\in P(LG)$ from every input bundle, i.e. the wires in $\{IN_k^b(\widehat{\Gamma})|1 \leq b \leq i, k \in P(LG)\}$ to simulate the leakage from $\widehat{\Gamma}$ perfectly (without knowing the values of the other input wires). Secondly, for every output wire w in $\widehat{\Gamma}$ that is k-th wire in any output bundle, i.e. $w \in OUT_k(\widehat{\Gamma})$, we have $P(w) = \{k\}$.

Consider an experiment where each of the wires in the gadget $\widehat{\Gamma}$ leaks independently with probability p. Let us call a set of leaking wires LR. Induced projection of the leakage $P(LR)$ defines a probability distribution over the subsets of $[n]$. We denote this *leakage projection distribution* with $D_p(\widehat{\Gamma})$. In the security proof it will be convenient to consider only the gadgets $\widehat{\Gamma}$ with the following property: if every wire of $\widehat{\Gamma}$ leaks independently with probability p then projection of the leakage contains every number $i \in [n]$ with some probability q *independently*. However it is not the case, e.g. for the product gadget. For that reason, we introduce *a projection probability function* describing a particular gadget. Essentially, it expresses with what probability do we need to add every particular number $\in [n]$ to the projection in order to make these events independent. More precisely, we will say that a function $f : [0,1] \to \mathbb{R}$ is *a projection probability function* for a gadget $\widehat{\Gamma}$ if the leakage projection distribution $D_p(\widehat{\Gamma})$ is covered by the standard distribution $D_{f(p)}([n])$ (as in Definition 1). Note that the function f may depend on the security parameter n, like in the case of product gadget $\widehat{\otimes}$.

4.2 The Gadgets Used in Our Construction

In this section we present all the gadgets used in our construction.

ISW Product Gadget. As the product gadget $\widehat{\otimes}$ in our construction we use the gadget proposed in [18]. Here we recall their scheme and prove that it satisfies the general gadget definition.

1. **Input:** 2 bundles $\overrightarrow{x} = (x_1, \dots x_n)$ and $\overrightarrow{y} = (y_1, \dots y_n)$
2. For $1 \leq i < j \leq n$ sample $z_{i,j} \leftarrow \mathbb{F}$
3. For $1 \leq i < j \leq n$ compute $z_{j,i} = (z_{i,j} \oplus x_i \otimes y_j) \oplus x_j \otimes y_i$
4. Compute the output encoding $(t_1, \dots t_n)$ as $t_i = x_i \otimes y_i \oplus \bigoplus_{j \neq i} z_{i,j}$
5. *Output:* a bundle $(t_1, \dots, t_n)$

We define the projection function P for this gadget as follows: For every wire w of the form x_i, y_i, $x_i \otimes y_i$, $z_{i,j}$ (for any $j \neq i$) or a sum of values of the above form (with t_i as a special case), $P(w) = \{i\}$. For the remaining wires w, which are of the form $x_i \otimes y_j$ or $z_{i,j} \oplus x_i \otimes y_j$, we define $P(w) = \{i, j\}$. The following lemmas are proven in the full version of this paper [15].

Lemma 1. *The ISW product gadget with its projection function satisfies a general gadget description (given in the Sect. 4.1) for the multiplication function $g(x, y) = x \cdot y$.*

Lemma 2. *The function $f(p) = n(8p + \sqrt{3p})$ is a projection probability function for the ISW product gadget.*

Other Gadgets. We already described the addition gadget $\widehat{\oplus}$ in Sect. 2.3. Besides of this, we use a *copy gadget* $\widehat{CP}$ that takes one input bundle $\overrightarrow{x} = (x_1, \dots, x_n)$. Then it applies the copy gate CP to each wire $x_1, \dots, x_n$ obtaining n respective pairs $(y_1, z_1), \dots, (y_n, z_n)$. We define two output bundles as $\overrightarrow{y} = (y_1, \dots, y_n)$ and $\overrightarrow{z} = (z_1, \dots, z_n)$. The *negation gadget* $\widehat{NEG}$. takes one input bundle $\overrightarrow{x} = (x_1, \dots, x_n)$ and it applies the negation gate NEG to each of n wires $x_1, \dots, x_n$ obtaining wires $y_1, \dots, y_n$. We define the output bundle of the gadget as $\overrightarrow{y} = (y_1, \dots, y_n)$. *Constant gadget* $\widehat{Const_\alpha}$ has zero input bundles and one output bundle carrying n constant values: $(\alpha, 0, \dots, 0)$. Finally, the *Identity gadget* $\widehat{ID}$ is a special gadget has one input and one output bundle, and simply outputs the input.

Properties of Gadgets Other Than ISW Product Gadget. It is clear that all the gadgets described above correctly implement the desired functions. Also, it is easy to see that for each gadget the leakage projection function P can be defined as follows: for any input or output wire w in the gadget, define $P(w) = \{i\}$, where i is an index of w in its (input or output) bundle. Clearly for each of these gadgets the function $f(p) = 3p$ is a projection probability function. For the gadgets $\widehat{Const_\alpha}$ and $\widehat{ID}$ even smaller function $f(p) = p$ is a projection probability function. We omit the proofs in these cases, as they are very straightforward.

4.3 Refreshing Gadget Properties

In this section we describe properties of the refreshing gadget $\widehat{R}$. (see Fig. 1a on p. 8) that are crucial to the security of the construction, and are used in the privacy proof. Below, by *refreshing bundle* $B_{\widehat{R}}$ we mean the wires that are used to generate the fresh encoding Enc(0) in the refreshing gadget $\widehat{R}$, i.e., wires carrying $b_1^j, \ldots, b_n^j$ and $c_1^j, \ldots, c_{n-1}^j$ on Fig. 1.

Refreshing Bundle Leakage Projection. Consider a refreshing bundle $B_{\widehat{R}}$. Suppose that LR is a set of leaking wires in $B_{\widehat{R}}$. We define a subset $S(LR)$ of $\{0, \ldots, n\}$ representing the leakage LR as follows: we start with the set $S = \{0, n\}$. For every wire of the form $c_k^j = b_1^j \oplus b_2^j \oplus \ldots \oplus b_k^j$ in LR, where $1 \leq k < n$, add k to S. For every wire of the form b_k^j in LR , where $1 < k \leq n$, add k and $k - 1$ to S.

One may think of the function $S(\cdot)$ as an analogue of the leakage projection function (introduced in Sect. 4.1) in case of a refreshing gadget. The difference is, however, that $S(\cdot)$ codomain size is $n + 1$ instead of n, and that 2 elements (0 and n) belong to $S(LR)$ "by default".

Leakage Projection Coverage. Here we show a random subset of $\{0, \ldots, n\}$ that covers the projection of the refreshing bundle leakage. Let us define a random subset R_q of $\{0, \ldots, n\}$ as follows: R_q contains 0 and n with probability 1, and for any other number $i \in \{0, \ldots, n\}$ R_q contains i with probability q, independently. The proof of the following lemma appears in the full version of this paper [15].

Lemma 3. *Let LR be a subset of leaking wires of a refreshing bundle $B_{\widehat{R}}$ when each wire leaks independently with probability p. Then the random subset $S(LR) \subset \{0, \ldots, n\}$ is covered by $R_{p+2\sqrt{3p}}$.*

Leakage Diagrams. The main technical concept of this work is a *leakage diagram* (already introduced informally in Sect. 2). Consider a transformed circuit $\widehat{C}$, as described in previous sections. Suppose that LW is the set of leaking wires in $\widehat{C}$. The leakage diagram is a representation of the set LW. As explained in Sect. 2, in the security proof the leakage diagram is used to determine whether the leakage compromises the secret or not. This is because of the property that if the *leftmost* and *rightmost* sides of the leakage diagram are disconnected then the privacy is preserved.

We first define the leakage diagram as a subgraph of $G = G(\widehat{C})$ - a graph associated with the transformed circuit $\widehat{C}$. The leakage diagram inherits all nodes of G and some of its edges, depending on the set of leaking wires LW. The exact construction of graph $G(\widehat{C})$ and the leakage diagram are described in the following paragraphs. Let C be any circuit and $\widehat{C}$ its transformation as described in Sect. 4. We define an associated undirected graph $G = G(\widehat{C})$ as follows. For each general gadget $\widehat{\Gamma}$ in $\widehat{C}$ (every gadget except the refreshing gadgets) $G(\widehat{C})$ contains *a crosswise path* of length n, where n is the security parameter of the construction. We denote the nodes of this path $N_0^{\widehat{\Gamma}}, \ldots, N_n^{\widehat{\Gamma}}$. Moreover, for every pair $\widehat{\Gamma}_1, \widehat{\Gamma}_2$ of connected gadgets in $\widehat{C}$, we add to the graph $G(\widehat{C})$ a *vertical matching* consisting of the following $n+1$ edges: $(N_0^{\widehat{\Gamma}_1}, N_0^{\widehat{\Gamma}_2}), \ldots, (N_n^{\widehat{\Gamma}_1}, N_n^{\widehat{\Gamma}_2})$. We call all the nodes of the form $N_0^{\widehat{\Gamma}}$, for some gadget $\widehat{\Gamma}$ in $\widehat{C}$, together with the edges between these nodes *a leftmost of* G. Analogically, we define *a rightmost of* G as all the nodes of the form $N_n^{\widehat{\Gamma}}$ with all the edges between them. The construction of $G(\widehat{C})$ can be naturally decomposed into separate subsets of edges - its crosswise paths and vertical matchings. We will call it a *decomposition* of $G(\widehat{C})$.

While the computation is executed on circuit $\widehat{C}$ some wires will leak the carried values. Let LW denote the set of all the leaking wires. We will be representing this set with *a leakage diagram* H - a subgraph of $G(\widehat{C})$. The leakage diagram inherits all the nodes from $G(\widehat{C})$ and some of its edges as in the following construction. Each leaking wire $w \in LW$ that belongs to some general gadget $\widehat{\Gamma}$ is *projected* onto the respective crosswise path in G. More precisely, if $P_{\widehat{\Gamma}}$ is leakage projection function for the gadget $\widehat{\Gamma}$ then we add to the leakage diagram H the edges in the crosswise path of order in $P_{\widehat{\Gamma}}(w)$, i.e. edges $\{(N_{i-1}^{\widehat{\Gamma}}, N_i^{\widehat{\Gamma}}) | i \in P_{\widehat{\Gamma}}(w)\}$.

The rest of the leaking wires in the set LW are part of some refreshing bundle $B_{\widehat{R}}$, where the refreshing gadget $\widehat{R}$ connects some gadgets $\widehat{\Gamma}_1$ and $\widehat{\Gamma}_2$. Let $LR \subset LW$ be a set of leaking wires in this refreshing bundle. It is represented in the leakage diagram H by the subset of the vertical matching between two respective crosswise paths, namely $\{(N_i^{\widehat{\Gamma}_1}, N_i^{\widehat{\Gamma}_2}) | i \in S(LR)\}$. An example of a leakage diagram is illustrated on Fig. 7.

Modification Vectors. In the security proof we use a sequence of hybrid experiments that produce exactly the same leakage. One of the hybrids requires to assign every gadget in $\widehat{C}$ with a *basic modification vector*. They were already informally introduced in Sect. 2. Let us now present their formal definition. A *basic modification vector* is a vector $\overrightarrow{m} = (m_1, \ldots, m_n)$ of length n whose coordinates are in the set $\{-1, 0, 1\}$ and additionally $\sum_{i=1}^{n} m_i = 1$. We assign a gadget $\widehat{\Gamma}$ with the basic modification vector $\overrightarrow{m}^{\widehat{\Gamma}}$ based on the leakage diagram H. Let LS be the connected component of the leftmost side of H. Let I be the set of nodes indices from $\{N_0^{\widehat{\Gamma}}, \ldots, N_n^{\widehat{\Gamma}}\}$ that belong to LS. Now, based on the set I we assign the modification vector $\overrightarrow{m}^{\widehat{\Gamma}}$ according to the following rule: the i-th coordinate of $\overrightarrow{m}^{\widehat{\Gamma}}$ equals 1 if $i-1 \in I$ and $i \notin I$, equals -1 if $i-1 \notin I$ and $i \in I$, and equals 0 in other cases.

We generalize the definition of a basic modification vector to a *modification vector*. We will say that a vector $\overrightarrow{w}$ is *a modification vector* if it can be written in the form $\overrightarrow{w} = v \cdot \overrightarrow{m}$ for some scalar value $v \in \mathbb{F}$ and a basic modification vector $\overrightarrow{m}$. Moreover, we will say that a modification vector $\overrightarrow{m} = (m_1, \ldots, m_n)$ is *disjoint* with a set $A \subset [n]$ if $m_a = 0$ for all $a \in A$. Let S be a subset of $\{0, \ldots, n\}$ and let $\overrightarrow{m}^1, \overrightarrow{m}^2$ be any modification vectors of length n. We will say that $\overrightarrow{m}^1$ and $\overrightarrow{m}^2$ are *indistinguishable* under S if for every $k \in S$ we have that $\sum_{i=1}^{k} m_i^1 = \sum_{i=1}^{k} m_i^2$.

Leakage and Extended Leakage from a Gadget. In this section we give the formal definitions of leakages from a gadget. Here, we consider only gadgets other than refreshing gadget.

Extended Leakage. In order to express the desired property of a gadget we define a random variable that we call *extended leakage*. It is a leakage from a subset of wires in $\widehat{\Gamma}$ together with values carried by *all* the output wires of $\widehat{\Gamma}$, including the non-leaking wires (a more restrictive definition that does not include these wires is given in the full version of this paper [15].

Definition 6. *Let $\widehat{\Gamma}$ be a gadget with i input bundles and o output bundles and let LG be a subset of its wires. We define a function $ExtLeak_{\widehat{\Gamma}}^{LG}(\overrightarrow{x_1}, \ldots, \overrightarrow{x_i})$ as the output of the following experiment:*

1. *The gadget $\widehat{\Gamma}$ is fed with input $(\overrightarrow{x_1}, \ldots, \overrightarrow{x_i})$ resulting with some assignment of the wires of $\widehat{\Gamma}$.*
2. *Let $\overrightarrow{y}_1, \ldots, \overrightarrow{y}_o$ be the produced output of $\widehat{\Gamma}$.*
3. Output: *(values assigned to wires in LG, values assigned to* all *the output wires $\overrightarrow{y_1}, \ldots, \overrightarrow{y_o}$).*

Extended Leakage Shiftability. Recall that in Sect. 2.1 one of the main technical tricks was to show that the experiments Exp_C and Exp_B^0 are indistinguishable from the point of view of the adversary. This was done by showing that the vectors encoding the secret can be "shifted" (i.e. a certain vector can be added to it) in way that is not noticeable to the adversary. This idea is formalized and generalized to gadgets below.

Definition 7. *Let $\overrightarrow{v_1}, \ldots \overrightarrow{v_k}$ and $\overrightarrow{m}$ be vectors of the same length. and let $T = (T_1, \ldots, T_k)$ be a sequence of k field elements. We define a $shift_{\overrightarrow{m}}^{T}(\overrightarrow{v_1}, \ldots, \overrightarrow{v_k})$ as follows: it is a sequence of vectors $\overrightarrow{w_1}, \ldots, \overrightarrow{w_k}$, with $\overrightarrow{w_j}$ being a modified vector $\overrightarrow{v_j}$:*

$$\overrightarrow{w_j} = \overrightarrow{v_j} + T_j \cdot \overrightarrow{m}.$$

Also, when applicable, we treat values $v_1, \ldots, v_l$ as vectors of length 1. Then we assume a default basic modification vector (1) and write $shift^T(v_1, \ldots, v_k)$ instead of $shift_{(1)}^T(v_1, \ldots, v_k)$.

Recall that the informal description in Sect. 2 was simplified, since it was focusing on the multi-round refreshing circuits only. Making this idea work for arbitrary circuits requires some extra work. In particular, we need to ensure that nothing goes wrong in the (non-refreshing) gadgets if their input is shifted. Let LG denote a fixed subset of leaking wires in the gadget $\widehat{\Gamma}$. Informally speaking, *the extended leakage shiftability property* says that shifting the value of the wires of index $w \notin P(LG)$ in the input bundles of $\widehat{\Gamma}$ results in shifting the extended leakage only on the index w in the output bundles. This is formalized below.

Definition 8. *Let $\widehat{\Gamma}$ be a gadget with i input bundles and o output bundles implementing a function g, and let P be its leakage projection function. We say that a pair $(\widehat{\Gamma}, P)$ satisfies an* extended leakage shiftability *property if the following holds: Let $x_1, \ldots, x_i$ be any input to g and suppose that $g(shift^S(x_1, \ldots, x_i)) = shift^T(g(x_1, \ldots, x_i))$ for some sequences S and T of lengths i and o, respectively. For any fixed encodings $\overrightarrow{x_1}, \ldots, \overrightarrow{x_i}$ of $x_1, \ldots, x_i$, any subset of leaking wires LG and any basic modification vector $\overrightarrow{m}$ that is disjoint with the set $P(LG)$ we have*

$$ExtLeak_{\widehat{\Gamma}}^{LG}(shift_{\overrightarrow{m}}^{S}(\overrightarrow{x_1}, \ldots, \overrightarrow{x_i})) = shift_{\overrightarrow{m}}^{T}(ExtLeak_{\widehat{\Gamma}}^{LG}(\overrightarrow{x_1}, \ldots, \overrightarrow{x_i})).$$

Here, when the function shift is applied to the output of the ExtLeak experiment, it is applied only to the second part of the experiment output i.e. values assigned to the output bundles of a $\widehat{\Gamma}$.

Based on the following lemma, whose proof appears in the full version of this paper [15], every gadget used in our construction satisfies the extended leakage shiftability property.

Lemma 4. *Every general gadget $\widehat{\Gamma}$ with its leakage projection function, as described in Sect. 4.1, satisfies the extended leakage shiftability property.*

In the proof of Theorem 1 we also use a concept of *refreshed gadget reconstruction* that is presented in the full version of this paper [15].

4.4 Privacy of the Construction

Here we present and prove a central theorem of our work.

Theorem 1. *Let C be any arithmetic circuit and $\widehat{C}$ its transformation as described in Sect. 4. Assume that for all gadgets used in $\widehat{C}$ the projection probability functions are upper-bounded by a function $q : [0,1] \to \mathbb{R}$, which also upper-bounds the function $f(p) = p + 2\sqrt{3p}$. Then $\widehat{C}$ is sound implementation of C and $\widehat{C}$ is $(p, |C| \cdot (4q(p))^n)$-private for any probability p.*

This theorem is proven along the lines of the intuitions presented in Sect. 2. Due to space limitation we give only a proof overview. Below we present in a formal way some tools that are used in the proof (and that were already informally discussed in Sect. 2). These tools are used in the proof of Theorem 1, which appears in the full version of this paper [15].

Proof Overview. To prove the privacy of our construction, we will show that any two inputs X_1, X_2 to circuit $\widehat{C}$ induce leakages that are close in terms of statistical distance. We compare these two leakages conditioned on the set of leaking wires being some *fixed* set LW. Let H be a leakage diagram induced by LW. We show that if the left and right sides of the graph H are not connected then the two leakages are actually identical. To this end, we use a hybrid argument, with the set of leaking wires being fixed to LW. We define a sequence of experiments, called hybrids, and show that every two consecutive experiments produce identical output. Here we briefly describe them:

Hybrid_1 (this corresponds to experiment Exp_A^0 in Sect. 2.1): simply outputs the leakage when $\widehat{C}$ is fed with X_1.

Hybrid_2 (this corresponds to experiment Exp_B^0): in this experiment each gadget in $\widehat{C}$ is evaluated separately, and the assignment of the refreshing bundles between the gadgets are derived from there. To this end, we consider the evaluation of the original circuit C when fed with X_1. If a particular wire w in C, which is an input to a gate Γ, is assigned with a value v then the respective input bundle in the gadget $\widehat{\Gamma}$ in $\widehat{C}$ is fed with a freshly chosen random encoding $\overrightarrow{v} \leftarrow \text{Enc}(v)$. Then each gadget in $\widehat{C}$ is evaluated accordingly to the chosen inputs. This determines the assignment of all the refreshing bundles in $\widehat{C}$. The output of the experiment consists of the values assigned to wires in LW.

Hybrid_3 (this corresponds to experiment Exp_C): this experiment is the same as Experiment 2, except for the random vectors that are assigned to the input bundles of each individual gadget. Here, after choosing a random encoding $\overrightarrow{v} \leftarrow \text{Enc}(v)$ just as in Experiment 2, we shift it by carefully chosen modification vector $\overrightarrow{m}$. As a result, we feed the particular input bundle with $\overrightarrow{v} + \overrightarrow{m}$. The modification vector for the input bundles of each gadget is constructed based on inputs X_1 and X_2, and the leakage diagram H. At this point we use the fact that the left and right sides of the leakage diagram H are not connected. The details of the construction for modification vectors are given in the Sect. 4.3.

Based on the properties of the refreshed gadgets subcircuits in $\widehat{C}$ and taking into account the construction of the modification vectors, we argue that shifting values that are fed to each gadget actually does not change the leakage. Hence this experiment outputs the same random variable as Experiment 2.

Hybrid_4 (this corresponds to experiment Exp_B^1): this experiment is analogous to the Experiment 2, with input X_2 instead of X_1. We argue that the random vectors assigned to the input bundles of each individual gadget in are actually the same in this experiment and in Experiment 3. Hence, the two experiments produce identical outputs.

Hybrid_5 (this corresponds to experiment Exp_A^1): this experiment is analogous to the Experiment 1, with input X_2 instead of X_1. Also the transition between Experiment 4 and this experiment is analogous to the transition for Experiments 1 and 2.

The hybrid argument above essentially shows that unless the left and right sides of the leakage diagram H are connected, the leakage is the same independently of the input X fed to the transformed circuit $\widehat{C}$. Now, to complete the privacy proof, it is enough to upper-bound the probability of the left and right sides of H being connected. This is a pure probability theory exercise, given that $q(p)$ upper-bounds the leakage projection function of used gadgets which means that each edge will be included to the leakage diagram independently with probability at most $q(p)$.

4.5 Concrete Results

In this section we present the concrete results implied by Theorem 1. These are immediate consequences of the theorem. For affine circuits we obtain the following.

Proposition 1. *Assume that a circuit C is an affine circuit. Our transformation $\widehat{C}$, as described in Sect. 4, is $(p, |C| \cdot (4p + 8\sqrt{3p})^n)$-private for any probability p.*

Proof. As stated in the Sect. 4.2, for every gadget used in $\widehat{C}$ its projection probability function is upper-bounded by $3p$ and hence by $p + 2\sqrt{3p}$. Thus, the Proposition is a consequence of the Theorem 1 for the function $q(p) = p + 2\sqrt{3p}$. □

For the general circuits we have the following.

Proposition 2. *Assume that a circuit C is an arithmetic circuit. Our transformation $\widehat{C}$, as described in Sect. 4, is $(p, |C| \cdot (32np + 4n\sqrt{3p})^n)$-private for any probability p.*

Proof. From the Sect. 4.2 we conclude that for every gadget used in $\widehat{C}$ its projection probability function is upper-bounded by $n(8p+\sqrt{3p})$. Assuming $n \geq 2$, this function also upper-bounds $p+2\sqrt{3p}$. Thus, the Proposition is a consequence of the Theorem 1 for the function $q(p) = n(8p + \sqrt{3p})$. □

Finally, let us state the result for the multi-round simple refreshing circuits.

Proposition 3. *Consider a k-round refreshing circuit (see Sect. 2). This circuit is $(p, k \cdot (4p + 8\sqrt{3p})^n)$-private for any probability p.*

Proof. As stated in the Sect. 4.2, the projection probability function of the identity gadgets $\widehat{ID}$ used in the circuit equals p and hence is upper-bounded by $p+2\sqrt{3p}$. Thus, the Proposition is a consequence of the Theorem 1 for the function $q(p) = p + 2\sqrt{3p}$. □

5 Conclusion

In this work we introduce a new method to analyze the security of masking schemes in the noisy leakage model of Prouff and Rivain [21]. Our approach enables us to show the security of a simple refreshing scheme which is optimal in terms of randomness complexity (it requires only $n-1$ random values), and uses a small number of arithmetic operations. Our results are achieved by introducing a new technique for analyzing masked circuits against noisy leakages, which is of independent interest.

We believe that our results are of practical importance to the analysis of side-channel resistant masking schemes. The reason for this are twofold. First, our refreshing scheme is very simple and efficient, and reduces the overheads of the masking countermeasure significantly – in particular, for certain types of computation. For example in the case of a secure key update mechanism as used in any cryptocraphic scheme, we can reduce randomness and circuit complexity from $O(n^2)$ using ISW-like refreshing to $O(n)$, where the asymptotic in the later is with nearly optimal constants. Second, while in [5] it was shown how to construct a very simple refreshing scheme (similar to the one used in our work), the security analysis was in a more restricted model (the bounded moment model), and carried out only for small n. In our case, the analysis works for any n and in the standard noisy model that is well accepted in practice.

Interesting questions for future research include to extend our analysis to other masking schemes [4], to explore the tightness of our bounds and to verify our results experimentally in practice (e.g., by providing simulations on the practical resistance of the countermeasure and its efficiency).

Acknowledgements. The authors thank Sonia Belaïd and the anonymous reviewers for their constructive comments. Sebastian Faust received funding from the German Federal Ministery of Education and Research and the Hessen State Ministry for Higher Education, Research and the Arts within their joint support of the National Research Center for Applied Cybersecurity (CRISP). Additionally, he received funding from the Emmy Noether Program FA 1320/1-1 of the German Research Foundation (DFG) and by the VeriSec project 16KIS0634 from the Federal Ministry of Education and Research (BMBF). Stefan Dziembowski and Karol Żebrowski received funding from the Foundation for Polish Science (grant agreement TEAM/2016-1/4) co-financed with the support of the EU Smart Growth Operational Programme (PO IR).

References

1. Ajtai, M.: Secure computation with information leaking to an adversary. In: 43rd Annual ACM Symposium on Theory of Computing, pp. 715–724. ACM Press (2011)
2. Ananth, P., Ishai, Y., Sahai, A.: Private circuits: a modular approach. In: Shacham, H., Boldyreva, A. (eds.) CRYPTO 2018. LNCS. Part III, vol. 10993, pp. 427–455. Springer, Cham (2018). https://doi.org/10.1007/978-3-319-96878-0_15
3. Andrychowicz, M., Dziembowski, S., Faust, S.: Circuit compilers with $O(1/\log(n))$ leakage rate. In: Fischlin, M., Coron, J.-S. (eds.) EUROCRYPT 2016. LNCS. Part II, vol. 9666, pp. 586–615. Springer, Heidelberg (2016). https://doi.org/10.1007/978-3-662-49896-5_21
4. Balasch, J., Faust, S., Gierlichs, B., Verbauwhede, I.: Theory and practice of a leakage resilient masking scheme. In: Wang, X., Sako, K. (eds.) ASIACRYPT 2012. LNCS, vol. 7658, pp. 758–775. Springer, Heidelberg (2012). https://doi.org/10.1007/978-3-642-34961-4_45

5. Barthe, G., Dupressoir, F., Faust, S., Grégoire, B., Standaert, F.-X., Strub, P.-Y.: Parallel implementations of masking schemes and the bounded moment leakage model. In: Coron, J.-S., Nielsen, J.B. (eds.) EUROCRYPT 2017. LNCS. Part I, vol. 10210, pp. 535–566. Springer, Cham (2017). https://doi.org/10.1007/978-3-319-56620-7_19
6. Barthe, G., et al.: Strong non-interference and type-directed higher-order masking. In: ACM CCS 2016: 23rd Conference on Computer and Communications Security, pp. 116–129. ACM Press (2016)
7. Barthe, G., Belaïd, S., Dupressoir, F., Fouque, P.-A., Grégoire, B., Strub, P.-Y.: Verified proofs of higher-order masking. In: Oswald, E., Fischlin, M. (eds.) EUROCRYPT 2015. LNCS. Part I, vol. 9056, pp. 457–485. Springer, Heidelberg (2015). https://doi.org/10.1007/978-3-662-46800-5_18
8. Belaïd, S., Goudarzi, D., Rivain, M.: Tight private circuits: achieving probing security with the least refreshing. In: Peyrin, T., Galbraith, S. (eds.) ASIACRYPT 2018. LNCS. Part II, vol. 11273, pp. 343–372. Springer, Cham (2018). https://doi.org/10.1007/978-3-030-03329-3_12
9. Coron, J.-S.: Formal verification of side-channel countermeasures via elementary circuit transformations. In: Preneel, B., Vercauteren, F. (eds.) ACNS 2018. LNCS, vol. 10892, pp. 65–82. Springer, Cham (2018). https://doi.org/10.1007/978-3-319-93387-0_4
10. Coron, J.-S., Prouff, E., Rivain, M., Roche, T.: Higher-order side channel security and mask refreshing. In: Moriai, S. (ed.) FSE 2013. LNCS, vol. 8424, pp. 410–424. Springer, Heidelberg (2014). https://doi.org/10.1007/978-3-662-43933-3_21
11. Duc, A., Dziembowski, S., Faust, S.: Unifying leakage models: from probing attacks to noisy leakage. In: Nguyen, P.Q., Oswald, E. (eds.) EUROCRYPT 2014. LNCS, vol. 8441, pp. 423–440. Springer, Heidelberg (2014). https://doi.org/10.1007/978-3-642-55220-5_24
12. Duc, A., Faust, S., Standaert, F.-X.: Making masking security proofs concrete. In: Oswald, E., Fischlin, M. (eds.) EUROCRYPT 2015. LNCS. Part I, vol. 9056, pp. 401–429. Springer, Heidelberg (2015). https://doi.org/10.1007/978-3-662-46800-5_16
13. Dziembowski, S., Faust, S.: Leakage-resilient cryptography from the inner-product extractor. In: Lee, D.H., Wang, X. (eds.) ASIACRYPT 2011. LNCS, vol. 7073, pp. 702–721. Springer, Heidelberg (2011). https://doi.org/10.1007/978-3-642-25385-0_38
14. Dziembowski, S., Faust, S., Skorski, M.: Noisy leakage revisited. In: Oswald, E., Fischlin, M. (eds.) EUROCRYPT 2015. LNCS. Part II, vol. 9057, pp. 159–188. Springer, Heidelberg (2015). https://doi.org/10.1007/978-3-662-46803-6_6
15. Dziembowski, S., Faust, S., Żebrowski, K.: Simple re-freshing in the noisy leakage model. Cryptology ePrint Archive. Extended version of this paper (2019)
16. Goudarzi, D., Joux, A., Rivain, M.: How to securely compute with noisy leakage in quasilinear complexity. In: Peyrin, T., Galbraith, S. (eds.) ASIACRYPT 2018. LNCS. Part II, vol. 11273, pp. 547–574. Springer, Cham (2018). https://doi.org/10.1007/978-3-030-03329-3_19
17. Goudarzi, D., Martinelli, A., Passelèegue, A., Prest, T.: Unifying leakage models on a Réenyi day. Cryptology ePrint Archive, Report 2019/138 (2019). https://eprint.iacr.org/2019/138
18. Ishai, Y., Sahai, A., Wagner, D.: Private circuits: securing hardware against probing attacks. In: Boneh, D. (ed.) CRYPTO 2003. LNCS, vol. 2729, pp. 463–481. Springer, Heidelberg (2003). https://doi.org/10.1007/978-3-540-45146-4_27

19. Kalai, Y.T., Reyzin, L.: A survey of leakage-resilient cryptography. IACR Cryptology ePrint Archive 2019, p. 302 (2019). https://eprint.iacr.org/2019/302
20. Nikova, S., Rechberger, C., Rijmen, V.: Threshold implementations against side-channel attacks and glitches. In: Ning, P., Qing, S., Li, N. (eds.) ICICS 2006. LNCS, vol. 4307, pp. 529–545. Springer, Heidelberg (2006). https://doi.org/10.1007/11935308_38
21. Prouff, E., Rivain, M.: Masking against side-channel attacks: a formal security proof. In: Johansson, T., Nguyen, P.Q. (eds.) EUROCRYPT 2013. LNCS, vol. 7881, pp. 142–159. Springer, Heidelberg (2013). https://doi.org/10.1007/978-3-642-38348-9_9
22. Rivain, M., Prouff, E.: Provably secure higher-order masking of AES. In: Mangard, S., Standaert, F.-X. (eds.) CHES 2010. LNCS, vol. 6225, pp. 413–427. Springer, Heidelberg (2010). https://doi.org/10.1007/978-3-642-15031-9_28

Symmetric Cryptography (2)

The Exchange Attack: *How to Distinguish Six Rounds of AES with $2^{88.2}$ Chosen Plaintexts*

Navid Ghaedi Bardeh(✉) and Sondre Rønjom

Department of Informatics, University of Bergen, 5020 Bergen, Norway
{navid.bardeh,sondre.ronjom}@uib.no

Abstract. In this paper we present *exchange-equivalence attacks* which is a new cryptanalytic attack technique suitable for SPN-like block cipher designs. Our new technique results in the first secret-key chosen plaintext distinguisher for 6-round AES. The complexity of the distinguisher is about $2^{88.2}$ in terms of data, memory and computational complexity. The distinguishing attack for AES reduced to six rounds is a straight-forward extension of an exchange attack for 5-round AES that requires 2^{30} in terms of chosen plaintexts and computation. This is also a new record for AES reduced to five rounds. The main result of this paper is that AES up to at least six rounds is biased when restricted to *exchange-invariant sets* of plaintexts.

Keywords: SPN · AES · Exchange-equivalence attacks · Exchange-invariant sets · Exchange-equivalence class · Secret-key model · Differential cryptanalysis

1 Introduction

Block ciphers are typically designed by iterating an efficiently computable round function many times in the hope that the resulting composition behaves like a randomly drawn permutation. The designer is typically constrained by various practical criterion, e.g. security target, implementation boundaries, and specialized applications, that might lead the designer to introduce symmetries and structures into the round function as a compromise between efficiency and security. In the compromise, a round function is iterated enough times to make sure that any symmetries and structural properties that might exist in the round function vanish. Thus, a round function is typically designed to increasingly de-correlate with structure and symmetries after several rounds. However, what actually constitutes *structure* is an open question which requires continuous investigation as long as using randomly drawn codebooks is out of reach.

Low data- and computational-complexity distinguishers and key-recovery attacks on round-reduced block ciphers have recently gained renewed interest in the literature. There are several reasons for this. In one direction cryptanalysis of block ciphers has focused on maximizing the number of rounds that can

S. D. Galbraith and S. Moriai (Eds.): ASIACRYPT 2019, LNCS 11923, pp. 347–370, 2019.
https://doi.org/10.1007/978-3-030-34618-8_12

be broken without exhausting the full codebook and key space. This often leads to attacks marginally close to that of pure brute-force. These are attacks that typically have been improved over time based on many years of cryptanalysis. The most successful attacks often become de-facto standard methods of cryptanalysis for a particular block cipher and might discourage anyone from pursuing new directions in cryptanalysis that do not reach the same number of rounds. This in itself might hinder new breakthroughs, thus it can be important to investigate new promising ideas that might not have reached its full potential yet. New methods of cryptanalysis that break or distinguish fewer rounds faster but with lower complexity than established cryptanalysis is therefore interesting in this process. Many constructions employ reduced round AES as part of their design. On the other hand, reduced versions of AES have nice and well-studied properties that can be favorable as components of larger designs (see for instance Simpira [10]).

The security of Rijndael-type block cipher designs is believed to be a well-studied topic and has been in the focus of a large group of cryptanalysts during the last 20 years (e.g. see [1–3,5,6,8,12,13]). Thus, it is rather surprising that new and quite fundamental results continuously appear for 2–4 rounds of AES that enables completely new types of more efficient attacks for an increasing number of rounds of AES. At Crypto 2016, the authors of [15] presented the very first secret-key 5-round distinguisher for AES. Secret-key (or key-independent) means that the attack does not care about the particular round keys (e.g. in contrast to related-key attacks). They extend a 4-round integral property to 5-rounds by exploiting properties of the AES MixColumn matrix. Although their distinguisher requires the whole codebook, it spawned a series of new fundamental results for AES. It was later improved to $2^{98.2}$ chosen plaintexts with 2^{107} computations by extending a 4-round impossible differential property to a 5-round property. Then, at Eurocrypt 2017, the authors of [9] proposed the first 5-round secret-key chosen plaintext distinguisher which requires 2^{32} chosen texts with a computational cost of $2^{35.6}$ look-ups into memory of size 2^{36} bytes. They showed that by encrypting cosets of certain subspaces of the plaintext space the number of times the difference of ciphertext pairs lie in a particular subspace of the state space always is a multiple of 8.

Later, at Asiacrypt 2017, the authors of [14] presented new fundamental properties for Rijndael-type block cipher designs leading to new types of 3- to 6-round secret-key distinguishers for AES that beats all previous records. The authors introduced a new deterministic 4-round property in AES, which states that sets of pairs of plaintexts that are equivalent by exchange of any subset of diagonals encrypts to a set of pairs of ciphertexts after four rounds that all have a difference of zero in exactly the same columns before the final linear layer. This was further explored in [7] under the name "mixture cryptanalysis".

1.1 Our Contribution

The first 5-round secret-key chosen-plaintext distinguisher for AES was introduced at Crypto 2016, *almost 20 years after Rinjdael was first proposed as a can-*

didate in the AES-competition, and required the whole codebook. In this paper, *only three years later*, we introduce the first 6-round secret-key distinguisher for AES that has complexity of about $2^{88.2}$ computations and ciphertexts. This is a giant leap for cryptanalysis of AES. Our distinguishers are based on simple techniques which are easy to verify theoretically and in practice. Moreover, we prove that *AES up to at least 6 rounds is biased on exchange-invariant sets.* The 5-round distinguisher has been practically verified on a scaled down version in C/C++ on a standard laptop[1].

1.2 Overview of This Paper and Main Results

In Sect. 2 we briefly describe results and notation that makes up the machinery for the rest of this paper. In particular, we describe what we call exchange operators, exchange-invariant sets and exchange-equivalence classes, and their relations to AES. In Sect. 3, we prove that five full rounds of AES is biased on exchange-invariant sets and in Sects. 4 and 5 we turn this result into simple distinguishers for AES reduced to five and six rounds.

The currently best secret-key distinguishers for 5- and 6-round AES are given in Table 1. We adopt that data complexity is measured in a minimum number of chosen plaintexts/ciphertexts CP/CC or adaptively chosen plaintexts/ciphertexts ACP/ACC. Time complexity is measured in equivalent number of AES encryptions (E), memory accesses (M) and/or XOR operations (XOR) - adopting that $20M \approx 1$ round of AES.

Table 1. *Secret-key distinguishers for AES*

Property	Rounds	Data	Cost	Ref.
Impossible Diff	5	2^{128} CP	$2^{129.6}$ XORs	[15]
Multiple-8	5	2^{32} CP	$2^{35.6}$ M	[9]
Exchange Attack	5	2^{30} CP	2^{30}E	**Sect. 4**
Zero difference	6	$2^{122.8}$ ACC	$2^{121.8}$ XOR	[14]
Exchange Attack	6	$2^{88.2}$ CP	$2^{88.2}$E	**Sect. 5**

2 Preliminaries

The *Advanced Encryption Standard* (AES) [4] is the most widely adopted block cipher in the world today and is a critical component in protecting information in both commercial and high-assurance communication. The AES internal state is typically represented by a 4 by 4 matrix in $\mathbb{F}_{2^8}^{4\times 4}$. The matrix representation is for the most part purely representational as the actual properties of the matrix (e.g. rank, order etc.) are not actually exploited for anything. One full round of AES consists of SubBytes (SB), ShiftRows (SR), MixColumns (MC) and

[1] https://github.com/Symmetric-crypto/ExchangeAttack.git.

AddKey (AK). The SB-layer applies a fixed permutation over $\mathbb{F}_{2^8}$ independently to each byte of the state, the SR-layer cyclically shifts the i-th row by i positions, while the MC-layer applies a fixed linear transformation to each column. The key addition adds a secret round-dependent value to the state. One full round is composed as $R = AK \circ MC \circ SR \circ SB$. We follow standard convention and simplify notation by writing $R^t(x)$ to mean t rounds of AES where each round key is fixed to some random value.

In this section we recall some basic results and introduce necessary notation. We begin by defining what we call *column exchange differences.*

Definition 1. *For a vector $v \in \mathbb{F}_2^4$ and a pair of states $\alpha, \beta \in \mathbb{F}_{2^8}^{4\times 4}$ define the column exchange difference $\Delta_v^{\alpha,\beta} \in \mathbb{F}_{2^8}^{4\times 4}$ where the i-th column is defined by*

$$(\Delta_v^{\alpha,\beta})_i = (\alpha_i \oplus \beta_i)v_i$$

where α_i and β_i are the i-th columns of α and β.

A pair of states define a set of $2^{wt_c(\alpha\oplus\beta)}$ possible column exchange differences where $wt_c(x)$ denotes the number of non-zero columns of x. We can now define three related operators that exchange diagonal, column and mixed values between a pair of AES states.

Definition 2 (*Column exchange*). *For a vector $v \in \mathbb{F}_2^4$ and a pair of states $\alpha, \beta \in \mathbb{F}_{2^8}^{4\times 4}$, define column exchange according to v as*

$$\rho_c^v(\alpha, \beta) = \alpha \oplus \Delta_v^{\alpha,\beta}.$$

It is easy to see that the pair of states $(\rho_c^v(\alpha, \beta), \rho_c^v(\beta, \alpha)) = (\alpha \oplus \Delta_v^{\alpha,\beta}, \beta \oplus \Delta_v^{\alpha,\beta})$ are formed by exchanging individual columns between α and β according to the binary coefficients of v. Thus, for any v it is easy to see that

$$\alpha \oplus \beta = \rho_c^v(\alpha, \beta) \oplus \rho_c^v(\beta, \alpha).$$

From the definition of column exchange, we may define *diagonal exchange* as follows.

Definition 3 (*Diagonal exchange*). *For a vector $v \in \mathbb{F}_2^4$ and a pair of states $\alpha, \beta \in \mathbb{F}_{2^8}^{4\times 4}$, define diagonal exchange according to v as*

$$\rho_d^v(\alpha, \beta) = \alpha \oplus SR^{-1}(\Delta_v^{SR(\alpha),SR(\beta)}).$$

The new pair $(\rho_d^v(\alpha, \beta), \rho_d^v(\beta, \alpha))$ is formed by exchanging individual diagonals between α and β according to the binary coefficients of v. The relationship between exchange of diagonals and exchange of columns is intuitively straightforward.

Lemma 1. *From the definition of ρ_d^v and ρ_c^v it follows that*

$$R(\rho_d^v(\alpha, \beta)) = \rho_c^v(R(\alpha), R(\beta)).$$

Proof. By definition of diagonal exchange, it follows that

$$MC \circ SR(\rho_d^v(\alpha, \beta)) = \rho_c^v(MC \circ SR(\alpha), MC \circ SR(\beta))$$

and since both ρ_d and ρ_c commute with SB, it follows that

$$R(\rho_d^v(\alpha, \beta)) = \rho_c^v(R(\alpha), R(\beta)).\square$$

The last exchange operation involves exchanging more general looking subspace components belonging to the subspaces formed by applying SR and MC to single columns.

Definition 4 (***Mixed exchange***)**.** *For a vector* $v \in \mathbb{F}_2^4$ *and a pair of states* $\alpha, \beta \in \mathbb{F}_{2^8}^{4\times 4}$ *define mixed exchange according to* v *as*

$$\rho_m^v(\alpha, \beta) = a \oplus L(\Delta_v^{L^{-1}(\alpha), L^{-1}(\beta)})$$

where $L = MC \circ SR$.

Lemma 2. *From the definition of* ρ_c^v *and* ρ_m^v *it follows that*

$$R(\rho_c^v(\alpha, \beta)) = \rho_m^v(R(\alpha), R(\beta)).$$

Proof. By definition of ρ_m^v, let $L = MC \circ SR$, we have that

$$\begin{aligned}\rho_m^v(R(\alpha), R(\beta)) &= R(a) \oplus L(\Delta_v^{L^{-1}(R(\alpha)), L^{-1}(R(\beta))}) \\ &= L \circ (SB(\alpha) \oplus \Delta_v^{SB(\alpha), SB(\beta)}) \\ &= L \circ SB(\alpha \oplus \Delta_v^{\alpha,\beta}) \\ &= R(\rho_c^v(\alpha, \beta)).\square\end{aligned}$$

Although the following trivial two-round property in AES is straight-forward, we add it as a simple theorem to summarise the exchange operators.

Theorem 1. *For two random states* α, β *and some non-zero vector* $v \in \mathbb{F}_2^4$*, we have that*

$$R^2(\rho_d^v(\alpha, \beta)) = \rho_m^v(R^2(\alpha), R^2(\beta)).$$

Proof. Follows by combining Lemmas 1 and 2. $\square$

The exchange operators are related to a type of sets called *exchange-invariant sets.*

Definition 5. *A set* $A \subset \mathbb{F}_{2^8}^{4\times 4}$ *is called exchange-invariant if it satisfies*

$$A = \{\rho^v(a, b) \mid a, b \in A,\ v \in \mathbb{F}_2^4\}$$

where ρ *is either of the three exchange operators.*

Diagonal exchange-invariant sets have the following form. Let $A = A_0 \oplus A_1 \oplus A_2 \oplus A_3$ where A_i corresponds to a subset of $\mathbb{F}_{2^8}^{4\times 4}$ of matrix states where only the i-th diagonal is non-zero. It then follows from the definition of the diagonal exchange operator that

$$A = \{\rho_d^v(a,b) \mid a,b \in A,\ v \in \mathbb{F}_2^4\}.$$

Similarly, we have that a *column exchange-invariant set* B has the form

$$\begin{aligned} B &= SR(A) \\ &= SR(A_0) \oplus SR(A_1) \oplus SR(A_2) \oplus SR(A_3) \\ &= B_0 \oplus B_1 \oplus B_2 \oplus B_3 \end{aligned}$$

and similarly, a *mixed exchange-invariant set* has the form

$$\begin{aligned} C &= SR \circ MC(B) \\ &= SR \circ MC(B_0) \oplus SR \circ MC(B_1) \oplus SR \circ MC(B_2) \oplus SR \circ MC(B_3) \\ &= C_0 \oplus C_1 \oplus C_2 \oplus C_3. \end{aligned}$$

Then from the definition of exchange-invariant sets and the definition of the exchange operator, it follows that two rounds of AES maps a diagonal exchange-invariant set $A = A_0 \oplus A_1 \oplus A_2 \oplus A_3$ to a mixed exchange-invariant set $C = C_0 \oplus C_1 \oplus C_2 \oplus C_3$ where $|C_i| = |A_i|$. The adversary may predict the exact size of each set C_i (since they are equal to the size of A_i's), but he may even predict new plaintext/ciphertext pairs over two rounds. For instance, let $A = A_0 \oplus A_1$ with $A_0 = \{a_0, a_1\}$ and $A_1 = \{b_0, b_1\}$ (i.e. $|A| = 4$). Then the adversary may encrypt two out of four plaintexts from the set A, say $a_0 \oplus b_0$ and $a_1 \oplus b_1$, for two rounds to a pair of ciphertexts c^0 and c^1 that provides him with a minimal set of generators (relative to the mixed exchange operator) which allows him to predict the remaining ciphertexts corresponding to the remaining two plaintexts in A, i.e.

$$C = \{\rho_m^v(c^0, c^1) \mid v \in \mathbb{F}_2^4\}.$$

For a state $s \in \mathbb{F}_{2^8}^{4\times 4}$, define $L^{-1}(s) = SR^{-1} \circ MC^{-1}(s)$ and let $\nu(s)$ denote the binary indicator vector which is 1 in position i if the i-th column of $L^{-1}(s)$ is non-zero and 0 otherwise. We use this notation to simplify the results and to avoid working with more complicated state spaces. Thus, $\nu(s)$ simply indicates the non-zero columns of the state *before* the last linear layer. For a subset $I \subset \{0,1,2,3\}$, we write $v^I \in \mathbb{F}_2^4$ to mean the indicator vector which has value $v_i^I = 1$ if $i \in I$ and 0 otherwise.

Definition 6. *Let α, β be a pair of states that are different in diagonals indicated by $H \subset \{0,1,2,3\}$ and let $H^* \subset H$ denote the set formed by removing one element from H. Then we define the exchange-equivalence class relative to (α, β) as*

$$S_{\alpha,\beta} = \{(\rho_d^{v^I}(\alpha,\beta), \rho_d^{v^I}(\beta,\alpha)) \mid I \subseteq H^*\}.$$

All pairs in $S_{\alpha,\beta}$ are exchange-equivalent to each other. Since 2^{t-1} of the 2^t possible exchange-equivalent pairs are unique (e.g. $\rho_d^{v^I+(1,1,1,1)}(\alpha,\beta) = \rho_d^{v^I}(\beta,\alpha)$ when $|H| = 4$), we fix one index in H in all pairs (i.e. we do not exchange it) and call it H^*.

Theorem 2. *Let $A = A_0 \oplus A_1 \oplus A_2 \oplus A_3$ be a diagonal exchange-invariant set and assume $|A_i| = m_i$ such that $|A| = m_0 \cdot m_1 \cdot m_2 \cdot m_3$. Then there are exactly*

$$L_t(m_0, m_1, m_2, m_3) = \sum_{\substack{I \subset \{0,1,2,3\} \\ wt(I)=t}} \prod_{i \in I} \binom{m_i}{2} \prod_{j \in \{0,1,2,3\} \setminus I} m_j$$

representative pairs $\alpha, \beta \in A$ which are different in exactly t diagonals and where each define a unique exchange-equivalence class $S_{\alpha,\beta}$ of size 2^{t-1}. It follows that

$$\sum_{t=1}^{4} L_t(m_0, m_1, m_2, m_3) 2^{t-1} = \binom{\prod_{i=0}^{3} m_i}{2}$$

is the total number of pairs in A and

$$\sum_{t=1}^{4} L_t(m_0, m_1, m_2, m_3)$$

is the number of distinct exchange-equivalence classes in A.

Proof. The number of pairs in A that are different in t diagonals I and equal in the remaining $(4-t)$ diagonals J, is given by $\prod_{i \in I} \binom{m_i}{2} \prod_{j \in J} m_j$. Each such combination corresponds to one unique exchange-equivalence class $S_{a,b}$ of size 2^{t-1}. By inspecting the terms in the sums over the L_t, it can also easily be seen that it is equivalent to $\binom{m_0 \cdot m_1 \cdot m_2 \cdot m_3}{2}$. □

Thus, the space of $\binom{|A|}{2}$ pairs can be grouped into $\sum_{t=1}^{4} L_t(m_0, m_1, m_2, m_3)$ exchange-equivalence classes, which provides us with a fine grained view of the exchange-equivalence structure of the sets.

We may write a pair in terms of their exchange indicators v, e.g. $a_v = (\rho_d^v(\alpha,\beta), \rho_d^v(\beta,\alpha))$ where v is drawn from a $(t-1)$-dimensional vector space (to ensure that we generate only unique pairs) defined by fixing one of the active diagonals in all exchanged pairs. Taking the $\binom{2^{t-1}}{2}$ combinations of all possible pairs a_v, a_u can be viewed as combining $(t-1)$-dimensional vectors u and v. We are interested in determining the number of combinations of pairs from a set $S_{\alpha,\beta}$ in which the first pair can be derived from the other by exchanging exactly t diagonals. Thus we will need the following.

Lemma 3. *The number of distinct pairs of vectors in $\mathbb{F}_2^n$ whose difference has Hamming weight t, is given by*

$$c(n,t) = \binom{n}{t} 2^{n-1}.$$

Proof. There are $\binom{n}{t}$ vectors of weight t. For each such vector, we need to identify the unique pairs that sum to this vector. For the t positions where the vector is 1, the two vectors can be set to 2^{t-1} unique combinations such that those positions sum one. The remaining positions in the two vectors must be identical, thus there are 2^{n-t} choices for this part. The proof follows. □

We can generate $\binom{2^{t-1}}{2}$ unique combinations of pairs (a_v, a_u) from $S_{a,b}$ where $c(t-1, j)$ counts the number of combinations of pairs (a_u, a_v) in $S_{a,b}$ which are exchange-equivalent if one pair can be obtained from the other by exchanging exactly j diagonals. In other words, $c(t-1, j)$ of the combinations of pairs in $S_t(\alpha, \beta)$ are equivalent if exactly j diagonals are exchanged between them. Moreover, it follows that

$$\begin{aligned}\sum_{j=1}^{t-1} c(t-1, j) &= \sum_{i=1}^{t-1} \binom{t-1}{j} 2^{t-2} \\ &= 2^{t-2} \cdot (2^{t-1} - 1) \\ &= \frac{2^{t-1} \cdot (2^{t-1} - 1)}{2} \\ &= \binom{2^{t-1}}{2}.\end{aligned}$$

We will need the following modified theorem from [14], which states an exchange-difference relation over 4 rounds of AES. Let R^4 denotes 4 full rounds of AES with randomly fixed round keys. Then Theorem 1 of [14] is equal to the following (slightly re-formulated) theorem.

Theorem 3 (***4-round exchange-difference relation***). *Let* $\alpha, \beta \in \mathbb{F}_{2^8}^{4\times 4}$ *and* $\alpha' = \rho_d^v(\alpha, \beta), \beta' = \rho_d^v(\beta, \alpha)$ *for any* $v \in \mathbb{F}_2^4$*, then*

$$\nu(R^4(\alpha) \oplus R^4(\beta)) = \nu(R^4(\alpha') \oplus R^4(\beta')).$$

In other words, the pattern of non-zero and zero columns in the difference $L^{-1}(R^4(\alpha) \oplus R^4(\beta))$ is preserved by diagonal exchange of plaintext pairs α and β, i.e. on exchange-equivalence classes $S_{\alpha,\beta}$. Figure 1 depicts this relation for the case when the exchanged pair of plaintexts is formed by exchanging the first diagonal and the first pair is zero in the last column before the last linear layer.

If we let $\mathrm{wt}(x)$ denote the ordinary Hamming weight of a binary vector, then one last property of AES will be important in this paper.

Theorem 4. *Assume a pair of states* α *and* β *with* $\mathrm{wt}(\nu(\alpha \oplus \beta)) = w_1$*. Then*

$$P(\mathrm{wt}(\nu(R(\alpha) \oplus R(\beta))) = w_2) = \binom{4}{4-w_2} (2^{-8})^{w_1(4-w_2)}.$$

Proof. If $SR^{-1} \circ MC^{-1}(\alpha) \oplus SR^{-1} \circ MC^{-1}(\beta)$ has w_1 active columns, then each column of $\alpha \oplus \beta$ can be written as a linear function of w_1 independent bytes. E.g.

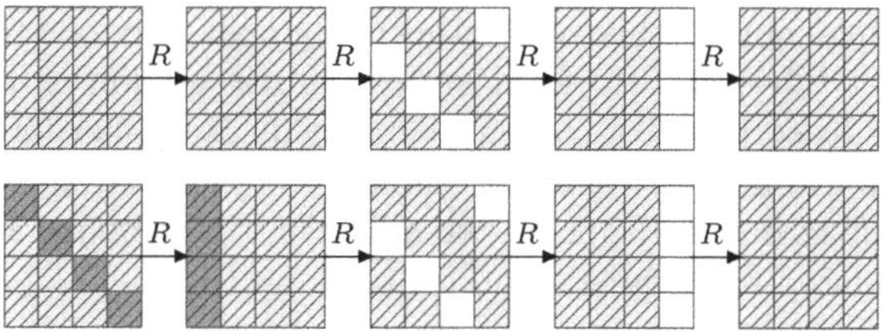

Fig. 1. 4-round exchange trail.

the probability that one column is zero, is thus exactly $(2^8)^{-w_1}$. Moreover, the probability that exactly w_2 of the columns are non-zero (i.e. $4-w_2$ of the columns are zero) is thus exactly $\binom{4}{4-w_2}(2^8)^{-w_1(4-w_2)}$. Since the s-box layer preserves zero differences in bytes (and thus columns), it follows that $\nu(R(\alpha) \oplus R(\beta))$ has the desired probability. □

2.1 Collision and Multicollision in a Set

The expected number of collision and multicollision in a set are computed in [11]. Suppose m objects taken uniformly at random from a given set (with replacement), of size N. Then by using a heuristic method, the expected number of collision is:

$$\frac{m \cdot (m-1)}{2N}$$

assuming independence between pairs of objects. In general, the expected number of collisions in two subsets, of cardinality m_0 and m_1, which obtained by drawing at random without replacement from a large set of size N is:

$$\frac{m_0 \cdot m_1}{N}$$

Multicollision can be considered in a set as have l different elements with the same value or in l different sets and search for an element common to all. Using the same heuristic method, the expected number of multicollisions in a subset of size m drawn from a set of size N is:

$$s(m, N, l) = \frac{\prod_{i=1}^{l} m + 1 - i}{l! N^{l-1}}. \qquad (1)$$

3 When Column Exchange Equals Diagonal Exchange

In the previous section we showed that exchanging diagonals between plaintexts is the same as exchanging column values after one round. In this section we describe the intersection of column exchange and diagonal exchange, i.e. the probabilistic case when exchange of some diagonals between a pair of plaintexts is equal to exchange of (possibly some other) diagonals after one round. We then combine this with Theorem 3 to form a probabilistic version of Theorem 3 that instructs us how to construct a chosen-plaintext distinguisher for five rounds of AES. For this we will need to count the number of bytes that are simultaneously active in both a fixed set of diagonals and a fixed set of columns. Thus, we define sets of indices related to diagonals and columns

Definition 7. *For a set $I \subset \{0,1,2,3\}$, let D_I denote the set of indices $D_I = \{(k, k+i) \mod 4) \,|\, 0 \le k < 4,\, i \in I\}$ where $(i,j) \in D_I$ if the byte at index (i,j) is activated by any of the diagonals indicated by I.*

Definition 8. *For a $J \subset \{0,1,2,3\}$, let $C_J = \{(k,i) \,|\, 0 \le k < 4\,,\, i \in J\}$ denote the set of indices (i,j), where the byte at position (i,j) is activated by any of the columns indicated by J.*

It is easy to see that the number of bytes that are simultaneously in a set of diagonals I and set of columns J is equal to $|D_I \cap C_J| = |I| \cdot |J|$. Thus, it follows that $|D_I \cup C_J| = 4(|I| + |J|) - |I| \cdot |J|$ bytes are activated in total by the diagonals I and by the columns J.

Assume we have a pair of plaintexts (p^0, p^1) that we encrypt one round to a pair of ciphertexts (c^0, c^1). Then assume that we make a new pair of plaintexts

$$(p'^0, p'^1) = (\rho_d^{(1000)}(p^0, p^1), \rho_d^{(1000)}(p^1, p^0)),$$

by exchanging the first diagonal such that the new pair of ciphertexts satisfy

$$\begin{aligned}(R(p'^0), R(p'^1)) &= (\rho_c^{(1000)}(c^0, c^1), \rho_c^{(1000)}(c^1, c^0)) \\ &= (c'^0, c'^1).\end{aligned}$$

We have a new pair of ciphertexts (c'^0, c'^1) formed by exchanging the first column between c^0 and c^1. Now let $I = \{0\}$ such that C_I contains the indices of the first column and imagine that there exists a set J such that the difference $c^0 \oplus c^1$ is zero in all indices in $C_I \cup D_J$ except exactly the indices in the intersection $C_I \cap D_J$, where it can be random. Then, certainly, if the column bytes indicated by C_I were exchanged between the ciphertexts (c^0, c^1) to get (c'^0, c'^1), then certainly we must also have had that diagonal bytes indicated by D_J, and thus the diagonals indicated by J, were exchanged too. Hence, the pair of states are in a configuration where exchanging columns and diagonals means the same thing. The following theorem summarizes the probability of this event.

Theorem 5. *Let $I, J, K \subset \{0,1,2,3\}$ and $\alpha, \beta \in \mathbb{F}_{2^8}^{4\times4}$ be two random states. Then the probability that a set of diagonals J are exchanged, given that a set of columns I are exchanged when the difference $\alpha \oplus \beta$ is zero in columns indicated by K, i.e.*

$$P((\rho_d^{v^J}(\alpha,\beta), \rho_d^{v^J}(\beta,\alpha)) = (\rho_c^{v^I}(\alpha,\beta), \rho_c^{v^I}(\beta,\alpha)))$$

is given by

$$P(|I|, |J|, |K|) = (2^{-8})^{4(|I|+|J|)-|K||J|-2|I|\cdot|J|}.$$

Proof. We restrict the state difference $\alpha \oplus \beta$ to bytes indicated by indices $C_I \cup D_J$ and require that all byte differences in this restriction is zero except for the bytes in the intersection $C_I \cap D_J$. Since $|C_I \cap D_J| = |I| \cdot |J|$ and $|C_I \cup D_J| = 4(|I| + |J|) - |I| \cdot |J|$, and since the bytes take on 2^8 values, it follows that with a probability

$$(2^8)^{|I| \cdot |J|} / (2^8)^{4(|I|+|J|)-|I| \cdot |J|} = (2^{-8})^{4(|I|+|J|)-2|I| \cdot |J|}$$

we have that exchanging columns I is equivalent to exchanging diagonals J (and vice versa). If columns K are equal this means that I can not take on values from K (else the relation become trivial), but only take on values $I \subset \{0, 1, 2, 3\} \setminus K$ not in K such that $|I| < 4 - |K|$. Thus, if the states are equal in K columns, then since the restriction of C_K to $C_I \cup D_J$ is equal to $C_K \cap D_J = |K| \cdot |J|$ bytes, the probability is increased by a factor of $(2^8)^{|K| \cdot |J|}$ to $(2^{-8})^{4(|I|+|J|)-|K||J|-2|I||J|}$. □

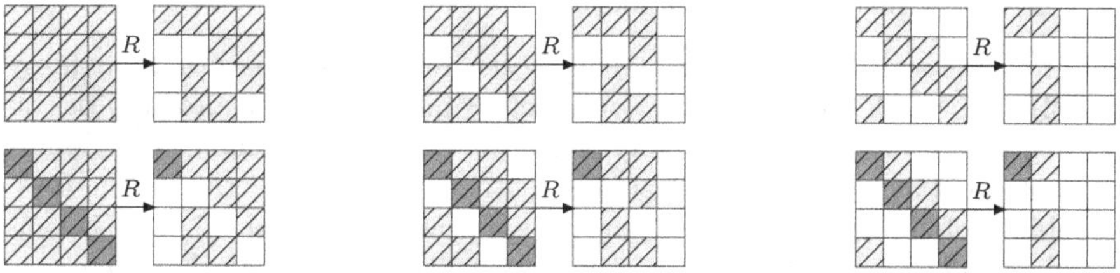

(a) $|K| = 0$ (b) $|K| = 1$ (c) $|K| = 2$

Fig. 2. Example conditions for column/diagonal exchange equivalence.

In other words we have that with some fixed probability, exchanging diagonals between plaintexts is the same as exchanging (possibly some other) diagonals between the intermediate states after one round. And if some diagonals are exchanged after one round, then with probability 1 we also have that Theorem 3 applies. For instance, suppose two random plaintexts verify the differential characteristic of one of the examples in Fig. 2. Then if we exchange the first diagonal between these two plaintexts, then after one round encryption only the first byte is exchanged between the intermediate states. As a consequence, both the first column and the first diagonal are exchanged between the intermediate pair after one round, and thus Theorem 3 can be extended to 5 rounds. This is summarized as follows.

Theorem 6. *Let $\alpha, \beta \in \mathbb{F}_{2^8}^{4\times 4}$ denote two plaintexts equal in $|K|$ diagonals indicated by $K \subset \{0, 1, 2, 3\}$ and assume $0 < \mathrm{wt}(\nu(R^5(\alpha) \oplus R^5(\beta))) < 4$. Then for a non-trivial choice of $I \subset \{0, 1, 2, 3\} \setminus K$ the relation*

$$\nu(R^5(\alpha) \oplus R^5(\beta))) = \nu(R^5(\rho_d^{v^I}(\alpha, \beta)) \oplus R^5(\rho_d^{v^I}(\beta, \alpha)))$$

holds with probability

$$P_5(|I|,|K|) = \sum_{d=1}^{3} \binom{4}{d} P(|I|,d,|K|)$$

Proof. The relation follows trivially by combining Theorems 3 and 5. Theorem 3 states that

$$\nu(R^4(\rho_d^v(R(\alpha),R(\beta))) \oplus R^4(\rho_d^v(R(\beta),R(\alpha)))) = \nu(R^5(\alpha) \oplus R^5(\beta))$$

for any non-zero $v \in \mathbb{F}_2^4$. Theorem 5 states that, if diagonals indicated by I are exchanged between the plaintexts α and β, then there is a probability $P(|I|,|J|,|K|) = (2^{-8})^{4(|I|+|J|)-|K||J|-2|I|\cdot|J|}$ that this equals exchanging diagonals J after one round, i.e.

$$(\rho_d^{v^J}(R(\alpha),R(\beta)), \rho_d^{v^J}(R(\beta),R(\alpha))) = (R(\rho_d^{v^I}(\alpha,\beta)), R(\rho_d^{v^I}(\beta,\alpha))).$$

Then by summing over the probabilities for each possible choice of J, for a fixed I and K, gives the desired expression. □

For instance, if $|K| = 2$ and $|I| = 1$, the relation holds with probability $P_5(1,2) = 2^{-28.19}$. Note that we could set $\geq$ in front of the probabilities in Theorems 5 and 6 instead of equality, since the case when exchange of columns does not equal an exchange of diagonals contributes a tiny fraction to the total probability of the event. However, for our applications this contribution is vanishingly small, thus we may think of it as equality. This will in the worst case mean that our attack analysis is pessimistic since a higher probability will only decrease the complexity of all of our attacks.

Assume an diagonal exchange-invariant set $A = A_0 \oplus A_1 \oplus A_2 \oplus A_3$. We then have the following result.

Theorem 7. *For a diagonal exchange-invariant set $A = A_0 \oplus A_1 \oplus A_2 \oplus A_3$ where $|A_i| = m_i$, the expected number of combinations of pairs $(a,b),(c,d) = (\rho_d^v(a,b), \rho_d^v(b,a))$ that satisfy*

$$(\rho_d^u(R(a),R(b)), \rho_d^u(R(b),R(a))) = (R(c),R(d))$$

for any u and v is expected to be

$$G(m_1,m_2,m_3,m_4) = \sum_{t=1}^{4} L_t(m_1,m_2,m_3,m_4) \cdot \sum_{j=1}^{t-1} c(t-1,j) \cdot P_5(j,4-t). \quad (2)$$

Proof. Let $S_{\alpha,\beta}$ denotes one of the exchange-equivalence classes in A of size 2^{t-1}. Then there are $c(t-1,j)$ combinations of two pairs a_u, a_v from $S_{\alpha,\beta}$ such that $\rho_d^z(a_u) = a_v$ for a vector z of weight j. For each of those combinations, the probability is $P_5(j,4-t)$ that the relation holds, and thus

$$\sum_{j=1}^{t-1} c(t-1,j) \cdot P_5(j,4-t)$$

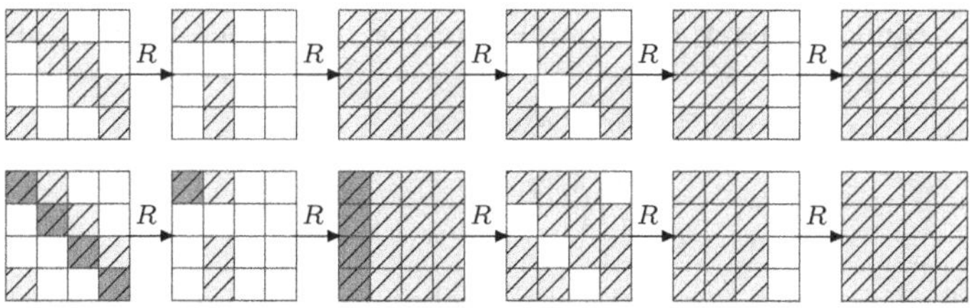

Fig. 3. 5-round exchange trail.

is the expected number of combinations of pairs from one such set $S_{\alpha,\beta}$ of size 2^{t-1} that satisfy the condition. Then since there are $L_t(m_1, m_2, m_3, m_4)$ exchange-equivalence classes of size 2^{t-1}, the expression follows (Fig. 3). □

4 The Exchange Attack on Five Rounds AES

Theorem 6 can be used directly to show that *AES limited to five full rounds is biased when plaintexts are closed under the action of diagonal exchange operations, i.e. diagonal invariant sets.* We show this using the following approach. Assume $f(x)$ is a random permutation acting on the same state space as AES and two random plaintexts p^i, p^j together with the exchanged plaintexts $p'^i = \rho_d^v(p^i, p^j), p'^j = \rho_d^v(p^j, p^i)$. We assume that p^i and p^j are different in at least two diagonal positions or else the exchanged pair will be equivalent to the original pair. Then let $c^i = f(p^i), c^j = f(p^j), c'^i = f(p'^i)$ and $c'^j = f(p'^j)$. Then we ask the question; what is the probability that

$$0 < \mathrm{wt}(\nu(c^i \oplus c^j)) = d < 4$$

and simultaneously

$$\nu(c^i \oplus c^j) = \nu(c'^i \oplus c'^j)?$$

In other words, what is the probability that $SR^{-1} \circ MC^{-1}(c^i \oplus c^j)$ is zero in $4-d$ columns *and* that $SR^{-1} \circ MC^{-1}(c'^i \oplus c'^j)$ is zero in exactly the same columns?

For a single combination of pairs, the probability that c^i and c^j satisfy $\mathrm{wt}(\nu(c^i \oplus c^j)) = d$ (i.e. $SR^{-1} \circ MC^{-1}(c^i)$ and $SR^{-1} \circ MC^{-1}(c^j)$ collide in $4-d$ columns) is given by

$$P_{first} = \binom{4}{4-d}(2^{32})^{-(4-d)}.$$

For instance, the probability that two ciphertexts satisfy $\mathrm{wt}(\nu(c^i \oplus c^j)) = 3$ is given by $\binom{4}{1} \cdot (2^{32})^{-1} = 2^{-30}$, and the probability that the pair has $\mathrm{wt}(\nu(c^i \oplus c^j)) = 2$ is given by $\binom{4}{2} \cdot (2^{32})^{-2} = 6 \cdot 2^{-64}$, and so on. The probability that the second pair is zero in the exact same columns as the first is then in general

$$P_{second} = (2^{32})^{-(4-d)}.$$

Thus, in the random case the probability of the two events is given by

$$P_{rand} = P_{first} \cdot P_{second} \tag{3}$$

$$= \binom{4}{d}(2^{32})^{-(4-d)} \cdot (2^{32})^{-(4-d)}. \tag{4}$$

But for AES, Theorem 6 states that the probability is $P_5(|I|, |K|)$ for the second event, where I is the set of exchanged diagonals while K is the set of diagonals that are equal in the initial plaintext pair. Thus, the total probability for AES becomes instead

$$P_{AES} = P_{first} \cdot P_5(|I|, |K|). \tag{5}$$

For instance, if we set $|I| = 1$ (i.e. one diagonal is exchanged) and $|K| = 2$ (i.e. the difference of the plaintexts is zero in two diagonals), we get that $P_5(1, 2) > 2^{-28.2}$, while it is 2^{-32} for random, the probability that second pair is zero in the exact same columns as the first one is. Notice that the second term of P_{rand}, P_{second}, is a function of the ciphertext collision event while the second term of P_{AES}, $P_5(|I|, |K|)$, is fixed and independent of this ciphertext collision condition. Thus, while the second term in the probability for the random case depends on the size of the space that the difference $c'^i \oplus c'^j$ is required to collide in (e.g. has probability 2^{-96} for the second event), the second term of P_{AES} is fixed and independent of this (e.g. has always probability $2^{-28.2}$ for the second event). Thus, since the probability of the exchange-equivalence condition of Theorem 5 can easily be made higher than the random collision condition for the second pair, a distinguishing condition follows.

For five rounds, the adversary bases the distinguisher more concretely on the following question with respect to a larger subset of plaintexts:

For a subset of plaintexts A, what is the probability that there exist two distinct pairs a_u, a_v from any of the exchange-equivalence classes $S_{a,b}$ in A that satisfy

$$\nu(R^5(a) \oplus R^5(b)) = \nu(R^5(c) \oplus R^5(d))$$

and

$$0 < \mathrm{wt}(\nu(R^5(a) \oplus R^5(b))) = d < 4.$$

Theorems 6 and 7 in the previous section can be used directly to set up a straight-forward 5-round chosen plaintext distinguisher for AES. If the adversary observes a pair of plaintexts (p^0, p^1) corresponding to a pair of ciphertexts that satisfy $0 < \mathrm{wt}(\nu(c^0, c^1)) < 4$, then Theorem 6 states that for any other pair $(p'^0, p'^1) \in S_{p^0,p^1}$, the probability of the event $\nu(c'^0 \oplus c'^1) = \nu(c^0 \oplus c^1)$ is significantly higher than for the random case. In Theorem 7 we showed that the expected number of combinations of diagonal exchange-equivalent pairs $(a, b), (c, d)$ from the diagonal exchange-invariant set $A = A_0 \oplus A_1 \oplus A_2 \oplus A_3$ that are also diagonal exchange equivalent after one round, is given by $G(m_1, m_2, m_3, m_4)$ where $m_i = |A_i|$. In this case, the combination of pairs also obey the additional 4 round exchange difference relation of Theorem 3, thus

$G(m_1, m_2, m_3, m_4)$ is also the expected number of combinations of exchange equivalent plaintext pairs that enjoy the 4-round exchange-difference relation for five rounds. Since $G(m_1, m_2, m_3, m_4)$ is the expected number of combinations of exchange-equivalent pairs a_u, a_v that satisfy Theorem 6, it follows that

$$E_{AES} = G(m_1, m_2, m_3, m_4) \cdot \binom{4}{d} (2^{-32})^{(4-d)}$$

is the expected number of combinations of pairs from the exchange-equivalence classes whose ciphertexts satisfy

$$\nu(c'^0 \oplus c'^1) = \nu(c^0 \oplus c^1)$$

when $\mathrm{wt}(\nu(c^0 \oplus c^1)) = d$. For the random case, the same probability becomes

$$E_{rand} = H(m_1, m_2, m_3, m_4) \cdot \binom{4}{d} (2^{-32})^{(4-d)} \cdot (2^{-32})^{(4-d)}$$

where

$$H(m_1, m_2, m_3, m_4) = \sum_{t=1}^{4} L_t(m_1, m_2, m_3, m_4) \binom{2^{t-1}}{2}$$

follows from Theorem 7 and is the total number of combinations of two pairs from each possible exchange-equivalence class $S_{a,b}$.

An algorithm for the 5-round distinguisher is presented in Algorithm 1. In our distinguisher for five rounds, we pick two random subsets A_0 and A_1 of $\mathbb{F}_{2^8}^4$, each of size m, and encrypt the resulting diagonal exchange-invariant set of m^2 plaintexts A formed by spanning the first diagonal with the possible elements from A_0 and the second diagonal with elements from A_1, while setting the remaining bytes to random constants. The structure of the plaintext subset is determined by optimizing E_{AES} relative to E_{rand} with the condition that $d = 3$, i.e. the two pairs of ciphertexts must collide in the same column before the last linear layer.

In this particular case, when only two diagonals are active in the plaintexts, the set of plaintexts contains exchange-equivalence classes of size 1 and 2 (note that we can not draw pairs from an exchange-equivalence class of size 1). If we set $m = 2^{15}$, we get that

$$\begin{aligned} E_{AES} &= G(m, m, 1, 1) \cdot 2^{-30} \\ &\approx 1 \end{aligned}$$

while

$$\begin{aligned} E_{rand} &= H(m, m, 1, 1) \cdot 2^{-62} \\ &\approx 2^{-4}. \end{aligned}$$

Thus, by encrypting a plaintext set $A = A_0 \oplus A_1 \oplus A_2 \oplus A_3$ where $|A_0| = |A_1| = m$ and $|A_2| = |A_3| = 1$ (i.e. $|A| \approx 2^{30}$), we are able to distinguish AES. An unoptimized algorithm for the distinguisher is presented in Algorithm 1.

Algorithm 1. Pseudo-code for 5-round distinguisher.

```
Input: m = 2^15, D = 2^30
Result: 1 if AES, −1 otherwise.
L^{-1} ← SR^{-1} ∘ MC^{-1}
Choose m random values A = {a_0, a_1, ..., a_{m-1}} ⊂ F^4_{2^8}
Choose m random values B = {b_0, b_1, ..., b_{m-1}} ⊂ F^4_{2^8}
Choose random constants z_2, z_3 ∈ F^4_{2^8}
C ← {}
T_0, T_1, T_2, T_3 = {} // empty hash tables containing unordered sets(e.g.
unordered multisets)
/* Encrypt and order 2^30 plaintexts                                        */
for i from 0 to m − 1 do
    for j from 0 to m − 1 do
        l ← i · m + j
        p^l ← (a_i, b_j, z_2, z_3) // a_i is the first diagonal value and b_j is
            the second diagonal value and so forth.
        c^l ← E_K(p^l)
        /* Add (i, j) to T_k[z] according to value z of column k of
            L^{-1}(c^l)                                                     */
        for k from 0 to 3 do
            z ← |L^{-1}(c^l)_k| // |L^{-1}(c^l)_k| is integer value of k-th column
            T_k[z] ← T_k[z] ∪ {(i, j)}
        end
        C ← C ∪ {c^l}
    end
end
/* Search for double collisions                                             */
for each c^i in C do
    /* coeffs(i) returns coefficients a, b s.t. a · m + b = i               */
    i_1, j_1 ← coeffs(i)
    for j from 0 to 3 do
        for i_2, j_2 ∈ T_j[|c^i_j|] do
            if i_1 ≠ i_2 and j_1 ≠ j_2 then
                if L^{-1}(c^{(i_2·m+j_1)} ⊕ c^{(i_1·m+j_2)})_k equals 0 then
                    /* Two pairs forming double collision found             */
                    return 1
                end
            end
        end
    end
end
return −1
```

4.1 Complexity of Distinguisher

The algorithm consists of two parts. In the first part, the adversary encrypts $D = m^2 = 2^{30}$ plaintexts and inserts the index (i, j) into each of the four tables

T_k according to the integer column values of $L^{-1}(c^{i \cdot m+j})$, i.e. the index (i, j) is inserted into $T_k[|L^{-1}(c^{i \cdot m+j})_k|]$ where $|L^{-1}(c^{i \cdot m+j})_k|$ is the integer value of the k-th column of $L^{-1}(c^{i \cdot m+j})$. The complexity of this part is roughly D encryptions plus D ciphertext-lookups times four insertions to the hash tables T_k, which is roughly about

$$\begin{aligned} C_{part_1} &= D + \frac{(4 \cdot D)}{80} \\ &\approx D \end{aligned}$$

if we use the convention that one encryption (i.e. $5 \cdot 16 = 80$ s-box lookups) corresponds to one unit of computation. To determine the complexity of the second part, we need to estimate approximately the expected number of entries that contains 0 values, 1 values, 2 values etc. using the formula for the expected number of multicollisions.

For $D = 2^{30}$ and $N = 2^{32}$, by using the formula 1, we do not expect any multicollisions involving more than seven ciphertexts (i.e. $s(D, N, 8) \approx 0.7$) and thus $T_k[r]$ contains at most seven values such that the complexity of testing each combination of ciphertexts related to an index entry of $T_k[r]$ takes at most $\binom{7}{2} = 21$. But to get a more accurate complexity estimate, we may iteratively compute the expected number of sets $T_k[r]$ which contains $l = 7$ elements (i.e. correspond to a 7-multicollision), which contains $l = 6$ elements, and so forth. To do this, we let $s_7 = s(D, N, 7)$. Then the number of 6-multicollisions not already inside a 7-multicollision is given by

$$s_6 = s(D, N, 6) - s_7 \cdot \binom{7}{6}.$$

Then the number of 5-multicollisions that are not already inside a 6-multicollision, is given by

$$s_5 = s(D, N, 5) - s_6 \cdot \binom{6}{5}$$

and so forth, obeying the recurrence

$$s_t = s(D, N, t) - s_{t+1} \cdot \binom{t+1}{t}$$

until we arrive at s_1, which is the expected number of entries which contains only one element. Moreover, we should have that $\sum_{t=1}^{7} s_t \cdot t \approx D$, and indeed we get that

$$\sum_{t=1}^{7} s_t \cdot t \approx 2^{30}$$

as expected. From this, we can compute the complexity corresponding to finding collisions in one of the tables T_k in part 2 as

$$\begin{aligned} C'_{part_2} &= s_1 + \sum_{t=2}^{7} s_t t^2 \\ &\approx (2^{29.7} + 2^{29}) \cdot C \\ &\approx 2^{30} \cdot C \end{aligned}$$

such that the total complexity of part 2 roughly becomes

$$\begin{aligned} C_{part_2} &= \frac{4 \cdot C'_{part_2}}{80} \\ &\approx 2^{25.7} \end{aligned}$$

if we adopt the convention that one operation equals one encryption, which can be viewed as $16 \cdot 5 = 80$ s-box lookups, where C is the number of ciphertext-lookups we do for each combinations of pairs. Hence, five rounds of AES can be distinguished using a chosen plaintext distinguisher with $D = 2^{30}$ data and about the same computational complexity. In the next section, we show that with a change of parameters the same distinguisher can be used to distinguish fix rounds of AES.

5 The Exchange Attack on Six Rounds AES

In this section we present the first 6-round secret-key chosen plaintext distinguisher for AES, which follows from a straight-forward extension of Theorem 6. Imagine a setup similar to the 5-round distinguisher, but where we encrypt two random plaintexts p^i and p^j which are non-zero in all bytes except the last diagonal. As before, let $p'^i = \rho_d^v(p^i, p^j)$ and $p'^j = \rho_d^v(p^j, p^i)$.

Now assume the following two conditions, where the first one is given by

$$\mathrm{wt}(\nu(R^5(p^i) \oplus R^5(p^j))) = 1 \tag{6}$$

and second one is given by

$$\mathrm{wt}(\nu(R^6(p^i) \oplus R^6(p^j))) = 1. \tag{7}$$

At random, the first condition happens with probability 2^{-94} and thus the second condition happens with probability $\binom{4}{3}(2^{-8})^3 = 2^{-22}$ by Theorem 4 conditioned on the first event. By symmetry of Theorem 4, the same condition applies in the reverse direction. If we observe that the second condition (7) holds, which happens with probability 2^{-94} at random, then the first condition (6) holds with probability 2^{-22} by Theorem 4. So assume that we observe a pair of ciphertexts $c^i = R^6(p^i)$ and $c^j = R^6(p^j)$ that happens to satisfy the second condition (7), i.e. $SR^{-1} \circ MC^{-1}(c^i \oplus c^j)$ contains exactly one active column. Such an event happens at random with probability

$$P_{R6} = 2^{-94}.$$

Then by Theorem 4 applied in the reverse direction, the probability of the first condition (6) conditioned on the event (7) is given by

$$P_{R5} = 2^{-22},$$

i.e. $SR^{-1} \circ MC^{-1}(R^5(p^i) \oplus R^5(p^j))$ contains exactly one active column too. But if the first event (6) occurs, then by Theorem 6 we also have that the event

$$\nu(R^5(p'^i) \oplus R^5(p'^j)) = \nu(R^5(p^i) \oplus R^5(p^j))$$

happens with probability probability $P_{R1} = P_5(3,1) \approx 2^{-38}$ for the exchanged pair. Hence, both pairs satisfy

$$\begin{aligned} \mathrm{wt}(\nu(R^5(p'^i) \oplus R^5(p'^j))) &= \mathrm{wt}(\nu(R^5(p^i) \oplus R^5(p^j))) \\ &= 1 \end{aligned}$$

and thus it follows from Theorem 4 applied to the exchanged pair in the fifth round that the probability of the two simultaneous events (conditioned on the previous events)

$$\mathrm{wt}(\nu(R^6(p'^i) \oplus R^6(p'^j))) = \mathrm{wt}(\nu(R^6(p^i) \oplus R^6(p^j))) = 1$$

is given by $P_{R6'} = 2^{-22}$. Hence, if the adversary observes a pair of ciphertexts that satisfy (7), then the probability that the event $\mathrm{wt}(\nu(R^6(p'^i) \oplus R^6(p'^j))) = 1$ occur (i.e. the same event happens for the exchanged pair too) is given by

$$\begin{aligned} P_{second} &= P_{R5} \cdot P_5(3,1) \cdot P_{R6'} \\ &\approx 2^{-44} \cdot 2^{-38} \\ &= 2^{-82}. \end{aligned}$$

In the random case, however, the probability that the second ciphertext pair satisfies the last condition (7) is 2^{-94} for both pairs of ciphertexts. Thus, for a random plaintext/ciphertext pair $(p^i, p^j) \to (c^i, c^j)$ and an exchanged pair $(p'^i, p'^j) \to (c'^i, c'^j)$, the probability that

$$\mathrm{wt}(\nu(c^i \oplus c^j)) = \mathrm{wt}(\nu(c'^i \oplus c'^j)) = 1$$

is for a random permutation given by

$$\begin{aligned} P_{rand} &= 2^{-94} \cdot 2^{-94} \\ &= 2^{-188} \end{aligned}$$

while it is equal to

$$\begin{aligned} P_{AES} &= 2^{-94} \cdot 2^{-82} \\ &= 2^{-176} \end{aligned}$$

for AES. We may summarize the result as follows.

Theorem 8. *Let $A = A_0 \oplus A_1 \oplus A_2 \oplus A_3$ with $|A_0| = |A_1| = |A_2| = 2^{29.4}$ and $|A_3| = 1$ such that $|A| = 2^{88.2}$, then the expected number of combinations of pairs $(a, b), (c, d)$ from the exchange-equivalence classes in A whose ciphertexts satisfy*

$$\Pr(\mathrm{wt}(\nu(R^6(a) \oplus R^6(b))) = \mathrm{wt}(\nu(R^6(c) \oplus R^6(d))) = 1)$$

is given by

$$\begin{aligned} E_{AES} &= G(m, m, m, 1) \cdot 2^{-44} \cdot 2^{-94} \\ &\approx 1 \end{aligned}$$

while

$$\begin{aligned} E_{rand} &= H(m, m, m, 1) \cdot 2^{-94} \cdot 2^{-94} \\ &\approx 2^{-11} \end{aligned}$$

for random.

Proof. Proof follows straight forwardly by combining Theorems 6, 7 and 4. Assume that two exchange-equivalent pairs satisfy the 5 round exchange relation and assume that one of them satisfy relation 6. Then the other pair must satisfy this relation. The probability that both pairs satisfy this condition is therefore 2^{-94} (in comparison to 2^{-94-94} for random). Then due to Theorem 4, it follows that the probability that both pairs of ciphertexts obey relation 7, is $2^{-22} \cdot 2^{-22} = 2^{-44}$. □

Hence, if the adversary encrypts a set of $D = (2^{29.4})^3 = 2^{88.2}$ plaintexts, we expect to find a combination of pairs that satisfy our condition, while we expect to find 2^{-10} double collisions for a random permutation. Thus, we have the basis for a distinguisher which can distinguish 6 full rounds of AES that requires $2^{88.2}$ chosen plaintexts.

5.1 Distinguishing Attack Algorithm for Six Rounds

Similar to the 5-round distinguisher, we pick three sets A_0, A_1 and A_2, each of size $\approx 2^{29.4}$ such that we may generate a diagonal exchange-invariant set of $D = 2^{88.2}$ plaintexts in such a way that the i-th diagonal of each plaintext is spanned by the possible elements of A_i while the last diagonal is set to a random constant. In fact, the algorithm is exactly the same as the 5-round distinguisher, except for a change of parameters and collision condition. Moreover, this time each of our hash-tables may in the worst case contain up to $2^{88.2}$ values. The algorithm for six rounds AES is presented in Algorithm 2 and it can be readily seen that ciphertexts are essentially the same as for five rounds. However, if we observe a pair of ciperhtexts c^i and c^j that have our desired collision property and which stems from a plaintext pair p^i and p^j which differ in all three diagonals, then we need to generate the remaining three possible exchanges of those to test the secondary condition. Due to our use of indices for locating ciphertexts in the algorithm, exchanging a pair of plaintexts corresponds to exchanging indices between the corresponding ciphertexts. However, in the algorithm we are more explicit for ease of understanding (Fig. 4).

Algorithm 2. Pseudo-code for 6-round distinguisher.

```
Input: m = 2^{29.4}, D = 2^{88.2}
Result: 1 if AES, −1 otherwise.
L^{-1} ← SR^{-1} ∘ MC^{-1}
Choose m random values A = {a_0, a_1, ..., a_{m-1}} ⊂ F_{2^8}^4
Choose m random values B = {b_0, b_1, ..., b_{m-1}} ⊂ F_{2^8}^4
Choose m random values C = {c_0, c_1, ..., c_{m-1}} ⊂ F_{2^8}^4
Choose random constants z_3 ∈ F_{2^8}^4
C ← {}
T_0, T_1, T_2, T_3 = {} // empty hash tables containing unordered sets(e.g.
    unordered multisets)
/* Encrypt 2^{88.2} plaintexts                                          */
for i from 0 to m − 1 do
    for j from 0 to m − 1 do
        for k from 0 to m − 1 do
            l ← i · m^2 + j · m + k
            p^l ← (a_i, b_j, c_k, z_3) // a_i is the first diagonal value and b_j
                is the second diagonal value and so forth.
            c^l ← E_K(p^l)
            /* T_r[z] contains indices (i, j, k) for ciphertext c^{i·m^2+j·m+k}
                with value z in the r-th column of L^{-1}(c^{i·m^2+j·m+k})   */
            for r from 0 to 3 do
                z ← |L^{-1}(c^l)_r| // |L^{-1}(c^l)_r| is integer value of r-th
                    column
                T_r[z] ← T_r[z] ∪ {(i, j, k)}
            end
            C ← C ∪ {c^l}
        end
    end
end
/* Search for double collisions                                         */
for each c^i in C do
    /* coeffs(i) returns coefficients a, b, c s.t. a · m^2 + b · m + c = i */
    i_1, j_1, k_1 ← coeffs(i)
    for j from 0 to 3 do
        for i_2, j_2, k_2 ∈ T_j[|c_j^i|] do
            /* G_{i,j} is the set of ciphertexts corresponding to
                exchange-equivalence class S_{p^i,p^j}                      */
            S ← G_{(i_1,j_1,k_1),(i_2,j_2,k_2)} //|G| ≤ 4
            for each pair (a, b) ∈ S do
                if wt(ν(R^6(a)) ⊕ R^6(b)) equals 1 then
                    /* Two pairs forming double collision found         */
                    return 1
                end
            end
        end
    end
end
return −1
```

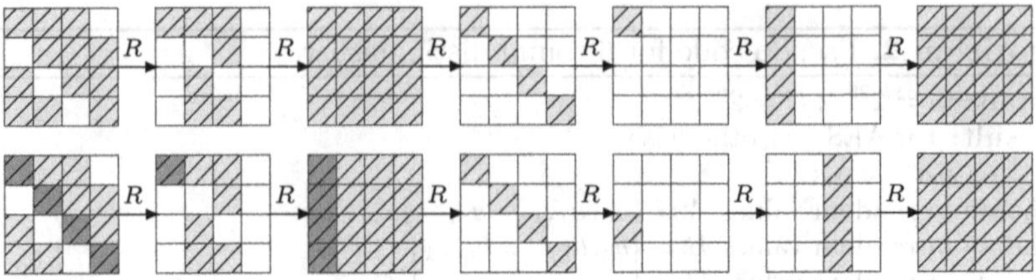

Fig. 4. 6-round exchange trail.

5.2 Complexity of Distinguisher

The analysis of the 6-round distinguisher pretty much follows the same line as the 5-round distinguisher. The distinguisher consists of two parts; first the adversary populates the tables T_k followed by a collision search. The first part is estimated in the same way as for the five rounds, i.e. we get roughly

$$\begin{aligned} C_{part_1} &= D + \frac{(4 \cdot D)}{96} \\ &\approx D. \end{aligned}$$

where we adopt the convention that one unit equals one encryption, where one encryption equals 96 s-box look-ups for six full rounds. For part 2, searching for collisions, the analysis is the same as for five rounds. Again, by using the recurrence (1) for multicollisions, we find that more than 10 collisions in the same entry $T_k[r]$ is unlikely (i.e. $s(2^{88.2}, 2^{96}, 10) \approx 2^{-4}$). Thus, again we may apply the recursion

$$s_t = s(D, N, t) - s_{t+1} \cdot \binom{t+1}{t}$$

to compute s_8 down to s_1 given that s_9 is expected to be $s(D, N, 9)$. This way, we find the expected number of entries in each table which has nine elements, eight elements, and so on. If the computation is correct, we should have that

$$\sum_{t=1}^{9} s_t \cdot t \approx 2^{88.2}$$

which we indeed get. From this, we can compute the complexity corresponding to finding collisions in one of the tables T_k in part 2 as

$$\begin{aligned} C'_{part_2} &= (s_1 + \sum_{t=2}^{9} s_t t^2) \cdot C \\ &\approx (2^{88.19} + 2^{81.4}) \cdot C \\ &\approx 2^{88.2} \cdot C \end{aligned}$$

where C is the number of ciphertext-lookups we do for each combinations of pairs. Note that the algorithm will spend most of the time detecting entries with no collision. In the second term above, each table index of size t is visited t times (once for each ciphertext in it, which is not optimal). We can do at most three additional exchanges between the observed pairs, or else we can do one or zero exchanges, depending on the size of the exchange-equivalence class these pairs belong to (either size one, two or four). In any case, this last term does not contribute to the final complexity C'_{part_2}. Thus, the expected complexity of evaluating the four tables then roughly becomes

$$\begin{aligned} C_{part_2} &= \frac{4 \cdot C'_{part_2}}{96} \\ &\approx 2^{83.6} \end{aligned}$$

where we adopt the convention that one unit equals one encryption, where one encryption equals 96 s-box look-ups for six full rounds. Thus, the total complexity of the algorithm is dominated by the number of required ciphertexts, $D = 2^{88.2}$, in terms of data, memory and computation.

6 Conclusion

In this paper we have introduced the first 6-round secret-key chosen-plaintext distinguisher for AES using a new type of attack called *exchange-equivalence attacks* (or simply, exchange attacks). The distinguisher has data and computational complexity of only $2^{88.2}$ and can thus be viewed as a giant leap in the cryptanalysis of AES when one considers that the first 5-round secret-key distinguisher for AES appeared nearly 20 years after the publication of Rijndael. All of our attacks can easily be turned into chosen ciphertext attacks on the inverted block cipher due to the inherent symmetry of the properties we are using. Our results are easily generalized to any SPN-like cipher, and in particular, we note that the theory in this paper can be generalized to extend the attacks for more rounds for ciphers with slower diffusion (e.g. lightweight designs). We are confident that our results lead the way to further breakthroughs on ciphers such as AES.

Acknowledgments. We thank the anonymous reviewers for their valuable comments and suggestions. This research was supported by the Norwegian Research Council.

References

1. Biham, E., Keller, N.: Cryptanalysis of reduced variants of Rijndael. In: 3rd AES Conference, vol. 230 (2000)
2. Bouillaguet, C., Derbez, P., Dunkelman, O., Fouque, P.A., Keller, N., Rijmen, V.: Low-data complexity attacks on AES. IEEE Trans. Inf. Theory **58**(11), 7002–7017 (2012)

3. Daemen, J., Rijmen, V.: Plateau characteristics. IET Inf. Secur. **1**, 11–17 (2007)
4. Daemen, J., Rijmen, V.: The Design of Rijndael: AES - The Advanced Encryption Standard. Springer, Heidelberg (2002). https://doi.org/10.1007/978-3-662-04722-4
5. Daemen, J., Rijmen, V.: Understanding two-round differentials in AES. In: De Prisco, R., Yung, M. (eds.) SCN 2006. LNCS, vol. 4116, pp. 78–94. Springer, Heidelberg (2006). https://doi.org/10.1007/11832072_6
6. Derbez, P., Fouque, P.-A.: Automatic search of meet-in-the-middle and impossible differential attacks. In: Robshaw, M., Katz, J. (eds.) CRYPTO 2016. LNCS. Part II, vol. 9815, pp. 157–184. Springer, Heidelberg (2016). https://doi.org/10.1007/978-3-662-53008-5_6
7. Grassi, L.: Mixture differential cryptanalysis: a new approach to distinguishers and attacks on round-reduced AES. IACR Trans. Symmetric Cryptol. **2018**(2), 133–160 (2018)
8. Grassi, L., Rechberger, C., Rønjom, S.: Subspace trail cryptanalysis and its applications to AES. IACR Trans. Symmetric Cryptol. **2016**(2), 192–225 (2016)
9. Grassi, L., Rechberger, C., Rønjom, S.: A new structural-differential property of 5-round AES. In: Coron, J.-S., Nielsen, J.B. (eds.) EUROCRYPT 2017. LNCS. Part II, vol. 10211, pp. 289–317. Springer, Cham (2017). https://doi.org/10.1007/978-3-319-56614-6_10
10. Gueron, S., Mouha, N.: Simpira v2: a family of efficient permutations using the AES round function. In: Cheon, J.H., Takagi, T. (eds.) ASIACRYPT 2016. LNCS. Part I, vol. 10031, pp. 95–125. Springer, Heidelberg (2016). https://doi.org/10.1007/978-3-662-53887-6_4
11. Joux, A.: Algorithmic Cryptanalysis, 1st edn. Chapman & Hall/CRC, Boca Raton (2009)
12. Knudsen, L.R., Wagner, D.: Integral cryptanalysis. In: Daemen, J., Rijmen, V. (eds.) FSE 2002. LNCS, vol. 2365, pp. 112–127. Springer, Heidelberg (2002). https://doi.org/10.1007/3-540-45661-9_9
13. Rijmen, V.: Cryptanalysis and design of iterated block ciphers. Doctoral dissertation, K.U. Leuven (1997)
14. Rønjom, S., Bardeh, N.G., Helleseth, T.: Yoyo tricks with AES. In: Takagi, T., Peyrin, T. (eds.) ASIACRYPT 2017. LNCS. Part I, vol. 10624, pp. 217–243. Springer, Cham (2017). https://doi.org/10.1007/978-3-319-70694-8_8
15. Sun, B., Liu, M., Guo, J., Qu, L., Rijmen, V.: New insights on AES-like SPN ciphers. In: Robshaw, M., Katz, J. (eds.) CRYPTO 2016. LNCS. Part I, vol. 9814, pp. 605–624. Springer, Heidelberg (2016). https://doi.org/10.1007/978-3-662-53018-4_22

Algebraic Cryptanalysis of STARK-Friendly Designs: Application to MARVELlous and MiMC

Martin R. Albrecht[1], Carlos Cid[1,2], Lorenzo Grassi[5,6], Dmitry Khovratovich[3,4,7], Reinhard Lüftenegger[5(✉)], Christian Rechberger[5], and Markus Schofnegger[5]

[1] Information Security Group, Royal Holloway, University of London, London, UK
{martin.albrecht,carlos.cid}@rhul.ac.uk
[2] Simula UiB, Bergen, Norway
[3] Dusk Network, Amsterdam, The Netherlands
khovratovich@gmail.com
[4] ABDK Consulting, Tallinn, Estonia
[5] IAIK, Graz University of Technology, Graz, Austria
{lorenzo.grassi,reinhard.lueftenegger,christian.rechberger,
markus.schofnegger}@iaik.tugraz.at
[6] Know-Center, Graz, Austria
[7] Evernym Inc., Salt Lake City, USA

Abstract. The block cipher Jarvis and the hash function Friday, both members of the MARVELlous family of cryptographic primitives, are among the first proposed solutions to the problem of designing symmetric-key algorithms suitable for transparent, post-quantum secure zero-knowledge proof systems such as ZK-STARKs. In this paper we describe an algebraic cryptanalysis of Jarvis and Friday and show that the proposed number of rounds is not sufficient to provide adequate security. In Jarvis, the round function is obtained by combining a finite field inversion, a full-degree affine permutation polynomial and a key addition. Yet we show that even though the high degree of the affine polynomial may prevent some algebraic attacks (as claimed by the designers), the particular algebraic properties of the round function make both Jarvis and Friday vulnerable to Gröbner basis attacks. We also consider MiMC, a block cipher similar in structure to Jarvis. However, this cipher proves to be resistant against our proposed attack strategy. Still, our successful cryptanalysis of Jarvis and Friday does illustrate that block cipher designs for "algebraic platforms" such as STARKs, FHE or MPC may be particularly vulnerable to algebraic attacks.

Keywords: Gröbner basis · MARVELlous · Jarvis · Friday · MiMC · ZK-STARKs · Algebraic cryptanalysis · Arithmetic circuits

S. D. Galbraith and S. Moriai (Eds.): ASIACRYPT 2019, LNCS 11923, pp. 371–397, 2019.
https://doi.org/10.1007/978-3-030-34618-8_13

1 Introduction

Background. Whenever a computation on sensitive data is outsourced to an untrusted machine, one has to ensure that the result is correct. Examples are database updates, user authentications, and elections. The underlying problem, formally called *computational integrity*, has been theoretically solved since the 1990s with the emergence of the PCP theorem. But the performance of actual implementations was too poor to handle any computation of practical interest. Only recently a few proof systems have appeared where the proving time is quasi-linear in the computation length (which is typically represented as an arithmetic circuit), e.g. ZK-SNARKs [Par+13], Bulletproofs [Bun+18], and ZK-STARKs [Ben+18]. While they all share the overall structure, these proof systems differ in details such as the need of a trusted setup, proof size, verifier scalability, and post-quantum resistance.

The cryptographic protocols that make use of such systems for zero-knowledge proofs often face the problem that whenever a hash function is involved, the associated circuit is typically long and complex, and thus the hash computation becomes a bottleneck in the proof. An example is the Zerocash cryptocurrency protocol [Ben+14]: in order to spend a coin anonymously, one has to present a zero-knowledge proof that the coin is in the set of all valid coins, represented by a Merkle tree with coins as leaves. When a traditional hash function such as SHA-256 is used in the Merkle tree, the proof generation takes almost a minute for 28-level trees such as in Zcash [Hop+19], which represents a real obstacle to the widespread use of privacy-oriented cryptocurrencies.

The demand for symmetric-key primitives addressing the needs of specific proof systems has been high, but only a few candidates have been proposed so far: a hash function based on Pedersen commitments [Hop+19], MPC-oriented LowMC [Alb+15], and big-field MiMC [Alb+16, Alb+19]. Even worse, different ZK proof systems use distinct computation representations. Concretely, ZK-SNARKs prefer pairing-friendly curves over prime scalar fields, Bulletproofs uses a fast curve over a scalar field, whereas ZK-STARKs are most comfortable operating over binary fields. Hence, the issue of different representations further limits the design space of ZK-friendly primitives.

STARKs. ZK-STARKs [Ben+18] is a novel proof system which, in contrast to SNARKs, does not need a trusted setup phase and whose security relies only on the existence of collision-resistant hash functions. The computation is represented as an execution trace, with polynomial relations among the trace elements. Concretely, the trace registers must be elements of some large binary field, and the polynomials should have low degree. The proof generation time is approximately[1] $O(S \log S)$, where

$$S \approx (\text{Maximum polynomial degree} \times \text{Trace length}) .$$

[1] We omit optimisations related to the trace layout.

The STARK paper came with a proposal to use Rijndael-based hash functions, but as these have been shown to be insecure [KBN09], custom designs are clearly needed.

JARVIS *and* FRIDAY. Ashur and Dhooghe recently addressed this need with the proposal of the block cipher JARVIS and the hash function FRIDAY [AD18]. The primitives were immediately endorsed by the ZK-STARK authors as possible solutions to reduce the STARK generation cost in many applications[2]. The new hash function was claimed to offer up to a 20-fold advantage over Pedersen hashes and an advantage by a factor of 2.5 over MiMC-based hash functions, regarding the STARK proof generation time [BS18].

Albeit similar in spirit to MiMC, JARVIS comes with novel design elements in order to considerably reduce the number of rounds, while still aiming to provide adequate security. In the original proposal several types of algebraic attacks were initially ruled out, and security arguments from RIJNDAEL/AES were used to inform the choice of the number of rounds, leading to a statement that attacks were expected to cover up to three rounds only. An extra security margin was added, leading to a recommendation of 10 rounds for the variant with an expected security of 128 bits. Variants with higher claims of security were also specified.

Algebraic Attacks. This class of attacks aims to utilise the algebraic properties of a construction. One example is the Gröbner basis attack, which proceeds by modelling the underlying primitive as a multivariate system of equations which is then solved using off-the-shelf Gröbner basis algorithms [Buc65, CLO97, Fau99, Fau02]. After some initial success against certain stream cipher constructions [Cou03b, Cou03a], algebraic attacks were also considered against block ciphers [MR02, CB07], albeit with limited success. Even approaches combining algebraic and statistical techniques [AC09] were later shown not to outperform known cryptanalytic techniques [Wan+11]. As a result algebraic attacks are typically not considered a major concern for new block ciphers. We note however that Gröbner basis methods have proven fruitful for attacking a number of public-key schemes [Fau+10, AG11, Alb+14, FPP14, Fau+15].

Contribution. In this paper we show that, while the overall design approach of JARVIS and FRIDAY seems sound, the choice for the number of rounds is not sufficient to offer adequate security. We do this by mounting algebraic attacks on the full-round versions of the primitives with the help of Gröbner bases. Our results show that designers of symmetric-key constructions targeting "algebraic platforms" – such as STARKs, FHE and MPC – must pay particular attention to the algebraic structure of their ciphers, and that algebraic attacks should receive renewed attention from the cryptographic community.

Organisation. The remainder of this work is organised as follows. In Sect. 2 we briefly describe the block cipher JARVIS and the hash function FRIDAY. Follow-

[2] The ciphers were announced among high anticipation of the audience at the prime Ethereum conference DevCon4, held in November 2018 [BS18].

ing, we discuss various algebraic attacks in Sect. 3, including higher-order differential attacks, interpolation attacks, and in particular attacks using Gröbner bases. In the following sections, we describe our attacks, including key-recovery attacks on JARVIS in Sect. 4 and preimage attacks on FRIDAY in Sect. 5. In Sect. 6, we describe our experimental results from running the attacks and discuss our findings. Finally, in Sect. 7 we analyse the S-box layer of JARVIS and compare it to the AES.

2 MARVELlous

MARVELLOUS [AD18, Aly+19] is a family of cryptographic primitives specifically designed for STARK applications. It includes the block cipher JARVIS as well as FRIDAY, a hash function based on this block cipher. We briefly describe the two primitives in this section.

As usual, we identify functions on $\mathbb{F}_{2^n}$ with elements in the quotient ring

$$\mathcal{R} := \mathbb{F}_{2^n}[X] / \langle X^{2^n} - X \rangle.$$

Whenever it is clear from the context, we refer to the corresponding polynomial representation in the above quotient ring when we speak of a function on $\mathbb{F}_{2^n}$ and use the notation $F(X)$, or just F, for the coset $F(X) + \langle X^{2^n} - X \rangle \in \mathcal{R}$.

2.1 JARVIS

JARVIS is a family of block ciphers operating on a state and a key of n bits, thus working entirely over the finite field $\mathbb{F}_{2^n}$. The construction is based on ideas used by the AES, most prominently the *wide-trail design* strategy, which guarantees security against differential and linear (statistical) attacks. However, where AES uses multiple small S-boxes in every round, JARVIS applies a single nonlinear transformation to the whole state, essentially using one large n-bit S-box. The S-box of JARVIS is defined as the generalised inverse function $S : \mathbb{F}_{2^n} \to \mathbb{F}_{2^n}$ with

$$S(x) := \begin{cases} x^{-1} & x \neq 0 \\ 0 & x = 0, \end{cases}$$

which corresponds to the element

$$S(X) := X^{2^n - 2} \in \mathcal{R}.$$

We note that this specific S-box makes the construction efficient in the STARK setting, because verifying it uses only one quadratic constraint (note that for non-zero $x \in \mathbb{F}_{2^n}$ the equality $\frac{1}{x} = y$ is equivalent to the equality $x \cdot y = 1$, and the constraint for the full S-box can be written as $x^2 \cdot y + x = 0$). We refer to [Ben+18, AD18] for more details.

The linear layer of JARVIS is composed by evaluating a high-degree affine polynomial

$$A(X) := L(X) + \hat{c} \in \mathcal{R},$$

where $\hat{c} \in \mathbb{F}_{2^n}$ is a constant and

$$L(X) := \sum_{i=0}^{n-1} l_{2^i} \cdot X^{2^i} \in \mathcal{R}$$

is a linearised permutation polynomial. Note that the set of all linearised permutation polynomials in $\mathcal{R}$ forms a group under composition modulo $X^{2^n} - X$, also known as the *Betti-Mathieu* group [LN96].

In JARVIS, the polynomial A is built from two affine monic permutation polynomials B, C of degree 4, that is

$$B(X) := L_B(X) + b_0 := X^4 + b_2X^2 + b_1X + b_0 \in \mathcal{R}$$

and

$$C(X) := L_C(X) + c_0 := X^4 + c_2X^2 + c_1X + c_0 \in \mathcal{R}$$

satisfying the equation

$$A = C \circ B^{-1}.$$

The operator $\circ$ indicates composition modulo $X^{2^n} - X$ and B^{-1} denotes the compositional inverse of B (with respect to the operator $\circ$) given by

$$B^{-1}(X) := {L_B}^{-1}(X) + {L_B}^{-1}(b_0).$$

Here, ${L_B}^{-1}$ denotes the inverse of L_B under composition modulo $X^{2^n} - X$, or in other words, the inverse of L_B in the Betti-Mathieu group. We highlight that the inverse B^{-1} shares the same affine structure with B, i.e. it is composed of a linearised permutation polynomial ${L_B}^{-1}$ and a constant term in $\mathbb{F}_{2^n}$, but has a much higher degree.

One round of JARVIS is shown in Fig. 1. Additionally, a whitening key k_0 is applied before the first round.

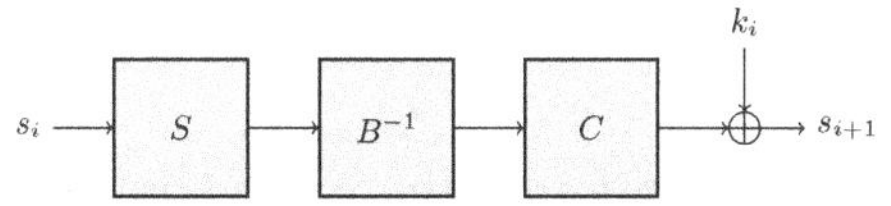

Fig. 1. One round of the JARVIS block cipher. For simplicity, the addition of the whitening key is omitted.

Key Schedule. The key schedule of JARVIS shares similarities with the round function itself, the main difference being that the affine transformations are omitted. In the key schedule, the first key k_0 is the master key and the next round key k_{i+1} is calculated by adding a round constant c_i to the (generalised) inverse $S(k_i)$ of the previous round key k_i. One round of the key schedule is depicted in Fig. 2.

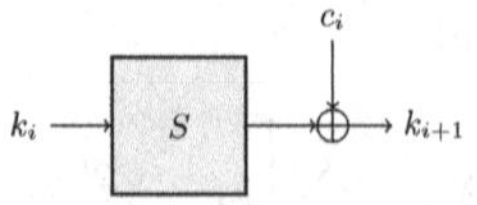

Fig. 2. The key schedule used by the JARVIS block cipher.

The first round constant c_0 is randomly selected from $\mathbb{F}_{2^n}$, while subsequent round constants c_i, $1 \leq i \leq r$, are calculated using the relation

$$c_i := a \cdot c_{i-1} + b$$

for random elements $a, b \in \mathbb{F}_{2^n}$.

Instantiations. The authors of [AD18] propose four instances of JARVIS-n, where $n \in \{128, 160, 192, 256\}$. For each of these instances the values c_1, a, b, and the polynomials B and C are specified. Table 1 presents the recommended number of rounds r for each instance, where the claimed security level is equal to the key size (and state size) n. We will use $r \in \mathbb{N}$ throughout this paper to denote the number of rounds of a specific instance.

Table 1. Instances of the JARVIS block cipher [AD18].

Instance	n	# of rounds r
JARVIS-128	128	10
JARVIS-160	160	11
JARVIS-192	192	12
JARVIS-256	256	14

2.2 FRIDAY

FRIDAY is a hash function based on a Merkle-Damgård construction, where the block cipher JARVIS is transformed into a compression function using the Miyaguchi-Preneel scheme. In this scheme, a (padded) message block m_i, $1 \leq i \leq t$, serves as input m to a block cipher $E(m, k)$ and the respective previous hash value h_{i-1} serves as key k. The output of the block cipher is then added to the sum of m_i and h_{i-1}, resulting in the new hash value h_i. The first hash value h_0 is an initialization vector and taken to be the zero element in $\mathbb{F}_{2^n}$ in the case of FRIDAY. The final state h_t is the output of the hash function. The hash function FRIDAY is thus defined by the following iterative formula

$$h_0 := IV := 0,$$
$$h_i := E(m_i, h_{i-1}) + h_{i-1} + m_i,$$

for $1 \leq i \leq t$, as illustrated in Fig. 3.

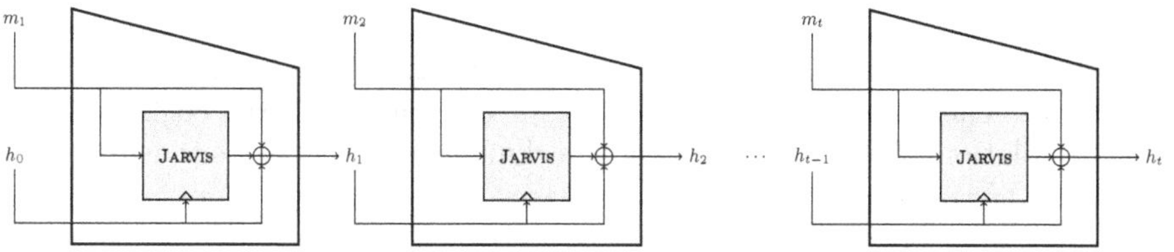

Fig. 3. The FRIDAY hash function.

3 Overview of Algebraic Attacks on JARVIS and FRIDAY

From an algebraic point of view, JARVIS offers security mainly by delivering a high degree for its linear transformations and for the S-box. In the original proposal, the authors analyse the security against various algebraic attack vectors, such as higher-order differential attacks and interpolation attacks.

3.1 Higher-Order Differential Attacks

Higher-order differential attacks [Knu95] can be regarded as algebraic attacks that exploit the low algebraic degree of a nonlinear transformation. If this degree is low enough, an attack using multiple plaintexts and their corresponding ciphertexts can be mounted. In more detail, if the algebraic degree of a Boolean function f is d, then when applying f to all elements of an affine vector space $\mathcal{V} \oplus c$ of dimension $> d$ and taking the sum of these values, the result is 0, i.e.

$$\bigoplus_{v \in \mathcal{V} \oplus c} v = \bigoplus_{v \in \mathcal{V} \oplus c} f(v) = 0.$$

Finding such a distinguisher possibly allows the attacker to recover the secret key.

However, higher-order differential attacks pose no threat to JARVIS. Indeed, the algebraic degree of $S(X) = X^{2^n-2}$ is the Hamming weight of $2^n - 2$, which is equal to $n - 1$ and thus maximal (note that the S-box is a permutation). This makes higher-order differential attacks and zero-sum distinguishers infeasible after only one round of JARVIS.

3.2 Interpolation Attacks

Interpolation attacks were introduced in 1997 [JK97] and are another type of algebraic attack where the attacker constructs the polynomial corresponding to the encryption (or decryption) function without knowing the secret key. The basis of interpolation attacks is a consequence of the Fundamental Theorem of Algebra: given $d + 1$ pairs $(x_0, y_0), \ldots, (x_d, y_d)$ of elements in a certain field $\mathbb{F}$, there is a *unique* polynomial $P(X) \in \mathbb{F}[X]$ of degree at most d which satisfies

$$P(x_i) = y_i$$

for all $0 \leq i \leq d$. To put it another way, the polynomial $P(X)$ *interpolates* the given pairs (x_i, y_i), which is why it deserves the denotation *interpolation polynomial.* There are several approaches for calculating all the coefficients of the interpolation polynomial. A classical technique is to choose Lagrange's basis $(L_0, L_1, \ldots, L_d)$, with

$$L_i(X) := \prod_{\substack{j=0 \\ j \neq i}}^{d} \frac{X - x_j}{x_i - x_j} \in \mathbb{F}[X],$$

as a basis for the $\mathbb{F}$-vector space $\mathbb{F}[X]$ and read off the solution $(p_0, \ldots, p_d)$ from the resulting system of equations

$$y_i = P(x_i) = p_0 L_0(x_i) + p_1 L_1(x_i) + \ldots + p_d L_d(x_i),\ 0 \leq i \leq d.$$

Lagrange's basis leads to a complexity of $\mathcal{O}(d^2)$ field operations and so does Newton's basis $\{N_0, N_1, \ldots, N_d\}$ with

$$N_i(X) := \prod_{j=0}^{i-1} (X - x_j) \in \mathbb{F}[X].$$

A different approach uses the fact that polynomial interpolation can be reduced to polynomial evaluation, as discussed by Horowitz [Hor72] and Kung [Kun73], leading to a complexity of $\mathcal{O}(d \log^2 d)$ field operations. In essence, this approach relies on the Fast Fourier Transform for polynomial multiplication.

From the above complexity estimates, it is thus desirable that the polynomial representation of the encryption function reaches a high degree and forces all possible monomials to appear. In Jarvis, a high word-level degree is already reached after only one round; additionally the polynomial expression of the encryption function is also dense after only two rounds. It follows that interpolation attacks pose no threat to Jarvis.

3.3 Gröbner Basis Attacks

The first step in a Gröbner basis attack is to describe the primitive by a system of polynomial equations. Subsequently, a Gröbner basis [Buc65, CLO97] for the ideal defined by the corresponding polynomials is calculated and finally used to solve for specified variables. In more detail, Gröbner basis attacks consist of three phases:

1. Set up an equation system and compute a Gröbner basis (typically for the *degrevlex* term order for performance reasons) using an algorithm such as Buchberger's algorithm [Buc65], F4 [Fau99], or F5 [Fau02].
2. Perform a change of term ordering for the computed Gröbner basis (typically going from the *degrevlex* term order to the *lex* one, which facilitates computing elimination ideals and hence eliminating variables) using an algorithm such as FGLM [Fau+93]. Note that in our applications all systems of algebraic equations result in zero-dimensional ideals, i.e. the systems have only finitely many solutions.

3. Solve the univariate equation for the last variable using a polynomial factoring algorithm, and substitute into other equations to obtain the full solution of the system.

Cost of Gröbner Basis Computation. For a generic system of n_e polynomial equations

$$F_1(x_1,\ldots,x_{n_v}) = F_2(x_1,\ldots,x_{n_v}) = \cdots = F_{n_e}(x_1,\ldots,x_{n_v}) = 0$$

in n_v variables $x_1,\ldots,x_{n_v}$, the complexity of computing a Gröbner basis [BFP12] is

$$\mathcal{C}_{\text{GB}} \in \mathcal{O}\left(\binom{n_v + D_{\text{reg}}}{D_{\text{reg}}}^{\omega}\right), \tag{1}$$

where $2 \leq \omega < 3$ is the linear algebra exponent representing the complexity of matrix multiplication and D_{reg} is the degree of regularity. The constants hidden by $\mathcal{O}(\cdot)$ are relatively small, which is why $\binom{n_v+D_{\text{reg}}}{D_{\text{reg}}}^{\omega}$ is typically used directly. In general, computing the degree of regularity is a hard problem. However, the degree of regularity for "regular sequences" [Bar+05] is given by

$$D_{\text{reg}} = 1 + \sum_{i=1}^{n_e}(d_i - 1), \tag{2}$$

where d_i is the degree of F_i. Regular sequences have $n_e = n_v$. More generally, for "semi-regular sequences" (the generalisation of regular sequences to $n_e > n_v$) the degree of regularity can be computed as the index of the first non-positive coefficient in

$$H(z) = \frac{1}{(1-z)^{n_v}} \times \prod_{i=1}^{n_e}(1 - z^{d_i}).$$

It is conjectured that most sequences are semi-regular [Fro85]. Indeed, experimental evidence suggests random systems behave like semi-regular systems with high probability. Hence, assuming our target systems of equations behave like semi-regular sequences, i.e. they have no additional structure, the complexity of computing a Gröbner basis depends on (a) the number of equations n_e, (b) the degrees $d_1, d_2, \ldots, d_{n_e}$ of the equations, and (c) the number of variables n_v. Crucially, our experiments described later in the paper indicate that the systems considered in this work do not behave like regular sequences.

Cost of Gröbner Basis Conversion. The complexity of the FGLM algorithm [Fau+93] is

$$\mathcal{C}_{\text{FGLM}} \in \mathcal{O}\left(n_v \cdot \deg(\mathcal{I})^3\right), \tag{3}$$

where $\deg(\mathcal{I})$ is called the *degree of the ideal* and defined as the dimension of the quotient ring $\mathbb{F}[X_1, X_2, \ldots, X_n]/\mathcal{I}$ as an $\mathbb{F}$-vector space. For the systems we are considering in this paper – which are expected to have a unique solution in $\mathbb{F}$ – the dimension of $R/\mathcal{I}$ corresponds to the degree of the unique univariate polynomial equation in the reduced Gröbner basis with respect to the canonical lexicographic order [KR00, Theorem 3.7.25]. Again, the hidden constants are small, permitting to use $n_v \cdot \deg(\mathcal{I})^3$ directly. A sparse variant of the algorithm also exists [FM11] with complexity $\mathcal{O}(\deg(\mathcal{I})(N_1 + n_v \log \deg(\mathcal{I})))$, where N_1 is the number of nonzero entries of a multiplication matrix, which is sparse even if the input system spanning $\mathcal{I}$ is dense. Thus, the key datum to establish for estimating the cost of this step is $\deg(\mathcal{I})$.

Cost of Factoring. Finally, we need to solve for the last variable using the remaining univariate polynomial equation obtained by computing all necessary elimination ideals. This can be done by using a factorisation algorithm. For example, the complexity of a modified version of the Berlekamp algorithm [Gen07] to factorise a polynomial P of degree D over $\mathbb{F}_{2^n}$ is

$$\mathcal{C}_{\text{Sol}} \in \mathcal{O}\left(D^3 n^2 + D n^3\right). \tag{4}$$

In our context, we can however reduce the cost of this step by performing the first and second steps of the attack for two (or more) (plaintext, ciphertext) pairs and then considering the GCD of the resulting univariate polynomials, which are univariate in the secret key variable k_0. Computing polynomial GCDs is quasi-linear in the degree of the input polynomials. In particular, we expect

$$\mathcal{C}_{\text{Sol}} \in \mathcal{O}\left(D(\log(D))^2\right). \tag{5}$$

We will again drop the $\mathcal{O}(\cdot)$ and use the expressions directly.

Our Algebraic Attacks on MARVELLOUS. All attacks on MARVELLOUS presented in this paper are inherently Gröbner basis attacks which, on the one hand, are based on the fact that the S-box $S(X) = X^{2^n-2}$ of JARVIS can be regarded as the function $S: \mathbb{F}_{2^n} \to \mathbb{F}_{2^n}$, where

$$S(x) = x^{-1}$$

for all elements *except* the zero element in $\mathbb{F}_{2^n}$. As a consequence, the relation

$$y = S(x) = x^{-1}$$

can be rewritten as an equation of degree 2 in two variables, namely

$$x \cdot y = 1,$$

which holds everywhere except for the zero element in $\mathbb{F}_{2^n}$. We will use this relation in our attacks, noting that $x = 0$ occurs with a negligibly small probability for $n \geq 128$.

On the other hand, we exploit the fact that the decomposition of the affine polynomial A originates from two low-degree polynomials B and C. When setting up the associated equations for JARVIS, we introduce intermediate variables in such a way that the low degree of B and C comes into effect, and then show that the particular combination of the inverse S-box $S(X) = X^{2^n-2}$ with the affine layer in JARVIS is vulnerable to Gröbner basis attacks.

Based on the above observations, we describe in the next sections:

- a key-recovery attack on reduced-round JARVIS and an optimised key-recovery attack on full-round JARVIS;
- its extension to a (two-block) preimage attack on full-round FRIDAY;
- a more efficient direct preimage attack on full-round FRIDAY.

4 Gröbner Basis Computation for JARVIS

We first describe a straightforward approach, followed by various optimisations which are necessary to extend the attack to all rounds.

4.1 Reduced-Round JARVIS

Let $B, C \in \mathcal{R}$ be the polynomials of the affine layer in JARVIS. Furthermore, in round i of JARVIS let us denote the intermediate state between the application of B^{-1} and C as x_i, for $1 \leq i \leq r$ (see Fig. 4).

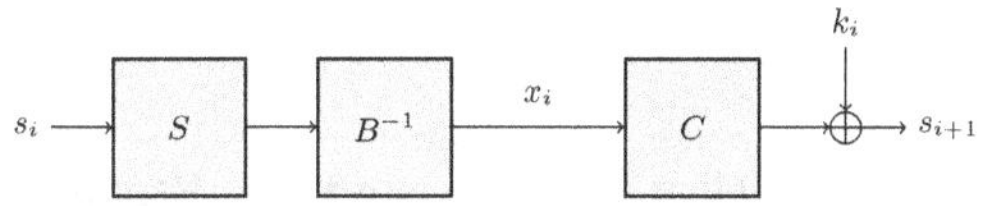

Fig. 4. Intermediate state x_i in one round of the encryption path.

As a result, two consecutive rounds of JARVIS can be related by the equation

$$(C(x_i) + k_i) \cdot B(x_{i+1}) = 1 \tag{6}$$

for $1 \leq i \leq r - 1$. As both polynomials B and C have degree 4, Eq. (6) yields a system of $r - 1$ polynomial equations, each of degree 8, in the variables $x_1, \ldots, x_r$ and $k_0, \ldots, k_r$. To make the system dependent on the plaintext p and the ciphertext c, we add the two equations

$$B(x_1) \cdot (p + k_0) = 1, \tag{7}$$

$$C(x_r) = c + k_r \tag{8}$$

to this system. Additionally, two successive round keys are connected through the equation

$$(k_{i+1} + c_i) \cdot k_i = 1 \tag{9}$$

for $0 \leq i \leq r-1$. In total, the above description of JARVIS amounts to $2 \cdot r + 1$ equations in $2 \cdot r + 1$ variables, namely:

- $r - 1$ equations of degree 8 (Eq. (6)),
- one equation of degree 5 (Eq. (7)),
- one equation of degree 4 (Eq. (8)),
- r equations of degree 2 (Eq. (9)),

in the $2 \cdot r + 1$ variables $x_1, \ldots, x_r$ and $k_0, \ldots, k_r$. Since the number of equations is equal to the number of variables, we can estimate the complexity of a Gröbner basis attack by using Eq. (2). According to this estimate, the computation of a Gröbner basis for the above system of equations is prohibitively expensive for full-round JARVIS. For example, Eq. (2) predicts a complexity of ≈ 120 bits (when setting $\omega = 2.8$) for computing a Gröbner basis for $r = 6$. However, we note that we were able to compute such a basis in practice (Sect. 6), which indicates that the above estimate is too pessimistic.

4.2 Optimisations for an Attack on Full-Round JARVIS

In order to optimise the computation from the previous section and extend it to full-round JARVIS, we introduce two main improvements. First, we reduce the number of variables and equations used for intermediate states. Secondly, we relate all round keys to the master key, which helps to further reduce the number of variables.

A More Efficient Description of Intermediate States. The main idea is to reduce the number of equations and variables for intermediate states at the expense of an increased degree in some of the remaining equations. By relating a fixed intermediate state x_i to the respective preceding and succeeding intermediate states x_{i-1} and x_{i+1}, we obtain the equations

$$B(x_i) = \frac{1}{C(x_{i-1}) + k_{i-1}}, \tag{10}$$

$$C(x_i) = \frac{1}{B(x_{i+1})} + k_i \tag{11}$$

for $2 \leq i \leq r-1$. Since both B and C are *monic affine* polynomials of degree 4, we claim that it is possible to find *monic affine* polynomials

$$D(X) := X^4 + d_2X^2 + d_1X + d_0$$

and

$$E(X) := X^4 + e_2X^2 + e_1X + e_0,$$

also of degree 4, such that

$$D(B) = E(C).$$

Indeed, comparing corresponding coefficients of $D(B)$ and $E(C)$ yields a system of 5 linear equations in the 6 unknown coefficients $d_0, d_1, d_2, e_0, e_1, e_2$, which can then be solved. We explain the construction of D and E in more detail in Appendix A.

From now on let us assume we have already found appropriate polynomials D and E. After applying D and E to Eqs. (10) and (11), respectively, we equate the right-hand side parts of the resulting equations and get

$$D\left(\frac{1}{C(x_{i-1}) + k_{i-1}}\right) = E\left(\frac{1}{B(x_{i+1})} + k_i\right) \tag{12}$$

for $2 \leq i \leq r - 1$. Eventually we obtain a system of polynomial equations of degree 36 by clearing denominators in Eq. (12).

The crucial point is that variables for every second intermediate state may now be dropped out of the description of JARVIS. This is because we can consider either only evenly indexed states or only odd ones, and by doing so, we have essentially halved the number of equations and variables needed to describe intermediate states. We note that in all optimised versions of our attacks we only work with *evenly* indexed intermediate states, as this choice allows for a more efficient description of JARVIS compared to working with odd ones.

Finally we relate the plaintext p and the ciphertext c to the appropriate intermediate state x_2 and x_r, respectively, and set

$$D\left(\frac{1}{p + k_0}\right) = E\left(\frac{1}{B(x_2)} + k_1\right), \tag{13}$$

$$C(x_r) + k_r = c. \tag{14}$$

Here, the degree of Eq. (13) is 24, while Eq. (14) has degree 4.

Remarks. It is worth pointing out that the above description uses several implicit assumptions. First, it may happen that some intermediate states become zero, with the consequence that our approach will not find a solution. However, this case only occurs with a negligibly small probability, in particular when considering instances with $n \geq 128$. If this event occurs we can use another plaintext-ciphertext pair. Secondly, when we solve the optimised system of equations (i.e. the system we obtain after applying D and E), not all of the solutions we find for this system are guaranteed to be valid solutions for the original system of equations. Lastly, Eq. (14) implicitly assumes an even number of rounds. If we wanted to attack an odd number of rounds instead, this equation had to be adjusted accordingly.

Relating Round Keys to the Master Key. Two consecutive round keys in JARVIS are connected by the relation

$$k_{i+1} = \frac{1}{k_i} + c_i$$

if $k_i \neq 0$, which is true with high probability for large state sizes n. As a consequence, each round key is a rational function of the master key k_0 of degree 1, i.e.

$$k_{i+1} = \frac{\alpha_i \cdot k_0 + \beta_i}{\gamma_i \cdot k_0 + \delta_i}.$$

We provide the exact values for α_i, β_i, γ_i, and δ_i in Appendix B. Expressing k_i as a rational function of k_0 in Eqs. (12) and (14) raises the total degree of these equations to 40 and 5, respectively. On the other hand, the degree of Eq. (13) remains unchanged.

4.3 Complexity Estimates of Gröbner Basis Computation for Jarvis

Assuming the number of rounds r to be even, the aforementioned two improvements yield

- $\frac{r}{2} - 1$ equations of degree 40 (Eq. (12)),
- one equation of degree 24 (Eq. (13)),
- one equation of degree 5 (Eq. (14)),

in $\frac{r}{2} + 1$ variables (the intermediate states $x_2, x_4, \ldots, x_r$ and the master key k_0). Since the number of equations equals the number of variables, we may calculate the degree of regularity using Eq. (2), again assuming the system behaves like a regular sequence.

Our results for the degree of regularity, and thus also for the complexity of computing a Gröbner basis, are listed in Table 2. Note that we assume $\omega = 2.8$. However, this is possibly a pessimistic choice, as the regarded systems are sparse. We therefore also give the complexities for $\omega = 2$ in parentheses.

Table 2. Complexity estimates of Gröbner basis computations for r-round Jarvis.

r	n_v	D_{reg}	Complexity in bits
6	4	106	63 (45)
8	5	145	82 (58)
10 (Jarvis-128)	6	184	100 (72)
12 (Jarvis-192)	7	223	119 (85)
14 (Jarvis-256)	8	262	138 (98)
16	9	301	156 (112)
18	10	340	175 (125)
20	11	379	194 (138)

These values show that we are able to compute Gröbner bases for, and therefore successfully attack, all full-round versions of Jarvis. We note that, even when pessimistically assuming that the memory complexity of a Gröbner basis computation is asymptotically the same as its time complexity (the memory complexity of any algorithm is bounded by its time complexity) and when considering the time-memory product (which is highly pessimistic from an attacker's point of view), our attacks against Jarvis-256 are still valid.

5 Gröbner Basis Computation for FRIDAY

In this section, we let $F : \mathbb{F}_{2^n} \times \mathbb{F}_{2^n} \to \mathbb{F}_{2^n}$ indicate the application of one block of FRIDAY.

5.1 Extending the Key-Recovery Attack on JARVIS to a Preimage Attack on FRIDAY

Using the same equations as for JARVIS described in Sect. 4, a preimage attack on FRIDAY may also be mounted. At its heart, the attack on FRIDAY with r rounds is an attack on JARVIS with $r-1$ rounds.

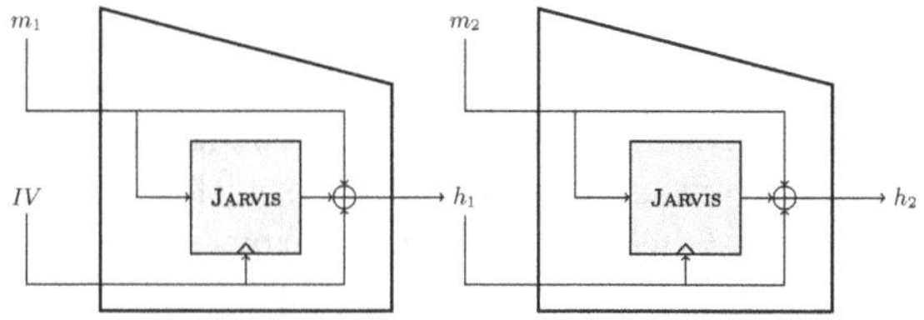

Fig. 5. Two blocks of FRIDAY.

We work with two blocks of FRIDAY, hence a message m is the concatenation

$$m = m_1 \mid\mid m_2$$

of two message blocks $m_1, m_2 \in \mathbb{F}_{2^n}$. The output of the first block is denoted by h_1 and the known (final) hash value of the second block is denoted by h_2. The hash values h_1 and h_2 can be expressed as

$$h_1 = F(m_1, IV)$$

and

$$h_2 = F(m_2, h_1).$$

The initialization vector IV is just the zero element in $\mathbb{F}_{2^n}$. We refer to Fig. 5 for an illustration of the introduced notation.

Our preimage attack proceeds as follows: in the first part, we use random values $\hat{m}_1$ for the input to the first block to populate a table T_1 in which each entry contains a pair $(\hat{m}_1, \hat{h}_1)$, where $\hat{h}_1$ denotes the corresponding intermediate hash value

$$\hat{h}_1 := F(\hat{m}_1, IV).$$

In the second part, we find pairs (m_2', h_1') with

$$F(m_2', h_1') = h_2,$$

or in other words, pseudo preimages for the known hash value h_2. To find such a pseudo preimage, we fix the sum $m_2 + h_1$ to an arbitrary value $v_0 \in \mathbb{F}_{2^n}$, i.e. we set

$$v_0 := m_2 + h_1.$$

This has two effects:

1. In the second block, the value v_1 entering the first round of JARVIS is fixed and known until the application of the second round key. Essentially, this means that one round of JARVIS can be skipped.
2. Since $v_0 = m_2 + h_1$ is fixed and known, the final output v_2 of JARVIS is defined by

$$v_2 := v_0 + h_2$$

and thus also known.

In the current scenario, the intermediate hash value h_1 serves as master key for the r round keys $k_1, k_2, \ldots, k_r$ applied in the second block. Using v_1 as plaintext and v_2 as ciphertext, an attack on JARVIS with $r-1$ rounds is sufficient to reveal these round keys. Once one of the round keys is recovered, we calculate the second part h_1' of a pseudo preimage (m_2', h_1') by applying the inverse key schedule to the recovered key. Finally, we set

$$m_2' := h_1' + v_0$$

and thereby obtain the remaining part of a pseudo preimage. How the presented pseudo preimage attack on r-round FRIDAY reduces to a key-recovery attack on $(r-1)$-round JARVIS is outlined in Fig. 6.

Conceptually, we repeat the pseudo preimage attack many times (for different values of v_0) and store the resulting pairs (m_2', h_1') in a table T_2. The aim is to produce matching entries $(\hat{m}_1, \hat{h}_1)$ and (m_2', h_1') in T_1 and T_2 such that

$$\hat{h}_1 = h_1',$$

which implies

$$F(m_2', F(\hat{m}_1, IV)) = F(m_2', \hat{h}_1) = F(m_2', h_1') = h_2,$$

giving us the preimage $(\hat{m}_1, m_2')$ we are looking for.

Remark. The (input, output) pairs (v_1, v_2) we use for the underlying key-recovery attack on JARVIS are *not* proper pairs provided by, e.g., an encryption oracle for JARVIS. Thus, it may happen that for some pairs (v_1, v_2) the key-recovery attack does not succeed, i.e. there is no key h_1' which maps v_1 to v_2. The probability for such an event is

$$P_{\text{fail}} = \left(\frac{2^n - 1}{2^n}\right)^{2^n} = \left(1 - \frac{1}{2^n}\right)^{2^n} \approx \lim_{k \to \infty} \left(1 - \frac{1}{k}\right)^k = \frac{1}{e}$$

for large n.

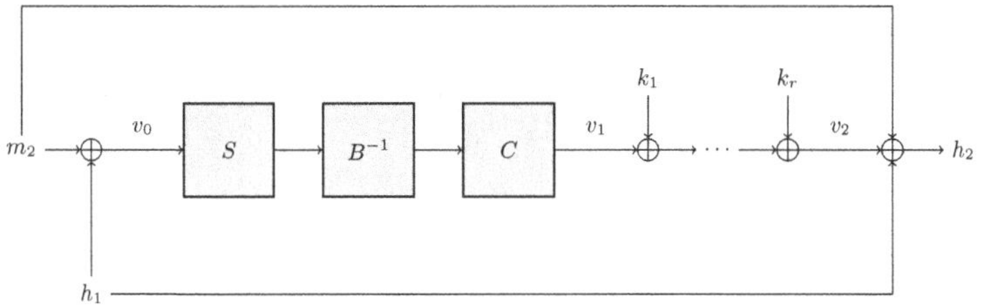

Fig. 6. Internals of the second block of FRIDAY. The values v_0, v_1 and v_2 are known.

5.2 Complexity of Generating Pseudo Preimages

The cost of generating pseudo preimages is not negligible. Hence, we cannot afford to generate tables T_1 and T_2, each with $2^{\frac{n}{2}}$ entries, and then look for a collision. However, given the attack complexities for JARVIS in Table 2, an attack on 9-round JARVIS has a complexity of around 83 bits (assuming $\omega = 2.8$). Considering JARVIS-128, for example, this means we can generate up to 2^{45} pseudo preimages.

Let us assume we calculate 2^{10} pseudo preimages $(\hat{m}_1, m_1')$ and $2^{\frac{n}{2}}$ intermediate pairs $(\hat{m}_1, \hat{h}_1)$, in both cases for FRIDAY instantiated with JARVIS-128. This leaves us with a table T_1 containing $2^{\frac{n}{2}}$ $(\hat{m}_1, \hat{h}_1)$ pairs and a table T_2 containing 2^{10} (m_2', h_1') pairs.

Assuming that all hash values in T_1 are pairwise distinct and that also all hash values in T_2 are pairwise distinct, the probability that we find at least one hash collision between a pair in T_1 and a pair in T_2 is

$$P = 1 - \prod_{i=0}^{|T_2|-1} \left(1 - \frac{|T_1|}{2^{128} - i}\right), \tag{15}$$

which is, unfortunately, too low for $|T_1| = 2^{\frac{n}{2}}$. However, we can increase this probability by generating more entries for T_1. Targeting a total complexity of, e.g., ≈ 120 bits, we can generate 2^{118} such entries. Note that the number of expected collisions in a table of m random n-bit entries is

$$N_c = m - 2^n + 2^n \cdot \left(\frac{2^n - 1}{2^n}\right)^m.$$

Therefore, the expected number of unique values in such a table is

$$N_u = \left(1 - \frac{N_c}{m}\right) \cdot m = m - N_c = 2^n - 2^n \cdot \left(\frac{2^n - 1}{2^n}\right)^m.$$

We want that $N_u \geq 2^{118}$, and by simple computation it turns out that 2^{119} hash evaluations are sufficient with high probability. Using these values in Eq. (15) yields a success probability of around 63%.

5.3 Direct Preimage Attack on FRIDAY

The preimage attack we present in this section works with *one* block of FRIDAY, as shown in Fig. 7.

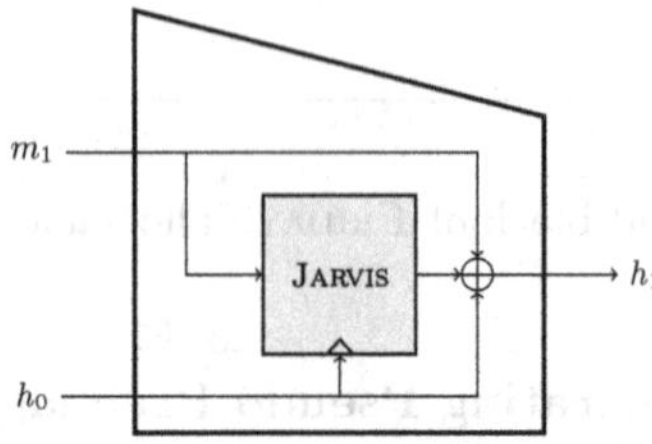

Fig. 7. Preimage attack on FRIDAY using one message block.

The description of the intermediate states $x_1, \ldots, x_r$ yields the same system of equations as before; however, in contrast to the optimised attack on JARVIS described in Sect. 4.2, in the current preimage attack on FRIDAY the master key k_0 and thus all subsequent round keys $k_1, \ldots, k_r$ are known. As an effect, we do not need to express round keys as a rational function of k_0 anymore. For the sake of completeness, we give Eq. (12) once more and note that the degree now decreases to 32 (from formerly 40). It holds that

$$D\left(\frac{1}{C(x_{i-1}) + k_{i-1}}\right) = E\left(\frac{1}{B(x_{i+1})} + k_i\right)$$

for $2 \leq i \leq r - 1$. Moreover, an additional equation is needed to describe the structure of the Miyaguchi-Preneel compression function (see Fig. 6), namely

$$B(x_1) \cdot (C(x_r) + k_r + h_1) = 1.$$

Again, we assume an even number of rounds r and work with intermediate states $x_2, x_4, \ldots, x_r$, which is why we need to apply the transformations D and E to cancel out the state x_1 in the above equation. Thus, eventually we have

$$D\left(\frac{1}{C(x_r) + k_r + h_1}\right) = E\left(\frac{1}{B(x_2)} + k_1\right). \tag{16}$$

Here, h_1 denotes the hash value for which we want to find a preimage m_1' such that

$$F(m_1', h_0) = h_1.$$

To obtain m_1' we solve for the intermediate state x_r and calculate

$$m_1' := C(x_r) + k_r + h_1 + h_0.$$

The value $h_0 = k_0$ can be regarded as the initialisation vector and is the zero element in $\mathbb{F}_{2^n}$. The above attack results in:

- $\frac{r}{2} - 1$ equations of degree 32 coming from Eq. (12) when considering even intermediate states, and
- one equation of degree 32 coming from Eq. (16),

in the $\frac{r}{2}$ variables $x_2, x_4, \ldots, x_r$. The number of equations is the same as the number of variables, and we can again use Eq. (2) to estimate the degree of regularity. The complexities of the Gröbner basis computations are summarised in Table 3, where we pessimistically assume $\omega = 2.8$, but also give the complexities for $\omega = 2$ in parentheses.

Table 3. Complexity estimates for the Gröbner basis step in preimage attacks on FRIDAY using r-round JARVIS.

r	n_v	D_{reg}	Complexity in bits
6	3	94	48 (34)
8	4	125	65 (47)
10 (JARVIS-128)	5	156	83 (59)
12 (JARVIS-192)	6	187	101 (72)
14 (JARVIS-256)	7	218	118 (85)
16	8	249	136 (97)
18	9	280	154 (110)
20	10	311	172 (123)

6 Behaviour of the Attacks Against JARVIS and FRIDAY

Recall that our attack has three steps:

1. Set up an equation system and compute a Gröbner basis using, e.g., the F4 algorithm [Fau99], with cost $\mathcal{C}_{\text{GB}}$.
2. Perform a change of term ordering for the computed Gröbner basis using the FGLM algorithm [Fau+93], with cost $\mathcal{C}_{\text{FGLM}}$.
3. Solve the remaining univariate equation for the last variable using a polynomial factoring algorithm, substitute into other equations, with cost $\mathcal{C}_{\text{Sol}}$.

For the overall cost of the attack we have[3]:

$$\mathcal{C} := 2\,\mathcal{C}_{\text{GB}} + 2\,\mathcal{C}_{\text{FGLM}} + \mathcal{C}_{\text{Sol}},$$

$$\mathcal{C} := 2\left(\binom{n_v + D}{D}^{\omega}\right) + 2\left(n_v \cdot {D_u}^3\right) + \left(D_u \log^2 D_u\right).$$

We can estimate $\mathcal{C}_{\text{GB}}$ if we assume that our systems behave like regular sequences. For the $\mathcal{C}_{\text{FGLM}}$ and $\mathcal{C}_{\text{Sol}}$ we need to establish the degree D_u of the

[3] As suggested in Sect. 3.3, our attack proceeds by running steps 1 and 2 twice, and recovering the last variable via the GCD computation, thus reducing the complexity of step 3.

univariate polynomial recovered, for which however we do not have an estimate. We have therefore implemented our attacks on JARVIS and FRIDAY using Sage v8.6 [Ste+19] with Magma v2.20–5 [BCP97] as the Gröbner basis engine. In particular, we implemented both the unoptimised and the optimised variants of the attacks from Sects. 4.2 and 5.3.

We observed that our attacks performed significantly better in our experiments than predicted. On the one hand, our Gröbner basis computations reached significantly lower degrees D than the (theoretically) expected D_{reg}. Furthermore, the degrees of the univariate polynomials seem to grow as $\approx 2 \cdot 5^r$ (JARVIS) and as $\approx 2 \cdot 4^r$ (FRIDAY), respectively, suggesting the second and third steps of our attack are relatively cheap.

We therefore conclude that the complexities given in Tables 2 and 3 are conservative upper bounds for our attacks on JARVIS and FRIDAY. We summarise our findings in Table 4, and we provide the source code of our attacks on MARVELLOUS as supplementary material. We summarise our findings in Table 4, and the source code of our attacks on MARVELLOUS is available on GitHub[4].

6.1 Comparison with MiMC

We note that the same attack strategy – direct Gröbner basis computation to recover the secret key – also applies, in principle, to MiMC, as pointed out by [Ash19]. In particular, it is easy to construct a multivariate system of equations for MiMC with degree 3 that is already a Gröbner basis by introducing a new state variable per round[5]. This makes the first step of a Gröbner basis attack free.[6] However, then the change of ordering has to essentially undo the construction to recover a univariate polynomial of degree $D_u \approx 3^r$. Performing this step twice produces two such polynomials from which we can recover the key by applying the GCD algorithm with complexity $\tilde{\mathcal{O}}(3^r)$. In [Alb+16], the security analysis implicitly assumes that steps 1 and 2 of our attack are free by constructing the univariate polynomial directly and costing only the third and final step of computing the GCD.

The reason our Gröbner basis attacks are so effective against FRIDAY and JARVIS is that the particular operations used in the ciphers – finite field inversion and low-degree linearised polynomials – allow us to construct a polynomial system with a relatively small number of variables, which can in turn be efficiently solved using our three-step attack strategy. We have not been able to construct such amenable systems for MiMC.

---

[4] https://github.com/IAIK/marvellous-attacks.

[5] This property was observed by Tomer Ashur and Alan Szepieniec and shared with us during personal communication.

[6] We note that this situation is somewhat analogous to the one described in [BPW06].

Table 4. Experimental results using Sage.

r	n_v	D_{reg}	$2\log_2\binom{n_v+D_{\text{reg}}}{D_{\text{reg}}}$	D	$2\log_2\binom{n_v+D}{D}$	$D_u=\deg(\mathcal{I})$	Time
JARVIS (optimised)							
3	2	47	20	26	17	256	0.3s
4	3	67	31	40	27	1280	9.4s
5	3	86	34	40	27	6144	891.4s
6	4	106	45	41	34	28672	99989.0s
JARVIS (unoptimised)							
3	4	25	29	10	20	256	0.5s
4	5	33	38	11	24	1280	23.9s
5	6	41	47	13	29	6144	2559.8s
6	7	47	55	14	34	28672	358228.6s
FRIDAY							
3	2	39	19	32	18	128	3.6s
4	2	63	22	36	19	512	0.5s
5	3	70	32	36	26	2048	36.5s
6	3	94	34	48	29	8192	2095.2s

In the table, r denotes the number of rounds, D_{reg} is the expected degree of regularity under the assumption that the input system is regular, n_v is the number of variables, $2 \cdot \log_2\binom{n_v+D_{\text{reg}}}{D_{\text{reg}}}$ is the expected bit security for $\omega = 2$ under the regularity assumption, D is the highest degree reached during the Gröbner basis computation, and $2 \cdot \log_2\binom{n_v+D}{D}$ is the expected bit security for $\omega = 2$ based on our experiments. The degree of the recovered univariate polynomial used for solving the system is denoted as D_u.

7 Comparing the S-Boxes of JARVIS and the AES

The non-linear operation in JARVIS shows similarities with the AES S-box $S_{\text{AES}}(X)$. In particular, $S_{\text{AES}}(X)$ is the composition of an $\mathbb{F}_2$-affine function A_{AES} and the multiplicative inverse of the input in $\mathbb{F}_{2^8}$, i.e.

$$S_{\text{AES}}(X) = A_{\text{AES}}(X^{254}),$$

where

$$A_{\text{AES}}(X) = \texttt{0x8F} \cdot X^{128} + \texttt{0xB5} \cdot X^{64} + \texttt{0x01} \cdot X^{32} + \texttt{0xF4} \cdot X^{16} + \texttt{0x25} \cdot X^{8} + \texttt{0xF9} \cdot X^{4} + \texttt{0x09} \cdot X^{2} + \texttt{0x05} \cdot X + \texttt{0x63}.$$

In JARVIS, we can also view the S-box as

$$S(X) = A(X^{254}),$$

where

$$A(X) = (C \circ B^{-1})(X)$$

and both B and C are of degree 4. In this section we show that A_{AES} *cannot* be split into

$$A_{\text{AES}}(X) = (\hat{C} \circ \hat{B}^{-1})(X),$$

with both $\hat{B}$ and $\hat{C}$ of low degree. To see this, first note that above decomposition implies

$$\hat{B}(X) = A_{\text{AES}}^{-1}(\hat{C}(X)),$$

where

$$A_{\text{AES}}^{-1}(X) = \texttt{0x6E} \cdot X^{128} + \texttt{0xDB} \cdot X^{64} + \texttt{0x59} \cdot X^{32} + \texttt{0x78} \cdot X^{16} + \texttt{0x5A} \cdot X^{8} + \texttt{0x7F} \cdot X^{4} + \texttt{0xFE} \cdot X^{2} + \texttt{0x5} \cdot X + \texttt{0x5}$$

is the compositional inverse polynomial of A_{AES} satisfying the relation

$$A_{\text{AES}}^{-1}(A_{\text{AES}}(x)) = x,$$

for every $x \in \mathbb{F}_{2^8}$. Hence, to show that at least one of $\hat{B}, \hat{C}$ is of degree > 4, it suffices to compute $A_{\text{AES}}^{-1}(\hat{C})$ assuming a degree 4 for $\hat{C}$, and to show that then the corresponding $\hat{B}$ has degree > 4.

Remark. First of all, note that since A_{AES} has degree 128, it is always possible to find polynomials $\hat{C}$ and $\hat{B}$ of degree 8 such that the equality $A_{\text{AES}}(X) = \hat{C}(\hat{B}^{-1}(X))$ is satisfied. Indeed, if both $\hat{C}$ and $\hat{B}$ have degree 8, then each one of them have all monomials of degrees $1, 2, 4$ and 8. The equality $A_{\text{AES}}(X) = \hat{C}(\hat{B}^{-1}(X))$ is then satisfied if 8 equations (one for each monomial of A_{AES}) in 8 variables (both $\hat{C}$ and $\hat{B}$ have 4 monomials each) are satisfied. Hence, a random polynomial A_{AES} satisfies the equality $A_{\text{AES}}(x) = \hat{C}(\hat{B}^{-1}(x))$ with negligible probability if both $\hat{C}$ and $\hat{B}$ have degree at most 4.

Property of A_{AES}. Let us assume a degree-4 polynomial

$$\hat{C}(X) = \hat{c}_4 X^4 + \hat{c}_2 X^2 + \hat{c}_1 X + \hat{c}_0.$$

We can now write down $A_{\text{AES}}^{-1}(\hat{C}(X))$, which results in $\hat{B}(X)$. However, we want $\hat{B}$ to be of degree at most 4, so we set all coefficients for the degrees $8, 16, 32, 64, 128$ to 0. This results in a system of five equations in the three variables $\hat{c}_1, \hat{c}_2, \hat{c}_4$, given in Appendix C. We tried to solve this system and confirmed that no solutions exist. Thus, the affine part of the AES S-box cannot be split into $\hat{C}(\hat{B}^{-1}(X))$ such that both $\hat{B}$ and $\hat{C}$ are of degree at most 4, whereas in Jarvis this is possible.

As a result, from this point of view, the main difference between AES and Jarvis/Friday is that the linear polynomial used to construct the AES S-box does not have the splitting property used in our attacks, while the same is not true for the case of Jarvis/Friday. In this latter case, even if $B(C^{-1})$ has high degree, it depends only on 9 variables instead of $n+1$ as expected by a linearised polynomial of degree 2^n (where $n \geq 128$). Thus, a natural question to ask is what happens if we replace B and C with other polynomials of higher degree.

8 Conclusion and Future Work

We have demonstrated that Jarvis and Friday are insecure against Gröbner basis attacks, mainly due to the algebraic properties of concatenating the finite field inversion with a function that is defined by composing two low-degree affine polynomials. In our attacks we modelled both designs as a system of polynomial equations in several variables. Additionally, we bridged equations over two rounds, with the effect of significantly reducing the number of variables needed to describe the designs.

Following our analysis, the area sees a dynamic development. Authors of Jarvis and Friday have abandoned their design. Their new construction [Aly+19] is substantially different, although it still uses basic components which we were able to exploit in our analysis. Whether our particular method of bridging internal state equations can be applied to the new hash functions is subject to future work. A broader effort is currently underway to identify designs practically useful for a range of modern proof systems. A noteable competition compares three new designs (Marvelous [Aly+19], Poseidon/Starkad [Gra+19], and GMiMC [Alb+19]) with the more established MiMC.

Acknowledgements. We thank Tomer Ashur for fruitful discussions about Jarvis, Friday, and a preliminary version of our analysis. The research described in this paper was supported by the Royal Society International Exchanges grant "Domain Specific Ciphers" (IES\R2\170211) and the "Lightest" project, which is partially funded by the European Commission as an Innovation Act as part of the Horizon 2020 program under grant agreement number 700321.

References

[AC09] Albrecht, M., Cid, C.: Algebraic techniques in differential cryptanalysis. In: Dunkelman, O. (ed.) FSE 2009. LNCS, vol. 5665, pp. 193–208. Springer, Heidelberg (2009). https://doi.org/10.1007/978-3-642-03317-9_12

[AD18] Ashur, T., Dhooghe, S.: MARVELlous: A STARKFriendly Family of Cryptographic Primitives. Cryptology ePrint Archive, Report 2018/1098. https://eprint.iacr.org/2018/1098 (2018)

[AG11] Arora, S., Ge, R.: New algorithms for learning in presence of errors. In: Aceto, L., Henzinger, M., Sgall, J. (eds.) ICALP 2011, Part I. LNCS, vol. 6755, pp. 403–415. Springer, Heidelberg (2011). https://doi.org/10.1007/978-3-642-22006-7_34

[Alb+14] Albrecht, M.R., Cid, C., Faugère, J.-C., Perret, L.: Algebraic Algorithms for LWE. Cryptology ePrint Archive, Report 2014/1018. http://eprint.iacr.org/2014/1018 (2014)

[Alb+15] Albrecht, M.R., Rechberger, C., Schneider, T., Tiessen, T., Zohner, M.: Ciphers for MPC and FHE. In: Oswald, E., Fischlin, M. (eds.) EUROCRYPT 2015, Part I. LNCS, vol. 9056, pp. 430–454. Springer, Heidelberg (2015). https://doi.org/10.1007/978-3-662-46800-5_17

[Alb+16] Albrecht, M., Grassi, L., Rechberger, C., Roy, A., Tiessen, T.: MiMC: efficient encryption and cryptographic hashing with minimal multiplicative complexity. In: Cheon, J.H., Takagi, T. (eds.) ASIACRYPT 2016, Part I. LNCS, vol. 10031, pp. 191–219. Springer, Heidelberg (2016). https://doi.org/10.1007/978-3-662-53887-6_7

[Alb+19] Albrecht, M.R., Grassi, L., Perrin, L., Ramacher, S., Rechberger, C., Rotaru, D. et al.: Feistel Structures for MPC, and More. Cryptology ePrint Archive, Report 2019/397, to appear in ESORICS 2019. https://eprint.iacr.org/2019/397 (2019)

[Aly+19] Aly, A., Ashur, T., Ben-Sasson, E., Dhooghe, S., Szepieniec, A.: Design of Symmetric-Key Primitives for Advanced Cryptographic Protocols. Cryptology ePrint Archive, Report 2019/426. https://eprint.iacr.org/2019/426 (2019)

[Ash19] Ashur, T.: Private Communication, March 2019

[Bar+05] Bardet, M., Faugere, J.C., Salvy, B., Yang, B.Y.: Asymptotic behaviour of the index of regularity of quadratic semi-regular polynomial systems. In: The Effective Methods in Algebraic Geometry Conference (MEGA), pp. 1–14 (2005)

[BCP97] Bosma, W., Cannon, J., Playoust, C.: The MAGMA algebra system I: the user language. J. Symbolic Comput. **24**, 235–265 (1997)

[Ben+14] Ben-Sasson, E., Chiesa, A., Garman, C., Green, M., Miers, I., Tromer, E., et al.: Zerocash: Decentralized Anonymous Payments from Bitcoin. Cryptology ePrint Archive, Report 2014/349 (2014). http://eprint.iacr.org/2014/349

[Ben+18] Ben-Sasson, E., Bentov, I., Horesh, Y., Riabzev, M.: Scalable, transparent, and post-quantum secure computational integrity. Cryptology ePrint Archive, Report 2018/046. https://eprint.iacr.org/2018/046 (2018)

[BFP12] Bettale, L., Faugère, J.-C., Perret, L.: Solving polynomial systems over finite fields: improved analysis of the hybrid approach. In: International Symposium on Symbolic and Algebraic Computation, ISSAC 2012, pp. 67–74. ACM (2012)

[BPW06] Buchmann, J., Pyshkin, A., Weinmann, R.-P.: A zero-dimensional Gröbner basis for AES-128. In: Robshaw, M. (ed.) FSE 2006. LNCS, vol. 4047, pp. 78–88. Springer, Heidelberg (2006). https://doi.org/10.1007/11799313_6

[BS18] Ben-Sasson, E.: State of the STARK, November 2018. https://drive.google.com/file/d/1Osa0MXu-04dfwn1YOSgN6CXOgWnsp-Tu/view

[Buc65] Buchberger, B.: Ein Algorithmus zum Auffinden der Basiselemente des Restklassenringes nach einem nulldimensionalen Polynomideal. Ph.D. thesis, University of Innsbruck (1965)

[Bun+18] Bünz, B., Bootle, J., Boneh, D., Poelstra, A., Wuille, P., Maxwell, G.: Bulletproofs: short proofs for confidential transactions and more. In: 2018 IEEE Symposium on Security and Privacy, pp. 315–334. IEEE Computer Society Press, May 2018. https://doi.org/10.1109/SP.2018.00020

[CB07] Courtois, N.T., Bard, G.V.: Algebraic cryptanalysis of the data encryption standard. In: Galbraith, S.D. (ed.) Cryptography and Coding 2007. LNCS, vol. 4887, pp. 152–169. Springer, Heidelberg (2007). https://doi.org/10.1007/978-3-540-77272-9_10

[CLO97] Cox, D.A., Little, J., O'Shea, D.: Ideals, Varieties, and Algorithms - An Introduction to Computational Algebraic Geometry and Commutative Algebra. Undergraduate Texts in Mathematics, 2nd edn. Springer, Heidelberg (1997). https://doi.org/10.1007/978-3-319-16721-3

[Cou03a] Courtois, N.T.: Fast algebraic attacks on stream ciphers with linear feedback. In: Boneh, D. (ed.) CRYPTO 2003. LNCS, vol. 2729, pp. 176–194. Springer, Heidelberg (2003). https://doi.org/10.1007/978-3-540-45146-4_11

[Cou03b] Courtois, N.T.: Higher order correlation attacks, XL algorithm and cryptanalysis of toyocrypt. In: Lee, P.J., Lim, C.H. (eds.) ICISC 2002. LNCS, vol. 2587, pp. 182–199. Springer, Heidelberg (2003). https://doi.org/10.1007/3-540-36552-4_13

[Fau+10] Faugère, J.-C., Gauthier, V., Otmani, A., Perret, L., Tillich, J.-P.: A Distinguisher for High Rate McEliece Cryptosystems. Cryptology ePrint Archive, Report 2010/331. http://eprint.iacr.org/2010/331 (2010)

[Fau+15] Faugère, J.-C., Gligoroski, D., Perret, L., Samardjiska, S., Thomae, E.: A polynomial-time key-recovery attack on MQQ cryptosystems. In: Katz, J. (ed.) PKC 2015. LNCS, vol. 9020, pp. 150–174. Springer, Heidelberg (2015). https://doi.org/10.1007/978-3-662-46447-2_7

[Fau+93] Faugère, J.-C., Gianni, P.M., Lazard, D., Mora, T.: Efficient computation of zero-dimensional Gröbner bases by change of ordering. J. Symb. Comput. **16**(4), 329–344 (1993)

[Fau02] Faugère, J.-C.: A new efficient algorithm for computing Gröbner bases without reduction to zero (F5). In: Mora, T. (ed.) Proceedings of the 2002 International Symposium on Symbolic and Algebraic Computation ISSAC, pp. 75-83. ACM Press, July 2002. ISBN 1-58113-484-3

[Fau99] Faugere, J.-C.: A new efficient algorithm for computing Gröbner bases (F4). J. Pure Appl. Algebra **139**(1–3), 61–88 (1999)

[FM11] Faugère, J.-C., Mou, C.: Fast algorithm for change of ordering of zero-dimensional Gröbner bases with sparse multiplication matrices. In: Schost, É., Emiris, I.Z. (eds.) Symbolic and Algebraic Computation, International Symposium, ISSAC 2011, pp. 115–122. ACM (2011). https://doi.org/10.1145/1993886.1993908

[FPP14] Faugère, J.-C., Perret, L., de Portzamparc, F.: Algebraic attack against variants of mceliece with goppa polynomial of a special form. In: Sarkar, P., Iwata, T. (eds.) ASIACRYPT 2014, Part I. LNCS, vol. 8873, pp. 21–41. Springer, Heidelberg (2014). https://doi.org/10.1007/978-3-662-45611-8_2

[Fro85] Fröberg, R.: An inequality for Hilbert series of graded algebras. Mathematica Scandinavica **56**, 117–144 (1985)

[Gen07] Genovese, G.: Improving the algorithms of Berlekamp and Niederreiter for factoring polynomials over finite fields. J. Symb. Comput. **42**(1–2), 159–177 (2007)

[Gra+19] Grassi, L., Kales, D., Khovratovich, D., Roy, A., Rechberger, C., Schofnegger, M.: Starkad and Poseidon: New Hash Functions for Zero Knowledge Proof Systems. Cryptology ePrint Archive, Report 2019/458. https://eprint.iacr.org/2019/458 (2019)

[Hop+19] Hopwood, D., Bowe, S., Hornby, T., Wilcox, N.: Zcash protocol specification: version 2019.0-beta-37 [Overwinter+Sapling]. Technical report, Zerocoin Electric Coin Company (2019). https://github.com/zcash/zips/blob/master/protocol/protocol.pdf

[Hor72] Horowitz, E.: A fast method for interpolation using preconditioning. Inf. Process. Lett. (IPL) **1**(4), 157–163 (1972)

[JK97] Jakobsen, T., Knudsen, L.R.: The interpolation attack on block ciphers. In: Biham, E. (ed.) FSE 1997. LNCS, vol. 1267, pp. 28–40. Springer, Heidelberg (1997). https://doi.org/10.1007/BFb0052332

[KBN09] Khovratovich, D., Biryukov, A., Nikolic, I.: Speeding up collision search for byte-oriented hash functions. In: Fischlin, M. (ed.) CT-RSA 2009. LNCS, vol. 5473, pp. 164–181. Springer, Heidelberg (2009). https://doi.org/10.1007/978-3-642-00862-7_11

[Knu95] Knudsen, L.R.: Truncated and higher order differentials. In: Preneel, B. (ed.) FSE 1994. LNCS, vol. 1008, pp. 196–211. Springer, Heidelberg (1995). https://doi.org/10.1007/3-540-60590-8_16
[KR00] Kreuzer, M., Robbiano, L.: Computational Commutative Algebra, 1st edn. Springer, New York (2000)
[Kun73] Kung, H.-T.: Fast Evaluation and Interpolation. Technical report, Department of Computer Science, Carnegie-Mellon University, January 1973
[LN96] Lidl, R., Niederreiter, H.: Finite Fields. Encyclopedia of Mathematics and its Applications, 2nd edn. Cambridge University Press (1996)
[MR02] Murphy, S., Robshaw, M.J.B.: Essential algebraic structure within the AES. In: Yung, M. (ed.) CRYPTO 2002. LNCS, vol. 2442, pp. 1–16. Springer, Heidelberg (2002). https://doi.org/10.1007/3-540-45708-9_1
[Par+13] Parno, B., Howell, J., Gentry, C., Raykova, M.: Pinocchio: nearly practical verifiable computation. In: 2013 IEEE Symposium on Security and Privacy, pp. 238–252. IEEE Computer Society Press, May 2013. https://doi.org/10.1109/SP.2013.47
[Ste+19] Stein, W., et al.: Sage Mathematics Software Version 8.6. The Sage Development Team (2019). http://www.sagemath.org
[Wan+11] Wang, M., Sun, Y., Mouha, N., Preneel, B.: Algebraic techniques in differential cryptanalysis revisited. In: Parampalli, U., Hawkes, P. (eds.) ACISP 2011. LNCS, vol. 6812, pp. 120–141. Springer, Heidelberg (2011). https://doi.org/10.1007/978-3-642-22497-3_9

A Polynomials of Section 4.2

In Sect. 4.2, we search for monic affine polynomials D, E such that the equality

$$D(B) = E(C)$$

is satisfied, where B, C are monic affine polynomials of degree 4. In particular, given

$$B(X) = X^4 + b_2X^2 + b_1X + b_0 \quad \text{and} \quad C(X) = X^4 + c_2X^2 + c_1X + c_0$$

the goal is to find

$$D(X) = X^4 + d_2X^2 + d_1X^1 + d_0 \quad \text{and} \quad E(X) = X^4 + e_2X^2 + e_1X + e_0$$

such that $D(B) = E(C)$.

By comparing the corresponding coefficients of $D(B)$ and $E(C)$, we obtain a system of 5 linear equations in the 6 variables $d_0, d_1, d_2, e_0, e_1, e_2$:

$$\begin{aligned} d_2 + e_2 &= b_2^4 + c_2^4, \\ d_1 + b_2^2 \cdot d_2 + e_1 + c_2^2 \cdot e_2 &= b_1^4 + c_1^4, \\ b_2 \cdot d_1 + b_1^2 \cdot d_2 + c_2 \cdot e_1 + c_1^2 \cdot e_2 &= 0, \\ b_1 \cdot d_1 + c_1 \cdot e_1 &= 0, \\ d_0 + b_0 \cdot d_1 + b_0^2 \cdot d_2 + e_0 + c_0 \cdot e_1 + c_0^2 \cdot e_2 &= b_0^4 + c_0^4. \end{aligned}$$

This system can be solved to recover D and E.

B Constants α_i, β_i, γ_i, and δ_i for the Round Keys

Each round key $k_{i+1} = \frac{1}{k_i} + c_i$ in Jarvis can be written as

$$k_{i+1} = \frac{\alpha_i \cdot k_0 + \beta_i}{\gamma_i \cdot k_0 + \delta_i},$$

where α_i, β_i, γ_i, and δ_i are constants. By simple computation, note that:

- $i = 0$:
$$k_1 = \frac{1}{k_0} + c_0 = \frac{c_0 k_0 + 1}{k_0},$$
and $\alpha_0 = c_0, \beta_0 = 1, \gamma_0 = 1, \delta_0 = 0$;
- $i = 1$:
$$k_2 = \frac{1}{k_1} + c_1 = \frac{(c_0 c_1 + 1)k_0 + c_1}{c_0 k_0 + 1},$$
and $\alpha_1 = 1 + c_0 c_1, \beta_1 = c_1, \gamma_1 = c_0, \delta_1 = 1$;
- $i = 2$:
$$k_3 = \frac{1}{k_2} + c_2 = \frac{(c_0 c_1 c_2 + c_0 + c_2)k_0 + c_1 c_2 + 1}{(c_0 c_1 + 1)k_0 + c_1},$$
and $\alpha_2 = c_0 c_1 c_2 + c_0 + c_2, \beta_2 = c_1 c_2 + 1, \gamma_2 = c_0 c_1 + 1, \delta_2 = c_1$;

and so on. Thus, we can derive recursive formulas to calculate the remaining values for generic $i \geq 0$:

$$\begin{aligned}
\alpha_{i+1} &= \alpha_i \cdot c_{i+1} + \gamma_i,\\
\beta_{i+1} &= \beta_i \cdot c_{i+1} + \delta_i,\\
\gamma_{i+1} &= \alpha_i,\\
\delta_{i+1} &= \beta_i.
\end{aligned}$$

C System of Equations from Section 7

The system of equations is constructed by symbolically computing $A_{\text{AES}}^{-1}(\hat{C}(x))$, as described in Sect. 7, and setting all coefficients for degrees $8, 16, 32, 64, 128$ to 0. These are five possible degrees and the following equations are the sum of all coefficients belonging to each of these degrees:

$$\begin{aligned}
\texttt{0x5a} \cdot \hat{c}_1^8 + \texttt{0x7f} \cdot \hat{c}_2^4 + \texttt{0xfe} \cdot \hat{c}_4^2 &= 0,\\
\texttt{0x78} \cdot \hat{c}_1^{16} + \texttt{0x5a} \cdot \hat{c}_2^8 + \texttt{0x7f} \cdot \hat{c}_4^4 &= 0,\\
\texttt{0x59} \cdot \hat{c}_1^{32} + \texttt{0x78} \cdot \hat{c}_2^{16} + \texttt{0x5a} \cdot \hat{c}_4^8 &= 0,\\
\texttt{0xdb} \cdot \hat{c}_1^{64} + \texttt{0x59} \cdot \hat{c}_2^{32} + \texttt{0x78} \cdot \hat{c}_4^{16} &= 0,\\
\texttt{0x6e} \cdot \hat{c}_1^{128} + \texttt{0xdb} \cdot \hat{c}_2^{64} + \texttt{0x59} \cdot \hat{c}_4^{32} &= 0.
\end{aligned}$$

By practical tests we found that no (nontrivial) coefficients $\hat{c}_1, \hat{c}_2, \hat{c}_4$ satisfy all previous equalities, which means that there are no polynomials $\hat{B}$ and $\hat{C}$ both of degree 4 that satisfy $A_{\text{AES}}(X) = (\hat{C} \circ \hat{B}^{-1})(X)$.

MILP-aided Method of Searching Division Property Using Three Subsets and Applications

Senpeng Wang(✉), Bin Hu, Jie Guan, Kai Zhang, and Tairong Shi

PLA SSF Information Engineering University, Zhengzhou, China
wsp2110@126.com

Abstract. Division property is a generalized integral property proposed by Todo at EUROCRYPT 2015, and then conventional bit-based division property (CBDP) and bit-based division property using three subsets (BDPT) were proposed by Todo and Morii at FSE 2016. At the very beginning, the two kinds of bit-based division properties once couldn't be applied to ciphers with large block size just because of the huge time and memory complexity. At ASIACRYPT 2016, Xiang *et al.* extended Mixed Integer Linear Programming (MILP) method to search integral distinguishers based on CBDP. BDPT can find more accurate integral distinguishers than CBDP, but it couldn't be modeled efficiently.

This paper focuses on the feasibility of searching integral distinguishers based on BDPT. We propose the pruning techniques and fast propagation of BDPT for the first time. Based on these, an MILP-aided method for the propagation of BDPT is proposed. Then, we apply this method to some block ciphers. For SIMON64, PRESENT, and RECTANGLE, we find more balanced bits than the previous longest distinguishers. For LBlock, we find a better 16-round integral distinguisher with less active bits. For other block ciphers, our results are in accordance with the previous longest distinguishers.

Cube attack is an important cryptanalytic technique against symmetric cryptosystems, especially for stream ciphers. And the most important step in cube attack is superpoly recovery. Inspired by the CBDP based cube attack proposed by Todo at CRYPTO 2017, we propose a method which uses BDPT to recover the superpoly in cube attack. We apply this new method to round-reduced Trivium. To be specific, the time complexity of recovering the superpoly of 832-round Trivium at CRYPTO 2017 is reduced from 2^{77} to practical, and the time complexity of recovering the superpoly of 839-round Trivium at CRYPTO 2018 is reduced from 2^{79} to practical. Then, we propose a theoretical attack which can recover the superpoly of Trivium up to 841 round.

Keywords: Integral distinguisher · Division property · MILP · Block cipher · Cube attack · Stream cipher

S. D. Galbraith and S. Moriai (Eds.): ASIACRYPT 2019, LNCS 11923, pp. 398–427, 2019.
https://doi.org/10.1007/978-3-030-34618-8_14

1 Introduction

Division property, a generalization of integral property [11], was proposed by Todo at EUROCRYPT 2015 [22]. It can exploit the algebraic structure of block ciphers to construct integral distinguishers even if the block ciphers have non-bijective, bit-oriented, or low-degree structures. Then, at CRYPTO 2015 [20], Todo applied this new technique to MISTY1 and achieved the first theoretical cryptanalysis of the full-round MISTY1. Sun *et al.* [18], revisited division property, and they studied the property of a set (multiset) satisfying certain division property. At CRYPTO 2016 [4], Boura and Canteaut introduced a new notion called *parity set* to exploit division property. They formulated and characterized the division property of S-box and found better integral distinguisher of PRESENT [3]. But it required large time and memory complexity. To solve this problem, Xie and Tian [28] proposed another concept called *term set*, based on which they found a 9-round distinguisher of PRESENT with 22 balanced bits.

In order to exploit the concrete structure of round function, Todo and Morii [21] proposed bit-based division property at FSE 2016. There are two kinds of bit-based division property: conventional bit-based division property (CBDP) and bit-based division property using three subsets (BDPT). CBDP focuses on that the parity $\bigoplus_{\boldsymbol{x}\in\mathbb{X}} \boldsymbol{x}^{\boldsymbol{u}}$ is 0 or unknown, while BDPT focuses on that the parity $\bigoplus_{\boldsymbol{x}\in\mathbb{X}} \boldsymbol{x}^{\boldsymbol{u}}$ is 0, 1, or unknown. Therefore, BDPT can find more accurate integral characteristics than CBDP. For example, CBDP proved the existence of the 14-round integral distinguisher of SIMON32 while BDPT found the 15-round integral distinguisher of SIMON32 [21].

Although CBDP and BDPT could find accurate integral distinguishers, the huge complexity once restricted their wide applications. At ASIACRYPT 2016, Xiang *et al.* [27] applied MILP method to search integral distinguishers based on CBDP, which allowed them to analyze block ciphers with large sizes. But there was still no MILP method to model the propagation of BDPT.

Cube attack, proposed by Dinur and Shamir [6] at EUROCRYPT 2009, is one of the general cryptanalytic techniques against symmetric cryptosystems. For a cipher with n secret variables $\boldsymbol{x} = (x_0, x_1, \ldots, x_{n-1})$ and m public variables $\boldsymbol{v} = (v_0, v_1, \ldots, v_{m-1})$, the output bit can be denoted as a polynomial $f(\boldsymbol{x}, \boldsymbol{v})$. The core idea of cube attack is to simplify $f(\boldsymbol{x}, \boldsymbol{v})$ by summing the output of cryptosystem over a subset of public variables, called *cube*. And the target of cube attack is to recover secret variables from the simplified polynomial called *superpoly*. In the original paper of cube attack [6], the authors regarded stream cipher as a blackbox polynomial and introduced a linearity test to recover superpoly. Recently, many variants of cube attacks were put forward such as dynamic cube attacks [7], conditional cube attacks [14], correlation cube attacks [15], CBDP based cube attacks [23,26], and deterministic cube attacks [30].

At EUROCRYPT 2018 [15], Liu *et al.* proposed *correlation cube attack*, which could mount to 835-round Trivium using small dimensional cubes. Then, in [30], Ye *et al.* proposed a new variant of cube attack, named *deterministic cube attacks*. Their attacks were developed based on degree evaluation method proposed by Liu *et al.* at CRYPTO 2017 [16]. They proposed a special type of cube that the numeric degree of every term was always less than or equal to the cube size, called *useful cube*. With a 37-dimensional useful cube, they recovered the corresponding exact superpoly for up to 838-round Trivium. However, as the authors wrote in their paper, it seemed hard to increase the number of attacking round when the cube size increased. Namely, their methods didn't work well for large cube size. Moreover, at CRYPTO 2018 [9], Fu *et al.* proposed a key recovery attack on 855-round Trivium which somewhat resembled dynamic cube attacks. For the attack in [9], the paper [12] pointed out that there was possibility that the correct key guesses and the wrong ones shared the same zero-sum property. It means that the key recovery attack may degenerate to distinguish attack.

It is noticeable that, at CRYPTO 2017 [23], Todo *et al.* treated the polynomial as non-blackbox and applied CBDP to the cube attack on stream ciphers. Due to the MILP-aided CBDP, they could evaluate the algebraic normal form (ANF) of the superpoly with large cube size. By using a 72-dimensional cube, they proposed a theoretical cube attack on 832-round Trivium. Then, at CRYPTO 2018 [26], Wang *et al.* improve the CBDP based cube attack and gave a key recovery attack on 839-round Trivium. For CBDP based cube attacks, the superpolies of large cubes can be recovered by theoretical method. But the theory of CBDP cannot ensure that the superpoly of a cube is non-constant. Hence the key recovery attack may be just a distinguish attack. BDPT can exploit the integral distinguisher whose sum is 1, which means BDPT may show a determined key recovery attack. However, compared with the propagation of CBDP, the propagation of BDPT is more complicated and cannot be modeled by MILP method directly. An automatically searching for a variant three-subset division property with STP solver was proposed in [13], but the variant is weaker than the original BDPT. How to trace the propagation of BDPT is an open problem.

1.1 Our Contributions

In this paper, we propose an MILP-aided method for BDPT. Then, we apply it to search integral distinguishers of block ciphers and recover superpolies of stream ciphers.

1.1.1 MILP-aided Method for BDPT

Pruning Properties of BDPT. When we evaluate the propagation of BDPT, there may be some vectors that have no impact on the BDPT of output bit. So we show the pruning properties when the vectors of BDPT can be removed.

Fast Propagation and Stopping Rules. Inspired by the "lazy propagation" in [21], we propose the notion of "fast propagation" which can translate BDPT into CBDP and show some bits are balanced. Then, based on "lazy propagation" and "fast propagation", we obtain three stopping rules. Finally, an MILP-aided method for the propagation of BDPT is proposed.

1.1.2 Searching Integral Distinguishers of Block Ciphers

We apply our MILP-aided method to search integral distinguishers of some block ciphers. The main results are shown in Table 1.

ARX Ciphers. For SIMON32, we find the 15-round integral distinguisher that cannot be found by CBDP. For 18-round SIMON64, we find 23 balanced bits which has one more bit than the previous longest integral distinguisher.

SPN Ciphers. For PRESENT, when the input data is 2^{60}, our method can find 3 more balanced bits than the previous longest integral distinguisher. Moreover, when the input data is 2^{63}, the integral distinguisher we got has 6 more balanced bits than that got by term set in the paper [28]. For RECTANGLE, when the input data is 2^{60}, our method can also obtain 11 more balanced bits than the previous longest 9-round integral distinguisher.

Generalized Feistel Cipher. For LBlock, we obtain a 17-round integral distinguisher which is the same with the previous longest integral distinguisher. Moreover, a better 16-round integral distinguisher with less active bits can also be obtained.

1.1.3 Recovering Superpoly of Stream Cipher

Using BDPT to Recover the ANF Coefficient of Superpoly. Inspired by the CBDP based cube attack in [23,26], our new method is based on the propagation of BDPT which can find integral distinguisher whose sum is 0 or 1. But it's nontrivial to recover the superpoly by integral distinguishers based on BDPT. Therefore, we proposed the notion of *similar polynomial.* We can recover the ANF coefficient of superpoly by researching the BDPT propagation of corresponding similar polynomial. In order to analyze the security of ciphers better, we divide ciphers into two categories: public-update ciphers and secret-update ciphers. For public-update ciphers, we proved that the exact ANF of superpoly can be fully recovered by BDPT.

Application to Trivium. In order to verify the correctness and effectiveness of our method, we apply BDPT to recover the superpoly of round-reduced Trivium which is a public cipher. To be specific, the time complexity of recovering the superpoly of 832-round Trivium at CRYPTO 2017 is reduced from 2^{77} to practical, and the time complexity of recovering the superpoly of 839-round Trivium at CRYPTO 2018 is reduced from 2^{79} to practical. Then, we propose a theoretical attack which can recover the superpoly of Trivium up to 841 round. The detailed information is shown in Table 2. And the time complexity in the table means the time complexity of recovering superpoly. And c is the average computational complexity of tracing the propagation of BDPT using MILP-aided method.

Table 1. Summarization of integral distinguishers

Cipher	Data	Round	Number of balanced bits	Time	Reference
SIMON32	2^{31}	15	3		[21]
		15	3	2m	Sect. 5.1
SIMON64	2^{63}	18	22	6.7m	[27]
		18	**23**	1h41m	Sect. 5.1
PRESENT	2^{60}	9	1	3.4m	[27]
		9	**4**	56m	Sect. 5.2
	2^{63}	9	22		[28]
		9	**28**	10m	Sect. 5.2
RECTANGLE	2^{60}	9	16	4.1m	[27]
		9	**27**	10m	Sect. 5.2
LBlock	2^{63}	16	32	4.9m	[27]
		17	4		[8]
		17	4	10h25m	Sect. 5.3
	$\mathbf{2^{62}}$	16	18	6h49m	Sect. 5.3

Table 2. Superpoly recovery of Trivium

Rounds	Cube size	Exact superpoly	Complexity	Reference
832	72	yes	2^{77}	[23]
			$2^{76.7}$	[26]
			practical	Sect. 7.3
835	36/37	no		[15]
838	37	yes	practical	[30]
839	78	yes	2^{79}	[26]
			practical	Sect. 7.3
841	78	yes	$2^{41} \cdot c$	Sect. 7.4

1.2 Outline of the Paper

This paper is organized as follows: Sect. 2 provides the background of MILP, division property, and cube attacks etc. In Sect. 3, some new propagation properties of BDPT are given. In Sect. 4, we propose an MILP-aided method for BDPT. Section 5 shows applications to block ciphers. In Sect. 6, we use BDPT to recover the superpoly in cube attack. Section 7 shows the application to Trivium. Section 8 concludes the paper. Some auxiliary materials are supplied in Appendix.

2 Preliminaries

2.1 Notations

Let $\mathbb{F}_2$ denote the finite field $\{0,1\}$ and $\boldsymbol{a} = (a_0, a_1, \dots, a_{n-1}) \in \mathbb{F}_2^n$ be an n-bit vector, where a_i denotes the i-th bit of $\boldsymbol{a}$. For n-bit vectors $\boldsymbol{x}$ and $\boldsymbol{u}$, define $\boldsymbol{x}^{\boldsymbol{u}} = \prod_{i=0}^{n-1} x_i^{u_i}$. Then, for any $\boldsymbol{k} \in \mathbb{F}_2^n$ and $\boldsymbol{k}' \in \mathbb{F}_2^n$, define $\boldsymbol{k} \succeq \boldsymbol{k}'$ if $k_i \geq k_i'$ holds for all $i = 0, 1, \ldots, n-1$ and define $\boldsymbol{k} \succ \boldsymbol{k}'$ if $k_i > k_i'$ holds for all $i = 0, 1, \ldots, n-1$. For a subset $I \subset \{0, 1, \ldots, n-1\}$, $\boldsymbol{u}_I$ denotes an n-dimensional bit vector $(u_0, u_1, \ldots, u_{n-1})$ satisfying $u_i = 1$ if $i \in I$ and $u_i = 0$ otherwise. We simply write $\mathbb{K} \leftarrow \boldsymbol{k}$ when $\mathbb{K} := \mathbb{K} \cup \{\boldsymbol{k}\}$ and $\mathbb{K} \rightarrow \boldsymbol{k}$ when $\mathbb{K} := \mathbb{K} \setminus \{\boldsymbol{k}\}$. And $|\mathbb{K}|$ denotes the number of elements in the set $|\mathbb{K}|$.

2.2 Mixed Integer Linear Programming

MILP is a kind of optimization or feasibility program whose objective function and constraints are linear, and the variables are restricted to be integers. Generally, an MILP model $\mathcal{M}$ consists of variables $\mathcal{M}.var$, constrains $\mathcal{M}.con$, and the objective function $\mathcal{M}.obj$. MILP models can be solved by solver like Gurobi [10]. If there is no feasible solution, the solver will returns *infeasible*. When there is no objective function in $\mathcal{M}$, the MILP solver will only return whether $\mathcal{M}$ is feasible or not.

2.3 Bit-Based Division Property

Two kinds of bit-based division property (CBDP and BDPT) were introduced by Todo and Morii at FSE 2016 [21]. In this subsection, we will briefly introduce them and their propagation rules.

Definition 1 (CBDP [21]**).** *Let $\mathbb{X}$ be a multiset whose elements take a value of $\mathbb{F}_2^n$. When the multiset $\mathbb{X}$ has the CBDP $\mathcal{D}_{\mathbb{K}}^{1^n}$, where $\mathbb{K}$ denotes a set of n-dimensional vectors whose i-th element takes a value between 0 and 1, it fulfills the following conditions:*

$$\bigoplus_{\boldsymbol{x} \in \mathbb{X}} \boldsymbol{x}^{\boldsymbol{u}} = \begin{cases} \text{unknown}, & \textit{if there exists } \boldsymbol{k} \in \mathbb{K} \textit{ satisfying } \boldsymbol{u} \succeq \boldsymbol{k}, \\ 0, & \textit{otherwise}. \end{cases}$$

Definition 2 (BDPT [21]**).** *Let $\mathbb{X}$ be a multiset whose elements take a value of $\mathbb{F}_2^n$. Let $\mathbb{K}$ and $\mathbb{L}$ be two sets whose elements take n-dimensional bit vectors. When the multiset $\mathbb{X}$ has the BDPT $\mathcal{D}_{\mathbb{K},\mathbb{L}}^{1^n}$, it fulfills the following conditions:*

$$\bigoplus_{\boldsymbol{x} \in \mathbb{X}} \boldsymbol{x}^{\boldsymbol{u}} = \begin{cases} \text{unknown}, & \textit{if there is } \boldsymbol{k} \in \mathbb{K} \textit{ satisfying } \boldsymbol{u} \succeq \boldsymbol{k}, \\ 1, & \textit{else if there is } \boldsymbol{\ell} \in \mathbb{L} \textit{ satisfying } \boldsymbol{u} = \boldsymbol{\ell}, \\ 0, & \textit{otherwise}. \end{cases}$$

According to [21], if there are $\boldsymbol{k} \in \mathbb{K}$ and $\boldsymbol{k}' \in \mathbb{K}$ satisfying $\boldsymbol{k} \succeq \boldsymbol{k}'$, $\boldsymbol{k}$ can be removed from $\mathbb{K}$ because the vector $\boldsymbol{k}$ is redundant. We denote this progress as ***Reduce0*** $(\mathbb{K})$. If there are $\boldsymbol{\ell} \in \mathbb{L}$ and $\boldsymbol{k} \in \mathbb{K}$ satisfying $\boldsymbol{\ell} \succeq \boldsymbol{k}$, the vector $\boldsymbol{\ell}$ can also be removed from $\mathbb{L}$. We denote this progress as ***Reduce1*** $(\mathbb{K}, \mathbb{L})$. For any $\boldsymbol{u}$, the redundant vectors in $\mathbb{K}$ and $\mathbb{L}$ will not affect the value of $\bigoplus_{\boldsymbol{x} \in \mathbb{X}} \boldsymbol{x}^{\boldsymbol{u}}$.

The propagation rules of $\mathbb{K}$ in CBDP are the same with BDPT. So here we only show the propagation rules of BDPT. For more details, please refer to [21].

BDPT Rule 1 (Copy [21]). *Let* $\boldsymbol{y} = f(\boldsymbol{x})$ *be a copy function, where* $\boldsymbol{x} = (x_0, x_1, \ldots, x_{n-1}) \in \mathbb{F}_2^n$, *and the output is calculated as* $\boldsymbol{y} = (x_0, x_0, x_1, \ldots, x_{n-1})$. *Assuming the input multiset* $\mathbb{X}$ *has* $\mathcal{D}^{1^n}_{\mathbb{K},\mathbb{L}}$, *then the output multiset* $\mathbb{Y}$ *has* $\mathcal{D}^{1^{n+1}}_{\mathbb{K}',\mathbb{L}'}$, *where*

$$\mathbb{K}' \leftarrow \begin{cases} (0,0,k_1,\ldots,k_{n-1}), & \text{if } k_0 = 0 \\ (1,0,k_1,\ldots,k_{n-1}), (0,1,k_1,\ldots,k_{n-1}), & \text{if } k_0 = 1 \end{cases},$$

$$\mathbb{L}' \leftarrow \begin{cases} (0,0,\ell_1,\ldots,\ell_{n-1}), & \text{if } \ell_0 = 0 \\ (1,0,\ell_1,\ldots,\ell_{n-1}), (0,1,\ell_1,\ldots,\ell_{n-1}), (1,1,\ell_1,\ldots,\ell_{n-1}), & \text{if } \ell_0 = 1 \end{cases},$$

are computed from all $\boldsymbol{k} \in \mathbb{K}$ *and all* $\boldsymbol{\ell} \in \mathbb{L}$, *respectively.*

BDPT Rule 2 (And [21]). *Let* $\boldsymbol{y} = f(\boldsymbol{x})$ *be a function compressed by an And, where the input* $\boldsymbol{x} = (x_0, x_1, \ldots, x_{n-1}) \in \mathbb{F}_2^n$, *and the output is calculated as* $\boldsymbol{y} = (x_0 \wedge x_1, x_2, \ldots, x_{n-1}) \in \mathbb{F}_2^{n-1}$. *Assuming the input multiset* $\mathbb{X}$ *has* $\mathcal{D}^{1^n}_{\mathbb{K},\mathbb{L}}$, *then the output multiset* $\mathbb{Y}$ *has* $\mathcal{D}^{1^{n-1}}_{\mathbb{K}',\mathbb{L}'}$, *where* $\mathbb{K}'$ *is computed from all* $\boldsymbol{k} \in \mathbb{K}$ *as*

$$\mathbb{K}' \leftarrow \left(\left\lceil \frac{k_0 + k_1}{2} \right\rceil, k_2, \ldots, k_{n-1} \right),$$

and $\mathbb{L}'$ *is computed from all* $\boldsymbol{\ell} \in \mathbb{L}$ *satisfying* $(\ell_0, \ell_1) = (0,0)$ *or* $(1,1)$ *as*

$$\mathbb{L}' \leftarrow \left(\left\lceil \frac{\ell_0 + \ell_1}{2} \right\rceil, \ell_2, \ldots, \ell_{n-1} \right).$$

BDPT Rule 3 (Xor [21]). *Let* $\boldsymbol{y} = f(\boldsymbol{x})$ *be a function compressed by an Xor, where the input* $\boldsymbol{x} = (x_0, x_1, \ldots, x_{n-1}) \in \mathbb{F}_2^n$, *and the output is calculated as* $\boldsymbol{y} = (x_0 \oplus x_1, x_2, \ldots, x_{n-1}) \in \mathbb{F}_2^{n-1}$. *Assuming the input multiset* $\mathbb{X}$ *has* $\mathcal{D}^{1^n}_{\mathbb{K},\mathbb{L}}$, *then the output multiset* $\mathbb{Y}$ *has* $\mathcal{D}^{1^{n-1}}_{\mathbb{K}',\mathbb{L}'}$, *where* $\mathbb{K}'$ *is computed from all* $\boldsymbol{k} \in \mathbb{K}$ *satisfying* $(k_0, k_1) = (0,0), (1,0)$, *or* $(0,1)$ *as*

$$\mathbb{K}' \leftarrow (k_0 + k_1, k_2, \ldots, k_{n-1}),$$

$\mathbb{L}'$ *is computed from all* $\boldsymbol{\ell} \in \mathbb{L}$ *satisfying* $(\ell_0, \ell_1) = (0,0)$, $(1,0)$, *or* $(0,1)$ *as*

$$\mathbb{L}' \xleftarrow{x} (\ell_0 + \ell_1, \ell_2, \ldots, \ell_{n-1}).$$

And $\mathbb{L} \xleftarrow{x} \boldsymbol{\ell}$ *means*

$$\mathbb{L} := \begin{cases} \mathbb{L} \cup \{\boldsymbol{\ell}\} & \text{if the original } \mathbb{L} \text{ does not include } \boldsymbol{\ell}, \\ \mathbb{L} \setminus \{\boldsymbol{\ell}\} & \text{if the original } \mathbb{L} \text{ includes } \boldsymbol{\ell}. \end{cases}$$

BDPT Rule 4 (Xor with Secret Key [21]). *Let* $\mathbb{X}$ *be the input multiset satisfying* $\mathcal{D}^{1^n}_{\mathbb{K},\mathbb{L}}$. *For the input* $\boldsymbol{x} \in \mathbb{X}$, *the output* $\boldsymbol{y} \in \mathbb{Y}$ *is computed as* $\boldsymbol{y} = (x_0, \ldots, x_{i-1}, x_i \oplus r_k, x_{i+1}, \ldots, x_{n-1})$, *where* r_k *is the secret key. Then, the output multiset* $\mathbb{Y}$ *has* $\mathcal{D}^{1^n}_{\mathbb{K}',\mathbb{L}'}$, *where* $\mathbb{K}'$ *and* $\mathbb{L}'$ *are computed as*

$$\begin{aligned} &\mathbb{L}' \leftarrow \boldsymbol{\ell}, \text{ for } \boldsymbol{\ell} \in \mathbb{L}, \\ &\mathbb{K}' \leftarrow \boldsymbol{k}, \text{ for } \boldsymbol{k} \in \mathbb{K}, \\ &\mathbb{K}' \leftarrow (\ell_0, \ell_1, \ldots, \ell_i \vee 1, \ldots, \ell_{n-1}), \text{ for } \boldsymbol{\ell} \in \mathbb{L} \text{ satisfying } \ell_i = 0. \end{aligned}$$

CBDP Rule 5 (S-box [4,27]**).** *Let* $\boldsymbol{y} = f(\boldsymbol{x})$ *be a function of S-box, where the input* $\boldsymbol{x} = (x_0, x_1, \ldots, x_{n-1}) \in \mathbb{F}_2^n$*, and the output* $\boldsymbol{y} = (y_0, y_1, \ldots, y_{m-1}) \in \mathbb{F}_2^m$*. Then, every* $y_i, i \in \{0, 1, \ldots, m-1\}$ *can be expressed as a Boolean function of* $(x_0, \ldots, x_{n-1})$*. For the input CBDP* $\mathbb{K}$*, the output CBDP* $\mathbb{K}'$ *is a set of vectors as follows:*

$$\mathbb{K}' = \{\boldsymbol{u}' \in \mathbb{F}_2^m |\ \textit{for any } \boldsymbol{u} \in \mathbb{K},\ \textit{if } \boldsymbol{y}^{\boldsymbol{u}'} \textit{ contains any term } \boldsymbol{x}^{\boldsymbol{v}} \textit{ satisfying } \boldsymbol{v} \succeq \boldsymbol{u}\}.$$

When there was no effective way to model the propagation of BDPT, Todo and Morii [21] proposed the notion of 'lazy propagation" to give the provable security of SIMON family against BDPT.

Definition 3 (Lazy Propagation [21]**).** *Let* $D^{1^n}_{\mathbb{K}_i, \mathbb{L}_i}$ *be the input BDPT of the i-th round function and* $D^{1^n}_{\overline{\mathbb{K}}_{i+1}, \overline{\mathbb{L}}_{i+1}}$ *be the BDPT from the lazy propagation. Then,* $\overline{\mathbb{K}}_{i+1}$ *is computed from only a part of vectors in* $\mathbb{K}_i$*, and* $\overline{\mathbb{L}}_{i+1}$ *always becomes the empty set* $\emptyset$*. Therefore, if the lazy propagation creates* $\mathcal{D}^{1^n}_{\overline{\mathbb{K}}_r, \emptyset}$*, where* $\overline{\mathbb{K}}_r$ *has n distinct vectors whose Hamming weight is one, the accurate propagation also creates the same n distinct vectors in the same round.*

2.4 The MILP Representation of CBDP

For an r-round iterative cipher of size n, attackers determine indices set $I = \{i_0, i_1, \ldots, i_{|I|-1},\} \subset \{0, 1, \ldots, n-1\}$ and prepare $2^{|I|}$ chosen plaintexts where variables indexed by I are taking all possible combinations of values and the other variables are set to constants. The CBDP of such chosen plaintexts is $\mathcal{D}^{1^n}_{\mathbb{K}_0 = \{\boldsymbol{k}_I\}}$. Based on the propagation rules, the propagation of CBDP from $\mathcal{D}^{1^n}_{\{\boldsymbol{k}_I\}}$ can be evaluated as $\{\boldsymbol{k}_I\} \stackrel{def}{=} \mathbb{K}_0 \rightarrow \mathbb{K}_1 \rightarrow \cdots \rightarrow \mathbb{K}_r$, where $\mathcal{D}^{1^n}_{\mathbb{K}_r}$ is the CBDP after r-round propagation. If the set $\mathbb{K}_r$ doesn't have the unit vector $\boldsymbol{e}_m \in \mathbb{F}_2^n$ whose only m-th element is 1, the m-th output bit of r-round ciphertexts is balanced. At ASIACRYPT 2016, Xiang *et al.* [27] applied MILP method to the propagation of CBDP. They first introduced the concept of CBDP trail, which is defined as follows.

Definition 4 (CBDP Trail [27]**).** *Let us consider the propagation of the CBDP* $\{\boldsymbol{k}_I\} \stackrel{def}{=} \mathbb{K}_0 \rightarrow \mathbb{K}_1 \rightarrow \cdots \rightarrow \mathbb{K}_r$*. For any vector* $\boldsymbol{k}_{i+1} \in \mathbb{K}_{i+1}$*, there must exist a vector* $\boldsymbol{k}_i \in \mathbb{K}_i$ *such that* $\boldsymbol{k}_i$ *can propagate to* $\boldsymbol{k}_{i+1}$ *by the propagation rules of CBDP. Furthermore, for* $(\boldsymbol{k}_0, \boldsymbol{k}_1, \ldots, \boldsymbol{k}_r) \in \mathbb{K}_0 \times \mathbb{K}_1 \times \cdots \times \mathbb{K}_r$*, if* $\boldsymbol{k}_i$ *can propagate to* $\boldsymbol{k}_{i+1}$ *for all* $i \in \{0, 1, \ldots r-1\}$*, we call* $\boldsymbol{k}_0 \rightarrow \boldsymbol{k}_1 \rightarrow \cdots \rightarrow \boldsymbol{k}_r$ *an r-round CBDP trail.*

In [27], the authors modeled CBDP propagations of basic operations (Copy, Xor, And) and S-box by linear inequalities. Therefore, they could build an MILP model to cover all the possible CBDP trails generated from a given initial CBDP. Here, we introduce the MILP models for **Copy**, **Xor**, **And** and S-box.

Model 1 (Copy [27]). *Let* $a \xrightarrow{Copy} (b_0, b_1, \ldots, b_{n-1})$ *be a CBDP trail of Copy. The following inequalities are sufficient to describe its CBDP propagation*

$$\begin{cases} \mathcal{M}.var \leftarrow a, b_0, b_1, \ldots, b_{n-1} \text{ as binary}, \\ \mathcal{M}.con \leftarrow a = b_0 + b_1 + \cdots + b_{n-1}. \end{cases}$$

Model 2 (Xor [27]). *Let* $(a_0, a_1, \ldots, a_{n-1}) \xrightarrow{Xor} b$ *be a division trail of Xor. The following inequalities are sufficient to describe its CBDP propagation*

$$\begin{cases} \mathcal{M}.var \leftarrow a_0, a_1, \ldots, a_{n-1}, b \text{ as binary}, \\ \mathcal{M}.con \leftarrow b = a_0 + a_1 + \cdots + a_{n-1}. \end{cases}$$

Model 3 (And [27]). *Let* $(a_0, a_1, \ldots, a_{n-1}) \xrightarrow{And} b$ *be a division trail of And. The following inequalities are sufficient to describe its CBDP propagation*

$$\begin{cases} \mathcal{M}.var \leftarrow a_0, a_1, \ldots, a_{n-1}, b \text{ as binary}, \\ \mathcal{M}.con \leftarrow b \geq a_i \text{ for all } i \in \{0, 1, \ldots, n-1\}. \end{cases}$$

Model 4 (S-box [27]). The CBDP Rule 5 in Sect. 2.3 can generate the CBDP propagation property of S-box. Then, we can using the **inequality_generator()** function in Sage software [17] to get a set of linear inequalities. Sometimes the number of linear inequalities in the set is large. Thus, some Greedy Algorithms [1,19] were proposed to reduced this set.

2.5 Cube Attack

Cube attack was proposed by Dinur and Shamir at EUROCRYPT 2009 [6]. For a cipher with n secret variables $\boldsymbol{x} = (x_0, x_1, \ldots, x_{n-1})$ and m public variables $\boldsymbol{v} = (v_0, v_1, \ldots, v_{m-1})$, the output bit can be represented as $f(\boldsymbol{x}, \boldsymbol{v})$. Attackers determine an indices subset $I_v = \{i_0, i_1, \ldots, i_{|I_v|-1}\} \subset \{0, 1, \ldots, m-1\}$, then $f(\boldsymbol{x}, \boldsymbol{v})$ can be uniquely represented as

$$f(\boldsymbol{x}, \boldsymbol{v}) = \boldsymbol{v}^{\boldsymbol{u}_{I_v}} \cdot p(\boldsymbol{x}, \boldsymbol{v}) \oplus q(\boldsymbol{x}, \boldsymbol{v}),$$

where $p(\boldsymbol{x}, \boldsymbol{v})$ is called the *superpoly* of C_{I_v,J_v,K_v} in $f(\boldsymbol{x}, \boldsymbol{v})$, and every term in $q(\boldsymbol{x}, \boldsymbol{v})$ misses at least one variable from $\{v_{i_0}, v_{i_1}, \ldots, v_{i_{|I_v|-1}}\}$.

Attackers can prepare a cube set denoted as C_{I_v,J_v,K_v}, where public variables indexed by I_v are taking all possible combinations of values, public variables indexed by $J_v \subset \{0, 1, \ldots, m-1\} - I_v$ are set to constant 1, and public variables indexed by $K_v = \{0, 1, \cdots, m-1\} - I_v - J_v$ are set to constant 0. Just as follows

$$C_{I_v,J_v,K_v} = \{\boldsymbol{v} \in \mathbb{F}_2^m | v_i \in \mathbb{F}_2 \text{ if } i \in I_v, v_j = 1 \text{ if } j \in J_v, v_k = 0 \text{ if } k \in K_v\} \quad (1)$$

What's more, the sum of $f(\boldsymbol{x}, \boldsymbol{v})$ over the cube set C_{I_v,J_v,K_v} is

$$\bigoplus_{\boldsymbol{v} \in C_{I_v,J_v,K_v}} f(\boldsymbol{x}, \boldsymbol{v}) = p_{I_v,J_v,K_v}(\boldsymbol{x}). \quad (2)$$

If $p_{I_v,J_v,K_v}(\boldsymbol{x})$ is not a constant polynomial, attackers can query the encryption oracle with the chosen cube set C_{I_v,J_v,K_v} to get the equation with secret variables.

2.6 The Cube Attack Based on CBDP

At CRYPTO 2017 [23], Todo *et al.* successfully applied CBDP to cube attack. They use CBDP to analyze the ANF coefficients of superpoly.

Lemma 1. [23] *Let* $f(\boldsymbol{x}) = \bigoplus_{\boldsymbol{u}\in\mathbb{F}_2^n} a_{\boldsymbol{u}}^f \cdot \boldsymbol{x}^{\boldsymbol{u}}$ *be a polynomial from* $\mathbb{F}_2^n$ *to* $\mathbb{F}_2$ *and* $a_{\boldsymbol{u}}^f \in \mathbb{F}_2$ *be the ANF coefficients. Let* $\boldsymbol{k}$ *be an* n*-dimensional bit vector. If there is no CBDP trail such that* $\boldsymbol{k} \xrightarrow{f} 1$*, then* $a_{\boldsymbol{u}}^f$ *is always 0 for* $\boldsymbol{u} \succeq \boldsymbol{k}$*.*

Proposition 1. [23] *Let* $f(\boldsymbol{x}, \boldsymbol{v})$ *be a polynomial, where* $\boldsymbol{x} \in \mathbb{F}_2^n$ *and* $\boldsymbol{v} \in \mathbb{F}_2^m$ *denote the secret and public variables, respectively. For a cube set* C_{I_v,J_v,K_v} *defined as Eq. (1), let* $\boldsymbol{e}_i$ *be an* n*-bit unit vector whose only* i*-th element is 1. If there is no CBDP trail such that* $(\boldsymbol{e}_i, \boldsymbol{u}_{I_v}) \xrightarrow{f} 1$*, then* x_i *is not involved in the superpoly of the cube* C_{I_v,J_v,K_v}*.*

When $f(\boldsymbol{x}, \boldsymbol{v})$ represents the output bit of target cipher, we can use MILP method to identify the involved keys set I by checking whether there is division trial $\{(\boldsymbol{e}_i, \boldsymbol{u}_{I_v})\} \xrightarrow{f} 1$ for $i = 0, 1, \cdots, n-1$. Then, at CRYPTO 2018 [26], Wang *et al.* proposed the degree bounding and term enumeration techniques to further reduce the complexity of recovering superpoly. The degree evaluation of superpoly is based on the following proposition.

Proposition 2. [26] *For a set* $I_x = \{i_0, i_1, \ldots, i_{|I_x|-1}\} \subset \{0, 1, \ldots, n-1\}$*, if there is no CBDP trail such that* $(\boldsymbol{u}_{I_x}, \boldsymbol{u}_{I_v}) \xrightarrow{f} 1$*, then* $\boldsymbol{x}^{\boldsymbol{u}_{I_x}}$ *is not involved in the superpoly of cube* C_{I_v,J_v,K_v}*.*

After getting the involved keys set I and the degree d of superpoly, the superpoly can be represented with $\sum_{i=0}^{d}\binom{|I|}{i}$ coefficients. Therefore, by selecting $\sum_{i=0}^{d}\binom{|I|}{i}$ different $\boldsymbol{x}$, a linear system with $\sum_{i=0}^{d}\binom{|I|}{i}$ variables can be constructed. Then, the whole ANF of $p_{I_v,J_v,K_v}(\boldsymbol{x})$ can be recovered by solving such a linear system. So the complexity of recovering the superpoly of cube C_{I_v,J_v,K_v} is $2^{|I_v|} \times \sum_{i=0}^{d}\binom{|I|}{i}$.

3 The Propagation Properties of BDPT

In this section, we will explore some new propagation properties of BDPT.

3.1 The BDPT Propagation of S-Box

In the Sect. 2.3, we have introduced the existing BDPT propagation rules of Copy, And, and Xor. Although any Boolean function can be evaluated by using these three rules, the propagation requires large time and memory complexity when the Boolean function is complex. Here, we propose a generalized method to calculate the BDPT propagation of S-box.

Theorem 1. *For an S-box:* $\mathbb{F}_2^n \rightarrow \mathbb{F}_2^m$, *let* $\boldsymbol{x} = (x_0, x_1, \ldots, x_{n-1})$ *and* $\boldsymbol{y} = (y_0, y_1, \ldots, y_{m-1})$ *denote the input and output. Every* $y_i, i \in \{0, 1, \ldots, m-1\}$ *can be expressed as a boolean function of* $(x_0, x_1, \ldots, x_{n-1})$. *If the input BDPT of S-box is* $\mathcal{D}^{1^n}_{\mathbb{K},\mathbb{L}=\{\boldsymbol{\ell}\}}$, *then the output BDPT of S-box can be calculated by* $\mathcal{D}^{1^m}_{\boldsymbol{Reduce0}(\underline{\mathbb{K}}),\boldsymbol{Reduce1}(\underline{\mathbb{K}},\underline{\mathbb{L}})}$, *where*

$$\underline{\mathbb{K}} = \{\boldsymbol{u}' \in \mathbb{F}_2^m | \text{ for any } \boldsymbol{u} \in \mathbb{K}, \text{ if } \boldsymbol{y}^{\boldsymbol{u}'} \text{ contain any term } \boldsymbol{x}^{\boldsymbol{v}} \text{ satisfying } \boldsymbol{v} \succeq \boldsymbol{u}\}.$$
$$\underline{\mathbb{L}} = \{\boldsymbol{u} \in \mathbb{F}_2^m | \boldsymbol{y}^{\boldsymbol{u}} \text{ contains the term } \boldsymbol{x}^{\boldsymbol{\ell}}\}.$$

Proof. Let $\mathbb{K}'$ be the set of output BDPT that has no redundant vectors. According to the CBDP rules 5 in Sect. 2.3, we know that $\mathbb{K}' = \boldsymbol{Reduce0}\,(\underline{\mathbb{K}})$.

Let $\mathbb{L}'$ be the set of output BDPT that has no redundant vectors. For any $\boldsymbol{u} \in \mathbb{L}'$, we have $\bigoplus_{\boldsymbol{y}\in\mathbb{Y}} \boldsymbol{y}^{\boldsymbol{u}} = 1$. Since there is only one vector $\boldsymbol{\ell}$ in the input $\mathbb{L}$, the ANF of $\boldsymbol{y}^{\boldsymbol{u}}$ must has the monomial $\boldsymbol{x}^{\boldsymbol{\ell}}$. Thus, we get $\mathbb{L}' \subset \underline{\mathbb{L}}$. Because the function ***Reduce1*** only removes the vectors satisfying $\bigoplus_{\boldsymbol{y}\in\mathbb{Y}} \boldsymbol{y}^{\boldsymbol{u}} = \textit{unknown}$, we have $\mathbb{L}' \subset \boldsymbol{Reduce1}\,(\underline{\mathbb{K}}, \underline{\mathbb{L}})$.

On the other hand, if $\boldsymbol{y}^{\boldsymbol{u}}$ contains the monomial $\boldsymbol{x}^{\boldsymbol{\ell}}$, we have $\bigoplus_{\boldsymbol{x}\in\mathbb{X}} \boldsymbol{y}^{\boldsymbol{u}}$ equals unknown or 1. For the set $\underline{\mathbb{L}}$, the function ***Reduce1*** will remove all the vectors satisfying $\bigoplus_{\boldsymbol{y}\in\mathbb{Y}} \boldsymbol{y}^{\boldsymbol{u}} = \textit{unknown}$. So all the remaining vectors satisfying $\bigoplus_{\boldsymbol{y}\in\mathbb{Y}} \boldsymbol{y}^{\boldsymbol{u}} = 1$. Then, we get $\boldsymbol{Reduce1}\,(\underline{\mathbb{K}}, \underline{\mathbb{L}}) \subset \mathbb{L}'$.

Altogether, we obtain $\mathbb{L}' = \boldsymbol{Reduce1}\,(\underline{\mathbb{K}}, \underline{\mathbb{L}})$. □

We apply Theorem 1 to the core operation of SIMON family, the obtained BDPT propagation rules are in accordance with that in [21]. Note that Theorem 1 can get the BDPT propagation rules when the input $\mathbb{L}$ has only one vector. If there are more vectors in $\mathbb{L}$, the paper [21] has showed an example on how to get its BDPT propagation rules. Let $\mathcal{D}^{1^n}_{\mathbb{K},\mathbb{L}=\{\boldsymbol{\ell}_0,\boldsymbol{\ell}_1,\ldots,\boldsymbol{\ell}_{r-1}\}}$ and $\mathcal{D}^{1^m}_{\mathbb{K}',\mathbb{L}'}$ be the input and output BDPT of S-box, respectively. According to Theorem 1, we can get the output BDPT $\mathcal{D}^{1^m}_{\mathbb{K}',\mathbb{L}'_i}$ from the corresponding input BDPT $\mathcal{D}_{\mathbb{K},\mathbb{L}=\{\boldsymbol{\ell}_i\}}$, where $i = 0, 1, \ldots, r-1$. Then,

$$\mathbb{L}' = \{\boldsymbol{\ell} | \boldsymbol{\ell} \text{ appears odd times in sets } \mathbb{L}'_0, \mathbb{L}'_1, \ldots, \mathbb{L}'_{r-1}\}.$$

And we also give an example in Sect. 5.1 to help readers understand the propagation of BDPT.

3.2 Pruning Techniques of BDPT

The previous works often divide ciphers into r rounds, and investigate the CBDP or BDPT of round functions. Round functions often have too many operations which will generate many redundant intermediate vectors of division property. When the round number or block size grows, it will make propagation impossible just because of complexity. In order to solve this problem, we divide the ciphers into small parts. And after getting the BDPT propagation of a part, we will use the pruning techniques to remove the redundant vectors. Then, the remaining vectors in BDPT can continue to propagate efficiently.

Let Q_i be the i-th round function of an r-round cipher $E = Q_r \circ Q_{r-1} \circ \cdots \circ Q_1$, then we divide Q_i into l_i parts $Q_i = Q_{i,l_i-1} \circ Q_{i,l_i-2} \circ \cdots \circ Q_{i,0}$. Let $E_{i,j} = (Q_{i,j-1} \circ Q_{i,j-2} \circ \cdots \circ Q_{i,0}) \circ (Q_{i-1} \circ Q_{i-2} \circ \cdots \circ Q_1)$ and $\overline{E_{i,j}} = (Q_r \circ Q_{r-1} \circ \cdots \circ Q_{i+1})(Q_{i,l_i-1} \circ Q_{i,l_i-2} \circ \cdots \circ Q_{i,j})$, then $E = \overline{E_{i,j}} \circ E_{i,j}$, where $1 \le i \le r, 0 \le j \le l_i - 1$ and $E_{1,0}$ is identity function.

Theorem 2 (Prune $\mathbb{K}$). *For r-round cipher $E = Q_r \circ Q_{r-1} \circ \cdots \circ Q_1$, let $\mathcal{D}^{1^n}_{\mathbb{K}_{i,j},\mathbb{L}_{i,j}}$ be the input BDPT of $\overline{E_{i,j}}$. For any vector $\boldsymbol{k} \in \mathbb{K}_{i,j}$, if there is no CBDP trail such that $\boldsymbol{k} \xrightarrow{\overline{E_{i,j}}} \boldsymbol{e}_m$, the BDPT propagation of $\mathcal{D}^{1^n}_{\mathbb{K}_{i,j},\mathbb{L}_{i,j}}$ is equivalent to that of $\mathcal{D}^{1^n}_{\mathbb{K}_{i,j}\to\boldsymbol{k},\mathbb{L}_{i,j}}$ on whether $\boldsymbol{e}_m \in \mathbb{K}_{r+1,0}$ and $\boldsymbol{e}_m \in \mathbb{L}_{r+1,0}$ or not.*

Proof. In Sect. 2.3, we know that for public function, the BDPT propagation of $\mathbb{K}_{i,j}$ and $\mathbb{L}_{i,j}$ is independent. Only when the secret round key is Xored, some vectors of $\mathbb{L}_{i,j}$ will affect $\mathbb{K}_{i,j}$, but they only adds some vectors into $\mathbb{K}_{i,j}$. Because every vector $\boldsymbol{k} \in \mathbb{K}_{i,j}$ is propagated independently based on CBDP, if there is no CBDP trail such that $\boldsymbol{k} \xrightarrow{\overline{E_{i,j}}} \boldsymbol{e}_m$, then removing it from $\mathbb{K}_{i,j}$ doesn't have any impact on whether $\mathbb{K}_{r+1,0}$ includes $\boldsymbol{e}_m$ or not. That means $\mathcal{D}^{1^n}_{\mathbb{K}_{i,j},\mathbb{L}_{i,j}}$ has the same result with $\mathcal{D}^{1^n}_{\mathbb{K}_{i,j}\to\boldsymbol{k},\mathbb{L}_{i,j}}$ on whether $\mathbb{K}_{r+1,0}$ includes $\boldsymbol{e}_m$ or not.

Because all the vectors of $\mathbb{L}_{r+1,0}$ are generated from $\mathbb{L}_{i,j}$, that is, removing $\boldsymbol{k}$ from $\mathbb{K}_{i,j}$ has no impact on the generation of $\boldsymbol{e}_m \in \mathbb{L}_{r+1,0}$. On the other hand, we have got that removing $\boldsymbol{k}$ from $\mathbb{K}_{i,j}$ doesn't have any impact on whether $\mathbb{K}_{r+1,0}$ includes $\boldsymbol{e}_m$ or not. So it has no impact on the reduction of $\boldsymbol{e}_m \in \mathbb{L}_{r+1,0}$. That means $\mathcal{D}^{1^n}_{\mathbb{K}_{i,j},\mathbb{L}_{i,j}}$ has the same result with $\mathcal{D}^{1^n}_{\mathbb{K}_{i,j}\to\boldsymbol{k},\mathbb{L}_{i,j}}$ on whether $\mathbb{L}_{r+1,0}$ includes $\boldsymbol{e}_m$ or not. □

Theorem 3 (Prune $\mathbb{L}$). *For r-round cipher $E = Q_r \circ Q_{r-1} \circ \cdots \circ Q_1$, let $\mathcal{D}^{1^n}_{\mathbb{K}_{i,j},\mathbb{L}_{i,j}}$ be the input BDPT of $\overline{E_{i,j}}$. For any vector $\boldsymbol{\ell} \in \mathbb{L}_{i,j}$, if there is no CBDP trail such that $\boldsymbol{\ell} \xrightarrow{\overline{E_{i,j}}} \boldsymbol{e}_m$, the BDPT propagation of $\mathcal{D}^{1^n}_{\mathbb{K}_{i,j},\mathbb{L}_{i,j}}$ is equivalent to that of $\mathcal{D}^{1^n}_{\mathbb{K}_{i,j},\mathbb{L}_{i,j}\to\boldsymbol{\ell}}$ on whether $\boldsymbol{e}_m \in \mathbb{K}_{r+1,0}$ and $\boldsymbol{e}_m \in \mathbb{L}_{r+1,0}$ or not.*

Proof. For any vector $\boldsymbol{\ell} \in \mathbb{L}_{i,j}$, if there is no CBDP trail such that $\boldsymbol{\ell} \xrightarrow{\overline{E_{i,j}}} \boldsymbol{e}_m$, according to Theorem 2, the BDPT propagation of $\mathcal{D}^{1^n}_{\mathbb{K}_{i,j},\mathbb{L}_{i,j}}$ is equivalent to that of $\mathcal{D}^{1^n}_{\mathbb{K}_{i,j}\leftarrow\boldsymbol{\ell},\mathbb{L}_{i,j}}$ on whether $\boldsymbol{e}_m \in \mathbb{K}_{r+1,0}$ and $\boldsymbol{e}_m \in \mathbb{L}_{r+1,0}$ or not.

Because $\mathbb{K}_{i,j} \leftarrow \boldsymbol{\ell}$, the vector $\boldsymbol{\ell}$ can be removed from $\mathbb{L}_{i,j}$ according to the definition of BDPT. So the BDPT $\mathcal{D}^{1^n}_{\mathbb{K}_{i,j}\leftarrow\boldsymbol{\ell},\mathbb{L}_{i,j}}$ is completely equivalent to $\mathcal{D}^{1^n}_{\mathbb{K}_{i,j}\leftarrow\boldsymbol{\ell},\mathbb{L}_{i,j}\to\boldsymbol{\ell}}$.

According to Theorem 2 again, the BDPT propagation of $\mathcal{D}^{1^n}_{\mathbb{K}_{i,j}\leftarrow\boldsymbol{\ell},\mathbb{L}_{i,j}\to\boldsymbol{\ell}}$ is equivalent to that of $\mathcal{D}^{1^n}_{\mathbb{K}_{i,j},\mathbb{L}_{i,j}\to\boldsymbol{\ell}}$ on whether $\boldsymbol{e}_m \in \mathbb{K}_{r+1,0}$ and $\boldsymbol{e}_m \in \mathbb{L}_{r+1,0}$ or not. □

The propagation of CBDP can be efficiently solved by MILP model. Therefore, the meaning of Theorems 2 and 3 is that we can use CBDP method to reduce the BDPT sets $\mathbb{K}_{i,j}$ and $\mathbb{L}_{i,j}$.

3.3 Fast Propagation

Inspired by the notion of "lazy propagation", we propose a notion called "fast propagation" to show the balanced information of output bits.

Definition 5 (Fast Propagation). *For r-round cipher* $E = Q_r \circ Q_{r-1} \circ \cdots \circ Q_1$, *let* $\mathcal{D}^{1^n}_{\mathbb{K}_{i,j},\mathbb{L}_{i,j}}$ *be the input BDPT of* $\overline{E_{i,j}}$. *Under fast propagation, we translate the BDPT into CBDP* $\mathcal{D}^{1^n}_{\overline{\mathbb{K}}_{i,j}}$, *where* $\overline{\mathbb{K}}_{i,j} = \mathbb{K}_{i,j} \cup \mathbb{L}_{i,j}$. *The output CBDP of* $\overline{E_{i,j}}$ *is computed from* $\mathcal{D}^{1^n}_{\overline{\mathbb{K}}_{i,j}}$.

The "fast propagation" removes all vectors from $\mathbb{L}_{i,j}$, and get the union set $\mathbb{K}_{i,j} \cup \mathbb{L}_{i,j}$. By its nature, "fast propagation" translate BDPT into CBDP. We can use the MILP method to solve the CBDP propagation of $\mathcal{D}^{1^n}_{\mathbb{K}_{i,j} \cup \mathbb{L}_{i,j}}$. Let us consider the meaning of "fast propagation". Assuming the input set of $\overline{E_{i,j}}$ has BDPT $\mathcal{D}^{1^n}_{\mathbb{K}_{i,j},\mathbb{L}_{i,j}}$, according to the definition of BDPT and CBDP, this set must also has CBDP $\mathcal{D}^{1^n}_{\mathbb{K}_{i,j} \cup \mathbb{L}_{i,j}}$. If for any $\boldsymbol{k} \in \mathbb{K}_{i,j} \cup \mathbb{L}_{i,j}$, there is no CBDP trial such that $\boldsymbol{k} \xrightarrow{\overline{E_{i,j}}} \boldsymbol{e}_m$, then the m-th output bit of $\overline{E_{i,j}}$ is balanced.

4 The MILP-aided Method for BDPT

Based on the work of [27], we first simplify the MILP algorithm of searching integral distinguishers based on CBDP to improve efficiency. Then, we show three stopping rules and propose an algorithm to search integral distinguishers based on BDPT.

4.1 Simplify the MILP Method of CBDP

Using the method in the paper [27], we can get a linear inequality set which describes the r-round CBDP division trails with the given initial CBDP $\mathcal{D}^{1^n}_{\{k\}}$. The former CBDP method will return a set of balanced bits. Because only one bit's balanced information is needed, our MILP model has no objective function which is added into the constrains. We can use the solver Gurobi [10] to determine whether the MILP model has feasible solutions or not. If it has feasible solutions, it shows that the m-th bit of the output is unknown. Otherwise, the m-th bit is balanced. The detail information is shown in Algorithm 1.

4.2 Stopping Rules

Based on "lazy propagation" and "fast propagation", in this subsection, we propose three stopping rules in searching integral distinguishers based on BDPT.

Algorithm 1. *SCBDP*$(E, \boldsymbol{k}, m)$

Input: The cipher E, the initial CBDP vector $\boldsymbol{k}$, and the number m
Output: Whether the m-th bit of the output is balanced or not based on CBDP
begin
  $\mathcal{L}$ is a linear inequality set which describe the CBDP division trails such that $\boldsymbol{k} \xrightarrow{E} \boldsymbol{e}_m$
  if $\mathcal{L}$ has feasible solutions **do**
    return unknown
  else
    return 0
end

Stopping Rule 1. *For an r-round cipher $E = Q_r \circ Q_{r-1} \circ \cdots \circ Q_1$, let $\mathcal{D}^{1^n}_{\mathbb{K}_{i,j},\mathbb{L}_{i,j}}$ be the input BDPT of $\overline{E_{i,j}}$. For any vector $\boldsymbol{k} \in \mathbb{K}_{i,j}$, if there is CBDP trail such that $\boldsymbol{k} \xrightarrow{\overline{E_{i,j}}} \boldsymbol{e}_m$, according to "lazy propagation", we stop the process and obtain that the m-th output bit of E is unknown.*

After Stopping Rule 1, if the searching procedure doesn't stop, all the vectors in $\mathbb{K}_{i,j}$ will be removed according to the pruning technique in Theorem 2. Then, we consider the following Stopping Rule 2.

Stopping Rule 2. *After removing the redundant vectors in the set $\mathbb{L}_{i,j}$ by the pruning technique in Theorem 3, if there is still vector $\boldsymbol{\ell} \in \mathbb{L}_{i,j}$, we cannot stop the procedure and $\boldsymbol{\ell}$ should be propagated to next part based on BDPT. If there is no vector in $\mathbb{L}_{i,j}$, according to "fast propagation", we can get that the m-th output bit of E is balanced.*

Different from Stopping Rule 1 which shows the m-th bit is unknown, Stopping Rule 2 can show the m-th bit is balanced based on BDPT. If the process doesn't stop even we get the output BDPT of E, Stopping Rule 3 can explain this situation.

Stopping Rule 3. *If $\mathbb{K}_{r+1,0} = \emptyset$ and $\mathbb{L}_{r+1,0} = \{\boldsymbol{e}_m\}$, then we find an integral distinguisher whose sum of the m-th output bit is 1.*

4.3 The MILP-aided Method of Searching Integral Distinguishers Based on BDPT

The algorithm of searching integral distinguishers often has a given initial BDPT $\mathcal{D}^{1^n}_{\mathbb{K}_{1,0},\mathbb{L}_{1,0}}$. For an indices set $I = \{i_0, i_1, \ldots, i_{|I|-1}\} \subset \{0, 1, \ldots, n-1\}$, attackers prepare $2^{|I|}$ chosen plaintexts where variables indexed by I are taking all possible combinations of values and the other variables are set to constants. The CBDP of such chosen plaintexts is $\mathcal{D}^{1^n}_{\{\boldsymbol{u}_I\}}$. Then, the BDPT of such chosen plaintexts is $\mathcal{D}_{\mathbb{K}_{1,0},\mathbb{L}_{1,0}}$, where $\mathbb{K}_{1,0} = \{\boldsymbol{u}' \in \mathbb{F}_2^n | \boldsymbol{u}' \succ \boldsymbol{u}_I\}$ and $\mathbb{L}_{1,0} = \{\boldsymbol{u}_I\}$. We illustrate the whole framework in Algorithm 2.

Algorithm 2. *BDPT*$(E, \mathbb{L}_{1,0}, \mathbb{K}_{1,0}, m)$

Input: The cipher E, the input BDPT $\mathcal{D}_{\mathbb{K}_{1,0},\mathbb{L}_{1,0}}$, and the number m
Output: The balanced information of the m-th output bit based on BDPT

1 **begin**
2   **for** $(i = 1; i \leq r; i++)$ **do**
3   **for** $(j = 0; j \leq l_i - 1; j++)$ **do**
4   **for** $\boldsymbol{k}$ **in** $\mathbb{K}_{i,j}$
5   **if** SCBDP$(\overline{E_{i,j}}, \boldsymbol{k}, m)$ is unknown
6   **return** unknown
7   else
8                     $\mathbb{K}_{i,j} \rightarrow \boldsymbol{k}$
9   **end**
10             $\mathbb{L}'_{i,j} = \emptyset$
11   **for** $\boldsymbol{\ell}$ in $\mathbb{L}_{i,j}$ **do**
12   **if** SCBDP $(\overline{E_{i,j}}, \boldsymbol{\ell}, m)$ is unknown
13                     $\mathbb{L}'_{i,j} = \mathbb{L}'_{i,j} \cup \boldsymbol{\ell}$
14   **end**
15   **end**
16   **if** $\mathbb{L}'_{i,j} = \emptyset$
17   **return** 0
18   **end**
19             $\mathcal{D}_{\mathbb{K}_{i+\lfloor (j+1)/l_i \rfloor,(j+1) \bmod l_i}, \mathbb{L}_{i+\lfloor (j+1)/l_i \rfloor,(j+1) \bmod l_i}} = BDPTP\left(Q_{i,j}, \mathcal{D}_{\emptyset,\mathbb{L}'_{i,j}}\right)$
20   **end**
21   **end**
22   **return** 1
23 **end**

We explain **Algorithm 2** line by line:

Line 2–3 The cipher E is divided into small parts.

Line 4–9 For every $\boldsymbol{k} \in \mathbb{K}_{i,j}$, if *SCBDP*$(\overline{E_{i,j}}, \boldsymbol{k}, m)$ is unknown (Algorithm 1), according to Stopping Rule 1, we know that the m-th output bit is unknown based on BDPT. Otherwise, we remove it from $\mathbb{K}_{i,j}$ according to the pruning technique in Theorem 2.

Line 10 Initialize $\mathbb{L}'_{i,j}$ to be an empty set.

Line 11–15 For any vector $\boldsymbol{\ell} \in \mathbb{L}_{i,j}$, if *SCBDP*$(\overline{E_{i,j}}, \boldsymbol{\ell}, m)$ can generate the unit vector $\boldsymbol{e}_m$, we store all these vectors in $\mathbb{L}'_{i,j}$.

Line 16–18 If the set $\mathbb{L}'_{i,j}$ is empty set, it satisfies Stopping Rule 2, that is, the m-th output bit is balanced.

Line 19 If we don't get the balanced information of the m-th bit, we should use the propagation rules of BDPT to get the input BDPT of the next part.

Line 22 It triggers Stopping Rules 3, and the sum of the m-th output bit is 1.

The principle of dividing the round function Q_i is that the vectors of BDPT don't expand too much. Only in this way can we run the searching algorithm efficiently. Algorithm 2 can show the balanced information of any output bit. Therefore, we can search the integral distinguishers of cipher in parallel.

5 Applications to Block Ciphers

In this section, we apply our algorithm to SIMON, SIMECK, PRESENT, RECTANGLE, and LBlock. All the experiments are conducted on the platform: Intel Core i5-4590 CPU @3,3 GHz, 8.00G RAM. And the optimizer we used to solve MILP models is Gurobi 8.1.0 [10]. For the integral distinguishers, what needs to be explained is that "a" denotes active bit, "c" denotes constant bit, "?" denotes the balanced information is unknown, and "b" denotes the balanced bit.

5.1 Applications to SIMON and SIMECK

SIMON is a lightweight block cipher family [2] based on Feistel structure which only involves bit-wise And, Xor, and Circular shift operations. Let SIMON2n be the SIMON cipher with $2n$-bit block length, where $n \in \{16, 24, 32, 48, 64\}$. And the left part of Fig. 1 shows the round structure of SIMON2n. The core operation of round function is represented by the right part of Fig. 1.

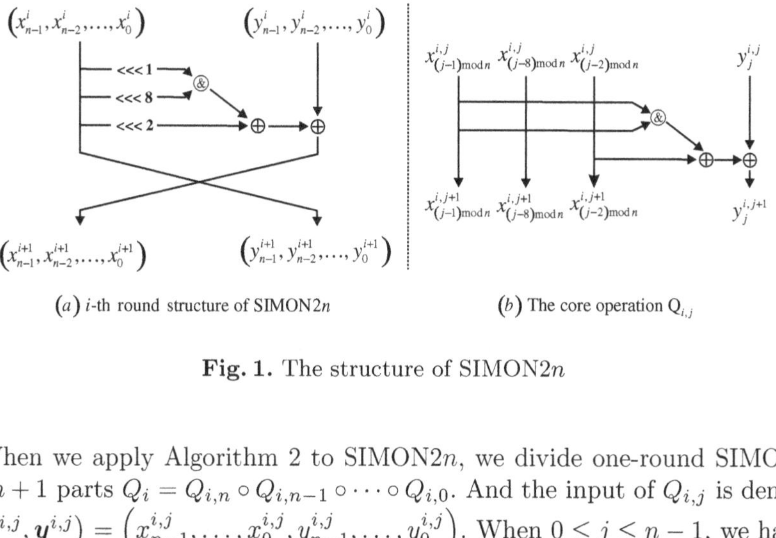

Fig. 1. The structure of SIMON2n

When we apply Algorithm 2 to SIMON2n, we divide one-round SIMON2n into $n+1$ parts $Q_i = Q_{i,n} \circ Q_{i,n-1} \circ \cdots \circ Q_{i,0}$. And the input of $Q_{i,j}$ is denoted as $(\boldsymbol{x}^{i,j}, \boldsymbol{y}^{i,j}) = \left(x^{i,j}_{n-1}, \ldots, x^{i,j}_0, y^{i,j}_{n-1}, \ldots, y^{i,j}_0\right)$. When $0 \le j \le n-1$, we have

$$Q_{i,j}\left(\boldsymbol{x}^{i,j}, \boldsymbol{y}^{i,j}\right) = \left(\boldsymbol{x}^{i,j}, y^{i,j}_{n-1}, \ldots, y^{i,j}_{j+1}, Y^{i,j}_j, y^{i,j}_{j-1}, \ldots, y^{i,j}_0\right),$$

where $Y^{i,j}_j = \left(x^{i,j}_{(j-1)\mathrm{mod}n} \& x^{i,j}_{(j-8)\mathrm{mod}n}\right) \bigoplus x^{i,j}_{(j-2)\mathrm{mod}n}$.

Moreover, $Q_{i,n}\left(\boldsymbol{x}^{i,n}, \boldsymbol{y}^{i,n}\right) = \left(\boldsymbol{y}^{i,n} \oplus \boldsymbol{k}^i, \boldsymbol{x}^{i,n}\right)$, where $\boldsymbol{k}^i$ is the i-th round key of SIMON2n.

For $Q_{i,j}, 0 \le j \le n-1$, when we consider the BDPT propagation rules of the function $BDPTP\left(Q_{i,j}, \mathcal{D}_{\emptyset, \mathbb{L}'_{i,j}}\right)$, $(2n-4)$ bits remain unchanged. Thus, only 4-bit $\left(x^{i,j}_{(j-1)\mathrm{mod}\ n}, x^{i,j}_{(j-2)\mathrm{mod}n}, x^{i,j}_{(j-8)\mathrm{mod}n}, y^{i,j}_{(j)\mathrm{mod}n}\right)$ of the BDPT vectors will be changed. We can view it as 4-bit S-box and use Theorem 1 to get its accurate BDPT propagation rules which are in accordance with that in the paper [21]. We show it in Appendix Table 7.

When we use Algorithm 2 to search the integral distinguishers of SIMON2n based on BDPT, we should call Algorithm 1 to build the MILP model based on CBDP. The paper [27] has showed us how to model CBDP division trails of 1-round SIMON2n. We introduce it as follows.

1-round Description of SIMON2n. Denote 1-round CBDP trail of SIMON2n by $(a^i_{n-1},\dots,a^i_0,b^i_{n-1},\dots,b^i_0) \rightarrow (a^{i+1}_{n-1},\dots,a^{i+1}_0,b^{i+1}_{n-1},\dots,b^{i+1}_0)$. In order to get a linear description of all CBDP trails of 1-round SIMON2n, we introduce four vectors of auxiliary variables which are $(u^i_{n-1},\dots,u^i_0)$, $(v^i_{n-1},\dots,v^i_0)$, $(w^i_{n-1},\dots,w^i_0)$ and $(t^i_{n-1},\dots,t^i_0)$. We denote $(u^i_{n-1},\dots,u^i_0)$ the input CBDP of the left circular shift by 1 bit. Similarly, denote $(v^i_{n-1},\dots,v^i_0)$ and $(w^i_{n-1},\dots,w^i_0)$ the input CBDP of the left circular shift by 8 bits and 2 bits, respectively. Let $(t^i_{n-1},\dots,t^i_0)$ denote the output CBDP of bit-wise And operation. The following inequalities are sufficient to model the Copy operation used in SIMON2n:

$$\mathcal{L}_1 : a^i_j - u^i_j - v^i_j - w^i_j - b^{i+1}_j = 0 \text{ for } j \in \{0,1,\dots,n-1\}.$$

Then, the bit-wise And operation used in SIMON2n can be modeled by:

$$\mathcal{L}_2 = \begin{cases} t^i_j - u^i_{(j-1)\text{mod}n} \geq 0, & \text{for } j \in \{0,1,\dots,n-1\}, \\ t^i_j - v^i_{(j-8)\text{mod}n} \geq 0, & \text{for } j \in \{0,1,\dots,n-1\}, \\ t^i_j - u^i_{(j-1)\text{mod}n} - v^i_{(j-8)\text{mod}n} \leq 0, & \text{for } j \in \{0,1,\dots,n-1\}. \end{cases}$$

At last, the Xor operation in SIMON2n can be modeled by:

$$\mathcal{L}_3 : a^{i+1}_j - b^i_j - t^i_j - w^i_{(j-2)\text{mod}n} = 1 \text{ for } j \in \{0,1,\dots,n-1\}.$$

So far, we get a description $\{\mathcal{L}_1,\mathcal{L}_2,\mathcal{L}_3\}$ of 1-round CBDP trails.

How to Describe the CBDP Propagation of Partial Round. For $\overline{E_{i,j}}$, the first round maybe a partial round $Q_{i,l_i-1} \circ Q_{i,l_i-2} \circ \dots \circ Q_{i,j}$. When considering the CBDP propagation of $Q_{i,j}$, if add constrain $b^{i+1,j}_j = b^{i,j}_j$, the output vector is the same with the input vector. Namely, $Q_{i,j}$ is transformed into identity function.

For 1-round SIMON2n, by adding the following constrains

$$\mathcal{L}_4 : a^{i+1}_j - b^i_j = 0 \text{ for } j \in \{0,1,\dots,j-1\},$$

we obtain a description $\{\mathcal{L}_1,\mathcal{L}_2,\mathcal{L}_3,\mathcal{L}_4\}$ of partial round $Q_{i,l_i-1} \circ Q_{i,l_i-2} \circ \cdots \circ Q_{i,j}$. Then, by repeating the constrains of 1-round $(r-i)$ times, we can get a linear inequality system $\mathcal{L}$ for $\overline{E_{i,j}}$.

How to Obtain the Output BDPT of $Q_{i,j}$. After the pruning techniques and stopping rules, if Algorithm 2 doesn't stop, we know that $\mathbb{K}_{i,j} = \emptyset$ and $\mathbb{L}_{i,j} \neq \emptyset$. In order to help readers understand our algorithm, we show an example of the propagation of BDPT.

For SIMON32, if the input BDPT of $Q_{1,15}$ is $\mathcal{D}_{\mathbb{K}_{1,15}=\emptyset, \mathbb{L}_{1,15}=\{\boldsymbol{\ell}_1, \boldsymbol{\ell}_2\}}$, where $\boldsymbol{\ell}_1 = (1, \mathbf{0}, \mathbf{1}, 1, 1, 1, 1, 1, \mathbf{1}, 1, 1, 1, 1, 1, 1, 1, \mathbf{0}, 1, 1, 1, 1, 1, 1, 1, 1, 1, 1, 1, 1, 1, 1, 1)$, $\boldsymbol{\ell}_2 = (1, \mathbf{0}, \mathbf{0}, 1, 1, 1, 1, 1, \mathbf{1}, 1, 1, 1, 1, 1, 1, 1, \mathbf{1}, 1, 1, 1, 1, 1, 1, 1, 1, 1, 1, 1, 1, 1, 1, 1)$. The 4 bits of $\boldsymbol{\ell}_1$ that may be updated by $Q_{1,15}$ is $(0, 1, 1, 0)$. Then, according to the BDPT propagation rules of core operation in Table 7. The output vector set is $\mathbb{L}' = \{[0, 1, 1, 0], [0, 1, 0, 1], [0, 1, 1, 1]\}$. So $\boldsymbol{\ell}_1$ generates three vectors as:

$$\begin{aligned}
&(1, \mathbf{0}, \mathbf{1}, 1, 1, 1, 1, 1, \mathbf{1}, 1, 1, 1, 1, 1, 1, 1, \mathbf{0}, 1, 1, 1, 1, 1, 1, 1, 1, 1, 1, 1, 1, 1, 1, 1)\\
&(1, \mathbf{0}, \mathbf{0}, 1, 1, 1, 1, 1, \mathbf{1}, 1, 1, 1, 1, 1, 1, 1, \mathbf{1}, 1, 1, 1, 1, 1, 1, 1, 1, 1, 1, 1, 1, 1, 1, 1)\\
&(1, \mathbf{0}, \mathbf{1}, 1, 1, 1, 1, 1, \mathbf{1}, 1, 1, 1, 1, 1, 1, 1, \mathbf{1}, 1, 1, 1, 1, 1, 1, 1, 1, 1, 1, 1, 1, 1, 1, 1)
\end{aligned}$$

In the same way, we can obtain that $\boldsymbol{\ell}_2$ generates only one vector as

$$(1, \mathbf{0}, \mathbf{0}, 1, 1, 1, 1, 1, \mathbf{1}, 1, 1, 1, 1, 1, 1, 1, \mathbf{1}, 1, 1, 1, 1, 1, 1, 1, 1, 1, 1, 1, 1, 1, 1, 1).$$

According to **BDPT Rule** 3, the vector $(1, \mathbf{0}, \mathbf{0}, 1, 1, 1, 1, 1, \mathbf{1}, 1, 1, 1, 1, 1, 1, 1, \mathbf{1}, 1, 1, 1, 1, 1, 1, 1, 1, 1, 1, 1, 1, 1, 1, 1)$ should be canceled because it is propagated from $\boldsymbol{\ell}_1$ and $\boldsymbol{\ell}_2$ twice. The output BDPT of $Q_{1,15}$ is $\mathcal{D}_{\mathbb{K}_{1,16}=\emptyset, \mathbb{L}_{1,16}=\{\boldsymbol{\ell}_3, \boldsymbol{\ell}_4\}}$, where

$$\begin{aligned}
\boldsymbol{\ell}_3 &= (1, \mathbf{0}, \mathbf{1}, 1, 1, 1, 1, 1, \mathbf{1}, 1, 1, 1, 1, 1, 1, 1, \mathbf{0}, 1, 1, 1, 1, 1, 1, 1, 1, 1, 1, 1, 1, 1, 1, 1),\\
\boldsymbol{\ell}_4 &= (1, \mathbf{0}, \mathbf{1}, 1, 1, 1, 1, 1, \mathbf{1}, 1, 1, 1, 1, 1, 1, 1, \mathbf{1}, 1, 1, 1, 1, 1, 1, 1, 1, 1, 1, 1, 1, 1, 1, 1).
\end{aligned}$$

Then, $Q_{1,16}$ has round keys Xored operation. So a new vector is generated from $\boldsymbol{\ell}_3$ and inserted into $\mathbb{K}_{1,16}$ according to the **BDPT Rule** 4. Moreover, a vector in $\mathbb{L}_{1,16}$ becomes redundant because of the new vector of $\mathbb{K}_{1,16}$. After the swapping, the output BDPT of $Q_{1,16}$ is $\mathcal{D}_{\mathbb{K}_{2,0}=\{\boldsymbol{k}\}, \mathbb{L}_{2,0}=\{\boldsymbol{\ell}_5\}}$, where

$$\begin{aligned}
\boldsymbol{k} &= (1, 1, 1, 1, 1, 1, 1, 1, 1, 1, 1, 1, 1, 1, 1, 1, 0, 1, 1, 1, 1, 1, 1, 1, 1, 1, 1, 1, 1, 1, 1, 1),\\
\boldsymbol{\ell}_5 &= (0, 1, 1, 1, 1, 1, 1, 1, 1, 1, 1, 1, 1, 1, 1, 1, 0, 1, 1, 1, 1, 1, 1, 1, 1, 1, 1, 1, 1, 1, 1, 1).
\end{aligned}$$

The High Efficiency of Our Algorithm. For 14-round SIMON32, we prepare chosen plaintexts such that the leftmost bit is constant and the others are active. Then, the BDPT of chosen plaintexts is $\mathcal{D}_{\mathbb{K}=\{(1,1,1,\ldots,1)\}, \mathbb{L}=\{(0,1,1,\ldots,1)\}}$. Table 3 shows the sizes of $|\mathbb{K}|$ and $|\mathbb{L}|$ in every round. The sizes in the paper [21] are obtained after removing redundant vectors according to the definition of BDPT, while the sizes in this paper are obtained after the pruning techniques. From Table 3, we find that $|\mathbb{L}|$ of the 5-th round in this paper becomes 0, it triggers Stopping Rule 2, and we obtain that the rightmost bit is balanced. Our pruning techniques can reduce the size of BDPT greatly.

Integral Distinguishers. SIMECK is a family of lightweight block cipher proposed at CHES 2015 [29], and its round function is very similar to that of SIMON except the rotation constants. We use Algorithm 2 to search the integral distinguishers of SIMON and SIMECK family based on BDPT. For SIMON32,

Table 3. Sizes of $\mathcal{D}_{\mathbb{K},\mathbb{L}}$ in obtaining balanced information of the rightmost output bit

Reference	BDPT	Size in every round															
		0	1	2	3	4	5	6	7	8	9	10	11	12	13	14	15
[21]	$\lvert\mathbb{L}\rvert$	1	1	5	19	138	2236	89878	4485379	47149981	2453101	20360	168	8	0	0	0
	$\lvert\mathbb{K}\rvert$	1	1	1	6	43	722	23321	996837	9849735	2524718	130724	7483	852	181	32	32
This paper	$\lvert\mathbb{L}\rvert$	1	1	1	2	2	0	0	0	0	0	0	0	0	0	0	0
	$\lvert\mathbb{K}\rvert$	1	0	0	0	0	0	0	0	0	0	0	0	0	0	0	0

our MILP algorithm finds the 14-round integral distinguisher that found in [21] by going through all the BDPT division trails. For 17-round SIMON64, we find an integral distinguisher with 23 balanced bits which has one more bit than the previous longest integral distinguisher. For SIMON48/96/128 and SIMECK32/48/64, the distinguishers we find are in accordance with the previous longest distinguishers that found in [27]. The detailed integral distinguishers of SIMON32 and SIMON64 are listed in Table 4. And all the integral distinguishers in Table 4 can be extended one more round by the technique in [25].

Table 4. Integral distinguishers of SIMON32 and SIMON64

Cipher	Distinguisher
14-SIMON32	In: (caaaaaaaaaaaaaaa, aaaaaaaaaaaaaaaa) Out: (????????????????, ?b??????b??????b)
17-SIMON64	In: (caaaaaaaaaaaaaaaaaaaaaaaaaaaaaaa,aaaaaaaaaaaaaaaaaaaaaaaaaaaaaaaa) Out: (????????????????????????????????,bbbbbbbbbbb??b??b?????bbbbbbbbbb)

5.2 Applications to PRESENT and RECTANGLE

PRESENT [3] has an SPN structure and uses 80- and 128-bit keys with 64-bit blocks through 31 rounds. In order to improve the hardware efficiency, it use a fully wired diffusion layer. Figure 2 illustrates one-round structure of PRESENT.

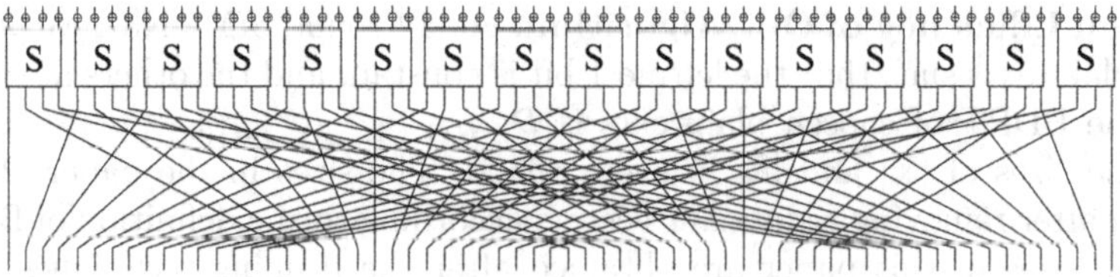

Fig. 2. One-round SPN structure of PRESENT

We divide one-round PRESENT into 17 parts $Q_i = Q_{i,16} \circ \cdots \circ Q_{i,0}$. When $0 \leq j \leq 15$, we have $Q_{i,j}\left(x_0^{i,j}, \ldots, x_{63}^{i,j}\right) = \left(x_0^{i,j}, \ldots, S\left(x_{4j}^{i,j}, \ldots, x_{4j+3}^{i,j}\right), \ldots, x_{63}^{i,j}\right)$, where $S\left(x_{4j}^{i,j}, \ldots, x_{4j+3}^{i,j}\right)$ is the S-box of PRESENT.

Moreover, $Q_{i,16}\left(x_0^{i,16},\ldots,x_{63}^{i,16}\right) = P\left(x_0^{i,16},x_1^{i,16},\ldots,x_{63}^{i,16}\right) \oplus \boldsymbol{k}^i$, where P is the linear permutation of PRESENT and $\boldsymbol{k}^i$ is the i-th round key.

RECTANGLE [31] is very like PRESENT. We apply Algorithm 2 to PRESENT and RECTANGLE, and the results are listed in Table 5.

Table 5. Integral distinguishers of PRESENT and RECTANGLE

Cipher	Distinguisher
9-PRESENT	In: (aaaaaaaaaaaaaaaaaaaaaaaaaaaaaaaa,aaaaaaaaaaaaaaaaaaaaaaaaaaaacccc) Out: (????????????????????????????????,???????????????????b???b???b???b)
9-PRESENT	In: (aaaaaaaaaaaaaaaaaaaaaaaaaaaaaaaa,aaaaaaaaaaaaaaaaaaaaaaaaaaaaaaac) Out: (???b???b???bbbbb???b???b???bbbbb, ???b???b???bbbbb???b???b???bbbbb)
9-RECTANGLE	In: (caaaaaaaaaaaaaaa,caaaaaaaaaaaaaaa,caaaaaaaaaaaaaaa,caaaaaaaaaaaaaaa) Out: (bbbbbbbbbbbbbbbb,bbbb??bb???bbbbb,????????????????,????????????????)

5.3 Applications to LBlock

LBlock is a lightweitht block cipher proposed by Wu and Zhang [24]. The block size is 64 bits and the key size is 80 bits. It employs a variant Feistel structure and consists of 32 rounds. One-round structure of LBlock is given in Fig. 3.

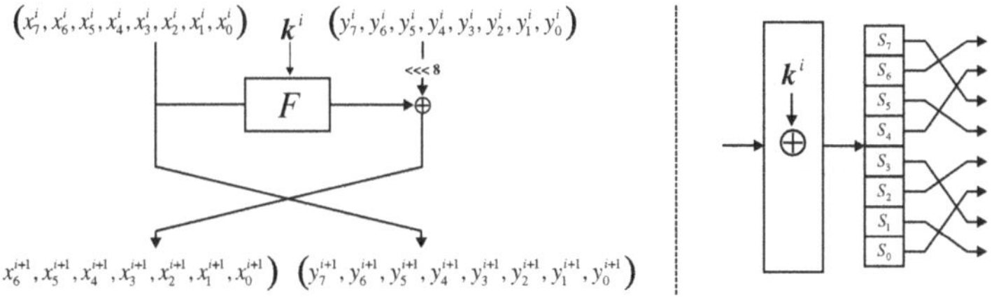

(*a*) *i*-th round structure of LBlock (b) The structure of F-function

Fig. 3. Round structure of LBlock

We divide one-round LBlock into 9 parts $Q_i = Q_{i,8} \circ \cdots \circ Q_{i,0}$. And the input of $Q_{i,j}$ is denoted as $\left(\boldsymbol{x}^{i,j},\boldsymbol{y}^{i,j}\right) = \left(\boldsymbol{x}_7^{i,j},\ldots,\boldsymbol{x}_0^{i,j},\boldsymbol{y}_7^{i,j},\ldots,\boldsymbol{y}_0^{i,j}\right)$. When $0 \le j \le 7$, we have $Q_{i,j}\left(\boldsymbol{x}^{i,j},\boldsymbol{y}^{i,j}\right) = \left(\boldsymbol{x}^{i,j},\boldsymbol{y}_7^{i,j},\ldots,\boldsymbol{y}_{P(j)+1}^{i,j},Y_{P(j)}^{i,j},\boldsymbol{y}_{P(j)-1}^{i,j},\ldots,\boldsymbol{y}_0^{i,j}\right)$, where $Y_{P(j)}^{i,j} = S_j\left(\boldsymbol{x}_j^{i,j} \bigoplus k_{i,j}\right) \bigoplus \boldsymbol{y}_{(P(j)-2)\bmod 8}^{i,j}$, S_j is the j-th S-box of LBlock, and $P(x)$ is the nibble diffusion function. Moreover, $Q_{i,8}\left(\boldsymbol{x}^{i,8},\boldsymbol{y}^{i,8}\right) = \left(\boldsymbol{y}^{i,8},\boldsymbol{x}^{i,8}\right)$.

Using Algorithm 2, we find a 17-round integral distinguisher of LBlock which is in accordance with the previous longest integral distinguisher [8], and a better 16-round integral distinguisher with less active bits. The detail forms of the integral distinguishers are shown in Table 6.

Table 6. Integral distinguishers of LBlock

Cipher	Distinguisher
17-LBlock	In: (caaaaaaaaaaaaaaaaaaaaaaaaaaaaaaa,aaaaaaaaaaaaaaaaaaaaaaaaaaaaaaaa) Out: (????????????????????????????????,??bb??????????????????????????bb)
16-LBlock	In: (aaccaaaaaaaaaaaaaaaaaaaaaaaaaaaa,aaaaaaaaaaaaaaaaaaaaaaaaaaaaaaaa) Out: (????????????????????????????????,??bbbbbbbbbb?b?bb?b?bbbb????????)

6 Using BDPT to Recover the Superpoly in Cube Attack

In this section, we analyze the ANF coefficients of non-blackbox polynomial and superpoly in cube attack. Then, we show an MILP-aided method based on BDPT to recover the ANF coefficients of superpoly.

6.1 Analyze the ANF Coefficients of Polynomial

Let $f(\boldsymbol{x}, \boldsymbol{v})$ be a polynomial, where $\boldsymbol{x} \in \mathbb{F}_2^n$ and $\boldsymbol{v} \in \mathbb{F}_2^m$ denote the secret and public variables, respectively. In cube attack, $f_{I_v,J_v,K_v}(\boldsymbol{x}, \boldsymbol{v})$ denotes a function that the public variables indexed by $I_v \subset \{0, 1, \cdots, m-1\}$ are chosen as cube variables, the public variables indexed by $J_v \subset \{0, 1, \cdots, m-1\} - I_v$ are set to 1, and the remaining public variables $K_v = \{0, 1, \cdots, m-1\} - I_v - J_v$ are set to 0. Then, the ANF of $f_{I_v,J_v,K_v}(\boldsymbol{x}, \boldsymbol{v})$ can be represented as follows

$$f_{I_v,J_v,K_v}(\boldsymbol{x}, \boldsymbol{v}) = \bigoplus_{\boldsymbol{u}_x \in \mathbb{F}_2^n, \boldsymbol{u}_v \preceq \boldsymbol{u}_I} a_{(\boldsymbol{u}_x, \boldsymbol{u}_v)}^{f_{I_v,J_v,K_v}} \cdot (\boldsymbol{x}, \boldsymbol{v})^{(\boldsymbol{u}_x, \boldsymbol{u}_v)}.$$

where $a_{(\boldsymbol{u}_x, \boldsymbol{u}_v)}^{f_{I_v,J_v,K_v}}$ is the ANF coefficient of term $(\boldsymbol{x}, \boldsymbol{v})^{(\boldsymbol{u}_x, \boldsymbol{u}_v)}$ in $f_{I_v,J_v,K_v}(\boldsymbol{x}, \boldsymbol{v})$.

For polynomial $f_{I_v,J_v,K_v}(\boldsymbol{x}, \boldsymbol{v})$ and an index subset $I_x \subset \{0, 1, \cdots, n-1\}$, if fixing all the secret variables $\{x_k | k \in \{0, 1, \cdots, n-1\} - I_x\}$ to 0, we can get a new polynomial denoted as $f_{I_x,I_v,J_v,K_v}(\boldsymbol{x}, \boldsymbol{v})$.

Definition 6. (Similar Polynomial). *For subsets of indices $I'_x \subset I_x$, the polynomial $f_{I'_x,I_v,J_v,K_v}(\boldsymbol{x}, \boldsymbol{v})$ is called the similar polynomial of $f_{I_x,I_v,J_v,K_v}(\boldsymbol{x}, \boldsymbol{v})$.*

Lemma 2. *If $f_{I'_x,I_v,J_v,K_v}(\boldsymbol{x}, \boldsymbol{v})$is the similar polynomial of $f_{I_x,I_v,J_v,K_v}(\boldsymbol{x}, \boldsymbol{v})$, then the value of ANF coefficient $a_{\left(\boldsymbol{u}_{I'_x}, \boldsymbol{u}_{I_v}\right)}^{f_{I'_x,I_v,J_v,K_v}}$ in $f_{I'_x,I_v,J_v,K_v}(\boldsymbol{x}, \boldsymbol{v})$ is equal to the value of ANF coefficients $a_{\left(\boldsymbol{u}_{I'_x}, \boldsymbol{u}_{I_v}\right)}^{f_{I_x,I_v,J_v,K_v}}$ in $f_{I_x,I_v,J_v,K_v}(\boldsymbol{x}, \boldsymbol{v})$.*

Proof. For $f_{I_x,I_v,J_v,K_v}(\boldsymbol{x},\boldsymbol{v})$, if all the variables of $\{x_i|i \in I_x - I'_x\}$ are assigned 0, it becomes the function $f_{I'_x,I_v,J_v,K_v}(\boldsymbol{x},\boldsymbol{v})$. Compared with the ANF of $f_{I_x,I_v,J_v,K_v}(\boldsymbol{x},\boldsymbol{v})$, the ANF of $f_{I'_x,I_v,J_v,K_v}(\boldsymbol{x},\boldsymbol{v})$ only misses terms that contain any variables of $\{x_i|i \in I_x - I'_x\}$. Moreover, $\boldsymbol{x}^{\boldsymbol{u}_{I'_x}}$ doesn't contain any variables of $\{x_i|i \in I_x - I'_x\}$, so $a^{f_{I'_x,I_v,J_v,K_v}}_{\left(\boldsymbol{u}_{I'_x},\boldsymbol{u}_{I_v}\right)} = a^{f_{I_x,I_v,J_v,K_v}}_{\left(\boldsymbol{u}_{I'_x},\boldsymbol{u}_{I_v}\right)}$.

6.2 Analyze the ANF Coefficients of Superpoly

The most important part of cube attack is recovering the superpoly. Once the superpoly is recovered, attackers can compute the sum of encryptions over the cube and get one equation about secret variables.

Let C_{I_v,J_v,K_v} be a cube set defined as Eq. (1) in Sect. 2.5. For polynomial $f_{I_v,J_v,K_v}(\boldsymbol{x},\boldsymbol{v})$, where $\boldsymbol{x} \in \mathbb{F}_2^n$ and $\boldsymbol{v} \in \mathbb{F}_2^m$, it can be unique represented as

$$f_{I_v,J_v,K_v}(\boldsymbol{x},\boldsymbol{v}) = \boldsymbol{v}^{\boldsymbol{u}_{I_v}} \cdot p_{I_v,J_v,K_v}(\boldsymbol{x}) \oplus q_{I_v,J_v,K_v}(\boldsymbol{x},\boldsymbol{v}). \tag{3}$$

where $p_{I_v,J_v,K_v}(\boldsymbol{x})$ does not contain any variable in $\{v_i|i \in I_v\}$, and each term of $q_{I_v,J_v,K_v}(\boldsymbol{x},\boldsymbol{v})$ is not divisible by $\boldsymbol{v}^{\boldsymbol{u}_{I_v}}$. Then, $p_{I_v,J_v,K_v}(\boldsymbol{x})$ is called the superpoly of C_{I_v,J_v,K_v} in $f_{I_v,J_v,K_v}(\boldsymbol{x},\boldsymbol{v})$.

Definition 7. *Let C_{I_x,I_v,J_v,K_v} be the set of $(\boldsymbol{x},\boldsymbol{v})$ satisfying secret variables $\{x_i|i \in I_x\}$ are taking all possible combinations of values, secret variables $\{x_i|i \in \{0,1,\ldots,n-1\} - I_x\}$ are set to constant 0, public variables $\{v_i|i \in I_v\}$ are taking all possible combinations of values, public variables $\{v_j|j \in J_v\}$ are set to constant 1, and public variables $\{v_k|k \in K_v\}$ are set to constant 0.*

Here, we propose a method to calculate the ANF coefficient of superpoly.

Proposition 3. *For any index subset $I_x \subset \{0,1,\ldots,n-1\}$, the ANF coefficient of term $\boldsymbol{x}^{\boldsymbol{u}_{I_x}}$ in the superpoly $p_{I_v,J_v,K_v}(\boldsymbol{x})$ can be calculated as*

$$a^{p_{I_v,J_v,K_v}}_{\boldsymbol{u}_{I_x}} = \bigoplus_{(\boldsymbol{x},\boldsymbol{v}) \in C_{I_x,I_v,J_v,K_v}} f_{I_x,I_v,J_v,K_v}(\boldsymbol{x},\boldsymbol{v}).$$

Proof. The ANF of $p_{I_v,J_v,K_v}(\boldsymbol{x})$ can be presented as

$$p_{I_v,J_v,K_v}(\boldsymbol{x}) = \bigoplus_{\boldsymbol{u} \in \mathbb{F}_2^n} a^{p_{I_v,J_v,K_v}}_{\boldsymbol{u}} \cdot \boldsymbol{x}^{\boldsymbol{u}}.$$

Then, the ANF of $\boldsymbol{v}^{\boldsymbol{u}_{I_v}} \cdot p_{I_v,J_v,K_v}(\boldsymbol{x})$ can be presented as

$$\boldsymbol{v}^{\boldsymbol{u}_{I_v}} \cdot p_{I_v,J_v,K_v}(\boldsymbol{x}) = \bigoplus_{\boldsymbol{u} \in \mathbb{F}_2^n} a^{p_{I_v,J_v,K_v}}_{\boldsymbol{u}} \cdot (\boldsymbol{x},\boldsymbol{v})^{(\boldsymbol{u},\boldsymbol{u}_{I_v})}.$$

So, the ANF coefficient of $(\boldsymbol{x},\boldsymbol{v})^{(\boldsymbol{u}_{I_x},\boldsymbol{u}_{I_v})}$ in $\boldsymbol{v}^{\boldsymbol{u}_{I_v}} \cdot p_{I_v,J_v,K_v}(\boldsymbol{x},\boldsymbol{v})$ is also $a^{p_{I_v,J_v,K_v}}_{\boldsymbol{u}_{I_x}}$.

Because $f_{I_v,J_v,K_v}(\boldsymbol{x},\boldsymbol{v})$ can be unique represented as Eq. (3) and every term in $q_{I_v,J_v,K_v}(\boldsymbol{x},\boldsymbol{v})$ misses at least one variable from $\{v_i|i \in I_v\}$, the term

$(\boldsymbol{x},\boldsymbol{v})^{(\boldsymbol{u}_{I_x},\boldsymbol{u}_{I_v})}$ doesn't exist in $q_{I_v,J_v,K_v}(\boldsymbol{x},\boldsymbol{v})$. According to Eq. (3), we obtain that the ANF coefficient of term $(\boldsymbol{x},\boldsymbol{v})^{\boldsymbol{u}_{I_x},\boldsymbol{u}_{I_v}}$ in $f_{I_v,J_v,K_v}(\boldsymbol{x},\boldsymbol{v})$ is $a_{\boldsymbol{u}_{I_x}}^{p_{I_v,J_v,K_v}}$. Namely,

$$a_{\boldsymbol{u}_{I_x}}^{p_{I_v,J_v,K_v}} = a_{(\boldsymbol{u}_{I_x},\boldsymbol{u}_{I_v})}^{f_{I_v,J_v,K_v}}. \tag{4}$$

From the Definition 6, we know that f_{I_x,I_v,J_v,K_v} is the similar polynomial of f_{I_v,J_v,K_v}. And according to Lemma 2, we obtain that

$$a_{\boldsymbol{u}_{I_x}}^{p_{I_v,J_v,K_v}} = a_{(\boldsymbol{u}_{I_x},\boldsymbol{u}_{I_v})}^{f_{I_v,J_v,K_v}} = a_{(\boldsymbol{u}_{I_x},\boldsymbol{u}_{I_v})}^{f_{I_x,I_v,J_v,K_v}}. \tag{5}$$

Then, we have

$$\begin{aligned}
&\bigoplus_{(\boldsymbol{x},\boldsymbol{v})\in C_{I_x,I_v,J_v,K_v}} f_{I_x,I_v,J_v,K_v}(\boldsymbol{x},\boldsymbol{v}) \\
&= \bigoplus_{(\boldsymbol{x},\boldsymbol{v})\in C_{I_x,I_v,J_v,K_v}} \bigoplus_{\boldsymbol{u}_x\preceq\boldsymbol{u}_{I_x},\boldsymbol{u}_v\preceq\boldsymbol{v}_{I_v}} a_{(\boldsymbol{u}_x,\boldsymbol{u}_v)}^{f_{I_x,I_v,J_v,K_v}}\cdot(\boldsymbol{x},\boldsymbol{v})^{(\boldsymbol{u}_x,\boldsymbol{u}_v)} \\
&= a_{(\boldsymbol{u}_{I_x},\boldsymbol{u}_{I_v})}^{f_{I_x,I_v,J_v,K_v}} = a_{\boldsymbol{u}_{I_x}}^{p_{I_v,J_v,K_v}}.
\end{aligned}$$

□

6.3 The Algorithm to Recover Superpoly

The set C_{I_x,I_v,J_v,K_v} can be viewed as a cube set, according to the definition of BDPT, we know that the BDPT of C_{I_x,I_v,J_v,K_v} is $\mathcal{D}_{\mathbb{K},\mathbb{L}}^{1^n}$, where $\mathbb{K}=\emptyset$, and $\mathbb{L}=\{(\boldsymbol{u}_{I_x},\boldsymbol{u}_v)\,|\boldsymbol{u}_{I_v}\preceq\boldsymbol{u}_v\preceq\boldsymbol{u}_{I_v}\oplus\boldsymbol{u}_{J_v}\}$. Then, we can use MILP-aided method (Algorithm 2) to research the propagation of $\mathcal{D}_{\mathbb{K},\mathbb{L}}^{1^n}$. The integral distinguisher got by BDPT recover the ANF coefficient of $\boldsymbol{x}^{\boldsymbol{u}_{I_x}}$ in superpoly $p_{I_v,J_v,K_v}(\boldsymbol{x})$. For example, if Algorithm 2 $BDPT(f_{I_x,I_v,J_v,K_v},\mathbb{K},\mathbb{L},0)$ return 1, it means that $\bigoplus_{(\boldsymbol{x},\boldsymbol{v})\in C_{I_x,I_v,J_v,K_v}} f_{I_x,I_v,J_v,K_v}(\boldsymbol{x},\boldsymbol{v})=1$. According to Proposition 3, we know that the ANF coefficient of $\boldsymbol{x}^{\boldsymbol{u}_{I_x}}$ in superpoly $p_{I_v,J_v,K_v}(\boldsymbol{x})$ equals 1. We illustrate the whole framework in Algorithm 3.

In order to analyze the ciphers better, we divide them into two categories: public-update ciphers and secret-update ciphers.

Definition 8. *For a function $f:\mathbb{F}_2^n\rightarrow\mathbb{F}_2^m$, if the ANF of f is definite, we call it public function. Let $E=Q_r\circ Q_{r-1}\circ\cdots\circ Q_1(\boldsymbol{x},\boldsymbol{v})$ be an r-round cipher, where Q_i is the i-th round update function, $\boldsymbol{x}$ denotes the secret variables, and $\boldsymbol{v}$ denotes the public variables. If all the round update functions Q_i, $i\in\{1,2,\cdots,r\}$ are public functions, the cipher E is public-update cipher. Otherwise we call it secret-update cipher.*

Proposition 4. *For a public-update cipher $f_{I_v,J_v,K_v}(\boldsymbol{x},\boldsymbol{v})$ and cube set C_{I_v,J_v,K_v}, the superpoly $p_{I_v,J_v,K_v}(\boldsymbol{x})$ can be fully recovered by the propagation of BDPT.*

Algorithm 3. Recover the ANF coefficient of $\boldsymbol{x}^{\boldsymbol{u}_{I_x}}$ in superpoly $p_{I_v,J_v,K_v}(\boldsymbol{x})$

procedure *RecoverCoefficient*(I_x, I_v, J_v, K_v)
  Initial $\mathbb{K} = \emptyset$, $\mathbb{L} = \{(\boldsymbol{u}_{I_x}, \boldsymbol{u}_v) \,|\, \boldsymbol{u}_{I_v} \preceq \boldsymbol{u}_v \preceq \boldsymbol{u}_{I_v} \oplus \boldsymbol{u}_{J_v}\}$
  if $BDPT(f_{I_x,I_v,J_v,K_v}, \mathbb{K}, \mathbb{L}, 0)$ return unknown
    return unknown
  else if $BDPT(f_{I_x,I_v,J_v,K_v}, \mathbb{K}, \mathbb{L}, 0)$ return 1
    return 1
  else
    return 0
end procedure

Proof. The superpoly $p_{I_v,J_v,K_v}(\boldsymbol{x})$ is a function of secret variables $\boldsymbol{x}$. If for arbitrary term $\boldsymbol{x}^{\boldsymbol{u}_{I_x}}$, we can determine its ANF coefficient. Then, the exact superpoly can be obtained.

Because $f_{I_v,J_v,K_v}(\boldsymbol{x}, \boldsymbol{v})$ is a public-update cipher, $f_{I_x,I_v,J_v,K_v}(\boldsymbol{x}, \boldsymbol{v})$ is also a public-update cipher. Then, for arbitrary term $\boldsymbol{x}^{\boldsymbol{u}_{I_x}}$, we research the propagation of BDPT $\mathcal{D}^{1^{n+m}}_{\mathbb{K},\mathbb{L}}$, where $\mathbb{K} = \emptyset$ and $\mathbb{L} = \{(\boldsymbol{u}_{I_x}, \boldsymbol{u}_v) \,|\, \boldsymbol{u}_{I_v} \preceq \boldsymbol{u}_v \preceq \boldsymbol{u}_{I_v} \oplus \boldsymbol{u}_{J_v}\}$. Let the output BDPT of $f_{I_x,I_v,J_v,K_v}(\boldsymbol{x}, \boldsymbol{v})$ be $\mathcal{D}^{1^{n+m}}_{\mathbb{K}',\mathbb{L}'}$. The initial $\mathbb{K} = \emptyset$ means that there is no division trail from $\mathbb{K} = \emptyset$ to $\mathbb{K}'$. From Sect. 2.3, we know that for public function, the BDPT propagation of $\mathbb{K}$ and $\mathbb{L}$ is independent. Only when the secret round key is involved, some vectors of $\mathbb{L}$ will affect $\mathbb{K}$. That means, there is no division trail from $\mathbb{L}$ to $\mathbb{K}'$ when all the update functions are public. The output set $\mathbb{K}' = \emptyset$ and the return value of Algorithm 3 is constant (0 or 1). So the ANF coefficient of arbitrary term $\boldsymbol{x}^{\boldsymbol{u}_{I_x}}$ can be recovered by BDPT. □

According to Sect. 2.6, for polynomial $f_{I_v,J_v,K_v}(\boldsymbol{x}, \boldsymbol{v})$ and cube set C_{I_v,J_v,K_v}, we can use MILP method to evaluate the secret variables involved in the superpoly and the upper bounding degree of superpoly. We denote the involved secret variables indices set as I and the upper bounding degree as d. Then, in order to recover the superpoly, we only need to determine the coefficients $a_{\boldsymbol{u}}^{p_{I_v,J_v,K_v}}$ satisfying $\boldsymbol{u} \preceq \boldsymbol{u}_I$ and $hw(\boldsymbol{u}) \leq d$.

Analysis of Public-Update Cipher. According to Proposition 4, we can query the Algorithm 3 $\sum_{i=0}^{d} \binom{|I|}{i}$ times to recover all the ANF coefficients of superpoly. The complexity is $c \cdot \sum_{i=0}^{d} \binom{|I|}{i}$, where c is the average computational complexity of Algorithm 3. Compared with CBDP based cube attack in Sect. 2.6, we can know that when $c < 2^{|I_v|}$, our method can obtain better results.

Analysis of Secret-Update Cipher. Due to the influence of secret keys in the intermediate rounds, new vectors may be generated from $\mathbb{L}_i$ and added to $\mathbb{K}_i$. Therefore, the condition that the output BDPT set $\mathbb{K}' = \emptyset$ may not hold. Namely, only a part of the ANF coefficients in superpoly $p_{I_v,J_v,K_v}(\boldsymbol{x}, \boldsymbol{v})$ can be obtained by BDPT. If there are N ANF coefficients that cannot be determined by BDPT, we have to get their ANF coefficients by the method used in the

CBDP based cube attack. Therefore, the complexity of recovering superpoly is $\left\{c \cdot \sum_{i=0}^{d} \binom{|I|}{i} + N \cdot 2^{|I_v|}\right\}$.

7 Application to Trivium

In order to verify the correctness and effectiveness of our method, we apply it to Trivium [5] which is a public-update cipher.

7.1 Descriptions of Trivium

Trivium [5] is a bit-oriented stream cipher with 288-bit internal state denoted by $\boldsymbol{s} = (s_0, s_1, \ldots, s_{287})$. To outline our technique more conveniently, we describe Trivium using the following expression. Let $\boldsymbol{x} = (x_0, x_1, \cdots, x_{79})$ denote the secret variables (80-bit Key), and $\boldsymbol{v} = (v_0, v_1, \cdots, v_{207})$ denote the public variables. For public variables, $v_{13}, v_{14}, \cdots, v_{92}$ are the IV variables whose values can be chosen by attackers (80-bit IV), $\{v_{205}, v_{206}, v_{207}\}$ are set to 1, and others are set to 0. Then, the algorithm would not output any keystream bit until the internal state is updated 1152 rounds. A complete description of Trivium is given by the following simple pseudo-code.

$(s_0, s_1, \ldots, s_{92}) \leftarrow (x_0, \ldots, x_{79}, v_0, \ldots, v_{12})$
$(s_{93}, s_{94}, \ldots, s_{176}) \leftarrow (v_{13}, \ldots, v_{96})$
$(s_{177}, s_{178}, \ldots, s_{287}) \leftarrow (v_{97}, \ldots, v_{207})$
for $i = 1$ **to** N **do**
  if $i > 1152$ **then**
    $z_{i-1152} \leftarrow s_{65} \oplus s_{92} \oplus s_{161} \oplus s_{176} \oplus s_{242} \oplus s_{287}$
  end if
  $t_1 \leftarrow s_{65} \oplus s_{90} \cdot s_{91} \oplus s_{92} \oplus s_{170}$
  $t_2 \leftarrow s_{161} \oplus s_{174} \cdot s_{175} \oplus s_{176} \oplus s_{263}$
  $t_3 \leftarrow s_{242} \oplus s_{285} \cdot s_{286} \oplus s_{287} \oplus s_{68}$
  $(s_0, s_1, \ldots, s_{92}) \leftarrow (t_2, s_0, \ldots, s_{91})$
  $(s_{93}, s_{94}, \ldots, s_{176}) \leftarrow (t_0, s_{93}, \ldots, s_{175})$
  $(s_{177}, s_{178}, \ldots, s_{287}) \leftarrow (t_1, s_{177}, \ldots, s_{286})$
end for

7.2 The MILP-aided Algorithm for Trivium

Because Trivium is a public-update cipher, during the progress of recovering the ANF coefficients of superpoly, the set $\mathbb{K}$ is always empty. The papers [23,26] have showed the method on how to build the CBDP model of Trivium. Here, we propose Algorithm 4 to get the $\mathbb{L}$'s propagation of Trivium's round function. The

---

Algorithm 4. The propagation of $\mathbb{L}$ for the round function

---

1 **procedure** $CorePropagation(\mathbb{L}, i_0, i_1, i_2, i_3, i_4)$
2   Let $\boldsymbol{x} = (x_0, x_1, x_2, x_3, x_4)$ be the variables
3   Let $\boldsymbol{y}$ be the function of $\boldsymbol{x}$, and $\boldsymbol{y} = (x_0, x_1, x_2, x_3, x_0x_1 + x_2 + x_3 + x_4)$
4      $\mathbb{L}' = \emptyset$
5   **for** ℓ in $\mathbb{L}$
6   **for all** $\boldsymbol{u} = (u_0, u_1, u_2, u_3, u_4) \in \mathbb{F}_2^5$ **do**
7   **if** $\boldsymbol{y}^{\boldsymbol{u}}$ contains the term $\boldsymbol{x}^{\left(\ell_{i_0}, \ell_{i_1}\ell_{i_2}, \ell_{i_3}, \ell_{i_4}\right)}$ **then**
8                  $\boldsymbol{\ell}' = \boldsymbol{\ell}$
9                  $\ell'_{i_0} = u_0,\ \ell'_{i_1} = u_1,\ \ell'_{i_2} = u_2,\ \ell'_{i_3} = u_3,\ \ell'_{i_4} = u_4$
10                 $\mathbb{L}' \xleftarrow{x} \boldsymbol{\ell}'$
11   **end if**
12   **end for**
13   **end for**
14   **return** $\mathbb{L}'$
15 **end procedure**

1 **procedure** $RoundPropagation(\mathbb{L}_r)$
2   **initial** $\mathbb{L}' = \emptyset,\ \mathbb{L}'' = \emptyset,\ \mathbb{L}''' = \emptyset,\ \mathbb{L}_{r+1} = \emptyset$
3      $\mathbb{L}' = CorePropagation(\mathbb{L}_r, 65, 170, 90, 91, 92)$
4      $\mathbb{L}'' = CorePropagation(\mathbb{L}', 161, 163, 174, 175, 176)$
5      $\mathbb{L}''' = CorePropagation(\mathbb{L}'', 242, 68, 285, 286, 287)$
6   **for** all ℓ in $\mathbb{L}'''$ **do**
7          $\mathbb{L}_{r+1} = \mathbb{L}_{r+1} \bigcup \{\ell \ggg 1\}$
8   **end for**
9   **return** $\mathbb{L}_{r+1}$
10 **end procedure**

---

input of procedure *RoundPropagation* in Algorithm 4 is the r-th round BDPT set $\mathbb{L}_r$, and the outputs is the $(r + 1)$-th round BDPT set $\mathbb{L}_{r+1}$.

At CRYPTO 2017 [23], Todo *et al.* proposed a CBDP based cube attack on the 832-round Trivium. Then, at CRYPTO 2018 [26], Wang *et al.* improved the result and presented a CBDP based cube attack on 839-round Trivium. But both methods cannot ensure whether the cube attacks are key recovery attacks or not. After applying Algorithm 3 to the 832-round and 839-round Trivium, we have the following results.

Result 1. *For cube set C_{I_v, J_v, K_v}, where $I_v = \{13, \ldots, 45, 47, \ldots, 58, 60, \ldots, 92\}$, no matter what the assignment to the non-cube IVs $\{46, 59\}$ is, the corresponding superpoly of 839-round Trivium in the paper [26] is constant. So the cube attack based on CBDP in the paper [26] is not key recovery attack.*

Result 2. *For the cube set C_{I_v, J_v, K_v}, where $I_v = \{13, 14, \ldots, 77, 79, 81, \ldots, 91\}$, the superpolies of some assignments are constant. For example, when $J_v = \{205, 206, 207\}$ and $K_v = \{0, 1, \ldots, 207\} - I_v - J_v$, the superpoly recovered is*

$p_{I_v,J_v,K_v}(\boldsymbol{x}) = 0$. *And the superpolies of some assignments are non-constant. For example, when* $J_v = \{80, 90, 205, 206, 207\}$ *and* $K_v = \{0, 1, \ldots, 207\} - I_v - J_v$, *the superpoly recovered is* $p_{I_v,J_v,K_v}(\boldsymbol{x}) = x_{56}x_{57}x_{58} + x_{32}x_{56} + x_{56}x_{59}$. *In a word, the assignment to the non-cube IVs will affect whether the cube attack on 832-round Trivium in the paper [23] is key recovery attack or not.*

7.3 Theoretical Result

Result 3. *Let* C_{I_v,J_v,K_v} *be a cube set, where* $I_v = \{13, 14, \ldots, 89, 91\}$, $J_v = \{205, 206, 207\}$, *and* $K_v = \{0, 1, \ldots, 204\} - I_v$. *Using the degree bounding technique in the paper [26], we can get that the degree of superpoly in 841-round Trivium is not larger than 10. Then, we have* $\sum_{i=0}^{d} \binom{|I|}{i} \leq \sum_{i=0}^{10} \binom{80}{i} \leq 2^{41}$. *That means we can use no more than* 2^{41} *MILP-aided propagation of BDPT to recover the exact superpoly of 841-round Trivium.*

Because our computing resources are limited, the exact superpoly of 841-round Trivium cannot be recovered in practical time. On our common PC (Intel Core i5-4590 CPU @3.3 GHz, 8.00G RAM), it takes about 18 days to complete the MILP-aided propagation of BDPT 100 times.

8 Conclusions

This paper is committed to solve the complexity problem of searching integral distinguishers based on BDPT. In order to make the propagation of BDPT efficient, we show the pruning techniques which can removing redundant vectors in time. Then, an algorithm is designed to estimate whether the m-th output bit is balanced or not based on BDPT. We apply the searching algorithm to some blocks, and the obtained integral distinguishers are the same or better than the previous longest integral distinguishers. It should be noted that the absence of integral distinguishers based on BDPT doesn't imply the absence of integral distinguishers. Any improvement on the accuracy of BDPT propagation may obtain better integral distinguishers. Moreover, our searching algorithm supposes that all round keys are chosen randomly. If consider the key scheduling algorithm, we may obtain better integral distinguishers.

Moreover, we apply BDPT to recover the superpoly in cube attack. As far as we know, this is the first application of BDPT to stream ciphers. For public-update ciphers, the exact ANF of superpoly can be fully recovered by exploring the propagation of BDPT. To verify the correctness and effectiveness of our method, we apply it to Trivium. For the cube attack on the 832-round Trivium [23], we obtain that only some proper non-cube IV assignments can obtain non-constant superpolies. For the cube attack on 839-round Trivium [26], our result shows that the superpoly is always constant. Because our method can determine the ANF coefficients of superpoly in practical time, we propose a theoretical superpoly recovery of 841-round Trivium.

For secret-update ciphers, due to the influence of intermediate round keys, not all the ANF coefficients can be obtained by BDPT. From this perspective, when we design stream ciphers, the secret-update ciphers are more secure. How to recover the superpoly of secret-update ciphers is our future work.

Acknowledgement. The authors would like to thank the anonymous reviewers for their detailed comments and suggestions. This work was supported by the National Natural Science Foundation of China [Grant No. 61572516, 61802437].

Appendix

Table 7. The $\mathbb{L}$ propagation of BDPT for the core operation of SIMON

Input $\mathcal{D}^{1^4}_{\mathbb{K},\{\boldsymbol{\ell}\}}$	Output $\mathcal{D}^{1^4}_{\mathbb{K}',\mathbb{L}'}$
$\boldsymbol{\ell}=[0,0,0,0]$	$\mathbb{L}'=\{[0,0,0,0]\}$
$\boldsymbol{\ell}=[1,0,0,0]$	$\mathbb{L}'=\{[1,0,0,0]\}$
$\boldsymbol{\ell}=[0,1,0,0]$	$\mathbb{L}'=\{[0,1,0,0]\}$
$\boldsymbol{\ell}=[1,1,0,0]$	$\mathbb{L}'=\{[1,1,0,0],[0,0,0,1],[1,0,0,1],[0,1,0,1],[1,1,0,1]\}$
$\boldsymbol{\ell}=[0,0,1,0]$	$\mathbb{L}'=\{[0,0,1,0],[0,0,0,1],[0,0,1,1]\}$
$\boldsymbol{\ell}=[1,0,1,0]$	$\mathbb{L}'=\{[1,0,1,0],[1,0,0,1],[1,0,1,1]\}$
$\boldsymbol{\ell}=[0,1,1,0]$	$\mathbb{L}'=\{[0,1,1,0],[0,1,0,1],[0,1,1,1]\}$
$\boldsymbol{\ell}=[1,1,1,0]$	$\mathbb{L}'=\{[1,1,1,0],[0,0,1,1],[1,0,1,1],[0,1,1,1],[1,1,0,1]\}$
$\boldsymbol{\ell}=[\ell_0,\ell_1,\ell_2,1]$	$\mathbb{L}'=\{[\ell_0,\ell_1,\ell_2,1]\}$

Experimental Verification

Example 1. For 591-round Trivium and cube set C_{I_v,J_v,K_v}, where $I_v=\{13,23,33,43,53,63,73,83\}$, $J_v=\{14,29,32,205,206,207\}$ and $K_v=\{0,1,\cdots,207\}-I_v-J_v$, we can get that the involved secret variables are $\{x_{22},x_{23},x_{24},x_{66}\}$, the degree of superpoly is not larger than 2. Then, we use Algorithm 3 to recover all the ANF coefficients of the superpoly, which is in accordance with the practically recovered superpoly as follows:

$$p_{I_v,J_v,K_v}(\boldsymbol{x})=x_{66}+x_{24}+x_{23}x_{22}+1.$$

Example 2. For 591-round Trivium and cube set C_{I_v,J_v,K_v}, where $I_v=\{13,23,33,43,53,63,73,83\}$, $J_v=\{29,32,82,205,206,207\}$, and $K_v=\{0,1,\cdots,207\}-I_v-J_v$, we can get that the involved secret variables are $\{x_{22},x_{23},x_{24},x_{65},x_{66}\}$, the degree of superpoly is not larger than 3. Then, we use Algorithm 3 to recover the superpoly, which is in accordance with the practically recovered superpoly as follows:

$$p_{I_v,J_v,K_v}(\boldsymbol{x})=x_{65}x_{23}x_{22}+x_{65}x_{24}+x_{66}x_{65}+x_{65}.$$

References

1. Abdelkhalek, A., Sasaki, Y., Todo, Y., Tolba, M., Youssef, M.: MILP modeling for (large) S-boxes to optimize probability of differential characteristics. IACR Trans. Symmetric Cryptol. **2017**(4), 99–129 (2017)
2. Beaulieu, R., Shors, D., Smith, J., Treatman-Clark, S., Weeks, B., Wingers, L.: The SIMON and SPECK families of lightweight block ciphers. IACR Cryptology ePrint Archive 2013:404 (2013). http://eprint.iacr.org/2013/404
3. Bogdanov, A., et al.: PRESENT: an ultra-lightweight block cipher. In: Paillier, P., Verbauwhede, I. (eds.) CHES 2007. LNCS, vol. 4727, pp. 450–466. Springer, Heidelberg (2007). https://doi.org/10.1007/978-3-540-74735-2_31
4. Boura, C., Canteaut, A.: Another view of the division property. In: Robshaw, M., Katz, J. (eds.) CRYPTO 2016. LNCS, vol. 9814, pp. 654–682. Springer, Heidelberg (2016). https://doi.org/10.1007/978-3-662-53018-4_24
5. De Cannière, C., Preneel, B.: Trivium. In: Robshaw, M., Billet, O. (eds.) New Stream Cipher Designs. LNCS, vol. 4986, pp. 244–266. Springer, Heidelberg (2008). https://doi.org/10.1007/978-3-540-68351-3_18
6. Dinur, I., Shamir, A.: Cube attacks on tweakable black box polynomials. In: Joux, A. (ed.) EUROCRYPT 2009. LNCS, vol. 5479, pp. 278–299. Springer, Heidelberg (2009). https://doi.org/10.1007/978-3-642-01001-9_16
7. Dinur, I., Shamir, A.: Breaking grain-128 with dynamic cube attacks. In: Joux, A. (ed.) FSE 2011. LNCS, vol. 6733, pp. 167–187. Springer, Heidelberg (2011). https://doi.org/10.1007/978-3-642-21702-9_10
8. Eskandari, Z., Kidmose, A.B., Kölbl, S., Tiessen, T.: Finding integral distinguishers with ease. In: Cid, C., Jacobson Jr., M. (eds.) SAC 2018. Lecture Notes in Computer Science, vol. 11349, pp. 115–138. Springer, Cham (2019). https://doi.org/10.1007/978-3-030-10970-7_6
9. Fu, X., Wang, X., Dong, X., Meier, W.: A key-recovery attack on 855-round Trivium. In: Shacham, H., Boldyreva, A. (eds.) CRYPTO 2018. LNCS, vol. 10992, pp. 160–184. Springer, Cham (2018). https://doi.org/10.1007/978-3-319-96881-0_6
10. Gurobi: http://www.gurobi.com/
11. Knudsen, L., Wagner, D.: Integral cryptanalysis. In: Daemen, J., Rijmen, V. (eds.) FSE 2002. LNCS, vol. 2365, pp. 112–127. Springer, Heidelberg (2002). https://doi.org/10.1007/3-540-45661-9_9
12. Hao, Y., Jiao, L., Li, C., Meier, W., Todo, Y., Wang, Q.: Observations on the dynamic cube attack of 855-Round TRIVIUM from Crypto 2018. IACR Cryptology ePrint Archive 2018:972 (2018). https://eprint.iacr.org/2018/972.pdf
13. Hu, K., Wang, M.: Automatic search for a variant of division property using three subsets. In: Matsui, M. (ed.) CT-RSA 2019. LNCS, vol. 11405, pp. 412–432. Springer, Cham (2019). https://doi.org/10.1007/978-3-030-12612-4_21
14. Huang, S., Wang, X., Xu, G., Wang, M., Zhao, J.: Conditional cube attack on reduced round keccak sponge function. In: Coron, J.-S., Nielsen, J.B. (eds.) EUROCRYPT 2017. LNCS, vol. 10211, pp. 259–288. Springer, Cham (2017). https://doi.org/10.1007/978-3-319-56614-6_9
15. Liu, M., Yang, J., Wang, W., Lin, D.: Correlation cube attacks: from weak-key distinguisher to key recovery. In: Nielsen, J.B., Rijmen, V. (eds.) EUROCRYPT 2018. LNCS, vol. 10821, pp. 715–744. Springer, Cham (2018). https://doi.org/10.1007/978-3-319-78375-8_23
16. Liu, M.: Degree evaluation of NFSR-based cryptosystems. In: Katz, J., Shacham, H. (eds.) CRYPTO 2017. LNCS, vol. 10403, pp. 227–249. Springer, Cham (2017). https://doi.org/10.1007/978-3-319-63697-9_8

17. Sage: http://www.sagemath.org/
18. Sun, B., Hai, X., Zhang, W., Cheng, L., Yang, Z.: New observation on division property. Sci. Chin. (Inf. Sci.) **2017**(09), 274–276 (2017)
19. Sun, S., Hu, L., Wang, P., Qiao, K., Ma, X., Song, L.: Automatic security evaluation and (related-key) differential characteristic search: application to SIMON, PRESENT, LBlock, DES(L) and other bit-oriented block ciphers. In: Sarkar, P., Iwata, T. (eds.) ASIACRYPT 2014. LNCS, vol. 8873, pp. 158–178. Springer, Heidelberg (2014). https://doi.org/10.1007/978-3-662-45611-8_9
20. Todo, Y.: Integral cryptanalysis on full MISTY1. In: Gennaro, R., Robshaw, M. (eds.) CRYPTO 2015. LNCS, vol. 9215, pp. 413–432. Springer, Heidelberg (2015). https://doi.org/10.1007/978-3-662-47989-6_20
21. Todo, Y., Morii, M.: Bit-based division property and application to Simon family. In: Peyrin, T. (ed.) FSE 2016. LNCS, vol. 9783, pp. 357–377. Springer, Heidelberg (2016). https://doi.org/10.1007/978-3-662-52993-5_18
22. Todo, Y.: Structural evaluation by generalized integral property. In: Oswald, E., Fischlin, M. (eds.) EUROCRYPT 2015. LNCS, vol. 9056, pp. 287–314. Springer, Heidelberg (2015). https://doi.org/10.1007/978-3-662-46800-5_12
23. Todo, Y., Isobe, T., Hao, Y., Meier, W.: Cube attacks on non-blackbox polynomials based on division property. In: Katz, J., Shacham, H. (eds.) CRYPTO 2017. LNCS, vol. 10403, pp. 250–279. Springer, Cham (2017). https://doi.org/10.1007/978-3-319-63697-9_9
24. Wu, W., Zhang, L.: LBlock: a lightweight block cipher. In: Lopez, J., Tsudik, G. (eds.) ACNS 2011. LNCS, vol. 6715, pp. 327–344. Springer, Heidelberg (2011). https://doi.org/10.1007/978-3-642-21554-4_19
25. Wang, Q., Liu, Z., Varıcı, K., Sasaki, Y., Rijmen, V., Todo, Y.: Cryptanalysis of reduced-round SIMON32 and SIMON48. In: Meier, W., Mukhopadhyay, D. (eds.) INDOCRYPT 2014. LNCS, vol. 8885, pp. 143–160. Springer, Cham (2014). https://doi.org/10.1007/978-3-319-13039-2_9
26. Wang, Q., Hao, Y., Todo, Y., Li, C., Isobe, T., Meier, W.: Improved division property based cube attacks exploiting algebraic properties of superpoly. In: Shacham, H., Boldyreva, A. (eds.) CRYPTO 2018. LNCS, vol. 10991, pp. 275–305. Springer, Cham (2018). https://doi.org/10.1007/978-3-319-96884-1_10
27. Xiang, Z., Zhang, W., Bao, Z., Lin, D.: Applying MILP method to searching integral distinguishers based on division property for 6 lightweight block ciphers. In: Cheon, J.H., Takagi, T. (eds.) ASIACRYPT 2016. LNCS, vol. 10031, pp. 648–678. Springer, Heidelberg (2016). https://doi.org/10.1007/978-3-662-53887-6_24
28. Xie, X., Tian, T.: Improved distinguisher search techniques based on parity sets. Sci. Chin. Inf. Sci. **55**, 2712 (2018)
29. Yang, G., Zhu, B., Suder, V., Aagaard, M.D., Gong, G.: The Simeck family of lightweight block ciphers. In: Güneysu, T., Handschuh, H. (eds.) CHES 2015. LNCS, vol. 9293, pp. 307–329. Springer, Heidelberg (2015). https://doi.org/10.1007/978-3-662-48324-4_16
30. Ye, C., Tian, T.: Deterministic cube attacks. IACR Cryptology ePrint Archive, 2018:1028 (2018). https://eprint.iacr.org/2018/1082.pdf
31. Zhang, W., Bao, Z., Lin, D., Rijmen, V., Yang, B., Verbauwhede, I.: Rectangle: a bit-slice lightweight block cipher suitable for multiple platforms. Sci. Chin. Inf. Sci. **58**(12), 1–15 (2015)

Cryptanalysis of GSM Encryption in 2G/3G Networks Without Rainbow Tables

Bin Zhang[1,2,3,4](✉)

[1] TCA, SKLCS, Institute of Software, Chinese Academy of Sciences, Beijing, China
martin_zhangbin@hotmail.com

[2] State Key Laboratory of Cryptology, P.O.Box 5159, Beijing 100878, China

[3] University of Chinese Academy of Sciences, Beijing 100049, China

[4] State Key Laboratory of Information Security, Institute of Information Engineering, Chinese Academy of Sciences, Beijing, China

Abstract. The GSM standard developed by ETSI for 2G networks adopts the A5/1 stream cipher to protect the over-the-air privacy in cell phone and has become the de-facto global standard in mobile communications, though the emerging of subsequent 3G/4G standards. There are many cryptanalytic results available so far and the most notable ones share the need of a heavy pre-computation with large rainbow tables or distributed cracking network. In this paper, we present a fast near collision attack on GSM encryption in 2G/3G networks, which is completely new and more threatening compared to the previous best results. We adapt the fast near collision attack proposed at Eurocrypt 2018 with the concrete irregular clocking manner in A5/1 to have a state recovery attack with a low complexity. It is shown that if the first 64 bits of one keystream frame are available, the secret key of A5/1 can be reliably found in $2^{31.79}$ cipher ticks, given around 1 MB memory and after the pre-computation of $2^{20.26}$ cipher ticks. Our current implementation clearly certified the validity of the suggested attack. Due to the fact that A5/3 and GPRS share the same key with A5/1, this can be converted into attacks against any GSM network eventually.

Keywords: Cryptanalysis · GSM · A5/1 · Near collision

1 Introduction

The GSM standard, developed by ETSI for 2G networks used by mobile phones, specifies the A5/1 stream cipher to protect the over-the-air privacy all over the world. As of today, A5/1 has become the de-facto global standard for mobile communications with more than 4 billion customers and over 90% market share, operating in over 219 countries and territories.

A5/1 is the strong version of the encryption algorithm in GSM standard and A5/2 is the weak version free of export limitations. A5/1 was designed in the 80's

S. D. Galbraith and S. Moriai (Eds.): ASIACRYPT 2019, LNCS 11923, pp. 428–456, 2019.
https://doi.org/10.1007/978-3-030-34618-8_15

of last century as a typical LFSR-based stream cipher with an irregular clocking mechanism. The exact design was reverse-engineered in [6] in 1999 and confirmed by the relevant authorities subsequently. A newly standardized version in GSM networks is A5/3, which is based on the block cipher KASUMI. In practice, the algorithm is frequently re-synchronized and a GSM conversation consists of a series of frames sent every 4.6 ms. From a 64-bit secret key and a 22-bit publicly known frame counter, only 228 bits keystream are generated in each frame, and after that a new frame counter is mixed with the same key again in initialization to generate another frame.

There are lots of attacks published so far against A5/1 and the GSM encryption, to name but a few [1–5, 7–9, 12–14]. Different cryptanalytic strategies, e.g., time/memory/data (TMD) tradeoff attacks, guess-and-determine attacks and (conditional) correlation attacks have been tried, resulting in more and more powerful attacks in terms of complexity cost. But from both the theoretical and practical viewpoints, only a few results [2, 13] have given new theoretical insights and have the direct and important consequence on the GSM encryption itself. At Crypto 2003, a practical ciphertext-only attack on A5/2 was depicted in [2], which could be extended to more complex and more expensive attacks on A5/1, of which the cheapest pre-computation required 35 PCs (that were available around 2003) to work a few years with about 600 GB disks. Given the protocol flaws of the GSM networks and with the assumption that the targeted mobile phone supports A5/2, various active attacks were launched accordingly. In January 2007, the Hacker's Choice started the A5/1 cracking project with plans to use FPGAs that allow A5/1 to be broken with a rainbow table attack. Then in 2010, K. Nohl announced a new attack in [13] without the reveal of many details, using a thick rainbow table with the distinguished points technique. It had an online attack time of about 10 s on a general-purpose GPU with the success probability of 87%, given 8 known keystream frames and 30 pre-computation tables of about 1.7 Terabytes. The most recent attack on A5/1 was presented in [11], where a unified rainbow table cryptanalytic method was introduced and applied to A5/1. To have a comparable success probability and online attack time to the relevant attacks, two pre-computation tables of 984 GB were needed, each has to be prepared in $55 \cdot 2 = 110$ days.

In this paper, we take an entirely different cryptanalytic approach to break A5/1 used in the GSM networks, free of the extremely heavy rainbow tables and long time pre-computation. While all the previous TMD tradeoff attacks regard the internal state of A5/1 as a whole and try to restore it in one shot immediately, we adopt the fast near collision attack (FNCA) strategy in [16] and make a divide-and-conquer partition of the full internal state. Precisely, we regard the internal state as the union of the crucial part (CP) and the rest part (RP), which could be retrieved easily given the corresponding CP and the keystream prefix. Thus, the security of the primitive mainly depends on the intractability of restoring the CP and the efficiency of restoring the RP accordingly. We first launch a fast near collision attack to recover the CP part of the internal state in A5/1 according to the irregular clocking mechanism, based on which the RP part could be recovered later by a dynamic guess-and-determine attack similar to [9]. It is surprising that the irregular clocking mechanism in A5/1 actually facilitates the list merging procedure in a fast near collision attack. Due to the event that not all the three registers moves simultaneously happens with a probability of 0.75, two more overlapping bits are gained for free for each such step. Further in A5/1, it is found that the parameter configuration in a fast near collision attack can be easily tuned to have the desirable non-random behaviour between the resultant list size and a good existence probability of each correct restricted internal state in FNCA. As a result of these findings, it is shown that if the first 64 bits of one keystream frame are intercepted, the internal state, thus the secret key of A5/1 can be reliably found in $2^{31.79}$ cipher ticks, given around 1 MB memory and after a pre-computation of $2^{20.26}$ cipher ticks. Our current implementation in C language on a single core of a PC clearly certified the correctness of this new attack. It takes tens of seconds on average to find the targeted internal state of A5/1, and we feel that further optimization of the code will reduce the time to several seconds. This is the best known attack on A5/1 so far and it is worthy noting that due to the fact that A5/3 and GPRS share the same key with A5/1 in GSM, our attack can be leveraged into attacks against any GSM 2G/3G network eventually.

This paper is organized as follows. A brief description of the A5/1 stream cipher and the GSM encryption scheme is presented in Sect. 2. Some basic definitions and preliminaries of the fast near collision attack relevant to our analysis are provided in Sect. 3. Then a high-level description of our attack against A5/1 and the technical details are presented in Sect. 4, followed by the experimental results in Sect. 5. We leverage our attack to any GSM network in Sect. 6 and finally, some conclusions are drawn in Sect. 7.

2 Description of A5/1 and the GSM Encryption

Let us first present the algorithmic details of the A5/1 stream cipher and the GSM encryption scheme that are relevant to our analysis. A5/1 consists of 3 short linear feedback shift registers (LFSR), denoted by R1, R2 and R3, which are of length 19, 22 and 23 bits respectively. Each register has a primitive feedback polynomial and thus generates a maximum-length binary sequence. The

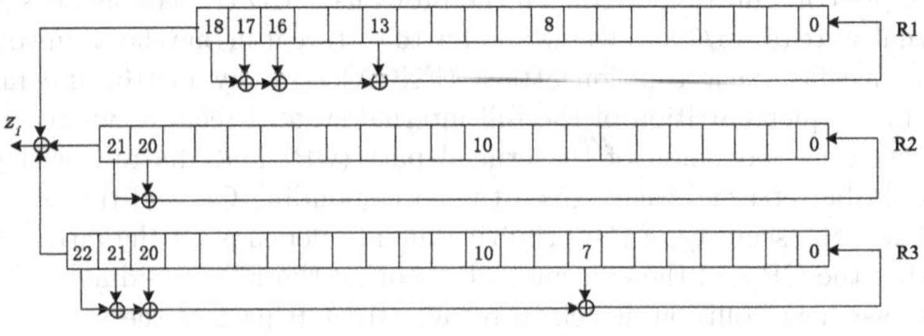

Fig. 1. The A5/1 stream cipher.

rightmost bit in each register is labelled as bit 0. The taps of R1 are at the 13th, 16th, 17th and 18th bit positions; the taps of R2 are at the 20th, 21st bits and the taps of R3 are at the 7th, 20th, 21st and 22nd bits, as depicted in Fig. 1.

Besides, each register has another clocking tap, i.e., bit 8 for R1, bit 10 for R2 and bit 10 for R3, to feed a majority function defined as $maj(ct_1, ct_2, ct_3) = ct_1 \cdot ct_2 \oplus ct_2 \cdot ct_3 \oplus ct_3 \cdot ct_1$, where $ct_1 = \text{R1}[8]$, $ct_2 = \text{R2}[10]$ and $ct_3 = \text{R3}[10]$. The three registers are clocked in a stop/go fashion using the majority rule: at each clock, take the majority function of the clocking taps and only run those registers whose clocking taps agree with the computed majority bit. It is easy to see that at each step, either two or three registers are clocked, and that each register moves with a probability of 3/4 and stops with a probability of 1/4.

In the initialization phase, the three registers are all first set to zero. The secret session key K and a publicly known frame number FN are first injected and then mixed, after that 228 keystream bits are generated in each frame.

One Session of GSM Encryption

Input Parameters:
1: $K = (K_{63}, \cdots, K_1, K_0)$ is the 64-bit secret key
2: $FN = (FN_{21}, \cdots, FN_0)$ is a 22-bit frame counter
3: **for** $i = 0$ to 63 **do**
4: regularly clock R1, R2 and R3
5: $\text{R1}[0] = \text{R1}[0] \oplus K_i$
6: $\text{R2}[0] = \text{R2}[0] \oplus K_i$
7: $\text{R3}[0] = \text{R3}[0] \oplus K_i$
8: **for** $i = 0$ to 21 **do**
9: regularly clock R1, R2 and R3
10: $\text{R1}[0] = \text{R1}[0] \oplus FN_i$
11: $\text{R2}[0] = \text{R2}[0] \oplus FN_i$
12: $\text{R3}[0] = \text{R3}[0] \oplus FN_i$
13: **for** $i = 0$ to 99 **do**
14: clock R1, R2 and R3 in the specified stop/go fashion without producing any output
15: **for** $i = 0$ to 227 **do**
16: clock R1, R2 and R3 in the specified stop/go fashion
17: generate $z_i = \text{R1}[18] \oplus \text{R2}[21] \oplus \text{R3}[22]$
Output: the keystream segment $\{z_i\}_{i=0}^{227}$

In the initialization phase (Step 3 to Step 14), the content of the three registers after step 12 is called the initial state, which is denoted by $S(0)$. The internal state before Step 15 is denoted by $S(100)$. In one session of a GSM conversation, one 114-bit keystream segment is generated for one direction communication and the other 114-bit segment for the opposite direction. Thus, only 228 keystream bits are available in each frame.

3 Preliminaries

In this section, some notations and basic definitions relevant to the fast near collision attack and our work are presented.

We start with the notions of keystream prefix, keystream segment difference (KSD) and internal state difference (ISD).

Definition 1. *For a specified cipher such as A5/1, a keystream prefix* $\mathbf{z} = (z_0, z_1, \cdots, z_{l-1})$ *of* l*-bit length is the keystream vector generated consecutively and directly from the corresponding internal state* $\mathbf{x}$.

Definition 2. *For a specified cipher such as A5/1 and two keystream prefixes* $\mathbf{z}$ *and* $\mathbf{z}'$*, if* $\mathbf{z}$ *is generated from the internal state* $\mathbf{x}$ *and* $\mathbf{z}'$ *is generated from the internal state* $\mathbf{x}'$*, the keystream segment difference (KSD) is* $\Delta\mathbf{z} = \mathbf{z} \oplus \mathbf{z}'$ *and the internal difference (ISD) is defined as* $\Delta\mathbf{x} = \mathbf{x} \oplus \mathbf{x}'$.

In a fast near collision attack, the full internal state which usually contains a large number of variables is not targeted directly; instead the adversary only considers the subset of the full internal state which is directly associated with a specified keystream prefix, as shown in the following definition of restricted internal state.

Definition 3. *For a specified cipher such as A5/1, the subset* $\mathbf{x} = (x_{i_0}, x_{i_1}, \cdots, x_{i_{n-1}})$ *of the full internal state which is directly associated with the keystream prefix* $\mathbf{z} = (z_0, z_1, \cdots, z_{l-1})$ *is called the restricted internal state of* $\mathbf{z}$.

Note that Definition 3 has a subtle difference with the corresponding Definition 1 in [16]. Here we only consider the keystream prefix which is generated consecutively from a given internal state, while Definition 1 in [16] covers the general cases that the keystream bits under consideration are non-consecutive, which is called the keystream vector.

We also need the notions of two kinds of sampling resistance of the primitive, which characterizes the enumeration procedure in a fast near collision attack.

Definition 4. *For a specified cipher such as A5/1, if there exists some efficient method to enumerate directly some subset of the full internal state which produces a special keystream prefix of* l *bits, e.g., a string of* 0*s, without trying and discarding the other states, then it has a BSW sampling resistance of* l *bits, or equivalently its BSW sampling resistance is* 2^{-l}.

It was shown in [5] that when targeting the full 64-bit internal state, A5/1 has a BSW sampling resisitance of 2^{-16} due to the poor choice of the clocking taps, which makes the register bits that affect the clock control and those affecting the keystream bits unrelated for about 16 clock cycles, so the adversary can independently choose them. This technique can be transformed as follows when only considering the subset of the full internal state.

Definition 5. *For a specified cipher such as A5/1, let* $\mathbf{z} = (z_0, z_1, \cdots, z_{l-1})$ *be the keystream prefix whose restricted internal state is* $\mathbf{x} = (x_{i_0}, x_{i_1}, \cdots, x_{i_{n-1}})$*, if* l *bits in* $\mathbf{x}$ *could be derived explicitly by* $\mathbf{z}$ *and the other bits in* $\mathbf{x}$*, then* l *is the restricted BSW sampling resistance corresponding to* $(\mathbf{x}, \mathbf{z})$.

Definition 4 is generalized in [16] to deal with the restricted internal states, as described in Definition 5. Now we only use very short keystream prefix once a time, e.g., $l = 2$ bits in a fast near collision attack, the time complexity to fulfill the enumeration step becomes negligible compared to the other procedures of the attack.

Let $\varnothing$ be the empty set, the partition of the full internal state into the CP and RP parts is formally introduced in the following definition.

Definition 6. *For a specified cipher such as A5/1, the subset* $\mathbf{x}^*$ *of the full internal state* $\mathbf{x}_{\mathbf{full}}$ *that is crucial for the security of the primitive under our framework is the CP part, the rest of the internal state* $\bar{\mathbf{x}^*}$*, where* $\mathbf{x}^* \cap \bar{\mathbf{x}^*} = \varnothing$ *and* $\mathbf{x}^* \cup \bar{\mathbf{x}^*} = \mathbf{x}_{\mathbf{full}}$ *is the RP part.*

Note that it is the freedom of the adversary to determine how to choose the CP part of the internal state. Usually, the CP part is selected in such a way that once retrieved, the rest internal state named RP could be relatively easy to restore, provided the corresponding keystream segment. The partition of CP and RP may be non-unique in a fast near collision attack, we do not exclude other possible choices.

Now we come to the theoretical foundation of a fast near collision attack. Let GF(2) be the binary field and its n-dimensional vector space is denoted by $\mathrm{GF}(2)^n$. First comes the definition of d-near-collision in [15,16].

Definition 7. *Two bit strings* $s_1 \in GF(2)^n$ *and* $s_2 \in GF(2)^n$ *are d-near-collision, if* $w_H(s_1 \oplus s_2) \leq d$*, where* $w_H(\cdot)$ *is the Hamming weight of the input.*

The basic near collision lemma is as follows, as stated in [16].

Lemma 1. *Let A and B be two random subsets of* $GF(2)^n$ *and D is a condition set, then there exist a pair* $(a, b) \in A \times B$ *satisfying one of the conditions in D if*

$$|A| \cdot |B| \geq c \cdot \frac{2^n}{|D|}$$

holds, where $|A|$, $|B|$ *and* $|D|$ *are the cardinalities of sets A, B and D respectively; c is the constant that determines the actual existence probability of the good pair* (a, b)*. In particular, if* $D = \{\Delta x \in GF(2)^n | w_H(\Delta x) \leq d\}$*, then* $|D| = v(n, d) = \sum_{i=0}^{d} \binom{n}{i}$ *is the total number of ISDs with* $w_H(\Delta x) \leq d$ *and* $(a, b) \in A \times B$ *is a d-near-collision pair.*

Lemma 1 has a very large connotation in the sense that it does not restrict what kind of condition is defined in D and has nothing to do with any secret information. The only premise is the randomness of the two involved sets. As in [16], we have the following statements as the corollary of Lemma 1.

Corollary 1. *For a specified cipher such as A5/1 and a constant c, if we choose* $|A| = 1$ *and* $|B| = c \cdot \frac{2^n}{|D|}$ *with A and B being the n-bit restricted internal states associated with a l-bit keystream prefix, and* $D = \{\Delta x \in GF(2)^n | w_H(\Delta x) \leq d\}$*, then there exists an element* $b_i \in B$ *such that the pair* (a, b_i) *with the unique element* $a \in A$ *forms a d-near collision pair with a probability dependent on c.*

Corollary 1 implies that we can carefully choose the constant c and the parameter d to control the existence probability of the desirably good near collision pair.

3.1 The Fast Near Collision Attack

At Eurocrypt 2018, a new cryptanalytic method called fast near collision attack was introduced in [16] to analyze modern stream ciphers with a large internal state. The idea is to combine a near collision property with the divide-and-conquer strategy to restore several partial internal states first, merge the recovered partial states together according to the concrete internal structure and finally recover the full large internal state based on the merged part.

While the original near collision attack in [15] tried to collect two keystream vector sets and to identify a near collision in the whole internal state at different time instants, a fast near collision attack only targets several well-chosen partial restricted internal states based on the refined self-contained method first, whose idea is depicted in Fig. 2. The observation here is that the adversary could simply collect only one keystream prefix set A and virtualize the other set B by directly computing it by himself/herself, i.e., B is not captured. Since we are playing with the internal states directly, the adversary could randomly produce the internal state $\mathbf{x}$ so that it will have the keystream prefix $\mathbf{z} = \texttt{prefix}$. As stated before, the premise in Lemma 1 is just the randomness of the two sets A and B, it does not restrict the way how the adversary gets them. Thus, the adversary could generate one keystream prefix set by himself/herself *without* knowing any secret key. Given the pre-computed table T mapping from a specified $\texttt{ksd}$ to all the possible ISDs, each partner state $\mathbf{x}' = \mathbf{x} \oplus \Delta\mathbf{x}$ will be checked and stored in the list L only if it would generate $\mathbf{z}' = \mathbf{z} \oplus \texttt{ksd}$. The following Algorithm a shows this technique in the general form, i.e., it could deal with both the full and the restricted internal state cases. This method will be integrated into the A5/1 setting with the corresponding chosen parameters and formally introduced in Sect. 4.4.

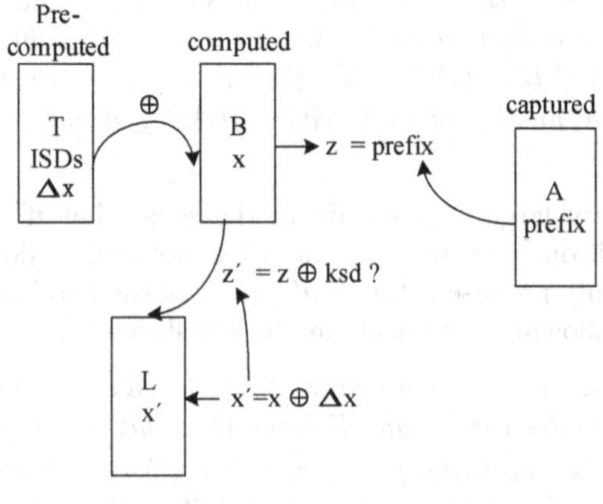

Fig. 2. The idea of the refined self_contained method.

Algorithm a. The refined self-contained method in general [10, 16]

for each `prefix` in A **do**
  Let $i = 0$
  while $i \leq c \cdot 2^n/|T|$ **do**
    randomly generate a new (full or restricted) internal state $\mathbf{x}$
    such that $\mathbf{x}$ produces $\mathbf{z} =$ `prefix`
      for each Δx in $|\mathrm{T}|$ **do**
        if $\mathbf{x}' = \mathbf{x} \oplus \Delta x$ generates $\mathbf{z}' =$ `prefix` $\oplus$ `ksd` **then**
          store $\mathbf{x}'$ into the list L indexed by `prefix`
    $i = i + 1$
Output: The list L

In general, there are two phases in a fast near collision attack: the offline and online phases. The pre-computation phase is quite efficient due to the divide-and-conquer strategy, in which the adversary tries to construct some relatively small tables, instead of one large table, mapping from a fixed `ksd` to all the possible ISDs of the corresponding restricted internal state. Now the adversary does not need to exhaustively search through all the possible ISDs over the full internal state; instead, he/she just search through all the possible ISDs over a specified restricted internal state associated with a given keystream `prefix`, which could be much smaller than the whole internal state. In the online phase, the adversary uses the above self-contained method to get a list of the candidates of the targeted restricted internal state. There is a distilling procedure afterwards to get a smaller list of the candidates with a higher existence probability of the correct partial internal state, which consists of some set intersection and union operations with carefully selected parameters. Then merge the restored restricted internal states together to cover the chosen CP portion of the full internal state, which is used later as the starting point for the full recovery of the internal state.

We will go into the details of the refined self-contained method, the distilling and merge procedures, and the final retrieval of the targeted internal state in the context of A5/1 in the following sections.

4 Our New Attack

In this section, we will present our new attack against A5/1 in a step-by-step manner following the above cryptanalytic principles.

4.1 A General Description of the Attack

We first present a high-level overview of our attack. As stated in Sect. 3, the goal is to first recover the CP part of the internal state, and then the RP part at a fixed time instance which is consistent with the captured keystream.

The main idea is as follows. In A5/1, the size of the internal state is only 64 bits, thus in the assumed fast near collision attack, if we target a well-chosen CP part of the internal state, the pre-computation and the memory complexities will be considerably reduced compared to the previous best attacks. What we need to do is just to prepare the partial differential tables which record the mappings from the l-bit KSDs to all the possible ISDs of the restricted internal state with the sorted occurring probabilities. In the online phase, we exploit the special internal structure of A5/1 to partition the 64-bit internal state into the CP and the RP part, and its irregular clocking mechanism to launch the concrete fast near collision attack against it. Formally, a high-level description of our attack is depicted in Algorithm 1.

Algorithm 1. Fast near collision attack on A5/1

Parameters: l, α, γ
Offline: Prepare the tables T[`ksd`, `prefix`]
1: **for** each possible value of (`ksd`, `prefix`) **do**
2: use the method in section 4.3 to construct T[`ksd`, `prefix`]
Input: A keystream segment $\mathbf{z} = (z_0, z_1, \ldots, z_{\gamma-1})$
Online: Recover the full internal state matching with $\mathbf{z}$
3: Divide $\mathbf{z}$ into α overlapping prefixes $\mathbf{z}_i$ $(0 \le i \le \alpha - 1)$ and a suffix $\mathbf{z}_\mu$
4: **for** $i = 0$ to $\alpha - 1$ **do**
5: derive the partial state list L_i for $\mathbf{z}_i$ in section 4.4 and 4.5
6: Merge L_is to get a candidate list for the CP part in section 4.6
7: Recover the RP in section 4.7 and check the consistency with $\mathbf{z}_\mu$

In the following, we will embed Algorithm 1 into the concrete attack scenario of A5/1 to demonstrate the attack in details.

4.2 Basic Facts of A5/1

As described in Sect. 2, A5/1 adopts the stop/go clocking fashion according to a majority function defined over the 3 taps from R1, R2 and R3, respectively. Thus, for one keystream bit z_i, it actually depends on 9 internal state bits: the 2 leftmost bits from each register, i.e., R1[18], R1[17], R2[21], R2[20], R3[22] and R3[21], and the three clock control bits R1[8], R2[10] and R3[10].

Further, as discussed in Sect. 3.1, the adversary needs to randomly generate the keystream prefix without knowing the secret key in a fast near collision attack. In this case, we could use either the oracle which generates the corresponding keystream prefix directly or the following Algebraic Norm Form (ANF) of z_i, which could be verified by enumerating all the possible values of $x = (x_0, x_1, x_2, \ldots, x_8)$:

$$\begin{aligned} f(x) &= x_3 + x_4 + x_5 + x_0x_6 + x_3x_6 + x_1x_7 + x_4x_7 + x_2x_8 + x_5x_8 \\ &+ x_0x_6x_7 + x_1x_6x_7 + x_2x_6x_7 + x_3x_6x_7 + x_4x_6x_7 + x_5x_6x_7 \\ &+ x_0x_6x_8 + x_1x_6x_8 + x_2x_6x_8 + x_3x_6x_8 + x_4x_6x_8 + x_5x_6x_8 \\ &+ x_0x_7x_8 + x_1x_7x_8 + x_2x_7x_8 + x_3x_7x_8 + x_4x_7x_8 + x_5x_7x_8, \end{aligned}$$

where x_0 = R1[18], x_1 = R2[21], x_2 = R3[22], x_3 = R1[17], x_4 = R2[20], x_5 = R3[21], x_6 = R1[8], x_7 = R2[10] and x_8 = R3[10]. As in [16], in order to have an efficient fast near collision attack, we have also tried to use the 2-bit keystream prefix to recover the corresponding restricted internal state, which contains 6 more variables than the single bit case, as depicted in Fig. 3. Due to the action of the majority function, there are now 15 input variables involved for a 2-bit keystream prefix. Precisely, in addition to the above 9 state bits in the ANF of z_i, we have $x_9 = \text{R1}[16]$, $x_{10} = \text{R2}[19]$, $x_{11} = \text{R3}[20]$, $x_{12} = \text{R1}[7]$, $x_{13} = \text{R2}[9]$ and $x_{14} = \text{R3}[9]$. For a 2-bit keystream prefix, we just directly exploit the keystream generation oracle to generate it in the refined self-contained method.

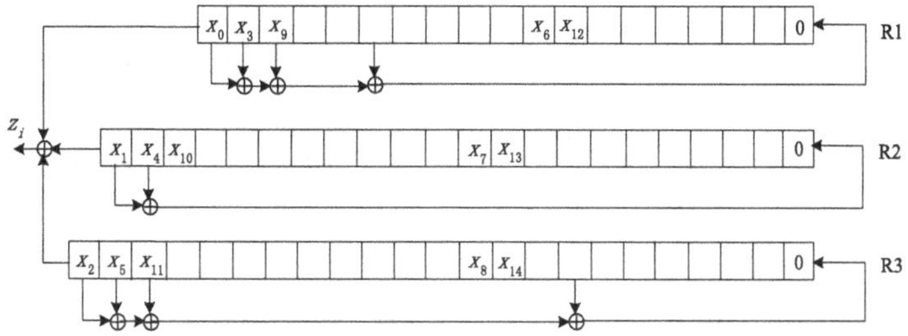

Fig. 3. The 15-bit restricted internal state that a 2-bit keystream prefix depends on.

4.3 Pre-computing the Partial Differential Tables

Now we come to the offline phase of Algorithm 1. Since there are only $n = 15$ or $n = 9$ variables involved for the 2-bit keystream prefix or 1-bit keystream, given a small value of d, we could fully enumerate all the possible values of the corresponding restricted internal state and all the $v(n, d)$ ISDs to *accurately* compute the occurring probabilities of each ISD.

Algorithm 2. The offline algorithm

Parameters: n, d and `ksd`, `prefix`

for each possible value of (`ksd`, `prefix`) **do**
  Initialize the table T[`ksd`, `prefix`]
  for each of the $v(n, d)$ Δx **do**
    Initialize $t = 0$ and $cc = 0$
    for each $i \in \{0, 1, \cdots, 2^n - 1\}$ **do**
      check **if** $\mathbf{x} = i$ generates the `prefix`
      if yes **then**
        $t = t + 1$
        generate the partial internal state $\mathbf{x}' = \mathbf{x} \oplus \Delta x$
        compute the prefix $\mathbf{z}'$ generated by $\mathbf{x}'$
        if $\mathbf{z}' =$ `prefix` $\oplus$ `ksd` **then** store Δx and $cc = cc + 1$
  Sort the ISDs according to the occurring rates cc/t

Algorithm 2 describes how to generate a series of pre-computed tables, which record the mappings from a specified `ksd` to all the possible ISDs Δx, each with a sorted occurring probability. The core point here is that the only premise in Lemma 1, i.e., the randomness, enables the adversary to freely compute the second set B by himself/herself. Thus, the adversary would know the matching between the partial internal state $\mathbf{x}'$ and the output prefix $\mathbf{z}' = \texttt{prefix} \oplus \texttt{ksd}$.

Table 1. The complete differential table when $ksd = 0x3$ and $d = 2$

ISD	Prob.	ISD	Prob.	ISD	Prob.	ISD	Prob.
(x_3, x_9)	0.75	(x_6, x_{11})	0.25	$(x_6, -)$	0.25	(x_0, x_5)	0.125
(x_5, x_{11})	0.75	(x_2, x_6)	0.25	(x_1, x_7)	0.25	(x_0, x_4)	0.125
(x_4, x_{10})	0.75	(x_6, x_{13})	0.25	$(x_7, -)$	0.25	(x_1, x_{14})	0.125
(x_0, x_3)	0.4375	(x_6, x_{14})	0.25	$(x_8, -)$	0.25	(x_1, x_{11})	0.125
(x_2, x_5)	0.4375	(x_2, x_7)	0.25	(x_6, x_{12})	0.234375	(x_2, x_4)	0.125
(x_1, x_4)	0.4375	(x_7, x_9)	0.25	(x_7, x_8)	0.234375	(x_2, x_{12})	0.125
(x_3, x_{12})	0.375	(x_2, x_8)	0.25	(x_6, x_8)	0.234375	(x_0, x_{11})	0.125
(x_5, x_{14})	0.375	(x_7, x_{10})	0.25	(x_8, x_{14})	0.234375	(x_2, x_{13})	0.125
(x_4, x_{13})	0.375	(x_0, x_7)	0.25	(x_7, x_{13})	0.234375	(x_1, x_{12})	0.125
(x_3, x_{14})	0.3125	(x_7, x_{11})	0.25	(x_6, x_7)	0.234375	(x_1, x_5)	0.125
(x_4, x_9)	0.3125	(x_1, x_6)	0.25	(x_4, x_8)	0.21875	(x_2, x_3)	0.125
(x_4, x_{12})	0.3125	(x_7, x_{12})	0.25	(x_3, x_7)	0.21875	(x_0, x_{13})	0.125
(x_4, x_{14})	0.3125	(x_1, x_8)	0.25	(x_5, x_6)	0.21875	(x_1, x_2)	0.125
(x_5, x_9)	0.3125	(x_7, x_{14})	0.25	(x_5, x_7)	0.21875	(x_0, x_{14})	0.125
(x_3, x_{11})	0.3125	(x_6, x_{10})	0.25	(x_3, x_8)	0.21875	(x_0, x_1)	0.125
(x_5, x_{10})	0.3125	(x_8, x_9)	0.25	(x_4, x_6)	0.21875	(x_1, x_{10})	0.0625
(x_3, x_{13})	0.3125	(x_4, x_5)	0.25	$(x_3, -)$	0.1875	(x_0, x_{12})	0.0625
(x_5, x_{12})	0.3125	(x_8, x_{10})	0.25	$(x_4, -)$	0.1875	(x_2, x_{11})	0.0625
(x_3, x_{10})	0.3125	(x_0, x_6)	0.25	$(x_5, -)$	0.1875	$(x_1, -)$	0.0625
(x_5, x_{13})	0.3125	(x_8, x_{11})	0.25	(x_0, x_{10})	0.125	(x_2, x_{14})	0.0625
(x_4, x_{11})	0.3125	(x_3, x_5)	0.25	(x_2, x_9)	0.125	(x_1, x_{13})	0.0625
(x_3, x_6)	0.28125	(x_8, x_{12})	0.25	(x_1, x_9)	0.125	(x_0, x_9)	0.0625
(x_4, x_7)	0.28125	(x_6, x_9)	0.25	(x_2, x_{10})	0.125	$(x_2, -)$	0.0625
(x_5, x_8)	0.28125	(x_8, x_{13})	0.25	(x_1, x_3)	0.125	$(x_0, -)$	0.0625
(x_0, x_8)	0.25	(x_3, x_4)	0.25	(x_0, x_2)	0.125		

Now we choose $d = 2$ in Algorithm 2, i.e., we consider the ISDs of Hamming weight less than or equal to 2-bit, there are $v(15, 2) = \sum_{i=0}^{2} \binom{15}{i} = 121$ ISDs for the 15-bit restricted internal state under consideration. For each ISD Δx satisfying $w_H(\Delta x) \leq d$, we enumerate all the possible values of the 15-bit restricted internal states $\mathbf{x}$. If $\mathbf{x}$ can generate the considered $\texttt{prefix}$, we xor it with the Δx and save Δx into the pre-computed table if and only if the xored state $\mathbf{x}'$ could generate $\texttt{prefix} \oplus \texttt{ksd}$. In Algorithm 2, t is the number of times that $\mathbf{x}$ generates the $\texttt{prefix}$, and cc is the number of occurrences of the event that $\mathbf{z}' = \texttt{prefix} \oplus \texttt{ksd}$ under the condition that $\mathbf{x}$ generates the $\texttt{prefix}$. Finally, the table is sorted according to the occurring probabilities of each ISD. We have computed $\mathrm{T}[\texttt{ksd}, \texttt{prefix}]$ for each combination of $(\texttt{ksd}, \texttt{prefix})$ when $l = 1$ and 2 with a low complexity of $2^{2l} \cdot v(n, d) \cdot 2^n \cdot l$. We list the complete table when $\texttt{ksd} = \texttt{0x3}$ with an arbitrary 2-bit keystream $\texttt{prefix}$ when $d = 2$ in Table 1.

Note that in Table 1, (x_i, x_j) for $0 \leq i, j \leq 14$ means the 2-bit differences are at the positions x_i and x_j in Fig. 3, respectively. For $(x_i, -)$ with $0 \leq i \leq 14$, it means the 1-bit difference is at the position x_i only. Given all the pre-computed tables T[`ksd`, `prefix`], it is easy to see that for a fixed `ksd`, the average reduction effect of each table indexed by `prefix` is almost the same, which is measured by the diversified probability defined below and in [16].

Definition 8. *The diversified probability is defined as* $P_{divs} = \frac{\sum_{\Delta x \in T} Pr_{\Delta x}}{|T|}$, *where* Δx *ranges over all the* $|T|$ *possible ISDs in the table T[ksd,prefix].*

This probability measures the average reducing effect of T[`ksd`, `prefix`] so that for a random restricted internal state $\mathbf{x}$ generating `prefix`, if the bits in $\mathbf{x}$ are flipped according to a $\Delta x \in T$ to get $\mathbf{x}'$, then with the probability P_{divs}, $\mathbf{x}'$ could produce `prefix` $\oplus$ `ksd`. The observation that P_{divs} is mainly determined by `ksd` further reduces the memory requirement of the pre-computation. Thus, given the fixed `ksd=0x3`, Table 1 actually works for an arbitrary keystream `prefix`. This property is crucial for the partial restricted state recovery, since the recovery procedure will be the same even if the keystream `prefix` under investigation has varied along the captured keystream, which is shown in Algorithm 3 in Sect. 4.4. It is interesting to see that this small differential table only has $99 < 121$ possible ISDs when `ksd` $=$ `0x3` and $d = 2$. Note that in the previous TMD tradeoff attacks, the adversary is expected to search through all the possible ISDs over the full internal state, which will result in a huge pre-computation complexity and memory consumption. Now we can just try the 99 possible ISDs over the 15-bit restricted internal state in the online phase. Further, the occurring probabilities of the ISDs in Table 1 ranges from 0.75 to 0.0625 and their distribution is heavily biased. We have tested all the cases for $l = 1$ and $l = 2$ with $2 \leq d \leq 4$ to find the optimal choice of the pre-computed table T[`ksd`, `prefix`]. The results are listed in Table 2.

Table 2 clearly shows that for the case $l = 2$ and $n = 15$, we have $P_{divs} = 0.234848$, which is the minimum value among all the empirical results. Besides, there are some more reasons for choosing this configuration, which are briefly discussed below. The cost of merging the candidates list in the following Sect. 4.6 accounts for some proportion of our attack, thus we expect to have less candidates for the involved restricted internal state. Note that the candidates are generated by xoring all the possible ISDs in the table T[`ksd`, `prefix`] in the refined self-contained method, thus we prefer a small number of possible ISDs in the selected pre-computed table. Further, we expect the average reduction effect, measured by P_{divs}, to be as low as possible for the efficiency reasons.

In Table 2, given $2 \leq d \leq 4$ and $0 \leq$ `ksd` ≤ 3, we have enumerated all the ISDs Δx such that $w_H(\Delta x) \leq d$. We have tested the 2^9 restricted internal states for the 1-bit keystream case and the 2^{15} restricted internal states for the 2-bit keystream prefix case, respectively. The experimental results in Table 2 imply that it seems to be a good choice to restore the involved 15-bit restricted internal state for $l = 2$ with (`ksd`, d) $=$ (`0x3`, 2). There are only 99 possible ISDs involved in the corresponding pre-computed table and the diversified probability is 0.234848. Note that each ISD in the table has 15 bits and can be stored in 2 bytes, which means that this table only requires about 198 bytes memory complexity.

Table 2. The empirical results for $l = 1$ and $l = 2$ with $2 \leq d \leq 4$

l	d	ksd	|T|	P_{divs}	l	d	ksd	|T|	P_{divs}
1	2	0x0	43	0.511628	2	2	0x0	118	0.313559
		0x1	45	0.533333			0x1	96	0.289063
		–	–	–			0x2	111	0.297297
		–	–	–			0x3	99	0.234848
	3	0x0	127	0.507874		3	0x0	555	0.269876
		0x1	128	0.511719			0x1	547	0.272795
		–	–	–			0x2	521	0.269494
		–	–	–			0x3	533	0.256273
	4	0x0	253	0.509881		4	0x0	1875	0.258133
		0x1	251	0.505976			0x1	1864	0.261266
		–	–	–			0x2	1853	0.261603
		–	–	–			0x3	1874	0.258938

4.4 Determining the Candidates List of the Involved Restricted Internal State

Now we are ready to enter the online phase of Algorithm 1, whose aim is to recover the restricted internal states for the targeted 2-bit keystream prefixes by using the pre-computed tables prepared in Sect. 4.3.

Algorithm 3. The refined self-contained method in online phase

Parameters: $l = 2$, $n = 15$, $d = 2$ and $\texttt{ksd} = \texttt{0x3}$
$\texttt{iter\_num} = 4 \cdot 2^{15}/|\mathrm{T}[\texttt{ksd}, \texttt{prefix}]| = 4 \cdot 2^{15}/99$

1: Initialize $i = 0$
2: **while** $i \leq \texttt{iter\_num}$ **do**
3: randomly generate a new restricted internal state $\mathbf{x}$
4: such that $\mathbf{x}$ produces $\mathbf{z} = \texttt{prefix}$
5: **for** each of the $|\mathrm{T}|$ possible ISDs Δx **do**
6: **if** $\mathbf{x}' = \mathbf{x} \oplus \Delta x$ generates $\mathbf{z}' = \texttt{prefix} \oplus \texttt{ksd}$ **then**
7: store $\mathbf{x}'$ into the list L
8: $i = i + 1$
9: **Output:** The list L

According to Corollary 1, given the pre-computed table $\mathrm{T}[\texttt{ksd}, \texttt{prefix}]$, let A with $|A| = 1$ be the 15-bit restricted internal state associated with a 2-bit keystream prefix and B with $|B| = c \cdot \frac{2^{15}}{|T|}$ be the virtualized 15-bit state in the refined self-contained method, then there should exist an element $b_i \in B$ such that the pair (a, b_i) with the unique element $a \in A$ forms a d-near collision pair with a probability dependent on c. This is the theoretical basis of Algorithm 3, where $c = 4$.

Recall that in the previous near collision attack in [15], the adversary needs to collect two random keystream vector sets, A and B, and tries to identify a d-near-collision state pair at two different time instants from the corresponding keystream segments. In this process, a strong wrong-candidate filter with a low complexity is needed, while in its form in [15], the reduction effect is not as good as expected. In [10], the self-contained method is introduced as a generic approach to obtain the candidates of the involved internal state. As briefly mentioned in Sect. 3.1, this method only requires one keystream prefix set, and allows the adversary to freely compute the other one based on the pre-computed table. Precisely, let A be the collected keystream prefix set, the adversary just randomly generates the other set B by himself/herself. For $\mathbf{x}' \in A$ and $\mathbf{x} \in B$, if the corresponding keystream prefixes satisfying $\mathbf{z}' \oplus \mathbf{z} = \mathtt{ksd}$ with $\mathtt{ksd}$ being the concrete value of the considered KSD, since the adversary knows the matching between the internal state $\mathbf{x}$ and its corresponding output $\mathbf{z}$, he/she could restore the targeted internal state $\mathbf{x}'$ by trying $\mathbf{x}' = \mathbf{x} \oplus \Delta x$. Thus the adversary could generate the candidates list of the targeted restricted internal state $\mathbf{x}'$ in this way. Taking into account the divide-and-conquer strategy on the partition of the full internal state, we have Algorithm 3.

Note that Steps 3 and 4 in Algorithm 3 can be fulfilled by a method similar to the BSW sampling enumeration technique in [5], but with a very small $l = 2$. Besides, Corollary 1 further implies that there is an inherent relationship between c, d and the existence probability of one good pair. For A5/1, we list this correspondence when 2-bit keystream prefix is considered with $d = 2$ in Table 3, which is achieved from 10^6 times repetition of empirical simulations. We have tried all the meaningful combinations of parameter configuration from the practical implementation point of view, and determined to choose $c = 4$ in our attack. Note that in this case, when $l = 2$ and $d = 2$, we have $n = 15$ and $v(n, d) = 121$ and the existence probability in one invoking of the refined self-contained method is 0.9835. As can be seen in the following sections, this choice provides a very good balance between the complexity aspects and the success probability.

Table 3. The correspondence between the constant c, the list size and the existence probability of one good pair for A5/1 when $d = 2$

c	List size r	Prob.
2	6903	0.8475
3	7654	0.9510
4	7963	0.9835

Further, Let us have a closer look at Algorithm 3 and Table 3 together. There is a magic fact here that though the adversary has iterated $4 \cdot 2^{15}/|\mathrm{T}[\mathtt{ksd}, \mathtt{prefix}]| = 4 \cdot 2^{15}/99 \doteq 1324$ times and there are 99 ISDs to be xored and checked the consistency with $\mathbf{z}'$ in each iteration in Algorithm 3, the number of hit values stored in the list L is much less than $\frac{2^{15}}{2^2} = 8192$ with a high existence probability of the correct candidate being in L, as can be seen from Table 3. All these facts can be well explained through the following Theorem 1 and are illustrated in *Example* 1.

Theorem 1 ([16]). *Let b be the number of all the values that can be hit and $a = c \cdot \frac{2^n}{|D|} \cdot |T| \cdot P_{divs}$, then after one invocation of the refined self-contained method, the expectation of the final number r of hitting values in the candidates list is*

$$E[r] = \sum_{r=1}^{a} \frac{\binom{b}{r} \cdot r! \cdot \left\{ {a \atop r} \right\} \cdot r}{b^a}, \tag{1}$$

where $\left\{ {a \atop r} \right\}$ is the Stirling number of the second kind, $\binom{b}{r}$ is the binomial coefficient and $r!$ is the factorial.

Example 1. Let $c = 4$, from Eq.(1), the mathematical expectation of the list size can be calculated as $E[r] = \sum_{r=1}^{a} \frac{\binom{b}{r} \cdot r! \cdot \left\{ {a \atop r} \right\} \cdot r}{b^a} \doteq 2^{12.96} \doteq 7963$, where $l = 2$, $n = 15$, $b = \frac{2^{15}}{2^2} = 8192$, $a = 4 \cdot \frac{2^{15}}{|T|} \cdot |T| \cdot P_{divs} = 4 \cdot 2^{15} \cdot P_{divs} = 2^{17} \cdot 0.234848 = 30782$. Here $P_{divs} = 0.234848$ is the diversified probability of the selected pre-computation table T[ksd, prefix] defined in Definition 8 in Sect. 4.3. □

4.5 Distilling Phase: Enhancing the Existence Probability of the Correct Candidate

In the real online attack, the adversary usually wants to have a candidate list of smaller size, while at the same time still containing the correct restricted internal state with a reasonably good probability. That is, we want to efficiently filter out the wrong candidates of the involved restricted internal state got from Algorithm 3. Note that due to the partition of the full internal state, now we cannot run A5/1 from the involved partial state forwards and backwards to check the consistency with the available keystream. This is the motivation of the distilling phase in this section, i.e., we need an efficient wrong-candidate filter without knowing the full internal state. For consistency with the work in [16], we follow the conventional notations and descriptions in this domain hereafter.

Algorithm 4. The Distilling Procedure 1: Intersection [16]

Parameters: $L = (L_1, \ldots, L_\beta)$: β candidate lists of the targeted partial state from Algorithm 3
1: **for** $i = 1$ **to** $\beta - 1$ **do**
2: $\quad L_{i+1} \leftarrow L_i \cap L_{i+1}$

To improve the hitting rate of the correct restricted internal state while keeping a smaller list size, we continuously modify the involved candidate lists got from Algorithm 3 by set intersection and union, depicted in Algorithm 4 and 5, respectively. Note that the candidates in list L are generated randomly in Algorithm 3 with the following property: the correct one has an existence probability determined by the constant c, which is usually higher than the random wrong ones, which appear in the list randomly. Some candidate of the restricted internal state may be hit several times by xoring different ISD Δx with different values of $\mathbf{x}$ in Algorithm 3.

Example 2. Let $\mathbf{x}$ = 111000010111110 be the correct restricted internal state, where the leftmost bit is x_0 and the rightmost bit x_{14}. It is observed that $\mathbf{x}$ is hit 30 times in one invocation of Algorithm 3 with the following ISDs: $(x_5, -)$, (x_3, x_4), $(x_5, -)$, (x_5, x_{13}), (x_3, x_4), (x_1, x_7), (x_7, x_{11}), (x_1, x_7), (x_1, x_9), (x_2, x_7), (x_2, x_5), (x_2, x_7), (x_2, x_6), (x_6, x_8), (x_5, x_{14}), (x_4, x_8), (x_6, x_8), (x_7, x_{12}), (x_1, x_9), (x_7, x_{11}), (x_5, x_{14}), (x_5, x_{13}), $(x_7, -)$, (x_0, x_5), (x_5, x_{10}), (x_1, x_{14}), (x_1, x_9), (x_2, x_6), (x_4, x_8), (x_3, x_{14}). □

Thus, it is natural to intersect the resultant lists from different independent invokes of Algorithm 3 to reduce the number of candidates, while still containing the correct one with a reasonable probability. During the intersection process, many wrong candidates can be removed from the list. Algorithm 4 depicts this distilling process. As proved in [16], the expected number of candidates after $\beta - 1$ steps of intersection can be estimated by $|L_1| \cdot (\frac{E[r]}{b})^{\beta-1}$, where $|L_1|$ is the number of candidates in the first generated list L_1, b and $E[r]$ are defined and listed in Theorem 1. We have found in practical experiments that $\beta = 6$ is a good choice for the reduction purpose when attacking A5/1, with an appropriate choice of the following parameter γ in Algorithm 5 at the same time.

On the other hand, we have also observed in the experiments that for a single invoking of Algorithm 3, there is some missing probability of the event that the correct partial state is not in the candidate list. If such an event happens, we may miss the correct partial state through the intersection process. This fact indicates that we should take some action to remedy such a situation and the existence probability of the correct restricted internal state in this case. The following Algorithm 5 resolves this problem by the set union operation. Precisely, after getting some intersected lists, we adopt Algorithm 5 to generate a candidate list for the considered restricted internal state. The union operation can well mitigate the influence of the missing event in Algorithm 3, i.e., as long as the correct restricted internal state survives in one of the γ lists obtained from Algorithm 4, it will be in the final list after Algorithm 5. Once Algorithm 5 is executed, we can get a candidate list for the n-bit restricted internal state associated with the l-bit keystream prefix.

Algorithm 5. The distilling procedure 2: Union [16]

Parameters: $U = (U_1, \ldots, U_\gamma)$: γ lists obtained from Algorithm 4

1: **for** $i = 1$ **to** $\gamma - 1$ **do**
2: $\quad U_{i+1} \leftarrow U_i \cup U_{i+1}$

Further, it is also proved in [16] that after the preparation of γ resultant lists from Algorithm 4, the expected number of candidates $|F_{i+1}|$ after i steps of union can be recursively derived as

$$|F_{i+1}| = |F_i| + |U_{i+1}| - \sum_{j=0}^{|U_{i+1}|} \frac{\binom{|F_i|}{j} \cdot \binom{|F_{i+1}|-|F_i|}{|U_{i+1}|-j}}{\binom{|F_{i+1}|}{|U_{i+1}|}} \cdot j \,,\ 1 \leq i \leq \gamma - 1, \qquad (2)$$

where F_i is the resultant list after $i-1$ steps of union. Equation (2) is derived from the fact that $|F_{i+1}| = |F_i| + |U_{i+1}| - |F_i \cap U_{i+1}|$ and the $|F_i \cap U_{i+1}|$ term follows the hypergeometric distribution. Finally, the following corollary provides the existence probability of the correct restricted internal state after Algorithm 5.

Corollary 2. *Let Pr_{Alg5} be the probability that the correct restricted internal state will exist in the final list generated by Algorithm 5, we have $Pr_{Alg5} = 1-(1-(p_1)^{\beta})^{\gamma}$, where p_1 is the existence probability of the correct restricted internal state in Table 3.*

Proof. It suffices to consider the opposite event that the correct restricted internal state does not exist after Algorithm 5, which means that it does not exist in any of the γ lists, each going through $\beta-1$ intersections. This completes the proof. □

In order to have a good existence probability of the correct restricted internal state with a smaller size of candidate list, we have tried $\beta = \{2,3,4,5,6,7,8,9\}$ and $\gamma = \{2,3,4,5\}$ in the experiments to identify the optimal parameter configuration for attacking A5/1. Here we only present some of the simulation results in Table 4.

Table 4. Some results for different parameter configurations with $c = 4$

γ	β	$\vert U\vert$	Prob.
2	3	8065	0.9940
2	4	7989	0.9927
2	5	7934	0.9912
2	6	7835	0.9903

From this table, the configuration that $\beta = 6$ and $\gamma = 2$ is selected to be used in our attack against A5/1. The success probability can be calculated as $Pr_{Alg5} = 1-(1-0.9835^{\beta})^{\gamma} = 0.9909$ according to Corollary 2, which is quite close to the result 0.9903 obtained from practical experiments, while the expected list size should be $2^{12.95}$ in theory according to the Eq. (2), which is also very close to the simulation result $7835 \doteq 2^{12.94}$. The following Lemma is the basis of our non-randomness observation after Algorithm 5.

Lemma 2. *(Chebyshev's Inequality) Let X be a random variable with the finite expected value μ and finite non-zero variance σ^2. Then for any real number $k > 0$, we have $Pr(|X-\mu| \geq k\sigma) \leq \frac{1}{k^2}$ and only the case $k > 1$ is useful.*

Based on Chebyshev's Inequality, we have the following statements on the non-randomness observation of the resultant candidate list after Algorithm 5.

Theorem 2. *Let $\beta = 6$ and $\gamma = 2$ in Algorithm 4 and 5, if $c = 4$, $\mathtt{ksd} = 0x3$, and $d = 2$ for a 2-bit keystream prefix, then the candidates list generated after Algorithm 5 has an averaged size of 7835 with the existence probability 0.9903 for the correct candidate being in the list, which is a non-random case.*

Proof. Here we follow the classic way of distinguishing two distributions in theory to show the non-randomness. Below we will investigate the distribution in the pure random case and that obtained in our attack respectively.

For a specified 2-bit keystream prefix, the candidate space has a size of $\frac{2^{15}}{2^2} = 2^{13} = 8192$ in the pure random case. The probability that a candidate 15-bit restricted internal state will generate the specified keystream prefix is $p = \frac{1}{4}$ and $q = 1 - p = \frac{3}{4}$ otherwise. We regard the list size as a sum of random variables which follow the binomial distribution with the corresponding parameters in each case and approximate it with the normal distribution. Then the standard deviation in the pure random case is $\sigma = \sqrt{2^{13} \cdot p \cdot q} \approx 39.19$. Further, the expectation of the list size in the pure random case with the existence probability 0.9903 should be $\mu_{union} = 2^{13} \cdot 0.9903 \approx 8113$.

On the other side, in our attack against A5/1 after Algorithm 5, the averaged list size is $\mu'_{union} = 7835$ with the existence probability 0.9903. Then according to the Chebyshev's inequality in Lemma 2, we can compute the coefficient k as

$$k = \frac{|\mu_{union} - \mu'_{union}|}{\sigma} \approx \frac{8113 - 7835}{39.19} \approx 7.09.$$

Thus, we can conclude that the resultant list size in our attack is non-random with a probability greater than 99%. This completes the proof. □

This non-random phenomenon is the basis of our new attack against A5/1. Note that the probability in theory 0.9909 and empirical rate 0.9903 from Table 4 are very close to each other, we have chosen to use the 'worse' value 0.9903 in the proof of Theorem 2 to demonstrate the strong validity of our attack.

4.6 Merging Phase: Restoring the CP Part of the Internal State

Now we show how to restore the CP part of the full internal state through the merging phase in Algorithm 1. The main difference between A5/1 and the target Grain v1 in [16] is that A5/1 executes according to the specified irregular clocking mechanism in a stop/go manner, while Grain v1 runs regularly. Though it is commonly believed that the irregular clocking mechanism improves the security of the primitive by introducing certain implicit non-linearity, we have found in the mounted fast near collision attack that the stop/go irregular clocking in A5/1 actually facilitates the list merging procedure in the attack. Let us first introduce a notion that is used in our attack.

Definition 9. *According to the stop/go clocking rule in A5/1, the intersection set of the two partial states at the time instants* t *and* $t+1$ *is called the check-state in the merging phase.*

From the irregular clocking rule in A5/1, the check-state in the merging phase has the following property.

Proposition 1. *The check-state in the merging phase has a cardinality of* 9 *if all the three registers are clocked once and has a cardinality of* 11 *if two registers are clocked once and one register stops at the corresponding time instant.*

Proof. There are 8 possible values for the three clock control bits, R1[8], R2[10] and R3[10]. It suffices to check the 8 cases one-by-one to conclude that only when the three bits take the value pattern $(0, 0, 0)$ or $(1, 1, 1)$, there will be 9 bits in the check-state as if all the registers are clocked regularly; otherwise there is one register unchanged which will offer 5 bits, while each of the two clocked registers will offer 3 bits. This completes the proof. □

Example 3. Let `111011011010010` be the starting restricted internal state, where the leftmost bit is x_0 and the rightmost bit is x_{14}. From Fig. 3, consider the following state transmission chain `111011011010010` → `111010010000000` → `010010010100101` → `010110111100101` → `110100101010001` to see the validity of the statements in Proposition 1. For example, when `111010010000000` → `010010010100101`, R2 stops with the clock control pattern `010`, thus the 5-bit state `11010` in R2 remains unchanged. □

Note that the event that not all the three registers move simultaneously happens with a probability of 0.75, in which case two more overlapping bits are gained for free from Proposition 1. This implies that in most cases, the check-state in the merging phase has an reduction effect which is unexpected from the cryptanalyst's point of view. Based on the check-state, we have Algorithm 6 for the merging phase, whose basic idea is to combine the candidates which are coincident on the identified check-state according to the clock control bits, and finally derive the candidates for the CP part.

Algorithm 6. Merging the lists from Algorithm 5 to restore CP

Parameters: $L_{z_i z_{i+1} \cdots z_{i+m-1}}$: one list from Algorithm 5
$L_{z_{i+1} z_{i+2} \cdots z_{i+m}}$: the other list to be merged
$L_{z_i z_{i+1} \cdots z_{i+m-1} z_{i+m}}$: the resultant list
Find the clock control bits set J between the two lists
for each value pattern in J **do**
  Form the subgroup of $L_{z_i \cdots z_{i+m-1}}$ according to the pattern
  Sort the subgroup according to the check-state values
for each value pattern in J **do**
  Form the subgroup of $L_{z_{i+1} \cdots z_{i+m}}$ according to the pattern
  Sort the subgroup according to the check-state values
for each pattern in J **do**
  for each value pattern of the corresponding check-state **do**
    Merge each pair of elements and save the result into $L_{z_i \cdots z_{i+m}}$
Output: $L_{z_i z_{i+1} \cdots z_{i+m-1} z_{i+m}}$

Algorithm 6 actually split the first list $L_{z_i \cdots z_{i+m-1}}$ into many sublists according to the value pattern of the clock control bits and the value of the corresponding check-state under each clock control pattern. Similarly, the second list $L_{z_{i+1} \cdots z_{i+m}}$ will be regrouped into sublists according to the value of the overlapping bits determined by each pattern of clock control bits. Then, select an element from each sublist with the same check-state pattern under the same clock control pattern to form a candidate of the merged state.

In our attack, we first generate 4 candidate lists[1], $L_{z_0 z_1}, L_{z_1 z_2}, L_{z_2 z_3}, L_{z_3 z_4}$, for the corresponding restricted internal states of the 5-bit keystream prefix $\mathbf{z} = (z_0, z_1, z_2, z_3, z_4)$, after that we run Algorithm 6 to merge the involved restricted internal states. The merged candidate list for the CP part of the internal state is used as the starting point for the following retrieval of the RP part in a guess-and-determine like manner. In our experiments, each list $L_{z_i z_{i+1}}$ for $0 \leq i \leq 3$ contains around 7835 candidates on average, thus storing the four lists costs about $\frac{7835 \cdot 15 \cdot 4}{2^{10} \cdot 8} \approx 58$ KB. The whole list merging process is depicted in Fig. 4.

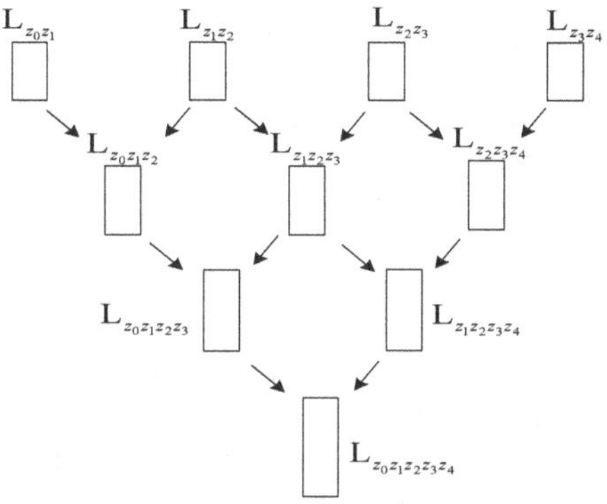

Fig. 4. The list merging process.

Precisely, in order to take the full advantage of the list size reduction during the merging process in Fig. 4, we adopt the merging routine as follows: let the input lists be denoted by $L_{z_0 z_1}, L_{z_1 z_2}, L_{z_2 z_3}, L_{z_3 z_4}$. We first merge every two consecutive candidate lists, and get the new candidate lists $L_{z_0 z_1 z_2}, L_{z_1 z_2 z_3}, L_{z_2 z_3 z_4}$. Now the consecutive lists will have more overlapping state bits, and can further reduce the memory complexity to a large extent. We repeat the similar procedure to get $L_{z_0 z_1 z_2 z_3}$ and $L_{z_1 z_2 z_3 z_4}$. Finally, we merge the last two candidate lists to get the final list for the CP part of the internal state.

Before running the merging procedure, we have already generated 4 candidate lists in the online phase from the 5-bit keystream prefix. The expected size of the check-state between two consecutive lists at the most upper level in Fig. 4 is $9 \cdot \frac{1}{4} + 11 \cdot \frac{3}{4} = 10.5$ bits. The averaged size of the input list is 7835, so it takes about

$$\frac{4 \cdot 8 \cdot 7835 + 3 \cdot 2^{12.3733} \cdot 2^{21}}{\Omega} \approx 2^{28.3}$$

cipher ticks to fulfill the merging procedure, where $\Omega = 2^{6.66}$ is the number of CPU-cycles to generate 1 bit keystream in A5/1. The detailed computation

[1] Other choices are also possible, e.g., we can get the first 6 keystream bits to launch the attack. For simplicity of description, we take the first 5 keystream bits here.

process of Ω is presented in Sect. 5. The expected number of merged candidates is $2^{16.6}$, each candidate can be stored in 5 bytes at most, so the memory requirement is $2^{18.92}$ bytes, approximately 486 KB. Note that we usually need to store 2 lists in memory when merging, thus the total memory cost is around 1 MB.

Note that we can recover the correct merged partial internal state if and only if the two candidate lists contain the correct candidate associated with their keystream prefixes. Taking into account the tree-like merging procedure in Fig. 4, the probability that the correct CP part will survive in the resultant list is $Pr_{merge} = (0.9903)^{\eta} = 0.9618$ where $\eta = 4$ is the number of the candidate lists $L_{z_i z_{i+1}}$. In other words, if we carry out the attack $\lceil \frac{1}{Pr_{merge}} \rceil \approx 2$ times, we are expected to find the actual CP part of the internal state from the resultant list. In our experiments, we have found that the probability p_1 defined in Corollary 2 is not stable sometimes, thus we multiply $\lceil \frac{1}{Pr_{merge}} \rceil$ by a small constant $\lambda = 4$ to recover the actual CP with a high probability.

Remarks. From the A5/1 specification and Fig. 3, for an i-bit keystream prefix with $i \geq 2$, the associated restricted internal state is of size $15 + 6(i - 2)$ bits. That is, for a 2-bit keystream prefix, the restricted internal state has 15 bits; for a 3-bit keystream prefix, the restricted internal state has $15 + 6 = 21$ bits; for a 4-bit keystream prefix, the restricted internal state has $21 + 6 = 27$ bits and for a 5-bit keystream prefix, the restricted internal state has $27 + 6 = 33$ bits. So far, we have already obtained a candidate list of the restricted internal state of 33 bits associated with the first 5 keystream bits $(z_0, z_1, z_2, z_3, z_4)$ in a probabilistic way. This is accomplished by independently treat with the 4 overlapping 2-bit keystream prefixes (z_i, z_{i+1}) for $i = 0, 1, 2, 3$. Precisely, for each keystream prefix, we derive the corresponding candidate list by the method in Sects. 4.4 and 4.5. By carefully choosing the attack parameters β and γ, we can guarantee that the corresponding correct restricted internal state for (z_i, z_{i+1}) is indeed in the candidate list with a reasonably good probability. Then we combine these 4 candidate lists together by the merging procedure in Sect. 4.6 based on the *randomness and independence* assumptions to have the larger partial internal state corresponding to the first 5 keystream bits. Once the 4 correct restricted internal states, corresponding to the 4 keystream prefixes (z_i, z_{i+1}) for $i = 0, 1, 2, 3$, are restored successfully in each case, the merging procedure will definitely retrieve the correct union larger state corresponding to the first 5 keystream bits with probability 1, in which process the merging operation will also massacre lots of candidates when the two adjacent restricted internal states have a number of overlapping bits.

4.7 Restoring the RP Part of the Internal State

We have already derived the CP part of the internal state in A5/1 after the merging phase, the RP part will be retrieved in this section. We decided to take a dynamic guess-and-determine like method to restore the RP part, similar to the approach in [9]. The difference is that in [9], all the possibilities of the guessed part of the internal state are tried, while now we have already obtained some subset of the CP part in the internal state *without* trying all the possibilities, which will reduce the overall complexity to a large extent.

Let $S(t)$ be the internal state of A5/1 at time t, then the state recovered after the merging phase is a subset of $S(100)$, while $S(101)$ is the targeted state. Since the clock control taps have already been recovered, it is easy to get $S(101)$ by clocking $S(100)$ one step forwards. As analyzed in [9][2], the state-transition function of A5/1 is not one-to-one and there are less than or equal to $5 \cdot 2^{61} \approx 2^{63.32}$ reachable internal states for $S(101)$. Thus, we take 63.32 instead of 64 to analyze the complexity after one step forwards, and the unreachable states can be easily distinguished by a set of linear equations.

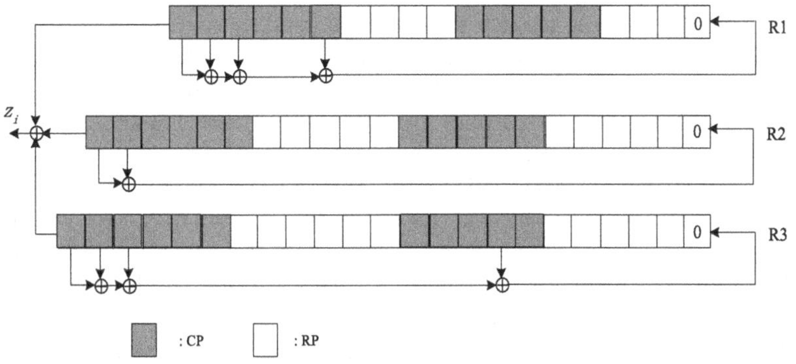

Fig. 5. The CP and RP parts of the internal state.

Our attack have exploited 5 keystream bits with $\eta = 4$ so far, as shown in Sect. 4.6. Due to the irregular clocking rule in A5/1, each register moves $\frac{3}{4}$ of the time, which means that each register is expected to clock $5 \times \frac{3}{4} \approx 4$ times after the merging phase. From each 2-bit keystream prefix, we can recover 15 bits of the corresponding restricted internal state, of which 11 bits are overlapped by the next restricted internal state with a probability of $\frac{3}{4}$, and 9 bits are overlapped with a probability of $\frac{1}{4}$. Hence, there are around $15 + 3 \cdot 6 = 33$ known bits after the merging phase, and the total entropy of the full internal state will be reduced to $63.32 - 33 = 30.32$ bits *on average*, which is depicted in Fig. 5.

The dynamic guess-and-determine attack is as follows. Note that in Figs. 1 and 5, the three registers are shifted towards the left direction and the newest feedback bits are injected into the state at the rightmost cells. The adversary could first guess y bits in each register from the first unknown clock control tap on to the right direction. One can thus obtain $3y$ linearly independent equations for the unknown bits in $S(101)$ if y is a small integer, e.g., $y = 4$ as shown in the register R1 in Fig. 5 whose rightmost 4 cells are unknown. As in [9], the adversary

[2] Though the clock control taps defined in [9] are different from here, this issue does not have any effect on the analysis of the state-transition properties under the condition that $\min(ct_1, ct_2, ct_3) \geq 2$.

can get one linear equation for free due to the fact that $S(101)$ is the targeted state and A5/1 first makes the stop/go clocking when producing the keystream bit. Corresponding to the directly guessed $3y$ clock control tap bits, there will be some number of linear equations obtained from the generated keystream bits for the adversary. Since on average each register clocks 3 times in 4 steps, the y guessed bits in each register can determine about $\frac{4y}{3}$ values of the clocking control taps in the register under consideration. Thus, the attacker can obtain around $1 + \frac{4y}{3}$ additional linear equations derived from the keystream bits on $S(101)$. When $y = 4$, each equation contains at least two new unknown bits that never appeared before, and these additional equations will be linearly independent to each other accordingly. The additional equations are linearly independent to the above $3y$ equations if and only if there is at least one bit not guessed in each of them. This is true when y is a small integer, e.g., $y = 4$. On the other side, if there are indeed some linear equations linearly dependent on the $3y$ single variable equations, these equations can be used as the linear consistency test of the guesses. Now there are around $3y + 1 + \frac{4y}{3} = 18.33$ linearly independent equations on average, and there are around $\tau = 30.32 - y = 12$ unknown bits which can be uniformly averaged among the three registers. Since there are some number of candidates left for the CP part when we only exploit the first 5 keystream bits, we need some very efficient procedure for the recovery of RP part. To fulfill this task, we then build a tree structure in the same manner as in [9] to derive the unknown bits dynamically and sequentially. Precisely, this tree structure sequentially record all the possibilities for the next coming bits consistent with the linear equations derived from the keystream bits. In each node of the tree, we store the next 3 clock taps of the three registers, representing the clocking which are consistent with the $1 + \frac{4y}{3}$ linear equations. As a result of the action of the majority function, there are 8 types of nodes, two of which have 8 possible successors and six of which have 4 possible successors. Thus, each node has $\frac{3}{4} \cdot 4 + \frac{1}{4} \cdot 8 = 5$ children in the tree on average. When checked with the corresponding linear equations, this number can be reduced to $5 \cdot \frac{1}{2} = 2.5$. Hence, to retrieve the τ unknown bits, the length of the tree should be $\frac{4}{3} \cdot \frac{\tau}{3}$ on average, and we have to check $2.5^{\frac{4}{3} \cdot \frac{\tau}{3}} \approx 2^{7.22}$ possibilities on average, instead of $2^{\tau} \approx 2^{12}$ possibilities under the independence assumption of the supercritical branching process, analyzed in the Appendix of [9]. In summary, the initial state $S(100)$ can be restored with a low complexity as shown above, the same as the state $S(101)$. Once we get the candidates of the full state, we can run A5/1 forwards for some ticks to check the consistency with the available keystream to decide whether the recovered state is correct or not.

5 Experimental Results

To check the validity and the actual performance of our attack, we have implemented the crucial steps of the suggested attack on a single core of a PC, running with Windows 7, Intel 3.4 GHz CPU and 16 GB RAM. In general, the experimental results verified the correctness of the crucial steps in the new method, and matched the theoretical prediction of each procedure quite well.

Since we take cipher ticks as the complexity unit in complexity analysis, we have first determined the constant Ω used in our attack. The source code of A5/1 for testing is a modification of a pedagogical implementation in [6]. We found that the average time to generate 1-bit keystream for A5/1 is $2.97 \cdot 10^{-8}$ seconds. Thus, one cipher tick of the A5/1 stream cipher accounts for $\Omega = 2.97 \cdot 10^{-8} \cdot 3.4 \cdot 10^{9} \doteq 2^{6.66}$ CPU cycles.

Since the basis of a fast near collision attack is the randomness, we use the RC4 stream cipher with a 128-bit secret key by discarding the first 8192 keystream words as the random source in our experiments. We adopted a random seed key dependent on the system time when invoking RC4, thus could assure that we have used random different sources among different calls[3].

To see the correctness of the suggested attack, there are essentially two points to check and verify. The first one is to assure that the correct restricted internal state associated with a 2-bit keystream prefix is indeed included in the candidates list generated by Algorithm 3 in Sect. 4.4. The other is to assure that with the suggested attack parameters, the attack will indeed behave as predicted in theory, which actually means that the 4 correct restricted internal states are *simultaneously* included in $L_{z_0 z_1}$, $L_{z_1 z_2}$, $L_{z_2 z_3}$ and $L_{z_3 z_4}$, respectively.

To check the first statement, following the analysis in Sect. 4.3, we have prepared the small differential tables indexed by $(\mathtt{ksd}, d) = (\mathtt{0x3}, 2)$ with all the possible ISDs in the pre-processing phase, requiring 198 bytes memory. Each pre-computed table is sorted according to the occurring probabilities of each involved ISD, which takes about $P = \frac{(2 \cdot 2^4 \cdot 2^{15} \cdot 121 + 2^4 \cdot 99 \cdot \log_2 99)}{\Omega} \approx 2^{20.26}$ cipher ticks.

Then in the online phase, we have tried all the 99 ISDs in the pre-computed table $c \cdot \frac{2^n}{|T|} = 4 \cdot \frac{2^{15}}{99} \doteq 1324$ times to make sure that the generated list contains the correct restricted internal state under consideration, so it takes $\frac{4 \cdot \frac{2^{15}}{99} \cdot 99}{\Omega} \approx 2^{10.34}$ cipher ticks for $n = 2^{15}$ and $|\mathtt{T}| = 99$. Then we enhance the existence probability of the correct restricted internal state by intersecting $\beta = 6$ candidate lists and unifying $\gamma = 2$ such resultant lists, as suggested in Algorithm 4 and 5. The complexity of the intersection operation in Algorithm 4 is $2^{12.93}$ cipher ticks, and the union operation takes about $2^{13.93}$ cipher ticks. Next, we could check whether the correct restricted internal states are indeed contained in the corresponding candidate lists generated by Algorithm 3. Precisely, we just set a flag value 1 when the check is successful, and expect to see 4 consecutive `1,1,1,1` when running the above steps a few times. In the experiments, we indeed saw the expected pattern `1,1,1,1` after running the routine a few times. Thus, we have verified the essential point for verifying the correctness of the suggested attack. This fact, together with the following merging step, also explained the surprising phenomenon that we actually restored around 33 internal state bits from only 5 keystream bits in a probabilistic way.

---

[3] The seed key in our c implementation is derived from the system time via some arithmetic operations such as modulo addition. We have also tried AES as the random source and obtained almost the same results as the RC4 case.

Note that with the correct restricted internal states being in the candidate lists, the merging procedure will preserve them, thus the correct union partial state with probability 1. Precisely, we could merge the restored restricted internal states to get the CP part of the internal state from the 4 unified lists from $L_{z_0 z_1}$ to $L_{z_3 z_4}$. The preparation for the merging procedure takes about $T_1 = 2^2 \cdot 2^{13.93} \doteq 2^{15.93}$ cipher ticks. Further, we adopt the list merging process depicted in Fig. 4 to restore the CP part of the internal state, whose time complexity is reduced to $T_2 = 2^{28.3}$ cipher ticks with approximately 1 MB memory, as discussed in Sect. 4.6.

Next, the RP part of the internal state is restored by the dynamic guess-and-determine attack in Sect. 4.7, which needs $2^{29.16}$ cipher ticks on average, which can be seen from the following deductions. First, the complexity of solving the $3y$ linear equations is at most 2^{12} when $y = 4$, and the subsequent steps take $2^{8.22}$ cipher ticks. The total complexity of the guess-and-determine attack is $\frac{2^{12+8.22}}{\Omega} = 2^{13.56}$ cipher ticks. The averaged number of trials to find the correct $S(101)$ is reduced to $2^{12.56}$ cipher ticks. Note that there are about $2^{16.6}$ candidates resultant from the merged CP state, so the complexity of the RP recovery is finally increased to $T_3 = 2^{16.6} \cdot 2^{12.56} = 2^{29.16}$ cipher ticks. The complexity of our attack is $T_1 + T_2 + T_3 = 2^{15.93} + 2^{28.3} + 2^{29.16} \approx 2^{29.79}$ cipher ticks. Since we choose $\lambda = 4$ to stabilize the success probability, the total complexity of our attack increase to $2^{31.79}$ cipher ticks.

We have run the above attack routines many times with different keys and frame numbers generated randomly, and the experimental results are always consistent with the theoretical analysis. One run of our attack is as follows. We use RC4 to generate the key `0xeab64598b32b32c4` and the frame number `0x3efd4c`, respectively. The produced keystream frame is `0x4a01e459770cdf81af52e70706a4c0` for one direction communication and `0xdefc02c7d0697294be821ae0f7adc0` for the other direction. The first 64 keystream bits are `0x4a01e459770cdf81`, where the first keystream bit z_0 is at the most significant bit position of a byte and z_7 is at the least significant bit position. Then we have $z_0 z_1 z_2 z_3 z_4 =$ `01001`. When one $l = 2$-bit keystream prefix $z_i z_{i+1}$ $(0 \leq i \leq 3)$ is used in our attack, it takes around more than one second to generate the pre-computed tables so far, stored in 198 bytes in RAM. Once constructed, these pre-computation tables can be used when the key or the frame numbers have changed. In the online phase, with the determined parameters `ksd=0x3`, $c = 4$, $\beta = 6$ and $\gamma = 2$, we could mount our attack step-by-step as presented above. The average size of the generated candidate list is 7835, while the theoretical size is $7880 \doteq 2^{12.95}$, which seems to imply that the practical implementation is more efficient to restore the correct restricted internal state than predicted in theory. We then ran the merging routine to restore the CP, which takes tens of seconds for the current non-optimized C implementation. The 4 correct restricted internal state `01` $\leftharpoondown$ `111000111010000`, `10` $\leftharpoondown$ `000010000011010`, `00` $\leftharpoondown$ `010011010001111` and `01` $\leftharpoondown$ `011011111101010` were simultaneously restored at the first invocation, which means that after the merging phase, the correct CP state `1110001110100000110100011111111010` would be recovered with probability 1. Finally, we retrieve the full internal state following the steps in Sect. 4.7. To improve the success probability, we set $\lambda = 4$ to retrieve the correct target internal state with a high probability. Note that the current implementation results are from the non-optimized code, and we think the attack can actually find the correct internal state, thus the secret key in a few seconds after future optimizations. The C language codes for verifying the validity of our attack are available via https://github.com/martinzhangbin.

We end this section by the comparison with other best known TMD attacks on A5/1, shown in Table 5. This table clearly shows the advantage of the new attack. Note that the (conditional) correlation attacks in [1,7,12] need much more keystream frames than our attack, though no long-term storage and no preprocessing are required.

Table 5. Comparison with the previous best known TMD attacks on A5/1

Attack	Data	Memory	Pre-comp.	Online time	Success rate
Nohl's Attack [13]	8 frames	1.7 TB	−weeks	10 seconds	87%
TMD Attack [11]	8 frames	1.968 TB	110days	9 seconds	81%
BB Attack [5]	$2^{14.67}$ frames	146 GB	$>$ 5 years	1 second	–
KP Attack [2]	4 frames	17.6 TB	$>$ 2300 years	5 minutes	$>$ 60%
Our Attack	$<$ 1 frame	1 MB	$2^{20.26}$ ticks	$2^{31.79}$ ticks	$\approx$ 99%

Note that the pre-computation phase of the Biased Birthday (BB) attack in [5] was extensively sampled rather than completely executed, i.e., there are non-trivial cases that the online time needed to find the correct internal state is much more than the claimed time.
KP means known plaintext.

Remarks. Since our attack is a known plaintext attack, we need first capture the first 64 keystream bits in 1 frame, which is always feasible in general in GSM networks. As shown in section 7 of [1], each traffic channel between the handset and the network is accompanied by a slower control channel, which is referred to as the Slow Associated Control CHannel (SACCH). The mobile uses the SACCH channel (on the uplink) to report its reception of adjacent cells. The network uses this channel (on the downlink) to send (general) system messages to the mobile, as well as to control the power and timing of the current conversation. The contents of the downlink SACCH can be inferred by passive eavesdropping. Besides, an attacker would still need to cope with the Frequency Hoping (FH) used by GSM, of which the hopping sequence can be determined through a quick exhaustive search due to the fact that given n, GSM defines only $64n$ hopping sequences (n cannot be large since the total number of frequencies in GSM is only about 1000, of which only 124 belong to GSM 900).

6 Leveraging the Attack to Any GSM Network

For completeness, we leverage our attack on A5/1 to other algorithms used in the GSM network in this section. It exploits the already detected flaw of the

GSM protocols that the same key is used in different encryption algorithms. The key only depends on RAND in GSM, thus once we have restored the secret key of A5/1 by our method with ease, we are able to use this key to encrypt or decrypt for A5/3 or GPRS.

Precisely, the adversary can carry out a man-in-the-middle attack by impersonating the base station to the victim and the customer to the base station. In the initialization of a conversation, there is an authentication phase, which is a basic challenge-response scheme between the mobile phone and the network. The network sends a RAND to the attacker, who will transmit it to the victim then. The victim computes the SRES and return it to the attacker. In GSM protocols, the customer phone only reports the list of ciphers that it supports, while the network chooses which encryption algorithm is to be used. Hence, the attacker asks the victim to encrypt with A5/1 so that he could retrieve the key of A5/1 by our attack to encrypt in the conversation of the base station. Then the attacker returns the SRES computed by the victim to the base station, and the authentication is finished. When the base station asks the attacker to encrypt with A5/3 or GPRS, the attacker has already recovered the same secret key as used by A5/1, and has the ability to encrypt or decrypt using the restored key. Due to the fact that the authentication phase may not be initialized frequently, the key recovered from the previous conversation may be used in future conversations as well.

There is also a similar attack against GPRS, whose security is based on the same mechanisms as of GSM except that it uses GPRS-RAND/GPRS-SRES in the authentication and key agreement, and the GPRS cipher, referred to as GPRS-A5 or GPRS Encryption Algorithm (GEA), is different from A5/1 or A5/2 and is never made public so far. The attacker can take advantage of the symmetry in the key agreement of GPRS and GSM by performing an active attack on the customer phone via a fake base station. That is, he impersonates the network and starts a radio session with the mobile phone victim that is protected by A5/1. Accordingly, the resulting key is identical to the one that is used in GPRS and the attacker can recover it using our attack on A5/1. This means that the attacker can encrypt or decrypt the corresponding GPRS traffic successfully.

7 Conclusion

In this paper, we have proposed an entirely different cryptanalytic approach to break the A5/1 stream cipher without the need of large rainbow tables and a huge pre-computation phase. We have taken the new viewpoint of the fast near collision attack to restore the internal state of A5/1 in a divide-and-conquer manner. Based on the refined self-contained method, we could efficiently recover the restricted internal state of a given keystream prefix and merge several restored partial states together according to the irregular clocking mechanism in A5/1. Now the pre-computation becomes quite lightweight and cheap for individual cryptanalysts. It is shown that the irregular stop/go mechanism in A5/1 does not frustrate our attack as expected in the merging phase and we can reliably find the initial internal state, thus the secret key of A5/1 from only 1 keystream frame in $2^{31.79}$ cipher ticks. The total storage of our attack is about 1 MB, which is much less than all the previous best known TMD attacks. Due to the fact that A5/3 and GPRS share the same key as A5/1 in GSM, our attack can be converted into attacks against any GSM network eventually. It is well known that the analysis of stream ciphers based on irregularly clocked shift registers is a long standing problem in theory, our results shed some light on this issue and open new possibilities to deal with irregularly clocked shift registers.

Acknowledgements. The author would like to thank the anonymous reviewers for very helpful comments and Yanyi Liu, Hui Peng and Di Zhai for the discussions on the topic. This work is supported by the National Key R&D Research programm (Grant No. 2017YFB0802504), the program of the National Natural Science Foundation of China (Grant No. 61572482), National Cryptography Development Fund (Grant No. MMJJ20170107) and National Grand Fundamental Research 973 Programs of China (Grant No. 2013CB338002).

References

1. Barkan, E., Biham, E.: Conditional estimators: an effective attack on A5/1. In: Preneel, B., Tavares, S. (eds.) SAC 2005. LNCS, vol. 3897, pp. 1–19. Springer, Heidelberg (2006). https://doi.org/10.1007/11693383_1
2. Barkan, E., Biham, E., Keller, N.: Instant ciphertext-only cryptanalysis of GSM encrypted communication. In: Boneh, D. (ed.) CRYPTO 2003. LNCS, vol. 2729, pp. 600–616. Springer, Heidelberg (2003). https://doi.org/10.1007/978-3-540-45146-4_35
3. Biham, E., Dunkelman, O.: Cryptanalysis of the A5/1 GSM stream cipher. In: Roy, B., Okamoto, E. (eds.) INDOCRYPT 2000. LNCS, vol. 1977, pp. 43–51. Springer, Heidelberg (2000). https://doi.org/10.1007/3-540-44495-5_5
4. Biryukov, A., Shamir, A.: Cryptanalytic time/memory/data tradeoffs for stream ciphers. In: Okamoto, T. (ed.) ASIACRYPT 2000. LNCS, vol. 1976, pp. 1–13. Springer, Heidelberg (2000). https://doi.org/10.1007/3-540-44448-3_1
5. Biryukov, A., Shamir, A., Wagner, D.: Real time cryptanalysis of A5/1 on a PC. In: Goos, G., Hartmanis, J., van Leeuwen, J., Schneier, B. (eds.) FSE 2000. LNCS, vol. 1978, pp. 1–18. Springer, Heidelberg (2001). https://doi.org/10.1007/3-540-44706-7_1
6. Briceno, M., Goldberg, I., Wagner, D.: A pedagogical implementation of A5/1, May 1999. http://www.scard.org
7. Ekdahl, P., Johansson, T.: Another attack on A5/1. IEEE Trans. Inf. Theory **49**(1), 284–289 (2003)
8. Gendrullis, T., Novotný, M., Rupp, A.: A real-world attack breaking A5/1 within hours. In: Oswald, E., Rohatgi, P. (eds.) CHES 2008. LNCS, vol. 5154, pp. 266–282. Springer, Heidelberg (2008). https://doi.org/10.1007/978-3-540-85053-3_17
9. Golić, J.D.: Cryptanalysis of alleged A5 stream cipher. In: Fumy, W. (ed.) EUROCRYPT 1997. LNCS, vol. 1233, pp. 239–255. Springer, Heidelberg (1997). https://doi.org/10.1007/3-540-69053-0_17
10. Koch, P.C.: Cryptanalysis of stream ciphers-analysis and application of the near collision attack for stream ciphers. Technical University of Denmark, Master thesis supervisor, Christian Rechberger, pp. 111–122, November 2013

11. Lu, J., Li, Z., Henricksen, M.: Time–memory trade-off attack on the GSM A5/1 stream cipher using commodity GPGPU. In: Malkin, T., Kolesnikov, V., Lewko, A.B., Polychronakis, M. (eds.) ACNS 2015. LNCS, vol. 9092, pp. 350–369. Springer, Cham (2015). https://doi.org/10.1007/978-3-319-28166-7_17
12. Maximov, A., Johansson, T., Babbage, S.: An improved correlation attack on A5/1. In: Handschuh, H., Hasan, M.A. (eds.) SAC 2004. LNCS, vol. 3357, pp. 1–18. Springer, Heidelberg (2004). https://doi.org/10.1007/978-3-540-30564-4_1
13. Nohl, K.: Attacking phone privacy. In: Black Hat USA 2010 Lecture Notes (2010). https://srlabs.de/decrypting-gsm/
14. Pornin, T., Stern, J.: Software-hardware trade-offs: application to A5/1 cryptanalysis. In: Koç, Ç.K., Paar, C. (eds.) CHES 2000. LNCS, vol. 1965, pp. 318–327. Springer, Heidelberg (2000). https://doi.org/10.1007/3-540-44499-8_25
15. Zhang, B., Li, Z., Feng, D., Lin, D.: Near collision attack on the grain v1 stream cipher. In: Moriai, S. (ed.) FSE 2013. LNCS, vol. 8424, pp. 518–538. Springer, Heidelberg (2014). https://doi.org/10.1007/978-3-662-43933-3_27
16. Zhang, B., Xu, C., Meier, W.: Fast near collision attack on the grain v1 stream cipher. In: Nielsen, J.B., Rijmen, V. (eds.) EUROCRYPT 2018. LNCS, vol. 10821, pp. 771–802. Springer, Cham (2018). https://doi.org/10.1007/978-3-319-78375-8_25

Functional Encryption

Tightly Secure Inner Product Functional Encryption: Multi-input and Function-Hiding Constructions

Junichi Tomida(✉)

NTT, Tokyo, Japan
junichi.tomida.vw@hco.ntt.co.jp

Abstract. Tightly secure cryptographic schemes have been extensively studied in the fields of chosen-ciphertext secure public-key encryption, identity-based encryption, signatures and more. We extend tightly secure cryptography to inner product functional encryption (IPFE) and present the first tightly secure schemes related to IPFE.

We first construct a new IPFE scheme that is tightly secure in the multi-user and multi-challenge setting. In other words, the security of our scheme does not degrade even if an adversary obtains many ciphertexts generated by many users. Our scheme is constructible on a pairing-free group and secure under the matrix decisional Diffie-Hellman (MDDH) assumption, which is the generalization of the decisional Diffie-Hellman (DDH) assumption. Applying the known conversions by Lin (CRYPTO 2017) and Abdalla et al. (CRYPTO 2018) to our scheme, we can obtain the first tightly secure function-hiding IPFE scheme and multi-input IPFE (MIPFE) scheme respectively.

Our second main contribution is the proposal of a new generic conversion from function-hiding IPFE to function-hiding MIPFE, which was left as an open problem by Abdalla et al. (CRYPTO 2018). We obtain the first tightly secure function-hiding MIPFE scheme by applying our conversion to the tightly secure function-hiding IPFE scheme described above.

Finally, the security reductions of all our schemes are fully tight, which means that the security of our schemes is reduced to the MDDH assumption with a constant security loss.

Keywords: Functional encryption · Inner product · Tight security

1 Introduction

(Multi-input) Inner Product Functional Encryption. Functional encryption (FE) [13,37] is a relatively novel cryptographic notion that has a crucially different feature from traditional encryption schemes. Specifically, FE schemes allow us to obtain computation results from encrypted data without revealing any other information about the underlying data. This is in contrast to traditional encryption schemes, in which only owners of legitimate keys can learn

S. D. Galbraith and S. Moriai (Eds.): ASIACRYPT 2019, LNCS 11923, pp. 459–488, 2019.
https://doi.org/10.1007/978-3-030-34618-8_16

entire underlying data from ciphertexts while others can learn nothing. An FE scheme supports a certain function class $\mathcal{F}$ and in which an owner of a master secret can issue a secret key sk_f for any function $f \in \mathcal{F}$. Decryption of a ciphertext ct_x of message x with sk_f yields the computation result $f(x)$ and nothing else.

Multi-input functional encryption (MIFE) [28] is a natural extension of FE, which can handle a function class that takes multiple inputs. Roughly speaking, an owner of sk_f can learn the computation result $f(x_1, \dots, x_\mu)$ from ciphertexts $\mathsf{ct}_{x_1}, \dots, \mathsf{ct}_{x_\mu}$ of messages $x_1, \dots, x_\mu$ for some natural number $\mu \geq 2$.

Known (MI)FE schemes can be classified into two categories with respect to their function classes.

General functionalities: This category consists of (MI)FE schemes for general circuits, e.g., [8,23,24,28,39]. Although they are powerful enough to handle all functions computable in polynomial time, known schemes are built on quite heavy cryptographic primitives such as indistinguishability obfuscation [23] or multi-linear maps [22]. Thus, they are captured as rather feasibility results.

Specific functionalities: The second category covers (MI)FE schemes for specific functions such as inner product and quadratic function, e.g., [2,4,6,9]. They are aimed at obtaining more practical features, namely, efficiency and concrete security, with sacrificing the generality. Therefore, they have simple constructions, and their security is based on standard assumptions.

Inner product functional encryption (IPFE) [2] and multi-input IPFE (MIPFE) [4], categorized into the latter, are FE and MIFE respectively, whose function classes are inner product. More precisely, in an (M)IPFE scheme, a secret key $\mathsf{sk}_{\mathbf{y}_1,\dots,\mathbf{y}_\mu}$ is associated with vectors $\mathbf{y}_1, \dots, \mathbf{y}_\mu$, and decrypting ciphertexts $\mathsf{ct}_{\mathbf{x}_1}, \dots, \mathsf{ct}_{\mathbf{x}_\mu}$ of vectors $\mathbf{x}_1, \dots, \mathbf{x}_\mu$ with $\mathsf{sk}_{\mathbf{y}_1,\dots,\mathbf{y}_\mu}$ reveals the summation of the inner products $\sum_{i \in [\mu]} \langle \mathbf{x}_i, \mathbf{y}_i \rangle$. When $\mu = 1$, the above description corresponds to an IPFE scheme. Inner product is a simple but powerful functionality, and many practical applications of IPFE have been suggested, e.g, biometric authentication, nearest-neighbor search and statistical analysis [2,32].

Function Privacy. In (MI)FE, we can consider two types of privacy: message privacy and function privacy. Message privacy, which is essential for standard (MI)FE schemes, is the property that ciphertexts do not reveal any information about underlying data. On the other hand, function privacy is an additional but important property for (MI)FE schemes, which indicates that secret keys also hide the information of the corresponding function. Function privacy is essential for some applications such as delegation of sensitive computation [15]. We often refer to (MI)FE with function privacy as function-hiding (MI)FE. Function-hiding (MI)FE schemes have also been studied for both general functionalities [14,15] and specific functionalities [12,18,32,38].

Tight Security. When we try to prove the security of a cryptographic scheme, we often construct a reduction algorithm that solves a problem assumed to be hard by utilizing a PPT adversary that breaks the security of the scheme. Then, breaking the security of the scheme immediately implies solving the hard problem. It is both theoretically and practically important to evaluate how difficult breaking the scheme is compared with solving the problem. More formally, when the reduction algorithm equipped with an adversary that breaks the scheme with probability ϵ in time t solves the underlying problem with probability ϵ/L in roughly the same time t, it is important to evaluate the security loss L. This is because we need to set the parameter size of the scheme large enough to negate the effect of L for the security guarantee. Thus, the smaller the security loss L, the more desirable the security reduction. We say that the security reduction is tight if the security loss is constant, i.e., $L = O(1)$.

When we consider public-key primitives such as public-key encryption (PKE) or identity-based encryption (IBE), we usually prove their security in the single-challenge setting. This is because the security of public-key primitives in the single-challenge setting normally implies that in the multi-user and multi-challenge setting via hybrid argument, which is more realistic setting where an adversary can make polynomially many challenge queries against multiple users. However, such a hybrid argument increases the security loss by the factor of μq, where μ is the number of users and q is the maximum number of challenge queries for each users [11]. Since the numbers of users and ciphertexts are quite large in practice, we strongly desire cryptographic schemes whose security is guaranteed independently of those numbers.

Motivated by the above reason, (almost) tightly secure cryptographic schemes have been extensively studied in various fields, especially on chosen-ciphertext secure PKE (CCA-secure PKE), IBE, and signature, e.g, [7,17,25,26,29–31,33]. In spite of such a great deal of effort, tightly secure schemes in the context of advanced encryption are known only for IBE except the very recent result on broadcast encryption by Gay et al. [27]. Hence, it is an important and interesting task to explore what kind of cryptographic schemes can achieve tight security.

Tight Security for IPFE. We would like to discuss the importance of tightly secure IPFE in more detail. We consider that the most significant situation where we need a tightly secure IPFE scheme is when a function-hiding scheme is needed. This is because the only way that we know to realize function-hiding IPFE schemes requires bilinear groups, which is relatively susceptible to security loss. One solution to compensate for security loss caused by loose reduction is to increase the parameter size of underlying primitives, e.g., bilinear groups, which will reinforce the difficulty of underlying problems, e.g., the matrix Diffie-Hellman problem. As observed by Abe et al. [5], however, this is not an easy task for bilinear groups because there are many factors that involve the security and efficiency of them such as the choice of curves, pairings, and various parameters like embedding degrees. Hence, we typically adopt one from existing well-studied settings, which are investigated only for standard parameters such as 128, 192, and 256-bit security. The main problem of this fact is that there is no intermediate instantiation among these parameters, and one have to hop to the next standard level if stronger security is necessary. A pairing computation is especially influenced by this hop; for instance, they state that a pairing in the

192-bit security takes 6 to 7 times more time than in the 128-bit security on ordinary personal computers [10,20].

Additionally, it is not unrealistic that an adversary obtains a large amount of ciphertexts so that we need to consider the security loss of IPFE schemes. Let us consider the case to use a function-hiding IPFE scheme for DNA analysis. Suppose a national institution holds a database consisting of a certain part of the human's DNA sequence. It is rational to assume that the part consists of 2^{13} bases and the number of the samples is 2^{20}; actually, GenBank operated by the National Center for Biotechnology Information has more than 2^{27} sequences [1]. Each sample is encoded to a binary vector setting as A = (1, 0, 0, 0), T = (0, 1, 0, 0), and so on, and stored in a cloud server with an encrypted form. We can check the number of the same bases between encrypted sequences and a target sequence by decrypting with a secret key for the target sequence. Because DNA sequences have a correlation with phenotypes, the DNA similarity check will be useful for genetical research, medical diagnosis, etc. We need the function-hiding property because target sequences are also personal data and thus sensitive. In this situation, the possibly untrusted server has $q = 2^{20}$ ciphertexts, large enough to consider the security loss of the scheme. Decryption of all known schemes involves the same number of pairings as the order of the vector length: $m = 2^{15}$ per one sample in our case. Thus, the choice of the security level significantly affects the efficiency of the system, and we can conclude that tight security is a very important concept in the context of IPFE as well as other cryptosystems.

1.1 Our Contributions

We extend the realm of tightly secure cryptography to IPFE and present a series of the first tightly secure (M)IPFE schemes. Our first main contribution is to construct the first tightly secure public-key IPFE scheme in the multi-user and multi-challenge setting. Note that previous IPFE schemes are tightly reduced to underlying assumptions in the single-challenge setting [6], which means that their security is independent from the number of secret key queries. To our knowledge, however, there are no results on tight security of IPFE in the multi-user and multi-challenge setting. Our tightly secure IPFE scheme is constructible from a pairing-free group and its security is based on the matrix decisional Diffie-Hellman (MDDH) assumption, which is a generalization of the well-studied decisional Diffie-Hellman (DDH) assumption, with a small constant security loss.

Our result can be easily extended to the multi-input setting. Recently, Abdalla et al. proposed a generic conversion from an IPFE scheme into a MIPFE scheme [3,4]. Their conversion employs parallel execution of μ instances of the underlying IPFE scheme that is secure in the multi-challenge setting. By this construction, their conversion incurs a security loss of $O(\mu q)$ if we apply it to an IPFE scheme that is secure in the single challenge setting, where μ is the number of slots of the converted scheme and q is the maximum number of ciphertext queries for each slot. Interestingly, this construction is precisely compatible with an IPFE scheme that is secure in the multi-user and multi-challenge setting. In other words, the security of the converted MIPFE scheme is tightly reduced to that of the underlying IPFE scheme if the underlying scheme is secure in the multi-user and multi-challenge setting. Thus, we can obtain the first tightly secure MIPFE scheme.

Another important issue is the realization of tightly secure function-hiding (M)IPFE schemes. All previous function-hiding schemes suffer from a security loss of $L = O(q_{\mathsf{ct}} + q_{\mathsf{sk}})$, where q_{ct} (resp. q_{sk}) refers to the total number of ciphertext (resp. secret key) queries [12,18,34,38]. To achieve tight security, we utilize Lin's technique, who presented a simple paradigm to construct a function-hiding (private-key) IPFE scheme from a (public-key) IPFE scheme [34]. Applying her paradigm to our IPFE scheme, we can obtain the first tightly secure function-hiding IPFE scheme that is based on bilinear groups. However, the naive application of her paradigm to our scheme results in a redundant scheme. Thus, we optimize the scheme by reducing the unnecessary part.

The final target is to construct a tightly secure function-hiding MIPFE scheme. Unfortunately, there is no known generic technique to achieve a function-hiding MIPFE scheme. In fact, Abdalla et al. mention that a powerful conversion to achieve a function-hiding MIPFE scheme is a very interesting open problem [3]. Furthermore, the techniques used in the rather specific constructions of known function-hiding MIPFE schemes [3,19] are not applicable to our case. This is because our scheme requires the selective setting in a certain step of the proof, if we naively try to prove the security similarly to [3,19].

Our second main contribution is overcoming this problem by solving the open problem posed by Abdalla et al., that is, we introduce a new powerful and generic conversion. It converts a (weakly) function-hiding IPFE scheme into a (fully) function-hiding MIPFE scheme. Our conversion is as general as that for constructing non-function-hiding MIPFE by Abdalla et al. [3]: the requirements for an underlying scheme are essentially the same. Hence, if new function-hiding IPFE schemes are proposed in the future, e.g., based on lattices, we may utilize our conversion to obtain new function-hiding MIPFE schemes though some modification will be necessary. Additionally, we can obtain (non-tightly-secure) function-hiding MIPFE schemes in a more modular way than the previous ones [3,19] by utilizing our conversion to function-hiding IPFE schemes, e.g., the scheme from AGRW17 [4] + Lin17 [34]. Applying our conversion to our tightly secure function-hiding IPFE scheme, we can finally achieve the first tightly secure function-hiding MIPFE scheme.

Similarly to all previous IPFE schemes based on a cyclic group or bilinear groups, the decryption algorithms of our schemes require to solve the discrete logarithm problem on a decryption value. As pointed out in [2,32], however, this step is not so problematic in many cases. This is mainly because decryption values will not become exponentially large in real applications. Additionally, although there are some IPFE schemes that allow exponentially large outputs, they are either inefficient due to the large modulus [6] or based on a non-standard assumption [16].

We summarize the comparison of our schemes with previous ones in Tables 1, 2, 3 and 4. In these tables, we count the numbers of elements assuming that a matrix distribution $\mathcal{D}_k$ is a uniform one over $\mathbb{Z}_p^{(k+1)\times k}$. Some readers may be concerned about the increase of the key and ciphertext sizes, which may slow the efficiency of the system even after the compensation of security loss. However, we would like to emphasize that our contribution is the first step toward more efficient tightly secure schemes. Furthermore, our schemes may outperform previous ones in some situations. For example, when we instantiate our function-hiding IPFE scheme from the SXDH, it takes almost 5 times more pairings in decryption than the state-of-the-art scheme (Table 3). As discussed in the previous subsection, the difference of security level possibly affects pairings by the factor of 6 to 7 in practice, and thus there is a possibility that the decryption, the most important process of IPFE, of our scheme is faster than those of previous ones in the same security level. We leave constructing more compact tightly secure IPFE schemes as an interesting open problem.

2 Technical Overview

2.1 Tightly Secure IPFE

Our scheme is secure in the multi-user and multi-challenge setting under the MDDH assumption, but here we describe our scheme based on the DDH assumption in the single-user and multi-challenge setting to ease the exposition. Our starting point is the adaptively secure IPFE scheme by Agrawal et al. [6]. We briefly describe their scheme below. Let m be a vector length in the scheme.

$\mathsf{Setup}(1^\lambda, 1^m)$: $a \xleftarrow{\mathsf{U}} \mathbb{Z}_p$, $\mathbf{W} \xleftarrow{\mathsf{U}} \mathbb{Z}_p^{m\times 2}$, $\mathbf{a} := (a, 1)$, $\mathsf{pk} := ([\mathbf{a}], [\mathbf{W}\mathbf{a}])$, $\mathsf{msk} := \mathbf{W}$.
$\mathsf{Enc}(\mathsf{pk}, \mathbf{x})$: $s \xleftarrow{\mathsf{U}} \mathbb{Z}_p$, $\mathsf{ct} := ([s\mathbf{a}], [s\mathbf{W}\mathbf{a} + \mathbf{x}])$.
$\mathsf{KeyGen}(\mathsf{pk}, \mathsf{msk}, \mathbf{y})$: $\mathsf{sk} := (-\mathbf{W}^\top \mathbf{y}, \mathbf{y})$.
$\mathsf{Dec}(\mathsf{pk}, \mathsf{ct}, \mathsf{sk})$: $-\mathbf{y}^\top \mathbf{W}[s\mathbf{a}] + \mathbf{y}^\top[s\mathbf{W}\mathbf{a} + \mathbf{x}] = [\langle \mathbf{x}, \mathbf{y}\rangle]$.

Next, we explain the security proof of this scheme by Abdalla et al. [4], which is somewhat different from the original proof by Agrawal et al. and roughly goes as follows. First, the form of the challenge ciphertext is changed from $\mathsf{ct} := ([s\mathbf{a}], [s\mathbf{W}\mathbf{a} + \mathbf{x}^\beta])$ to $\mathsf{ct} := ([s\mathbf{a} + s'\mathbf{b}], [\mathbf{W}(s\mathbf{a} + s'\mathbf{b}) + \mathbf{x}^\beta])$, where $s' \xleftarrow{\mathsf{U}} \mathbb{Z}_p$, $\mathbf{b} := (1, 0)$, and $\beta \xleftarrow{\mathsf{U}} \{0, 1\}$. This change is computationally indistinguishable under the DDH assumption. At this point, we redefine $\mathbf{W}$ as

$$\mathbf{W} := \widetilde{\mathbf{W}} + u(\mathbf{x}^1 - \mathbf{x}^0)\mathbf{a}^{\perp\top}, \tag{2.1}$$

where $u \xleftarrow{\mathsf{U}} \mathbb{Z}_p$, $\widetilde{\mathbf{W}} \xleftarrow{\mathsf{U}} \mathbb{Z}_p^{m\times 2}$, and $\mathbf{a}^\perp := (1, -a)$, and note that $\mathbf{a}^{\perp\top}\mathbf{b} = 1$. In fact, $\mathbf{x}^0$ and $\mathbf{x}^1$ may depend on $\widetilde{\mathbf{W}}$ because the information of $\widetilde{\mathbf{W}}$ is leaked to the adversary from the public key and queried secret keys. However, we can assume that $\mathbf{x}^0$ and $\mathbf{x}^1$ do not depend on $\widetilde{\mathbf{W}}$ (and formally we use complexity leveraging to argue that). Then, redefined $\mathbf{W}$ is also a random element in $\mathbb{Z}_p^{m\times 2}$ and we have

Table 1. Comparison of adaptively secure IPFE schemes in the multi-user and multi-challenge setting. The columns $|\mathsf{pk}|$ and $|\mathsf{ct}|$ refer to the number of group elements. The columns $|\mathsf{msk}|$ and $|\mathsf{sk}|$ refer to the number of $\mathbb{Z}_p$ elements. The number m refers to the vector length. The number q_{ct} refers to the total number of ciphertext queries by an adversary. Note that we omit the group description from $|\mathsf{pk}|$.

IPFE schemes						
scheme	$\|\mathsf{pk}\|$	$\|\mathsf{msk}\|$	$\|\mathsf{ct}\|$	$\|\mathsf{sk}\|$	sec. loss	assumption
ALS16 [6]	$m+1$	$2m$	$m+2$	$m+2$	$O(q_{\mathsf{ct}})$	DDH
AGRW17 [4]	$km+k^2+k$	$(k+1)m$	$m+k+1$	$m+k+1$	$O(q_{\mathsf{ct}})$	$\mathcal{D}_k$-MDDH
Ours	m^2+1	$2m^2$	$3m$	$3m$	$O(1)$	DDH
	$k^2m^2+k^2+k$	$(k^2+k)m^2$	$(k^2+k+1)m$	$(k^2+k+1)m$	$O(1)$	$\mathcal{D}_k$-MDDH

Table 2. Comparison of MIPFE schemes based on a pairing-free group. The columns $|\mathsf{msk}|$ and $|\mathsf{sk}|$ refer to the number of $\mathbb{Z}_p$ elements. The column $|\mathsf{ct}|$ refers to the number of group elements. The number m refers to the vector length. The number μ refers to the number of slots. The number q_{ct} refers to the total number of ciphertext queries for all slots by an adversary.

MIPFE schemes					
scheme	$\|\mathsf{msk}\|$	$\|\mathsf{ct}\|$	$\|\mathsf{sk}\|$	sec. loss	assumption
ACFGU18[3]	$\{k^2+k+(k+1)m\}\mu$	$m+k+1$	$(m+k+1)\mu+1$	$O(q_{\mathsf{ct}})$	$\mathcal{D}_k$-MDDH
Ours	$(k^2m+km+1)m\mu$	$(k^2+k+1)m$	$(k^2+k+1)m\mu+1$	$O(1)$	$\mathcal{D}_k$-MDDH

Table 3. Comparison of fully function-hiding IPFE schemes in the standard model. Lin17 [34] refers to the scheme obtained by applying her paradigm to the IPFE scheme AGRW17 [4]. The column $|\mathsf{msk}|$ refers to the number of $\mathbb{Z}_p$ elements. The columns $|\mathsf{ct}|$ and $|\mathsf{sk}|$ refer to the number of group elements in G_1 and G_2 respectively. The number m refers to the vector length. The numbers q_{ct} and q_{sk} refer to the total numbers of ciphertext queries and secret key queries by an adversary respectively.

function-hiding IPFE schemes					
scheme	$\|\mathsf{msk}\|$	$\|\mathsf{ct}\|$	$\|\mathsf{sk}\|$	sec. loss	assumption
DDM16 [18]	$8m^2+12m+28$	$4m+8$	$4m+8$	$O(q_{\mathsf{ct}}+q_{\mathsf{sk}})$	SXDH
TAO16 [38]	$4m^2+18m+20$	$2m+5$	$2m+5$	$O(q_{\mathsf{ct}}+q_{\mathsf{sk}})$	XDLIN
Lin17 [34]	$(k+1)(4m+3k+1)$	$2m+2k+2$	$2m+2k+2$	$O(q_{\mathsf{ct}}+q_{\mathsf{sk}})$	$\mathcal{D}_k$-MDDH
Ours	$32m^2$	$10m$	$10m$	$O(1)$	SXDH
	$(4k^4+8k^3+12k^2+8k)m^2$	$(4k^2+4k+2)m$	$(4k^2+4k+2)m$	$O(1)$	$\mathcal{D}_k$-MDDH

Table 4. Comparison of fully function-hiding MIPFE schemes. The column $|\mathsf{msk}|$ refers to the number of $\mathbb{Z}_p$ elements. The columns $|\mathsf{ct}|$ and $|\mathsf{sk}|$ refer to the number of group elements in G_1 and G_2 respectively. The number m refers to the vector length. The number μ refers to the number of slots. The numbers q_{ct} and q_{sk} refer to the total numbers of ciphertext queries for all slots and secret key queries by an adversary respectively.

function-hiding MIPFE schemes			
scheme	$\|\mathsf{msk}\|$	$\|\mathsf{ct}\|$	$\|\mathsf{sk}\|$
DOT18 [19]	$(2m+2k+1)^2\mu$	$2m+2k+1$	$(2m+2k+1)\mu$
ACFGU18 [3]	$\{(k+1)(4m+5k+1)+k\}\mu$	$2m+3k+2$	$(2m+3k+2)\mu(+\|G_T\|)$
Ours	$\{(k^4+2k^3+3k^2+2k)(2m+1)^2+m\}\mu$	$(2k^2+2k+1)(2m+1)$	$(2k^2+2k+1)(2m+1)\mu$
scheme	sec. loss	assumption	
DOT18 [19]	$O(q_{\mathsf{ct}}+q_{\mathsf{sk}})$	k-Lin	
ACFGU18 [3]	$O(q_{\mathsf{ct}}+\mu q_{\mathsf{sk}})$	$\mathcal{D}_k$-MDDH	
Ours	$O(1)$	$\mathcal{D}_k$-MDDH	

$$\mathbf{W}\mathbf{a} = \widetilde{\mathbf{W}}\mathbf{a}, \tag{2.2}$$
$$\mathbf{W}^\top \mathbf{y}_\ell = \widetilde{\mathbf{W}}^\top \mathbf{y}_\ell \quad (\ell \text{ is an index for the query number}), \tag{2.3}$$
$$\begin{aligned}\mathbf{W}(s\mathbf{a} + s'\mathbf{b}) + \mathbf{x}^\beta &= \widetilde{\mathbf{W}}(s\mathbf{a} + s'\mathbf{b}) + us'(\mathbf{x}^1 - \mathbf{x}^0) + \mathbf{x}^\beta \\ &= \widetilde{\mathbf{W}}(s\mathbf{a} + s'\mathbf{b}) + (us' + \beta)(\mathbf{x}^1 - \mathbf{x}^0) + \mathbf{x}^0.\end{aligned} \tag{2.4}$$

In the indistinguishability-based security game, we impose a query condition on the adversary to avoid a trivial attack. That is, for all secret key queries, we have $\mathbf{x}^0\mathbf{y}_\ell = \mathbf{x}^1\mathbf{y}_\ell$. Equation (2.3) follows from this condition. Finally, from Eq. (2.4), we can argue that the information of β is hidden from the adversary by the term us' unless $s' = 0$, because u is a fresh randomness from the viewpoint of the adversary. Thus, the scheme is secure under the DDH assumption. In the multi-challenge setting, however, this proof strategy needs a hybrid argument for each challenge and incurs the security loss of $O(q_{\mathsf{ct}})$, where q_{ct} is the number of the ciphertext challenges. Intuitively, this is because the matrix $\mathbf{W}$ is shared in all challenge ciphertexts and we cannot redefine $\mathbf{W}$ suitable for all challenge ciphertexts simultaneously in Eq. (2.1).

The first attempt to obtain a tight reduction is setting $\mathbf{W}$ in Eq. (2.1) as

$$u_1, \ldots, u_L \xleftarrow{\mathsf{U}} \mathbb{Z}_p, \quad \mathbf{W} := \widetilde{\mathbf{W}} + \sum_{\iota \in [L]} u_\iota \mathbf{x}_\iota \mathbf{a}^{\perp\top},$$

where $L(\leq m)$ is the dimension of the space V spanned by $\mathbf{x}_j^1 - \mathbf{x}_j^0 \in \mathbb{Z}_p^m$ for all $j \in [q_{\mathsf{ct}}]$, and $\{\mathbf{x}_\iota\}_{\iota \in [L]}$ are a basis of V. In this case, Eqs. (2.2) and (2.3) do not change and Eq. (2.4) becomes

$$\mathbf{W}(s_j\mathbf{a} + s_j'\mathbf{b}) + \mathbf{x}_j^\beta = \widetilde{\mathbf{W}}(s_j\mathbf{a} + s_j'\mathbf{b}) + s_j' \sum_{\iota \in [L]} u_\iota \mathbf{x}_\iota + \beta(\mathbf{x}_j^1 - \mathbf{x}_j^0) + \mathbf{x}_j^0,$$

where j is the index of challenge queries. If we can say that $\{[s_j' u_\iota]\}_{j \in [q_{\mathsf{ct}}], \iota \in [L]}$ are indistinguishable from $\{[r_{j,\iota}]\}_{j \in [q_{\mathsf{ct}}], \iota \in [L]}$, which are $q_{\mathsf{ct}} L$ random elements in G, we can conclude that the term $s_j' \sum_{\iota \in [L]} u_\iota \mathbf{x}_\iota$ hides the information of β. This is because $\mathbf{x}_j^1 - \mathbf{x}_j^0 \in V$ for all $j \in [q_{\mathsf{ct}}]$, and each $\sum_{\iota \in [L]} r_{j,\iota} \mathbf{x}_\iota$ is a completely random element in V. Fortunately, it is well known that $\{s_j' u_\iota\}_{j \in [q_{\mathsf{ct}}], \iota \in [L]}$ on the exponent forms a synthesizer [36], and they are computationally indistinguishable from $q_{\mathsf{ct}} L$ random group elements with the security loss being either q_{ct} or L. Thus, we can prove the security of the scheme by Agrawal et al. with the security loss of $O(m)$, which is independent from the adversaries' behavior.

However, the above proof contains two deficiencies. The first is that the security reduction is still not tight. The second is that the above strategy is useful against only selective adversaries. This is because the reduction algorithm needs to know about V to simulate each challenge ciphertext, but V depends on all challenge queries that the adversary makes. Thus, we have to overcome these two problems.

Toward Tight Security. The solution for the first problem (and partly for the second problem as a result) is to increase the column of the part $\mathbf{a}$, which allows us to embed more randomness into ciphertexts. That is, we modify the scheme as

$\mathsf{Setup}(1^\lambda, 1^m)$:

$$a \xleftarrow{\mathsf{U}} \mathbb{Z}_p, \quad \mathbf{W} \xleftarrow{\mathsf{U}} \mathbb{Z}_p^{m\times 2m}, \quad \mathbf{a} := (a, 1),$$

$$\mathbf{A} := \mathbf{I}_m \otimes \mathbf{a} = \overbrace{\begin{pmatrix} \mathbf{a} & & & \\ & \mathbf{a} & & \\ & & \ddots & \\ & & & \mathbf{a} \end{pmatrix}}^{m \text{ vectors}} \in \mathbb{Z}_p^{2m\times m},$$

$$\mathsf{pk} := ([\mathbf{a}], [\mathbf{WA}]), \quad \mathsf{msk} := \mathbf{W}.$$

$\mathsf{Enc}(\mathsf{pk}, \mathbf{x})$: $\mathbf{s} := (s_1, \ldots, s_m) \xleftarrow{\mathsf{U}} \mathbb{Z}_p^m$, $\mathsf{ct} := ([\mathbf{As}], [\mathbf{WAs} + \mathbf{x}])$.
$\mathsf{KeyGen}(\mathsf{pk}, \mathsf{msk}, \mathbf{y})$: $\mathsf{sk} := (-\mathbf{W}^\top \mathbf{y}, \mathbf{y})$.
$\mathsf{Dec}(\mathsf{pk}, \mathsf{ct}, \mathsf{sk})$: $-\mathbf{y}^\top \mathbf{W}[\mathbf{As}] + \mathbf{y}^\top[\mathbf{WAs} + \mathbf{x}] = [\langle \mathbf{x}, \mathbf{y}\rangle]$.

The security proof goes as follows. First, the form of all challenge ciphertexts is changed to

$$\begin{aligned} &\mathbf{B} := \mathbf{I}_m \otimes (1, 0) \in \mathbb{Z}_p^{2m\times m}, \quad \mathbf{s}'_j := (s'_{j,1}, \ldots, s'_{j,m}) \xleftarrow{\mathsf{U}} \mathbb{Z}_p^m, \\ &\mathsf{ct} := ([\mathbf{As}_j + \mathbf{Bs}'_j], [\mathbf{W}(\mathbf{As}_j + \mathbf{Bs}'_j) + \mathbf{x}_j^\beta]). \end{aligned} \tag{2.5}$$

The DDH problem is tightly reduced to the problem of distinguishing this change by the random self-reducibility. Next, we redefine $\mathbf{W}$ as

$$u \xleftarrow{\mathsf{U}} \mathbb{Z}_p, \quad \mathbf{W} := \widetilde{\mathbf{W}} + u \sum_{\iota\in[L]} \mathbf{x}_\iota {\mathbf{a}_\iota^\perp}^\top, \tag{2.6}$$

where $\mathbf{a}_\iota^\perp \in \mathbb{Z}_p^{2m}$ is the ι-th column of $\mathbf{A}^\perp := \mathbf{I}_m \otimes \mathbf{a}^\perp$. Then, we have

$$\begin{aligned} &\mathbf{WA} = \widetilde{\mathbf{W}}\mathbf{A}, \\ &\mathbf{W}^\top \mathbf{y}_\ell = \widetilde{\mathbf{W}}^\top \mathbf{y}_\ell, \\ &\mathbf{W}(\mathbf{As}_j + \mathbf{Bs}'_j) + \mathbf{x}_j^\beta = \widetilde{\mathbf{W}}(\mathbf{As}_j + \mathbf{Bs}'_j) + u \sum_{\iota\in[L]} s'_{j,\iota}\mathbf{x}_\iota + \beta(\mathbf{x}_j^1 - \mathbf{x}_j^0) + \mathbf{x}_j^0. \end{aligned} \tag{2.7}$$

In this case, we can see that $\{[us'_{j,\iota}]\}_{j\in[q_{\mathsf{ct}}],\iota\in[L]}$ are computationally indistinguishable from $\{[r_{j,\iota}]\}_{j\in[q_{\mathsf{ct}}],\iota\in[L]}$, which are $q_{\mathsf{ct}} L$ random elements in G, and this indistinguishability is tightly reduced to the DDH assumption by the random self-reducibility. Then, the information of β is completely hidden by the same argument as before in the selective security model.

Toward Adaptive Security. In this paragraph, we refer to the computational change from $\mathbf{A}\mathbf{s}_j$ to $\mathbf{A}\mathbf{s}_j + \mathbf{B}\mathbf{s}'_j$ as the first step and that from $\{[us'_{j,\iota}]\}_{j\in[q_{\mathrm{ct}}],\iota\in[L]}$ to $\{[r_{j,\iota}]\}_{j\in[q_{\mathrm{ct}}],\iota\in[L]}$ as the second step. The main obstacle to achieve the adaptive security is that the reduction algorithm needs to know about the space V before seeing all challenge queries in the second step. Our observation is that we do not need a random element in V to hide the information of β in each ciphertext. Let V_j be a space spanned by $\mathbf{x}^1_\iota - \mathbf{x}^0_\iota \in \mathbb{Z}_p^m$ for all $\iota \in [j]$. Then, a random element in V_j suffices to hide the information of β in the j-th ciphertext. Fortunately, the reduction algorithm knows about V_j when it simulates the j-th ciphertext because it already receives vectors that span V_j.

To do so, we modify the first step. In particular, we change the way of choosing $\mathbf{s}'_j$ in Eq. (2.5) as

$$s'_{j,1},\ldots,s'_{j,\phi(j)} \xleftarrow{\mathsf{U}} \mathbb{Z}_p,\;\; \mathbf{s}'_j := (s'_{j,1},\ldots,s'_{j,\phi(j)},0^{m-\phi(j)}) \in \mathbb{Z}_p^m,$$

where $\phi(j) := \dim V_j$. Next, we modify the definition of $\mathbf{x}_\iota$ as $\mathbf{x}_\iota := \mathbf{x}^1_{\rho(\iota)} - \mathbf{x}^0_{\rho(\iota)} \in \mathbb{Z}_p^m$ for all $\iota \in [L]$, where $\rho(\iota) := \min \phi^{-1}(\iota)$. It is not difficult to confirm that $\{\mathbf{x}_\iota\}_{\iota\in[\phi(j)]}$ form a basis of V_j. Then, Eq. (2.7) is changed to

$$\mathbf{W}(\mathbf{A}\mathbf{s}_j + \mathbf{B}\mathbf{s}'_j) + \mathbf{x}^\beta_j = \widetilde{\mathbf{W}}(\mathbf{A}\mathbf{s}_j + \mathbf{B}\mathbf{s}'_j) + u \sum_{\iota\in[\phi(j)]} s'_{j,\iota}\mathbf{x}_\iota + \beta(\mathbf{x}^1_j - \mathbf{x}^0_j) + \mathbf{x}^0_j.$$

Observe that the reduction algorithm can compute $\mathbf{x}_\iota$ for $\iota \in [\phi(j)]$ when it simulates the j-th ciphertext. As explained in the previous paragraph, $\{[us'_{j,\iota}]\}_{j\in[q_{\mathrm{ct}}],\iota\in[\phi(j)]}$ are computationally indistinguishable from $\{[r_{j,\iota}]\}_{j\in[q_{\mathrm{ct}}],\iota\in[\phi(j)]}$, and the term $\sum_{\iota\in[\phi(j)]} r_{j,\iota}\mathbf{x}_\iota$ hides the information of β in the j-th ciphertext. Thus, we can achieve the adaptive security.

2.2 Conversion from Function-Hiding IPFE to Function-Hiding MIPFE

Similarly to previous MIPFE schemes, our conversion utilizes parallel execution of an underlying function-hiding IPFE scheme. The construction of our conversion can be seen as the combination of the non-function-hiding MIPFE scheme by Abdalla et al. [3] and the function-hiding MIPFE scheme by Datta et al. [19]. For simplicity, we consider the IPFE scheme over $\mathbb{Z}_n$ for some integer n, which means that the functionality of FE is inner product over $\mathbb{Z}_n$. Let m be a vector length and μ be a number of slots of the converted scheme, and $\mathsf{IPFE} := (\mathsf{Setup}', \mathsf{Enc}', \mathsf{KeyGen}', \mathsf{Dec}')$ be an underlying weakly function-hiding IPFE scheme. Then, our conversion invokes Setup' with setting the vector length as $2m + 1$ and generates μ master secret keys $\mathsf{msk}'_1,\ldots,\mathsf{msk}'_\mu$ (we omit public parameters here). In addition, it chooses μ random vectors $\mathbf{u}_1,\ldots,\mathbf{u}_\mu \xleftarrow{\mathsf{U}} \mathbb{Z}_n^m$ and sets a master secret key of the converted scheme as $\mathsf{msk} := (\mathsf{msk}'_1,\ldots,\mathsf{msk}'_\mu,\mathbf{u}_1,\ldots,\mathbf{u}_\mu)$. To encrypt a vector $\mathbf{x}_i$ for the index i, it encrypts $\tilde{\mathbf{x}}_i := (\mathbf{x}_i + \mathbf{u}_i, 0^m, 1)$ as $\mathsf{ct}'_i \leftarrow \mathsf{Enc}'(\mathsf{msk}_i, \tilde{\mathbf{x}}_i)$ and outputs ct'_i. To generate a secret key for $\{\mathbf{y}_i\}_{i\in[\mu]}$, it first generates secret shares of $-\sum_{i\in[\mu]}\langle \mathbf{y}_i, \mathbf{u}_i\rangle$

as $r_1, \ldots, r_\mu \xleftarrow{\mathsf{U}} \mathbb{Z}_n$ such that $\sum_{i\in[\mu]} r_i = -\sum_{i\in[\mu]} \langle \mathbf{y}_i, \mathbf{u}_i \rangle \pmod{n}$. These shares prevent the leakage of partial inner product values. Then, our conversion generates a secret key for $\tilde{\mathbf{y}}_i := (\mathbf{y}_i, 0^m, r_i)$ as $\mathsf{sk}'_i \leftarrow \mathsf{KeyGen}'(\mathsf{msk}'_i, \tilde{\mathbf{y}}_i)$ for all $i \in [\mu]$. Finally, it sets the secret key for converted scheme as $\mathsf{sk} := (\mathsf{sk}'_1, \ldots, \mathsf{sk}'_\mu)$. The decryption algorithm simply computes $\sum_{i\in[\mu]} \mathsf{Dec}'(\mathsf{ct}'_i, \mathsf{sk}'_i) \pmod{n}$. The correctness of the converted scheme is not difficult to confirm because $\sum_{i\in[\mu]} \langle \tilde{\mathbf{x}}_i, \tilde{\mathbf{y}}_i \rangle = \sum_{i\in[\mu]} \langle \mathbf{x}_i, \mathbf{y}_i \rangle$.

Although our conversion is as simple as that by Abdalla et al. [3], the security proof needs a more ingenious technique. To see this, we briefly recall the proof strategy of their conversion and show that the naive application of their strategy to our conversion does not work. Here, we assume that the converted MIPFE scheme is weakly function-hiding, meaning that an adversary against the converted scheme has the following condition on the queries in the security game. Let $q_{\mathsf{ct},i}$ be the total number of ciphertext queries for index i and q_{sk} be the total number of secret key queries. Then, for all $(j_1, \ldots, j_\mu) \in [q_{\mathsf{ct},1}] \times \cdots \times [q_{\mathsf{ct},\mu}]$, and $\ell \in [q_{\mathsf{sk}}]$, we have

$$\sum_{i\in[\mu]} \langle \mathbf{x}^0_{i,j_i}, \mathbf{y}^0_{i,\ell} \rangle = \sum_{i\in[\mu]} \langle \mathbf{x}^0_{i,j_i}, \mathbf{y}^1_{i,\ell} \rangle = \sum_{i\in[\mu]} \langle \mathbf{x}^1_{i,j_i}, \mathbf{y}^1_{i,\ell} \rangle. \tag{2.8}$$

The proof employs a series of games, and the goal is that the adversary does not obtain any information about a random bit β in the final game. The first step is to redefine $\mathbf{u}_i := \tilde{\mathbf{u}}_i + \mathbf{x}^0_{i,1} - \mathbf{x}^\beta_{i,1}$, where $\tilde{\mathbf{u}}_i \xleftarrow{\mathsf{U}} \mathbb{Z}_n$. This information-theoretic change does not affect secret keys because $\sum_{i\in[\mu]} \langle \mathbf{x}^0_{i,1} - \mathbf{x}^\beta_{i,1}, \mathbf{y}^\beta_{i,\ell} \rangle = 0$ from Eq. (2.8). The second step is to change $\tilde{\mathbf{x}}_{i,j_i}$ from $(\mathbf{x}^\beta_{i,j_i} + \tilde{\mathbf{u}}_i + \mathbf{x}^0_{i,1} - \mathbf{x}^\beta_{i,1}, 0^m, 1)$ to $(\mathbf{x}^0_{i,j_i} + \tilde{\mathbf{u}}_i, 0^m, 1)$. This change is justified by the security of the underlying IPFE scheme because $\langle \mathbf{x}^\beta_{i,j_i} - \mathbf{x}^\beta_{i,1}, \mathbf{y}^\beta_{i,\ell} \rangle = \langle \mathbf{x}^0_{i,j_i} - \mathbf{x}^0_{i,1}, \mathbf{y}^\beta_{i,\ell} \rangle$ for all $i \in [\mu]$, which can be derived from Eq. (2.8). Finally, we want to change $\tilde{\mathbf{y}}_{i,\ell}$ from $(\mathbf{y}^\beta_{i,\ell}, 0^m, r_{i,\ell})$ to $(\mathbf{y}^0_{i,\ell}, 0^m, r'_{i,\ell})$ to hide the information of β. However, we cannot make this change in the adaptive setting. The reason is that the reduction algorithm needs to set $r'_{i,\ell} := r_{i,\ell} + \Delta_{i,\ell}$, where $\Delta_{i,\ell} := \langle \mathbf{x}^0_{i,j_i} + \mathbf{u}_i, \mathbf{y}^\beta_{i,\ell} - \mathbf{y}^0_{i,\ell} \rangle = \langle \mathbf{x}^0_{i,1} + \mathbf{u}_i, \mathbf{y}^\beta_{i,\ell} - \mathbf{y}^0_{i,\ell} \rangle$ (the second equality follows from Eq. (2.8)), to keep the inner product value when it simulates the ℓ-th secret key. If the adversary makes a secret key query before it makes the first ciphertext query for some index i, the reduction algorithm cannot simulate a secret key because it does not know the value $\langle \mathbf{x}^0_{i,1}, \mathbf{y}^\beta_{i,\ell} - \mathbf{y}^0_{i,\ell} \rangle$. Hence, this strategy does not work.

To circumvent this problem, we introduce another proof strategy. Recall that this problem occurs in the second step, where $\mathbf{y}^\beta_{i,\ell}$ is changed to $\mathbf{y}^0_{i,\ell}$, whereas the first step goes well, where $\mathbf{x}^\beta_{i,j_i}$ is changed to $\mathbf{x}^0_{i,j_i}$. Intuitively, our solution for this problem is to make both changes in one-shot in the same manner as the first step. That is, we do not take the intermediate step where the inner product values of queried vectors are $\sum_{i\in[\mu]} \langle \mathbf{x}^0_{i,j_i}, \mathbf{y}^\beta_{i,\ell} \rangle$, and we change the replies such that the inner product values of queried vectors are directly changed from $\sum_{i\in[\mu]} \langle \mathbf{x}^\beta_{i,j_i}, \mathbf{y}^\beta_{i,\ell} \rangle$ to $\sum_{i\in[\mu]} \langle \mathbf{x}^0_{i,j_i}, \mathbf{y}^0_{i,\ell} \rangle$. This means that our conversion allows us to directly achieve a fully function-hiding MIPFE scheme. This is possible if we prepare $2n+1$ dimensions for the underlying scheme and use the similar technique to that by Tomida et al. [38]. To do so, we want to create a situation where $\tilde{\mathbf{x}}_{i,j_i} := (\mathbf{x}^{\beta}_{i,j_i} + \tilde{\mathbf{u}}_i - \mathbf{x}^{\beta}_{i,1}, \mathbf{x}^0_{i,1}, 1)$ and $\tilde{\mathbf{y}}_{i,\ell} := (\mathbf{y}^{\beta}_{i,\ell}, \mathbf{y}^0_{i,\ell}, r'_{i,\ell})$. This is because if we have the above situation, we can change $\tilde{\mathbf{x}}_{i,j_i}$ to $(\tilde{\mathbf{u}}_i, \mathbf{x}^0_{i,j_i} - \mathbf{x}^0_{i,1} + \mathbf{x}^0_{i,1}, 1) = (\tilde{\mathbf{u}}_i, \mathbf{x}^0_{i,j_i}, 1)$ by the security of the underlying scheme and the relation $\langle \mathbf{x}^{\beta}_{i,j_i} - \mathbf{x}^{\beta}_{i,1}, \mathbf{y}^{\beta}_{i,\ell} \rangle = \langle \mathbf{x}^0_{i,j_i} - \mathbf{x}^0_{i,1}, \mathbf{y}^0_{i,\ell} \rangle$, which also can be derived from Eq. (2.8).

To reach the situation starting from the real game, however, we need one more trick. This is because the reduction algorithm needs to compute the value $\Delta_{i,\ell} := \langle \mathbf{x}^0_{i,1}, \mathbf{y}^0_{i,\ell} \rangle$ to adjust inner products with the term $r'_{i,\ell}$ when it simulates the ℓ-th secret key. Thus, the same problems as above occurs. To solve this problem, we take the intermediate step where $\tilde{\mathbf{x}}_{i,j_i} := (\mathbf{x}^{\beta}_{i,j_i} + \mathbf{u}_i, \mathbf{v}_i, 1)$ and $\tilde{\mathbf{y}}_{i,\ell} := (\mathbf{y}^{\beta}_{i,\ell}, \mathbf{y}^0_{i,\ell}, r_{i,\ell})$, where $\mathbf{v}_i \xleftarrow{\mathsf{U}} \mathbb{Z}_n^m$ is randomly chosen at the beginning of the game. This is possible because computing $\Delta_{i,\ell} := \langle \mathbf{v}_i, \mathbf{y}^0_{i,\ell} \rangle$ suffices for the reduction algorithm to reach the step. After the step, we redefine $\mathbf{u}_i := \tilde{\mathbf{u}}_i - \mathbf{x}^{\beta}_{i,1}$ and $\mathbf{v}_i := \tilde{\mathbf{v}}_i + \mathbf{x}^0_{i,1}$ where $\tilde{\mathbf{u}}_i, \tilde{\mathbf{v}}_i \xleftarrow{\mathsf{U}} \mathbb{Z}_n^m$. This change is information-theoretic and we do not need to care about when the adversary makes the first ciphertext query. By these steps, our proof strategy goes well since there are no steps where reduction algorithms need to compute values related to $\mathbf{x}^0_{i,1}$ when it simulates secret keys.

The interesting points of our technique are to crucially utilize the blank space, namely the $n+1$ to $2n$-th dimensions, and directly construct a fully function-hiding MIPFE scheme from a weakly function-hiding IPFE scheme. This is in contrast to the function-hiding scheme in [3], where they first construct a weakly function-hiding MIPFE scheme, setting a vector length of an underlying IPFE scheme as almost n. Then, they convert it into a fully function-hiding scheme by doubling the vector length of the scheme.

3 Preliminary

3.1 Notation

For a natural number $n \in \mathbb{N}$, $\mathbb{Z}_n$ denotes a ring $\mathbb{Z}/n\mathbb{Z}$ and $[n]$ denotes a set $\{1, \ldots, n\}$. For a set S, $s \xleftarrow{\mathsf{U}} S$ denotes that s is uniformly chosen from S. We treat vectors as column vectors. For a vector $\mathbf{x}$, $||\mathbf{x}||_\infty$ denotes its infinity norm. For vectors $\mathbf{v}_1, \mathbf{v}_2, \ldots, \mathbf{v}_n$, $(\mathbf{v}_1, \mathbf{v}_2, \ldots, \mathbf{v}_n)$ denotes a vector generated by the vertical concatenation of these vectors. For matrices (including vectors) with the same number of rows $\mathbf{A}_1, \mathbf{A}_2, \ldots, \mathbf{A}_n$, $(\mathbf{A}_1||\mathbf{A}_2||\cdots||\mathbf{A}_n)$ denotes a matrix generated by the horizontal concatenation of these matrices. For a generator g_i of a cyclic group G_i of order p and $a \in \mathbb{Z}_p$, $[a]_i$ denotes g_i^a. Furthermore, for a matrix $\mathbf{A} := (a_{j,\ell})_{j,\ell}$ over $\mathbb{Z}_p$, $[\mathbf{A}]_i$ denotes a matrix over G_i whose (i,j) entry is $g_i^{a_{j,\ell}}$. For vectors $\mathbf{x} := (x_1, \ldots, x_n)$ and $\mathbf{y} := (y_1, \ldots, y_n) \in \mathbb{Z}_p^n$, let $e([\mathbf{x}]_1, [\mathbf{y}]_2) := e(g_1, g_2)^{\langle \mathbf{x}, \mathbf{y} \rangle}$ be a function that computes the inner product on the exponent by $\prod_{i\in[n]} e([x_i]_1, [y_i]_2)$. A matrix $\mathbf{I}_n$ denotes the $n \times n$ identity matrix. A matrix $\mathbf{O}_{m\times n}$ denotes the $m \times n$ zero matrix. A function $f : \mathbb{N} \to \mathbb{R}$ is called negligible if $f(\lambda) = \lambda^{-\omega(1)}$ and denotes $f(\lambda) \leq \mathsf{negl}(\lambda)$. For families of distributions $X := \{X_\lambda\}_{\lambda\in\mathbb{N}}$ and $Y := \{Y_\lambda\}_{\lambda\in\mathbb{N}}$, $X \approx_c Y$ means that they are computationally indistinguishable.

3.2 Basic Tools and Assumption

Definition 3.1 (Cyclic Group). A description of a cyclic group $\mathbb{G}_{\mathsf{CG}} := (p, G, g)$ consists of a prime p, a cyclic group G of order p, and a generator g. A cyclic group generator $\mathcal{G}_{\mathsf{CG}}(1^\lambda)$ takes a security parameter 1^λ and outputs a description of a cyclic group $\mathbb{G}_{\mathsf{CG}}$ with a λ-bit prime p.

Definition 3.2 (Bilinear Groups). A description of bilinear groups $\mathbb{G}_{\mathsf{BG}} := (p, G_1, G_2, G_T, g_1, g_2, e)$ consist of a prime p, cyclic groups G_1, G_2, G_T of order p, generators g_1 and g_2 of G_1 and G_2 respectively, and a bilinear map $e : G_1 \times G_2 \to G_T$, which has two properties.

- (Bilinearity): $\forall h_1 \in G_1, h_2 \in G_2, a, b \in \mathbb{Z}_p, e(h_1^a, h_2^b) = e(h_1, h_2)^{ab}$.
- (Non-degeneracy): For generators g_1 and g_2, $g_T := e(g_1, g_2)$ is a generator of G_T.

A bilinear group generator $\mathcal{G}_{\mathsf{BG}}(1^\lambda)$ takes a security parameter 1^λ and outputs a description of bilinear groups $\mathbb{G}_{\mathsf{BG}}$ with a λ-bit prime p.

Definition 3.3 ($\mathcal{D}_k$-MDDH Assumption [21]). Let $\mathcal{D}_k$ be a matrix distribution over full rank matrices in $\mathbb{Z}_p^{(k+1)\times k}$. We can assume that, wlog, the first k rows of a matrix $\mathbf{A}$ chosen from $\mathcal{D}_k$ forms an invertible matrix. We consider the following distribution:

$$\mathbb{G}_{\mathsf{CG}} \leftarrow \mathcal{G}_{\mathsf{CG}}(1^\lambda), \quad \mathbb{G}_{\mathsf{BG}} \leftarrow \mathcal{G}_{\mathsf{BG}}(1^\lambda),$$
$$\mathbf{A} \leftarrow \mathcal{D}_k, \quad \mathbf{v} \xleftarrow{\mathsf{U}} \mathbb{Z}_p^k, \quad \mathbf{t}_0 := \mathbf{A}\mathbf{v}, \quad \mathbf{t}_1 \xleftarrow{\mathsf{U}} \mathbb{Z}_p^{k+1}.$$

We say that the $\mathcal{D}_k$-MDDH assumption holds with respect to $\mathcal{G}_{\mathsf{CG}}$ if the advantage of any PPT adversary $\mathcal{A}$ defined below is negligible,

$$\mathsf{Adv}_{\mathcal{A},\mathsf{CG}}^{\mathcal{D}_k\text{-}\mathsf{MDDH}}(\lambda) := |\mathsf{Pr}[1 \leftarrow \mathcal{A}(\mathbb{G}_{\mathsf{CG}}, [\mathbf{A}], [\mathbf{t}_0])] - \mathsf{Pr}[1 \leftarrow \mathcal{A}(\mathbb{G}_{\mathsf{CG}}, [\mathbf{A}], [\mathbf{t}_1])]|,$$

and with respect to $\mathcal{G}_{\mathsf{BG}}$ if the advantage of any PPT adversary $\mathcal{A}$ for both $i \in \{1, 2\}$ defined below is negligible,

$$\mathsf{Adv}_{\mathcal{A},\mathsf{BG},i}^{\mathcal{D}_k\text{-}\mathsf{MDDH}}(\lambda) := |\mathsf{Pr}[1 \leftarrow \mathcal{A}(\mathbb{G}_{\mathsf{BG}}, [\mathbf{A}]_i, [\mathbf{t}_0]_i)] - \mathsf{Pr}[1 \leftarrow \mathcal{A}(\mathbb{G}_{\mathsf{BG}}, [\mathbf{A}]_i, [\mathbf{t}_1]_i)]|.$$

Random Self-reducibility. By the random self-reducibility, we can obtain arbitrarily many instances of the $\mathcal{D}_k$-MDDH problem without additional security loss. For any $n \in \mathbb{N}$, we additionally define the following distribution:

$$\mathbf{V} \xleftarrow{\mathsf{U}} \mathbb{Z}_p^{k\times n}, \quad \mathbf{T}_0 := \mathbf{A}\mathbf{V}, \quad \mathbf{T}_1 \xleftarrow{\mathsf{U}} \mathbb{Z}_p^{(k+1)\times n}.$$

The advantages of $\mathcal{A}$ against n-fold $\mathcal{D}_k$-MDDH assumption with respect to $\mathcal{G}_{\mathsf{CG}}$ and $\mathcal{G}_{\mathsf{BG}}$ are defined as:

$$\mathsf{Adv}_{\mathcal{A},\mathsf{CG}}^{\mathsf{n}\text{-}\mathcal{D}_\mathsf{k}\text{-}\mathsf{MDDH}}(\lambda) := |\Pr[1 \leftarrow \mathcal{A}(\mathbb{G}_{\mathsf{CG}}, [\mathbf{A}], [\mathbf{T}_0])] - \Pr[1 \leftarrow \mathcal{A}(\mathbb{G}_{\mathsf{CG}}, [\mathbf{A}], [\mathbf{T}_1])]|,$$
$$\mathsf{Adv}_{\mathcal{A},\mathsf{BG},i}^{\mathsf{n}\text{-}\mathcal{D}_\mathsf{k}\text{-}\mathsf{MDDH}}(\lambda) := |\Pr[1 \leftarrow \mathcal{A}(\mathbb{G}_{\mathsf{BG}}, [\mathbf{A}]_i, [\mathbf{T}_0]_i)] - \Pr[1 \leftarrow \mathcal{A}(\mathbb{G}_{\mathsf{BG}}, [\mathbf{A}]_i, [\mathbf{T}_1]_i)]|.$$

Then, for any PPT adversaries $\mathcal{A}_1, \mathcal{A}_2$ and both $i \in \{1,2\}$, there exist PPT adversaries $\mathcal{B}_1, \mathcal{B}_2$ and we have

$$\mathsf{Adv}_{\mathcal{A}_1,\mathsf{CG}}^{\mathsf{n}\text{-}\mathcal{D}_\mathsf{k}\text{-}\mathsf{MDDH}}(\lambda) \leq \mathsf{Adv}_{\mathcal{B}_1,\mathsf{CG}}^{\mathcal{D}_\mathsf{k}\text{-}\mathsf{MDDH}}(\lambda) + 2^{-\Omega(\lambda)},$$
$$\mathsf{Adv}_{\mathcal{A}_2,\mathsf{BG},i}^{\mathsf{n}\text{-}\mathcal{D}_\mathsf{k}\text{-}\mathsf{MDDH}}(\lambda) \leq \mathsf{Adv}_{\mathcal{B}_2,\mathsf{BG},i}^{\mathcal{D}_\mathsf{k}\text{-}\mathsf{MDDH}}(\lambda) + 2^{-\Omega(\lambda)},$$
$$\mathsf{Time}(\mathcal{B}_j) \approx \mathsf{Time}(\mathcal{A}_j) + n\mathsf{poly}_j(\lambda) \text{ for both } j \in \{1,2\},$$

where $\mathsf{poly}_j(\lambda)$ is independent from $\mathsf{Time}(\mathcal{A}_j)$.

3.3 Definitions of Inner Product Functional Encryption

In this paper, we treat both single-input inner product functional encryption (IPFE) and multi-input IPFE. In both cases, the inner product functionality is defined over $\mathbb{Z}$ and its domain is limited depending on the infinity norms of the input vectors. We formally define the functionality called bonded-norm inner product.

Definition 3.4 (Bounded-Norm Inner Product over $\mathbb{Z}$). This function family $\mathcal{F}$ consists of functions $f_{\mathbf{y}_1,\ldots,\mathbf{y}_\mu}^{X,Y} : \mathbb{Z}^m \times \cdots \times \mathbb{Z}^m \rightarrow \mathbb{Z}$ where $m, \mu, X, Y \in \mathbb{N}$, $\mathbf{y}_i \in \mathbb{Z}^m$ s.t. $||\mathbf{y}_i||_\infty \leq Y$. For all $(\mathbf{x}_1, \ldots, \mathbf{x}_\mu) \in (\mathbb{Z}^m)^\mu$ s.t. $\forall i \in [\mu], ||\mathbf{x}_i||_\infty \leq X$, we define the function as

$$f_{\mathbf{y}_1,\ldots,\mathbf{y}_\mu}^{X,Y}(\mathbf{x}_1, \ldots, \mathbf{x}_\mu) := \sum_{i \in [\mu]} \langle \mathbf{x}_i, \mathbf{y}_i \rangle.$$

We call μ a number of slots. We refer to the function as single-input inner product when $\mu = 1$, and multi-input inner product when $\mu > 1$.

With respect to single-input IPFE, there are two types of IPFE: public-key IPFE and private-key IPFE. To achieve the function privacy, we need the private-key setting as defined below. Roughly speaking, this is because an adversary can learn the information of functions embedded in secret keys by decrypting ciphetexts generated by itself with the secret keys in the public-key setting.

Definition 3.5 (Public-Key Inner Product Functional Encryption). Let $\mathcal{X} := \{X_\lambda\}_{\lambda \in \mathbb{N}}, \mathcal{Y} := \{Y_\lambda\}_{\lambda \in \mathbb{N}}$ be ensembles of norm-bounds. Public-key inner product functional encryption (Pub-IPFE) consists of five algorithms.

$\mathsf{Par}(1^\lambda)$: It takes a security parameter 1^λ and outputs a public parameter pp.
$\mathsf{Setup}(1^m, \mathsf{pp})$: It takes a vector length 1^m and pp and outputs a public key pk and a master secret key msk.

$\mathcal{O}_{\mathsf{ct}}(\beta \in \{0,1\}, i \in [\mu], (\mathbf{x}^0, \mathbf{x}^1) \in (\mathbb{Z}^m)^2)$	$\mathcal{O}_{\mathsf{sk}}(i \in [\mu], \mathbf{y} \in \mathbb{Z}^m)$
$\mathsf{ct}_i \overset{\mathsf{U}}{\leftarrow} \mathsf{Enc}(\mathsf{pk}_i, \mathbf{x}^\beta)$	$\mathsf{sk}_i \overset{\mathsf{U}}{\leftarrow} \mathsf{KeyGen}(\mathsf{pk}_i, \mathsf{msk}_i, \mathbf{y})$
return ct_i	return sk_i

Fig. 1. The description of oracles in the security game for Pub-IPFE.

$\mathsf{Enc}(\mathsf{pk}, \mathbf{x})$**:** It takes pk and a vector $\mathbf{x} := (x_1, \ldots, x_m) \in \mathbb{Z}^m$ and outputs a ciphertext ct.

$\mathsf{KeyGen}(\mathsf{pk}, \mathsf{msk}, \mathbf{y})$**:** It takes $\mathsf{pk}, \mathsf{msk}$, and a vector $\mathbf{y} := (y_1, \ldots, y_m) \in \mathbb{Z}^m$ and outputs a secret key sk.

$\mathsf{Dec}(\mathsf{pk}, \mathsf{ct}, \mathsf{sk})$**:** It takes pk, ct and sk and outputs a decrypted value $d \in \mathbb{Z}$ or a symbol $\perp$.

Correctness. Pub-IPFE is *correct* if it satisfies the following condition. For any $\lambda, m \in \mathbb{N}$ and for any $\mathbf{x}, \mathbf{y} \in \mathbb{Z}^m$ s.t. $||\mathbf{x}||_\infty \leq X_\lambda$ and $||\mathbf{y}||_\infty \leq Y_\lambda$, we have

$$\Pr\left[d = \langle \mathbf{x}, \mathbf{y} \rangle \;\middle|\; \begin{array}{l} \mathsf{pp} \leftarrow \mathsf{Par}(1^\lambda) \\ (\mathsf{pk}, \mathsf{msk}) \leftarrow \mathsf{Setup}(1^m, \mathsf{pp}) \\ \mathsf{ct} \leftarrow \mathsf{Enc}(\mathsf{pk}, \mathbf{x}) \\ \mathsf{sk} \leftarrow \mathsf{KeyGen}(\mathsf{pk}, \mathsf{msk}, \mathbf{y}) \\ d := \mathsf{Dec}(\mathsf{pk}, \mathsf{ct}, \mathsf{sk}) \end{array} \right] = 1.$$

Security. Let $\mu \in \mathbb{N}$ be a natural number that represents the number of users. Pub-IPFE is *adaptively secure in the multi-user and multi-challenge setting* if it satisfies the following condition. That is, the advantage of $\mathcal{A}$ against Pub-IPFE defined as follows is negligible in λ for any constant $m, \mu \in \mathbb{N}$, and PPT adversary $\mathcal{A}$,

$$\mathsf{Adv}_{\mathcal{A}}^{\mathsf{Pub\text{-}IPFE}}(\lambda) := \left| 2\Pr\left[\beta = \beta' \;\middle|\; \begin{array}{l} \beta \overset{\mathsf{U}}{\leftarrow} \{0,1\}, \;\; \mathsf{pp} \leftarrow \mathsf{Par}(1^\lambda) \\ \{(\mathsf{pk}_i, \mathsf{msk}_i)\}_{i \in [\mu]} \leftarrow \mathsf{Setup}(1^m, \mathsf{pp}) \\ \beta' \leftarrow \mathcal{A}^{\mathcal{O}_{\mathsf{ct}}(\beta,\cdot,\cdot), \mathcal{O}_{\mathsf{sk}}(\cdot,\cdot)}(1^\lambda, \{\mathsf{pk}_i\}_{i \in [\mu]}) \end{array} \right] - 1 \right|.$$

The description of the oracles $\mathcal{O}_{\mathsf{ct}}$ and $\mathcal{O}_{\mathsf{sk}}$ is presented in Fig. 1. We refer to queries to $\mathcal{O}_{\mathsf{ct}}$ and $\mathcal{O}_{\mathsf{sk}}$ as a ciphertext query and a secret key query respectively. To avoid a trivial attack of $\mathcal{A}$, we have the following condition on $\mathcal{A}$'s queries. Let $q_{\mathsf{ct},i}$ and $q_{\mathsf{sk},i}$ be the total number of ciphertext queries and secret key queries for index i respectively. Then, for all $i \in [\mu]$, $j_i \in [q_{\mathsf{ct},i}]$, and $\ell_i \in [q_{\mathsf{sk},i}]$, we have

$$\langle \mathbf{x}^0_{i,j_i}, \mathbf{y}_{i,\ell_i} \rangle = \langle \mathbf{x}^1_{i,j_i}, \mathbf{y}_{i,\ell_i} \rangle. \tag{3.1}$$

Definition 3.6 (Private-Key Inner Product Functional Encryption). Let $\mathcal{X} := \{X_\lambda\}_{\lambda \in \mathbb{N}}, \mathcal{Y} := \{Y_\lambda\}_{\lambda \in \mathbb{N}}$ be ensembles of norm-bounds. Private-key inner product functional encryption (Priv-IPFE) consists of five algorithms.

$\mathsf{Par}(1^\lambda)$**:** It takes a security parameter 1^λ and outputs a public parameter pp.

$\mathsf{Setup}(1^m, \mathsf{pp})$: It takes a vector length 1^m and pp and outputs a master secret key msk.
$\mathsf{Enc}(\mathsf{pp}, \mathsf{msk}, \mathbf{x})$: It takes pp, msk, and a vector $\mathbf{x} := (x_1, \ldots, x_m) \in \mathbb{Z}^m$ and outputs a ciphertext ct.
$\mathsf{KeyGen}(\mathsf{pp}, \mathsf{msk}, \mathbf{y})$: It takes pp, msk, and a vector $\mathbf{y} := (y_1, \ldots, y_m) \in \mathbb{Z}^m$ and outputs a secret key sk.
$\mathsf{Dec}(\mathsf{pp}, \mathsf{ct}, \mathsf{sk})$: It takes pp, ct and sk and outputs a decrypted value $d \in \mathbb{Z}$ or a symbol $\perp$.

Correctness. Priv-IPFE is *correct* if it satisfies the following condition. For any $\lambda, m \in \mathbb{N}$ and for any $\mathbf{x}, \mathbf{y} \in \mathbb{Z}^m$ s.t. $||\mathbf{x}||_\infty \leq X_\lambda$ and $||\mathbf{y}||_\infty \leq Y_\lambda$, we have

$$\Pr\left[d = \langle \mathbf{x}, \mathbf{y} \rangle \;\middle|\; \begin{array}{l} \mathsf{pp} \leftarrow \mathsf{Par}(1^\lambda) \\ \mathsf{msk} \leftarrow \mathsf{Setup}(1^m, \mathsf{pp}) \\ \mathsf{ct} \leftarrow \mathsf{Enc}(\mathsf{pp}, \mathsf{msk}, \mathbf{x}) \\ \mathsf{sk} \leftarrow \mathsf{KeyGen}(\mathsf{pp}, \mathsf{msk}, \mathbf{y}) \\ d := \mathsf{Dec}(\mathsf{pp}, \mathsf{ct}, \mathsf{sk}) \end{array} \right] = 1.$$

Security. Let $\mu \in \mathbb{N}$ be a natural number that represents the number of users. Priv-IPFE is *fully function-hiding in the multi-user setting* if it satisfies the following condition. That is, the advantage of $\mathcal{A}$ against Priv-IPFE defined as follows is negligible in λ for any constant $m, \mu \in \mathbb{N}$ and any PPT adversary $\mathcal{A}$,

$$\mathsf{Adv}^{\mathsf{Priv\text{-}IPFE}}_{\mathcal{A},\mathsf{f\text{-}fh}}(\lambda) := \left| \begin{array}{l} \Pr\left[\beta' = 1 \;\middle|\; \begin{array}{l} \mathsf{pp} \leftarrow \mathsf{Par}(1^\lambda) \\ \{\mathsf{msk}_i\}_{i\in[\mu]} \leftarrow \mathsf{Setup}(1^m, \mathsf{pp}) \\ \beta' \leftarrow \mathcal{A}^{\mathcal{O}_{\mathsf{ct}}(0,\cdot,\cdot), \mathcal{O}_{\mathsf{sk}}(0,\cdot,\cdot)}(\mathsf{pp}) \end{array} \right] \\ -\Pr\left[\beta' = 1 \;\middle|\; \begin{array}{l} \mathsf{pp} \leftarrow \mathsf{Par}(1^\lambda) \\ \{\mathsf{msk}_i\}_{i\in[\mu]} \leftarrow \mathsf{Setup}(1^m, \mathsf{pp}) \\ \beta' \leftarrow \mathcal{A}^{\mathcal{O}_{\mathsf{ct}}(1,\cdot,\cdot), \mathcal{O}_{\mathsf{sk}}(1,\cdot,\cdot)}(\mathsf{pp}) \end{array} \right] \end{array} \right|.$$

It is convenient for our paper to define the advantage on Priv-IPFE as above rather than the form like $|2\Pr[\beta = \beta'] - 1|$, and both formulations are equivalent. The description of the oracles $\mathcal{O}_{\mathsf{ct}}$ and $\mathcal{O}_{\mathsf{sk}}$ is presented in Fig. 2. We refer to queries to $\mathcal{O}_{\mathsf{ct}}$ and $\mathcal{O}_{\mathsf{sk}}$ as a ciphertext query and a secret key query respectively. To avoid a trivial attack of $\mathcal{A}$, we have the following condition on $\mathcal{A}$'s queries. Let $q_{\mathsf{ct},i}$ and $q_{\mathsf{sk},i}$ be the total numbers of ciphertext queries and secret key queries for index i respectively. Then, for all $i \in [\mu]$, $j_i \in [q_{\mathsf{ct},i}]$, and $\ell_i \in [q_{\mathsf{sk},i}]$, we have

$$\langle \mathbf{x}^0_{i,j_i}, \mathbf{y}^0_{i,\ell_i} \rangle = \langle \mathbf{x}^1_{i,j_i}, \mathbf{y}^1_{i,\ell_i} \rangle. \tag{3.2}$$

We say that Priv-IPFE is *weakly function-hiding in the multi-user setting* if it satisfies the above definition except that the query condition of $\mathcal{A}$ is more restricted as follows. That is, for all $i \in [\mu]$, $j_i \in [q_{\mathsf{ct},i}]$, and $\ell_i \in [q_{\mathsf{sk},i}]$, we have

$$\langle \mathbf{x}^0_{i,j_i}, \mathbf{y}^0_{i,\ell_i} \rangle = \langle \mathbf{x}^1_{i,j_i}, \mathbf{y}^0_{i,\ell_i} \rangle = \langle \mathbf{x}^1_{i,j_i}, \mathbf{y}^1_{i,\ell_i} \rangle. \tag{3.3}$$

We denote the advantage of $\mathcal{A}$ in weakly function-hiding game in the multi-user setting by $\mathsf{Adv}^{\mathsf{Priv\text{-}IPFE}}_{\mathcal{A},\mathsf{w\text{-}fh}}(\lambda)$.

$\mathcal{O}_{\mathsf{ct}}(\beta \in \{0,1\}, i \in [\mu], (\mathbf{x}^0, \mathbf{x}^1) \in (\mathbb{Z}^m)^2)$	$\mathcal{O}_{\mathsf{sk}}(\beta \in \{0,1\}, i \in [\mu], (\mathbf{y}^0, \mathbf{y}^1) \in (\mathbb{Z}^m)^2)$
$\mathsf{ct}_i \xleftarrow{\mathsf{U}} \mathsf{Enc}(\mathsf{pp}, \mathsf{msk}_i, \mathbf{x}^\beta)$ return ct_i	$\mathsf{sk}_i \xleftarrow{\mathsf{U}} \mathsf{KeyGen}(\mathsf{pp}, \mathsf{msk}_i, \mathbf{y}^\beta)$ return sk_i

Fig. 2. The description of oracles in the security game for Priv-IPFE.

As pointed out by Abdalla et al. [4], public-key multi-input IPFE (MIPFE) is almost meaningless because it inherently leaks the same amount of information as parallel execution of single-input IPFE. Therefore, following them, we only consider private-key MIPFE in this paper.

Definition 3.7 (Multi-input Inner Product Functional Encryption). Let $\mathcal{X} := \{X_\lambda\}_{\lambda\in\mathbb{N}}, \mathcal{Y} := \{Y_\lambda\}_{\lambda\in\mathbb{N}}$ be ensembles of norm-bound. Multi-input inner product functional encryption (MIPFE) consists of four algorithms.

$\mathsf{Setup}(1^\lambda, 1^m, 1^\mu)$**:** It takes a security parameter 1^λ, a vector length 1^m, and a number of slots 1^μ. Then, it outputs a public parameter pp and a master secret key msk.

$\mathsf{Enc}(\mathsf{pp}, \mathsf{msk}, i, \mathbf{x})$**:** It takes pp, msk, an index $i \in [\mu]$, and a vector $\mathbf{x} := (x_1, \ldots, x_m) \in \mathbb{Z}^m$ and outputs a ciphertext ct_i.

$\mathsf{KeyGen}(\mathsf{pp}, \mathsf{msk}, \{\mathbf{y}_i\}_{i\in[\mu]})$**:** It takes $\mathsf{pp}, \mathsf{msk}$, and vectors $\{\mathbf{y}_i := (y_{i,1}, \ldots, y_{i,m})\}_{i\in[\mu]} \in (\mathbb{Z}^m)^\mu$, and outputs a secret key sk.

$\mathsf{Dec}(\mathsf{pp}, \mathsf{ct}_1, \ldots, \mathsf{ct}_\mu, \mathsf{sk})$**:** It takes $\mathsf{pp}, \mathsf{ct}_1, \ldots, \mathsf{ct}_\mu$ and sk and outputs a decrypted value $d \in \mathbb{Z}$ or a symbol $\perp$.

Correctness. MIPFE is *correct* if it satisfies the following condition. For any $\lambda, m, \mu \in \mathbb{N}$ and for any $\{\mathbf{x}_i\}_{i\in[\mu]}, \{\mathbf{y}_i\}_{i\in[\mu]} \in (\mathbb{Z}^m)^\mu$ s.t. $\forall i, ||\mathbf{x}_i||_\infty \leq X_\lambda$ and $||\mathbf{y}_i||_\infty \leq Y_\lambda$, we have

$$\Pr\left[d = \sum_{i\in[\mu]} \langle \mathbf{x}_i, \mathbf{y}_i \rangle \;\middle|\; \begin{array}{l} \mathsf{pp}, \mathsf{msk} \leftarrow \mathsf{Setup}(1^\lambda, 1^m, 1^\mu) \\ \mathsf{ct}_i \leftarrow \mathsf{Enc}(\mathsf{pp}, \mathsf{msk}, i, \mathbf{x}_i) \quad \text{for all } i \in [\mu] \\ \mathsf{sk} \leftarrow \mathsf{KeyGen}(\mathsf{pp}, \mathsf{msk}, \{\mathbf{y}_i\}_{i\in[\mu]}) \\ d := \mathsf{Dec}(\mathsf{pp}, \mathsf{ct}, \mathsf{sk}) \end{array} \right] = 1.$$

Security. MIPFE is *fully function-hiding* if it satisfies the following condition. That is, the advantage of $\mathcal{A}$ against MIPFE defined as follows is negligible in λ for any constant $m, \mu \in \mathbb{N}$ and any PPT adversary $\mathcal{A}$,

$$\mathsf{Adv}^{\mathsf{MIPFE}}_{\mathcal{A},\mathsf{f\text{-}fh}}(\lambda) := \left| 2\Pr\left[\beta = \beta' \;\middle|\; \begin{array}{l} \beta \xleftarrow{\mathsf{U}} \{0,1\}, \\ (\mathsf{pp}, \mathsf{msk}) \leftarrow \mathsf{Setup}(1^\lambda, 1^m, 1^\mu) \\ \beta' \leftarrow \mathcal{A}^{\mathcal{O}_{\mathsf{ct}}(\beta,\cdot,\cdot), \mathcal{O}_{\mathsf{sk}}(\beta,\cdot)}(\mathsf{pp}) \end{array} \right] - 1 \right|.$$

The description of the oracles $\mathcal{O}_{\mathsf{ct}}$ and $\mathcal{O}_{\mathsf{sk}}$ is presented in Fig. 3. We refer to queries to $\mathcal{O}_{\mathsf{ct}}$ and $\mathcal{O}_{\mathsf{sk}}$ as a ciphertext query and a secret key query respectively. To avoid a trivial attack of $\mathcal{A}$, we have the following condition on $\mathcal{A}$'s queries.

$\mathcal{O}_{\mathsf{ct}}(\beta \in \{0,1\}, i \in [\mu], (\mathbf{x}^0, \mathbf{x}^1) \in (\mathbb{Z}^m)^2)$
$\mathsf{ct}_i \xleftarrow{\mathsf{U}} \mathsf{Enc}(\mathsf{pp}, \mathsf{msk}, i, \mathbf{x}^\beta)$ return ct_i
$\mathcal{O}_{\mathsf{sk}}(\beta \in \{0,1\}, (\{\mathbf{y}_i^0\}_{i\in[\mu]}, \{\mathbf{y}_i^1\}_{i\in[\mu]}) \in ((\mathbb{Z}^m)^\mu)^2)$
$\mathsf{sk} \xleftarrow{\mathsf{U}} \mathsf{KeyGen}(\mathsf{pp}, \mathsf{msk}, \{\mathbf{y}_i^\beta\}_{i\in[\mu]})$ return sk

Fig. 3. The description of oracles in the security game for MIPFE.

Let $q_{\mathsf{ct},i}$ be the total number of ciphertext queries for index i and q_{sk} be the total number of secret key queries. Then, for all $(j_1, \ldots, j_\mu) \in [q_{\mathsf{ct},1}] \times \cdots \times [q_{\mathsf{ct},\mu}]$, and $\ell \in [q_{\mathsf{sk}}]$,

$$\sum_{i\in[\mu]} \langle \mathbf{x}_{i,j_i}^0, \mathbf{y}_{i,\ell}^0 \rangle = \sum_{i\in[\mu]} \langle \mathbf{x}_{i,j_i}^1, \mathbf{y}_{i,\ell}^1 \rangle. \tag{3.4}$$

In this paper, we assume that $q_{\mathsf{ct},i} \geq 1$ for all $i \in [\mu]$ and $q_{\mathsf{sk}} \geq 1$. Note that this condition can be easily removed by simply utilizing symmetric key encryption [4,19].

We say that MIPFE is just *adaptively secure* if it satisfies the above definition except that $\mathcal{O}_{\mathsf{sk}}(\beta, \cdot)$ is replaced to $\mathsf{KeyGen}(\mathsf{pp}, \mathsf{msk}, \cdot)$, and $\mathbf{y}_{i,\ell}^0$ and $\mathbf{y}_{i,\ell}^1$ are changed to $\mathbf{y}_{i,\ell}$ in Eq. (3.4). This security definition captures only the message privacy of MIPFE schemes, i.e., the scheme is non-function-hiding. We denote the advantage of $\mathcal{A}$ in the adaptive-security game by $\mathsf{Adv}_{\mathcal{A},\mathsf{ad}}^{\mathsf{MIPFE}}(\lambda)$. Note that we do not explicitly use the word "adaptive" in the definitions of function-hiding because it seems wordy, but we consider only the adaptive security for function-hiding schemes in this paper.

4 Tightly Secure (Multi-input) Inner Product Functional Encryption

In this section, we present our tightly secure Pub-IPFE scheme and non-function-hiding MIPFE scheme, the latter is obtained by applying the conversion by Abdalla et al. [3] to our IPFE scheme.

4.1 Construction

Let $\mathcal{D}_k$ be a matrix distribution over full rank matrices in $\mathbb{Z}_p^{(k+1)\times k}$ and norm bounds X_λ and Y_λ be polynomials in λ.

$\mathsf{Par}(1^\lambda)$: It takes a security parameter 1^λ and outputs pp as follows.

$$\mathbb{G}_{\mathsf{CG}} \leftarrow \mathcal{G}_{\mathsf{CG}}(1^\lambda), \;\; \tilde{\mathbf{A}} \leftarrow \mathcal{D}_k, \;\; \mathsf{pp} := (\mathbb{G}_{\mathsf{CG}}, [\tilde{\mathbf{A}}])$$

$\mathsf{Setup}(1^m, \mathsf{pp})$: It takes a vector length 1^m and a public parameter pp. Then, it outputs a public key pk and a master secret key msk as follows.

$$\mathbf{W} \xleftarrow{\mathsf{U}} \mathbb{Z}_p^{m \times k(k+1)m}, \quad \mathbf{A} := \overbrace{\begin{pmatrix} \tilde{\mathbf{A}} & & & \\ & \tilde{\mathbf{A}} & & \\ & & \ddots & \\ & & & \tilde{\mathbf{A}} \end{pmatrix}}^{km \text{ matrices}} \in \mathbb{Z}_p^{k(k+1)m \times k^2 m}, \tag{4.1}$$

$$\mathsf{pk} := (\mathbb{G}_{\mathsf{CG}}, [\tilde{\mathbf{A}}], [\mathbf{WA}]), \quad \mathsf{msk} := \mathbf{W}.$$

$\mathsf{Enc}(\mathsf{pk}, \mathbf{x})$: It takes pk and $\mathbf{x} \in \mathbb{Z}^m$ and outputs a ciphertext ct as follows.

$$\mathbf{s} \xleftarrow{\mathsf{U}} \mathbb{Z}_p^{k^2 m}, \quad \mathbf{c}_1 := \mathbf{As} \in \mathbb{Z}_p^{k(k+1)m}, \quad \mathbf{c}_2 := \mathbf{WAs} + \mathbf{x} \in \mathbb{Z}_p^m, \quad \mathsf{ct} := ([\mathbf{c}_1], [\mathbf{c}_2]).$$

$\mathsf{KeyGen}(\mathsf{pk}, \mathsf{msk}, \mathbf{y})$: It takes pp, msk, and $\mathbf{y} \in \mathbb{Z}^m$ and outputs a secret key sk as follows.

$$\mathbf{k}_1 := -\mathbf{W}^\top \mathbf{y} \in \mathbb{Z}_p^{k(k+1)m}, \quad \mathbf{k}_2 := \mathbf{y} \in \mathbb{Z}_p^m, \quad \mathsf{sk} := (\mathbf{k}_1, \mathbf{k}_2).$$

$\mathsf{Dec}(\mathsf{pk}, \mathsf{ct}, \mathsf{sk})$: It takes pk, ct, and sk. Then it computes $[d] := [\mathbf{k}_1^\top \mathbf{c}_1 + \mathbf{k}_2^\top \mathbf{c}_2]$ and searches for d exhaustively in the range of $-mX_\lambda Y_\lambda$ to $mX_\lambda Y_\lambda$. If such d is found, it outputs d. Otherwise, it outputs $\perp$.

Correctness. Observe that if ct is an encryption of $\mathbf{x}$ and sk is a secret key of $\mathbf{y}$,

$$d = -\mathbf{y}^\top \mathbf{WAs} + \mathbf{y}^\top \mathbf{WAs} + \mathbf{y}^\top \mathbf{x} = \langle \mathbf{x}, \mathbf{y} \rangle.$$

Therefore, if $||\mathbf{x}||_\infty \le X_\lambda$ and $||\mathbf{y}||_\infty \le Y_\lambda$, the output of the decryption algorithm is $d = \langle \mathbf{x}, \mathbf{y} \rangle$.

4.2 Security

Theorem 4.1. *Assume that the $\mathcal{D}_k$-MDDH assumption holds with respect to $\mathcal{G}_{\mathsf{CG}}$, then our Pub-IPFE scheme is adaptively secure in the multi-user and multi-challenge setting. More formally, let μ be a number of users, $q_{\mathsf{ct}} := \sum_{i \in [\mu]} q_{\mathsf{ct},i}$ be the total number of the ciphertext queries by $\mathcal{A}$, $q_{\mathsf{sk}} := \sum_{i \in [\mu]} q_{\mathsf{sk},i}$ be the total number of the secret key queries by $\mathcal{A}$, and m be a vector length. Then, for any PPT adversary $\mathcal{A}$ and security parameter λ, there exist PPT adversaries $\mathcal{B}_1$ and $\mathcal{B}_2$ for the $\mathcal{D}_k$-MDDH and we have*

$$\mathsf{Adv}_{\mathcal{A}}^{\mathsf{Pub\text{-}IPFE}}(\lambda) \le 2\mathsf{Adv}_{\mathcal{B}_1,\mathsf{CG}}^{\mathcal{D}_k\text{-}\mathsf{MDDH}}(\lambda) + 2\mathsf{Adv}_{\mathcal{B}_2,\mathsf{CG}}^{\mathcal{D}_k\text{-}\mathsf{MDDH}}(\lambda) + 2^{-\Omega(\lambda)},$$
$$\max\{\mathsf{Time}(\mathcal{B}_1), \mathsf{Time}(\mathcal{B}_2)\} \approx \mathsf{Time}(\mathcal{A}) + (\mu + q_{\mathsf{ct}} + q_{\mathsf{sk}})\mathsf{poly}(\lambda, m),$$

where $\mathsf{poly}(\lambda, m)$ *is independent from* $\mathsf{Time}(\mathcal{A})$.

Proof. We employ a series of games and evaluate the advantage of the adversary in each game. In the overveiw, we used the variable i to denote the index of users and j_i (resp. ℓ_i) to denote the index of ciphertext (resp. secret key) queries for user i. For example, a vector $\mathbf{s}$ in j_i-th ciphertext for user i will be denoted by $\mathbf{s}_{i,j_i}$. In the security proof, however, we change the forms of ciphertexts and secret keys for every user in the same way simultaneously. Thus, we do not need to specify users when we consider adversary's queries. For conciseness, we omit the index i from (i, j_i) and (i, ℓ_i), and just use j and ℓ to denote the indices of queries (but j and ℓ are implicitly associated with i).

Game 0: This game is the same as the real game. Then, for all $j \in [q_{\mathsf{ct},i}]$, the j-th ciphertext that $\mathcal{A}$ obtains from the oracle corresponds to

$$\mathbf{s}_j \xleftarrow{\mathsf{U}} \mathbb{Z}_p^{k^2 m}, \quad \mathbf{c}_{j,1} := \mathbf{A}\mathbf{s}_j, \quad \mathbf{c}_{j,2} := \mathbf{W}_i \mathbf{A}\mathbf{s}_j + \mathbf{x}_j^{\beta}.$$

Game 1: The reply for ciphertext queries is changed as follows. For $j \in [q_{\mathsf{ct},i}]$, we define $\mathbf{x}_j := \mathbf{x}_j^1 - \mathbf{x}_j^0 \in \mathbb{Z}_p^m$. Let $\phi_i : [q_{\mathsf{ct},i}] \to [m]$ be a map such that $\phi_i(j) := \mathsf{rank}(\mathbf{x}_1 || \cdots || \mathbf{x}_j)$. Then, for all $j \in [q_{\mathsf{ct},i}]$, the j-th ciphertext that $\mathcal{A}$ obtains from the oracle corresponds to

$$\mathbf{b} \xleftarrow{\mathsf{U}} \mathbb{Z}_p^{k+1} \backslash \mathsf{span}(\tilde{\mathbf{A}}), \quad \mathbf{B} := \overbrace{\begin{pmatrix} \mathbf{b} & & & \\ & \mathbf{b} & & \\ & & \ddots & \\ & & & \mathbf{b} \end{pmatrix}}^{km \text{ vectors}} \in \mathbb{Z}_p^{k(k+1)m \times km}, \tag{4.2}$$

$$\tilde{\mathbf{s}}_{j,1}, \ldots, \tilde{\mathbf{s}}_{j,\phi_i(j)} \xleftarrow{\mathsf{U}} \mathbb{Z}_p^k, \quad \mathbf{s}'_j := (\tilde{\mathbf{s}}_{j,1}, \ldots, \tilde{\mathbf{s}}_{j,\phi_i(j)}, 0^{k(m-\phi_i(j))}) \in \mathbb{Z}_p^{km},$$
$$\mathbf{c}_{j,1} := \mathbf{A}\mathbf{s}_j + \boxed{\mathbf{B}\mathbf{s}'_j}, \quad \mathbf{c}_{j,2} := \mathbf{W}_i(\mathbf{A}\mathbf{s}_j + \boxed{\mathbf{B}\mathbf{s}'_j}) + \mathbf{x}_j^{\beta}.$$

Game 2: The reply for ciphertext queries is changed as follows. Let $\rho_i : [\phi_i(q_{\mathsf{ct},i})] \to [q_{\mathsf{ct},i}]$ be a map such that $\rho_i(\iota) := \min \phi_i^{-1}(\iota)$. In other words, on an input ι, ρ_i returns the first query number j such that the rank of the matrix $(\mathbf{x}_1 || \cdots || \mathbf{x}_j)$ equals ι. Then, for all $j \in [q_{\mathsf{ct},i}]$, the j-th ciphertext that $\mathcal{A}$ obtains from the oracle corresponds to

$$\mathbf{u} \xleftarrow{\mathsf{U}} \mathbb{Z}_p^k,$$
$$\mathbf{c}_{j,1} := \mathbf{A}\mathbf{s}_j + \mathbf{B}\mathbf{s}'_j, \quad \mathbf{c}_{j,2} := \mathbf{W}_i(\mathbf{A}\mathbf{s}_j + \mathbf{B}\mathbf{s}'_j) + \mathbf{x}_j^{\beta} + \boxed{\sum_{\iota \in [\phi_i(j)]} \langle \mathbf{u}, \tilde{\mathbf{s}}_{j,\iota} \rangle \mathbf{x}_{\rho_i(\iota)}}.$$

Note that $\tilde{\mathbf{s}}_{j,\iota}$ is defined in Game 1.

Game 3: The reply for ciphertext queries is changed as follows. For all $j \in [q_{\mathsf{ct},i}]$, the j-th ciphertext that $\mathcal{A}$ obtains from the oracle corresponds to

$$r_{j,1}, \ldots, r_{j,\phi_i(j)} \xleftarrow{\mathsf{U}} \mathbb{Z}_p,$$
$$\mathbf{c}_{j,1} := \mathbf{A}\mathbf{s}_j + \mathbf{B}\mathbf{s}'_j, \quad \mathbf{c}_{j,2} := \mathbf{W}_i(\mathbf{A}\mathbf{s}_j + \mathbf{B}\mathbf{s}'_j) + \mathbf{x}_j^{\beta} + \boxed{\sum_{\iota \in [\phi_i(j)]} r_{j,\iota} \mathbf{x}_{\rho_i(\iota)}}.$$

Game 4: The reply for ciphertext queries is changed as follows. For all $j \in [q_{\mathsf{ct},i}]$, the j-th ciphertext that $\mathcal{A}$ obtains from the oracle corresponds to

$$r_{j,1}, \ldots, r_{j,\phi_i(j)} \xleftarrow{\mathsf{U}} \mathbb{Z}_p,$$
$$\mathbf{c}_{j,1} := \mathbf{A}\mathbf{s}_j + \mathbf{B}\mathbf{s}'_j, \;\; \mathbf{c}_{j,2} := \mathbf{W}_i(\mathbf{A}\mathbf{s}_j + \mathbf{B}\mathbf{s}'_j) + \boxed{\mathbf{x}_j^0} + \sum_{\iota \in [\phi_i(j)]} r_{j,\iota}\mathbf{x}_{\rho_i(\iota)}.$$

We present proofs of the indistinguishability among these games in the full version of this paper. □

4.3 Application to Multi-input Inner Product Functional Encryption

We can obtain an adaptively secure MIPFE scheme whose security is tightly reduced to the $\mathcal{D}_k$ - MDDH assumption by applying the generic conversion by Abdalla et al. [3] to our scheme. Let Pub-IPFE be a Pub-IPFE scheme that is adaptively secure in the multi-user and multi-challenge setting. It is not difficult to see that the security of the MIPFE scheme obtained by applying the conversion to Pub-IPFE is reduced to that of Pub-IPFE with the security loss being 1. Thus, we obtain the following corollary.

Corollary 4.1. *Let MIPFE be the MIPFE scheme obtained by applying the conversion in [3] to our Pub-IPFE scheme. Then MIPFE is adaptively secure. More formally, let μ be a number of slots, $q_{\mathsf{ct}} := \sum_{i \in [\mu]} q_{\mathsf{ct},i}$ be the total number of the ciphertext queries by $\mathcal{A}$, q_{sk} be the total number of the secret key queries by $\mathcal{A}$, and m be a vector length. Then, for any PPT adversary $\mathcal{A}$ and security parameter λ, there exist PPT adversaries $\mathcal{B}_1$ and $\mathcal{B}_2$ for the $\mathcal{D}_k$-MDDH and we have*

$$\mathsf{Adv}^{\mathsf{MIPFE}}_{\mathcal{A},\mathsf{ad}}(\lambda) \le 2\mathsf{Adv}^{\mathcal{D}_k\text{-}\mathsf{MDDH}}_{\mathcal{B}_1}(\lambda) + 2\mathsf{Adv}^{\mathcal{D}_k\text{-}\mathsf{MDDH}}_{\mathcal{B}_2}(\lambda) + 2^{-\Omega(\lambda)},$$
$$\max\{\mathsf{Time}(\mathcal{B}_1), \mathsf{Time}(\mathcal{B}_2)\} \approx \mathsf{Time}(\mathcal{A}) + (\mu + q_{\mathsf{ct}} + \mu q_{\mathsf{sk}})\mathsf{poly}(\lambda, m),$$

where $\mathsf{poly}(\lambda, m)$ is independent from $\mathsf{Time}(\mathcal{A})$.

5 Function-Hiding Inner Product Functional Encryption

Lin proposed a simple framework that allows us to construct a function-hiding IPFE scheme from a public key IPFE scheme [34]. We can apply her framework to our scheme and obtain a tightly function-hiding IPFE scheme in the multi-user setting. Informally, her framework is as follows.

First, we can see that a ciphertext and a secret key in our IPFE scheme consist of vectors, and decryption involves inner product of these vectors. That is, a ciphertext of a vector $\mathbf{x}$ corresponds to a vector $\mathbf{c}_{\mathsf{in}} := (\mathbf{c}_{\mathsf{in},1}, \mathbf{c}_{\mathsf{in},2}) := (\mathbf{A}\mathbf{s}, \mathbf{W}\mathbf{A}\mathbf{s} + \mathbf{x}) \in \mathbb{Z}_p^{(k^2+k+1)m}$ and a secret key of a vector $\mathbf{y}$ corresponds to a vector $\mathbf{k}_{\mathsf{in}} :=$

$(\mathbf{k}_{\mathsf{in},1}, \mathbf{k}_{\mathsf{in},2}) := (-\mathbf{W}^\top \mathbf{y}, \mathbf{y}) \in \mathbb{Z}_p^{(k^2+k+1)m}$. Decryption just computes $\langle \mathbf{c}_{\mathsf{in}}, \mathbf{k}_{\mathsf{in}} \rangle$. We call the scheme described above an inner scheme.

To ensure the confidentiality of secret keys, we "encrypt" secret keys in the same way as ciphertexts in our IPFE scheme. That is, a secret key of the function-hiding IPFE scheme is generated as $\mathsf{sk} := (\mathbf{c}_{\mathsf{out},1}, \mathbf{c}_{\mathsf{out},2}) := \left(\mathbf{Dr} \in \mathbb{Z}_p^{k(k+1)(k^2+k+1)m}, \mathbf{VDr} + \mathbf{k}_{\mathsf{in}} \in \mathbb{Z}_p^{(k^2+k+1)m}\right)$, where $\mathbf{V}$, $\mathbf{D}$, and $\mathbf{r}$ correspond to $\mathbf{W}$, $\mathbf{A}$, and $\mathbf{s}$ respectively in our scheme presented in Sect. 4.1. We call the scheme utilized to encrypt secret keys an outer scheme. We also need to transform ciphertexts to make them compatible with sk, which can be done by "generating a secret key" of $\mathbf{c}_{\mathsf{in}}$ in the outer scheme. That is, we define a ciphertext of the function-hiding IPFE scheme as $\mathsf{ct} := (\mathbf{k}_{\mathsf{out},1}, \mathbf{k}_{\mathsf{out},2}) := \left(-\mathbf{V}^\top \mathbf{c}_{\mathsf{in}} \in \mathbb{Z}_p^{k(k+1)(k^2+k+1)m}, \mathbf{c}_{\mathsf{in}} \in \mathbb{Z}_p^{(k^2+k+1)m}\right)$. Observe that $\langle \mathsf{ct}, \mathsf{sk} \rangle = \langle \mathbf{c}_{\mathsf{in}}, \mathbf{k}_{\mathsf{in}} \rangle = \langle \mathbf{x}, \mathbf{y} \rangle$.

To achieve the security, of course we need to encode both ct and sk on the exponent of group elements. We employ bilinear groups that allow us to compute inner product over the group elements, which is necessary for decryption. Then, the confidentiality of ciphertexts is assured by the inner scheme and that of secret keys is assured by the outer scheme.

5.1 Actual Scheme and Optimization

As described above, if we directly apply Lin's framework to our scheme, the first components of a ciphertext and a secret key will consist of $k(k+1)(k^2+k+1)m$ group elements. Recall the reason we need $k(k+1)m$ group elements in the first components of a ciphertext and a secret key in the original scheme. That is, the maximum dimension of the space spanned by the vectors $\mathbf{x}_j = \mathbf{x}_j^1 - \mathbf{x}_j^0$ is m, and this fact directly affects the number of group elements in the first components. Because the vector length handled in the outer scheme is $(k^2+k+1)m$, the first components seem to require $k(k+1)(k^2+k+1)m$ group elements. However, observe that the maximum dimension of the space spanned by the vectors $\mathbf{k}_{\mathsf{out},\ell} := \mathbf{k}_{\mathsf{out},\ell}^1 - \mathbf{k}_{\mathsf{out},\ell}^0 := (-\mathbf{W}^\top \mathbf{y}_\ell^1, \mathbf{y}_\ell^1) - (-\mathbf{W}^\top \mathbf{y}_\ell^0, \mathbf{y}_\ell^0)$ for all $\ell \in [q_{\mathsf{sk}}]$ is m, not $(k^2+k+1)m$. Hence, we can reduce the number of group elements in the first components to $k(k+1)m$, and the resulting scheme is given as follows.

Let $\mathcal{D}_k$ be a matrix distribution over full rank matrices in $\mathbb{Z}_p^{(k+1)\times k}$ and norm bounds X_λ and Y_λ be polynomials in λ.

$\mathsf{Par}(1^\lambda)$: It takes a security parameter 1^λ and outputs pp as follows.

$$\mathbb{G}_{\mathsf{BG}} \leftarrow \mathcal{G}_{\mathsf{BG}}(1^\lambda), \quad \tilde{\mathbf{A}}, \tilde{\mathbf{D}} \leftarrow \mathcal{D}_k, \quad \mathsf{pp} := (\mathbb{G}_{\mathsf{BG}}, [\tilde{\mathbf{A}}]_1, [\tilde{\mathbf{D}}]_2).$$

$\mathsf{Setup}(1^m, \mathsf{pp})$: It takes a vector length 1^m and a public parameter pp. Then, it outputs a master secret key msk as follows.

$$\mathbf{W} \xleftarrow{\mathsf{U}} \mathbb{Z}_p^{m\times k(k+1)m}, \quad \mathbf{V} \xleftarrow{\mathsf{U}} \mathbb{Z}_p^{(k^2+k+1)m\times k(k+1)m}, \quad \mathsf{msk} := (\mathbf{W}, \mathbf{V}).$$

$\mathsf{Enc}(\mathsf{pp}, \mathsf{msk}, \mathbf{x})$: It takes pp, msk, and $\mathbf{x} \in \mathbb{Z}^m$ and outputs a ciphertext ct as follows.

$$\mathbf{A} := \overbrace{\begin{pmatrix} \tilde{\mathbf{A}} & & & \\ & \tilde{\mathbf{A}} & & \\ & & \ddots & \\ & & & \tilde{\mathbf{A}} \end{pmatrix}}^{km \text{ matrices}} \in \mathbb{Z}_p^{k(k+1)m \times k^2 m},$$

$$\mathbf{s} \xleftarrow{\mathsf{U}} \mathbb{Z}_p^{k^2 m}, \quad \mathbf{c}_{\mathsf{in}} := (\mathbf{As}, \mathbf{WAs} + \mathbf{x}) \in \mathbb{Z}_p^{(k^2+k+1)m},$$
$$\mathbf{k}_{\mathsf{out},1} := -\mathbf{V}^\top \mathbf{c}_{\mathsf{in}} \in \mathbb{Z}_p^{k(k+1)m}, \quad \mathbf{k}_{\mathsf{out},2} := \mathbf{c}_{\mathsf{in}}, \quad \mathsf{ct} := ([\mathbf{k}_{\mathsf{out},1}]_1, [\mathbf{k}_{\mathsf{out},2}]_1).$$

$\mathsf{KeyGen}(\mathsf{pp}, \mathsf{msk}, \mathbf{y})$: It takes pp, msk, and $\mathbf{y} \in \mathbb{Z}^m$ and outputs a secret key sk as follows.

$$\mathbf{D} := \overbrace{\begin{pmatrix} \tilde{\mathbf{D}} & & & \\ & \tilde{\mathbf{D}} & & \\ & & \ddots & \\ & & & \tilde{\mathbf{D}} \end{pmatrix}}^{km \text{ matrices}} \in \mathbb{Z}_p^{k(k+1)m \times k^2 m},$$

$$\mathbf{r} \xleftarrow{\mathsf{U}} \mathbb{Z}_p^{k^2 m}, \quad \mathbf{k}_{\mathsf{in}} := (-\mathbf{W}^\top \mathbf{y}, \mathbf{y}) \in \mathbb{Z}_p^{(k^2+k+1)m},$$
$$\mathbf{c}_{\mathsf{out},1} := \mathbf{Dr} \in \mathbb{Z}_p^{k(k+1)m}, \quad \mathbf{c}_{\mathsf{out},2} := \mathbf{VDr} + \mathbf{k}_{\mathsf{in}} \in \mathbb{Z}_p^{(k^2+k+1)m},$$
$$\mathsf{sk} := ([\mathbf{c}_{\mathsf{out},1}]_2, [\mathbf{c}_{\mathsf{out},2}]_2).$$

$\mathsf{Dec}(\mathsf{pp}, \mathsf{ct}, \mathsf{sk})$: It takes pp, ct, and sk. Then it computes $[d]_T := e([\mathbf{k}_{\mathsf{out},1}]_1, [\mathbf{c}_{\mathsf{out},1}]_2) e([\mathbf{k}_{\mathsf{out},2}]_1, [\mathbf{c}_{\mathsf{out},2}]_2)$ and searches for d exhaustively in the range of $-mX_\lambda Y_\lambda$ to $mX_\lambda Y_\lambda$. If such d is found, it outputs d. Otherwise, it outputs $\perp$.

Correctness. Observe that if ct is an encryption of $\mathbf{x}$ and sk is a secret key of $\mathbf{y}$,

$$d = -\mathbf{c}_{\mathsf{in}}^\top \mathbf{VDr} + \mathbf{c}_{\mathsf{in}}^\top \mathbf{VDr} + \mathbf{c}_{\mathsf{in}}^\top \mathbf{k}_{\mathsf{in}} = \langle \mathbf{c}_{\mathsf{in}}, \mathbf{k}_{\mathsf{in}} \rangle = \langle \mathbf{x}, \mathbf{y} \rangle.$$

Therefore, if $||\mathbf{x}||_\infty \leq X_\lambda$ and $||\mathbf{y}||_\infty \leq Y_\lambda$, the output of the decryption algorithm is $d = \langle \mathbf{x}, \mathbf{y} \rangle$.

5.2 Security

Theorem 5.1. *Assume that the $\mathcal{D}_k$-MDDH assumption holds with respect to $\mathcal{G}_{\mathsf{BG}}$, then our Priv-IPFE scheme is weakly function-hiding in the multi-user setting. More formally, let μ be a number of users, $q_{\mathsf{ct}} := \sum_{i \in [\mu]} q_{\mathsf{ct},i}$ be the total number of the ciphertext queries by $\mathcal{A}$, $q_{\mathsf{sk}} := \sum_{i \in [\mu]} q_{\mathsf{sk},i}$ be the total number of the secret key queries by $\mathcal{A}$, and m be a vector length. Then, for any PPT* *adversary $\mathcal{A}$ and security parameter λ, there exist PPT adversaries $\mathcal{B}_1, \ldots, \mathcal{B}_4$ for the $\mathcal{D}_k$-MDDH, and we have*

$$\mathsf{Adv}^{\mathsf{Priv\text{-}IPFE}}_{\mathcal{A},\mathsf{w\text{-}fh}}(\lambda) \leq 2 \sum_{\iota \in \{1,2\}} \mathsf{Adv}^{\mathcal{D}_k\text{-}\mathsf{MDDH}}_{\mathcal{B}_\iota,\mathsf{BG},1}(\lambda) + 2 \sum_{\iota \in \{3,4\}} \mathsf{Adv}^{\mathcal{D}_k\text{-}\mathsf{MDDH}}_{\mathcal{B}_\iota,\mathsf{BG},2}(\lambda) + 2^{-\Omega(\lambda)},$$

$$\max_{\iota \in [4]}\{\mathsf{Time}(\mathcal{B}_\iota)\} \approx \mathsf{Time}(\mathcal{A}) + (\mu + q_{\mathsf{ct}} + q_{\mathsf{sk}})\mathsf{poly}(\lambda, m),$$

where $\mathsf{poly}(\lambda, m)$ *is independent from* $\mathsf{Time}(\mathcal{A})$.

Theorem 5.1 follows from Theorem 4.1 and Lin's observation [34]. That is, the following relations hold:

$$\left\{\{\mathsf{ct}^0_j\}_{j\in[q_{\mathsf{ct},i}]}, \{\mathsf{sk}^0_\ell\}_{\ell\in[q_{\mathsf{sk},i}]}\right\}_{i\in[\mu]} \approx_c \left\{\{\mathsf{ct}^1_j\}_{j\in[q_{\mathsf{ct},i}]}, \{\mathsf{sk}^0_\ell\}_{\ell\in[q_{\mathsf{sk},i}]}\right\}_{i\in[\mu]}$$
$$\approx_c \left\{\{\mathsf{ct}^1_j\}_{j\in[q_{\mathsf{ct},i}]}, \{\mathsf{sk}^1_\ell\}_{\ell\in[q_{\mathsf{sk},i}]}\right\}_{i\in[\mu]}.$$

The first indistinguishability follows from the security of the inner scheme and Eq. (3.3), and the second indistinguishability follows from the security of the outer scheme and Eq. (3.3). More precisely, we use the relations $\langle \mathbf{x}^0_{i,j_i}, \mathbf{y}^0_{i,\ell_i}\rangle = \langle \mathbf{x}^1_{i,j_i}, \mathbf{y}^0_{i,\ell_i}\rangle$ for the inner scheme and $\langle \mathbf{c}^1_{\mathsf{in},i,j_i}, \mathbf{k}^0_{\mathsf{in},i,\ell_i}\rangle = \langle \mathbf{c}^1_{\mathsf{in},i,j_i}, \mathbf{k}^1_{\mathsf{in},i,\ell_i}\rangle$ for the outer scheme. Both relations can be derived from Eq. (3.3). Note that because our scheme is adaptively secure, the above relations hold even if ciphertexts and secret keys are queried by an adversary adaptively.

Remark 5.1. Although the above scheme is weakly function-hiding in the multi-user setting, we can easily convert it into one that is fully function-hiding in the multi-user setting by the conversion proposed by Lin and Vaikuntanathan [35]. The conversion is very simple and works by only doubling vector lengths. When encrypting $\mathbf{x} \in \mathbb{Z}^m$, we just encrypt $(\mathbf{x}, 0^m)$ in the original scheme. Key generation is also done in the same way. In addition, this conversion is tight. That is, for any PPT adversary $\mathcal{A}$ and security parameter λ, there exist PPT adversaries $\mathcal{B}_1, \mathcal{B}_2, \mathcal{B}_3$ and we have

$$\mathsf{Adv}^{\mathsf{Priv\text{-}IPFE}}_{\mathcal{A},\mathsf{f\text{-}fh}}(\lambda) \leq \sum_{\iota\in[3]} \mathsf{Adv}^{\mathsf{Priv\text{-}IPFE}}_{\mathcal{B}_\iota,\mathsf{w\text{-}fh}}(\lambda),$$

$$\max_{\iota \in [3]}\{\mathsf{Time}(\mathcal{B}_\iota)\} \approx \mathsf{Time}(\mathcal{A}) + (\mu + q_{\mathsf{ct}} + q_{\mathsf{sk}})\mathsf{poly}(\lambda, m),$$

where $\mathsf{poly}(\lambda, m)$ is independent from $\mathsf{Time}(\mathcal{A})$.

6 From Single to Multi-input Function-Hiding Inner Product Functional Encryption

In this section, we present a generic conversion from weakly function-hiding single-input IPFE to fully function-hiding multi-input IPFE. Because all known function-hiding single-input IPFE schemes are based on bilinear groups, we design the conversion to be compatible with group based schemes. As in [3], however, we believe that our conversion is so generic that we can easily modify it to be suitable to schemes based on other primitives.

6.1 Conversion

Property. Let $\mathsf{Priv\text{-}IPFE} := (\mathsf{Par}, \mathsf{Setup}, \mathsf{Enc}, \mathsf{KeyGen}, \mathsf{Dec})$ be a Priv-IPFE scheme (Definition 3.6). In our conversion, we require that an underlying scheme has the following properties.

1. Priv-IPFE is weakly function-hiding in the multi-user setting.
2. A public parameter pp defines an order n, a group G of order n with group law $\circ$, and an encoding function $E : \mathbb{Z}_n \to G$.
3. A decryption algorithm Dec can be divided into the two algorithms Dec_1 and Dec_2 with the following properties. For any $\lambda, m \in \mathbb{N}$, any $\mathbf{x}, \mathbf{y} \in \mathbb{Z}^m$, and any $z \in \mathbb{Z}_n$ such that $|z| \leq mX_\lambda Y_\lambda$, we have

$$\Pr\left[d = E(\langle \mathbf{x}, \mathbf{y} \rangle \bmod n) \;\middle|\; \begin{array}{l} \mathsf{pp} \leftarrow \mathsf{Par}(1^\lambda) \\ \mathsf{msk} \leftarrow \mathsf{Setup}(1^m, \mathsf{pp}) \\ \mathsf{ct} \leftarrow \mathsf{Enc}(\mathsf{pp}, \mathsf{msk}, \mathbf{x}) \\ \mathsf{sk} \leftarrow \mathsf{KeyGen}(\mathsf{pp}, \mathsf{msk}, \mathbf{y}) \\ d := \mathsf{Dec}_1(\mathsf{pp}, \mathsf{ct}, \mathsf{sk}) \end{array} \right] = 1,$$

$$\mathsf{Dec}_2(\mathsf{pp}, E(z)) = z.$$

4. For any $a, b \in \mathbb{Z}_n$, we have $E(a) \circ E(b) = E(a+b)$.

Conversion. Let $\mathsf{Priv\text{-}IPFE} := (\mathsf{Par}', \mathsf{Setup}', \mathsf{Enc}', \mathsf{KeyGen}', \mathsf{Dec}' := (\mathsf{Dec}_1', \mathsf{Dec}_2'))$ be a Priv-IPFE scheme with the property defined above. Let $\mathsf{MIPFE} := (\mathsf{Setup}, \mathsf{Enc}, \mathsf{KeyGen}, \mathsf{Dec})$ be a converted MIPFE scheme. Let $X_\lambda := X_\lambda'/\mu$ be a norm bound of MIPFE, where X_λ' is a norm bound of Priv-IPFE. Our conversion is performed as follows.

$\mathsf{Setup}(1^\lambda, 1^m, 1^\mu)$: It takes a security parameter 1^λ, a vector length 1^m, and a number of slots 1^μ. Then, it outputs a public parameter pp and a master secret key msk as follows.

$$\mathsf{pp}' \leftarrow \mathsf{Par}'(1^\lambda), \;\; \{\mathsf{msk}_i'\}_{i \in [\mu]} \leftarrow \mathsf{Setup}'(1^{2m+1}, \mathsf{pp}'), \;\; \{\mathbf{u}_i\}_{i \in [\mu]} \xleftarrow{\mathsf{U}} \mathbb{Z}_n^m,$$
$$\mathsf{pp} := \mathsf{pp}', \;\; \mathsf{msk} := (\{\mathsf{msk}_i'\}_{i \in [\mu]}, \{\mathbf{u}_i\}_{i \in [\mu]}).$$

$\mathsf{Enc}(\mathsf{pp}, \mathsf{msk}, i, \mathbf{x})$: It takes pp, msk, $i \in [\mu]$ and $\mathbf{x} \in \mathbb{Z}^m$ and outputs a ciphertext ct_i as follows.

$$\tilde{\mathbf{x}} := (\mathbf{x} + \mathbf{u}_i, 0^m, 1) \in \mathbb{Z}_n^{2m+1}, \;\; \mathsf{ct}_i' \leftarrow \mathsf{Enc}'(\mathsf{pp}', \mathsf{msk}_i', \tilde{\mathbf{x}}), \;\; \mathsf{ct}_i := \mathsf{ct}_i'.$$

$\mathsf{KeyGen}(\mathsf{pp}, \mathsf{msk}, \{\mathbf{y}_i\}_{i \in [\mu]})$: It takes pp, msk, and $\{\mathbf{y}_i\}_{i \in [\mu]} \in \mathbb{Z}^m$ and outputs a secret key sk as follows.

$$\{r_i\}_{i \in [\mu-1]} \xleftarrow{\mathsf{U}} \mathbb{Z}_n, \;\; r_\mu := -\left(\sum_{i \in [\mu-1]} r_i + \sum_{i \in [\mu]} \langle \mathbf{y}_i, \mathbf{u}_i \rangle \right) \in \mathbb{Z}_n,$$
$$\tilde{\mathbf{y}}_i := (\mathbf{y}_i, 0^m, r_i) \in \mathbb{Z}_n^{2m+1}, \;\; \mathsf{sk}_i' \leftarrow \mathsf{KeyGen}'(\mathsf{pp}', \mathsf{msk}_i', \tilde{\mathbf{y}}_i) \text{ for all } i \in [\mu],$$
$$\mathsf{sk} := \{\mathsf{sk}_i'\}_{i \in [\mu]}.$$

$\mathsf{Dec}(\mathsf{pp}, \{\mathsf{ct}_i\}_{i\in[\mu]}, \mathsf{sk})$: It takes pp, $\{\mathsf{ct}_i\}_{i\in[\mu]}$, and sk. Then, it computes decryption value d as follows.

$$d_i := \mathsf{Dec}'_1(\mathsf{pp}', \mathsf{ct}'_i, \mathsf{sk}'_i) \in \mathbb{G} \text{ for all } i \in [\mu], \quad d := \mathsf{Dec}'_2(\mathsf{pp}', d_1 \circ \cdots \circ d_\mu).$$

Correctness. From property 3, we have

$$d_i = E(\langle \mathbf{x}_i + \mathbf{u}_i, \mathbf{y}_i \rangle + r_i \mod n).$$

From property 4, we have

$$d_1 \circ \cdots \circ d_\mu = E\left(\sum_{i\in[\mu]} (\langle \mathbf{x}_i + \mathbf{u}_i, \mathbf{y}_i \rangle + r_i) \mod n\right) = E\left(\sum_{i\in[\mu]} \langle \mathbf{x}_i, \mathbf{y}_i \rangle \mod n\right).$$

Then, from property 3 and the correctness of Priv-IPFE, we have $d := \mathsf{Dec}'_2(d_1 \circ \cdots \circ d_\mu) = \sum_{i\in[\mu]} \langle \mathbf{x}_i, \mathbf{y}_i \rangle$.

Remark 6.1. Typically, we define Priv-IPFE as consisting of four algorithms (Setup, Enc, KeyGen, Dec) and Setup outputs pp and msk when we consider Priv-IPFE in the single-user setting. To apply our conversion to such a Priv-IPFE scheme, just setting $\mathsf{pp} := \mathsf{pp}'_1, \ldots, \mathsf{pp}'_\mu$ suffices in the setup algorithm. In the security proof, however, we need a hybrid argument for each slot similarly to [3]. Thus, the security reduction will not become tight.

6.2 Security

Theorem 6.1. *Let Priv-IPFE be a Priv-IPFE scheme that satisfies the properties described above. Then converted scheme, MIPFE, is a fully function-hiding MIPFE scheme. More formally, let μ be a number of slots, $q_{\mathsf{ct}} := \sum_{i\in[\mu]} q_{\mathsf{ct},i}$ be the total number of the ciphertext queries by $\mathcal{A}$, q_{sk} be the total number of the secret key queries by $\mathcal{A}$, and m be a vector length. Then, for any PPT adversary $\mathcal{A}$ and security parameter λ, there exist PPT adversaries $\mathcal{B}_1, \mathcal{B}_2$ for Priv-IPFE and we have*

$$\mathsf{Adv}^{\mathsf{MIPFE}}_{\mathcal{A},\mathsf{f\text{-}fh}}(\lambda) \le 2 \sum_{\iota\in[2]} \mathsf{Adv}^{\mathsf{Priv\text{-}IPFE}}_{\mathcal{B}_\iota,\mathsf{w\text{-}fh}}(\lambda),$$

$$\max_{\iota\in[2]}\{\mathsf{Time}(\mathcal{B}_\iota)\} \approx \mathsf{Time}(\mathcal{A}) + (\mu + q_{\mathsf{ct}} + \mu q_{\mathsf{sk}})\mathsf{poly}(\lambda, m),$$

where $\mathsf{poly}(\lambda, m)$ *is independent from* $\mathsf{Time}(\mathcal{A})$.

Proof. We employ a series of games and evaluate the advantage of the adversary in each game. For ease of exposition, we first consider six games: Games 0 to 5, and show that the each transition of games is justified by the security of the underlying scheme (or an information-theoretical argument). Then, we explain that the transition from Game 0 to 2 and that from Game 3 to 5 can be done in one-shot. We summarize forms of ciphertexts and secret keys in each game in Table 5. Similarly to in Sect. 4.2, we omit index i from index j_i and just denote it by j. We present formal proof in the full version of this paper. □

Table 5. Overview of the game change.

game	$\tilde{\mathbf{x}}_{i,j}$ in ct	$\tilde{\mathbf{y}}_{i,\ell}$ in sk	$-\sum r_{i,\ell}$
0 (real)	$(\mathbf{x}^{\beta}_{i,j}+\mathbf{u}_i, 0^m, 1)$	$(\mathbf{y}^{\beta}_{i,\ell}, 0^m, r_{i,\ell})$	$\sum\langle\mathbf{y}^{\beta}_{i,\ell}, \mathbf{u}_i\rangle$
1	$(\mathbf{x}^{\beta}_{i,j}+\mathbf{u}_i, \boxed{\mathbf{v}_i}, 1)$	$(\mathbf{y}^{\beta}_{i,\ell}, 0^m, r_{i,\ell})$	$\sum\langle\mathbf{y}^{\beta}_{i,\ell}, \mathbf{u}_i\rangle$
2	$(\mathbf{x}^{\beta}_{i,j}+\mathbf{u}_i, \mathbf{v}_i, 1)$	$(\mathbf{y}^{\beta}_{i,\ell}, \boxed{\mathbf{y}^{0}_{i,\ell}}, r_{i,\ell})$	$\boxed{\sum(\langle\mathbf{y}^{\beta}_{i,\ell}, \mathbf{u}_i\rangle+\langle\mathbf{y}^{0}_{i,\ell}, \mathbf{v}_i\rangle)}$
3	$(\mathbf{x}^{\beta}_{i,j}\boxed{-\mathbf{x}^{\beta}_{i,1}}+\mathbf{u}_i, \boxed{\mathbf{x}^{0}_{i,1}}+\mathbf{v}_i, 1)$	$(\mathbf{y}^{\beta}_{i,\ell}, \mathbf{y}^{0}_{i,\ell}, r_{i,\ell})$	$\sum(\langle\mathbf{y}^{\beta}_{i,\ell}, \mathbf{u}_i\rangle+\langle\mathbf{y}^{0}_{i,\ell}, \mathbf{v}_i\rangle)$
4	$(\mathbf{u}_i, \boxed{\mathbf{x}^{0}_{i,j}}+\mathbf{v}_i, 1)$	$(\mathbf{y}^{\beta}_{i,\ell}, \mathbf{y}^{0}_{i,\ell}, r_{i,\ell})$	$\sum(\langle\mathbf{y}^{\beta}_{i,\ell}, \mathbf{u}_i\rangle+\langle\mathbf{y}^{0}_{i,\ell}, \mathbf{v}_i\rangle)$
5 (final)	$(\mathbf{u}_i, \mathbf{x}^{0}_{i,j}+\mathbf{v}_i, 1)$	$(\boxed{0^m}, \mathbf{y}^{0}_{i,\ell}, r_{i,\ell})$	$\boxed{\sum\langle\mathbf{y}^{0}_{i,\ell}, \mathbf{v}_i\rangle}$

6.3 Application to Our Scheme

Applying the conversion to our scheme presented in Sect. 5.1, we can obtain a tightly secure fully function-hiding MIPFE scheme. First, we confirm that our scheme satisfies the property presented in Sect. 6.1.

1. Theorem 5.1 says that our scheme is weakly function-hiding.
2. We can define that $n := p$, $G := G_T$, and $E : a \in \mathbb{Z}_p \to [a]_T \in G_T$. The group law $\circ$ corresponds to the multiplication over G_T.
3. We can define that Dec_1 computes $[d]_T$ and Dec_2 searches for the discrete logarithm of $[d]_T$.
4. It is obvious that $g_T^a \cdot g_T^b = g_T^{a+b}$.

Then, from Theorems 5.1 and 6.1, we obtain the following corollary.

Corollary 6.1. *Let* MIPFE *be the MIPFE scheme obtained by applying the conversion in Sect. 6.1 to our weakly function-hiding Priv-IPFE scheme. Then* MIPFE *is fully function-hiding. More formally, let* μ *be a number of slots,* $q_{\mathsf{ct}} := \sum_{i\in[\mu]} q_{\mathsf{ct},i}$ *be the total number of the ciphertext queries by* $\mathcal{A}$*,* q_{sk} *be the total number of the secret key queries by* $\mathcal{A}$*, and* m *be a vector length. Then, for any PPT adversary* $\mathcal{A}$ *and security parameter* λ*, there exist PPT adversaries* $\mathcal{B}_1, \ldots, \mathcal{B}_4$ *for the* $\mathcal{D}_k$*-MDDH and we have*

$$\mathsf{Adv}^{\mathsf{MIPFE}}_{\mathcal{A},\mathsf{f\text{-}fh}}(\lambda) \le 8 \sum_{\iota\in\{1,2\}} \mathsf{Adv}^{\mathcal{D}_k\text{-}\mathsf{MDDH}}_{\mathcal{B}_\iota,\mathsf{BG},1}(\lambda) + 8 \sum_{\iota\in\{3,4\}} \mathsf{Adv}^{\mathcal{D}_k\text{-}\mathsf{MDDH}}_{\mathcal{B}_\iota,\mathsf{BG},2}(\lambda) + 2^{-\Omega(\lambda)},$$

$$\max_{\iota\in[4]}\{\mathsf{Time}(\mathcal{B}_\iota)\} \approx \mathsf{Time}(\mathcal{A}) + (\mu + q_{\mathsf{ct}} + \mu q_{\mathsf{sk}})\mathsf{poly}(\lambda, m),$$

where $\mathsf{poly}(\lambda, m)$ *is independent from* $\mathsf{Time}(\mathcal{A})$.

References

1. GenBank and WGS statistics. https://www.ncbi.nlm.nih.gov/genbank/statistics/
2. Abdalla, M., Bourse, F., De Caro, A., Pointcheval, D.: Simple functional encryption schemes for inner products. In: Katz, J. (ed.) PKC 2015. LNCS, vol. 9020, pp. 733–751. Springer, Heidelberg (2015). https://doi.org/10.1007/978-3-662-46447-2_33

3. Abdalla, M., Catalano, D., Fiore, D., Gay, R., Ursu, B.: Multi-input functional encryption for inner products: function-hiding realizations and constructions without pairings. In: Shacham, H., Boldyreva, A. (eds.) CRYPTO 2018, Part I. LNCS, vol. 10991, pp. 597–627. Springer, Cham (2018). https://doi.org/10.1007/978-3-319-96884-1_20
4. Abdalla, M., Gay, R., Raykova, M., Wee, H.: Multi-input inner-product functional encryption from pairings. In: Coron, J.-S., Nielsen, J.B. (eds.) EUROCRYPT 2017, Part I. LNCS, vol. 10210, pp. 601–626. Springer, Cham (2017). https://doi.org/10.1007/978-3-319-56620-7_21
5. Abe, M., Hofheinz, D., Nishimaki, R., Ohkubo, M., Pan, J.: Compact structure-preserving signatures with almost tight security. In: Katz, J., Shacham, H. (eds.) CRYPTO 2017, Part II. LNCS, vol. 10402, pp. 548–580. Springer, Cham (2017). https://doi.org/10.1007/978-3-319-63715-0_19
6. Agrawal, S., Libert, B., Stehlé, D.: Fully secure functional encryption for inner products, from standard assumptions. In: Robshaw, M., Katz, J. (eds.) CRYPTO 2016, Part III. LNCS, vol. 9816, pp. 333–362. Springer, Heidelberg (2016). https://doi.org/10.1007/978-3-662-53015-3_12
7. Attrapadung, N., Hanaoka, G., Yamada, S.: A framework for identity-based encryption with almost tight security. In: Iwata, T., Cheon, J.H. (eds.) ASIACRYPT 2015, Part I. LNCS, vol. 9452, pp. 521–549. Springer, Heidelberg (2015). https://doi.org/10.1007/978-3-662-48797-6_22
8. Badrinarayanan, S., Gupta, D., Jain, A., Sahai, A.: Multi-input functional encryption for unbounded arity functions. In: Iwata, T., Cheon, J.H. (eds.) ASIACRYPT 2015, Part I. LNCS, vol. 9452, pp. 27–51. Springer, Heidelberg (2015). https://doi.org/10.1007/978-3-662-48797-6_2
9. Baltico, C.E.Z., Catalano, D., Fiore, D., Gay, R.: Practical functional encryption for quadratic functions with applications to predicate encryption. In: Katz, J., Shacham, H. (eds.) CRYPTO 2017, Part I. LNCS, vol. 10401, pp. 67–98. Springer, Cham (2017). https://doi.org/10.1007/978-3-319-63688-7_3
10. Barreto, P.S.L.M., Costello, C., Misoczki, R., Naehrig, M., Pereira, G.C.C.F., Zanon, G.: Subgroup security in pairing-based cryptography. In: Lauter, K., Rodríguez-Henríquez, F. (eds.) LATINCRYPT 2015. LNCS, vol. 9230, pp. 245–265. Springer, Cham (2015). https://doi.org/10.1007/978-3-319-22174-8_14
11. Bellare, M., Boldyreva, A., Micali, S.: Public-key encryption in a multi-user setting: security proofs and improvements. In: Preneel, B. (ed.) EUROCRYPT 2000. LNCS, vol. 1807, pp. 259–274. Springer, Heidelberg (2000). https://doi.org/10.1007/3-540-45539-6_18
12. Bishop, A., Jain, A., Kowalczyk, L.: Function-hiding inner product encryption. In: Iwata, T., Cheon, J.H. (eds.) ASIACRYPT 2015, Part I. LNCS, vol. 9452, pp. 470–491. Springer, Heidelberg (2015). https://doi.org/10.1007/978-3-662-48797-6_20
13. Boneh, D., Sahai, A., Waters, B.: Functional encryption: definitions and challenges. In: Ishai, Y. (ed.) TCC 2011. LNCS, vol. 6597, pp. 253–273. Springer, Heidelberg (2011). https://doi.org/10.1007/978-3-642-19571-6_16
14. Brakerski, Z., Komargodski, I., Segev, G.: Multi-input functional encryption in the private-key setting: stronger security from weaker assumptions. In: Fischlin, M., Coron, J.-S. (eds.) EUROCRYPT 2016, Part II. LNCS, vol. 9666, pp. 852–880. Springer, Heidelberg (2016). https://doi.org/10.1007/978-3-662-49896-5_30
15. Brakerski, Z., Segev, G.: Function-private functional encryption in the private-key setting. In: Dodis, Y., Nielsen, J.B. (eds.) TCC 2015, Part II. LNCS, vol. 9015, pp. 306–324. Springer, Heidelberg (2015). https://doi.org/10.1007/978-3-662-46497-7_12

16. Castagnos, G., Laguillaumie, F., Tucker, I.: Practical fully secure unrestricted inner product functional encryption modulo p. In: Peyrin, T., Galbraith, S. (eds.) ASIACRYPT 2018, Part II. LNCS, vol. 11273, pp. 733–764. Springer, Cham (2018). https://doi.org/10.1007/978-3-030-03329-3_25
17. Chen, J., Wee, H.: Fully, (almost) tightly secure ibe and dual system groups. In: Canetti, R., Garay, J.A. (eds.) CRYPTO 2013, Part II. LNCS, vol. 8043, pp. 435–460. Springer, Heidelberg (2013). https://doi.org/10.1007/978-3-642-40084-1_25
18. Datta, P., Dutta, R., Mukhopadhyay, S.: Functional encryption for inner product with full function privacy. In: Cheng, C.-M., Chung, K.-M., Persiano, G., Yang, B.-Y. (eds.) PKC 2016, Part I. LNCS, vol. 9614, pp. 164–195. Springer, Heidelberg (2016). https://doi.org/10.1007/978-3-662-49384-7_7
19. Datta, P., Okamoto, T., Tomida, J.: Full-hiding (unbounded) multi-input inner product functional encryption from the k-linear assumption. In: Abdalla, M., Dahab, R. (eds.) PKC 2018, Part II. LNCS, vol. 10770, pp. 245–277. Springer, Cham (2018). https://doi.org/10.1007/978-3-319-76581-5_9
20. Enge, A., Milan, J.: Implementing cryptographic pairings at standard security levels. In: Chakraborty, R.S., Matyas, V., Schaumont, P. (eds.) SPACE 2014. LNCS, vol. 8804, pp. 28–46. Springer, Cham (2014). https://doi.org/10.1007/978-3-319-12060-7_3
21. Escala, A., Herold, G., Kiltz, E., Ràfols, C., Villar, J.L.: An algebraic framework for Diffie-Hellman assumptions. J. Cryptol. **30**(1), 242–288 (2017)
22. Garg, S., Gentry, C., Halevi, S.: Candidate multilinear maps from ideal lattices. In: Johansson, T., Nguyen, P.Q. (eds.) EUROCRYPT 2013. LNCS, vol. 7881, pp. 1–17. Springer, Heidelberg (2013). https://doi.org/10.1007/978-3-642-38348-9_1
23. Garg, S., Gentry, C., Halevi, S., Raykova, M., Sahai, A., Waters, B.: Candidate indistinguishability obfuscation and functional encryption for all circuits. In: 54th FOCS, pp. 40–49. IEEE Computer Society Press, October 2013
24. Garg, S., Gentry, C., Halevi, S., Zhandry, M.: Functional encryption without obfuscation. In: Kushilevitz, E., Malkin, T. (eds.) TCC 2016, Part II. LNCS, vol. 9563, pp. 480–511. Springer, Heidelberg (2016). https://doi.org/10.1007/978-3-662-49099-0_18
25. Gay, R., Hofheinz, D., Kiltz, E., Wee, H.: Tightly CCA-secure encryption without pairings. In: Fischlin, M., Coron, J.-S. (eds.) EUROCRYPT 2016, Part I. LNCS, vol. 9665, pp. 1–27. Springer, Heidelberg (2016). https://doi.org/10.1007/978-3-662-49890-3_1
26. Gay, R., Hofheinz, D., Kohl, L.: Kurosawa-Desmedt meets tight security. In: Katz, J., Shacham, H. (eds.) CRYPTO 2017, Part III. LNCS, vol. 10403, pp. 133–160. Springer, Cham (2017). https://doi.org/10.1007/978-3-319-63697-9_5
27. Gay, R., Kowalczyk, L., Wee, H.: Tight adaptively secure broadcast encryption with short ciphertexts and keys. In: Catalano, D., De Prisco, R. (eds.) SCN 2018. LNCS, vol. 11035, pp. 123–139. Springer, Cham (2018). https://doi.org/10.1007/978-3-319-98113-0_7
28. Goldwasser, S., et al.: Multi-input functional encryption. In: Nguyen, P.Q., Oswald, E. (eds.) EUROCRYPT 2014. LNCS, vol. 8441, pp. 578–602. Springer, Heidelberg (2014). https://doi.org/10.1007/978-3-642-55220-5_32
29. Hofheinz, D.: Algebraic partitioning: fully compact and (almost) tightly secure cryptography. In: Kushilevitz, E., Malkin, T. (eds.) TCC 2016, Part I. LNCS, vol. 9562, pp. 251–281. Springer, Heidelberg (2016). https://doi.org/10.1007/978-3-662-49096-9_11

30. Hofheinz, D.: Adaptive partitioning. In: Coron, J.-S., Nielsen, J.B. (eds.) EUROCRYPT 2017, Part III. LNCS, vol. 10212, pp. 489–518. Springer, Cham (2017). https://doi.org/10.1007/978-3-319-56617-7_17
31. Hofheinz, D., Jager, T.: Tightly secure signatures and public-key encryption. In: Safavi-Naini, R., Canetti, R. (eds.) CRYPTO 2012. LNCS, vol. 7417, pp. 590–607. Springer, Heidelberg (2012). https://doi.org/10.1007/978-3-642-32009-5_35
32. Kim, S., Lewi, K., Mandal, A., Montgomery, H., Roy, A., Wu, D.J.: Function-hiding inner product encryption is practical. In: Catalano, D., De Prisco, R. (eds.) SCN 2018. LNCS, vol. 11035, pp. 544–562. Springer, Cham (2018). https://doi.org/10.1007/978-3-319-98113-0_29
33. Libert, B., Peters, T., Joye, M., Yung, M.: Compactly hiding linear spans. In: Iwata, T., Cheon, J.H. (eds.) ASIACRYPT 2015, Part I. LNCS, vol. 9452, pp. 681–707. Springer, Heidelberg (2015). https://doi.org/10.1007/978-3-662-48797-6_28
34. Lin, H.: Indistinguishability obfuscation from SXDH on 5-linear maps and locality-5 PRGs. In: Katz, J., Shacham, H. (eds.) CRYPTO 2017, Part I. LNCS, vol. 10401, pp. 599–629. Springer, Cham (2017). https://doi.org/10.1007/978-3-319-63688-7_20
35. Lin, H., Vaikuntanathan, V.: Indistinguishability obfuscation from DDH-like assumptions on constant-degree graded encodings. In: Dinur, I. (ed.) 57th FOCS, pp. 11–20. IEEE Computer Society Press, October 2016
36. Naor, M., Reingold, O.: Synthesizers and their application to the parallel construction of pseudo-random functions. J. Comput. Syst. Sci. **58**(2), 336–375 (1999)
37. O'Neill, A.: Definitional issues in functional encryption. Cryptology ePrint Archive, Report 2010/556 (2010). http://eprint.iacr.org/2010/556
38. Tomida, J., Abe, M., Okamoto, T.: Efficient functional encryption for inner-product values with full-hiding security. In: Bishop, M., Nascimento, A.C.A. (eds.) ISC 2016. LNCS, vol. 9866, pp. 408–425. Springer, Cham (2016). https://doi.org/10.1007/978-3-319-45871-7_24
39. Waters, B.: A punctured programming approach to adaptively secure functional encryption. In: Gennaro, R., Robshaw, M. (eds.) CRYPTO 2015, Part II. LNCS, vol. 9216, pp. 678–697. Springer, Heidelberg (2015). https://doi.org/10.1007/978-3-662-48000-7_33

Public-Key Function-Private Hidden Vector Encryption (and More)

James Bartusek[1(✉)], Brent Carmer[2], Abhishek Jain[3], Zhengzhong Jin[3], Tancrède Lepoint[4], Fermi Ma[5], Tal Malkin[6], Alex J. Malozemoff[2], and Mariana Raykova[4]

[1] UC Berkeley, Berkeley, USA
bartusek.james@gmail.com
[2] Galois, Portland, USA
{bcarmer,amaloz}@galois.com
[3] Johns Hopkins University, Baltimore, USA
abhishek@cs.jhu.edu, zjin12@jhu.edu
[4] Google, Mountain View, USA
tancrede@google.com, mariana@google.com
[5] Princeton University, Princeton, USA
fermima@alum.mit.edu
[6] Columbia University, New York, USA
tal@cs.columbia.edu

Abstract. We construct *public-key function-private* predicate encryption for the "small superset functionality," recently introduced by Beullens and Wee (PKC 2019). This functionality captures several important classes of predicates:

- Point functions. For point function predicates, our construction is equivalent to public-key function-private anonymous identity-based encryption.
- Conjunctions. If the predicate computes a conjunction, our construction is a public-key function-private hidden vector encryption scheme. This addresses an open problem posed by Boneh, Raghunathan, and Segev (ASIACRYPT 2013).
- d-CNFs and read-once conjunctions of d-disjunctions for constant-size d.

Our construction extends the group-based obfuscation schemes of Bishop et al. (CRYPTO 2018), Beullens and Wee (PKC 2019), and Bartusek et al. (EUROCRYPT 2019) to the setting of public-key function-private predicate encryption. We achieve an average-case notion of function privacy, which guarantees that a decryption key sk_f reveals nothing about f as long as f is drawn from a distribution with sufficient entropy. We formalize this security notion as a generalization of the (enhanced) real-or-random function privacy definition of Boneh, Raghunathan, and Segev (CRYPTO 2013). Our construction relies on bilinear groups, and we prove security in the generic bilinear group model.

Research conducted at Princeton University.

S. D. Galbraith and S. Moriai (Eds.): ASIACRYPT 2019, LNCS 11923, pp. 489–519, 2019.
https://doi.org/10.1007/978-3-030-34618-8_17

1 Introduction

Predicate encryption [BW07,KSW08] is a powerful tool which enables fine-grained access to encrypted information. Roughly speaking, a sender can encrypt a message m (commonly referred to as a payload) with respect to an *attribute* x, while each decryption key sk_f is tied to a specific predicate f in some function class $\mathcal{F}$; the key sk_f correctly decrypts a ciphertext if and only if the associated attribute x satisfies $f(x) = 1$.

Generally, predicate encryption schemes simultaneously achieve *payload-hiding* and *attribute-hiding* security. At a high level, payload-hiding guarantees that an encryption of m with respect to attribute x reveals nothing about m to an adversary who does not possess a decryption key sk_f where $f(x) = 1$. Attribute-hiding guarantees that ciphertexts hide any information about x beyond what is leaked from successful decryption. That is, an adversary holding decryption keys $\mathsf{sk}_{f_1}, \ldots, \mathsf{sk}_{f_n}$ may learn the 0/1 evaluations of $f_1, \ldots, f_n$ on x, but should not be able to learn anything else about x.

For certain applications, however, these two security guarantees may not be enough. Suppose an email user wants to set up a gateway that routes encrypted emails differently depending on whether or not they are spam. The user would like to avoid giving the gateway full access to the content of their emails; instead the user may have some list of potential spam email addresses, and would prefer the gateway only apply its spam filtering algorithm (which requires reading the email plaintext) if the email is sent from this set of addresses.[1]

The predicate encryption-based solution treats the contents of the email as the "payload," and the sender email address as the "attribute." The user generates some decryption key sk_f for their filtering predicate f (in this scenario, the predicate would output one if the email address belongs to a list of potential spammers), and sends this to the gateway. It is easy to imagine that the user may want to hide its particular choice of f from the gateway, since after all the user views the gateway as an untrusted party. But given sk_f, the standard payload-hiding and attribute-hiding definitions say nothing about whether one can learn the description of f.

Public-Key Function Privacy. Boneh, Raghunathan, and Segev (BRS) [BRS13a] address this problem by defining public-key *function-private* predicate encryption, which requires that sk_f leak nothing about f beyond what is leaked through honest decryption.[2] They demonstrate that this notion is achievable with a new construction of function-private anonymous identity-based encryption (i.e., predicate encryption for equality predicates). We note that here, "function private" means that the identity embedded in the decryption key is hidden, "anonymous" means that the intended recipient of the ciphertext (the attribute) is

[1] Note that we would require a public-key predicate encryption scheme for this scenario, with the assumption that an email client would encrypt any email to the user under the user's public key.

[2] Function privacy had been studied before the work of BRS [BRS13a], albeit in the private key setting [SWP00,OS07,BSW09,SSW09].

hidden; and finally the message being encrypted also stays secret. In follow-up work, BRS [BRS13b] extended public-key function privacy to a significantly larger class of *subspace-membership* predicates. We stress that in both works, BRS present function privacy as an average-case definition, which is essentially inherent in the public-key setting (see Sect. 1.2 for further discussion).

While BRS [BRS13a, BRS13b] laid the groundwork for the study of function privacy in the public-key setting, a number of important questions remained unanswered. In particular, BRS explicitly identified three important directions for further exploration [BRS13b]:

1. **Computational Function Privacy.** In both works, BRS construct *statistically* function-private schemes. They conjectured, however, that it might be possible to leverage group-based assumptions to achieve more powerful/expressive *computationally* function-private predicate encryption schemes.
2. **Hidden Vector Encryption.** The seminal work of Boneh and Waters [BW07] introduced *hidden vector encryption* (HVE) as a general approach to performing equality, comparison, and subset queries on encrypted data. In HVE, predicates are specified by a vector $\mathbf{v} \in \Sigma^k$, where $\Sigma = \mathbb{Z}_s \cup \{*\}$. We refer to s as the alphabet size and $*$ as a *wildcard* character. A message m encrypted under attribute $\mathbf{x} \in \mathbb{Z}_s^k$ can be decrypted under key f_v if x_i matches v_i at each i where $v_i \neq *$. Follow-up work by Katz, Sahai, and Waters [KSW13] introduced *inner product encryption* (i.e., predicate encryption for inner product predicates) as a generalization of HVE. In turn, inner product predicates are a subclass of more general *subspace-membership predicates*. Therefore, predicate encryption for subspace-membership trivially implies inner product encryption and HVE.
   However, BRS [BRS13b] observe that these implications crucially do not preserve function privacy. That is, their function-private subspace-membership encryption construction is *not* a function-private HVE. In fact, BRS remark that even *defining* function privacy for HVE is not straightforward, and they leave defining and constructing function-private HVE as an open problem.
3. **Enhanced Function Privacy.** The plain definition of function privacy given by BRS [BRS13a] comes with a serious drawback. At a high level, the definition assumes that the adversary holding decryption key sk_f will never encounter a ciphertext with a matching attribute x (i.e., where $f(x) = 1$). The authors argue that such an assumption is necessary in many settings, since if an adversary could generate such matching ciphertexts, it must know some x where $f(x) = 1$. For equality predicates, this amounts to learning f entirely.
   In almost any natural application, however, we should expect that the party in possession of sk_f will encounter "matching" ciphertexts; the crucial point is that they would not be generating these ciphertexts themselves. To capture this, BRS define a stronger notion called *enhanced function privacy* where the adversary is given access to an "encryption oracle" that outputs matching ciphertexts.
   Unfortunately, the only known construction of a public-key scheme achieving enhanced function-privacy is the anonymous identity-based encryption

construction presented by BRS [BRS13a]. Therefore, constructing enhanced-function-private predicate encryption schemes for any class of predicates beyond equality predicates has remained open since.

1.1 Our Contributions

In this work, we make substantial progress on all three fronts. Compared to BRS [BRS13a, BRS13b], our results come from using a qualitatively different high-level approach. In particular, BRS construct public-key function-private predicate encryption by starting from schemes that satisfy only data privacy (a definition combining attribute-hiding and payload-hiding), and transforming them to achieve data privacy and function privacy simultaneously.

We take the opposite approach. We begin with constructions that satisfy function privacy but not data privacy and transform them to achieve both data privacy and function privacy. In more standard terminology, our high-level approach is to think of an *obfuscated program* [BGI+01] as a decryption key within a "predicate encryption" scheme that has no data privacy whatsoever (since obfuscated programs are run directly on non-encrypted inputs). We then show that several obfuscation schemes from the literature can be appropriately transformed to achieve public-key function-private predicate encryption.

Our starting point is the recent line of work [BKM+18, BW19, BLMZ19] that constructs simple, group-based obfuscation schemes for what Beullens and Wee refer to as the "big subset" predicate [BW19].[3] For our work, we re-interpret these predicates as "small superset" predicates, a notion we find slightly more natural for our applications. A "small superset" predicate $f_{n,t,X}$ is parameterized by a target set $X \subseteq [n]$, an integer size bound $t \leq n$, and takes as input any set $Y \subseteq [n]$. $f_{n,t,X}(Y)$ outputs 1 if and only if $X \subseteq Y$ and $|Y| \leq t$ (that is, Y is a small superset of X). We show that "small superset" predicates capture several natural and expressive predicate classes, including large-alphabet conjunctions, functions in conjunctive normal form with a constant number of inputs per conjunct (a.k.a., d-CNFs for $d = O(1)$), and read-once conjunctions of d-disjunctions for $d = O(1)$.

Our primary contributions are the following:

1. We draw upon a correspondence between program obfuscation and function privacy to formulate new and versatile simulation-based definitions of average-case function-privacy and enhanced function-privacy.[4] While our definitions incorporate elements of the *distributional virtual black box* notion from obfuscation [BGI+01, BR17], we view our (enhanced) function privacy definition as a natural extension of the definition of BRS [BRS13a]. Unlike

[3] We remark that [BKM+18, BLMZ19] framed their results as obfuscation for conjunctions. Beullens and Wee [BW19] were the first to notice that these techniques are in fact obfuscating a more general class of "big subset" predicates, which in particular encompass conjunctions.

[4] While our definitions are new, we are not the first to observe the connection between program obfuscation and function-privacy. See also [AAB+15, ITZ16, ABF16].

these prior function-privacy notions [BRS13a, BRS13b], which are tailored to specific classes of predicates, our definition is completely agnostic to the predicate class.[5] For the special case of HVE (i.e., large-alphabet conjunctions), we demonstrate that constructions achieving our function privacy definitions hide strictly more information about the underlying predicate than constructions achieving other recently proposed HVE function-privacy definitions (e.g., [PM18, PMR19]).

2. We leverage bilinear maps to construct a public-key predicate encryption scheme for small superset predicates. At a very high level, our construction works by embedding the group-based constructions developed in [BKM+18, BW19, BLMZ19] in group $\mathbb{G}_1$, encoding messages/attributes in group $\mathbb{G}_2$, and decrypting using the bilinear map. We prove that our construction achieves enhanced function privacy in the generic bilinear group model. We note that generic analysis is somewhat unavoidable in our setting, as the underlying obfuscation constructions we build on are not known to be secure under any falsifiable assumption [Nao03, GW11].
3. We show that our general construction of public-key enhanced function-private predicate encryption for "small superset" immediately yields the following:
   - Anonymous IBE achieving enhanced function privacy as long as the underlying distribution on points has super-logarithmic min-entropy.
   - Public-key enhanced-function-private HVE whenever the underlying distribution meets a certain entropy threshold.
   - Public-key enhanced-function-private predicate encryption for d-CNFs and read-once conjunctions of d-disjunctions, subject to certain entropy requirements.

1.2 Technical Overview

Our Approach: From Obfuscation to Function-Private Predicate Encryption. We begin by recalling the notion of *program obfuscation* [BGI+01], which is the starting point for all of the constructions in this work. Roughly speaking, a program obfuscator takes in a description of some program P and outputs an obfuscated program $\mathsf{Obf}(P)$ that is functionally equivalent to P, but hides all of the implementation details. A natural approach to formalizing obfuscation security is the notion of a virtual black box (VBB), which asks that anything (precisely, any one-bit predicate) one can learn given $\mathsf{Obf}(P)$ can also be learned from black-box access to an oracle for P. While VBB obfuscation for general programs is known to be impossible [BGI+01], there have been a number of positive results that achieve (average-case or worst-case) VBB security for

[5] We note that we are not the first to give a public-key function-private definition that is agnostic to the predicate class. In particular, this is also achieved by the definition of [ITZ16]. However, their definition does not extend to enhanced function privacy, and furthermore they do not give any constructions achieving their definition except under a strengthening of indistinguishability obfuscation due to [BCKP14].

limited classes of functionalities, such as point functions [Can97,LPS04,Wee05], conjunctions [BR13,BR17,BKM+18,BLMZ19,BW19], Hamming balls [DS05], hyperplanes [CRV10], "compute-and-compare" functions [WZ17,GKW17], etc.

As mentioned in Sect. 1.1, there is a strong intuitive connection between program obfuscation and function-private predicate-encryption in the public-key setting (this has also been observed in prior work [AAB+15,ITZ16,ABF16]). In both settings, the goal is to allow evaluation of a specific functionality without leaking anything else about the functionality itself. The difference is that an obfuscated program runs on an arbitrary public input, while in function-private predicate encryption, function evaluation occurs when applying a decryption key for some predicate f to a ciphertext whose hidden attribute is the input to the function.

Moreover, we can imagine defining function privacy so that a decryption key sk_f for some function f leaks no more than a VBB obfuscation of f. In this case, public-key function-private predicate encryption for some function class $\mathcal{F}$ is a strictly stronger primitive than program obfuscation for $\mathcal{F}$. This follows trivially from the fact that anyone holding a decryption key sk_f for $f \in \mathcal{F}$ can use it as an obfuscated program: to learn whether $f(x)$ outputs 0 or 1, use the public key to encrypt a message payload under attribute x and check if decryption succeeds.

In this work, we leverage this intuitive connection to build public-key function-private predication encryption schemes by transforming simple obfuscators [Can97,BKM+18,BLMZ19,BW19] that have appeared in the literature. Our core construction will be based on an obfuscator for the "small superset" functionality, which is essentially equivalent to the "big subset" functionality introduced by Beullens and Wee [BW19]. We define our "small superset" function $f_{n,t,X}$, parameterized by a target set $X \subseteq [n]$, a positive integer n, and an integer size bound $t \leq n$, to output 1 on input $Y \subseteq [n]$ if $X \subseteq Y$ and $|Y| \leq t$, and 0 otherwise.[6]

A simple group-based obfuscator for the "small superset" functionality follows easily from prior work [BLMZ19,BW19] (which are inspired by the construction of Bishop et al. [BKM+18]). The obfuscation achieves an average-case notion of security (i.e., VBB holds if the set X is drawn from a distribution with appropriate entropy). From this, we build a public-key (enhanced) function-private predicate encryption scheme supporting the class of small superset predicates.

In the remainder of this technical overview, we describe a slightly simplified version of our construction to highlight the main ideas. Instead of starting with the "small superset" functionality, we use the simpler obfuscator of Canetti [Can97] for point functions. This yields a public-key function-private predicate encryption scheme for the equality predicate, or equivalently public-key anonymous IBE. We then provide an extensive discussion on our new definitions of function privacy and enhanced function privacy. Finally, we demonstrate how

[6] The "big subset" function of Beullens and Wee [BW19] is also parameterized by the same n, t, X, but it outputs 1 if and only if $Y \subseteq X$ and $|Y| \geq t$. The functionalities are seen to be equivalent by associating each input set Y with its complement $[n] \setminus Y$.

our predicate encryption for "small supersets" naturally captures hidden vector encryption.

Remark on Presentation. After the technical overview, we will not return to the construction of public-key anonymous IBE based on Canetti's obfuscator [Can97]; the construction described in this technical overview follows trivially from our full-fledged "small superset" obfuscator in Sect. 5. Details and definitions for our extensions to d-CNFs and read-once conjunctions of d-disjunctions (for constant d) can be found in the full version; we note that these constructions follow from a straightforward generalization of our main techniques.

Function-Private Anonymous IBE from Point Obfuscation. We start with Canetti's point function obfuscator [Can97]. Recall that a point function I_x is a boolean-valued function that outputs 1 on input x, and 0 elsewhere. Fix a cryptographic group with order p and generator g. Given x, we obfuscate I_x by drawing a uniformly random $r \leftarrow \mathbb{F}_p$ and outputting

$$\mathsf{Obf}(I_x) = (g^r, g^{rx}).$$

Anyone can evaluate I_x on arbitrary input y by computing $(g^r)^y$ and comparing with g^{rx}. Moreover, Canetti proves that if x is drawn from any distribution with super-logarithmic min-entropy, the above construction hides x under a strengthening of the Decisional Diffie-Hellman (DDH) assumption [Can97].

Handling Encrypted Inputs: A First Attempt. A natural idea to upgrade Canetti's obfuscator to work for encrypted inputs y is to use a bilinear map, and to "obfuscate" the input y in a similar manner. Consider groups $\mathbb{G}_1, \mathbb{G}_2, \mathbb{G}_T$ with associated generators g_1, g_2, g_T equipped with a bilinear map $e\colon \mathbb{G}_1 \times \mathbb{G}_2 \to \mathbb{G}_T$. To generate the public key, we draw a uniformly random $r \leftarrow \mathbb{F}_p$ and give out $g_2^{r^{-1}}$. We treat r as a secret key which is given to the obfuscator. To encrypt a plaintext y, the user computes $(g_2^{r^{-1}})^{y^{-1}}$. A function decryption key for I_x is simply $\mathsf{Obf}(I_x) = g_1^{rx}$.

A user holding an encryption $g_2^{r^{-1}y^{-1}}$ of y and a function decryption key g_1^{rx} for I_x can easily verify whether $I_x(y) = 1$ (i.e. $y = x$) by using the bilinear map and checking whether

$$e(g_1^{rx}, g_2^{r^{-1}y^{-1}}) \stackrel{?}{=} g_T.$$

However, this simple method of "encrypting" y fails to achieve even semantic security for ciphertexts since the encryption algorithm is deterministic. That is, an attacker trying to distinguish between an encryption of y_0 and an encryption of y_1 can easily encrypt both and compare to the challenge ciphertext. A natural approach to randomizing the encoding procedure would be to encode y as $g_2^{r^{-1}y^{-1}\alpha}$ for a random $\alpha \leftarrow \mathbb{F}_p$. However, for evaluation to work, the ciphertext would have to include g_T^α, and essentially the same limitation would arise since the attacker can request decryption keys of their choice.[7]

Handling Encrypted Inputs Securely. Our goal now is to modify the scheme so that we can introduce randomness into the encryption procedure without disturbing correctness. The idea is to generalize the above procedure to first encode x as a 2-dimensional vector $[x\ x^2]$, and to replace the role of y^{-1} with a uniformly random vector orthogonal to $[y\ y^2]$. Now if we compute the dot product of these vectors, we get 0 if $y = x$ and a non-zero value otherwise. Note that a random vector orthogonal to $[y\ y^2]$ can be written as $[-\beta y\ \beta]^\top$ where $\beta \leftarrow \mathbb{F}_p$ is uniformly random. The role of the random scalar r in the previous scheme can be replaced by a uniformly random invertible 2×2 matrix $\mathbf{R} \leftarrow \mathbb{F}_p^{2\times 2}$. We are also free to introduce independent randomness α during obfuscation/secret key generation. The resulting scheme is as follows.

- Setup. Draw random invertible $\mathbf{R} \leftarrow \mathbb{F}_p^{2\times 2}$, and output $\mathsf{pk} = g_2^{\mathbf{R}^{-1}}$, $\mathsf{sk} = \mathbf{R}$.[8]
- $\mathsf{KeyGen}(\mathsf{sk}, x)$. Parse sk as $\mathbf{R}$. Draw random $\alpha \leftarrow \mathbb{F}_p$ and output $\mathsf{sk}_x = g_1^{[\alpha x\ \alpha x^2]\mathbf{R}}$.
- $\mathsf{Enc}(\mathsf{pk}, y)$. Parse pk as $g_2^{\mathbf{R}^{-1}}$. Draw random $\beta \leftarrow \mathbb{F}_p$ and output $g_2^{\mathbf{R}^{-1}[-\beta y\ \beta]^\top}$.
- $\mathsf{Dec}(\mathsf{sk}_x, c)$. Parse sk_x as $g_1^{[v_1\ v_2]}$ and c as $g_2^{[u_1\ u_2]^\top}$. Use the bilinear map e to compute $g_T^{[v_1\ v_2]\cdot\begin{bmatrix}u_1\\ u_2\end{bmatrix}}$ and output 1 if this equals g_T^0.

Adding Payloads for Function-Private Anonymous IBE. At the moment, the above scheme corresponds to an IBE scheme without message payloads; if we interpret x and y as user identities, currently a user only learns whether or not they were the correct recipient of a ciphertext. To obtain full IBE, we need to modify the encryption algorithm to incorporate a message payload μ. To enable this, we extend $\mathbf{R}$ to a 3×3 matrix, and extend the obfuscated row vector to $[1\ x\ x^2]$. During encryption we choose more randomness γ, extend the encrypted column vector to $[\gamma\ k_1\ k_2]^\top$, and additionally release $\mu \cdot g_T^\gamma$. An accepting input will now decrypt to g_T^γ rather than the identity, which can be divided out from $\mu \cdot g_T^\gamma$ to recover μ.

---

[7] In more detail, an attacker trying to distinguish between an encryption of y_0 and an encryption of y_1 (for y_0, y_1 of their choice) is free to request decryption keys corresponding to any function I_x provided that I_x does not trivially allow the attacker to distinguish between y_0 and y_1. The attacker can therefore request g_1^{rx} for any x that does not equal y_0 or y_1. Given challenge $g_2^{\alpha r^{-1} y_b^{-1}}$, g_T^α and decryption key g_1^{rx}, the attacker can use the fact that they know x, y_0, y_1 in the clear to determine b as follows. The attacker raise g_T^α to the exponent xy_0^{-1} to obtain $g_T^{\alpha x y_0^{-1}}$, and then computes $e(g_1^{rx}, g_2^{\alpha r^{-1} y_b^{-1}})$. If $b = 0$, these quantities match, and otherwise they do not.

[8] We use the shorthand $g^{\mathbf{V}}$ where $\mathbf{V} = (v_{i,j})_{i\in[k],j\in[\ell]}$ to denote the matrix of group elements $(g^{v_{i,j}})_{i\in[k],j\in[\ell]}$.

On Function-Private Identity Based Encryption. A construction of function-private anonymous IBE appears in Boneh, Raghunathan, and Segev [BRS13a]. Their approach starts with an existing (anonymous) IBE scheme and "upgrades" it to statistically hide the function using a randomness extractor. As outlined earlier, our approach and construction differ in several important dimensions. First, our approach starts with an existing point obfuscation scheme, "upgrades" it to encrypt the inputs, and then subsequently introduces the ability to encrypt a message payload. Second, our construction achieves computational function privacy for any distribution with super-logarithmic min-entropy, rather than λ min-entropy as required in [BRS13a] (this requirement was also relaxed in [PMR19]). However, the drawback of our approach is that we can only prove security in the generic (bilinear) group model [Nec94, Sho97, Mau05], whereas [BRS13a] is proven secure in the standard model.

Building Public-Key Function-Private Predicate Encryption for "Small Supersets". We now briefly describe how to extend the above function-private anonymous IBE to handle the significantly more expressive "small superset" functionality, described earlier. First, we describe how to generate a function decryption key for $f_{n,t,X}$, where $X \subseteq [n]$. Now, $\mathbf{R}$ is a uniformly random width $t+1$ matrix (instead of width 2). We now follow essentially the same procedure as before for each $x \in X$. That is, for each $x \in X$ we form the row vector $[x\ x^2\ \dots\ x^{t+1}]$ and compute the row vector $[x\ x^2\ \dots\ x^{t+1}] \cdot \mathbf{R}$. We collect the row vectors resulting from this process into a matrix $\mathbf{M}_X$ where the rows are indexed by elements $x \in X$.

A set $Y \subseteq [n]$ (corresponding to a set that will be given as input to the "small superset" functionality) can be encrypted as follows. We assemble a matrix $\mathbf{W}_Y$ whose rows are indexed by elements $y \in Y$. The row corresponding to y is simply $[y\ y^2\ \dots\ y^{t+1}]$. Draw a uniformly random vector $\mathbf{v}$ in the right kernel of $\mathbf{W}_Y$, and output $\mathbf{v}_Y = \mathbf{R}^{-1} \cdot \mathbf{v}$. Note that this is only possible if $|Y| \leq t$.

To decrypt, compute the matrix-vector product $\mathbf{M}_X \cdot \mathbf{v}_Y$ in the exponent, which will be the all-zeros vector if and only if $X \subseteq Y$. To minimize the size of the obfuscation, we can collapse the matrix $\mathbf{M}_X$ to a vector $\mathbf{u}_X^\top$, by left multiplying by a uniformly random vector of the appropriate dimension. Then decryption simply computes the (dot) product $\mathbf{u}_X^\top \cdot \mathbf{v}_Y$ in the exponent. In the body, we describe these obfuscation and encryption procedures in the language of linear codes, which results in a cleaner presentation.

Note that our construction is efficient as long as t is polynomial in the security parameter, since the vector and matrix dimensions are all determined by t. In particular, the universe size n could be exponential. On the other hand, the obfuscation construction given by Beullens and Wee [BW19] for large subset is only efficient for polynomial sized universe.

Finally, we remark that it is also easy to extend this to function-private predicate encryption for small superset by adding a payload in the same manner as for identity based encryption.

Function-Private Hidden Vector Encryption. We now describe how function-private predicate encryption for small superset gives rise to function-private hidden vector encryption [BW07]. Consider a vector $\mathbf{v} = (v_i)_{i\in[k]} \in (\mathbb{Z}_s \cup \{*\})^k$. Hidden vector encryption corresponds to predicate encryption for the predicate

$$P_{\mathbf{v}}(\mathbf{u}) = \begin{cases} 1 & \text{if for all } i \in [k] : (v_i = u_i \text{ or } v_i = *), \\ 0 & \text{otherwise.} \end{cases}$$

Let the universe size of the small superset instance be $n = ks$ and the threshold value be k. Let the set X corresponding to $\mathbf{v}$ be defined as $X := \{(i-1)s+v_i\}_{i\in B}$, where B denotes the non-wildcard positions of $\mathbf{v}$. Then an input vector $\mathbf{u} = (u_i)_{i\in[k]} \in \mathbb{Z}_s^k$ corresponds to the set $Y_{\mathbf{u}} := \{(i-1)s + u_i\}_{i\in[k]}$, which has size exactly k. Finally, we have $P_{\mathbf{v}}(\mathbf{u}) = 1 \iff X \subseteq Y_{\mathbf{u}}$. Since hidden vector encryption is most generally defined over exponentially sized alphabets, we would like to take s and thus n to be exponential. Thus, we crucially rely on the fact that the universe size of our small superset instance is allowed to be exponential.

Function-Privacy Definitions. When considering public-key function-private predicate encryption, the appropriate notion of function privacy is somewhat tricky to define. We choose to generalize the original notion of "real-or-random" function privacy of Boneh et al. [BRS13a]. This definition was originally stated just for point functions and was later extended to inner products [BRS13b]. Roughly, the definition considers an oracle which is set to be in either "real" or "random" mode, and which accepts a distribution over points. If it is in real mode, it produces a key for a point drawn from the queried distribution, and if it is in random mode, it produces a key for a uniformly random point. Security is parameterized by a class of allowed distributions for which the adversary can query its oracle, and requires that an adversary cannot determine which mode its oracle is in.

Extending this definition to a larger class $\mathcal{C}$ of functions would require a natural notion of a uniformly random function from $\mathcal{C}$. We choose to instead view the random mode as a "simulated" mode, where the behavior of the oracle is independent of the queried distribution, but otherwise arbitrary. This definition now naturally extends to any class of functions, and captures the same intuition that an adversary learns nothing about the function that it has a key for, as long as it is drawn from a particular class of distributions. We refer to this oracle now as the Real-or-Sim oracle.

We note here that although our predicate encryption constructions are inspired by and built from existing obfuscation constructions (in particular, those that already satisfy distributional virtual black box security), this notion of function privacy is incomparable to distributional VBB. In particular, distributional VBB is defined relative to a distribution $\mathcal{D}$ over functions in $\mathcal{C}$, and essentially requires that no adversary, given the obfuscation of a function f drawn from $\mathcal{D}$, can guess the value of any predicate $\mathcal{P}$ applied to f. On the other hand, our definition of function privacy is defined relative to an entire class of distributions $\mathbb{D}$, and does not consider predicates on functions drawn from individual distributions $\mathcal{D}$. Instead, we require that the class of distributions $\mathbb{D}$ is simulatable in the sense described above. Note that our constructions also satisfy distributional VBB, but we focus on this function-private predicate encryption style of definition, as it aligns more closely with previous work.

Enhanced Function Privacy. As in prior work [BRS13a], we will be concerned with evasive distributions over functions, where it is difficult to find an accepting input given oracle access to a function drawn from the distribution. However, it is crucial for applications that given a decryption key for an unknown function, the key can be used to successfully decrypt payloads without sacrificing function privacy. This means in particular that an adversary should not be able to produce accepting inputs to its decryption key, *even given* encryptions of arbitrary accepting inputs.

This is captured by Boneh et al. [BRS13a] by the notion of *enhanced* function privacy, where in the real-or-random game, the adversary is additionally given an encryption oracle. The adversary can query this oracle to obtain encryptions of arbitrary accepting inputs to the unknown functions corresponding to the decryption keys in its possession. Enhanced function privacy requires that the adversary still cannot determine what mode its Real-or-Sim oracle is in. We prove that our predicate encryption scheme for small superset satisfies this enhanced function privacy notion, which implies that our hidden vector encryption construction does as well.

Secure Distributions for Function-Private HVE (and More). We determine which distributions over HVE instances induce an evasive distribution over small superset instances, under the mapping defined above. We parameterize HVE distributions by an alphabet size s, and an input length k. For a particular distribution $\mathcal{D}_{k,s}$, let $H_\infty(\mathcal{D}_{k,s})$ be the min-entropy of the vector $\mathbf{v} \leftarrow \mathcal{D}_{k,s}$. Following the proof strategy from [BW19, Lemma 2], we show enhanced function-private hidden vector encryption for the set of distributions containing any $\mathcal{D}_{k,s}$ such that $H_\infty(\mathcal{D}_{k,s}) \geq k + \omega(\log k)$.

Note that the min-entropy requirement scales with the input length, but not with the alphabet size. Thus as the alphabet size increases, we obtain security for a larger and larger class of distributions. If we instead had a polynomial limit on the universe size of our small superset instances (like in [BW19]), then to support exponentially large alphabets $\mathbb{Z}_s$, we would be forced to first write each element as a bitstring (or more generally a string over a polynomially sized alphabet), increasing the input length. This would cause the min-entropy requirement to scale with the size of the alphabet.

On the other hand, this result severely restricts the possible distributions when s is a small constant. Thus we give an additional set of secure distributions (that also appear in [BW19] in the context of conjunction obfuscation) over vectors with a fixed number of wildcards w. We obtain enhanced function-private hidden vector encryption for the set of distributions containing any

$$\mathcal{D}_{k,s} \text{ such that } H_\infty(\mathcal{D}_{k,s}) = \log \binom{k}{w} + \omega(\log(k)),$$

and where $\mathcal{D}_{k,s}$ is supported on vectors with exactly w wildcards. Note that for some values of w, this min-entropy bound is much less than the input length k, and thus supports a large and interesting class of distributions even for small alphabet size s.

Extentions to d-CNF and Read-once Conjunction of d-disjunctions. We also extend the enhanced-function-private predicate encryption of "small supersets" to d-CNF and conjunction of d-disjunctions for $d = O(1)$.
d-CNFs for $d = O(1)$. The underlying technique in the BKMPRS construction is to translate the evaluation of the conjunction functionality into a polynomial interpolation, which is successful if and only if all input values (one per comparison clause) are valid points on the underling polynomial. This is achieved by evaluating the comparison functionality as a *lookup table* which contains either valid shares for matching input, or random values, otherwise (all encoded in the exponent for security). Our observation is that we can use a similar lookup table approach to implement *any circuit functionality* besides comparisons, and this technique is polynomially efficient as long as the underlying circuits have constant input length.

A d-CNF for k-bit input is a circuit $C = C_1 \wedge C_2 \wedge \cdots \wedge C_m$ where for each $i \in [m]$, C_i is a boolean circuit which depends only on the inputs bits with indices in a subset, denoted as $I_i \subseteq [k]$. We now show how to reduce the d-CNF to the "small superset" functionality.

Given a d-CNF $C = C_1 \wedge C_2 \wedge \cdots \wedge C_m$, denote $K = \binom{k}{d}$ and $D = 2^d$. We create a universe of $n = KD$ elements. Then we reform the set $[KD]$ into a $K \times D$ matrix. The rows of the matrix corresponds to subsets of size d in $[k]$. The columns of the matrix corresponds to the input strings of k-bits. Now we specify a subset X of $[KD]$. For $I \in \binom{[k]}{d}$ and $v \in \{0,1\}^k$, X contains the elements in I-th row and v-th column, if there exists a $C_i, i \in [m]$ such that C_i only depends on I and $C_i(v) = 0$. On input $x \in \{0,1\}^k$, we specify a subset $Y \in [KD]$. For every $I \in \binom{[n]}{d}$, Y contains all elements in I-th row, except the one in x_I-th column. Since Y contains $K(D-1)$ elements, we simply set the threshold $t = K(D-1)$. Then, $C(x) = 1$ if and only if $X \subseteq Y$. This is because, $C(x) = 1$ if and only if the following condition holds: for every $I \in \binom{[k]}{d}$, either there exists a C_i such that C_i only depends on I and $C_i(x_I) = 1$, or such C_i doesn't appear in C. The above condition is equivalent to $X \subseteq Y$. We prove security for some special distributions over d-CNF.

Function Distribution. We prove the security for two distributions. The first distribution essentially corresponds to the "uniform" case. Here, we achieve the *best possible* parameter, namely, $m = \omega(\log k)$.[9] Our proof in this case is a natural extension of the BKMPRS proof. The second distribution is useful for obtaining obfuscation of conjunctions of d-disjunctions via the mapping discussed earlier. Crucially, in this distribution, we do *not* require the distribution over C_i to be independent. Consequently, the proof of security for this distribution is more involved. Specifically, since C_i's may be dependent, in order to use a combinatorial argument similar to BKMPRS, we first need to "break" the dependence.

[9] Indeed, m must be $\omega(\log k)$ in order to make the function family evasive.

We address this by choosing a subset of sets, say $\mathcal{I}$, such that the sets in $\mathcal{I}$ are disjoint. Clearly, $\mathcal{I}$ has the necessary independence. To choose such a subset, we build a graph, where each vertex of the graph represents a set, and draw an edge between the two vertices if and only if the intersection of two vertex is non-empty. We then bound the degree of this graph and argue that the number of color used for coloring the graph is also bounded. Finally, we use the pigeonhole principle to pick such a subset $\mathcal{I}$. Due to lack of space, we refer the reader to the full version for details.

Read Once Conjunctions of d-Disjunctions. We also consider a class of functionalities that directly generalizes the conjunctions functionality but in a different way from d-CNFs. While the conjunctions functionality constrains the value of each input bit independently, in our generalization we constrain the values of several consecutive input bits together. More precisely, our functionality is defined as

$$\left(p_1^{(1)} \vee \cdots \vee p_d^{(1)}\right) \wedge \cdots \wedge \left(p_1^{(\ell)} \vee \cdots \vee p_d^{(\ell)}\right),$$

where $p_j^{(i)}$ is a length k_i string over alphabet $\{0, 1, ?\}$, and $\sum_i k_i = k$. It evaluates to one on input string $x = x^{(1)} \| \cdots \| x^{(\ell)} \in \{0,1\}^n$ if and only if for every $i \in [\ell]$, it holds that $|x^{(i)}| = k_i$ and $x^{(i)}$ matches one of $\{p_j^{(i)}\}_{j\in[d]}$.

One direct way to achieve the above functionality using the d-CNF construction is by considering each $\left(p_1^{(i)} \vee \cdots \vee p_d^{(i)}\right)$ as the functionality of the clause C_i. However, this will impose a restriction that each $k_i = O(1)$. Instead, We provide a different mapping to the d-CNF class with the only restriction that $\sum_i k_i = k$ when $k = O(1)$. This mapping transforms the conjunction of disjunction over strings into a conjunctions of disjunctions over bits by representing the matching $y \stackrel{?}{=} x$ of a longer string $x = x_1 \ldots x_t$ as the conjunction over bit comparisons $y_1 \stackrel{?}{=} x_1 \wedge \cdots \wedge y_t \stackrel{?}{=} x_t$.

1.3 Outline

The rest of the paper is structure as follows. In Sect. 2 we define notation and provide background definitions. In Sect. 3 we present our construction for obfuscating small supersets. In Sect. 4 we formally define our security notions of data privacy and (enhanced) function privacy, and in Sect. 5 we present our construction for function-private predicate encryption for small supersets. In the full version we present applications of our construction to hidden vector encryption, d-CNFs for $d = O(1)$, and read-once conjunctions of d-disjunctions for $d = O(1)$.

2 Preliminaries

We use the standard Landau notations. A function $\epsilon(\lambda)$ is written as $\mathsf{negl}(\lambda)$ if for all positive integers c, $\epsilon(\lambda) = o(\frac{1}{\lambda^c})$. For a positive integer n, we let $[n]$ denote the set $\{1, 2, \ldots, n\}$. For a finite set S, $x \leftarrow S$ denotes a uniformly random sample.

We write scalars as lowercase unbolded letters (e.g., α or a), vectors as lowercase bold letters (e.g., $\mathbf{v}$) and matrices as uppercase bold letters (e.g. $\mathbf{M}$). We use the shorthand $g^{\mathbf{v}}$ where $\mathbf{v} = (v_1, \ldots, v_n)$ to denote the vector of group elements $g^{v_1}, \ldots, g^{v_n}$, and naturally extend this notation to matrices $\mathbf{V}$. To distinguish between the case where x refers to a specific value and the case where x is used as a formal variable, we will explicitly write $\hat{x}$ if it is a formal variable. This notation will also extend to vectors $\hat{\mathbf{v}} = (\hat{v}_1, \ldots, \hat{v}_n)$ where each entry is itself a formal variable, as well as to matrices $\hat{\mathbf{M}}$ where each entry is a formal variable.

2.1 Bilinear Groups

We briefly recall the definition of an asymmetric bilinear group [Jou04,BF01]. Let $\mathbb{G}_1, \mathbb{G}_2, \mathbb{G}_T$ be distinct groups, all of prime order q, and let $e \colon \mathbb{G}_1 \times \mathbb{G}_2 \to \mathbb{G}_T$ be a mapping from $\mathbb{G}_1 \times \mathbb{G}_2$ onto the target group $\mathbb{G}_T$. Let g_1, g_2 be generators for $\mathbb{G}_1$ and $\mathbb{G}_2$, respectively. We say that $(\mathbb{G}_1, \mathbb{G}_2, \mathbb{G}_T, e)$ is an asymmetric bilinear group if the following conditions are met:

- (Efficiency) The group operations in $\mathbb{G}_1, \mathbb{G}_2, \mathbb{G}_T$ as well as the mapping $e(\cdot, \cdot)$ are all efficiently computable.
- (Non-degeneracy) $e(g_1, g_2) = g_T$, where g_T is a generator of $\mathbb{G}_T$.
- (Bilinearity) $e(g_1^a, g_2^b) = g_T^{ab}$ for all $a, b \in \mathbb{Z}_q$.

2.2 Generic Bilinear Group Model

We use an extension of the generic group model [Nec94,Sho97] adapted to bilinear groups. The following definition is taken verbatim from [KLM+18].

Definition 1 (Generic Bilinear Group Oracle). *A generic bilinear group oracle is a stateful oracle* BG *that responds to queries as follows:*

- *On a query* $\mathsf{BG.Setup}(1^\lambda)$*, the oracle generates two fresh nonces* $\mathsf{pp}, \mathsf{sp} \leftarrow \{0,1\}^\lambda$ *and a prime* p*. It outputs* $(\mathsf{pp}, \mathsf{sp}, p)$*. It stores the generated values, initializes an empty table* $T \leftarrow \{\}$*, and sets the internal state so subsequent invocations of* BG.Setup *fail.*
- *On a query* $\mathsf{BG.Encode}(k, x, i)$ *where* $k \in \{0,1\}^\lambda, x \in \mathbb{Z}_p$ *and* $i \in \{1, 2, T\}$*, the oracle checks that* $k = \mathsf{sp}$ *(returning* $\perp$ *otherwise). The oracle then generates a fresh nonce* $h \leftarrow \{0,1\}^\lambda$*, adds the entry* $h \mapsto (x, i)$ *to the table* T*, and outputs* h*.*
- *On a query* $\mathsf{BG.Add}(k, h_1, h_2)$ *where* $k, h_1, h_2 \in \{0,1\}^\lambda$*, the oracle checks that (1)* $k = \mathsf{pp}$*, and (2) the handles* h_1, h_2 *are present in its internal table* T *and are mapped to the values* (x_1, i_1) *and* (x_2, i_2)*, respectively, with* $i_1 = i_2$ *(returning* $\perp$ *otherwise). The oracle then generates a fresh handle* $h \leftarrow \{0,1\}^\lambda$*, computes* $x = x_1 + x_2 \in \mathbb{Z}_p$*, adds the entry* $h \mapsto (x, i_1)$ *to* T*, and outputs* h*.*
- *On a query* $\mathsf{BG.Pair}(k, h_1, h_2)$ *where* $k, h_1, h_2 \in \{0,1\}^\lambda$*, the oracle checks that (1)* $k = \mathsf{pp}$*, and (2) the handles* h_1, h_2 *are present in* T *and are mapped to values* $(x_1, 1)$ *and* $(x_2, 2)$*, respectively (returning* $\perp$ *otherwise). The oracle then generates a fresh handle* $h \leftarrow \{0,1\}^\lambda$*, computes* $x = x_1 x_2 \in \mathbb{Z}_p$*, adds the entry* $h \mapsto (x, T)$ *to* T*, and outputs* h*.*

- *On a query* $\mathsf{BG.ZeroTest}(k, x)$ *where* $k, x \in \{0,1\}^\lambda$, *the oracle checks that (1)* $k = \mathsf{pp}$, *and (2) the handle* h *is present in* T *and it maps to some value* (x, i) *(returning* $\perp$ *otherwise). The oracle then outputs "zero" if* $x = 0 \in \mathbb{Z}_p$ *and "non-zero" otherwise.*

2.3 Virtual Black Box Obfuscation

We recall the definition of a distributional virtual black-box (VBB) obfuscator. We roughly follow the definition of Brakerski and Rothblum [BR13].

Definition 2 (Distributional VBB Obfuscation). *Let* $\mathcal{C} = \{\mathcal{C}_n\}_{n\in\mathbb{N}}$ *be a family of polynomial-size circuits, where* $\mathcal{C}_n$ *is a set of boolean circuits operating on inputs of length* n, *and let* Obf *be a PPT algorithm which takes as input an input length* $n \in \mathbb{N}$ *and a circuit* $C \in \mathcal{C}_n$ *and outputs a boolean circuit* $\mathsf{Obf}(C)$ *(not necessarily in* $\mathcal{C}$*). Let* $\mathcal{D} = \{\mathcal{D}_n\}_{n\in\mathbb{N}}$ *be an ensemble of distribution families* $\mathcal{D}_n$ *where each* $D \in \mathcal{D}_n$ *is a distribution over* $\mathcal{C}_n$.

Obf *is a distributional VBB obfuscator for the distribution class* $\mathcal{D}$ *over the circuit family* $\mathcal{C}$ *if it has the following properties:*

1. *(Strong) Functionality Preservation: For every* $n \in \mathbb{N}$, $C \in \mathcal{C}_n$, *there exists a negligible function* μ *such that*
$$\Pr[\mathsf{Obf}(C, 1^n)(x) = C(x)\ \forall x \in \{0,1\}^n] = 1 - \mu(n)\,.$$
2. *Polynomial Slowdown: For every* $n \in \mathbb{N}$ *and* $C \in \mathcal{C}_n$, *the evaluation of* $\mathsf{Obf}(C, 1^n)$ *can be performed in time poly*$(|C|, n)$.
3. *Distributional Virtual Black-Box: For every PPT adversary* $\mathcal{A}$, *there exists a (non-uniform) polynomial size simulator* $\mathcal{S}$ *such that for every* $n \in \mathbb{N}$, *every distribution* $D \in \mathcal{D}_n$ *(a distribution over* $\mathcal{C}_n$*), and every predicate* $\mathcal{P}\colon \mathcal{C}_n \to \{0,1\}$, *there exists a negligible function* μ *such that*
$$\left|\Pr_{C\leftarrow\mathcal{D}_n}[\mathcal{A}(\mathsf{Obf}(C,1^n)) = \mathcal{P}(C)] - \Pr_{C\leftarrow\mathcal{D}_n}[\mathcal{S}^C(1^{|C|}, 1^n) = \mathcal{P}(C)]\right| = \mu(n)\,.$$

2.4 Predicate Encryption

Let $\mathcal{F} = \{\mathcal{F}_\lambda\}_\lambda$ be a function class, where $\mathcal{F}_\lambda = \{f : \mathcal{X}_\lambda \to \{0,1\}\}$. Let $\mathcal{M} = \{\mathcal{M}_\lambda\}_\lambda$ be a message space.

Definition 3 (Public-key Predicate Encryption). *A* public-key predicate encryption *scheme* $\Pi = (\mathsf{Setup}, \mathsf{KeyGen}, \mathsf{Enc}, \mathsf{Dec})$ *for a function class* $\mathcal{F}$ *and message space* $\mathcal{M}$ *is a tuple of* PPT *algorithms defined as follows:*

- $\mathsf{Setup}(1^\lambda)$: *On input security parameter* $\lambda \in \mathbb{N}$ *provided in unary, output master secret key* msk *and public key* pk.
- $\mathsf{KeyGen}(\mathsf{msk}, f)$: *On input master secret key* msk *and function* $f \in \mathcal{F}_\lambda$, *output decryption key* sk_f.
- $\mathsf{Enc}(\mathsf{pk}, x, \mu)$: *On input public key* pk *an attribute* $x \in \mathcal{X}_\lambda$, *and a payload* $\mu \in \mathcal{M}_\lambda$, *output ciphertext* ct.

- $\mathsf{Dec}(\mathsf{sk}_f, \mathsf{ct})$: *On input decryption key* sk_f *for function* $f \in \mathcal{F}_\lambda$ *and ciphertext* ct, *output an element of* $\mathcal{M}_\lambda \cup \{\bot\}$.

A public-key predicate encryption scheme Π for function class $\mathcal{F} = \{\mathcal{F}_\lambda\}_\lambda$ and message space $\mathcal{M} = \{\mathcal{M}_\lambda\}_\lambda$ is correct if for all $\lambda \in \mathbb{N}$, $f \in \mathcal{F}_\lambda, x \in \mathcal{X}_\lambda$, and $\mu \in \mathcal{M}_\lambda$, it holds that:

$$\Pr\left[\begin{array}{l}(\mathsf{msk}, \mathsf{pk}) \leftarrow \mathsf{Setup}(1^\lambda)\\ \mathsf{sk}_f \leftarrow \mathsf{KeyGen}(\mathsf{msk}, f)\\ \mathsf{ct} \leftarrow \mathsf{Enc}(\mathsf{pk}, x, \mu)\end{array} \;\middle|\; \mathsf{Dec}(\mathsf{sk}_f, \mathsf{ct}) = \begin{cases}\mu & \text{if } f(x) = 1\\ \bot & \text{if } f(x) = 0\end{cases}\right] = 1 - \nu(\lambda),$$

where the probability is taken over the internal randomness of the algorithms and $\nu(\cdot)$ is a negligible function.

We defer the security notions for predicate encryption to Sect. 4.

3 Obfuscating Small Supersets

We define the "small superset" functionality. As mentioned earlier, this is an alternative but virtually identical view of the "big subset" functionality proposed by Beullens and Wee [BW19]. However, we find the "small superset" formulation to be significantly more intuitive for our applications. The small superset functionality $f_{n,t,X}$ is parameterized by a universe size n, a threshold value t, and a set $X \subseteq [n]$. $f_{n,t,X}$ takes as input a set $Y \subseteq [n]$, and accepts if $|Y| \leq t$ and $X \subseteq Y$. While Beullens and Wee [BW19] limit n to be polynomial-size, we integrate the approach of Bartusek, Lepoint, Ma, and Zhandry [BLMZ19] for large-alphabet conjunctions to handle exponential size n, provided $t = \mathsf{poly}(\lambda)$.

Definition 4. *Let* $X \subseteq \mathbb{F}_q$ *consist of elements* $x_1, \ldots, x_k$. *Let* $\mathbf{B}_{t,X,q} \in \mathbb{F}_q^{k\times(t+1)}$ *be defined as*

$$\mathbf{B}_{t,X,q} := \begin{pmatrix} x_1 & x_1^2 & \cdots & x_1^{t+1}\\ x_2 & x_2^2 & \cdots & x_2^{t+1}\\ \vdots & \vdots & \ddots & \vdots\\ x_k & x_k^2 & \cdots & x_k^{t+1}\end{pmatrix}.$$

We also define the following helper functionalities.

- $\mathsf{SampCodeword}(\mathbf{B} \in \mathbb{F}_q^{k\times(t+1)})$. Output a random codeword in the code generated by $\mathbf{B}$ by sampling uniformly random $\mathbf{e} \in \mathbb{F}_q^k$ and outputting $\mathbf{e}^\top \cdot \mathbf{B}$.
- $\mathsf{SampDualCodeword}(\mathbf{B} \in \mathbb{F}_q^{k\times(t+1)})$. Output a uniformly random vector $\mathbf{w} \in \mathbb{F}_q^{t+1}$ in the right kernel of $\mathbf{B}$, i.e., $\mathbf{w}$ such that $\mathbf{B} \cdot \mathbf{w} = 0$.

3.1 Small Superset Obfuscation Construction

In this section, we define a small superset obfuscator using the above helper functionalities. The following construction is similar to the generic group constructions in [BW19,BLMZ19] (which build on [BKM+18]), though the presentation is tailored to fit the scope of this work. We assume that global parameters (λ, n, t, q) are set in advance, where λ is the security parameter, $n = 2^{\mathsf{poly}(\lambda)}$ is the universe size, $t = \mathsf{poly}(\lambda)$ is the threshold size, and q is a prime larger than n (for strong functionality preservation, we will require that $q \geq 2^\lambda \binom{n'}{t} = 2^{\mathsf{poly}(\lambda)}$, where $n' = \max\{n, 2t\}$). Let $\mathbb{G}$ be a group of order q with generator g.

- $\mathsf{Obf}((n, t, q), X \subseteq [n])$. $\mathbf{c}^\top \leftarrow \mathsf{SampCodeword}(\mathbf{B}_{t,X,q})$. Output $g^{\mathbf{c}^\top}$ (interpreted as $g^{c_1}, \ldots, g^{c_{t+1}}$).
- $\mathsf{Eval}((n, t, q), g^{\mathbf{c}^\top} \in \mathbb{G}^{t+1}, Y \subseteq [n])$. Let $\mathbf{w} \leftarrow \mathsf{SampDualCodeword}(\mathbf{B}_{t,Y,q})$. Accept if and only if $g^{\mathbf{c}^\top \cdot \mathbf{w}} = g^0$.

3.2 Functionality Preservation

We rely on the following fact (also stated in [BLMZ19]).

Lemma 1. *For any $t+1$ values of $x_1, \ldots, x_{t+1} < q$, the corresponding set of $t+1$ vectors $\{(x_i\ x_i^2 \cdots x_i^{t+1})\}_{i \in [t+1]}$ are linearly independent over $\mathbb{F}_q$.*

Functionality preservation now follows almost immediately from the following.

Lemma 2. *Let $X, Y \subseteq \mathbb{Z}_q$ be such that $|X|, |Y| \leq t$ and $X \not\subseteq Y$. Let $\mathbf{c}^\top \leftarrow \mathsf{SampCodeword}(\mathbf{B}_{t,X,q})$ and $\mathbf{w} \leftarrow \mathsf{SampDualCodeword}(\mathbf{B}_{t,Y,q})$. Then $\Pr[\mathbf{c}^\top \cdot \mathbf{w} = 0] \leq 2/q$.*

Proof. We first show that $\Pr[\mathbf{B}_{t,X,q} \cdot \mathbf{w} = \mathbf{0}] \leq 1/q$. By definition, there must be some element $x \in X$ such that $x \notin Y$. By Lemma 1, the row vector $(x\ x^2\ \ldots\ x^{t+1})$ is not in the row span of $\mathbf{B}_{t,Y,q}$. Thus, only a $1/q$ fraction of the vectors in the kernel of $\mathbf{B}_{t,Y,q}$ are orthogonal to $(x\ x^2\ \ldots\ x^{t+1})$. Noting that $\mathbf{w}$ is a uniformly random vector in the kernel of $\mathbf{B}_{t,Y,q}$, and that $(x\ x^2\ \ldots\ x^{t+1})$ is a row of $\mathbf{B}_{t,X,q}$ establishes the claim. Finally, note that if $\mathbf{B}_{t,X,q} \cdot \mathbf{w} \neq 0$, then the uniform randomness of the vector $\mathbf{c}$ chosen by $\mathsf{SampCodeword}$ implies that $\Pr[\mathbf{c}^\top \cdot \mathbf{w} = 0] = 1/q$. Thus $\Pr[\mathbf{c}^\top \cdot \mathbf{w} = 0] \leq 1/q + ((q-1)/q)(1/q) \leq 2/q$. □

Now, if $X \subseteq Y$, then $\mathbf{w}$ is in the kernel of $\mathbf{B}_{t,X,q}$, so $\mathbf{c}^\top \cdot \mathbf{w} = 0$ with probability 1. Otherwise, the above lemma shows that $\mathbf{c}^\top \cdot \mathbf{w} \neq 0$ except with probability $2/q$. Now let $q \geq 2^\lambda \binom{n'}{t}$, where $n' = \max\{n, 2t\}$, and consider all sets Y such that $|Y| \leq t$ and $X \not\subseteq Y$. The number of such sets is at most $\sum_{i \in [t]} i \binom{n}{i} < t \binom{n'}{t}$. A union bound shows that strong functionality is preserved except with probability at most $t/2^\lambda = \mathsf{negl}(\lambda)$.

3.3 Security

Definition 5. *Let $n(\cdot)$ and $t(\cdot)$ be functions of the security parameter. We say that a family of distributions $\{\mathcal{D}_{n,t,\lambda}\}_\lambda$ where each $\mathcal{D}_{n,t,\lambda}$ is a distribution over subsets $X \subseteq [n(\lambda)]$ such that $|X| \leq t(\lambda)$, is an evasive distribution for the small-superset functionality, if for all fixed $Y \subseteq [n(\lambda)], |Y| \leq t(\lambda)$,*

$$\Pr[X \subseteq Y \mid X \leftarrow \mathcal{D}_{n,t,\lambda}] = \mathsf{negl}(\lambda).$$

Theorem 1. *For any functions $n(\cdot)$ and $t(\cdot)$, and evasive family of distributions $\{\mathcal{D}_{n,t,\lambda}\}_\lambda$ for the small superset functionality, the above construction is a distributional-VBB secure obfuscator in the generic group model.*

The proof is similar to the generic group proofs given in prior work [BLMZ19, BW19].

4 Function-Private Predicate Encryption Security Definitions

4.1 Data Privacy

Our data privacy definition is standard and captures the property that an adversary should not be able to tell the difference between two encrypted attributes x_0 and x_1 or payloads μ_0 and μ_1, provided that it does not have decryption keys that allow it to distinguish trivially. We allow the adversary access to a key generation oracle, allowing it to produce decryption keys for functions f of its choice (from a specified function class), subject to the usual requirement that $f(x_0) = f(x_1)$, and if $f(x_0) = f(x_1) = 1$ for some queried f, then $\mu_0 = \mu_1$.

Definition 6 (Data Privacy). *Let Π be a public-key predicate encryption scheme for function class $\mathcal{F} = \{\mathcal{F}_\lambda\}_\lambda$ where $\mathcal{F}_\lambda = \{f : \mathcal{X}_\lambda \to \{0,1\}\}$ and message space $\mathcal{M} = \{\mathcal{M}_\lambda\}_\lambda$, and let $\mathcal{A}$ be a stateful adversary. We define the* data privacy (DP) *advantage as*

$$\mathbf{Adv}^{\mathsf{DP}}_{\Pi,\mathcal{A}}(\lambda) \stackrel{\text{def}}{=} \left|\Pr\left[\mathsf{Expt}^{\mathsf{DP}}_{\Pi,\mathcal{A}}(\lambda, 0) = 1\right] - \Pr\left[\mathsf{Expt}^{\mathsf{DP}}_{\Pi,\mathcal{A}}(\lambda, 1) = 1\right]\right|,$$

where for $\lambda \in \mathbb{N}$ and $b \in \{0,1\}$, we define experiment $\mathsf{Expt}^{\mathsf{DP}}_{\Pi,\mathcal{A}}(\lambda, b)$ as on the right, where $x_0, x_1 \in \mathcal{X}_\lambda$, and $\mu_0, \mu_1 \in \mathcal{M}_\lambda$. We additionally require that $\mathcal{A}$ is admissible *in the following sense: for all* KeyGen *queries $f \in \mathcal{F}_\lambda$ made by $\mathcal{A}$ we have that $f(x_0) = f(x_1)$, and if there exists an f such that $f(x_0) = f(x_1) = 1$, then $\mu_0 = \mu_1$.*

$\mathsf{Expt}^{\mathsf{DP}}_{\Pi,\mathcal{A}}(\lambda, n, b)$
$(\mathsf{msk}, \mathsf{pk}) \leftarrow \mathsf{Setup}(1^\lambda)$ $(x_0, x_1, \mu_0, \mu_1) \leftarrow \mathcal{A}^{\mathsf{KeyGen}(\mathsf{msk},\cdot)}(1^\lambda, \mathsf{pk})$ $\mathsf{ct} \leftarrow \mathsf{Enc}(\mathsf{pk}, x_b, \mu_b)$ **return** $\mathcal{A}^{\mathsf{KeyGen}(\mathsf{msk},\cdot)}(\mathsf{ct})$

We say Π is a data-private predicate encryption scheme if for all admissible PPT adversaries $\mathcal{A}$, there exists a negligible function $\nu(\cdot)$ such that for all $\lambda \in \mathbb{N}$, $\mathbf{Adv}^{\mathsf{DP}}_{\Pi,\mathcal{A}}(\lambda) \le \nu(\lambda)$.

4.2 Function Privacy

Now consider a set of distribution ensembles over functions, where for each choice of λ, we have a set of distributions $\mathbb{D}_\lambda$. Our first function privacy notion states that function keys for functions drawn from any distribution $\mathcal{D} \in \mathbb{D}_\lambda$ can be simulated even without the description of $\mathcal{D}$. We consider two experiments. In the first, the adversary has access to a "distributional key generation oracle" that takes as input some $\mathcal{D} \in \mathbb{D}_\lambda$ and outputs a decryption key for a boolean function f drawn from $\mathcal{D}$. In the second, we replace the distributional key generation oracle by a simulator. This simulator has no access to the input $\mathcal{D}$, and thus must produce "fake" decryption keys that are indistinguishable to any PPT adversary from real decryption keys produced by the key generation oracle.

Since we consider public-key schemes, an adversary essentially has oracle access to the function corresponding to any decryption key in its possession. Thus it may be easy for the adversary to distinguish these two experiments if for $f \leftarrow \mathcal{D}$, it can find an attribute x such that $f(x) = 1$. So this notion is only realizable for carefully chosen sets of distributions $\mathbb{D}_\lambda$, which in particular must consist solely of *evasive* distributions $\mathcal{D}$ [BBC+14]. That is, for $f \leftarrow \mathcal{D}$, finding x such that $f(x) = 1$ given oracle access to f is computationally intractable.

Definition 7 (Function Privacy). *Let Π be a public-key predicate encryption scheme for function class $\mathcal{F}$ and message space $\mathcal{M}$, let $\mathcal{A}$ be a stateful adversary, and let $\mathcal{S}$ be an explicit* PPT *algorithm simulating* KeyGen. *We define the* function privacy (FP) *advantage for set of distribution ensembles $\mathbb{D} = \{\mathbb{D}_\lambda\}_\lambda$ as*

$$\mathbf{Adv}^{\mathsf{FH}}_{\Pi,\mathcal{S},\mathcal{A}}(\lambda, \mathbb{D}) \stackrel{\text{def}}{=} \left|\Pr\left[\mathsf{Expt}^{\mathsf{FH}}_{\Pi,\mathcal{S},\mathcal{A}}(\lambda, \mathbb{D}, 0) = 1\right] - \Pr\left[\mathsf{Expt}^{\mathsf{FH}}_{\Pi,\mathcal{S},\mathcal{A}}(\lambda, \mathbb{D}, 1) = 1\right]\right|,$$

where for $\lambda \in \mathbb{N}$ and $b \in \{0, 1\}$, we define experiment $\mathsf{Expt}^{\mathsf{FH}}_{\Pi,\mathcal{S},\mathcal{A}}(\lambda, \mathbb{D}, b)$ as:

$\mathsf{Expt}^{\mathsf{FH}}_{\Pi,\mathcal{S},\mathcal{A}}(\lambda, \mathbb{D}, 0)$
$(\mathsf{msk}, \mathsf{pk}) \leftarrow \mathsf{Setup}(1^\lambda)$ **return** $\mathcal{A}^{\mathcal{O}_{\mathsf{DKeyGen}}(\mathsf{msk},\cdot)}(1^\lambda, \mathsf{pk})$ $\underline{\mathcal{O}_{\mathsf{DKeyGen}}(\mathsf{msk}, \mathcal{D})}$ If $\mathcal{D} \in \mathbb{D}_\lambda$, $f \leftarrow \mathcal{D}$, **return** $\mathsf{KeyGen}(\mathsf{msk}, f)$ Else **return** $\perp$

$\mathsf{Expt}^{\mathsf{FH}}_{\Pi,\mathcal{S},\mathcal{A}}(\lambda, \mathbb{D}, 1)$
$(\mathsf{msk}, \mathsf{pk}) \leftarrow \mathsf{Setup}(1^\lambda)$ **return** $\mathcal{A}^{\mathcal{S}(\mathsf{msk})}(1^\lambda, \mathsf{pk})$

We say Π is a $\mathbb{D}$-function-private predicate encryption scheme if there exists a simulator $\mathcal{S}$ such that for all PPT adversaries $\mathcal{A}$, there exists a negligible function $\nu(\cdot)$ such that for all $\lambda \in \mathbb{N}$, $\mathbf{Adv}^{\mathsf{FH}}_{\Pi,\mathcal{S},\mathcal{A}}(\lambda, \mathbb{D}) \leq \nu(\lambda)$.

4.3 Enhanced Function Privacy

In the standard notion of function privacy described above, the fact that the adversary only receives decryption keys sk_{f_j} for functions f_j (where f_j denotes the jth output of $\mathcal{O}_{\mathsf{DKeyGen}}$ during the course of the experiment) drawn from an evasive distribution $\mathcal{D}$ implies it will not be able to generate a ciphertext c encrypting (x, μ) such that $\mathsf{Dec}(\mathsf{sk}_{f_j}, c) \rightarrow \mu$ (except with negligible probability). We now describe a strictly stronger notion of function privacy known as *enhanced function privacy*, where we provide the adversary with an oracle $\mathcal{O}_{\mathsf{Enc}}$

that generates ciphertexts of (x, μ) such that $f_j(x) = 1$ for some f_j. More precisely, $\mathcal{O}_{\mathsf{Enc}}$ takes pk and an index j as input, and outputs a ciphertext c of some arbitrary (x, μ) such that $\mathsf{Dec}(\mathsf{sk}_{f_j}, c) \to \mu$.

Note that normal (non-enhanced) function privacy does not guarantee any security the moment a function decryption key holder receives a ciphertext of (x, μ) such that $f(x) = 1$. This renders the standard function privacy notion almost useless in many settings, since as soon as a user is able to use its decryption key to decrypt any payload μ, all function privacy may be lost. We give our formal definition below, which generalizes the enhanced function privacy notion proposed by Boneh et al. [BRS13a, §3.2] in the context of identity-based encryption (IBE).

Definition 8 (Enhanced Function Privacy). *Let Π be a public-key predicate encryption scheme for function class $\mathcal{F}$ and message space $\mathcal{M}$, let $\mathcal{A}$ be a stateful adversary, and let $\mathcal{S} = (\mathcal{S}_{\mathsf{DKeyGen}}, \mathcal{S}_{\mathsf{Enc}})$ be an explicit* PPT *algorithm simulating* KeyGen *and* Enc. *We define the* enhanced function privacy (eFP) *advantage for distribution ensemble $\mathbb{D} = \{\mathbb{D}_\lambda\}_\lambda$ as*

$$\mathbf{Adv}^{\mathsf{eFP}}_{\Pi,\mathcal{S},\mathcal{A}}(\lambda, \mathbb{D}) \stackrel{\text{def}}{=} \left| \Pr\left[\mathsf{Expt}^{\mathsf{eFP}}_{\Pi,\mathcal{S},\mathcal{A}}(\lambda, \mathbb{D}, 0) = 1\right] - \Pr\left[\mathsf{Expt}^{\mathsf{eFP}}_{\Pi,\mathcal{S},\mathcal{A}}(\lambda, \mathbb{D}, 1) = 1\right] \right|,$$

where for $\lambda \in \mathbb{N}$ and $b \in \{0, 1\}$, we define experiment $\mathsf{Expt}^{\mathsf{eFH}}_{\Pi,\mathcal{S},\mathcal{A}}(\lambda, \mathbb{D}, b)$ as:

$\mathsf{Expt}^{\mathsf{eFH}}_{\Pi,\mathcal{S},\mathcal{A}}(\lambda, \mathbb{D}, 0)$	$\mathsf{Expt}^{\mathsf{eFH}}_{\Pi,\mathcal{S},\mathcal{A}}(\lambda, \mathbb{D}, 1)$
$(\mathsf{msk}, \mathsf{pk}) \leftarrow \mathsf{Setup}(1^\lambda)$ $j := 1$ **return** $\mathcal{A}^{\mathcal{O}_{\mathsf{DKeyGen}}(\mathsf{msk},\cdot), \mathcal{O}_{\mathsf{Enc}}(\mathsf{pk},\cdot)}(1^\lambda, \mathsf{pk})$ $\underline{\mathcal{O}_{\mathsf{DKeyGen}}(\mathsf{msk}, \mathcal{D})}$ If $\mathcal{D} \in \mathbb{D}_\lambda$, $f_j \leftarrow \mathcal{D}$, **return** $\mathsf{KeyGen}(\mathsf{msk}, f)$ Else **return** $\perp$ $(j := j + 1)$ $\underline{\mathcal{O}_{\mathsf{Enc}}(\mathsf{pk}, j)}$ **choose** any $(x, \mu) \in \mathcal{X}_\lambda \times \mathcal{M}_\lambda$ **such that** $f_j(x) = 1$ **return** $\mathsf{Enc}(\mathsf{pk}, x, \mu)$	$(\mathsf{msk}, \mathsf{pk}) \leftarrow \mathsf{Setup}(1^\lambda)$ **return** $\mathcal{A}^{\mathcal{S}_{\mathsf{DKeyGen}}(\mathsf{msk}), \mathcal{S}_{\mathsf{Enc}}(\mathsf{pk},\cdot)}(1^\lambda, \mathsf{pk})$

We say Π is a $\mathbb{D}$-enhanced function-private predicate encryption scheme if there exists a simulator $\mathcal{S} = (\mathcal{S}_{\mathsf{DKeyGen}}, \mathcal{S}_{\mathsf{Enc}})$ such that for all PPT adversaries $\mathcal{A}$, there exists a negligible function $\nu(\cdot)$ such that for all $\lambda \in \mathbb{N}$, $\mathbf{Adv}^{\mathsf{eFH}}_{\Pi,\mathcal{S},\mathcal{A}}(\lambda, \mathbb{D}) \leq \nu(\lambda)$.

4.4 Discussion

We view our enhanced function privacy definition as a direct generalization of the "real-or-random" enhanced function privacy definition considered by Boneh et al. [BRS13a]. Boneh et al. give their definition in the context of identity-based encryption (IBE), where an adversary is given an oracle that accepts distributions $\mathcal{D}$ over identities. The guarantee is that the adversary cannot determine whether the oracle is in "real" or "random" mode, where real mode means that it will return the secret key for an identity $\mathcal{I}$ drawn from the input distribution $\mathcal{D}$, and random mode means that it will return the secret key for a uniformly random identity. When attempting to generalize this definition to more expressive function classes (note that IBE corresponds to predicate encryption for point functions), it is not necessarily clear what the behavior of the random mode oracle should be.

We instead view the random mode oracle as a simulator which does not get to see the input distribution $\mathcal{D}$. In the case of IBE, one possible simulator could be defined to return a secret key for a uniformly random identity. But in general, we can allow the simulator's behavior to be arbitrary, as long as it does not depend on the queried distribution.

We note that our definition is weaker than the Boneh et al. definition [BRS13a] for IBE in one sense: we no longer provide the adversary with an explicit KeyGen oracle, which can be used to obtain secret keys for arbitrary functions of the adversary's choice (our only key generation oracle outputs functions drawn from evasive distributions). This is because such a definition is trivially unachievable when considering general functionalities such as small superset.

Indeed, since the behavior of the Enc oracle is arbitrary, assume that given an index j corresponding to a secret key for hidden subset X_j, it encrypts using the attribute X_j itself, a valid accepting input. Assume further that the universe size n is polynomial. Now an adversary can use the KeyGen oracle n times to receive a secret key for each of the subsets $\{i\}$ for $i \in [n]$. Then it simply tries to decrypt the encryption with attribute X_j with each of the keys, and can figure out exactly what X_j is, breaking function privacy. This style of attack does not exist when considering IBE where the functions encrypted are simply point functions.

As a final note about our definitions, we compare to those of Patranabis et al. [PMR19], which to the best of our knowledge is the only previous work proposing function private hidden vector encryption. They consider a notion of "left-or-right" security, where the adversary queries two distributions at a time to its oracle, and the oracle chooses which one to draw the function from depending on whether it is in "left" or "right" mode. They do not consider the enhanced version where an Enc oracle is provided, but they do provide a KeyGen oracle as in the original Boneh et al. [BRS13a] definition.

In the full version, we augment our basic (non-enhanced) function privacy definition to include a KeyGen oracle, sketch a proof that our small superset construction obtains this definition, and then show that this definition implies the left-or-right definition considered by Patranabis et al. [PMR19]. By going through our HVE-to-small-superset compiler and comparing the class of distributions considered in this work with those of Patranabis et al. [PMR19] (which, in particular, reveal the positions of the wildcards), we demonstrate that the security of our function private HVE construction generalizes that of Patranabis et al. [PMR19].

5 Function-Private Predicate Encryption for Small Superset

The following construction Π relies on an asymmetric bilinear map $e\colon \mathbb{G}_1 \times \mathbb{G}_2 \to \mathbb{G}_T$. We let $[a]_1, [b]_2, [c]_T$ denote encodings of a, b, c in groups $\mathbb{G}_1, \mathbb{G}_2, \mathbb{G}_T$ respectively. For a vector $\mathbf{v}$ or matrix $\mathbf{M}$, we use the shorthand $[\mathbf{v}]$ or $[\mathbf{M}]$ (for any of the three groups) to denote the group elements obtained by encoding each entry of $\mathbf{v}$ or $\mathbf{M}$ respectively. Let $n(\cdot), t(\cdot)$ be functions of the security parameter. Let the message space $\mathcal{M} := \mathcal{M}_\lambda$ be a subset of the target group $\mathbb{G}_T$ such that $|\mathcal{M}|/|\mathbb{G}_T| = \mathsf{negl}(\lambda)$.[10]

- $\mathsf{Setup}(1^\lambda)$. Set $n := n(\lambda), t := t(\lambda)$. Pick a prime $q > \max\{n, 2^\lambda\}$. Sample a uniformly random matrix $\mathbf{R} \in \mathbb{F}_q^{(t+2)\times(t+2)}$ and compute $\mathbf{R}^{-1}$. Output $\mathsf{msk} := \mathbf{R}^{-1}$ and $\mathsf{pk} := [\mathbf{R}]_2$.
- $\mathsf{KeyGen}(\mathsf{msk}, X)$. Parse msk as $\mathbf{R}^{-1}$. To encrypt a subset $X \subseteq [n]$, draw $\mathbf{c}^\top \leftarrow \mathsf{SampCodeword}(\mathbf{B}_{t,X,q})$, sample $\alpha \leftarrow \mathbb{F}_q$, and output
$$[(1 \mid \alpha \cdot \mathbf{c}^\top) \cdot \mathbf{R}^{-1}]_1.$$
- $\mathsf{Enc}(\mathsf{pk}, Y, \mu)$. Parse pk as $[\mathbf{R}]_2$. To encrypt a message $Y \subseteq [n]$ such that $|Y| \leq t$, let $\mathbf{w} \leftarrow \mathsf{SampDualCodeword}(\mathbf{B}_{t,Y,q})$, sample $\beta, \gamma \leftarrow \mathbb{F}_q$, and output
$$[\mathbf{R}]_2 \cdot (\gamma \mid \mathbf{w}^\top \cdot \beta)^\top, \quad \mu \cdot [\gamma]_T.$$
- $\mathsf{Dec}(\mathsf{sk}_X, \mathsf{ctxt}_Y)$. Parse sk_X as $[\mathbf{v}^\top]_1$ and ctxt_Y as $([\mathbf{w}]_2, h)$. Compute
$$\mu := h/[\mathbf{v}^\top \cdot \mathbf{w}]_T,$$
where the dot product in the target group is computed using the bilinear operation. If $\mu \in \mathcal{M}$, output μ, otherwise output $\perp$.

Correctness. If $X \subseteq Y$, note that the $\mathbf{w}$ vector associated with ctxt_Y is a codeword in the dual of the code from which the $\mathbf{c}^\top$ vector from sk_X was drawn. This follows since every row in the generator matrix of $\mathbf{c}^\top$'s code is one of the rows in the generator matrix of $\mathbf{w}$'s dual code. Thus the dot product computed during decryption will be equal to γ, and dividing h by $[\gamma]_T$ will give the encrypted payload μ.

If $X \not\subseteq Y$, a straightforward application of Lemma 2 shows that with overwhelming probability, $h/[\mathbf{v}^\top \cdot \mathbf{w}]_T$ will be a uniformly random group element, and will thus be an element of $\mathcal{M}$ with negligible probability. Thus, decryption will output $\perp$ with overwhelming probability.

[10] In [BW07], it is noted that this restriction on the size of the message space can be avoided in practice by essentially setting the payload to be the key of a symmetric key encryption scheme, and releasing an encryption of the actual message under this key (along with a consistency check). This technique can easily be applied in our setting.

5.1 Security

We make use of a variant [BGMZ18] of a lemma by Badrinarayanan et al. [BMSZ16]. In fact, we only need a particular special case of the lemma, stated below.

Lemma 3 ([BMSZ16,BGMZ18]). *Let* $\hat{\mathbf{R}}$ *be an* $n \times n$ *matrix of distinct formal variables* $\hat{r}_{i,j}$*, and* $\mathbf{u}, \mathbf{v} \in \mathbb{F}_q^n$ *be two arbitrary vectors. Let* $\hat{\mathbf{u}} = \mathbf{u}^\top \cdot \hat{\mathbf{R}}^{-1}$ *and* $\hat{\mathbf{v}} = \hat{\mathbf{R}} \cdot \mathbf{v}$ *be two vectors of rational functions over the* $\hat{r}_{i,j}$ *formal variables. Let* P *be a polynomial over the entries of* $\hat{\mathbf{u}}$ *and* $\hat{\mathbf{v}}$ *such that each monomial contains exactly one entry from* $\hat{\mathbf{u}}$ *and one from* $\hat{\mathbf{v}}$*. Then if* P *is identically a constant over the* $\hat{r}_{i,j}$ *variables, it must be a constant multiple of the inner product of* $\hat{\mathbf{u}}$ *and* $\hat{\mathbf{v}}$*.*

Theorem 2. *The above construction* Π *is a* data-private *predicate encryption scheme for small superset.*

Proof. Consider any $Y_0, Y_1 \subseteq [n]$ such that $|Y_0|, |Y_1| \leq t$, and $\mu_0, \mu_1 \in \mathbb{G}_T$. The adversary $\mathcal{A}$ receives the public key and an encryption of (Y_b, μ_b) where $b \leftarrow \{0,1\}$. For convenience, we will let μ_0' and μ_1' be the discrete logs of μ_0, μ_1. $\mathcal{A}$ is free to request keys for sets X_i such that X_i is not contained in either Y_0 or Y_1. If $\mu_0 = \mu_1$, it is also free to request keys for sets X_i such that X_i is contained in both Y_0 and Y_1.

Thus, $\mathcal{A}$ has access to the handles of the elements

$$[\mathbf{R}]_2, \{[(1 \mid \alpha_i \cdot \mathbf{c}_{X_i}^\top) \cdot \mathbf{R}^{-1}]_1\}_i, [\mathbf{R} \cdot (\gamma \mid \mathbf{w}_{Y_b}^\top \cdot \beta)^\top]_2, [\mu_b' + \gamma]_T.$$

Recall that the only distinguishing information the adversary can obtain in the generic group model is the responses to zero-test queries in the target group. We first imagine replacing $\{\alpha_i\}_i, \beta, \gamma$, and the entries of $\mathbf{R}$ with formal variables. So we let $\hat{\mathbf{R}}$ be a $(t+2) \times (t+2)$ matrix of formal variables $\hat{r}_{i,j}$ for $i, j \in [t+2]$. We would like to apply Schwartz-Zippel to every zero-test query submitted by $\mathcal{A}$ in order to conclude that $\mathcal{A}$ cannot distinguish this switch except with negligible probability. However, the resulting zero-test expressions are rational functions of the above formal variables. We instead imagine taking each zero-test query and multiplying through by $\det(\hat{\mathbf{R}})$, which does not change whether it is identically zero or not. By construction, this results in a polynomial of degree at most $t + 5 = \mathsf{poly}(\lambda)$ over the formal variables. Thus applying Schwartz-Zippel and union bounding over the polynomially many zero-test queries submitted by $\mathcal{A}$ establishes that $\mathcal{A}$ cannot distinguish this switch except with negligible probability.

Now, define $(\hat{\mathbf{u}}^{(i)})^\top := (\hat{\alpha}_i^{-1} \mid \mathbf{c}_{X_i}^\top) \cdot \hat{\mathbf{R}}^{-1}$, and $\hat{\mathbf{v}} := \hat{\mathbf{R}} \cdot (\hat{\beta}^{-1}\hat{\gamma} \mid \mathbf{w}_{Y_b}^\top)^\top$. Using this notation, we will write down a general expression for any zero-test query submitted by the adversary. We consider all the possible ways that $\mathcal{A}$ can produce elements in the target group: pairing its ciphertext, secret key, or public key elements with a constant in the other group, or pairing its secret key elements with public key or ciphertext elements. Then we write a general linear combination of such elements, where $\kappa_{i,j}$, τ_k, $\delta_{k,\ell}$, $\eta_{i,j,k}$, $\rho_{i,j,k,\ell}$, and ν represent coefficients submitted by $\mathcal{A}$. This results in the following expression.

$$\sum_{i,j} \kappa_{i,j}\hat{\alpha}_i\hat{\mathbf{u}}_j^{(i)} + \sum_k \tau_k \hat{\mathbf{v}}_k\hat{\beta} + \sum_{k,\ell} \delta_{k,\ell}\hat{r}_{k,\ell} + \sum_{i,j,k} \eta_{i,j,k}\hat{\alpha}_i(\hat{\mathbf{u}}_j^{(i)})^\top \hat{\mathbf{v}}_k\hat{\beta}$$
$$+ \sum_{i,j,k,\ell} \rho_{i,j,k,\ell}\hat{\alpha}_i\hat{\mathbf{u}}_j^{(i)}\hat{r}_{k,\ell} + \nu(\mu'_b + \hat{\gamma})$$
$$= \hat{\beta}\left(\sum_k \tau_k\hat{\mathbf{v}}_k + \sum_i \hat{\alpha}_i\left(\sum_{j,k}\eta_{i,j,k}(\hat{\mathbf{u}}_j^{(i)})^\top\hat{\mathbf{v}}_k\right)\right) + \sum_{i,j}\kappa_{i,j}\hat{\alpha}_i\hat{\mathbf{u}}_j^{(i)} + \sum_{k,\ell}\delta_{k,\ell}\hat{r}_{k,\ell}$$
$$+ \sum_{i,j,k,\ell} \rho_{i,j,k,\ell}\hat{\alpha}_i\hat{\mathbf{u}}_j^{(i)}\hat{r}_{k,\ell} + \nu(\mu'_b + \hat{\gamma})$$

Now any potentially distinguishing zero-test query must result in an identically zero rational function for at least one setting of $b \in \{0,1\}$, and thus must set the coefficient on $\hat{\beta}$ to some scaling of $\hat{\beta}^{-1}$ for one of these settings (since $\hat{\beta}$ does not appear in the other terms). This implies a few things about the adversary's coefficients. First, for each k, $\tau_k = 0$, since each entry of $\hat{\mathbf{v}}$ is a sum over distinct formal variables from $\hat{\mathbf{R}}$ which cannot be canceled out elsewhere in the coefficient on $\hat{\beta}$. Next, for each i, the coefficient on $\hat{\alpha}_i$ within this $\hat{\beta}$ coefficient must be some scaling of $\hat{\alpha}_i^{-1}$. Then by Lemma 3, for each i, the coefficients $\{\eta_{i,j,k}\}_{j,k}$ must be set to induce a scaling of the inner product of $\hat{\mathbf{u}}^{(i)}$ and $\hat{\mathbf{v}}$. Let z_i denote this scaling. We can rewrite the above expression as follows.

$$\hat{\beta}\left(\sum_i \hat{\alpha}_i\left(z_i(\hat{\alpha}_i^{-1}\hat{\beta}^{-1}\hat{\gamma} + \mathbf{c}_{X_i}^\top \cdot \mathbf{w}_{Y_b})\right)\right) + \sum_{i,j}\kappa_{i,j}\hat{\alpha}_i\hat{\mathbf{u}}_j^{(i)} + \sum_{k,\ell}\delta_{k,\ell}\hat{r}_{k,\ell}$$
$$+ \sum_{i,j,k,\ell} \rho_{i,j,k,\ell}\hat{\alpha}_i\hat{\mathbf{u}}_j^{(i)}\hat{r}_{k,\ell} + \nu(\mu'_b + \hat{\gamma})$$
$$= \hat{\gamma}\left(\sum_i z_i + \nu\right) + \hat{\beta}\left(\sum_i \hat{\alpha}_i z_i \mathbf{c}_{X_i}^\top \cdot \mathbf{w}_{Y_b}\right) + \sum_{i,j}\kappa_{i,j}\hat{\alpha}_i\hat{\mathbf{u}}_j^{(i)} + \sum_{k,\ell}\delta_{k,\ell}\hat{r}_{k,\ell}$$
$$+ \sum_{i,j,k,\ell} \rho_{i,j,k,\ell}\hat{\alpha}_i\hat{\mathbf{u}}_j^{(i)}\hat{r}_{k,\ell} + \nu\mu'_b$$

Now observe that we need the coefficient on $\hat{\gamma}$ to be zero in order to obtain a successful zero-test. We consider two cases. First, if $z_i = 0$ for all i, then the coefficient on $\hat{\gamma}$ is zero only if $\nu = 0$. But in this case, the remaining term is

$$\sum_{i,j}\kappa_{i,j}\hat{\alpha}_i\hat{\mathbf{u}}_j^{(i)} + \sum_{k,\ell}\delta_{k,\ell}\hat{r}_{k,\ell} + \sum_{i,j,k,\ell}\rho_{i,j,k,\ell}\hat{\alpha}_i\hat{\mathbf{u}}_j^{(i)}\hat{r}_{k,\ell},$$

which is independent of the bit b. Thus, such a zero-test cannot be used to distinguish.

Otherwise, let S be the set of i such that $z_i \neq 0$. If the coefficient on $\hat{\beta}$ is zero for some b, this implies that $\mathbf{c}_{X_i}^\top \cdot \mathbf{w}_{Y_b} = 0$ for each $i \in S$, and thus by correctness, $X_i \subseteq Y_b$ for each $i \in S$. Then by admissibility, $X_i \subseteq Y_0, Y_1$ for each $i \in S$, meaning that the coefficient on $\hat{\beta}$ is zero regardless of b. But again by admissibility, this also implies that $\mu'_0 = \mu'_1$. Then it is clear that the remaining expression

$$\sum_{i,j} \kappa_{i,j} \hat{\alpha}_i \hat{\mathbf{u}}_j^{(i)} + \sum_{k,\ell} \delta_{k,\ell} \hat{r}_{k,\ell} + \sum_{i,j,k,\ell} \rho_{i,j,k,\ell} \hat{\alpha}_i \hat{\mathbf{u}}_j^{(i)} \hat{r}_{k,\ell} + \nu \mu'_b$$

is independent of the bit b, completing the proof. □

Definition 9. *Let $n(\cdot), t(\cdot)$ be functions of the security parameter. Let $\mathcal{E}_{n,t}$ be the entire set of families of evasive small superset distributions $\{\mathcal{D}_{n,t,\lambda}\}_\lambda$. Write $\mathcal{E}_{n,t} = \{\mathcal{E}_{n,t,\lambda}\}_\lambda$.*

Theorem 3. *For any $n(\cdot), t(\cdot)$, the above construction Π is an $\mathcal{E}_{n,t}$-enhanced function-private predicate encryption scheme for small superset.*

Proof. In $\mathsf{Expt}^{\mathsf{eFP}}_{\Pi,\mathcal{S},\mathcal{A}}(1^\lambda, \mathcal{E}_{n,t}, 0)$ from Definition 8, $\mathcal{A}$ interacts with an honest implementation of the construction Π in the generic bilinear group model. We prove through a series of hybrid experiments that $\mathcal{A}$'s view in the honest world is indistinguishable from its view in $\mathsf{Expt}^{\mathsf{eFP}}_{\Pi,\mathcal{S},\mathcal{A}}(1^\lambda, \mathcal{E}_{n,t}, 1)$, in which the oracles $\mathcal{O}_{\mathsf{DKeyGen}}$ and $\mathcal{O}_{\mathsf{Enc}}$ are implemented by the simulator $\mathcal{S}$ with no knowledge of the queried distributions in $\mathcal{E}_{n,t,\lambda}$. Note that the oracles $\mathcal{O}_{\mathsf{DKeyGen}}$ and $\mathcal{O}_{\mathsf{Enc}}$ are allowed to share state.

First, we make explicit the following generic group instantiation of $\mathsf{Expt}^{\mathsf{eFP}}_{\Pi,\mathcal{S},\mathcal{A}}(1^\lambda, \mathcal{E}_{n,t}, 0)$. Note that the adversary $\mathcal{A}$ in the below experiment and all following hybrid experiments also implicitly has access to generic group bilinear map operations described in Definition 1. Since $\mathcal{A}$ is PPT, we'll say that $\mathcal{A}$ makes $J = \mathsf{poly}(\lambda)$ queries to $\mathcal{O}_{\mathsf{DKeyGen}}$ and $K = \mathsf{poly}(\lambda)$ queries to $\mathcal{O}_{\mathsf{Enc}}$.

$\mathsf{Expt}^{\mathsf{eFP}}_{\Pi,\mathcal{S},\mathcal{A}}(1^\lambda, \mathcal{E}_{n,t}, 0)$:

1. Set $n := n(\lambda), t := t(\lambda)$, $q > \max\{n, 2^\lambda\}$.
2. Sample $\mathbf{R} \leftarrow \mathbb{F}_q^{(t+2)\times(t+2)}$ and set $\mathsf{msk} := \mathbf{R}^{-1}$.
3. Generate fresh handles in group 2 for each entry of $\mathbf{R}$, letting pk consist of this set of handles.
4. Output $\mathcal{A}^{\mathcal{O}_{\mathsf{DKeyGen}}(\mathsf{msk},\cdot),\mathcal{O}_{\mathsf{Enc}}(\mathsf{pk},\cdot)}(1^\lambda, \mathsf{pk})$.

$\mathcal{O}_{\mathsf{DKeyGen}}(\mathsf{msk}, \mathcal{D})$:

This oracle maintains an internal counter j, initialized at $j = 1$. After each oracle call, increment j. On each oracle call:

1. Sample $X_j \leftarrow \mathcal{D}$ and set $(\mathbf{c}^{(j)})^\top \leftarrow \mathsf{SampCodeword}(\mathbf{B}_{t,X_j,q})$.
2. Sample $\alpha_j \leftarrow \mathbb{F}_q$.
3. Set $(\mathbf{u}^{(j)})^\top := (1 \mid \alpha_j \cdot (\mathbf{c}^{(j)})^\top) \cdot \mathbf{R}^{-1}$.
4. Generate and return fresh handles in group 1 for $(\mathbf{u}^{(j)})^\top$.

$\mathcal{O}_{\mathsf{Enc}}(\mathsf{msk}, j)$:

On the kth oracle call, do the following:

1. Let $Y_k \subseteq [n]$ be any set satisfying $|Y_k| \leq t$ and $X_j \subseteq Y_k$.
2. Let $\mu'_k \in \mathbb{F}_q$.
3. Sample:
   - $\mathbf{w}^{(k)} \leftarrow \mathsf{SampDualCodeword}(\mathbf{B}_{t,Y_k,q})$
   - $\beta_k, \gamma_k \leftarrow \mathbb{F}_q$
   - $\mathbf{v}^{(k)} := \mathbf{R} \cdot (\gamma_k \mid (\mathbf{w}^{(k)})^\top \cdot \beta_k)^\top$
4. Generate and return fresh handles in group 2 for $\mathbf{v}^{(k)}$, and a fresh handle in group T for $\mu'_k + \gamma_k$.

Now, we present a series of hybrid experiments, beginning with the above experiment and ending with a generic group instantiation of $\mathsf{Expt}^{\mathsf{eFP}}_{\Pi,\mathcal{S},\mathcal{A}}(1^\lambda, \mathcal{E}_{n,t}, 1)$.

- Expt_0 is exactly $\mathsf{Expt}^{\mathsf{eFP}}_{\Pi,\mathcal{S},\mathcal{A}}(1^\lambda, \mathcal{E}_{t,n}, 0)$.
- Expt_1 is obtained from Expt_0 by modifying $\mathcal{O}_{\mathsf{Enc}}(\mathsf{msk}, \cdot)$ to the following:
  $\mathcal{O}_{\mathsf{Enc}}(\mathsf{msk}, j)$:
  On the kth oracle call, do the following:
  1. Let $\mu'_k \in \mathbb{F}_q$.
  2. Sample $\beta_k, \gamma_k \leftarrow \mathbb{F}_q$.
  3. Define t new formal variables $\hat{w}_{k,1}, \ldots, \hat{w}_{k,t}$.
  4. Define $\mathbf{w}^{(k)} := \left[\hat{w}_{k,1}, \ldots, \hat{w}_{k,t} - \frac{1}{\mathbf{c}^{(j)}_{t+1}} \sum_{i=1}^{t} \mathbf{c}^{(j)}_i \hat{w}_{k,i}\right]$.
  5. Set $\hat{\mathbf{v}}^{(k)} := \mathbf{R} \cdot (\gamma_k \mid (\mathbf{w}^{(k)})^\top \cdot \beta_k)^\top$.
  6. Generate and return fresh handles in group 2 for $\hat{\mathbf{v}}^{(k)}$, and a fresh handle in group T for $\mu'_k + \gamma_k$.

  Note that the generic bilinear group operations are now performed over the ring $\mathbb{Z}[\{\hat{w}_{k,i}\}_{k\in[K],i\in[t]}]$. Also note that $\mathcal{O}_{\mathsf{Enc}}$ and $\mathcal{O}_{\mathsf{DKeyGen}}$ are sharing state, in particular the set of $\mathbf{c}^{(j)}$ vectors.
- $\mathsf{Expt}_{2,\ell}$ (for $\ell = 0, \ldots, J$) is obtained from Expt_1, except $\mathcal{O}_{\mathsf{DKeyGen}}(\mathsf{msk}, \cdot)$ is modified to the following:
  $\mathcal{O}_{\mathsf{DKeyGen}}(\mathsf{msk}, \mathcal{D})$:
  The oracle maintains an internal counter j, initialized at $j = 1$. After each oracle call, increment j. On each oracle call:
  1. If $j \leq \ell$, sample uniformly random $(\mathbf{c}^{(j)})^\top \leftarrow \mathbb{F}_q^{t+1}$. If $j > \ell$, sample $X_j \leftarrow \mathcal{D}$ and set $(\mathbf{c}^{(j)})^\top \leftarrow \mathsf{SampCodeword}(\mathbf{B}_{t,X_j,q})$.
  2. Sample $\alpha_j \leftarrow \mathbb{F}_q$.
  3. Set $(\hat{\mathbf{u}}^{(j)})^\top := (1 \mid \alpha_j \cdot (\mathbf{c}^{(j)})^\top) \cdot \mathbf{R}^{-1}$.
  4. Generate and return fresh handles in group 1 for $(\hat{\mathbf{u}}^{(j)})^\top$.

  Observe that $\mathsf{Expt}_1 = \mathsf{Expt}_{2,0}$, and that $\mathsf{Expt}_{2,J}$ is a generic group instantiation of $\mathsf{Expt}^{\mathsf{eFP}}_{\Pi,\mathcal{S},\mathcal{A}}(1^\lambda, \mathcal{E}_{n,t}, 1)$. This follows since the input $\mathcal{D}$ is not used by $\mathcal{O}_{\mathsf{DKeyGen}}$ at any point during the course of the experiment, so $\mathcal{O}_{\mathsf{DKeyGen}}$ can be simulated by $\mathcal{S}$.

Claim. $\mathcal{A}$ cannot distinguish between Expt_0 and Expt_1 except with $\mathsf{negl}(\lambda)$ advantage.

Proof. Let j_k denote the index input to $\mathcal{O}_{\mathsf{Enc}}(\mathsf{pk}, \cdot)$ on the kth query. We condition on the event that for each Y_k, $X_{j'} \not\subseteq Y_k$ for all $j' \neq j_k$. This occurs with overwhelming probability due to the definition of $\mathcal{E}_{n,t}$ and a union bound over $J, K = \mathsf{poly}(\lambda)$. We further condition on the event that in Expt_0, for all k, j, $(\mathbf{c}^{(j)})^\top \cdot \mathbf{w}^{(k)} = 0$ if and only if $j = j_k$, which follows from Lemma 2 and a union bound.

In both games, consider replacing all the entries of $\mathbf{R}$ and all $\alpha_j, \beta_k, \gamma_k$ with formal variables, and call the resulting games $\mathsf{Sim\text{-}Real'}$ and $\mathsf{Sim\text{-}Enc'}$. By a similar argument as in the proof of Theorem 2, $\mathcal{A}$ notices this switch with negligible probability. Now fix any zero-test query that $\mathcal{A}$ submits. We claim that it evaluates to identically zero in $\mathsf{Sim\text{-}Real'}$ if and only if it does so in $\mathsf{Sim\text{-}Enc'}$.

As in the proof of Theorem 2, we first write explicitly the form of a zero-test query in $\mathsf{Sim\text{-}Real'}/\mathsf{Sim\text{-}Enc'}$. Define $(\hat{\mathbf{u}}'^{(j)})^\top := (\hat{\alpha}_j^{-1} \mid (\mathbf{c}^{(j)})^\top) \cdot \hat{\mathbf{R}}^{-1}$, and $\hat{\mathbf{v}}'^{(k)} := \hat{\mathbf{R}} \cdot (\hat{\beta}_k^{-1}\hat{\gamma}_k \mid (\mathbf{w}^{(k)})^\top)^\top$. Letting $\kappa_{j,m}, \tau_{k,\ell}, \delta_{k,\ell}, \eta_{j,m,k,\ell}, \rho_{j,m,k,\ell}, \nu_k$ refer to the coefficients submitted by $\mathcal{A}$, the general form of a zero-test query is

$$\sum_{j,m} \kappa_{j,m}\hat{\alpha}_j\hat{\mathbf{u}}'^{(j)}_m + \sum_{k,\ell} \tau_{k,\ell}\hat{\mathbf{v}}'^{(k)}_\ell\hat{\beta}_k + \sum_{k,\ell} \delta_{k,\ell}\hat{r}_{k,\ell} + \sum_{j,m,k,\ell} \eta_{j,m,k,\ell}\hat{\alpha}_j(\hat{\mathbf{u}}'^{(j)}_m)^\top\hat{\mathbf{v}}'^{(k)}_\ell\hat{\beta}_k$$
$$+ \sum_{j,m,k,\ell} \rho_{j,m,k,\ell}\hat{\alpha}_j\hat{\mathbf{u}}'^{(j)}_m\hat{r}_{k,\ell} + \sum_k \nu_k(\mu'_k + \hat{\gamma}_k).$$

First, notice that all but the second and fourth terms are identical between the two games $\mathsf{Sim\text{-}Real'}$ and $\mathsf{Sim\text{-}Enc'}$, since the only difference lies in the $\mathbf{v}'^{(k)}$ vectors. Note further that an adversary can only hope to obtain a successful zero-test in either game by setting $\tau_{k,\ell} = 0$ for all k, ℓ. This follows from a similar argument as in the proof of Theorem 2, where the entire expression is stratified by the $\hat{\beta}_k$ variables. Looking at each $\hat{\beta}_k$ term, it is clear that the formal variables from $\hat{\mathbf{R}}$ in the elements of the $\hat{\mathbf{v}}'^{(k)}$ vectors cannot be canceled out.

Thus we focus on the fourth term, and stratify by the $\hat{\alpha}_j$ and $\hat{\beta}_k$ variables to obtain

$$\sum_{j,k} \hat{\alpha}_j\hat{\beta}_k \left(\sum_{m,\ell} \eta_{j,m,k,\ell}(\hat{\mathbf{u}}'^{(j)}_m)^\top\hat{\mathbf{v}}'^{(k)}_\ell \right).$$

$\mathcal{A}$ can only hope to obtain a successful zero-test if the coefficient on each $\hat{\alpha}_j\hat{\beta}_k$ is a constant multiple of $\hat{\alpha}_j^{-1}\hat{\beta}_k^{-1}$. So by Lemma 3, for this to happen, it must be the case that for each (j, k), the coefficients $\{\eta_{j,m,k,\ell}\}_{m,\ell}$ induce a scaling of the inner product between $\hat{\mathbf{u}}'^{(j)}$ and $\hat{\mathbf{v}}'^{(k)}$. For each (j, k), let $z_{j,k}$ be this scaling. Now we can re-write this term as

$$\sum_{j,k} \hat{\alpha}_j\hat{\beta}_k z_{j,k}(\hat{\alpha}_j^{-1}\hat{\beta}_k^{-1}\hat{\gamma}_k + (\mathbf{c}^{(j)})^\top \cdot \mathbf{w}^{(k)}) =$$
$$\sum_k \hat{\gamma}_k \left(\sum_j z_{j,k} \right) + \sum_{j,k} \hat{\alpha}_j\hat{\beta}_k z_{j,k}(\mathbf{c}^{(j)})^\top \cdot \mathbf{w}^{(k)}.$$

Again notice that the first term will be identical in both games, so focus attention on the second. We see that the term will be zero if and only, for each (j,k) such that $z_{j,k} \neq 0$, $(\mathbf{c}^{(j)})^\top \cdot \mathbf{w}^{(k)} = 0$. Finally, we see that $(\mathbf{c}^{(j)})^\top \cdot \mathbf{w}^{(k)} = 0$ under the exact same conditions in both games, namely, if and only $j = j_k$ (due to the conditioning at the beginning of this proof). This completes the proof of the claim. □

Claim. For $\ell = 1, \ldots, J$, $\mathcal{A}$ cannot distinguish between $\mathsf{Expt}_{2,\ell-1}$ and $\mathsf{Expt}_{2,\ell}$ except with $\mathsf{negl}(\lambda)$ advantage.

Proof. This follows from a straightforward reduction to the generic group security of small superset obfuscation with the simulator specified in the proof of Theorem 1 (which initializes the adversary with $t+1$ uniformly random group elements). Let $\hat{W}$ refer to the set of formal variables $\{\hat{w}_{k,i}\}_{k \in [K], i \in [t]}$. Notice that the only difference between $\mathsf{Expt}_{2,\ell-1}$ and $\mathsf{Expt}_{2,\ell}$ is whether $\mathbf{c}^{(\ell)}$ is a uniformly random vector or an obfuscation of X_ℓ. Consider a reduction $\mathcal{B}$ interacting with the generic group model game for small superset obfuscation. $\mathcal{B}$ associates the $t+1$ handles it receives with $\mathbf{c}^{(\ell)}$, which it sets to be formal variables $\hat{c}_1, \ldots, \hat{c}_{t+1}$. Let $\hat{C}$ refer to this set of formal variables. It can now simulate $\mathsf{Expt}_{2,\ell-1}$ or $\mathsf{Expt}_{2,\ell}$ for $\mathcal{A}$, maintaining its table with polynomials over $\hat{C}$ and $\hat{W}$. Whenever $\mathcal{A}$ make a zero-test query, $\mathcal{B}$ stratifies the resulting polynomial by the $\hat{W}$ variables, considering separately each coefficient on $\hat{w}_{k,i}$. Note that by the restrictions imposed by the bilinear generic group model, each such coefficient must be a linear polynomial over the $\hat{C}$ variables. Therefore, $\mathcal{B}$ can determine whether it is zero via a zero-test query to its own generic group oracle. Combining the results, $\mathcal{B}$ can respond appropriately to $\mathcal{A}$. If $\mathcal{B}$'s generic group oracle is implementing the valid obfuscation, then $\mathcal{A}$ sees exactly $\mathsf{Expt}_{2,\ell-1}$. If $\mathcal{B}$'s generic group oracle is initializing $\mathcal{B}$ with $t+1$ random elements, then $\mathcal{A}$ sees exactly $\mathsf{Expt}_{2,\ell}$. This completes the proof of the claim. □

Acknowledgements. This research was supported in part by ARO and DARPA Safeware under contracts W911NF-15-C-0227, W911NF-15-C-0236, W911NF-16-1-0389, W911NF-15-C-0213, and by NSF grants CNS-1633282, 1562888, 1565208, and 1814919. Any opinions, findings and conclusions or recommendations expressed in this material are those of the authors and do not necessarily reflect the views of the ARO and DARPA.

References

[AAB+15] Agrawal, S., Agrawal, S., Badrinarayanan, S., Kumarasubramanian, A., Prabhakaran, M., Sahai, A.: On the practical security of inner product functional encryption. In: Katz, J. (ed.) PKC 2015. LNCS, vol. 9020, pp. 777–798. Springer, Heidelberg (2015). https://doi.org/10.1007/978-3-662-46447-2_35

[ABF16] Arriaga, A., Barbosa, M., Farshim, P.: Private functional encryption: indistinguishability-based definitions and constructions from obfuscation. In: Dunkelman, O., Sanadhya, S.K. (eds.) INDOCRYPT 2016. LNCS, vol. 10095, pp. 227–247. Springer, Cham (2016). https://doi.org/10.1007/978-3-319-49890-4_13

[BBC+14] Barak, B., Bitansky, N., Canetti, R., Kalai, Y.T., Paneth, O., Sahai, A.: Obfuscation for evasive functions. In: Lindell, Y. (ed.) TCC 2014. LNCS, vol. 8349, pp. 26–51. Springer, Heidelberg (2014). https://doi.org/10.1007/978-3-642-54242-8_2

[BCKP14] Bitansky, N., Canetti, R., Kalai, Y.T., Paneth, O.: On virtual grey box obfuscation for general circuits. In: Garay, J.A., Gennaro, R. (eds.) CRYPTO 2014. LNCS, vol. 8617, pp. 108–125. Springer, Heidelberg (2014). https://doi.org/10.1007/978-3-662-44381-1_7

[BF01] Boneh, D., Franklin, M.K.: Identity-based encryption from the Weil pairing. In: Kilian, J. (ed.) CRYPTO 2001. LNCS, vol. 2139, pp. 213–229. Springer, Heidelberg (2001). https://doi.org/10.1007/3-540-44647-8_13

[BGI+01] Barak, B., et al.: On the (im)possibility of obfuscating programs. In: Kilian, J. (ed.) CRYPTO 2001. LNCS, vol. 2139, pp. 1–18. Springer, Heidelberg (2001). https://doi.org/10.1007/3-540-44647-8_1

[BGMZ18] Bartusek, J., Guan, J., Ma, F., Zhandry, M.: Return of GGH15: provable security against zeroizing attacks. In: Beimel, A., Dziembowski, S. (eds.) TCC 2018. LNCS, vol. 11240, pp. 544–574. Springer, Cham (2018). https://doi.org/10.1007/978-3-030-03810-6_20

[BKM+18] Bishop, A., Kowalczyk, L., Malkin, T., Pastro, V., Raykova, M., Shi, K.: A simple obfuscation scheme for pattern-matching with wildcards. In: Shacham, H., Boldyreva, A. (eds.) CRYPTO 2018. LNCS, vol. 10993, pp. 731–752. Springer, Cham (2018). https://doi.org/10.1007/978-3-319-96878-0_25

[BLMZ19] Bartusek, J., Lepoint, T., Ma, F., Zhandry, M.: New techniques for obfuscating conjunctions. In: Ishai, Y., Rijmen, V. (eds.) EUROCRYPT 2019. LNCS, vol. 11478, pp. 636–666. Springer, Cham (2019). https://doi.org/10.1007/978-3-030-17659-4_22

[BMSZ16] Badrinarayanan, S., Miles, E., Sahai, A., Zhandry, M.: Post-zeroizing obfuscation: new mathematical tools, and the case of evasive circuits. In: Fischlin, M., Coron, J.-S. (eds.) EUROCRYPT 2016. LNCS, vol. 9666, pp. 764–791. Springer, Heidelberg (2016). https://doi.org/10.1007/978-3-662-49896-5_27

[BR13] Brakerski, Z., Rothblum, G.N.: Obfuscating conjunctions. In: Canetti, R., Garay, J.A. (eds.) CRYPTO 2013. LNCS, vol. 8043, pp. 416–434. Springer, Heidelberg (2013). https://doi.org/10.1007/978-3-642-40084-1_24

[BR17] Brakerski, Z., Rothblum, G.N.: Obfuscating conjunctions. J. Crypt. **30**(1), 289–320 (2017)

[BRS13a] Boneh, D., Raghunathan, A., Segev, G.: Function-private identity-based encryption: hiding the function in functional encryption. In: Canetti, R., Garay, J.A. (eds.) CRYPTO 2013. LNCS, vol. 8043, pp. 461–478. Springer, Heidelberg (2013). https://doi.org/10.1007/978-3-642-40084-1_26

[BRS13b] Boneh, D., Raghunathan, A., Segev, G.: Function-private subspace-membership encryption and its applications. In: Sako, K., Sarkar, P. (eds.) ASIACRYPT 2013. LNCS, vol. 8269, pp. 255–275. Springer, Heidelberg (2013). https://doi.org/10.1007/978-3-642-42033-7_14

[BSW09] Bethencourt, J., Song, D., Waters, B.: New techniques for private stream searching. ACM Trans. Inf. Syst. Secur. (TISSEC) **12**(3), 16 (2009)

[BW07] Boneh, D., Waters, B.: Conjunctive, subset, and range queries on encrypted data. In: Vadhan, S.P. (ed.) TCC 2007. LNCS, vol. 4392, pp. 535–554. Springer, Heidelberg (2007). https://doi.org/10.1007/978-3-540-70936-7_29

[BW19] Beullens, W., Wee, H.: Obfuscating simple functionalities from knowledge assumptions. In: Lin, D., Sako, K. (eds.) PKC 2019. LNCS, vol. 11443, pp. 254–283. Springer, Cham (2019). https://doi.org/10.1007/978-3-030-17259-6_9

[Can97] Canetti, R.: Towards realizing random oracles: hash functions that hide all partial information. In: Kaliski, B.S. (ed.) CRYPTO 1997. LNCS, vol. 1294, pp. 455–469. Springer, Heidelberg (1997). https://doi.org/10.1007/BFb0052255

[CRV10] Canetti, R., Rothblum, G.N., Varia, M.: Obfuscation of hyperplane membership. In: Micciancio, D. (ed.) TCC 2010. LNCS, vol. 5978, pp. 72–89. Springer, Heidelberg (2010). https://doi.org/10.1007/978-3-642-11799-2_5

[DS05] Dodis, Y., Smith, A.: Correcting errors without leaking partial information. In: 37th ACM STOC (2005)

[GKW17] Goyal, R., Koppula, V., Waters, B.: Lockable obfuscation. In: 58th FOCS (2017)

[GW11] Gentry, C., Wichs, D.: Separating succinct non-interactive arguments from all falsifiable assumptions. In: 43rd ACM STOC (2011)

[ITZ16] Iovino, V., Tang, Q., Zebrowski, K.: On the power of public-key function-private functional encryption. In: CANS 2016 (2016)

[Jou04] Joux, A.: A one round protocol for tripartite Diffie-Hellman. J. Cryptol. **17**(4), 263–276 (2004)

[KLM+18] Kim, S., Lewi, K., Mandal, A., Montgomery, H., Roy, A., Wu, D.J.: Function-hiding inner product encryption is practical. In: Catalano, D., De Prisco, R. (eds.) SCN 2018. LNCS, vol. 11035, pp. 544–562. Springer, Cham (2018). https://doi.org/10.1007/978-3-319-98113-0_29

[KSW08] Katz, J., Sahai, A., Waters, B.: Predicate encryption supporting disjunctions, polynomial equations, and inner products. In: Smart, N. (ed.) EUROCRYPT 2008. LNCS, vol. 4965, pp. 146–162. Springer, Heidelberg (2008). https://doi.org/10.1007/978-3-540-78967-3_9

[KSW13] Katz, J., Sahai, A., Waters, B.: Predicate encryption supporting disjunctions, polynomial equations, and inner products. J. Cryptol. **26**(2), 191–224 (2013)

[LPS04] Lynn, B., Prabhakaran, M., Sahai, A.: Positive results and techniques for obfuscation. In: Cachin, C., Camenisch, J.L. (eds.) EUROCRYPT 2004. LNCS, vol. 3027, pp. 20–39. Springer, Heidelberg (2004). https://doi.org/10.1007/978-3-540-24676-3_2

[Mau05] Maurer, U.M.: Abstract models of computation in cryptography. In: Smart, N.P. (ed.) Cryptography and Coding 2005. LNCS, vol. 3796, pp. 1–12. Springer, Heidelberg (2005). https://doi.org/10.1007/11586821_1

[Nao03] Naor, M.: On cryptographic assumptions and challenges. In: Boneh, D. (ed.) CRYPTO 2003. LNCS, vol. 2729, pp. 96–109. Springer, Heidelberg (2003). https://doi.org/10.1007/978-3-540-45146-4_6

[Nec94] Nechaev, V.I.: Complexity of a determinate algorithm for the discrete logarithm. Math. Notes **55**(2), 165–172 (1994)

[OS07] Ostrovsky, R., Skeith, W.E.: Private searching on streaming data. J. Cryptol. **20**(4), 397–430 (2007)

[PM18] Patranabis, S., Mukhopadhyay, D.: New lower bounds on predicate entropy for function private public-key predicate encryption. Cryptology ePrint Archive, Report 2018/190 (2018). https://eprint.iacr.org/2018/190

[PMR19] Patranabis, S., Mukhopadhyay, D., Ramanna, S.C.: Function private predicate encryption for low min-entropy predicates. In: Lin, D., Sako, K. (eds.) PKC 2019. LNCS, vol. 11443, pp. 189–219. Springer, Cham (2019). https://doi.org/10.1007/978-3-030-17259-6_7

[Sho97] Shoup, V.: Lower bounds for discrete logarithms and related problems. In: Fumy, W. (ed.) EUROCRYPT 1997. LNCS, vol. 1233, pp. 256–266. Springer, Heidelberg (1997). https://doi.org/10.1007/3-540-69053-0_18

[SSW09] Shen, E., Shi, E., Waters, B.: Predicate privacy in encryption systems. In: Reingold, O. (ed.) TCC 2009. LNCS, vol. 5444, pp. 457–473. Springer, Heidelberg (2009). https://doi.org/10.1007/978-3-642-00457-5_27

[SWP00] Song, D.X., Wagner, D., Perrig, A.: Practical techniques for searches on encrypted data. In: 2000 IEEE Symposium on Security and Privacy (2000)

[Wee05] Wee, H.: On obfuscating point functions. In: 37th ACM STOC (2005)

[WZ17] Wichs, D., Zirdelis, G.: Obfuscating compute-and-compare programs under LWE. In: 58th FOCS (2017)

Multi-Client Functional Encryption for Linear Functions in the Standard Model from LWE

Benoît Libert[1,2(✉)] and Radu Ţiţiu[2,3]

[1] CNRS, Laboratoire LIP, Lyon, France
[2] ENS de Lyon, Laboratoire LIP (U. Lyon, CNRS, ENSL, INRIA, UCBL), Lyon, France
benoit.libert@ens-lyon.fr, radu.titiu@gmail.com
[3] Bitdefender, Bucharest, Romania

Abstract. Multi-client functional encryption (MCFE) allows ℓ clients to encrypt ciphertexts $(\mathbf{C}_{t,1}, \mathbf{C}_{t,2}, \ldots, \mathbf{C}_{t,\ell})$ under some label. Each client can encrypt his own data X_i for a label t using a private encryption key ek_i issued by a trusted authority in such a way that, as long as all $\mathbf{C}_{t,i}$ share the same label t, an evaluator endowed with a functional key dk_f can evaluate $f(X_1, X_2, \ldots, X_\ell)$ without learning anything else on the underlying plaintexts X_i. Functional decryption keys can be derived by the central authority using the master secret key. Under the Decision Diffie-Hellman assumption, Chotard *et al.* (Asiacrypt 2018) recently described an adaptively secure MCFE scheme for the evaluation of linear functions over the integers. They also gave a decentralized variant (DMCFE) of their scheme which does not rely on a centralized authority, but rather allows encryptors to issue functional secret keys in a distributed manner. While efficient, their constructions both rely on random oracles in their security analysis. In this paper, we build a standard-model MCFE scheme for the same functionality and prove it fully secure under adaptive corruptions. Our proof relies on the Learning-With-Errors (LWE) assumption and does not require the random oracle model. We also provide a decentralized variant of our scheme, which we prove secure in the static corruption setting (but for adaptively chosen messages) under the LWE assumption.

Keywords: Multi-client functional encryption · Inner product evaluation · LWE · Standard model · Decentralization

1 Introduction

Functional encryption (FE) [19,62] is a modern paradigm that overcomes the all-or-nothing nature of ordinary encryption schemes. In FE, the master secret key msk allows deriving a sub-key dk_f associated with a specific function f. If a ciphertext C encrypts a message X under the master public key mpk, when dk_f

S. D. Galbraith and S. Moriai (Eds.): ASIACRYPT 2019, LNCS 11923, pp. 520–551, 2019.
https://doi.org/10.1007/978-3-030-34618-8_18

is used to decrypt C, the decryptor only obtains $f(X)$ and nothing else about X. Functional encryption is an extremely general concept as it subsumes identity-based encryption [17,29], searchable encryption [16], attribute-based encryption [44,62], broadcast encryption [32] and many others.

As formalized by Boneh, Sahai and Waters [19], FE only allows evaluating a function f over data provided by a single sender whereas many natural applications require to compute over data coming from distinct distrustful parties. A straightforward solution to handle multiple senders is to distribute the generation of ciphertexts by means of a multi-party computation (MPC) protocol. Unfortunately, jointly generating a ciphertext incurs potentially costly interactions between the senders who should be online at the same time and have their data ready to be submitted. Ideally, the participants should be able to supply their input without interacting with one another and go off-line immediately after having sent their contribution. This motivates the concepts of multi-input [37,38] and multi-client [37,43] functional encryption, which support the evaluation of multivariate functions over data coming from distinct sources.

1.1 (Decentralized) Multi-Client FE

Multi-client functional encryption. As defined in [37,43], multi-client functional encryption (MCFE) allows computing over input vectors $(X_1, \ldots, X_\ell)$ of which each coordinate X_i may be sent by a different client. Each ciphertext C_i is associated with a client index i and a tag t (also called "label"): on input of a vector of ciphertexts $(C_1 = \mathsf{Encrypt}(1, X_1, t), \ldots, C_\ell = \mathsf{Encrypt}(\ell, X_\ell, t))$, where C_i is generated by client i using a secret encryption key ek_i for each $i \in [\ell]$, anyone holding a functional decryption key dk_f for an ℓ-ary function can compute $f(X_1, \ldots, X_\ell)$ as long as all C_i are labeled with the same tag t (which may be a time-specific information or a dataset name). No further information than $f(X_1, \ldots, X_\ell)$ is revealed about individual inputs X_i and nothing can be inferred by combining ciphertexts generated for different tags. MCFE can thus be seen as a multi-party computation (MPC) where each ciphertext C_i can be generated independently of others and no communication is needed between data providers.

Decentralized multi-client functional encryption. Most FE flavors involve a single central authority that should not only be trusted by all users, but also receives the burden of generating all functional secret keys. In decentralized FE systems [24,52], multiple authorities can operate independently without even being aware of one another.

Like its single-client counterpart, multi-client FE requires a trusted entity, which is assigned the task of generating a master key msk as well as handing out encryption keys ek_i to all clients and functional decryption keys dk_f to all decryptors. In some applications, clients may be reluctant to rely on a single point of trust. This motivates the design of a decentralized version of MCFE, as introduced by Chotard *et al.* [27]. Decentralized multi-client functional encryption (DMCFE) obviates the need for a centralized authority by shifting the task of generating functional secret keys to the clients themselves. In a setup phase, the clients $\mathcal{S}_1, \ldots, \mathcal{S}_\ell$ first generate public parameters by running an interactive protocol but no further interaction is needed among clients when it comes to generating functional secret keys later on. When a decryptor wishes to obtain a functional secret key for an ℓ-ary function f, it interacts with each client i independently so as to obtain partial functional decryption keys $\mathsf{dk}_{f,i}$. The decryptor can then fold $\{\mathsf{dk}_{f,i}\}_{i=1}^{\ell}$ into a functional decryption key dk_f for f. By doing so, each client has full control over his individual data and the functions for which secret keys are given out. Importantly, no interaction among senders is required beyond the setup phase, where public parameters are generated.

As a motivating example, Chotard *et al.* [27] consider the use-case of a financial analyst that is interested in mining several companies' private data so as to better understand the dynamics of an economical sector. These companies have some incentives to collaborate, but they do not want their clients' data to be abused (in which case, they would risk heavy fines owing to the EU General Data Protection Regulation). After having interactively set up DMCFE parameters, each company can encrypt its own data with respect to a time-stamp. Then, the analyst can contact each company to obtain partial functional keys and reconstruct a key that only reveals a weighted aggregate of companies' private inputs provided they are labeled with the same time-stamp.

Chotard *et al.* [27] described a DMCFE scheme that allows evaluating linear functions over encrypted data: namely, if $(X_1, \ldots, X_\ell) \in \mathbb{Z}^\ell$ are the individual contributions sent by ℓ senders, a functional secret key dk_f for the integer vector $\boldsymbol{y} = (y_1, \ldots, y_\ell) \in \mathbb{Z}^\ell$ allows computing $\sum_{i=1}^{\ell} y_i \cdot X_i$ from $\{C_i = \mathsf{Encrypt}(i, X_i, t)\}_{i=1}^{\ell}$, where C_i is generated by the i-th sender. In the decentralized setting, each sender can also generate a partial functional secret key $\mathsf{dk}_{f,i}$ for $\boldsymbol{y} = (y_1, \ldots, y_\ell) \in \mathbb{Z}^\ell$ using their secret encryption key ek_i.

1.2 Our Contributions

The MCFE scheme of Chotard *et al.* [27] was proved fully secure (as opposed to selectively secure) in the random oracle model under the standard Decision Diffie-Hellman assumption in groups without a bilinear maps. Its decentralized variant was proved secure under the Symmetric eXternal Diffie-Hellman (SXDH) assumption in groups endowed with an asymmetric bilinear map. While efficient, the schemes of [27] both require the random oracle model. Chotard *et al.* thus left open the problem of designing a (D)MCFE system under well-studied hardness assumptions without using random oracles: even in the centralized setting, the only known MCFE candidates in the standard model [37,43] rely on indistinguishability obfuscation. They also left open the problem of instantiating their schemes under the LWE assumption or any other assumption than DDH.

In this paper, we address both problems. For linear functions over the integers (i.e., the same functionality as [27]), we construct the first MCFE scheme in the standard model and prove it fully secure under the Learning-With-Errors assumption [60] in the adaptive corruption setting (note that only static corruptions were considered in [43, Section 2.3]). This construction turns out to be the first standard-model realization of an MCFE system with labels – albeit for a restricted functionality – that does not require obfuscation. Next, we extend our centralized system to obtain the first labeled DMCFE scheme without random oracles. Like [27], our decentralized solution is only proved secure in the static corruption setting although we can handle adaptive corruptions in its centralized version. Both constructions are proved secure under the LWE assumption with sub-exponential approximation factors. Our security proofs stand in the standard model in the sense of the same security definitions as those considered in [27].

We leave it as an open problem to achieve security under an LWE assumption with polynomial approximation factor. Another natural open question is the feasibility of (D)MCFE beyond linear functions under standard assumptions.

1.3 Challenges and Techniques

We start from the observation that the DDH-based MCFE scheme of Chotard *et al.* [27] can be interpreted as relying on (a variant of) the key-homomorphic pseudorandom function [18] of Naor, Pinkas and Reingold [58]. Namely, the scheme of [27] encrypts $x_i \in \mathbb{Z}_q$ for the tag t by computing $C_i = g^{x_i} \cdot H_{t,1}^{s_i} \cdot H_{t,2}^{t_i}$, where $(s_i, t_i) \in \mathbb{Z}_q^2$ is the i-th sender's secret key and $(H_{t,1}, H_{t,2}) = H(t) \in \mathbb{G}^2$ is derived from a random oracle in a DDH-hard group $\mathbb{G} = \langle g \rangle$.

The security proof of [27] crucially exploits the entropy of the secret key (s_i, t_i) in a hybrid argument over all encryption queries. To preserve this entropy, they need to prevent the encryption oracle from leaking too much about uncorrupted users' secret keys $\{(s_i, t_i)\}_i$. For this purpose, they rely on the DDH assumption to modify the random oracle $H : \{0,1\}^* \to \mathbb{G}^2$ in such a way that, in all encryption queries but one, the hash value $H(t) \in \mathbb{G}^2$ lives in a one-dimensional subspace. In order to transpose this technique in the standard model, we would need a programmable hash function [46] that ranges over a one-dimensional subspace of $\mathbb{G}^2$ on polynomially-many inputs while mapping an extra input outside this subspace with noticeable probability. The results of Hanaoka *et al.* [45] hint that such programmable hash functions are hardly instantiable in prime-order DDH groups. While the multi-linear setting [33] allows bypassing the impossibility results of [45], it is not known to enable standard assumptions.

A natural idea is to replace the random-oracle-based key-homomorphic PRF of [58] by an LWE-based key-homomorphic PRF [11,18]. However, analogously to Chotard *et al.* [27],[1] we aim at an MCFE system that can be proved secure in a game where the adversary is allowed to corrupt senders adaptively. In order to deal with the adaptive corruption of senders, we thus turn to the adaptively secure distributed PRF proposed by Libert, Stehlé and Titiu [55]. The latter can be seen as instantiating the programmable hash function of Freire *et al.* [33] in the context of homomorphic encryption (FHE). Their PRF maps an input x

[1] While their decentralized scheme is only proved secure under static corruptions, its centralized version is proved secure under adaptive corruptions.

to $\lfloor \mathbf{A}(x)^\top \cdot \mathbf{s} \rfloor_p$, where[2] $\mathbf{s} \in \mathbb{Z}^n$ is the secret key and $\mathbf{A}(x) \in \mathbb{Z}_q^{n \times m}$ is derived from public matrices using the Gentry-Sahai-Waters FHE [36]. More precisely, the matrix $\mathbf{A}(x)$ is obtained as the product of GSW ciphertexts dictated by the output of an admissible hash function [15] applied to the PRF input. The security proof of [55] uses the property that, with noticeable probability, the input-dependent matrix $\mathbf{A}(x)$ is a GSW encryption of 1 for the challenge input $x^\star$: namely, $\mathbf{A}(x^\star)$ is a matrix of of the form $\mathbf{A}(x^\star) = \mathbf{A} \cdot \mathbf{R}^\star + \mathbf{G}$, where $\mathbf{G} \in \mathbb{Z}_q^{n \times m}$ is the gadget matrix of Micciancio and Peikert [57] and $\mathbf{R}^\star \in \mathbb{Z}^{m \times m}$ is a small-norm matrix. At the same time, all evaluation queries are associated with a matrix $\mathbf{A}(x)$ consisting of a GSW encryption of 0 (i.e., a matrix $\mathbf{A}(x) = \mathbf{A} \cdot \mathbf{R}$, for a small-norm $\mathbf{R} \in \mathbb{Z}^{m \times m}$). Then, the proof of [55] appeals to the lossy mode of LWE [39] and replaces the uniform matrix $\mathbf{A}^\top \in \mathbb{Z}_q^{m \times n}$ by a lossy matrix of the form $\hat{\mathbf{A}}^\top \cdot \mathbf{C} + \mathbf{E}$, where $\mathbf{E} \in \mathbb{Z}^{m \times n}$ is a short integer matrix with Gaussian entries, $\mathbf{C} \in \mathbb{Z}_q^{n_1 \times n}$ is random, and $\hat{\mathbf{A}} \in \mathbb{Z}_q^{n_1 \times m}$ has rank $n_1 \ll n$. In all evaluation queries, the smallness of $\mathbf{s} \in \mathbb{Z}^n$ then ensures that the values $\lfloor \mathbf{A}(x)^\top \cdot \mathbf{s} \rfloor_p$ always reveal the same information about $\mathbf{s}$, which amounts to the product $\mathbf{C} \cdot \mathbf{s} \in \mathbb{Z}_q^{n_1}$. Since $\mathbf{A}(x^\star)$ depends on $\mathbf{G}$ for the challenge input $x^\star$, the function $\lfloor \mathbf{A}(x^\star)^\top \cdot \mathbf{s} \rfloor_p$ is in fact an injective function of $\mathbf{s}$, meaning that it has high min-entropy.

Our MCFE scheme relies on the lossy mode of LWE in a similar way to [55], except that we add a Gaussian noise instead of using the Learning-With-Rounding technique [12]. The i-th sender uses his secret key $\mathbf{s}_i \in \mathbb{Z}^n$ to encrypt a short integer vector as $\boldsymbol{x}_i \in \mathbb{Z}^{n_0}$ as $\mathbf{C}_i = \mathbf{G}_0^\top \cdot \boldsymbol{x}_i + \mathbf{A}(t)^\top \cdot \mathbf{s}_i + \mathsf{noise} \in \mathbb{Z}_q^m$, where $\mathbf{A}(t) \in \mathbb{Z}_q^{n \times m}$ is a tag-dependent matrix derived as a product of GSW ciphertexts indexed by the bits of t and $\mathbf{G}_0 \in \mathbb{Z}_q^{n_0 \times m}$ is a gadget matrix for which the lattice $\Lambda^\perp(\mathbf{G}_0)$ has a short public basis. A functional secret key for the vector $\boldsymbol{y} = (y_1, \ldots, y_\ell)^\top$ consists of $\mathsf{dk}_{\boldsymbol{y}} = \sum_{i=1}^{\ell} y_i \cdot \mathbf{s}_i \in \mathbb{Z}^n$ and allows computing $\mathbf{G}_0^\top \cdot (\sum_{i=1} y_i \cdot \boldsymbol{x}_i) + \mathsf{small} \in \mathbb{Z}_q^m$ from $\sum_{i=1}^{\ell} y_i \cdot \mathbf{C}_i \in \mathbb{Z}_q^m$ and eventually recovering the linear function $\sum_{i=1} y_i \cdot \boldsymbol{x}_i \in \mathbb{Z}^{n_0}$ of $\mathbf{X} = [\boldsymbol{x}_1 \mid \ldots \mid \boldsymbol{x}_\ell] \in \mathbb{Z}_q^{n_0 \times \ell}$.

At this point, adapting the security proof of [55] is non-trivial. We cannot rely on the DPRF of [55] in a modular way as it would require a DPRF where partial evaluations are themselves pseudorandom so long as the adversary does not obtain the underlying secret key shares: in our setting, a challenge ciphertext contains a bunch of partial evaluations (one for each message slot) rather than a threshold recombination of such evaluations. We emphasize that, in the LWE-based DPRF of [55], partial evaluations are not proven pseudorandom: [55] only proves – via a deterministic randomness extraction argument – the pseudorandomness of the final PRF value obtained by combining partial evaluations. They cannot apply (and neither can we) a randomness extractor to individual partial DPRF evaluations as it would destroy their key homomorphic property. Instead of relying on the pseudorandomness of partial evaluations, we actually prove a milder indistinguishability property which suffices for our purposes.

[2] Introduced in [12], the notation $\lfloor x \rfloor_p$ stands for the rounded value $\lfloor (p/q) \cdot x \rfloor \in \mathbb{Z}_p$, where $x \in \mathbb{Z}_q$, and $p < q$.

The first step is to make sure that all encryption queries will involve a lossy matrix $\mathbf{A}(t)^\top = \mathbf{R}_t \cdot \hat{\mathbf{A}}^\top \cdot \mathbf{C} + \mathbf{E}_t$, for small-norm $\mathbf{R}_t \in \mathbb{Z}^{m \times m}$ and $\mathbf{E}_t \in \mathbb{Z}^{m \times n}$, so that honest senders' ciphertexts are of the form $\mathbf{C}_i = \mathbf{G}_0^\top \cdot \boldsymbol{x}_i + \mathbf{R}_t \cdot \hat{\mathbf{A}}^\top \cdot \mathbf{C} \cdot \mathbf{s}_i + \mathsf{noise}$ and thus leak nothing about $\mathbf{s}_i \in \mathbb{Z}^n$ beyond $\mathbf{C} \cdot \mathbf{s}_i \in \mathbb{Z}_q^{n_1}$. The difficulty arises in the challenge queries $(i, t^\star, \boldsymbol{x}_{0,i}^\star, \boldsymbol{x}_{1,i}^\star)$, where $\mathbf{A}(t^\star) \in \mathbb{Z}_q^{n \times m}$ is not a lossy matrix and we must find a way to replace $\mathbf{C}_i^\star = \mathbf{G}_0^\top \cdot \boldsymbol{x}_{0,i}^\star + \mathbf{A}(t^\star)^\top \cdot \mathbf{s}_i + \mathsf{noise}$ by $\mathbf{C}_i^\star = \mathbf{G}_0^\top \cdot \boldsymbol{x}_{1,i}^\star + \mathbf{A}(t^\star)^\top \cdot \mathbf{s}_i + \mathsf{noise}$ without the adversary noticing. In [55], the proof relies on a deterministic randomness extraction[3] argument to extract statistically uniform bits from $\lfloor \mathbf{A}(x^\star)^\top \cdot \mathbf{s} \rfloor_p$, which has high min-entropy when $\mathbf{A}(x^\star)$ is of the form $\mathbf{A} \cdot \mathbf{R}^\star + \mathbf{G}$. Here, we do not see how to apply deterministic extractors in the proof while preserving the functionality of the MCFE scheme.

Our solution is to program the public parameters in such a way that, with noticeable probability, the challenge ciphertexts are generated for a matrix $\mathbf{A}(t^\star) \in \mathbb{Z}_q^{n \times m}$ of the form

$$\mathbf{A}(t^\star)^\top = \mathbf{R}^\star \cdot \mathbf{A}^\top + \mathbf{G}_0^\top \cdot \mathbf{V} = \mathbf{R}^\star \cdot \hat{\mathbf{A}}^\top \cdot \mathbf{C} + \mathbf{G}_0^\top \cdot \mathbf{V} + \mathsf{noise}, \tag{1}$$

for a statistically random matrix $\mathbf{V} \in \mathbb{Z}_q^{n_0 \times n}$ included in the public parameters. In the proof, the simulator generates a statistically uniform matrix $\mathbf{U} = \left[\begin{smallmatrix}\mathbf{V}\\ \mathbf{C}\end{smallmatrix}\right]$, where $\mathbf{C} \in \mathbb{Z}_q^{n_1 \times n}$ is used to build the lossy matrix $\mathbf{A}^\top = \hat{\mathbf{A}}^\top \cdot \mathbf{C} + \mathbf{E}$, together with a trapdoor $\mathbf{T}_\mathbf{U}$ for $\Lambda^\perp(\mathbf{U})$. (The idea of embedding a trapdoor in the LWE secret of a lossy matrix is borrowed from [54]). Using $\mathbf{T}_\mathbf{U}$, the simulator can sample a short matrix $\mathbf{T} \in \mathbb{Z}^{n \times n_0}$ satisfying $\mathbf{U} \cdot \mathbf{T} = \left[\begin{smallmatrix}\mathbf{I}_{n_0}\\ \mathbf{0}\end{smallmatrix}\right] \bmod q$, allowing it to define an alternative secret key $\mathbf{s}_i' = \mathbf{s}_i + \mathbf{T} \cdot (\boldsymbol{x}_{0,i}^\star - \boldsymbol{x}_{1,i}^\star) \in \mathbb{Z}^n$. As long as $\mathbf{s}_i$ is sampled from a Gaussian distribution with sufficiently large standard deviation, $\mathbf{s}_i'$ and $\mathbf{s}_i$ are negligibly far apart in terms of statistical distance (note that, as in [13,67], the simulator can guess $\boldsymbol{x}_{0,i}^\star - \boldsymbol{x}_{1,i}^\star$ upfront without affecting the polynomial running time of the reduction since we are in the middle of a purely statistical argument). The alternative secret keys $\{\mathbf{s}_i'\}_{i=1}^\ell$ further satisfy $\sum_{i=1}^\ell y_i \cdot \mathbf{s}_i' = \sum_{i=1}^\ell y_i \cdot \mathbf{s}_i$ for all legal functional key queries $\boldsymbol{y} = (y_1, \ldots, y_\ell)$ made by the adversary. The definition of $\mathbf{s}_i'$ finally ensures that $\mathbf{C} \cdot \mathbf{s}_i' = \mathbf{C} \cdot \mathbf{s}_i \bmod q$, meaning that $\mathbf{s}_i'$ is compatible with all encryption queries for which $\mathbf{A}(t)$ is lossy. From (1), the condition $\mathbf{V} \cdot \mathbf{T} = \mathbf{I}_{n_0} \bmod q$ then implies that the challenge ciphertext can be interpreted as an encryption of $\boldsymbol{x}_{1,i}^\star$ since $\mathbf{C}_i^\star = \mathbf{G}_0^\top \cdot \boldsymbol{x}_{1,i}^\star + \mathbf{A}(t^\star)^\top \cdot \mathbf{s}_i' + \mathsf{noise}$ is statistically close to $\mathbf{C}_i^\star = \mathbf{G}_0^\top \cdot \boldsymbol{x}_{0,i}^\star + \mathbf{A}(t^\star)^\top \cdot \mathbf{s}_i + \mathsf{noise}$.

We insist that our construction and proof are not merely obtained by plugging the DPRF of [55] into the high-level design principle of [27]. In particular, we do not rely on the pseudorandomness of partial PRF evaluations, but rather prove a milder indistinguishability property in some transition in our sequence of games. To do this, we need to modify the proof of [55], by introducing a matrix $\mathbf{V}$ and embedding a trapdoor in the matrix $\mathbf{U}$ obtained by stacking up $\mathbf{V}$ and the secret matrix $\mathbf{C}$ of the lossy mode of LWE.

[3] The standard Leftover Hash Lemma cannot be applied since the source $\lfloor \mathbf{A}(x^\star)^\top \cdot \mathbf{s} \rfloor_p$ is not guaranteed to be independent of the seed. A deterministic extractor based on k-wise independent functions [31] is thus needed in [55].

In order to build a DMCFE system, we proceed analogously to [27] and combine two instances of our centralized MCFE scheme. The first one is only used to generate partial functional secret keys whereas the second one is used exactly as in the centralized system. As in [27], we first have the senders run an interactive protocol allowing them to jointly generate public parameters for the two MCFE instances. At the end of this protocol (which may involve costly MPC operations, but is only executed once), each sender holds an encryption key $\mathsf{ek}_i = (\mathbf{s}_i, \mathbf{t}_i)$ consisting of encryption keys for the two underlying instances. In order to have the i-th sender $\mathcal{S}_i$ generate a partial functional secret key $\mathsf{dk}_{f,i}$ for a vector $\boldsymbol{y} = (y_1, \ldots, y_\ell)^\top$, we exploit the fact that our centralized scheme allows encrypting vectors. Namely, the decryptor obtains from $\mathcal{S}_i$ an MCFE encryption of the vector $y_i \cdot \mathbf{s}_i \in \mathbb{Z}^n$ under the encryption key $\mathbf{t}_i$ of the first instance.

1.4 Related Work

Functional encryption was implicitly introduced by Sahai and Waters in [62], where they also constructed a scheme for threshold functions. Constructions of FE for point functions (known as identity-based encryption) [17,29] existed already, but were not viewed through the lens of FE until later. Subsequent works saw constructions for several more advanced functionalities such as inner product functions [7,50], Boolean formulas [44,51,53,59,65], membership checking [20] and even finite state automaton [66]. Recently, the landscape of functional encryption improved considerably. Gorbunov *et al.* [42] and Garg *et al.* [34] provided the first constructions of attribute-based encryption for all circuits; Goldwasser *et al.* [41] constructed succinct simulation-secure single-key FE scheme for all circuits and also obtained FE for Turing machines [40]. In a breakthrough result, Garg *et al.* [34] designed indistinguishability-secure multi-key FE schemes for all circuits. However, while the constructions of [41,42] rely on standard assumptions, the assumptions underlying the other constructions [34,40] are still ill-understood and have not undergone much cryptanalytic effort.

FE FOR SIMPLE CIRCUITS. Abdalla, Bourse, De Caro and Pointcheval [3] considered the question of building FE for linear functions (a functionality dubbed IPFE for "inner product functional encryption"). Here, a ciphertext C encrypts a vector $\boldsymbol{y} \in \mathcal{D}^\ell$ over some ring $\mathcal{D}$, a secret key for the vector $\boldsymbol{x} \in \mathcal{D}^\ell$ allows computing $\langle \boldsymbol{x}, \boldsymbol{y} \rangle$ and nothing else about $\boldsymbol{y}$. Abdalla *et al.* [3] described two constructions under the Decision Diffie-Hellman (DDH) and Learning-With-Errors (LWE) assumptions, respectively. On the downside, Abdalla *et al.* [3] only proved their schemes to be secure against selective adversaries. Namely, in the security game, the adversary chooses two vectors $\boldsymbol{x}_0, \boldsymbol{x}_1 \in \mathcal{D}^\ell$ and expects to receive an encryption of one of these in the challenge phase. Selective security forces the adversary to declare $\boldsymbol{x}_0, \boldsymbol{x}_1$ before seeing the public key and before obtaining any private key. Agrawal, Libert and Stehlé subsequently upgraded the constructions of [3] so as to prove security against adaptive adversaries, which may choose $\boldsymbol{x}_0, \boldsymbol{x}_1$ after having seen the public key and obtained a number of private keys. Agrawal *et al.* [8] described several IPFE schemes under well-established assumptions which include the standard Decision Diffie-Hellman (DDH) assumption, the Decision Composite Residuosity (DCR) assumption and the LWE assumption. Under the DCR and LWE assumptions, the schemes of [8] can evaluate both inner products over the integers and modulo a prime or composite number. The IPFE constructions of [3,8] served as building blocks for FE schemes handling general functionalities [9] in the bounded collusion setting [42,61]. Quite recently, the IPFE functionality [3,8] was extended into FE schemes supporting the evaluation of quadratic functions over encrypted data [10,56]. The schemes of [10,56] are only proved secure against selective adversaries and they can only compute functions which have their output confined in a small interval. For the time being, the only known FE schemes that support the evaluation of more general functions than quadratic polynomials either require fancy tools like obfuscation [34], or are restricted to bounded collusions [9,42].

MULTI-INPUT AND MULTI-CLIENT FUNCTIONAL ENCRYPTION. Goldwasser *et al.* [37,38] introduced the concept of multi-input functional encryption (MIFE). MIFE and MCFE are both more interesting in the secret-key setting than in the public-key setting, where much more information inevitably leaks about the data (see, e.g., [5,27,38]). Similarly to MCFE, MIFE operates over input vectors $(X_1, \ldots, X_\ell)$ comprised of messages sent by distinct parties, but without assigning a tag to ciphertexts: each user i can encrypt X_i as $C_i = \mathsf{Encrypt}(X_i)$ in such a way that anyone equipped with a functional secret key dk_f for an ℓ-argument function f can compute $f(X_1, \ldots, X_n)$ given multiple ciphertexts $\{C_i = \mathsf{Encrypt}(X_i)\}_{i=1}^{\ell}$. Brakerski *et al.* [22] gave a transformation for constructing adaptively secure general-purpose MIFE schemes for a constant n from any general-purpose private-key single-input scheme. Like MCFE, MIFE for general functionalities necessarily rely on indistinguishability obfuscation or multilinear maps, so that instantiations under standard assumptions are currently lacking. Under the SXDH assumption, Abdalla *et al.* [5] managed to construct a MIFE scheme for the inner product functionality. In their scheme, each input slot encrypts a vector $\boldsymbol{x}_i \in \mathbb{Z}_p^m$ while each functional secret key $\mathsf{sk}_{\boldsymbol{y}}$ corresponds to a vector $\boldsymbol{y} \in \mathbb{Z}_p^{\ell \cdot m}$, where ℓ is the total number of slots. On input of encrypted data $\boldsymbol{X} = (\boldsymbol{x}_1, \ldots, \boldsymbol{x}_\ell)$ such that $\boldsymbol{x}_i$ is encrypted by sender i in the i-th slot, their multi-input inner product functionality computes $\langle \boldsymbol{X}, \boldsymbol{y} \rangle$ using $\mathsf{sk}_{\boldsymbol{y}}$. Function-hiding MIFE schemes were described in [4,30]. Abdalla *et al.* [4] notably gave a generic single-input to multi-input transformation, which yields MIFE constructions for the inner product functionality under the DDH, LWE and DCR assumptions.

Besides syntactical differences, MCFE departs from MIFE in the amount of information leaked about plaintexts. The MIFE model [37,38] allows any slot of any ciphertext to be combined with any other slot of any other ciphertext. As soon as senders encrypt more than one ciphertext per slot, a given functional secret key can thus compute a much larger number of values. As discussed in [27], this feature incurs a much more important information leakage, especially when many functional secret keys are given out. In contrast, the multi-client setting only allows functional secret keys to operate over ciphertexts that share the same tag. As long as tags are single-use (e.g., a timestamp), this allows clients to retain a more accurate control over the information leaked about their data.

The first MCFE realization was proposed in [37,43] and relies on the DDH assumption and on indistinguishability obfuscation to handle general circuits. The notion of aggregator-oblivious encryption (AOE) [14,23,48,64] allows an untrusted aggregator to compute sums of encrypted values without learning anything else about individual inputs. As such, AOE can be seen as a form of MCFE with single-key security (namely, the only key revealed to the aggregator is for the vector $(1, 1, \ldots, 1)^\top$) for the evaluation of inner products. So far, all non-interactive AOE constructions [14,48,64] rely on the random oracle model.

The first efficient MCFE scheme with multi-key security was described by Chotard *et al.* [27] who also introduced the concept of decentralized MCFE. Their schemes both rely on DDH-like assumptions in the random oracle model. At the time of writing, we are not aware of any (D)MCFE construction based on a well-studied assumption in the standard model.

DECENTRALIZED FUNCTIONAL ENCRYPTION. The first examples of decentralized FE schemes were given in the context of attribute-based encryption (ABE) [25,26]. Lewko and Waters [52] gave the first ciphertext-policy ABE where users' attributes may be certified by completely independent authorities. Boneh and Zhandry [21] suggested distributed broadcast encryption systems, which dispense with the need for an authority handing out keys to registered users. Chandran *et al.* [24] considered decentralized general-purpose FE using obfuscation. The decentralization of multi-client FE was first considered by Chotard *et al.* [27] in a model where all clients run an interactive protocol to generate public parameters, but eliminate any interaction beyond the setup phase.

Abdalla *et al.* [2] described generic transformations providing DMCFE schemes from any MCFE system satisfying extra properties. While applying their compilers to [4] yields DMCFE schemes in the standard model, the resulting ciphertexts are not labeled. Without labels, the functionality leaks much more information about encrypted messages for a given functional key since there is no restriction on the way slots from different ciphertexts can be combined together (any slot from any ciphertext can be combined with any other slot from any other ciphertext). In this paper, our goal is to support labels, which is significantly more challenging and was only achieved in the random oracle model so far.

Chotard *et al.* [28] gave a technique to remove the restriction that forces the adversary to make challenge queries for all uncorrupted ciphertext slots. Their technique upgrades any MCFE scheme satisfying our definition (which is the definition introduced in [27] and called "pos-IND" security in [2]) so as to prove security under a stronger definition where the adversary can obtain incomplete ciphertexts. Their technique builds on a "secret-sharing layer" (SSL) primitive which is only known to exist assuming pairings and random oracles as their SSL scheme [28, Section 4.2] is implicitly based on the Boneh-Franklin IBE [17]. Abdalla *et al.* [2] suggested a different technique to handle incomplete ciphertexts without using pairings, but they either require random oracles or they do not support labels (except in a model with static corruptions and selective security).

Chotard *et al.* [28] also showed how to transform the ROM-based scheme from [27] in such a way that users are allowed multiple encryption queries for each slot-label pair. Their technique is not generic and only works for their DDH-based construction (as they mention in Sect. 6.2). Finally, [2,28] both give generic compilers from MCFE to DMCFE. Abdalla *et al.* [2] obtain DMCFE under adaptive corruptions, but they need to start from an MCFE which computes inner products modulo an integer L (instead of inner products over $\mathbb{Z}$). Hence, their compiler does not imply DMCFE from LWE in the standard model. As it turns out, neither [2,28] implies MCFE with labels in the standard model from LWE (nor any standard assumption), even for the security definition of [27]. In a concurrent and independent work [1], Abdalla *et al.* provide a solution to this problem via a generic construction of labeled MCFE from single-input IPFE schemes evaluating modular inner products. While their construction satisfies a stronger security notion than ours (which allows multiple encryption queries for the same slot-label pair), their scheme of [1, Section 3] requires longer ciphertext than ours as each slot takes a full IPFE ciphertext of linear size in ℓ if ℓ is the number of slots.

In their construction and in ours, handling incomplete ciphertexts expands partial ciphertexts by a factor $O(\ell)$. In our most efficient schemes, we still need to assume that the adversary obtains challenge ciphertexts for all clients as in [27]. In the full version of the paper, we show that a variant of the compiler of Abdalla *et al.* [2] allows proving security in the standard model, even when the adversary is allowed to obtain incomplete challenge ciphertexts. Our compiler relies on pseudorandom functions satisfying a specific security definition in the multi-instance setting. The concurrent work of Abdalla *et al.* [1] achieves a similar result using any PRF satisfying a standard security definition.

2 Background

2.1 Lattices

For any $q \geq 2$, we let $\mathbb{Z}_q$ denote the ring of integers modulo q. For a vector $\mathbf{x} \in \mathbb{R}^n$ denote $\|\mathbf{x}\| = \sqrt{x_1^2 + x_2^2 + \cdots x_n^2}$ and $\|\mathbf{x}\|_\infty = \max_i |x_i|$. If $\mathbf{M}$ is a matrix over $\mathbb{R}$, then $\|\mathbf{M}\| := \sup_{\mathbf{x}\neq 0} \frac{\|\mathbf{M}\mathbf{x}\|}{\|\mathbf{x}\|}$ and $\|\mathbf{M}\|_\infty := \sup_{\mathbf{x}\neq 0} \frac{\|\mathbf{M}\mathbf{x}\|_\infty}{\|\mathbf{x}\|_\infty}$. For a finite set S, we let $U(S)$ denote the uniform distribution over S. If X and Y are distributions over the same domain, then $\Delta(X, Y)$ denotes their statistical distance. Let $\mathbf{\Sigma} \in \mathbb{R}^{n\times n}$ be a symmetric positive-definite matrix, and $\mathbf{c} \in \mathbb{R}^n$. We define the Gaussian function on $\mathbb{R}^n$ by $\rho_{\mathbf{\Sigma},\mathbf{c}}(\mathbf{x}) = \exp(-\pi(\mathbf{x}-\mathbf{c})^\top \mathbf{\Sigma}^{-1}(\mathbf{x}-\mathbf{c}))$ and if $\mathbf{\Sigma} = \sigma^2 \cdot \mathbf{I}_n$ and $\mathbf{c} = \mathbf{0}$ we denote it by ρ_σ. For an n dimensional lattice $\Lambda \subset \mathbb{R}^n$ and for any lattice vector $\mathbf{x} \in \Lambda$ the discrete gaussian is defined $\rho_{\Lambda,\mathbf{\Sigma},\mathbf{c}}(\mathbf{x}) = \frac{\rho_{\mathbf{\Sigma},\mathbf{c}}}{\rho_{\mathbf{\Sigma},\mathbf{c}}(\Lambda)}$. For an n-dimensional lattice Λ, we define $\eta_\varepsilon(\Lambda)$ as the smallest $r > 0$ such that $\rho_{1/r}(\widehat{\Lambda} \setminus \mathbf{0}) \leq \varepsilon$ with $\widehat{\Lambda}$ denoting the dual of Λ, for any $\varepsilon \in (0, 1)$. For a matrix $\mathbf{A} \in \mathbb{Z}_q^{n\times m}$, we define $\Lambda^\perp(\mathbf{A}) = \{\mathbf{x} \in \mathbb{Z}^m : \mathbf{A} \cdot \mathbf{x} = \mathbf{0} \bmod q\}$ and

$\Lambda(\mathbf{A}) = \mathbf{A}^\top \cdot \mathbb{Z}^n + q\mathbb{Z}^m$. For an arbitrary vector $\mathbf{u} \in \mathbb{Z}_q^n$, we also define the shifted lattice $\Lambda^{\mathbf{u}}(\mathbf{A}) = \{\mathbf{x} \in \mathbb{Z}^m : \mathbf{A} \cdot \mathbf{x} = \mathbf{u} \bmod q\}$.

Definition 2.1 (LWE). *Let $m \geq n \geq 1$, $q \geq 2$ and $\alpha \in (0,1)$ be functions of a security parameter λ. The* LWE *problem consists in distinguishing between the distributions $(\mathbf{A}, \mathbf{A}\mathbf{s} + \mathbf{e})$ and $U(\mathbb{Z}_q^{m\times n} \times \mathbb{Z}_q^m)$, where $\mathbf{A} \sim U(\mathbb{Z}_q^{m\times n})$, $\mathbf{s} \sim U(\mathbb{Z}_q^n)$ and $\mathbf{e} \sim D_{\mathbb{Z}^m,\alpha q}$. For an algorithm $\mathcal{A} : \mathbb{Z}_q^{m\times n} \times \mathbb{Z}_q^m \to \{0,1\}$, we define:*

$$\mathbf{Adv}^{\mathsf{LWE}}_{q,m,n,\alpha}(\mathcal{A}) = |\Pr[\mathcal{A}(\mathbf{A}, \mathbf{A}\mathbf{s} + \mathbf{e}) = 1] - \Pr[\mathcal{A}(\mathbf{A}, \mathbf{u}) = 1|,$$

where the probabilities are over $\mathbf{A} \sim U(\mathbb{Z}_q^{m\times n})$, $\mathbf{s} \sim U(\mathbb{Z}_q^n)$, $\mathbf{u} \sim U(\mathbb{Z}_q^m)$ and $\mathbf{e} \sim D_{\mathbb{Z}^m,\alpha q}$ and the internal randomness of $\mathcal{A}$. We say that $\mathsf{LWE}_{q,m,n,\alpha}$ is hard if, for any ppt *algorithm $\mathcal{A}$, the advantage $\mathbf{Adv}^{\mathsf{LWE}}_{q,m,n,\alpha}(\mathcal{A})$ is negligible.*

Micciancio and Peikert [57] described a trapdoor mechanism for LWE. Their technique uses a "gadget" matrix $\mathbf{G} \in \mathbb{Z}_q^{n\times w}$, with $w = n \log q$, for which anyone can publicly sample short vectors $\mathbf{x} \in \mathbb{Z}^w$ such that $\mathbf{G} \cdot \mathbf{x} = \mathbf{0}$.

Lemma 2.2 ([57, Section 5]). *Let $m \geq 3n \log q$. There exists a* ppt *algorithm* GenTrap *that outputs a statistically uniform matrix $\mathbf{A} \in \mathbb{Z}_q^{n\times m}$, together with a trapdoor $\mathbf{T}_{\mathbf{A}} \in \mathbb{Z}^{m\times m}$ for $\Lambda^\perp(\mathbf{A})$, such that $\max_j \|\tilde{\mathbf{t}}_j\| \leq O(\sqrt{n \log q})$, where $\tilde{\mathbf{t}}_j$ are the corresponding Gram-Schmidt vectors.*

It is known [57] that, for any $\mathbf{u} \in \mathbb{Z}_q^n$, a trapdoor for $\mathbf{A} \in \mathbb{Z}_q^{n\times m}$ allows sampling from $D_{\Lambda^{\mathbf{u}}(\mathbf{A}),s\cdot\omega(\sqrt{\log m})}$ for $s = O(\sqrt{n \log q})$. Since

$$\eta_{2^{-m}}\left(\Lambda^\perp(\mathbf{A})\right) \leq \max_j \|\tilde{\boldsymbol{t}}_j\| \cdot \omega(\sqrt{\log m}) \leq s \cdot \omega(\sqrt{\log m})$$

for large enough $s = O(\sqrt{n \log q})$, the magnitude of a vector $\mathbf{x}$ sampled from $D_{\Lambda^{\mathbf{u}}(\mathbf{A}),s\cdot\omega(\sqrt{\log m})}$, is bounded by $\|\mathbf{x}\| \leq s\sqrt{m} \cdot \omega(\sqrt{\log m})$.

Remark 2.3. For $m \geq 3n \log q$, we can thus sample a statistically uniform matrix $\mathbf{A}$ from $\mathbb{Z}_q^{n\times m}$ together with a trapdoor, which allows finding small solutions of $\mathbf{A} \cdot \mathbf{x} = \mathbf{u} \bmod q$, with $\|\mathbf{x}\| \leq s\sqrt{m} \cdot \omega(\sqrt{\log m}) = O(\sqrt{mn \log q}) \cdot \omega(\sqrt{\log m})$.

We sometimes rely on the so-called "noise flooding" technique via the next lemma.

Lemma 2.4 ([39, Lemma 3]). *Let $\boldsymbol{y} \in \mathbb{Z}^m$. The statistical distance between $D_{\mathbb{Z}^m,\sigma}$ and $\boldsymbol{y} + D_{\mathbb{Z}^m,\sigma}$ is at most $\Delta\left(D_{\mathbb{Z}^m,\sigma}, \boldsymbol{y} + D_{\mathbb{Z}^m,\sigma}\right) \leq m \cdot \frac{\|\boldsymbol{y}\|_\infty}{\sigma}$.*

Lemma 2.5 ([35, Theorem 4.1]). *There is a* ppt *algorithm that, given a basis $\mathbf{B}$ of an n-dimensional lattice $\Lambda = \mathcal{L}(\mathbf{B})$, a parameter $s > \|\tilde{\mathbf{B}}\| \cdot \omega(\sqrt{\log n})$, and a center $\mathbf{c} \in \mathbb{R}^n$, outputs a sample from a distribution statistically close to $D_{\Lambda,s,\mathbf{c}}$.*

2.2 Admissible Hash Functions

Admissible hash functions were introduced by Boneh and Boyen [15] as a combinatorial tool for partitioning-based security proofs for which Freire *et al.* [33] gave a simplified definition. Jager [47] considered the following generalization in order to simplify the analysis of reductions under decisional assumption.

Definition 2.6 ([47]). *Let $\ell(\lambda), L(\lambda) \in \mathbb{N}$ be functions of a security parameter $\lambda \in \mathbb{N}$. Let $\mathsf{AHF} : \{0,1\}^\ell \to \{0,1\}^L$ be an efficiently computable function. For every $K \in \{0,1,\perp\}^L$, let the partitioning function $P_K : \{0,1\}^\ell \to \{0,1\}$ such that*

$$P_K(X) := \begin{cases} 0 & \text{if} \quad \forall i \in [L] \quad (\mathsf{AHF}(X)_i = K_i) \ \vee \ (K_i = \perp) \\ 1 & \text{otherwise} \end{cases}$$

We say that AHF *is a* **balanced admissible hash function** *if there exists an efficient algorithm* $\mathsf{AdmSmp}(1^\lambda, Q, \delta)$ *that takes as input $Q \in \mathsf{poly}(\lambda)$ and a non-negligible $\delta(\lambda) \in (0,1]$ and outputs a key $K \in \{0,1,\perp\}^L$ such that, for all $X^{(1)}, \ldots, X^{(Q)}, X^\star \in \{0,1\}^\ell$ such that $X^\star \notin \{X^{(1)}, \ldots, X^{(Q)}\}$, we have*

$$\gamma_{\max}(\lambda) \geq \Pr_K \left[P_K(X^{(1)}) = \cdots = P_K(X^{(Q)}) = 1 \ \wedge \ P_K(X^\star) = 0 \right] \geq \gamma_{\min}(\lambda),$$

where $\gamma_{\max}(\lambda)$ and $\gamma_{\min}(\lambda)$ are functions such that

$$\tau(\lambda) = \gamma_{\min}(\lambda) \cdot \delta(\lambda) - \frac{\gamma_{\max}(\lambda) - \gamma_{\min}(\lambda)}{2}$$

is a non-negligible function of λ.

Intuitively, the condition that $\tau(\lambda)$ be non-negligible requires $\gamma_{\min}(\lambda)$ to be noticeable and the difference of $\gamma_{\max}(\lambda) - \gamma_{\min}(\lambda)$ to be small.

It is known [47] that balanced admissible hash functions exist for $\ell, L = \Theta(\lambda)$.

Theorem 2.7 ([47, Theorem 1]). *Let $(C_\ell)_{\ell \in \mathbb{N}}$ be a family of codes $C_\ell : \{0,1\}^\ell \to \{0,1\}^L$ with minimal distance $c \cdot L$ for some constant $c \in (0,1/2)$. Then, $(C_\ell)_{\ell \in \mathbb{N}}$ is a family of balanced admissible hash functions. Furthermore,* $\mathsf{AdmSmp}(1^\lambda, Q, \delta)$ *outputs a key $K \in \{0,1,\perp\}^L$ for which $\eta = \lfloor \frac{\ln(2Q + Q/\delta)}{-\ln((1-c))} \rfloor$ components are not $\perp$ and $\gamma_{\max} = 2^{-\eta}$, $\gamma_{\min} = \left(1 - Q(1-c)\right)^\eta \cdot 2^{-\eta}$, so that $\tau = (2\delta - (2\delta + 1) \cdot Q \cdot (1-c)^\eta)/2^{\eta+1}$ is a non-negligible function of λ.*

Lemma 2.8 ([49, Lemma 8],[6, Lemma 28]). *Let $K \leftarrow \mathsf{AdmSmp}(1^\lambda, Q, \delta)$, an input space $\mathcal{X}$ and the mapping γ that maps a $(Q+1)$-uple $(X^\star, X_1, \ldots, X_Q)$ in $\mathcal{X}^{Q+1}$ to a probability value in $[0,1]$, given by:*

$$\gamma(X^\star, X_1, \ldots, X_Q) := \Pr_K \left[P_K(X^{(1)}) = \cdots = P_K(X^{(Q)}) = 1 \ \wedge \ P_K(X^\star) = 0 \right].$$

We consider the following experiment where we first execute the PRF security game, in which the adversary eventually outputs a guess $\hat{b} \in \{0,1\}$ of the challenger's bit $b \in \{0,1\}$ and wins with advantage ε. We denote by $X^\star \in \mathcal{X}$ the *challenge input and* $X_1, \ldots, X_Q \in \mathcal{X}$ *the evaluation queries. At the end of the game, we flip a fair random coin* $b'' \hookleftarrow U(\{0,1\})$. *If the condition* $P_K(X^{(1)}) = \cdots = P_K(X^{(Q)}) = 1 \wedge P_K(X^\star) = 0$ *is satisfied we define* $b' = \hat{b}$. *Otherwise, we define* $b' = b''$. *Then, we have* $|\Pr[b' = b] - 1/2| \geq \gamma_{\min} \cdot \varepsilon - \frac{\gamma_{\max} - \gamma_{\min}}{2}$, *where* $\gamma_{\min}$ *and* $\gamma_{\max}$ *are the maximum and minimum of* $\gamma(\mathbb{X})$ *for any* $\mathbb{X} \in \mathcal{X}^{Q+1}$.

2.3 Randomness Extraction

The Leftover Hash Lemma was used by Agrawal *et al.* [6] to re-randomize matrices over $\mathbb{Z}_q$ by multiplying them with small-norm matrices.

Lemma 2.9 ([6]). *Let integers* m, n *such that* $m > 2n \cdot \log q$, *for some prime* $q > 2$. *Let* $\mathbf{A}, \mathbf{U} \hookleftarrow U(\mathbb{Z}_q^{n \times m})$ *and* $\mathbf{R} \hookleftarrow U(\{-1,1\}^{m \times m})$. *The distributions* $(\mathbf{A}, \mathbf{AR})$ *and* $(\mathbf{A}, \mathbf{U})$ *are within* $2^{-\Omega(n)}$ *statistical distance.*

2.4 Multi-Client Functional Encryption

We recall the syntax of multi-client functional encryption as introduced in [43].

Definition 2.10. *A* **multi-client functional encryption** *(MCFE) scheme for a message space* $\mathcal{M}$ *and tag space* $\mathcal{T}$ *is a tuple* (Setup, Encrypt, DKeygen, Decrypt) *of efficient algorithm with the following specifications:*

Setup$(\mathsf{cp}, 1^\ell)$:*Takes in global parameters* cp *and a pre-determined number of users* 1^ℓ, *where* cp *specifies a security parameter* 1^λ. *It outputs a set of public parameters* mpk, *a master secret key* msk, *and a set of encryption keys* $\{\mathsf{ek}_i\}_{i=1}^{\ell}$. *We assume that* mpk *is included in all encryption keys* ek_i.

Encrypt(ek_i, x_i, t) : *Takes as input the encryption key* ek_i *of user* $i \in [\ell]$, *a message* x_i *and a tag* $t \in \mathcal{T}$. *It output a ciphertext* $C_{t,i}$.

DKeygen(msk, f) : *Takes as input the master secret key* msk *and an* ℓ*-argument function* $f : \mathcal{M}^\ell \to \mathcal{R}$. *It outputs a functional decryption key* dk_f.

Decrypt$(\mathsf{dk}_f, t, \mathbf{C})$: *Takes as input a functional decryption key* dk_f, *a tag* t, *and an* ℓ*-vector of ciphertexts* $\mathbf{C} = (C_{t,1}, \ldots, C_{t,\ell})$. *It outputs a function evaluation* $f(\boldsymbol{x}) \in \mathcal{R}$ *or an error message* $\perp$.

Correctness. For any set of public parameters cp, any $(\mathsf{mpk}, \mathsf{msk}, \{\mathsf{ek}_i\}_{i=1}^{\ell}) \leftarrow \mathsf{Setup}(\mathsf{cp}, 1^\ell)$, any vector $\boldsymbol{x} \in \mathcal{M}^n$ any tag $t \in \mathcal{T}$ and any function $f : \mathcal{M}^\ell \to \mathcal{R}$, if $C_{t,i} \leftarrow \mathsf{Encrypt}(\mathsf{ek}_i, x_i, t)$ for all $i \in [\ell]$ and $\mathsf{dk}_f \leftarrow \mathsf{DKeygen}(\mathsf{msk}, f)$, we have $\mathsf{Decrypt}(\mathsf{dk}_f, t, \mathbf{C}_t = (C_{t,1}, \ldots, C_{t,\ell})) = f(\boldsymbol{x})$ with overwhelming probability.

We now recall the security definition given in [43] for an adaptively secure MCFE, and then we will give the definition that we use in this work. These two definitions are in fact equivalent.

Definition 2.11. (IND-sec). *For an MCFE scheme with* ℓ *senders, consider the following game between an adversary* $\mathcal{A}$ *and a challenger* $\mathcal{C}$. *The game involves a set* $\mathcal{HS}$ *of honest senders (initialized to* $\mathcal{HS} := [\ell]$*) and a set* $\mathcal{CS}$ *(initialized to* $\mathcal{CS} := \emptyset$*) of corrupted senders.*

Initialization: *The challenger $\mathcal{C}$ chooses* cp *and runs* $(\mathsf{mpk}, \mathsf{msk}, \{\mathsf{ek}_i\}_{i=1}^{\ell}) \leftarrow \mathsf{Setup}(\mathsf{cp}, 1^{\ell})$. *Then, it chooses a random bit* $b \leftarrow \{0,1\}$ *and gives the master public key* mpk *to the adversary.*

Encryption queries: *The adversary $\mathcal{A}$ can adaptively make encryption queries* $\mathsf{QEncrypt}(i, x^0, x^1, t)$, *to which the challenger replies with* $\mathsf{Encrypt}(\mathsf{ek}_i, x^b, t)$. *For any given pair* (i,t), *only one query is allowed and subsequent queries involving the same* (i,t) *are ignored.*

Functional decryption key queries: *The adversary can adaptively obtain functional decryption keys by making queries of the form* $\mathsf{QDKeygen}(f)$. *The challenger returns* $\mathsf{dk}_f \leftarrow \mathsf{DKeygen}(\mathsf{msk}, f)$.

Corruption queries: *For any user* $i \in \mathcal{HS}$, *the adversary can adaptively make queries* $\mathsf{QCorrupt}(i)$, *to which the challenger replies with* ek_i *and updates* $\mathcal{HS}$ *and* $\mathcal{CS}$ *by setting* $\mathcal{CS} := \mathcal{CS} \cup \{i\}$ *and* $\mathcal{HS} := \mathcal{HS} \setminus \{i\}$.

Finalize: *The adversary makes its guess* $b' \in \{0,1\}$; $\mathcal{A}$ *wins the game if* $\beta = b$, *where* β *is defined to be* $\beta := b'$ *except in the following situations.*

1. *An encryption query* $\mathsf{QEncrypt}(i, x^0, x^1, t)$ *has been made for an index* $i \in \mathcal{CS}$ *with* $x^0 \neq x^1$.
2. *For some label* t, *an encryption query* $\mathsf{QEncrypt}(i, x_i^0, x_i^1, t)$ *has been asked for* $i \in \mathcal{HS}$, *but encryption queries* $\mathsf{QEncrypt}(j, x_j^0, x_j^1, t)$ *have not been asked for all* $j \in \mathcal{HS}$.
3. *For a label* t *and some function* f *queried to* $\mathsf{QDKeygen}$, *there exists a pair of vectors* $(\boldsymbol{x}^0, \boldsymbol{x}^1)$ *such that* $f(\boldsymbol{x}^0) \neq f(\boldsymbol{x}^1)$, *where*
   - $x_i^0 = x_i^1$ *for all* $i \in \mathcal{CS}$;
   - $\mathsf{QEncrypt}(i, x_i^0, x_i^1, t)$ *have been asked for all* $i \in \mathcal{HS}$.

In any of the above cases, $\mathcal{A}$*'s output is replaced by a random* $\beta \leftarrow U(\{0,1\})$.

An MCFE scheme provides IND *security if, for any efficient adversary* $\mathcal{A}$, *we have* $\mathbf{Adv}^{\mathsf{IND}}(\mathcal{A}) := |\Pr[\beta = 1 \mid b = 1] - \Pr[\beta = 1 \mid b = 0]| \in \mathsf{negl}(\lambda)$.

In the following, it will be convenient to work with the following security definition, which is equivalent to Definition 2.11.

Definition 2.12. (1-challenge IND-sec). *For an MCFE scheme with ℓ senders, we consider the following game between an adversary $\mathcal{A}$ and a challenger $\mathcal{C}$. The game involves a set $\mathcal{HS}$ (initialized to $\mathcal{HS} := [\ell]$), of honest senders and a set $\mathcal{CS}$ (initialized to $\mathcal{CS} := \emptyset$), of corrupted senders.*

Initialization: *The challenger $\mathcal{C}$ generates* cp *and runs* $(\mathsf{mpk}, \mathsf{msk}, \{\mathsf{ek}_i\}_{i=1}^{\ell}) \leftarrow \mathsf{Setup}(\mathsf{cp}, 1^{\ell})$. *Then, it chooses a random bit* $b \leftarrow \{0,1\}$ *and gives the master public key* mpk *to the adversary* $\mathcal{A}$.

Encryption queries: *The adversary can adaptively make encryption queries* $\mathsf{QEncrypt}(i, x, t)$, *to which the challenger replies with* $\mathsf{Encrypt}(\mathsf{ek}_i, x, t)$. *Any further query involving the same pair* (i,t) *is ignored.*

Challenge queries: *The adversary adaptively makes challenge queries of the form* $\mathsf{CQEncrypt}(i, x_i^{\star 0}, x_i^{\star 1}, t^\star)$. *The challenger replies with* $\mathsf{Encrypt}(\mathsf{ek}_i, x_i^{\star b}, t^\star)$. *Only one tag* $t^\star$ *can be involved in a challenge query. If* $t^\star$ *denotes the tag of the first query, the challenger only replies to subsequent challenge queries for the same label* $t^\star$. *Moreover, only one query* $(i, t^\star)$ *is allowed for each* $i \in [\ell]$ *and subsequent queries involving the same* $i \in [\ell]$ *are ignored.*

Functional decryption key queries: *The adversary can adaptively obtain functional decryption keys via queries* $\mathsf{QDKeygen}(f)$. *At each query, the challenger returns* $\mathsf{dk}_f \leftarrow \mathsf{DKeygen}(\mathsf{msk}, f)$.

Corruption queries: *For any user* $i \in \mathcal{HS}$, *the adversary can adaptively make queries* $\mathsf{QCorrupt}(i)$, *to which the challenger replies with* ek_i *and updates* $\mathcal{HS}$ *and* $\mathcal{CS}$ *by setting* $\mathcal{CS} := \mathcal{CS} \cup \{i\}$ *and* $\mathcal{HS} := \mathcal{HS} \setminus \{i\}$.

Finalize: *The adversary outputs a bit* $b' \in \{0,1\}$. *The adversary* $\mathcal{A}$ *wins if* $\beta = b$, *where* β *is defined as* $\beta := b'$, *unless of the situations below occurred.*

1. *A challenge query* $\mathsf{CQEncrypt}(i, x_i^{\star 0}, x_i^{\star 1}, t^\star)$ *has been made for an index* $i \in \mathcal{CS}$ *with* $x_i^{\star 0} \neq x_i^{\star 1}$.
2. *An encryption query* $\mathsf{QEncrypt}(i, x, t^\star)$ *has been made for the challenge tag* $t^\star$ *for some index* $i \in [\ell]$.
3. *For the challenge tag* $t^\star$, *a challenge query* $\mathsf{CQEncrypt}(i, x_i^{\star 0}, x_i^{\star 1}, t^\star)$ *has been asked for some* $i \in \mathcal{HS}$, *but challenge queries* $\mathsf{CQEncrypt}(j, x_j^{\star 0}, x_j^{\star 1}, t^\star)$ *have not been asked for all* $j \in \mathcal{HS}$.
4. *For the challenge tag* $t^\star$ *and some function* f *queried to* $\mathsf{QDKeygen}$, *there exists a pair of vectors* $(\boldsymbol{x}^{\star 0}, \boldsymbol{x}^{\star 1})$ *such that* $f(\boldsymbol{x}^{\star 0}) \neq f(\boldsymbol{x}^{\star 1})$, *where*
   - $x_i^{\star 0} = x_i^{\star 1}$ *for all* $i \in \mathcal{CS}$;
   - $\mathsf{CQEncrypt}(i, x_i^{\star 0}, x_i^{\star 1}, t^\star)$ *have been asked for all* $i \in \mathcal{HS}$.

If any of these events occurred, $\mathcal{A}$*'s output is overwritten by* $\beta \leftarrow U(\{0,1\})$.

We say that an MCFE scheme provides 1Ch-IND *security if, for any efficient adversary* $\mathcal{A}$, *we have* $\mathbf{Adv}^{\mathsf{1Ch\text{-}IND}}(\mathcal{A}) := \left|\Pr[\beta = b] - \frac{1}{2}\right| \in \mathsf{negl}(\lambda)$.

In the full version of the paper, we show that 1Ch-IND security implies IND security. We also note that condition 2 of "Finalize" could be:

2'. Both $\mathsf{QEncrypt}(i, x, t^\star)$ and $\mathsf{CQEncrypt}(i, x_i^{\star 0}, x_i^{\star 1}, t^\star)$ have been made for an index i and the challenge label $t^\star$, such that $x_i^{0\star} \neq x_i^{1\star}$

This allows the adversary to make both an encryption query $\mathsf{QEncrypt}(i, x, t^\star)$ and a challenge query $\mathsf{CQEncrypt}(i, x_i^{\star 0}, x_i^{\star 1}, t^\star)$ where $x_i^{\star 0} = x_i^{\star 1}$. In the full version of the paper, we show that replacing condition 2 by condition $2'$ does not make the adversary any stronger.

Our first construction is proven secure under Definition 2.12. Abdalla *et al.* [2] and Chotard *et al.* [28] independently showed constructions that can be proven secure in the sense of a stronger definition which eliminates restriction 3 from the "Finalize" stage. In the full version of the paper, we show that a variant of the compiler of [2, Section 4.2] is secure in the standard model. Recently, Abdalla *et al.* [1] independently obtained a similar result. While their PRF-based compiler [1] can rely on any PRF, we obtain a tighter reduction using a specific PRF described in [55]. Chotard *et al.* [28] additionally show how to enable repetitions by allowing multiple encryption queries for the same pair (i, t). However, they need random oracles for this purpose.

2.5 Decentralized Multi-Client Functional Encryption

We use the same syntax as Chotard *et al.* [27] with the difference that we explicitly assume common public parameters cp. As in [27], we assume that each function f can be injectively encoded as a tag t_f (called "label" in [27]) taken as input by the partial functional key generation algorithm.

Definition 2.13. *For a message space $\mathcal{M}$ and tag space $\mathcal{T}$, a* **decentralized multi-client functional encryption** *(DMCFE) scheme between ℓ senders $\{\mathcal{S}_i\}_{i=1}^{\ell}$ and a functional decryptor $\mathcal{FD}$ is specified by the following components.*

Setup$(\mathsf{cp}, 1^\ell)$: *This is an interactive protocol between the senders $\{\mathcal{S}_i\}_{i=1}^{\ell}$, which allows them to generate their own secret keys sk_i and encryption keys ek_i, for $i \in [\ell]$, as well as a set of public parameters mpk.*

Encrypt(ek_i, x_i, t) : *Takes as input the encryption key ek_i of user $i \in [\ell]$, a message x_i and a tag $t \in \mathcal{T}$. It output a ciphertext $C_{t,i}$.*

DKeygenShare(sk_i, t_f) : *Takes as input a user's secret key sk_i and the label t_f of a function $f : \mathcal{M}^\ell \to \mathcal{R}$. It outputs a partial functional decryption key $\mathsf{dk}_{f,i}$ for the function described by t_f.*

DKeygenComb$(\{\mathsf{dk}_{f,i}\}_i, t_f)$: *Takes as input a set of partial functional decryption keys $\{\mathsf{dk}_{f,i}\}_i$ and the label t_f of a function $f : \mathcal{M}^\ell \to \mathcal{R}$. It outputs a full functional decryption key dk_f for the function f described by t_f*

Decrypt$(\mathsf{dk}_f, t, \mathbf{C})$: *Takes as input a functional decryption key dk_f, a tag t, and an ℓ-vector of ciphertexts $\mathbf{C} = (C_{t,1}, \ldots, C_{t,\ell})$. It outputs a function evaluation $f(\boldsymbol{x}) \in \mathcal{R}$ or a message $\perp$ indicating a decryption failure.*

For simplicity, we assume that mpk is included in all secret keys and encryption keys, as well as in (partial) functional decryption keys. We also assume that a description of f is included in (partial) functional decryption keys.

Correctness. For any $\lambda \in \mathbb{N}$, any $(\mathsf{mpk}, \{\mathsf{sk}_i\}_{i=1}^{\ell}, \{\mathsf{ek}_i\}_{i=1}^{\ell}) \leftarrow \mathsf{Setup}(\mathsf{cp}, 1^\ell)$, any $\boldsymbol{x} \in \mathcal{M}^n$, any tag $t \in \mathcal{T}$ and any function $f : \mathcal{M}^\ell \to \mathcal{R}$, if $C_{t,i} \leftarrow \mathsf{Encrypt}(\mathsf{ek}_i, x_i, t)$ for all $i \in [\ell]$ and $\mathsf{dk}_f \leftarrow \mathsf{DKeyComb}(\{\mathsf{DKeyGenShare}(\mathsf{sk}_i, t_f)\}_i, t_f)$, with overwhelming probability, we have $\mathsf{Decrypt}(\mathsf{dk}_f, t, \mathbf{C}_t = (C_{t,1}, \ldots, C_{t,\ell})) = f(\boldsymbol{x})$.

Definition 2.14. (IND-sec **for DMCFE**). *For a DMCFE scheme with ℓ senders, we consider the following game between an adversary and a challenger. It involves a set $\mathcal{HS}$ of honest senders (initialized to $\mathcal{HS} := [\ell]$) and a set $\mathcal{CS}$ (initialized to $\mathcal{CS} := \emptyset$) of the corrupted senders.*

Initialization: *The challenger $\mathcal{C}$ generates cp and runs $(\mathsf{mpk}, \{\mathsf{sk}_i\}_{i=1}^{\ell}, \{\mathsf{ek}_i\}_{i=1}^{\ell}) \leftarrow \mathsf{Setup}(\mathsf{cp}, 1^\ell)$. Then, it flips a fair coin $b \leftarrow \{0, 1\}$ and gives the master public key mpk to the adversary $\mathcal{A}$.*

Encryption queries: *The adversary $\mathcal{A}$ can adaptively make encryption queries $\mathsf{QEncrypt}(i, x^0, x^1, t)$, to which the challenger replies with $\mathsf{Encrypt}(\mathsf{ek}_i, x^b, t)$. For any given pair (i, t), only one query is allowed and subsequent queries involving the same (i, t) are ignored.*

Functional decryption key queries: *Via queries* $\mathsf{QDKeygen}(i, f)$, $\mathcal{A}$ *can adaptively obtain partial functional decryption keys on behalf of uncorrupted senders. At each query, the challenger returns* $\mathsf{dk}_f \leftarrow \mathsf{DKeygenShare}(\mathsf{sk}_i, t_f)$ *if* $i \in \mathcal{HS}$ *(if* $i \in \mathcal{CS}$*, the oracle returns* $\perp$*).*

Corruption queries: *For any user* $i \in \mathcal{HS}$*, the adversary can adaptively make queries* $\mathsf{QCorrupt}(i)$ *and the challenger replies by returning* $(\mathsf{sk}_i, \mathsf{ek}_i)$*. It also updates the sets* $\mathcal{HS}$ *and* $\mathcal{CS}$ *by setting* $\mathcal{CS} := \mathcal{CS} \cup \{i\}$ *and* $\mathcal{HS} := \mathcal{HS} \setminus \{i\}$.

Finalize: *The adversary outputs a bit* $b' \in \{0,1\}$*. The adversary* $\mathcal{A}$ *wins if* $\beta = b$*, where* β *is defined as* $\beta := b'$*, unless of the situations below occurred.*

1. *An encryption query* $\mathsf{QEncrypt}(i, x_i^0, x_i^1, t)$ *has been made for an index* $i \in \mathcal{CS}$ *with* $x_i^0 \neq x_i^1$.
2. *For some label* t*, an encryption query* $\mathsf{QEncrypt}(i, x_i^0, x_i^1, t)$ *has been asked for* $i \in \mathcal{HS}$*, but encryption queries* $\mathsf{QEncrypt}(j, x_j^0, x_j^1, t)$ *have not been asked for all* $j \in \mathcal{HS}$.
3. *For a tag* t *and some function* f *queried to* $\mathsf{QDKeygen}(i, .)$ *for all* $i \in \mathcal{HS}$*, there exists a pair of vectors* $(\boldsymbol{x}^0, \boldsymbol{x}^1)$ *such that* $f(\boldsymbol{x}^0) \neq f(\boldsymbol{x}^1)$*, where*
   - $x_i^0 = x_i^1$ *for all* $i \in \mathcal{CS}$;
   - $\mathsf{QEncrypt}(i, x_i^0, x_i^1, t)$ *have been asked for all* $i \in \mathcal{HS}$.

If any of these events occurred, $\mathcal{A}$*'s output is overwritten by* $\beta \leftarrow U(\{0,1\})$.

We say that a DMCFE scheme provides IND *security if, for any efficient adversary* $\mathcal{A}$*, we have* $\mathbf{Adv}^{\mathsf{IND}}(\mathcal{A}) := |\Pr[\beta = 1 \mid b = 1] - \Pr[\beta = 1 \mid b = 0]| \in \mathsf{negl}(\lambda)$.

The above definition captures adaptive corruptions in that the $\mathsf{QCorrupt}(\cdot)$ oracle may be invoked at any time during the game. In the static corruption setting, all queries to $\mathsf{QCorrupt}(\cdot)$ should be made at once before the initialization phase. In this case, the sets $\mathcal{HS}$ and $\mathcal{CS}$ are thus determined before the generation of $(\mathsf{mpk}, \{\mathsf{sk}_i\}_{i=1}^{\ell}, \{\mathsf{ek}_i\}_{i=1}^{\ell})$. We denote by sta-IND-sec the latter security game.

Our scheme of Sect. 4 will be proven secure under static corruptions. We insist that only corruptions are static: the encryption oracle can be queried on adaptively chosen messages (x^0, x^1), which is stronger than the selective security game, where the challenge messages have to be declared upfront.

3 Our MCFE Scheme for Linear Functions

The scheme encrypts $\boldsymbol{x}_i \in \mathbb{Z}^{n_0}$ as a vector $\mathbf{C}_{t,i} = \mathbf{G}_0^\top \cdot \boldsymbol{x}_i + \mathbf{A}(\tau)^\top \cdot \mathbf{s}_i + \mathbf{e}_i$, where $\mathbf{G}_0$ is a gadget matrix; $\tau = \mathsf{AHF}(t) \in \{0,1\}^L$ is an admissible hash of the tag t; and $\mathbf{e}_i$ is a Gaussian noise. This is done in a way that a functional secret key $\mathbf{s}_y = \sum_{i=1}^{\ell} y_i \cdot \mathbf{s}_i \in \mathbb{Z}^n$ allows computing $\sum_{i=1}^{\ell} y_i \cdot \boldsymbol{x}_i$ from $\{\mathbf{C}_{t,i}\}_{i=1}^{\ell}$ by using the public trapdoor of the lattice $\Lambda^\perp(\mathbf{G}_0)$.

We derive $\mathbf{A}(\tau)$ from a set of $2L$ public matrices $\{\mathbf{A}_{i,0}, \mathbf{A}_{i,1}\}_{i=1}^{L}$ and an additional matrix $\mathbf{V} \in \mathbb{Z}_q^{n_0 \times n}$. Like [55], our proof interprets each $\mathbf{A}_{i,b} \in \mathbb{Z}_q^{n \times m}$ as a GSW ciphertext $\mathbf{A}_{i,b} = \mathbf{A} \cdot \mathbf{R}_{i,b} + \mu_{i,b} \cdot \mathbf{G}$, where $\mathbf{R}_{i,b} \in \{-1,1\}^{m \times m}$, $\mu_{i,b} \in \{0,1\}$ and $\mathbf{G} \in \mathbb{Z}_q^{n \times m}$ is the gadget matrix of [57]. Then, we homomorphically compute $\mathbf{A}(\tau)$ as an FHE ciphertext $\mathbf{A} \cdot \mathbf{R}'_\tau + (\prod_{i=1}^{L} \mu_{i,\tau[i]}) \cdot \mathbf{G}$, for some small-norm

$\mathbf{R}'_\tau \in \mathbb{Z}^{m\times m}$, which is in turn multiplied by $\mathbf{G}^{-1}(\mathbf{V}^\top \cdot \mathbf{G}_0)$ in such a way that $\mathbf{A}(\tau) = \mathbf{A}\cdot\mathbf{R}_\tau + (\prod_{i=1}^{L} \mu_{i,\tau[i]})\cdot(\mathbf{V}^\top \cdot \mathbf{G}_0)$. Via a careful choice of $\{\mu_{i,b}\}_{i\in[L],b\in\{0,1\}}$, the properties of admissible hash functions imply that $\prod_{i=1}^{L} \mu_{i,x[i]}$ vanishes in all encryption queries but evaluates to 1 on the challenge tag $\tau^\star$. In order to prevent the encryption oracle from leaking too much about $\mathbf{s}_i \in \mathbb{Z}^n$, we proceed as in [55] and replace the random $\mathbf{A} \in \mathbb{Z}_q^{n\times m}$ by a lossy matrix $\mathbf{A}^\top = \hat{\mathbf{A}}^\top \cdot \mathbf{C} + \mathbf{E}$, where $\hat{\mathbf{A}} \hookleftarrow U(\mathbb{Z}_q^{n_1\times m})$, $\mathbf{C} \hookleftarrow U(\mathbb{Z}_q^{n_1\times n})$ and for a small-norm $\mathbf{E} \in \mathbb{Z}^{m\times n}$.

Our construction and proof depart from [55] in that we use an additional multiplication by $\mathbf{G}^{-1}(\mathbf{V}^\top \cdot \mathbf{G}_0)$ in order to introduce a matrix $\mathbf{V} \in \mathbb{Z}_q^{n_0\times n}$ in the expression of $\mathbf{A}(\tau^\star)$. In addition, unlike [55], we do not rely on a randomness extraction argument to exploit the entropy of $\mathbf{A}(\tau^\star)^\top \cdot \mathbf{s}_i + \mathbf{e}_i$ in the challenge phase. Instead, we use a trapdoor for the matrix $\mathbf{U} = [\begin{smallmatrix}\mathbf{V}\\ \mathbf{C}\end{smallmatrix}]$ to "equivocate" the challenge ciphertexts and explain them as an encryption of $\boldsymbol{x}_{1,i}^\star$ instead of $\boldsymbol{x}_{0,i}^\star$.

Another difference with [55] is that the product $\mathbf{A}(\tau)$ of GSW ciphertexts $\{\mathbf{A}_{i,\tau[i]}\}_{i=1}^{L}$ is evaluated in a sequential manner[4] (as in the "right-spine" PRF construction of [11]) in order for the noise matrix $\mathbf{R}_\tau$ to retain small entries.

3.1 Description

In the following description, we assume public parameters

$$\mathsf{cp} := \Big(\lambda,\ \ell_{\max},\ X,\ Y,\ n_0,\ n_1,\ n,\ m,\ \alpha,\ \alpha_1,\ \sigma,\ \ell_t,\ L,\ q,\ \mathsf{AHF}\Big),$$

consisting of a security parameter λ and the following quantities:

- $(X, Y, \ell_{\max}, n_0, n_1, n, m)$, which are all in $\mathsf{poly}(\lambda)$ $X = 1$, $n_1 = \lambda^d$, $q = 2^{\lambda^{d-1}}$, $\alpha = 2^{-\sqrt{\lambda}}$, $\alpha_1 = 2^{-\lambda^{d-1}+d\log\lambda}$, $n_0 = o(\lambda^{d-2})$, $n = O(\lambda^{2d-1})$, $\sigma = 2^{\lambda^{d-1}-2\lambda}$ and $n_0 \cdot \ell_{max} = O(\lambda^{d-2})$ where d is a constant; for instance $d = 3$ works asymptotically.
- The description of a tag space $\mathcal{T} = \{0,1\}^{\ell_t}$, for some $\ell_t \in \mathsf{poly}(\lambda)$, such that tags may be arbitrary strings (e.g., time period numbers or dataset names).
- The description of a balanced admissible hash function $\mathsf{AHF} : \{0,1\}^{\ell_t} \to \{0,1\}^L$, for a suitable $L \in \Theta(\lambda)$.
- The message space will be $\mathcal{M} = [-X, X]^{n_0}$, for some $n_0 \in \mathsf{poly}(\lambda)$.
- Integers $n, n_0, n_1, m \in \mathsf{poly}(\lambda)$ satisfying the conditions $m > 2n \cdot \lceil \log q \rceil$ and $n > 3 \cdot (n_0 + n_1) \cdot \lceil \log q \rceil$.
- A real $\alpha > 0$ and a Gaussian parameter $\sigma > 0$, which specifies an interval $[-\beta, \beta] = [-\sigma\sqrt{n}, \sigma\sqrt{n}]$ where the coordinates of users' secret keys will live (with probability exponentially close to 1).

Letting $\ell \in \mathsf{poly}(\lambda)$, with $\ell \leq \ell_{max}$, be the number of users, our function space is the set of all functions $f_{\boldsymbol{y}} : \mathbb{Z}^{n_0\times \ell} \to \mathbb{Z}^{n_0}$ indexed by an integer vector $\boldsymbol{y} \in \mathbb{Z}^\ell$ of infinity norm $\|\boldsymbol{y}\|_\infty < Y$.

[4] In [55], the multiplication of ciphertexts $\{\mathbf{A}_{i,\tau[i]}\}_{i=1}^{L}$ was computed in a parallel fashion $\mathbf{A}_0 \cdot \prod_{i=1}^{L} \mathbf{G}^{-1}(\mathbf{A}_{i,\tau[i]})$ because their initial proof required the matrices $\{\mathbf{A}_{i,b}\}_{i,b}$ to be generated in such a way that $\mathbf{G}^{-1}(\mathbf{A}_{i,b})$ was invertible over $\mathbb{Z}_q$.

We define $\mathbf{G}_0 \in \mathbb{Z}_q^{n_0 \times m}$ to be the gadget matrix

$$\mathbf{G}_0 = [\mathbf{I}_{n_0} \otimes (1, 2, 4, \ldots, 2^{\lceil \log q \rceil}) \mid \mathbf{0}^{n_0} \mid \ldots \mid \mathbf{0}^{n_0}] \in \mathbb{Z}_q^{n_0 \times m}$$

where the product $\mathbf{I}_{n_0} \otimes (1, 2, 4, \ldots, 2^{\lceil \log q \rceil})$ is padded with $m - n_0 \cdot \lceil \log q \rceil$ zero columns. We similarly denote by $\mathbf{G} \in \mathbb{Z}_q^{n \times m}$ the gadget matrix of rank n:

$$\mathbf{G} = [\mathbf{I}_n \otimes (1, 2, 4, \ldots, 2^{\lceil \log q \rceil}) \mid \mathbf{0}^n \mid \ldots \mid \mathbf{0}^n] \in \mathbb{Z}_q^{n \times m}.$$

Our MCFE construction goes as follows.

Setup$(\mathsf{cp}, 1^\ell)$: On input of cp and a number of users ℓ, do the following.
1. Choose random matrices $\mathbf{A}_{i,b} \hookleftarrow U(\mathbb{Z}_q^{n \times m})$, for each $i \in [L]$, $b \in \{0, 1\}$.
2. Choose a uniformly random matrix $\mathbf{V} \hookleftarrow U(\mathbb{Z}_q^{n_0 \times n})$.
3. For each $i \in [\ell]$, sample $\mathbf{s}_i \hookleftarrow D_{\mathbb{Z}^n, \sigma}$ and define $\mathsf{ek}_i = \mathbf{s}_i \in \mathbb{Z}^n$.

Output the master secret key $\mathsf{msk} := \{\mathsf{ek}_i\}_{i=1}^{\ell}$ and the public parameters

$$\mathsf{mpk} := \Big(\mathsf{cp},\ \mathbf{V},\ \{\mathbf{A}_{i,0}, \mathbf{A}_{i,1}\ \in \mathbb{Z}_q^{n \times m}\ \}_{i=1}^{L}\Big).$$

DKeygen$(\mathsf{msk}, f_{\boldsymbol{y}})$: Given the master secret key $\mathsf{msk} := \{\mathsf{ek}_i\}_{i=1}^{\ell}$ and a linear function $f_{\boldsymbol{y}} : \mathbb{Z}^{n_0 \times \ell} \to \mathbb{Z}^{n_0}$ defined by an integer vector $\boldsymbol{y} = (y_1, \ldots, y_\ell)^\top \in \mathbb{Z}^\ell$ which maps an input $\mathbf{X} = [\boldsymbol{x}_1 \mid \ldots \mid \boldsymbol{x}_\ell] \in \mathbb{Z}^{n_0 \times \ell}$ to $f_{\boldsymbol{y}}(\mathbf{X}) = \mathbf{X} \cdot \boldsymbol{y} \in \mathbb{Z}^{n_0}$, parse each ek_i as a vector $\mathbf{s}_i \in \mathbb{Z}^n$. Then, compute and output the functional secret key $\mathsf{dk}_{\boldsymbol{y}} := (\boldsymbol{y}, \mathbf{s}_{\boldsymbol{y}})$, where $\mathbf{s}_{\boldsymbol{y}} = \sum_{i=1}^{\ell} \mathbf{s}_i \cdot y_i \in \mathbb{Z}^n$.

Encrypt$(\mathsf{ek}_i, \boldsymbol{x}_i, t)$: Given $\mathsf{ek}_i = \mathbf{s}_i \in \mathbb{Z}^n$, $\boldsymbol{x}_i \in [-X, X]^{n_0}$, and $t \in \{0, 1\}^{\ell_t}$,
1. Compute $\tau = \mathsf{AHF}(t) \in \{0, 1\}^L$ and parse it as $\tau = \tau[1] \ldots \tau[L]$.
2. Define $\mathbf{W} = \mathbf{G}_0^\top \cdot \mathbf{V} \in \mathbb{Z}_q^{m \times n}$ and compute

$$\mathbf{A}(\tau) = \mathbf{A}_{L,\tau[L]} \cdot \mathbf{G}^{-1}\Big(\mathbf{A}_{L-1,\tau[L-1]} \cdot \mathbf{G}^{-1}(\ldots \mathbf{A}_{2,\tau[2]} \cdot \mathbf{G}^{-1}(\mathbf{A}_{1,\tau[1]}))\Big) \cdot \mathbf{G}^{-1}(\mathbf{W}^\top) \in \mathbb{Z}_q^{n \times m}. \quad (2)$$

3. Sample a noise vector $\mathbf{e}_i \hookleftarrow D_{\mathbb{Z}^m, \alpha q}$. Then, compute and output

$$\mathbf{C}_{t,i} = \mathbf{G}_0^\top \cdot \boldsymbol{x}_i + \mathbf{A}(\tau)^\top \cdot \mathbf{s}_i + \mathbf{e}_i \ \in \mathbb{Z}_q^m.$$

Decrypt$(\mathsf{dk}_{\boldsymbol{y}}, t, \mathbf{C}_t)$: On input of a functional secret key $\mathsf{dk}_{\boldsymbol{y}} = (\boldsymbol{y}, \mathbf{s}_{\boldsymbol{y}})$ for a vector $\boldsymbol{y} = (y_1, \ldots, y_\ell)^\top \in [-Y, Y]^\ell$, a tag $t \in \{0, 1\}^{\ell_t}$, and an ℓ-vector of ciphertexts $\mathbf{C}_t = (\mathbf{C}_{t,1}, \ldots, \mathbf{C}_{t,\ell}) \in (\mathbb{Z}_q^m)^\ell$, conduct the following steps.
1. Compute $\tau = \mathsf{AHF}(t) \in \{0, 1\}^L$ and parse it as $\tau = \tau[1] \ldots \tau[L]$.
2. Compute $\mathbf{A}(\tau) \in \mathbb{Z}_q^{n \times m}$ as per (2).
3. Compute $\mathbf{f}_{t,\boldsymbol{y}} = \sum_{i=1}^{\ell} y_i \cdot \mathbf{C}_{t,i} - \mathbf{A}(\tau)^\top \cdot \mathbf{s}_{\boldsymbol{y}} \mod q$.
4. Interpret $\mathbf{f}_{t,\boldsymbol{y}} \in \mathbb{Z}_q^m$ as a vector of the form $\mathbf{f}_{t,\boldsymbol{y}} = \mathbf{G}_0^\top \cdot \boldsymbol{z} + \tilde{\mathbf{e}} \mod q$, for some error vector $\tilde{\mathbf{e}} \in [-B, B]^m$. Using the public trapdoor of $\Lambda^\perp(\mathbf{G}_0)$, compute and output the underlying vector $\boldsymbol{z} \in [-\ell \cdot X \cdot Y, \ell \cdot X \cdot Y]^{n_0}$.

The following lemma is proved in the full version of the paper.

Lemma 3.1. (Correctness). *Assume that* $\alpha q = \omega(\sqrt{\log \ell})$, $Y \cdot \ell \cdot \alpha q \cdot \log q < q/2$ *and* $\ell \cdot X \cdot Y < q/2$. *Then, for any* $(\mathsf{mpk}, \mathsf{msk}, \{\mathsf{ek}_i\}_{i=1}^{\ell}) \leftarrow \mathsf{Setup}(\mathsf{cp}, 1^\lambda)$, *any message* $\mathbf{X} = [\boldsymbol{x}_1 | \cdots | \boldsymbol{x}_\ell] \in [-X, X]^{n_0 \times \ell}$, *any* $\boldsymbol{y} \in [-Y, Y]^\ell$, *any tag* $t \in \{0, 1\}^{\ell_t}$, *algorithm* $\mathsf{Decrypt}(\mathsf{dk}_{\boldsymbol{y}}, t, \mathbf{C}_t)$ *outputs* $\mathbf{X} \cdot \boldsymbol{y} \in \mathbb{Z}^{n_0}$ *with probability exponentially close to* 1, *where* $\mathbf{C}_{t,i} \leftarrow \mathsf{Encrypt}(\mathsf{ek_i}, \boldsymbol{x_i}, \mathrm{t})$ *and* $\mathsf{dk}_y \leftarrow \mathsf{DKeygen}(\mathsf{msk}, f_{\boldsymbol{y}})$.

3.2 Security

We now prove the security of the scheme in the sense of Definition 2.12 (and thus Definition 2.11 modulo some loss of tightness in the reduction).

For the current parameters $n_1 = \lambda^d$, $q = 2^{\lambda^{d-1}}$, and $\alpha_1 = 2^{-\lambda^{d-1}+d\log\lambda}$, $\alpha_1 q = \Omega(\sqrt{n_1})$, we know from [60] that $\mathsf{LWE}_{q,n_1,\alpha_1}$ is at least as hard as GapSVP_γ, with $\gamma = \tilde{O}(n_1/\alpha_1) = \tilde{O}(2^{\lambda^{d-1}})$. The best known algorithms [63] for solving GapSVP_γ run in $2^{\tilde{O}(\frac{n_1}{\log\gamma})}$, which for our parameters is $2^{\tilde{O}(\lambda)}$.

Theorem 3.2. *The above MCFE schemes provides adaptive security under the* $\mathsf{LWE}_{q,m,n_1,\alpha_1}$ *assumption.*

Proof. The proof considers a sequence of games. In each game, we denote by W_i the event that $b' = b$. For each i, the adversary's advantage function in Game_i is $\mathbf{Adv}_i(\mathcal{A}) := |\Pr[b' = b] - 1/2| = \frac{1}{2} \cdot |\Pr[b' = 1 \mid b = 1] - \Pr[b' = 1 \mid b = 0]|$.

Game_0: This is the real security game. We denote by $t^\star$ the tag of the challenge phase while $t^{(1)}, \ldots, t^{(Q)}$ are the tags involved in encryption queries. Namely, for each $j \in [Q]$, $t^{(j)}$ stands for the j-th distinct tag involved in an encryption query. Since up to ℓ encryption queries $(i, \boldsymbol{x}_i, t)$ are allowed for each tag t, the adversary can make a total of $\ell \cdot Q$ encryption queries. The game begins with the challenger initially choosing encryption keys $\{\mathsf{ek}_i\}_{i=1}^{\ell}$ by sampling $\mathsf{ek}_i = \mathbf{s}_i \hookleftarrow D_{\mathbb{Z}^n,\sigma}$ for each $i \in [\ell]$. In addition, the challenger flips a fair coin $b \hookleftarrow U(\{0,1\})$ which will determine the response to challenge queries. At each corruption query $i \in [\ell]$, the adversary obtains ek_i and the challenger updates a set $\mathcal{CS} := \mathcal{CS} \cup \{i\}$, which is initially empty. At each encryption query $(i, \boldsymbol{x}_i^{(j)}, t^{(j)})$, the challenger samples $\mathbf{e}_i^{(j)} \hookleftarrow D_{\mathbb{Z}^m,\alpha q}$ and returns

$$\mathbf{C}_{t,i} = \mathbf{G}_0^\top \cdot \boldsymbol{x}_i^{(j)} + \mathbf{A}(\tau^{(j)})^\top \cdot \mathbf{s}_i + \mathbf{e}_i^{(j)} \in \mathbb{Z}_q^m,$$

where $\tau^{(j)} = \mathsf{AHF}(t^{(j)})$. In the challenge phase, the adversary $\mathcal{A}$ chooses a fresh tag $t^\star$ and two vectors of messages $\mathbf{X}_0^\star = [\boldsymbol{x}_{0,1}^\star \mid \ldots \mid \boldsymbol{x}_{0,\ell}^\star] \in [-X, X]^{n_0 \times \ell}$ and $\mathbf{X}_1^\star = [\boldsymbol{x}_{1,1}^\star \mid \ldots \mid \boldsymbol{x}_{1,\ell}^\star] \in [-X, X]^{n_0 \times \ell}$ subject to the constraint that, for any private key query $\boldsymbol{y} \in [-Y, Y]^\ell$ made by $\mathcal{A}$, we must have $\mathbf{X}_0^\star \cdot \boldsymbol{y} = \mathbf{X}_1^\star \cdot \boldsymbol{y}$ over $\mathbb{Z}$. In addition, the invariant that $\boldsymbol{x}_{0,i}^\star = \boldsymbol{x}_{1,i}^\star$ for any $i \in \mathcal{CS}$ must be satisfied at any time during the game. In response to a challenge query $(i, \boldsymbol{x}_{0,i}^\star, \boldsymbol{x}_{1,i}^\star, t^\star)$, the challenger generates a challenge ciphertext $\mathbf{C}_{t^\star,i}$, where

$$\mathbf{C}_{t^\star,i} = \mathbf{G}_0^\top \cdot \boldsymbol{x}_{b,i}^\star + \mathbf{A}(\tau^\star)^\top \cdot \mathbf{s}_i + \mathbf{e}_i^\star, \tag{3}$$

where $\tau^\star = \mathsf{AHF}(t^\star)$ and $\mathbf{e}_i^\star \hookleftarrow D_{\mathbb{Z}^m,\alpha q}$ for all $i \in [\ell]$.

When $\mathcal{A}$ halts, it outputs $\hat{b} \in \{0,1\}$ and the challenger defines $b' := \hat{b}$. We have $\mathbf{Adv}(\mathcal{A}) := |\Pr[W_0] - 1/2|$, where W_0 is event that $b' = b$.

Game_1: This game is identical to Game_0 except for the following changes. First, the challenger runs $K \leftarrow \mathsf{AdmSmp}(1^\lambda, Q, \delta)$ to generate a key $K \in \{0, 1, \perp\}^L$

for a balanced admissible hash function $\mathsf{AHF} : \{0,1\}^{\ell_t} \to \{0,1\}^L$. When the adversary halts and outputs $\hat{b} \in \{0,1\}$, the challenger checks if the conditions

$$P_K(t^{(1)}) = \cdots = P_K(t^{(Q)}) = 1 \;\wedge\; P_K(t^\star) = 0 \tag{4}$$

are satisfied. If conditions (4) do not hold, the challenger ignores $\mathcal{A}$'s output $\hat{b} \in \{0,1\}$ and overwrites it with a random bit $b'' \hookleftarrow \{0,1\}$ to define $b' = b''$. If conditions (4) are satisfied, the challenger sets $b' = \hat{b}$. By Lemma 2.8,

$$|\Pr[W_1] - 1/2| \geq \gamma_{\min} \cdot \mathbf{Adv}(\mathcal{A}) - \frac{1}{2} \cdot (\gamma_{\max} - \gamma_{\min}) = \tau,$$

where $\tau(\lambda)$ is a noticeable function.

Game_2: In this game, we modify the generation of mpk in the following way. Initially, the challenger samples a uniformly random matrix $\mathbf{A} \hookleftarrow U(\mathbb{Z}_q^{n \times m})$. Next, for each $i \in [L]$, it samples $\mathbf{R}_{i,0}, \mathbf{R}_{i,1} \hookleftarrow U(\{-1,1\})^{m \times m}$ and defines $\{\mathbf{A}_{i,0}, \mathbf{A}_{i,1}\}_{i=1}^L$ as follows for all $i \in [L]$ and $j \in \{0,1\}$:

$$\mathbf{A}_{i,j} := \begin{cases} \mathbf{A} \cdot \mathbf{R}_{i,j} & \text{if} \;\; (j \neq K_i) \wedge (K_i \neq \perp) \\ \mathbf{A} \cdot \mathbf{R}_{i,j} + \mathbf{G} & \text{if} \;\; (j = K_i) \vee (K_i = \perp) \end{cases} \tag{5}$$

Since $\mathbf{A} \in \mathbb{Z}_q^{n \times m}$ was chosen uniformly, the Leftover Hash Lemma ensures that $\{\mathbf{A}_{i,0}, \mathbf{A}_{i,1}\}_{i=1}^L$ are statistically independent and uniformly distributed over $\mathbb{Z}_q^{n \times m}$. It follows that $|\Pr[W_2] - \Pr[W_1]| \leq L \cdot 2^{-\lambda}$.

We note that, at each encryption query $(i, \boldsymbol{x}_i^{(j)}, t^{(j)})$, the admissible hash function maps $t^{(j)}$ to $\tau^{(j)} = \mathsf{AHF}(t^{(j)})$, which is itself mapped to a GSW encryption

$$\mathbf{A}(\tau^{(j)}) = \mathbf{A} \cdot \mathbf{R}_{\tau^{(j)}} + (\prod_{i=1}^{L} \mu_i) \cdot \mathbf{W}^\top, \tag{6}$$

of a product $\prod_{i=1}^L \mu_i$, for some small norm matrix $\mathbf{R}_{\tau^{(j)}} \in \mathbb{Z}^{m \times m}$, where

$$\mu_i := \begin{cases} 0 & \text{if} \;\; (\mathsf{AHF}(t^{(j)}))_i \neq K_i) \wedge (K_i \neq \perp) \\ 1 & \text{if} \;\; (\mathsf{AHF}(t^{(j)}))_i = K_i) \vee (K_i = \perp) \end{cases}$$

If conditions (4) are satisfied, at each encryption query $(i, x_i^{(j)}, t^{(j)})$, the admissible hash function ensures that $\tau^{(j)} = \mathsf{AHF}(t^{(j)})$ satisfies

$$\mathbf{A}(\tau^{(j)}) = \mathbf{A} \cdot \mathbf{R}_{\tau^{(j)}} \qquad \forall j \in [Q], \tag{7}$$

for some small norm $\mathbf{R}_{\tau^{(j)}} \in \mathbb{Z}^{m \times m}$. Moreover, the challenge tag $t^\star$ is mapped to an L-bit string $\tau^\star = \mathsf{AHF}(t^\star)$ such that

$$\mathbf{A}(\tau^\star) = \mathbf{A} \cdot \mathbf{R}_{\tau^\star} + \mathbf{W}^\top = \mathbf{A} \cdot \mathbf{R}_{\tau^\star} + \mathbf{V}^\top \cdot \mathbf{G}_0 \tag{8}$$

Game$_3$: In this game, we modify the distribution of mpk and replace the uniform matrix $\mathbf{A} \in \mathbb{Z}_q^{n \times m}$ by a lossy matrix such that

$$\mathbf{A}^\top = \hat{\mathbf{A}}^\top \cdot \mathbf{C} + \mathbf{E} \ \in \mathbb{Z}_q^{m \times n}, \tag{9}$$

where $\hat{\mathbf{A}} \hookleftarrow U(\mathbb{Z}_q^{n_1 \times m})$, $\mathbf{C} \hookleftarrow U(\mathbb{Z}_q^{n_1 \times n})$ and $\mathbf{E} \hookleftarrow D_{\mathbb{Z}^{m \times n}, \alpha_1 q}$, for $n_1 \ll n$. The matrix (9) is thus "close" to a matrix $\hat{\mathbf{A}}^\top \cdot \mathbf{C}$ of much lower rank than n. Under the LWE assumption in dimension n_1 with error rate α_1, this change should not significantly affect $\mathcal{A}$'s behavior and a straightforward reduction $\mathcal{B}$ shows that $|\Pr[W_3] - \Pr[W_2]| \leq n \cdot \mathbf{Adv}_{\mathcal{B}}^{\mathsf{LWE}_{q,m,n_1,\alpha_1}}(\lambda)$, where the factor n comes from the use of an LWE assumption with n secrets.

Game$_4$: In this game, we modify the encryption oracle. At each encryption query $(i, \boldsymbol{x}_i^{(j)}, t^{(j)})$, the challenger generates the ciphertext by computing:

$$\mathbf{C}_{t,i} = \mathbf{G}_0^\top \cdot \boldsymbol{x}_i^{(j)} + \mathbf{R}_{\tau^{(j)}}^\top \cdot \hat{\mathbf{A}}^\top \cdot \mathbf{C} \cdot \mathbf{s}_i + \mathbf{e}_i^{(j)} \ \in \mathbb{Z}_q^m, \tag{10}$$

and for each challenge query $(i, \boldsymbol{x}_{0,i}^\star, \boldsymbol{x}_{1,i}^\star, t^\star)$ the challenger replies with:

$$\mathbf{C}_{t^\star,i} = \mathbf{G}_0^\top \cdot \boldsymbol{x}_{b,i}^\star + \left(\mathbf{R}_{\tau^\star}^\top \cdot \hat{\mathbf{A}}^\top \cdot \mathbf{C} + \mathbf{G}_0^\top \cdot \mathbf{V}\right) \cdot \mathbf{s}_i + \mathbf{e}_i^\star \ \in \mathbb{Z}_q^m \tag{11}$$

where $\mathbf{e}_i^{(j)} \hookleftarrow D_{\mathbb{Z}^m, \alpha q}$ and $\mathbf{e}_i^\star \hookleftarrow D_{\mathbb{Z}^m, \alpha q}$. The only difference between Game$_3$ and Game$_4$ is thus that the terms $\mathbf{R}_{\tau^{(j)}}^\top \cdot \mathbf{E} \cdot \mathbf{s}_i + \mathbf{e}_i^{(j)}$ and $\mathbf{R}_{\tau^\star}^\top \cdot \mathbf{E} \cdot \mathbf{s}_i + \mathbf{e}_i^\star$ are replaced by $\mathbf{e}_i^{(j)}$ and $\mathbf{e}_i^\star$ respectively, at each encryption or challenge query. However, the smudging lemma (Lemma 2.4) ensures that the two distributions are statistically close as long as α is sufficiently large with respect to α_1 and σ. Concretely, Lemma 3.3 implies $|\Pr[W_4] - \Pr[W_3]| \leq \ell \cdot (Q+1) \cdot 2^{-\Omega(\lambda)}$.

Game$_5$: This game is like Game$_4$ but we modify the challenge oracle. Instead of encrypting $\mathbf{X}_b^\star = [\boldsymbol{x}_{b,1}^\star \mid \ldots \mid \boldsymbol{x}_{b,\ell}^\star]$ as in (11), the challenger encrypts a linear combination of $\mathbf{X}_0^\star$ and $\mathbf{X}_1^\star$. It initially chooses a uniformly random $\gamma \hookleftarrow U(\mathbb{Z}_q)$ and, at each challenge query $(i, \boldsymbol{x}_{0,i}^\star, \boldsymbol{x}_{1,i}^\star, t^\star)$, computes $\mathbf{C}_{t^\star,i}$ as

$$\mathbf{C}_{t^\star,i} = \mathbf{G}_0^\top \cdot \left((1-\gamma) \cdot \boldsymbol{x}_{b,i}^\star + \gamma \cdot \boldsymbol{x}_{1-b,i}^\star\right) + \left(\mathbf{R}_{\tau^\star}^\top \cdot \hat{\mathbf{A}}^\top \cdot \mathbf{C} + \mathbf{G}_0^\top \cdot \mathbf{V}\right) \cdot \mathbf{s}_i + \mathbf{e}_i^\star,$$

with $\mathbf{e}_i^\star \hookleftarrow D_{\mathbb{Z}^m, \alpha q}$, for all $i \in [\ell]$. Lemma 3.4 shows that Game$_4$ and Game$_5$ are negligibly far part as $|\Pr[W_5] - \Pr[W_4]| \leq 2^{-\Omega(\lambda)}$.

In Game$_5$, we clearly have $\Pr[W_5] = 1/2$ since the challenge ciphertexts $(\mathbf{C}_{t,1}^\star, \ldots, \mathbf{C}_{t,\ell}^\star)$ reveal no information about $b \in \{0,1\}$. □

Lemma 3.3. *Let $\mathbf{R}_\tau \in \mathbb{Z}^{m \times m}$ be as in equation (6). Let $\mathbf{E} \hookleftarrow D_{\mathbb{Z}^{m \times n}, \alpha_1 q}$ and $\mathbf{s} \hookleftarrow D_{\mathbb{Z}^n, \sigma}$. If $\alpha_1 q = \omega(\sqrt{\log n})$, $\sigma = \omega(\sqrt{\log n})$ and $\alpha \geq 2^\lambda \cdot L \cdot m^4 \cdot n^{3/2} \cdot \alpha_1 \cdot \sigma$, we have the statistical distance upper bound $\Delta\left(D_{\mathbb{Z}^m, \alpha q},\ \mathbf{R}_\tau^\top \cdot \mathbf{E} \cdot \mathbf{s} + D_{\mathbb{Z}^m, \alpha q}\right) \leq 2^{-\lambda}$.* (The proof is given in the full version of the paper.)

Lemma 3.4. *We have* $|\Pr[W_5] - \Pr[W_4]| \le 2^{-\Omega(\lambda)}$.

Proof. To prove the result, we resort to a technique of guessing in advance the difference $\mathbf{X}^\star_{1-b} - \mathbf{X}^\star_b$, which was previously used in [13,67] and can be seen as complexity leveraging with respect to a statistical argument. We consider the following variants of Game_4 and Game_5, respectively.

We define Game'_4 and Game'_5 simultaneously by using an index $k \in \{4,5\}$:

Game'_k: This game is like Game_k with one difference in the setup phase. To generate mpk, the challenger $\mathcal{B}$ generates a statistically uniform $\mathbf{U} \in \mathbb{Z}_q^{(n_0+n_1)\times n}$ with a trapdoor $\mathbf{T}_U$ for the lattice $\Lambda^\perp(\mathbf{U})$. Then, $\mathcal{B}$ parses $\mathbf{U}$ as

$$\mathbf{U} = \left[\frac{\mathbf{V}}{\mathbf{C}}\right] \in \mathbb{Z}_q^{(n_0+n_1)\times n},$$

where $\mathbf{V} \in \mathbb{Z}_q^{n_0\times n}$ and $\mathbf{C} \in \mathbb{Z}_q^{n_1\times n}$ are statistically independent and uniform over $\mathbb{Z}_q$. Next, it computes

$$\mathbf{A}^\top = \hat{\mathbf{A}}^\top \cdot \mathbf{C} + \mathbf{E} \in \mathbb{Z}_q^{m\times n},$$

where $\hat{\mathbf{A}} \hookleftarrow U(\mathbb{Z}_q^{n_1\times m})$ and $\mathbf{E} \hookleftarrow D_{\mathbb{Z}^{m\times n},\alpha_1 q}$. The obtained matrix $\mathbf{A} \in \mathbb{Z}_q^{n\times m}$ is then used to generate $\{\mathbf{A}_{i,j}\}_{i\in[L],j\in\{0,1\}}$ as per (5). The upper part $\mathbf{V} \in \mathbb{Z}_q^{n_0\times n}$ of $\mathbf{U}$ is included in mpk, the distribution of which is statistically close to that of Game_k: we indeed have $|\Pr[W'_k] - \Pr[W_k]| \le 2^{-\Omega(\lambda)}$.

We do the same as above and define Game''_4 and Game''_5 simultaneously by using an index $k \in \{4,5\}$:

Game''_k: This game is identical to Game'_k with the following difference. At the outset of the game, the challenger randomly chooses $\mathbf{\Delta X} \hookleftarrow U([-2X,2X]^{n_0\times \ell})$ as a guess for the difference $\mathbf{X}^\star_{1-b} - \mathbf{X}^\star_b$ between the challenge messages $\mathbf{X}^\star_0, \mathbf{X}^\star_1$. In the challenge phase, the challenger checks if $\mathbf{\Delta X} = \mathbf{X}^\star_{1-b} - \mathbf{X}^\star_b$. If not, it aborts and replaces $\mathcal{A}$'s output $\hat{b}$ with a random bit $b'' \hookleftarrow U(\{0,1\})$. If the guess for $\mathbf{X}^\star_{1-b} - \mathbf{X}^\star_b$ was successful (we call Guess this event), the challenger proceeds exactly as it did in Game'_k.

Since the choice of $\mathbf{\Delta X} \hookleftarrow U([-2X,2X]^{n_0\times\ell})$ is completely independent of $\mathcal{A}$'s view, we clearly have $\Pr[\mathsf{Guess}] = 1/(4X)^{n_0\ell}$. Since Game''_4 is identical to Game'_4 when Guess occurs, this implies $\mathbf{Adv}_{4'}(\mathcal{A}) = (4X)^{n_0\ell} \cdot \mathbf{Adv}_{4''}(\mathcal{A})$. Indeed,

$$\begin{aligned}\mathbf{Adv}_{4''}(\mathcal{A}) := \frac{1}{2} \cdot |\Pr[b' = 1 \mid b = 1, \mathsf{Guess}] \cdot \Pr[\mathsf{Guess}] + \frac{1}{2} \cdot \Pr[\neg\mathsf{Guess}] \\ - \Pr[b' = 1 \mid b = 0, \mathsf{Guess}] \cdot \Pr[\mathsf{Guess}] - \frac{1}{2} \cdot \Pr[\neg\mathsf{Guess}]| \\ = \frac{1}{2} \cdot \Pr[\mathsf{Guess}] \cdot |\Pr[b' = 1 \mid b = 1, \mathsf{Guess}] - \Pr[b' = 1 \mid b = 0, \mathsf{Guess}]| \\ = \Pr[\mathsf{Guess}] \cdot \mathbf{Adv}_{4'}(\mathcal{A}) = \frac{1}{(4X)^{n_0\ell}} \cdot \mathbf{Adv}_{4'}(\mathcal{A})\end{aligned}$$

and we can similarly show that $\mathbf{Adv}_{5'}(\mathcal{A}) = (4X)^{n_0\ell} \cdot \mathbf{Adv}_{5''}(\mathcal{A})$.

Game$_5'''$: This game is identical to Game$_4''$ except that encryption keys $\{\mathsf{ek}_i\}_{i=1}^{\ell}$ are replaced by alternative encryption keys $\{\mathsf{ek}'_i\}_{i=1}^{\ell}$, which are generated as follows. After having sampled $\mathsf{ek}_i = \mathbf{s}_i \hookleftarrow D_{\mathbb{Z}^n,\sigma}$ for all $i \in [\ell]$, the challenger $\mathcal{B}$ chooses $\gamma \hookleftarrow U(\mathbb{Z}_q)$ and uses the trapdoor $\mathbf{T}_{\mathbf{U}}$ for $\Lambda^{\perp}(\mathbf{U})$ to sample a small-norm matrix $\mathbf{T} \in \mathbb{Z}^{n \times n_0}$ satisfying

$$\mathbf{U} \cdot \mathbf{T} = \begin{bmatrix} \gamma \cdot \mathbf{I}_{n_0} \\ \hline \mathbf{0}^{n_1 \times n_0} \end{bmatrix} \mod q, \tag{12}$$

so that $\mathbf{V} \cdot \mathbf{T} = \gamma \cdot \mathbf{I}_{n_0} \bmod q$ and $\mathbf{C} \cdot \mathbf{T} = \mathbf{0}^{n_1 \times n_0} \bmod q$. For each $i \in [\ell]$, $\mathcal{B}$ then defines the alternative key $\mathsf{ek}'_i = \mathbf{s}'_i$ of user i to be

$$\mathbf{s}'_i = \mathbf{s}_i + \mathbf{T} \cdot \boldsymbol{\Delta x}_i \ \in \mathbb{Z}^n \qquad \forall i \in [\ell], \tag{13}$$

where $\boldsymbol{\Delta x}_i$ is the i-th column of $\boldsymbol{\Delta}\mathbf{X}$ (i.e., the guess for $\boldsymbol{x}^{\star}_{1-b,i} - \boldsymbol{x}^{\star}_{b,i}$). These modified encryption keys $\{\mathsf{ek}'_i = \mathbf{s}'_i\}_{i=1}^{\ell}$ are used to answer all encryption queries and to generate the challenge ciphertext. At each corruption query i, the adversary is also given ek'_i instead of ek_i.

We first claim that, conditionally on Guess, Game$_5'''$ is statistically close to Game$_4''$. To see this, we first argue that trading $\{\mathsf{ek}_i\}_{i=1}^{\ell}$ for $\{\mathsf{ek}'_i\}_{i=1}^{\ell}$ has no incidence on queries made by a legitimate adversary:

- We have $\mathbf{C} \cdot \mathbf{s}'_i = \mathbf{C} \cdot \mathbf{s}_i \bmod q$, so that encryption queries obtain the same responses no matter which key set is used among $\{\mathsf{ek}_i\}_{i=1}^{\ell}$ and $\{\mathsf{ek}'_i\}_{i=1}^{\ell}$.
- We have $\sum_{i=1}^{\ell} \mathbf{s}'_i \cdot \boldsymbol{y}_i = \sum_{i=1}^{\ell} \mathbf{s}_i \cdot \boldsymbol{y}_i$ so long as the adversary only obtains private keys for vectors $\boldsymbol{y} \in \mathbb{Z}^{\ell}$ such that $(\mathbf{X}^{\star}_0 - \mathbf{X}^{\star}_1) \cdot \boldsymbol{y} = \mathbf{0}$ (over $\mathbb{Z}$).
- For any corrupted user $i \in \mathcal{CS}$, it should be the case that $\boldsymbol{x}^{\star}_{0,i} = \boldsymbol{x}^{\star}_{1,i}$, meaning that $\mathbf{s}'_i = \mathbf{s}_i$ as long as Guess occurs.

This implies that Game$_5'''$ is identical to Game$_4''$, except that users' secret keys are defined via (13) and thus have a slightly different distribution. Lemma 3.5 shows that the statistical distance between the distributions of $\{\mathbf{s}'_i\}_{i=1}^{\ell}$ and $\{\mathbf{s}_i\}_{i=1}^{\ell}$ is at most $2^{-\lambda} \cdot (4X)^{-n_0\ell}$. This implies that Game$_4''$ and Game$_5'''$ are statistically close assuming that Guess occurs. When Guess does not occur, both games output a random $b' \hookleftarrow U(\{0,1\})$, so that $|\Pr[W_5'''] - \Pr[W_4'']| \leq 2^{-\lambda} \cdot (4X)^{-n_0\ell}$.

We finally claim that, from the adversary's view Game$_5'''$ is identical to Game$_5''$. Indeed, our choice of $\mathbf{T}$ ensures that $\mathbf{V} \cdot \mathbf{T} = \gamma \cdot \mathbf{I}_{n_0} \bmod q$, so that we have $\mathbf{V} \cdot \mathbf{s}'_i = \mathbf{V} \cdot \mathbf{s}_i + \gamma \cdot (\boldsymbol{x}^{\star}_{1-b,i} - \boldsymbol{x}^{\star}_{b,i}) \bmod q$. This implies

$$\begin{aligned} \mathbf{C}_{t^{\star},i} &= \mathbf{G}_0^{\top} \cdot \boldsymbol{x}^{\star}_{b,i} + (\mathbf{R}_{\tau^{\star}}^{\top} \cdot \hat{\mathbf{A}}^{\top} \cdot \mathbf{C} + \mathbf{G}_0^{\top} \cdot \mathbf{V}) \cdot \mathbf{s}'_i + \mathbf{e}_i{}^{\star} \\ &= \mathbf{G}_0^{\top} \cdot \big((1-\gamma) \cdot \boldsymbol{x}^{\star}_{b,i} + \gamma \cdot \boldsymbol{x}^{\star}_{1-b,i}\big) + (\mathbf{R}_{\tau^{\star}}^{\top} \cdot \hat{\mathbf{A}}^{\top} \cdot \mathbf{C} + \mathbf{G}_0^{\top} \cdot \mathbf{V}) \cdot \mathbf{s}_i + \mathbf{e}_i{}^{\star} \end{aligned}$$

which is exactly the distribution from Game$_5''$.

Putting the above altogether, we find $|\Pr[W_4''] - \Pr[W_5'']| \leq 2^{-\Omega(\lambda)} \cdot (4X)^{-n_0\ell}$, which in turn implies $|\Pr[W_4] - \Pr[W_5]| \leq 2^{-\Omega(\lambda)}$, as claimed.

□

Lemma 3.5. *If $\sigma \geq 2^{\lambda} \cdot n_0 \cdot (4X)^{n_0\ell+1} \cdot \omega(n^2\sqrt{\log n})$, then we have the inequality* $\Delta\,(D_{\mathbb{Z}^n,\sigma}, \mathbf{T} \cdot \boldsymbol{\Delta x_i} + D_{\mathbb{Z}^n,\sigma}) \leq 2^{-\lambda} \cdot (4X)^{-n_0\ell}$. (The proof is in the full version.)

4 A DMCFE Scheme for Linear Functions

As in [27], our DMCFE scheme combines two instances of the underlying centralized scheme of Sect. 3. While the second instance is used exactly in the same way as in the centralized construction, the first instance is used for the sole purpose of generating partial functional secret keys without having the senders communicate with one another. As in [27], the senders have to initially run an interactive protocol in order to jointly generate public parameters for the two schemes. Note that this protocol is the only step that requires interaction among senders and it is only executed once. This interactive step ends with each sender holding an encryption key $\mathsf{ek}_i = (\mathbf{s}_i, \mathbf{t}_i)$ comprised of encryption keys for the two MCFE instances. The distributed protocol also ensures that a functional secret key $\mathbf{t} = \sum_{i=1}^{\ell} \mathbf{t}_i$ for the all-one vector $(1, 1, \ldots, 1)^\top \in \mathbb{Z}^\ell$ be made publicly available for the first MCFE instance. Later on, when a decryptor wishes to obtain a partial functional secret key $\mathsf{dk}_{f,i}$ for a vector $\boldsymbol{y} = (y_1, \ldots, y_\ell)^\top$ from the i-th sender $\mathcal{S}_i$, the latter can generate an MCFE encryption of the vector $y_i \cdot \mathbf{s}_i \in \mathbb{Z}^n$ under his secret key $\mathbf{t}_i$. Having obtained partial functional secret keys $\mathsf{dk}_{f,i}$ from all senders $\{\mathcal{S}_i\}_{i=1}^{\ell}$, the decryptor can then use the functional secret key $\mathbf{t} = \sum_{i=1}^{\ell} \mathbf{t}_i$ to compute $\mathbf{s}_y = \sum_{i=1}^{\ell} y_i \cdot \mathbf{s}_i \in \mathbb{Z}^n$.

4.1 Description

We assume global public parameters

$$\mathsf{cp} := \Big(\lambda,\ \ell_{\max},\ X,\ \bar{X},\ Y,\ \bar{Y},\ n_0,\ n_1,\ \bar{n}_1,\ n,\ \bar{n},,\ m,\ \bar{m},\ \alpha,\ \alpha_1,\ \bar{\alpha}_1,\ \sigma,\ \bar{\sigma},\ \ell_t,\ \ell_f,\ L,\ q,\ \bar{q},\ \mathsf{AHF}_t,\ \mathsf{AHF}_f \Big),$$

which specify a security parameter λ and the following quantities

- Let $\ell_{max} = \lambda^k$, $n_1 = \lambda^d$, $\bar{d} = 3d + k - 1$, $q = 2^{\lambda^{d-1}+\lambda}$, $\bar{q} = 2^{\lambda^{\bar{d}-1}+\lambda}$, $\bar{n}_1 = \lambda^{\bar{d}}$, $\alpha_1 = 2^{-\lambda^{d-1}+d\log\lambda}$, $\bar{\alpha}_1 = 2^{-\lambda^{\bar{d}-1}+\bar{d}\log\lambda}$, $\alpha = 2^{-\sqrt{\lambda}}$, $n_0 \cdot \ell_{max} = O(\lambda^{d-2})$, $n_0 = O(\lambda^{d-2})$, $n = O(\lambda^{2d-1})$, $\bar{n} = O(\lambda^{4d+k-2})$, $X = 1$, $\bar{Y} = 1$, $\sigma = 2^{\lambda^{d-1}-2\lambda}$, $\bar{\sigma} = 2^{\lambda^{\bar{d}-1}-2\lambda}$, $\bar{X} = 2\ell \cdot Y \cdot \sigma\sqrt{n}$ and the rest of the parameters $Y, m, \bar{m}$ are all in $\mathsf{poly}(\lambda)$
- A tag length $\ell_t \in \Theta(\lambda)$ and a length $\ell_f \in \Theta(\lambda)$ of function labels.
- Dimensions $n, m, n_0, n_1, \bar{n}, \bar{m} \in \mathsf{poly}(\lambda)$ such that $n > 3 \cdot (n_0 + n_1) \cdot \lceil \log q \rceil$, $m > 2 \cdot n \cdot \lceil \log q \rceil$, $\bar{n} > 3 \cdot (n + \bar{n}_1) \cdot \lceil \log \bar{q} \rceil$ and $\bar{m} > 2 \cdot \bar{n} \cdot \lceil \log \bar{q} \rceil$.
- The description of balanced admissible hash functions $\mathsf{AHF}_t : \{0,1\}^{\ell_t} \to \{0,1\}^L$ and $\mathsf{AHF}_f : \{0,1\}^{\ell_f} \to \{0,1\}^L$, for a suitable $L \in \Theta(\lambda)$.
- A real $\alpha > 0$ and a Gaussian parameter $\sigma > 0$, which will specify an interval $[-\beta, \beta] = [-\sigma\sqrt{n}, \sigma\sqrt{n}]$ where the coordinates of the secret will live (with probability exponentially close to 1).

We define $\bar{\mathbf{G}} \in \mathbb{Z}_{\bar{q}}^{n \times \bar{m}}$ to be the gadget matrix

$$\bar{\mathbf{G}} = [\mathbf{I}_n \otimes (1, 2, 4, \ldots, 2^{\lceil \log \bar{q} \rceil}) \mid \mathbf{0}^n \mid \ldots \mid \mathbf{0}^n] \ \in \mathbb{Z}_{\bar{q}}^{n \times \bar{m}}$$

where $\mathbf{I}_n \otimes (1, 2, 4, \ldots, 2^{\lceil \log \bar{q} \rceil})$ is padded with $\bar{m} - n \cdot \lceil \log \bar{q} \rceil$ zero columns.

Setup$(\mathsf{cp}, 1^\ell)$: On input of a number of users $\ell < \ell_{\max}$, the senders $\{\mathcal{S}_i\}_{i=1}^{\ell}$ run an interactive protocol at the end of which the following quantities are made publicly available.
- Random matrices $\mathbf{A}_{i,b} \hookleftarrow U(\mathbb{Z}_q^{n \times m})$, for each $i \in [L]$, $b \in \{0,1\}$.
- Random matrices $\mathbf{B}_{i,b} \hookleftarrow U(\mathbb{Z}_{\bar{q}}^{\bar{n} \times \bar{m}})$, for each $i \in [L]$, $b \in \{0,1\}$.
- Random matrices $\mathbf{V} \hookleftarrow U(\mathbb{Z}_q^{n_0 \times n})$, $\bar{\mathbf{V}} \hookleftarrow U(\mathbb{Z}_{\bar{q}}^{n \times \bar{n}})$.
- The sum $\mathbf{t} = \sum_{i=1}^{\ell} \mathbf{t}_i \in \mathbb{Z}^{\bar{n}}$ of Gaussian vectors $\mathbf{t}_i \hookleftarrow D_{\mathbb{Z}^{\bar{n}},\bar{\sigma}}$ for $i \in [\ell]$.

In addition, for each $i \in [\ell]$, the i-th sender $\mathcal{S}_i$ privately obtains the following:
- The i-th term $\mathbf{t}_i \in \mathbb{Z}^{\bar{n}}$ of the sum $\mathbf{t} = \sum_{i=1}^{\ell} \mathbf{t}_i$.
- A Gaussian vector $\mathbf{s}_i \hookleftarrow D_{\mathbb{Z}^n,\sigma}$, which is used to define $\mathcal{S}_i$'s encryption key $\mathsf{ek}_i = \mathbf{s}_i \in \mathbb{Z}^n$ and the corresponding secret key $\mathsf{sk}_i = (\mathbf{s}_i, \mathbf{t}_i) \in \mathbb{Z}^n \times \mathbb{Z}^{\bar{n}}$.

The master public key is defined to be

$$\mathsf{mpk} := \Big(\mathsf{cp},\ \mathbf{V},\ \bar{\mathbf{V}}, \{\mathbf{A}_{i,0}, \mathbf{A}_{i,1} \ \in \mathbb{Z}_q^{n \times m}\ \}_{i=1}^{L},\\ \{\mathbf{B}_{i,0}, \mathbf{B}_{i,1} \ \in \mathbb{Z}_{\bar{q}}^{\bar{n} \times \bar{m}}\ \}_{i=1}^{L},\ \mathbf{t}\Big),$$

while $\mathcal{S}_i$ obtains $\mathsf{ek}_i = \mathbf{s}_i \in \mathbb{Z}^n$ and $\mathsf{sk}_i = (\mathbf{s}_i, \mathbf{t}_i) \in \mathbb{Z}^n \times \mathbb{Z}^{\bar{n}}$ for each $i \in [\ell]$.

DKeygenShare(sk_i, t_f) : Given the secret key $\mathsf{sk}_i = (\mathbf{s}_i, \mathbf{t}_i) \in \mathbb{Z}^n \times \mathbb{Z}^{\bar{n}}$ and the label t_f of a linear function $f_{\boldsymbol{y}} : \mathbb{Z}^{n_0 \times \ell} \to \mathbb{Z}^{n_0}$ described by a vector $\boldsymbol{y} = (y_1, \ldots, y_\ell)^\top \in [-Y, Y]^\ell$, conduct the following steps.
1. Compute $\tau_f = \tau_f[1] \ldots \tau_f[L] = \mathsf{AHF}_f(t_f) \in \{0,1\}^L$ as well as

$$\mathbf{B}(\tau_f) = \mathbf{B}_{L,\tau_f[L]} \cdot \bar{\mathbf{G}}^{-1}\Big(\mathbf{B}_{L-1,\tau_f[L-1]} \cdot \bar{\mathbf{G}}^{-1}\big(\ldots \mathbf{B}_{2,\tau_f[2]} \cdot \bar{\mathbf{G}}^{-1}\big(\mathbf{B}_{1,\tau_f[1]}\big)\big)\Big)\\ \cdot \bar{\mathbf{G}}^{-1}(\bar{\mathbf{W}}^\top) \ \in \mathbb{Z}_{\bar{q}}^{\bar{n} \times \bar{m}}, \quad (14)$$

   where $\bar{\mathbf{W}} = \bar{\mathbf{G}}^\top \cdot \bar{\mathbf{V}} \in \mathbb{Z}_{\bar{q}}^{\bar{m} \times \bar{n}}$.
2. Sample a noise vector $\mathbf{e}_{f,i} \hookleftarrow D_{\mathbb{Z}^{\bar{m}},\alpha\bar{q}}$. Then, compute

$$\mathsf{dk}_{f,i} = \bar{\mathbf{G}}^\top \cdot (y_i \cdot \boldsymbol{s}_i) + \mathbf{B}(\tau_f)^\top \cdot \mathbf{t}_i + \mathbf{e}_{f,i} \ \in \mathbb{Z}_{\bar{q}}^{\bar{m}}. \quad (15)$$

   Output the partial functional decryption key $\mathsf{dk}_{f,i} \in \mathbb{Z}_{\bar{q}}^{\bar{m}}$.

DKeygenComb$(\{\mathsf{dk}_{f,i}\}_i, t_f)$: Given the label of a function described by a vector $\boldsymbol{y} = (y_1, \ldots, y_\ell) \in [-Y, Y]^\ell$ and ℓ partial functional keys $\{\mathsf{dk}_{f,i}\}_{i=1}^{\ell}$ where $\mathsf{dk}_{f,i} \in \mathbb{Z}_{\bar{q}}^{\bar{m}}$ for each $i \in [\ell]$, conduct the following steps.
1. Compute $\tau_f = \mathsf{AHF}_f(t_f) \in \{0,1\}^L$ and parse it as $\tau_f = \tau_f[1] \ldots \tau_f[L]$.
2. Compute $\mathbf{B}(\tau_f) \in \mathbb{Z}_{\bar{q}}^{\bar{n} \times \bar{m}}$ as per (14).
3. Compute $\mathbf{d}_{t_f} = \sum_{i=1}^{\ell} \mathsf{dk}_{f,i} - \mathbf{B}(\tau_f)^\top \cdot \mathbf{t} \mod \bar{q}$, where $\mathbf{t} \in \mathbb{Z}^{\bar{n}}$ is taken from mpk.

4. Interpret $\mathbf{d}_{t_f} \in \mathbb{Z}_{\bar{q}}^{\bar{m}}$ as a vector of the form $\mathbf{d}_{t_f} = \bar{\mathbf{G}}^\top \cdot \mathbf{s}_{\mathbf{y}} + \tilde{\mathbf{e}}_f \mod \bar{q}$, for some error vector $\tilde{\mathbf{e}}_f \in [-\bar{B}, \bar{B}]^{\bar{m}}$. Using the public trapdoor of $\Lambda^\perp(\bar{\mathbf{G}})$, compute the underlying $\mathbf{s}_{\mathbf{y}} \in [-\ell \cdot \beta \cdot Y, \ell \cdot \beta \cdot Y]^n$.
Output the functional secret key $\mathsf{dk}_{\boldsymbol{y}} = (\boldsymbol{y}, \mathbf{s}_y)$.

Encrypt$(\mathsf{ek}_i, \boldsymbol{x}_i, t)$: Given $\mathsf{ek}_i = \mathbf{s}_i \in \mathbb{Z}^n$, $\boldsymbol{x}_i \in [-X, X]^{n_0}$, and $t \in \{0,1\}^{\ell_t}$,
1. Compute $\tau = \mathsf{AHF}(t) \in \{0,1\}^L$ and parse it as $\tau = \tau[1] \ldots \tau[L]$.
2. Letting $\mathbf{W} = \mathbf{G}_0^\top \cdot \mathbf{V} \in \mathbb{Z}_q^{m \times n}$, compute

$$\mathbf{A}(\tau) = \mathbf{A}_{L,\tau[L]} \cdot \mathbf{G}^{-1}\Big(\mathbf{A}_{L-1,\tau[L-1]} \cdot \mathbf{G}^{-1}\big(\ldots \mathbf{A}_{2,\tau[2]} \cdot \mathbf{G}^{-1}(\mathbf{A}_{1,\tau[1]})\big)\Big) \cdot \mathbf{G}^{-1}(\mathbf{W}^\top) \in \mathbb{Z}_q^{n \times m}. \quad (16)$$

3. Sample a noise vector $\mathbf{e}_i \hookleftarrow D_{\mathbb{Z}^m, \alpha q}$. Then, compute and output

$$\mathbf{C}_{t,i} = \mathbf{G}_0^\top \cdot \boldsymbol{x}_i + \mathbf{A}(\tau)^\top \cdot \mathbf{s}_i + \mathbf{e}_i \ \in \mathbb{Z}_q^m.$$

Decrypt$(\mathsf{dk}_{\boldsymbol{y}}, t, \mathbf{C}_t)$: On input of a functional secret key $\mathsf{dk}_{\boldsymbol{y}} = (\boldsymbol{y}, \mathbf{s}_{\boldsymbol{y}})$ for a vector $\boldsymbol{y} = (y_1, \ldots, y_\ell)^\top \in [-Y, Y]^\ell$, a tag $t \in \{0,1\}^{\ell_t}$, and an ℓ-vector of ciphertexts $\mathbf{C}_t = (\mathbf{C}_{t,1}, \ldots, \mathbf{C}_{t,\ell}) \in (\mathbb{Z}_q^m)^\ell$, conduct the following steps.
1. Compute $\tau = \mathsf{AHF}(t) \in \{0,1\}^L$ and parse it as $\tau = \tau[1] \ldots \tau[L]$.
2. Compute $\mathbf{A}(\tau) \in \mathbb{Z}_q^{n \times m}$ as per (16).
3. Compute $\mathbf{f}_{t,\boldsymbol{y}} = \sum_{i=1}^{\ell} y_i \cdot \mathbf{C}_{t,i} - \mathbf{A}(\tau)^\top \cdot \mathbf{s}_{\boldsymbol{y}} \mod q$.
4. Interpret $\mathbf{f}_{t,\boldsymbol{y}} \in \mathbb{Z}_q^m$ as a vector of the form $\mathbf{f}_{t,\boldsymbol{y}} = \mathbf{G}_0^\top \cdot \boldsymbol{z} + \tilde{\mathbf{e}} \mod q$, for some error vector $\tilde{\mathbf{e}} \in [-B, B]^m$. Using the public trapdoor of $\Lambda^\perp(\mathbf{G}_0)$, compute and output the underlying vector $\boldsymbol{z} \in [-\ell \cdot X \cdot Y, \ell \cdot X \cdot Y]^{n_0}$.

The scheme's correctness is implied by that of the two underlying centralized schemes. In turn, these are correct by Lemma 3.1 and the choice of parameters.

4.2 Security

The proof of Theorem 4.1 is given in the full version of the paper. In order to reduce the security of the centralized scheme to that of its decentralized variant, the proof first moves to a game where the partial functional key generation oracle of Definition 2.14 can be simulated using the functional key generation oracle of Definition 2.11. To this end, it relies on the security of the first MCFE instance. The next step is to move to a game where encryption queries $(i, \boldsymbol{x}_{i,0}, \boldsymbol{x}_{i,1}, t)$ are answered by returning encryptions of $\boldsymbol{x}_{i,1}$ instead of $\boldsymbol{x}_{i,0}$. To this end, we rely on the security of the second MCFE instance, which is possible since the partial key generation oracle can be simulated using the centralized key generation oracle. The final transition restores the partial key generation oracle of Definition 2.14 to its original output distribution. To this end, we invoke again the security of the first MCFE instance and reverse the transition of the first step.

Theorem 4.1. *The above DMCFE scheme provides* sta-IND-sec *security under the* LWE *assumption.*

Acknowledgements. Part of this work was funded by the French ANR ALAMBIC project (ANR-16-CE39-0006) and by BPI-France in the context of the national project RISQ (P141580). This work was also supported by the European Union PROMETHEUS project (Horizon 2020 Research and Innovation Program, grant 780701).

References

1. Abdalla, M., Benhamouda, F., Gay, R.: From single-input to multi-client inner product functional encryption. In: Galbraith, S., Moriai, S. (eds.) ASIACRYPT 2019, LNCS, vol. 11923, pp. 552–582. Springer, Heidelberg (2019)
2. Abdalla, M., Benhamouda, F., Kohlweiss, M., Waldner, H.: Decentralizing inner-product functional encryption. In: Lin, D., Sako, K. (eds.) PKC 2019. LNCS, vol. 11443, pp. 128–157. Springer, Cham (2019). https://doi.org/10.1007/978-3-030-17259-6_5
3. Abdalla, M., Bourse, F., De Caro, A., Pointcheval, D.: Simple functional encryption schemes for inner products. In: Katz, J. (ed.) PKC 2015. LNCS, vol. 9020, pp. 733–751. Springer, Heidelberg (2015). https://doi.org/10.1007/978-3-662-46447-2_33
4. Abdalla, M., Catalano, D., Fiore, D., Gay, R., Ursu, B.: Multi-input functional encryption for inner products: function-hiding realizations and constructions without pairings. In: Shacham, H., Boldyreva, A. (eds.) CRYPTO 2018. LNCS, vol. 10991, pp. 597–627. Springer, Cham (2018). https://doi.org/10.1007/978-3-319-96884-1_20
5. Abdalla, M., Gay, R., Raykova, M., Wee, H.: Multi-input inner-product functional encryption from pairings. In: Coron, J.-S., Nielsen, J.B. (eds.) EUROCRYPT 2017. LNCS, vol. 10210, pp. 601–626. Springer, Cham (2017). https://doi.org/10.1007/978-3-319-56620-7_21
6. Agrawal, S., Boneh, D., Boyen, X.: Efficient lattice (H)IBE in the standard model. In: Gilbert, H. (ed.) EUROCRYPT 2010. LNCS, vol. 6110, pp. 553–572. Springer, Heidelberg (2010). https://doi.org/10.1007/978-3-642-13190-5_28
7. Agrawal, S., Freeman, D.M., Vaikuntanathan, V.: Functional encryption for inner product predicates from learning with errors. In: Lee, D.H., Wang, X. (eds.) ASIACRYPT 2011. LNCS, vol. 7073, pp. 21–40. Springer, Heidelberg (2011). https://doi.org/10.1007/978-3-642-25385-0_2
8. Agrawal, S., Libert, B., Stehlé, D.: Fully secure functional encryption for inner products, from standard assumptions. In: Robshaw, M., Katz, J. (eds.) CRYPTO 2016. LNCS, vol. 9816, pp. 333–362. Springer, Heidelberg (2016). https://doi.org/10.1007/978-3-662-53015-3_12
9. Agrawal, S., Rosen, A.: Functional encryption for bounded collusions, revisited. In: Kalai, Y., Reyzin, L. (eds.) TCC 2017. LNCS, vol. 10677, pp. 173–205. Springer, Cham (2017). https://doi.org/10.1007/978-3-319-70500-2_7
10. Baltico, C., Catalano, D., Fiore, D., Gay, R.: Practical functional encryption for quadratic functions with applications to predicate encryption. In: Katz, J., Shacham, H. (eds.) CRYPTO 2017. LNCS, vol. 10401, pp. 67–98. Springer, Cham (2017). https://doi.org/10.1007/978-3-319-63688-7_3
11. Banerjee, A., Peikert, C.: New and improved key-homomorphic pseudorandom functions. In: Garay, J.A., Gennaro, R. (eds.) CRYPTO 2014. LNCS, vol. 8616, pp. 353–370. Springer, Heidelberg (2014). https://doi.org/10.1007/978-3-662-44371-2_20

12. Banerjee, A., Peikert, C., Rosen, A.: Pseudorandom functions and lattices. In: Pointcheval, D., Johansson, T. (eds.) EUROCRYPT 2012. LNCS, vol. 7237, pp. 719–737. Springer, Heidelberg (2012). https://doi.org/10.1007/978-3-642-29011-4_42
13. Benhamouda, F., Bourse, F., Lipmaa, H.: CCA-secure inner-product functional encryption from projective hash functions. In: Fehr, S. (ed.) PKC 2017. LNCS, vol. 10175, pp. 36–66. Springer, Heidelberg (2017). https://doi.org/10.1007/978-3-662-54388-7_2
14. Benhamouda, F., Joye, M., Libert, B.: A framework for privacy-preserving aggregation of time-series data. ACM Trans. Inf. Syst. Secur. (ACM-TISSEC) **18**(3), 10:1–10:21 (2016)
15. Boneh, D., Boyen, X.: Secure identity based encryption without random oracles. In: Franklin, M. (ed.) CRYPTO 2004. LNCS, vol. 3152, pp. 443–459. Springer, Heidelberg (2004). https://doi.org/10.1007/978-3-540-28628-8_27
16. Boneh, D., Di Crescenzo, G., Ostrovsky, R., Persiano, G.: Public key encryption with keyword search. In: Cachin, C., Camenisch, J.L. (eds.) EUROCRYPT 2004. LNCS, vol. 3027, pp. 506–522. Springer, Heidelberg (2004). https://doi.org/10.1007/978-3-540-24676-3_30
17. Boneh, D., Franklin, M.: Identity-based encryption from the Weil pairing. In: Kilian, J. (ed.) CRYPTO 2001. LNCS, vol. 2139, pp. 213–229. Springer, Heidelberg (2001). https://doi.org/10.1007/3-540-44647-8_13
18. Boneh, D., Lewi, K., Montgomery, H., Raghunathan, A.: Key homomorphic PRFs and their applications. In: Canetti, R., Garay, J.A. (eds.) CRYPTO 2013. LNCS, vol. 8042, pp. 410–428. Springer, Heidelberg (2013). https://doi.org/10.1007/978-3-642-40041-4_23
19. Boneh, D., Sahai, A., Waters, B.: Functional encryption: definitions and challenges. In: Ishai, Y. (ed.) TCC 2011. LNCS, vol. 6597, pp. 253–273. Springer, Heidelberg (2011). https://doi.org/10.1007/978-3-642-19571-6_16
20. Boneh, D., Waters, B.: Conjunctive, subset, and range queries on encrypted data. In: Vadhan, S.P. (ed.) TCC 2007. LNCS, vol. 4392, pp. 535–554. Springer, Heidelberg (2007). https://doi.org/10.1007/978-3-540-70936-7_29
21. Boneh, D., Zhandry, M.: Multiparty key exchange, efficient traitor tracing, and more from indistinguishability obfuscation. In: Garay, J.A., Gennaro, R. (eds.) CRYPTO 2014. LNCS, vol. 8616, pp. 480–499. Springer, Heidelberg (2014). https://doi.org/10.1007/978-3-662-44371-2_27
22. Brakerski, Z., Komargodski, I., Segev, G.: Multi-input functional encryption in the private-key setting: stronger security from weaker assumptions. In: Fischlin, M., Coron, J.-S. (eds.) EUROCRYPT 2016. LNCS, vol. 9666, pp. 852–880. Springer, Heidelberg (2016). https://doi.org/10.1007/978-3-662-49896-5_30
23. Chan, T., Shi, E., Song, D.: Privacy-preserving stream aggregation with fault tolerance. In: Keromytis, A.D. (ed.) FC 2012. LNCS, vol. 7397, pp. 200–214. Springer, Heidelberg (2012). https://doi.org/10.1007/978-3-642-32946-3_15
24. Chandran, N., Goyal, V., Jain, A., Sahai, A.: Functional encryption: decentralised and delegatable. Cryptology ePrint Archive: Report 2015/1017
25. Chase, M.: Multi-authority attribute based encryption. In: Vadhan, S.P. (ed.) TCC 2007. LNCS, vol. 4392, pp. 515–534. Springer, Heidelberg (2007). https://doi.org/10.1007/978-3-540-70936-7_28
26. Chase, M., Chow, S.: Improving privacy and security in multi-authority attribute-based encryption. In: ACM-CCS (2009)

27. Chotard, J., Dufour Sans, E., Gay, R., Phan, D.-H., Pointcheval, D.: Decentralized multi-client functional encryption for inner product. In: Peyrin, T., Galbraith, S. (eds.) ASIACRYPT 2018. LNCS, vol. 11273, pp. 703–732. Springer, Cham (2018). https://doi.org/10.1007/978-3-030-03329-3_24
28. Chotard, J., Dufour Sans, E., Gay, R., Phan, D.-H., Pointcheval, D.: Multi-client functional encryption with repetition for inner product. Cryptology ePrint Archive: Report 2018/1021 (2018)
29. Cocks, C.: An identity based encryption scheme based on quadratic residues. In: Honary, B. (ed.) Cryptography and Coding 2001. LNCS, vol. 2260, pp. 360–363. Springer, Heidelberg (2001). https://doi.org/10.1007/3-540-45325-3_32
30. Datta, P., Okamoto, T., Tomida, J.: Full-hiding (Unbounded) multi-input inner product functional encryption from the k-linear assumption. In: Abdalla, M., Dahab, R. (eds.) PKC 2018. LNCS, vol. 10770, pp. 245–277. Springer, Cham (2018). https://doi.org/10.1007/978-3-319-76581-5_9
31. Dodis, Y.: Exposure-resilient cryptography. Ph.D. thesis, MIT (2000)
32. Fiat, A., Naor, M.: Broadcast encryption. In: Stinson, D.R. (ed.) CRYPTO 1993. LNCS, vol. 773, pp. 480–491. Springer, Heidelberg (1994). https://doi.org/10.1007/3-540-48329-2_40
33. Freire, E., Hofheinz, D., Paterson, K., Striecks, C.: Programmable hash functions in the multilinear setting. In: Canetti, R., Garay, J.A. (eds.) CRYPTO 2013. LNCS, vol. 8042, pp. 513–530. Springer, Heidelberg (2013). https://doi.org/10.1007/978-3-642-40041-4_28
34. Garg, S., Gentry, C., Halevi, S., Raykova, M., Sahai, A., Waters, B.: Candidate indistinguishability obfuscation and functional encryption for all circuits. In: FOCS (2013)
35. Gentry, C., Peikert, C., Vaikuntanathan, V.: Trapdoors for hard lattices and new cryptographic constructions. In: STOC (2008)
36. Gentry, C., Sahai, A., Waters, B.: Homomorphic encryption from learning with errors: conceptually-simpler, asymptotically-faster, attribute-based. In: Canetti, R., Garay, J.A. (eds.) CRYPTO 2013. LNCS, vol. 8042, pp. 75–92. Springer, Heidelberg (2013). https://doi.org/10.1007/978-3-642-40041-4_5
37. Goldwasser, S., et al.: Multi-input functional encryption. In: Nguyen, P.Q., Oswald, E. (eds.) EUROCRYPT 2014. LNCS, vol. 8441, pp. 578–602. Springer, Heidelberg (2014). https://doi.org/10.1007/978-3-642-55220-5_32
38. Goldwasser, S., Goyal, V., Jain, A., Sahai, A.: Multi-input functional encryption. Cryptology ePrint Archive: Report 2013/727 (2013)
39. Goldwasser, S., Kalai, Y., Peikert, C., Vaikuntanathan, V.: Robustness of the learning with errors assumption. In: ICS (2010)
40. Goldwasser, S., Tauman Kalai, Y., Popa, R., Vaikuntanathan, V., Zeldovich, N.: How to run turing machines on encrypted data. In: Canetti, R., Garay, J.A. (eds.) CRYPTO 2013. LNCS, vol. 8043, pp. 536–553. Springer, Heidelberg (2013). https://doi.org/10.1007/978-3-642-40084-1_30
41. Goldwasser, S., Tauman Kalai, Y., Popa, R., Vaikuntanathan, V., Zeldovich, N.: Reusable garbled circuits and succinct functional encryption. In: STOC (2013)
42. Gorbunov, S., Vaikuntanathan, V., Wee, H.: Functional encryption with bounded collusions via multi-party computation. In: Safavi-Naini, R., Canetti, R. (eds.) CRYPTO 2012. LNCS, vol. 7417, pp. 162–179. Springer, Heidelberg (2012). https://doi.org/10.1007/978-3-642-32009-5_11
43. Gordon, S., Katz, J., Liu, F.-H., Shi, E., Zhou, H.-S.: Multi-input functional encryption. Cryptology ePrint Archive: Report 2013/774 (2014)

44. Goyal, V., Pandey, O., Sahai, A., Waters, B.: Attribute-based encryption for fine-grained access control of encrypted data. In: ACM-CCS (2006)
45. Hanaoka, G., Matsuda, T., Schuldt, J.: On the impossibility of constructing efficient key encapsulation and programmable hash functions in prime order groups. In: Safavi-Naini, R., Canetti, R. (eds.) CRYPTO 2012. LNCS, vol. 7417, pp. 812–831. Springer, Heidelberg (2012). https://doi.org/10.1007/978-3-642-32009-5_47
46. Hofheinz, D., Kiltz, E.: Programmable hash functions and their applications. In: Wagner, D. (ed.) CRYPTO 2008. LNCS, vol. 5157, pp. 21–38. Springer, Heidelberg (2008). https://doi.org/10.1007/978-3-540-85174-5_2
47. Jager, T.: Verifiable random functions from weaker assumptions. In: Dodis, Y., Nielsen, J.B. (eds.) TCC 2015. LNCS, vol. 9015, pp. 121–143. Springer, Heidelberg (2015). https://doi.org/10.1007/978-3-662-46497-7_5
48. Joye, M., Libert, B.: A scalable scheme for privacy-preserving aggregation of time-series data. In: Sadeghi, A.-R. (ed.) FC 2013. LNCS, vol. 7859, pp. 111–125. Springer, Heidelberg (2013). https://doi.org/10.1007/978-3-642-39884-1_10
49. Katsumata, S., Yamada, S.: Partitioning via non-linear polynomial functions: more compact IBEs from ideal lattices and bilinear maps. In: Cheon, J.H., Takagi, T. (eds.) ASIACRYPT 2016. LNCS, vol. 10032, pp. 682–712. Springer, Heidelberg (2016). https://doi.org/10.1007/978-3-662-53890-6_23
50. Katz, J., Sahai, A., Waters, B.: Predicate encryption supporting disjunctions, polynomial equations, and inner products. In: Smart, N. (ed.) EUROCRYPT 2008. LNCS, vol. 4965, pp. 146–162. Springer, Heidelberg (2008). https://doi.org/10.1007/978-3-540-78967-3_9
51. Lewko, A., Okamoto, T., Sahai, A., Takashima, K., Waters, B.: Fully secure functional encryption: attribute-based encryption and (hierarchical) inner product encryption. In: Gilbert, H. (ed.) EUROCRYPT 2010. LNCS, vol. 6110, pp. 62–91. Springer, Heidelberg (2010). https://doi.org/10.1007/978-3-642-13190-5_4
52. Lewko, A., Waters, B.: Decentralizing attribute-based encryption. In: Paterson, K.G. (ed.) EUROCRYPT 2011. LNCS, vol. 6632, pp. 568–588. Springer, Heidelberg (2011). https://doi.org/10.1007/978-3-642-20465-4_31
53. Lewko, A., Waters, B.: New proof methods for attribute-based encryption: achieving full security through selective techniques. In: Safavi-Naini, R., Canetti, R. (eds.) CRYPTO 2012. LNCS, vol. 7417, pp. 180–198. Springer, Heidelberg (2012). https://doi.org/10.1007/978-3-642-32009-5_12
54. Libert, B., Sakzad, A., Stehlé, D., Steinfeld, R.: All-but-many lossy trapdoor functions and selective opening chosen-ciphertext security from LWE. In: Katz, J., Shacham, H. (eds.) CRYPTO 2017. LNCS, vol. 10403, pp. 332–364. Springer, Cham (2017). https://doi.org/10.1007/978-3-319-63697-9_12
55. Libert, B., Stehlé, D., Titiu, R.: Adaptively secure distributed PRFs from LWE. In: Beimel, A., Dziembowski, S. (eds.) TCC 2018. LNCS, vol. 11240, pp. 391–421. Springer, Cham (2018). https://doi.org/10.1007/978-3-030-03810-6_15
56. Lin, H.: Indistinguishability obfuscation from SXDH on 5-linear maps and locality-5 PRGs. In: Katz, J., Shacham, H. (eds.) CRYPTO 2017. LNCS, vol. 10401, pp. 599–629. Springer, Cham (2017). https://doi.org/10.1007/978-3-319-63688-7_20
57. Micciancio, D., Peikert, C.: Trapdoors for lattices: simpler, tighter, faster, smaller. In: Pointcheval, D., Johansson, T. (eds.) EUROCRYPT 2012. LNCS, vol. 7237, pp. 700–718. Springer, Heidelberg (2012). https://doi.org/10.1007/978-3-642-29011-4_41
58. Naor, M., Pinkas, B., Reingold, O.: Distributed pseudo-random functions and KDCs. In: Stern, J. (ed.) EUROCRYPT 1999. LNCS, vol. 1592, pp. 327–346. Springer, Heidelberg (1999). https://doi.org/10.1007/3-540-48910-X_23

59. Okamoto, T., Takashima, K.: Fully secure functional encryption with general relations from the decisional linear assumption. In: Rabin, T. (ed.) CRYPTO 2010. LNCS, vol. 6223, pp. 191–208. Springer, Heidelberg (2010). https://doi.org/10.1007/978-3-642-14623-7_11
60. Regev, O.: On lattices, learning with errors, random linear codes, and cryptography. In: STOC (2005)
61. Sahai, A., Seyalioglu, H.: Worry-free encryption: functional encryption with public keys. In: ACM-CCS (2010)
62. Sahai, A., Waters, B.: Fuzzy identity-based encryption. In: Cramer, R. (ed.) EUROCRYPT 2005. LNCS, vol. 3494, pp. 457–473. Springer, Heidelberg (2005). https://doi.org/10.1007/11426639_27
63. Schnorr, C.P.: A hierarchy of polynomial time lattice basis reduction algorithms. Theor. Comput. Sci. **53**(2–3), 201–224 (1987)
64. Shi, E., Chan, T., Rieffel, E., Chow, R., Song, D.: Privacy-preserving aggregation of time-series data. In: NDSS (2011)
65. Waters, B.: Ciphertext-policy attribute-based encryption: an expressive, efficient, and provably secure realization. In: Catalano, D., Fazio, N., Gennaro, R., Nicolosi, A. (eds.) PKC 2011. LNCS, vol. 6571, pp. 53–70. Springer, Heidelberg (2011). https://doi.org/10.1007/978-3-642-19379-8_4
66. Waters, B.: Functional encryption for regular languages. In: Safavi-Naini, R., Canetti, R. (eds.) CRYPTO 2012. LNCS, vol. 7417, pp. 218–235. Springer, Heidelberg (2012). https://doi.org/10.1007/978-3-642-32009-5_14
67. Wee, H.: Dual system encryption via predicate encodings. In: Lindell, Y. (ed.) TCC 2014. LNCS, vol. 8349, pp. 616–637. Springer, Heidelberg (2014). https://doi.org/10.1007/978-3-642-54242-8_26

From Single-Input to Multi-client Inner-Product Functional Encryption

Michel Abdalla[1,2](✉), Fabrice Benhamouda[3], and Romain Gay[4]

[1] DIENS, École normale supérieure, CNRS, PSL University, Paris, France
michel.abdalla@ens.fr
[2] INRIA, Paris, France
[3] Algorand Foundation, New York, NY, USA
fabrice.benhamouda@normalesup.org
[4] University of California, Berkeley, CA, USA
rgay@berkeley.edu

Abstract. We present a new generic construction of multi-client functional encryption (MCFE) for inner products from single-input functional inner-product encryption and standard pseudorandom functions. In spite of its simplicity, the new construction supports labels, achieves security in the standard model under adaptive corruptions, and can be instantiated from the plain DDH, LWE, and Paillier assumptions. Prior to our work, the only known constructions required discrete-log-based assumptions and the random-oracle model. Since our new scheme is not compatible with the compiler from Abdalla et al. (PKC 2019) that decentralizes the generation of the functional decryption keys, we also show how to modify the latter transformation to obtain a decentralized version of our scheme with similar features.

1 Introduction

Functional encryption [11,18] is a generalization of standard encryption which allows for a more fine-grained control over the decryption capabilities of third parties. In these schemes, the owner of a master secret key can derive secret keys for specific functions via a key derivation algorithm. Then, given the encryption of a message x, the holder of a secret decryption key sk_f for a function f can compute $f(x)$ using the decryption algorithm. Informally, a FE scheme is deemed secure if it is infeasible for an adversary to learn any information about x other than what it can be computed using the secret keys at its disposal.

Multi-input functional encryption [16] is an extension of the functional encryption in which the function can be computed over several different inputs that can be encrypted independently. More precisely, the decryption algorithm of such schemes takes as input a secret key sk_f for a function f together with n different ciphertexts $\mathsf{Enc}(x_1), \ldots, \mathsf{Enc}(x_n)$ and outputs the value of the function f applied to underlying plaintexts $(x_1, \ldots, x_n)$.

In the setting in which each ciphertext of a multi-input functional encryption scheme is generated by a different party or client P_i. we often refer to these

S. D. Galbraith and S. Moriai (Eds.): ASIACRYPT 2019, LNCS 11923, pp. 552–582, 2019.
https://doi.org/10.1007/978-3-030-34618-8_19

schemes as multi-client functional encryption (MCFE) schemes [13,16]. In this setting, it is natural to assume that the adversary can corrupt these parties and learn their secret encryption keys. The master secret key, however, is still assumed to be owned by a trusted third party.

Another important property of multi-client functional encryption considered by Chotard et al. [13] is the inclusion of labels in the encryption process. More precisely, in a labeled MCFE scheme, the individual encryption algorithms each take a label as an additional parameter and decryption should only be possible when using ciphertexts generated with respect to the same label. That is, labels allow the users to have more control over the mix-and-match capabilities, as opposed to MCFE without labels, where the owner of a functional decryption key can mix and match all the ciphertexts.

Note that labels can be obtained without loss of generality for MCFE for all functions; however, this is not the case of the practical constructions for restricted classes of functions, such as inner products, which is the focus of this paper. Reciprocally, any MCFE with labels can be turned into a label-free MCFE for the same functionality, simply by setting the labels used by the encryption algorithm to be always a fixed value $\perp$. Put simply, labels are an extra feature that offers a better control over the information leaked by each generated functional decryption key.

For instance, suppose we want to use MCFE to allow teachers to grade their students in a way that the students can use these grades in different college applications and that colleges only learn the average grades of the students with weights of their choice. In this scenario, each teacher would encrypt the grade of each student for their subject. Each college would have a functional decryption key to compute the weighted average of all the grades of each student. It is very important that the teachers use the student ID as a label, otherwise colleges would be able to compute weighted average of a mix of multiple students (like Maths from student A and Physics from student B), which significantly hinders privacy.

Prior Work. As remarked in [5], most of the prior work in the multi-input setting are either feasibility results for general functionalities (e.g,. [8,9,12,16]) or efficient constructions for particular functionalities (e.g.,[2,4,5,13–15]). In the latter case, which is the setting in which we are interested in this paper, the main functionality under consideration is the inner-product functionality, in which functions are associated to a collection $\boldsymbol{y}$ of n vectors $\boldsymbol{y}_1, \ldots, \boldsymbol{y}_n$. In particular, on input a collection $\boldsymbol{x}$ of n vectors $\boldsymbol{x}_1, \ldots, \boldsymbol{x}_n$, it outputs $f_{\boldsymbol{y}}(\boldsymbol{x}) = \sum_{i=1}^{n} \langle \boldsymbol{x}_i, \boldsymbol{y}_i \rangle = \langle \boldsymbol{x}, \boldsymbol{y} \rangle$. As noted in prior works [3,5,13], inner-product functionalities can be quite useful for computing statistics or performing data mining on encrypted databases.

Among the constructions of multi-input functional inner-product encryption schemes without labels, the work of Abdalla et al. in [4] is the one requiring the weakest assumptions since it can be built from any single-input functional encryption scheme satisfying some mild properties (recalled in Sect. 3). In particular, by instantiating it with the public-key functional inner-product encryption schemes in [7], one can obtain constructions based on the DDH, Paillier, and LWE assumptions. Moreover, as recently shown in [2], their schemes remain secure even when the secret encryption keys can be adaptively corrupted by the adversary. Unfortunately, as we further discuss below, we do not know how to generalize the ACFGU scheme to the labeled setting. In fact, the construction from [4] relies on an information-theoretic multi-input FE (as they put it, the functional encryption equivalent of a one-time pad) to obtain security in the restricted context of one challenge ciphertext per input slot. Then, they bootstrap security to many challenge ciphertexts using an extra layer of single-input FE. That information-theoretic approach cannot be emulated, since we need to hide messages for arbitrarily many labels in our case. Thus, an entropy argument can be used to show that we need to resort to a computational assumption, even for proving security in the context of one challenge ciphertext per input slot and label. In our case, we use PRFs.

Among the constructions of multi-input functional inner-product encryption schemes with labels, the works of Abdalla et al. [2] and Chotard et al. [14] currently represent the state of the art in this area. In particular, both schemes provide labeled MCFE schemes in the random-oracle model in discrete-log-based groups. The main advantage of the work of Chotard et al. is that its ciphertexts are shorter and that it allows for multiple ciphertexts under the same label. However, it requires pairing groups. The main advantage of the work of Abdalla et al. is that it can be instantiated in pairing-free groups. However, its ciphertexts are longer and it only allows for one ciphertext per label, a restriction inhereted from [13]. As in the case of other discrete-log-based constructions of functional inner-product encryption schemes (e.g, [3,5,7,10]), the size of supported messages is restricted for both schemes since the decryption algorithm needs to compute discrete logarithms.

Contributions. In order to address the shortcomings of previous labeled MCFE schemes, the main contribution of this paper is to provide the first construction of labeled MCFE schemes in the standard model from more general assumptions than discrete-logarithm-based ones. As in the work of Abdalla et al. in [4], our constructions can be built from any single-input public-key functional encryption scheme satisfying some mild properties (recalled in Sect. 3). In particular, by instantiating it with the schemes in [7], one can obtain constructions based on the DDH, Paillier, and LWE assumptions. Our constructions have no restriction on the number of ciphertexts per label and are proven secure with respect to adaptive corruptions.

In order to achieve our main result, our security proof proceeds in two parts. First, we prove the security of our MCFE scheme in a setting in which the adversary is required to query the encryption oracle in all n positions for each label. Then, in a second step, we apply the compiler suggested in [2] to remove this requirement. Since the proof for the latter transformation given in [2] is in the random-oracle model, an additional contribution of our work is to provide an alternative proof for it in Sect. 4 which does not require random oracles.

Finally, since our main construction is not compatible with the transformation from [2] that decentralizes the generation of the functional decryption keys, we also show how to modify the latter to obtain a decentralized version of our scheme with similar features. As a result, we obtain the first decentralized labeled MCFE schemes in the standard model based on the DDH, Paillier, and LWE assumptions.

Independent Work. In a recent work [6], the authors define multi-input functional encryption schemes with decentralized key generation and setup, in which users can join the system dynamically. They give a feasibility result for general functions, and also provide a construction for inner products, from a standard assumption (LWE). However, their construction does not handle labels.

Overview of Our Construction. Following the proof strategy first used in [5] in the context of multi-input FE for inner products, we start with a scheme whose security only holds when there is only one challenge ciphertext per input slot. The novelty compared to multi-input FE is that we have to handle arbitrarily many labels, even if there is only one challenge ciphertext per slot and label.

One-time Security With Labels. We modify the scheme from [4], where the one-time secure MIFE is simply obtained using a one-time pad of the messages. The functional decryption keys are simply the linear combination of these pads. Namely, for any input slot i, we have $\mathsf{ct}_i := \boldsymbol{x}_i + \boldsymbol{t}_i$, and for $\mathsf{sk}_{\boldsymbol{y}} := \sum_{i=1}^{n} \langle \boldsymbol{t}_i, \boldsymbol{y}_i \rangle$, where $\boldsymbol{t}_i \leftarrow \mathbb{Z}_L^m$, m denotes the dimension of individual messages $\boldsymbol{x}_i$, and everything is computed modulo L, for some specified integer L. Here, we write $\boldsymbol{y} := (\boldsymbol{y}_1 \| \dots \| \boldsymbol{y}_n)$, the concatenation of n vectors, each of dimension m. To decrypt the set of ciphertext $\{\mathsf{ct}_i\}_i$, one simply compute $\sum_i \langle \mathsf{ct}_i, \boldsymbol{y}_i \rangle$, and subtract by the key $\mathsf{sk}_{\boldsymbol{y}}$ to get $\sum_i \langle \boldsymbol{x}_i, \boldsymbol{y}_i \rangle$. Security follows by a perfect statistical argument.

The technical challenge is to emulate this idea to a setting where ciphertexts can be generated for many labels. Since the number of label is not a priori bounded, we cannot resort to a perfectly statistical argument: the master secret key (which in the previous scheme contains all the vectors $\boldsymbol{t}_i$) is simply too small to contain all possible pads $\boldsymbol{t}_{i,\ell}$ for all labels $\ell \in \mathsf{Labels}$ that would required to perform such an argument. We must resort to a computation argument. A natural but flawed idea would to generate the pads $\boldsymbol{t}_{i,\ell}$ using a PRF applied on a label $\ell \in \mathsf{Labels}$. This approach faces two issues: first, if one slot is corrupted, then the security of the entire system is compromised, since each input slot needs the PRF key to encrypt. Second, since the labels are only known at encryption time, the generation of functional decryption keys is unable to produce the value $\sum_i \langle \boldsymbol{y}_i, \boldsymbol{t}_{i,\ell} \rangle$.

To circumvent these issues we generate the pads $\boldsymbol{t}_{i,\ell} := \sum_{j \neq i} (-1)^{j<i} \mathsf{PRF}_{\mathsf{K}_{i,j}}(\ell)$, where for all $i < j \in [n]$, the keys $\mathsf{K}_{i,j} \leftarrow \{0,1\}^\lambda$, and $\mathsf{K}_{j,i} = \mathsf{K}_{i,j}$, and $(-1)^{j<i}$ denotes -1 if $j < i$, 1 otherwise. This construction has first been used in [17] to decentralize the computation of the sum of private values in a non-interactive way. Each input slot $i \in [n]$ needs the set of keys $\{\mathsf{K}_{i,j}\}_{j\in[n]}$ to encrypt. Assuming the security of the PRF, it produces pseudorandom pads, which will be able to mask the messages $\boldsymbol{x}_i$ simultaneously for all used label $\ell \in \mathsf{Labels}$. Thus, we prove that this holds even when some users $i \in [n]$ are corrupted (in fact, up to $n-2$ can be corrupted). This solves the first issue mentioned above. To solve the second issue, namely, ensuring correctness holds for all possible labels, we use the structure property that holds for all label $\ell \in \mathsf{Labels}$: $\sum_{i\in[n]} \boldsymbol{t}_{i,\ell} = \mathbf{0}$, where $\mathbf{0}$ denotes the zero vector. Otherwise stated, these pads are shares of a perfect n out of n secret sharing of $\mathbf{0}$. We use this by setting the ciphertext for slot $i \in [n]$ and label $\ell \in \mathsf{Labels}$ to be an encryption of the vector $\boldsymbol{w}_{i,\ell} := (\mathbf{0}\|\ldots\|\mathbf{0}\|\boldsymbol{x}_i\|\mathbf{0}\|\ldots\|\mathbf{0}) + \boldsymbol{t}_{i,\ell} \in \mathbb{Z}_L^{mn}$. This way, we have $\langle \boldsymbol{w}_{i,\ell}, \boldsymbol{y}\rangle = \langle \boldsymbol{x}_i, \boldsymbol{y}_i\rangle + \langle \boldsymbol{t}_{i,\ell}, \boldsymbol{y}\rangle$ for all slots $i \in [n]$, therefore: $\sum_{i\in[n]}\langle \boldsymbol{w}_{i,\ell}, \boldsymbol{y}\rangle = \sum_{i\in[n]}\langle \boldsymbol{x}_i, \boldsymbol{y}_i\rangle$. The last step is to encrypt the vector $\boldsymbol{w}_{i,\ell}$ using any single-input, public-key FE for inner products. The functional decryption key is simply the functional decryption key of the single-input inner-product FE for the associated vector $\boldsymbol{y}$. Correctness is preserved, since the decryption only needs to compute the inner product between $\boldsymbol{w}_{i,\ell}$ and $\boldsymbol{y}$.

Full-fledged Security. To obtain security with many challenge ciphertexts per input slot and label, we use similar techniques to those used in [4] in the context of multi-input inner-product FE. However, these can only be applied when the adversary does not make use of the information revealed by partial ciphertexts $\{\mathsf{ct}_{i,\ell}\}_{i\in[n]\setminus\{\text{missing}\}}$, where {missing} denotes the set of missing slots for label ℓ. Prior works [2,14] provides generic compilers that precisely avoid partial ciphertexts to leak any information about the underlying plaintext (decryption is only successful when ciphertexts for all slots are present), but they are only proven secure in the random oracle model, and for [14], use additional assumptions (pairings). Since our focus it to build simple MCFE schemes from weak assumptions, we give a new generic transformation (in Sect. 4) that avoids the leakage of information of partial ciphertexts, with no extra assumption (only PRFs, in the standard model), and that handles adaptive corruptions.

Decentralizing MCFE. In order to decentralize the generation of functional decryption keys, we adapt the construction from [2]. The main idea is to secret share the master secret key, since computing the functional secret key is a linear operation, it can be done non-interactively from these shares.

Outline. The rest of the paper is organized as follows. After giving the relevant technical preliminaries and definitions in Sect. 2, we give our new construction of MCFE from single-input FE for inner products in Sect. 3. In Sect. 4, we show how to generically strengthen the security of our MCFE construction, thereby removing any artificial restrictions on the security model. Finally, in Sect. 5, we show how to decentralize our MCFE to obtain a DMCFE.

2 Definitions and Security Models

Notation. We use $[n]$ to denote the set $\{1, \ldots, n\}$. We write $\boldsymbol{x}$ for vectors and x_i for the i-th element. For security parameter λ and additional parameters n, we denote the winning probability of an adversary $\mathcal{A}$ in a game or experiment G as $\mathsf{Win}^{\mathsf{G}}_{\mathcal{A}}(\lambda, n)$, which is $\Pr[\mathsf{G}(\lambda, n, \mathcal{A}) = 1]$. The probability is taken over the random coins of G and $\mathcal{A}$.

2.1 Multi-Client Functional Encryption

In this section, we recall the definition of MCFE [16]. It is taken almost verbatim from [2], with the following differences: the use of a stronger security definition (see Remark 2.3) and the introduction of a master public key mpk, so that *public-key* functional encryption becomes a particular case of MCFE.

Definition 2.1. ***(Multi-Client Functional Encryption)*** *Let $\mathcal{F} = \{\mathcal{F}_\rho\}_\rho$ be a family (indexed by ρ) of sets $\mathcal{F}_\rho$ of functions $f\colon \mathcal{X}_{\rho,1} \times \cdots \times \mathcal{X}_{\rho,n_\rho} \to \mathcal{Y}_\rho$.*[1] *Let $\mathsf{Labels} = \{0,1\}^*$ or $\{\perp\}$ be a set of labels. A multi-client functional encryption scheme (MCFE) for the function family $\mathcal{F}$ and the label set Labels is a tuple of five algorithms $\mathsf{MCFE} = (\mathsf{Setup}, \mathsf{KeyGen}, \mathsf{KeyDer}, \mathsf{Enc}, \mathsf{Dec})$:*

$\mathsf{Setup}(1^\lambda, 1^n)$: *Takes as input a security parameter λ and the number of parties n, and generates public parameters pp. The public parameters implicitly define an index ρ corresponding to a set $\mathcal{F}_\rho$ of n-ary functions (i.e., $n = n_\rho$).*

$\mathsf{KeyGen}(\mathsf{pp})$: *Takes as input the public parameters pp and outputs n secret keys $\{\mathsf{sk}_i\}_{i\in[n]}$, a master secret key msk, and a master public key mpk.*

$\mathsf{KeyDer}(\mathsf{pp}, \mathsf{msk}, f)$: *Takes as input the public parameters pp, the master secret key msk and a function $f \in \mathcal{F}_\rho$, and outputs a functional decryption key sk_f.*

$\mathsf{Enc}(\mathsf{pp}, \mathsf{mpk}, \mathsf{sk}_i, x_i, \ell)$: *Takes as input the public parameters pp, a master public key mpk, a secret key sk_i, a message $x_i \in \mathcal{X}_{\rho,i}$ to encrypt, a label $\ell \in \mathsf{Labels}$, and outputs ciphertext $\mathsf{ct}_{i,\ell}$.*

$\mathsf{Dec}(\mathsf{pp}, \mathsf{sk}_f, \mathsf{ct}_{1,\ell}, \ldots, \mathsf{ct}_{n,\ell})$: *Takes as input the public parameters pp, a functional key sk_f and n ciphertexts under the same label ℓ and outputs a value $y \in \mathcal{Y}_\rho$.*

A scheme MCFE is correct, if for all $\lambda, n \in \mathbb{N}$, $\mathsf{pp} \leftarrow \mathsf{Setup}(1^\lambda, 1^n)$, $f \in \mathcal{F}_\rho$, $\ell \in \mathsf{Labels}$, $x_i \in \mathcal{X}_{\rho,i}$, when $(\{\mathsf{sk}_i\}_{i\in[n]}, \mathsf{msk}, \mathsf{mpk}) \leftarrow \mathsf{KeyGen}(\mathsf{pp})$ and $\mathsf{sk}_f \leftarrow \mathsf{KeyDer}(\mathsf{pp}, \mathsf{msk}, f)$, we have for $\boldsymbol{x} = (x_1, \ldots, x_n)$:

$$\Pr\left[\mathsf{Dec}(\mathsf{pp}, \mathsf{sk}_f, \mathsf{Enc}(\mathsf{pp}, \mathsf{mpk}, \mathsf{sk}_1, x_1, \ell), \ldots, \mathsf{Enc}(\mathsf{pp}, \mathsf{mpk}, \mathsf{sk}_n, x_n, \ell)) = f(\boldsymbol{x})\right] = 1.$$

When ρ is clear from context, the index ρ is omitted. Note that the case of (single-input) functional encryption as defined in [11,18] corresponds to the case $n = 1$, and $\mathsf{Labels} = \{\perp\}$. For such schemes, we also consider the *public-key* variant, where $\mathsf{sk}_1 = \perp$, that is, the encryption algorithm only requires the public

[1] All the functions inside the same set $\mathcal{F}_\rho$ have the same domain and the same range.

parameters pp and the master public key mpk to encrypt the message x_1. In this setting, sk_1 is omitted.

Except for public-key single-input functional encryption, the master public-key can be included in each secret key sk_i and we omit it.

We follow the notation of [2] here, where the algorithm Setup only generates public parameters that determine the set of functions for which functional decryption keys can be created, and the secret/encryption keys and the master secret keys are generated by another algorithm KeyGen, while the functional decryption keys are generated by KeyDer.

In the following, we define security as adaptive left-or-right indistinguishability under both static (sta), and adaptive (adt) corruption. We also consider two variants of these notions (any, pos$^+$) related to the number of encryption queries asked by the adversary for each slot.

Definition 2.2. ***(Security of MCFE)*** *Let* MCFE *be an MCFE scheme,* $\mathcal{F} = \{\mathcal{F}_\rho\}_\rho$ *a function family indexed by* ρ *and* Labels *a label set. For* $\text{xx} \in \{\text{sta}, \text{adt}\}$, $\text{yy} \in \{\text{any}, \text{pos}^+\}$, *and* $\beta \in \{0, 1\}$, *we define the experiment* $\text{xx-yy-IND}_\beta^{\mathsf{MCFE}}$ *in Fig. 1, where the oracles are defined as:*

Corruption oracle $\mathsf{QCor}(i)$: *Outputs the encryption key* sk_i *of slot* i. *We denote by* $\mathcal{CS}$ *the set of corrupted slots at the end of the experiment.*

Left-Right oracle $\mathsf{QLeftRight}(i, x_i^0, x_i^1, \ell)$: *Outputs* $\mathsf{ct}_{i,\ell} = \mathsf{Enc}(\mathsf{pp}, \mathsf{sk}_i, x_i^\beta, \ell)$ *on a query* (i, x_i^0, x_i^1, ℓ). *We denote by* $Q_{i,\ell}$ *the number of queries of the form* $\mathsf{QLeftRight}(i, \cdot, \cdot, \ell)$.

Encryption oracle $\mathsf{QEnc}(i, x_i, \ell)$: *outputs* $\mathsf{ct}_{i,\ell} = \mathsf{Enc}(\mathsf{pp}, \mathsf{mpk}, \mathsf{sk}_i, x_i, \ell)$ *on a query* (i, x_i, ℓ).

Key derivation oracle $\mathsf{QKeyD}(f)$: *Outputs* $\mathsf{sk}_f = \mathsf{KeyDer}(\mathsf{pp}, \mathsf{msk}, f)$.

and where Condition (*) *holds if all the following conditions hold:*

- *If* $i \in \mathcal{CS}$ *(i.e., slot* i *is corrupted): for any query* $\mathsf{QLeftRight}(i, x_i^0, x_i^1, \ell)$, $x_i^0 = x_i^1$.[2]
- *For any label* $\ell \in$ Labels, *for any family of queries* $\{\mathsf{QLeftRight}(i, x_i^0, x_i^1, \ell)$ *or* $\mathsf{QEnc}(i, x_i, \ell)\}_{i \in [n] \setminus \mathcal{CS}}$, *for any family of inputs* $\{x_i \in \mathcal{X}_{\rho,i}\}_{i \in \mathcal{CS}}$, *for any query* $\mathsf{QKeyD}(f)$, *we define* $x_i^0 := x_i$ *and* $x_i^1 := x_i$ *for any slot* $i \in \mathcal{CS}$ *and any slot queried to* $\mathsf{QEnc}(i, x_i, \ell)$, *and we require that:*

$$f(\boldsymbol{x}^0) = f(\boldsymbol{x}^1) \qquad \textit{where } \boldsymbol{x}^b = (x_1^b, \ldots, x_n^b) \textit{ for } b \in \{0,1\}\ .$$

We insist that if one index $i \notin \mathcal{CS}$ *is not queried for the label* ℓ, *there is no restriction.*

---

[2] We could define a stronger security notion without this restriction. However, in this paper, as in the prior works on MCFE, we add this restriction. In particular, we allow the secret key for the slot i to decrypt ciphertexts for the slot i. We leave achieving stronger security as an interesting open problem.

- *When* yy = pos$^+$*: for any slot* $i \in [n]$ *and* $\ell \in$ Labels*, if* $Q_{i,\ell} > 0$*, then for any slot* $j \in [n] \setminus \mathcal{CS}$*,* $Q_{j,\ell} > 0$*. In other words, for any label, either the adversary makes no left-right encryption query or makes at least one left-right encryption query for each slot* $i \in [n] \setminus \mathcal{CS}$*.*

We define the advantage of an adversary $\mathcal{A}$ *in the following way:*

$$\mathsf{Adv}^{\text{xx-yy-IND}}_{\mathsf{MCFE},\mathcal{A}}(\lambda, n) = \Big| \Pr[\text{xx-yy-IND}^{\mathsf{MCFE}}_0(\lambda, n, \mathcal{A}) = 1] \\ - \Pr[\text{xx-yy-IND}^{\mathsf{MCFE}}_1(\lambda, n, \mathcal{A}) = 1] \Big| \ .$$

A multi-client functional encryption scheme MCFE *is* xx-yy-*IND secure, if for any* n*, for any polynomial-time adversary* $\mathcal{A}$*, there exists a negligible function* negl *such that:* $\mathsf{Adv}^{\text{xx-yy-IND}}_{\mathsf{MCFE},\mathcal{A}}(\lambda, n) \leq \text{negl}(\lambda)$.

We omit n when it is clear from the context. We also often omit $\mathcal{A}$ from the parameter of experiments or games when it is clear from the context.

Remark 2.3 (The role of the oracle QEnc*).* The security definitions we give are slightly stronger than those given in [2], since the oracle QEnc gives out information that is not captured by Condition (*), for pos$^+$, hence the use of the notation pos$^+$ instead of pos in [2]. For any, this addition of QEnc has no effect, as QEnc queries can be simulated using QLeftRight. But for pos$^+$/pos, there is no equivalence in general between the security definition with and without the encryption oracle. We add this oracle QEnc so that we can reduce the security with respect to one label to the security with respect to multiple queried labels, via a simple hybrid argument (which would not be valid without the QEnc oracle), as done in [14]. This will be used in our generic compiler from pos$^+$ to any security, in Sect. 4.

Now we define a seemingly weaker security notion than xx-yy-IND, which we call xx-yy-IND-1-label, since the adversary is restricted to query the oracle QLeftRight on at most one label, and it cannot query the oracle QEnc oracle on that label. Using a standard hybrid argument (cf Lemma 2.5), we show that this is equivalent to the original xx-yy-IND security defined above. These restrictions will make the proofs easier in the rest of the paper.

Definition 2.4 ***(1-label Security)*** *Let* MCFE *be an MCFE scheme,* $\mathcal{F} = \{\mathcal{F}_\rho\}_\rho$ *a function family indexed by* ρ *and* Labels *a label set. For* xx $\in$ {sta, adt}*,* yy $\in$ {any, pos$^+$}*, and* $\beta \in \{0, 1\}$*, we define the experiment* xx-yy-IND$^{\mathsf{MCFE}}_\beta$ *exactly as in Fig. 1, where the oracles are defined as for Definition 2.2, except:*

Left-Right oracle $\mathsf{QLeftRight}(i, x_i^0, x_i^1, \ell)$**:** *Outputs* $\mathsf{ct}_{i,\ell} = \mathsf{Enc}(\mathsf{pp}, \mathsf{sk}_i, x_i^\beta, \ell)$ *on a query* (i, x_i^0, x_i^1, ℓ)*. This oracle can be queried at most on one label. Further queries with distinct labels will be ignored.*

Encryption oracle $\mathsf{QEnc}(i, x_i, \ell)$**:** *outputs* $\mathsf{ct}_{i,\ell} = \mathsf{Enc}(\mathsf{pp}, \mathsf{mpk}, \mathsf{sk}_i, x_i, \ell)$ *on a query* (i, x_i, ℓ)*. If this oracle is queried on the same label that is queried to* QLeftRight*, the game ends and return* 0.

Condition (*) *is defined as for Definition 2.2.*

We define the advantage of an adversary $\mathcal{A}$ in the following way:

$$\mathsf{Adv}^{\text{xx-yy-IND-1-label}}_{\mathsf{MCFE},\mathcal{A}}(\lambda, n) = \big|\Pr[\text{xx-yy-IND-1-label}^{\mathsf{MCFE}}_0(\lambda, n, \mathcal{A}) = 1] - \Pr[\text{xx-yy-IND-1-label}^{\mathsf{MCFE}}_1(\lambda, n, \mathcal{A}) = 1]\big| .$$

Lemma 2.5 (From one to many labels). *Let* MCFE *be a scheme that is* xx-yy-*IND-1-label secure, for* $\text{xx} \in \{\text{sta}, \text{adt}\}$ *and* $\text{yy} \in \{\text{pos}^+, \text{any}\}$. *Then it is also secure against PPT adversaries that query* QLeftRight *on many distinct labels (*xx-yy-*IND security). Namely, for any PPT adversary $\mathcal{A}$, there exists a PPT adversary $\mathcal{B}$ such that:*

$$\mathsf{Adv}^{\text{xx-yy-IND}}_{\mathsf{MCFE},\mathcal{A}}(\lambda, n) \le q_{\mathsf{Enc}} \cdot \mathsf{Adv}^{\text{xx-yy-IND-1-label}}_{\mathsf{MCFE},\mathcal{B}}(\lambda, n),$$

where $\mathsf{Adv}^{\text{xx-yy-IND-1-label}}_{\mathsf{MCFE},\mathcal{B}}(\lambda, n)$ *denotes the advantage of $\mathcal{B}$ against an experiment defined as above, except* QLeftRight *can be queried on at most one label and* QEnc *must not be queried on that label. By q_{Enc} we denote the number of distinct labels queried by $\mathcal{A}$ to* QLeftRight *in the original security game.*

Proof (Sketch).

First, let us consider the case of yy = any security. The proof uses a hybrid argument which goes over all the labels $\ell_1, ...\ell_Q$ queried to both the oracles QEnc and QLeftRight. In the k'th hybrid, the queries for the first k'th labels to the QLeftRight oracle are answered with the right plaintext, and the the last $Q-k$ labels are answered with the left plaintext. To go from hybrid $k-1$ to k, $\mathcal{B}$ uses its own QEnc oracle to answer $\mathcal{A}$'s queries to QLeftRight for labels ℓ_j for $j < k$, and $j > k$ (using the right and left plaintext respectively), and uses its own oracle QLeftRight for label ℓ_k. The queries made by $\mathcal{A}$ to QEnc and QCor are answered straightforwardly by $\mathcal{B}$ from its own oracles. Note that the queries made by $\mathcal{B}$ satisfy the 1-label restriction, since QLeftRight is only queried on ℓ_k, and QEnc is not queried on ℓ_k.

For the case of yy = pos$^+$ security, to go from hybrid $k-1$ to k, $\mathcal{B}$ uses the QEnc oracle to answer QLeftRight queries for labels ℓ_j for $j < k$ and $j > k$ (using the right and left plaintext respectively). For the label ℓ_k, $\mathcal{B}$ uses its own oracle QLeftRight to answer $\mathcal{A}$'s queries to both QLeftRight and QEnc. So far, the reduction works as for the case of yy = any security. However, the difference is yy = pos$^+$ security requires additional conditions on the queries made to QLeftRight, in particular, if one honest slot is queried to QLeftRight for ℓ_k, then all honest slots should be queried. Thus, we need to distinguish two cases: case (1) ℓ_k is queried to QEnc, but never on QLeftRight, in which case $\mathcal{B}$ uses its own QEnc oracle; case (2) ℓ_k is queried to QLeftRight at some point (and by definition of pos$^+$ security, that means it's queried to all honest slots). In case 2, the queries of $\mathcal{B}$ to QLeftRight will satisfy the condition required by the yy = pos$^+$ security game, namely, if QLeftRight is queried on ℓ_k for some honest input slot, then it has to be queried on the same label ℓ_k for all honest input slots. Note that this restriction doesn't apply to the queries made to QEnc. In case 1, we use the

sta-yy-IND$_\beta^{\mathsf{MCFE}}(\lambda, n, \mathcal{A})$

$\mathcal{CS} \leftarrow \mathcal{A}(1^\lambda, 1^n)$
$\mathsf{pp} \leftarrow \mathsf{Setup}(1^\lambda, 1^n)$
$(\{\mathsf{sk}_i\}_{i\in[n]}, \mathsf{mpk}, \mathsf{msk}) \leftarrow \mathsf{KeyGen}(\mathsf{pp})$
$\alpha \leftarrow \mathcal{A}^{\mathsf{QEnc}(\cdot,\cdot,\cdot),\mathsf{QLeftRight}(\cdot,\cdot,\cdot,\cdot),\mathsf{QKeyD}(\cdot)}(\mathsf{pp}, \mathsf{mpk}, \{\mathsf{sk}_i\}_{i\in\mathcal{CS}})$
Output: α if Condition (*) is satisfied, 0 otherwise.

adt-yy-IND$_\beta^{\mathsf{MCFE}}(\lambda, n, \mathcal{A})$

$\mathsf{pp} \leftarrow \mathsf{Setup}(1^\lambda, 1^n)$
$(\{\mathsf{sk}_i\}_{i\in[n]}, \mathsf{msk}, \mathsf{mpk}) \leftarrow \mathsf{KeyGen}(\mathsf{pp})$
$\alpha \leftarrow \mathcal{A}^{\mathsf{QCor}(\cdot),\mathsf{QEnc}(\cdot,\cdot,\cdot),\mathsf{QLeftRight}(\cdot,\cdot,\cdot,\cdot),\mathsf{QKeyD}(\cdot)}(\mathsf{pp}, \mathsf{mpk})$
Output: α if Condition (*) is satisfied, 0 otherwise.

Fig. 1. Security games for MCFE

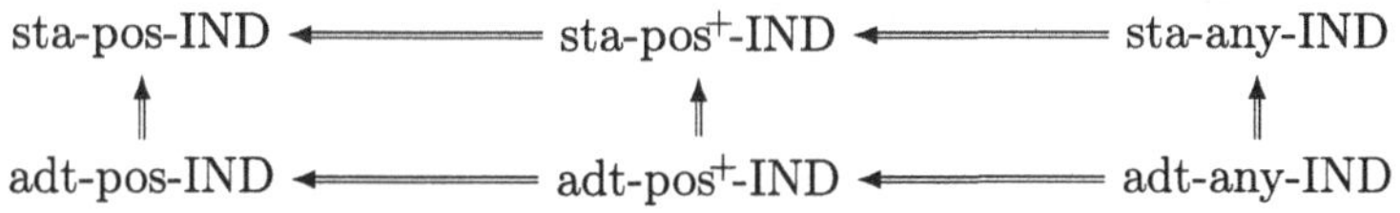

Fig. 2. Relations between the MCFE security notions (arrows indicate implication or being "a stronger security notion than")

fact that the two hybrid games $k-1$ and k are exactly the same. Therefore, at the end of the simulation, $\mathcal{B}$ checks whether case 1 occurs, and if it does, simply outputs 0 to its own experiment, ignoring $\mathcal{A}$'s output. Otherwise, it means it is case 2, and $\mathcal{B}$ forwards the output from $\mathcal{A}$ to its own experiment. □

We summarize the relations between the six security notions in Fig. 2, where xx-pos-IND is the notion defined in [2] (i.e., it is like xx-pos$^+$-IND without the QEnc oracle).

2.2 Decentralized Multi-Client Functional Encryption

Now, we introduce the definition of decentralized multi-client functional encryption (DMCFE) [13]. As for our definition of MCFE, we separate the algorithm Setup which generates public parameters defining in particular the set of functions, from the algorithm KeyGen. We do not consider public-key variants of DMCFE and hence completely omit the master public key mpk.

Definition 2.6. ***(Decentralized Multi-Client Functional Encryption)*** *Let $\mathcal{F} = \{\mathcal{F}_\rho\}_\rho$ be a family (indexed by ρ) of sets $\mathcal{F}_\rho$ of functions $f: \mathcal{X}_{\rho,1} \times \cdots \times \mathcal{X}_{\rho,n_\rho} \to \mathcal{Y}_\rho$. Let* $\mathsf{Labels} = \{0,1\}^*$ *or* $\{\perp\}$ *be a set of labels. A decentralized multi-client functional encryption scheme (DMCFE) for the function family $\mathcal{F}$ and the label set* Labels *is a tuple of six algorithms* DMCFE = (Setup, KeyGen, KeyDerShare, KeyDerComb, Enc, Dec)*:*

$\mathsf{Setup}(1^\lambda, 1^n)$ *is defined as for MCFE in Definition 2.1.*

$\mathsf{KeyGen}(\mathsf{pp})$: *Takes as input the public parameters* pp *and outputs n secret keys* $\{\mathsf{sk_i}\}_{i\in[n]}$.

$\mathsf{KeyDerShare}(\mathsf{pp}, \mathsf{sk}_i, f)$: *Takes as input the public parameters* pp, *a secret key* sk_i *from position i and a function $f \in \mathcal{F}_\rho$, and outputs a partial functional decryption key* $\mathsf{sk}_{i,f}$.

$\mathsf{KeyDerComb}(\mathsf{pp}, \mathsf{sk}_{1,f}, \ldots, \mathsf{sk}_{n,f})$: *Takes as input the public parameters* pp, *n partial functional decryption keys* $\mathsf{sk}_{1,f}, \ldots, \mathsf{sk}_{n,f}$ *and outputs the functional decryption key* sk_f.

$\mathsf{Enc}(\mathsf{pp}, \mathsf{sk}_i, x_i, \ell)$ *is defined as for MCFE in Definition 2.1.*

$\mathsf{Dec}(\mathsf{pp}, \mathsf{sk}_f, \mathsf{ct}_{1,\ell}, \ldots, \mathsf{ct}_{n,\ell})$ *is defined as for MCFE in Definition 2.1.*

A scheme DMCFE *is correct, if for all $\lambda, n \in \mathbb{N}$,* $\mathsf{pp} \leftarrow \mathsf{Setup}(1^\lambda, 1^n)$, $f \in \mathcal{F}_\rho$, $\ell \in \mathsf{Labels}$, $x_i \in \mathcal{X}_{\rho,i}$, *when* $\{\mathsf{sk}_i\}_{i\in[n]} \leftarrow \mathsf{KeyGen}(\mathsf{pp})$, $\mathsf{sk}_{i,f} \leftarrow \mathsf{KeyDerShare}(\mathsf{sk}_i, f)$ *for $i \in [n]$, and* $\mathsf{sk}_f \leftarrow \mathsf{KeyDerComb}(\mathsf{pp}, \mathsf{sk}_{1,f}, \ldots, \mathsf{sk}_{n,f})$, *we have*

$$\Pr\left[\mathsf{Dec}(\mathsf{pp}, \mathsf{sk}_f, \mathsf{Enc}(\mathsf{pp}, \mathsf{sk}_1, x_1, \ell), \ldots, \mathsf{Enc}(\mathsf{pp}, \mathsf{sk}_n, x_n, \ell)) = f(x_1, \ldots, x_n)\right] = 1 \ .$$

We remark that there is no master secret key msk. Furthermore, similarly to [13], our definition does not explicitly ask the setup to be decentralized. Our DMCFE construction based on DDH (Sect. 5) however has a setup which can be easily decentralized.

We consider a similar security definition for the decentralized multi-client scheme. We point out that contrary to [13], we do not differentiate encryption keys from secret keys. This is without loss of generality, as corruptions in [13] only allow to corrupt both keys at the same time.

Definition 2.7. ***(Security of DMCFE)*** *The* xx-yy-*IND security notion of an DMCFE scheme* (xx $\in$ {sta, adt} *and* yy $\in$ {any, pos$^+$}) *is similar to the one of an MCFE (Definition 2.2), except that there is no master secret key* msk *and the key derivation oracle is now defined as:*

Key derivation oracle $\mathsf{QKeyD}(f)$: *Computes* $\mathsf{sk}_{i,f} := \mathsf{KeyDerShare}(\mathsf{pp}, \mathsf{sk}_i, f)$ *for $i \in [n]$ and outputs* $\{\mathsf{sk}_{i,f}\}_{i\in[n]}$.

2.3 Inner-Product Functionality

We describe the functionalities supported by the constructions in this paper. The index of the family is defined as $\rho = (\mathcal{R}, n, m, X, Y)$ where $\mathcal{R}$ is either $\mathbb{Z}$ or $\mathbb{Z}_L$ for some integer L, and n, m, X, Y are positive integers. If X, Y are omitted, then $X = Y = L$ is used (i.e., no constraint).

This defines $\mathcal{F}_\rho^{\mathsf{ip}} = \{f_{\boldsymbol{y}_1,\ldots,\boldsymbol{y}_n} : (\mathcal{R}^m)^n \to \mathcal{R}\}$ where

$$f_{\boldsymbol{y}_1,\ldots,\boldsymbol{y}_n}(\boldsymbol{x}_1,\ldots,\boldsymbol{x}_n) = \sum_{i=1}^{n} \langle \boldsymbol{x}_i, \boldsymbol{y}_i \rangle = \langle \boldsymbol{x}, \boldsymbol{y} \rangle ,$$

where the vectors satisfy the following bounds: $\|\boldsymbol{x}_i\|_\infty < X$, $\|\boldsymbol{y}_i\|_\infty < Y$ for $i \in [n]$, and where $\boldsymbol{x} \in \mathcal{R}^{mn}$ and $\boldsymbol{y} \in \mathcal{R}^{mn}$ are the vectors corresponding to the concatenation of the n vectors $\boldsymbol{x}_1, \ldots, \boldsymbol{x}_n$ and $\boldsymbol{y}_1, \ldots, \boldsymbol{y}_n$ respectively.

2.4 Pseudorandom Functions (PRF)

We make use of a pseudorandom function $\mathsf{PRF}_\mathsf{K}(\ell)$, indexed by a key $\mathsf{K} \in \{0,1\}^\lambda$, that takes as input a label $\ell \in \mathsf{Labels}$, and outputs a value in the output space $\mathcal{Z}$. For a uniformly random key $\mathsf{K} \leftarrow \{0,1\}^\lambda$, this function is computationally indistinguishable from a truly random function from Labels to $\mathcal{Z}$.

We define the advantage of an adversary $\mathcal{A}$ in the following way:

$$\mathsf{Adv}_{\mathsf{PRF},\mathcal{A}}(\lambda) = \left| \Pr[\mathrm{IND}_0^{\mathsf{PRF}}(\lambda,\mathcal{A}) = 1] - \Pr[\mathrm{IND}_1^{\mathsf{PRF}}(\lambda,\mathcal{A}) = 1] \right| ,$$

where $\mathrm{IND}_0^{\mathsf{PRF}}(\lambda,\mathcal{A})$ is the experiment where $\mathcal{A}$ has an oracle access to $\mathsf{PRF}_\mathsf{K}(\cdot)$, whereas $\mathrm{IND}_1^{\mathsf{PRF}}(\lambda,\mathcal{A})$ is the experiment where $\mathcal{A}$ has an oracle access to a truly random function instead.

A PRF is secure, if for any any polynomial-time adversary $\mathcal{A}$, there exists a negligible function negl such that: $\mathsf{Adv}_{\mathsf{PRF},\mathcal{A}}(\lambda) \leq \mathrm{negl}(\lambda)$.

2.5 Symmetric-Key Encryption (SE)

A symmetric encryption with key space $\mathcal{K}$ consists of the following PPT algorithms:

- $\mathsf{Enc}(K, m)$: given a symmetric key K and a message m, outputs a ciphertext.
- $\mathsf{Dec}(K, \mathsf{ct})$: given a symmetric key K and a ciphertext ct, outputs a message (or $\perp$ if it fails to decrypt).

For all message in the message space, we have $\Pr[\mathsf{Dec}(k, \mathsf{Enc}(k, m)) = m] = 1$, where the probability is taken over the random choice of $K \leftarrow \mathcal{K}$. We say a symmetric-key encryption with key space $\mathcal{K}$ is compatible with a PRF with output space $\mathcal{Z}$ if $\mathcal{K} = \mathcal{Z}$.

Definition 2.8 (SE). *For any* SE *with key space* $\mathcal{K}$*, any bit* $\beta \in \{0,1\}$*, any security parameter* λ*, and any adversary* $\mathcal{A}$*, we define the experiment* $\mathbf{IND}^{\mathsf{PRF}}{}_\beta$ *as follows.*

We define the advantage of an adversary $\mathcal{A}$ *in the following way:*

$$\mathsf{Adv}_{\mathsf{SE},\mathcal{A}}(\lambda, n) = \left| \Pr[\mathrm{IND}_0^{\mathsf{PRF}}(\lambda,\mathcal{A}) = 1] - \Pr[\mathrm{IND}_1^{\mathsf{SE}}(\lambda,\mathcal{A}) = 1] \right| .$$

A SE is secure, if for any any polynomial-time adversary $\mathcal{A}$*, there exists a negligible function* negl *such that:* $\mathsf{Adv}_{\mathsf{SE},\mathcal{A}}(\lambda) \leq \mathrm{negl}(\lambda)$.

$\mathbf{IND}^{\mathsf{SE}}_{\beta}(\lambda, \mathcal{A})$
$\mathsf{K} \leftarrow \mathcal{K}$
$\alpha \leftarrow \mathcal{A}^{\mathcal{O}_{\mathsf{SE}}(\cdot)}(1^\lambda)$
Output: α

Fig. 3. Security games for SE. The oracle $\mathcal{O}_{\mathsf{SE}}(m_0, m_1)$ returns $\mathsf{Enc}(K, m_\beta)$.

3 MCFE from Public-Key Single-Input FE

In this section, we build a multi-client FE for inner products generically from any public-key single-input FE and a standard PRF.

3.1 Construction

The construction resembles the multi-input FE from [4], where an inner layer of information-theoretic one-time FE is combined with an outer layer of single-input FE. We manage to extend this paradigm to the setting where the encryption additionally takes a label as input: the one-time pads are replaced by pads which are pseudorandom for all used labels ℓ, using techniques similar to those used in [2] to decentralize the generation of functional secret keys.

The underlying single-input FE is required to satisfy simple structural properties, originally defined in [4] and recalled below (converted to the public-key setting), which are satisfied by all known existing single-input FE for inner products.

Definition 3.1 (Two-step decryption [4]**).** *A public-key FE scheme* $\mathsf{FE} = (\mathsf{Setup}, \mathsf{KeyGen}, \mathsf{KeyDer}, \mathsf{Enc}, \mathsf{Dec})$ *for the function ensemble* $\mathcal{F}^{\mathsf{ip}}_{\rho}, \rho = (\mathbb{Z}, 1, m, X, Y)$ *satisfies the* two-step decryption *property if it admits PPT algorithms* $\mathsf{Setup}^\star$, $\mathsf{Dec}_1, \mathsf{Dec}_2$ *and an encoding function* $\mathcal{E}$ *such that:*

1. *For all* $\lambda \in \mathbb{N}, \mathsf{Setup}^\star(1^\lambda, 1^n)$ *outputs* pp *where* pp *includes* $\rho = (\mathbb{Z}, 1, m, X, Y)$ *and a bound* $B \in \mathbb{R}^+$*, as well as the description of a group* $\mathbb{G}$ *(with group law* $\circ$*) of order* $L > n \cdot m \cdot X \cdot Y$*, which defines the encoding function* $\mathcal{E} : \mathbb{Z}_L \times \mathbb{Z} \to \mathbb{G}$.
2. *For all* $(\mathsf{msk}, \mathsf{mpk}) \leftarrow \mathsf{KeyGen}(\mathsf{pp}), \boldsymbol{x} \in \mathbb{Z}^m, \mathsf{ct} \leftarrow \mathsf{Enc}(\mathsf{pp}, \mathsf{mpk}, \boldsymbol{x}), \boldsymbol{y} \in \mathbb{Z}^m$*, and* $\mathsf{sk} \leftarrow \mathsf{KeyDer}(\mathsf{msk}, \boldsymbol{y})$*, we have*

$$\mathsf{Dec}_1(\mathsf{pp}, \mathsf{sk}, \mathsf{ct}) = \mathcal{E}(\langle \boldsymbol{x}, \boldsymbol{y} \rangle \bmod L, \mathsf{noise}) \ ,$$

for some $\mathsf{noise} \in \mathbb{Z}$ *that depends on* ct *and* sk*. Furthermore, it holds that* $Pr[|\mathsf{noise}| < B] = 1 - \mathrm{negl}(\lambda)$*, where the probability is taken over the random coins of* KeyGen *and* KeyDer*. Note that there is no restriction on the norm of* $\langle \boldsymbol{x}, \boldsymbol{y} \rangle$ *here.*

$\underline{\mathsf{Setup}(1^\lambda, 1^n):}$

$\mathsf{pp}_{\mathsf{ipfe}} \leftarrow \mathsf{Setup}^\star_{\mathsf{ipfe}}(1^\lambda, 1^n)$, with L implicitly defined from $\mathsf{pp}_{\mathsf{ipfe}}$

Return $\mathsf{pp} = \mathsf{pp}_{\mathsf{ipfe}}$

$\underline{\mathsf{KeyGen}(\mathsf{pp}):}$

$(\mathsf{msk}_{\mathsf{ipfe}}, \mathsf{mpk}_{\mathsf{ipfe}}) \leftarrow \mathsf{KeyGen}_{\mathsf{ipfe}}(\mathsf{pp}_{\mathsf{ipfe}})$; $\mathsf{msk} := \mathsf{msk}_{\mathsf{ipfe}}$

For $i \in [n]$, $j > i : \mathsf{K}_{i,j} = \mathsf{K}_{j,i} \leftarrow \{0,1\}^\lambda$

Return $\{\mathsf{sk}_i = (\mathsf{mpk}, \{\mathsf{K}_{i,j}\}_{j \in [n]})\}_{i \in [n]}$ and msk

$\underline{\mathsf{Enc}(\mathsf{pp}, \mathsf{sk}_i, \boldsymbol{x}_i \in \mathcal{R}^m, \ell \in \mathsf{Labels}):}$

Parse $\mathsf{sk}_i = (\mathsf{mpk}_{\mathsf{ipfe}}, \{\mathsf{K}_{i,j}\}_{j \in [n]})$

$\boldsymbol{t}_{i,\ell} := \sum_{j \neq i} (-1)^{j<i} \mathsf{PRF}_{\mathsf{K}_{i,j}}(\ell) \in \mathbb{Z}_L^{mn}$

$\boldsymbol{w}_i := (\mathbf{0} \| \ldots \| \mathbf{0} \| \boldsymbol{x}_i \| \mathbf{0} \| \ldots \| \mathbf{0}) + \boldsymbol{t}_{i,\ell} \bmod L$

$\mathsf{ct}_i \leftarrow \mathsf{Enc}_{\mathsf{ipfe}}(\mathsf{pp}_{\mathsf{ipfe}}, \mathsf{mpk}_{\mathsf{ipfe}}, \boldsymbol{w}_i)$

Return ct_i

$\underline{\mathsf{KeyDer}(\mathsf{pp}, \mathsf{msk}, \boldsymbol{y} \in \mathcal{R}^{mn}):}$

Return $\mathsf{sk}_{\boldsymbol{y}} \leftarrow \mathsf{KeyDer}_{\mathsf{ipfe}}(\mathsf{pp}_{\mathsf{ipfe}}, \mathsf{msk}_{\mathsf{ipfe}}, \boldsymbol{y})$

$\underline{\mathsf{Dec}(\mathsf{pp}, \mathsf{sk}_{\boldsymbol{y}}, \{\mathsf{ct}_i\}_{i \in [n]}):}$

For $i \in [n]$, $\mathcal{E}(\langle \boldsymbol{w}_i, \boldsymbol{y} \rangle \bmod L, \mathsf{noise}_i) \leftarrow \mathsf{Dec}_{\mathsf{ipfe},1}(\mathsf{pp}_{\mathsf{ipfe}}, \mathsf{sk}_{\boldsymbol{y}}, \mathsf{ct}_i)$

Return $\mathsf{Dec}_{\mathsf{ipfe},2}(\mathsf{pp}_{\mathsf{ipfe}}, \mathcal{E}(\langle \boldsymbol{w}_1, \boldsymbol{y} \rangle \bmod L, \mathsf{noise}_1)) \circ \cdots \circ \mathcal{E}(\langle \boldsymbol{w}_n, \boldsymbol{y} \rangle \bmod L, \mathsf{noise}_n))$

Fig. 4. Inner-Product MCFE for $\mathcal{F}_\rho, \rho = (\mathbb{Z}, n, m, X, Y)$ built from a public-key FE $\mathsf{FE} := (\mathsf{Setup}_{\mathsf{ipfe}}, \mathsf{Enc}_{\mathsf{ipfe}}, \mathsf{KeyDer}_{\mathsf{ipfe}}, \mathsf{Dec}_{\mathsf{ipfe}})$ for $\mathcal{F}_{\rho_{\mathsf{ipfe}}}, \rho_{\mathsf{ipfe}} = (\mathbb{Z}, 1, n \cdot m, 2X, Y)$. We assume FE satisfies the two-step decryption property (see Definition 3.1), hence the existence of PPT algorithms $\mathsf{Setup}^\star_{\mathsf{ipfe}}$, $\mathsf{Dec}_{\mathsf{ipfe},1}$ and $\mathsf{Dec}_{\mathsf{ipfe},2}$. Here, for any $\mathsf{K} \in \{0,1\}^\lambda$, $\mathsf{PRF}_{\mathsf{K}} : \mathsf{Labels} \rightarrow \mathbb{Z}_L^{mn}$ is a pseudorandom function (see Sect. 2.4).

3. *The encoding $\mathcal{E}$ is linear, that is: for all $\gamma, \gamma' \in \mathbb{Z}_L$, $\mathsf{noise}, \mathsf{noise}' \in \mathbb{Z}$, we have*

$$\mathcal{E}(\gamma, \mathsf{noise}) \circ \mathcal{E}(\gamma', \mathsf{noise}') = \mathcal{E}(\gamma + \gamma' \bmod L, \mathsf{noise} + \mathsf{noise}') \ .$$

4. *For all $\gamma < n \cdot m \cdot X \cdot Y$, and $|\mathsf{noise}| < n \cdot B$, $\mathsf{Dec}_2(\mathsf{pp}, \mathcal{E}(\gamma, \mathsf{noise})) = \gamma$.*

Definition 3.2 (Linear encryption [4]**).** *A secret-key FE scheme* $\mathsf{FE} = (\mathsf{Setup}, \mathsf{KeyGen}, \mathsf{KeyDer}, \mathsf{Enc}, \mathsf{Dec})$ *is said to satisfy the linear encryption property if there exists a deterministic algorithm* Add *that takes as input a ciphertext* *and a message, such that for all* $\boldsymbol{x}, \boldsymbol{x}' \in \mathbb{Z}^m$*, the following are identically distributed:*

$$\mathsf{Add}(\mathsf{Enc}(\mathsf{pp}, \mathsf{msk}, \boldsymbol{x}), \boldsymbol{x}'), \quad \textit{and} \quad \mathsf{Enc}\big(\mathsf{pp}, \mathsf{msk}, (\boldsymbol{x} + \boldsymbol{x}' \bmod L)\big) .$$

Recall that the value $L \in \mathbb{N}$ *is defined as part of the output of the algorithm* $\mathsf{Setup}^\star$ *(see the two-step decryption property above).*

Correctness. The correctness of the scheme in Fig. 4 follows from (i) the correctness and Definition 3.1 (two-step decryption) of the single-input scheme, and (ii) the fact that for all $\ell \in \mathsf{Labels}$, $\sum_{i \in [n]} \boldsymbol{t}_{i,\ell} = \boldsymbol{0}$, by definition of the vectors $\boldsymbol{t}_{i,\ell}$. Thus, writing $\boldsymbol{w}_i := (\boldsymbol{0}\| \dots \|\boldsymbol{0}\|\boldsymbol{x}_i\|\boldsymbol{0}\| \dots \|\boldsymbol{0}) + \boldsymbol{t}_{i,\ell} \bmod L$, we have $\sum_{i \in [n]} \boldsymbol{w}_i \bmod L = \boldsymbol{x} \bmod L \in \mathbb{Z}_L^{mn}$, where $\boldsymbol{x} \in \mathcal{R}^{nm}$ denotes the concatenation of the n vectors $\boldsymbol{x}_1, \dots, \boldsymbol{x}_n$.

More precisely, consider any vector $\boldsymbol{x} := (\boldsymbol{x}_1\| \cdots \|\boldsymbol{x}_n) \in (\mathbb{Z}^m)^n$, $\boldsymbol{y} \in \mathbb{Z}^{mn}$, such that $\|\boldsymbol{x}\|_\infty < X$, $\|\boldsymbol{y}\|_\infty < Y$ and let $\mathsf{pp} \leftarrow \mathsf{Setup}(1^\lambda)$, $(\{\mathsf{sk}_i\}_{i \in [n]}, \mathsf{msk}) \leftarrow \mathsf{KeyGen}(\mathsf{pp})$, $\mathsf{sk}_{\boldsymbol{y}} \leftarrow \mathsf{KeyDer}(\mathsf{pp}, \mathsf{msk}, \boldsymbol{y})$, and $\mathsf{ct}_i \leftarrow \mathsf{Enc}(\mathsf{pp}, \mathsf{sk}_i, \boldsymbol{x}_i, \ell)$ for all $i \in [n]$.

By (2) of Definition 3.1, the decryption algorithm $\mathsf{Dec}(\mathsf{pp}, \mathsf{sk}_{\boldsymbol{y}}, \{\mathsf{ct}_i\}_{i \in [n]})$ computes $\mathcal{E}(\langle \boldsymbol{w}_i, \boldsymbol{y} \rangle \bmod L, \mathsf{noise}_i) \leftarrow \mathsf{Dec}_{\mathsf{ipfe},1}(\mathsf{pp}, \mathsf{sk}_i, \mathsf{ct}_i)$ where for all $i \in [n]$, $|\mathsf{noise}_i| < B$ with probability $1 - \mathrm{negl}(\lambda)$, where $B \in \mathbb{R}^+$ is the bound output by $\mathsf{Setup}^\star_{\mathsf{ipfe}}$.

By (3) of Definition 3.1 (linearity of $\mathcal{E}$) we have:

$$\begin{aligned} &\mathcal{E}(\langle \boldsymbol{w}_1, \boldsymbol{y} \rangle \bmod L, \mathsf{noise}_1) \circ \cdots \circ \mathcal{E}(\langle \boldsymbol{w}_n, \boldsymbol{y} \rangle \bmod L, \mathsf{noise}_n) \\ &= \mathcal{E}\left(\langle \sum_{i \in [n]} \boldsymbol{w}_i, \boldsymbol{y} \rangle, \sum_{i \in [n]} \mathsf{noise}_i \right) = \mathcal{E}\left(\langle \boldsymbol{x}, \boldsymbol{y} \rangle \bmod L, \sum_{i \in [n]} \mathsf{noise}_i \right) . \end{aligned}$$

Since $\langle \boldsymbol{x}, \boldsymbol{y} \rangle < n \cdot m \cdot X \cdot Y < L$ and $\left| \sum_{i \in [n]} \mathsf{noise}_i \right| < n \cdot B$, we have

$$\mathsf{Dec}_{\mathsf{ipfe},2}\big(\mathcal{E}(\langle \boldsymbol{x}, \boldsymbol{y} \rangle \bmod L, \sum_{i \in [n]} \mathsf{noise}_i)\big) = \langle \boldsymbol{x}, \boldsymbol{y} \rangle,$$

by (4) of Definition 3.1.

3.2 Static Security

Now we proceed to prove the sta-pos$^+$-IND-security of the scheme, that is, security with static corruption, which serves as a warm up to the more complicated proof of adt-pos$^+$-IND-security, that we give later. Using the generic transformation in Sect. 4, we can remove the pos$^+$ restriction, and obtain adt-any-IND security.

Theorem 3.3 (adt-pos$^+$-IND-security). *If the FE scheme* $\mathsf{FE} = (\mathsf{Setup}_{\mathsf{ipfe}}, \mathsf{KeyGen}_{\mathsf{ipfe}}, \mathsf{KeyDer}_{\mathsf{ipfe}}, \mathsf{Enc}_{\mathsf{ipfe}}, \mathsf{Dec}_{\mathsf{ipfe}})$ *is an* any*-IND-secure FE scheme for the inner product functionality defined as* $\mathcal{F}^{\mathsf{ip}}_{\rho_{\mathsf{ipfe}}}, \rho_{\mathsf{ipfe}} = (\mathbb{Z}, 1, m, 2X, Y)$*, and* PRF *is secure, then* MCFE *from Fig. 4 is* sta-pos$^+$*-IND-secure for the functionality defined as* $\mathcal{F}^{\mathsf{ip}}_{\rho}, \rho = (\mathbb{Z}, n, m, X, Y)$*. Namely, for any PPT adversary* $\mathcal{A}$*, there exist PPT adversaries* $\mathcal{B}$ *and* $\mathcal{B}'$ *such that:*

$$\mathsf{Adv}^{sta\text{-}pos^+\text{-}IND}_{\mathsf{MCFE},\mathcal{A}}(\lambda, n) \leq 2q_{\mathsf{Enc}} \cdot \mathsf{Adv}^{any\text{-}IND}_{\mathsf{FE},\mathcal{B}}(\lambda) + 2(n-1)q_{\mathsf{Enc}} \cdot \mathsf{Adv}_{\mathsf{PRF},\mathcal{B}'}(\lambda),$$

where q_{Enc} *denotes the number of distinct labels queried to* QLeftRight.

Proof. For simplicity, we consider the case where $\mathcal{A}$ only queries QLeftRight on one label $\ell^\star$, and never queries QEnc on $\ell^\star$. We build PPT adversaries $\mathcal{B}$ and $\mathcal{B}'$ such that: $\mathsf{Adv}^{\text{sta-pos}^+\text{-IND-1-label}}_{\mathsf{MCFE},\mathcal{A}}(\lambda, n) \leq 2 \cdot \mathsf{Adv}^{\text{any-IND}}_{\mathsf{FE},\mathcal{B}}(\lambda) + 2(n-1) \cdot \mathsf{Adv}_{\mathsf{PRF},\mathcal{B}'}(\lambda)$, where $\mathsf{Adv}^{\text{sta-pos}^+\text{-IND-1-label}}_{\mathsf{MCFE},\mathcal{A}}(\lambda, n)$ is defined as $\mathsf{Adv}^{\text{sta-pos}^+\text{-IND}}_{\mathsf{MCFE},\mathcal{A}}(\lambda, n)$, except with the limitations mentioned above, namely, $\mathcal{A}$ can query QLeftRight on at most one label, which cannot be queried to QEnc. Then we use Lemma 2.5 to obtain the theorem.

First, consider the case where there is only one honest user. In this case, the security follows directly from the any-IND security of FE. Namely, in that case we build a PPT adversary $\mathcal{B}$ such that $\mathsf{Adv}^{\text{sta-pos}^+\text{-IND-1-label}}_{\mathsf{MCFE},\mathcal{A}}(\lambda, n) \leq \mathsf{Adv}^{\text{any-IND}}_{\mathsf{FE},\mathcal{B}}(\lambda)$. Given $\mathsf{pp}_{\mathsf{ipfe}}$, $\mathcal{B}$ first samples the keys $\mathsf{K}_{i,j}$ for all $i, j \in [n]$, thanks to which it can compute pp, $\{\mathsf{sk}_i\}_{i \in [n]}$, and send $(\mathsf{pp}, \{\mathsf{sk}_i\}_{i \in \mathcal{CS}})$ to $\mathcal{A}$. $\mathcal{B}$ can answer all queries to $\mathsf{QEnc}(i, \boldsymbol{x}^j_i, \ell)$, by returning $\mathsf{Enc}(\mathsf{pp}, \mathsf{sk}_i, \boldsymbol{x}^j_i, \ell)$, since it know sk_i for all $i \in [n]$. Call $i^\star$ the only honest slot. $\mathcal{B}$ can answer all queries to $\mathsf{QEnc}(i, \cdot, \cdot, \cdot)$ and $\mathsf{QLeftRight}(i, \cdot, \cdot, \cdot)$ for $i \neq i^\star$, using pp and $\{\mathsf{sk}_i\}_{i \in [n]}$. Whenever $\mathcal{A}$ queries $\mathsf{QLeftRight}(i^\star, \boldsymbol{x}^{j,0}_{i^\star}, \boldsymbol{x}^{j,1}_{i^\star}, \ell^\star)$, $\mathcal{B}$ queries its own left right oracle on $(\mathbf{0}\|\ldots\|\mathbf{0}\|\boldsymbol{x}^{j,0}_{i^\star}\|\mathbf{0}\|\ldots\|\mathbf{0}), (\mathbf{0}\|\ldots\|\boldsymbol{x}^{j,1}_{i^\star}\|\mathbf{0}\|\ldots\|\mathbf{0})$, to receive $\mathsf{ct}_i := \mathsf{Enc}_{\mathsf{ipfe}}(\mathsf{pp}_{\mathsf{ipfe}}, \mathsf{mpk}_{\mathsf{ipfe}}, \mathsf{sk}_{i^\star}, (\mathbf{0}\|\ldots\|\boldsymbol{x}^{j,\beta}_{i^\star}\|\mathbf{0}\|\ldots\|\mathbf{0}))$, where $\beta \in \{0, 1\}$, depending on the experiment $\mathcal{B}$ is interacting with. Then, $\mathcal{B}$ computes $\boldsymbol{t}_{i^\star,\ell^\star}$ as described in Fig. 4, and returns $\mathsf{Add}(\mathsf{ct}_{i^\star}, \boldsymbol{t}_{i^\star,\ell^\star})$ to $\mathcal{A}$, which, according to the property from Definition 3.2 (linear encryption), is identically distributed to $\mathsf{Enc}_{\mathsf{ipfe}}(\mathsf{pp}_{\mathsf{ipfe}}, \mathsf{mpk}_{\mathsf{ipfe}}, (\mathbf{0}\|\ldots\|\boldsymbol{x}^{j,\beta}_{i^\star}\|\mathbf{0}\|\ldots\|\mathbf{0}) + \boldsymbol{t}_{i^\star,\ell^\star} \bmod L)$. Whenever $\mathcal{A}$ queries QKeyD on input $\boldsymbol{y}$, $\mathcal{B}$ queries its own QKeyD on the same input, and forwards the output to $\mathcal{A}$. For all $\boldsymbol{y}$ queried to QKeyD, we have $\langle(\mathbf{0}\|\ldots\|\boldsymbol{x}^{j,0}_{i^\star}\|\mathbf{0}\|\ldots\|\mathbf{0}), \boldsymbol{y}\rangle = \langle(\mathbf{0}\|\ldots\|\boldsymbol{x}^{j,1}_{i^\star}\|\mathbf{0}\|\ldots\|\mathbf{0}), \boldsymbol{y}\rangle$, by Condition (*). Moreover, for all $\beta \in \{0, 1\}$, $\|(\mathbf{0}\|\ldots\|\boldsymbol{x}^{j,\beta}_{i^\star}\|\mathbf{0}\|\ldots\|\mathbf{0})\|_\infty < 2X$. Thus, the queries $\mathcal{B}$ sends to its left-right oracle are legitimate. This concludes the case where there is only one honest user.

Second, we consider the case where there is more than one honest user. For this case, we proceed via a hybrid argument, using the games described in Fig. 5. Note that G_0 corresponds to sta-pos$^+$-$\text{IND}^{\mathsf{MCFE}}_0(\lambda, n, \mathcal{A})$, and G_4 corresponds to sta-pos$^+$-$\text{IND}^{\mathsf{MCFE}}_1(\lambda, n, \mathcal{A})$, with the one label restriction. Thus, we have:

$$\mathsf{Adv}^{\text{sta-pos}^+\text{-IND-1-label}}_{\mathsf{MCFE},\mathcal{A}}(\lambda, n) = \left|\mathsf{Win}^{\mathsf{G}_0}_{\mathcal{A}}(\lambda, n) - \mathsf{Win}^{\mathsf{G}_4}_{\mathcal{A}}(\lambda, n)\right|.$$

G_0, G_1, G_2, G_3, G_4:

$\mathcal{CS} \leftarrow \mathcal{A}(1^\lambda, 1^n)$
$(\{\mathsf{sk}_i\}_{i\in[n]}, \mathsf{msk}) \leftarrow \mathsf{KeyGen}(\mathsf{pp})$
$\alpha \leftarrow \mathcal{A}^{\mathsf{QLeftRight}(\cdot,\cdot,\cdot,\cdot),\mathsf{QEnc}(\cdot,\cdot,\cdot),\mathsf{QKeyD}(\cdot)}(\mathsf{pp}, \{\mathsf{sk}_i\}_{i\in\mathcal{CS}})$
Output: α if Condition (*) is satisfied, or 0 otherwise.

$\mathsf{QKeyD}(\boldsymbol{y})$:
Return $\mathsf{sk}_{\boldsymbol{y}} \leftarrow \mathsf{KeyDer}(\mathsf{pp}, \mathsf{msk}, \boldsymbol{y})$

$\mathsf{QEnc}(i, \boldsymbol{x}_i^j, \ell)$:
$\boldsymbol{t}_{i,\ell} \leftarrow \mathsf{Gen}(i, \ell)$
$\boldsymbol{w}_i := (\mathbf{0}\|\ldots\|\mathbf{0}\|\boldsymbol{x}_i^j\|\mathbf{0}\|\ldots\|\mathbf{0}) + \boldsymbol{t}_{i,\ell} \bmod L$
$\mathsf{ct}_i \leftarrow \mathsf{Enc}_{\mathsf{ipfe}}(\mathsf{pp}_{\mathsf{ipfe}}, \mathsf{mpk}_{\mathsf{ipfe}}, \boldsymbol{w}_i)$
Return ct_i

$\mathsf{QLeftRight}(i, \boldsymbol{x}_i^{j,0}, \boldsymbol{x}_i^{j,1}, \ell^\star)$:
$\boldsymbol{t}_{i,\ell^\star} \leftarrow \mathsf{Gen}(i, \ell^\star)$
$\boldsymbol{w}_i := (\mathbf{0}\|\ldots\|\mathbf{0}\|\boldsymbol{x}_i^{j,0} + \boldsymbol{x}_i^{1,1} - \boldsymbol{x}_i^{1,0}\|\mathbf{0}\|\ldots\|\mathbf{0}) + \boldsymbol{t}_{i,\ell^\star} \bmod L$
$\boldsymbol{w}_i := (\mathbf{0}\|\ldots\|\mathbf{0}\|\boldsymbol{x}_i^{j,1}\|\mathbf{0}\|\ldots\|\mathbf{0}) + \boldsymbol{t}_{i,\ell^\star} \bmod L$
$\mathsf{ct}_i \leftarrow \mathsf{Enc}_{\mathsf{ipfe}}(\mathsf{pp}_{\mathsf{ipfe}}, \mathsf{mpk}_{\mathsf{ipfe}}, \boldsymbol{w}_i)$
Return ct_i

$\mathsf{Gen}(i, \ell)$:
Parse $\mathsf{sk}_i = \{\mathsf{K}_{i,j}\}_{j\in[n]}$
$\boldsymbol{t}_{i,\ell} := \sum_{j\neq i}(-1)^{j<i}\mathsf{PRF}_{\mathsf{K}_{i,j}}(\ell) \in \mathbb{Z}_L^{mn}$

If $i \in \mathcal{HS} := \{i_1, \ldots, i_h\}$, then:
- If $i = i_1$, $\boldsymbol{t}_{i,\ell} := \sum_{j\in\mathcal{CS}}(-1)^{j<i}\mathsf{PRF}_{\mathsf{K}_{i,j}}(\ell) + \sum_{t=2}^{h}\mathsf{RF}(t, \ell)$.
- If $i = i_t$, for $t \in [2, \ldots, h]$, $\boldsymbol{t}_{i,\ell} := \sum_{j\in[n]\setminus\{i_t, i_1\}}(-1)^{j<i}\mathsf{PRF}_{\mathsf{K}_{i,j}}(\ell) - \mathsf{RF}(t, \ell)$.

Return $\boldsymbol{t}_{i,\ell}$

Fig. 5. Games for the proof of Theorem 3.3. Here, $\mathcal{HS} := [n] \setminus \mathcal{CS}$. Condition (*) is given in Definition 2.1. Here, RF denotes a random function that is computed on the fly. WLOG, QLeftRight is only queried on label $\ell^\star$, and QEnc isn't queried on $\ell^\star$.

Game G_1. In game G_1, we change the way the vectors $\boldsymbol{t}_{i,\ell}$ used by QEnc and QLeftRight are generated, switching the values $\mathsf{PRF}_{K_{i_1,i_t}}(\ell)$ to $\mathsf{RF}(t,\ell)$, for all $t \in [2, h]$, where we write the set of honest users $\mathcal{HS} := \{i_1, \ldots, i_h\}$, and RF denotes a random function, computed on the fly (see Fig. 5). The transition from G_0 to G_1 is justified by the security of the PRF. Namely, in Lemma 3.4, we exhibit a PPT adversary $\mathcal{B}_0$ such that:

$$\left|\mathsf{Win}_{\mathcal{A}}^{\mathsf{G}_0}(\lambda, n) - \mathsf{Win}_{\mathcal{A}}^{\mathsf{G}_1}(\lambda, n)\right| \leq (h-1) \cdot \mathsf{Adv}_{\mathsf{PRF},\mathcal{B}_0}(\lambda),$$

where $h \leq n$ denotes the number of honest users.

Game G_2. In game G_2, the vectors $\boldsymbol{w}_i$ used to generate the challenge ciphertexts contain an additional vector $(\mathbf{0}\|\ldots\|\mathbf{0}\|\boldsymbol{x}_i^{1,1}-\boldsymbol{x}_i^{1,0}\|\mathbf{0}\|\ldots\|\mathbf{0})$. The transition from G_1 to G_2 is justified by the any-IND security of FE. Namely, in Lemma 3.5, we exhibit a PPT adversary $\mathcal{B}_1$ such that:

$$\left|\mathsf{Win}_{\mathcal{A}}^{\mathsf{G}_1}(\lambda,n)-\mathsf{Win}_{\mathcal{A}}^{\mathsf{G}_2}(\lambda,n)\right|\leq \mathsf{Adv}_{\mathsf{FE},\mathcal{B}_1}^{\text{any-IND}}(\lambda).$$

Game G_3. In game G_3, the vectors $\boldsymbol{w}_i$ used in the challenge ciphertexts are of the form: $\boldsymbol{w}_i:=(\mathbf{0}\|\ldots\|\mathbf{0}\|\boldsymbol{x}_i^{j,1}\|\mathbf{0}\|\ldots\|\mathbf{0})$. The transition from G_2 to G_3 is justified by the any-IND security of FE. Namely, in Lemma 3.6, we exhibit a PPT adversary $\mathcal{B}_2$ such that:

$$\left|\mathsf{Win}_{\mathcal{A}}^{\mathsf{G}_2}(\lambda,n)-\mathsf{Win}_{\mathcal{A}}^{\mathsf{G}_3}(\lambda,n)\right|\leq \mathsf{Adv}_{\mathsf{FE},\mathcal{B}_2}^{\text{any-IND}}(\lambda).$$

Game G_4. This game is sta-pos^{+}-$\text{IND}_1^{\mathsf{MCFE}}(\lambda,n,\mathcal{A})$. The transition from G_3 to G_4 is symmetric to the transition from G_0 to G_1, justified by the security of the PRF. Namely, it can be proven as in Lemma 3.4 that there exists a PPT adversary $\mathcal{B}_3$ such that:

$$\left|\mathsf{Win}_{\mathcal{A}}^{\mathsf{G}_3}(\lambda,n)-\mathsf{Win}_{\mathcal{A}}^{\mathsf{G}_4}(\lambda,n)\right|\leq (h-1)\cdot \mathsf{Adv}_{\mathsf{PRF},\mathcal{B}_3}(\lambda),$$

where $h\leq n$ denotes the number of honest users. We defer to the proof of Lemma 3.4 for further details.

Putting everything together, we obtain the theorem. □

Lemma 3.4 (Transition from G_0 *to* G_1**).** *There exists a PPT adversary* $\mathcal{B}'$ *such that* $\left|\mathsf{Win}_{\mathcal{A}}^{\mathsf{G}_0}(\lambda,n)-\mathsf{Win}_{\mathcal{A}}^{\mathsf{G}_1}(\lambda,n)\right|\leq (h-1)\cdot \mathsf{Adv}_{\mathsf{PRF},\mathcal{B}'}(\lambda)$.

Proof. We can use the security of the PRF on all keys $\mathsf{K}_{i,j}$ where $i,j\in\mathcal{HS}$, since these are hidden from the adversary $\mathcal{A}$. We show that using the security of the PRF on $h-1$ carefully chosen such keys is sufficient to transition from G_0 to G_1. Namely, if we write $\mathcal{HS}:=\{i_1,\ldots,i_h\}$, where the indices $i_1<i_2<\cdots<i_h$ are ordered, we use the security of the PRF on keys of the form $\mathsf{K}_{i_1,j}$ for all $j\in\mathcal{HS}\setminus\{i_1\}$.

We build the adversary $\mathcal{B}'$ as follows. Given $\mathcal{CS}$ sent by $\mathcal{A}$, it samples $\mathsf{pp}_{\mathsf{ipfe}}\leftarrow\mathsf{Setup}^{\star}_{\mathsf{ipfe}}(1^\lambda,1^n)$ and $\mathsf{msk}_{\mathsf{ipfe}}\leftarrow\mathsf{KeyGen}_{\mathsf{ipfe}}(\mathsf{pp}_{\mathsf{ipfe}})$. For all $i\in[n]\setminus\{i_1\}$, for all $j>i$, $\mathcal{B}'$ samples $\mathsf{K}_{i,j}=\mathsf{K}_{j,i}\leftarrow\{0,1\}^\lambda$, thanks to which it can compute $\mathsf{sk}_i:=\{\mathsf{K}_{i,j}\}_{j\in[n]}$ for all $i\in\mathcal{CS}$ and send them to $\mathcal{A}$. $\mathcal{B}'$ can simulate the oracle QKeyD using $\mathsf{msk}_{\mathsf{ipfe}}$, and answers the queries to $\mathsf{QEnc}(i,\boldsymbol{x}_i^j,\ell)$ for $i\in\mathcal{CS}$, and $\mathsf{QLeftRight}(i,\boldsymbol{x}_i^{j,0},\boldsymbol{x}_i^{j,1},\ell^\star)$ for $i\in\mathcal{CS}$ using sk_i.

To answer $\mathsf{QEnc}(i_1,\boldsymbol{x}_{i_1}^j,\ell)$ or $\mathsf{QLeftRight}(i_1,\boldsymbol{x}_{i_1}^{j,0},\boldsymbol{x}_{i_1}^{j,1},\ell^\star)$, $\mathcal{B}'$ computes

$$t_{i_1,\ell}:=\sum_{j\in\mathcal{CS}}(-1)^{j<i_1}\mathsf{PRF}_{\mathsf{K}_{i_1,j}}(\ell)+\sum_{t=2}^{h}\mathsf{RF}(t,\ell).$$

To answer $\mathsf{QEnc}(i_t, \boldsymbol{x}_{i_t}^j, \ell)$ or $\mathsf{QLeftRight}(i_t, \boldsymbol{x}_{i_t}^{j,0}, \boldsymbol{x}_{i_t}^{j,1}, \ell^\star)$, for $t \in [2, \dots, h]$, $\mathcal{B}'$ computes

$$\boldsymbol{t}_{i_t,\ell} := \sum_{j \in [n] \setminus \{i_t, i_1\}} (-1)^{j<i_t} \mathsf{PRF}_{\mathsf{K}_{i_t,j}}(\ell) - \mathsf{RF}(t, \ell).$$

Here, $\mathsf{RF}(t, \ell)$ is either a truly random function, or $\mathsf{PRF}_{\mathsf{K}_{i_1,i_t}}(\ell)$, depending on the experiment $\mathcal{B}'$ is interacting with. In fact, we implicitly use a hybrid argument which goes over all $t \in [2, \dots, h]$ here, in order to switch the values $\mathsf{PRF}_{\mathsf{K}_{i_1,i_t}}(\ell)$ to $\mathsf{RF}(t, \ell)$. Thus, we obtain $\left|\mathsf{Win}_{\mathcal{A}}^{\mathsf{G}_0}(\lambda, n) - \mathsf{Win}_{\mathcal{A}}^{\mathsf{G}_1}(\lambda, n)\right| \le (h-1) \cdot \mathsf{Adv}_{\mathsf{PRF},\mathcal{B}'}(\lambda)$. □

Lemma 3.5 (Transition from G_1 to G_2). *There exists a PPT adversary $\mathcal{B}_1$ such that* $\left|\mathsf{Win}_{\mathcal{A}}^{\mathsf{G}_1}(\lambda, n) - \mathsf{Win}_{\mathcal{A}}^{\mathsf{G}_2}(\lambda, n)\right| \le \mathsf{Adv}_{\mathsf{FE},\mathcal{B}_1}^{\textit{any-IND}}(\lambda)$.

Proof. The adversary $\mathcal{B}_1$ works as follows. Given $\mathcal{CS}$ sent by $\mathcal{A}$, and $\mathsf{pp}_{\mathsf{ipfe}}$ from its own experiment, $\mathcal{B}_1$ samples $\mathsf{K}_{i,j} = \mathsf{K}_{j,i} \leftarrow \{0,1\}^\lambda$ for all $i < j \in [n]$, thanks to which it can send the sk_i for all $i \in \mathcal{CS}$, together with $\mathsf{pp}_{\mathsf{ipfe}}$ to $\mathcal{A}$. Since $\mathcal{B}_1$ knows the sk_i for all $i \in [n]$, it can answer the oracle QEnc as described in Fig. 5.

Whenever $\mathcal{A}$ queries QKeyD on input $\boldsymbol{y}$, $\mathcal{B}_1$ queries its own oracle on the same input, and forwards the answer to $\mathcal{A}$.

Since we are considering pos$^+$-IND security, we know $\mathcal{A}$ queries all honest slots on $\mathsf{QLeftRight}(\cdot,\cdot,\cdot,\ell^\star)$ and we denote by $i_{t^\star}$ the last honest slot queried on $\mathsf{QLeftRight}(\cdot,\cdot,\cdot,\ell^\star)$. We call $\Delta_{\boldsymbol{x}} := (\boldsymbol{x}_1^{1,1} - \boldsymbol{x}_1^{1,0}, \dots, \boldsymbol{x}_n^{1,1} - \boldsymbol{x}_n^{1,0})$, where for all $i \in \mathcal{HS}$, $(i, \boldsymbol{x}_i^{1,0}, \boldsymbol{x}_i^{1,1}, \ell^\star)$ is the first query of the form $\mathsf{QLeftRight}(i,\cdot,\cdot,\ell^\star)$, and for all $i \in \mathcal{CS}$, we define $\boldsymbol{x}_i^{1,1} - \boldsymbol{x}_i^{1,0} := \mathbf{0} \in \mathbb{Z}^m$ (note that $\mathsf{QLeftRight}$ can be queried on a corrupted slot, but by Condition (*), that means the query is of the form $(i, \boldsymbol{x}_i^{1,0}, \boldsymbol{x}_i^{1,1}, \ell^\star)$).

Whenever $\mathcal{A}$ queries $\mathsf{QLeftRight}(i, \boldsymbol{x}_i^{j,0}, \boldsymbol{x}_i^{j,1}, \ell^\star)$, $\mathcal{B}_1$ computes the vectors $\boldsymbol{t}_{i,\ell^\star}$ for all $i \in [n]$, using sk_i and computing the random function RF on the fly, as described in Fig. 5. Then, if $i \ne i_{t^\star}$, it computes $\boldsymbol{w}_i := (\mathbf{0}\|\dots\|\mathbf{0}\|\boldsymbol{x}_i^{j,0}\|\mathbf{0}\|\dots\|\mathbf{0}) + \boldsymbol{t}_{i,\ell^\star} \bmod L$, and returns $\mathsf{Enc}_{\mathsf{ipfe}}(\mathsf{pp}_{\mathsf{ipfe}}, \mathsf{mpk}_{\mathsf{ipfe}}, \boldsymbol{w}_i)$ to $\mathcal{A}$. If $i = i_{t^\star}$, then $\mathcal{B}_1$ queries its left-right oracle on input $(\mathbf{0}, \Delta_{\boldsymbol{x}})$ to get $\mathsf{ct}_i := \mathsf{Enc}_{\mathsf{ipfe}}(\mathsf{pp}_{\mathsf{ipfe}}, \mathsf{mpk}_{\mathsf{ipfe}}, \mathbf{0})$ or $\mathsf{ct}_i := \mathsf{Enc}_{\mathsf{ipfe}}(\mathsf{pp}_{\mathsf{ipfe}}, \mathsf{mpk}_{\mathsf{ipfe}}, \Delta_{\boldsymbol{x}})$, depending on the experiment $\mathcal{B}_1$ is interacting with. Note that at this point, $\Delta_{\boldsymbol{x}}$ is entirely known to $\mathcal{B}_1$, since $i_{t^\star}$ is the last honest slot to be queried to $\mathsf{QLeftRight}(\cdot,\cdot,\cdot,\ell^\star)$. Then, $\mathcal{B}_1$ computes $\boldsymbol{w}_i := (\mathbf{0}\|\dots\|\mathbf{0}\|\boldsymbol{x}_i^{j,0}\|\mathbf{0}\|\dots\|\mathbf{0}) + \boldsymbol{t}_{i,\ell^\star} \bmod L$ and returns $\mathsf{ct}'_i := \mathsf{Add}(\mathsf{ct}_i, \boldsymbol{w}_i)$, which, according to the property from Definition 3.2 (linear encryption), is identically distributed to $\mathsf{Enc}_{\mathsf{ipfe}}(\mathsf{pp}_{\mathsf{ipfe}}, \mathsf{mpk}_{\mathsf{ipfe}}, \boldsymbol{w}_i \bmod L)$ or $\mathsf{Enc}_{\mathsf{ipfe}}(\mathsf{pp}_{\mathsf{ipfe}}, \mathsf{mpk}_{\mathsf{ipfe}}, \boldsymbol{w}_i + \Delta_{\boldsymbol{x}} \bmod L)$, (again, depending on which experiment $\mathcal{B}_1$ is interacting with). For all $\boldsymbol{y}$ queried to QKeyD, we have $\langle \Delta_{\boldsymbol{x}}, \boldsymbol{y} \rangle = 0$, by Condition (*). Moreover, $\|\Delta_{\boldsymbol{x}}\|_\infty < 2X$. Thus, the queries $\mathcal{B}_1$ sends to its left-right oracle are legitimate. Finally, $\mathcal{B}_1$ returns ct'_i to $\mathcal{A}$.

To conclude, we show that when $\mathcal{B}_1$ is interacting with $\textbf{any-IND}_0^{\mathsf{FE}}(\lambda, 1, \mathcal{A})$, then it simulates the game G_1, whereas it simulates the game G_2 when it is interacting with $\textbf{any-IND}_1^{\mathsf{FE}}(\lambda, 1, \mathcal{A})$. It is clear for the case $\textbf{any-IND}_0^{\mathsf{FE}}(\lambda, 1, \mathcal{A})$. For the case **any-IND**$_1^{\mathsf{FE}}(\lambda, 1, \mathcal{A})$, we consider the vectors $\{\boldsymbol{u}_t\}_{t\in[h]}$, where we write $\mathcal{HS} := \{i_1, \ldots, i_h\}$ and we denote by $\boldsymbol{u}_1 := -\sum_{t=2}^{h} \mathsf{RF}(t, \ell^\star)$ and $\boldsymbol{u}_t := \mathsf{RF}(t, \ell^\star)$, for all $t \in [2, \ldots, n]$. These are shares of a perfect h out of h secret sharing of $\mathbf{0}$, that is, they are uniformly random conditioned on $\sum_{t\in[h]} \boldsymbol{u}_t = \mathbf{0}$. Thus, $\{\boldsymbol{u}_t\}_{t\in[t]\setminus\{t^\star\}} \cup \{\boldsymbol{u}_{t^\star} + \Delta_{\boldsymbol{x}}\}$ is a set of shares for a secret sharing of the vector $\Delta_{\boldsymbol{x}}$. Thus, the following distributions are identical:

$$\{\boldsymbol{u}_t\}_{t\in[h]\setminus\{t^\star\}} \cup \{\boldsymbol{u}_{t^\star} + \Delta_{\boldsymbol{x}}\}$$

and

$$\{\boldsymbol{u}_t + (\mathbf{0}\|\ldots\|\boldsymbol{x}_{i_t}^{1,1} - \boldsymbol{x}_{i_t}^{1,0}\|\mathbf{0}\|\ldots\|\mathbf{0})\}_{t\in[h]},$$

where for all $t \in [h]$, $\boldsymbol{u}_t \leftarrow \mathbb{Z}_L^{mn}$ such that $\sum_{t\in[h]} \boldsymbol{u}_t = \mathbf{0}$. The uppermost distribution corresponds to the simulation by $\mathcal{B}_1$ when it is interacting with **any-IND**$_1^{\mathsf{FE}}(\lambda, 1, \mathcal{A})$, while the lowermost distribution corresponds to the game $\mathsf{G}_{1.\rho}$. This concludes the proof. □

Lemma 3.6 (Transition from G_2 *to* G_3**).** *There exists a PPT adversary* $\mathcal{B}_2$ *such that* $\left|\mathsf{Win}_{\mathcal{A}}^{\mathsf{G}_2}(\lambda, n) - \mathsf{Win}_{\mathcal{A}}^{\mathsf{G}_3}(\lambda, n)\right| \leq \mathsf{Adv}_{\mathsf{FE},\mathcal{B}}^{\textit{any-IND}}(\lambda)$.

Proof. We build an adversary $\mathcal{B}_2$ against the any-IND security of FE as follows.

Given $\mathcal{CS}$ sent by $\mathcal{A}$, and $\mathsf{pp}_{\mathsf{ipfe}}$ from its own experiment, $\mathcal{B}_2$ samples $\mathsf{K}_{i,j} = \mathsf{K}_{j,i} \leftarrow \{0,1\}^\lambda$ for all $i < j \in [n]$, thanks to which it can send the sk_i for all $i \in \mathcal{CS}$, together with $\mathsf{pp}_{\mathsf{ipfe}}$ to $\mathcal{A}$, and answer the oracle queries to QEnc as described in Fig. 5.

Then, whenever $\mathcal{A}$ queries QKeyD on input $\boldsymbol{y}$, $\mathcal{B}_2$ queries its own oracle on the same input, and forwards the answer to $\mathcal{A}$. Whenever $\mathcal{A}$ queries $\mathsf{QLeftRight}(i, \boldsymbol{x}_i^{j,0}, \boldsymbol{x}_i^{j,1}, \ell^\star)$, $\mathcal{B}_2$ computes $\boldsymbol{t}_{i,\ell^\star}$ using sk_i and computing the random function RF on the fly, as described in Fig. 5. Then, $\mathcal{B}_2$ queries its left-right oracle on input $(\mathbf{0}\|\ldots\|\mathbf{0}\|\boldsymbol{x}_i^{j,0} - \boldsymbol{x}_i^{1,0}\|\mathbf{0}\|\ldots\|\mathbf{0}), (\mathbf{0}\|\ldots\|\mathbf{0}\|\boldsymbol{x}_i^{j,1} - \boldsymbol{x}_i^{1,1}\|\mathbf{0}\|\ldots\|\mathbf{0})$ to get

$$\mathsf{ct}_i := \mathsf{Enc}_{\mathsf{ipfe}}(\mathsf{pp}_{\mathsf{ipfe}}, \mathsf{mpk}_{\mathsf{ipfe}}(\mathbf{0}\|\ldots\|\mathbf{0}\|\boldsymbol{x}_i^{j,\beta} - \boldsymbol{x}_i^{1,\beta}\|\mathbf{0}\|\ldots\|\mathbf{0})),$$

where $\beta \in \{0,1\}$, depending on the experiment $\mathcal{B}_2$ is interacting with. Finally, $\mathcal{B}_2$ computes $\boldsymbol{v}_i := (\mathbf{0}\|\ldots\|\mathbf{0}\|\boldsymbol{x}_i^{1,1}\|\mathbf{0}\|\ldots\|\mathbf{0}) + \boldsymbol{t}_{i,\ell^\star} \bmod L$, and returns $\mathsf{ct}'_i := \mathsf{Add}(\mathsf{ct}_i, \boldsymbol{v}_i)$ to $\mathcal{A}$, which, according to the property from Definition 3.2, is identically distributed to $\mathsf{Enc}_{\mathsf{ipfe}}(\mathsf{pp}_{\mathsf{ipfe}}, \mathsf{mpk}_{\mathsf{ipfe}}, (\mathbf{0}\|\ldots\|\mathbf{0}\|\boldsymbol{x}_i^{j,\beta} - \boldsymbol{x}_i^{1,\beta} + \boldsymbol{x}_i^{1,1}\|\mathbf{0}\|\ldots\|\mathbf{0}) + \boldsymbol{t}_{i,\ell^\star} \bmod L)$. For all $\boldsymbol{y}$ queried to QKeyD, Condition (*) implies that $\langle(\mathbf{0}\|\ldots\|\mathbf{0}\|\boldsymbol{x}_i^{j,0} - \boldsymbol{x}_i^{1,0}\|\mathbf{0}\|\ldots\|\mathbf{0}), \boldsymbol{y}\rangle = \langle(\mathbf{0}\|\ldots\|\mathbf{0}\|\boldsymbol{x}_i^{j,1} - \boldsymbol{x}_i^{1,1}\|\mathbf{0}\|\ldots\|\mathbf{0}), \boldsymbol{y}\rangle$ for all queries $(i, \boldsymbol{x}_i^{j,0}, \boldsymbol{x}_i^{j,1}, \ell^\star)$ to $\mathsf{QLeftRight}$. Moreover, for all $\beta \in \{0,1\}$, we have $\|(\mathbf{0}\|\ldots\|\mathbf{0}\|\boldsymbol{x}_i^{j,\beta} - \boldsymbol{x}_i^{1,\beta}\|\mathbf{0}\|\ldots\|\mathbf{0})\|_\infty < 2X$. Thus, the queries $\mathcal{B}_2$ sends to its left-right oracle are legitimate. □

3.3 Adaptive Security

Now we proceed to prove the adt-pos$^+$-IND-security of the scheme, that is, security with adaptive corruption. As before, using the generic transformation in Sect. 4, we can remove the pos$^+$ restriction, and obtain adt-any-IND security.

Theorem 3.7 (adt-pos$^+$-IND-security). *If the FE scheme* $\mathsf{FE} = (\mathsf{Setup_{ipfe}}, \mathsf{KeyGen_{ipfe}}, \mathsf{KeyDer_{ipfe}}, \mathsf{Enc_{ipfe}}, \mathsf{Dec_{ipfe}})$ *is an* any*-IND-secure FE scheme for the inner product functionality defined as* $\mathcal{F}^{\mathsf{ip}}_{\rho_{\mathsf{ipfe}}}, \rho_{\mathsf{ipfe}} = (\mathbb{Z}, 1, m, 2X, Y)$, *and* PRF *is secure, then* MCFE *from Fig. 4 is* adt-pos$^+$*-IND-secure for the functionality defined as* $\mathcal{F}^{\mathsf{ip}}_{\rho}, \rho = (\mathbb{Z}, n, m, X, Y)$. *Namely, for any PPT adversary* $\mathcal{A}$, *there exist PPT adversaries* $\mathcal{B}$ *and* $\mathcal{B}'$ *such that:*

$$\mathsf{Adv}^{\textit{adt-pos}^+\textit{-IND}}_{\mathsf{MCFE},\mathcal{A}}(\lambda, n) \leq 2(n+1)n(n-1)^2 q_{\mathsf{Enc}} \cdot \mathsf{Adv}_{\mathsf{PRF},\mathcal{B}}(\lambda) + 2(n+1)q_{\mathsf{Enc}} \cdot \mathsf{Adv}^{\textit{any-IND}}_{\mathsf{FE},\mathcal{B}'}(\lambda) \;,$$

where q_{Enc} *denotes the number of distinct labels queried to* QLeftRight.

Proof. WLOG, we can assume that adversary $\mathcal{A}$ only queries QLeftRight on one label $\ell^\star$, that isn't queried to QEnc. Namely, we show that there exist PPT adversaries $\mathcal{B}$ and $\mathcal{B}'$ such that:

$$\mathsf{Adv}^{\text{adt-pos}^+\text{-IND-1-label}}_{\mathsf{MCFE},\mathcal{A}}(\lambda, n) \leq 2(n+1)n(n-1)^2 \cdot \mathsf{Adv}_{\mathsf{PRF},\mathcal{B}}(\lambda) + 2(n+1) \cdot \mathsf{Adv}^{\text{pos}^+\text{-IND}}_{\mathsf{FE},\mathcal{B}'}(\lambda) \;.$$

The theorem then follows from Lemma 2.5.

We proceed via a hybrid argument, using the games described in Fig. 6. The lemmas from the transitions are provided in the full version [1].

Game $\mathsf{G}^{\star}_0$: is as xx-yy-IND-1-label$_0$, except the size of $\mathcal{Q}_{\ell^\star}$, which denotes the set of slots queried to $\mathsf{QLeftRight}(\cdot,\cdot,\cdot,\ell^\star)$, is initially guessed by the experiment, by choosing a uniformly random $\kappa^\star \leftarrow \{0,\ldots,n\}$. The game behaves exactly as xx-yy-IND-1-label$_0$, except it ignores the $\mathcal{A}$'s output α, and outputs 0 instead, in case the guess $\kappa^\star$ was incorrect. Since this guess is correct with probability $\frac{1}{n+1}$, we have

$$\mathsf{Win}^{\mathsf{G}^\star_0}_{\mathcal{A}}(\lambda, n) = \frac{1}{n+1} \cdot \mathsf{Win}^{\text{xx-yy-IND-1-label}_0}_{\mathcal{A}}(\lambda, n) \;.$$

Game $\mathsf{G}^\star_1$: in this game, we change the distribution of the ciphertexts output QLeftRight, for the case $\kappa^\star \geq 2$. For these, the vector $(\mathbf{0}\|\ldots\|\mathbf{0}\|\boldsymbol{x}^{j,0}_i\|\mathbf{0}\|\ldots\|\mathbf{0})$ to be encrypted is added a share of a perfect $\kappa^\star$ out of $\kappa^\star$ secret sharing of $\mathbf{0}$. This game is similar to the game G_1 from Fig. 5 for the proof of Theorem 3.3. We justify this transition using the security of the PRF, as in Lemma 3.5, with the crucial difference that corruptions are adaptive here. Thus, the set of slots $\mathcal{Q}_{\ell^\star}$ queried to QLeftRight is not known

$G_0^\star$, $\boxed{G_1^\star, G_2^\star, G_3^\star}$, $G_4^\star$:

$\kappa^\star \leftarrow \{0, \ldots, n\}$, $\beta \leftarrow \{0,1\}$, $\boxed{\text{for all } t \in [2, \ldots, \kappa^\star],\ \boldsymbol{u}_t \leftarrow \mathbb{Z}_L^{mn}}$
$(\{\mathsf{sk}_i\}_{i\in[n]}, \mathsf{msk}) \leftarrow \mathsf{KeyGen}(\mathsf{pp})$
$\alpha \leftarrow \mathcal{A}^{\mathsf{QEnc}(\cdot,\cdot,\cdot),\mathsf{QKeyD}(\cdot),\mathsf{QCor}(\cdot)}(\mathsf{pp})$
Output α if Condition (*) is satisfied AND the guess $\kappa^\star$ is correct; 0 otherwise.

$\mathsf{QEnc}(i, \boldsymbol{x}_i^j, \ell)$:
Return $\mathsf{Enc}(\mathsf{pp}, \mathsf{sk}_i, \boldsymbol{x}_i^j, \ell)$

$\mathsf{QKeyD}(\boldsymbol{y})$:
Return $\mathsf{sk}_{\boldsymbol{y}} \leftarrow \mathsf{KeyDer}(\mathsf{pp}, \mathsf{msk}, \boldsymbol{y})$

$\mathsf{QCor}(i)$:
Return sk_i

$\mathsf{QLeftRight}(i, \boldsymbol{x}_i^{j,0}, \boldsymbol{x}_i^{j,1}, \ell^\star)$:
Parse $\mathsf{sk}_i := \{\mathsf{K}_{i,j}\}_{j\in[n]}$, $\boldsymbol{v}_{i,\ell} := \sum_{j\neq i}(-1)^{j<i}\mathsf{PRF}_{\mathsf{K}_{i,j}}(\ell) \in \mathbb{Z}_L^{mn}$, $\boldsymbol{t}_{i,\ell} := \boldsymbol{v}_{i,\ell}$.

We write $\{i_1, \ldots, i_\kappa\}$ the set of slots queried to QLeftRight, in the order they are queried (that is, i_1 is the first queried, i_2 is the second, and so forth). If $\kappa^\star \geq 2$, then do the following.

- If $i = i_1$, then $\boldsymbol{t}_{i,\ell} := \boldsymbol{v}_{i,\ell} + \sum_{t=2}^{\kappa^\star} \boldsymbol{u}_t$.
- If $i = i_t$, for $t \in [2, \ldots, \kappa^\star]$, then $\boldsymbol{t}_{i,\ell} := \boldsymbol{v}_{i,\ell} - \boldsymbol{u}_t$.
- If $i = i_t$, for $t > \kappa^\star$, that means $\kappa > \kappa^\star$, the guess was incorrect.

Ends the game and output 0.

$\boldsymbol{w}_i := (\mathbf{0}\|\ldots\|\mathbf{0}\|\boldsymbol{x}_i^{j,0}\|\mathbf{0}\|\cdots\|\mathbf{0}) + \boldsymbol{t}_{i,\ell} \bmod L$
If $\kappa^\star \geq 2$: $\boldsymbol{w}_i := (\mathbf{0}\|\ldots\|\mathbf{0}\|\boldsymbol{x}_i^{j,0} + \boldsymbol{x}_i^{1,1} - \boldsymbol{x}_i^{1,0}\|\mathbf{0}\|\cdots\|\mathbf{0}) + \boldsymbol{t}_{i,\ell} \bmod L$
$\boldsymbol{w}_i := (\mathbf{0}\|\ldots\|\mathbf{0}\|\boldsymbol{x}_i^{j,1}\|\mathbf{0}\|\ldots\|\mathbf{0}) + \boldsymbol{t}_{i,\ell} \bmod L$
$\mathsf{ct}_i \leftarrow \mathsf{Enc}^{\mathsf{ipfe}}(\mathsf{pp}_{\mathsf{ipfe}}, \boldsymbol{w}_i)$
Return ct_i

Fig. 6. Games for the proof of Theorem 3.7. We say the guess $\kappa^\star$ is correct if the size of $\mathcal{Q}_{\ell^\star}$ is $\kappa^\star$.

in advance by the reduction. Since guessing the entire set would incur an exponential security loss, we introduce gradually the shares, starting with a 2 out of 2 perfect secret sharing, then 3 out of 3, and so forth, via a hybrid argument, until we reach the $\kappa^\star$ out of $\kappa^\star$ secret sharing among all queried slots. To go from one hybrid to another, we only require to guess a pair of users (i, j) (as opposed to guessing the entire set of honest users) to use the security of the PRF on the key $\mathsf{K}_{i,j}$. Namely, in the full version [1], we show that there exists a PPT adversary $\mathcal{B}_0$ such that:

$$\left|\mathsf{Win}_{\mathcal{A}}^{G_0^\star}(\lambda, n) - \mathsf{Win}_{\mathcal{A}}^{G_1^\star}(\lambda, n)\right| \leq n(n-1)^2 \cdot \mathsf{Adv}_{\mathsf{PRF},\mathcal{B}_0}(\lambda)$$

Game $\mathsf{G}_2^\star$**:** in this game, the vectors $\boldsymbol{w}_i$ used to generate the ciphertexts output by $\mathsf{QLeftRight}$ contain an additional vector $(\mathbf{0}\|\ldots\|\mathbf{0}\|\boldsymbol{x}_i^{1,1} - \boldsymbol{x}_i^{1,0}\|\mathbf{0}\|\ldots\|\mathbf{0})$. The transition from $\mathsf{G}_1^\star$ to $\mathsf{G}_2^\star$ is justified by the any-IND security of FE, similarly than the transition from G_1 to G_2 in Fig. 5 for the proof of Theorem 3.3. Namely, in the full version [1], we exhibit a PPT adversary $\mathcal{B}_1$ such that:

$$\left|\mathsf{Win}_{\mathcal{A}}^{\mathsf{G}_1^\star}(\lambda, n) - \mathsf{Win}_{\mathcal{A}}^{\mathsf{G}_2^\star}(\lambda, n)\right| \leq \mathsf{Adv}_{\mathsf{FE},\mathcal{B}_1}^{\text{any-IND}}(\lambda).$$

Game $\mathsf{G}_3^\star$**:** in this game, the vectors $\boldsymbol{w}_i$ used in the ciphertexts output by $\mathsf{QLeftRight}$ are of the form: $\boldsymbol{w}_i := (\mathbf{0}\|\ldots\|\mathbf{0}\|\boldsymbol{x}_i^{j,1}\|\mathbf{0}\|\ldots\|\mathbf{0}) + \boldsymbol{t}_{i,\ell^\star} \bmod L$. The transition from $\mathsf{G}_{\rho-1.2}^\star$ to $\mathsf{G}_{\rho-1.3}^\star$ is justified by the pos$^+$-IND security of FE, similarly than the transition from G_2 to G_3 in Fig. 5 for the proof of Theorem 3.3. Namely, in the full version [1], we build a PPT adversary $\mathcal{B}_2$ such that:

$$\left|\mathsf{Win}_{\mathcal{A}}^{\mathsf{G}_2^\star}(\lambda, n) - \mathsf{Win}_{\mathcal{A}}^{\mathsf{G}_3^\star}(\lambda, n)\right| \leq \mathsf{Adv}_{\mathsf{FE},\mathcal{B}_2}^{\text{any-IND}}(\lambda).$$

Game $\mathsf{G}^{\star_4}$**.** The transition from $\mathsf{G}_3^\star$ to $\mathsf{G}_4^\star$ is symmetric to the transition from $\mathsf{G}_0^\star$ to $\mathsf{G}_1^\star$, justified by the security of the PRF. Namely, we prove in in the full version [1] that there exists a PPT adversary $\mathcal{B}_3$ such that:

$$\left|\mathsf{Win}_{\mathcal{A}}^{\mathsf{G}_3^\star}(\lambda, n) - \mathsf{Win}_{\mathcal{A}}^{\mathsf{G}_4^\star}(\lambda, n)\right| \leq n(n-1)^2 \cdot \mathsf{Adv}_{\mathsf{PRF},\mathcal{B}_3}(\lambda).$$

We defer to the full version [1] for further details. Since $\mathsf{G}_4^\star$ is exactly as the game xx-yy-$\text{IND}_0^{\mathsf{MCFE}}$ except it it guesses $\kappa^\star \leftarrow \{0, \ldots, n\}$, we have

$$\mathsf{Win}_{\mathcal{A}}^{\mathsf{G}_4^\star}(\lambda, n) = \frac{1}{n+1} \cdot \mathsf{Win}_{\mathcal{A}}^{\text{xx-yy-IND-1-label}_1}(\lambda, n).$$

Putting everything together, we obtain the theorem. □

4 From pos$^+$-IND to any-IND Security

In this section, we give a compiler that generically transforms any adt-pos$^+$-IND secure (D)MCFE into an adt-any-IND secure (D)MCFE. Our construction builds up from the compiler from [2, Section 4.1], which does not support labels. Our technical contribution is to handle multiple labels, many challenge ciphertexts per label and input slots, and adaptive corruptions, without resorting to the random oracle model, as opposed to [2, Section 4.2]. This is the first generic transformation to support such features, and when combined with our MCFE from Sect. 3, it gives the first MCFE for inner products whose adt − any-IND security is proven in the standard model. Our construction is given in Fig. 7. I've added this last sentence. It is stated in terms of DMCFE, but a similar transformation works for MCFE.

$\underline{\mathsf{Setup}'(1^\lambda, 1^n):}$	$\underline{\mathsf{KeyDerShare}'(\mathsf{pp}, \mathsf{sk}'_i, f):}$
Return $\mathsf{pp} \leftarrow \mathsf{Setup}(1^\lambda, 1^n)$	Parse $\mathsf{sk}'_i = (\mathsf{sk}_i, \{\mathsf{k}_{i,j}, \mathsf{k}_{j,i}\}_{j\in[n]})$ Return $\mathsf{sk}'_{i,f} \leftarrow \mathsf{KeyDerShare}(\mathsf{sk}_i, f)$
$\underline{\mathsf{KeyGen}'(\mathsf{pp}):}$ $\{\mathsf{sk}_i\}_{i\in[n]} \leftarrow \mathsf{KeyGen}(\mathsf{pp})$ For $i \in [n]$: $\mathsf{k}_{i,1}, \dots, \mathsf{k}_{i,n} \leftarrow \{0,1\}^\lambda$ Return $\{\mathsf{sk}'_i = (\mathsf{sk}_i, \{\mathsf{k}_{i,j}, \mathsf{k}_{j,i}\}_{j\in[n]})\}_{i\in[n]}$	$\underline{\mathsf{KeyDerComb}'(\mathsf{pp}, \{\mathsf{sk}'_{i,f}\}_{i\in[n]}):}$ $\mathsf{sk}_f := \mathsf{KeyDerComb}(\mathsf{pp}, \{\mathsf{sk}'_{i,f}\}_{i\in[n]})$ Return sk_f
$\underline{\mathsf{Enc}'(\mathsf{pp}, \mathsf{sk}'_i, x_i, \ell):}$ Parse $\mathsf{sk}'_i = (\mathsf{sk}_i, \{\mathsf{k}_{i,j}, \mathsf{k}_{j,i}\}_{j\in[n]})$ $\mathsf{ct}_i \leftarrow \mathsf{Enc}(\mathsf{pp}, \mathsf{sk}_i, x_i)$ For all $j \in [n]$: $\mathsf{k}_{i,j}(\ell) := \mathsf{PRF}_{\mathsf{k}_{i,j}}(\ell)$ $\mathsf{K}_i(\ell) := \oplus_{j\in[n]} \mathsf{k}_{i,j}(\ell)$ $\mathsf{ct}'_i \leftarrow \mathsf{Enc}_{\mathsf{SE}}(\mathsf{K}_i(\ell), \mathsf{ct}_i)$ Return $(\mathsf{ct}'_i, \{\mathsf{k}_{j,i}(\ell)\}_{j\in[n]})$	$\underline{\mathsf{Dec}'(\mathsf{pp}, \mathsf{sk}_f, \mathsf{ct}''_1, \dots, \mathsf{ct}''_n):}$ Parse $\{\mathsf{ct}''_i = (\mathsf{ct}'_i, \{\mathsf{k}_{j,i}(\ell)\}_{j\in[n]})\}_{i\in[n]}$ For $i \in [n]$: $\quad \mathsf{K}_i(\ell) = \oplus_{j\in[n]} \mathsf{k}_{i,j}(\ell)$ $\quad \mathsf{ct}_i \leftarrow \mathsf{Dec}_{\mathsf{SE}}(\mathsf{K}_i(\ell), \mathsf{ct}'_i)$ Return $\mathsf{Dec}(\mathsf{pp}, \mathsf{sk}_f, \mathsf{ct}_1, \dots, \mathsf{ct}_n)$.

Fig. 7. Compiler from an xx-pos$^+$-IND DMCFE DMCFE into an xx-any-IND DMCFE DMCFE' using an IND-CPA symmetric-key encryption scheme SE.

Theorem 4.1. (Security). *Let the tuple* $\mathsf{DMCFE} = (\mathsf{Setup}, \mathsf{KeyGen}, \mathsf{KeyDerShare}, \mathsf{KeyDerComb}, \mathsf{Enc}, \mathsf{Dec})$ *be an* adt-pos$^+$*-IND-secure DMCFE scheme for a family of functions* $\mathcal{F}$*. Let* $\mathsf{SE} = (\mathsf{Enc}_{\mathsf{SE}}, \mathsf{Dec}_{\mathsf{SE}})$ *be an IND-CPA symmetric-key encryption scheme. Let* PRF *be a pseudorandom function. Then the DMCFE scheme* $\mathsf{DMCFE}' = (\mathsf{Setup}', \mathsf{KeyGen}', \mathsf{KeyDerShare}', \mathsf{KeyDerComb}', \mathsf{Enc}', \mathsf{Dec}')$ *described in Fig. 7 is* adt-any*-IND secure. Namely, for any PPT adversary* $\mathcal{A}$*, there exist PPT adversaries* $\mathcal{B}$*,* $\mathcal{B}'$*, and* $\mathcal{B}''$ *such that:*

$$\begin{aligned}\mathsf{Adv}^{\textit{adt-any-IND}}_{\mathsf{DMCFE}',\mathcal{A}}(\lambda, n) \leq q_{\mathsf{Enc}} \cdot \mathsf{Adv}^{\textit{adt-pos}^+\textit{-IND}}_{\mathsf{DMCFE},\mathcal{B}}(\lambda, n) \\ + q_{\mathsf{Enc}} n^2 \cdot \mathsf{Adv}^{\textit{IND-CPA}}_{\mathsf{SE},\mathcal{B}'}(\lambda) + 2 q_{\mathsf{Enc}} n^2 \cdot \mathsf{Adv}_{\mathsf{PRF},\mathcal{B}''}(\lambda),\end{aligned}$$

where q_{Enc} *denotes the number of distinct labels queried to* $\mathsf{QLeftRight}$.

Proof. WLOG, we can consider the security where only one label is queried to $\mathsf{QLeftRight}$, and that label is not queried to QEnc. Namely, we show there exist PPT adversaries $\mathcal{B}$, $\mathcal{B}'$ and $\mathcal{B}''$ such that $\mathsf{Adv}^{\text{adt-any-IND-1-label}}_{\mathsf{DMCFE}',\mathcal{A}}(\lambda, n) \leq \mathsf{Adv}^{\text{adt-pos}^+\text{-IND}}_{\mathsf{DMCFE},\mathcal{B}}(\lambda, n) + n \cdot \mathsf{Adv}^{\text{IND-CPA}}_{\mathsf{SE},\mathcal{B}'}(\lambda) + 2n \cdot \mathsf{Adv}_{\mathsf{PRF},\mathcal{B}''}(\lambda)$. The theorem follows from Lemma 2.5 (from one to many labels). We call $\ell^\star$ the unique label queried to $\mathsf{QLeftRight}$ (if $\mathsf{QLeftRight}$ is not queried, the security follows trivially).

Intuitively, the proof uses the adt-pos$^+$-IND security of DMCFE for the case where all honest slots are queried to $\mathsf{QLeftRight}(\cdot, \cdot, \cdot, \ell^\star)$, and the security of the PRF together witht the IND-CPA security of SE for the case where not all honest slots are queried to $\mathsf{QLeftRight}(\cdot, \cdot, \cdot, \ell^\star)$.

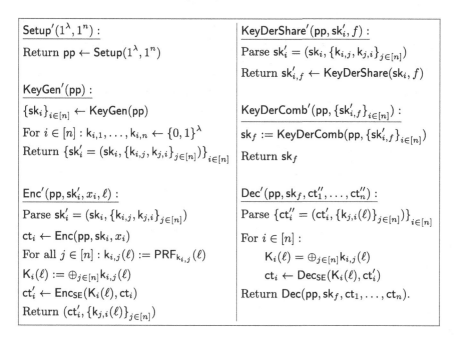

Formally, for all $b \in \{0,1\}$, we define $\mathsf{G}_b^\star$ as adt-yy-IND$_1^{\mathsf{DMCFE}'}(\lambda, n, \mathcal{A})$, except the game guesses an honest slot that is not going to be queried to $\mathsf{QLeftRight}(\cdot,\cdot,\cdot,\ell^\star)$, by sampling uniformly at random $i^\star \leftarrow \{0,...,n\}$, where $i^\star = 0$ means that all honest slots are queried to $\mathsf{QLeftRight}(\cdot,\cdot,\cdot,\ell^\star)$. The output of $\mathsf{G}_b^\star$ is the same adt-yy-IND$_1^{\mathsf{DMCFE}'}(\lambda, n, \mathcal{A})$, unless the guess is unsuccessful, in which case, $\mathsf{G}_b^\star$ outputs 0. Clearly, we have $\Pr[\mathsf{G}_b^\star(\lambda, n, \mathcal{A}) = 1] = \frac{1}{n+1} \cdot \Pr[\text{adt-yy-IND}_b^{\mathsf{DMCFE}'}(\lambda, n, \mathcal{A}) = 1]$.

When $i^\star = 0$, we can rely on the adt-pos$^+$-IND security of DMCFE. Namely, we have a PPT adversary $\mathcal{B}$ such that:

$$\Big| \Pr[\mathsf{G}_0^\star(\lambda, n, \mathcal{A}) = 1 | i^\star = 0] - \Pr[\mathsf{G}_1^\star(\lambda, n, \mathcal{A}) = 1 | i^\star = 0] \Big| \leq \mathsf{Adv}_{\mathsf{DMCFE},\mathcal{B}}^{\text{adt-pos}^+\text{-IND}}(\lambda, n).$$

For all $j \in [n]$, we prove that there exist PPT adversaries $\mathcal{B}'$ and $\mathcal{B}''$ such that:

$$\Big| \Pr[\mathsf{G}_0^\star(\lambda, n, \mathcal{A}) = 1 | i^\star = j] - \Pr[\mathsf{G}_1^\star(\lambda, n, \mathcal{A}) = 1 | i^\star = j] \Big| \leq n \cdot \mathsf{Adv}_{\mathsf{SE},\mathcal{B}'}^{\text{IND-CPA}}(\lambda, n) + 2n \cdot \mathsf{Adv}_{\mathsf{PRF},\mathcal{B}''}(\lambda, n).$$

To prove the statement above, we use the fact that if there is a query $\mathsf{QLeftRight}(i, \boldsymbol{x}_i^{j,0}, \boldsymbol{x}_i^{j,1}, \ell^\star)$ with $\boldsymbol{x}_i^{j,0} \neq \boldsymbol{x}_i^{j,1}$, then the slot $i \in [n]$ cannot be corrupted without violating the Condition (*) from the security definition given in Definition 2.2. We call such a slot explicitly honest, and such a query explicitly honest. We define hybrid games H_ρ for all $\rho \in \{0, \ldots, n\}$, defined as $\mathsf{G}_0^\star$, except that every explicitly honest query $\mathsf{QLeftRight}(i, \boldsymbol{x}_i^{j,0}, \boldsymbol{x}_i^{j,1}, \ell^\star)$ is answered by $\mathsf{Enc}'(\mathsf{pp}, \mathsf{sk}'_i, \boldsymbol{x}_i^{j,1}, \ell^\star)$ if $i \leq \rho$, and is answered by $\mathsf{Enc}'(\mathsf{pp}, \mathsf{sk}'_i, \boldsymbol{x}_i^{j,0}, \ell^\star)$ if $i > \rho$. The game H_0 is the same as $\mathsf{G}_0^\star$, and H_n is the same as $\mathsf{G}_1^\star$. We prove that for all $j \in [n]$, for all $\rho \in [n]$, there exist PPT adversaries $\mathcal{B}_\rho$ and $\mathcal{B}'_\rho$ such that:

$$\Big| \Pr[\mathsf{H}_{\rho-1}(\lambda, n, \mathcal{A}) = 1 | i^\star = j] - \Pr[\mathsf{H}_\rho(\lambda, n, \mathcal{A}) = 1 | i^\star = j] \Big| \leq \mathsf{Adv}_{\mathsf{SE},\mathcal{B}_\rho}^{\text{IND-CPA}}(\lambda, n) + 2 \cdot \mathsf{Adv}_{\mathsf{PRF},\mathcal{B}'_\rho}(\lambda, n).$$

The transition from $\mathsf{H}_{\rho-1}^\star$ and $\mathsf{H}_{\rho-1}^\star$ is justified as follows. If ρ is not an explicitly honest slot, then the two games are the same by definition. Otherwise, we use the security of the PRF to switch the key $\mathsf{k}_{\rho,i^\star}(\ell^\star)$ to uniformly random (note that we can do so since the slots ρ and $i^\star$ are known beforehand by the reduction). If the guess $i^\star$ is correct (i.e $i^\star$ is honest but never queried to $\mathsf{QLeftRight}$), then the key $\mathsf{k}_{\rho,i^\star}(\ell^\star) := \mathsf{PRF}_{\mathsf{k}_{\rho,i^\star}}(\ell^\star)$ only appears in the output $\mathsf{QLeftRight}(\rho,\cdot,\cdot,\ell^\star)$. So, for these challenge ciphertexts, we have a uniformly random key $\mathsf{K}_\rho(\ell^\star)$, which allows us to use the IND-CPA security of SE, and changes encryption of $\boldsymbol{x}_\rho^{j,0}$ as in $\mathsf{G}_{\rho-1}^\star$ into encryption of $\boldsymbol{x}_\rho^{j,1}$, as in $\mathsf{G}_\rho^\star$. Then we switch back the key $\mathsf{k}_{\rho,i^\star}$ from uniformly random to pseudo-random, using the security of the PRF once again. Summarizing, we have:

$$\Pr[\mathsf{H}_{\rho-1}^\star(\lambda, n, \mathcal{A}) = 1 | i^\star = j] - \Pr[\mathsf{H}_\rho^\star(\lambda, n, \mathcal{A}) = 1 | i^\star = j] = \mathsf{Adv}_{\mathsf{SE},\mathcal{B}_\rho}^{\text{IND-CPA}}(\lambda, n) + 2 \cdot \mathsf{Adv}_{\mathsf{PRF},\mathcal{B}'_\rho}(\lambda, n).$$

Summing up for all $\rho \in [n]$, we obtain the following for all $j \in [n]$:

$$\begin{aligned}&\left|\Pr[\mathsf{G}_0^\star(\lambda,n,\mathcal{A})=1|i^\star=j]-\Pr[\mathsf{G}_1^\star(\lambda,n,\mathcal{A})=1|i^\star=j]\right|\\&\leq n\cdot\mathsf{Adv}^{\text{IND-CPA}}_{\mathsf{SE},\mathcal{B}'}(\lambda,n)+2n\cdot\mathsf{Adv}_{\mathsf{PRF},\mathcal{B}''}(\lambda,n).\end{aligned}$$

Thus, we have:

$$\begin{aligned}&\left|\Pr[\mathsf{G}_0^\star(\lambda,n,\mathcal{A})=1]-\Pr[\mathsf{G}_1^\star(\lambda,n,\mathcal{A})=1]\right|\\&\leq\frac{1}{n+1}\mathsf{Adv}^{\text{adt-pos}^+\text{-IND}}_{\mathsf{DMCFE},\mathcal{B}}(\lambda,n)\\&\qquad+\frac{n^2}{n+1}\cdot\mathsf{Adv}^{\text{IND-CPA}}_{\mathsf{SE},\mathcal{B}'}(\lambda,n)+\frac{2n^2}{n+1}\cdot\mathsf{Adv}_{\mathsf{PRF},\mathcal{B}''}(\lambda,n)\ .\end{aligned}$$

Therefore, we obtain:

$$\begin{aligned}&\left|\Pr[\text{adt-yy-IND}_0^{\mathsf{DMCFE}'}(\lambda,n,\mathcal{A})=1]-\Pr[\text{adt-yy-IND}_1^{\mathsf{DMCFE}'}(\lambda,n,\mathcal{A})=1]\right|\\&\leq\mathsf{Adv}^{\text{adt-pos}^+\text{-IND}}_{\mathsf{DMCFE},\mathcal{B}}(\lambda,n)+n^2\cdot\mathsf{Adv}^{\text{IND-CPA}}_{\mathsf{SE},\mathcal{B}'}(\lambda,n)+2n^2\cdot\mathsf{Adv}_{\mathsf{PRF},\mathcal{B}''}(\lambda,n)\ .\end{aligned}$$

□

5 Decentralized Multi-Client Function Encryption

In this section, we modify the generic construction of Sect. 3 to make it decentralized. We cannot use directly the transformation from [2], because the master secret key msk may be arbitrary, and not necessarily the concatenation of the parties' secret keys sk_i (for $i \in [n]$), as required by [2]. Moreover, the functional decryption keys sk_f may not be computed just from sk_i. Instead, we additively secret share the master secret key of the underlying single-input FE. For key derivation to be possible in a decentralized way, we require an extra property on the single-input FE, that is fulfilled by most known constructions of single-input inner FE for inner products. This property is called special key derivation, and is very similar to special key derivation for MCFE defined in [2].

Definition 5.1 (FE with Special Key Derivation). *Let* $\mathsf{FE} = (\mathsf{Setup}, \mathsf{KeyGen}, \mathsf{KeyDer}, \mathsf{Enc}, \mathsf{Dec})$ *be a public-key FE scheme for the inner product functionality* $\mathcal{F}_\rho^{\mathsf{ip}}$*, where* $\rho = (\mathcal{R}, 1, n \cdot m, X, Y)$ *where* $\mathcal{R}$ *is either* $\mathbb{Z}$ *or* $\mathbb{Z}_L$ *for some integer* L*, and* n, m, X, Y *are positive integers.* FE *is said to have special key derivation modulo* M *if:*

- *The algorithm* $\mathsf{KeyGen}(\mathsf{pp})$ *generates a master secret key of the form* $\mathsf{msk} := \mathbf{U} \in \mathbb{Z}_M^{\kappa \times mn}$*, for some constant* κ *(which can depend on* pp*).*
- $\mathsf{sk}_{\boldsymbol{y}} \leftarrow \mathsf{KeyDer}(\mathsf{pp}, \mathsf{msk}, \boldsymbol{y})$ *outputs* $\mathsf{sk}_{\boldsymbol{y}} = (\boldsymbol{y}, \mathbf{U} \cdot \boldsymbol{y} \in \mathbb{Z}_M^\kappa)$*.*

For our security proof, we require M to be a prime number.

Instantiations. All the stateless[3] IPFE constructions in [7] satisfy the special key derivation property. More precisely, the DDH construction has special key derivation modulo p, the prime order of the used cyclic group, and $\kappa = 2$ (using notations from [7], the matrix $\mathbf{U}$ is defined by $U_{1,i} = s_i$ and $U_{2,i} = t_i$). The Paillier and LWE constructions have special key derivation modulo any large enough prime number M so that $\mathbf{U} \cdot \boldsymbol{y}$ is the same modulo M and over the integers with overwhelming probability over the generation of msk. For Paillier, $\kappa = 1$ and $U_{1,i} = s_i$, while for LWE, $\kappa = m$ and $\mathbf{U} = \mathbf{Z}$ (using notations from [7]).

Construction. The construction is provided in Fig. 8.

When instantiated with the DDH construction from [7], KeyGen can be decentralized non-interactively. Let G be the underlying cyclic group of order p and g and h be two generators of G. Each party i independently generates $\mathbf{U}_i \leftarrow \mathbb{Z}_p^{2\times mn}$ and $\mathsf{K}'_{i,j} \leftarrow \{0,1\}^\lambda$, computes

$$h_{k,i} := g^{U_{i,1,k}} \cdot h^{U_{i,2,k}} \qquad \text{for } k \in [mn] \ .$$

It then sends $(\{h_{k,i}\}_{k\in[mn]}, \mathsf{K}'_{i,j})$ to party j, for each $j \in [n]$. After receiving all the messages from the other parties, each party i computes and sets:

$$\begin{aligned} \mathsf{mpk}_{\mathsf{ipfe}} &:= \{h_k := \textstyle\prod_{i=1}^n h_{k,i}\}_{k\in[mn]} \ , \\ \mathsf{K}_{i,j} &:= \mathsf{K}_{j,i} := \mathsf{K}'_{i,j} \oplus \mathsf{K}'_{j,i} \qquad \text{for } j \in [n] \ , \\ \mathsf{sk}_i &:= (\mathsf{mpk}_{\mathsf{ipfe}}, \mathbf{U}_i, \{\mathsf{K}_{i,j}\}_{j\in[n]}) \ . \end{aligned}$$

When instantiated with the Paillier or DDH construction from [7], we do not know how to decentralize KeyGen this way. The issue is that in these constructions, $\mathbf{U}$ is not uniform in $\mathbb{Z}_M^{\kappa\times mn}$ but is sampled according to some Gaussian distribution.

Correctness. The only remaining part of correctness to be proven for the scheme in Fig. 8 is to show that the key computed by the algorithms $\mathsf{KeyDerShare}$ and $\mathsf{KeyDerComb}$ corresponds to the one that would have been computed by KeyDer. This follows from the following fact:

$$\mathsf{sk}_{\boldsymbol{y}} = \sum_{i=1}^n \mathsf{sk}_{i,\boldsymbol{y}} = \sum_{i=1}^n \mathbf{U}_i \cdot \boldsymbol{y} = \mathbf{U} \cdot \boldsymbol{y} \ .$$

Theorem 5.2 (adt-pos$^+$-IND-security). *If the FE scheme* $\mathsf{FE} = (\mathsf{Setup}_{\mathsf{ipfe}}, \mathsf{KeyGen}_{\mathsf{ipfe}}, \mathsf{KeyDer}_{\mathsf{ipfe}}, \mathsf{Enc}_{\mathsf{ipfe}}, \mathsf{Dec}_{\mathsf{ipfe}})$ *is an any-IND-secure FE scheme for the inner product functionality defined as* $\mathcal{F}^{\mathsf{ip}}_{\rho_{\mathsf{ipfe}}}$, $\rho_{\mathsf{ipfe}} = (\mathbb{Z}, 1, m, 2X, Y)$, *if* FE *has the special key derivation property modulo the prime number* M, *and if* PRF

[3] In this paper, our definitions do not allow for the encryption to be stateful.

$\underline{\mathsf{KeyGen}(\mathsf{pp}):}$

$(\mathsf{msk}_{\mathsf{ipfe}}, \mathsf{mpk}_{\mathsf{ipfe}}) \leftarrow \mathsf{KeyGen}^{\mathsf{ipfe}}(\mathsf{pp}_{\mathsf{ipfe}})$; $\mathsf{msk} := \mathsf{msk}_{\mathsf{ipfe}} := \mathbf{U} \in \mathbb{Z}_M^{\kappa \times mn}$

For $i \in [n]$, $j > i$: $\mathsf{K}_{i,j} = \mathsf{K}_{j,i} \leftarrow \{0,1\}^\lambda$

For $i \in [n-1]$: $\mathbf{U}_i \leftarrow \mathbb{Z}_M^{\kappa \times mn}$

$$\mathbf{U}_n := \mathbf{U} - \sum_{i=1}^{n-1} \mathbf{U}_i \in \mathbb{Z}_M^{\kappa \times mn}$$

Return $\{\mathsf{sk}_i = (\mathsf{mpk}_{\mathsf{ipfe}}, \mathbf{U}_i, \{\mathsf{K}_{i,j}\}_{j\in[n]})\}_{i\in[n]}$ and msk

$\underline{\mathsf{KeyDerShare}(\mathsf{pp}, \mathsf{sk}_i, \boldsymbol{y} \in \mathcal{R}^{mn}):}$

Return $\mathsf{sk}_{i,\boldsymbol{y}} := \mathbf{U}_i \cdot \boldsymbol{y} \in \mathbb{Z}_M^\kappa$

$\underline{\mathsf{KeyDerComb}(\mathsf{pp}, \mathsf{sk}_{1,\boldsymbol{y}}, \ldots, \mathsf{sk}_{mn,\boldsymbol{y}}):}$

$$\text{Return } \mathsf{sk}_{\boldsymbol{y}} := \sum_{i=1}^{n} \mathsf{sk}_{i,\boldsymbol{y}} \in \mathbb{Z}_M^\kappa$$

Fig. 8. Algorithms KeyGen, KeyDerShare and KeyDerComb making the inner-product MCFE from Fig. 4 a DMCFE, assuming that $\mathsf{FE} := (\mathsf{Setup}^{\mathsf{ipfe}}, \mathsf{Enc}^{\mathsf{ipfe}}, \mathsf{KeyDer}^{\mathsf{ipfe}}, \mathsf{Dec}^{\mathsf{ipfe}})$ has the special key derivation property modulo a prime number M.

is secure, then DMCFE *from Fig. 8 is* adt-pos$^+$*-IND-secure for the functionality defined as* $\mathcal{F}_\rho^{\mathsf{ip}}, \rho = (\mathbb{Z}, n, m, X, Y)$. *Namely, for any PPT adversary* $\mathcal{A}$, *there exist PPT adversaries* $\mathcal{B}$ *and* $\mathcal{B}'$ *such that:*

$$\mathsf{Adv}_{\mathsf{MCFE},\mathcal{A}}^{adt\text{-}pos^+\text{-}IND}(\lambda, n) \le 2n^2(n-1)q_{\mathsf{Enc}} \cdot \mathsf{Adv}_{\mathsf{PRF},\mathcal{B}}(\lambda) + 2q_{\mathsf{Enc}} \cdot \mathsf{Adv}_{\mathsf{FE},\mathcal{B}'}^{\text{any-}IND}(\lambda),$$

where q_{Enc} *denotes the number of distinct labels queried to* QLeftRight.

Proof. Let $\mathcal{A}$ be a PPT adversary against the security of MCFE. We proceed via a hybrid argument, using the games described in Fig. 9. Note that G_0 corresponds to the game adt-pos$^+$-$\text{IND}_0^{\mathsf{DMCFE}}(\lambda, n, \mathcal{A})$, and G_3 corresponds to the game adt-pos$^+$-$\text{IND}_1^{\mathsf{DMCFE}}(\lambda, n, \mathcal{A})$. Thus, we have: $\mathsf{Adv}_{\mathsf{DMCFE},\mathcal{A}}^{\text{adt-pos}^+\text{-IND}}(\lambda, n) = \left|\mathsf{Win}_{\mathcal{A}}^{\mathsf{G}_0}(\lambda, n) - \mathsf{Win}_{\mathcal{A}}^{\mathsf{G}_3}(\lambda, n)\right|$.

Game G_1. In game G_1, we change the way the oracles QCor and QKeyD answer: instead of using each individual share $\mathbf{U}_i$, they generate their answers on-the-fly to be consistent with previous answers and $\mathsf{KeyDer}^{\mathsf{ipfe}}(\mathsf{pp}_{\mathsf{ipfe}}, \mathsf{msk}_{\mathsf{ipfe}}, \boldsymbol{y})$ in the case of QKeyD. The transition from G_0 to G_1 is justified by linear algebra: the two games are perfectly indistinguishable. A formal proof can be derived from [2, Lemma A.2] (for $\kappa = 1$, the lemma applies directly, while for $\kappa \ge 2$, we just need to apply for each row of $\mathbf{U}$.

G_0, $\boxed{\mathsf{G}_3}$:

$(\{\mathsf{sk}_i\}_{i\in[n]}, \mathsf{msk}) \leftarrow \mathsf{KeyGen}(\mathsf{pp})$
$\alpha \leftarrow \mathcal{A}^{\mathsf{QCor}(\cdot)\mathsf{QEnc}(\cdot,\cdot,\cdot,\cdot),\mathsf{QKeyD}(\cdot)}(\mathsf{pp})$
Output: α if Condition (*) is satisfied, or a 0 otherwise.

$\mathsf{QCor}(i)$:
Return $\mathsf{sk}_i = (\mathbf{U}_i, \{\mathsf{K}_{i,j}\}_{j\in[n]})$

$\mathsf{QKeyD}(\boldsymbol{y})$:
For any $i \in [n]$, $\mathsf{sk}_{i,\boldsymbol{y}} := \mathbf{U}_i \cdot \boldsymbol{y} \in \mathbb{Z}_M^{\kappa}$
Return $\{\mathsf{sk}_{i,\boldsymbol{y}}\}_{i\in[n]}$

G_0, G_1, G_2, $\boxed{\mathsf{G}_3}$:

$\mathsf{QEnc}(i, \boldsymbol{x}_i^{j,0}, \boldsymbol{x}_i^{j,1}, \ell)$:
$\boldsymbol{x}_i^j := \boldsymbol{x}_i^{j,0}$
$\boxed{\boldsymbol{x}_i^j := \boldsymbol{x}_i^{j,1}}$
Return $\mathsf{Enc}(\mathsf{pp}, \mathsf{sk}_i, \boldsymbol{x}_i^j, \ell)$

G_1, $\boxed{\mathsf{G}_2}$:

$(\{\mathsf{sk}_i\}_{i\in[n]}, \mathsf{msk}) \leftarrow \mathsf{KeyGen}(\mathsf{pp})$ except $\mathbf{U}_i$ not generated
$\alpha \leftarrow \mathcal{A}^{\mathsf{QCor}(\cdot),\mathsf{QEnc}(\cdot,\cdot,\cdot,\cdot),\mathsf{QKeyD}(\cdot)}(\mathsf{pp})$
$S := \emptyset$
Output: α if Condition (*) is satisfied, or 0 otherwise.

$\mathsf{QCor}(i)$:
Pick $\mathbf{U}_i$ uniformly under the constraint $\forall \boldsymbol{y} \in S$, $\mathsf{sk}_{i,\boldsymbol{y}} = \mathbf{U}_i \cdot \boldsymbol{y}$
Return $\mathsf{sk}_i := (\mathbf{U}_i, \{\mathsf{K}_{i,j}\}_{j\in[n]})$

$\mathsf{QKeyD}(\boldsymbol{y})$:
$\mathsf{sk}_{\boldsymbol{y}} := \mathbf{U} \cdot \boldsymbol{y} = \mathsf{KeyDer}^{\mathsf{ipfe}}(\mathsf{pp}_{\mathsf{ipfe}}, \mathsf{msk}_{\mathsf{ipfe}}, \boldsymbol{y})$
For any $i \in \mathcal{CS}$, $\mathsf{sk}_{i,\boldsymbol{y}} := \mathbf{U}_i \cdot \boldsymbol{y} \in \mathbb{Z}_M^{\kappa}$
If $\boldsymbol{y} \in \mathrm{Vect}(S)$,
  Pick $\{\mathsf{sk}_{i,\boldsymbol{y}}\}_{i\notin\mathcal{CS}}$ uniformly under the constraint $\sum_{i\in[n]} \mathsf{sk}_{i,\boldsymbol{y}} = \mathsf{sk}_{\boldsymbol{y}}$
  Add $\boldsymbol{y}$ to the set S
If $\boldsymbol{y} \notin \mathrm{Vect}(S)$
  Find $\{\mu_{\boldsymbol{y}'}\}_{\boldsymbol{y}'\in S}$ s.t. $\boldsymbol{y} = \sum_{\boldsymbol{y}'\in S} \mu_{\boldsymbol{y}'} \cdot \boldsymbol{y}'$
  Set $\mathsf{sk}_{i,\boldsymbol{y}} := \sum_{\boldsymbol{y}'\in S} \mu_{\boldsymbol{y}'} \cdot \mathsf{sk}_{i,\boldsymbol{y}'}$
Return $\{\mathsf{sk}_{i,\boldsymbol{y}}\}_{i\in[n]}$

Fig. 9. Games for the proof of Theorem 5.2. Condition (*) is given in Definition 2.1.

Game G_2**.** In game G_2, the challenge ciphertexts encrypts $\boldsymbol{x}_i^{j,1}$ instead of $\boldsymbol{x}_i^{j,0}$. The transition from G_1 to G_2 is justified by the adt-pos$^+$-IND security of MCFE proven in Theorem 3.7.

Game G_3**.** In game G_3, we change back the way the oracles QCor and QKeyD answer to match adt-pos$^+$-$\mathrm{IND}_1^{\mathsf{DMCFE}}(\lambda, n, \mathcal{A})$. The transition from G_2 to G_3 is similar to the one from G_1 to G_0: G_3 and G_2 are perfectly indistinguishable.

Putting everything together, we obtain the theorem. □

Acknowledgments. This work was supported in part by the European Union's Horizon 2020 Research and Innovation Programme under grant agreement 780108 (FENTEC), by the ERC Project aSCEND (H2020 639554), by the French *Programme d'Investissement d'Avenir* under national project RISQ P141580, and by the French FUI project ANBLIC. The third author was partially supported by a Google PhD Fellowship in Privacy and Security. Part of this work was done while the second author was at IBM Research, Yorktown Heights, USA, and the third author was at École normale supérieure, Paris, France.

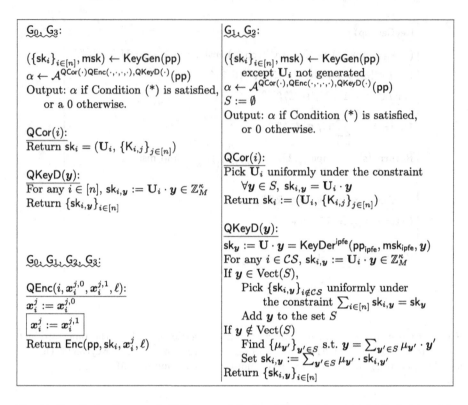

References

1. Abdalla, M., Benhamouda, F., Gay, R.: From single-input to multi-client inner-product functional encryption. Cryptology ePrint Archive, Report 2019/487 (2019). https://eprint.iacr.org/2019/487
2. Abdalla, M., Benhamouda, F., Kohlweiss, M., Waldner, H.: Decentralizing inner-product functional encryption. In: Lin, D., Sako, K. (eds.) PKC 2019, Part II. LNCS, vol. 11443, pp. 128–157. Springer, Cham (2019). https://doi.org/10.1007/978-3-030-17259-6_5
3. Abdalla, M., Bourse, F., De Caro, A., Pointcheval, D.: Simple functional encryption schemes for inner products. In: Katz, J. (ed.) PKC 2015. LNCS, vol. 9020, pp. 733–751. Springer, Heidelberg (2015). https://doi.org/10.1007/978-3-662-46447-2_33
4. Abdalla, M., Catalano, D., Fiore, D., Gay, R., Ursu, B.: Multi-input functional encryption for inner products: function-hiding realizations and constructions without pairings. In: Shacham, H., Boldyreva, A. (eds.) CRYPTO 2018, Part I. LNCS, vol. 10991, pp. 597–627. Springer, Cham (2018). https://doi.org/10.1007/978-3-319-96884-1_20
5. Abdalla, M., Gay, R., Raykova, M., Wee, H.: Multi-input inner-product functional encryption from pairings. In: Coron, J.-S., Nielsen, J.B. (eds.) EUROCRYPT 2017, Part I. LNCS, vol. 10210, pp. 601–626. Springer, Cham (2017). https://doi.org/10.1007/978-3-319-56620-7_21
6. Agrawal, S., Clear, M., Frieder, O., Garg, S., O'Neill, A., Thaler, J.: Ad hoc multi-input functional encryption. Cryptology ePrint Archive, Report 2019/356 (2019). https://eprint.iacr.org/2019/356
7. Agrawal, S., Libert, B., Stehlé, D.: Fully secure functional encryption for inner products, from standard assumptions. In: Robshaw, M., Katz, J. (eds.) CRYPTO 2016, Part III. LNCS, vol. 9816, pp. 333–362. Springer, Heidelberg (2016). https://doi.org/10.1007/978-3-662-53015-3_12
8. Ananth, P., Jain, A.: Indistinguishability obfuscation from compact functional encryption. In: Gennaro, R., Robshaw, M. (eds.) CRYPTO 2015, Part I. LNCS, vol. 9215, pp. 308–326. Springer, Heidelberg (2015). https://doi.org/10.1007/978-3-662-47989-6_15
9. Badrinarayanan, S., Gupta, D., Jain, A., Sahai, A.: Multi-input functional encryption for unbounded arity functions. In: Iwata, T., Cheon, J.H. (eds.) ASIACRYPT 2015, Part I. LNCS, vol. 9452, pp. 27–51. Springer, Heidelberg (2015). https://doi.org/10.1007/978-3-662-48797-6_2
10. Bishop, A., Jain, A., Kowalczyk, L.: Function-hiding inner product encryption. In: Iwata, T., Cheon, J.H. (eds.) ASIACRYPT 2015, Part I. LNCS, vol. 9452, pp. 470–491. Springer, Heidelberg (2015). https://doi.org/10.1007/978-3-662-48797-6_20
11. Boneh, D., Sahai, A., Waters, B.: Functional encryption: definitions and challenges. In: Ishai, Y. (ed.) TCC 2011. LNCS, vol. 6597, pp. 253–273. Springer, Heidelberg (2011). https://doi.org/10.1007/978-3-642-19571-6_16
12. Brakerski, Z., Komargodski, I., Segev, G.: Multi-input functional encryption in the private-key setting: stronger security from weaker assumptions. J. Cryptol. **31**(2), 434–520 (2018). https://doi.org/10.1007/s00145-017-9261-0
13. Chotard, J., Dufour Sans, E., Gay, R., Phan, D.H., Pointcheval, D.: Decentralized multi-client functional encryption for inner product. In: Peyrin, T., Galbraith, S. (eds.) ASIACRYPT 2018, Part II. LNCS, vol. 11273, pp. 703–732. Springer, Cham (2018). https://doi.org/10.1007/978-3-030-03329-3_24

14. Chotard, J., Dufour Sans, E., Gay, R., Phan, D.H., Pointcheval, D.: Multi-client functional encryption with repetition for inner product. Cryptology ePrint Archive, Report 2018/1021 (2018). http://eprint.iacr.org/2018/1021
15. Datta, P., Okamoto, T., Tomida, J.: Full-hiding (unbounded) multi-input inner product functional encryption from the k-linear assumption. In: Abdalla, M., Dahab, R. (eds.) PKC 2018, Part II. LNCS, vol. 10770, pp. 245–277. Springer, Cham (2018). https://doi.org/10.1007/978-3-319-76581-5_9
16. Goldwasser, S., et al.: Multi-input functional encryption. In: Nguyen, P.Q., Oswald, E. (eds.) EUROCRYPT 2014. LNCS, vol. 8441, pp. 578–602. Springer, Heidelberg (2014). https://doi.org/10.1007/978-3-642-55220-5_32
17. Kursawe, K., Danezis, G., Kohlweiss, M.: Privacy-friendly aggregation for the smart-grid. In: Fischer-Hübner, S., Hopper, N. (eds.) PETS 2011. LNCS, vol. 6794, pp. 175–191. Springer, Heidelberg (2011). https://doi.org/10.1007/978-3-642-22263-4_10
18. O'Neill, A.: Definitional issues in functional encryption. Cryptology ePrint Archive, Report 2010/556 (2010). http://eprint.iacr.org/2010/556

Public Key Encryption (2)

Rate-1 Trapdoor Functions from the Diffie-Hellman Problem

Nico Döttling[1(✉)], Sanjam Garg[2], Mohammad Hajiabadi[2], Kevin Liu[2], and Giulio Malavolta[3]

[1] CISPA Helmholtz Center for Information Security, Saarbrücken, Germany
nico.doettling@gmail.com
[2] University of California, Berkeley, USA
[3] Simons Institute for the Theory of Computing, Berkeley, USA

Abstract. Trapdoor functions (TDFs) are one of the fundamental building blocks in cryptography. Studying the underlying assumptions and the efficiency of the resulting instantiations is therefore of both theoretical and practical interest. In this work we improve the input-to-image rate of TDFs based on the Diffie-Hellman problem. Specifically, we present:

(a) A rate-1 TDF from the computational Diffie-Hellman (CDH) assumption, improving the result of Garg, Gay, and Hajiabadi [EUROCRYPT 2019], which achieved linear-size outputs but with large constants. Our techniques combine non-binary alphabets and high-rate error-correcting codes over large fields.

(b) A rate-1 deterministic public-key encryption satisfying block-source security from the decisional Diffie-Hellman (DDH) assumption. While this question was recently settled by Döttling et al. [CRYPTO 2019], our scheme is conceptually simpler and *concretely more efficient*. We demonstrate this fact by implementing our construction.

1 Introduction

Trapdoor functions (TDFs) are the public-key variant of the notion of one-way functions. Informally, TDFs are (families of) one-to-one functions, where each function can be computed in the forward direction using the *index key*,

S. Garg—Supported in part from DARPA/ARL SAFEWARE Award W911NF15C0210, AFOSR Award FA9550-15-1-0274, AFOSR Award FA9550-19-1-0200, AFOSR YIP Award, NSF CNS Award 1936826, DARPA and SPAWAR under contract N66001-15-C-4065, a Hellman Award and research grants by the Okawa Foundation, Visa Inc., and Center for Long-Term Cybersecurity (CLTC, UC Berkeley). The views expressed are those of the author and do not reflect the official policy or position of the funding agencies.
G. Malavolta—"Part of this work was done while the author was at Carnegie Mellon University.

S. D. Galbraith and S. Moriai (Eds.): ASIACRYPT 2019, LNCS 11923, pp. 585–606, 2019.
https://doi.org/10.1007/978-3-030-34618-8_20

and in the backward direction using a corresponding *trapdoor key.* Moreover, without knowledge of a trapdoor, a randomly chosen function should be one-way. Trapdoor functions, or extensions thereof such as lossy TDFs or deterministic public-key encryption, have important applications in the construction of primitives with CCA security, selective-opening security, and more recently in the context of designated-verifier non-interactive zero knowledge [BFOR08, BHY09, BBN+09, MY10, BCPT13, LQR+19].

A series of works, some quite recent, have shown how to build TDFs and related primitives based on almost any specific assumptions from which public-key encryption (PKE) is known [PW08, FGK+10, PW11, Wee12, GH18, GGH19]. However, all these constructions are less efficient than those of PKE from the corresponding assumptions, in particular with respect to the sizes of public-keys and ciphertexts. For instance, we have constructions of PKE for which ciphertext expansion factors are small constants, sometimes even approaching 1. Yet, the situation for TDFs is different: All TDFs either have quadratic ciphertext expansions or linear expansions with large constants.

In this work we build TDFs and deterministic-encryption schemes with rates approaching 1 based on standard assumptions in cyclic groups, specifically the Computational Diffie Hellman (CDH) and Decisional Diffie Hellman (DDH) assumptions. Concretely, for an image y of an input x, the ratio $|\mathsf{x}|/|\mathsf{y}|$ approaches 1 as $|\mathsf{x}|$ grows. The first TDF constructions based on DDH [PW08, FGK+10] resulted in schemes in which the size of the image is quadratic in the input size. In a nutshell, the optimized TDF construction of Peikert and Waters, due to [FGK+10], computes a linear function *in the exponent* on a binary encoding of the input. In particular, recall that in a group with a generator g, if we have an encoding $[\mathbf{M}] = g^{\mathbf{M}}$ of an invertible matrix $\mathbf{M}$ of exponents, then we can encode any column vector X of bits by computing $\mathbf{M} \cdot \mathsf{X}$ in the exponent. This will allow for inversion if one possesses $\mathbf{M}^{-1}$. We can argue lossiness in a very elegant way by making the matrix $\mathbf{M}$ rank-deficient. On the downside however, we need to spend an entire group element in the output for each input bit, resulting in an expansion factor of $\Omega(\lambda)$.

A recent result of Garg, Gay and Hajiabadi [GGH19] shows how to construct linearly-expanding TDFs and DE schemes based on CDH or DDH. In particular, they give schemes in which the image expansion ratio is $O(1)$. However, this linear expansion hides big constants—a rough estimate of the constant is at least 20. At a high level, the constructions of [GGH19] achieve linear-expansion rates via the following two steps:

(a) For some constant c, first build a so-called local TDF, in which the inversion algorithm for every coordinate of the input either manages to recover the underlying bit correctly or outputs $\perp$, the latter happening with probability at most $1/2^c$.
(b) Boost correctness of local TDFs by applying erasure-correcting codes.

Their local TDFs from step (a) already incur an expansion factor of at least $2c$. Also, since erasure corrections for strings over $\mathbb{F}_2$ can tolerate only relatively small erasure rates (i.e., the ratio between the maximum number of tolerated erasures and the total length), they have to choose the constant c bit enough in Step (a)—at least 10.

The problem of rate-1 (lossy) TDFs from DDH was recently resolved in the work of Döttling et al. [DGI+19], who presented a construction based on the interplay of [GGH19] and techniques developed in the context of homomorphic secret sharing [BGI16]. Their approach however results in large index keys and does not appear to extend to the more challenging CDH settings.

Our Results. In this work, we show how to build rate-1 TDFs based on CDH or DDH, satisfying stronger properties such as block-source deterministic-encryption security in the sense of [BBO07,BFO08,BFOR08].[1] This notion of security requires that the deterministic encryptions of any two distributions each having high min-entropy (more than a threshold k) should be computationally indistinguishable. Ideally, we want $k << n$, where n is the bit length of the input.

At a high-level, our CDH-based construction deviates from the paradigm of [GGH19] by parsing the input into elements from a poly-sized field $\mathbb{F}$ (i.e., $|\mathbb{F}| = \mathsf{poly}(\lambda)$). Then for every block $\mathsf{B}_i \in \mathbb{F}$ of the input, we provide a corresponding "hinting" block O_i in the output of almost equal size. We then show how to perform inversion in a way which allows us to recover all except a $1/\mathsf{poly}_1(\lambda)$ fraction of the input blocks, for some polynomial poly_1. By choosing an appropriate error-correcting code over $\mathbb{F}$ and by choosing poly_1 appropriately based on $|\mathbb{F}|$, we are able to achieve rate 1. The main technical novelty of our work lies in providing the hints in a succinct way. See Sect. 1.1 for more details.

Under the DDH assumptions, we give a more direct rate-1 construction without the need of relying on error-correction techniques. For an input $\mathsf{x} \in \{0,1\}^n$, the output of the TDF contains only one group element plus exactly n bits. The construction has perfect correctness (i.e., can be inverted with probability 1), is conceptually simple, and is *concretely efficient.* We show this by providing a proof-concept-implementation in Python. Our implementation confirms our expectation of having short ciphertexts and relatively fast encryption/decryption times. Both encryptions and decryption times take less than a second on inputs of 128 Bytes (1024 bits).

Comparison with [DGI+19]**.** The work of [DGI+19] also shows how to build lossy TDFs (and deterministic encryption) based on DDH achieving rate 1 as in our construction. However, our construction achieves shorter public keys, saving an additive factor of at least $3n^2$ group elements, and is much simpler. In particular, the construction of [DGI+19] relies on non-trivial techniques such as

[1] We mention that building rate-1 TDFs satisfying one-wayness alone is trivial. If a TDF TDF maps n-bit inputs to n^c-bit outputs, then define a second TDF whose input is of the form $(\mathsf{x} \in \{0,1\}^n, \mathsf{x}' \in \{0,1\}^{n^{c+1}})$, and the output is $(\mathsf{TDF}(\mathsf{x}), \mathsf{x}')$. While this trivial construction achieves rate-1, it destroys stronger properties such as deterministic-encryption security.

those developed in the context of homomorphic secret sharing [BGI16] as well as error-correcting code type techniques. We rely on neither of these tools.

Open Problems. Our rate-1 primitives only provide CPA security. It would be interesting to see if techniques from [GGH19], along with those developed in this work, yield a rate-1 CCA primitive. One challenge is that in [GGH19] the (constant) multiplicative overhead of ciphertexts in the CCA case is much larger than the CPA case. In particular, our current techniques do not appear to naturally yield a rate-1 CCA primitive. We leave this as an open problem.

1.1 Technical Overview

In the following we provide an informal overview of the techniques developed in this work. We first discuss how to construct a CDH-based trapdoor function with rate 1, then we turn our attention to the DDH-based settings.

The Basic Building Block. The starting point of this work is the following group-based hash function, which maps $\{0,1\}^n$ into a group $\mathbb{G}$

$$\mathsf{Hash}(\mathsf{k},\mathsf{x}) := \prod_{j=1}^{n} g_{j,\mathsf{x}_j}$$

where the key

$$\mathsf{k} := \begin{pmatrix} g_{1,1}, g_{2,1}, \dots, g_{n,1} \\ g_{1,2}, g_{2,2}, \dots, g_{n,2} \end{pmatrix} \xleftarrow{\$} \mathbb{G}^{2\times n}.$$

is chosen uniformly at random and $\mathsf{x} \in \{0,1\}^n$ is the input. By choosing n larger than the representation size of a group element in $\mathbb{G}$, this function becomes compressing. This surprisingly powerful function plays a central role in recent constructions of identity based encryption [DG17b], trapdoor functions [GH18], deterministic encryption and lossy trapdoor functions [GGH19].

In a first step, we increase the alphabet size of the input x, i.e instead of taking x from $\{0,1\}^n$, we take it from Σ^n for an alphabet $\Sigma := \{1,\dots,\sigma\}$ of size $\sigma = \mathsf{poly}(\lambda)$. While the definition of the function Hash is unchanged, we need to account for the increased alphabet size by sampling the key as

$$\mathsf{k} := \begin{pmatrix} g_{1,1}, g_{2,1}, \dots, g_{n,1} \\ g_{1,2}, g_{2,2}, \dots, g_{n,2} \\ \dots, \dots, \dots, \dots \\ g_{1,\sigma}, g_{2,\sigma}, \dots, g_{n,\sigma} \end{pmatrix} \xleftarrow{\$} \mathbb{G}^{\sigma\times n}.$$

The main effect of this modification for now is that the size of the key is increased by a σ factor. While this modification seems insignificant at first, it will be instrumental in achieving rate 1.

Adding the Encryption. We now show how this function can be augmented with the encryption functionality, using techniques of [GGH19]. Let $\mathsf{y} := \mathsf{Hash}(\mathsf{k}, \mathsf{x})$, for a certain input $\mathsf{x} \in \Sigma^n$, our objective is to design an encryption algorithm such that a ciphertext encrypted under an index $i \in [n]$, a symbol $f \in \Sigma$ and y, can be decrypted with the knowledge of x only if $\mathsf{x}_i = f$. This is done by sampling a uniform $\rho \xleftarrow{\$} \mathbb{Z}_p$ and publishing

$$\mathsf{ct}_{i,f} := \begin{pmatrix} g_{1,1}^{\rho}, & g_{2,1}^{\rho}, & \ldots, & \bot, & \ldots, & g_{n,1}^{\rho} \\ \ldots, & \ldots, & \ldots, & \ldots, & \ldots, & \ldots \\ g_{1,f}^{\rho}, & g_{2,f}^{\rho}, & \ldots, & g_{i,f}^{\rho}, & \ldots, & g_{n,f}^{\rho} \\ \ldots, & \ldots, & \ldots, & \ldots, & \ldots, & \ldots \\ g_{1,\sigma}^{\rho}, & g_{2,\sigma}^{\rho}, & \ldots, & \bot, & \ldots, & g_{n,\sigma}^{\rho} \end{pmatrix},$$

as the ciphertext, and letting y^{ρ} be the underlying (secret) encapsulated value. Given x, anyone can recover y^{ρ} by simply computing

$$\mathsf{y}^{\rho} := \prod_{j=1}^{m} g_{j,\mathsf{x}_j}^{\rho}.$$

It is not hard to show that recovering the y^{ρ} if $\mathsf{x}_j \neq f$ is as hard as solving the Diffie-Hellman problem.

Constructing Trapdoor Functions. The key observation of [GH18] (later improved in [GGH19]) is that the same value can be recovered from y using the trapdoor ρ, without the knowledge of x. This allows us to use the above structure to construct a trapdoor function by sampling the trapdoor as a matrix

$$\mathsf{tk} := \begin{pmatrix} \rho_{1,1}, & \rho_{2,1}, & \ldots, & \rho_{n,1} \\ \rho_{1,2}, & \rho_{2,2}, & \ldots, & \rho_{n,2} \\ \ldots, & \ldots, & \ldots, & \ldots \\ \rho_{1,\sigma}, & \rho_{2,\sigma}, & \ldots, & \rho_{n,\sigma} \end{pmatrix} \xleftarrow{\$} \mathbb{Z}_p^{\sigma \times n}$$

and setting the index key as

$$\mathsf{ik} := \mathsf{k}, \begin{pmatrix} \mathsf{ct}_{1,1}, & \mathsf{ct}_{2,1}, & \ldots, & \mathsf{ct}_{n,1} \\ \mathsf{ct}_{1,2}, & \mathsf{ct}_{2,2}, & \ldots, & \mathsf{ct}_{n,2} \\ \ldots, & \ldots, & \ldots, & \ldots \\ \mathsf{ct}_{1,\sigma}, & \mathsf{ct}_{2,\sigma}, & \ldots, & \mathsf{ct}_{n,\sigma} \end{pmatrix}, \begin{pmatrix} a_{1,1}, & a_{2,1}, & \ldots, & a_{n,1} \\ a_{1,2}, & a_{2,2}, & \ldots, & a_{n,2} \\ \ldots, & \ldots, & \ldots, & \ldots \\ a_{1,\sigma}, & a_{2,\sigma}, & \ldots, & a_{n,\sigma} \end{pmatrix}$$

where k and $\mathsf{ct}_{i,f}$ are defined as above and each $a_{i,f} \xleftarrow{\$} \mathbb{G}$ is a random group elements. The purpose of these random elements is to shift the (negligible) inversion error from the random choice of x to the random choice of ik (see [GGH19] for a detailed discussion). Given an input x, the output of the trapdoor function is defined to be

$$\mathsf{u} := (\mathsf{y}, v_1 := \mathsf{y}^{\rho_{1,\mathsf{x}_1}} \oplus a_{1,\mathsf{x}_1}, \ldots, v_n := \mathsf{y}^{\rho_{n,\mathsf{x}_n}} \oplus a_{n,\mathsf{x}_n}).$$

Note that, as discussed before, this computation can be performed without the trapdoor tk. On the other hand, the function can be easily inverted with the knowledge of tk (and without x) by simply recomputing each $\mathsf{y}^{\rho_{i,f}} \oplus a_{i,f}$ and comparing it with v_i. If it matches, then the i-th symbol is set to f. While this gives us a trapdoor function, its rate is far from 1: To encode one symbol $\mathsf{x}_i \in \Sigma$, we need to spend one group element $v_i \in \mathbb{G}$.

Boosting the Rate. However, we can improve the rate of this construction with a surprisingly simple idea. Namely, we will use a hardcore function (in the sense of [GL89]) H to hash the element $\mathsf{y}^{\rho_{i,\mathsf{x}_i}}$ into a polynomial-size domain $\{0,1\}^w$ and sample $a_{i,f}$ from the same domain. Image values of the function now look as follows

$$\mathsf{u} := (\mathsf{y}, v_1 := \mathsf{H}(\mathsf{y}^{\rho_{1,\mathsf{x}_1}}) \oplus a_{1,\mathsf{x}_1}, \ldots, v_n := \mathsf{H}(\mathsf{y}^{\rho_{n,\mathsf{x}_n}}) \oplus a_{n,\mathsf{x}_n}).$$

Inversion is done as before: Given y and the trapdoor key tk one can check whether $v_i \stackrel{?}{=} \mathsf{H}(\mathsf{y}^{\rho_{i,f}}) \oplus a_{i,f}$ for all possible $f \in \Sigma$. If one finds a unique $f \in \Sigma$ with this property, then it must hold that $\mathsf{x}_i = f$. However, as $\{0,1\}^w$ is a domain of polynomial size, collisions can and will occur. That is, there can occur *false positives* $f' \neq \mathsf{x}_i$ which satisfy the above condition. Given that such collisions are not too frequent, we can protect against them by pre-processing x with a suitable code which also has high rate. Our analysis shows that the number of indices i at which such collisions occur is at most $2n \cdot \sigma/2^w$, where $\sigma = |\Sigma|$ is the size of the alphabet.

Achieving Rate 1. The crucial observation now is that we can choose Σ and $\{0,1\}^w$ in such a way that $\sigma/2^w$ is sublinear, but at the same time $\log(\sigma)/w$ approaches 1. Note that $\log(\sigma)/w$ is the rate at which we encode a symbol $\mathsf{x}_i \in \Sigma$ by $\mathsf{H}(\mathsf{y}^{\rho_{i,\mathsf{x}_i}}) \oplus a_{i,\mathsf{x}_i}$. This is e.g. achieved by choosing $\sigma \geq \lambda$ and $2^w = \sigma \cdot \log(\lambda)$. This choice gives us $\sigma/2^w \leq 1/\log(\lambda)$ and

$$\frac{\log(\sigma)}{w} = \frac{\log(\sigma)}{\log(\sigma) + \log\log(\lambda)} = 1 - \frac{\log\log(\lambda)}{\log(\sigma) + \log\log(\lambda)} \geq 1 - \frac{\log\log(\lambda)}{\log(\lambda)},$$

which approaches 1. Finally, we can pre-process the input x with a code which can handle a $2 \cdot \sigma/2^w = 2/\log(\lambda)$ fraction of erasures, such as a $[n, n - 2n/\log(\lambda), 2/\log(\lambda) + 1]$ Reed Solomon code over a field Σ of size $\sigma \geq n$. This code has rate $1 - 2/\log(\lambda)$. Concluding, the image (ignoring the group element y which causes only an additive overhead) encodes a message x at rate

$$\left(1 - \frac{2}{\log(\lambda)}\right) \cdot \left(1 - \frac{\log\log(\lambda)}{\log(\lambda)}\right) \geq 1 - \Omega\left(\frac{\log\log(\lambda)}{\log(\lambda)}\right).$$

The last question to address is how to instantiate H to extract enough randomness from a CDH instance. By our choice of parameters above, $w = O(\log(\lambda))$ random bits suffice, which allows us to use the standard Goldreich-Levin [GL89] hardcore function.

DDH-Based Deterministic Encryption (DE). Recall that we say a TDF with input space $\{0,1\}^n$ has (k,n)-CPA-security if the evaluations of any two distributions with min-entropy at least k result in computationally-indistinguishable distributions. We show how to realize this notion in a very simple and ciphertext-compact way using DDH.

The index key of our TDF consists of a random vector $\mathbf{g} \in \mathbb{G}^n$ together with n vectors $\{\mathbf{g}_i \in \mathbb{G}^n\}_{i\in[n]}$, where each $\mathbf{g}_i$ is an element-wise exponentiation of $\mathbf{g}$ to a random power ρ_i. To evaluate an input $\mathsf{x} \in \{0,1\}^n$, we return a group element $g' := \mathsf{x} \cdot \mathbf{g}$ (where $\cdot$ denotes the hash $\prod_{j=1}^{n} g_j^{\mathsf{x}_j}$), as well as an encoded bit $b_i := \mathsf{BL}(\mathsf{x} \cdot \mathbf{g}_i) \oplus \mathsf{x}_i$ for the i-th bit of the input. Here $\mathsf{BL}\colon \mathbb{G} \to \{0,1\}$ is a balanced function, meaning that the output of $\mathsf{BL}(g_u)$ on a uniformly-random g_u is a uniformly-random bit. Inversion can be performed by knowing all the exponents ρ_i's.

We show if $k \geq \log p + \omega(\log \lambda)$—where p is the size of the group—then we have (k,n)-CPA security. To argue this, first recall that an index key is of the form $(\mathbf{g}, \mathbf{g}_1, \ldots, \mathbf{g}_n)$, where each $\mathbf{g}_i$ is an exponentiation of $\mathbf{g}$. Say two x and x' are siblings if $\mathsf{x} \cdot \mathbf{g} = \mathsf{x}' \cdot \mathbf{g}$. (That is, if both result in the same group element in the output.) We show that for any $\mathsf{x} \in \{0,1\}^n$, one may sample the $\mathbf{g}_i$ components of the index key in a manner correlated with x to get a correlated ik^* in such a way that:

1. $(\mathsf{x}, \mathsf{ik}^*) \stackrel{c}{\equiv} (\mathsf{x}, \mathsf{ik})$, where ik is a real index key; and
2. ik^* will lose information w.r.t. all siblings of x. That is, if x' is a sibling of x, then $\mathsf{TDF.F}(\mathsf{ik}^*, \mathsf{x}) = \mathsf{TDF.F}(\mathsf{ik}^*, \mathsf{x}')$.
3. The joint distribution $(\mathsf{ik}^*, \mathsf{TDF.F}(\mathsf{ik}^*, \mathsf{x}))$ can be formed just by knowing $g' := \mathsf{x} \cdot \mathbf{g}$, and especially without knowing x.

Let us first sketch why the above properties imply DE security. Let $\mathcal{D}_0$ and $\mathcal{D}_1$ be the underlying high-entropy distributions. Let $\mathsf{x}_b \stackrel{\$}{\leftarrow} \mathcal{D}_b$, $\mathbf{g} \stackrel{\$}{\leftarrow} \mathbb{G}^n$ and $g'_b = \mathsf{x}_b \cdot \mathbf{g}$. Also, let ik_b^* be the corresponding correlated index key which by Item 3 can be formed just by knowing g'_b. By Item 1 we have $(\mathsf{ik}, \mathsf{TDF.F}(\mathsf{ik}, \mathsf{x}_b)) \stackrel{c}{\equiv} (\mathsf{ik}_b^*, \mathsf{TDF.F}(\mathsf{ik}_b^*, \mathsf{x}_b))$. Now since by Item 3 the joint distribution $(\mathsf{ik}_b^*, \mathsf{TDF.F}(\mathsf{ik}_b^*, \mathsf{x}_b))$ can be sampled just by knowing g'_b and since we have $g'_0 \stackrel{s}{\equiv} g'_1$ (by the leftover hash lemma), we have $(\mathsf{ik}_0^*, \mathsf{TDF.F}(\mathsf{ik}_0^*, \mathsf{x}_0)) \stackrel{s}{\equiv} (\mathsf{ik}_1^*, \mathsf{TDF.F}(\mathsf{ik}_1^*, \mathsf{x}_1))$, establishing the desired security.

Now let us explain how to sample such "lossy" index key ik^* for x just by knowing $g_c = \mathsf{x} \cdot \mathbf{g}$. We form $\mathsf{ik}^* := (\mathbf{g}, \mathbf{g}_1^*, \ldots, \mathbf{g}_n^*)$, where each $\mathbf{g}_i^*$ is formed exactly as in $\mathbf{g}_i$, except that we multiply the i-th element of the vector $\mathbf{g}_i := \mathbf{g}^{\rho_i}$ with a random group element g'_i which satisfies $\mathsf{BL}(g_c^{\rho_i}) = 1 \oplus \mathsf{BL}(g_c^{\rho_i} \cdot g'_i)$. Namely, each $\mathbf{g}_i^*$ is an "almost" exponentiation of $\mathbf{g}$ in that we tamper with the i-th element of the resulting exponentiated vector.

Using simple inspection we can verify that Property 2 follows by the particular way in which ik^* is sampled. Also, the way ik^* is defined allows us to sample the joint distribution $(\mathsf{ik}^*, \mathsf{TDF.F}(\mathsf{ik}^*, \mathsf{x}))$ just by knowing g_c and ρ_i's, establishing Property 3. Finally, via a sequence of hybrids, we show how to establish Property 1 based on DDH.

2 Preliminaries

We denote the security parameter by λ. We use $\stackrel{c}{\equiv}$ to denote computational indistinguishability between two distributions and use $\equiv$ to denote two distributions are identical. We write $\stackrel{s}{\equiv}$ for statistical indistinguishability and we write $\approx_\varepsilon$ to denote that two distributions are statistically close, within statistical distance ε. For a distribution $\mathcal{S}$ we use $x \stackrel{\$}{\leftarrow} \mathcal{S}$ to mean x is sampled according to $\mathcal{S}$ and use $y \in \mathcal{S}$ to mean $y \in \sup(\mathcal{S})$, where sup denotes the support of a distribution. For a set S we overload the notation to use $x \stackrel{\$}{\leftarrow} \mathsf{S}$ to indicate that x is chosen uniformly at random from S. The set $\{1, \ldots, n\}$ is often abbreviated as $[n]$. We say that a machine is PPT if it runs in probabilistic polynomial-time.

The min-entropy of a distribution $\mathcal{S}$ is $\mathsf{H}_\infty(\mathcal{S}) \triangleq -\log(\max_{\mathsf{x}} \Pr[\mathcal{S} = \mathsf{x}])$. For a finite alphabet Σ, we say a distribution $\mathcal{S}$ is a *k-source* over Σ^n if $\mathsf{H}_\infty(\mathcal{S}) \geq k$. When the alphabet Σ is clear from context, we say $\mathcal{S}$ is a (k, n)-source.

Lemma 1 (Chernoff Inequality). *Let X be binomially distributed with parameters $n \in \mathbb{N}$ and $p \in [0,1]$. Let $p' > p$. Then*

$$\Pr[X > 2p'n] < e^{-p'n/3}.$$

Lemma 2 (Leftover Hash Lemma [ILL89]). *Let $\mathcal{X}$ be a random variable over X and $h : \mathsf{S} \times \mathsf{X} \to \mathsf{Y}$ be a 2-universal hash function, where $|\mathsf{Y}| \leq 2^m$ for some $m > 0$. If $m \leq \mathsf{H}_\infty(\mathcal{X}) - 2\log\left(\frac{1}{\varepsilon}\right)$, then $(h(\mathcal{S}, \mathcal{X}), \mathcal{S}) \approx_\varepsilon (\mathcal{U}, \mathcal{S})$, where $\mathcal{S}$ is uniform over S and $\mathcal{U}$ is uniform over Y.*

Lemma 3 (Log-Many Bits Hardcore Functions [GL89]). *Let $\mathsf{f}\colon \{0,1\}^n \to \{0,1\}^m$ be an OWF with respect to a distribution $\mathcal{D}$. Let $\mathsf{B}_i\colon \{0,1\}^n \times \{0,1\}^{2n} \to \{0,1\}$ be a function defined as $\mathsf{B}_i(\mathsf{x}, \mathsf{s}) := \langle \mathsf{x}, \mathsf{s}[i, i+n-1] \rangle \mod 2$, where $\mathsf{s}[i, i+n-1] := (\mathsf{s}_i, \ldots, \mathsf{s}_{i+n-1})$. Then for any constant $c > 0$, the function $\mathsf{H}\colon \{0,1\}^n \times \{0,1\}^{2n} \to \{0,1\}^{c\lceil \log n \rceil}$ defined as*

$$\mathsf{H}(\mathsf{x}, \mathsf{s}) := (\mathsf{B}_1(\mathsf{x}, \mathsf{s}), \ldots, \mathsf{B}_{c\lceil \log n \rceil}(\mathsf{x}, \mathsf{s}))$$

is a hardcore function for f. That is, $(\mathsf{s}, \mathsf{f}(\mathsf{x}), \mathsf{H}(\mathsf{x}, \mathsf{s})) \stackrel{c}{\equiv} (\mathsf{s}, \mathsf{f}(\mathsf{x}), \mathsf{w})$, where $\mathsf{s} \stackrel{\$}{\leftarrow} \{0,1\}^{2n}$, $\mathsf{x} \stackrel{\$}{\leftarrow} \mathcal{D}$ and $\mathsf{w} \stackrel{\$}{\leftarrow} \{0,1\}^{c\lceil \log n \rceil}$.

2.1 Error Correcting Codes

For our constructions we will rely on efficiently correctable error correcting block codes. Fix a finite alphabets Σ, and two parameters k and m. We will represent codes by two efficient algorithms Encode and Decode, where Encode takes as input a message $\mathsf{x} = (\mathsf{x}_1, \ldots, \mathsf{x}_k) \in \Sigma^k$ and outputs a codeword $\mathsf{c} = (\mathsf{c}_1, \ldots, \mathsf{c}_m) \in \Sigma^m$. We refer to the support of the algorithm Encode as the code C. The algorithm Decode takes as input a string $\hat{\mathsf{c}} \in (\Sigma \cup \{\perp\})^m$ and outputs a message $\mathsf{x} \in \Sigma^k$ or $\perp$. We say that such a code jointly corrects r errors and s erasures, if it holds for every $\hat{\mathsf{c}} \in (\Sigma \cup \{\perp\})^m$ which can be obtained from a codeword $\mathsf{c} \in \mathsf{C}$ by changing at most r positions and erasing at most s positions that $\mathsf{Decode}(\hat{\mathsf{c}}) = \mathsf{Decode}(\mathsf{c})$.

The specific class of codes which we use in our constructions are Reed Solomon (RS) codes [RS60]. The alphabets of Reed Solomon codes are finite fields $\mathbb{F}_q$, and an $[m,k]$-RS code exists whenever $m \leq q$. The encoding procedure Encode of a $[m,k]$-RS code represents the message $\mathsf{x} \in \mathbb{F}_q^k$ as a polynomial P of degree $k-1$ over $\mathbb{F}_q$ via the coefficient embedding, and computes and outputs $(P(\xi_1), \ldots, P(\xi_m))$, where $\xi_1, \ldots, \xi_m$ are pairwise distinct elements of $\mathbb{F}_q$. There exists an efficient decoding algorithm, the so-called Berlekamp-Welch decoder [WB86], which can jointly decode r errors and s erasures given that $2r + s \leq m - k$. We say this RS code has minimum-distance $m - k + 1$.

In abuse of notation, we will provide as input to the encoding algorithm a binary string, i.e. an element from $\{0,1\}^n$. For $\Sigma = \mathbb{F}_{2^\kappa}$, such a string $x \in \{0,1\}^n$ can be mapped to a string $\mathsf{x} \in \mathbb{F}_{2^k}$ by chopping x into blocks of length κ and letting each block represent an element of $\mathbb{F}_{2^\kappa}$. Finally, we will always assume that there is a canonical enumeration of the elements in Σ. This lets us identify each element in Σ with a corresponding element in the set $\{1, \ldots, \Sigma\}$.

2.2 Trapdoor Functions

We recall the definition of trapdoor function (TDFs).

Definition 4 (Trapdoor Functions). *Let $n = n(\lambda)$ be a polynomial. A family of trapdoor functions* TDF *with domain $\{0,1\}^n$ consists of three PPT algorithms* TDF.KG, TDF.F *and* $\mathsf{TDF.F}^{-1}$ *with the following syntax and security properties.*

- $\mathsf{TDF.KG}(1^\lambda)$: *Takes the security parameter 1^λ and outputs a pair* $(\mathsf{ik}, \mathsf{tk})$ *of index/trapdoor keys.*
- $\mathsf{TDF.F}(\mathsf{ik}, \mathsf{x})$: *Takes an index key* ik *and a domain element* $\mathsf{x} \in \{0,1\}^n$ *and deterministically outputs an image element* u.
- $\mathsf{TDF.F}^{-1}(\mathsf{tk}, \mathsf{u})$: *Takes a trapdoor key* tk *and an image element* u *and outputs a value* $\mathsf{x} \in \{0,1\}^n \cup \{\perp\}$.

We require the following properties.

- ***Correctness:***

$$\Pr_{(\mathsf{ik},\mathsf{tk})}\left[\exists \mathsf{x} \in \{0,1\}^n \ s.t.\ \mathsf{TDF.F}^{-1}(\mathsf{tk}, \mathsf{TDF.F}(\mathsf{ik}, \mathsf{x})) \neq \mathsf{x}\right] = \mathsf{negl}(\lambda), \quad (1)$$

where the probability is taken over $(\mathsf{ik}, \mathsf{tk}) \xleftarrow{\$} \mathsf{TDF.KG}(1^\lambda)$.
- ***One-wayness:*** *For any PPT adversary $\mathcal{A}$:* $\Pr[\mathcal{A}(\mathsf{ik}, \mathsf{u}) = \mathsf{x}] = \mathsf{negl}(\lambda)$, *where* $(\mathsf{ik}, *) \xleftarrow{\$} \mathsf{TDF.KG}(1^\lambda)$, $\mathsf{x} \xleftarrow{\$} \{0,1\}^n$ *and* $\mathsf{u} := \mathsf{TDF.F}(\mathsf{ik}, \mathsf{x})$.

We also define a stronger security property, called CPA block-source security.

- ***CPA-deterministic security:*** *We say* TDF *is (k,n)-CPA-secure if for any two (k,n)-sources $\mathcal{D}_0$ and $\mathcal{D}_1$:* $(\mathsf{ik}, \mathsf{TDF.F}(\mathsf{ik}, \mathcal{D}_0)) \stackrel{c}{\equiv} (\mathsf{ik}, \mathsf{TDF.F}(\mathsf{ik}, \mathcal{D}_1))$, *where* $(\mathsf{ik}, *) \xleftarrow{\$} \mathsf{TDF.KG}(1^\lambda)$.

We give the definition of rate for a TDF, which captures the asymptotic input-to-image ratio.

Definition 5 (Rate). *A TDF* $(\mathsf{TDF.KG}, \mathsf{TDF.F}, \mathsf{TDF.F}^{-1})$ *has rate* ρ *if for all* $\lambda \in \mathbb{N}$, *all polynomials* $n(\lambda)$, *all* ik *in the support of* $\mathsf{TDF.KG}(1^\lambda)$, *all inputs in* $\mathsf{x} \in \{0,1\}^{n(\lambda)}$*:*

$$\liminf_{\lambda \to \infty} \frac{n(\lambda)}{|\mathsf{TDF.F}(\mathsf{ik}, \mathsf{x})|} = \rho.$$

2.3 The Diffie-Hellman Problems

We recall the classical Diffie-Hellman problem [DH76] both in its search and decisional version. Let G be a group-generator scheme, which on input 1^λ outputs $(\mathbb{G}, p, g)$, where $\mathbb{G}$ is the description of a group, p is the order of the group which is always a prime number and g is a generator for the group. In favor of a simpler analysis, we consider groups such that $\log(p) = |g| = \lambda$.

Definition 6 (Diffie-Hellman Assumptions). *We say* G *is CDH-hard if for any PPT adversary* $\mathcal{A}$ $\Pr[\mathcal{A}(\mathbb{G}, p, g, g^{a_1}, g^{a_2}) = g^{a_1 a_2}] = \mathsf{negl}(\lambda)$ *where* $(\mathbb{G}, p, g) \xleftarrow{\$} \mathsf{G}(1^\lambda)$ *and* $(a_1, a_2) \xleftarrow{\$} \mathbb{Z}_p^2$*. We say* G *is DDH-hard if* $(\mathbb{G}, p, g, g^{a_1}, g^{a_2}, g^{a_3}) \overset{c}{\equiv} (\mathbb{G}, p, g, g^{a_1}, g^{a_2}, g^{a_1 a_2})$*, where* $(\mathbb{G}, p, g) \xleftarrow{\$} \mathsf{G}(1^\lambda)$ *and* $(a_1, a_2, a_3) \xleftarrow{\$} \mathbb{Z}_p^3$*.*

3 Smooth Recyclable OWFE

We recall the definition of recyclable one-way function with encryption (OWFE) from [GH18]. The following definitions are taken almost in verbatim from [GGH19], except that we consider a generalized version of the primitive over any finite alphabets Σ. The notion of OWFE in turn builds on related notions known in the literature as (chameleon) hash encryption and its variants [DG17b, DG17a, BLSV18, DGHM18].

Definition 7 (Recyclable one-way function with encryption). *Let* $\Sigma = \{1, \ldots, \sigma\}$ *for some integer* σ*. A* w*-bit recyclable* (k, n)*-OWFE scheme consists of the PPT algorithms* Gen, Hash, Enc_1, Enc_2 *and* Dec *with the following syntax.*

- $\mathsf{Gen}(1^\lambda)$*: Takes the security parameter* 1^λ *and outputs a public parameter* k *(by tossing coins) for a function* $\mathsf{Hash}(\mathsf{k}, \cdot)$ *from* n *bits to* ν *bits.*
- $\mathsf{Hash}(\mathsf{k}, \mathsf{x})$*: Takes a public parameter* k *and a preimage* $\mathsf{x} \in \Sigma^n$*, and deterministically outputs an element* y*.*
- $\mathsf{Enc}_1(\mathsf{k}, (i, z); \rho)$*: Takes a public parameter* k*, an index* $i \in [n]$*, a word* $z \in \Sigma$ *and randomness* ρ*, and outputs a ciphertext* ct*. We implicitly assume that* ct *contains* (i, z)*.*
- $\mathsf{Enc}_2(\mathsf{k}, \mathsf{y}, (i, z); \rho)$*: Takes a public parameter* k*, a value* y*, an index* $i \in [n]$*, a word* $z \in \Sigma$ *and randomness* ρ*, and outputs a string* $\mathsf{e} \in \{0,1\}^w$*. Notice that unlike* Enc_1*, which does not take* y *as input, the algorithm* Enc_2 *does take* y *as input.*

- $\mathsf{Dec}(\mathsf{k}, \mathsf{ct}, \mathsf{x})$: *Takes a public parameter* k, *a ciphertext* ct *and a preimage* $\mathsf{x} \in \Sigma^n$, *and deterministically outputs a string* $\mathsf{e} \in \{0,1\}^w$.

We require the following properties.

- ***Correctness.*** *For any choice of* $\mathsf{k} \in \mathsf{Gen}(1^\lambda)$, *any index* $i \in [n]$, *any preimage* $\mathsf{x} \in \Sigma^n$ *and any randomness value* ρ,

$$\Pr[\mathsf{Enc}_2(\mathsf{k}, \mathsf{y}, (i, \mathsf{x}_i); \rho) = \mathsf{Dec}(\mathsf{k}, \mathsf{ct}, \mathsf{x})] = 1$$

where $\mathsf{y} := \mathsf{Hash}(\mathsf{k}, \mathsf{x})$ *and* $\mathsf{ct} := \mathsf{Enc}_1(\mathsf{k}, (i, \mathsf{x}_i); \rho)$.
- (k, n)***-One-wayness:*** *For any* k*-source* $\mathcal{S}$ *over* Σ^n *and any PPT adversary* $\mathcal{A}$:

$$\Pr[\mathsf{Hash}(\mathsf{k}, \mathcal{A}(\mathsf{k}, \mathsf{y})) = \mathsf{y}] = \mathsf{negl}(\lambda),$$

where $\mathsf{k} \xleftarrow{\$} \mathsf{Gen}(1^\lambda)$, $\mathsf{x} \xleftarrow{\$} \mathcal{S}$ *and* $\mathsf{y} := \mathsf{Hash}(\mathsf{k}, \mathsf{x})$. *If* $k = n$, *then we simply refer to an OWFE scheme (without specifying the parameters).*
- (k, n)***-Smoothness:*** *For any two* (k, n)*-sources* $\mathcal{S}_1$ *and* $\mathcal{S}_2$:

$$(\mathsf{k}, \mathsf{Hash}(\mathsf{k}, \mathsf{x}_1)) \stackrel{c}{\equiv} (\mathsf{k}, \mathsf{Hash}(\mathsf{k}, \mathsf{x}_2))$$

where $\mathsf{k} \xleftarrow{\$} \mathsf{Gen}(1^\lambda)$, $\mathsf{x}_1 \xleftarrow{\$} \mathcal{S}_1$ *and* $\mathsf{x}_2 \xleftarrow{\$} \mathcal{S}_2$.
- ***Security for Encryption:*** *For any* $i \in [n]$, *any* $\mathsf{x} \in \Sigma^n$, *and any* $f \in \Sigma \setminus \{\mathsf{x}_i\}$:

$$(\mathsf{x}, \mathsf{k}, \mathsf{ct}, \mathsf{e}) \stackrel{c}{\equiv} (\mathsf{x}, \mathsf{k}, \mathsf{ct}, \mathsf{e}')$$

where $\mathsf{k} \xleftarrow{\$} \mathsf{Gen}(1^\lambda)$, $\rho \xleftarrow{\$} \{0,1\}^*$, $\mathsf{ct} := \mathsf{Enc}_1(\mathsf{k}, (i, f); \rho)$, $\mathsf{e} := \mathsf{Enc}_2\big(\mathsf{k}, \mathsf{Hash}(\mathsf{k}, \mathsf{x}), (i, f); \rho\big)$ *and* $\mathsf{e}' \xleftarrow{\$} \{0,1\}^w$.

3.1 Smooth Recyclable OWFE from CDH

We generalize the recyclable OWFE from [GH18] to any finite alphabet Σ. Although this modification might look insignificant, it will be our main leverage to construct a rate-1 trapdoor function.

Construction 8 (Smooth recyclable OWFE from CDH). *Let* G *be a CDH-hard group-generator scheme and let* $\Sigma := \{1, \ldots, \sigma\}$ *be a finite alphabet.*

- $\mathsf{Gen}(1^\lambda)$: *Sample* $(\mathbb{G}, p, g) \xleftarrow{\$} \mathbb{G}(1^\lambda)$. *For each* $j \in [n]$ *and* $f \in \Sigma$, *choose* $g_{j,f} \xleftarrow{\$} \mathbb{G}$. *Output*

$$\mathsf{k} := \begin{pmatrix} g_{1,1}, & g_{2,1}, & \ldots, & g_{n,1} \\ g_{1,2}, & g_{2,2}, & \ldots, & g_{n,2} \\ \ldots, & \ldots, & \ldots, & \ldots \\ g_{1,\sigma}, & g_{2,\sigma}, & \ldots, & g_{n,\sigma} \end{pmatrix}. \tag{2}$$

- $\mathsf{Hash}(\mathsf{k}, \mathsf{x})$: *Parse* k *as in Eq. 2, and output* $\mathsf{y} := \prod_{j \in [n]} g_{j, \mathsf{x}_j}$.

- $\mathsf{Enc}_1(\mathsf{k}, (i, z); \rho)$: *Parse* k *as in Eq. 2. Given the randomness* $\rho \xleftarrow{\$} \mathbb{Z}_p$, *proceed as follows:*
  - *For every* $j \in [n] \setminus \{i\}$, *and every* $f \in \Sigma$ *set* $c_{j,f} := g_{j,z}^{\rho}$.
  - *For every* $f \in \Sigma \setminus \{z\}$ *set* $c_{i,f} := \bot$, *then set* $c_{i,z} := g_{i,z}^{\rho}$.
  - *Output*

$$\mathsf{ct} := \begin{pmatrix} c_{1,1}, c_{2,1}, \ldots, c_{n,1} \\ c_{1,2}, c_{2,2}, \ldots, c_{n,2} \\ \ldots, \ldots, \ldots, \ldots \\ c_{1,\sigma}, c_{2,\sigma}, \ldots, c_{n,\sigma} \end{pmatrix}. \tag{3}$$

- $\mathsf{Enc}_2(\mathsf{k}, (\mathsf{y}, i, z); \rho)$: *Given the randomness* $\rho \xleftarrow{\$} \mathbb{Z}_p$, *output* $\mathsf{H}(\mathsf{y}^{\rho})$, *where* $\mathsf{H} : \mathbb{G} \to \{0,1\}^w$ *denotes a hardcore function (e.g., the function from Lemma 3).*
- $\mathsf{Dec}(\mathsf{k}, \mathsf{ct}, \mathsf{x})$: *Parse* ct *as in Eq. 3, and output* $\mathsf{H}\left(\prod_{j\in[n]} c_{j,\mathsf{x}_j}\right)$.

Correctness of the scheme is immediate. We now show that the construction satisfies all of the required security properties properties.

Theorem 9 (One-Wayness). *Let* G *generate a CDH-hard group, then for all* $n \geq \omega(\log(p))$, *Construction 8 is one-way.*

Proof. This is shown with a reduction to the discrete logarithm problem. On input a challenge random element $h \in \mathbb{G}$, sample a random pair of indices $i^* \xleftarrow{\$} [n]$ and $f^* \xleftarrow{\$} \Sigma$ and set $g_{i^*,f^*} := h$. For all $i \xleftarrow{\$} [n]$ and $f \xleftarrow{\$} \Sigma$, except for the pair (i^*, f^*), set $g_{i,f} := g^{r_{i,f}}$, for a uniform $r_{i,f} \xleftarrow{\$} \mathbb{Z}_p$. Define the public key as

$$\mathsf{k} := \begin{pmatrix} g_{1,1}, g_{2,1}, \ldots, g_{n,1} \\ g_{1,2}, g_{2,2}, \ldots, g_{n,2} \\ \ldots, \ldots, \ldots, \ldots \\ g_{1,\sigma}, g_{2,\sigma}, \ldots, g_{n,\sigma} \end{pmatrix}.$$

Then sample a uniform $\mathsf{x} \xleftarrow{\$} \Sigma^n$ such that $\mathsf{x}_{i^*} \neq f^*$ and compute $\mathsf{y} := \prod_{j\in[n]} g_{j,\mathsf{x}_j}$. Give (k, y) to the adversary and receive some x'. By Lemma 2, $\mathsf{x}'_{i^*} = f^*$ with probability close to $1/\sigma$, which allows us to compute the discrete logarithm of h.

Theorem 10 ((k, n)-Smoothness). *Let* G *generate a CDH-hard group and* $k \geq \log p + \omega(\log \lambda)$, *then Construction 8 is* (k, n)*-smooth.*

Proof. Let $\mathcal{S}_1$ and $\mathcal{S}_2$ be two (k, n) sources. The smoothness is a direct consequence of Lemma 2. Namely, assuming $\mathsf{k} \xleftarrow{\$} \mathsf{Gen}(1^{\lambda})$, $\mathsf{x}_1 \xleftarrow{\$} \mathcal{S}_1$ and $\mathsf{x}_2 \xleftarrow{\$} \mathcal{S}_2$, since Hash is a 2-universal hash function, by Lemma 2 we know that the outputs of both $\mathsf{Hash}(\mathsf{k}, \mathsf{x}_1)$ and $\mathsf{Hash}(\mathsf{k}, \mathsf{x}_2)$ are statistically $\frac{1}{2^{\omega(\log \lambda)}}$ close to the uniform over $\mathbb{G}$, and hence negligibly close (statistically) to each other.

Theorem 11 (Security for Encryption). *Let* G *generate a CDH-hard group, then Construction 8 is secure for encryption.*

Proof. Assume towards contradiction that there exists some i^*, x, and $f^* \neq \mathsf{x}_{i^*}$ such that an adversary can successfully distinguish on those input. Let (g, h_1, h_2, s) be a CDH challenge, where $s \in \{0,1\}^w$ is either a random string or the output of the hardcore function. For all $i \in [n] \setminus i^*$ and all $f \in \Sigma$ set $g_{i,f} := g^{r_{i,f}}$, for a uniform $r_{i,f} \xleftarrow{\$} \mathbb{Z}_p$. Similarly, for all $f \in \Sigma \setminus \mathsf{x}_{i^*}$ set $g_{i^*,f} := g^{r_{i^*,f}}$, for a uniform $r_{i^*,f} \xleftarrow{\$} \mathbb{Z}_p$. Finally set

$$g_{i^*,\mathsf{x}_{i^*}} := \frac{h_1}{\prod_{j\in[n]\setminus i^*} g_{j,\mathsf{x}_j}}$$

and define k accordingly. Define

$$\mathsf{ct} := \begin{pmatrix} c_{1,1}, & c_{2,1}, & \dots, & c_{n,1} \\ c_{1,2}, & c_{2,2}, & \dots, & c_{n,2} \\ \dots, & \dots, & \dots, & \dots \\ c_{1,\sigma}, & c_{2,\sigma}, & \dots, & c_{n,\sigma} \end{pmatrix}$$

where for $i \in [n] \setminus i^*$ and all $f \in \Sigma$ we have $c_{i,f} := h_2^{r_{i,f}}$ and $c_{i^*,f^*} := h_2^{r_{i^*,f^*}}$, whereas the other terms are set to $\perp$. The adversary is given $(\mathsf{x}, \mathsf{k}, \mathsf{ct}, s)$ and the reduction returns whatever the adversary returns. Security follows from the fact that

$$\mathsf{Hash}(\mathsf{k}, \mathsf{x}) = \prod_{j\in[n]} g_{j,\mathsf{x}_j} = h_1.$$

4 Rate-1 CDH-Based Trapdoor Function

In this section we give a construction of rate-1 TDFs based on CDH, satisfying deterministic-encryption security. The result of [GGH19] gives CDH-based TDF constructions with rates $1/c$ for a constant c. A rough estimate of the constant c is at least 20. The main reason behind the large constant is that [GGH19] first builds an intermediate *local TDF* which (1) outputs two bits for every bit of the input (i.e., a rate less than 1/2) and (2) the TDF has a local property in that for each bit of the input, the inversion algorithm either recovers the bit or gives up for that particular bit, each happening with probability 1/2. The construction of [GGH19] then performs error correction over bitstrings to boost correctness. This results in another constant blowup.

At a high level our approach for achieving rate 1 proceeds as follows. We encode the input to the TDF block-by-block, instead of bit-by-bit. Each block is a symbol of an alphabet over a field for which erasure correction with better rates can be done. We then show how to provide an almost equally-sized *hint* for every block of the input, achieving a rate 1 at the end. The main technical novelty of our work relies on how to form the hint in a succinct way.

Construction 12 (Rate-1 TDF from CDH). *Let* $\Sigma = \{1, \dots, \sigma\}$ *be a finite alphabet, let* $(\mathsf{Gen}, \mathsf{Hash}, \mathsf{Enc}_1, \mathsf{Enc}_2, \mathsf{Dec})$ *be a* w*-bit OWFE, and let* $(\mathsf{Encode}, \mathsf{Decode})$ *be an error-correcting code, where* $\mathsf{Encode} : \{0,1\}^n \to \Sigma^m$. *We define our TDF construction* $(\mathsf{TDF.KG}, \mathsf{TDF.F}, \mathsf{TDF.F}^{-1})$ *as follows.*

- $\mathsf{TDF.KG}(1^\lambda)$:
  1. *Sample* $\mathsf{k} := \mathsf{Gen}(1^\lambda)$.
  2. *For all* $i \in [m]$ *and all* $f \in \Sigma$:
     (a) *Sample* $\rho_{i,f} \xleftarrow{\$} \{0,1\}^\lambda$ *and* $a_{i,f} \xleftarrow{\$} \{0,1\}^w$.
     (b) *Compute* $\mathsf{ct}_{i,f} := \mathsf{Enc}_1(\mathsf{k}, (i, f); \rho_{i,f})$.
  3. *Set the trapdoor key as*

$$\mathsf{tk} := \mathsf{k}, \begin{pmatrix} \rho_{1,1}, \rho_{2,1}, \ldots, \rho_{m,1} \\ \rho_{1,2}, \rho_{2,2}, \ldots, \rho_{m,2} \\ \ldots, \ldots, \ldots, \ldots \\ \rho_{1,\sigma}, \rho_{2,\sigma}, \ldots, \rho_{m,\sigma} \end{pmatrix} \tag{4}$$

and the index key as

$$\mathsf{ik} := \mathsf{k}, \begin{pmatrix} \mathsf{ct}_{1,1}, \mathsf{ct}_{2,1}, \ldots, \mathsf{ct}_{m,1} \\ \mathsf{ct}_{1,2}, \mathsf{ct}_{2,2}, \ldots, \mathsf{ct}_{m,2} \\ \ldots, \ldots, \ldots, \ldots \\ \mathsf{ct}_{1,\sigma}, \mathsf{ct}_{2,\sigma}, \ldots, \mathsf{ct}_{m,\sigma} \end{pmatrix}, \begin{pmatrix} a_{1,1}, a_{2,1}, \ldots, a_{m,1} \\ a_{1,2}, a_{2,2}, \ldots, a_{m,2} \\ \ldots, \ldots, \ldots, \ldots \\ a_{1,\sigma}, a_{2,\sigma}, \ldots, a_{m,\sigma} \end{pmatrix}. \tag{5}$$

- $\mathsf{TDF.F}(\mathsf{ik}, \mathsf{x} \in \{0,1\}^n)$:
  1. *Parse* ik *as in Eq. 5.*
  2. *Let* $\mathsf{z} := \mathsf{Encode}(\mathsf{x}) \in \Sigma^m$ *and* $\mathsf{y} := \mathsf{Hash}(\mathsf{k}, \mathsf{z})$.
  3. *For all* $i \in [m]$:
     (a) *Let* $h_i := \mathsf{Dec}(\mathsf{k}, \mathsf{ct}_{i,\mathsf{z}_i}, \mathsf{z})$.
     (b) *Set* $v_i := h_i \oplus a_{i,\mathsf{z}_i} \in \{0,1\}^w$.
  4. *Return* $\mathsf{u} := (\mathsf{y}, v_1, \ldots, v_m)$.
- $\mathsf{TDF.F}^{-1}(\mathsf{tk}, \mathsf{u})$:
  1. *Parse* tk *as in Eq. 4 and* $\mathsf{u} := (\mathsf{y}, v_1, \ldots, v_m)$.
  2. *Retrieve* z' *element-by-element as follows. For* $i \in [m]$, *to retrieve the i-th element:*
     (a) *If there exists one and only one index* $f \in \Sigma$ *such that*

$$\mathsf{Enc}_2(\mathsf{k}, \mathsf{y}, (i, f); \rho_{i,f}) = a_{i,f} \oplus v_i, \tag{6}$$

     then set $\mathsf{z}'_i = f$.
     (b) *Otherwise, set* $\mathsf{z}'_i = \perp$.
  3. *Return* $\mathsf{Decode}(\mathsf{z}')$.

4.1 Analysis

In the following we show that our construction is a correct and secure TDF.

Theorem 13 (Correctness). *Let* $(\mathsf{Gen}, \mathsf{Hash}, \mathsf{Enc}_1, \mathsf{Enc}_2, \mathsf{Dec})$ *be a* w*-bit OWFE, where* $2^w \geq 2\sigma/\eta$, *for some* $\eta \in (0,1]$. *Let* $(\mathsf{Encode}, \mathsf{Decode})$ *be an error-correcting code resilient against a* η*-fraction of erasures, where* $\mathsf{Encode} : \{0,1\}^n \to \Sigma^m$ *and* $m \geq 6\lambda/\eta$. *Then Construction 12 is correct except with probability* $e^{-\lambda}$: $\Pr[\exists \mathsf{x} \in \{0,1\}^n : \mathsf{TDF.F}^{-1}(\mathsf{tk}, \mathsf{TDF.F}(\mathsf{ik}, \mathsf{x})) \neq \mathsf{x}] \leq e^{-\lambda}$, *where* $(\mathsf{ik}, \mathsf{tk}) \xleftarrow{\$} \mathsf{TDF.KG}(1^\lambda)$.

Proof. Let $(\mathsf{ik}, \mathsf{tk}) := \mathsf{TDF.KG}(1^\lambda)$, let $\mathsf{u} := \mathsf{TDF.F}(\mathsf{ik}, \mathsf{x})$ for a uniform $\mathsf{x} \xleftarrow{\$} \{0,1\}^n$, and let $\mathsf{z} := \mathsf{Encode}(\mathsf{x})$. For all $i \in [m]$, by the correctness of the OWFE we have

$$v_i = h_i \oplus a_{i,\mathsf{z}_i} = \mathsf{Dec}(\mathsf{k}, \mathsf{ct}_{i,\mathsf{z}_i}, \mathsf{z}) \oplus a_{i,\mathsf{z}_i} = \mathsf{Enc}_2(\mathsf{k}, \mathsf{y}, (i, \mathsf{z}_i); \rho_{i,\mathsf{z}_i}) \oplus a_{i,\mathsf{z}_i}.$$

Thus, the index z_i satisfies Eq. 6. Now we consider the probability that some $f \neq \mathsf{z}_i$ satisfies the same condition. Since $a_{i,f}$ is chosen uniformly and independently at random, the two values $\mathsf{Enc}_2(\mathsf{k}, (\mathsf{y}, i, f); \rho_{i,f}) \oplus v_i$ and $a_{i,f}$ are independent. Thus, the probability that the index $f \neq \mathsf{z}_i$ satisfies Eq. 6 is

$$\Pr\left[\mathsf{Enc}_2(\mathsf{k}, \mathsf{y}, (i, f); \rho_{i,f}) \oplus v_i = a_{i,f}\right] \leq \frac{1}{2^w}.$$

By a union bound, the probability that such an $f \neq \mathsf{z}_i$ exists is at most

$$\Pr\left[\exists f \in \Sigma \setminus \{\mathsf{z}_i\} : \mathsf{Enc}_2(\mathsf{k}, \mathsf{y}, (i, f); \rho_{i,f}) \oplus v_i = a_{i,f}\right] \leq \frac{\sigma}{2^w} \leq \frac{\eta}{2}$$

as $2^w \geq 2\sigma/\eta$. Applying the Chernoff bound (Lemma 1) yields that at most $\eta \cdot m$ of the indices contain a non unique decoding and therefore z'_i is set to $\perp$, except with probability $e^{-\frac{\eta \cdot m}{6}} \leq e^{-\lambda}$. Let $S \subseteq [m]$ be the set of indices for which there exists a $\mathsf{z}'_i = \perp$. By the above, $|S| \leq \eta \cdot m$, except with probability $e^{-\lambda}$. As we assume the code $(\mathsf{Encode}, \mathsf{Decode})$ is capable of handling a η-fraction of erasures, $\mathsf{Decode}(\mathsf{z}')$ will output x with overwhelming probability.

We show that our TDF is one-way.

Theorem 14 (One-Wayness and DE security). *Assuming* $(\mathsf{Gen}, \mathsf{Hash}, \mathsf{Enc}_1, \mathsf{Enc}_2, \mathsf{Dec})$ *is an* (n, m)*-OWFE scheme, then Construction 12 is one-way. Moreover, if* $(\mathsf{Gen}, \mathsf{Hash}, \mathsf{Enc}_1, \mathsf{Enc}_2, \mathsf{Dec})$ *is* (k, m)*-smooth, the resulting TDF is* (k, m)*-CPA indistinguishable.*

Proof. Let $\mathsf{x} \in \{0,1\}^n$ be the random input to the TDF, and let $\mathsf{z} := \mathsf{Encode}(\mathsf{x})$. Also, let $\mathsf{y} := \mathsf{Hash}(\mathsf{k}, \mathsf{z})$. We first construct a simulator $\mathsf{Sim}(\mathsf{k}, \mathsf{y})$, which—without knowledge of x—samples a simulated index key ik_{sim} together with a corresponding u_{sim} as follows.

- $\mathsf{Sim}(\mathsf{k}, \mathsf{y})$:
  1. For all $i \in [m]$: Sample $a_i \xleftarrow{\$} \{0,1\}^w$.
  2. For all $i \in [m]$ and $f \in \Sigma$:
     (a) Sample $\rho_{i,f} \xleftarrow{\$} \{0,1\}^*$.
     (b) Compute $\mathsf{ct}_{i,f} := \mathsf{Enc}_1(\mathsf{k}, (i, f); \rho_{i,f})$.
     (c) Compute $a_{i,f} := a_i \oplus \mathsf{Enc}_2(\mathsf{k}, \mathsf{y}, (i, f); \rho_{i,f})$.
  3. Set the index key as

$$\mathsf{ik}_{\text{sim}} := \mathsf{k}, \begin{pmatrix} \mathsf{ct}_{1,1}, & \mathsf{ct}_{2,1}, & \ldots, & \mathsf{ct}_{m,1} \\ \mathsf{ct}_{1,2}, & \mathsf{ct}_{2,2}, & \ldots, & \mathsf{ct}_{m,2} \\ \ldots, & \ldots, & \ldots, & \ldots \\ \mathsf{ct}_{1,\sigma}, & \mathsf{ct}_{2,\sigma}, & \ldots, & \mathsf{ct}_{m,\sigma} \end{pmatrix}, \begin{pmatrix} a_{1,1}, & a_{2,1}, & \ldots, & a_{m,1} \\ a_{1,2}, & a_{2,2}, & \ldots, & a_{m,2} \\ \ldots, & \ldots, & \ldots, & \ldots \\ a_{1,\sigma}, & a_{2,\sigma}, & \ldots, & a_{m,\sigma} \end{pmatrix}$$

  and the image as $\mathsf{u} := (\mathsf{y}, a_1, \ldots, a_m)$.

We now show that for any distribution $\mathcal{S}$ over $\{0,1\}^n$

$$(\mathsf{x}, \mathsf{ik}, \mathsf{TDF.F}(\mathsf{ik}, \mathsf{x})) \stackrel{c}{\equiv} (\mathsf{x}, \mathsf{Sim}(\mathsf{k}, \mathsf{y})) \tag{7}$$

where $\mathsf{x} \xleftarrow{\$} \mathcal{S}$, $(\mathsf{ik}, *) \xleftarrow{\$} \mathsf{TDF.KG}(1^\lambda)$, $\mathsf{k} \xleftarrow{\$} \mathsf{Gen}(1^\lambda)$, and $\mathsf{y} := \mathsf{Hash}(\mathsf{k}, \mathsf{Encode}(\mathsf{x}))$. This will yield both the one-wayness and deterministic-encryption security claims of the lemma.

We define $\mathsf{Sim}'(\mathsf{k}, \mathsf{x}, \mathsf{y})$ as follows. For all $i \in [m]$ and $f \in \Sigma$, sample $\mathsf{ct}_{i,f}$ exactly as in $\mathsf{Sim}(\mathsf{k}, \mathsf{y})$, and letting $\mathsf{z} := \mathsf{Encode}(\mathsf{x})$, sample $a_{i,j}$ as follows:

- If $f = \mathsf{z}_i$, then set $a_{i,f} := a_i \oplus \mathsf{Enc}_2(\mathsf{k}, \mathsf{y}, (i, f); \rho_{i,f})$, exactly as in $\mathsf{Sim}(\mathsf{k}, \mathsf{y})$.
- If $f \neq \mathsf{z}_i$, then sample $a_{i,f} \xleftarrow{\$} \{0,1\}^w$.

By the security-for-encryption requirement of the underlying OWFE

$$(\mathsf{x}, \mathsf{Sim}(\mathsf{k}, \mathsf{y})) \stackrel{c}{\equiv} (\mathsf{x}, \mathsf{Sim}'(\mathsf{k}, \mathsf{x}, \mathsf{y})).$$

By simple inspection we can see that the distribution $(\mathsf{x}, \mathsf{Sim}'(\mathsf{k}, \mathsf{x}, \mathsf{y}))$ is identically distributed to $(\mathsf{x}, \mathsf{ik}, \mathsf{TDF}(\mathsf{ik}, \mathsf{x}))$, where $(\mathsf{ik}, *) \xleftarrow{\$} \mathsf{TDF.KG}(1^\lambda)$. The proof is now complete.

4.2 Parameters

We analyze the rate of our scheme and we discuss possible instantiations for the underlying building blocks.

Theorem 15 (Rate). *Let* $\sigma \geq \lambda$ *and let* $(\mathsf{Encode}, \mathsf{Decode})$ *be an error correcting code for an alphabet* Σ *of size* σ *that can correct a fraction of erasure* $\eta = 1/\log(\lambda)$ *and has rate* $1 - 1/\log(\lambda)$. *Let* $2^w = 2 \cdot \sigma/\eta$ *and* $m \geq 6\lambda/\eta$. *Then Construction 12 has rate* 1.

Proof. By definition we have that

$$2^w = 2 \cdot \sigma/\eta = 2 \cdot \sigma \cdot \log(\lambda).$$

Then the value $v_i \in \{0,1\}^w$ encodes the codeword symbol $\mathsf{z}_i \in \Sigma$ of the codeword x_i. Thus, each codeword symbol is encoded at rate

$$\frac{\log(\sigma)}{w} = \frac{\log(\sigma)}{\log(\sigma) + \log\log(\lambda) + 1} \geq 1 - 2\frac{\log\log(\lambda)}{\log(\sigma)} \geq 1 - 2\frac{\log\log(\lambda)}{\log(\lambda)}.$$

Recall that $(\mathsf{Encode}, \mathsf{Decode})$ has rate $1 - 1/\log(\lambda)$ and can efficiently decode from a $\eta = 1/\log(\lambda)$ fraction of errors. Taking into account that the sender message includes an additional group element $\mathsf{y} \in \mathbb{G}$ and assuming that $\log(|\mathbb{G}|) = \lambda$, this accounts for a decrease of the rate by a factor

$$\frac{m \cdot w}{m \cdot w + \log(|\mathbb{G}|)} = 1 - \frac{\log(|\mathbb{G}|)}{m \cdot w + \log(|\mathbb{G}|)} \geq 1 - \frac{1}{m \cdot \log(\lambda)}.$$

Consequently, the total rate of our scheme is lower bounded by

$$\left(1 - \frac{1}{m \cdot \log(\lambda)}\right)\left(1 - \frac{1}{\log(\lambda)}\right)\left(1 - 2\frac{\log\log(\lambda)}{\log(\lambda)}\right) \geq 1 - 3\frac{\log\log(\lambda)}{\log(\lambda)},$$

which approaches 1.

Note that the constraints $2^w = 2\sigma \cdot \log(\lambda)$ and $\sigma \geq \lambda$ require us to instantiate a hardcore function H that extracts $O(\log(\lambda))$ random bits from a CDH instance. This is well in reach of the function given in Lemma 3. What is left to be shown is a code that handles a $\eta = 1/\log(\lambda)$ fraction of errors. A natural choice is a Reed Solomon code over the alphabet Σ^2, specifically a $[m, m - m/\log(\lambda), m/\log(\lambda) + 1]$ Reed Solomon code. For this code, we can efficiently decode $m/\log(\lambda) = \eta m$ erasures, ensuring correctness of our scheme. This code has rate $1 - 1/\log(\lambda)$.

5 Rate-1 DDH-Based Deterministic Encryption

In this section we show how to build TDFs satisfying DE security with the following two properties: (a) the index key contains $(n^2 + 1)$ group elements and (b) the image contains one group element plus exactly n bits. We mention that a recent result of Döttling et al. [DGI+19] achieve the same image size, but at the cost of bigger index keys, containing at least $4n^2$ group elements. Moreover, the construction of [DGI+19] is highly non-trivial, using techniques from [BGI16] as well as error-correcting codes. In contrast, our construction is fairly elementary and does not need ECC-based techniques.

We will make use of a balanced predicate during our construction, defined as follows.

Definition 16 (Balanced predicates). *We say a predicate* $\mathsf{P} : \mathsf{S} \times \{0,1\}^* \rightarrow \{0,1\}$ *is balanced over a set* S *if for all* $b_1, b_2 \in \{0,1\}$*:* $\Pr[\mathsf{P}(x_1; r) = b_1 \wedge \mathsf{P}(x_2; r) = b_2] = 1/4$*, where* $x_1, x_2 \xleftarrow{\$} \mathsf{S}$ *and* $r \xleftarrow{\$} \{0,1\}^*$*.*

An obvious example of a balanced predicate is the inner-product function mod 2. However, in some situations one may be able to give a more direct (and sometimes a deterministic) construction. For example, if the underlying set S is $\{0,1\}^n$, then we may simply define $\mathsf{P}(x) = x_1$.

Notation. For $\mathsf{x} \in \{0,1\}^n$ and a vector $\mathbf{g} := (g_1, \dots, g_n)$ we define $\mathsf{x} \cdot \mathbf{g} = \Pi_{i \in [n]} g_i^{\mathsf{x}_i}$.

Construction 17 (Linear-image TDF). *Let* G *be a group scheme and let* BL *be a balanced predicate for the underlying group (Definition 16).*

We define our TDF construction $(\mathsf{TDF.KG}, \mathsf{TDF.F}, \mathsf{TDF.F}^{-1})$ *as follows.*

[2] By increasing the the size of Σ to e.g. the next power of 2, the bit representation of each symbol in Σ grows by at most one bit, i.e., the rate of such an encoding is $1 - 1/\lambda$.

- $\mathsf{TDF.KG}(1^\lambda)$:
  1. *Sample* $(\mathbb{G}, p, g) \xleftarrow{\$} \mathsf{G}(1^\lambda)$ *and* $\mathbf{g} := (g_1, \ldots, g_n) \xleftarrow{\$} \mathbb{G}^n$.
  2. *For all* $i \in [n]$, *sample* $\rho_i \xleftarrow{\$} \mathbb{Z}_p$ *and set* $\mathbf{g}_i := \mathbf{g}^{\rho_i}$, *where* $\mathbf{g}^{\rho_i}$ *denotes element-wise exponentiation to the power of* ρ_i.
  3. *For each* $i \in [n]$ *sample random coins* $r_i \xleftarrow{\$} \{0,1\}^*$ *for* BL.[3]
  4. *Set* $\mathsf{tk} := (\rho_1, \ldots, \rho_n, \{r_i\})$ *as the trapdoor key and* $\mathsf{ik} := (\mathbf{g}, \mathbf{g}_1, \ldots, \mathbf{g}_n, (r_i)_{i \in [n]})$ *as the index key.*
- $\mathsf{TDF.F}(\mathsf{ik}, \mathsf{x} \in \{0,1\}^n)$: *Parse* $\mathsf{ik} := (\mathbf{g}, \mathbf{g}_1, \ldots, \mathbf{g}_n, (r_i)_{i \in [n]})$. *Return*

$$\mathsf{u} := (\mathsf{x} \cdot \mathbf{g}, \mathsf{BL}(\mathsf{x} \cdot \mathbf{g}_1; r_1) \oplus \mathsf{x}_1, \ldots, \mathsf{BL}(\mathsf{x} \cdot \mathbf{g}_n; r_n) \oplus \mathsf{x}_n) \in \mathbb{G} \times \{0,1\}^n. \quad (8)$$

- $\mathsf{TDF.F}^{-1}(\mathsf{tk}, \mathsf{u})$:
  1. *Parse* $\mathsf{tk} := (\rho_1, \ldots, \rho_n, (r_i)_{i \in [n]})$ *and* $\mathsf{u} := (g_c, b'_1, \ldots, b'_n)$.
  2. *Return* $(\mathsf{BL}(g_c^{\rho_1}; r_1) \oplus b'_1, \ldots, \mathsf{BL}(g_c^{\rho_n}; r_n) \oplus b'_n)$.

5.1 Analysis

The correctness of the scheme is immediate.

Lemma 18 (Deterministic-encryption security). *Assuming the underlying group is DDH-hard, then for any* $k \leq n$ *such that* $k \geq \log p + \omega(\log \lambda)$, *the TDF given in construction 17 provides* (k, n)*-CPA security.*

Proof. For any two (k, n)-sources $\mathcal{D}_0$ and $\mathcal{D}_1$ we need to show $(\mathsf{ik}, \mathsf{TDF.F}(\mathsf{ik}, \mathcal{D}_0)) \stackrel{c}{\equiv} (\mathsf{ik}, \mathsf{TDF.F}(\mathsf{ik}, \mathcal{D}_1))$, where $(\mathsf{ik}, *) \xleftarrow{\$} \mathsf{TDF.KG}(1^\lambda)$. We do this via a series of hybrids, where in each hybrid we sample $(\mathsf{ik}, \mathsf{u})$ as follows.

- Hyb_b [Real game for $\mathcal{D}_b$]: sample $\mathbf{g} \xleftarrow{\$} \mathbb{G}^n$ and set $\mathbf{g}_i := \mathbf{g}^{\rho_i}$, for $\rho_i \xleftarrow{\$} \mathbb{Z}_p$. Set $\mathsf{ik} := (\mathbf{g}, \mathbf{g}_1, \ldots, \mathbf{g}_n)$. Sample $\mathsf{x} \xleftarrow{\$} \mathcal{D}_b$ and return $(\mathsf{ik}, \mathsf{TDF.F}(\mathsf{ik}, \mathsf{x}))$.
- Hyb'_b:
  1. Sample $\mathbf{g} \xleftarrow{\$} \mathbb{G}^n$, $\mathsf{x} \xleftarrow{\$} \mathcal{D}_b$, and set $g_c := \mathsf{x} \cdot \mathbf{g}$.
  2. For $i \in [n]$ sample $\rho_i \xleftarrow{\$} \mathbb{Z}_p$ and set $\mathbf{g}'_i := \mathbf{g}^{\rho_i}$.
  3. Set $\mathbf{g}_i := \mathbf{g}'_i \cdot \mathbf{v}_i$, where

$$\mathbf{v}_i := (1, \ldots, 1, \underbrace{g'_i}_{i\text{th position}}, 1, \ldots, 1) \quad (9)$$

  and
     (a) sample $g'_i \xleftarrow{\$} \mathbb{G}$ in such a way that $1 \oplus \mathsf{BL}(g''_i \cdot g'_i; r_i) = \mathsf{BL}(g''_i; r_i)$, where $g''_i := \mathsf{x} \cdot \mathbf{g}'_i$.
  4. Set $\mathsf{ik} := (\mathbf{g}, \mathbf{g}_1, \ldots, \mathbf{g}_n)$ and $\mathsf{u} := \mathsf{TDF.F}(\mathsf{ik}, \mathsf{x})$.

[3] We can also prove security just by sampling a single r, but the proof will be more complicated.

Note 1 about Hyb'_b. For x and ik sampled as in Hyb'_b, we have the following relation: $\mathsf{TDF.F}(\mathsf{ik}, \mathsf{x}) = (g_c, \mathsf{BL}(g_c^{\rho_1}; r_1), \ldots, \mathsf{BL}(g_c^{\rho_n}; r_n))$. In particular, the output of $\mathsf{TDF.F}(\mathsf{ik}, \mathsf{x})$ can be sampled without knowing x, and just by knowing g_c and ρ_i's. Notice that the value of $g''_i := \mathsf{x} \cdot \mathbf{g}'_i$ (Item 3a above) can alternatively be computed as $g''_i = g_c^{\rho_i}$, without knowing x. We will make use of this fact in our proofs below.

Indistinguishability in Hyb': We have $\mathsf{Hyb}'_0 \stackrel{c}{\equiv} \mathsf{Hyb}'_1$. This follows from two facts. First, by Note 1 above, Hyb'_b can be sampled by just knowing g_c and ρ_i's and especially without knowing x. The second fact is that the distributions of g_c in Hyb'_0 and Hyb'_1 are statistically indistinguishable, by the leftover hash lemma and the fact that $\mathsf{H}_\infty(\mathcal{D}_b) \geq k$. These two facts together imply $\mathsf{Hyb}'_0 \stackrel{c}{\equiv} \mathsf{Hyb}'_1$.

Proof of $\mathsf{Hyb}_b \stackrel{c}{\equiv} \mathsf{Hyb}'_b$ for both $b \in \{0, 1\}$: We prove this for $b = 0$, and the proof for the other case is the same. To prove $\mathsf{Hyb}_0 \stackrel{c}{\equiv} \mathsf{Hyb}''_0$, define the following two hybrids:

- HybRnd_0: same as Hyb_0 except for every i we replace $\mathbf{g}_i$ with a random vector chosen uniformly from $\mathbb{G}^n$.
- HybRnd'_0: same as Hyb'_0 except for every i we replace $\mathbf{g}'_i$ with a random vector chosen uniformly from $\mathbb{G}^n$.

We will now show $\mathsf{Hyb}_0 \stackrel{c}{\equiv} \mathsf{HybRnd}_0 \stackrel{c}{\equiv} \mathsf{HybRnd}'_0 \stackrel{c}{\equiv} \mathsf{Hyb}'_0$, and this will complete the proof.

Proof for $\mathsf{Hyb}_0 \stackrel{c}{\equiv} \mathsf{HybRnd}_0$. The proof follows from DDH, by considering the fact that either hybrid can be simulated just by knowing $\mathbf{g}_i$ and that in one hybrid we have $\mathbf{g}_i := \mathbf{g}^{\rho_i}$ for a random exponent ρ_i, and in the other exponent $\mathbf{g}_i \stackrel{\$}{\leftarrow} \mathbb{G}^n$.

Proof for $\mathsf{HybRnd}_0 \stackrel{c}{\equiv} \mathsf{HybRnd}'_0$. These two distributions are identical, because in either distribution $\mathbf{g}_i$ is uniformly random.

Proof for $\mathsf{HybRnd}'_0 \stackrel{c}{\equiv} \mathsf{Hyb}'_0$. The proof follows from DDH, by considering the fact that either hybrid can be simulated just by knowing x and $\mathbf{g}'_i$'s, and that in one hybrid we have $\mathbf{g}'_i \stackrel{\$}{\leftarrow} \mathbb{G}^n$, and in the other hybrid $\mathbf{g}'_i := \mathbf{g}^{\rho_i}$ for a random exponent ρ_i.

6 Experimental Results

In this section we report proof-of-concept implementations of our DDH-based TDF construction (Construction 17) using Python. We report the resulting parameters of the scheme in Table 1.

Our group is an elliptic curve group on Ed25519 and the size of a group element in our implementation is 32 Bytes (B) = 256 bits. The encryption and decryption algorithms take less than a second. The table shows the growth of ciphertext size based on input size. We have not optimized our code for achieving more compact ciphertexts. Essentially, we used a serialization package (Pickle) which resulted in extra overhead in ciphertext size. As expected, the key-generation algorithm is main bottleneck in our implementation, together with the resulting index/trapdoor keys.

The machine specifications are as follows.

```
Architecture:            x86_64
CPU op-mode(s):          32-bit, 64-bit
Byte Order:              Little Endian
CPU(s):                  4
On-line CPU(s) list:     0-3
Thread(s) per core:      1
Core(s) per socket:      1
Socket(s):               4
NUMA node(s):            1
Vendor ID:               GenuineIntel
CPU family:              6
Model:                   6
Model name:              QEMU Virtual CPU version 2.5+
Stepping:                3
CPU MHz:                 2599.998
BogoMIPS:                5199.99
Hypervisor vendor:       KVM
Virtualization type:     full
L1d cache:               32K
L1i cache:               32K
L2 cache:                4096K
L3 cache:                16384K
NUMA node0 CPU(s):       0-3
```

Table 1. Experimental results of our TDF construction. Here B denotes bytes (8 bits). The size of the group element is 32 B.

msg	ct	tk	ik	KG time	Enc time	Dec time
64 B	274 B	18 KB	19 MB	15.6 s	0.11 s	0.04 s
128 B	338 B	35 KB	75 MB	62.3 s	0.42 s	0.07 s

References

[BBN+09] Bellare, M., et al.: Hedged public-key encryption: how to protect against bad randomness. In: Matsui, M. (ed.) ASIACRYPT 2009. LNCS, vol. 5912, pp. 232–249. Springer, Heidelberg (2009). https://doi.org/10.1007/978-3-642-10366-7_14

[BBO07] Bellare, M., Boldyreva, A., O'Neill, A.: Deterministic and efficiently searchable encryption. In: Menezes, A. (ed.) CRYPTO 2007. LNCS, vol. 4622, pp. 535–552. Springer, Heidelberg (2007). https://doi.org/10.1007/978-3-540-74143-5_30

[BCPT13] Birrell, E., Chung, K.-M., Pass, R., Telang, S.: Randomness-dependent message security. In: Sahai, A. (ed.) TCC 2013. LNCS, vol. 7785, pp. 700–720. Springer, Heidelberg (2013). https://doi.org/10.1007/978-3-642-36594-2_39

[BFO08] Boldyreva, A., Fehr, S., O'Neill, A.: On notions of security for deterministic encryption, and efficient constructions without random oracles. In: Wagner, D. (ed.) CRYPTO 2008. LNCS, vol. 5157, pp. 335–359. Springer, Heidelberg (2008). https://doi.org/10.1007/978-3-540-85174-5_19

[BFOR08] Bellare, M., Fischlin, M., O'Neill, A., Ristenpart, T.: Deterministic encryption: definitional equivalences and constructions without random oracles. In: Wagner, D. (ed.) CRYPTO 2008. LNCS, vol. 5157, pp. 360–378. Springer, Heidelberg (2008). https://doi.org/10.1007/978-3-540-85174-5_20

[BGI16] Boyle, E., Gilboa, N., Ishai, Y.: Breaking the circuit size barrier for secure computation under DDH. In: Robshaw, M., Katz, J. (eds.) CRYPTO 2016, Part I. LNCS, vol. 9814, pp. 509–539. Springer, Heidelberg (2016). https://doi.org/10.1007/978-3-662-53018-4_19

[BHY09] Bellare, M., Hofheinz, D., Yilek, S.: Possibility and impossibility results for encryption and commitment secure under selective opening. In: Joux, A. (ed.) EUROCRYPT 2009. LNCS, vol. 5479, pp. 1–35. Springer, Heidelberg (2009). https://doi.org/10.1007/978-3-642-01001-9_1

[BLSV18] Brakerski, Z., Lombardi, A., Segev, G., Vaikuntanathan, V.: Anonymous IBE, leakage resilience and circular security from new assumptions. In: Nielsen, J.B., Rijmen, V. (eds.) EUROCRYPT 2018, Part I. LNCS, vol. 10820, pp. 535–564. Springer, Cham (2018). https://doi.org/10.1007/978-3-319-78381-9_20

[DG17a] Döttling, N., Garg, S.: From selective IBE to full IBE and selective HIBE. In: Kalai, Y., Reyzin, L. (eds.) TCC 2017, Part I. LNCS, vol. 10677, pp. 372–408. Springer, Cham (2017). https://doi.org/10.1007/978-3-319-70500-2_13

[DG17b] Döttling, N., Garg, S.: Identity-based encryption from the diffie-hellman assumption. In: Katz, J., Shacham, H. (eds.) CRYPTO 2017, Part I. LNCS, vol. 10401, pp. 537–569. Springer, Cham (2017). https://doi.org/10.1007/978-3-319-63688-7_18

[DGHM18] Döttling, N., Garg, S., Hajiabadi, M., Masny, D.: New constructions of identity-based and key-dependent message secure encryption schemes. In: Abdalla, M., Dahab, R. (eds.) PKC 2018, Part I. LNCS, vol. 10769, pp. 3–31. Springer, Cham (2018). https://doi.org/10.1007/978-3-319-76578-5_1

[DGI+19] Döttling, N., Garg, S., Ishai, Y., Malavolta, G., Mour, T., Ostrovsky, R.: Trapdoor hash functions and their applications. In: Boldyreva, A., Micciancio, D. (eds.) CRYPTO 2019, Part III. LNCS, vol. 11694, pp. 3–32. Springer, Cham (2019). https://doi.org/10.1007/978-3-030-26954-8_1

[DH76] Diffie, W., Hellman, M.E.: New directions in cryptography. IEEE Trans. Inf. Theory **22**(6), 644–654 (1976)

[FGK+10] Freeman, D.M., Goldreich, O., Kiltz, E., Rosen, A., Segev, G.: More constructions of lossy and correlation-secure trapdoor functions. In: Nguyen, P.Q., Pointcheval, D. (eds.) PKC 2010. LNCS, vol. 6056, pp. 279–295. Springer, Heidelberg (2010). https://doi.org/10.1007/978-3-642-13013-7_17

[GGH19] Garg, S., Gay, R., Hajiabadi, M.: New techniques for efficient trapdoor functions and applications. In: Ishai, Y., Rijmen, V. (eds.) EUROCRYPT 2019, Part III. LNCS, vol. 11478, pp. 33–63. Springer, Cham (2019). https://doi.org/10.1007/978-3-030-17659-4_2

[GH18] Garg, S., Hajiabadi, M.: Trapdoor functions from the computational diffie-hellman assumption. In: Shacham, H., Boldyreva, A. (eds.) CRYPTO 2018, Part II. LNCS, vol. 10992, pp. 362–391. Springer, Cham (2018). https://doi.org/10.1007/978-3-319-96881-0_13

[GL89] Goldreich, O., Levin, L.A.: A hard-core predicate for all one-way functions. In: 21st ACM STOC, Seattle, WA, USA, 15–17 May 1989, pp. 25–32. ACM Press (1989)

[ILL89] Impagliazzo, R., Levin, L.A., Luby, M.: Pseudo-random generation from one-way functions (extended abstracts). In: 21st ACM STOC, Seattle, WA, USA, 15–17 May 1989, pp. 12–24. ACM Press (1989)

[LQR+19] Lombardi, A., Quach, W., Rothblum, R.D., Wichs, D., Wu, D.J.: New constructions of reusable designated-verifier NIZKs. In: Boldyreva, A., Micciancio, D. (eds.) CRYPTO 2019, Part III. LNCS, vol. 11694, pp. 670–700. Springer, Cham (2019). https://doi.org/10.1007/978-3-030-26954-8_22

[MY10] Mol, P., Yilek, S.: Chosen-ciphertext security from slightly lossy trapdoor functions. In: Nguyen, P.Q., Pointcheval, D. (eds.) PKC 2010. LNCS, vol. 6056, pp. 296–311. Springer, Heidelberg (2010). https://doi.org/10.1007/978-3-642-13013-7_18

[PW08] Peikert, C., Waters, B.: Lossy trapdoor functions and their applications. In: 40th ACM STOC, Victoria, BC, Canada, 17–20 May 2008, pp. 187–196. ACM Press (2008)

[PW11] Peikert, C., Waters, B.: Lossy trapdoor functions and their applications. SIAM J. Comput. **40**(6), 1803–1844 (2011)

[RS60] Reed, I.S., Solomon, G.: Polynomial codes over certain finite fields. J. Soc. Ind. Appl. Math. **8**(2), 300–304 (1960)

[WB86] Welch, L.R., Berlekamp, E.R.: Error correction for algebraic block codes, December 30 1986. US Patent 4,633,470

[Wee12] Wee, H.: Dual Projective hashing and its applications—lossy trapdoor functions and more. In: Pointcheval, D., Johansson, T. (eds.) EUROCRYPT 2012. LNCS, vol. 7237, pp. 246–262. Springer, Heidelberg (2012). https://doi.org/10.1007/978-3-642-29011-4_16

The Local Forking Lemma and Its Application to Deterministic Encryption

Mihir Bellare[1(✉)], Wei Dai[1], and Lucy Li[2]

[1] University of California San Diego, La Jolla, USA
{mihir,weidai}@eng.ucsd.edu
[2] Cornell University, Ithaca, USA
lucy@cs.cornell.edu

Abstract. We bypass impossibility results for the deterministic encryption of public-key-dependent messages, showing that, in this setting, the classical Encrypt-with-Hash scheme provides message-recovery security, across a broad range of message distributions. The proof relies on a new variant of the forking lemma in which the random oracle is reprogrammed on just a single fork point rather than on all points past the fork.

1 Introduction

Deterministic Encryption. In a scheme DE for Deterministic Public-Key Encryption (D-PKE) [2], the encryption algorithm DE.Enc takes public encryption key ek and message m to deterministically return a ciphertext c. The standard privacy goal is most easily understood as the same as for randomized public-key encryption—IND-CPA, asking for indistinguishability of encryptions of different messages—but with two restrictions: (1) That messages not depend on the public key, and (2) that messages be unpredictable, meaning have high min entropy. We will use the IND formalism of [5], shown by the latter to be equivalent to the PRIV formalization of [2] as well as to several other formalizations. A canonical and practical construction is EwH (Encrypt with Hash) [2]. It encrypts message m under a (any) randomized IND-CPA scheme RE with the coins set to a hash of $ek \| m$, and is proven IND-secure if the hash function is a random oracle [2]. Further schemes and considerations can be found in [6,13,14,19,20,27,30].

Why D-PKE? Determinism allows sorting of ciphertexts, enabling fast search on encrypted data, the motivating application in BBO's introduction of D-PKE [2]. Determinism also closes the door to vulnerabilities arising from poor randomness [15,28]. Understood to be a threat already when its causes were inadvertent system errors [31], poor randomness is now even more a threat when we see that it can be intentional, arising from the subversion of RNGs happening as part of mass-surveillance activities [11].

Narrowing the Gap. We benefit, in light of the above motivations for D-PKE, from the latter providing privacy as close to IND-CPA as possible. We can't expect of course to entirely close the gap—no D-PKE scheme can achieve IND-CPA—but

S. D. Galbraith and S. Moriai (Eds.): ASIACRYPT 2019, LNCS 11923, pp. 607–636, 2019.
https://doi.org/10.1007/978-3-030-34618-8_21

we'll narrow it. Our target will be the first of the two limitations of IND noted above, namely that it guarantees no privacy when messages depend on the public key. In particular, for all we know, in this case, one could recover the entire message from an EwH-ciphertext. This is the gap we will close, showing EwH message recovery is not possible across a broad range of public-key-dependent message distributions. We'll explain how this bypasses, rather than contradicts, prior impossibility results that have inhibited progress on the question, while also contributing new, more fine-grained impossibility results to indicate that our own possibility results will not extend much beyond the message distributions for which we establish them. Underlying our possibility result is a new variant of the forking lemma [1,7,29], that we call the Local Forking Lemma, of independent interest. We now look at all this in more detail.

Prior Work. Given that the public key is, as the name indicates, public, messages depending on it are a possibility in practice, and IND-CPA provides privacy even for such messages. But, for D-PKE, the literature says that security for public-key-dependent messages is impossible [2,5]. The argument supporting this claim is that the following attack violates IND-security of *any* D-PKE scheme DE. In its message finding stage, the adversary, given the public key ek, picks m_0, m_1 at random—of length, say, equal to the security parameter—subject to the constraint that the first bits of $h_0 \leftarrow \textsc{Hash}(\mathsf{DE.Enc}(ek, m_0))$ and $h_1 \leftarrow \textsc{Hash}(\mathsf{DE.Enc}(ek, m_1))$ are 0, 1, respectively, where $\textsc{Hash}$ is a random oracle. The messages m_0, m_1 are unpredictable, but given a ciphertext $c \leftarrow \mathsf{DE.Enc}(ek, m_b)$ encrypting m_b, the adversary can determine b as the first bit of the hash $\textsc{Hash}(c)$ of the ciphertext.

That IND cannot be achieved for public-key dependent messages doesn't mean no security is possible in this setting; perhaps guarantees can be provided under some other, meaningful metric (definition) of security X. Raghunathan, Segev and Vadhan (RSV) [30] were the first to pursue this, making a choice of X that we'll refer to as PDIND. In X = PDIND, security is parameterized by the number $N(\cdot)$ of (public-key dependent) distributions from which the message may be drawn. RSV [30] show that, if one first fixes an upper bound $N(\cdot) = 2^{p(\cdot)}$ on the number of allowed message distributions, then one can build a PDIND secure D-PKE scheme, with the scheme and its parameters depending on $N(\cdot)$. While theoretically interesting, this result has limitations from a practical perspective. The scheme is expensive, with key size and computation time growing polynomially with p, and this is inherent. Security is fragile: If the number of message distributions exceeds the bound $N(\cdot)$, security may—and in some of their schemes, will—fail. There is difficulty of use: it is not clear how a designer or implementer can, with confidence, pick $N(\cdot)$ a priori, but they must have $N(\cdot)$ in hand to build the scheme.

PDMR Security. Our target is a simple, meaningful security guarantee (when deterministically encrypting public-key dependent messages) that we can establish for *practical* schemes. We reach this by making a different choice of X above. We formalize and target X = PDMR, *message recovery* security for public-key-dependent messages. The definition, in Sect. 4, considers a *source* S that, given the public key *ek* and access to the random oracle HASH, returns a sequence of unpredictable messages. Encryptions under *ek* of these messages are then provided to the adversary A, who, continuing to have *ek* and access to HASH, must, to win, recover (in full) one of the messages. Unlike PDIND [30], there is no a priori restriction on the number of message distributions (here, sources).

One might object that message recovery security is a weak security guarantee, in response to which we note the following. *First*, in practice adversaries benefit more by recovering the full message from a ciphertext than by merely distinguishing the encryptions of two messages. So, even when distinguishing attacks are possible, a scheme preventing message recovery can add significant security. *Second*, right now, practical schemes like EwH are not proven to provide *any* security for public-key dependent messages, so if we can show PDMR is present, we have improved security guarantees without increasing cost. *Third*, in providing PDMR, we will insist that IND be maintained, so that overall security only goes up, not down. In other words, for messages not depending on the public key, we continue to provide the guarantee that is standard and viewed as best possible (IND), supplementing this with a meaningful guarantee (PDMR) for messages that do depend on the public key.

It is useful to define $n(\cdot)$-PDMR security as PDMR security for sources that output n messages. We will establish PDMR first for $n = 1$ and then boost to more messages.

One-Message PDMR Security of EwH. The core possibility result of this paper is that EwH is 1-PDMR secure, meaning provides message-recovery security for the encryption of one unpredictable message *even when the latter depends arbitrarily on the public key*. The underlying randomized public-key encryption scheme RE is assumed, only and correspondingly, to itself provide security against message-recovery. (This is implied by IND-CPA and hence true for EwH [2], but strictly weaker.) The hash function HASH continues, as in [2], to be modeled as a random oracle.

The proof requires new techniques. Let m denote the challenge message produced by the source, and let $c_1 \leftarrow \mathsf{RE.Enc}(ek, m; r_1)$ where $r_1 \leftarrow \mathrm{HASH}(ek\|m)$. The approach of [2] would replace c_1 with a ciphertext $c_0 \leftarrow \mathsf{RE.Enc}(ek, m; r_0)$ for random r_0, allowing a reduction to the assumed message-recovery security of RE. This requires that neither the source nor the adversary make query $ek\|m$ to HASH, for otherwise they can differentiate c_0 from c_1. But this in turn requires that the source not have *ek*. Indeed, in our setting, where it does have *ek*, it *can* query $ek\|m$ to HASH, and we must assume that it does so. The prior argument now breaks down entirely and it is not clear how to do the reduction. We obtain our result, instead, via a novel rewinding argument. Two executions are forked at the crucial hash query, one corresponding to response r_1 and the other to response r_0, but with a twist. In the classical rewinding technique [7,29], *all* answers to random-oracle queries after the fork are random and independent in the two forks. This fails to work in our case. Instead we are able to re-program the random oracle at just one point in the rewinding and argue that the two executions both result in correct guesses by the adversary.

The analysis relies on what we call the Local Forking lemma, a (new) variant of the forking lemmas of [1,7,29] that we give and prove. As with the General Forking Lemma of BN [7], our Local Forking Lemma is a purely probabilistic result, knowing or saying nothing about encryption. Handing off to our Local Forking Lemma the core probabilistic analysis in the above-discussed proof of 1-PDMR security not only makes the latter more modular but allows an extension to security against chosen-ciphertext attacks.

Many-Message PDMR Security of EwH. We show that EwH provides PDMR security for all sources (distributions on message sequences) that are what we call resampling indistinguishible (RI). Very roughly—the formal definition is in Sect. 5—RI asks that different messages in the sequence, although all allowed to depend on the public key in different ways, are themselves almost independently distributed.

Our first step is a general result showing that if a D-PKE scheme is 1-PDMR then it provides PDMR for *any* RI source. That is, once we have PDMR security when encrypting just one, single message, we also have it when encrypting *any* polynomial number of RI messages. This is a general result, holding for *any* D-PKE scheme. An interesting element of this result is that the public-key dependence of messages is a plus, exploited crucially in the proof.

To put this in context, for IND, security for one message does not, in general—that is, for arbitrary message distributions—imply security for multiple messages [2]. It has been shown to do so for particular message distributions, namely block sources, by Boldyreva, Fehr and O'Neill [13]. But they do not consider public-key-dependent messages, and block sources and RI distributions are not the same.

That EwH provides PDMR security for all RI sources now of course follows directly from the general reduction just mentioned and our above-discussed result establishing 1-PDMR security of EwH.

That these results are for EwH rather than some other scheme is important for two reasons. The first is that EwH is efficient and practical. The second is that we know that EwH already achieves IND for messages that do not depend on the public key [2]. As discussed above it is important that PDMR be provided while maintaining IND so that we augment, not reduce, existing guarantees.

CCA Too. All the above considered security under chosen-plaintext attack (CPA). This is certainly the first and foremost goal, but one can ask also about security against chosen-ciphertext attack (CCA), particularly if our quest is parity (to the best extent possible) with randomized encryption, where motivated by applications [12], efficient IND-CCA schemes have been sought and provided [9,17,18,21,23,25].

Our results extend to CCAs. Namely, we show that, under chosen-ciphertext attack, EwH continues to provide 1-PPDMR, and PDMR for RI sources, assuming the underlying randomized public-key encryption scheme itself provides message recovery under CCA, which is implied by IND-CCA. Put another way, EwH promotes message-recovery security of RE to message-recovery security of the constructed DE, in both the CPA and the CCA cases, for public-key dependent messages produced by RI sources. In the body of the paper, we give unified definitions and a single, unified result that cover both CCA and CPA by viewing the latter as the special case of the former in which adversaries make no decryption queries, exploiting our Local Forking lemma to provide a modular proof.

Impossibility Results. Our possibility results show that PDMR security is achievable when messages in the sequence are somewhat independent of each other, formalized as RI. We complement these possibility results with negative ones, showing that, when messages in a sequence are closely related, PDMR security is not possible. Section 6 gives attacks to show that PDMR security can be violated even when encrypting just two, closely related messages, even though both messages are unpredictable. This is true for *any* D-PKE scheme. These attacks are novel; the above-mentioned attacks understood in the literature violate indistinguishability security for public-key-dependent messages, but do not recover messages and thus, unlike ours, do not violate PDMR. We believe that a contribution here is not just to give these attacks, but with rigorous and formal analyses (Theorems 4 and 5), which is unusual in the literature. The proof of unpredictability in Theorem 5 relies on techniques from the proof of the Leftover Hash Lemma [24].

Discussion and Further Directions. It is interesting to note that Goldwasser and Micali's original definition of semantic security for public-key encryption [22] only required privacy for messages not depending on the public key. This was pointed out by Micali, Rackoff and Sloan (MRS) [26], who strengthened the definition in this regard. (In their terminology, this corresponds to three pass versus one pass notions). Modern definitions of semantic security (IND-CPA) [4, 16] accordingly ask for privacy even for messages that depend on the public key, and modern public-key encryption schemes provide this privacy. Our work continues the quest, started by RSV [30], to bring D-PKE to parity as much as possible in this regard.

There is a great deal of work on D-PKE including many schemes without random oracles [6,13,14,19,20,27,30]. A direction for future work is to assess whether these schemes provide PDMR security, or give new schemes without random oracles that provide both IND and PDMR security.

The full and most current version of this paper is available as [3].

2 Preliminaries

Notation and Terminology. By $\lambda \in \mathbb{N}$ we denote the security parameter and by 1^λ its unary representation. We denote the number of coordinates of a vector $\mathbf{x}$ by $|\mathbf{x}|$, the length of a string $x \in \{0,1\}^*$ by $|x|$ and the size of a set S by $|S|$. If x is a string then $x[i]$ is its i-th bit. Algorithms are randomized unless otherwise indicated. Running time is worst case. "PT" stands for "polynomial-time", whether for randomized algorithms or deterministic ones. For integers $a \leq b$ we let $[a..b] = \{a, a+1, \ldots, b\}$. We let $y \leftarrow A^{\mathsf{O}_1,\ldots}(x_1,\ldots;r)$ denote executing algorithm A on inputs $x_1,\ldots$ and coins r with access to oracles $\mathsf{O}_1,\ldots$ and letting y be the result. We let $y \leftarrow\$ A^{\mathsf{O}_1,\ldots}(x_1,\ldots)$ be the resulting of picking r at random and letting $y \leftarrow A^{\mathsf{O}_1,\ldots}(x_1,\ldots;r)$. We let $[A^{\mathsf{O}_1,\ldots}(x_1,\ldots)]$ denote the set of all possible outputs of A when invoked with inputs $x_1,\ldots$ and oracles $\mathsf{O}_1,\ldots$. We use $q_A^{\mathsf{O}_i}$ to denote the number of queries that A makes to O_i in the worst case. We recall that a function $f\colon \mathbb{N} \to \mathbb{R}$ is negligible if for every positive polynomial p, there exists $n_p \in \mathbb{N}$ such that $f(n) < 1/p(n)$ for all $n > n_p$. An adversary is an algorithm or a tuple of algorithms. The running time of a tuple of algorithms is defined as the sum of the individual running times. We use t_A to denote the running time of an adversary A.

Games. We use the code based game playing framework of [10]. (See Fig. 4 for an example). By $\mathrm{G} \Rightarrow y$ we denote the event that the execution of game G results in output y, the game output being what is returned by the game. We write $\Pr[\mathrm{G}]$ as shorthand for $\Pr[\mathrm{G} \Rightarrow \mathsf{true}]$, the probability that the game returns true.

Random Oracle Model (ROM). In the ROM [8], we give parties a random oracle $\textsc{Hash}$ that on input a string $x \in \{0,1\}^*$ returns a an output y that is (conceptually at least) a random, infinite string. The caller will then read a prefix of y, of any length it wants, and be charged, in terms of computation, an amount proportional only to the number of bits read.

Let $\mathbf{T}$ denote the set of all functions $T\colon \{0,1\}^* \to \{0,1\}^\infty$. Then, mathematically, a random oracle $\textsc{Hash}$ is a function drawn at random from $\mathbf{T}$. We view each $T \in \mathbf{T}$ as a table so that values in it can be reprogrammed, and thus may write $T[\cdot]$ in place of $T(\cdot)$. $\textsc{Hash}$ could be a procedure in games, for example in Fig. 3, where return values are sampled lazily as they are needed. Alternatively, we also sample the table T that describes $\textsc{Hash}$ uniformly at random from $\mathbf{T}$ at the beginning of the game (and write T in place of $\textsc{Hash}$), for example in Fig. 1. We note that the above two ways of implementing the random oracle $\textsc{Hash}$ are equivalent.

It is sometimes useful to give parties a variable output length random oracle. This takes two inputs, $x \in \{0,1\}^*$ and $\ell \in \mathbb{N}$, and returns a random ℓ-bit string, and, even for a fixed x, the outputs for different lengths ℓ must be independent. We can implement such a variable output length RO in our model above, and now discuss how. First, what does *not* work is to query x and take the ℓ-bit prefix of the infinite-length string returned, since in this case the result for x, ℓ is a prefix of the result for x, ℓ' whenever $\ell' > \ell$, and so the two are not independent as required. However, one can first fix an efficient injective encoding of the form $\{0,1\}^* \times \mathbb{N} \to \{0,1\}^*$. Then, a query of the form x, ℓ to a variable-length RO can be simulated by quering encoding of the pair (x, ℓ) to our single-input random oracle $\textsc{Hash}$. With this understood, we will work in our model above.

Game $\mathrm{G}^{\mathrm{single}}_{\mathrm{SAMP},F}$	Game $\mathrm{G}^{\mathrm{double}}_{\mathrm{SAMP},F}$
$\pi \leftarrow\!\!\$\ \mathrm{SAMP}()$ $T \leftarrow\!\!\$\ \mathbf{T}$ $(\alpha, x) \leftarrow F^T(\pi)$ Return $(\alpha \geq 1)$	$\pi \leftarrow\!\!\$\ \mathrm{SAMP}()$ $T \leftarrow\!\!\$\ \mathbf{T}$ $(\alpha, x) \leftarrow F^T(\pi)$ $T' \leftarrow T$ $T'[x] \leftarrow\!\!\$\ \{0,1\}^\infty$ $(\alpha', x') \leftarrow F^{T'}(\pi)$ Return $((\alpha = \alpha') \wedge (\alpha \geq 1))$

Fig. 1. Games $\mathrm{G}^{\mathrm{single}}_{\mathrm{SAMP},F}$ (single run) and $\mathrm{G}^{\mathrm{double}}_{\mathrm{SAMP},F}$ (double run) associated with algorithms SAMP and F.

3 The Local Forking Lemma

We consider two algorithms SAMP and F. The first could be randomized but has no oracle. The second is deterministic and has access to a random oracle HASH as defined in Sect. 2. These algorithms work as follows.

Via $\pi \leftarrow\!\!\$\ \mathrm{SAMP}()$, algorithm SAMP returns a value π that we think of as parameters that are input to F. Via $(\alpha, x) \leftarrow F^T(\pi)$, algorithm F, with input π, and with access to oracle $T \in \mathbf{T}$, returns a pair, where $\alpha \geq 0$ is an integer and x is a string. We require that if $\alpha \geq 1$ then x must be the α-th query that F has made to its oracle. If $\alpha = 0$, there is no requirement on x. Think of $\alpha = 0$ as denoting rejection and $\alpha \geq 1$ as denoting acceptance. We let q denote maximum value that α can take. Furthermore, we require that the first q queries that F make must be distinct.

Consider the games $\mathrm{G}^{\mathrm{single}}_{\mathrm{SAMP},F}$ and $\mathrm{G}^{\mathrm{double}}_{\mathrm{SAMP},F}$ in Fig. 1. They are parameterized by algorithms SAMP and F. Game $\mathrm{G}^{\mathrm{single}}_{\mathrm{SAMP},F}$ is a "normal" execution, in which π is sampled via SAMP, then F is executed with oracle T, the game returned true if $\alpha \geq 1$ (acceptance) and false if $\alpha = 0$ (rejection). Game $\mathrm{G}^{\mathrm{double}}_{\mathrm{SAMP},F}$ begins with the same "normal" run. Then, it reruns F with a different oracle T'. The difference is in just one point, namely the reply to the α-th query. Otherwise, T' is the same as T. This "local", as opposed to "global" change in T' versus T is the main difference from the General Forking Lemma of [7]. Our Local Forking Lemma relates the probability of these games returning true. Our proof follows the template of [7].

Lemma 1 (Local Forking Lemma). *Let* SAMP, F *and* q *be as above. Then*

$$\Pr[\mathrm{G}^{\mathrm{double}}_{\mathrm{SAMP},F}] \geq \frac{1}{q} \cdot \Pr[\mathrm{G}^{\mathrm{single}}_{\mathrm{SAMP},F}]^2 \ . \tag{1}$$

Proof (Lemma 1). Consider the games of Fig. 2. They are like the corresponding games of Fig. 1 except that $\pi \in [\mathrm{SAMP}()]$ is fixed as a parameter of the game rather than chosen via SAMP in the game. Our main claim, that we will establish below, is that for every $\pi \in [\mathrm{SAMP}()]$ we have

GAME $\mathrm{G}^{\mathrm{single}}_{\pi,F}$	GAME $\mathrm{G}^{\mathrm{double}}_{\pi,F}$
$T \leftarrow_{\$} \mathbf{T}$	$T \leftarrow_{\$} \mathbf{T}$
$(\alpha, x) \leftarrow F^T(\pi)$	$(\alpha, x) \leftarrow F^T(\pi)$
Return $(\alpha \geq 1)$	$T' \leftarrow T$
	$T'[x] \leftarrow_{\$} \{0,1\}^\infty$
	$(\alpha', x') \leftarrow F^{T'}(\pi)$
	Return $((\alpha = \alpha') \wedge (\alpha \geq 1))$

Fig. 2. Games $\mathrm{G}^{\mathrm{single}}_{\pi,F}$ (single run) and $\mathrm{G}^{\mathrm{double}}_{\pi,F}$ (double run), with the parameter π now fixed.

$$\Pr[\mathrm{G}^{\mathrm{double}}_{\pi,F}] \geq \frac{1}{q} \cdot \Pr[\mathrm{G}^{\mathrm{single}}_{\pi,F}]^2 . \tag{2}$$

From this we obtain Eq. (1) as in [7]. Namely, define $\mathsf{Y}_1, \mathsf{Y}_2$: $[\textsc{Samp}()] \to [0,1]$ by $\mathsf{Y}_1(\pi) = \Pr[\mathrm{G}^{\mathrm{single}}_{\pi,F}]$ and $\mathsf{Y}_2(\pi) = \Pr[\mathrm{G}^{\mathrm{double}}_{\pi,F}]$, and regard these as random variables over the choice of $\pi \leftarrow_{\$} \textsc{Samp}()$. Then, from Eq. (2), we have

$$\begin{aligned} \Pr[\mathrm{G}^{\mathrm{double}}_{\textsc{Samp},F}] &= \mathbf{E}[\mathsf{Y}_2] \\ &\geq \mathbf{E}\left[\frac{1}{q} \cdot \mathsf{Y}_1^2\right] \\ &\geq \frac{1}{q}\mathbf{E}\left[\mathsf{Y}_1\right]^2 \qquad (3) \\ &= \frac{1}{q} \cdot \Pr[\mathrm{G}^{\mathrm{single}}_{\textsc{Samp},F}]^2 , \end{aligned}$$

Where Eq. (3) is by Jensen's inequality. This establishes Eq. (1). We proceed to the main task, namely to prove Eq. (2). Henceforth, regard $\pi \in [\textsc{Samp}()]$ as fixed.

Since F makes a finite number of oracle queries and has finite running time, we can fix an integer L such that any query x made by F has $|x| \leq L$ and also the maximum number of bits of any reply read by F is at most L. This allows us to work over a finite sample space. Namely, let $D = \{0,1\}^{\leq L}$ be the set of all strings of length at most L and let $R = \{0,1\}^L$ be shorthand for the set of strings of length L. Then let OS be the set of all functions T: $D \to R$. Now we can view T in the games as being sampled from the finite set OS.

We let $Q_1, Q_2, \ldots, Q_q$ denote the query functions of F, corresponding to the first q queries. Function Q_i: $R^{i-1} \to D$ takes a list $h_1, \ldots, h_{i-1}$ of answers to queries $1, \ldots, i-1$ and returns the query that F would make next. To be formal, the only possible input to Q_1 is the empty string ε, and it returns the first query made by F, which is uniquely defined since F is deterministic. On input a string $h_1 \in R$, function Q_2 returns the query that F would make if it received h_1 as the answer to its first query. And so on, so that function Q_i, given $h_1, \ldots, h_{i-1} \in R$, returns the i-th query that F would make had it received $h_1, \ldots, h_{i-1}$ as responses to its prior queries. We note again that the determinism of F is important for these (deterministic) query functions to be well defined.

For $i \in [1..q]$ we let $\mathbf{Q}(h_1, \ldots, h_{i-1}) = (Q_1(\varepsilon), Q_2(h_1), \ldots, Q_i(h_1, \ldots, h_{i-1}))$ be the vector consisting of the first i queries given responses $h_1, \ldots, h_{i-1}$. Note that by our assumptions on F, the i entries of this vector are always distinct.

We will be wanting to tinker with a function T, erasing it at some points, and then adding in new values. We now develop some language to facilitate this. If V is a vector, we let $[V]$ denote the set whose elements are the entries of V, for example $[(1, 7, 5)] = \{1, 7, 5\}$. For a vector $Q \in D^i$ of possible queries, we let OS_Q denote the set of all functions $S\colon\ D \setminus [Q] \to R$, meaning functions just like those in OS but undefined at inputs in $[Q]$. Now if $S \in \mathsf{OS}_Q$ and $H \in R^i$ is a vector of possible answers, we let $S[H]$ denote the function $T \in \mathsf{OS}$ that reprograms S on the query points, leaving it intact on others. In detail, for $1 \le j \le i$ we let $T(Q[j]) = H[j]$, and for $x \notin [Q]$, we let $T(x) = S(x)$.

Recall that F's output is a pair of the form (α, x) where $0 \le \alpha \le q$ is an integer. We are only interested in the first output α, and it is convenient to let F_1 denote the algorithm that returns this. Also if $i, \alpha \ge 0$ are integers, $\mathrm{Ind}_i(\alpha)$ is defined to be 1 if $\alpha = i$ and 0 otherwise. Now suppose $i \in [1..q]$. We let Ω_i be the set of all $(h_1, \ldots, h_{i-1}, S)$ such that $h_1, \ldots, h_{i-1} \in R$ and $S \in \mathsf{OS}_{\mathbf{Q}(h_1,\ldots,h_{i-1})}$, meaning S is undefined at the first i queries made by F. The function $\mathsf{X}_i\colon\ \Omega_i \to [0, 1]$ is then defined by

$$\begin{aligned}\mathsf{X}_i(h_1, \ldots, h_{i-1}, S) &= \Pr\left[\alpha = i \;:\; h \leftarrow\!\!\$\, R\,;\ \alpha \leftarrow F_1^{S[(h_1,\ldots,h_{i-1},h)]}(\pi)\right]\\ &= \frac{1}{|R|} \cdot \sum_{h \in R} \mathrm{Ind}_i\left(F_1^{S[(h_1,\ldots,h_{i-1},h)]}(\pi)\right) .\end{aligned}$$

This function fixes the answers to the first $i-1$ queries, which uniquely determines the i-th query, and also fixes, as S the answers to all but these i queries, taking the probability only over the answer h to the i-th query. Let I and I' be the random variables taking values α and α', respectively, in game $\mathrm{G}^{\mathrm{double}}_{\pi,F}$. Then

$$\begin{aligned}\Pr[\mathrm{G}^{\mathrm{double}}_{\pi,F}] &= \Pr[\mathsf{I} \ge 1 \wedge \mathsf{I}' = \mathsf{I}]\\ &= \sum_{i=1}^{q} \Pr[\mathsf{I} = i \wedge \mathsf{I}' = i]\\ &= \sum_{i=1}^{q} \Pr[\,\mathsf{I} = i\,] \cdot \Pr[\,\mathsf{I}' = i \mid \mathsf{I} = i\,]\\ &= \sum_{i=1}^{q} \frac{1}{|\Omega_i|} \sum_{(h_1,\ldots,h_{i-1},S) \in \Omega_i} \mathsf{X}_i(h_1, \ldots, h_{i-1}, S)^2\\ &= \sum_{i=1}^{q} \mathbf{E}[\mathsf{X}_i^2] \qquad (4)\\ &\ge \sum_{i=1}^{q} \mathbf{E}[\mathsf{X}_i]^2 \,. \qquad (5)\end{aligned}$$

In Eq. (4), we regard X_i as a random variable over Ω_i, and refer to its expectation. Eq. (5) is by Jensen's inequality. Now recall that if $q \geq 1$ is an integer and $x_1, \ldots, x_q \geq 0$ are real numbers, then

$$q \cdot \sum_{i=1}^{q} x_i^2 \geq \left(\sum_{i=1}^{q} x_i \right)^2 .$$

This can be shown via Jensen's inequality or the Cauchy-Schwartz inequality, and a proof is in [7]. Setting $x_i = \mathbf{E}[\mathsf{X}_i]$, we have

$$q \cdot \sum_{i=1}^{q} \mathbf{E}[\mathsf{X}_i]^2 \geq \left(\sum_{i=1}^{q} \mathbf{E}[\mathsf{X}_i] \right)^2 .$$

At this point, we would like to invoke linearity of expectation to say that $\mathbf{E}[\mathsf{X}_1] + \cdots + \mathbf{E}[\mathsf{X}_q] = \mathbf{E}[\mathsf{X}_1 + \cdots + \mathsf{X}_q]$, but there is a difficulty, namely that linearity of expectation only makes sense when the random variables are over the same sample space, and ours are not, so the sum is not really even defined. (This is glossed over in [7]). So instead we expand the expectations again,

$$\begin{aligned} \sum_{i=1}^{q} \mathbf{E}[\mathsf{X}_i] &= \sum_{i=1}^{q} \frac{1}{|\Omega_i|} \sum_{(h_1,\ldots,h_{i-1},S)\in\Omega_i} \mathsf{X}_i(h_1, \ldots, h_{i-1}, S) \\ &= \sum_{i=1}^{q} \Pr[\mathsf{I} = i] \\ &= \Pr[\mathsf{I} \geq 1] = \Pr[\mathrm{G}^{\mathrm{single}}_{\pi,F}] . \end{aligned}$$

Putting all the above together, we have Eq. (2). □

4 Public-Key-Dependent Message-Recovery Security

We start by recalling definitions for public-key encryption schemes.

Public-Key Encryption. A public-key encryption (PKE) scheme PKE defines PT algorithms $\mathsf{PKE.Kg}, \mathsf{PKE.Enc}, \mathsf{PKE.Dec}$, the last deterministic. Algorithm $\mathsf{PKE.Kg}$ takes as input 1^λ and outputs a public encryption key $ek \in \{0,1\}^{\mathsf{PKE.ekl}(\lambda)}$ and a secret decryption key dk, where $\mathsf{PKE.ekl}\colon \mathbb{N} \to \mathbb{N}$ is the public-key length of PKE. Algorithm $\mathsf{PKE.Enc}$ takes as input $1^\lambda, ek$ and a message m with $|m| \in \mathsf{PKE.IL}(\lambda)$ to return a ciphertext $c \in \{0,1\}^{\mathsf{PKE.cl}(\lambda,|m|)}$, where $\mathsf{PKE.IL}$ is the input-length function of PKE, so that $\mathsf{PKE.IL}(\lambda) \subseteq \mathbb{N}$ is the set of allowed input (message) lengths, and $\mathsf{PKE.cl}\colon \mathbb{N} \times \mathbb{N} \to \mathbb{N}$ is the ciphertext length function of PKE. Algorithm $\mathsf{PKE.Dec}$ takes $1^\lambda, dk, c$ and outputs $m \in \{0,1\}^* \cup \{\bot\}$. Correctness requires that $\mathsf{PKE.Dec}(1^\lambda, dk, c) = m$ for all $\lambda \in \mathbb{N}$, all $(ek, dk) \in [\mathsf{PKE.Kg}(1^\lambda)]$ all m with $|m| \in \mathsf{PKE.IL}(\lambda)$ and all $c \in [\mathsf{PKE.Enc}(1^\lambda, ek, m)]$. Let $\mathsf{PKE.rl}\colon \mathbb{N} \to \mathbb{N}$ denote the randomness-length

Game $\mathbf{G}^{\$\mathrm{ind}}_{\mathsf{PKE},A}(\lambda)$	$\mathrm{LR}(m_0, m_1)$
$(ek, dk) \leftarrow_{\$} \mathsf{PKE.Kg}(1^\lambda)$ $b \leftarrow_{\$} \{0,1\}$ $b' \leftarrow_{\$} A^{\mathrm{LR},\mathrm{Dec}}(1^\lambda, ek)$ Return $(b = b')$	If $(\lvert m_0\rvert \neq \lvert m_1\rvert)$ Return $\bot$ $c \leftarrow_{\$} \mathsf{PKE.Enc}(1^\lambda, ek, m_b)$ $S \leftarrow S \cup \{c\}$ Return c
$\mathrm{Hash}(x, \ell)$	$\mathrm{Dec}(c)$
If not $T[x,\ell]$ then $\quad T[x,\ell] \leftarrow_{\$} \{0,1\}^\ell$ Return $T[x,\ell]$	If $c \in S$ then return $\bot$ $m \leftarrow_{\$} \mathsf{PKE.Dec}(1^\lambda, dk, c)$ Return m

Fig. 3. Game $\mathbf{G}^{\$\mathrm{ind}}$ defining \$IND security of PKE.

function of PKE, meaning $\mathsf{PKE.Enc}(1^\lambda, \cdot, \cdot)$ draws its coins at random from the set $\{0,1\}^{\mathsf{PKE.rl}(\lambda)}$.

Via game $\mathbf{G}^{\$\mathrm{ind}}_{\mathsf{PKE},A}(\lambda)$ of Fig. 3, we recall the definition of what is usually called IND-CCA. We use the notation \$IND to emphasize that this is for randomized schemes and to avoid confusion with "IND" also being a notion for D-PKE schemes [5], and we cut the "CCA" for succinctness. We explicitly write the random oracle Hash as a variable-output-length one, so that it takes a string x and integer ℓ to return a random ℓ-bit string. (This can be implemented as discussed in Sect. 2 via a RO that, like in Lemma 1, takes one string input and returns strings of infinite length.) We let

$$\mathsf{Adv}^{\$\mathsf{ind}}_{\mathsf{PKE},A}(\lambda) = 2\Pr[\mathbf{G}^{\$\mathrm{ind}}_{\mathsf{PKE},A}(\lambda)] - 1 .$$

We say that PKE is \$IND-secure if the function $\mathsf{Adv}^{\$\mathsf{ind}}_{\mathsf{PKE},A}(\cdot)$ is negligible for every PT adversary A. We don't have to define what is conventionally called IND-CPA separately, but can recover it by saying that PKE is \$IND-CPA secure if the function $\mathsf{Adv}^{\$\mathsf{ind}}_{\mathsf{PKE},A}(\cdot)$ is negligible for every PT adversary A that makes zero queries to the Dec oracle.

We say that a PKE scheme PKE is a deterministic public-key encryption (D-PKE) [2] scheme if the encryption algorithm DE.Enc is deterministic. Formally, $\mathsf{PKE.rl}(\cdot) = 0$, so that the randomness can only be the empty string.

The EwH D-PKE Scheme. We recall the Encrypt-with-Hash D-PKE scheme (formally, a transform) [2]. Let PKE be a PKE scheme. Then $\mathsf{DE} = \mathsf{EwH}[\mathsf{PKE}]$ is a ROM scheme defined as follows. First, $\mathsf{DE.Kg} = \mathsf{PKE.Kg}$ and $\mathsf{DE.Dec} = \mathsf{PKE.Dec}$, meaning the key generation and decryption algorithms of DE are the same as those of PKE. We also have that $\mathsf{DE.IL}(\lambda) = \mathsf{PKE.IL}(\lambda)$ and $\mathsf{DE.cl}(\lambda, \ell) = \mathsf{PKE.cl}(\lambda, \ell)$, for all λ and message lengths ℓ. We let $\mathsf{DE.rl}(\lambda) = 0$ for all λ. The encryption algorithm of DE is as follows:

$\mathsf{DE.Enc}^{\mathrm{Hash}}(1^\lambda, ek, m)$

$r \leftarrow \mathrm{Hash}(ek \| m, \mathsf{PKE.rl}(\lambda))$; $c \leftarrow \mathsf{PKE.Enc}(1^\lambda, ek, m; r)$
Return c

Above, Hash is the variable output length random oracle as discussed previously.

Game $\mathbf{G}^{\text{pred}}_{\mathsf{PKE},\mathcal{S},P}(\lambda)$	Game $\mathbf{G}^{\text{pdmr}}_{\mathsf{PKE},S,A}(\lambda)$
$(ek, dk) \leftarrow_{\$} \mathsf{PKE.Kg}(1^\lambda)$	$(ek, dk) \leftarrow_{\$} \mathsf{PKE.Kg}(1^\lambda)$
$cc \leftarrow_{\$} S.\mathsf{cx}(1^\lambda)$	$cc \leftarrow_{\$} S.\mathsf{cx}(1^\lambda)$
$\mathbf{m} \leftarrow_{\$} S.\mathsf{msg}^{\text{Hash},\text{Dec}}(1^\lambda, cc, ek)$	$\mathbf{m} \leftarrow_{\$} S.\mathsf{msg}^{\text{Hash},\text{Dec}}(1^\lambda, cc, ek)$
For $i = 1, \ldots, \lvert\mathbf{m}\rvert$ do	For $i = 1$ to $\lvert\mathbf{m}\rvert$ do
$\quad \boldsymbol{\ell}[i] \leftarrow \lvert\mathbf{m}[i]\rvert$	$\quad \mathbf{c}[i] \leftarrow_{\$} \mathsf{PKE.Enc}^{\text{Hash}}(1^\lambda, ek, \mathbf{m}[i])$
$(m, i) \leftarrow_{\$} P^{\text{Hash},\text{Dec}}(1^\lambda, cc, ek, \lvert\mathbf{m}\rvert, \boldsymbol{\ell})$	$(m, i) \leftarrow_{\$} A^{\text{Hash},\text{Dec}}(1^\lambda, cc, ek, \mathbf{c})$
Return $(m = \mathbf{m}[i])$	Return $(m = \mathbf{m}[i])$
$\underline{\text{Hash}(x, 1^\ell)}$	$\underline{\text{Hash}(x, 1^\ell)}$
If not $T[x, \ell]$ then	If not $T[x, \ell]$ then
$\quad T[x, \ell] \leftarrow_{\$} \{0,1\}^\ell$	$\quad T[x, \ell] \leftarrow_{\$} \{0,1\}^\ell$
Return $T[x, \ell]$	Return $T[x, \ell]$
$\underline{\text{Dec}(c)}$	$\underline{\text{Dec}(c)}$
Return $\mathsf{DE.Dec}(1^\lambda, dk, c)$	If $(\exists i : c = \mathbf{c}[i])$ then Return $\bot$
	Return $\mathsf{PKE.Dec}(1^\lambda, dk, c)$

Fig. 4. Left: Game defining unpredictability of source S. Right: Game defining PDMR security of PKE scheme PKE with source S and PDMR adversary A.

PDMR. We know that D-PKE cannot provide indistinguishability-style security for messages that depend on the public key [2]. We ask whether, for public-key dependent messages, it could nonetheless provide a form of security that, although weaker, is desirable and meaningful in practice, namely security against message recovery. Here we give the necessary definitions, but in the general setting of PKE instead of restricting to D-PKE.

Let PKE be a PKE scheme. A source S for PKE specifies PT algorithms $S.\mathsf{cx}$ and $S.\mathsf{msg}$, the first called the *context sampler* and the second called the *message sampler*. A PDMR adversary for source S is an algorithm A. We associate to PKE, S, A, and $\lambda \in \mathbb{N}$ the game $\mathbf{G}^{\text{pdmr}}_{\mathsf{PKE},S,A}(\lambda)$ in the right panel of Fig. 4. Via $cc \leftarrow_{\$} S.\mathsf{cx}(1^\lambda)$, the game samples the *context*. Via $\mathbf{m} \leftarrow_{\$} S.\mathsf{msg}^{\text{Hash},\text{Dec}}(1^\lambda, cc, ek)$, the message sampler $S.\mathsf{msg}$ produces a target-message vector $\mathbf{m}$. We require that $\lvert\mathbf{m}[i]\rvert \in \mathsf{PKE.IL}(\lambda)$ for all i. *The fact that* $S.\mathsf{msg}$ *has ek as input means that target messages may depend on the public key.* For $i = 1, \ldots, \lvert\mathbf{m}\rvert$, the game then encrypts message $\mathbf{m}[i]$ to create target ciphertext $\mathbf{c}[i]$. Via $(m, i) \leftarrow_{\$} A^{\text{Hash},\text{Dec}}(1^\lambda, cc, ek, \mathbf{c})$, the adversary A produces a (guess) message m and an index i in the range $1 \leq i \leq \lvert\mathbf{c}\rvert$; it is guessing that $\mathbf{m}[i] = m$, and wins if this guess is correct. Note that A is not allowed to query Dec on any of the ciphertexts in the vector $\mathbf{c}$. The PDMR-advantage $\mathsf{Adv}^{\mathsf{pdmr}}_{\mathsf{PKE},S,A}(\lambda) = \Pr[\mathbf{G}^{\text{pdmr}}_{\mathsf{PKE},S,A}(\lambda)]$ of A is the probability that the game returns true. For convenience of notation, we omit writing Dec in the superscript if the source or adversary do not query it.

Classes of Sources. We define classes of sources (a set of message samplers) as a convenient way to state our results. For $n\colon \mathbb{N} \to \mathbb{N}$, we let $\mathcal{S}^n$ denote the class of

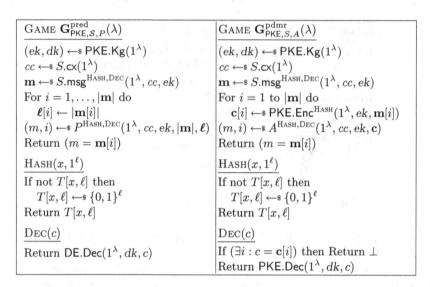

$\mathcal{S}^n$	Sources that output $n(\lambda)$ messages
$\mathcal{S}^{\mathsf{up}}$	Unpredictable sources
$\mathcal{S}^{\mathsf{ri}}$	Resampling-indistinguishable sources

Fig. 5. Classes of message samplers of interest. See text for explanations.

sources whose message sampler's output vector $\mathbf{m} \leftarrow S.\mathsf{msg}^{\textsc{Hash},\textsc{Dec}}(1^\lambda, \cdot, \cdot)$ has length $|\mathbf{m}| = n(\lambda)$. In some of our usage, n will be a constant and we will refer, for example to $\mathcal{S}^1$ or $\mathcal{S}^2$. Later we will define other classes as well. A summary is in Fig. 5.

Unpredictability. We cannot expect PDMR security for predictable target messages. Indeed, if, say, there are s known choices for $\mathbf{m}[1]$ then A can return one of them at random to get PDMR advantage $1/s$. Alternatively, A could encrypt all s candidates and return the one whose encryption equals $\mathbf{c}[1]$, getting an advantage of 1. We formalize unpredictability of a source S via game $\mathbf{G}^{\mathrm{pred}}_{\mathsf{PKE},S,P}$ specified in the left panel of Fig. 4, associated to D-PKE scheme PKE, source S and an adversary P that we call a predictor. Source S is run as in the message-recovery game. Next, instead of running A, predictor P is run and it tries to predict (guess) some component of $\mathbf{m}$. Unlike A, predictor P is not given $\mathbf{c}$. Instead it gets $|\mathbf{m}|$, the lengths of all component messages of this vector, and $1^\lambda, cc, ek$. Note that P gets the decryption oracle $\textsc{Dec}$, with no restrictions on querying it. Predictor P wins the game if $m = \mathbf{m}[i]$. For $\lambda \in \mathbb{N}$ we define the prediction advantage of P to be

$$\mathsf{Adv}^{\mathsf{pred}}_{\mathsf{PKE},S,P}(\lambda) = \Pr[\mathbf{G}^{\mathrm{pred}}_{\mathsf{PKE},S,P}(\lambda)] \; .$$

For $\lambda \in \mathbb{N}$ we also define

$$\mathsf{Adv}^{\mathsf{pred}}_{\mathsf{PKE},S}(\lambda) = \max_P \mathsf{Adv}^{\mathsf{pred}}_{\mathsf{PKE},S,P}(\lambda) \; .$$

where the maximum is over all predictors P, with no limit on their running time or the number of $\textsc{Hash}$ queries. We say that S is *unpredictable* if $\mathsf{Adv}^{\mathsf{pred}}_{\mathsf{PKE},S}(\cdot)$ is negligible. We let $\mathcal{S}^{\mathsf{up}}$ be the class of all unpredictable sources S.

Parameterized Security. We will see that achievability of PDMR security depends very much on the class (set) of sources. Let $\mathcal{S}$ be a class of sources. We say that PKE scheme PKE is PDMR-secure against $\mathcal{S}$ if $\mathsf{Adv}^{\mathsf{pdmr}}_{\mathsf{PKE},S,A}(\cdot)$ is negligible for all $S \in \mathcal{S}$ and all PT A. We say that PKE scheme PKE is PDMR-CPA-secure against $\mathcal{S}$ if $\mathsf{Adv}^{\mathsf{pdmr}}_{\mathsf{PKE},S,A}(\lambda)$ is negligible for all $S \in \mathcal{S}$ that make no $\textsc{Dec}$ queries and all PT A that make no $\textsc{Dec}$ queries.

\$IND Implies PDMR. We show that \$IND-security implies PDMR security for *randomized* PKE schemes. It is important that this *does not* apply to D-PKE schemes as these cannot achieve \$IND security. Let PKE be a PKE scheme, S be a source for PKE and A be a PDMR adversary for S. The following implies that if PKE is \$IND secure, then it is PDMR-secure against $\mathcal{S}^n \cap \mathcal{S}^{\mathsf{up}}$ for any polynomial n. Since the reduction preserves the number of decryption queries, the result holds in that case as well.

Proposition 1. *Let* PKE *be a PKE scheme, and n a polynomial. Let $S \in \mathcal{S}^n$ be a source for* PKE *and let A be a PDMR adversary. The proof gives \$IND adversary B and predictor P such that*

$$\mathsf{Adv}^{\$\mathsf{ind}}_{\mathsf{PKE},B}(\lambda) + \mathsf{Adv}^{\mathsf{pred}}_{\mathsf{PKE},S,P}(\lambda) \geq \mathsf{Adv}^{\mathsf{pdmr}}_{\mathsf{PKE},S,A}(\lambda) .$$

Furthermore, the resources of adversary B and predictor P relate to those of S and A as follows:

$$q^{\mathrm{LR}}_B = n, \quad q^{\mathrm{HASH}}_B = q^{\mathrm{HASH}}_S + q^{\mathrm{HASH}}_A, \quad q^{\mathrm{DEC}}_B = q^{\mathrm{DEC}}_S + q^{\mathrm{DEC}}_A, \quad t_B \approx t_S + t_A ,$$

and

$$q^{\mathrm{HASH}}_P = q^{\mathrm{HASH}}_A + n \cdot q^{\mathrm{HASH}}_{\mathsf{PKE.Enc}}, \quad q^{\mathrm{DEC}}_P = q^{\mathrm{DEC}}_A, \quad t_P \approx n \cdot t_{\mathsf{PKE.Enc}} + t_A .$$

Proof (of Proposition 1). \$IND adversary B and predictor P are as follows:

Adversary $B^{\mathrm{LR,HASH,DEC}}(1^\lambda, ek)$

$cc \leftarrow_\$ S.\mathsf{cx}(1^\lambda)$
$\mathbf{m} \leftarrow_\$ S.\mathsf{msg}^{\mathrm{HASH,DEC}}(1^\lambda, cc, ek)$
For $i = 1, \ldots |\mathbf{m}|$ do
  $\mathbf{m}'[i] \leftarrow_\$ \{0,1\}^{|\mathbf{m}[i]|}$
  $\mathbf{c}[i] \leftarrow_\$ \mathrm{LR}(\mathbf{m}'[i], \mathbf{m}[i])$
$(\overline{m}, i) \leftarrow_\$ A^{\mathrm{HASH,DEC}}(1^\lambda, cc, ek, \mathbf{c})$
Return $(\overline{m} = \mathbf{m}[i])$

Adversary $P^{\mathrm{HASH,DEC}}(1^\lambda, cc, ek, \boldsymbol{\ell})$

For $i = 1, \ldots, |\boldsymbol{\ell}|$ do
  $\mathbf{m}'[i] \leftarrow_\$ \{0,1\}^{\boldsymbol{\ell}[i]}$
  $\mathbf{c}[i] \leftarrow_\$ \mathsf{PKE.Enc}^{\mathrm{HASH}}(1^\lambda, ek, \mathbf{m}'[i])$
$(\overline{m}, i) \leftarrow_\$ A^{\mathrm{HASH,DECSIM}}(1^\lambda, cc, ek, \mathbf{c})$
Return $(\overline{m}, i)$

Algorithm $\mathrm{DECSIM}(x)$

If $(\exists i : x = \mathbf{c}[i])$ then return $\perp$
Return $\mathrm{DEC}(x)$

Adversary B uses $\mathbf{m}$ output by $S.\mathsf{msg}$ as well as $\mathbf{m}'$ that is sampled uniformly at random at each component i subjected to $|\mathbf{m}[i]| = |\mathbf{m}'[i]|$. Adversary B will query $\mathrm{LR}(m', m)$ to obtain ciphertext c. Adversary B then runs A on ciphertext $\mathbf{c}$ and checks if the guess of A matches message m. Predictor P obtains the encryption of a randomly sampled messages $\mathbf{m}'$ where component i has length $\boldsymbol{\ell}[i]$. Then it runs A and returns its output. We have

$$\begin{aligned}\mathsf{Adv}^{\$\mathsf{ind}}_{\mathsf{PKE},B}(\lambda) &= 2 \cdot \Pr[b = b'] - 1 \\ &= \Pr[b' = 1 \mid b = 1] - (1 - \Pr[b' = 0 \mid b = 0]) \\ &= \Pr[b' = 1 \mid b = 1] - \Pr[b' = 1 \mid b = 0] ,\end{aligned}$$

Where b' and b are random variables associated to game $\mathbf{G}^{\$\mathsf{ind}}_{\mathsf{PKE},B}(\lambda)$. It is standard to check that

$$\Pr[b' = 1 \mid b = 1] = \mathsf{Adv}^{\mathsf{pdmr}}_{\mathsf{PKE},S,A}(\lambda) , \tag{6}$$

and

$$\Pr[b' = 1 \mid b = 0] = \mathsf{Adv}^{\mathsf{pred}}_{\mathsf{PKE},S,P}(\lambda) . \tag{7}$$

Combining the above two equations, we obtain Proposition 1. □

5 Possibility Results

In this section, we show that when messages are not too strongly related to each other—more precisely when they are resampling-indistinguishable, to be defined shortly—PDMR security is possible. Furthermore this is not just in principle, but in practice: we show that such PDMR security is provided by the simple and efficient EwH scheme. Thus we can add, to the IND security for public-key independent messages we know this scheme already provides [2], a good privacy guarantee for messages that depend on the public key. This supports the security of existing or future uses of the scheme.

In more detail, our main technical result, Theorem 1, shows that $\mathsf{DE} = \mathsf{EwH}[\mathsf{PKE}]$ is PDMR-secure against $\mathcal{S}^1$ sources (namely, for the encryption of a single message) as long as the same is true for the randomized PKE. The proof relies crucially on Lemma 1. Note that this reduction does not need to assume unpredictability of the source. It follows from Proposition 1 that $\mathsf{DE} = \mathsf{EwH}[\mathsf{PKE}]$ is PDMR-secure against $\mathcal{S}^{\mathsf{up}} \cap \mathcal{S}^1$ sources as long as the randomized PKE is \$IND-secure.

The above is all for encryption of a single message. We will then turn to the encryption of multiple messages. We define a class of sources $\mathcal{S}^{\mathsf{ri}}$ that we call resampling indistinguishable. Such sources produce a polynomially-long vector of messages, reflecting that we are asking for privacy when encrypting many messages. Theorem 2 is a general result saying that *any* scheme that is PDMR-secure for $\mathcal{S}^1 \cap \mathcal{S}^{\mathsf{up}}$ is automatically PDMR security for $\mathcal{S}^{\mathsf{ri}} \cap \mathcal{S}^{\mathsf{up}}$, meaning PDMR-security for a single unpredictable message implies it for any polynomial number of unpredictable resampling-indistinguishable messages. Putting all this together, we get that $\mathsf{DE} = \mathsf{EwH}[\mathsf{PKE}]$ is PDMR-secure against $\mathcal{S}^{\mathsf{up}} \cap \mathcal{S}^{\mathsf{ri}}$ sources as long as the randomized PKE is \$IND-secure.

<u>Remark Regarding CPA.</u> All of the results in this section are stated in the presence of a decryption oracle. However, our reductions will preserve the number of decryption queries, so that analogous CPA-type result can be obtained simply by restricting the number of decryption queries q^{DEC} to be 0 for all sources and adversaries involved. Thus the statement "...PDMR(-CPA)-secure ...\$IND(-CPA)-secure ...", is read as two separate statements: "...PDMR-secure ...\$IND-secure" and "...PDMR-CPA-secure ...\$IND-CPA-secure ...".

<u>Remark Regarding PKE Schemes that Rely on a Random Oracle.</u> For simplicity we assume that the starting randomized PKE scheme is not a ROM scheme. However our result applies also to the case where it is in fact a ROM scheme like those of [9,18]. For this, we simply use domain separation, effectively making the RO used by EwH and the RO used by PKE independent.

5.1 Security of EwH for a Single Message

<u>Canonical 1-Sources and PDMR Adversaries.</u> Let $S \in \mathcal{S}^1$ be a source for DE, and A be a PDMR adversary for S. Since $S.\mathsf{msg}$ only produces one message, we can assume that the message index given by A is always 1. Hence, we can view the

$B^{\text{HASH,DEC}}(1^\lambda, cc, ek, c)$
$Q \leftarrow \emptyset$; $\mu \leftarrow\!\!\$\ A^{\text{HSIM,DEC}}(1^\lambda, cc, ek, c)$
$x \leftarrow\!\!\$\ Q$; $(ek\|m^*, \ell) \leftarrow x$
Return m^*

$\text{HSIM}(w)$
$Q \leftarrow Q \cup \{w\}$
Return $\text{HASH}(w)$

Algorithm SAMP
$(ek, dk) \leftarrow\!\!\$\ \mathsf{PKE.Kg}(1^\lambda)$
$cc \leftarrow\!\!\$\ S.\mathsf{cx}(1^\lambda)$
$\rho_S \leftarrow\!\!\$\ \{0,1\}^{S.\mathsf{msg.rl}(\lambda)}$
$\rho_A \leftarrow\!\!\$\ \{0,1\}^{A.\mathsf{rl}(\lambda)}$
Return $(ek, dk, cc, \rho_S, \rho_A)$

Subroutine $\text{DECSIM}(d)$
If $(d = c)$ then return $\perp$
Return $\mathsf{PKE.Dec}(1^\lambda, dk, d)$

Algorithm $F^{\text{HASH}}((ek, dk, cc, \rho_S, \rho_A))$
$j \leftarrow 0$; $Q \leftarrow \emptyset$; $\mathsf{ms} \leftarrow \mathsf{true}$
$m \leftarrow S.\mathsf{msg}^{\text{HSIM,DECSIM}}(1^\lambda, cc, ek; \rho_S)$
$x \leftarrow (ek\|m, \mathsf{PKE.rl})$; $r \leftarrow \text{HASH}(x)[1..\mathsf{PKE.rl}]$
$c \leftarrow \mathsf{PKE.Enc}(1^\lambda, ek, m; r)$; $\mathsf{ms} \leftarrow \mathsf{false}$
$\mu \leftarrow A^{\text{HASH,DECSIM}}(1^\lambda, cc, ek, c; \rho_A)$
If $(x \in Q)$ then $\alpha \leftarrow \text{Idx}(x)$ else $\alpha \leftarrow 0$
Return (α, x)

Subroutine $\text{HSIM}(w)$
If (ms) then $j \leftarrow j+1$; $\text{Idx}(w) \leftarrow j$
Else $Q \leftarrow Q \cup \{w\}$
Return $\text{HASH}(w)$

Fig. 6. Top is our PDMR adversary B against PKE in the proof of Theorem 1. It invokes a given PDMR adversary A against EwH[PKE]. Bottom are algorithms SAMP, F used in the analysis.

output of both $S.\mathsf{msg}$ and A as a single message. Next, we note that we can require $S.\mathsf{msg}$ and A to query HASH at $(ek\|m, \mathsf{PKE.rl}(\lambda))$ if they output m. This can always be done at the expense of one more query to HASH. For the following results, we shall assume canonical 1-sources and PDMR adversaries for them.

PDMR-Security of PKE Implies PDMR of EwH. The following says that if randomized scheme PKE is PDMR-secure for a source $S \in \mathcal{S}^1$, then so is deterministic scheme DE = EwH[PKE]. As noted above, the theorem itself does not assume unpredictability of the source. That will enter later.

Theorem 1. *Let* PKE *be a public-key encryption scheme. Let* DE = EwH[PKE] *be the associated deterministic public-key encryption scheme. Let* $S \in \mathcal{S}^1$ *be a 1-message source. Let* A *be a PDMR adversary for* S*, and let* B *be the PDMR adversary for* S *given in Fig. 6. Then*

$$\mathsf{Adv}^{\mathsf{pdmr}}_{\mathsf{PKE},S,B}(\lambda) \geq \frac{1}{(1+q_S^{\text{HASH}}) \cdot (1+q_A^{\text{HASH}})} \cdot \left(\mathsf{Adv}^{\mathsf{pdmr}}_{\mathsf{DE},S,A}(\lambda)\right)^2 .$$

Additionally $q_B^{\text{HASH}} \leq 1 + q_A^{\text{HASH}}$, $q_B^{\text{DEC}} = q_A^{\text{DEC}}$ *and* $t_B \approx t_A + \mathcal{O}(q_A^{\text{HASH}})$.

Proof (of Theorem 1). Let $\ell = \mathsf{PKE.rl}$. We assume that if $S^{\text{HASH,DEC}}(1^\lambda, cc, ek)$ outputs message m then it has always queried $(ek\|m, \ell)$ to HASH. Likewise, we assume that if $A^{\text{HASH,DEC}}(1^\lambda, cc, ek, c)$ outputs message μ then it has always queried $(ek\|\mu, \ell)$ to HASH. In both cases, as discussed above, this can be ensured

Games G_0, G_1

$(ek, dk, cc, \rho_S, \rho_A) \leftarrow\$ \textsc{Samp}()$
$j, j', j'' \leftarrow 0$; $Q, Q', Q'' \leftarrow \emptyset$; $c, c', c'' \leftarrow \bot$; $\mathsf{ms} \leftarrow \mathsf{true}$; $\ell \leftarrow \mathsf{PKE.rl}$

$T \leftarrow\$ \mathbf{T}$; $m \leftarrow S.\mathsf{msg}^{\textsc{HSim},\textsc{DecSim}}(1^\lambda, cc, ek; \rho_S)$; $x \leftarrow (ek \| m, \ell)$
$T' \leftarrow T$; $T'[x] \leftarrow\$ \{0,1\}^\infty$; $m' \leftarrow S.\mathsf{msg}^{\textsc{HSim}',\textsc{DecSim}'}(1^\lambda, cc, ek; \rho_S)$; $x' \leftarrow (ek \| m', \ell)$
$\alpha \leftarrow \mathrm{Idx}(x)$; $\alpha' \leftarrow \mathrm{Idx}'(x')$; $\mathsf{ms} \leftarrow \mathsf{false}$
$r \leftarrow T(x)[1..\ell]$; $r' \leftarrow T'(x')[1..\ell]$; $r'' \leftarrow T'(x)[1..\ell]$

$c \leftarrow \mathsf{PKE.Enc}(1^\lambda, ek, m; r)$; $\mu \leftarrow A^{\textsc{HSim},\textsc{DecSim}}(1^\lambda, cc, ek, c; \rho_A)$
$c' \leftarrow \mathsf{PKE.Enc}(1^\lambda, ek, m'; r')$; $\mu' \leftarrow A^{\textsc{HSim}',\textsc{DecSim}'}(1^\lambda, cc, ek, c'; \rho_A)$
$c'' \leftarrow \mathsf{PKE.Enc}(1^\lambda, ek, m; r'')$; $\mu'' \leftarrow A^{\textsc{HSim}'',\textsc{DecSim}''}(1^\lambda, cc, ek, c''; \rho_A)$

If $(x \notin Q)$ then $\alpha \leftarrow 0$
If $(x' \notin Q')$ then $\alpha' \leftarrow 0$

Return $(x \in Q'')$ // G_0
Return $((x \in Q'') \wedge (\alpha = \alpha') \wedge (\alpha \geq 1))$ // G_1

Procedure $\textsc{DecSim}(d)$
If $(d = c)$ then return $\bot$
Return $\mathsf{PKE.Dec}(1^\lambda, dk, d)$

Procedure $\textsc{DecSim}'(d)$
If $(d = c')$ then return $\bot$
Return $\mathsf{PKE.Dec}(1^\lambda, dk, d)$

Procedure $\textsc{DecSim}''(d)$
If $(d = c'')$ then return $\bot$
Return $\mathsf{PKE.Dec}(1^\lambda, dk, d)$

Procedure $\textsc{HSim}(w)$
If (ms) then $j \leftarrow j + 1$; $\mathrm{Idx}(w) \leftarrow j$
Else $Q \leftarrow Q \cup \{w\}$
Return $T(w)$

Procedure $\textsc{HSim}'(w)$
If (ms) then $j' \leftarrow j' + 1$; $\mathrm{Idx}'(w) \leftarrow j$
Else $Q' \leftarrow Q' \cup \{w\}$
Return $T'(w)$

Procedure $\textsc{HSim}''(w)$
If (not ms) then $Q'' \leftarrow Q'' \cup \{w\}$
Return $T(w)$

Fig. 7. Games G_0, G_1 for proof of Theorem 1, in the top box, differ only in their Return statements, and use the procedures in the bottom box.

by modifying the algorithm to make the required query if it did not already do so, increasing the number of Hash queries by at most one. So, letting $q_1 = q_S^{\textsc{Hash}}$ and $q_2 = q_A^{\textsc{Hash}}$, we now regard the number of Hash queries of S and A as $1 + q_1$ and $1 + q_2$, respectively. We assume that all Hash queries of S are distinct, and also that all Hash queries of A are distinct. Crucially, we do not, and cannot, assume distinctness across these queries, meaning A could repeat queries made by S.

Fix some $\lambda \in \mathbb{N}$. We start the analysis with the Samp algorithm of Fig. 6. (Ignore the rest of that Figure for now). It picks keys, common coins cc, coins ρ_S for the message-finding phase of sampler S, and coins ρ_A for A, so that these can be fixed and maintained across multiple executions of the algorithms. Now consider

Games G_2, G_3

$(ek, dk, cc, \rho_S, \rho_A) \leftarrow\$ \textsc{Samp}()$

$j, j', j'' \leftarrow 0$; $Q, Q', Q'' \leftarrow \emptyset$; $c, c', c'' \leftarrow \bot$; $\mathsf{ms} \leftarrow \mathsf{true}$; $\ell \leftarrow \mathsf{PKE.rl}$

$T \leftarrow\$ \mathbf{T}$; $m \leftarrow S.\mathsf{msg}^{\textsc{HSim},\textsc{DecSim}}(1^\lambda, cc, ek; \rho_S)$; $x \leftarrow (ek\|m, \ell)$

$T' \leftarrow T$; $T'[x] \leftarrow\$ \{0,1\}^\infty$; $m' \leftarrow S.\mathsf{msg}^{\textsc{HSim}',\textsc{DecSim}'}(1^\lambda, cc, ek; \rho_S)$; $x' \leftarrow (ek\|m', \ell)$

$\alpha \leftarrow \mathrm{Idx}(x)$; $\alpha' \leftarrow \mathrm{Idx}'(x')$; $\mathsf{ms} \leftarrow \mathsf{false}$

$r \leftarrow T(x)[1..\ell]$; $r' \leftarrow T'(x)[1..\ell]$

$c \leftarrow \mathsf{PKE.Enc}(1^\lambda, ek, m; r)$; $\mu \leftarrow A^{\textsc{HSim},\textsc{DecSim}}(1^\lambda, cc, ek, c; \rho_A)$

$c' \leftarrow \mathsf{PKE.Enc}(1^\lambda, ek, m; r')$; $\mu' \leftarrow A^{\textsc{HSim}',\textsc{DecSim}'}(1^\lambda, cc, ek, c'; \rho_A)$

$\mu'' \leftarrow A^{\textsc{HSim}'',\textsc{DecSim}'}(1^\lambda, cc, ek, c'; \rho_A)$ // G_2

$\mu'' \leftarrow \mu'$ // G_3

If $(x \notin Q)$ then $\alpha \leftarrow 0$

If $(x \notin Q')$ then $\alpha' \leftarrow 0$

Return $((x \in Q'') \wedge (\alpha = \alpha') \wedge (\alpha \geq 1))$ // G_2

Return $((x \in Q') \wedge (\alpha = \alpha') \wedge (\alpha \geq 1))$ // G_3

Fig. 8. Games $\mathrm{G}_2, \mathrm{G}_3$ for the proof of Theorem 1. They use the procedures at the bottom of Fig. 7.

games $\mathrm{G}_0, \mathrm{G}_1$ at the top of Fig. 7. They invoke Samp at the very beginning. They also invoke the procedures in the bottom of Fig. 7. We claim that

$$\mathsf{Adv}^{\mathsf{pdmr}}_{\mathsf{PKE},S,B}(\lambda) \geq \frac{1}{1+q_2} \cdot \Pr[\mathrm{G}_0] . \tag{8}$$

This is justified as follows. The message m in G_0 is created just as in game $\mathbf{G}^{\mathsf{pdmr}}_{\mathsf{PKE},S,A}(\lambda)$, the oracle Hash being set, by procedure HSim, to T. In game $\mathbf{G}^{\mathsf{pdmr}}_{\mathsf{PKE},S,A}(\lambda)$, ciphertext c is created by encryption of m under coins that are random and independent of Hash, captured in G_0 as $T'[x]$. However, B runs A with its own oracle Hash, here T, not T', captured in G_0 as HSim$''$. We have written HSim and HSim$''$ as two, separate, oracles, even though both reply simply via T, because they keep track of different things. In the message-sampling phase (flag $\mathsf{ms} = \mathsf{true}$) they store the index of each query, and when A is run (flag $\mathsf{ms} = \mathsf{false}$), they store the queries in a set. Note that in G_0, we are not concerned with $x', r', c', \mu, \mu', \alpha, \alpha'$, meaning all these quantities can be ignored in the context of Eq. (8). Game G_0 returns true if $x \in Q''$, meaning if A made query $x = (ek\|m, \ell)$ to HSim$''$. We have assumed that A always makes hash query $(ek\|\mu, \ell)$ on output μ, and we have $|Q''| \leq 1 + q_2$, yielding Eq. (8).
Games $\mathrm{G}_0, \mathrm{G}_1$ differ only in what they return, and the boolean returned by G_1 is the one returned by G_0 ANDed with more stuff. So, regardless of what is this stuff, we must have

$$\Pr[\mathrm{G}_0] \geq \Pr[\mathrm{G}_1] . \tag{9}$$

Suppose the winning condition of game G_1 is met, so that $\alpha = \alpha' \neq 0$. This implies $(x, m, r', c') = (x', m', r'', c'')$. To explain, we have assumed the hash queries of S are distinct, we have maintained the coins of S across the runs, and T, T' differ only at x, so until x is queried, the executions of S are the same, so $x = x'$. This implies $r' = r''$. From the definitions of x, x' we get $m = m'$, and thus we also get $c' = c''$. In game G_2 of Fig. 8—the procedures used continue to be those at the bottom of Fig. 7—we rewrite and simplify the code of G_1 under the assumption that $(x, m, r', c') = (x', m', r'', c'')$. Since G_2 maintains the winning condition of G_1, and we have seen this implies $(x, m, r', c') = (x', m', r'', c'')$, we have

$$\Pr[\mathrm{G}_1] = \Pr[\mathrm{G}_2]\ . \tag{10}$$

In game G_2, consider the computations of μ' and μ''. The only difference is that in the first A has oracle $\textsc{HSim}'$, and in the second, $\textsc{HSim}''$. However, the replies from these oracles differ only at query x, and the winning condition of G_2 depends only on x and other quantities determined prior to the reply to hash query x being obtained by A. This means that the winning condition of game G_3 is equivalent to that of G_2. (Game G_3 no longer computes μ'' as in game G_2 to ensure $\textsc{HSim}''$ is no longer used, and sets μ'' instead, correctly, to μ', but this quantity is not used in the winning condition). We have

$$\Pr[\mathrm{G}_2] = \Pr[\mathrm{G}_3]\ . \tag{11}$$

Now consider algorithm F of Fig. 6, and consider executing game $\mathrm{G}^{\mathrm{double}}_{\textsc{Samp},F}$ of Fig. 1. We have

$$\Pr[\mathrm{G}_3] \geq \Pr[\mathrm{G}^{\mathrm{double}}_{\textsc{Samp},F}] \tag{12}$$

$$\geq \frac{1}{1+q_2} \cdot \Pr[\mathrm{G}^{\mathrm{single}}_{\textsc{Samp},F}]^2\ , \tag{13}$$

where Eq. 13 is by Lemma 1. Now we observe that

$$\Pr[\mathrm{G}^{\mathrm{single}}_{\textsc{Samp},F}] \geq \mathsf{Adv}^{\mathsf{pdmr}}_{\mathsf{DE},S,A}(\lambda)\ . \tag{14}$$

Combining the equations above completes the proof. □

PDMR Security of EwH for Unpredictable One-Message Sources. An immediate corollary of Proposition 1 and Theorem 1 is that \$IND(-CPA) security of PKE implies PDMR(-CPA)-security of $\mathsf{EwH}[\mathsf{PKE}]$ against $\mathcal{S}^1 \cap \mathcal{S}^{\mathsf{up}}$.

Corollary 1. *Let* PKE *be a public-key encryption scheme. Let* $\mathsf{DE} = \mathsf{EwH}[\mathsf{PKE}]$ *be the associated deterministic public-key encryption scheme. Let* $S \in \mathcal{S}^1$ *be a 1-message source. Let* A *be a PDMR adversary for* S. *The proof specifies PDMR adversary* B *for* S, *and predictor* P, *such that*

$$\mathsf{Adv}^{\mathsf{pdmr}}_{\mathsf{DE},S,A}(\lambda) \leq \sqrt{(1+q_S^{\textsc{Hash}})(1+q_A^{\textsc{Hash}})\left(\mathsf{Adv}^{\mathsf{\$ind}}_{\mathsf{PKE},B}(\lambda) + \mathsf{Adv}^{\mathsf{pred}}_{\mathsf{PKE},S,P}(\lambda)\right)}\ .$$

Game $\mathbf{G}^{\mathrm{ri}}_{\mathsf{DE},S,D}(\lambda)$	$\mathrm{Hash}(x, 1^{\ell})$
$(ek, dk) \leftarrow \mathsf{DE.Kg}(1^\lambda)$	If not $T[x,\ell]$ then
$cc \leftarrow_{\$} S.\mathsf{cx}(1^\lambda)$; $b \leftarrow_{\$} \{0,1\}$	$T[x,\ell] \leftarrow_{\$} \{0,1\}^{\ell}$
$j \leftarrow_{\$} [n(\lambda)]$	Return $T[x,\ell]$
$\mathbf{m}_0 \leftarrow_{\$} S.\mathsf{msg}^{\mathrm{Hash},\mathrm{Dec}}(1^\lambda, cc, ek)$	$\mathrm{Dec}(c)$
$\mathbf{m}_1 \leftarrow \mathbf{m}_0$	Return $\mathsf{DE.Dec}(1^\lambda, dk, c)$
$\mathbf{m}_1[j] \leftarrow_{\$} S.\mathsf{msg}^{\mathrm{Hash},\mathrm{Dec}}(1^\lambda, cc, ek)[j]$	
$b' \leftarrow_{\$} D^{\mathrm{Hash},\mathrm{Dec}}(1^\lambda, cc, ek, \mathbf{m}_b, j)$	
Return $(b = b')$	

Fig. 9. Game defining resampling indistinguishability of source S for DE.

The resources of B and P are related to those of S and A as follows:

$$q_B^{\mathrm{Hash}} = q_S^{\mathrm{Hash}} + q_A^{\mathrm{Hash}}, \quad q_B^{\mathrm{Dec}} = q_S^{\mathrm{Dec}} + q_A^{\mathrm{Dec}}, \quad t_B \approx t_S + t_A ,$$

and

$$q_P^{\mathrm{Hash}} = q_A^{\mathrm{Hash}} + q_{\mathsf{Enc}}^{\mathrm{Hash}}, \quad q_P^{\mathrm{Dec}} = q_A^{\mathrm{Dec}}, \quad t_P \approx t_{\mathsf{Enc}} + t_A .$$

5.2 Resampling Indistinguishability

We define what it means for an adversary A to be resampling indistinguishable. At a high level, the condition is that, the distribution of the vector of messages produced by the adversary is not detectably changed by replacing one of the components of the vector with a component from another vector produced by a second run of the adversary using independent coins. This captures a weak form of independence of the components of the vector. We give accompanying examples after the precise definition.

Definition. Let DE be a D-PKE scheme. Consider the game $\mathbf{G}^{\mathrm{ri}}_{\mathsf{DE},S,D}$ given in Fig. 9, where S is a $n(\lambda)$-source for DE and D is an adversary called the resampling distinguisher. In this game, a message vector $\mathbf{m}_0$ is obtained by running $S.\mathsf{msg}$. Then $\mathbf{m}_1$ is created to be the same as $\mathbf{m}_0$ except at one, random, location j. The value it takes at j is the j-th component of a message vector obtained by running $S.\mathsf{msg}$ again, independently and with fresh coins, but on the same inputs $1^\lambda, cc, ek$. Finally, D takes input $(1^\lambda, cc, ek, \mathbf{m}_b, j)$ and attempts to guess the value of b. We let

$$\mathsf{Adv}^{\mathrm{ri}}_{\mathsf{DE},S,D}(\lambda) = 2\Pr[\mathbf{G}^{\mathrm{ri}}_{\mathsf{DE},S,D}(\lambda)] - 1 .$$

We say that S is resampling-indistinguishable if the function $\mathsf{Adv}^{\mathrm{ri}}_{\mathsf{DE},S,D}(\cdot)$ is negligible for any PT distinguisher D. We let $\mathcal{S}^{\mathrm{ri}}$ be the class of resampling-indistinguishable sources.

Examples of Message Samples in $\mathcal{S}^{\mathrm{ri}}$. We give some examples of RI sources. First, if each $\mathbf{m}[i]$ is sampled independently from some distribution depending on i, then S is RI *even when these distribution depends on the public key.* More precisely, suppose, for some PT algorithm X and polynomial $n(\cdot)$, sampler $S.\mathsf{msg}$ works as follows:

$$\begin{array}{l}\underline{\text{Adversary } S.\mathsf{msg}^{\textsc{Hash}}(1^\lambda, cc, ek)}\\ \text{For } i = 1,\ldots,n(\lambda) \text{ do } \mathbf{m}[i] \leftarrow\!\!\$\; X^{\textsc{Hash}}(1^\lambda, cc, ek, i)\\ \text{Return } \mathbf{m}\end{array}$$

Then, for any choices of X, n, sampler $S.\mathsf{msg}$ as above (together with any context sampler) is RI. Moreover, S is perfectly RI, i.e. $\mathsf{Adv}^{\mathsf{ri}}_{A,D}(\lambda) = 0$ for any distinguisher D. Note that the class of such adversaries, defined by all the choices of PT X and polynomials n, is too large for the constructions of RSV [30], so our positive results give schemes providing security for classes of message distributions for which their schemes do not provide security. This example extends naturally to sources S' such that the output of $S'.\mathsf{msg}$ is *indistinguishable* from the output of $S.\mathsf{msg}$ (for some choice of X and $n(\cdot)$). The notion of RI also allows us to capture correlation in $\mathbf{m}$ that cannot be efficiently detected. For example, consider S that does the following. It first generate a random string $r \leftarrow\!\!\$\; \{0,1\}^n$. Then, it sets $\mathbf{m}[i] \leftarrow \textsc{Hash}(r\|i, 1^n)$ for $i \in \{1,2\}$. Note that there is strong information-theoretic correlation between $\mathbf{m}[1]$ and $\mathbf{m}[2]$, given the entire function table of $\textsc{Hash}$. However, any distinguisher D making q queries to $\textsc{Hash}$ cannot detect this correlation with advantage more than $q/2^n$. Finally, we note that resampling-indistinguishability is independent of predictability. In particular, if X always returns a constant message (that is compatible with the message space of the encryption scheme), then the source constructed before is still RI, but it is trivially predictable.

Reduction to 1-PDMR Security. A useful property of RI adversaries is that their PDMR security reduces to the PDMR security of the encryption of just one message. This is formalized via the theorem below, which says that DE is PDMR-(CCA-)secure for $\mathcal{S}^{\mathsf{up}} \cap \mathcal{S}^1$, then it is PDMR-(CCA-)secure for $\mathcal{S}^{\mathsf{up}} \cap \mathcal{S}^{\mathsf{ri}}$.

Theorem 2. *Let* DE *be any D-PKE scheme. Let* S_1 *be any* $n(\lambda)$*-source and* A *be a PDMR adversary for D-PKE scheme* DE*. Consider the* 1*-source* S_2 *and PDMR adversary* B *given in Fig. 10. Then*

$$\mathsf{Adv}^{\mathsf{pdmr}}_{\mathsf{DE},S_1,A}(\lambda) \le n(\lambda) \cdot \left(\mathsf{Adv}^{\mathsf{pdmr}}_{\mathsf{DE},S_2,B}(\lambda) + \mathsf{Adv}^{\mathsf{ri}}_{\mathsf{DE},S_1,D}(\lambda)\right) . \tag{15}$$

Source S_2*, adversary* B*, and distinguisher* D *are efficient as long as* S_1 *and* A *are. In particular,*

$$q_B^{\textsc{Hash}} = q_{S_1}^{\textsc{Hash}} + n(\lambda) \cdot q_{\mathsf{DE.Enc}}^{\textsc{Hash}} + q_A^{\textsc{Hash}}, \quad q_B^{\textsc{Dec}} = q_{S_1}^{\textsc{Dec}} + q_A^{\textsc{Dec}} ,$$
$$t_B \approx t_S + n(\lambda) \cdot t_{\mathsf{DE.Enc}} + t_A ,$$
$$q_D^{\textsc{Hash}} = n(\lambda) \cdot q_{\mathsf{DE.Enc}}^{\textsc{Hash}} + q_A^{\textsc{Hash}}, \quad q_D^{\textsc{Dec}} = q_A^{\textsc{Dec}} ,$$
$$t_D \approx n(\lambda) \cdot t_{\mathsf{DE.Enc}} + t_A .$$

$S_2.\mathsf{cx}(1^\lambda)$

$cc \leftarrow\!\!\$\; S_1.\mathsf{cx}(1^\lambda)\;;\; j \leftarrow\!\!\$\; [n]$
Return (cc, j)

$S_2.\mathsf{msg}^{\textsc{Hash},\textsc{Dec}}(1^\lambda, \overline{cc}, ek)$

$(cc, j) \leftarrow \overline{cc}$
$\mathbf{m} \leftarrow S_1.\mathsf{msg}^{\textsc{Hash},\textsc{Dec}}(1^\lambda, cc, ek)$
Return $\mathbf{m}[j]$

$B^{\textsc{Hash},\textsc{Dec}}(1^\lambda, \overline{cc}, ek, c)$

$(cc, j) \leftarrow \overline{cc}$
$\mathbf{m} \leftarrow\!\!\$\; S_1.\mathsf{msg}^{\textsc{Hash},\textsc{Dec}}(1^\lambda, cc, ek)$
For $i \leftarrow 1, \ldots, n(\lambda)$ do
  $\mathbf{c}[i] \leftarrow\!\!\$\; \mathsf{DE.Enc}^{\textsc{Hash}}(1^\lambda, ek, \mathbf{m}[i])$
$\mathbf{c}[j] \leftarrow c$
$(m, i) \leftarrow A^{\textsc{Hash},\textsc{DecSim}}(1^\lambda, cc, ek, \mathbf{c})$
If $(i = j)$ then Return m
Else Return $\bot$

Algorithm $\textsc{DecSim}(x)$

If $(\exists i : x = \mathbf{c}[i])$ then return $\bot$
Return $\textsc{Dec}(x)$

$D^{\textsc{Hash},\textsc{Dec}}(1^\lambda, cc, ek, \mathbf{m}, j)$

For $i \leftarrow 1, \ldots, |\mathbf{m}|$ do $\mathbf{c}[i] \leftarrow \mathsf{DE.Enc}^{\textsc{Hash}}(1^\lambda, ek, \mathbf{m}[i])$
$(m, i) \leftarrow\!\!\$\; A^{\textsc{Hash},\textsc{DecSim}}(1^\lambda, cc, ek, \mathbf{c})$
Return $\neg((\mathbf{m}[i] = m) \text{ and } (j = i))$

Algorithm $\textsc{DecSim}(x)$

If $(\exists i : x = \mathbf{c}[i])$ then return $\bot$
Return $\textsc{Dec}(x)$

Fig. 10. Source S_2 (top left), adversary B (top right), and distinguisher D (bottom) used in Theorem 2.

Furthermore, S_2 is unpredictable if S_1 is. Given any predictor P_2 for S_2, the proof gives predictor P_1 such that

$$\mathsf{Adv}^{\mathsf{pred}}_{\mathsf{DE},S_2,P_2}(\lambda) \leq \mathsf{Adv}^{\mathsf{pred}}_{\mathsf{DE},S_1,P_1}(\lambda)\,, \tag{16}$$

and

$$q^{\textsc{Hash}}_{S_2} = q^{\textsc{Hash}}_{S_1}\,, \quad q^{\textsc{Dec}}_{S_2} = q^{\textsc{Dec}}_{S_1}\,, \quad t_{S_2} \approx t_{S_1}\,,$$
$$q^{\textsc{Hash}}_{P_2} = q^{\textsc{Hash}}_{P_1}\,, \quad q^{\textsc{Dec}}_{P_2} = q^{\textsc{Dec}}_{P_1}\,, \quad t_{P_2} \approx t_{P_1}\,.$$

The intuition behind the proof of Theorem 2 is straightforward—resampling-indistinguishability allows a PDMR adversary to *simulate* the ciphertext vector $\mathbf{c}$ in order to run *any* RI PDMR adversary. We give the details below.

Proof (of Theorem 2). Consider game G_0 and G_1 given in Fig. 11, where G_1 contains the boxed code, while G_0 does not. By construction,

$$\Pr[G_1] = \Pr[\mathbf{G}^{\mathrm{pdmr}}_{\mathsf{DE},S_2,B}(\lambda)]\,. \tag{17}$$

Next, we claim that

$$\Pr[G_0] = \frac{1}{n(\lambda)} \cdot \Pr[\mathbf{G}^{\mathrm{pdmr}}_{\mathsf{DE},S_1,A}(\lambda)]\,. \tag{18}$$

$$
\begin{array}{l}
\underline{\text{GAME } \mathrm{G}_0 \ \boxed{\mathrm{G}_1}} \\
ek \leftarrow_{\$} \mathsf{DE.Kg}(1^\lambda) \ ; \ cc \leftarrow_{\$} S_1.\mathsf{cx}(1^\lambda) \ ; \ \mathbf{m} \leftarrow_{\$} S_1.\mathsf{msg}^{\mathrm{HASH,DEC}}(1^\lambda, cc, ek) \\
j \leftarrow_{\$} [n] \ ; \ \boxed{\mathbf{m}[j] \leftarrow_{\$} S_1.\mathsf{msg}^{\mathrm{HASH}}(1^\lambda, cc, ek)[j]} \\
\text{For } i \leftarrow 1, \ldots, |\mathbf{m}| \text{ do } \mathbf{c}[i] \leftarrow \mathsf{DE.Enc}^{\mathrm{HASH}}(1^\lambda, ek, \mathbf{m}[i]) \\
(m, i) \leftarrow_{\$} A^{\mathrm{HASH,DEC}}(1^\lambda, cc, ek, \mathbf{c}) \ ; \ \text{Return } ((\mathbf{m}[i] = m) \text{ and } (j = i)) \\
\underline{\mathrm{HASH}(x, 1^\ell)} \\
\text{If not } T[x, \ell] \text{ then } T[x, \ell] \leftarrow_{\$} \{0,1\}^\ell \\
\text{Return } T[x, \ell] \\
\underline{\text{Algorithm } \mathrm{DEC}(c)} \\
\text{If } (\exists i : c = \mathbf{c}[i]) \text{ then return } \bot \\
\text{Return } \mathsf{DE.Dec}(1^\lambda, dk, c)
\end{array}
$$

Fig. 11. Games G_0 and G_1 used in the proof of Theorem 2.

This is because j is uniformly sampled and is not used any where in G_0 besides computing the return value. Finally, let us consider $\mathbf{G}^{\mathrm{ri}}_{S_1,D}$. We note that by construction of D, it holds for $i \in \{0,1\}$ that

$$\Pr[\mathrm{G}_i] = \Pr[D \text{ outputs } 0 \mid b = i] \ , \tag{19}$$

Where the second probability is taken over game $\mathbf{G}^{\mathrm{ri}}_{\mathsf{DE},S_1,D}$ and b is as sampled in the game. Hence,

$$\Pr[\mathrm{G}_0] - \Pr[\mathrm{G}_1] = \mathsf{Adv}^{\mathsf{ri}}_{\mathsf{DE},S_1,D}(\lambda) \ . \tag{20}$$

Combining Eqs. (17), (18) and (20), we obtain Eq. (15). Lastly, let P_2 be a predictor for S_2, consider the following predictor P_1 for S_1:

$$
\begin{array}{l}
\underline{P_1^{\mathrm{HASH,DEC}}(1^\lambda, \overline{cc}, ek, n, \boldsymbol{\ell})} \\
(cc, j) \leftarrow \overline{cc} \ ; \ m \leftarrow P_2^{\mathrm{HASH,DEC}}(1^\lambda, cc, ek, 1, \boldsymbol{\ell}[j]) \\
\text{Return } (m, j)
\end{array}
$$

It is easy to check that Eq. (16) holds.

5.3 Security of EwH Against $\mathcal{S}^{\mathsf{up}} \cap \mathcal{S}^{\mathsf{ri}}$

Combining Theorem 2 and Corollary 1, we obtain the following theorem, which says that if PKE is \$IND(-CPA) secure, then DE = EwH[PKE] is PDMR(-CPA)-secure against $\mathcal{S}^{\mathsf{ri}} \cap \mathcal{S}^{\mathsf{up}}$.

Theorem 3. *Let* PKE *be a public-key encryption scheme. Let* DE = EwH[PKE] *be the associated deterministic public-key encryption scheme. Let S be a $n(\lambda)$-source for* DE. *Let A be a PDMR adversary for S. \$IND adversary B for* PKE, *predictor P, and distinguisher D can be constructed such that*

$$\mathsf{Adv}^{\mathsf{pdmr}}_{\mathsf{DE},A}(\lambda) \le n(\lambda) \cdot \mathsf{Adv}^{\mathsf{ri}}_{S,D}(\lambda)$$
$$+ n(\lambda)\sqrt{\left(q_S^{\text{HASH}}+1\right)\left(q_S^{\text{HASH}}+q_A^{\text{HASH}}+n(\lambda)+1\right)\left(\mathsf{Adv}^{\$\mathsf{ind}}_{\mathsf{PKE},B}(\lambda)+\mathsf{Adv}^{\mathsf{pred}}_{S,P}(\lambda)\right)}.$$

Furthermore, D, B and P are efficient as long as S and A are. In particular,

$$q_B^{\text{HASH}} = 2 \cdot q_S^{\text{HASH}} + q_A^{\text{HASH}} + n(\lambda) + 1, \quad q_B^{\text{DEC}} = 2 \cdot q_S^{\text{DEC}} + q_A^{\text{DEC}}$$

$$t_B = 2 \cdot t_S + n(\lambda) \cdot t_{\mathsf{PKE.Enc}} + t_A \ ,$$

$$q_D^{\text{HASH}} = n(\lambda) + q_A^{\text{HASH}}, \ q_D^{\text{DEC}} = q_A^{\text{DEC}}, \ t_D \approx n(\lambda) \cdot t_{\mathsf{DE.Enc}} + t_A \ ,$$

and

$$q_P^{\text{HASH}} = q_S^{\text{HASH}} + q_A^{\text{HASH}} + n(\lambda) + 1, \ q_P^{\text{DEC}} = q_S^{\text{DEC}} + q_A^{\text{DEC}},$$

$$t_P \approx t_S + n(\lambda) \cdot t_{\mathsf{PKE.Enc}} + t_A \ .$$

The proof of Theorem 3 is straight forward given Theorem 2 and Corollary 1 and we only sketch it here. We first apply Theorem 2 to source S and adversary A to obtain a 1-source S', adversary A' and distinguisher D. Then, we can apply Corollary 1 to S' and A' to obtain adversary B and predictor P.

6 Impossibility Results

In this section, we explore what goes wrong when messages can have correlation. The known attacks showing IND-style security is unachievable [2,5] only distinguish between encryptions of unpredictable messages. Here we give attacks showing that public-key-dependent messages can in fact be recovered in full by the adversary—that is, PDMR security is violated—as long as two or more *closely related* messages are encrypted. In particular, we show that no D-PKE scheme is secure against $\mathcal{S}^{\mathsf{up}}$ (in particular $\mathcal{S}^{\mathsf{up}} \cap \mathcal{S}^2$). We start with a basic attack on schemes that can encrypt messages of any length, and then extend this to schemes that can only encrypt messages of a fixed length.

Basic Attack. The basic PDMR attack works when the D-PKE scheme allows the encryption of messages of arbitrary length, meaning $\mathsf{DE.IL}(\cdot) = \mathbb{N}$. The idea is simple. Since the message-choosing adversary A_1 has the public key, it can encrypt. It sets the second message to the encryption of a first, random message. The first challenge ciphertext is thus the second message. This requires that the scheme be able to encrypt messages of varying length because the ciphertext will not (usually) have the same length as the plaintext. For the attack to be valid, we must also show that the adversary is unpredictable. The following theorem formalizes this intuition. Here $\mu(\cdot)$ is a parameter representing the message length. The adversary is statistically unpredictable for $\mu(\cdot) = \omega(\log(\cdot))$, ruling out even weak PDMR security. The D-PKE scheme is *arbitrary* subject to being able to encrypt messages of arbitrary length.

$\underline{S.\mathsf{cx}^{\text{HASH}}(1^\lambda)}$	$\underline{S.\mathsf{cx}^{\text{HASH}}(1^\lambda)}$
Return ϵ	$hk \leftarrow_\$ \{0,1\}^{\mathsf{H.kl}}$
$\underline{S.\mathsf{msg}^{\text{HASH}}(1^\lambda, \varepsilon, \mathsf{pp})}$	Return hk
$\mathbf{m}[1] \leftarrow_\$ \{0,1\}^{\mu(\lambda)}$	$\underline{S.\mathsf{msg}^{\text{HASH}}(1^\lambda, hk, \mathsf{pp})}$
$\mathbf{m}[2] \leftarrow \mathsf{DE.Enc}^{\text{HASH}}(1^\lambda, \mathsf{pp}, \mathbf{m}[1])$	$\mathbf{m}[1] \leftarrow_\$ \{0,1\}^{\mu(\lambda)}$
Return $\mathbf{m}$	$c \leftarrow \mathsf{DE.Enc}^{\text{HASH}}(1^\lambda, \mathsf{pp}, \mathbf{m}[1])$
	$\mathbf{m}[2] \leftarrow \mathsf{H.Ev}(1^\lambda, hk, c)$
$\underline{A^{\text{HASH}}(1^\lambda, \varepsilon, \mathsf{pp}, \mathbf{c})}$	Return $\mathbf{m}$
Return $(\mathbf{c}[1], 2)$	$\underline{A^{\text{HASH}}(1^\lambda, hk, \mathsf{pp}, \mathbf{c})}$
	Return $(\mathsf{H.Ev}(1^\lambda, hk, \mathbf{c}[1]), 2)$

Fig. 12. Left: Source S and PDMR adversary A used in Theorem 4. **Right:** Source S and PDMR adversary A used in Theorem 5.

Theorem 4. *Let* DE *be a D-PKE scheme with* $\mathsf{DE.IL}(\lambda) = \mathbb{N}$ *for all* λ*. Then,* DE *is not PDMR-secure against* $\mathcal{S}^{\mathsf{up}}$ *message samplers. In particular, let* $\mu\colon \mathbb{N} \to \mathbb{N}$ *be any function, and* S, A *be the source and adversary given on the left in Fig. 12. Then, we claim that* $S \in \mathcal{S}^2 \cap \mathcal{S}^{\mathsf{up}}$*; in particular, for predictors* P *and all* λ*,*

$$\mathsf{Adv}^{\mathsf{pred}}_{\mathsf{DE},S,P}(\lambda) \leq 2^{-\mu(\lambda)} . \tag{21}$$

But for all λ*,*

$$\mathsf{Adv}^{\mathsf{pdmr}}_{\mathsf{DE},S,A}(\lambda) = 1 . \tag{22}$$

Proof (of Theorem 4). We first prove Eq. (22). Adversary A wins game $\mathbf{G}^{\mathrm{pdmr}}_{\mathsf{DE},A}(\lambda)$ as long as $\mathbf{m}[2]$, as computed by A_1, equals $\mathbf{c}[1]$, as computed by the game. Both are computed independently as $\mathsf{DE.Enc}^{\text{HASH}}(1^\lambda, \mathsf{pp}, \mathbf{m}[1])$, so they will always be equal, since DE.Enc is deterministic.

We move on to prove Eq. (21). Let P be any predictor, and consider game $\mathbf{G}^{\mathrm{pred}}_{\mathsf{DE},S,P}(\lambda)$. For $i = 1, 2$ let E_i be the event that in game $\mathbf{G}^{\mathrm{pred}}_{\mathsf{DE},S,P}(\lambda)$, predictor P outputs a guess of the form (m', i), for some string m'. The following inequalities, which complete the proof, are justified after they are stated:

$$\begin{aligned} \mathsf{Adv}^{\mathsf{pred}}_{\mathsf{DE},S,P}(\lambda) &= \sum_{i=1}^{2} \Pr[\, \mathbf{G}^{\mathrm{pred}}_{\mathsf{DE},S,P}(\lambda) \mid \mathrm{E}_i \,] \cdot \Pr[\mathrm{E}_i] \\ &\leq 2^{-\mu(\lambda)} \cdot \Pr[\mathrm{E}_1] + 2^{-\mu(\lambda)} \cdot \Pr[\mathrm{E}_2] && (23) \\ &\leq 2^{-\mu(\lambda)} . && (24) \end{aligned}$$

Since the first message $\mathbf{m}[1]$ is randomly chosen from $\{0,1\}^{\mu(\lambda)}$, the probability that $m' = \mathbf{m}[1]$ when P returns $(m', 1)$ is at most $2^{-\mu(\lambda)}$. The second message $\mathbf{m}[2]$ is the deterministic encryption of the first message, $\mathbf{m}[1]$. Since the function $\mathsf{DE.Enc}(1^\lambda, ek, \cdot)$ is injective, and there are $2^{\mu(\lambda)}$ possible values for $\mathbf{m}[1]$, there will also be $2^{\mu(\lambda)}$ possible values for $\mathbf{m}[2]$. So again, the probability that $m' = \mathbf{m}[2]$ when P returns $(m', 2)$ is at most $2^{-\mu(\lambda)}$. This justifies Eq. (23). Equation (24) holds simply because $\Pr[\mathrm{E}_1] + \Pr[\mathrm{E}_2] \leq 1$.

General Attack. The basic attack assumed the D-PKE scheme could encrypt messages of varying length. Many D-PKE schemes—and even definitions—in the literature restrict the space of allowed messages to ones of a single length. We now extend the basic attack to one that works in this case, showing that no D-PKE scheme is (even weakly) PDMR-secure for the encryption of two or more messages, even if these are of the same length.

Function Families. A family of functions (or function family) F specifies a deterministic PT evaluation algorithm $\mathsf{F.Ev}$ such that $\mathsf{F.Ev}(1^\lambda, \cdot, \cdot)\colon \{0,1\}^{\mathsf{F.kl}(\lambda)} \times \{0,1\}^{\mathsf{F.il}(\lambda)} \to \{0,1\}^{\mathsf{F.ol}(\lambda)}$ for all $\lambda \in \mathbb{N}$, where $\mathsf{F.kl}$, $\mathsf{F.il}$ and $\mathsf{F.ol}$ are the key, input and ouput length functions, respectively. Many security attributes may be defined and considered for such families.

Universal Hash Functions. As a tool we need a family of universal hash functions, so we start by recalling the definition. Let H be a family of functions. For $\lambda \in \mathbb{N}$, a key $hk \in \{0,1\}^{\mathsf{H.kl}(\lambda)}$ and inputs $x_1, x_2 \in \{0,1\}^{\mathsf{H.il}(\lambda)}$ we define the collision probabilities

$$\begin{aligned} \mathbf{cp}_{\mathsf{H}}(\lambda, x_1, x_2) &= \Pr[\mathsf{H.Ev}(1^\lambda, hk, x_1) = \mathsf{H.Ev}(1^\lambda, hk, x_2)] \\ \mathbf{cp}_{\mathsf{H}}(\lambda) &= \max \mathbf{cp}_{\mathsf{H}}(\lambda, x_1, x_2) \,, \end{aligned}$$

Where the probability is over $hk \leftarrow\!\!\$ \{0,1\}^{\mathsf{H.kl}(\lambda)}$ and the max is over all distinct $x_1, x_2 \in \{0,1\}^{\mathsf{H.il}(\lambda)}$. We say that H is *universal* if $\mathbf{cp}_{\mathsf{H}}(\lambda) = 2^{-\mathsf{H.ol}(\lambda)}$ for all $\lambda \in \mathbb{N}$.

Theorem 5. *Let* DE *be a D-PKE scheme. Let* $\mu\colon \mathbb{N} \to \mathbb{N}$ *be any function such that* $\mu(\lambda) \in \mathsf{DE.IL}(\lambda)$ *for all* $\lambda \in \mathbb{N}$. *We claim that* DE *is not PDMR-secure for* $\mathcal{S}^{\mathsf{up}}$ *message samplers. More precisely, let* H *be a universal family of functions with* $\mathsf{H.il}(\lambda) = \mathsf{DE.cl}(\lambda, \mu(\lambda))$ *and* $\mathsf{H.ol}(\lambda) = \mu(\lambda)$ *for all* $\lambda \in \mathbb{N}$. *Let* S, A *be the source and PDMR adversary for* DE *shown on the right in Fig. 12. Then* $S \in \mathcal{S}^2 \cap \mathcal{S}^{\mathsf{up}}$; *in particular, for all predictors* P *and all* λ,

$$\mathsf{Adv}^{\mathsf{pred}}_{\mathsf{DE},S,P}(\lambda) \leq \sqrt{2} \cdot 2^{-\mu(\lambda)/2} \,. \tag{25}$$

But for all λ,

$$\mathsf{Adv}^{\mathsf{pdmr}}_{\mathsf{DE},S,A}(\lambda) = 1 \,. \tag{26}$$

The adversary picks $\mathbf{m}[1]$ as before and hashes its encryption down to get $\mathbf{m}[2]$. Note that both these strings have the same length $\mu(\lambda)$, so the attack works even if there is just one allowed message length. The key hk for the hash function is shared using the common coins, so is available to both the message source and the adversary. The adversary continues to have PDMR advantage one. The more difficult task is to establish its unpredictability. The theorem shows that the prediction advantage has degraded (increased) relative to Theorem 4, being about the square root of what it was before, but this is still exponentially vanishing with $\mu(\cdot)$. The proof of this bound uses techniques from the proof of the Leftover Hash Lemma [24].

Proof (of Theorem 5). We first prove Eq. (26). Adversary A wins game $\mathbf{G}^{\mathrm{pdmr}}_{\mathsf{DE},S,A}(\lambda)$ as long as $\mathbf{m}[2]$ from S equals $\mathsf{H.Ev}(1^\lambda, hk, \mathbf{c}[1])$, as computed by the game. Both are calculated as $\mathsf{H.Ev}(1^\lambda, hk, \mathsf{DE.Enc}^{\mathrm{HASH}}(1^\lambda, \mathsf{pp}, \mathbf{m}[1]))$, so they will always be equal, since $\mathsf{DE.Enc}$ is deterministic.

Now we prove Eq. (25). Let P be any predictor, and consider game $\mathbf{G}^{\mathrm{pred}}_{\mathsf{DE},S,P}(\lambda)$. For $i = 1, 2$ let E_i be the event that in game $\mathbf{G}^{\mathrm{pred}}_{\mathsf{DE},S,P}(\lambda)$, predictor P outputs a guess of the form (m', i), for some string m'. We claim that

$$\Pr[\, \mathbf{G}^{\mathrm{pred}}_{A,P}(\lambda) \mid \mathrm{E}_1 \,] \le 2^{-\mu(\lambda)} \tag{27}$$

$$\Pr[\, \mathbf{G}^{\mathrm{pred}}_{A,P}(\lambda) \mid \mathrm{E}_2 \,] \le \sqrt{2} \cdot 2^{-\mu(\lambda)/2} \,. \tag{28}$$

Given the above, we can complete the proof via

$$\begin{aligned} \mathsf{Adv}^{\mathsf{pred}}_{\mathsf{DE},S,P}(\lambda) &= \sum_{i=1}^{2} \Pr[\, \mathbf{G}^{\mathrm{pred}}_{\mathsf{DE},S,P}(\lambda) \mid \mathrm{E}_i \,] \cdot \Pr[\mathrm{E}_i] \\ &\le 2^{-\mu(\lambda)} \cdot \Pr[\mathrm{E}_1] + \sqrt{2} \cdot 2^{-\mu(\lambda)/2} \cdot \Pr[\mathrm{E}_2] \\ &\le \sqrt{2} \cdot 2^{-\mu(\lambda)/2} \end{aligned}$$

Equation (27) is true for the same reason as in Theorem 4, namely that, since the first message $\mathbf{m}[1]$ is randomly chosen from $\{0,1\}^{\mu(\lambda)}$, the probability that $m' = \mathbf{m}[1]$ when P returns $(m', 1)$ is at most $2^{-\mu(\lambda)}$. The main issue is Eq. (28), which we now prove.

Let $(ek, dk) \in [\mathsf{DE.Kg}(1^\lambda)]$ and $hk \in \{0,1\}^{\mathsf{H.kl}(\lambda)}$. Define $\mathsf{X}_{ek,hk}\colon \{0,1\}^{\mu(\lambda)} \to \{0,1\}^{\mu(\lambda)}$ by

$$\mathsf{X}_{ek,hk}(m) = \mathsf{H.Ev}(1^\lambda, hk, \mathsf{DE.Enc}(1^\lambda, ek, m)) \,.$$

Regard this as a random variable over the random choice of $m \leftarrow_\$ \{0,1\}^{\mu(\lambda)}$. Now consider the guessing and collision probabilities of this random variable,

$$\begin{aligned} \mathbf{gp}(\mathsf{X}_{ek,hk}) &= \max_{h \in \{0,1\}^{\mathsf{H.ol}(\lambda)}} \Pr[\mathsf{X}_{ek,hk} = h] \\ \mathbf{cp}(\mathsf{X}_{ek,hk}) &= \sum_{h \in \{0,1\}^{\mathsf{H.ol}(\lambda)}} \Pr[\mathsf{X}_{ek,hk} = h]^2 \,. \end{aligned}$$

Further define $\mathsf{GP}_{ek}, \mathsf{CP}_{ek}\colon \{0,1\}^{\mathsf{H.kl}(\lambda)} \to [0,1]$ by

$$\mathsf{GP}_{ek}(hk) = \mathbf{gp}(\mathsf{X}_{ek,hk}) \quad \text{and} \quad \mathsf{CP}_{ek}(hk) = \mathbf{cp}(\mathsf{X}_{ek,hk}) \,,$$

and regard them as random variables over the random choice of $hk \leftarrow_{\$} \{0,1\}^{\mathsf{H.kl}(\lambda)}$. Below we will show that

$$\mathbf{E}[\,\mathsf{GP}_{ek}\,] \leq \sqrt{2} \cdot 2^{-\mu(\lambda)/2} \tag{29}$$

for every $(ek, dk) \in [\mathsf{DE.Kg}(1^\lambda)]$. Now, hk is an input to P, so

$$\begin{aligned}\Pr[\,\mathbf{G}^{\text{pred}}_{A,P}(\lambda) \mid \mathrm{E}_2\,] &\leq \max_{(ek,dk)\in[\mathsf{DE.Kg}(1^\lambda)]} \mathbf{E}[\,\mathsf{GP}_{ek}\,] \\ &\leq \sqrt{2} \cdot 2^{-\mu(\lambda)/2}\end{aligned}$$

Where the second equation is by Eq. (29). This proves Eq. (28).
Fixing $(ek, dk) \in [\mathsf{DE.Kg}(1^\lambda)]$, we now prove Eq. (29). It is clear (and a standard relation between guessing and collision probabilities of a random variable) that for all hk we have

$$\mathbf{gp}(\mathsf{X}_{ek,hk})^2 \leq \mathbf{cp}(\mathsf{X}_{ek,hk})\ .$$

Thus

$$\mathsf{GP}_{ek} \leq \sqrt{\mathsf{CP}_{ek}}\ .$$

By Jensen's inequality and concavity of the square-root function,

$$\mathbf{E}[\,\mathsf{GP}_{ek}\,] \leq \mathbf{E}\left[\sqrt{\mathsf{CP}_{ek}}\right] \leq \sqrt{\mathbf{E}[\,\mathsf{CP}_{ek}\,]}\ .$$

Now with the expectation over $hk \leftarrow_{\$} \{0,1\}^{\mathsf{H.kl}(\lambda)}$ and the probability over $m_1, m_2 \leftarrow_{\$} \{0,1\}^{\mu(\lambda)}$, we have

$$\mathbf{E}[\,\mathsf{CP}_{ek}\,] = \mathbf{E}[\,\Pr[\mathsf{X}_{ek,hk}(m_1) = \mathsf{X}_{ek,hk}(m_2)]\,] \leq 2^{-\mu(\lambda)} + \mathbf{cp}_{\mathsf{H}}(\lambda)\ .$$

This is by considering two cases. The first is that $m_1 = m_2$, which happens with probability $2^{-\mu(\lambda)}$. The second is that $m_1 \neq m_2$, in which case the inputs to $\mathsf{H.Ev}(1^\lambda, hk, \cdot)$ are different due to the injectivity of $\mathsf{DE.Enc}(1^\lambda, ek, \cdot)$, and we can exploit the universality of H. Now by assumption of universality of H, $\mathbf{cp}_{\mathsf{H}}(\lambda) = 2^{-\mu(\lambda)}$, so putting everything together we have Eq. (29).

Acknowledgments. The first and second authors are supported in part by NSF grants CNS-1526801 and CNS-1717640, ERC Project ERCC FP7/615074 and a gift from Microsoft. The second author is supported in part by a Powell fellowship. The third author was supported in part by NSF grant CNS-1564102.

We thank reviewers from Asiacrypt 2019 and Crypto 2019 for their detailed and extensive comments.

References

1. Bagherzandi, A., Cheon, J.H., Jarecki, S.: Multisignatures secure under the discrete logarithm assumption and a generalized forking lemma. In: Ning, P., Syverson, P.F., Jha, S. (eds.) ACM CCS 2008, pp. 449–458. ACM Press (October 2008)
2. Bellare, M., Boldyreva, A., O'Neill, A.: Deterministic and efficiently searchable encryption. In: Menezes, A. (ed.) CRYPTO 2007. LNCS, vol. 4622, pp. 535–552. Springer, Heidelberg (2007). https://doi.org/10.1007/978-3-540-74143-5_30
3. Bellare, M., Dai, W., Li, L.: The local forking lemma and its application to deterministic encryption. Cryptology ePrint Archive, Report 2019/1017 (2019). https://eprint.iacr.org/2019/1017
4. Bellare, M., Desai, A., Pointcheval, D., Rogaway, P.: Relations among notions of security for public-key encryption schemes. In: Krawczyk, H. (ed.) CRYPTO 1998. LNCS, vol. 1462, pp. 26–45. Springer, Heidelberg (1998). https://doi.org/10.1007/BFb0055718
5. Bellare, M., Fischlin, M., O'Neill, A., Ristenpart, T.: Deterministic encryption: definitional equivalences and constructions without random oracles. In: Wagner, D. (ed.) CRYPTO 2008. LNCS, vol. 5157, pp. 360–378. Springer, Heidelberg (2008). https://doi.org/10.1007/978-3-540-85174-5_20
6. Bellare, M., Hoang, V.T.: Resisting randomness subversion: fast deterministic and hedged public-key encryption in the standard model. In: Oswald, E., Fischlin, M. (eds.) EUROCRYPT 2015. LNCS, vol. 9057, pp. 627–656. Springer, Heidelberg (2015). https://doi.org/10.1007/978-3-662-46803-6_21
7. Bellare, M., Neven, G.: Multi-signatures in the plain public-key model and a general forking lemma. In: Juels, A., Wright, R.N., De Capitani di Vimercati, S. (eds.) ACM CCS 2006, pp. 390–399. ACM Press, October/November 2006
8. Bellare, M., Rogaway, P.: Random oracles are practical: A paradigm for designing efficient protocols. In: Denning, D.E., Pyle, R., Ganesan, R., Sandhu, R.S., Ashby, V. (eds.) ACM CCS 1993, pp. 62–73. ACM Press (November 1993)
9. Bellare, M., Rogaway, P.: Optimal asymmetric encryption. In: Santis, A.D. (ed.) EUROCRYPT 1994. LNCS, vol. 950, pp. 92–111. Springer, Heidelberg (1995). https://doi.org/10.1007/BFb0053428
10. Bellare, M., Rogaway, P.: The security of triple encryption and a framework for code-based game-playing proofs. In: Vaudenay, S. (ed.) EUROCRYPT 2006. LNCS, vol. 4004, pp. 409–426. Springer, Heidelberg (2006). https://doi.org/10.1007/11761679_25
11. Bernstein, D.J., Lange, T., Niederhagen, R.: Dual EC: a standardized back door. Cryptology ePrint Archive, Report 2015/767 (2015). http://eprint.iacr.org/2015/767
12. Bleichenbacher, D.: On the security of the KMOV public key cryptosystem. In: Kaliski, B.S. (ed.) CRYPTO 1997. LNCS, vol. 1294, pp. 235–248. Springer, Heidelberg (1997). https://doi.org/10.1007/BFb0052239
13. Boldyreva, A., Fehr, S., O'Neill, A.: On notions of security for deterministic encryption, and efficient constructions without random oracles. In: Wagner, D. (ed.) CRYPTO 2008. LNCS, vol. 5157, pp. 335–359. Springer, Heidelberg (2008). https://doi.org/10.1007/978-3-540-85174-5_19
14. Brakerski, Z., Segev, G.: Better security for deterministic public-key encryption: the auxiliary-input setting. In: Rogaway, P. (ed.) CRYPTO 2011. LNCS, vol. 6841, pp. 543–560. Springer, Heidelberg (2011). https://doi.org/10.1007/978-3-642-22792-9_31

15. Brown, D.R.L.: A weak-randomizer attack on RSA-OAEP with e = 3. Cryptology ePrint Archive, Report 2005/189 (2005). http://eprint.iacr.org/2005/189
16. Cramer, R., Shoup, V.: A practical public key cryptosystem provably secure against adaptive chosen ciphertext attack. In: Krawczyk, H. (ed.) CRYPTO 1998. LNCS, vol. 1462, pp. 13–25. Springer, Heidelberg (1998). https://doi.org/10.1007/BFb0055717
17. Cramer, R., Shoup, V.: Design and analysis of practical public-key encryption schemes secure against adaptive chosen ciphertext attack. SIAM J. Comput. **33**(1), 167–226 (2003)
18. Fujisaki, E., Okamoto, T.: Secure integration of asymmetric and symmetric encryption schemes. In: Wiener, M.J. (ed.) CRYPTO 1999. LNCS, vol. 1666, pp. 537–554. Springer, Heidelberg (1999). https://doi.org/10.1007/3-540-48405-1_34
19. Fuller, B., O'Neill, A., Reyzin, L.: A unified approach to deterministic encryption: new constructions and a connection to computational entropy. J. Cryptol. **28**(3), 671–717 (2015)
20. Garg, S., Gay, R., Hajiabadi, M.: New techniques for efficient trapdoor functions and applications. In: Ishai, Y., Rijmen, V. (eds.) EUROCRYPT 2019. LNCS, vol. 11478, pp. 33–63. Springer, Cham (2019). https://doi.org/10.1007/978-3-030-17659-4_2
21. Gay, R., Hofheinz, D., Kiltz, E., Wee, H.: Tightly CCA-secure encryption without pairings. In: Fischlin, M., Coron, J.-S. (eds.) EUROCRYPT 2016. LNCS, vol. 9665, pp. 1–27. Springer, Heidelberg (2016). https://doi.org/10.1007/978-3-662-49890-3_1
22. Goldwasser, S., Micali, S.: Probabilistic encryption. J. Comput. Syst. Sci. **28**(2), 270–299 (1984)
23. Hofheinz, D., Kiltz, E.: Practical chosen ciphertext secure encryption from factoring. In: Joux, A. (ed.) EUROCRYPT 2009. LNCS, vol. 5479, pp. 313–332. Springer, Heidelberg (2009). https://doi.org/10.1007/978-3-642-01001-9_18
24. Impagliazzo, R., Zuckerman, D.: How to recycle random bits. In: 30th FOCS, pp. 248–253. IEEE Computer Society Press, October/November 1989
25. Kurosawa, K., Desmedt, Y.: A new paradigm of hybrid encryption scheme. In: Franklin, M. (ed.) CRYPTO 2004. LNCS, vol. 3152, pp. 426–442. Springer, Heidelberg (2004). https://doi.org/10.1007/978-3-540-28628-8_26
26. Micali, S., Rackoff, C., Sloan, B.: The notion of security for probabilistic cryptosystems. SIAM J. Comput. **17**(2), 412–426 (1988). Special issue on cryptography
27. Mironov, I., Pandey, O., Reingold, O., Segev, G.: Incremental deterministic public-key encryption. In: Pointcheval, D., Johansson, T. (eds.) EUROCRYPT 2012. LNCS, vol. 7237, pp. 628–644. Springer, Heidelberg (2012). https://doi.org/10.1007/978-3-642-29011-4_37
28. Ouafi, K., Vaudenay, S.: Smashing SQUASH-0. In: Joux, A. (ed.) EUROCRYPT 2009. LNCS, vol. 5479, pp. 300–312. Springer, Heidelberg (2009). https://doi.org/10.1007/978-3-642-01001-9_17
29. Pointcheval, D., Stern, J.: Security arguments for digital signatures and blind signatures. J. Cryptol. **13**(3), 361–396 (2000)
30. Raghunathan, A., Segev, G., Vadhan, S.: Deterministic public-key encryption for adaptively chosen plaintext distributions. In: Johansson, T., Nguyen, P.Q. (eds.) EUROCRYPT 2013. LNCS, vol. 7881, pp. 93–110. Springer, Heidelberg (2013). https://doi.org/10.1007/978-3-642-38348-9_6
31. Yilek, S., Rescorla, E., Shacham, H., Enright, B., Savage, S.: When private keys are public: results from the 2008 Debian OpenSSL vulnerability. In: Proceedings of the 9th ACM SIGCOMM Conference on Internet Measurement, pp. 15–27. ACM (2009)

Fine-Grained Cryptography Revisited

Shohei Egashira[1], Yuyu Wang[2(✉)], and Keisuke Tanaka[1]

[1] Tokyo Institute of Technology, Tokyo, Japan
egashira.s.aa@m.titech.ac.jp, keisuke@is.titech.ac.jp
[2] University of Electronic Science and Technology of China, Chengdu, China
wangyuyu@uestc.edu.cn

Abstract. *Fine-grained cryptographic primitives* are secure against adversaries with bounded resources and can be computed by honest users with less resources than the adversaries. In this paper, we revisit the results by Degwekar, Vaikuntanathan, and Vasudevan in Crypto 2016 on fine-grained cryptography and show the constructions of three key fundamental fine-grained cryptographic primitives: *one-way permutations*, *hash proof systems* (which in turn implies a *public-key encryption scheme against chosen chiphertext attacks*), and *trapdoor one-way functions*. All of our constructions are computable in NC^1 and secure against (*non-uniform*) NC^1 circuits under the widely believed worst-case assumption $\mathsf{NC}^1 \subsetneq \oplus\mathsf{L/poly}$.

Keywords: Fine-grained cryptography · NC^1 circuit · One-way permutation · Hash proof system · Trapdoor one-way function

1 Introduction

1.1 Background

To prove the security of a cryptographic scheme, we typically reduce the security to some computational hardness assumption with a precise security definition. Due to the fact that most assumptions are unproven, it is desirable to make the underlying assumptions as weak as possible. However, it turns out to be very hard to construct a public-key cryptographic scheme without assuming the existence of one-way functions (OWF). Moreover, for a vast majority of primitives (including public-key encryption (PKE)), we further need to assume the hardness of specific problems such as factoring, discrete-logarithm, learning with errors, etc. It still remains open whether it is possible to construct even basic cryptographic primitives under no assumptions, or at least mild complexity-theoretic assumptions. For instance, the complexity-theoretic assumption $\mathsf{NP} \not\subseteq \mathsf{BPP}$, which is strictly weaker than the assumption of OWFs, has been proven to be insufficient for constructing even OWFs as shown by Akavia et al. [4].

Due to the difficulty of directly constructing cryptographic primitives against any polynomial probabilistic time adversaries based on mild complexity-theoretic assumptions such as $\mathsf{NP} \not\subseteq \mathsf{BPP}$, a line of beautiful works focused on fine-grained

S. D. Galbraith and S. Moriai (Eds.): ASIACRYPT 2019, LNCS 11923, pp. 637–666, 2019.
https://doi.org/10.1007/978-3-030-34618-8_22

cryptographic primitives [16], where (1) the resource of an adversary is a-prior bounded, (2) an honest party can run the algorithms with less resource than an adversary, and (3) the underlying assumption is extremely mild.

Merkle [35] initialized the study in this field by constructing a non-interactive key exchange scheme, which can be run in time $O(n)$ and adversaries running in time $o(n^2)$ cannot break the security. The construction only requires random functions (i.e., the random oracle). Subsequent to his work, Biham et al. [10] showed the existence of strong OWFs based on the same assumption.

While Merkle restricted adversaries in the term of running time, Maurer considered a model where adversaries have infinite computing power but only restricted storage [34]. Afterwards, he proposed a key exchange protocol in this model [36]. Following these works, Cachin and Maurer [13] constructed a symmetric-key encryption scheme and a key exchange protocol which can be run with storage $O(s)$ and are unconditionally secure against adversaries with storage $o(s^2)$. Besides, there have been many other works focusing on primitives in this model [8,9,18–20,42].

In the constant depth circuit model, Ajtai and Wigderson [3] constructed an unconditional secure pseudo-random generator. Then, Boppana and Lagarias [12] exploited the results by Ajitai [2] and Furst et al. [21], which shows that parity cannot be computed in size-bounded circuits, to achieve OWFs. The proposed OWF can be computed in AC^0 (constant-depth polynomial-sized) circuits consisting of AND, OR, and NOT gates of unbounded fan-in, while the inverse cannot. Afterwards, several works treating the same model have been proposed [6,28,43,44].

Recently, Degwekar et al. [16] proposed fine-grained cryptographic primitives against adversaries captured by two (non-uniform) classes of adversaries, which are AC^0 and NC^1 (logarithmic-depth polynomial-sized) circuits consisting of AND, OR, and NOT gates of fan-in 2. They first constructed an unconditionally secure pseudorandom generator with arbitrary polynomial stretch, a weak pseudorandom function, and a secret-key encryption scheme, all of which are computable in AC^0 and secure against adversaries that are AC^0 circuits. Then, under the widely believed separation assumption $\mathsf{NC}^1 \subsetneq \oplus\mathsf{L/poly}$, they constructed a OWF, a pseudorandom generator, a collision-resistant hash function, and a semantically secure PKE scheme that are computable in NC^1 and secure against NC^1 circuits.

Following the above work, Campanelli and Gennaro [14] constructed a somewhat homomorphic encryption and a verifiable computation against NC^1 circuits. As in [16], the underlying assumption is $\mathsf{NC}^1 \subsetneq \oplus\mathsf{L/poly}$.

While the above sequence of works have achieved amazing success, it still remains open whether it is possible to construct other fine-grained primitives, such as one-way permutation (OWP), PKE against chosen ciphertext attacks (CCA), and even trapdoor one-way function (TDF).

1.2 Our Results and Techniques

In this paper, we propose several fine-grained cryptographic primitives under the assumption $\mathsf{NC}^1 \subsetneq \oplus\mathsf{L/poly}$. Specifically, we propose a OWP, a hash proof system (HPS) (which in turn derives a CCA-secure PKE scheme), and a TDF. All of them are computable in NC^1 and secure against adversaries captured by the class of NC^1 circuits. Since a lot of results have been devoted to constructing advanced primitives from these fundamental ones, our results greatly alleviate the efforts to achieve more fine-grained primitives from scratch.

Our constructions rely on the fact shown in the papers by Appelebaum, Ishai, and Kushilevitz [5,29], that if $\mathsf{NC}^1 \subsetneq \oplus\mathsf{L/poly}$, there exist a distribution D_0^n over $n \times n$ matrices of rank $(n-1)$ and a distribution D_1^n over $n \times n$ matrices of rank n, which are indistinguishable for NC^1 circuits.

One-Way Permutation. As one of the most fundamental cryptographic primitives, OWP has been shown to be sufficient for constructing many primitives (e.g. pseudorandom generators [11] and universal one-way hash functions [37]). Compared with primitives built from OWFs which are not bijective (e.g., [26,40]), ones built from OWPs are usually more efficient [7,33].

In the previous work, Degwekar et al. [16] showed a construction of fine-grained OWFs in NC^1. Their construction relies on a randomized encoding of a boolean function f, which is a randomized function outputting the distribution related only to $f(x)$. Specifically, let $\hat{f} : \{0,1\}^n \times \{0,1\}^m \rightarrow \{0,1\}^{m+1} \in \mathsf{NC}^1$ be the randomized encoding of $f \in \oplus\mathsf{L/poly}$, where the existence of $\hat{f}$ is shown in [5]. Then, their construction of a OWF is $g(x) = \hat{f}(0^n, x)$.[1] However, the domain and range of g are $\{0,1\}^m$ and $\{0,1\}^{m+1}$ respectively, i.e., the domain and range of g are inconsistent. Thus their construction is not a permutation. Moreover, since they define OWFs using randomized encoding directly, it is difficult to make their construction a permutation, i.e., it is not clear how to further achieve OWPs under the same worst-case assumption.

In this work, we propose a collection of OWPs and extend it to a OWP, both of which are computable in NC^1 and secure against NC^1 circuits under the assumption $\mathsf{NC}^1 \subsetneq \oplus\mathsf{L/poly}$.

To achieve the goal, we exploit the two distributions D_0^n and D_1^n described above. Essentially, our idea is to construct a "lossy function family" $\{f_{\mathbf{M}}(\mathbf{x}) = \mathbf{M}\mathbf{x}\}_{\mathbf{M} \in D_1^n}$. We let $\mathbf{M} \leftarrow D_1^n$ and $\mathbf{M} \leftarrow D_0^n$ in the injective and lossy model respectively, and the indistinguishability between the two models can be reduced to the indistinguishability between D_1^n and D_0^n. Then we follow the Peikert-Waters [38] approach to prove that $f_{\mathbf{M}}$ in the injective model satisfies one-wayness. Furthermore, since a matrix $\mathbf{M} \leftarrow D_1^n$ is of full rank, it holds that $f_{\mathbf{M}}$ in the injective model is a permutation. Therefore, $\{f_{\mathbf{M}}(\mathbf{x}) = \mathbf{M}\mathbf{x}\}_{\mathbf{M} \in D_1^n}$ is a collection of OWPs. Next, we extend it to a OWP, i.e., we give a construction of OWP based on a collection of OWPs which satisfies the distribution of index sample algorithm is identical to the uniform distribution over index set as follows.

[1] The one-wayness of g is based on the indistinguishability of the output distributions of $\hat{f}$ conditioned on $f(x) = 0$ and $f(x) = 1$, which can be reduced to $\mathsf{NC}^1 \subsetneq \oplus\mathsf{L/poly}$.

For a collection of OWPs $\{f_i : D_i \to D_i\}_{i \in I}$ where I is an index set, define a function g with the domain $D := \bigcup_{i \in I}(\{i\} \times D_i)$ and $g((i,x) \in D) = (i, f_i(x))$. Since f_i is a permutation and one-way, g is a permutation and one-way as well, i.e., g is a OWP.

Hash Proof System and CCA Secure PKE Scheme. The notion of HPS, which can be treated as designated verifier non-interactive zero-knowledge proof system for a language, was first introduced by Cramer and Shoup [15] for the purpose of constructing a CCA secure PKE scheme. An HPS allows one to generate a valid proof π proving that a statement x is in a language L by using a witness w and a public key pk. Also, one can generate a valid proof for x (not necessarily in L) by using only a secret key sk. For $x \in L$, proofs generated in these two ways should be the same. Typically, L is required to be a hard subset membership one, i.e., statements sampled from inside and outside the language should be indistinguishable. Furthermore, an HPS usually satisfies universality and smoothness. Universality means that for fixed x outside L and pk, the entropy of π is high enough (due to the entropy of sk). Smoothness means that for x outside L, the distribution of π honestly generated with sk is close to the uniform distribution in the proof space. HPSs are very versatile. Besides the application of PKE schemes, they play important roles in constructing various primitives, such as password authenticated key exchange [24,32], oblivious transfer [1,31], and zero-knowledge arguments [30].

In previous works, there has been no known way to construct HPSs that is computable in NC^1 and secure against adversaries bounded in NC^1 yet. Note that HPS is a quite different primitive from the ones in [14,16], and its instantiation cannot be achieved via some simple extension. The main bottleneck is that it is not clear how to construct an HPS, where we can reduce the hardness of the subset membership to the indistinguishability between D_0^n and D_1^n. To overcome this problem, we define two sets L and L' that are identical to the supported language in a somewhat sophisticated way. The interesting part is that we can reduce the indistinguishability between L' and X/L to that between D_0^n and D_1^n. Also we did very careful analysis on the entropy of secret keys with respect to fixed public keys the to prove smoothness and universality. More details are given as follows.

In this work, we propose the first HPS that is computable in NC^1 and secure against NC^1 adversaries based on the worst-case assumption $\mathsf{NC}^1 \subsetneq \oplus\mathsf{L/poly}$.

Our idea is to let a proof in the HPS be of the form $\mathbf{sk}^\top\mathbf{M}^\top\mathbf{w}$, where $\mathbf{M} \leftarrow D_0^n$, $\mathbf{x} = \mathbf{M}^\top\mathbf{w}$ is the statement with witness $\mathbf{w}$, $\mathbf{sk}$ is the secret key, and $\mathbf{M}\,\mathbf{sk} = \mathbf{pk}$ is the public key. A proof can be generated as either $\mathbf{pk}^\top\mathbf{w}$ or $\mathbf{sk}^\top\mathbf{x}$. The language that our HPS supports is $\mathrm{Im}(\mathbf{M}^\top)$. To achieve the hardness of our subset membership problem, we exploit the fact that $\mathrm{Im}(\mathbf{M}^\top)$ is identical to both

$$L = \{\mathbf{x}|\mathbf{w} \in 1\times\{0,1\}^{n-1}, \mathbf{x} = \mathbf{M}^\top\mathbf{w}\} \text{ and } L' = \{\mathbf{x}|\mathbf{w}\in 0\times\{0,1\}^{n-1}, \mathbf{x} = \mathbf{M}^\top\mathbf{w}\}.$$

We prove that if we sample $\mathbf{M}$ as $\mathbf{M} \leftarrow D_1^n$ instead of $\mathbf{M} \leftarrow D_0^n$ for L', L' becomes exactly $X \setminus L$ where $X = \{0,1\}^n$. Then we reduce the indistinguishability between the uniform distributions over L and $X \setminus L$ to that between D_0^n and D_1^n. To prove universality and smoothness, we show that for one $\mathbf{pk}$, there exist different valid secret keys, which lead to different outputs for any statement not in the language. Hence, the entropy of the proof is high due to the entropy of the secret key for a fixed $\mathbf{pk}$ and statement. We refer the reader to Sect. 4 for further details.

The proof size of the above scheme is only one single bit, while we can extend it to an HPS with multi-bit proofs by running many HPSs in parallel and show that the extension is still computable in NC^1 and secure against NC^1 circuits.

We now can instantiate the generic constructions [15] of a CCA-secure PKE scheme with our HPSs. The resulting scheme is secure against NC^1 circuits allowed to make constant rounds of adaptive decryption queries, while in each round, it can make arbitrary polynomial number of queries. This restriction is natural and defined in the same way as the adversaries for the NC^1-verifiable computation scheme in [14].

As far as we know, this is the first PKE that is CCA secure against NC^1 circuits under a mild complexity-theoretic assumptions, and there is no known way to make the PKE in [16] and the somewhat homomorphic encryption scheme in [14], which are malleable, CCA secure.

Trapdoor One-Way Function. TDF is a fundamental primitive introduced by Diffie and Hellman [17]. Unlike PKE schemes, where the decryption algorithm only recovers the plaintext (not including the internal randomness used in the encryption procedure), the inversion algorithm of a TDF recovers the entire pre-image. The property of TDF mentioned above is useful in many applications, where proofs of well-formedness are required [22]. However, in the same time, it makes constructing TDFs very challenging.

In the previous works [14,16], the PKEs use randomness in the encrypting procedures and it is difficult to recover the randomness in the decrypting procedures since the constructions recover the plaintexts by canceling the randomness using the property of the kernel of $\mathbf{M} \leftarrow D_0^n$. Namely, it is not easy to extend their construction to achieve a TDF. In fact, it has been shown that a TDF cannot be built from a PKE scheme in a black-box way [25][2]. On the other hand, it seems that there is a naive approach to construct a TDF f by defining it in the same way as our OWP, i.e., $f(\mathbf{x}) = \mathbf{M}_1\mathbf{x}$ where $\mathbf{M}_1$ is a sampled from D_1^n, and sample the inverse $\mathbf{M}_1^{-1}$ or some other elements that can be used to solve linear equations efficiently as the trapdoor. However, there is no known way to perform such a sampling procedure in NC^1 circuits. Therefore, some more sophisticated approach has to be taken.

In this work, we propose a TDF that is computable in NC^1 and secure against NC^1 circuits based on $\mathsf{NC}^1 \subsetneq \oplus\mathsf{L/poly}$. The intuition is as follows.

We first change the domain to $\{0,1\}^t \times (L \times X \setminus L)^t$ where $\mathbf{M} \leftarrow D_0^n$, $L = \mathrm{Im}(\mathbf{M})$, and $X = \{0,1\}^n$. On input $(x, (\mathbf{c}_1, \mathbf{c}'_1), \cdots, (\mathbf{c}_t, \mathbf{c}'_t)) \in \{0,1\}^t \times (L \times X \setminus L)^t$, our TDF computes $y = f(\mathbf{x})$, and additionally outputs $(\mathbf{c}_i, \mathbf{c}'_i)$ if

[2] There is no rigorous proof showing that the separation holds for NC^1, while it is an evidence that TDF is not easy to achieve.

$x_i = 0$ and $(\mathbf{c}'_i, \mathbf{c}_i)$ otherwise for all i. Here, f is a OWF that is computable in NC^1 and secure against NC^1 and x_i denotes the ith bit of x. Then, if we have a non-zero vector $\mathbf{k}$ in the kernel of $\mathbf{M}$, which is samplable in NC^1 [16], we can determine whether $\mathbf{x} \in \{0,1\}^n$ is in $\mathrm{Im}(\mathbf{M}^\top)$ or $\{0,1\}^n \setminus \mathrm{Im}(\mathbf{M}^\top)$ and recover x_i by checking whether $\mathbf{c}_i$ and $\mathbf{c}'_i$ are swapped. This provides us an efficiently samplable trapdoor. Due to the subset membership problem for $L = \mathrm{Im}(\mathbf{M})$ we described before, the uniform distributions over $\mathrm{Im}(\mathbf{M}^\top)$ and $\{0,1\}^n \setminus \mathrm{Im}(\mathbf{M}^\top)$ are indistinguishable when $\mathbf{M}$ is a matrix sampled from D_0^n. Therefore, the adversary in the one-wayness game can only obtain information on $f(x)$ (which is one-way) and the additional pairs do little help to it.

The above technique of sampling additional pairs is called *bits planting* which was used by Garg et al. [23] to construct a TDF based on the computational Diffie-Hellman problem. Although both our construction and the one in [23] aim at constructing trapdoor TDFs, we use the *bits planting* in a different way. In [23], this technique is exploited to recover the randomness used in the computation procedure of the TDF (see [23] for details), while in our work, we use it to avoid sampling the inverse of $\mathbf{M}$ so that every operation can be performed in NC^1.

1.3 Possibility on the Extension from Our Proposed NC^1 Fine-Grained Primitives

As described above, the fundamental cryptographic primitives we considered play key roles in a great deal of applications. Hence, our results directly imply the existence of more advanced NC^1-fine-grained primitives. As a simple instance, besides CCA secure PKE schemes, our HPS immediately implies the existence of a non-interactive key exchange scheme according to the recent construction by [27]. However, some NC^1 primitives can not be directly derived from existing ones by adopting previous generic conversions in the polynomial-time world since the resulting primitive may not be in NC^1 any more. For example, although it is well known that pseudorandom functions can be constructed from OWF/OWPs, ones in NC^1 are neither implied by our NC^1-OWP nor the OWF in [16]. It remains open how to construct such fine-grained primitives, and we believe that our works will serve a good starting point.

2 Preliminaries

2.1 Notation

For a distribution D, we denote sampling x according to D by $x \leftarrow D$. For a set S, we denote sampling x from S uniformly at random by $x \leftarrow S$. We denote the set $\{1, \cdots, n\}$ by $[n]$ and the ith element of a vector $\boldsymbol{x}$ by x_i. For a vector $\mathbf{x} \in \{0,1\}^*$, $\mathbf{x}$ will be regarded by default as a column vector. For a matrix $\mathbf{M}$, we denote the sets $\{\mathbf{y} \mid \exists \mathbf{x} \;\; s.t. \;\; \mathbf{y} = \mathbf{M}\mathbf{x}\}$ and $\{\mathbf{x} \mid \mathbf{M}\mathbf{x} = 0\}$ by $\mathrm{Im}(\mathbf{M})$ and $\mathrm{Ker}(\mathbf{M})$ respectively. Let X and Y be random variables over a finite set S. The *statistical distance* between X and Y is defined to be

$$\mathrm{Dist}(X, Y) = \frac{1}{2} \sum_{s \in S} |\Pr[X = s] - \Pr[Y = s]|.$$

We say that X and Y are ϵ-close if $\mathrm{Dist}(X, Y) \leq \epsilon$.

We note that all arithmetic computations are over $GF(2)$ in this work. Namely, all arithmetic computations are performed with a modulus of 2. By negl we denote an unspecified negligible function.

2.2 Definitions

In this section, we recall the definitions of a function family, NC^1 circuits, and $\oplus\mathsf{L/poly}$.

Definition 1 (Function Family). *A function family is a family of (possibly randomized) functions $\mathcal{F} = \{f_\lambda\}_{\lambda \in \mathbb{N}}$, where for each λ, f_λ has a domain D^f_λ and a range R^f_λ.*

Definition 2 (NC^1). *The class of (non-uniform) NC^1 function families is the set of all function families $\mathcal{F} = \{f_\lambda\}$ for which there is a polynomial p and constant c such that for each λ, f_λ can be computed by a (randomized) circuit of size $p(\lambda)$, depth $c\log(\lambda)$ and fan-in 2 using* AND, OR, *and* NOT *gates.*

Definition 3 ($\oplus\mathsf{L/poly}$). *$\oplus\mathsf{L/poly}$ is the set of all boolean function families $\mathcal{F} = \{f_\lambda\}$ for which there is a constant c such that for each λ, there is a non-deterministic Turing machine M_λ such that for each input x with length λ, $M_\lambda(x)$ uses at most $c\log(\lambda)$ space, and $f_\lambda(x)$ is equal to the parity of the number of accepting paths of $M_\lambda(x)$.*

We now give the lemma about the number of solutions for the linear equations defined by a matrix. It is straightforwardly follows from the fact that the rank of $\mathcal{A}$ is $n - 1$.

Lemma 1. *For any $n \times n$ matrix $\mathbf{A}$, if the rank of $\mathbf{A}$ is $n - 1$ and all arithmetic computations are over $GF(2)$, then for any $\mathbf{y} \in \mathrm{Im}(\mathbf{A})$, there exist and only exist two different vectors $\mathbf{x}$ and $\mathbf{x}'$ such that $\mathbf{Ax} = \mathbf{Ax}' = \mathbf{y}$.*

2.3 Definitions in Fine-Grained Cryptography

In this section, we define several cryptographic primitives which are secure against restricted complexity classes of adversaries and easy to run for honest parties. In the following definitions, we denote the class of honest parties by $\mathcal{C}_1$ i.e., function families that compose the primitive are in the class $\mathcal{C}_1$ and the class of adversaries by $\mathcal{C}_2$, and the condition $\mathcal{C}_1 \subseteq \mathcal{C}_2$ is implicit in each definition and hence left unmentioned.

Definition 4 (One-Way Function [16]). *Let l be a polynomial in λ. Let $\mathcal{F} = \{f_\lambda : \{0,1\}^\lambda \to \{0,1\}^{l(\lambda)}\}$ be a function family. $\mathcal{F}$ is a $\mathcal{C}_1$-*one-way function (OWF) *against $\mathcal{C}_2$ if:*

- **Computability:** *For each* λ, f_λ *is deterministic.*
- **One-wayness:** *For any* $\mathcal{G} = \{g_\lambda : \{0,1\}^{l(\lambda)} \to \{0,1\}^\lambda\}$ *and any* $\lambda \in \mathbb{N}$*:*

$$\Pr\left[f_\lambda(g_\lambda(y)) = y \,\middle|\, \begin{array}{c} x \leftarrow \{0,1\}^\lambda \\ y = f_\lambda(x) \end{array}\right] \leq \mathsf{negl}(\lambda).$$

Definition 5 (One-Way Permutation). *Let* $\mathcal{F} = \{f_\lambda : D_\lambda \to D_\lambda\}$ *be a function family.* $\mathcal{F}$ *is a* $\mathcal{C}_1$-one-way permutation (OWP) *against* $\mathcal{C}_2$ *if:*

- **Permutation:** *For each* λ, f_λ *is a permutation.*
- **One-wayness:** *For any* $\mathcal{G} = \{g_\lambda : D_\lambda \to D_\lambda\}$ *and any* $\lambda \in \mathbb{N}$*:*

$$\Pr\left[g_\lambda(y) = x \,\middle|\, \begin{array}{c} x \leftarrow D_\lambda \\ y = f_\lambda(x) \end{array}\right] \leq \mathsf{negl}(\lambda).$$

Definition 6 (Collection of OWPs). *Let* $\mathcal{KeyGen} = \{\mathsf{KeyGen}_\lambda : \phi \to K_\lambda\}$ *and* $\mathcal{E}val = \{\mathsf{Eval}_\lambda : K_\lambda \times \{0,1\}^\lambda \to \{0,1\}^\lambda\}$ *be function families.* $(\mathcal{KeyGen}, \mathcal{E}val)$ *is a collection of* $\mathcal{C}_1$*-OWPs against* $\mathcal{C}_2$ *if:*

- **Permutation:** *For each* λ *and* $k \leftarrow \mathsf{KeyGen}_\lambda$, $\mathsf{Eval}_\lambda(k, \cdot) : D_{\lambda,k} \to D_{\lambda,k}$ *is a permutation where* $D_{\lambda,k} \subseteq \{0,1\}^\lambda$.
- **One-wayness:** *For any* $\mathcal{G} = \{g_\lambda : K_\lambda \times \{0,1\}^\lambda \to \{0,1\}^\lambda\}$ *and any* $\lambda \in \mathbb{N}$*:*

$$\Pr\left[g_\lambda(k, y) = x \,\middle|\, \begin{array}{c} k \leftarrow \mathsf{KeyGen}_\lambda \\ x \leftarrow D_{\lambda,k} \subseteq \{0,1\}^\lambda \\ y = \mathsf{Eval}_\lambda(k, x) \end{array}\right] \leq \mathsf{negl}(\lambda).$$

Definition 7 (Hash Proof System). *Let* $PP_\lambda = (X_\lambda, L_\lambda, W_\lambda, R_\lambda, SK_\lambda, PK_\lambda, \Pi_\lambda, H_\lambda, \alpha_\lambda, \mathsf{aux}_\lambda)$ *where* X_λ *is a finite non-empty set,* L_λ *is a subset of* X *such that* $x \in L_\lambda$ *iff there exists a witness* $w \in W_\lambda$ *with* $(x, w) \in R_\lambda \subset X_\lambda \times W_\lambda$, SK_λ *is a secret key space,* PK_λ *is a public key space,* Π_λ *is a proof space,* $H_\lambda : SK_\lambda \times X_\lambda \to \Pi_\lambda$ *is a hash function,* $\alpha_\lambda : SK_\lambda \to PK_\lambda$ *is a projective map, and* aux_λ *is an auxiliary information. Define the following function families.*

- $\mathcal{S}etup = \{\mathsf{Setup}_\lambda : \phi \to PP_\lambda\}$ *where* Setup_λ *outputs a public parameter* $\mathsf{pp} \in PP_\lambda$.
- $\mathcal{S}amp\mathcal{Y}es = \{\mathsf{SampYes}_\lambda : PP_\lambda \to R_\lambda\}$ *where* $\mathsf{SampYes}_\lambda$ *on input* $\mathsf{pp} \in PP_\lambda$ *outputs a random element* $x \in L_\lambda$ *with a witness* $w \in W_\lambda$*, i.e., a random element* $(x, w) \in R_\lambda$.
- $\mathcal{S}amp\mathcal{N}o = \{\mathsf{SampNo}_\lambda : PP_\lambda \to X_\lambda \backslash L_\lambda\}$ *where* SampNo_λ *on input* $\mathsf{pp} \in PP_\lambda$ *outputs a random element* $x \in X_\lambda \setminus L_\lambda$.
- $\mathcal{KeyGen} = \{\mathsf{KeyGen}_\lambda : PP_\lambda \to PK_\lambda \times SK_\lambda\}$ *where* KeyGen_λ *on input* $\mathsf{pp} \in PP_\lambda$ *outputs a public key* pk *and secret key* sk *such that* $pk = \alpha_\lambda(sk)$.
- $\mathcal{P}riv = \{\mathsf{Priv}_\lambda : PP_\lambda \times SK_\lambda \times X_\lambda \to \Pi_\lambda\}$ *where* Priv_λ *on input* $\mathsf{pp} \in PP_\lambda$, $sk \in SK_\lambda$*, and an instance* $x \in X_\lambda$ *outputs its proof* $\pi = H_\lambda(sk, x)$.
- $\mathcal{P}ub = \{\mathsf{Pub}_\lambda : PP_\lambda \times PK_\lambda \times R_\lambda \to \Pi_\lambda\}$ *where* Pub_λ *on input* $\mathsf{pp} \in PP_\lambda$, $pk \in PK_\lambda$*, and an instance with a witness* $(x, w) \in R_\lambda$ *outputs its proof* $\pi \in \Pi_\lambda$.

$(\mathcal{S}etup, \mathcal{S}amp\mathcal{Y}es, \mathcal{S}amp\mathcal{N}o, \mathcal{K}ey\mathcal{G}en, \mathcal{P}riv, \mathcal{P}ub)$ *is a* $\mathcal{C}_1$-hash proof system (HPS) against $\mathcal{C}_2$ *if for any* $\lambda \in \mathbb{N}$, *it holds that:*

- **Correctness:** *For any* $(x, w) \in R_\lambda$, *we have*

$$\mathsf{Priv}_\lambda(\mathsf{pp}, sk, x) = H_\lambda(sk, x) = \mathsf{Pub}_\lambda(\mathsf{pp}, pk, x, w)$$

  where $\mathsf{pp} \leftarrow \mathsf{Setup}_\lambda$ *and* $(pk, sk) \leftarrow \mathsf{KeyGen}_\lambda(\mathsf{pp})$.
- **Subset membership problem:**
  - *The distributions of* x *and* x' *are identical where* $\mathsf{pp} \leftarrow \mathsf{Setup}_\lambda$, $(x, w) \leftarrow \mathsf{SampYes}_\lambda(\mathsf{pp})$, *and* $x' \leftarrow L_\lambda$.
  - *The distributions of* x *and* x' *are identical where* $\mathsf{pp} \leftarrow \mathsf{Setup}_\lambda$, $x \leftarrow \mathsf{SampNo}_\lambda(\mathsf{pp})$, *and* $x' \leftarrow X_\lambda \setminus L_\lambda$.
  - *For any* $\mathcal{G} = \{g_\lambda\} \in \mathcal{C}_2$,

$$|\Pr[g_\lambda(\mathsf{pp}, x_0) = 1] - \Pr[g_\lambda(\mathsf{pp}, x_1) = 1]| \leq \mathsf{negl}(\lambda)$$

    where $\mathsf{pp} \leftarrow \mathsf{Setup}_\lambda$, $(x_0, w) \leftarrow \mathsf{SampYes}_\lambda(\mathsf{pp})$, *and* $x_1 \leftarrow \mathsf{SampNo}_\lambda(\mathsf{pp})$.

$(\mathcal{S}etup, \mathcal{S}amp\mathcal{Y}es, \mathcal{S}amp\mathcal{N}o, \mathcal{K}ey\mathcal{G}en, \mathcal{P}riv, \mathcal{P}ub)$ *is* perfectly smooth $\mathcal{C}_1$-*HPS against* $\mathcal{C}_2$ *if it satisfies the following property.*

- **Perfect smoothness:** *For any* $\mathsf{pp} \leftarrow \mathsf{Setup}_\lambda$, *the following random variables are identical, i.e., 0-close.*

$$(x, pk, \pi),\ (x, pk, \pi')$$

  where $\mathbf{x} \leftarrow \mathsf{SampNo}_\lambda(\mathsf{pp})$, $(pk, sk) \leftarrow \mathsf{KeyGen}_\lambda(\mathsf{pp})$, $\pi = \mathsf{Priv}_\lambda(\mathsf{pp}, sk, x)$, *and* $\pi' \leftarrow \Pi$.

$(\mathcal{S}etup, \mathcal{S}amp\mathcal{Y}es, \mathcal{S}amp\mathcal{N}o, \mathcal{K}ey\mathcal{G}en, \mathcal{P}riv, \mathcal{P}ub)$ *is* ϵ-universal$_1$ $\mathcal{C}_1$-HPS against $\mathcal{C}_2$ *if it satisfies the following property.*

- *ϵ-universality$_1$: For any* $\mathsf{pp} \leftarrow \mathsf{Setup}_\lambda$, $pk \in PK_\lambda$, $x \in X_\lambda \setminus L_\lambda$ *and* $\pi \in \Pi_\lambda$, *it holds that*

$$\Pr[\mathsf{Priv}_\lambda(\mathsf{pp}, sk, x) = \pi \mid \alpha_\lambda(sk) = pk] \leq \epsilon.$$

  If ϵ *is a negligible function, then* $(\mathcal{S}etup, \mathcal{S}amp\mathcal{Y}es, \mathcal{S}amp\mathcal{N}o, \mathcal{K}ey\mathcal{G}en, \mathcal{P}riv, \mathcal{P}ub)$ *is a* strong universal$_1$ $\mathcal{C}_1$-HPS against $\mathcal{C}_2$.

$(\mathcal{S}etup, \mathcal{S}amp\mathcal{Y}es, \mathcal{S}amp\mathcal{N}o, \mathcal{K}ey\mathcal{G}en, \mathcal{P}riv, \mathcal{P}ub)$ *is* ϵ-universal$_2$ $\mathcal{C}_1$-HPS against $\mathcal{C}_2$ *if it satisfies the following property.*

- *ϵ-universality$_2$: For any* $\mathsf{pp} \leftarrow \mathsf{Setup}_\lambda$, $pk \in PK_\lambda$, $x, x^* \in X_\lambda$ *and* $\pi, \pi^* \in \Pi_\lambda$ *with* $x \notin L_\lambda \cup \{x^*\}$, *it holds that*

$$\Pr[\mathsf{Priv}_\lambda(\mathsf{pp}, sk, x) = \pi \mid \mathsf{Priv}_\lambda(\mathsf{pp}, sk, x^*) = \pi^* \wedge \alpha_\lambda(sk) = pk] \leq \epsilon.$$

  If ϵ *is a negligible function, then* $(\mathcal{S}etup, \mathcal{S}amp\mathcal{Y}es, \mathcal{S}amp\mathcal{N}o, \mathcal{K}ey\mathcal{G}en, \mathcal{P}riv, \mathcal{P}ub)$ *is a* strong universal$_2$ $\mathcal{C}_1$-HPS against $\mathcal{C}_2$.

Definition 8 (Trapdoor One-Way Function). *Let* $\mathcal{KeyGen} = \{\mathsf{KeyGen}_\lambda : \phi \to EK_\lambda \times TK_\lambda\}$*,* $\mathcal{Eval} = \{\mathsf{Eval}_\lambda : EK_\lambda \times \mathcal{D}_\lambda \to \mathcal{R}_\lambda\}$ *and* $\mathcal{Inverse} = \{\mathsf{Inverse}_\lambda : TK_\lambda \times \mathcal{D}_\lambda \to \mathcal{R}_\lambda\}$ *be function families where* $\mathcal{D}_\lambda$ *and* $\mathcal{R}_\lambda$ *are determined by the key pair* (ek, tk) *generated by* KeyGen_λ*.* $(\mathcal{KeyGen}, \mathcal{Eval}, \mathcal{Inverse})$ *is a* $\mathcal{C}_1$*-*trapdoor one-way function (TDF) against $\mathcal{C}_2$ *if:*

- **Correctness:** *For any* $\lambda \in \mathbb{N}$*, any* $(ek, tk) \leftarrow \mathsf{KeyGen}_\lambda$*, and any* $X \in \mathcal{D}_\lambda$*:*

$$\mathsf{Inverse}_\lambda(tk, \mathsf{Eval}_\lambda(ek, X)) = X.$$

- **One-wayness:** *For any* $\mathcal{G} = \{g_\lambda\} \in \mathcal{C}_2$*, and any* $\lambda \in \mathbb{N}$*:*

$$\Pr\left[\mathsf{Eval}_\lambda(ek, g_\lambda(ek, Y)) = Y \;\middle|\; \begin{array}{c}(ek, tk) \leftarrow \mathsf{KeyGen}_\lambda \\ X \leftarrow \mathcal{D}_\lambda \\ Y = \mathsf{Eval}_\lambda(ek, X)\end{array}\right] \le \mathsf{negl}(\lambda).$$

2.4 Sampling Procedure

In this section, we recall the sampling procedure in [16], and then show several lemmas on the sampling procedure that will be used later in the security proofs.

Construction 1 (Sampling Procedure). *Let* $\mathbf{M}_0^n$ *and* $\mathbf{M}_1^n$ *be the following* $n \times n$ *matrices:*

$$\mathbf{M}_0^n = \begin{pmatrix} 0 & & \cdots & 0 & 0 \\ 1 & 0 & & & 0 \\ 0 & 1 & \ddots & & \vdots \\ \vdots & \vdots & & \ddots & 0 \\ 0 & \cdots & 0 & 1 & 0 \end{pmatrix}, \quad \mathbf{M}_1^n = \begin{pmatrix} 0 & & \cdots & 0 & 1 \\ 1 & 0 & & & 0 \\ 0 & 1 & \ddots & & \vdots \\ \vdots & \vdots & & \ddots & 0 \\ 0 & \cdots & 0 & 1 & 0 \end{pmatrix}.$$

- $\mathsf{LSamp}(n)$*:*
  1. *Output the following* $n \times n$ *upper triangular matrix:*

$$\begin{pmatrix} 1 & r_{1,2} & \cdots & r_{1,n-1} & r_{1,n} \\ 0 & 1 & r_{2,3} & \cdots & r_{2,n} \\ 0 & 0 & \ddots & & \vdots \\ \vdots & \vdots & \ddots & 1 & r_{n-1,n} \\ 0 & \cdots & 0 & 0 & 1 \end{pmatrix}$$

  where $r_{i,j} \leftarrow \{0, 1\}$*.*
- $\mathsf{RSamp}(n)$*:*
  1. *Output the following* $n \times n$ *matrix:*

$$\begin{pmatrix} 1 & & \cdots & 0 & r_1 \\ 0 & 1 & & & r_2 \\ 0 & 0 & \ddots & & \vdots \\ \vdots & \vdots & & \ddots\ 1 & r_{n-1} \\ 0 & \cdots & 0 & 0 & 1 \end{pmatrix}$$

  where $r_i \leftarrow \{0, 1\}$*.*

– ZeroSamp*(n):*
  1. *Sample* $\mathbf{R}_1 \leftarrow \mathsf{LSamp}(n)$ *and* $\mathbf{R}_2 \leftarrow \mathsf{RSamp}(n)$.
  2. *Output* $\mathbf{R}_1\mathbf{M}_0^n\mathbf{R}_2$.
– OneSamp*(n):*
  1. *Sample* $\mathbf{R}_1 \leftarrow \mathsf{LSamp}(n)$ *and* $\mathbf{R}_2 \leftarrow \mathsf{RSamp}(n)$.
  2. *Output* $\mathbf{R}_1\mathbf{M}_1^n\mathbf{R}_2$.

Here, the output of $\mathsf{ZeroSamp}(n)$ *is always a matrix of rank* $n-1$ *and the output of* $\mathsf{OneSamp}(n)$ *is always a matrix of full rank.*

Lemma 2 **([5,29]).** *If* $\mathsf{NC}^1 \subsetneq \oplus\mathsf{L/poly}$, *then there is a polynomial* n *such that for any family* $\mathcal{F} = \{f_\lambda\}$ *in* NC^1 *and any* $\lambda \in \mathbb{N}$, *we have*

$$|\Pr[f_\lambda(\mathbf{M}) = 1 \mid \mathbf{M} \leftarrow \mathsf{ZeroSamp}(n(\lambda))] - \Pr[f_\lambda(\mathbf{M}') = 1 \mid \mathbf{M}' \leftarrow \mathsf{OneSamp}(n(\lambda))]| \leq \mathsf{negl}(\lambda).$$

Lemma 3. *For any* $\mathbf{M} \leftarrow \mathsf{ZeroSamp}(n)$, *it holds that* $\mathrm{Ker}(\mathbf{M}) = \{\mathbf{0}, \mathbf{k}\}$ *where* $\mathbf{k}$ *is a vector such that* $\mathbf{k} \in \{0,1\}^{n-1} \times 1$.

Proof. $\mathbf{M}$ is a matrix sampled from $\mathsf{ZeroSamp}(n)$, i.e.,

$$\mathbf{M} = \mathbf{R}_1\mathbf{M}_1^n\mathbf{R}_2 = \mathbf{R}_1 \begin{pmatrix} 0 & & \cdots & 0 & 0 \\ 1 & 0 & & & 0 \\ 0 & 1 & \ddots & & \vdots \\ \vdots & \vdots & & \ddots & 0 \\ 0 & \cdots & & 0 & 1 & 0 \end{pmatrix} \begin{pmatrix} 1 & & \cdots & 0 & r_1 \\ 0 & 1 & & & r_2 \\ 0 & 0 & \ddots & & \vdots \\ \vdots & \vdots & & \ddots\ 1 & r_{n-1} \\ 0 & \cdots & & 0\ 0 & 1 \end{pmatrix}$$

where $\mathbf{R}_1 \leftarrow \mathsf{LSamp}(n)$ and $\mathbf{R}_2 \leftarrow \mathsf{RSamp}(n)$. Then, we have $\mathbf{k} = (r_1 r_2 \cdots 1)^\top \in \mathrm{Ker}(\mathbf{M})$ since

$$\mathbf{M} = \mathbf{R}_1\mathbf{M}_1^n\mathbf{R}_2\mathbf{k} = \mathbf{R}_1 \begin{pmatrix} 0 & & \cdots & 0 & 0 \\ 1 & 0 & & & 0 \\ 0 & 1 & \ddots & & \vdots \\ \vdots & \vdots & & \ddots & 0 \\ 0 & \cdots & & 0 & 1 & 0 \end{pmatrix} \begin{pmatrix} 1 & & \cdots & 0 & r_1 \\ 0 & 1 & & & r_2 \\ 0 & 0 & \ddots & & \vdots \\ \vdots & \vdots & & \ddots\ 1 & r_{n-1} \\ 0 & \cdots & & 0\ 0 & 1 \end{pmatrix} \begin{pmatrix} r_1 \\ r_2 \\ \vdots \\ r_{n-1} \\ 1 \end{pmatrix}$$

$$= \mathbf{R}_1 \begin{pmatrix} 0 & & \cdots & 0 & 0 \\ 1 & 0 & & & 0 \\ 0 & 1 & \ddots & & \vdots \\ \vdots & \vdots & & \ddots & 0 \\ 0 & \cdots & & 0 & 1 & 0 \end{pmatrix} \begin{pmatrix} 0 \\ 0 \\ \vdots \\ 0 \\ 1 \end{pmatrix} = \mathbf{R}_1\mathbf{0} = \mathbf{0}.$$

Moreover, according to Lemma 1, there are only two vectors $\boldsymbol{v}$ such that $\mathbf{M}\boldsymbol{v} = \mathbf{0}$. Therefore, we have $\mathrm{Ker}(\mathbf{M}) = \{\mathbf{0}, \mathbf{k}\}$, completing the proof of Lemma 3. □

Lemma 4. *For any* $\mathbf{M} \leftarrow \mathsf{ZeroSamp}(n)$*, it holds that* $\mathrm{Ker}(\mathbf{M}^\top) = \{\mathbf{0}, \mathbf{k}\}$ *where* $\mathbf{k}$ *is a vector such that* $\mathbf{k} \in 1 \times \{0,1\}^{n-1}$.

Proof. $\mathbf{M}$ is a matrix sampled from $\mathsf{ZeroSamp}(n)$ i.e., $\mathbf{M} = \mathbf{R}_1\mathbf{M}_0^n\mathbf{R}_2$, where $\mathbf{R}_1 \leftarrow \mathsf{LSamp}(n)$, $\mathbf{R}_2 \leftarrow \mathsf{RSamp}(n)$. Since $\mathbf{R}_1^\top$ has full rank, the equation $\mathbf{R}_1^\top\mathbf{x} = (1\ 0\ \cdots\ 0)^\top$ has a unique solution $\mathbf{x}^*$. $\mathbf{x}^*$ is in the kernel of $\mathbf{M}^\top$ since $\mathbf{R}_2^\top\mathbf{M}_0^{n\top}\mathbf{R}_1^\top\mathbf{x}^* = \mathbf{R}_2^\top\mathbf{M}_0^{n\top}(1\ 0\ \cdots\ 0)^\top = \mathbf{R}_2^\top\mathbf{0} = \mathbf{0}$. According to the following equation

$$\mathbf{R}_1^\top\mathbf{x}^* = \begin{pmatrix} 1 & 0 & \cdots & 0 & 0 \\ r_{2,1} & 1 & 0 & \cdots & 0 \\ r_{3,1} & r_{3,2} & \ddots & & \vdots \\ \vdots & \vdots & \ddots & 1 & 0 \\ r_{n,1} & \cdots & & r_{n,n-1} & 1 \end{pmatrix} \begin{pmatrix} x_1^* \\ x_2^* \\ x_3^* \\ \vdots \\ x_n^* \end{pmatrix} = \begin{pmatrix} 1 \\ 0 \\ 0 \\ \vdots \\ 0 \end{pmatrix},$$

we have $x_1^* = 1$, i.e., $\mathbf{x} \in 1 \times \{0,1\}^{n-1}$.

Moreover, according to Lemma 1 and the fact that the rank of $\mathbf{M}^\top$ is $n-1$, there are only two vectors $\boldsymbol{v}$ such that $\mathbf{M}^\top\boldsymbol{v} = \mathbf{0}$. Therefore, we have $\mathrm{Ker}(\mathbf{M}^\top) = \{\mathbf{0}, \mathbf{x}^*\}$, completing the proof of Lemma 4. □

Lemma 5. *For any* $\mathbf{M} \leftarrow \mathsf{ZeroSamp}(\lambda)$*, it holds that*

$$\mathrm{Im}(\mathbf{M}^\top) = \{\mathbf{x} | \mathbf{w} \in 0\times\{0,1\}^{\lambda-1}, \mathbf{x} = \mathbf{M}^\top\mathbf{w}\} = \{\mathbf{x} | \mathbf{w} \in 1\times\{0,1\}^{\lambda-1}, \mathbf{x} = \mathbf{M}^\top\mathbf{w}\}.$$

Proof. Let U be a set such that $U = \{\mathbf{x}|\mathbf{w} \in 0\times\{0,1\}^{\lambda-1}, \mathbf{x} = \mathbf{M}^\top\mathbf{w}\}$ and V be a set such that $V = \{\mathbf{x}|\mathbf{w} \in 1 \times \{0,1\}^{\lambda-1}, \mathbf{x} = \mathbf{M}^\top\mathbf{w}\}$. Let $\mathbf{k}$ be a non-zero vector such that $\mathbf{k} \in \mathrm{Ker}(\mathbf{M}^\top)$. According to Lemma 4, we have $\mathbf{k} \in 1 \times \{0,1\}^{\lambda-1}$. Therefore, for any $\mathbf{x} \in U$ such that $\mathbf{x} = \mathbf{M}^\top\mathbf{w}$ where $\mathbf{w} \in 0 \times \{0,1\}^{\lambda-1}$, we have $\mathbf{x} = \mathbf{M}^\top\mathbf{w} = \mathbf{M}^\top(\mathbf{w}+\mathbf{k}) \in V$ since $(\mathbf{w}+\mathbf{k}) \in 1 \times \{0,1\}^{\lambda-1}$. Moreover, for any $\mathbf{x} \in V$ such that $\mathbf{x} = \mathbf{M}^\top\mathbf{w}$ where $\mathbf{w} \in 1 \times \{0,1\}^{\lambda-1}$, we have $\mathbf{x} = \mathbf{M}^\top\mathbf{w} = \mathbf{M}^\top(\mathbf{w}+\mathbf{k}) \in U$ since $(\mathbf{w}+\mathbf{k}) \in 0 \times \{0,1\}^{\lambda-1}$. Therefore, we have $U = V$ and it follows that $\mathrm{Im}(\mathbf{M}^\top) = U \cup V = U \cup U = U = \{\mathbf{x}|\mathbf{w} \in 0 \times \{0,1\}^{\lambda-1}, \mathbf{x} = \mathbf{M}^\top\mathbf{w}\}$. In the same way, we have $\mathrm{Im}(\mathbf{M}^\top) = U \cup V = V \cup V = V = \{\mathbf{x}|\mathbf{w} \in 1 \times \{0,1\}^{\lambda-1}, \mathbf{x} = \mathbf{M}^\top\mathbf{w}\}$. As a result, we have

$$\mathrm{Im}(\mathbf{M}^\top) = \{\mathbf{x} | \mathbf{w} \in 0\times\{0,1\}^{\lambda-1}, \mathbf{x} = \mathbf{M}^\top\mathbf{w}\} = \{\mathbf{x} | \mathbf{w} \in 1\times\{0,1\}^{\lambda-1}, \mathbf{x} = \mathbf{M}^\top\mathbf{w}\},$$

completing the proof of Lemma 5. □

Lemma 6. *The distributions of* $\mathbf{M} + \mathbf{N}$ *and* $\mathbf{M}'$ *are identical, where* $\mathbf{M} \leftarrow \mathsf{ZeroSamp}(\lambda)$, $\mathbf{M}' \leftarrow \mathsf{OneSamp}(\lambda)$*, and* $\mathbf{N}$ *is the following matrix.*

$$\mathbf{N} = \begin{pmatrix} 0 & \cdots & 0 & 1 \\ 0 & 0 & \cdots & 0 \\ \vdots & & \ddots & \vdots \\ 0 & \cdots & 0 & 0 \end{pmatrix}.$$

Proof. For $\mathbf{R}_1 \leftarrow \mathsf{LSamp}(\lambda)$ and $\mathbf{R}_2 \leftarrow \mathsf{RSamp}(\lambda)$, we have

$$\begin{aligned}
\mathbf{M}' = \mathbf{R}_1\mathbf{M}_1^\lambda\mathbf{R}_2 &= \mathbf{R}_1 \begin{pmatrix} 0 & & \cdots & 0 & 1 \\ 1 & 0 & & \vdots & 0 \\ \vdots & \vdots & \ddots & 0 & \vdots \\ 0 & \cdots & 0 & 1 & 0 \end{pmatrix} \mathbf{R}_2 \\
&= \mathbf{R}_1 \begin{pmatrix} 0 & & \cdots & 0 & 0 \\ 1 & 0 & & \vdots & 0 \\ \vdots & \vdots & \ddots & 0 & \vdots \\ 0 & \cdots & 0 & 1 & 0 \end{pmatrix} \mathbf{R}_2 + \mathbf{R}_1 \begin{pmatrix} 0 & & \cdots & 0 & 1 \\ 0 & 0 & & \vdots & 0 \\ \vdots & \vdots & \ddots & 0 & \vdots \\ 0 & \cdots & 0 & 0 & 0 \end{pmatrix} \mathbf{R}_2 \\
&= \mathbf{R}_1 \begin{pmatrix} 0 & & \cdots & 0 & 0 \\ 1 & 0 & & \vdots & 0 \\ \vdots & \vdots & \ddots & 0 & \vdots \\ 0 & \cdots & 0 & 1 & 0 \end{pmatrix} \mathbf{R}_2 + \begin{pmatrix} 0 & & \cdots & 0 & 1 \\ 0 & 0 & & \vdots & 0 \\ \vdots & \vdots & \ddots & 0 & \vdots \\ 0 & \cdots & 0 & 0 & 0 \end{pmatrix} \\
&= \mathbf{R}_1\mathbf{M}_0^\lambda\mathbf{R}_2 + \mathbf{N} = \mathbf{M} + \mathbf{N}.
\end{aligned}$$

Hence, the distributions of $\mathbf{M} + \mathbf{N}$ and $\mathbf{M}'$ are identical for $\mathbf{M} \leftarrow \mathsf{ZeroSamp}(\lambda)$ and $\mathbf{M}' \leftarrow \mathsf{OneSamp}(\lambda)$, completing the proof of Lemma 6. □

3 Construction of NC^1-OWP Against NC^1

In this section, we first give our construction of a collection of NC^1-OWPs against NC^1 under the assumption $\mathsf{NC}^1 \subsetneq \oplus\mathsf{L/poly}$. Next, we extend it to a NC^1-OWP against NC^1 based on the same assumption.

Construction 2 (Collection of NC^1-OWPs). *Let λ be a security parameter. We define the families $\mathcal{KeyGen} = \{\mathsf{KeyGen}_\lambda\}$ with key spaces $\{K_\lambda = \{\mathbf{M} \mid \mathbf{M} \in \mathsf{OneSamp}(\lambda)\}\}$ and $\mathcal{Eval} = \{\mathsf{Eval}_\lambda\}$ as follows.*

- KeyGen_λ*:*
  1. *Sample* $\mathbf{M} \leftarrow \mathsf{OneSamp}(\lambda)$*.*
  2. *Output* $\mathbf{M}$ *(which defines the domain as $D_{\lambda,\mathbf{M}} := \{0,1\}^\lambda$).*
- $\mathsf{Eval}_\lambda(\mathbf{M}, \mathbf{x})$*:*
  1. *Compute* $\mathbf{y} := \mathbf{M}\mathbf{x}$ *and output* $\mathbf{y}$*.*

Theorem 1. *$(\mathcal{KeyGen}, \mathcal{Eval})$ defined as Construction 2 is a collection of NC^1-OWPs against NC^1 under the assumption $\mathsf{NC}^1 \subsetneq \oplus\mathsf{L/poly}$.*

Proof Sketch. As described in Introduction, our construction is essentially a "lossy function". More specifically, it is straightforward that our scheme is a permutation, since $\mathbf{M}$ is of full rank when $\mathbf{M} \leftarrow \mathsf{OneSamp}(\lambda)$. Moreover, when we generate $\mathbf{M} \leftarrow \mathsf{ZeroSamp}(\lambda)$ instead of $\mathbf{M} \leftarrow \mathsf{OneSamp}(\lambda)$ in KeyGen_λ, we can prove that an adversary $\mathcal{A}$ breaking the one-wayness of our construction with probability ϵ can also be used to find a second pre-image $\mathbf{x}'$ for $\mathsf{Eval}_\lambda(\mathbf{M}, \mathbf{x})$ such that $\mathbf{x} \neq \mathbf{x}'$ with probability $\frac{1}{2}\epsilon$. This is due to the fact that $\mathbf{M}$ is not of full rank in this case and $\mathcal{A}$ has no information on whether the pre-image is $\mathbf{x}$ or $\mathbf{x}'$. However, it is unlikely that $\mathcal{A}$ can find such a second pre-image, since this construction is indistinguishable with the original one, where $\mathbf{M}$ is generated as $\mathbf{M} \leftarrow \mathsf{OneSamp}(\lambda)$ and there exists no second pre-image for each $\mathbf{M}$. Therefore, we can conclude that this scheme is one-way, which immediately gives us the one-wayness of the original scheme (due to the indistinguishability between $\mathsf{OneSamp}(\lambda)$ and $\mathsf{ZeroSamp}(\lambda)$).

The formal proof is as follows.

Proof. First note that both $\mathcal{KeyGen}$ and $\mathcal{Eval}$ are computable in NC^1, since they only involve operations including multiplications of a constant number of matrices, inner products, and sampling random bits. We now show that $(\mathcal{KeyGen}, \mathcal{Eval})$ satisfies computability and one-wayness.

Permutation. Since for $\mathbf{M} \leftarrow \mathsf{OneSamp}_\lambda$, $\mathbf{M}$ is a full rank matrix, we have that $\mathsf{Eval}_\lambda(\mathbf{M}, \mathbf{x}) = \mathbf{M}\mathbf{x} \in D_{\lambda,\mathbf{M}} = \{0,1\}^\lambda$ is a permutation.

One-Wayness. Let $\mathcal{A} = \{a_\lambda\}$ be any adversary in NC^1. We give hybrid games to show that the advantage of $\mathcal{A}$ in breaking the one-wayness of Construction 2 is negligible.

Game 0: This is the original one-wayness game for $\mathcal{A} = \{a_\lambda\}$. $\mathcal{CH}$ runs $\mathbf{M} \leftarrow \mathsf{KeyGen}_\lambda$ and samples $\mathbf{x} \leftarrow \{0,1\}^\lambda$. Then, it runs $\mathbf{y} = \mathsf{Eval}_\lambda(\mathbf{M}, \mathbf{x})$ and sends $\mathbf{y}$ to a_λ. a_λ succeeds if it outputs $\tilde{\mathbf{x}}$ such that $\mathbf{x} = \tilde{\mathbf{x}}$. Otherwise, it fails.

Game 1: This game is the same as **Game 0** except that $\mathcal{CH}$ runs $\mathsf{ZeroSamp}(\lambda)$ instead of $\mathsf{OneSamp}(\lambda)$ in the key generation procedure.

Lemma 7. *If $\mathcal{A} = \{a_\lambda\}$ succeeds with advantage ϵ_0 (resp., ϵ_1) in* **Game 0** *(resp.,* **Game 1***), then $|\epsilon_0 - \epsilon_1| = \mathsf{negl}(\lambda)$.*

Proof. We now construct $\mathcal{B} = \{b_\lambda\} \in \mathsf{NC}^1$ that distinguishes $\mathbf{M} \leftarrow \mathsf{ZeroSamp}(\lambda)$ and $\mathbf{M} \leftarrow \mathsf{OneSamp}(\lambda)$ with advantage $|\epsilon_0 - \epsilon_1|$, which contradicts to Lemma 2.

b_λ takes as input $\mathbf{M}$, which is generated as $\mathbf{M} \leftarrow \mathsf{ZeroSamp}(\lambda)$ or $\mathbf{M} \leftarrow \mathsf{OneSamp}(\lambda)$ from its challenger. Then, it samples $\mathbf{x} \leftarrow \{0,1\}^\lambda$. Next, b_λ runs $\mathbf{y} = \mathsf{Eval}_\lambda(\mathbf{M}, \mathbf{x})$ and sends $\mathbf{y}$ to a_λ. When a_λ outputs $\tilde{\mathbf{x}}$, if $\mathbf{x} = \tilde{\mathbf{x}}$, b_λ outputs 1. Otherwise, it outputs 0.

Since all operations in b_λ are performed in NC^1, we have $\mathcal{B} = \{b_\lambda\} \in \mathsf{NC}^1$.

One can see that when $\mathbf{M} \leftarrow \mathsf{ZeroSamp}(\lambda)$ (resp., $\mathbf{M} \leftarrow \mathsf{OneSamp}(\lambda)$), the view of a_λ is identical to its view in **Game 0** (resp., **Game 1**), i.e., b_λ outputs 1 with probability ϵ_0 (resp., ϵ_1). Therefore, $\mathcal{B} = \{b_\lambda\}$ distinguishes $\mathbf{M} \leftarrow \mathsf{ZeroSamp}(\lambda)$ and $\mathbf{M} \leftarrow \mathsf{OneSamp}(\lambda)$ with advantage $|\epsilon_0 - \epsilon_1|$, which should be negligible according to Lemma 2, completing the proof of Lemma 7. □

Game 2: This game is the same as **Game 1** except that a_λ succeeds if $\mathbf{x} \neq \tilde{\mathbf{x}} \wedge \mathsf{Eval}_\lambda(\mathbf{M}, \mathbf{x}) = \mathsf{Eval}(\mathbf{M}, \tilde{\mathbf{x}})$.

Lemma 8. *If $\mathcal{A} = \{a_\lambda\}$ succeeds with advantage ϵ_1 (resp., ϵ_2) in* **Game 1** *(resp.,* **Game 2***), then $\epsilon_1 = \epsilon_2$.*

Proof. According to Lemma 1 and due to the fact that the rank of $\mathbf{M} \leftarrow \mathsf{ZeroSamp}(\lambda)$ is $\lambda - 1$, for any $\mathbf{y} \in \mathrm{Im}(\mathbf{M})$, there are two vectors $\mathbf{x}, \mathbf{x}'$ such that $\mathbf{M}\mathbf{x} = \mathbf{M}\mathbf{x}' = \mathbf{y} \wedge \mathbf{x} \neq \mathbf{x}'$, and we have

$$\begin{aligned}
\epsilon_1 &= \Pr\left[\tilde{\mathbf{x}} = \mathbf{x}^* \,\middle|\, \begin{array}{c} \mathbf{x}^* \leftarrow \{\mathbf{x}, \mathbf{x}'\} \\ \mathbf{y} = \mathbf{M}\mathbf{x}^* \\ \tilde{\mathbf{x}} \leftarrow a_\lambda(\mathbf{y}) \end{array}\right] \\
&= \frac{1}{2}\Pr\left[\tilde{\mathbf{x}} = \mathbf{x} \,\middle|\, \begin{array}{c} \mathbf{x}^* = \mathbf{x} \\ \mathbf{y} = \mathbf{M}\mathbf{x}^* \\ \tilde{\mathbf{x}} \leftarrow a_\lambda(\mathbf{y}) \end{array}\right] + \frac{1}{2}\Pr\left[\tilde{\mathbf{x}} = \mathbf{x}' \,\middle|\, \begin{array}{c} \mathbf{x}^* = \mathbf{x}' \\ \mathbf{y} = \mathbf{M}\mathbf{x}^* \\ \tilde{\mathbf{x}} \leftarrow a_\lambda(\mathbf{y}) \end{array}\right] \\
&= \frac{1}{2}\Pr\left[\tilde{\mathbf{x}} = \mathbf{x} \,\middle|\, \begin{array}{c} \mathbf{x}^* = \mathbf{x}' \\ \mathbf{y} = \mathbf{M}\mathbf{x}^* \\ \tilde{\mathbf{x}} \leftarrow a_\lambda(\mathbf{y}) \end{array}\right] + \frac{1}{2}\Pr\left[\tilde{\mathbf{x}} = \mathbf{x}' \,\middle|\, \begin{array}{c} \mathbf{x}^* = \mathbf{x} \\ \mathbf{y} = \mathbf{M}\mathbf{x}^* \\ \tilde{\mathbf{x}} \leftarrow a_\lambda(\mathbf{y}) \end{array}\right] \\
&= \Pr\left[\begin{array}{c} \tilde{\mathbf{x}} \neq \mathbf{x}^* \wedge \\ \mathsf{Eval}_\lambda(\mathbf{M}, \tilde{\mathbf{x}}) = \mathsf{Eval}_\lambda(\mathbf{M}, \mathbf{x}^*) \end{array} \,\middle|\, \begin{array}{c} \mathbf{x}^* \leftarrow \{\mathbf{x}, \mathbf{x}'\} \\ \mathbf{y} = \mathbf{M}\mathbf{x}^* \\ \tilde{\mathbf{x}} \leftarrow a_\lambda(\mathbf{y}) \end{array}\right] = \epsilon_2,
\end{aligned}$$

completing the proof of Lemma 8.

Lemma 9. *If $\mathcal{A} = \{a_\lambda\}$ succeeds with advantage ϵ_2 in* **Game 2***, then $\epsilon_2 = \mathsf{negl}(\lambda)$.*

Proof. We now construct $\mathcal{B} = \{b_\lambda\} \in \mathsf{NC}^1$ that distinguishes $\mathbf{M} \leftarrow \mathsf{ZeroSamp}(\lambda)$ and $\mathbf{M} \leftarrow \mathsf{OneSamp}(\lambda)$ with advantage ϵ_2, which contradicts to Lemma 2.

b_λ takes as input $\mathbf{M}$, which is generated as $\mathbf{M} \leftarrow \mathsf{ZeroSamp}(\lambda)$ or $\mathbf{M} \leftarrow \mathsf{OneSamp}(\lambda)$ from its challenger. Then, it samples $\mathbf{x} \leftarrow \{0,1\}^\lambda$. Next, b_λ runs $\mathbf{y} = \mathsf{Eval}_\lambda(\mathbf{M}, \mathbf{x})$ and send $\mathbf{y}$ to a_λ. When a_λ outputs $\tilde{\mathbf{x}}$, if $\mathbf{x} \neq \tilde{\mathbf{x}} \wedge \mathbf{y} = \mathsf{Eval}_\lambda(\mathbf{M}, \mathbf{x})$, b_λ outputs 1. Otherwise, it outputs 0.

Since all operations in b_λ are performed in NC^1, we have $\mathcal{B} = \{b_\lambda\} \in \mathsf{NC}^1$.

One can see that when $\mathbf{M} \leftarrow \mathsf{ZeroSamp}(\lambda)$, the view of a_λ is identical to its view in **Game 2**, i.e., b_λ outputs 1 with probability ϵ_2.

When $\mathbf{M} \leftarrow \mathsf{OneSamp}(\lambda)$, since $\mathsf{Eval}(\mathbf{M}, \tilde{\mathbf{x}})$ is permutation, there is no vector $\tilde{\mathbf{x}}$ such that $\mathbf{x} \neq \tilde{\mathbf{x}} \wedge \mathbf{y} = \mathsf{Eval}_\lambda(\mathbf{k}, \mathbf{x})$, i.e. b_λ outputs 1 with probability 0.

Therefore, $\mathcal{B} = \{b_\lambda\}$ distinguishes $\mathbf{M} \leftarrow \mathsf{ZeroSamp}(\lambda)$ and $\mathbf{M} \leftarrow \mathsf{OneSamp}(\lambda)$ with advantage ϵ_2, which should be negligible according to Lemma 2, completing the proof of Lemma 9. □

Since $|\epsilon_0 - \epsilon_1| = \mathsf{negl}(\lambda)$, $\epsilon_1 = \epsilon_2$, and $\epsilon_2 = \mathsf{negl}(\lambda)$, we have

$$\epsilon_0 \leq |\epsilon_0 - \epsilon_1| + \epsilon_1 = \mathsf{negl}(\lambda) + \epsilon_2 = \mathsf{negl}(\lambda),$$

i.e., Construction 2 satisfies one-wayness. This completes the proof of Theorem 1. □

Extension to NC^1-OWPs Against NC^1. We now show a transformation from collections of NC^1-OWPs against NC^1, where the output distributions of the key generation algorithms are uniformly random over key space, to NC^1-OWPs against NC^1. Specifically, given a collection of OWPs $\{f_k : D_k \to D_k\}_{k \in K}$ where K is the key space, we construct a OWP $g : D \to D$ where $D := \bigcup_{k \in K}(\{k\} \times D_k)$ and $g((k,x) \in D) = (k, f_k(x))$. This transformation can be applied in NC^1, and the properties of permutation and one-wayness of g hold due to those properties of f. Note that in [5], it is shown that $\mathsf{OneSamp}(\lambda)$ samples $\mathbf{M} \leftarrow \{\mathbf{M} \in \mathsf{OneSamp}(\lambda)\}$ uniformly. Thus, KeyGen_λ of our construction samples $k \leftarrow K_\lambda = \{\mathbf{M} \mid \mathbf{M} \in \mathsf{OneSamp}(\lambda)\}$ uniformly, and we can apply this transformation to our collection of NC^1-OWPs against NC^1. We refer the reader to the full paper for the details.

Computability in $\mathsf{AC}^0[2]$. Perhaps interestingly, our one-way permutation can be run by an even smaller class of circuits $\mathsf{AC}^0[2]$, which satisfies $\mathsf{AC}^0[2] \subsetneq \mathsf{NC}^1$ [39,41] and consists of constant-depth circuits with $\mathsf{MOD2}$ gates. The reason is that it only involves multiplications of a constant number of matrices, inner products, and sampling random bits. Due to the same reason, our constructions of single-bit HPS introduced later in Sect. 4 is also computable in $\mathsf{AC}^0[2]$.

4 Construction of NC^1-HPS Against NC^1

In this section, we start by giving a construction of perfectly smooth and $\frac{1}{2}$-universal$_1$ NC^1-HPS against NC^1 such that the proof space is one-bit. Next, we turn this construction into a perfectly smooth and strong universal$_1$ NC^1-HPS against NC^1 such that the proof space is multi-bit. Finally, we construct a strong universal$_2$ NC^1-HPS against NC^1 such that the language L supports $\{0,1\}^n$.

4.1 Perfectly Smooth and Universal$_1$ for One-Bit

In this section, we give our construction of perfectly smooth and $\frac{1}{2}$-universal$_1$ NC^1-HPS against NC^1 circuits under the assumption $\mathsf{NC}^1 \subsetneq \oplus\mathsf{L/poly}$.

Construction 3 (NC^1**-HPS**)**.** *Let λ be a security parameter. We define the families $\mathcal{S}etup = \{\mathsf{Setup}_\lambda\}$, $\mathcal{S}amp\mathcal{Y}es = \{\mathsf{SampYes}_\lambda\}$, $\mathcal{S}amp\mathcal{N}o = \{\mathsf{SampNo}_\lambda\}$, $\mathcal{K}ey\mathcal{G}en = \{\mathsf{KeyGen}_\lambda\}$, $\mathcal{P}riv = \{\mathsf{Priv}_\lambda\}$, and $\mathcal{P}ub = \{\mathsf{Pub}_\lambda\}$ as follows.*

- Setup_λ*:*
  1. *Sample* $\mathbf{M} \leftarrow \mathsf{ZeroSamp}(\lambda)$.
  2. *Output* $\mathsf{pp} = (X_\lambda, L_\lambda, W_\lambda, R_\lambda, SK_\lambda, PK_\lambda, \Pi_\lambda, H_\lambda, \alpha_\lambda, \mathsf{aux}_\lambda)$ *where*
     - $X_\lambda := \{0,1\}^\lambda$.
     - $L_\lambda := \{\mathbf{x} | \mathbf{w} \in 1 \times \{0,1\}^{\lambda-1}, \mathbf{x} = \mathbf{M}\mathbf{w}\} = \mathrm{Im}(\mathbf{M}^\top)$ *($\because$ Lemma 5).*
     - $W_\lambda := 1 \times \{0,1\}^{\lambda-1}$.
     - $R_\lambda := \{(\mathbf{x}, \mathbf{w}) | \ \mathbf{w} \in 1 \times \{0,1\}^{\lambda-1}, \mathbf{x} = \mathbf{M}\mathbf{w}\}$.
     - $SK_\lambda := \{0,1\}^\lambda$.
     - $PK_\lambda := \mathrm{Im}(\mathbf{M})$.

- $\Pi_\lambda := \{0,1\}$.
- $H_\lambda(\mathbf{sk}, \mathbf{x}) := \mathbf{sk}^\top \mathbf{x}$.
- $\alpha_\lambda(\mathbf{sk}) := \mathbf{M}\,\mathbf{sk}$.
- $\mathsf{aux}_\lambda := \mathbf{M}$.

– $\mathsf{SampYes}_\lambda(\mathsf{pp})$:
  1. *Parse* $\mathsf{pp} = (X_\lambda, L_\lambda, W_\lambda, R_\lambda, SK_\lambda, PK_\lambda, \Pi_\lambda, H_\lambda, \alpha_\lambda, \mathsf{aux}_\lambda)$ *and let* $\mathsf{aux}_\lambda = \mathbf{M}$.
  2. *Sample* $\mathbf{w} \leftarrow 1 \times \{0,1\}^{\lambda-1}$.
  3. *Compute* $\mathbf{x} := \mathbf{M}^\top \mathbf{w}$ *and output* $\mathbf{x}$.

– $\mathsf{SampNo}_\lambda(\mathsf{pp})$:
  1. *Parse* $\mathsf{pp} = (X_\lambda, L_\lambda, W_\lambda, R_\lambda, SK_\lambda, PK_\lambda, \Pi_\lambda, H_\lambda, \alpha_\lambda, \mathsf{aux}_\lambda)$ *and let* $\mathsf{aux}_\lambda = \mathbf{M}$.
  2. *Sample* $\mathbf{w} \leftarrow 1 \times \{0,1\}^{\lambda-1}$.
  3. *Compute* $\mathbf{M}'$ *as*

$$\mathbf{M}' = \mathbf{M} + \begin{pmatrix} 0 & \cdots & 0 & 1 \\ 0 & 0 & \cdots & 0 \\ \vdots & & \ddots & \vdots \\ 0 & \cdots & 0 & 0 \end{pmatrix}.$$

  4. *Compute* $\mathbf{x} := \mathbf{M}'^\top \mathbf{w}$ *and output* $\mathbf{x}$.

– $\mathsf{KeyGen}_\lambda(\mathsf{pp})$:
  1. *Parse* $\mathsf{pp} = (X_\lambda, L_\lambda, W_\lambda, R_\lambda, SK_\lambda, PK_\lambda, H_\lambda, \Pi_\lambda, \alpha_\lambda, \mathsf{aux}_\lambda)$.
  2. *Sample* $\mathbf{sk} \leftarrow SK_\lambda$.
  3. *Compute* $\mathbf{pk} := \alpha_\lambda(\mathbf{sk})$ *and output* $(\mathbf{pk}, \mathbf{sk})$.

– $\mathsf{Priv}_\lambda(\mathsf{pp}, \mathbf{sk}, \mathbf{x})$:
  1. *Parse* $\mathsf{pp} = (X_\lambda, L_\lambda, W_\lambda, R_\lambda, SK_\lambda, PK_\lambda, \Pi_\lambda, H_\lambda, \alpha_\lambda, \mathsf{aux}_\lambda)$.
  2. *Compute* $\pi := H_\lambda(\mathbf{sk}, \mathbf{x})$ *and output* π.

– $\mathsf{Pub}_\lambda(\mathsf{pp}, \mathbf{pk}, \mathbf{x}, \mathbf{w})$:
  1. *Compute* $\pi := \mathbf{pk}^\top \mathbf{w}$ *and output* π.

Theorem 2. *If* $\mathsf{NC}^1 \subsetneq \oplus\mathsf{L/poly}$, *then* ($\mathcal{S}etup$, $\mathcal{S}amp\mathcal{Y}es$, $\mathcal{S}amp\mathcal{N}o$, $\mathcal{K}ey\mathcal{G}en$, $\mathcal{P}riv$, $\mathcal{P}ub$) *defined as Construction 3 is a perfectly smooth and* $\frac{1}{2}$*-universal*$_1$ NC^1*-HPS against* NC^1 *circuits.*

Proof Sketch. It is straightforward that this HPS is correct.

To show the subset membership problem of our construction, we first give two observations: (1) for any $\mathbf{M}$ sampled from $\mathsf{ZeroSamp}(\lambda)$, the distribution of $\mathbf{M} + \mathbf{N}$ is identical to $\mathsf{OneSamp}(\lambda)$, where

$$\mathbf{N} = \begin{pmatrix} 0 & \cdots & 0 & 1 \\ 0 & 0 & \cdots & 0 \\ \vdots & & \ddots & \vdots \\ 0 & \cdots & 0 & 0 \end{pmatrix},$$

and (2) perhaps interestingly, for any $\mathbf{w} \in 0 \times \{0,1\}^{n-1}$ (respectively, $\mathbf{w} \in 1 \times \{0,1\}^{n-1}$), there is a vector $\mathbf{k}$ in the kernel of $\mathbf{M}$ such that $\mathbf{M}\mathbf{w}^\top =$

$\mathbf{M}(\mathbf{w}+\mathbf{k})^\top$ and $(\mathbf{w}+\mathbf{k}) \in 1 \times \{0,1\}^{n-1}$ (respectively, $(\mathbf{w}+\mathbf{k}) \in 0 \times \{0,1\}^{n-1}$), which implies $\mathrm{Im}(\mathbf{M}^\top) = \{\mathbf{x} | \mathbf{w} \in 0 \times \{0,1\}^{n-1}, \mathbf{x} = \mathbf{M}^\top \mathbf{w}\} = \{\mathbf{x} | \mathbf{w} \in 1 \times \{0,1\}^{n-1}, \mathbf{x} = \mathbf{M}^\top \mathbf{w}\}$. Since for any vector $\mathbf{w} \in 0 \times \{0,1\}^{n-1}$, it holds that $(\mathbf{M}+\mathbf{N})^\top \mathbf{w} = \mathbf{M}^\top \mathbf{w} + \mathbf{N}^\top \mathbf{w} = \mathbf{M}^\top \mathbf{w} + \mathbf{0} = \mathbf{M}^\top \mathbf{w}$, we have $L = \mathrm{Im}(\mathbf{M}^\top) = \{\mathbf{x} | \mathbf{w} \in 0 \times \{0,1\}^{n-1}, \mathbf{x} = (\mathbf{M}+\mathbf{N})^\top \mathbf{w}\}$ due to observation (2). Moreover, since $\mathbf{M}+\mathbf{N}$ is of full rank due to observation (1), we have $X = \{0,1\}^n = \{\mathbf{x} | \mathbf{w} \in \{0,1\}^n, \mathbf{x} = (\mathbf{M}+\mathbf{N})^\top \mathbf{w}\}$. Thus, we can conclude that $X \setminus L = \{\mathbf{x} | \mathbf{w} \in 1 \times \{0,1\}^{n-1}, \mathbf{x} = (\mathbf{M}+\mathbf{N})^\top \mathbf{w}\}$. Then, the subset membership problem follows from the fact that $\mathrm{Im}(\mathbf{M}^\top) = \{\mathbf{x} | \mathbf{w} \in 1 \times \{0,1\}^{n-1}, \mathbf{x} = \mathbf{M}^\top \mathbf{w}\}$ and the indistinguishability between the distributions over $\mathrm{Im}(\mathbf{M}^\top)$ and $X \setminus L$ can be reduced to the indistinguishability between $\mathsf{ZeroSamp}(\lambda)$ and $\mathsf{OneSamp}(\lambda)$.

We now explain the intuition of the proof of universal$_1$. Since the rank of $\mathbf{M}$ is $n-1$, when we fix the public key $\mathbf{pk}$, there are two different secret keys $\mathbf{sk}$ and $\mathbf{sk}'$ such that $\mathbf{pk} = \mathbf{M}\,\mathbf{sk} = \mathbf{M}\,\mathbf{sk}'$. As explained before, for any $\mathbf{x} \in X \setminus L$, there exists $\mathbf{w} \in 1 \times \{0,1\}^{n-1}$ such that $\mathbf{x} = (\mathbf{M}+\mathbf{N})^\top \mathbf{w}$, and $(\mathbf{M}+\mathbf{N})$ is a full rank matrix. Therefore, we have $(\mathbf{M}+\mathbf{N})\mathbf{sk} \neq (\mathbf{M}+\mathbf{N})\mathbf{sk}'$ which implies $\mathbf{N}\,\mathbf{sk} \neq \mathbf{N}\,\mathbf{sk}'$, i.e., either $\mathbf{N}\,\mathbf{sk}$ or $\mathbf{N}\,\mathbf{sk}'$ is zero-vector and the other is $(1\ 0 \cdots 0)^\top$. Therefore, when we let $\mathbf{N}\,\mathbf{sk} = (0 \cdots 0)^\top$ and $\mathbf{N}\,\mathbf{sk}' = (1\ 0 \cdots 0)^\top$, it holds that $H(\mathbf{sk}, \mathbf{x}) = \mathbf{sk}^\top (\mathbf{M}+\mathbf{N})^\top \mathbf{w} = \mathbf{sk}^\top \mathbf{M}^\top \mathbf{w} + (0 \cdots 0)\mathbf{w} = \mathbf{sk}^\top \mathbf{M}^\top \mathbf{w}$ and $H(\mathbf{sk}', \mathbf{x}) = \mathbf{sk}'^\top (\mathbf{M}+\mathbf{N})^\top \mathbf{w} = \mathbf{sk}'^\top \mathbf{M}^\top \mathbf{w} + (1\ 0 \cdots 0)\mathbf{w} = \mathbf{sk}'^\top \mathbf{M}^\top \mathbf{w} + 1$, which implies $H(\mathbf{sk}, \mathbf{x}) \neq H(\mathbf{sk}', \mathbf{x})$. As a result, for fixed $\mathbf{pk}$, one can guess the proof for an instance $\mathbf{x} \in X \setminus L$ with probability at most $\frac{1}{2}$ since there is no information on whether the secret key is $\mathbf{sk}$ or $\mathbf{sk}'$.

The formal proof is as follows.

Proof. First note that all of the algorithms $\mathcal{S}etup$, $\mathcal{S}amp\mathcal{Y}es$, $\mathcal{S}amp\mathcal{N}o$, $\mathcal{K}ey\mathcal{G}en$, $\mathcal{P}riv$, and $\mathcal{P}ub$ are in NC^1, since they only involve operations including multiplications of a constant number of matrices, inner products, and sampling random bits.

Next we prove that Construction 3 satisfies correctness, subset membership problem, perfect smoothness, and $\frac{1}{2}$-universality$_1$.

Correctness. Since $\mathsf{Priv}_\lambda(\mathsf{pp}, \mathbf{sk}, \mathbf{x}) = H_\lambda(\mathbf{sk}, \mathbf{x}) = \mathbf{sk}^\top \mathbf{x} = \mathbf{sk}^\top \mathbf{M}^\top \mathbf{w} = (\mathbf{M}\mathbf{sk})^\top \mathbf{w} = \mathbf{pk}^\top \mathbf{w} = \mathsf{Pub}_\lambda(\mathsf{pp}, \mathbf{pk}, \mathbf{x}, \mathbf{w})$, Construction 3 satisfies correctness.

Subset Membership Problem. We now propose and prove three propositions corresponding to the three properties in the definition of subset membership problem (see Definition 7) respectively.

Proposition 1. *The distributions of* $\mathbf{x}$ *and* $\mathbf{x}'$ *are identical where* $\mathsf{pp} \leftarrow \mathsf{Setup}_\lambda$, $\mathbf{x} \leftarrow \mathsf{SampYes}_\lambda(\mathsf{pp})$, *and* $\mathbf{x}' \leftarrow L_\lambda$.

Proof. Let $\mathbf{M} \leftarrow \mathsf{ZeroSamp}(\lambda)$ be a matrix generated in the procedure of Setup_λ. Let f be a map $f: 1 \times \{0,1\}^{\lambda-1} \to \mathrm{Im}(\mathbf{M}^\top)$ such that $f(\mathbf{w}) = \mathbf{M}^\top \mathbf{w}$. One can see that for any $\mathsf{pp} \leftarrow \mathsf{Setup}_\lambda$, the distributions of $\mathbf{x}$ and $\mathbf{x}'$ are identical where $\mathbf{x} \leftarrow \mathsf{SampYes}_\lambda(\mathsf{pp})$, $\mathbf{w}' \leftarrow 1 \times \{0,1\}^{\lambda-1}$, and $\mathbf{x}' = f(\mathbf{w}')$. Moreover, if f is bijective, the distributions of $\mathbf{x}'$ and $\mathbf{x}''$ are identical for $\mathbf{w}' \leftarrow 1 \times \{0,1\}^{\lambda-1}$, $\mathbf{x}' = f(\mathbf{w}')$, and $\mathbf{x}'' \leftarrow \text{Im}(\mathbf{M}^\top)$. Therefore, if f is bijective, the distributions of $\mathbf{x}$ and $\mathbf{x}''$ are identical. Namely, to show Proposition 1, we only have to show that f is bijective.

Injectivity. We now show that for any $\mathbf{w}, \mathbf{w}' \leftarrow 1 \times \{0,1\}^{\lambda-1}$ such that $\mathbf{w} \neq \mathbf{w}'$, we have $f(\mathbf{w}) \neq f(\mathbf{w}')$. We prove by contradiction, i.e. we show that if there are $\mathbf{w}, \mathbf{w}' \leftarrow 1 \times \{0,1\}^{\lambda-1}$ such that $\mathbf{w} \neq \mathbf{w}'$ and $f(\mathbf{w}) = f(\mathbf{w}')$, then it contradicts on Lemma 4.

Since $\mathbf{M}^\top \mathbf{w} = \mathbf{M}^\top \mathbf{w}'$, we have $\mathbf{M}^\top(\mathbf{w} - \mathbf{w}') = \mathbf{0}$. Moreover, since $\mathbf{w} \neq \mathbf{w}'$ and $\mathbf{w}, \mathbf{w}' \in 1 \times \{0,1\}^{\lambda-1}$, $\mathbf{w} - \mathbf{w}'$ is the non-zero vector in the kernel of $\mathbf{M}^\top$ and $\mathbf{w} - \mathbf{w}' \in 0 \times \{0,1\}^{\lambda-1}$. However, according to Lemma 4, we have $\text{Ker}(\mathbf{M}^\top) = \{\mathbf{0}, \mathbf{k}\}$ where $\mathbf{k} \in 1 \times \{0,1\}^\lambda$, which gives us the conflict.

Surjectivity. We now show that for any $\mathbf{x} \in \text{Im}(\mathbf{M}^\top)$, there exists a vector $\mathbf{w} \in 1 \times \{0,1\}^{\lambda-1}$ such that $\mathbf{x} = f(\mathbf{w})$, i.e., $\mathbf{x} = \mathbf{M}^\top \mathbf{w}$. According to Lemma 5, we have $\text{Im}(\mathbf{M}^\top) = \{\mathbf{x} | \mathbf{w} \in 1 \times \{0,1\}^\lambda, \mathbf{x} = \mathbf{M}^\top \mathbf{w}\}$. Therefore, it holds that for any $\mathbf{x} \in \text{Im}(\mathbf{M}^\top)$, there exists $\mathbf{w} \in 1 \times \{0,1\}^\lambda$ such that $\mathbf{x} = \mathbf{M}^\top \mathbf{w}$, i.e., $\mathbf{x} = f(\mathbf{w})$, completing the proof of surjectivity.

Putting all the above together, Proposition 1 immediately follows. □

Proposition 2. *The distributions of* $\mathbf{x}$ *and* $\mathbf{x}'$ *are identical for* $\mathsf{pp} \leftarrow \mathsf{Setup}_\lambda$, $\mathbf{x} \leftarrow \mathsf{SampNo}_\lambda(\mathsf{pp})$, *and* $\mathbf{x}' \leftarrow X_\lambda \setminus L_\lambda$.

Proof. Let $\mathbf{M} \leftarrow \mathsf{ZeroSamp}(\lambda)$ and $\mathbf{M}'$ be

$$\mathbf{M}' = \mathbf{M} + \begin{pmatrix} 0 & \cdots & 0 & 1 \\ 0 & 0 & \cdots & 0 \\ \vdots & & \ddots & \vdots \\ 0 & \cdots & 0 & 0 \end{pmatrix}.$$

We first show that for any $\mathbf{w} \in 1 \times \{0,1\}^{\lambda-1}$, we have $\mathbf{M}'^\top \mathbf{w} \in \{0,1\}^\lambda \setminus \text{Im}(\mathbf{M}^\top)$.

($\bigstar$) For any $\mathbf{w} \in 0 \times \{0,1\}^{\lambda-1}$, we have

$$\begin{aligned} \mathbf{M}'\mathbf{w} &= \left(\mathbf{M} + \begin{pmatrix} 0 & \cdots & 0 & 1 \\ 0 & 0 & \cdots & 0 \\ \vdots & & \ddots & \vdots \\ 0 & \cdots & 0 & 0 \end{pmatrix} \right)^\top \mathbf{w} \\ &= \mathbf{M}^\top \mathbf{w} + \begin{pmatrix} 0 & \cdots & & 0 \\ \vdots & 0 & \cdots & 0 \\ 0 & & \ddots & \vdots \\ 1 & 0 & \cdots & 0 \end{pmatrix} \begin{pmatrix} 0 \\ w_2 \\ \vdots \\ w_\lambda \end{pmatrix} = \mathbf{M}^\top \mathbf{w} + \mathbf{0} = \mathbf{M}^\top \mathbf{w}. \end{aligned}$$

Moreover, according to Lemma 5, we have $\text{Im}(\mathbf{M}^\top) = \{\mathbf{x} | \mathbf{w} \in 0 \times \{0,1\}^{\lambda-1}, \mathbf{x} = \mathbf{M}^\top \mathbf{w}\}$. Hence, we have

$$\begin{aligned} \{\mathbf{x} \mid \mathbf{w} \in 0 \times \{0,1\}^{\lambda-1},\ \mathbf{x} = \mathbf{M}'^\top \mathbf{w}\} &= \{\mathbf{x} \mid \mathbf{w} \in 0 \times \{0,1\}^{\lambda-1},\ \mathbf{x} = \mathbf{M}^\top \mathbf{w}\} \ (\bigstar) \\ &= \text{Im}(\mathbf{M}^\top). \end{aligned}$$

As a result, for all $\mathbf{x} \in \mathrm{Im}(\mathbf{M}^\top)$, there exists $\mathbf{w}' \in 0 \times \{0,1\}^{\lambda-1}$ such that $\mathbf{x} = \mathbf{M}'^\top \mathbf{w}'$. Moreover, according to Lemma 6, $\mathbf{M}'$ is a full rank matrix, which means that for any $\mathbf{w} \in 1 \times \{0,1\}^{\lambda-1}$ and any $\mathbf{x} \in \mathrm{Im}(\mathbf{M}^\top)$, we have $\mathbf{M}'^\top \mathbf{w} \neq \mathbf{x}$. Namely, for any $\mathbf{w} \in 1 \times \{0,1\}^{\lambda-1}$, we have $\mathbf{M}'^\top \mathbf{w} \in \{0,1\}^\lambda \setminus \mathrm{Im}(\mathbf{M}^\top)$.

It is straightforward that for any $\mathsf{pp} \leftarrow \mathsf{Setup}_\lambda$, the distributions of $\mathbf{x} \leftarrow \mathsf{SampNo}_\lambda(\mathsf{pp})$ and $\mathbf{x}' = \mathbf{M}'^\top \mathbf{w}'$ are identical where $\mathbf{w}' \leftarrow 1 \times \{0,1\}^{\lambda-1}$. Moreover, since $\mathbf{M}'^\top$ is of full rank, the map $f : 1 \times \{0,1\}^{n-1} \to \{0,1\}^\lambda \setminus \mathrm{Im}(\mathbf{M}^\top)$ such that $f(\mathbf{w}) = \mathbf{M}'^\top \mathbf{w}$ is bijective, i.e., the distributions of $\mathbf{x}' = f(\mathbf{w}')$ and $\mathbf{x}'' \leftarrow \{0,1\}^\lambda \setminus \mathrm{Im}(\mathbf{M}^\top)$ are identical for $\mathbf{w}' \leftarrow 1 \times \{0,1\}^{\lambda-1}$, completing the proof of Proposition 2. □

Proposition 3. *For any* $\mathcal{A} = \{a_\lambda\} \in \mathsf{NC}^1$,

$$|\Pr[a_\lambda(\mathsf{pp}, \mathbf{x}_0) = 1] - \Pr[a_\lambda(\mathsf{pp}, \mathbf{x}_1) = 1]| \leq \mathsf{negl}(\lambda)$$

where $\mathsf{pp} \leftarrow \mathsf{Setup}_\lambda$, $(\mathbf{x}_0, \mathbf{w}) \leftarrow \mathsf{SampYes}_\lambda(\mathsf{pp})$, *and* $\mathbf{x}_1 \leftarrow \mathsf{SampNo}_\lambda(\mathsf{pp})$.

Proof. Let $\mathcal{A} = \{a_\lambda\}$ be any adversary in NC^1. We give hybrid games to show that the advantage of $\mathcal{A}$ in breaking the hardness of subset membership problem is negligible.

Game 0: This is the original $\mathsf{SampYes}$ game for $\mathcal{A}$. $\mathcal{CH}$ runs $\mathsf{pp} \leftarrow \mathsf{Setup}_\lambda$, $(\mathbf{x}, \mathbf{w}) \leftarrow \mathsf{SampYes}_\lambda(\mathsf{pp})$. Then it sends $(\mathsf{pp}, \mathbf{x})$ to a_λ. a_λ succeeds if a_λ outputs 1. Otherwise, it fails.

Game 1: This game is the same as **Game 0** except that $\mathcal{CH}$ runs $\mathbf{M} \leftarrow \mathsf{OneSamp}_\lambda$ in the procedure of Setup_λ.

Lemma 10. *If* $\mathcal{A} = \{a_\lambda\}$ *succeeds with advantage* ϵ_0 *(resp.,* ϵ_1*) in* **Game 0** *(resp.,* **Game 1***), then* $|\epsilon_0 - \epsilon_1| = \mathsf{negl}(\lambda)$.

Proof. We now construct $\mathcal{B} = \{b_\lambda\} \in \mathsf{NC}^1$ that distinguishes $\mathbf{M} \leftarrow \mathsf{ZeroSamp}(\lambda)$ and $\mathbf{M} \leftarrow \mathsf{OneSamp}(\lambda)$ with advantage $|\epsilon_0 - \epsilon_1|$, which contradicts to Lemma 2.

b_λ takes as input $\mathbf{M}$, which is generated as $\mathbf{M} \leftarrow \mathsf{ZeroSamp}_\lambda$ or $\mathbf{M} \leftarrow \mathsf{OneSamp}_\lambda$ from its challenger. Then, it runs $\mathsf{pp} \leftarrow \mathsf{Setup}_\lambda$ using $\mathbf{M}$, samples $\mathbf{w} \leftarrow 1 \times \{0,1\}^{\lambda-1}$, and sets $x := \mathbf{M}^\top \mathbf{w}$. Next, b_λ gives $(\mathsf{pp}, \mathbf{x})$ to a_λ. When a_λ outputs b, then b_λ outputs b.

Since all operations in b_λ are performed in NC^1, we have $\mathcal{B} = \{b_\lambda\} \in \mathsf{NC}^1$.

One can see that when $\mathbf{M} \leftarrow \mathsf{ZeroSamp}(\lambda)$ (resp., $\mathbf{M} \leftarrow \mathsf{OneSamp}(\lambda)$), the view of a_λ is identical to its view in **Game 1** (resp., **Game 2**), i.e., b_λ outputs 1 with probability ϵ_0 (resp., ϵ_1). Therefore, $\mathcal{B} = \{b_\lambda\}$ distinguishes $\mathbf{M} \leftarrow \mathsf{ZeroSamp}(\lambda)$ and $\mathbf{M} \leftarrow \mathsf{OneSamp}(\lambda)$ with advantage $|\epsilon_0 - \epsilon_1|$, which should be negligible according to Lemma 2, completing the proof of Lemma 10. □

Game 2: This is the original SampNo game for $\mathcal{A}$, i.e., it is the same as **Game 1** except that $\mathcal{CH}$ runs $\mathbf{M}' \leftarrow \mathsf{ZeroSamp}(\lambda)$ and set $\mathbf{M} := \mathbf{M}' + \mathbf{N}$ in the procedure of Setup_λ, where

$$\mathbf{N} = \begin{pmatrix} 0 & \cdots & 0 & 1 \\ 0 & 0 & \cdots & 0 \\ \vdots & & \ddots & \vdots \\ 0 & \cdots & 0 & 0 \end{pmatrix}.$$

Lemma 11. *If $\mathcal{A} = \{a_\lambda\}$ succeeds with advantage ϵ_1 (resp., ϵ_2) in* **Game 1** *(resp.,* **Game 2***), then $\epsilon_1 = \epsilon_2$.*

Proof. Lemma 11 follows from the fact that the distributions of $\mathbf{M}_0 + \mathbf{N}$ and $\mathbf{M}_1$ are identical where $\mathbf{M}_0 \leftarrow \mathsf{ZeroSamp}(\lambda)$ and $\mathbf{M}_1 \leftarrow \mathsf{OneSamp}(\lambda)$ (according to Lemma 6). □

Note that, $\epsilon_0 = \Pr[a_\lambda(\mathsf{pp}, \mathbf{x}) = 1]$ and $\epsilon_2 = \Pr[a_\lambda(\mathsf{pp}, \mathbf{x}') = 1]$ where $\mathsf{pp} \leftarrow \mathsf{Setup}_\lambda$, $(\mathbf{x}, \mathbf{w}) \leftarrow \mathsf{SampYes}_\lambda(\mathsf{pp})$, and $\mathbf{x}' \leftarrow \mathsf{SampNo}_\lambda(\mathsf{pp})$. Moreover, since $|\epsilon_0 - \epsilon_1| = \mathsf{negl}(\lambda)$ and $\epsilon_1 = \epsilon_2$, we have

$$|\epsilon_0 - \epsilon_2| \leq |\epsilon_0 - \epsilon_1| + |\epsilon_1 - \epsilon_2| = \mathsf{negl}(\lambda).$$

□

According to Propositions 1, 2, and 3, Construction 3 satisfies the subset membership problem, completing this part of proof.

Perfect Smoothness. We now show that for any $\mathsf{pp} \leftarrow \mathsf{Setup}_\lambda$, the random variables $(\mathbf{x}, \mathbf{pk}, \pi)$ and $(\mathbf{x}, \mathbf{pk}, \pi')$ are identical where $\mathbf{x} \leftarrow X_\lambda \setminus L_\lambda$, $(\mathbf{pk}, \mathbf{sk}) \leftarrow \mathsf{KeyGen}_\lambda(\mathsf{pp})$, and $\pi' \leftarrow \Pi_\lambda$.

According to Lemma 1, for any $\mathbf{pk}^* \in PK_\lambda$, there are only two secret keys $\mathbf{sk}$ and $\mathbf{sk}'$ such that $\mathbf{pk}^* = \alpha_\lambda(\mathbf{sk}) = \mathbf{M}\mathbf{sk} = \alpha_\lambda(\mathbf{sk}') = \mathbf{M}\mathbf{sk}'$. Moreover, according to Lemma 3, we have $\mathbf{sk} = \mathbf{sk}' + \mathbf{k}$ where $\mathbf{k}$ is a vector such that $\mathbf{k} \in \mathrm{Ker}(\mathbf{M})$ and $\mathbf{k} \in \{0,1\}^{\lambda-1} \times 1$, i.e., the last elements in $\mathbf{sk}$ and $\mathbf{sk}'$ are different (one is 1 and other is 0). Therefore, for any $\mathbf{x}^* \in X_\lambda \setminus L_\lambda$ and $\mathbf{pk}^* \in PK_\lambda$, we have

$$\begin{aligned} \pi = \mathsf{Priv}_\lambda(\mathsf{pp}, \mathbf{sk}, \mathbf{x}^*) &= \mathbf{sk}^\top \mathbf{M}'^\top \mathbf{w}^* \\ &= \mathbf{sk}^\top \left(\mathbf{M} + \begin{pmatrix} 0 & \cdots & 0 & 1 \\ 0 & 0 & \cdots & 0 \\ \vdots & & \ddots & \vdots \\ 0 & \cdots & 0 & 0 \end{pmatrix} \right)^\top \mathbf{w}^* \\ &= \mathbf{sk}^\top \mathbf{M}^\top \mathbf{w}^* + (sk_1 \cdots sk_\lambda) \begin{pmatrix} 0 & \cdots & & 0 \\ \vdots & 0 & \cdots & 0 \\ 0 & & \ddots & \vdots \\ 1 & 0 & \cdots & 0 \end{pmatrix} \begin{pmatrix} 1 \\ w_2^* \\ \vdots \\ w_\lambda^* \end{pmatrix} \\ &= \mathbf{pk}^{*\top} \mathbf{w}^* + sk_\lambda, \end{aligned}$$

and it follows that for any $\mathbf{x}^* \in X_\lambda \setminus L_\lambda$ and $\mathbf{pk}^* \in PK_\lambda$, there are two secret key $\mathbf{sk}, \mathbf{sk}'$ such that $\mathbf{pk}^* = \alpha_\lambda(\mathbf{sk}) = \alpha_\lambda(\mathbf{sk}')$, $\mathsf{Priv}_\lambda(\mathbf{sk}, \mathbf{x}^*) = 0$, and $\mathsf{Priv}_\lambda(\mathsf{pp}, \mathbf{sk}', \mathbf{x}^*) = 1$. Namely, the number of secret keys satisfying $\mathbf{pk}^* = \alpha_\lambda(\mathbf{sk}) \wedge \pi^* = \mathsf{Priv}_\lambda(\mathbf{sk}, \mathbf{x}^*)$ is 1. Therefore, we have

$$\begin{aligned}\Pr[(\mathbf{x},\mathbf{pk},\pi)=(\mathbf{x}^*,\mathbf{pk}^*,\pi^*)] &= \Pr\left[\begin{array}{c}\mathbf{pk}=\mathbf{pk}^*\wedge\\ \pi=\pi^*\end{array}\middle|\ \mathbf{x}=\mathbf{x}^*\right]\Pr[\mathbf{x}=\mathbf{x}^*]\\ &= \Pr\left[\begin{array}{c}\mathbf{pk}^*=\alpha_\lambda(\mathbf{sk})\wedge\\ \pi^*=\mathsf{Priv}_\lambda(\mathsf{pp},\mathbf{sk},\mathbf{x}^*)\end{array}\right]\Pr[\mathbf{x}=\mathbf{x}^*]\\ &= \frac{1}{|SK_\lambda|}\Pr[\mathbf{x}=\mathbf{x}^*]\end{aligned}$$

where $\mathbf{sk}\leftarrow SK_\lambda$ and $\mathbf{x}\leftarrow X_\lambda\times L_\lambda$. Similarly, we have

$$\begin{aligned}\Pr[(\mathbf{x},\mathbf{pk},\pi')=(\mathbf{x}^*,\mathbf{pk}^*,\pi^*)] &= \Pr[\pi'=\pi^*]\Pr[\mathbf{pk}=\mathbf{pk}^*]\Pr[\mathbf{x}=\mathbf{x}^*]\\ &= \frac{1}{2}\frac{2}{|SK_\lambda|}\Pr[\mathbf{x}=\mathbf{x}^*]\\ &= \frac{1}{|SK_\lambda|}\Pr[\mathbf{x}=\mathbf{x}^*].\end{aligned}$$

Therefore, we have $\Pr[(\mathbf{x},\mathbf{pk},\pi)=(\mathbf{x}^*,\mathbf{pk}^*,\pi^*)]=\Pr[(\mathbf{x},\mathbf{pk},\pi')=(\mathbf{x}^*,\mathbf{pk}^*,\pi^*)]$ and it follows that Construction 3 satisfies perfect smoothness.

$\frac{1}{2}$-universality$_1$. $\frac{1}{2}$-universality$_1$ follows from the fact that for any $\mathsf{pp}\leftarrow\mathsf{Setup}_\lambda$, $\mathbf{x}\in X_\lambda\setminus L_\lambda$, $\mathbf{pk}\in PK_\lambda$, and $\pi\in\Pi_\lambda$, the number of secret keys such that $\mathbf{pk}=\alpha_\lambda(\mathbf{sk})$ is 2 and the number of secret keys such that $\mathbf{pk}=\alpha_\lambda(\mathbf{sk})\wedge\pi=\mathsf{Priv}_\lambda(\mathsf{pp},\mathbf{sk},\mathbf{x})$ is 1 as described above. Therefore, we have

$$\begin{aligned}\Pr[\mathsf{Priv}_\lambda(\mathsf{pp},\mathbf{sk},\mathbf{x})=\pi\wedge\alpha_\lambda(\mathbf{sk})=\mathbf{pk}] &= \frac{1}{|SK_\lambda|}=\frac{1}{2}\frac{2}{|SK_\lambda|}\\ &= \frac{1}{2}\Pr[\alpha_\lambda(\mathbf{sk})=\mathbf{pk}]\\ \Leftrightarrow \Pr[\mathsf{Priv}_\lambda(\mathsf{pp},\mathbf{sk},\mathbf{x})=\pi|\alpha_\lambda(\mathbf{sk})=\mathbf{pk}] &= \frac{1}{2}.\end{aligned}$$

Therefore, Construction 3 satisfies $\frac{1}{2}$-universality$_1$.

Putting all the above together, Theorem 2 immediately follows. □

Multi-bit NC^1-HPS. Notice that the size of proof space of Construction 3 is only one-bit, which makes it less useful. However, we can extend this construction with multi-bit proofs by running multiple HPS in parallel. We refer the reader to the full paper for the multi-bit version of our HPS and the security proof.

Universal$_2$ NC^1-HPS. By carefully adopting the technique by Cramer and Shoup [15], we achieve a universal$_2$ NC^1-HPS. The resulting scheme can be computed in NC^1 and it is secure against NC^1 circuits under the assumption $\mathsf{NC}^1 \subsetneq \oplus\mathsf{L/poly}$. We refer the reader to the full paper for the details.

4.2 Application: NC^1-CCA Secure PKE

As one of the most important application of HPSs, Cramer and shoup [15] constructed a CCA secure PKE scheme. Interestingly, by instantiating the underlying HPS with our construction, we immediately achieve an NC^1-CCA secure

PKE scheme against NC^1 circuits restricted in the same way as the ones defined for verifiable computation schemes by Campanelli and Gennaro [14], i.e., ones allowed to make constant rounds of adaptive queries to the decryption oracle, while in each round, they can make arbitrary polynomial number of queries. We refer the reader to the full paper for the details on this application.

5 Construction of NC^1-TDF Against NC^1

In this section, we give our construction of NC^1-TDF against NC^1 under the assumption $\mathsf{NC}^1 \subsetneq \oplus\mathsf{L/poly}$.

Construction 4 **(NC^1-TDF).** *Let λ be a security parameter and l be a polynomial in λ. Let $\mathcal{F} = \{f_\lambda : \{0,1\}^\lambda \to \{0,1\}^{l(\lambda)}\}$ be a NC^1-OWF against NC^1. We define the families $\mathcal{KeyGen} = \{\mathsf{KeyGen}_\lambda\}$ with key spaces $EK_\lambda = \{\mathbf{M} \mid \mathbf{M} \leftarrow \mathsf{ZeroSamp}(\lambda)\}$ and $TK_\lambda = \mathrm{Ker}(\mathbf{M})$, $\mathcal{Eval} = \{\mathsf{Eval}_\lambda\}$ and $\mathcal{Inverse} = \{\mathsf{Inverse}_\lambda\}$ as follows.*

- KeyGen_λ*:*
  1. *Run $\mathbf{R} \leftarrow \mathsf{LSamp}(\lambda)$, $\mathbf{R}' \leftarrow \mathsf{RSamp}(\lambda)$.*
  2. *Set $\mathbf{k} := (\boldsymbol{r}\ 1)^\top$ where $(\boldsymbol{r}\ 1)^\top$ is the last column of $\mathbf{R}'$.*
  3. *Compute $\mathbf{M} := \mathbf{R}\mathbf{M}_0^\lambda\mathbf{R}'$ where $\mathbf{M}_0^\lambda$ is defined as Construction 1.*
  4. *Set $ek := \mathbf{M}$ and $tk := \mathbf{k}$, and output (ek, tk) (according to the proof of Lemma 3, it holds that $\mathbf{k} \in \mathrm{Ker}(\mathbf{M})$).*

  The domain $\mathcal{D}_{\lambda,ek}$ and range $\mathcal{R}_{\lambda,ek}$ are defined as follows.

$$\mathcal{D}_{\lambda,ek} := \{0,1\}^\lambda \times \left(\mathrm{Im}(\mathbf{M}^\top) \times \{0,1\}^\lambda \setminus \mathrm{Im}(\mathbf{M}^\top)\right)^\lambda.$$
$$\mathcal{R}_{\lambda,ek} := \{0,1\}^{l(\lambda)+2\lambda^2}.$$

- $\mathsf{Eval}_\lambda(ek, X)$*:*
  1. *Parse $X := (x, (\mathbf{c}_{1,0}, \mathbf{c}_{1,1}), (\mathbf{c}_{2,0}, \mathbf{c}_{2,1}), \cdots, (\mathbf{c}_{\lambda,0}, \mathbf{c}_{\lambda,1})) \in \mathcal{D}_{\lambda,ek}$.*
  2. *For $x = x_1x_2\cdots x_\lambda \in \{0,1\}^\lambda$ and all $i \in [\lambda]$, if $x_i = 0$, set $(\mathbf{c}_i, \mathbf{c}_i') := (\mathbf{c}_{i,0}, \mathbf{c}_{i,1})$, otherwise set $(\mathbf{c}_i, \mathbf{c}_i') := (\mathbf{c}_{i,1}, \mathbf{c}_{i,0})$.*
  3. *Compute $y := f_\lambda(x)$.*
  4. *Set $Y := (y, (\mathbf{c}_1, \mathbf{c}_1'), (\mathbf{c}_2, \mathbf{c}_2'), \cdots, (\mathbf{c}_\lambda, \mathbf{c}_\lambda'))$ and output Y.*
- $\mathsf{Inverse}_\lambda(tk, Y)$*:*
  1. *Parse $tk := \mathbf{k}$ and $Y := (y, (\mathbf{c}_1, \mathbf{c}_1'), (\mathbf{c}_2, \mathbf{c}_2'), \cdots, (\mathbf{c}_\lambda, \mathbf{c}_\lambda')) \in \mathcal{R}_{\lambda,ek}$.*
  2. *For all $i \in [\lambda]$, if $\mathbf{k}^\top\mathbf{c}_i = 0 \wedge \mathbf{k}^\top\mathbf{c}_i' = 1$, then set $x_i := 0$ and $(\mathbf{c}_{i,0}, \mathbf{c}_{i,1}) := (\mathbf{c}_i, \mathbf{c}_i')$.*
  3. *Else if $\mathbf{k}^\top\mathbf{c}_i = 1 \wedge \mathbf{k}^\top\mathbf{c}_i' = 0$, then set $x_i := 1$ and $(\mathbf{c}_{i,0}, \mathbf{c}_{i,1}) := (\mathbf{c}_i', \mathbf{c}_i)$.*
  4. *Else output $\perp$ and halt.*
  5. *Set $X = (x, (\mathbf{c}_{1,0}, \mathbf{c}_{1,1}), \cdots, (\mathbf{c}_{\lambda,0}, \mathbf{c}_{\lambda,1}))$ and output X.*

Theorem 3. *If $\mathsf{NC}^1 \subsetneq \oplus\mathsf{L/poly}$ and there exists an NC^1-OWF against NC^1 circuits, Construction 4 is an NC^1-TDF against NC^1.*

Proof Sketch. Let $\mathbf{k}$ and $\mathbf{M}$ be the trapdoor key and evaluation key generated by KeyGen_λ respectively. For any $\mathbf{c} \in \mathrm{Im}(\mathbf{M}^\top)$, we must have $\mathbf{k}^\top \mathbf{c} = 0$ since $\mathbf{k} \in \mathrm{Ker}(\mathbf{M})$. Also, we prove that for any $\mathbf{c} \in \{0,1\}^\lambda \setminus \mathrm{Im}(\mathbf{M}^\top)$, there must exists $\mathbf{w}$ such that $\mathbf{w} \in 1 \times \{0,1\}^{\lambda-1}$ and $\mathbf{c} = \left(\mathbf{M} + \begin{pmatrix} 0 \cdots & 0 & 1 \\ 0 \; 0 & \cdots & 0 \\ \vdots & \ddots & \vdots \\ 0 \cdots & 0 & 0 \end{pmatrix}\right) \mathbf{w}$. Since

$\mathbf{k}^\top \mathbf{M}^\top \mathbf{w} = 0$ and $\mathbf{k}^\top \begin{pmatrix} 0 \cdots & 0 & 1 \\ 0 \; 0 & \cdots & 0 \\ \vdots & \ddots & \vdots \\ 0 \cdots & 0 & 0 \end{pmatrix}^\top \mathbf{w} = 1$, we must have $\mathbf{k}^\top \mathbf{c} = 1$ in this case.

Therefore, $\mathbf{k}$, which is samplable in NC^1, can be used to determine whether $\mathbf{c}$ is in $\mathrm{Im}(\mathbf{M}^\top)$ or $\{0,1\}^\lambda \setminus \mathrm{Im}(\mathbf{M}^\top)$ and recover x_i by checking whether $\mathbf{c}_i$ and $\mathbf{c}'_i$ are swapped in the inversing procedure, i.e., correctness holds.

Moreover, due to the subset membership problem for $L = \mathrm{Im}(\mathbf{M})$, the uniform distributions over $\mathrm{Im}(\mathbf{M}^\top)$ and $\{0,1\}^\lambda \setminus \mathrm{Im}(\mathbf{M}^\top)$ are indistinguishable when $\mathbf{M}$ is a correctly generated evaluation key, i.e., the distributions $(\mathbf{c}_{i,0}, \mathbf{c}_{i,1}) \leftarrow \mathrm{Im}(\mathbf{M}^\top) \times \{0,1\}^\lambda \setminus \mathrm{Im}(\mathbf{M}^\top)$ and $(\mathbf{c}_{i,0}, \mathbf{c}_{i,1}) \leftarrow \mathrm{Im}(\mathbf{M}^\top) \times \mathrm{Im}(\mathbf{M}^\top)$ are indistinguishable. Therefore, the adversary in the one-way game can only obtain information on $f_\lambda(x)$ (which is one-way), and the additional pairs $(\mathbf{c}_{i,0}, \mathbf{c}_{i,1})$ can be simulated by just sampling them from $\mathrm{Im}(\mathbf{M}^\top) \times \mathrm{Im}(\mathbf{M}^\top)$, i.e., they reveal little information on x.

The formal proof is as follows.

Proof. Note that $\mathcal{KeyGen}$, $\mathcal{E}val$, and $\mathcal{I}nverse$ only involve operations including multiplications of the constant number of matrices, inner products and sampling random bits. Since these operation can be performed in NC^1, we have $\mathcal{KeyGen}$, $\mathcal{E}val$, and $\mathcal{I}nverse$ can be computed in NC^1.

Next, we prove that Construction 4 satisfies correctness and one-wayness.

Correctness. For any $\mathbf{M} \leftarrow \mathsf{ZeroSamp}(\lambda)$ and any $\mathbf{c} \in \mathrm{Im}(\mathbf{M}^\top)$, we have

$$\mathbf{k}^\top \mathbf{c} = \mathbf{k}^\top \mathbf{M}^\top \mathbf{w} = (\mathbf{M}\mathbf{k})^\top \mathbf{w} = \mathbf{0}^\top \mathbf{w} = 0$$

where $\mathbf{k} \in \mathrm{Ker}(\mathbf{M})$ and $\mathbf{w}$ is a vector such that $\mathbf{c} = \mathbf{M}\mathbf{w}$.

Next we show that when $\mathbf{c} \in \{0,1\}^\lambda \setminus \mathrm{Im}(\mathbf{M}^\top)$ then $\mathbf{k}^\top \mathbf{c} = 1$. Before showing this, we first propose the following lemma, which is straightforwardly implied by Proposition 2 in Theorem 2.

Lemma 12. *For any* $\mathbf{M} \leftarrow \mathsf{ZeroSamp}(\lambda)$*, it holds that*

$$\{0,1\}^\lambda \setminus \mathrm{Im}(\mathbf{M}^\top) = \{\mathbf{x} \mid \exists \mathbf{w} \in 1 \times \{0,1\}^{\lambda-1}, \mathbf{x} = \mathbf{M}'^\top \mathbf{w}\}$$

where

$$\mathbf{M}' = \mathbf{M} + \begin{pmatrix} 0 \cdots & 0 & 1 \\ 0 \; 0 & \cdots & 0 \\ \vdots & \ddots & \vdots \\ 0 \cdots & 0 & 0 \end{pmatrix}.$$

According to Lemma 12, for any $\mathbf{M} \leftarrow \mathsf{ZeroSamp}(\lambda)$ and any $\boldsymbol{c} \in \{0,1\}^\lambda \setminus \mathrm{Im}(\mathbf{M}^\top)$, we have

$$\begin{aligned}\mathbf{k}^\top \boldsymbol{c} = \mathbf{k}^\top \mathbf{M}'^\top \mathbf{w} &= \mathbf{k}^\top \left(\mathbf{M} + \begin{pmatrix} 0 & \cdots & 0 & 1 \\ 0 & 0 & \cdots & 0 \\ \vdots & & \ddots & \vdots \\ 0 & \cdots & 0 & 0 \end{pmatrix} \right)^\top \mathbf{w} \\ &= (\mathbf{M}\mathbf{k})^\top \mathbf{w} + \left(\begin{pmatrix} 0 & \cdots & 0 & 1 \\ 0 & 0 & \cdots & 0 \\ \vdots & & \ddots & \vdots \\ 0 & \cdots & 0 & 0 \end{pmatrix} \mathbf{k} \right)^\top \mathbf{w} \\ &= \mathbf{0}^\top \mathbf{w} + (1\ 0\ \cdots\ 0)\mathbf{w} = 0 + 1 = 1\end{aligned}$$

where $\mathbf{k}$ is a vector in the kernel of $\mathbf{M}$ and $\mathbf{w}$ is a vector such that $\mathbf{w} \in 1 \times \{0,1\}^{\lambda-1} \wedge \boldsymbol{c} = \mathbf{M}'^\top \mathbf{w}$.

As a result, for all $i \in [\lambda]$, $\mathbf{k}$ generated by KeyGen_λ can be used to determine whether $\boldsymbol{c}_i$ (resp., $\boldsymbol{c}'_i$) generated by Eval_λ are in $\mathrm{Im}(\mathbf{M}^\top)$ or $\{0,1\}^\lambda \setminus \mathrm{Im}(\mathbf{M}^\top)$, and hence recover x_i.

One-Wayness. Let $\mathcal{A} = \{a_\lambda\}$ be any adversary in NC^1. We give hybrid games to show that the advantage of $\mathcal{A}$ in breaking the one-wayness of Construction 4 is negligible.

Game 0: This is the original one-wayness game for $\mathcal{A} = \{a_\lambda\}$. $\mathcal{CH}$ runs $(ek, tk) \leftarrow \mathsf{KeyGen}_\lambda$, samples $X \leftarrow \mathcal{D}_{\lambda,ek}$, and runs $Y = \mathsf{Eval}_\lambda(ek, X)$. Then it sends (ek, Y) to a_λ. a_λ succeeds if a_λ outputs X^* such that $\mathsf{Eval}_\lambda(ek, X^*) = Y$. Otherwise it fails.

Game 1 $\sim$ Game $\boldsymbol{\lambda}$: For $i \in [\lambda]$, **Game i** is the same as **Game i-1** except that $\mathcal{CH}$ samples $(\mathbf{c}_{i,0}, \mathbf{c}_{i,1}) \leftarrow \mathrm{Im}(\mathbf{M}^\top) \times \mathrm{Im}(\mathbf{M}^\top)$.

Lemma 13. *If $\mathcal{A} = \{a_\lambda\}$ succeeds with advantage ϵ_{i-1} (resp., ϵ_i) in* **Game i-1** *(resp.,* **Game i***), then $|\epsilon_{i-1} - \epsilon_i| = \mathsf{negl}(\lambda)$.*

Proof. According to the proof of the part of subset membership problem in Theorem 2, we have the following lemma.

Lemma 14. *For any $\mathcal{G} = \{g_\lambda\} \in \mathsf{NC}^1$,*

$$|\Pr[g_\lambda(\mathbf{M}, \mathbf{c}_0) = 1] - \Pr[g_\lambda(\mathbf{M}, \mathbf{c}_1) = 1]| \leq \mathsf{negl}(\lambda),$$

where $\mathbf{M} \leftarrow \mathsf{ZeroSamp}(\lambda)$, $\mathbf{c}_0 \leftarrow \mathrm{Im}(\mathbf{M}^\top)$, and $\mathbf{c}_1 \leftarrow \{0,1\}^\lambda \setminus \mathrm{Im}(\mathbf{M}^\top)$.

Proof. Let $(\mathcal{Setup}, \mathcal{SampYes}, \mathcal{SampNo}, \mathcal{KeyGen}, \mathcal{Priv}, \mathcal{Pub})$ be a strong smooth HPS defined as Construction 3. According to Propositions 1, 2, and 3, we have

- the distributions of $\mathbf{x}_0$ and $\mathbf{c}_0$ are identical where $\mathsf{pp} \leftarrow \mathsf{Setup}_\lambda$, $(\mathbf{x}_0, w) \leftarrow \mathsf{SampYes}_\lambda(\mathsf{pp})$, and $\mathbf{c} \leftarrow \mathrm{Im}(\mathbf{M}^\top)$ ($\because$ Proposition 1).

- the distributions of $\mathbf{x}_1$ and $\mathbf{c}_1$ are identical where $\mathsf{pp} \leftarrow \mathsf{Setup}_\lambda$, $(\mathbf{x}_0, w) \leftarrow \mathsf{SampNo}_\lambda(\mathsf{pp})$, and $\mathbf{c} \leftarrow \{0,1\}^\lambda \setminus \mathrm{Im}(\mathbf{M}^\top)$ ($\because$ Proposition 2).
- it holds that for any $\mathcal{G} = \{g_\lambda\} \in \mathcal{C}_2$,

$$|\Pr[g_\lambda(\mathsf{pp}, \mathbf{x}_0) = 1] - \Pr[g_\lambda(\mathsf{pp}, \mathbf{x}_1) = 1]| \leq \mathsf{negl}(\lambda)$$

where $\mathsf{pp} \leftarrow \mathsf{Setup}_\lambda$, $(\mathbf{x}_0, w) \leftarrow \mathsf{SampYes}_\lambda(\mathsf{pp})$, and $\mathbf{x}_1 \leftarrow \mathsf{SampNo}_\lambda(\mathsf{pp})$ ($\because$ Proposition 3).

Moreover, the distribution of $\mathsf{pp} \leftarrow \mathsf{Setup}_\lambda$ depends only on the distribution of $\mathbf{M} \leftarrow \mathsf{ZeroSamp}(\lambda)$. Therefore, for any $\mathcal{G} = \{g_\lambda\} \in \mathcal{C}_2$, we have

$$|\Pr[g_\lambda(\mathbf{M}, \mathbf{c}_0) = 1] - \Pr[g_\lambda(\mathbf{M}, \mathbf{c}_1) = 1]| \leq \mathsf{negl}(\lambda),$$

where $\mathbf{M} \leftarrow \mathsf{ZeroSamp}(\lambda)$, $\mathbf{c}_0 \leftarrow \mathrm{Im}(\mathbf{M}^\top)$, and $\mathbf{c}_1 \leftarrow \{0,1\}^\lambda \setminus \mathrm{Im}(\mathbf{M}^\top)$, completing the proof of Lemma 14. □

We now construct $\mathcal{B} = \{b_\lambda\} \in \mathsf{NC}^1$ that distinguishes $\mathbf{c} \leftarrow \mathrm{Im}(\mathbf{M}^\top)$ and $\mathbf{c} \leftarrow \{0,1\}^\lambda \setminus \mathrm{Im}(\mathbf{M}^\top)$ with advantage $|\epsilon_{i-1} - \epsilon_i|$, which contradicts to Lemma 14.

b_λ takes as input $(\mathbf{M}, \mathbf{c})$, which is generated as $\mathbf{M} \leftarrow \mathsf{ZeroSamp}(\lambda)$ and $\mathbf{c}$ sampled as $\mathbf{c} \leftarrow \mathrm{Im}(\mathbf{M}^\top)$ or $\mathbf{c} \leftarrow \{0,1\}^\lambda \setminus \mathrm{Im}(\mathbf{M}^\top)$ from its challenger. Then, it sets $ek := \mathbf{M}$ and $\mathbf{c}_{i,1} := \mathbf{c}$. Next, b_λ samples $x \leftarrow \{0,1\}^\lambda$, $(\mathbf{c}_{j,0}, \mathbf{c}_{j,1}) \leftarrow \mathrm{Im}(\mathbf{M}^\top) \times \mathrm{Im}(\mathbf{M}^\top)$ for all $j \in [i-1]$, $\mathbf{c}_{i,0} \leftarrow \mathrm{Im}(\mathbf{M}^\top)$, and $(\mathbf{c}_{j,0}, \mathbf{c}_{j,1}) \leftarrow \mathrm{Im}(\mathbf{M}^\top) \times \{0,1\}^\lambda \setminus \mathrm{Im}(\mathbf{M}^\top)$ for all $j \in \{i+1, \cdots, \lambda\}$. Next, b_λ sets $X := (x, (\mathbf{c}_{1,0}, \mathbf{c}_{1,1}), (\mathbf{c}_{2,0}, \mathbf{c}_{2,1}), \cdots, (\mathbf{c}_{\lambda,0}, \mathbf{c}_{\lambda,1}))$ and computes $Y = \mathsf{Eval}_\lambda(ek, X)$. Finally, b_λ gives (ek, Y) to a_λ. When a_λ output X^*, if $Y = \mathsf{Eval}_\lambda(ek, X^*)$, b_λ outputs 1. Otherwise, it outputs 0.

Since all operations in b_λ are performed in NC^1, we have $\mathcal{B} = \{b_\lambda\} \in \mathsf{NC}^1$.

One can see that when $\mathbf{c} \leftarrow \mathrm{Im}(\mathbf{M}^\top)$ (resp., $\mathbf{c} \leftarrow \{0,1\}^\lambda \setminus \mathrm{Im}(\mathbf{M}^\top)$), the view of a_λ is identical to its view in **Game i-1** (resp., **Game i**), i.e., b_λ outputs 1 with probability ϵ_{i-1} (resp., ϵ_i). Therefore, $\mathcal{B} = \{b_\lambda\}$ distinguishes $\mathbf{c} \leftarrow \mathrm{Im}(\mathbf{M}^\top)$ and $\mathbf{c} \leftarrow \{0,1\}^\lambda \setminus \mathrm{Im}(\mathbf{M}^\top)$ with advantage $|\epsilon_{i-1} - \epsilon_i|$, which should be negligible according to Lemma 14, completing the proof of Lemma 13. □

Lemma 15. *If $\mathcal{A} = \{a_\lambda\}$ succeeds with advantage ϵ_λ in* **Game $\boldsymbol{\lambda}$**, *then $\epsilon_\lambda = \mathsf{negl}(\lambda)$.*

Proof. We now construct $\mathcal{B} = \{b_\lambda\} \in \mathsf{NC}^1$ that breaks the one-wayness of $\mathcal{F} = \{f_\lambda\}$ with advantage ϵ_n.

b_λ takes as input y, which is generated as $x \leftarrow \{0,1\}^\lambda$ and $y = f_\lambda(y)$ from its challenger. Then, b_λ runs $(ek, tk) \leftarrow \mathsf{KeyGen}_\lambda$, parses $ek := \mathbf{M}$, and samples $((\mathbf{c}_{0,0}, \mathbf{c}_{0,1}), \cdots, (\mathbf{c}_{\lambda,0}, \mathbf{c}_{\lambda,1})) \leftarrow \mathrm{Im}(\mathbf{M}^\top)^{2\lambda}$. Next b_λ gives $(y, (\mathbf{c}_{0,0}, \mathbf{c}_{0,1}), \cdots, (\mathbf{c}_{\lambda,0}, \mathbf{c}_{\lambda,1}))$ to a_λ. When a_λ outputs X^*, b_λ parses $X^* := (x^*(\mathbf{c}^*_{0,0}, \mathbf{c}^*_{0,1}), \cdots, (\mathbf{c}^*_{\lambda,0}, \mathbf{c}^*_{\lambda,1}))$ and outputs x^*.

Since all operations in b_λ are performed in NC^1, we have $\mathcal{B} = \{b_\lambda\} \in \mathsf{NC}^1$.

One can see that the view of a_λ is identical to its view in **Game $\boldsymbol{\lambda}$**, i.e., b_λ outputs x^* such that $y = f_\lambda(x^*)$ with probability ϵ_λ. Therefore, $\mathcal{B} = \{b_\lambda\}$ breaks the one-wayness of $\mathcal{F} = \{f_\lambda\}$ with advantage ϵ_λ, which should be negligible, completing the proof of Lemma 15.

Since for $i \in [\lambda]$, $|\epsilon_{i-1} - \epsilon_i| = \mathsf{negl}(\lambda)$, $\epsilon_\lambda = \mathsf{negl}(\lambda)$, we have

$$\epsilon_0 \leq \sum_{i=1}^{\lambda} |\epsilon_{i-1} - \epsilon_i| + \epsilon_\lambda = \mathsf{negl}(\lambda).$$

Therefore, Construction 4 satisfies one-wayness. □

Putting all the above together, Theorem 3 immediately follows. □

6 Conclusion

In this paper, we formalize fine-grained OWPs, HPSs (which in turn derives a CCA-secure PKE), and TDFs, and show how to construct the NC^1 versions of them secure against NC^1 adversaries. Compared with traditional cryptographic primitives, our schemes treat restricted class of adversaries, while they can be run more efficiently and are only based on the mild worst case assumption $\mathsf{NC}^1 \subsetneq \oplus\mathsf{L/poly}$. It remains open how to construct more fine-grained primitives not implied by our results, such as pseudo-random functions and signature schemes, in the same model.

Acknowledgements. A part of this work was supported by NTT Secure Platform Laboratories, JST OPERA JPMJOP1612, JST CREST JPMJCR14D6, JSPS KAKENHI JP16H01705, JP17H01695, and the Sichuan Science and Technology Program under Grant 2017GZDZX0002 and 2018GZDZX0006.

References

1. Abdalla, M., Benhamouda, F., Blazy, O., Chevalier, C., Pointcheval, D.: SPHF-friendly non-interactive commitments. In: Sako, K., Sarkar, P. (eds.) ASIACRYPT 2013, Part I. LNCS, vol. 8269, pp. 214–234. Springer, Heidelberg (2013). https://doi.org/10.1007/978-3-642-42033-7_12
2. Ajtai, M.: Σ_1^1-formulae on finite structures. Ann. Pure Appl. Logic **24**(1), 1–48 (1983)
3. Ajtai, M., Wigderson, A.: Deterministic simulation of probabilistic constant depth circuits (preliminary version). In: 26th Annual Symposium on Foundations of Computer Science, pp. 11–19. IEEE Computer Society Press (October 1985)
4. Akavia, A., Goldreich, O., Goldwasser, S., Moshkovitz, D.: Erratum for: on basing one-way functions on NP-hardness. In: Schulman, L.J. (ed.) 42nd Annual ACM Symposium on Theory of Computing, pp. 795–796. ACM Press (June 2010)
5. Applebaum, B., Ishai, Y., Kushilevitz, E.: Cryptography in NC^0. In: 45th Annual Symposium on Foundations of Computer Science, pp. 166–175. IEEE Computer Society Press (October 2004)
6. Applebaum, B., Ishai, Y., Kushilevitz, E.: On pseudorandom generators with linear stretch in NC^0. Comput. Complex. **17**(1), 38–69 (2008)

7. Asharov, G., Segev, G.: On constructing one-way permutations from indistinguishability obfuscation. In: Kushilevitz, E., Malkin, T. (eds.) TCC 2016, Part II. LNCS, vol. 9563, pp. 512–541. Springer, Heidelberg (2016). https://doi.org/10.1007/978-3-662-49099-0_19
8. Aumann, Y., Ding, Y.Z., Rabin, M.O.: Everlasting security in the bounded storage model. IEEE Trans. Inf. Theory **48**(6), 1668–1680 (2002)
9. Aumann, Y., Rabin, M.O.: Information theoretically secure communication in the limited storage space model. In: Wiener, M. (ed.) CRYPTO 1999. LNCS, vol. 1666, pp. 65–79. Springer, Heidelberg (1999). https://doi.org/10.1007/3-540-48405-1_5
10. Biham, E., Goren, Y.J., Ishai, Y.: Basing weak public-key cryptography on strong one-way functions. In: Canetti, R. (ed.) TCC 2008. LNCS, vol. 4948, pp. 55–72. Springer, Heidelberg (2008). https://doi.org/10.1007/978-3-540-78524-8_4
11. Blum, M., Micali, S.: How to generate cryptographically strong sequences of pseudo-random bits. SIAM J. Comput. **13**(4), 850–864 (1984)
12. Boppana, R.B., Lagarias, J.C.: One- way functions and circuit complexity. In: Structure in Complexity Theory, Proceedings of the Conference hold at the University of California, Berkeley, California, USA, June 2–5, 1986, pp. 51–65 (1986)
13. Cachin, C., Maurer, U.: Unconditional security against memory-bounded adversaries. In: Kaliski, B.S. (ed.) CRYPTO 1997. LNCS, vol. 1294, pp. 292–306. Springer, Heidelberg (1997). https://doi.org/10.1007/BFb0052243
14. Campanelli, M., Gennaro, R.: Fine-grained secure computation. In: Beimel, A., Dziembowski, S. (eds.) TCC 2018, Part II. LNCS, vol. 11240, pp. 66–97. Springer, Cham (2018). https://doi.org/10.1007/978-3-030-03810-6_3
15. Cramer, R., Shoup, V.: Universal hash proofs and a paradigm for adaptive chosen ciphertext secure public-key encryption. In: Knudsen, L.R. (ed.) EUROCRYPT 2002. LNCS, vol. 2332, pp. 45–64. Springer, Heidelberg (2002). https://doi.org/10.1007/3-540-46035-7_4
16. Degwekar, A., Vaikuntanathan, V., Vasudevan, P.N.: Fine-grained cryptography. In: Robshaw, M., Katz, J. (eds.) CRYPTO 2016, Part III. LNCS, vol. 9816, pp. 533–562. Springer, Heidelberg (2016). https://doi.org/10.1007/978-3-662-53015-3_19
17. Diffie, W., Hellman, M.E.: New directions in cryptography. IEEE Trans. Inf. Theory **22**(6), 644–654 (1976)
18. Ding, Y.Z.: Oblivious transfer in the bounded storage model. In: Kilian, J. (ed.) CRYPTO 2001. LNCS, vol. 2139, pp. 155–170. Springer, Heidelberg (2001). https://doi.org/10.1007/3-540-44647-8_9
19. Ding, Y.Z., Harnik, D., Rosen, A., Shaltiel, R.: Constant-round oblivious transfer in the bounded storage model. In: Naor, M. (ed.) TCC 2004. LNCS, vol. 2951, pp. 446–472. Springer, Heidelberg (2004). https://doi.org/10.1007/978-3-540-24638-1_25
20. Dziembowski, S., Maurer, U.: On generating the initial key in the bounded-storage model. In: Cachin, C., Camenisch, J.L. (eds.) EUROCRYPT 2004. LNCS, vol. 3027, pp. 126–137. Springer, Heidelberg (2004). https://doi.org/10.1007/978-3-540-24676-3_8
21. Furst, M.L., Saxe, J.B., Sipser, M.: Parity, circuits, and the polynomial-time hierarchy. In: 22nd Annual Symposium on Foundations of Computer Science, Nashville, Tennessee, USA, 28–30 October 1981, pp. 260–270 (1981)
22. Garg, S., Gay, R., Hajiabadi, M.: New techniques for efficient trapdoor functions and applications. In: Ishai, Y., Rijmen, V. (eds.) EUROCRYPT 2019, Part III. LNCS, vol. 11478, pp. 33–63. Springer, Cham (2019). https://doi.org/10.1007/978-3-030-17659-4_2

23. Garg, S., Hajiabadi, M.: Trapdoor functions from the computational Diffie-Hellman assumption. In: Shacham, H., Boldyreva, A. (eds.) CRYPTO 2018, Part II. LNCS, vol. 10992, pp. 362–391. Springer, Cham (2018). https://doi.org/10.1007/978-3-319-96881-0_13
24. Gennaro, R., Lindell, Y.: A framework for password-based authenticated key exchange. In: Biham, E. (ed.) EUROCRYPT 2003. LNCS, vol. 2656, pp. 524–543. Springer, Heidelberg (2003). https://doi.org/10.1007/3-540-39200-9_33
25. Gertner, Y., Malkin, T., Reingold, O.: On the impossibility of basing trapdoor functions on trapdoor predicates. In: 42nd Annual Symposium on Foundations of Computer Science, pp. 126–135. IEEE Computer Society Press (October 2001)
26. Håstad, J., Impagliazzo, R., Levin, L.A., Luby, M.: A pseudorandom generator from any one-way function. SIAM J. Comput. **28**(4), 1364–1396 (1999)
27. Hesse, J., Hofheinz, D., Kohl, L.: On tightly secure non-interactive key exchange. In: Shacham, H., Boldyreva, A. (eds.) CRYPTO 2018, Part II. LNCS, vol. 10992, pp. 65–94. Springer, Cham (2018). https://doi.org/10.1007/978-3-319-96881-0_3
28. Impagliazzo, R., Naor, M.: Efficient cryptographic schemes provably as secure as subset sum. In: 30th Annual Symposium on Foundations of Computer Science, pp. 236–241. IEEE Computer Society Press, October/November 1989
29. Ishai, Y., Kushilevitz, E.: Randomizing polynomials: a new representation with applications to round-efficient secure computation. In: 41st Annual Symposium on Foundations of Computer Science, pp. 294–304. IEEE Computer Society Press (November 2000)
30. Jutla, C., Roy, A.: Relatively-sound NIZKs and password-based key-exchange. In: Fischlin, M., Buchmann, J., Manulis, M. (eds.) PKC 2012. LNCS, vol. 7293, pp. 485–503. Springer, Heidelberg (2012). https://doi.org/10.1007/978-3-642-30057-8_29
31. Kalai, Y.T.: Smooth projective hashing and two-message oblivious transfer. In: Cramer, R. (ed.) EUROCRYPT 2005. LNCS, vol. 3494, pp. 78–95. Springer, Heidelberg (2005). https://doi.org/10.1007/11426639_5
32. Katz, J., Vaikuntanathan, V.: Round-optimal password-based authenticated key exchange. In: Ishai, Y. (ed.) TCC 2011. LNCS, vol. 6597, pp. 293–310. Springer, Heidelberg (2011). https://doi.org/10.1007/978-3-642-19571-6_18
33. Matsuda, T.: On the impossibility of basing public-coin one-way permutations on trapdoor permutations. In: Lindell, Y. (ed.) TCC 2014. LNCS, vol. 8349, pp. 265–290. Springer, Heidelberg (2014). https://doi.org/10.1007/978-3-642-54242-8_12
34. Maurer, U.M.: Conditionally-perfect secrecy and a provably-secure randomized cipher. J. Cryptol. **5**(1), 53–66 (1992)
35. Merkle, R.C.: Secure communications over insecure channels. Commun. ACM (CACM) **21**(4), 294–299 (1978)
36. Mitchell, C.J.: A storage complexity based analogue of Maurer key establishment using public channels. In: Boyd, C. (ed.) Cryptography and Coding 1995. LNCS, vol. 1025, pp. 84–93. Springer, Heidelberg (1995). https://doi.org/10.1007/3-540-60693-9_11
37. Naor, M., Yung, M.: Universal one-way hash functions and their cryptographic applications. In: 21st Annual ACM Symposium on Theory of Computing, pp. 33–43. ACM Press (May 1989)
38. Peikert, C., Waters, B.: Lossy trapdoor functions and their applications. In: Ladner, R.E., Dwork, C. (eds.) 40th Annual ACM Symposium on Theory of Computing, pp. 187–196. ACM Press (May 2008)

39. Razborov, A.A.: Lower bounds on the size of bounded depth circuits over a complete basis with logical addition. Math. Notes Acad. Sci. USSR **41**(4), 333–338 (1987)
40. Rompel, J.: One-way functions are necessary and sufficient for secure signatures. In: 22nd Annual ACM Symposium on Theory of Computing, pp. 387–394. ACM Press (May 1990)
41. Smolensky, R.: Algebraic methods in the theory of lower bounds for Boolean circuit complexity. In: Proceedings of the Nineteenth Annual ACM Symposium on Theory of Computing, STOC 1987, pp. 77–82. ACM, New York (1987)
42. Vadhan, S.P.: On constructing locally computable extractors and cryptosystems in the bounded storage model. In: Boneh, D. (ed.) CRYPTO 2003. LNCS, vol. 2729, pp. 61–77. Springer, Heidelberg (2003). https://doi.org/10.1007/978-3-540-45146-4_4
43. Viola, E.: On constructing parallel pseudorandom generators from one-way functions. Cryptology ePrint Archive, Report 2005/159 (2005). http://eprint.iacr.org/2005/159
44. Viola, E.: The complexity of distributions. In: 51st Annual Symposium on Foundations of Computer Science, pp. 202–211. IEEE Computer Society Press (October 2010)

Zero Knowledge

Shorter QA-NIZK and SPS with Tighter Security

Masayuki Abe[1], Charanjit S. Jutla[2], Miyako Ohkubo[3], Jiaxin Pan[4], Arnab Roy[5], and Yuyu Wang[6](✉)

[1] NTT Corporation, Tokyo, Japan
abe.masayuki@lab.ntt.co.jp
[2] IBM T. J. Watson Research Center, Yorktown Heights, USA
csjutla@us.ibm.com
[3] Security Fundamentals Laboratories, CSR, NICT, Tokyo, Japan
m.ohkubo@nict.go.jp
[4] Department of Mathematical Sciences, NTNU – Norwegian University of Science and Technology, Trondheim, Norway
jiaxin.pan@ntnu.no
[5] Fujitsu Laboratories of America, Sunnyvale, USA
aroy@us.fujitsu.com
[6] University of Electronic Science and Technology of China, Chengdu, China
wangyuyu@uestc.edu.cn

Abstract. Quasi-adaptive non-interactive zero-knowledge proof (QA-NIZK) systems and structure-preserving signature (SPS) schemes are two powerful tools for constructing practical pairing-based cryptographic schemes. Their efficiency directly affects the efficiency of the derived advanced protocols.

We construct more efficient QA-NIZK and SPS schemes with tight security reductions. Our QA-NIZK scheme is the *first* one that achieves both tight simulation soundness and constant proof size (in terms of number of group elements) at the same time, while the recent scheme from Abe et al. (ASIACRYPT 2018) achieved tight security with proof size linearly depending on the size of the language and the witness. Assuming the hardness of the Symmetric eXternal Diffie-Hellman (SXDH) problem, our scheme contains only 14 elements in the proof and remains independent of the size of the language and the witness. Moreover, our scheme has tighter simulation soundness than the previous schemes.

Technically, we refine and extend a partitioning technique from a recent SPS scheme (Gay et al., EUROCRYPT 2018). Furthermore, we improve the efficiency of the tightly secure SPS schemes by using a relaxation of NIZK proof system for OR languages, called designated-prover NIZK system. Under the SXDH assumption, our SPS scheme

J. Pan—Research was conducted at KIT, Germany under the DFG grant HO 4534/4-1. Yuyu Wang—Research was conducted at Tokyo Institute of Technology. A part of this work was supported by the Sichuan Science and Technology Program under Grant 2017GZDZX0002 and 2018GZDZX0006, Input Output Cryptocurrency Collaborative Research Chair funded by IOHK, JST OPERA JPMJOP1612, JST CREST JPMJCR14D6, JSPS KAKENHI JP16H01705, JP17H01695.

S. D. Galbraith and S. Moriai (Eds.): ASIACRYPT 2019, LNCS 11923, pp. 669–699, 2019.
https://doi.org/10.1007/978-3-030-34618-8_23

contains 11 group elements in the signature, which is shortest among the tight schemes and is the same as an early non-tight scheme (Abe et al., ASIACRYPT 2012). Compared to the shortest known non-tight scheme (Jutla and Roy, PKC 2017), our scheme achieves tight security at the cost of 5 additional elements.

All the schemes in this paper are proven secure based on the Matrix Diffie-Hellman assumptions (Escala et al., CRYPTO 2013). These are a class of assumptions which include the well-known SXDH and DLIN assumptions and provide clean algebraic insights to our constructions. To the best of our knowledge, our schemes achieve the best efficiency among schemes with the same functionality and security properties. This naturally leads to improvement of the efficiency of cryptosystems based on simulation-sound QA-NIZK and SPS.

Keywords: Quasi-adaptive NIZK · simulation soundness · Structure-preserving signature · Tight reduction

1 Introduction

Bilinear pairing groups have enabled the construction of a plethora of rich cryptographic primitives in the last two decades, starting from the seminal works on three-party key exchange [30] and identity-based encryption (IBE) [11]. In particular, the Groth-Sahai non-interactive zero knowledge (NIZK) proof system [24] for proving algebraic statements over pairing groups has proven to be a powerful tool to construct more efficient advanced cryptographic protocols, such as group signatures [21], anonymous credentials [7], and UC-secure commitment [17] schemes.

Quasi-Adaptive NIZK for Linear Subspaces. There are many applications which require NIZK systems for proving membership in linear subspaces of group vectors. A couple of examples are CCA2-secure public-key encryption via the Naor-Yung paradigm [42], and publicly verifiable CCA2-secure IBE [29].

For proving linear subspace membership, the Groth-Sahai system has a proof size linear in the dimension of the language and the subspace, in terms of number of group elements. To achieve better efficiency, Jutla and Roy proposed a weaker notion [32] called quasi-adaptive NIZK arguments (QA-NIZK), where the common reference string (CRS) may depend on the linear subspace and the soundness is computationally adaptive. For computationally adaptive soundness, the adversary is allowed to submit a proof for its adaptively chosen invalid statement. Based on their work, further improvements [1,33,38] gave QA-NIZK systems with constant proof size. This directly led to KDM-CCA2-secure PKE and publicly verifiable CCA2-secure IBE with constant-size ciphertexts.

Structure-Preserving Signature. Structure-Preserving (SP) cryptography [3] has evolved as an important paradigm in designing modular protocols. In order to enable interoperability, it is required for SP primitives to support verification only by pairing product equations, which enable zero-knowledge proofs using Groth-Sahai NIZKs.

Structure-preserving signature (SPS) schemes are the most important building blocks in constructing anonymous credential [7], voting systems and mixnets [22], and privacy-preserving point collection [25]. In an SPS, all the public keys, messages, and signatures are group elements and verification is done by checking pairing-product equations. Constructing SPS is a very challenging task, as traditional group-based signatures use hash functions, which are not structure-preserving.

TIGHT SECURITY. The security of a cryptographic scheme is proven by constructing a reduction $\mathcal{R}$ which uses a successful adversary $\mathcal{A}$ against the security of the scheme to solve some hard problem. Concretely, this argument establishes the relation between the success probability of $\mathcal{A}$ (denoted by $\varepsilon_{\mathcal{A}}$) and that of $\mathcal{R}$ (denoted by $\varepsilon_{\mathcal{R}}$) as $\varepsilon_{\mathcal{A}} \leq \ell \cdot \varepsilon_{\mathcal{R}} + \mathsf{negl}(\lambda)$, where $\mathsf{negl}(\lambda)$ is negligible in the security parameter λ. The reduction $\mathcal{R}$ is called *tight* if ℓ is a small constant and the running time of $\mathcal{R}$ is approximately the same as that of $\mathcal{A}$. Most of the recent works consider a variant notion of tight security, called *almost* tight security, where the only difference is that ℓ may linearly (or, even better, logarithmically) depend on the security parameter λ. It is worth mentioning that the security loss in all our schemes is $O(\log Q)$, where Q is the number of $\mathcal{A}$'s queries. We note that $Q \ll 2^{\lambda}$ and thus our security loss is much less than $O(\lambda)$. In this paper, we do not distinguish tight security and almost tight security, but we do provide the concrete security bounds.

Tightly secure schemes are more desirable than their non-tight counterparts, since tightly secure schemes do not need to compensate much for their security loss and allow universal key-length recommendations independent of the envisioned size of an application. In recent years, there have been significant efforts in developing schemes with tight security, such as PKEs [18,19,26–28], IBEs [9,13,29], and signatures [4,8,20,28].

As discussed above, QA-NIZK and SPS are important building blocks for advanced protocols which are embedded in larger scale settings. Designing efficient QA-NIZK and SPS with tight security is very important, since non-tight schemes can result in much larger security loss in the derived protocols.

QA-NIZK: TIGHT SECURITY OR COMPACT PROOFS? Several of the aforementioned applications of QA-NIZK require a stronger security notion, called simulation soundness, where an adversary can adaptively query simulated proofs for vectors either inside or outside the linear subspace and in the end the adversary needs to forge a proof on a vector outside the subspace. We assume that the simulation oracle can be queried by the adversary up to Q times. If $Q > 1$, we call the QA-NIZK scheme unbounded simulation-sound and if $Q = 1$, we call it one-time simulation-sound. Many applications, such as multi-challenge (KDM-)CCA2-secure PKE and CCA2-secure IBE, require unbounded simulation soundness.

If we consider the tightness, CRS and proof sizes[1] of previous works, we have three different flavors of unbounded simulation-sound QA-NIZK schemes:

[1] We only count numbers of group elements.

(1) schemes with non-tight security, but compact CRS-es (which only depend on the dimension of the subspace) and constant-size proofs [37]; (2) schemes with tight security and constant-size proofs, but linear-size CRS-es (which are linearly in λ) [18,29]; and (3) schemes with tight security and compact CRS-es, but linear-size proofs (in the dimension of the language and the subspace) [5,6].

A few remarks are made for the tightly secure QA-NIZK scheme of Abe et al. [5,6]. Its proceedings version has a bug and the authors fix it in the ePrint version [6], but the proof size of the new scheme linearly depends on the dimension of the language and the subspace. To be more technical, the work of Abe et al. achieves tight simulation soundness via the (structure-preserving) adaptive partitioning of [4,31]. Due to its use of OR proofs (cf. Fig. 1 in their full version [6]), the QA-NIZK proof size ends up being linear in the size of the language and the subspace (in particular, $|\pi| = O(n_1 + n_2)$). Thus, it remained open and interesting to construct a tightly simulation-sound QA-NIZK with compact CRS-es and constant-size proofs.

SPS: Tightness with Shorter Signatures. In the past few years, substantial progress was made to improve the efficiency of SPS. So far the schemes with shortest signatures have 6 signature elements with non-tight reduction [34] by improving [36], or 12 elements with security loss $36\log(Q)$ [6], or 14 elements with security loss $6\log(Q)$ [20], where Q is the number of signing queries. Our goal is to construct tightly secure SPS with shorter signatures and less security loss.

1.1 Our Contributions

To make progress on the aforementioned two questions, we construct a QA-NIZK scheme with 14 proof elements and an SPS scheme with 11 signature elements, based on the Symmetric eXternal Diffie-Hellman (SXDH) assumption. The security of both schemes is proven with tight reduction to the Matrix Diffie-Hellman (MDDH) assumption [16], which is an algebraic generalization of Diffie-Hellman assumptions (including SXDH). The security proof gives us algebraic insights to our constructions and furthermore our constructions can be implemented by (possibly weaker) linear assumptions beyond SXDH.

Our QA-NIZK scheme is the *first* one that achieves tight simulation soundness, compact CRS-es and constant-size proofs at the same time. Even among the tightly simulation-sound schemes, our scheme has less security loss. Since it achieves better efficiency, using our scheme immediately improves the efficiency of the applications of QA-NIZK with unbounded simulation soundness, including publicly verifiable CCA2-secure PKE with multiple challenge ciphertexts.

In contrast to the Abe et al. framework [5], we use a simpler and elegant framework to achieve better efficiency. Technically, we make novel use of the recent core lemma from [20] to construct a designated-verifier QA-NIZK (DV-QA-NIZK) and then compile it to (publicly verifiable) QA-NIZK by using the bilinearity of pairings. As a by-product, we achieve a tightly secure DV-QA-NIZK, where the verifier holds a secret verification key.

Let $\mathcal{L}_{[\mathbf{M}]_1} := \{[\mathbf{y}]_1 \in \mathbb{G}_1^{n_1} : \exists \mathbf{w} \in \mathbb{Z}_p^{n_2} \text{ such that } \mathbf{y} = \mathbf{M}\mathbf{w}\}^2$ be a linear subspace, where $\mathbf{M} \in \mathbb{Z}_p^{n_1 \times n_2}$ and $n_1 > n_2$. We compare the efficiency and security loss of QA-NIZK schemes in Table 1. Here we instantiate our schemes (in both Tables 1 and 2) based on the SXDH assumption for a fair comparison.

Table 1. Comparison of unbounded simulation-sound QA-NIZK schemes for proving membership in $\mathcal{L}_{[\mathbf{M}]_1}$. $|\mathsf{crs}|$ and $|\pi|$ denote the size of CRS-es and proofs in terms of numbers of group elements. For asymmetric pairings, notation (x, y) means x elements in $\mathbb{G}_1$ and y elements in $\mathbb{G}_2$. Q denotes the number of simulated proofs and λ is the security parameter.

Scheme	Type	$\lvert\mathsf{crs}\rvert$	$\lvert\pi\rvert$	Sec. los.	Ass.
LPJY14 [38]	QA-NIZK	$2n_1 + 3(n_2 + \lambda) + 10$	20	$O(Q)$	DLIN
KW15 [37]	QA-NIZK	$(2n_2 + 6, n_1 + 6)$	$(4, 0)$	$O(Q)$	SXDH
LPJY15 [39]	QA-NIZK	$2n_1 + 3n_2 + 24\lambda + 55$	42	$3\lambda + 7$	DLIN
GHKW16 [18]	DV-QA-NIZK	$n_2 + \lambda$	4	$8\lambda + 2$	DDH
GHKW16 [18]	QA-NIZK	$(n_2 + 6\lambda + 1, n_1 + 2)$	$(3, 0)$	$4\lambda + 1$	SXDH
AJOR18 [5,6]	QA-NIZK	$(3n_2 + 15, n_1 + 12)$	$(n_1 + 16, 2(n_2 + 5))$	$36 \log(Q)$	SXDH
Ours (Sect. 3.1)	DV-QA-NIZK	$(2n_2 + 3, 4)$	$(7, 6)$	$6 \log(Q)$	SXDH
Ours (Sect. 3.2)	QA-NIZK	$(4n_2 + 4, 8 + 2n_1)$	$(8, 6)$	$6 \log(Q)$	SXDH

Our second contribution is a more efficient tightly secure SPS. It contains 11 signature elements and $n_1 + 15$ public key elements, while the scheme from [5] contains 12 and $3n_1 + 23$ elements respectively, where n_1 denotes the number of group elements in a message vector. We give a comparison between our scheme and previous ones in Table 2. Compared with GHKP18, our construction has shorter signatures and less pairing-product equations (PPEs) with the same level of security loss. Compared with AJOR18, our construction has shorter signature and tighter security, but slightly more PPEs. We leave constructing an SPS with the same signature size and security loss but less PPEs as an interesting open problem. As an important building block of our SPS, we propose the notion of designated-prover OR proof systems for a unilateral language, where a prover holds a secret proving key and the language is defined in one single group. We believe that it is of independent interest.

1.2 Our QA-NIZK: Technical Overview

The Kiltz-Wee Framework. In contrast to the work of Abe et al. [5], our construction is motivated by the simple Kiltz-Wee framework [37], where they implicitly constructed a simulation-sound DV-QA-NIZK and then compiled it to a simulation-sound QA-NIZK with pairings. However, their simulation-sound DV-QA-NIZK is not tight. In the following, we focus on constructing a tightly simulation-sound DV-QA-NIZK. By a similar "DV-QA-NIZK $\rightarrow$ QA-NIZK transformation as in [37], we derive our QA-NIZK with shorter proofs and tighter simulation soundness in the end.

The DV-QA-NIZK in [37] is essentially a simple hash proof system [14] for the linear language $\mathcal{L}_{[\mathbf{M}]_1}$: to prove that $[\mathbf{y}]_1 = [\mathbf{M}\mathbf{x}]_1$ for some $\mathbf{x} \in \mathbb{Z}_p$, the prover

[2] We follow the implicit notation of a group element. $[\cdot]_s$ ($s \in \{1, 2, T\}$) denotes the entry-wise exponentiation in $\mathbb{G}_s$.

Table 2. Comparison of structure-preserving signatures for message space $\mathbb{G}^{n_1}$ (in their most efficient variants). "|m|", "|σ|", and "|vk|" denote the size of messages, signatures, and public keys in terms of numbers of group elements. Q denotes the number of signing queries. "# PPEs" denotes the number of pairing-product equations. "NL" denotes the number of non-linear equations that includes signatures in both groups. "L1" denotes the number of linear equations in $\mathbb{G}_1$ group. "L2" denotes the number of linear equations in $\mathbb{G}_2$ group.

Scheme	|m|	|σ|	|vk|	**Sec. loss**	**Assumption**	**# PPEs**			
						Total	NL	L1	L2
HJ12 [28]	1	$10\ell + 6$	13	$O(1)$	DLIN	$6\ell + 3$			
ACDKNO16 [2]	$(n_1, 0)$	$(7, 4)$	$(5, n_1 + 12)$	$O(Q)$	SXDH, XDLIN	5	1	2	2
LPY15 [40]	$(n_1, 0)$	$(10, 1)$	$(16, 2n_1 + 5)$	$O(Q)$	SXDH, XDLINX	5	3	2	
KPW15 [36]	$(n_1, 0)$	$(6, 1)$	$(0, n_1 + 6)$	$O(Q^2)$	SXDH	3	2	1	
JR17 [34]	$(n_1, 0)$	$(5, 1)$	$(0, n_1 + 6)$	$O(Q \log Q)$	SXDH	2	1	1	
AHNOP17 [4]	$(n_1, 0)$	$(13, 12)$	$(18, n_1 + 11)$	$O(\lambda)$	SXDH	15	4	3	8
JOR18 [31]	$(n_1, 0)$	$(11, 6)$	$(7, n_1 + 16)$	$O(\lambda)$	SXDH	8	4	2	2
GHKP18 [20]	$(n_1, 0)$	$(8, 6)$	$(2, n_1 + 9)$	$6 \log(Q)$	SXDH	9	8	1	
AJOR18 [5,6]	$(n_1, 0)$	$(6, 6)$	$(n_1 + 11, 2n_1 + 12)$	$36 \log(Q)$	SXDH	6	4	1	1
Ours(unilateral)	$(n_1, 0)$	$(7, 4)$	$(2, n_1 + 11)$	$6 \log(Q)$	SXDH	7	6	1	

outputs a proof as $\pi := [\mathbf{x}^\top \mathbf{p}]_1$, where the projection $[\mathbf{p}]_1 := [\mathbf{M}^\top \mathbf{k}]_1$ is published in the CRS. With the vector $\mathbf{k}$ as the secret verification key, a designated verifier can check whether $\pi = [\mathbf{y}^\top \mathbf{k}]_1$. By using $\mathbf{k}$ as a simulation trapdoor, a zero-knowledge simulator can return the simulated proof as $\pi := [\mathbf{y}^\top \mathbf{k}]_1$, due to the following equation:

$$\mathbf{x}^\top \mathbf{p} = \mathbf{x}^\top (\mathbf{M}^\top \mathbf{k}) = \mathbf{y}^\top \mathbf{k}.$$

Soundness is guaranteed by the fact that the value $\mathbf{y}^{*\top} \mathbf{k}$ is uniformly random, given $\mathbf{M}^\top \mathbf{k}$, if $\mathbf{y}^*$ is outside the span of $\mathbf{M}$.

Affine MACs and Unbounded Simulation Soundness. To achieve unbounded simulation soundness, we need to hide the information of $\mathbf{k}$ in all the Q_s-many simulation queries, in particular for the information outside the span of $\mathbf{M}^\top$. The Kiltz-Wee solution is to blind the term $\mathbf{y}^\top \mathbf{k}$ with a 2-universal hash proof system. Via a non-tight reduction the hash proof system can be proved to be a pseudorandom affine message authentication code (MAC) scheme proposed by [9]. Technically, unbounded simulation soundness requires the underlying affine MAC to be pseudorandom against multiple challenge queries. This notion has been formally considered in [29] later and it is stronger than the original security in [9]. Because of that, the affine MAC based on the Naor-Reingold PRF in [9] cannot be directly used in constructing tightly simulation-sound QA-NIZK.

Gay et al. [18] constructed a tightly secure unbounded simulation-sound QA-NIZK[3]. Essentially, their tight PCA-secure PKE against multiple challenge ciphertexts is a pseudorandom affine MAC against multiple challenge queries. Then they use this MAC to blind the term $\mathbf{y}^\top \mathbf{k}$. However, this tight solution

[3] We note that the tight affine MAC in [29] can also be used to construct a DV-QA-NIZK and a QA-NIZK with tight unbounded simulation soundness. Their efficiency is slightly better than those in [18].

has a large CRS, namely, the number of group elements in the CRS is linear in the security parameter. That is because the number of $\mathbb{Z}_p$ elements in the underlying affine MAC secret keys is also linear in the security parameter. These $\mathbb{Z}_p$ elements are later converted as group elements in the CRS of QA-NIZK. To the best of our knowledge, current pairing-based affine MACs enjoy either tight security and linear size secret keys or constant size secret keys but non-tight security. Therefore, it may be more promising to develop a new method, other than affine MACs, to hide $\mathbf{y}^\top \mathbf{k}$ with compact CRS and tight security.

Our Solution. We solve the above dilemma by a novel use of the core lemma from [20]. To give more details, we fix some matrices $\mathbf{A}_0, \mathbf{A}_1 \in \mathbb{Z}_p^{2k \times k}$, choose a random vector $\mathbf{k}'$ and consider $\mu := ([\mathbf{t}]_1, [u']_1, \pi')$ that has the distribution:

$$\begin{aligned} &\mathbf{t} \xleftarrow{\$} \mathsf{Span}(\mathbf{A}_0) \cup \mathsf{Span}(\mathbf{A}_1) \\ &u' = \mathbf{t}^\top \mathbf{k}' \in \mathbb{Z}_p \\ &\pi' : \text{ proves that } \mathbf{t} \in \mathsf{Span}(\mathbf{A}_0) \cup \mathsf{Span}(\mathbf{A}_1) \end{aligned} \quad . \tag{1}$$

In a nutshell, the NIZK proof π' guarantees that $\mathbf{t}$ is from the disjunction space and, by introducing randomness in the "right" space, the core lemma shows that $[u']_1$ is pseudorandom with tight reductions. The core lemma itself is not a MAC scheme, since it does not have message inputs, although it has been used to construct a tightly secure (non-affine) MAC in [20].

A "naive" attempt: Using the Core Lemma. To have unbounded simulation soundness, our first attempt is to use the pseudorandom value $[u']_1$ to directly blind the term $\mathbf{y}^\top \mathbf{k}$ from the DV-QA-NIZK with only adaptive soundness in a straightforward way. Then the resulting DV-QA-NIZK outputs the proof $([\mathbf{t}]_1, [u]_1, \pi')$, which has the following distribution:

$$\begin{aligned} &\mathbf{t} \xleftarrow{\$} \mathsf{Span}(\mathbf{A}_0) \cup \mathsf{Span}(\mathbf{A}_1) \\ &u = \mathbf{y}^\top \mathbf{k} + \boxed{\mathbf{t}^\top \mathbf{k}'} \in \mathbb{Z}_p \\ &\pi' : \text{ proves that } \mathbf{t} \in \mathsf{Span}(\mathbf{A}_0) \cup \mathsf{Span}(\mathbf{A}_1) \end{aligned} \quad . \tag{2}$$

In order to publicly generate a proof for a valid statement $[\mathbf{y}]_1 = [\mathbf{M}\mathbf{x}]_1$ with witness $\mathbf{x} \in \mathbb{Z}_p^{n_2}$, we publish $[\mathbf{M}^\top \mathbf{k}]_1, [\mathbf{A}_0^\top \mathbf{k}']_1$ and CRS for generating π' in the CRS of our DV-QA-NIZK. Verification is done with designated verification key $(\mathbf{k}, \mathbf{k}')$. Zero knowledge can be proven using $(\mathbf{k}, \mathbf{k}')$.

However, when we try to prove the unbounded simulation soundness, we run into a problem. The core lemma shows the following two distributions are tightly indistinguishable:

$$\mathsf{REAL} := \{([\mathbf{t}_i]_1, [\mathbf{t}_i^\top \mathbf{k}']_1, \pi'_i)\} \approx_c \{([\mathbf{t}_i]_1, [\mathbf{t}_i^\top \mathbf{k}'_i]_1, \pi'_i)\} =: \mathsf{RAND},$$

where $\mathbf{k}', \mathbf{k}'_i \xleftarrow{\$} \mathbb{Z}_p^{2k}$ and $i = 1, ..., Q$. In the proof of unbounded simulation soundness, we switch from REAL to RAND and then we can argue that all our simulated proofs are random, since $\mathbf{y}^\top \mathbf{k}$ is blinded by the random value $\mathbf{t}_i^\top \mathbf{k}'_i$. Unfortunately, here we cannot use an information-theoretical argument to show that an adversary cannot compute a forgery for an invalid statement: An adversary can reuse the $\mathbf{k}_j$ in the j-th ($1 \leq j \leq Q$) simulation query on $[\mathbf{y}_j]_1 \in \mathsf{Span}([\mathbf{M}']_1)$ and $\mathsf{Span}([\mathbf{M}']_1) \cap \mathsf{Span}([\mathbf{M}]_1) = \{[\mathbf{0}]_1\}$ and given the additional information $\mathbf{M}'^\top \mathbf{k}$ from the j-th query an adversary can compute a valid proof for another invalid statement $\mathbf{y}^* \in \mathsf{Span}(\mathbf{M}')$.

Moreover, this straightforward scheme has an attack: An adversary can ask for a simulated proof $\pi := ([\mathbf{t}]_1, [u]_1, \pi')$ on an invalid $[\mathbf{y}]_1$. Then it computes $([2\mathbf{t}]_1, [2u]_1)$ and adapts the OR proof π' accordingly to $\hat{\pi}$. The proof $\pi^* := ([2\mathbf{t}]_1, [2u]_1, \hat{\pi})$ is a valid proof for an invalid statement $[\mathbf{y}^*]_1 := [2\mathbf{y}]_1 \notin \mathsf{Span}([\mathbf{M}]_1)$.

From Failure to Success via Pairwise Independence. The above problem happens due to the malleability in the "naive" attempt. We introduce non-malleability by using a pairwise independent function in $\mathbf{k}$. More precisely, let $\tau \in \mathbb{Z}_p$ be a tag and our DV-QA-NIZK proof is still $([\mathbf{t}]_1, [u]_1, \pi')$ with $([\mathbf{t}]_1, \pi')$ as in Eq. (2) but

$$u := \mathbf{y}^\top(\mathbf{k}_0 + \tau \mathbf{k}_1) + \mathbf{t}^\top \mathbf{k}'.$$

We assume that all the tags in the simulated proofs and forgery are distinct, which can be achieved by using a collision-resistant hash as $\tau := H([\mathbf{y}]_1, [\mathbf{t}]_1, \pi') \in \mathbb{Z}_p$. Given $\mathbf{k}_j$ the adversary can only see $\mathbf{y}_j^\top(\mathbf{k}_0 + \tau_j \mathbf{k}_1)$ from the j-th query and for all the other queries the random values $\mathbf{t}_i^\top \mathbf{k}_i$ ($i \neq j$) hide the information about $\mathbf{k}_0$ and $\mathbf{k}_1$. Given $\mathbf{k}_0 + \tau_j \mathbf{k}_1$ for a τ_j, the pairwise independence guarantees that even for a computationally unbounded adversary it is hard to compute $\mathbf{k}_0 + \tau^* \mathbf{k}_1$ for any $\tau^* \neq \tau_j$. Thus, the unbounded simulation soundness is concluded. Details are presented in Sect. 3.1. In a nutshell, we use the pseudorandom element $[u']_1$ from the core lemma to hide $[\mathbf{y}^\top(\mathbf{k}_0 + \tau \mathbf{k}_1)]_1$ from a one-time simulation sound DV-QA-NIZK.

From Designated to Public Verification. What is left to do is to convert our DV-QA-NIZK scheme into a QA-NIZK. Intuitively, we first make u publicly verifiable via the (tuned) Groth-Sahai proof technique, and then modify the QA-NIZK so that we can embed the secret key of our DV-QA-NIZK into it without changing the view of the adversary. Then we can extract a forgery for the USS experiment of the DV-QA-NIZK from the forgery by the adversary. Similar ideas have been used in many previous works [9,12,20,33,36,37].

1.3 Our SPS: Technical Overview

The recent SPS schemes exploit the adaptive partitioning paradigm [4,19,27] to achieve tight security. In this paradigm, NIZK for OR languages [23,43] plays an important role, while at the same time, it also incurs high cost. Our basic idea is to replace the full-fledged OR proof system proposed by Gay et al. [20] with one in the designated-prover setting, where a prover is allowed to use a secret proving key. Intuitively, it is easier to achieve an efficient scheme in such a setting since it suffers less restrictions. In fact, the previous SPS scheme in [5] has already exploited the designated-prover setting to reduce the proof size. However, it only works for bilateral OR language (i.e., one out of two words lies in the linear span of its corresponding space), while an OR-proof for unilateral language (i.e., a single word lies in the linear span of either one of two spaces) is required in the construction of [20]. Thus, some new technique is necessary for solving this problem.

For ease of exposition, we focus on the SXDH setting now, where the following OR-language is in consideration:

$$\mathcal{L}_1 := \{[\mathbf{y}]_1 \in \mathbb{G}_1^2 \mid \exists r \in \mathbb{Z}_p \colon [\mathbf{y}]_1 = [\mathbf{A}_0]_1 \cdot r \vee [\mathbf{y}]_1 = [\mathbf{A}_1]_1 \cdot r\}.$$

Let $\mathbf{A}_1 = (a,\ b)^\top$, we observe that it is equivalent to the following language.

$$\mathcal{L}_2 := \{[y_0, y_1]_1^\top \in \mathbb{G}_1^2 \mid \exists x, x' \in \mathbb{Z}_p \colon [y_1]_1 - [y_0]_1 \cdot \frac{b}{a} = [x]_1 \wedge [\mathbf{y}]_1 \cdot x = [\mathbf{A}_0]_1 \cdot x'\}.$$

Specifically, when $x = 0$, we have $[y_1]_1 - [y_0]_1 \cdot \frac{b}{a} = [0]_1$, i.e., $[y_0, y_1]_1^\top$ is in the span of $\mathbf{A}_1$. Otherwise, we have $[\mathbf{y}]_1 = [\mathbf{A}_0]_1 \cdot \frac{x'}{x}$, i.e., $[y_0, y_1]_1^\top$ is in the span of $\mathbf{A}_0$. Note that this language is an "AND-language" now. More importantly, a witness consists only of 2 scalars and a statement consists only of 3 equations. Hence, when applying the Groth-Sahai proof [15,24], the proof size will be only 7 (4 elements for committing the witness and 3 elements for equations), which is shorter than the well-known OR proof in [43] (10 elements). However, the statement contains $\frac{b}{a}$ now, which may leak information on a witness. To avoid this, we make $\frac{b}{a}$ part of the witness and store its commitment (which consists of 2 group elements) in the common reference string. By doing this, we can ensure that the information on $\frac{b}{a}$ will not be leaked and $\frac{b}{a}$ is always "fixed", due to the hiding and biding properties of commitments respectively. Also, notice that this does not increase the size of proofs at all. This scheme satisfies perfect soundness, and zero-knowledge can be tightly reduced to the SXDH assumption. Since the prover has to use $\frac{b}{a}$ to generate a witness for $\mathcal{L}_2$ given a witness for $\mathcal{L}_1$, this scheme only works in the designated-prover setting. However, notice that when simulating the proof, $\mathbf{A}_0$ and $\mathbf{A}_1$ are not necessary, which is a crucial property when applying to the partitioning paradigm.

We further generalize this scheme to one under the $\mathcal{D}_k$-MDDH assumptions for a fixed k. The size of proof will become $O(k^3)$, and the zero-knowledge property can be reduced to the $\mathcal{D}_k$-MDDH assumption with almost no security loss.

Replacing the OR-proof system of [20] with our designated-prover ones immediately derives the most efficient SPS by now. We refer the reader to Table 2 for the comparison between our scheme and the previous ones.

Additionally, we give another designated-prover OR proof scheme where the proof size is $O(k^2)$, which is smaller than the above scheme when $k > 1$. As a trade-off, it suffers a security loss of k. When $k = 1$, its efficiency is the same as that of our original designated-prover OR proof scheme described above. In symmetric groups, we adapt the designated-prover OR proof to provide the most efficient full NIZK (i.e., one with public prover and verifier algorithms) for OR languages based on the $\mathcal{D}_k$-MDDH assumptions by now.

2 Preliminaries

NOTATIONS. We denote an empty string as ϵ. We use $x \xleftarrow{\$} S$ to denote the process of sampling an element x from set S uniformly at random. For positive integers $k > 1, \eta \in \mathbb{Z}^+$ and a matrix $\mathbf{A} \in \mathbb{Z}_p^{(k+\eta)\times k}$, we denote the upper square matrix of $\mathbf{A}$ by $\overline{\mathbf{A}} \in \mathbb{Z}_p^{k\times k}$ and the lower η rows of $\mathbf{A}$ by $\underline{\mathbf{A}} \in \mathbb{Z}_p^{\eta\times k}$. Similarly, for a column vector $\mathbf{v} \in \mathbb{Z}_p^{k+\eta}$, we denote the upper k elements by $\overline{\mathbf{v}} \in \mathbb{Z}_p^k$ and the lower η elements of $\mathbf{v}$ by $\underline{\mathbf{v}} \in \mathbb{Z}_p^\eta$. For a bit string $m \in \{0,1\}^n$, m_i denotes the ith bit of m $(i \le n)$ and $m_{|i}$ denotes the first i bits of m.

All our algorithms are probabilistic polynomial time unless we stated otherwise. If $\mathcal{A}$ is a probabilistic polynomial time algorithm, then we write $a \xleftarrow{\$} \mathcal{A}(b)$ to denote the random variable that outputted by $\mathcal{A}$ on input b.

GAMES. We follow [9] to use code-based games for defining and proving security. A game G contains procedures INIT and FINALIZE, and some additional procedures $\mathrm{P}_1, \ldots, \mathrm{P}_n$, which are defined in pseudo-code. All variables in a game are initialized as 0, and all sets are empty (denote by $\emptyset$). An adversary $\mathcal{A}$ is executed in game G (denote by $\mathsf{G}^{\mathcal{A}}$) if it first calls INIT, obtaining its output. Next, it may make arbitrary queries to P_i (according to their specification) and obtain their output, where the total number of queries is denoted by Q. Finally, it makes one single call to FINALIZE$(\cdot)$ and stops. We use $\mathsf{G}^{\mathcal{A}} \Rightarrow d$ to denote that G outputs d after interacting with $\mathcal{A}$, and d is the output of FINALIZE.

2.1 Collision Resistant Hash Functions

Let $\mathcal{H}$ be a family of hash functions $H : \{0,1\}^* \to \{0,1\}^\lambda$. We assume that it is efficient to sample a function from $\mathcal{H}$, which is denoted by $H \xleftarrow{\$} \mathcal{H}$.

Definition 1 (Collision resistance). *We say a family of hash functions $\mathcal{H}$ is collision-resistant (*CR*) if for all adversaries $\mathcal{A}$*

$$\mathsf{Adv}^{\mathsf{cr}}_{\mathcal{H},\mathcal{A}}(\lambda) := \Pr[x \neq x' \wedge H(x) = H(x') \mid H \xleftarrow{\$} \mathcal{H}, (x, x') \xleftarrow{\$} \mathcal{A}(1^\lambda, H)]$$

is negligible.

2.2 Pairing Groups and Matrix Diffie-Hellman Assumptions

Let GGen be a probabilistic polynomial time (PPT) algorithm that on input 1^λ returns a description $\mathcal{G} := (\mathbb{G}_1, \mathbb{G}_2, \mathbb{G}_T, p, P_1, P_2, e)$ of asymmetric pairing groups where $\mathbb{G}_1$, $\mathbb{G}_2$, $\mathbb{G}_T$ are cyclic groups of order p for a λ-bit prime p, P_1 and P_2 are generators of $\mathbb{G}_1$ and $\mathbb{G}_2$, respectively, and $e : \mathbb{G}_1 \times \mathbb{G}_2 \to \mathbb{G}_T$ is an efficient computable (non-degenerated) bilinear map. Define $P_T := e(P_1, P_2)$, which is a generator in $\mathbb{G}_T$. In this paper, we only consider Type III pairings, where $\mathbb{G}_1 \neq \mathbb{G}_2$ and there is no efficient homomorphism between them.

We use implicit representation of group elements as in [16]. For $s \in \{1, 2, T\}$ and $a \in \mathbb{Z}_p$ define $[a]_s = aP_s \in \mathbb{G}_s$ as the implicit representation of a in $\mathbb{G}_s$.

Similarly, for a matrix $\mathbf{A} = (a_{ij}) \in \mathbb{Z}_p^{n\times m}$ we define $[\mathbf{A}]_s$ as the implicit representation of $\mathbf{A}$ in $\mathbb{G}_s$. $\mathsf{Span}(\mathbf{A}) := \{\mathbf{Ar}|\mathbf{r} \in \mathbb{Z}_p^m\} \subset \mathbb{Z}_p^n$ denotes the linear span of $\mathbf{A}$, and similarly $\mathsf{Span}([\mathbf{A}]_s) := \{[\mathbf{Ar}]_s|\mathbf{r} \in \mathbb{Z}_p^m\} \subset \mathbb{G}_s^n$. Note that it is efficient to compute $[\mathbf{AB}]_s$ given $([\mathbf{A}]_s, \mathbf{B})$ or $(\mathbf{A}, [\mathbf{B}]_s)$ with matching dimensions. We define $[\mathbf{A}]_1 \circ [\mathbf{B}]_2 := e([\mathbf{A}]_1, [\mathbf{B}]_2) = [\mathbf{AB}]_T$, which can be efficiently computed given $[\mathbf{A}]_1$ and $[\mathbf{B}]_2$.

Next we recall the definition of the Matrix Decisional Diffie-Hellman (MDDH) [16] and related assumptions [41].

Definition 2 (Matrix distribution). *Let $k, \ell \in \mathbb{N}$ with $\ell > k$. We call $\mathcal{D}_{\ell,k}$ a matrix distribution if it outputs matrices in $\mathbb{Z}_p^{\ell\times k}$ of full rank k in polynomial time. By $\mathcal{D}_k$ we denote $\mathcal{D}_{k+1,k}$.*

Without loss of generality, we assume the first k rows of $\mathbf{A} \stackrel{\$}{\leftarrow} \mathcal{D}_{\ell,k}$ form an invertible matrix. For a matrix $\mathbf{A} \stackrel{\$}{\leftarrow} \mathcal{D}_{\ell,k}$, we define the set of kernel matrices of $\mathbf{A}$ as

$$\mathsf{ker}(\mathbf{A}) := \{\mathbf{a}^\perp \in \mathbb{Z}_p^{(\ell-k)\times \ell} \mid \mathbf{a}^\perp \cdot \mathbf{A} = \mathbf{0} \in \mathbb{Z}_p^{(\ell-k)\times k} \text{ and } \mathbf{a}^\perp \text{ has rank } (\ell-k)\}.$$

Given a matrix $\mathbf{A}$ over $\mathbb{Z}_p^{\ell\times k}$, it is efficient to sample an $\mathbf{a}^\perp$ from $\mathsf{ker}(\mathbf{A})$.

The $\mathcal{D}_{\ell,k}$-Matrix Diffie-Hellman problem is to distinguish the two distributions $([\mathbf{A}], [\mathbf{Aw}])$ and $([\mathbf{A}], [\mathbf{u}])$ where $\mathbf{A} \stackrel{\$}{\leftarrow} \mathcal{D}_{\ell,k}$, $\mathbf{w} \stackrel{\$}{\leftarrow} \mathbb{Z}_p^k$ and $\mathbf{u} \stackrel{\$}{\leftarrow} \mathbb{Z}_p^\ell$.

Definition 3 ($\mathcal{D}_{\ell,k}$-matrix decisional Diffie-Hellman assumption). *Let $\mathcal{D}_{\ell,k}$ be a matrix distribution and $s \in \{1, 2, T\}$. We say that the $\mathcal{D}_{\ell,k}$-Matrix Diffie-Hellman ($\mathcal{D}_{\ell,k}$-MDDH) is hard relative to GGen in group $\mathbb{G}_s$ if for all PPT adversaries $\mathcal{A}$, it holds that*

$$\mathsf{Adv}^{\mathsf{mddh}}_{\mathbb{G}_s,\mathcal{D}_{\ell,k},\mathcal{A}}(\lambda) := |\Pr[1 \stackrel{\$}{\leftarrow} \mathcal{A}(\mathcal{G}, [\mathbf{A}]_s, [\mathbf{Aw}]_s)] - \Pr[1 \stackrel{\$}{\leftarrow} \mathcal{A}(\mathcal{G}, [\mathbf{A}]_s, [\mathbf{u}]_s)]|$$

is negligible in the security parameter λ, where the probability is taken over $\mathcal{G} \stackrel{\$}{\leftarrow} \mathsf{GGen}(1^\lambda)$, $\mathbf{A} \stackrel{\$}{\leftarrow} \mathcal{D}_{\ell,k}, \mathbf{w} \stackrel{\$}{\leftarrow} \mathbb{Z}_p^k$ and $\mathbf{u} \stackrel{\$}{\leftarrow} \mathbb{Z}_p^\ell$.

We define the Kernel Diffie-Hellman assumption $\mathcal{D}_k$-KerMDH [41] which is a natural search variant of the $\mathcal{D}_k$-MDDH assumption.

Definition 4 ($\mathcal{D}_k$-kernel Diffie-Hellman assumption, $\mathcal{D}_k$-KerMDH). *Let $\mathcal{D}_k$ be a matrix distribution and $s \in \{1, 2\}$. We say that the $\mathcal{D}_k$-kernel Matrix Diffie-Hellman ($\mathcal{D}_k$-KerMDH) is hard relative to GGen in group $\mathbb{G}_s$ if for all PPT adversaries $\mathcal{A}$, it holds that*

$$\mathsf{Adv}^{\mathsf{kmdh}}_{\mathbb{G}_s,\mathcal{D}_{\ell,k},\mathcal{A}}(\lambda) := \Pr[\mathbf{c}^\top \mathbf{A} = \mathbf{0} \wedge \mathbf{c} \neq \mathbf{0} | [\mathbf{c}]_{3-s} \stackrel{\$}{\leftarrow} \mathcal{A}(\mathcal{G}, [\mathbf{A}]_s)]$$

is negligible in security parameter λ, where the probability is taken over $\mathcal{G} \stackrel{\$}{\leftarrow} \mathsf{GGen}(1^\lambda)$, $\mathbf{A} \stackrel{\$}{\leftarrow} \mathcal{D}_k$.

The following lemma shows that the $\mathcal{D}_k$-KerMDH assumption is a relaxation of the $\mathcal{D}_k$-MDDH assumption since one can use a non-zero vector in the kernel of $\mathbf{A}$ to test membership in the column space of $\mathbf{A}$.

Lemma 1 **($\mathcal{D}_k$-MDDH $\Rightarrow$ $\mathcal{D}_k$-KerMDH [41]).** *For any matrix distribution $\mathcal{D}_k$, if $\mathcal{D}_k$-MDDH is hard relative to* GGen *in group $\mathbb{G}_s$, then $\mathcal{D}_k$-KerMDH is hard relative to* GGen *in group $\mathbb{G}_s$.*

For $Q > 1$, $\mathbf{W} \xleftarrow{\$} \mathbb{Z}_p^{k\times Q}$, $\mathbf{U} \xleftarrow{\$} \mathbb{Z}_p^{\ell\times Q}$, consider the Q-fold $\mathcal{D}_{\ell,k}$-MDDH problem which is distinguishing the distributions $([\mathbf{A}], [\mathbf{AW}])$ and $([\mathbf{A}], [\mathbf{U}])$. That is, the Q-fold $\mathcal{D}_{\ell,k}$-MDDH problem contains Q independent instances of the $\mathcal{D}_{\ell,k}$-MDDH problem (with the same $\mathbf{A}$ but different $\mathbf{w}_i$). The following lemma shows that the two problems are tightly equivalent and the reduction only loses a constant factor $\ell - k$.

Lemma 2 (Random self-reducibility [16]). *For $\ell > k$ and any matrix distribution $\mathcal{D}_{\ell,k}$, $\mathcal{D}_{\ell,k}$-MDDH is random self-reducible. In particular, for any $Q \geq 1$, if $\mathcal{D}_{\ell,k}$-MDDH is hard relative to* GGen *in group $\mathbb{G}_s$, then Q-fold $\mathcal{D}_{\ell,k}$-MDDH is hard relative to* GGen *in group $\mathbb{G}_s$, where $\mathsf{T}(\mathcal{B}) \approx \mathsf{T}(\mathcal{A}) + Q \cdot \mathsf{poly}(\lambda)$ and*

$$\mathsf{Adv}^{Q\text{-}\mathsf{mddh}}_{\mathbb{G}_s,\mathcal{D}_{\ell,k},\mathcal{A}}(\lambda) \leq (\ell - k)\mathsf{Adv}^{\mathsf{mddh}}_{\mathbb{G}_s,\mathcal{D}_{\ell,k},\mathcal{B}}(\lambda) + \frac{1}{p-1}.$$

The *boosting lemma* in [35] shows that the $\mathcal{D}_{2k,k}$-MDDH assumption reduces to the $\mathcal{D}_k$-MDDH assumption with a security loss of a factor of k.

2.3 Non-interactive Zero-Knowledge Proof

In this section, we follow [24,37] to recall the notion of a non-interactive zero-knowledge proof [10] and then an instantiation for an OR-language.

Let par be the public parameter and $\mathcal{L} = \{\mathcal{L}_{\mathsf{par}}\}$ be a family of languages with efficiently computable witness relation $\mathcal{R}_{\mathcal{L}}$. This definition is as follows .

Definition 5 (Non-interactive zero-knowledge proof [24]). *A non-interactive zero-knowledge proof (*NIZK*) for $\mathcal{L}$ consists of five PPT algorithms Π =* (Gen, TGen, Prove, Ver, Sim) *such that:*

- Gen(par) *returns a common reference string* crs.
- TGen(par) *returns* crs *and a trapdoor* td.
- Prove(crs, x, w) *returns a proof* π.
- Ver(crs, x, π) *returns* 1 *(accept) or* 0 *(reject). Here,* Ver *is deterministic.*
- Sim(crs, td, x) *returns a proof* π.

Perfect completeness is satisfied if for all crs $\in$ Gen $(1^\lambda, \mathsf{par})$, *all $x \in \mathcal{L}$, all witnesses w such that $\mathcal{R}_{\mathcal{L}}(x, w) = 1$, and all $\pi \in$* Prove(crs, x, w), *we have*

$$\mathsf{Ver}(\mathsf{crs}, x, \pi) = 1.$$

Zero-knowledge is satisfied if for all PPT adversaries $\mathcal{A}$ we have that

$$\mathsf{Adv}^{\mathsf{zk}}_{\Pi,\mathcal{A}}(\lambda) := \Big| \Pr[\mathcal{A}^{\mathsf{Prove}(\mathsf{crs},\cdot,\cdot)}(1^\lambda, \mathsf{crs}) = 1 \mid \mathsf{crs} \xleftarrow{\$} \mathsf{Gen}(1^\lambda, \mathsf{par})]$$
$$- \Pr[\mathcal{A}^{Sim(\mathsf{crs},\cdot,\cdot)}(1^\lambda, \mathsf{crs}) = 1 \mid (\mathsf{crs}, \mathsf{td}) \xleftarrow{\$} \mathsf{TGen}(1^\lambda, \mathsf{par})] \Big|$$

is negligible, where Sim(crs, x, w) *returns* $\pi \xleftarrow{\$} \mathsf{Sim}(\mathsf{crs}, \mathsf{td}, x)$ *if* $\mathcal{R}_{\mathcal{L}}(x, w) = 1$ *and aborts otherwise.*

Perfect soundness is satisified if for all $\mathsf{crs} \in \mathsf{Gen}(\mathsf{par})$*, for all words* $x \notin \mathcal{L}$ *and all proofs* π *it holds* $\mathsf{Ver}(\mathsf{crs}, x, \pi) = 0$.

Notice that Gay et al. [20] adopted a stronger notion of composable zero-knowledge. However, one can easily see that the standard we defined above is enough for their constructions, as well as ours introduced later. Also, we can define *perfect zero-knowledge*, which requires $\mathsf{Adv}^{\mathsf{zk}}_{\Pi,\mathcal{A}}(\lambda) = 0$, and *computational soundness*, which requires that for all for all words $x \notin \mathcal{L}$,

$$\mathsf{Adv}^{\mathsf{snd}}_{\Pi,\mathcal{A}} = \left| \Pr[\mathsf{Ver}(\mathsf{crs}, x, \pi) = 1 \mid \mathsf{crs} \xleftarrow{\$} \mathsf{Gen}(1^\lambda, \mathsf{par}), \pi \xleftarrow{\$} \mathcal{A}(1^\lambda, \mathsf{crs})] \right|$$

is negligible.

NIZK FOR AN OR-LANGUAGE. Let $\mathcal{G} \leftarrow \mathsf{GGen}(1^\lambda)$, $k \in \mathbb{N}$, $\mathbf{A}_0, \mathbf{A}_1 \xleftarrow{\$} \mathcal{D}_{2k,k}$, and $\mathsf{par} := (\mathcal{G}, [\mathbf{A}_0]_1, [\mathbf{A}_1]_1)$. We refer the reader to the full paper for a NIZK proof scheme, which was previously presented in [37] and also implicitly given in [23,43], for the OR-language

$$\mathcal{L}^{\vee}_{\mathbf{A}_0,\mathbf{A}_1} := \{[\mathbf{x}]_1 \in \mathbb{G}_1^{2k} \mid \exists \mathbf{r} \in \mathbb{Z}_p^k \colon [\mathbf{x}]_1 = [\mathbf{A}_0]_1 \cdot \mathbf{r} \vee [\mathbf{x}]_1 = [\mathbf{A}_1]_1 \cdot \mathbf{r}\}.$$

It will be used as a building block of our QANIZK proof.

2.4 Quasi-Adaptive Zero-Knowledge Argument

The notion of Quasi-Adaptive Zero-Knowledge Argument (QANIZK) was proposed by Jutla and Roy [32], where the common reference string CRS depends on the specific language for which proofs are generated. In the following, we recall the definition of QANIZK [18,37]. For simplicity, we only consider arguments for linear subspaces.

Let par be the public parameters for QANIZK and $\mathcal{D}_{\mathsf{par}}$ be a probability distribution over a collection of relations $R = \{R_{[\mathbf{M}]_1}\}$ parametrized by a matrix $[\mathbf{M}]_1 \in \mathbb{G}_1^{n_1 \times n_2}$ $(n_1 > n_2)$ with associated language $\mathcal{L}_{[\mathbf{M}]_1} = \{[\mathbf{t}]_1 : \exists \mathbf{w} \in \mathbb{Z}_q^t, \text{ s.t. } [\mathbf{t}]_1 = [\mathbf{M}\mathbf{w}]_1\}$. We consider witness sampleable distributions [32] where there is an efficiently sampleable distribution $\mathcal{D}'_{\mathsf{par}}$ outputs $\mathbf{M}' \in \mathbb{Z}_q^{n_1 \times n_2}$ such that $[\mathbf{M}']_1$ distributes the same as $[\mathbf{M}]_1$. We note that the matrix distribution in Definition 2 is sampleable.

We define the notions of QANIZK, designated-prover QANIZK (DPQANIZK), designated-verifier QANIZK (DVQANIZK), designated-prover-verifier QANIZK (DPVQANIZK) as follow.

Definition 6 (QANIZK). *Let* $\mathsf{X} \in \{\epsilon, \mathsf{DP}, \mathsf{DV}, \mathsf{DPV}\}$*. An* XQANIZK *for a language distribution* $\mathcal{D}_{\mathsf{par}}$ *consists of four PPT algorithms* $\Pi = (\mathsf{Gen}, \mathsf{Prove}, \mathsf{Ver}, \mathsf{Sim})$.

- $\mathsf{Gen}(\mathsf{par}, [\mathbf{M}]_1)$ *returns a common reference string* crs*, a prover key* prk*, a verifier key* vrk *and a simulation trapdoor* td*:*

- $X = \epsilon$ *iff* $\mathsf{prk} = \mathsf{vrk} = \epsilon$.
- $X = \mathsf{DP}$ *iff* $\mathsf{vrk} = \epsilon$.
- $X = \mathsf{DV}$ *iff* $\mathsf{prk} = \epsilon$.
- $X = \mathsf{DPV}$ *iff* $\mathsf{prk} \neq \epsilon$ *and* $\mathsf{vrk} \neq \epsilon$.

– $\mathsf{Prove}(\mathsf{crs}, \mathsf{prk}, [\mathbf{y}]_1, \mathbf{w})$ *returns a proof* π.
– $\mathsf{Ver}(\mathsf{crs}, \mathsf{vrk}, [\mathbf{y}]_1, \pi)$ *returns* 1 *(accept) or* 0 *(reject). Here,* Ver *is a deterministic algorithm.*
– $\mathsf{Sim}(\mathsf{crs}, \mathsf{td}, [\mathbf{y}]_1)$ *returns a simulated proof* π.

Perfect completeness is satisfied if for all λ*, all* $[\mathbf{M}]_1$*, all* $([\mathbf{y}]_1, \mathbf{w})$ *with* $[\mathbf{y}]_1 = [\mathbf{M}\mathbf{w}]_1$*, all* $(\mathsf{crs}, \mathsf{prk}, \mathsf{vrk}, \mathsf{td}) \in \mathsf{Gen}(\mathsf{par}, [\mathbf{M}]_1)$*, and all* $\pi \in \mathsf{Prove}(\mathsf{crs}, \mathsf{prk}, [\mathbf{y}]_1, \mathbf{w})$*, we have*

$$\mathsf{Ver}(\mathsf{crs}, \mathsf{vrk}, [\mathbf{y}]_1, \pi) = 1.$$

Perfect zero knowledge is satisfied if for all λ*, all* $[\mathbf{M}]_1$*, all* $([\mathbf{y}]_1, \mathbf{w})$ *with* $[\mathbf{y}]_1 = [\mathbf{M}\mathbf{w}]_1$*, and all* $(\mathsf{crs}, \mathsf{prk}, \mathsf{vrk}, \mathsf{td}) \in \mathsf{Gen}(\mathsf{par}, [\mathbf{M}]_1)$*, the following two distributions are identical:*

$$\mathsf{Prove}(\mathsf{crs}, \mathsf{prk}, [\mathbf{y}]_1, \mathbf{w}) \quad \textit{and} \quad \mathsf{Sim}(\mathsf{crs}, \mathsf{td}, [\mathbf{y}]_1).$$

We define the (unbounded) simulation soundness for all types of QANIZK.

Definition 7 (Unbounded simulation soundness). *Let* $X \in \{\epsilon, \mathsf{DP}, \mathsf{DV}, \mathsf{DPV}\}$*. An* XQANIZK $\Pi := (\mathsf{Gen}, \mathsf{Prove}, \mathsf{Ver}, \mathsf{Sim})$ *is unbounded simulation sound (*USS*) if for any adversary* $\mathcal{A}$*,*

$$\mathsf{Adv}^{\mathsf{uss}}_{\Pi,\mathcal{A}}(\lambda) := \Pr[\mathsf{USS}^{\mathcal{A}} \Rightarrow 1]$$

is negligible, where Game USS *is defined in Fig. 1.*

$\underline{\text{INIT}(\mathbf{M}):}$	$\underline{\text{FINALIZE}([\mathbf{y}^*]_1, \pi^*):}$
$(\mathsf{crs}, \mathsf{prk}, \mathsf{vrk}, \mathsf{td}) \xleftarrow{\$} \mathsf{Gen}(\mathsf{par}, [\mathbf{M}]_1)$	If $[\mathbf{y}^*]_1 \notin \mathcal{L}_{[\mathbf{M}]_1} \wedge ([\mathbf{y}^*]_1, \pi^*) \notin \mathcal{Q}_{\mathsf{sim}}$ then
Return crs.	return $\mathsf{Ver}(\mathsf{crs}, \mathsf{vrk}, [\mathbf{y}^*]_1, \pi^*)$
	Else return 0
$\underline{\text{SIM}([\mathbf{y}]_1):}$ // Q_s queries	
$\pi \xleftarrow{\$} \mathsf{Sim}(\mathsf{crs}, \mathsf{td}, [\mathbf{y}]_1)$	
$\mathcal{Q}_{\mathsf{sim}} := \mathcal{Q}_{\mathsf{sim}} \cup ([\mathbf{y}]_1, \pi)$	
Return π	

Fig. 1. USS security game for XQANIZK.

WEAK USS. We can also consider a weak notion of simulation soundness. in the sense that it is only required that $[\mathbf{y}^*]_1 \notin \mathcal{Q}_{\mathsf{sim}}$.[4]

[4] In [5], the defined security is this weak version. However, it is not sufficient for constructing a CCA2 secure encryption scheme, since it does not prevent an adversary from forging a new ciphertext for a challenge message and sending that it as a decryption query.

WITNESS-SAMPLABLE DISTRIBUTION. Here we define simulation soundness for witness-sampleable distributions, namely, INIT gets $\mathbf{M} \in \mathbb{Z}_p^{n_1 \times n_2}$ as input, proofs of our DVQANIZK and QANIZK schemes do not require the explicit $\mathbf{M}$ over $\mathbb{Z}_p$. In all the standard definitions of (simulation) soundness of QANIZK for linear subspaces, the challenger needs information on $\mathbf{M}$ in $\mathbb{Z}_p$ (not necessary the whole matrix) to check whether the target word $[\mathbf{y}^*]_1$ is inside the language $\mathsf{Span}([\mathbf{M}]_1)$. This information can be a non-zero kernel vector of $\mathbf{M}$ (either in $\mathbb{Z}_p$ or in $\mathbb{G}_2$). We can also define USS with respect to non-witness sampleable distributions while our security proofs (with straightforward modifications) introduced later also hold. In this case, we have to allow the challenger to use super polynomial computational power to check whether $[\mathbf{y}^*]_1 \in \mathsf{Span}(\mathbf{M})$, i.e., then the USS game becomes non-falsifiable. Otherwise, we have to assume that the attacker always gives $[\mathbf{y}^*]_1 \notin \mathsf{Span}(\mathbf{M})$ in USS. In fact, we note that many constructions and applications of simulation-sound QANIZKs consider witness-sampleable distributions (c.f., [18,29,32,38]).

2.5 Structure-Preserving Signature

We now recall the notion of structure-preserving signature (SPS) [3] and unforgeability against chosen message attacks (UF-CMA).

Definition 8 (Signature). *A signature scheme is a tuple of PPT algorithms* $\mathsf{SIG} := (\mathsf{Gen}, \mathsf{Sign}, \mathsf{Ver})$ *such that:*

- $\mathsf{Gen}(\mathsf{par})$ *returns a verification/signing key pair* $(\mathsf{vk}, \mathsf{sk})$.
- $\mathsf{Sign}(\mathsf{sk}, \mathsf{m})$ *returns a signature* σ *for* $\mathsf{m} \in \mathcal{M}$.
- $\mathsf{Ver}(\mathsf{vk}, \mathsf{m}, \sigma)$ *returns* 1 *(accept) or* 0 *(reject). Here* Ver *is deterministic.*

Correctness is satisfied if for all $\lambda \in \mathbb{N}$, *all* $\mathsf{m} \in \mathcal{M}$, *and all* $(\mathsf{vk}, \mathsf{sk}) \in \mathsf{Gen}(\mathsf{par})$,

$$\mathsf{Ver}(\mathsf{vk}, \mathsf{m}, \mathsf{Sign}(\mathsf{sk}, \mathsf{m})) = 1.$$

Definition 9 (Structure-preservation). *A signature scheme is said to be structure-preserving if its verification keys, signing messages, and signatures consist only of group elements and verification proceeds via only a set of pairing product equations.*

Definition 10 (UF-CMA security). *For a signature scheme* $\mathsf{SIG} := (\mathsf{Gen}, \mathsf{Sign}, \mathsf{Ver})$ *and any adversary* $\mathcal{A}$, *we define the following experiment:*

<u>INIT</u>:	<u>SIGNO(m)</u>:	<u>FINALIZE(m*, σ*)</u>:
$(\mathsf{vk}, \mathsf{sk}) \leftarrow \mathsf{Gen}(\mathsf{par})$	$\mathcal{Q}_{\mathsf{sign}} := \mathcal{Q}_{\mathsf{sign}} \cup \{\mathsf{m}\}$	If $\mathsf{m}^* \notin \mathcal{Q}_{\mathsf{sign}}$ and $\mathsf{Ver}(\mathsf{vk}, \mathsf{m}^*, \sigma^*) = 1$
Return vk	$\sigma \leftarrow \mathsf{Sign}(\mathsf{sk}, \mathsf{m})$	Return 1
	Return σ	Else return 0

Fig. 2. UF-CMA security game for SIG.

A signature scheme SIG *is unforgeable against chosen message attacks* (UF-CMA)*, if for all PPT adversaries* $\mathcal{A}$*,*

$$\mathsf{Adv}^{\mathsf{uf\text{-}cma}}_{\mathsf{SIG},\mathcal{A}}(\lambda) := \Pr[\mathsf{UF\text{-}CMA}^{\mathcal{A}} \Rightarrow 1]$$

is negligible, where Game UF-CMA *is defined in Fig. 2.*

3 Quasi-Adaptive NIZK

In this section, we construct a QANIZK with tight simulation soundness. As a stepping stone, we develop a DVQANIZK based on the Matrix Diffie-Hellman assumption. By using the Kernel Matrix Diffie-Hellman assumption and pairings, our DVQANIZK gives us a more efficient QANIZK. All the security reductions in this section are tight.

The Core Lemma. We recall the useful core lemma from [20], which can computationally introduce randomness. More precisely, it shows that moving from experiment Core_0 to Core_1 can (up to negligible terms) only increase the winning chances of an adversary.

$\underline{\textsc{Init}_{\mathsf{core}}}$:	$\underline{\textsc{Eval}_{\mathsf{core}}}$:	$\underline{\textsc{Finalize}_{\mathsf{core}}(\mu)}$:
$\mathsf{c} := 0$	$\mathsf{c} := \mathsf{c} + 1$	Parse $\mu =: ([\mathbf{t}]_1, [u']_1, \pi_{\mathsf{or}})$
$\mathbf{A}_0, \mathbf{A}_1 \xleftarrow{\$} \mathcal{D}_{2k,k}$	$\mathbf{s} \xleftarrow{\$} \mathbb{Z}_p^k, \mathbf{t} := \mathbf{A}_0\mathbf{s} \in \mathbb{Z}_p^{2k}$	If $\mathsf{Ver}_{\mathsf{or}}(\mathsf{crs}_{\mathsf{or}}, [\mathbf{t}]_1, \pi_{\mathsf{or}}) = 0$
$\mathsf{par}_{\mathsf{or}} := (\mathsf{par}, [\mathbf{A}_0]_1, [\mathbf{A}_1]_1)$	$u' := \mathbf{t}^\top(\mathbf{k} \boxed{+\mathbf{RF}(\mathsf{c})}) \in \mathbb{Z}_p$	then return 0
$\mathsf{crs}_{\mathsf{or}} \leftarrow \mathsf{Gen}_{\mathsf{or}}(\mathsf{par}_{\mathsf{or}}, 1^\lambda)$	$\pi_{\mathsf{or}} \xleftarrow{\$} \mathsf{Prove}_{\mathsf{or}}(\mathsf{crs}_{\mathsf{or}}, [\mathbf{t}]_1, \mathbf{s})$	If $[u']_1 = \mathbf{t}^\top(\mathbf{k} + \boxed{\mathbf{RF}(\mathsf{c}')})$
$\mathbf{k} \xleftarrow{\$} \mathbb{Z}_p^{2k}$	$\mu := ([\mathbf{t}]_1, [u']_1, \pi_{\mathsf{or}})$	and $0 \le \mathsf{c}' \le \mathsf{c}$ then
$\mathbf{p} := \mathbf{A}_0^\top(\mathbf{k} \boxed{+\mathbf{RF}(0)})$	Return μ	return 1
$\mathsf{crs} := (\mathsf{crs}_{\mathsf{or}}, [\mathbf{A}_0]_1, [\mathbf{p}]_1)$		Else return 0
Return crs		

Fig. 3. Security games Core_0 and Core_1 for the core lemma. $\mathbf{RF} : \mathbb{Z}_p \to \mathbb{Z}_p^{2k}$ is a random function. All the codes are executed in both games, except the boxed codes which are only executed in Core_1.

Lemma 3 (Core lemma). *If the* $\mathcal{D}_k$-MDDH *assumption holds in the group* $\mathbb{G}_2$*, and* $\Pi^{\mathsf{or}} = (\mathsf{Gen}_{\mathsf{or}}, \mathsf{TGen}_{\mathsf{or}}, \mathsf{Prove}_{\mathsf{or}}, \mathsf{Ver}_{\mathsf{or}}, \mathsf{Sim}_{\mathsf{or}})$ *is a* NIZK *for* $\mathcal{L}^{\vee}_{\mathbf{A}_0,\mathbf{A}_1}$ *with perfect completeness, perfect soundness, and zero-knowledge, then for any adversary* $\mathcal{A}$ *against the core lemma, there exist adversaries* $\mathcal{B}, \mathcal{B}'$ *with running time* $T(\mathcal{B}) \approx T(\mathcal{B}') \approx T(\mathcal{A}) + Q \cdot \mathsf{poly}(\lambda)$ *such that*

$$\begin{aligned}\mathsf{Adv}^{\mathsf{core}}_{\mathcal{A}}(\lambda) :=& \Pr[\mathsf{Core}_0^{\mathcal{A}} \Rightarrow 1] - \Pr[\mathsf{Core}_1^{\mathcal{A}} \Rightarrow 1] \\ \le& (4k\lceil \log Q \rceil + 2) \cdot \mathsf{Adv}^{\mathsf{mddh}}_{\mathbb{G}_2, \mathcal{D}_{2k,k}, \mathcal{B}}(\lambda) + (2\lceil \log Q \rceil + 2) \cdot \mathsf{Adv}^{\mathsf{zk}}_{\mathsf{NIZK},\mathcal{B}'}(\lambda) \\ &+ \lceil \log Q \rceil \cdot \Delta_{\mathcal{D}_{2k,k}} + \frac{4\lceil \log Q \rceil + 2}{p-1} + \frac{\lceil \log Q \rceil \cdot Q}{p},\end{aligned}$$

where $\Delta_{\mathcal{D}_{2k,k}}$ *is a statistically small term for* $\mathcal{D}_{2k,k}$.

In a slight departure from [20], we include the term $[\mathbf{A}_0^\top \mathbf{k}]_1$ in crs. We argue that the core lemma still holds by the following reasons (for notation, our $\mathbf{k}$ is their $\mathbf{k}_0$):

- The main purpose of $\mathbf{k}$ is to introduce the constant random function $\mathbf{F}_0(\epsilon)$ in the transition from G_2 to $\mathsf{G}_{3.0}$ in Lemma 4 in [20]. The same argument still holds, given $[\mathbf{A}_0^\top \mathbf{k}]_1$.
- The randomization of Lemma 5 in [20] is done by switching $[\mathbf{t}]_1$ into the right span, and this can be done independent of $\mathbf{k}$. Additionally, we note that, given $[\mathbf{A}_0^\top \mathbf{k}]_1$, one cannot efficiently compute $[\mathbf{t}^\top \mathbf{k}]_1$ without knowing $\mathbf{s} \in \mathbb{Z}_p^k$ s.t. $\mathbf{t} = \mathbf{A}_0 \mathbf{s}$.

We give some brief intuition about the proof of the lemma here. Similar to [20], we re-randomize $\mathbf{k}$ via a sequence of hybrid games. In the i-th hybrid game, we set $u = \mathbf{t}^\top(\mathbf{k} + \mathbf{RF}_i(\mathsf{c}_{|i}))$ where $\mathbf{RF}_i$ is a random function and $\mathsf{c}_{|i}$ denotes the first i-bit prefix of the counter c for queries to $\textsc{Eval}_{\mathsf{core}}$. To proceed from the i-th game to the $(i+1)$-th, we choose $\mathbf{t} \in \mathsf{Span}(\mathbf{A}_{\mathsf{c}_{i+1}})$ in $\textsc{Eval}_{\mathsf{core}}$ depending on the $(i+1)$-th bit of c. We note that the view of the adversary does not change due to the $\mathcal{D}_{2k,k}$-MDDH assumption. Then, as in [20], we can construct $\mathbf{RF}_i$ in the way that it satisfies $\mathbf{t}^\top \mathbf{RF}_{i+1}(\mathsf{c}_{|i+1}) = \mathbf{t}^\top \mathbf{RF}_i(\mathsf{c}_{|i})$. The main difference is that our $\mathbf{RF}_i$ additionally satisfies $\mathbf{A}_0^\top(\mathbf{k} + \mathbf{RF}_{i+1}(0^{i+1})) = \mathbf{A}_0^\top(\mathbf{k} + \mathbf{RF}_i(0^i))$, namely, it not only re-randomizes $\mathbf{k}$ but also ensures that the $\mathbf{A}_0^\top \mathbf{k}$ part in crs is always independent of all the u'-s generated by $\textsc{Eval}_{\mathsf{core}}$. We furthermore make consistent changes to $\textsc{Finalize}_{\mathsf{core}}$ as in [20]. We refer the reader to the full paper for the full proof.

3.1 Stepping Stone: Designated-Verifier QA-NIZK

Let $\mathcal{G} \leftarrow \mathsf{GGen}(1^\lambda)$, $\mathsf{par} := \mathcal{G}$, $k \in \mathbb{N}$, $\mathcal{H}$ be a collision-resistant hash function family, and $\Pi^{\mathsf{or}} := (\mathsf{Gen}_{\mathsf{or}}, \mathsf{Prove}_{\mathsf{or}}, \mathsf{Ver}_{\mathsf{or}})$ be a NIZK system for language $\mathcal{L}^{\vee}_{\mathbf{A}_0,\mathbf{A}_1}$. Our $\mathsf{DVQANIZK}$ $\Pi^{\mathsf{dv}} := (\mathsf{Gen}, \mathsf{Prove}, \mathsf{Ver}, \mathsf{Sim})$ is defined as in Fig. 4. We note that our scheme can be easily extended to a tag-based scheme by putting the label ℓ inside the hash function. Thus, our scheme can be used in all the applications that require tag-based $\mathsf{DVQANIZK}$.

Theorem 1 (Security of Π^{dv}). *Π^{dv} is a $\mathsf{DVQANIZK}$ with perfect zero-knowledge and (tightly) unbound simulation soundness. In particular, for any adversary $\mathcal{A}$, there exist adversaries $\mathcal{B}$ and $\mathcal{B}'$ with $\mathsf{T}(\mathcal{B}) \approx \mathsf{T}(\mathcal{A})$ and*

$$\begin{aligned}\mathsf{Adv}^{\mathsf{uss}}_{\Pi^{\mathsf{dv}},\mathcal{A}}(\lambda) \leq & \mathsf{Adv}^{\mathsf{cr}}_{\mathcal{H},\hat{\mathcal{B}}}(\lambda) + (4k\lceil \log Q \rceil + 2) \cdot \mathsf{Adv}^{\mathsf{mddh}}_{\mathbb{G}_1,\mathcal{D}_{2k,k},\mathcal{B}}(\lambda) \\ & + (2\lceil \log Q \rceil + 2) \cdot \mathsf{Adv}^{\mathsf{zk}}_{\Pi^{\mathsf{or}},\mathcal{B}'}(\lambda) + \lceil \log Q \rceil \cdot \Delta_{\mathcal{D}_{2k,k}} \\ & + \frac{4\lceil \log Q \rceil + 2}{p-1} + \frac{(\lceil \log Q \rceil + 1) \cdot Q + 1}{p}.\end{aligned}$$

Proof (of Theorem 1). Perfect completeness follows directly from the correctness of the OR proof system and the fact that for all $\mathbf{y} = \mathbf{M}\mathbf{w}$, $\mathbf{p} := \mathbf{A}_0^\top \mathbf{k}$, $\mathbf{p}_0 := \mathbf{M}^\top \mathbf{k}_0$, $\mathbf{p}_1 := \mathbf{M}^\top \mathbf{k}_1$, and $\mathbf{t} = \mathbf{A}_0 \mathbf{s}$, for any τ, we have

$\underline{\mathsf{Gen}(\mathsf{par}, [\mathbf{M}]_1 \in \mathbb{G}_1^{n_1 \times n_2})}$: $\mathbf{A}_0, \mathbf{A}_1 \xleftarrow{\$} \mathcal{D}_{2k,k},\ H \xleftarrow{\$} \mathcal{H}$ $\mathsf{par}_{\mathsf{or}} := (\mathsf{par}, [\mathbf{A}_0]_1, [\mathbf{A}_1]_1)$ $\mathsf{crs}_{\mathsf{or}} \leftarrow \mathsf{Gen}_{\mathsf{or}}(\mathsf{par}_{\mathsf{or}}, 1^\lambda)$ $\mathbf{k}_0, \mathbf{k}_1 \xleftarrow{\$} \mathbb{Z}_p^{n_1},\ \mathbf{k} \xleftarrow{\$} \mathbb{Z}_p^{2k}$ $[\mathbf{p}]_1 := [\mathbf{A}_0^\top \mathbf{k}]_1 \in \mathbb{G}_1^k$ $[\mathbf{p}_0]_1 := [\mathbf{M}^\top \mathbf{k}_0]_1 \in \mathbb{G}_1^{n_2}$ $[\mathbf{p}_1]_1 := [\mathbf{M}^\top \mathbf{k}_1]_1 \in \mathbb{G}_1^{n_2}$ $\mathsf{crs} := (\mathsf{crs}_{\mathsf{or}}, [\mathbf{A}_0]_1, [\mathbf{p}]_1, [\mathbf{p}_0]_1, [\mathbf{p}_1]_1, H)$ $\mathsf{td} := (\mathbf{k}_0, \mathbf{k}_1)$ $\mathsf{vk} := (\mathbf{k}, \mathbf{k}_0, \mathbf{k}_1)$ Return $(\mathsf{crs}, \mathsf{vk}, \mathsf{td})$ $\underline{\mathsf{Sim}(\mathsf{crs}, \mathsf{td}, [\mathbf{y}]_1)}$: $\mathbf{s} \xleftarrow{\$} \mathbb{Z}_p^k,\ \mathbf{t} := \mathbf{A}_0 \mathbf{s}$ $\pi_{\mathsf{or}} \xleftarrow{\$} \mathsf{Prove}_{\mathsf{or}}(\mathsf{crs}_{\mathsf{or}}, [\mathbf{t}]_1, \mathbf{s})$ $\tau := H([\mathbf{y}]_1, [\mathbf{t}]_1, \pi_{\mathsf{or}}) \in \mathbb{Z}_p$ $[u]_1 := [\mathbf{y}^\top(\mathbf{k}_0 + \tau \mathbf{k}_1)]_1 + [\mathbf{s}^\top \mathbf{p}]_1$ Return $\pi := ([\mathbf{t}]_1, [u]_1, \pi_{\mathsf{or}})$	$\underline{\mathsf{Prove}(\mathsf{crs}, [\mathbf{y}]_1, \mathbf{w})}$: $/\!/\ \mathbf{y} = \mathbf{M}\mathbf{w} \in \mathbb{Z}_p^{n_1}$ $\mathbf{s} \xleftarrow{\$} \mathbb{Z}_p^k,\ [\mathbf{t}]_1 := [\mathbf{A}_0]_1 \mathbf{s}$ $\pi_{\mathsf{or}} \xleftarrow{\$} \mathsf{Prove}_{\mathsf{or}}(\mathsf{crs}_{\mathsf{or}}, [\mathbf{t}]_1, \mathbf{s})$ $\tau := H([\mathbf{y}]_1, [\mathbf{t}]_1, \pi_{\mathsf{or}}) \in \mathbb{Z}_p$ $[u]_1 := [\mathbf{w}^\top(\mathbf{p}_0 + \tau \mathbf{p}_1) + \mathbf{s}^\top \mathbf{p}]_1$ Return $\pi := ([\mathbf{t}]_1, [u]_1, \pi_{\mathsf{or}})$ $\underline{\mathsf{Ver}(\mathsf{crs}, \mathsf{vk}, [\mathbf{y}]_1, \pi)}$: Parse $\pi = ([\mathbf{t}]_1, [u]_1, \pi_{\mathsf{or}})$ $\tau := H([\mathbf{y}]_1, [\mathbf{t}]_1, \pi_{\mathsf{or}}) \in \mathbb{Z}_p$ If $\mathsf{Ver}_{\mathsf{or}}(\mathsf{crs}_{\mathsf{or}}, [\mathbf{t}]_1, \pi_{\mathsf{or}}) = 0$ then return 0 If $[u]_1 = [\mathbf{y}^\top]_1(\mathbf{k}_0 + \tau \mathbf{k}_1) + [\mathbf{t}^\top]_1 \mathbf{k}$ then return 1 Else return 0

Fig. 4. Construction of $\Pi^{\mathsf{dv}} := (\mathsf{Gen}, \mathsf{Prove}, \mathsf{Ver}, \mathsf{Sim})$.

$$\begin{aligned}\mathbf{w}^\top(\mathbf{p}_0 + \tau \mathbf{p}_1) + \mathbf{s}^\top \mathbf{p} &= \mathbf{w}^\top(\mathbf{M}^\top \mathbf{k}_0 + \tau \mathbf{M}^\top \mathbf{k}_1) + \mathbf{s}^\top \mathbf{A}_0^\top \mathbf{k}\\ &= \mathbf{y}^\top(\mathbf{k}_0 + \tau \mathbf{k}_1) + \mathbf{t}^\top \mathbf{k}.\end{aligned}$$

Moreover, since

$$\begin{aligned}\mathbf{w}^\top(\mathbf{p}_0 + \tau \mathbf{p}_1) + \mathbf{s}^\top \mathbf{p} &= \mathbf{w}^\top(\mathbf{M}^\top \mathbf{k}_0 + \tau \mathbf{M}^\top \mathbf{k}_1) + \mathbf{s}^\top \mathbf{p}\\ &= \mathbf{y}^\top(\mathbf{k}_0 + \tau \mathbf{k}_1) + \mathbf{s}^\top \mathbf{p},\end{aligned}$$

proofs generated by Prove and Sim for the same $\mathbf{y} = \mathbf{M}\mathbf{w}$ are identical. Hence, perfect zero knowledge is also satisfied.

We now focus on the tight simulation soundness of Π^{dv}. Let $\mathcal{A}$ be an adversary against the unbounded simulation soundness of Π^{dv}. We bound the advantage of $\mathcal{A}$ via a sequence of games defined in Fig. 5.

G_0 is the real USS experiment for DVQANIZK as defined in Definition 7.

Lemma 4 (G_0). $\Pr[\mathsf{USS}^{\mathcal{A}} \Rightarrow 1] = \Pr[\mathsf{G}_0^{\mathcal{A}} \Rightarrow 1]$.

Lemma 5 (G_0 *to* G_1). *There is an adversary $\mathcal{B}$ breaking the collision resistance of $\mathcal{H}$ with* $\mathsf{T}(\mathcal{B}) \approx \mathsf{T}(\mathcal{A})$ *and* $\mathsf{Adv}^{\mathsf{cr}}_{\mathcal{H},\mathcal{B}}(\lambda) \geq |\Pr[\mathsf{G}_0^{\mathcal{A}} \Rightarrow 1] - \Pr[\mathsf{G}_1^{\mathcal{A}} \Rightarrow 1]|$.

Proof. We note that in G_0 and G_1 the value u is uniquely defined by $\mathbf{y}, \mathbf{t}$ and π_{or}. Thus, if $\mathcal{A}$ asks FINALIZE with $([\mathbf{y}^*]_1, [\mathbf{t}^*]_1, \pi^*_{\mathsf{or}})$ that appears from one of the SIM queries, then FINALIZE will output 0, since $([\mathbf{y}^*]_1, \pi^* := ([\mathbf{y}^*]_1, [\mathbf{t}^*]_1, [u^*]_1, \pi^*_{\mathsf{or}})) \in \mathcal{Q}_{\mathsf{sim}}$. Now if $([\mathbf{y}^*]_1, [\mathbf{t}^*]_1, \pi^*_{\mathsf{or}})$ has never appeared from one of the SIM queries, but $\tau^* = H([\mathbf{y}^*]_1, [\mathbf{t}^*]_1, \pi^*_{\mathsf{or}}) \in \mathcal{Q}_{\mathsf{tag}}$, the we can construct a straightforward reduction $\mathcal{B}$ to break the CR property of $\mathcal{H}$. □

$\underline{\text{INIT}(\mathbf{M})}$: $\boxed{\mathsf{G}_2}$

$i := 0$

$\mathbf{A}_0, \mathbf{A}_1 \xleftarrow{\$} \mathcal{D}_{2k,k}$, $H \xleftarrow{\$} \mathcal{H}$

$\mathsf{par}_{\mathsf{or}} := (\mathsf{par}, [\mathbf{A}_0]_1, [\mathbf{A}_1]_1)$

$\mathsf{crs}_{\mathsf{or}} \leftarrow \mathsf{Gen}_{\mathsf{or}}(\mathsf{par}_{\mathsf{or}}, 1^\lambda)$

$\mathbf{k}_0, \mathbf{k}_1 \xleftarrow{\$} \mathbb{Z}_p^{n_1}$, $\mathbf{k} \xleftarrow{\$} \mathbb{Z}_p^{2k}$

$[\mathbf{p}]_1 := [\mathbf{A}_0^\top(\mathbf{k} \boxed{+\mathbf{RF}(0)})]_1 \in \mathbb{G}_1^k$

$\mathbf{p}_0 := \mathbf{M}^\top \mathbf{k}_0 \in \mathbb{Z}_p^{n_2}$

$\mathbf{p}_1 := \mathbf{M}^\top \mathbf{k}_1 \in \mathbb{Z}_p^{n_2}$

$\mathsf{crs} := (\mathsf{crs}_{\mathsf{or}}, [\mathbf{A}_0]_1, [\mathbf{p}]_1, [\mathbf{p}_0]_1, [\mathbf{p}_1]_1, H)$

Return crs

$\underline{\text{SIM}([\mathbf{y}]_1)}$: // $\boxed{\mathsf{G}_2}$

$\mathsf{c} := \mathsf{c} + 1$

$\mathbf{s} \xleftarrow{\$} \mathbb{Z}_p^k$, $[\mathbf{t}]_1 := [\mathbf{A}_0]_1 \mathbf{s}$

$\pi_{\mathsf{or}} \xleftarrow{\$} \mathsf{Prove}_{\mathsf{or}}(\mathsf{crs}_{\mathsf{or}}, [\mathbf{t}]_1, \mathbf{s})$

$\tau := H([\mathbf{y}]_1, [\mathbf{t}]_1, \pi_{\mathsf{or}}) \in \mathbb{Z}_p$

$[u]_1 := [\mathbf{y}^\top(\mathbf{k}_0 + \tau \mathbf{k}_1) + \mathbf{t}^\top(\mathbf{k} \boxed{+\mathbf{RF}(\mathsf{c})})]_1$

$\pi := ([\mathbf{t}]_1, [u]_1, \pi_{\mathsf{or}})$

$\mathcal{Q}_{\mathsf{sim}} := \mathcal{Q}_{\mathsf{sim}} \cup \{([\mathbf{y}]_1, \pi)\}$, $\mathcal{Q}_{\mathrm{tag}} := \mathcal{Q}_{\mathrm{tag}} \cup \{\tau\}$

Return π

$\underline{\text{FINALIZE}([\mathbf{y}^*]_1, \pi^*)}$: // G_1-G_2, $\boxed{\mathsf{G}_2}$

Parse $\pi^* =: ([\mathbf{t}^*]_1, [u^*]_1, \pi_{\mathsf{or}}^*)$

$\tau^* := H([\mathbf{y}^*]_1, [\mathbf{t}^*]_1, \pi_{\mathsf{or}}^*) \in \mathbb{Z}_p$

If $\tau^* \in \mathcal{Q}_{\mathrm{tag}}$ then return 0

If $[\mathbf{y}^*]_1 \in \mathcal{L}_{[\mathbf{M}]_1}$ or $([\mathbf{y}^*]_1, \pi^*) \in \mathcal{Q}_{\mathsf{sim}}$ then return 0

If $\mathsf{Ver}_{\mathsf{or}}(\mathsf{crs}_{\mathsf{or}}, [\mathbf{t}^*]_1, \pi_{\mathsf{or}}^*) = 0$ then return 0

$\mathcal{S} := \{[\mathbf{y}^{*\top}(\mathbf{k}_0 + \tau^* \mathbf{k}_1) + \mathbf{t}^{*\top}(\mathbf{k} \boxed{+\mathbf{RF}(j^*)})]_1 : 0 \leq j^* \leq \mathsf{c}\}$

If $[u^*]_1 \in \mathcal{S}$ then return 1

Else return 0

Fig. 5. Games G_0, G_1 and G_2 for the proof of Theorem 1. $\mathbf{RF} : \mathbb{Z}_p \rightarrow \mathbb{Z}_p^{2k}$ is a random function. Given $\mathbf{M}$ over $\mathbb{Z}_p$, it is efficient to check whether $[\mathbf{y}^*]_1 \in \mathcal{L}_{[\mathbf{M}]_1}$.

Lemma 6 (G_1 *to* G_2). *There is an adversary $\mathcal{B}$ breaking the core lemma (cf. Lemma 3) with running time* $\mathsf{T}(\mathcal{B}) \approx \mathsf{T}(\mathcal{A})$ *and* $\mathsf{Adv}_{\mathcal{B}}^{\mathsf{core}}(\lambda) = \Pr[\mathsf{G}_1^{\mathcal{A}} \Rightarrow 1] - \Pr[\mathsf{G}_2^{\mathcal{A}} \Rightarrow 1]$.

Proof. We construct the reduction $\mathcal{B}$ defined in Fig. 6 to break the core lemma. Clearly, if $\mathcal{B}$'s oracle access is from Core_0, then $\mathcal{B}$ simulates G_1; and if $\mathcal{B}$'s oracle access is from Core_1 (which uses a random function $\mathbf{RF}$), then $\mathcal{B}$ simulates G_2. Thus, $\Pr[\mathsf{G}_1^{\mathcal{A}} \Rightarrow 1] - \Pr[\mathsf{G}_2^{\mathcal{A}} \Rightarrow 1] = \Pr[\mathsf{Core}_0^{\mathcal{B}} \Rightarrow 1] - \Pr[\mathsf{Core}_1^{\mathcal{B}} \Rightarrow 1] = \mathsf{Adv}_{\mathcal{B}}^{\mathsf{core}}(\lambda)$, which concludes the lemma. $\square$

Lemma 7 (G_2). $\Pr[\mathsf{G}_2^{\mathcal{A}} \Rightarrow 1] = \frac{Q}{p}$.

Proof. We apply the following information-theoretical arguments to show that even a computationally unbounded adversary $\mathcal{A}$ can win in G_2 only with negligible probability. If $\mathcal{A}$ wants to win in G_2, then $\mathcal{A}$ needs to output a fresh and valid $\pi^* := ([\mathbf{t}^*]_1, [u^*]_1, \pi_{\mathsf{or}}^*)$. According to the additional rejection rule introduced in G_2, $u = \mathbf{y}^{*\top}(\mathbf{k}_0 + \tau^* \mathbf{k}_1) + \mathbf{t}^{*\top}(\mathbf{k} + \mathbf{RF}(j^*))$ must hold for some $0 \leq j^* \leq Q$. Fix a $j^* \leq Q$, we show that $\mathcal{A}$ can compute such a u with probability at most $1/p$.

The argument is based on the information leak about $\mathbf{k}_0$ and $\mathbf{k}_1$:

- For the j-th SIM query ($j \neq j^*$), the term $\mathbf{t}^\top \mathbf{RF}(j)$ completely blinds the information about $\mathbf{k}_0$ and $\mathbf{k}_1$ as long as $\mathbf{t} \neq \mathbf{0}$.
- For the j^*-th SIM query, we cannot use the entropy from the term $(\mathbf{k}+\mathbf{RF}(j^*))$ to hide $\mathbf{k}_0$ and $\mathbf{k}_1$ anymore, but we make the following stronger argument.

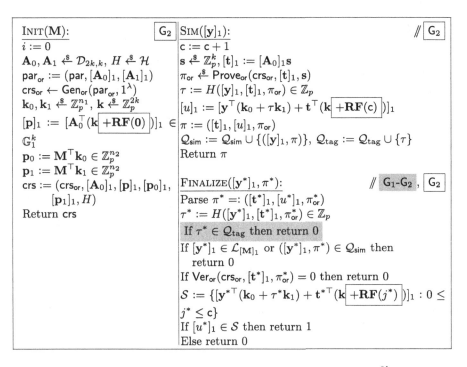

$\underline{\text{INIT}(\mathbf{M})\text{:}}$
$i := 0$
$\mathsf{crs}' \xleftarrow{\$} \text{INIT}_{\mathsf{core}}$
Parse $\mathsf{crs}' =: (\mathsf{crs}_{\mathsf{or}}, [\mathbf{A}_0]_1, [\mathbf{p}]_1)$
$\mathbf{k}_0, \mathbf{k}_1 \xleftarrow{\$} \mathbb{Z}_p^{n_1}$, $H \xleftarrow{\$} \mathcal{H}$
$[\mathbf{p}_0]_1 := [\mathbf{M}^\top \mathbf{k}_0]_1 \in \mathbb{G}_1^{n_2}$
$[\mathbf{p}_1]_1 := [\mathbf{M}^\top \mathbf{k}_1]_1 \in \mathbb{G}_1^{n_2}$
$\mathsf{crs} := (\mathsf{crs}', [\mathbf{p}_0]_1, [\mathbf{p}_1]_1, H)$
Return crs

$\underline{\text{SIM}([\mathbf{y}]_1)\text{:}}$
$\mathsf{c} := \mathsf{c} + 1$
$([\mathbf{t}]_1, [u']_1, \pi_{\mathsf{or}}) \xleftarrow{\$} \text{EVAL}_{\mathsf{core}}$
$\tau := H([\mathbf{y}]_1, [\mathbf{t}]_1, \pi_{\mathsf{or}}) \in \mathbb{Z}_p$
$[u]_1 := [\mathbf{y}^\top(\mathbf{k}_0 + \tau \mathbf{k}_1) + u']_1$
$\pi := ([\mathbf{t}]_1, [u]_1, \pi_{\mathsf{or}})$
$\mathcal{Q}_{\mathsf{sim}} := \mathcal{Q}_{\mathsf{sim}} \cup \{([\mathbf{y}]_1, \pi)\}$, $\mathcal{Q}_{\mathsf{tag}} := \mathcal{Q}_{\mathsf{tag}} \cup \{\tau\}$
Return π

$\underline{\text{FINALIZE}([\mathbf{y}^*]_1, \pi^*)\text{:}}$
Parse $\pi^* =: ([\mathbf{t}^*]_1, [u^*]_1, \pi^*_{\mathsf{or}})$
$\tau^* := H([\mathbf{y}^*]_1, [\mathbf{t}^*]_1, \pi^*_{\mathsf{or}}) \in \mathbb{Z}_p$
If $\tau^* \in \mathcal{Q}_{\mathsf{tag}}$ then return 0
If $[\mathbf{y}^*]_1 \in \mathcal{L}_{[\mathbf{M}]_1}$ or $([\mathbf{y}^*]_1, \pi^*) \in \mathcal{Q}_{\mathsf{sim}}$ then return 0
$[u'^*]_1 = [u^*]_1 - [\mathbf{y}^{*\top}(\mathbf{k}_0 + \tau^* \mathbf{k}_1)]_1$
Return $\text{FINALIZE}_{\mathsf{core}}([\mathbf{t}^*]_1, [u'^*]_1, \pi^*_{\mathsf{or}})$

Fig. 6. Reduction $\mathcal{B}$ for the proof of Lemma 6 with oracle $\text{INIT}_{\mathsf{core}}$, $\text{EVAL}_{\mathsf{core}}$, $\text{FINALIZE}_{\mathsf{core}}$ defined in Fig. 3. We highlight the oracle calls with grey.

We assume that $\mathcal{A}$ learns the term $\mathbf{t}^\top(\mathbf{k} + \mathbf{RF}(j^*))$, and thus $\mathbf{y}^\top(\mathbf{k}_0 + \tau \mathbf{k}_1)$ is also leaked to $\mathcal{A}$. However, since $\tau^* \neq \tau$, the terms $(\mathbf{k}_0 + \tau^* \mathbf{k}_1)$ and $(\mathbf{k}_0 + \tau \mathbf{k}_1)$ are pairwise independent.

Now together with the information leaked from $\mathbf{M}^\top \mathbf{k}_0$ and $\mathbf{M}^\top \mathbf{k}_1$ in crs, from $\mathcal{A}$'s view, the term $\mathbf{y}^{*\top}(\mathbf{k}_0 + \tau^* \mathbf{k}_1)$ is distributed uniformly at random, given $\mathbf{y}^\top(\mathbf{k}_0 + \tau \mathbf{k}_1)$ from the j^*-th SIM query ($[\mathbf{y}]_1$ may not be in $\mathcal{L}_{[\mathbf{M}]_1}$). Thus, $\mathcal{A}$ can compute the random term $\mathbf{y}^{*\top}(\mathbf{k}_0 + \tau^* \mathbf{k}_1)$ and make FINALIZE output 1 with probability at most $1/p$. By the union bound, $\mathcal{A}$ can win in G_2 with probability at most $(Q+1)/p$. □

From Lemmata 4 to 7, we have $\mathsf{Adv}^{\mathsf{uss}}_{\Pi^{\mathsf{dv}},\mathcal{A}}(\lambda) := \Pr[\mathsf{USS}^{\mathcal{A}}] \leq \mathsf{Adv}^{\mathsf{cr}}_{\mathcal{H},\hat{\mathcal{B}}}(\lambda) + \mathsf{Adv}^{\mathsf{core}}_{\mathcal{B}'}(\lambda) + \frac{(Q+1)}{p}$. By Lemma 3, we conclude Theorem 1 as

$$\begin{aligned}\mathsf{Adv}^{\mathsf{uss}}_{\Pi^{\mathsf{dv}},\mathcal{A}}(\lambda) \leq & \mathsf{Adv}^{\mathsf{cr}}_{\mathcal{H},\hat{\mathcal{B}}}(\lambda) + (4k\lceil\log Q\rceil + 2) \cdot \mathsf{Adv}^{\mathsf{mddh}}_{\mathbb{G}_1,\mathcal{D}_{2k,k},\mathcal{B}}(\lambda) \\ & + (2\lceil\log Q\rceil + 2) \cdot \mathsf{Adv}^{\mathsf{zk}}_{\mathsf{NIZK},\mathcal{B}'}(\lambda) + \lceil\log Q\rceil \cdot \Delta_{\mathcal{D}_{2k,k}} \\ & + \frac{4\lceil\log Q\rceil + 2}{p-1} + \frac{(\lceil\log Q\rceil + 1) \cdot Q + 1}{p}.\end{aligned}$$

□

3.2 QA-NIZK

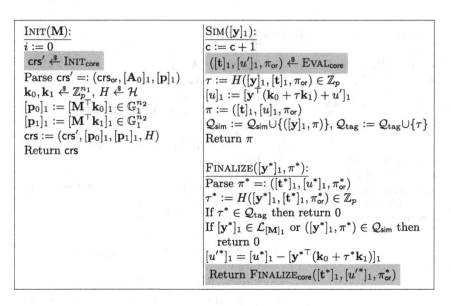

Let $\mathcal{G} \leftarrow \mathsf{GGen}(1^\lambda)$, $\mathsf{par} := \mathcal{G}$, $k \in \mathbb{N}$, $\mathcal{H}$ be a collision-resistant hash function family, and $\Pi^{\mathsf{or}} := (\mathsf{Gen}_{\mathsf{or}}, \mathsf{Prove}_{\mathsf{or}}, \mathsf{Ver}_{\mathsf{or}})$ be a NIZK system for language $\mathcal{L}^{\vee}_{\mathbf{A}_0,\mathbf{A}_1}$. Our (publicly verifiable) QANIZK $\Pi := (\mathsf{Gen}, \mathsf{Prove}, \mathsf{Ver}, \mathsf{Sim})$ is defined as in Fig. 7. The main idea behind our construction is to tightly compile the DVQANIZK Π^{dv} from Fig. 4 by using pairings. Again we note that our scheme can be easily extended to a tag-based scheme by putting the label ℓ inside the hash function. Thus, our scheme can be used in all the applications that require tag-based QANIZK.

$\underline{\mathsf{Gen}(\mathsf{par}, [\mathbf{M}]_1 \in \mathbb{G}_1^{n_1 \times n_2})}$:	$\underline{\mathsf{Prove}(\mathsf{crs}, [\mathbf{y}]_1, \mathbf{w})}$: $\quad /\!/ \mathbf{y} = \mathbf{M}\mathbf{w} \in \mathbb{Z}_p^{n_1}$
$\mathbf{A}_0, \mathbf{A}_1 \xleftarrow{\$} \mathcal{D}_{2k,k}$, $\mathbf{A} \xleftarrow{\$} \mathcal{D}_k$, $H \xleftarrow{\$} \mathcal{H}$ $\mathsf{par}_{\mathsf{or}} := (\mathsf{par}, [\mathbf{A}_0]_1, [\mathbf{A}_1]_1)$ $\mathsf{crs}_{\mathsf{or}} \leftarrow \mathsf{Gen}_{\mathsf{or}}(\mathsf{par}_{\mathsf{or}}, 1^\lambda)$ $\mathbf{K} \xleftarrow{\$} \mathbb{Z}_p^{2k \times (k+1)}$ $\mathbf{K}_0 \xleftarrow{\$} \mathbb{Z}_p^{n_1 \times (k+1)}$, $\mathbf{K}_1 \xleftarrow{\$} \mathbb{Z}_p^{n_1 \times (k+1)}$ $\mathbf{P} := \mathbf{A}_0^\top \mathbf{K} \in \mathbb{Z}_p^{k \times (k+1)}$ $[\mathbf{P}_0]_1 := [\mathbf{M}^\top \mathbf{K}_0]_1 \in \mathbb{G}_1^{n_2 \times (k+1)}$ $[\mathbf{P}_1]_1 := [\mathbf{M}^\top \mathbf{K}_1]_1 \in \mathbb{G}_1^{n_2 \times (k+1)}$ $\mathbf{C} := \mathbf{K}\mathbf{A} \in \mathbb{Z}_p^{2k \times k}$ $\mathbf{C}_0 := \mathbf{K}_0\mathbf{A} \in \mathbb{Z}_p^{n_1 \times k}$ $\mathbf{C}_1 := \mathbf{K}_1\mathbf{A} \in \mathbb{Z}_p^{n_1 \times k}$ $\mathsf{crs} := (\mathsf{crs}_{\mathsf{or}}, [\mathbf{A}_0]_1, [\mathbf{P}]_1, [\mathbf{P}_0]_1, [\mathbf{P}_1]_1,$ $\quad [\mathbf{A}]_2, [\mathbf{C}]_2, [\mathbf{C}_0]_2, [\mathbf{C}_1]_2, H)$ $\mathsf{td} := (\mathbf{K}_0, \mathbf{K}_1)$ Return $(\mathsf{crs}, \mathsf{td})$	$\mathbf{s} \xleftarrow{\$} \mathbb{Z}_p^k$, $[\mathbf{t}]_1 := [\mathbf{A}_0]_1\mathbf{s}$ $\pi_{\mathsf{or}} \xleftarrow{\$} \mathsf{Prove}_{\mathsf{or}}(\mathsf{crs}_{\mathsf{or}}, [\mathbf{t}]_1, \mathbf{s})$ $\tau := H([\mathbf{y}]_1, [\mathbf{t}]_1, \pi_{\mathsf{or}}) \in \mathbb{Z}_p$ $[\mathbf{u}]_1 := \mathbf{w}^\top([\mathbf{P}_0]_1 + \tau[\mathbf{P}_1]_1) + \mathbf{s}^\top[\mathbf{P}]_1 \in \mathbb{G}_1^{1 \times (k+1)}$ Return $\pi := ([\mathbf{t}]_1, [\mathbf{u}]_1, \pi_{\mathsf{or}})$ $\underline{\mathsf{Ver}(\mathsf{crs}, [\mathbf{y}]_1, \pi)}$: Parse $\pi = ([\mathbf{t}]_1, [\mathbf{u}]_1, \pi_{\mathsf{or}})$ $\tau := H([\mathbf{y}]_1, [\mathbf{t}]_1, \pi_{\mathsf{or}}) \in \mathbb{Z}_p$ If $\mathsf{Ver}_{\mathsf{or}}(\mathsf{crs}_{\mathsf{or}}, [\mathbf{t}]_1, \pi_{\mathsf{or}}) = 0$ then return 0 If $[\mathbf{u}]_1 \circ [\mathbf{A}]_2 = [\mathbf{y}^\top]_1 \circ [\mathbf{C}_0 + \tau\mathbf{C}_1]_2 + [\mathbf{t}^\top]_1 \circ [\mathbf{C}]_2$ then $\quad$ return 1 Else return 0 $\underline{\mathsf{Sim}(\mathsf{crs}, \mathsf{td}, [\mathbf{y}]_1)}$: $\mathbf{s} \xleftarrow{\$} \mathbb{Z}_p^k$, $\mathbf{t} := \mathbf{A}_0\mathbf{s}$ $\pi_{\mathsf{or}} \xleftarrow{\$} \mathsf{Prove}_{\mathsf{or}}(\mathsf{crs}_{\mathsf{or}}, [\mathbf{t}]_1, \mathbf{s})$ $\tau := H([\mathbf{y}]_1, [\mathbf{t}]_1, \pi_{\mathsf{or}}) \in \mathbb{Z}_p$ $[\mathbf{u}]_1 := [\mathbf{y}^\top(\mathbf{K}_0 + \tau\mathbf{K}_1)]_1 + [\mathbf{s}^\top\mathbf{P}]_1$ Return $\pi := ([\mathbf{t}]_1, [\mathbf{u}]_1, \pi_{\mathsf{or}})$

Fig. 7. Construction of Π.

Theorem 2 (Security of Π). *Π defined in Fig. 7 is a* QANIZK *with perfect zero-knowledge and (tight) unbounded simulation soundness if the $\mathcal{D}_k$-*KerMDH *assumption holds in $\mathbb{G}_2$ and the* DVQANIZK *Π^{dv} in Fig. 4 is unbounded simulation sound. In particular, for any adversary $\mathcal{A}$, there exist adversaries $\mathcal{B}$ and $\mathcal{B}'$ with $\mathsf{T}(\mathcal{B}) \approx \mathsf{T}(\mathcal{B}') \approx \mathsf{T}(\mathcal{A}) + Q \cdot \mathsf{poly}(\lambda)$, where Q is the number of queries to* Sim, poly *is independent of Q and*

$$\mathsf{Adv}^{\mathsf{uss}}_{\Pi,\mathcal{A}}(\lambda) \leq \mathsf{Adv}^{\mathsf{kmdh}}_{\mathbb{G}_1,\mathcal{D}_k,\mathcal{B}}(\lambda) + \mathsf{Adv}^{\mathsf{uss}}_{\Pi^{\mathsf{dv}},\mathcal{B}'}(\lambda).$$

Proof (of Theorem 2). Perfect completeness follows directly from the completeness of the OR proof system and the fact that for all $\mathbf{P} := \mathbf{A}_0^\top\mathbf{K}$, $\mathbf{P}_0 := \mathbf{M}^\top\mathbf{K}_0$, $\mathbf{P}_1 := \mathbf{M}^\top\mathbf{K}_1$, $\mathbf{C} := \mathbf{K}\mathbf{A}$, $\mathbf{C}_0 := \mathbf{K}_0\mathbf{A}$, $\mathbf{C}_1 := \mathbf{K}_1\mathbf{A}$, and any τ

$$\begin{aligned}
&[\mathbf{w}^\top(\mathbf{P}_0+\tau\mathbf{P}_1)+\mathbf{s}^\top\mathbf{P}]_1\circ[\mathbf{A}]_2\\
=&[\mathbf{w}^\top(\mathbf{M}^\top\mathbf{K}_0+\tau\mathbf{M}^\top\mathbf{K}_1)+\mathbf{s}^\top\mathbf{A}_0^\top\mathbf{K}]_1\circ[\mathbf{A}]_2\\
=&[\mathbf{w}^\top\mathbf{M}^\top]_1\circ[\mathbf{K}_0\mathbf{A}+\tau\mathbf{K}_1\mathbf{A}]_2+[\mathbf{s}^\top\mathbf{A}_0^\top]_1\circ[\mathbf{K}\mathbf{A}]_2\\
=&[\mathbf{y}^\top]_1\circ[\mathbf{C}_0+\tau\mathbf{C}_1]_2+[\mathbf{t}^\top]_1\circ[\mathbf{C}]_2.
\end{aligned}$$

Moreover, since

$$\begin{aligned}
\mathbf{w}^\top(\mathbf{P}_0+\tau\mathbf{P}_1)+\mathbf{s}^\top\mathbf{P}&=\mathbf{w}^\top(\mathbf{M}^\top\mathbf{K}_0+\tau\mathbf{M}^\top\mathbf{K}_1)+\mathbf{s}^\top\mathbf{P}\\
&=\mathbf{y}^\top(\mathbf{K}_0+\tau\mathbf{K}_1)+\mathbf{s}^\top\mathbf{P},
\end{aligned}$$

the output of Prove is identical to that of Sim for the same $\mathbf{y}=\mathbf{Mw}$. Hence, perfect zero knowledge is also satisfied.

We now focus on the tight simulation soundness of Π. We prove it by a sequence of games: G_0 is defined as the real experiment, USS (we omit the description here), G_1 and G_2 are defined as in Fig. 8.

$\underline{\text{INIT}(\mathbf{M})}$:
$\mathbf{A}_0,\mathbf{A}_1 \xleftarrow{\$} \mathcal{D}_{2k,k}$, $\mathbf{A} \xleftarrow{\$} \mathcal{D}_k$, $H \xleftarrow{\$} \mathcal{H}$
$\boxed{\mathbf{a}^\perp \xleftarrow{\$} \mathsf{ker}(\mathbf{A})}$
$\boxed{/\!/ \mathbf{a}^\perp \in \mathbb{Z}_p^{1\times(k+1)} \text{ and } \mathbf{a}^\perp\cdot\mathbf{A}=\mathbf{0}}$
$\mathsf{par}_{\mathsf{or}} := (\mathsf{par}, [\mathbf{A}_0]_1, [\mathbf{A}_1]_1)$
$\mathsf{crs}_{\mathsf{or}} \leftarrow \mathsf{Gen}_{\mathsf{or}}(\mathsf{par}_{\mathsf{or}}, 1^\lambda)$
$\mathbf{K}' \xleftarrow{\$} \mathbb{Z}_p^{2k\times(k+1)}$
$\mathbf{K}_0' \xleftarrow{\$} \mathbb{Z}_p^{n_1\times(k+1)}$, $\mathbf{K}_1' \xleftarrow{\$} \mathbb{Z}_p^{n_1\times(k+1)}$
$\mathbf{k}_0=\mathbf{k}_1 := \mathbf{0}\in\mathbb{Z}_p^{n_1}$, $\mathbf{k} := \mathbf{0}\in\mathbb{Z}_p^{2k}$
$\boxed{\mathbf{k}_0,\mathbf{k}_1 \xleftarrow{\$} \mathbb{Z}_p^{n_1},\ \mathbf{k} \xleftarrow{\$} \mathbb{Z}_p^{2k}}$
$\mathbf{K} := \mathbf{K}' + \mathbf{k}\cdot\mathbf{a}^\perp$
$\mathbf{K}_0 := \mathbf{K}_0' + \mathbf{k}_0'\cdot\mathbf{a}^\perp$
$\mathbf{K}_1 := \mathbf{K}_1' + \mathbf{k}_1'\cdot\mathbf{a}^\perp$
$\mathbf{P} := \mathbf{A}_0^\top\mathbf{K} \in \mathbb{Z}_p^{k\times(k+1)}$
$[\mathbf{P}_0]_1 := [\mathbf{M}^\top\mathbf{K}_0]_1 \in \mathbb{G}_1^{n_2\times(k+1)}$
$[\mathbf{P}_1]_1 := [\mathbf{M}^\top\mathbf{K}_1]_1 \in \mathbb{G}_1^{n_2\times(k+1)}$
$\mathbf{C} := \mathbf{K}\mathbf{A} \in \mathbb{Z}_p^{2k\times k}$
$\mathbf{C}_0 := \mathbf{K}_0\mathbf{A} \in \mathbb{Z}_p^{n_1\times k}$
$\mathbf{C}_1 := \mathbf{K}_1\mathbf{A} \in \mathbb{Z}_p^{n_1\times k}$
$\mathsf{crs} := (\mathsf{crs}_{\mathsf{or}}, [\mathbf{A}_0]_1, [\mathbf{P}]_1, [\mathbf{P}_0]_1, [\mathbf{P}_1]_1$
$\quad [\mathbf{A}]_2, [\mathbf{C}]_2, [\mathbf{C}_0]_2, [\mathbf{C}_1]_2, H)$
Return crs

$\underline{\text{SIM}([\mathbf{y}]_1)}$:
$\mathbf{s} \xleftarrow{\$} \mathbb{Z}_p^k$, $[\mathbf{t}]_1 := [\mathbf{A}_0]_1\mathbf{s}$
$\pi_{\mathsf{or}} \xleftarrow{\$} \mathsf{Prove}_{\mathsf{or}}(\mathsf{crs}_{\mathsf{or}}, [\mathbf{t}]_1, \mathbf{s})$
$\tau := H([\mathbf{y}]_1, [\mathbf{t}]_1, \pi_{\mathsf{or}}) \in \mathbb{Z}_p$
$[\Delta]_1 := [(\mathbf{y}^\top(\mathbf{k}_0+\tau\mathbf{k}_1)+\mathbf{t}^\top\mathbf{k})\cdot\mathbf{a}^\perp]_1$
$[\mathbf{u}]_1 := [\mathbf{y}^\top(\mathbf{K}_0'+\tau\mathbf{K}_1')+\mathbf{t}^\top\mathbf{K}'+\Delta]_1$
$\mathcal{Q}_{\mathsf{sim}} := \mathcal{Q}_{\mathsf{sim}} \cup \{([\mathbf{y}]_1, \pi)\}$
Return $\pi := ([\mathbf{t}]_1, [\mathbf{u}]_1, \pi_{\mathsf{or}})$

$\underline{\text{FINALIZE}([\mathbf{y}^*]_1, \pi^*)}$:
Parse $\pi = ([\mathbf{t}^*]_1, [\mathbf{u}^*]_1, \pi^*_{\mathsf{or}})$
$\tau^* := H([\mathbf{y}^*]_1, [\mathbf{t}^*]_1, \pi^*_{\mathsf{or}}) \in \mathbb{Z}_p$
If $([\mathbf{y}^*]_1, \pi^*) \in \mathcal{Q}_{\mathsf{sim}}$ or $[\mathbf{y}^*]_1 \in \mathcal{L}_{[\mathbf{M}]_1}$ or
$\mathsf{Ver}(\mathsf{crs}, [\mathbf{y}^*]_1, \pi^*) = 0$ then
  return 0
$[\Delta^*]_1 := [(\mathbf{y}^{*\top}(\mathbf{k}_0+\tau^*\mathbf{k}_1)+\mathbf{t}^{*\top}\mathbf{k})\cdot\mathbf{a}^\perp]_1$
If $[\mathbf{u}^*]_1 = [\mathbf{y}^{*\top}(\mathbf{K}_0'+\tau^*\mathbf{K}_1')+\mathbf{t}^{*\top}\mathbf{K}'+\Delta^*]_1$
then
  return 1
Else return 0

Fig. 8. Games G_1 and $\boxed{\mathsf{G}_2}$ for proving Theorem 2.

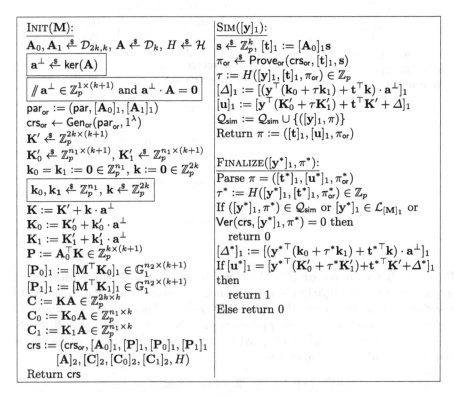

Lemma 8 (G_0). $\Pr[\mathsf{USS}^{\mathcal{A}} \Rightarrow 1] = \Pr[\mathsf{G}_0^{\mathcal{A}} \Rightarrow 1]$.

In G_1, FINALIZE additionally verifies the adversarial forgery with secret keys $\mathbf{K}$, $\mathbf{K}_0$, and $\mathbf{K}_1$ as in Fig. 8.

Lemma 9 (G_0 *to* G_1). *There is an adversary $\mathcal{B}$ breaking the $\mathcal{D}_k$-KerMDH assumption over $\mathbb{G}_2$ with $\mathsf{T}(\mathcal{B}) \approx \mathsf{T}(\mathcal{A}) + Q \cdot \mathsf{poly}(\lambda)$ and $\mathsf{Adv}^{\mathsf{kmdh}}_{\mathbb{G}_2,\mathcal{D}_k,\mathcal{B}}(\lambda) \geq |\Pr[\mathsf{G}_0^{\mathcal{A}} \Rightarrow 1] - \Pr[\mathsf{G}_1^{\mathcal{A}} \Rightarrow 1]|$.*

Proof. It is straightforward that a pair $([\mathbf{y}^*]_1, \pi^*)$ passing the FINALIZE in G_1 always passes the FINALIZE in G_0. We now bound the probability that $\mathcal{A}$ produces $([\mathbf{y}^*]_1, \pi^*)$ that passes the verification in G_0 but not that in G_1. For $\pi^* = ([\mathbf{t}^*]_1, [\mathbf{u}^*]_1, \pi^*_{\mathsf{or}})$, the verification equation in G_0 is:

$$[\mathbf{u}^*]_1 \circ [\mathbf{A}]_2 = [\mathbf{y}^{*\top}]_1 \circ [\mathbf{K}_0\mathbf{A} + \tau\mathbf{K}_1\mathbf{A}]_2 + [\mathbf{t}^\top]_1 \circ [\mathbf{K}\mathbf{A}]_2$$
$$\Leftrightarrow [\mathbf{u}^* - \mathbf{y}^{*\top}(\mathbf{K}_0 + \tau\mathbf{K}_1) - \mathbf{t}^\top\mathbf{K}]_1 \circ [\mathbf{A}]_2 = [\mathbf{0}]_T.$$

One can see that for any $([\mathbf{t}^*]_1, [\mathbf{u}^*]_1, \pi^*_{\mathsf{or}})$ that passes the verification equation in G_0 but not that in G_1, $\mathbf{u}^* - \mathbf{y}^*(\mathbf{K}_0 + \tau\mathbf{K}_1) - \mathbf{t}^\top\mathbf{K}$ is a non-zero vector in the kernel of $\mathbf{A}$.

We now construct an adversary $\mathcal{B}$ as follows. On receiving $(\mathcal{G}, [\mathbf{A}]_1)$ from the $\mathcal{D}_k$-KerMDH experiment, $\mathcal{B}$ samples all other parameters by itself and simulates G_0 for $\mathcal{A}$. When $\mathcal{A}$ outputs a tuple $([\mathbf{t}^*]_1, [\mathbf{u}^*]_1, \pi^*_{\mathsf{or}})$, $\mathcal{B}$ outputs $\mathbf{u}^* - \mathbf{y}^{*\top}(\mathbf{K}_0 + \tau\mathbf{K}_1) - \mathbf{t}^\top\mathbf{K}$. Since $\mathcal{B}$ succeeds in its experiment when $\mathcal{A}$ outputs a tuple such that $\mathbf{u}^* - \mathbf{y}^{*\top}(\mathbf{K}_0 + \tau\mathbf{K}_1) - \mathbf{t}^\top\mathbf{K}$ is a non-zero vector in the kernel of $\mathbf{A}$, we have $\mathsf{Adv}^{\mathsf{kmdh}}_{\mathbb{G}_1,\mathcal{D}_k,\mathcal{B}}(\lambda) \geq |\Pr[\mathsf{G}_0^{\mathcal{A}} \Rightarrow 1] - \Pr[\mathsf{G}_1^{\mathcal{A}} \Rightarrow 1]|$, completing the proof of this lemma. □

Lemma 10 (G_1 *to* G_2). $\Pr[\mathsf{G}_1^{\mathcal{A}} \Rightarrow 1] = \Pr[\mathsf{G}_2^{\mathcal{A}} \Rightarrow 1]$.

Proof. Now we finish the reduction to the KerMDH assumption and we can have $\mathbf{A}$ over $\mathbb{Z}_p$. In G_2, for $i \in \{0,1\}$ we replace $\mathbf{K}_i$ by $\mathbf{K}'_i + \mathbf{k}_i\mathbf{a}^\perp$ for $\mathbf{a}^\perp \in \ker(\mathbf{A})$, where $\mathbf{K}'_i \xleftarrow{\$} \mathbb{Z}_p^{n_1 \times (k+1)}$, and $\mathbf{k}_i \xleftarrow{\$} \mathbb{Z}_p^{n_1}$. Furthermore, we replace $\mathbf{K}$ by $\mathbf{K}' + \mathbf{k}\mathbf{a}^\perp$ for $\mathbf{K}' \xleftarrow{\$} \mathbb{Z}_p^{2k \times (k+1)}$ and $\mathbf{k} \xleftarrow{\$} \mathbb{Z}_p^{2k}$. Since $\mathbf{K}'$ and $\mathbf{K}'_i$ are uniformly random, $\mathbf{K}$ and $\mathbf{K}_i$ in G_2 are distributed at random and the same as in G_1. Thus, G_2 is distributed the same as G_1. □

Lemma 11 (G_2). *There is an adversary $\mathcal{B}'$ breaking the USS security of Π^{dv} defined in Fig. 4 with $\mathsf{T}(\mathcal{B}') \approx \mathsf{T}(\mathcal{A}) + Q \cdot \mathsf{poly}(\lambda)$ and $\Pr[\mathsf{G}_2^{\mathcal{A}} \Rightarrow 1] \leq \mathsf{Adv}^{\mathsf{uss}}_{\Pi^{\mathsf{dv}},\mathcal{B}'}(\lambda)$.*

Proof. We construct a reduction $\mathcal{B}'$ in Fig. 9 to break the USS security of Π^{dv} defined in Fig. 4.

We note that the $[\mathbf{p}]_1, [\mathbf{p}_i]_1$ $(i = 0, 1)$ from $\text{INIT}_{\mathsf{dv}}$ have the forms, $\mathbf{p} = \mathbf{A}_0^\top\mathbf{k}$ and $\mathbf{p}_i = \mathbf{M}^\top\mathbf{k}_i$ for some random $\mathbf{k} \in \mathbb{Z}_p^{2k}$ and $\mathbf{k}_i \in \mathbb{Z}_p^{n_1}$, and furthermore the value $[u]_1$ from SIM_{dv} has the form $u = \mathbf{y}^\top(\mathbf{k}_0 + \tau\mathbf{k}_1) + \mathbf{t}^\top\mathbf{k}$. Hence, essentially, $\mathcal{B}'$ simulate the security game with $\mathbf{K}$ and $\mathbf{K}_i$ that are implicitly defined as $\mathbf{K} := \mathbf{K}' + \mathbf{k} \cdot \mathbf{a}^\perp$ and $\mathbf{K}_i := \mathbf{K}'_i + \mathbf{k}_i \cdot \mathbf{a}^\perp$. The simulated INIT and SIM are identical to those in G_2.

$\underline{\text{INIT}(\mathbf{M})}$:	$\underline{\text{SIM}([\mathbf{y}]_1)}$:
$\mathbf{A} \xleftarrow{\$} \mathcal{D}_k$	$([\mathbf{t}]_1, [u]_1, \pi_{\mathsf{or}}) \xleftarrow{\$} \text{SIM}_{\mathsf{dv}}([\mathbf{y}]_1)$
$\mathbf{a}^\perp \xleftarrow{\$} \mathsf{ker}(\mathbf{A})$	$[\Delta]_1 := [u]_1 \cdot \mathbf{a}^\perp$
$/\!\!/ \mathbf{a}^\perp \in \mathbb{Z}_p^{1\times(k+1)}$ and $\mathbf{a}^\perp\mathbf{A} = \mathbf{0}$	$[\mathbf{u}]_1 := [\mathbf{y}^\top(\mathbf{K}_0' + \tau\mathbf{K}_1') + \mathbf{t}^\top\mathbf{K}' + \Delta]_1$
$\mathsf{crs}_{\mathsf{dv}} \xleftarrow{\$} \text{INIT}_{\mathsf{dv}}(\mathbf{M})$	$\mathcal{Q}_{\mathsf{sim}} := \mathcal{Q}_{\mathsf{sim}} \cup \{([\mathbf{y}]_1, \pi)\}$
Parse $\mathsf{crs}_{\mathsf{dv}} =: (\mathsf{crs}_{\mathsf{or}}, [\mathbf{A}_0]_1, [\mathbf{p}]_1, [\mathbf{p}_0]_1, [\mathbf{p}_1]_1, H)$	Return $\pi := ([\mathbf{t}]_1, [\mathbf{u}]_1, \pi_{\mathsf{or}})$
$\mathbf{K}' \xleftarrow{\$} \mathbb{Z}_p^{2k\times(k+1)}$	
$\mathbf{K}_0' \xleftarrow{\$} \mathbb{Z}_p^{n_1\times(k+1)}$, $\mathbf{K}_1' \xleftarrow{\$} \mathbb{Z}_p^{n_1\times(k+1)}$	$\underline{\text{FINALIZE}([\mathbf{y}^*]_1, \pi^*)}$:
$[\mathbf{P}]_1 := [\mathbf{A}_0]_1^\top\mathbf{K}' + [\mathbf{p}]_1\mathbf{a}^\perp$	Parse $\pi = ([\mathbf{t}^*]_1, [\mathbf{u}^*]_1, \pi_{\mathsf{or}}^*)$
$[\mathbf{P}_0]_1 := [\mathbf{M}]_1^\top\mathbf{K}_0' + [\mathbf{p}_0]_1\mathbf{a}^\perp$	If $([\mathbf{y}^*]_1, \pi^*) \in \mathcal{Q}_{\mathsf{sim}}$ or $[\mathbf{y}^*]_1 \in \mathcal{L}_{[\mathbf{M}]_1}$ or
$[\mathbf{P}_1]_1 := [\mathbf{M}]_1^\top\mathbf{K}_1' + [\mathbf{p}_1]_1\mathbf{a}^\perp$	$\mathsf{Ver}(\mathsf{crs}, [\mathbf{y}^*]_1, \pi^*) = 0$ then return 0
$\mathbf{C} := \mathbf{K}'\mathbf{A} \in \mathbb{Z}_p^{2k\times k}$	Compute $[v]_1$ such that
$\mathbf{C}_0 := \mathbf{K}_0'\mathbf{A} \in \mathbb{Z}_p^{n_1\times k}$	$[v]_1\mathbf{a}^\perp = [\mathbf{u}^* - \mathbf{y}^{*\top}(\mathbf{K}_0' + \tau^*\mathbf{K}_1') - \mathbf{t}^{*\top}\mathbf{K}']_1$
$\mathbf{C}_1 := \mathbf{K}_1'\mathbf{A} \in \mathbb{Z}_p^{n_1\times k}$	Return $\text{FINALIZE}_{\mathsf{dv}}([\mathbf{y}^*]_1, ([\mathbf{t}^*]_1, [v]_1, \pi_{\mathsf{or}}^*))$
$\mathsf{crs} := (\mathsf{crs}_{\mathsf{or}}, [\mathbf{A}_0]_1, [\mathbf{P}]_1, [\mathbf{P}_0]_1, [\mathbf{P}_1]_1, [\mathbf{A}]_2, [\mathbf{C}]_2, [\mathbf{C}_0]_2, [\mathbf{C}_1]_2, H)$	
Return crs	

Fig. 9. Reduction $\mathcal{B}'$ for the proof of Lemma 11 with oracle access to $\text{INIT}_{\mathsf{dv}}$, SIM_{dv} and $\text{FINALIZE}_{\mathsf{dv}}$ as defined in G_0 of Fig. 5. We highlight the oracle calls with grey.

In G_2, $\text{FINALIZE}([\mathbf{y}^*]_1, \pi^* := ([\mathbf{t}^*]_1, [\mathbf{u}^*]_1, \pi_{\mathsf{or}}^*))$ outputs 1 if

$$\mathbf{u}^* = \mathbf{y}^{*\top}(\mathbf{K}_0' + \tau^*\mathbf{K}_1') + \mathbf{t}^{*\top}\mathbf{K}' + \underbrace{(\mathbf{y}^{*\top}(\mathbf{k}_0 + \tau^*\mathbf{k}_1) + \mathbf{t}^{*\top}\mathbf{k})}_{=:v} \cdot \mathbf{a}^\perp$$

and $([\mathbf{y}^*]_1, \pi^*) \notin \mathcal{Q}_{\mathsf{sim}}$ and $[\mathbf{y}^*]_1 \notin \mathcal{L}_{[\mathbf{M}]_1}$ and $\mathsf{Ver}(\mathsf{crs}, [\mathbf{y}^*]_1, \pi^*) = 1$. Thus, if $\mathcal{A}$ can make $\text{FINALIZE}([\mathbf{y}^*]_1, \pi^*)$ output 1 then $\mathcal{B}'$ can extract the corresponding $[v]_1$ to break the USS security. We conclude the lemma. □

To sum up, we have $\Pr[\mathsf{USS}^{\mathcal{A}} \Rightarrow 1] \leq \mathsf{Adv}^{\mathsf{kmdh}}_{\mathbb{G}_1,\mathcal{D}_k,\mathcal{B}}(\lambda) + \mathsf{Adv}^{\mathsf{uss}}_{\Pi^{\mathsf{dv}},\mathcal{B}'}(\lambda)$ with $\mathcal{B}$ and $\mathcal{B}'$ as defined above. □

3.3 Application: Tightly IND-mCCA-Secure PKE

By instantiating the labeled (enhanced) USS-QA-NIZK in the generic construction in [5] with our construction in Sect. 3.2, we immediately obtain a more efficient publicly verifiable labeled public-key encryption (PKE) with tight IND-CCA2 security in the multi-user, multi-challenge setting (IND-mCCA). The security reduction is independent of the number of decryption-oracle requests of the CCA2 adversary. We refer the reader to the full paper for the definition of labeled IND-mCCA secure PKE and the construction.

4 Tightly Secure Structure-Preserving Signature

In this section, we present an SPS via a designated-prover NIZK for the OR-language, whose security can be tightly reduced to the $\mathcal{D}_{2k,k}$-MDDH and $\mathcal{D}_k$-MDDH assumptions.

4.1 Designated-Prover OR-Proof

In this section, we construct NIZKs in the designated-prover setting. In contrast to [5], we focus on the language $\mathcal{L}^{\vee}_{\mathbf{A}_0,\mathbf{A}_1}$ defined in Sect. 2.3, where a single word $\mathbf{y}$ is required to be in the linear span of either one of two spaces given by matrices $\mathbf{A}_0$ and $\mathbf{A}_1$.

While previous techniques [23,43] require ten group elements in a proof, our novel solution gives a QANIZK with only seven group elements under the SXDH hardness assumption, by leveraging the privacy of the prover CRS.

DEFINITION. For $\mathbf{A}_0, \mathbf{A}_1 \xleftarrow{\$} \mathcal{D}_{2k,k}$, we define the notion of designated-prover OR-proof for $\mathcal{L}^{\vee}_{\mathbf{A}_0,\mathbf{A}_1}$.

Definition 11 (Designated-Prover OR-Proof). *A designated-prover proof system for $\mathcal{L}^{\vee}_{\mathbf{A}_0,\mathbf{A}_1}$ is the same as that of* NIZK *for $\mathcal{L}^{\vee}_{\mathbf{A}_0,\mathbf{A}_1}$ (see Sect. 2.3), except that*

- Gen *takes* $(\mathsf{par}, \mathbf{A}_0, \mathbf{A}_1)$ *as input instead of* $(\mathsf{par}, [\mathbf{A}_0]_1, [\mathbf{A}_1]_1)$ *and outputs an additional prover key* prk.
- Prove *takes* prk *as additional input.*
- *In the soundness definition, the Adversary is given oracle access to* Prove *with* prk *instantiated by the one output by* Gen.

CONSTRUCTION. Let $\mathcal{G} \leftarrow \mathsf{GGen}(1^\lambda)$, $\mathsf{par} := \mathcal{G}$, and $k \in \mathbb{N}$. In Fig. 10 we present a Designated-Prover OR-proof system for $\mathcal{L}^{\vee}_{\mathbf{A}_0,\mathbf{A}_1}$.

Lemma 12. *If the $\mathcal{D}_k$-*MDDH *assumption holds in the group $\mathbb{G}_2$, then the proof system $\Pi^{\mathsf{or}} = (\mathsf{Gen}_{\mathsf{or}}, \mathsf{TGen}_{\mathsf{or}}, \mathsf{Prove}_{\mathsf{or}}, \mathsf{Ver}_{\mathsf{or}}, \mathsf{Sim}_{\mathsf{or}})$ as defined in Fig. 10 is a designated-prover or-proof system for $\mathcal{L}^{\vee}_{\mathbf{A}_0,\mathbf{A}_1}$ with perfect completeness, perfect soundness, and zero-knowledge. More precisely, for all adversaries $\mathcal{A}$ attacking the zero-knowledge property of Π^{or}, we obtain an adversary $\mathcal{B}$ with $T(\mathcal{B}) \approx T(\mathcal{A}) + Q \cdot \mathsf{poly}(\lambda)$ and $\mathsf{Adv}^{\mathsf{zk}}_{\Pi^{\mathsf{or}},\mathcal{A}}(\lambda) \leq \mathsf{Adv}^{\mathsf{mddh}}_{\mathcal{G},\mathbb{G}_2,\mathcal{D}_k,\mathcal{B}}(\lambda)$.*

We refer the reader to Introduction for the high-level idea of our construction. We refer the reader to the full paper for the full proof.

EXTENSIONS. For larger matrices $\mathbf{A}_0, \mathbf{A}_1$, and under $\mathcal{D}_k$-MDDH assumption for a fixed k, we improve our proof size so that it asymptotically approaches a factor of two. As a trade-off, it loses a factor of k.

Roughly, for some invertible matrix $\mathbf{U}$, we exploit the following language instead:

$$\mathcal{L}^{\vee}_{\mathbf{A}_0,\mathbf{A}_1} := \{[\mathbf{y}]_1 \in \mathbb{G}_1^{2k} \mid \exists \mathbf{x} \in \mathbb{Z}_p^{1\times k}, \mathbf{X} \in \mathbb{Z}_p^{k\times k} : \mathbf{A}_0\mathbf{X} = \mathbf{y}\mathbf{x} \vee \mathbf{y}^\top \mathbf{A}_1^\perp \mathbf{U} = \mathbf{x}\}.$$

One can see that it is also equal to $\mathcal{L}^{\vee}_{\mathbf{A}_0,\mathbf{A}_1}$, since $\mathbf{y}$ is in the span of $\mathbf{A}_0$ if $\mathbf{x} \neq \mathbf{0}$ and in the span of $\mathbf{A}_1$ otherwise. Instead of directly applying the Groth-Sahai proof to it as before, we make careful adjustment on the proof for $[\mathbf{y}]_1^\top \mathbf{A}_1^\perp \mathbf{U} = [\mathbf{x}]_1$ and commitment of the information on $\mathbf{A}_1^\perp$ in this case. We also extend it to an efficient OR-Proof in the symmetric pairing, which might be of independent interest. We refer the reader to the full paper for the constructions and security proofs.

$\mathsf{Gen}_{\mathsf{or}}(\mathsf{par}, \mathbf{A}_0 \in \mathbb{Z}_p^{2k\times k}, \mathbf{A}_1 \in \mathbb{Z}_p^{2k\times k})$:

$\mathbf{V} \xleftarrow{\$} \mathcal{D}_k\ \mathbf{u} \xleftarrow{\$} \mathbb{Z}_p^{k+1} \setminus \mathsf{Span}(\mathbf{V})$

$\begin{pmatrix}\mathbf{d}_1\\ \vdots \\ \mathbf{d}_k\end{pmatrix} := \underline{\mathbf{A}_1}\overline{\mathbf{A}_1}^{-1} \in \mathbb{Z}_p^{k\times k}$

For $i = 1, \cdots, k$:

$\quad \mathbf{S}_i \xleftarrow{\$} \mathbb{Z}_p^{k\times k}, \mathbf{D}_i := \mathbf{d}_i^\top \mathbf{u}^\top + \mathbf{S}_i\mathbf{V}^\top$

$\mathsf{crs}_{\mathsf{or}} := (\mathsf{par}, [\mathbf{A}_0]_1, [\mathbf{A}_1]_1, [\mathbf{u}]_2, [\mathbf{V}]_2, ([\mathbf{D}_i]_2)_{1\le i\le k})$

$\mathsf{sk}_{\mathsf{or}} := (\mathbf{A}_0, \mathbf{A}_1, (\mathbf{S}_i)_{1\le i\le k})$

Return $(\mathsf{crs}_{\mathsf{or}}, \mathsf{sk}_{\mathsf{or}})$

$\mathsf{Prove}_{\mathsf{or}}(\mathsf{crs}_{\mathsf{or}}, \mathsf{sk}_{\mathsf{or}}, [\mathbf{y}]_1, \mathbf{r})$:

Parse $\mathsf{sk}_{\mathsf{or}} =: (\mathbf{A}_0, \mathbf{A}_1, (\mathbf{S}_i)_{1\le i\le k})$

If $\neg(\exists j \in \{0,1\} : [\mathbf{y}]_1 = [\mathbf{A}_j\mathbf{r}]_1)$ then abort

$\mathbf{d} := \begin{pmatrix}\mathbf{d}_1\\ \vdots \\ \mathbf{d}_k\end{pmatrix} := \underline{\mathbf{A}_1}\overline{\mathbf{A}_1}^{-1} \in \mathbb{Z}_p^{k\times k}$

$(x_1, \cdots, x_k) := \overline{\mathbf{y}}^\top \mathbf{d}^\top - \underline{\mathbf{y}}^\top \in \mathbb{Z}_p^{1\times k}$

$(\mathbf{x}_1, \cdots, \mathbf{x}_k) := \mathbf{r}(x_1, \cdots, x_k) \in \mathbb{Z}_p^{k\times k}$

For $i = 1, \cdots, k$:

$\quad \mathbf{R}_i \xleftarrow{\$} \mathbb{Z}_p^{k\times k}$

$\quad [\mathbf{C}_i]_2 := \mathbf{x}_i[\mathbf{u}^\top]_2 + \mathbf{R}_i[\mathbf{V}^\top]_2$

$\quad \mathbf{r}_i \xleftarrow{\$} \mathbb{Z}_p^{1\times k}$

$\quad [\mathbf{c}_i]_2 := x_i[\mathbf{u}^\top]_2 + \mathbf{r}_i[\mathbf{V}^\top]_2$

$\quad \mathbf{\Pi}_i := \mathbf{A}_0\mathbf{R}_i - \mathbf{y}\mathbf{r}_i$

$\quad \pi_i := \overline{\mathbf{y}}^\top \mathbf{S}_i - \mathbf{r}_i$

Return $(([\mathbf{C}_i, \mathbf{c}_i]_2, [\mathbf{\Pi}_i, \pi_i]_1)_{1\le i\le k})$

$\mathsf{TGen}_{\mathsf{or}}(\mathsf{par}, [\mathbf{A}_0]_1, [\mathbf{A}_1]_1)$:

$\mathbf{V} \xleftarrow{\$} \mathcal{D}_k, \mathbf{z} \leftarrow \mathbb{Z}_p^k, \mathbf{u} := \mathbf{V}\mathbf{z}$

For $i = 1, \cdots, k$:

$\quad \mathbf{S}_i \xleftarrow{\$} \mathbb{Z}_p^{k\times k}, \mathbf{D}_i := \mathbf{S}_i\mathbf{V}^\top$

$\mathsf{crs}_{\mathsf{or}} := (\mathsf{par}, [\mathbf{A}_0]_1, [\mathbf{A}_1]_1, [\mathbf{u}]_2, [\mathbf{V}]_2, ([\mathbf{D}_i]_2)_{1\le i\le k})$,

$\mathsf{td}_{\mathsf{or}} := (\mathbf{z}, (\mathbf{S}_i)_{1\le i\le k})$

Return $(\mathsf{crs}_{\mathsf{or}}, \mathsf{td}_{\mathsf{or}})$

$\mathsf{Ver}_{\mathsf{or}}(\mathsf{crs}_{\mathsf{or}}, [\mathbf{y}]_1, ([\mathbf{C}_i, \mathbf{c}_i]_2, [\mathbf{\Pi}_i, \pi_i]_1)_{1\le i\le k})$:

Parse $\pi =: ([\mathbf{C}_i, \mathbf{c}_i]_2, [\mathbf{\Pi}_i, \pi_i]_1)_{1\le i\le k}$

$[\underline{\mathbf{y}}]_1 =: [(y_1, \cdots, y_k)^\top]_1$

For $i = 1, \cdots, k$:

$\quad$ If $[\mathbf{A}_0]_1 \circ [\mathbf{C}_i]_2 - [\mathbf{y}]_1 \circ [\mathbf{c}_i]_2 \neq [\mathbf{\Pi}_i]_1 \circ [\mathbf{V}^\top]_2$ then return 0

$\quad$ If $[\overline{\mathbf{y}}^\top]_1 \circ [\mathbf{D}_i]_2 - [y_i]_1 \circ [\mathbf{u}^\top]_2 - [1]_1 \circ [\mathbf{c}_i]_2 \neq [\pi_i]_1 \circ [\mathbf{V}^\top]_2$ then return 0

Else return 1

$\mathsf{Sim}_{\mathsf{or}}(\mathsf{crs}_{\mathsf{or}}, \mathsf{td}_{\mathsf{or}}, [\mathbf{y}]_1)$:

Parse $\mathsf{td}_{\mathsf{or}} =: (\mathbf{z}, (\mathbf{S})_{1\le i\le k})$

Parse $[\underline{\mathbf{y}}]_1 =: [(y_1, \cdots, y_k)]_1$

For $i = 1, \cdots, k$,

$\quad \mathbf{R}_i \xleftarrow{\$} \mathbb{Z}_p^{k\times k}, [\mathbf{C}_i]_2 := [\mathbf{R}_i\mathbf{V}^\top]_2$

$\quad \mathbf{r}_i \xleftarrow{\$} \mathbb{Z}_p^{1\times k}, [\mathbf{c}_i]_2 := [\mathbf{r}_i\mathbf{V}^\top]_2$

$\quad [\mathbf{\Pi}_i]_1 := [\mathbf{A}_0\mathbf{R}_i - \mathbf{y}\mathbf{r}_i]_1$

$\quad [\pi_i]_1 := [\overline{\mathbf{y}}^\top \mathbf{S}_i - \mathbf{r}_i - y_i\mathbf{z}^\top]_1$

Return $([\mathbf{C}_i, \mathbf{c}_i]_2, [\mathbf{\Pi}_i, \pi_i]_1)_{1\le i\le k}$

Fig. 10. Designated-prover OR-proof for $\mathcal{L}^{\vee}_{\mathbf{A}_0,\mathbf{A}_1}$.

4.2 Structure-Preserving Signature

By replacing the underlying OR-proof in the SPS in [20] with our designated-prover one, we immediately obtain a more efficient SPS. A signature consists only of 11 elements, which is the shortest known for tightly secure SPS-es.

Theorem 3 (Security of Σ). *If $\Pi^{\mathsf{or}} := (\mathsf{Gen}_{\mathsf{or}}, \mathsf{TGen}_{\mathsf{or}}, \mathsf{Ver}_{\mathsf{or}}, \mathsf{Sim}_{\mathsf{or}})$ is a non-interactive zero-knowledge proof system for $\mathcal{L}^{\vee}_{\mathbf{A}_0,\mathbf{A}_1}$, the signature scheme Σ described in Fig. 11 is* UF-CMA *secure under the $\mathcal{D}_{2k,k}$-*MDDH *and $\mathcal{D}_k$-*MDDH *assumptions. Namely, for any adversary $\mathcal{A}$, there exist adversaries $\mathcal{B}, \mathcal{B}'$ with running time $T(\mathcal{B}) \approx T(\mathcal{B}') \approx T(\mathcal{A}) + Q \cdot \mathsf{poly}(\lambda)$, where Q is the number of signing queries,* poly *is independent of Q, and*

$\underline{\mathsf{Gen}(\mathsf{par})}$:
$\mathbf{A}_0, \mathbf{A}_1 \xleftarrow{\$} \mathcal{D}_{2k,k}$
$(\mathsf{crs}_{\mathsf{or}}, \mathsf{sk}_{\mathsf{or}}) \leftarrow \mathsf{Gen}_{\mathsf{or}}(\mathsf{par}, \mathbf{A}_0, \mathbf{A}_1)$
$\mathbf{A} \xleftarrow{\$} \mathcal{D}_k$
$\mathbf{K}_0 \xleftarrow{\$} \mathbb{Z}_p^{2k\times(k+1)}$
$\mathbf{K} \xleftarrow{\$} \mathbb{Z}_p^{(n+1)\times(k+1)}$
$\mathbf{C}_0 = \mathbf{K}_0\mathbf{A} \in \mathbb{Z}_p^{2k\times k}$
$\mathbf{C} = \mathbf{K}\mathbf{A} \in \mathbb{Z}_p^{(n+1)\times k}$
$\mathsf{vk} := (\mathsf{crs}_{\mathsf{or}}, [\mathbf{A}_0]_1, [\mathbf{A}]_2, [\mathbf{C}_0]_2, [\mathbf{C}]_2)$
$\mathsf{sk} := (\mathbf{K}_0, \mathbf{K}, \mathsf{sk}_{\mathsf{or}})$
Return $(\mathsf{vk}, \mathsf{sk})$

$\underline{\mathsf{Sign}(\mathsf{vk}, \mathsf{sk}, [\mathbf{m}]_1 \in \mathbb{G}_1^n)}$:
$\mathbf{r} \xleftarrow{\$} \mathbb{Z}_p^k$, $[\mathbf{t}]_1 := [\mathbf{A}_0]_1\mathbf{r}$
$\pi_{\mathsf{or}} \leftarrow \mathsf{Prove}_{\mathsf{or}}(\mathsf{crs}_{\mathsf{or}}, [\mathbf{t}]_1, \mathbf{r})$
$[\mathbf{u}]_1 := \mathbf{K}_0^\top[\mathbf{t}]_1 + \mathbf{K}^\top \begin{bmatrix}\mathbf{m}\\ 1\end{bmatrix}_1$
Return $\sigma := ([\mathbf{t}]_1, \pi_{\mathsf{or}}, [\mathbf{u}]_1)$

$\underline{\mathsf{Ver}(\mathsf{vk}, \sigma, [\mathbf{m}]_1)}$:
Parse $\sigma := ([\mathbf{t}]_1, \pi_{\mathsf{or}}, [\mathbf{u}]_1)$
$b \leftarrow \mathsf{Ver}_{\mathsf{or}}(\mathsf{vk}, [\mathbf{t}]_1, \pi_{\mathsf{or}})$
If $b = 1$ and
  $[\mathbf{u}^\top]_1 \circ [\mathbf{A}]_2 = [\mathbf{t}^\top]_1 \circ [\mathbf{C}_0]_2 + [\mathbf{m}^\top, 1]_1 \circ [\mathbf{C}]_2$
  return 1
Else return 0

Fig. 11. Tightly UF-CMA secure structure-preserving signature scheme Σ with message space $\mathbb{G}_1^n$. $k \in \mathbb{N}$ and the public parameter is $\mathsf{par} = \mathcal{G}$ where $\mathcal{G} \leftarrow \mathsf{GGen}(1^\lambda)$.

$$\begin{aligned}\mathsf{Adv}^{\mathsf{uf\text{-}cma}}_{\mathsf{SPS},\mathcal{A}} \leq &(4k\lceil\log Q\rceil + 2)\cdot \mathsf{Adv}^{\mathsf{mddh}}_{\mathbb{G}_1,\mathcal{D}_{2k,k},\mathcal{B}} \\ &+ (2\lceil\log Q\rceil + 3)\cdot \mathsf{Adv}^{\mathsf{mddh}}_{\mathbb{G}_2,\mathcal{D}_k,\mathcal{B}'} + \lceil\log Q\rceil \cdot \Delta_{\mathcal{D}_{2k,k}} \\ &+ \frac{4\lceil\log Q\rceil + 2}{p-1} + \frac{(Q+1)\lceil\log Q\rceil + Q}{p} + \frac{Q}{p^k}.\end{aligned}$$

We omit the proof of the above theorem since it is exactly the same as the security proof of the SPS in [20] except that we adopt the notion of standard zero knowledge instead of the composable one and the OR-proof system is a designated-prover one now, which does not affect the validity of the proof at all. We refer the reader to [20] for the details. Notice that in the MDDH games of the security proof, the reduction algorithm is not allowed to see $\mathbf{A}_0$ and $\mathbf{A}_1$ so that it cannot run the honest generation algorithm $\mathsf{Gen}_{\mathsf{or}}(\mathsf{par}, \mathbf{A}_0, \mathbf{A}_1)$. However, it does not have to, since in all the MDDH games, common reference strings are always switched to simulated ones, namely, the reduction algorithms only have to run $\mathsf{TGen}_{\mathsf{or}}(\mathsf{par}, [\mathbf{A}_0]_1, [\mathbf{A}_1]_1)$.

4.3 DPQANIZK and Black-Box Construction

We can also use our designated-or-proof system to construct a structure-preserving DPQANIZK with weak USS, which might be of independent interest. We refer the reader to the full paper for the construction and security proof of it.

On the other hand, as shown in [5,6], there is an alternative approach for constructing SPS directly from DPQANIZK. It is just mapping a message to an invalid instance out of the language and simulating a proof with a trapdoor behind a common reference string published as a public key. In the concrete construction in [5,6], $n_0 + 1$ extra elements are included in a public key so that they are used to make sure that messages consisting of n_0 elements are certainly mapped to invalid instances. We can take the same approach but with improved mapping that requires only one extra element assuming the hardness of the computational Diffie-Hellman problem. The resulting signature size is exactly the same as that of proofs of DPQANIZK and the public-key size is that of a common-reference string plus one element.

References

1. Abdalla, M., Benhamouda, F., Pointcheval, D.: Disjunctions for hash proof systems: new constructions and applications. In: Oswald, E., Fischlin, M. (eds.) EUROCRYPT 2015. LNCS, vol. 9057, pp. 69–100. Springer, Heidelberg (2015). https://doi.org/10.1007/978-3-662-46803-6_3
2. Abe, M., Chase, M., David, B., Kohlweiss, M., Nishimaki, R., Ohkubo, M.: Constant-size structure-preserving signatures: generic constructions and simple assumptions. J. Cryptol. **29**(4), 833–878 (2016)
3. Abe, M., Fuchsbauer, G., Groth, J., Haralambiev, K., Ohkubo, M.: Structure-preserving signatures and commitments to group elements. J. Cryptol. **29**(2), 363–421 (2016)
4. Abe, M., Hofheinz, D., Nishimaki, R., Ohkubo, M., Pan, J.: Compact structure-preserving signatures with almost tight security. In: Katz, J., Shacham, H. (eds.) CRYPTO 2017. LNCS, vol. 10402, pp. 548–580. Springer, Cham (2017). https://doi.org/10.1007/978-3-319-63715-0_19
5. Abe, M., Jutla, C.S., Ohkubo, M., Roy, A.: Improved (Almost) tightly-secure simulation-sound QA-NIZK with applications. In: Peyrin, T., Galbraith, S. (eds.) ASIACRYPT 2018. LNCS, vol. 11272, pp. 627–656. Springer, Cham (2018). https://doi.org/10.1007/978-3-030-03326-2_21
6. Abe, M., Jutla, C.S., Ohkubo, M., Roy, A.: Improved (almost) tightly-secure simulation-sound QA-NIZK with applications. IACR Cryptology ePrint Archive 2018/849 (2018)
7. Belenkiy, M., Chase, M., Kohlweiss, M., Lysyanskaya, A.: P-signatures and non-interactive anonymous credentials. In: Canetti, R. (ed.) TCC 2008. LNCS, vol. 4948, pp. 356–374. Springer, Heidelberg (2008). https://doi.org/10.1007/978-3-540-78524-8_20
8. Blazy, O., Kakvi, S.A., Kiltz, E., Pan, J.: Tightly-secure signatures from chameleon hash functions. In: Katz, J. (ed.) PKC 2015. LNCS, vol. 9020, pp. 256–279. Springer, Heidelberg (2015). https://doi.org/10.1007/978-3-662-46447-2_12
9. Blazy, O., Kiltz, E., Pan, J.: (Hierarchical) identity-based encryption from affine message authentication. In: Garay, J.A., Gennaro, R. (eds.) CRYPTO 2014. LNCS, vol. 8616, pp. 408–425. Springer, Heidelberg (2014). https://doi.org/10.1007/978-3-662-44371-2_23
10. Blum, M., Feldman, P., Micali, S.: Non-interactive zero-knowledge and its applications (extended abstract). In: 20th ACM STOC, pp. 103–112. ACM Press, May 1988
11. Boneh, D., Franklin, M.: Identity-based encryption from the weil pairing. In: Kilian, J. (ed.) CRYPTO 2001. LNCS, vol. 2139, pp. 213–229. Springer, Heidelberg (2001). https://doi.org/10.1007/3-540-44647-8_13
12. Chen, J., Gay, R., Wee, H.: Improved dual system ABE in prime-order groups via predicate encodings. In: Oswald, E., Fischlin, M. (eds.) EUROCRYPT 2015. LNCS, vol. 9057, pp. 595–624. Springer, Heidelberg (2015). https://doi.org/10.1007/978-3-662-46803-6_20

13. Chen, J., Wee, H.: Fully, (almost) tightly secure IBE and dual system groups. In: Canetti, R., Garay, J.A. (eds.) CRYPTO 2013. LNCS, vol. 8043, pp. 435–460. Springer, Heidelberg (2013). https://doi.org/10.1007/978-3-642-40084-1_25
14. Cramer, R., Shoup, V.: Universal hash proofs and a paradigm for adaptive chosen ciphertext secure public-key encryption. In: Knudsen, L.R. (ed.) EUROCRYPT 2002. LNCS, vol. 2332, pp. 45–64. Springer, Heidelberg (2002). https://doi.org/10.1007/3-540-46035-7_4
15. Escala, A., Groth, J.: Fine-tuning Groth-Sahai proofs. In: Krawczyk, H. (ed.) PKC 2014. LNCS, vol. 8383, pp. 630–649. Springer, Heidelberg (2014). https://doi.org/10.1007/978-3-642-54631-0_36
16. Escala, A., Herold, G., Kiltz, E., Ràfols, C., Villar, J.: An algebraic framework for Diffie-Hellman assumptions. In: Canetti, R., Garay, J.A. (eds.) CRYPTO 2013. LNCS, vol. 8043, pp. 129–147. Springer, Heidelberg (2013). https://doi.org/10.1007/978-3-642-40084-1_8
17. Fischlin, M., Libert, B., Manulis, M.: Non-interactive and re-usable universally composable string commitments with adaptive security. In: Lee, D.H., Wang, X. (eds.) ASIACRYPT 2011. LNCS, vol. 7073, pp. 468–485. Springer, Heidelberg (2011). https://doi.org/10.1007/978-3-642-25385-0_25
18. Gay, R., Hofheinz, D., Kiltz, E., Wee, H.: Tightly CCA-secure encryption without pairings. In: Fischlin, M., Coron, J.-S. (eds.) EUROCRYPT 2016. LNCS, vol. 9665, pp. 1–27. Springer, Heidelberg (2016). https://doi.org/10.1007/978-3-662-49890-3_1
19. Gay, R., Hofheinz, D., Kohl, L.: Kurosawa-Desmedt meets tight security. In: Katz, J., Shacham, H. (eds.) CRYPTO 2017. LNCS, vol. 10403, pp. 133–160. Springer, Cham (2017). https://doi.org/10.1007/978-3-319-63697-9_5
20. Gay, R., Hofheinz, D., Kohl, L., Pan, J.: More efficient (almost) tightly secure structure-preserving signatures. In: Nielsen, J.B., Rijmen, V. (eds.) EUROCRYPT 2018. LNCS, vol. 10821, pp. 230–258. Springer, Cham (2018). https://doi.org/10.1007/978-3-319-78375-8_8
21. Groth, J.: Fully anonymous group signatures without random oracles. In: Kurosawa, K. (ed.) ASIACRYPT 2007. LNCS, vol. 4833, pp. 164–180. Springer, Heidelberg (2007). https://doi.org/10.1007/978-3-540-76900-2_10
22. Groth, J., Lu, S.: A non-interactive shuffle with pairing based verifiability. In: Kurosawa, K. (ed.) ASIACRYPT 2007. LNCS, vol. 4833, pp. 51–67. Springer, Heidelberg (2007). https://doi.org/10.1007/978-3-540-76900-2_4
23. Groth, J., Ostrovsky, R., Sahai, A.: New techniques for non-interactive zero-knowledge. J. ACM **59**(3), 11:1–11:35 (2012). https://doi.org/10.1145/2220357.2220358
24. Groth, J., Sahai, A.: Efficient non-interactive proof systems for bilinear groups. In: Smart, N. (ed.) EUROCRYPT 2008. LNCS, vol. 4965, pp. 415–432. Springer, Heidelberg (2008). https://doi.org/10.1007/978-3-540-78967-3_24
25. Hartung, G., Hoffmann, M., Nagel, M., Rupp, A.: BBA+: improving the security and applicability of privacy-preserving point collection. In: Thuraisingham, B.M., Evans, D., Malkin, T., Xu, D. (eds.) ACM CCS 2017, pp. 1925–1942. ACM Press (2017)
26. Hofheinz, D.: Algebraic partitioning: fully compact and (almost) tightly secure cryptography. In: Kushilevitz, E., Malkin, T. (eds.) TCC 2016. LNCS, vol. 9562, pp. 251–281. Springer, Heidelberg (2016). https://doi.org/10.1007/978-3-662-49096-9_11

27. Hofheinz, D.: Adaptive partitioning. In: Coron, J.-S., Nielsen, J.B. (eds.) EUROCRYPT 2017. LNCS, vol. 10212, pp. 489–518. Springer, Cham (2017). https://doi.org/10.1007/978-3-319-56617-7_17
28. Hofheinz, D., Jager, T.: Tightly secure signatures and public-key encryption. In: Safavi-Naini, R., Canetti, R. (eds.) CRYPTO 2012. LNCS, vol. 7417, pp. 590–607. Springer, Heidelberg (2012). https://doi.org/10.1007/978-3-642-32009-5_35
29. Hofheinz, D., Jia, D., Pan, J.: Identity-based encryption tightly secure under chosen-ciphertext attacks. In: Peyrin, T., Galbraith, S. (eds.) ASIACRYPT 2018. LNCS, vol. 11273, pp. 190–220. Springer, Cham (2018). https://doi.org/10.1007/978-3-030-03329-3_7
30. Joux, A.: A one round protocol for tripartite Diffie-Hellman. In: Proceedings of the Algorithmic Number Theory, 4th International Symposium, ANTS-IV, Leiden, The Netherlands, 2–7 July 2000, pp. 385–394 (2000)
31. Jutla, C.S., Ohkubo, M., Roy, A.: Improved (almost) tightly-secure structure-preserving signatures. In: Abdalla, M., Dahab, R. (eds.) PKC 2018. LNCS, vol. 10770, pp. 123–152. Springer, Cham (2018). https://doi.org/10.1007/978-3-319-76581-5_5
32. Jutla, C.S., Roy, A.: Shorter quasi-adaptive NIZK proofs for linear subspaces. In: Sako, K., Sarkar, P. (eds.) ASIACRYPT 2013. LNCS, vol. 8269, pp. 1–20. Springer, Heidelberg (2013). https://doi.org/10.1007/978-3-642-42033-7_1
33. Jutla, C.S., Roy, A.: Switching lemma for bilinear tests and constant-size NIZK proofs for linear subspaces. In: Garay, J.A., Gennaro, R. (eds.) CRYPTO 2014. LNCS, vol. 8617, pp. 295–312. Springer, Heidelberg (2014). https://doi.org/10.1007/978-3-662-44381-1_17
34. Jutla, C.S., Roy, A.: Improved structure preserving signatures under standard bilinear assumptions. In: Fehr, S. (ed.) PKC 2017. LNCS, vol. 10175, pp. 183–209. Springer, Heidelberg (2017). https://doi.org/10.1007/978-3-662-54388-7_7
35. Jutla, C.S., Roy, A.: Smooth NIZK arguments. In: Beimel, A., Dziembowski, S. (eds.) TCC 2018. LNCS, vol. 11239, pp. 235–262. Springer, Cham (2018). https://doi.org/10.1007/978-3-030-03807-6_9
36. Kiltz, E., Pan, J., Wee, H.: Structure-preserving signatures from standard assumptions, revisited. In: Gennaro, R., Robshaw, M. (eds.) CRYPTO 2015. LNCS, vol. 9216, pp. 275–295. Springer, Heidelberg (2015). https://doi.org/10.1007/978-3-662-48000-7_14
37. Kiltz, E., Wee, H.: Quasi-adaptive NIZK for linear subspaces revisited. In: Oswald, E., Fischlin, M. (eds.) EUROCRYPT 2015. LNCS, vol. 9057, pp. 101–128. Springer, Heidelberg (2015). https://doi.org/10.1007/978-3-662-46803-6_4
38. Libert, B., Peters, T., Joye, M., Yung, M.: Non-malleability from malleability: simulation-sound quasi-adaptive NIZK proofs and CCA2-secure encryption from homomorphic signatures. In: Nguyen, P.Q., Oswald, E. (eds.) EUROCRYPT 2014. LNCS, vol. 8441, pp. 514–532. Springer, Heidelberg (2014). https://doi.org/10.1007/978-3-642-55220-5_29
39. Libert, B., Peters, T., Joye, M., Yung, M.: Compactly hiding linear spans. In: Iwata, T., Cheon, J.H. (eds.) ASIACRYPT 2015. LNCS, vol. 9452, pp. 681–707. Springer, Heidelberg (2015). https://doi.org/10.1007/978-3-662-48797-6_28
40. Libert, B., Peters, T., Yung, M.: Short group signatures via structure-preserving signatures: standard model security from simple assumptions. In: Gennaro, R., Robshaw, M. (eds.) CRYPTO 2015. LNCS, vol. 9216, pp. 296–316. Springer, Heidelberg (2015). https://doi.org/10.1007/978-3-662-48000-7_15

41. Morillo, P., Ràfols, C., Villar, J.L.: The kernel matrix Diffie-Hellman assumption. In: Cheon, J.H., Takagi, T. (eds.) ASIACRYPT 2016. LNCS, vol. 10031, pp. 729–758. Springer, Heidelberg (2016). https://doi.org/10.1007/978-3-662-53887-6_27
42. Naor, M., Yung, M.: Public-key cryptosystems provably secure against chosen ciphertext attacks. In: 22nd ACM STOC, pp. 427–437. ACM Press, May 1990
43. Ràfols, C.: Stretching Groth-Sahai: NIZK proofs of partial satisfiability. In: Dodis, Y., Nielsen, J.B. (eds.) TCC 2015. LNCS, vol. 9015, pp. 247–276. Springer, Heidelberg (2015). https://doi.org/10.1007/978-3-662-46497-7_10

Efficient Noninteractive Certification of RSA Moduli and Beyond

Sharon Goldberg[1(✉)], Leonid Reyzin[1], Omar Sagga[1], and Foteini Baldimtsi[2]

[1] Boston University, Boston, MA, USA
{goldbe,reyzin}@cs.bu.edu, osagga@bu.edu
[2] George Mason University, Fairfax, VA, USA
foteini@gmu.edu

Abstract. In many applications, it is important to verify that an RSA public key (N, e) specifies a permutation over the entire space $\mathbb{Z}_N$, in order to prevent attacks due to adversarially-generated public keys. We design and implement a simple and efficient noninteractive zero-knowledge protocol (in the random oracle model) for this task. Applications concerned about adversarial key generation can just append our proof to the RSA public key without any other modifications to existing code or cryptographic libraries. Users need only perform a one-time verification of the proof to ensure that raising to the power e is a permutation of the integers modulo N. For typical parameter settings, the proof consists of nine integers modulo N; generating the proof and verifying it both require about nine modular exponentiations.

We extend our results beyond RSA keys and also provide efficient noninteractive zero-knowledge proofs for other properties of N, which can be used to certify that N is suitable for the Paillier cryptosystem, is a product of two primes, or is a Blum integer. As compared to the recent work of Auerbach and Poettering (PKC 2018), who provide two-message protocols for similar languages, our protocols are more efficient and do not require interaction, which enables a broader class of applications.

1 Introduction

Many applications use an RSA public key (N, e) that is chosen by a party who may be adversarial. In such applications, it is often necessary to ensure that the public key defines a permutation over $\mathbb{Z}_N$: that is, raising to the power e modulo N must be bijective, or, equivalently, every integer between 0 and $N-1$ must have an eth root modulo N. An attacker who deliberately generates a bad key pair may subvert the security of other users—see for example, [MRV99, CMS99, MPS00, LMRS04]. In particular, our work was motivated by TumbleBit [HAB+17], a transaction-anonymizing system deployed [Str17] on top of Bitcoin, in which a bad key pair can lead to a devastating attack (see footnote 2 in Sect. 5 for the attack specifics).

Interactive proofs for correctness of RSA keys are available (see, for example, [AP18] and references therein), but interaction with the key owner is often not

S. D. Galbraith and S. Moriai (Eds.): ASIACRYPT 2019, LNCS 11923, pp. 700–727, 2019.
https://doi.org/10.1007/978-3-030-34618-8_24

possible in the application. Thus, the folklore solution, used, for example, in [MRV99,CMS99,MPS00,LMRS04], is to choose the public RSA exponent e such that e is prime and larger than N. This solution has two major drawbacks.

First, because the folklore solution requires $e > N$, e is not in the set of standard values typically used for e in RSA implementations *e.g.*, $e \in \{3, 17, 2^{16}+1\}$. Unless a large prime value for e is standardized, before using the public key, one would have to perform a one-time primality test on e, to ensure that e really is prime. This primality test is quite expensive (see Sect. 5).

Second, most RSA implementations choose a small value for e, typically from a set of standard values $e \in \{3, 17, 2^{16} + 1\}$. Choosing a small e significantly reduces the cost of performing RSA public key operations. However, this efficiency advantage is eliminated in the folklore solution, which requires $e > N$. Unlike the previous drawback, which results in a one-time cost for each public key used, this drawback makes *every* public-key operation about two orders of magnitude more expensive.

In addition, this solution is not compatible with existing RSA standards and off-the-shelf implementations. This is because the folklore solution does not ensure that the public key operation is a permutation over $\mathbb{Z}_N$, where $\mathbb{Z}_N = \{0, 1, ..., N-1\}$. Instead, it ensures only that the public key operation defines a permutation over the set $\mathbb{Z}_N^*$, where $\mathbb{Z}_N^*$ is the set of values in $\mathbb{Z}_N$ that are relatively prime with N. Thus, there are no assurances about the values in the set $\mathbb{Z}_N - \mathbb{Z}_N^*$, *i.e.,* the set of values that are less than N but *not* relatively prime with N. (To see this, consider the example $N = 9$ and $e = 11$.) If the RSA public key is generated honestly, this is not a problem, because the set $\mathbb{Z}_N - \mathbb{Z}_N^*$ contains only a negligible fraction of $\mathbb{Z}_N$. However, if an adversary chooses the RSA public key (N, e) maliciously, then it could choose N so that the set $\mathbb{Z}_N - \mathbb{Z}_N^*$ is a large fraction of $\mathbb{Z}_N$.[1] To address this attack, the folklore solution additionally requires a gcd check along with *every* RSA public-key operation, to ensure that the exponentiated value is relatively prime with N.

1.1 Our Contributions

Proving that an RSA Key Specifies a Permutation over all of $\mathbb{Z}_N$ We present a simple noninteractive zero-knowledge proof (NIZK) in the random oracle model, that allows the holder of an RSA secret key to prove that the corresponding public key defines a permutation over all of $\mathbb{Z}_N$, without leaking information about the corresponding secret key. Our NIZK can be used even when the RSA exponent e is small, which is useful for applications that require fast RSA public key operations. In addition to the NIZK algorithm and a concrete security proof,

[1] It has been observed that such an N could be detected by checking if N has small divisors. However, the risk of being detected is not usually an adequate deterrent, unless implemented and deployed as part of a protocol. But if such a check is deployed, then the adversary, knowing what check has been deployed, could set divisors of N to be just slightly larger than the limits of the check, and thus still ensure that $\mathbb{Z}_N - \mathbb{Z}_N^*$ is a nonnegligible fraction of $\mathbb{Z}_N$.

we present a detailed specification of the prover and verifier algorithms, as well as production-quality implementation and an analysis of its performance. Because our NIZK is for all values in $\mathbb{Z}_N$, it is compliant with existing cryptographic specifications of RSA (*e.g.*, RFC8017 [MKJR16]).

For typical parameter settings, our NIZK consists of 9 elements of $\mathbb{Z}_N$. Generating the NIZK costs roughly 9 full-length RSA exponentiations modulo N. Meanwhile, each verifier pays the one-time cost of verifying our NIZK, which is also roughly equal to 9 full-length exponentiations. When compared to the folklore solution we described earlier, our solution (1) avoids the more expensive one-time primality test and (2) allows the verifier to continue using a small value of e, resulting in better performance for every public-key verification.

We view this result as of most immediate practical applicability (in fact, it has already been deployed). We therefore present not only a high-level explanation of this protocol (Sect. 3.3), but also its detailed specification (Appendix C) and implementation results (Sect. 5 and code at [cod]).

Suitability for Paillier and Other Properties of N. We also present simple NIZK proofs for several other properties of N, such as ensuring that N is square-free (Sect. 3.2), is suitable for Paillier encryption (required in [Lin17] and [HMRT12]; see Sect. 3.2), is a product of exactly two primes (Sect. 3.4), or is a Blum integer (i.e., product of two primes that are each 3 modulo 4; see Sect. 3.5). Most of these problems have been addressed only via interactive protocols in prior literature [AP18]. Noninteractive proofs have considerably broader applicability than interactive ones, because the owner of the public key can simply generate a nonineractive proof once and publish it once together with the public key, whereas in the interactive, the owner needs to be online, handle potentially high query loads, and be subject to denial of service attacks.

Our proofs for square-freeness and suitability for Paillier are of similar efficiency to the permutation proof, requiring only 8 elements in Z_N for typical parameter settings and 8 full-length modular exponentiations. Our proofs for products of two primes and Blum integers require the proof of square-freeness and a test for prime powers (same as in [AP18]), plus one more component, which is less efficient for the prover, but more efficient for the verifier. For 128-bit security, this additional component requires about 1420 square root operations $\mathbb{Z}_N$ by the prover (note, however, that this is done one-time during key generation). The verifier, on the other hand, needs to perform only Jacobi-symbol computations and modular squarings, which are much more efficient, making the verifier cost comparable to the cost of just a few full-length modular exponentiation. This additional component requires the publication of 1420 elements of Z_N.

All of our protocols are presented first as two-message public-coin honest-verifier protocols. We then convert them to noninteractive using the Fiat-Shamir heuristics, by obtaining the verifier's public-coin message through an application of the random oracle to the protocol input (see Sect. 4). They all have perfect completeness, perfect honest-verifier zero-knowledge, and statistical soundness, with the exception of the protocol for showing that N is a product of two primes, which has computational honest-verifier zero-knowledge under the quadratic residuosity assumption.

1.2 Related Work

Auerbach and Poettering [AP18] present two-message interactive protocols in the random oracle model for the same problems as we consider, with the exception of proving that (N, e) specifies a permutation over $\mathbb{Z}_N$ (they prove only that (N, e) specifies a permutation over $\mathbb{Z}_N^*$, which, would require users to modify their RSA implementations to add a gcd computation to every public-key operation). As already mentioned, noninteractive protocols have broader applicability than interactive ones. It is much more appealing to be able to post, say an RSA public key along with a NIZK proof of being well formed, as opposed to be expected to run an online, interactive protocol with each verifier. Their protocols for proving that (N, e) specifies a permutation, N is square-free, or is suitable for Paillier, are all considerably less efficient than ours, requiring 81–128 modular exponentiations for 128-bit security level. Their protocols for proving that N is a product of exactly two primes or is a Blum integer are also less efficient for the verifier (because the first step in those protocols is proving square-freeness); they are about 10 times more efficient for the prover if we consider only one-time use, but, because they are interactive, they must be run repeatedly by the prover, while in our noninteractive case, the prover needs to run them only once.

Kakvi, Kiltz, and May [KKM12] show how to verify that RSA is a permutation by providing only the RSA public key (N, e) and no additional information, as long as $e > N^{1/4}$. They also show that when e is small, it is impossible, under reasonable complexity assumptions, to verify that (N, e) is a permutation without any additional information [KKM12, Section 1]. Thus, their approach cannot be used when e is small. We circumvent their impossibility by having the prover additionally provide our NIZK (rather than just (N, e)) to the verifier.

Wong, Chan, and Zhu [WCZ03, Section 3.2] and Catalano, Pointcheval, and Pornin [CPP07, Appendix D.2] present interactive protocols (using techniques similar to ours) that, like the protocols of [AP18], also work only over $\mathbb{Z}_N^*$ rather than the entire $\mathbb{Z}_N$.

The protocols given by Camenisch and Michels [CM99, Section 5.2] and Benhamouda et al. [BFGN17] achieve much stronger goals. The former proves $N = pq$ is a product of two safe primes (*i.e.*, $p, q, (p-1)/2$, and $(q-1)/2$ are all prime); the second can prove that any prespecified procedure for generating the primes p and q was followed. These protocols can be used to prove that (N, e) specifies a permutation by imposing mild additional conditions on e (and the prime generation procedure for [BFGN17]). However, these stronger goals are not necessary for our purposes. Our protocol is considerably simpler and more efficient, and does not restrict p and q in any way.

Our protocol for showing that (N, e) specifies a permutation over $\mathbb{Z}_N$ builds on the protocol of Bellare and Yung [BY96], who showed how to prove that any function is "close" to a permutation. However, "close" is not good enough for our purposes, because the adversary may be able to force the honest parties to use the few values in $\mathbb{Z}_N$ at which the permutation property does not hold. Thus, additional work is required for our setting. This additional work is accomplished with the help of a simple sub-protocol from Gennaro, Micciancio, and Rabin [GMR98, Section 3.1] for showing the square-freeness of N (a similar sub-protocol in the interactive setting was discovered earlier by Boyar, Friedl, and Lund [BFL89, Section 2.2]). We demonstrate how to combine the ideas of [BY96] and [GMR98] to prove that (N, e) specifies a permutation over $\mathbb{Z}_N$.

2 Preliminaries

Some number-theoretic preliminaries are presented in Appendix A.

Here, we first recall the standard notion of honest-verifier zero-knowledge (HVZK).

Definition 1. *(Honest-Verifier Zero Knowledge (HVZK)) An interactive proof system between a prover and verifier (P, V) for a NP language L is said to be* honest-verifier zero knowledge *if the following properties hold:*

1. *(perfect) Completeness. For every $x \in L$ and every NP-witness w for x,*

$$Pr[\langle P(x, w), V(x)\rangle = 1] = 1.$$

2. *(statistical) Soundness. For every $x \notin L$ and every interactive algorithm P^**

$$Pr[\langle P^*(x), V(x)\rangle = 1] = \text{negl}(|x|)$$

3. *HVZK. There exists a probabilistic polynomial-time simulator S such that for all $x \in L$ and all PPT distinguishers D we have:*

$$\mathtt{view}D^{\langle P(x,w),V(x)\rangle} \approx \mathtt{view}D^{S(x)}.$$

We say (P, V) is *public coin* if all the messages sent by verifier V to prover P are random coin tosses.

Promise Problems. We also recall the notion of a promise problem, which is a generalization of the notion of a language. A promise problem consists of two disjoint sets: $\mathrm{L_{yes}}$ and $\mathrm{L_{no}}$. In a language, $\mathrm{L_{no}} = \overline{\mathrm{L_{yes}}}$, but in a promise problem, there may be strings that are neither in $\mathrm{L_{yes}}$ nor $\mathrm{L_{no}}$, and we generally do not care what happens if such a string is input. Thus, in a ZK proof for a promise problem, completeness and zero-knowledge need to hold for inputs in $\mathrm{L_{yes}}$, while soundness needs to hold for inputs in $\mathrm{L_{no}}$.

3 HVZK Proofs for Properties of N and e

3.1 HVZK Proof for a Permutation over $\mathbb{Z}_N^*$

Bellare and Yung [BY96] showed how to certify that any function is close to a permutation. The idea is to simply ask the prover to invert the permutation on random points. It is a standard fact from number theory (Lemma 8) that raising to eth power is either a permutation of $\mathbb{Z}_N^*$ or very far from one—in fact, it is either a permutation or an e'-to-1 function, where e' is the smallest prime divisor of e. Here, we adapt the protocol of [BY96] to show that the RSA function is not just close to a permutation, but is actually a permutation over $\mathbb{Z}_N^*$: if we check that e' is high enough, then not many random points will be needed.

It is also a standard fact (recalled in Lemma 2) that raising to the power e defines a permutation over $\mathbb{Z}_N^*$ if and only if e is relatively prime to $\phi(N)$. Thus, let

$$\mathrm{L}_{\mathsf{perm}\mathbb{Z}_N^*} = \{(N,e) \mid N, e > 0 \text{ and } \gcd(e, \phi(N)) = 1\}.$$

Let

$$\mathrm{L}_{e'} = \{(N,e) \mid N, e > 0 \text{ and no prime less than } e' \text{ divides } e\}.$$

(In typical RSA implementations, e is a fixed small prime, such as 3, 17, or 65537, and one would use $e' = e$.)

The following is an HVZK protocol for $\mathrm{L}_{\mathsf{perm}\mathbb{Z}_N^*} \cap \mathrm{L}_{e'}$ with perfect completeness, perfect zero-knowledge, and statistical soundness error $2^{-\kappa}$. We emphasize that, while the protocol is similar to that of [BY96], it is not identical. Specifically, the addition of $\mathrm{L}_{e'}$ and the verifier check in Step 4a allow us to guarantee that the RSA function is a permutation over $\mathbb{Z}_N^*$ much more efficiently than the protocol of [BY96].

Protocol $\mathcal{P}_{\mathsf{perm}\mathbb{Z}_N^*}$

1. Both prover and verifier let $m = \lceil \kappa / \log_2 e' \rceil$.
2. The verifier chooses m random values $\rho_i \in \mathbb{Z}_N^*$ and sends them to prover.
3. The prover sends back eth roots of ρ_i modulo N:

   $$\sigma_i = (\rho_i)^{e^{-1} \bmod \phi(N)} \bmod N$$

   for $i = 1 \ldots m$.
4. The verifier accepts that $N \in \mathrm{L}_{\mathsf{perm}\mathbb{Z}_N^*} \cap \mathrm{L}_{e'}$ if all of the following checks pass.
   (a) Check that N, e, and σ_i for $i = 1 \ldots m$ are positive integers, and that e not divisible by all the primes less than e' (if e is a fixed prime as in typical RSA implementations, this check simply involves checking that $e = e'$).
   (b) Verify that $\rho_i = (\sigma_i)^e \bmod N$ for $i = 1 \ldots m$.

Theorem 1. *$\mathcal{P}_{\mathsf{perm}\mathbb{Z}_N^*}$ is a 2-message public-coin protocol with perfect completeness, perfect honest-verifier zero-knowledge, and statistical soundness error $2^{-\kappa}$ for the language $\mathrm{L}_{\mathsf{perm}\mathbb{Z}_N^*} \cap \mathrm{L}_{e'}$.*

Proof. It is a standard fact that raising to the power N is a permutation of $\mathbb{Z}_N^*$ whenever $e \in \mathrm{L}_{\mathsf{perm}\mathbb{Z}_N^*}$, and the inverse of this permutation is raising to the power $(e^{-1} \bmod \phi(N))$ (see Lemma 2). This fact gives perfect completeness and a perfect HVZK simulator who simply chooses σ_i and computes $\rho_i = (\sigma_i)^e \bmod N$ for $i = 1 \ldots m$ (recall that by definition, completeness and HVZK apply only to $(N, e) \in \mathrm{L}_{\mathsf{perm}\mathbb{Z}_N^*} \cap \mathrm{L}_{e'}$). Statistical soundness with error $2^{-\kappa}$ follows from the fact that if $(N, e) \notin \mathrm{L}_{\mathsf{perm}\mathbb{Z}_N^*}$ but $(N, e) \in \mathrm{L}_{e'}$, then size the image of the map $\sigma \mapsto \sigma^e$ is at most $|\mathbb{Z}_N^*|/e'$ by Lemma 8. Thus, the probability that a σ_i exists for every ρ_i is at most $1/(e')^m = 2^{-m \log_2 e'} \leq 2^{-\kappa}$. □

3.2 HVZK Proofs for Paillier and Square-Free N

The Paillier cryptosystem requires a modulus N that is relatively prime with $\phi(N)$. Thus, let

$$\mathrm{L}_{\mathsf{pailler}\text{-}N} = \{N > 0 \,|\, \gcd(N, \phi(N)) = 1\}\,.$$

We emphasize that, unlike [AP18], we do not verify the properties of the generator g in the Paillier cryptosystem—but since the common choice is to use $g = N + 1$ per [DJ01], verifying that $N \in \mathrm{L}_{\mathsf{pailler}\text{-}N}$ is sufficient for the common case.

Let

$$\mathrm{L}_{\mathsf{square\text{-}free}} = \{N > 0 \,|\, \text{there is no prime } p \text{ such that } p^2 \text{ divides } N\}\,.$$

Note that to be in $\mathrm{L}_{\mathsf{pailler}\text{-}N}$, N has to be in $\mathrm{L}_{\mathsf{square\text{-}free}}$ and also have no prime divisors p, q such that $p \,|\, q-1$ (by definition of $\phi(N)$, as recalled in Appendix A), so $\mathrm{L}_{\mathsf{pailler}\text{-}N} \subset \mathrm{L}_{\mathsf{square\text{-}free}}$ (see Lemma 3).

Thus, letting

$$\mathrm{L}_{\mathsf{gap}} = \{N \in \mathrm{L}_{\mathsf{square\text{-}free}} | N \text{ has two prime divisors } p, q \text{ such that } p \text{ divides } q-1\},$$

we know that $\mathrm{L}_{\mathsf{square\text{-}free}} - \mathrm{L}_{\mathsf{gap}} = \mathrm{L}_{\mathsf{pailler}\text{-}N}$.

Our protocols for proving suitability for Paillier or square-freeness will depend on a parameter α and the corresponding language

$$\mathrm{L}_{\alpha} = \{N > 0 \,|\, \text{no prime less than } \alpha \text{ divides } N\}\,.$$

We now describe the protocol $\mathcal{P}_{\mathsf{pailler}\text{-}N}$, an HVZK protocol for $\mathrm{L}_{\alpha} \cap \mathrm{L}_{\mathsf{pailler}\text{-}N}$ with perfect completeness, perfect zero-knowledge, and statistical soundness error $2^{-\kappa}$. This protocol builds on the protocol from [GMR98, Section 3.1], but is not identical to it: specifically, the addition of L_{α} and verifier's Step 4a gives better performance. Setting $\alpha = 2$ gives a protocol for $\mathrm{L}_{\mathsf{pailler}\text{-}N}$, but a higher setting of α will improve efficiency (see Sect. 5 for a discussion of how to pick α).

The idea of the protocol is to ask the prover to take Nth roots of random points—they will not exist for many points if $N \notin \mathrm{L}_{\mathsf{pailler}\text{-}N}$, because raising to the power N will be far from a permutation. The protocol is the same as the protocol $\mathrm{L}_{\mathsf{perm}\mathbb{Z}_N^*}$ described in Sect. 3.1, replacing e with N and e' with α.

Protocol $\mathcal{P}_{\mathsf{pailler}\text{-}N}$:

1. Both prover and verifier let $m = \lceil \kappa / \log_2 \alpha \rceil$.
2. The verifier chooses m random values $\rho_i \in \mathbb{Z}_N^*$ and sends them to prover.
3. The prover sends back Nth roots of ρ_i modulo N:

$$\sigma_i = (\rho_i)^{N^{-1} \bmod \phi(N)} \bmod N$$

   for $i = 1 \ldots m$.
4. The verifier accepts that $N \in \mathrm{L}_{\mathsf{pailler}\text{-}N} \cap \mathrm{L}_\alpha$ if all of the following checks pass.
   (a) Check that N is a positive integer and is not divisible by all the primes less than α.
   (b) Check that σ_i is a positive integer for $i = 1 \ldots m$.
   (c) Verify that $\rho_i = (\sigma_i)^N \bmod N$ for $i = 1 \ldots m$.

Theorem 2 (GMR98). *$\mathcal{P}_{\mathsf{pailler}\text{-}N}$ is a 2-message public-coin proof with perfect completeness, perfect honest-verifier zero-knowledge, and statistical soundness error $2^{-\kappa}$ for the language $\mathrm{L}_\alpha \cap \mathrm{L}_{\mathsf{pailler}\text{-}N}$.*

Note that choosing elements in $\mathbb{Z}_N^*$ in step 2 of the protocol requires a gcd computation by the verifier (because the verifier cannot be sure that the difference between $\mathbb{Z}_N$ and $\mathbb{Z}_N^*$ is negligible). To avoid this computation, the verifier can choose values in $\mathbb{Z}_N$ instead. Then the verifier may have a lower probability of rejecting inputs outside of $\mathrm{L}_\alpha \cap \mathrm{L}_{\mathsf{pailler}\text{-}N}$, but is still guaranteed to reject inputs outside of $\mathrm{L}_\alpha \cap \mathrm{L}_{\mathsf{square}\text{-}\mathsf{free}}$ with probability $1 - 2^{-\kappa}$, as we show in Lemma 6. Perfect completeness and zero-knowledge still hold for $\mathrm{L}_\alpha \cap \mathrm{L}_{\mathsf{pailler}\text{-}N}$, and thus for an honestly generated RSA modulus. Let us call this modified protocol $\mathcal{P}_{\mathsf{square}\text{-}\mathsf{free}}$.

Protocol $\mathcal{P}_{\mathsf{square}\text{-}\mathsf{free}}$: Same as the protocol $\mathcal{P}_{\mathsf{pailler}\text{-}N}$ described above, replacing Z_N^* with Z_N in step 2 and $N \in \mathrm{L}_{\mathsf{pailler}\text{-}N} \cap \mathrm{L}_\alpha$ with $N \in \mathrm{L}_{\mathsf{square}\text{-}\mathsf{free}} \cap \mathrm{L}_\alpha$ in step 4. Specifically,

1. Both prover and verifier let $m = \lceil \kappa / \log_2 \alpha \rceil$.
2. The verifier chooses m random values $\rho_i \in \mathbb{Z}_N$ and sends them to prover.
3. The prover sends back Nth roots of ρ_i modulo N:

$$\sigma_i = (\rho_i)^{N^{-1} \bmod \phi(N)} \bmod N$$

   for $i = 1 \ldots m$.
4. The verifier accepts that $N \in \mathrm{L}_{\mathsf{square}\text{-}\mathsf{free}} \cap \mathrm{L}_\alpha$ if all of the following checks pass.
   (a) Check that N is a positive integer and is not divisible by all the primes less than α.
   (b) Check that σ_i is a positive integer for $i = 1 \ldots m$.
   (c) Verify that $\rho_i = (\sigma_i)^N \bmod N$ for $i = 1 \ldots m$.

Theorem 3. *$\mathcal{P}_{\text{square-free}}$ is a 2-message public-coin proof with perfect completeness, perfect honest-verifier zero-knowledge, and statistical soundness error $2^{-\kappa}$ for the promise problem* $(\mathrm{L}_{\text{yes}} = \mathrm{L}_\alpha \cap \mathrm{L}_{\text{pailler-}N}, \mathrm{L}_{\text{no}} = \overline{\mathrm{L}_\alpha \cap \mathrm{L}_{\text{square-free}}})$.

Proof. It is a standard fact that raising to the power N is a permutation of $\mathbb{Z}_N$ whenever $N \in \mathrm{L}_{\text{pailler-}N}$, and the inverse of this permutation is raising to the power $(N^{-1} \bmod \phi(N))$ (see Lemmas 3 and 4, setting $f = N$). This fact gives perfect completeness and a perfect HVZK simulator who simply chooses σ_i and computes $\rho_i = (\sigma_i)^N \bmod N$ for $i = 1 \dots m$ (recall that by definition, completeness and HVZK apply only to $N \in \mathrm{L}_{\text{yes}}$). Statistical soundness with error $2^{-\kappa}$ follows from the fact that if $p \geq \alpha$ is a prime such that $p^2 | N$, then the map $\sigma \mapsto \sigma^N$ is at least α-to-1 over Z_N (per Lemma 6); thus, the probability that a σ_i exists for every ρ_i is at most $1/\alpha^m = 2^{-m \log_2 \alpha} \leq 2^{-\kappa}$. □

3.3 HVZK Proof for Permutation over Entire $\mathbb{Z}_N$

As explained in the introduction, ensuring that raising to the power e is a permutation over the entire $\mathbb{Z}_N$ is more desirable than ensuring only that it is a permutation over $\mathbb{Z}_N^*$. In this section, we show that a careful combination of protocols $\mathcal{P}_{\text{square-free}}$ and $\mathcal{P}_{\text{perm}\mathbb{Z}_N^*}$ gives an efficient two-message public-coin HVZK protocol for proving that an RSA public key defines a permutation over the entire $\mathbb{Z}_N$.

Let $\mathrm{L}_{\text{perm}\mathbb{Z}_N^*}$ and $\mathrm{L}_{e'}$ be as in Sect. 3.1, and L_α, $\mathrm{L}_{\text{square-free}}$, $\mathrm{L}_{\text{pailler-}N}$, and L_{gap} be as in Sect. 3.2, except defined not just on integers N, but on pairs (N, e) for an arbitrary $e > 0$.

Let $\mathrm{L}_{\text{perm}\mathbb{Z}_N} = \{(N, e) \mid N, e > 0$ and raising to the power e is a permutation over $\mathbb{Z}_N\}$.

Note that

$$\left(\mathrm{L}_{\text{pailler-}N} \cap \mathrm{L}_{\text{perm}\mathbb{Z}_N^*}\right) \subset \left(\mathrm{L}_{\text{square-free}} \cap \mathrm{L}_{\text{perm}\mathbb{Z}_N^*}\right) \subset \mathrm{L}_{\text{perm}\mathbb{Z}_N} .$$

(the first $\subset$ property follows from Lemma 3; the second $\subset$ property follows from Lemma 4). Note that the only pairs (N, e) in $\mathrm{L}_{\text{perm}\mathbb{Z}_N} - \left(\mathrm{L}_{\text{square-free}} \cap \mathrm{L}_{\text{perm}\mathbb{Z}_N^*}\right)$ are those for which $e = 1$ and N is not square-free (per Lemma 5).

We want to design a protocol for $\mathrm{L}_{\text{perm}\mathbb{Z}_N}$. For efficiency reasons, we will focus instead on $\mathrm{L}_{\text{perm}\mathbb{Z}_N} \cap \mathrm{L}_\alpha \cap \mathrm{L}_{e'}$, i.e., require N and e to not have divisors smaller than α and e', respectively. Moreover, just like in protocol $\mathcal{P}_{\text{square-free}}$ of Sect. 3.2, we will consider slightly weaker completeness: if N is square-free, but has two prime divisors p, q such that $p \mid (q-1)$ (i.e., falls into L_{gap}), the verifier will be permitted to reject N. Thus, let

$$\mathrm{L}_{\text{yes}} = \mathrm{L}_{\text{pailler-}N} \cap \mathrm{L}_{\text{perm}\mathbb{Z}_N^*} \cap \mathrm{L}_\alpha \cap \mathrm{L}_{e'}$$
$$\mathrm{L}_{\text{no}} = \overline{\mathrm{L}_{\text{perm}\mathbb{Z}_N}} \cup \overline{\mathrm{L}_\alpha} \cup \overline{\mathrm{L}_{e'}}$$

The gap between L_{yes} and L_{no} (i.e., the only pairs (N, e) not in $\mathrm{L}_{\text{yes}} \cup \mathrm{L}_{\text{no}}$) is almost the same as in Theorem 3: namely, $\mathrm{L}_{\text{gap}} \cap \mathrm{L}_{e'} \cap \mathrm{L}_\alpha$, as well as some pairs

(N, e) with $e = 1$. Naturally occurring RSA moduli should never fall into this gap. Every (N, e) in the gap still defines a permutation over the entire $\mathbb{Z}_N$, but the prover may be unable to show this fact.

We now present a protocol $\mathcal{P}_{\mathsf{perm}\mathbb{Z}_N}$ for the promise problem $(\mathrm{L_{yes}}, \mathrm{L_{no}})$. The protocol $\mathcal{P}_{\mathsf{perm}\mathbb{Z}_N}$ is not simply a combination of $\mathcal{P}_{\mathsf{square\text{-}free}}$ and $\mathcal{P}_{\mathsf{perm}\mathbb{Z}_N^*}$: we save space by using the same ρ_i for both eth roots and Nth roots. Because any value that has an (eN)th root also has an eth root and an Nth root, we combine the two protocols simply by checking the ρ_i values have eNth roots.

The protocol $\mathcal{P}_{\mathsf{perm}\mathbb{Z}_N}$ depends on two parameters α and e', which are both primes, at most about 16 bits long. The verifier will reject any N that is divisible by a prime less than α and any e that is divisible by a prime less than e'. Any setting of α and e' is valid for security; varying these parameters affects only efficiency. An optimal setting of these parameters is implementation-dependent, since larger e' and α will result in some additional work for the verifier, but will also reduce work for the prover and verifier since m_1 and m_2 in Eq. (1) below become smaller. When e is a fixed prime like 3, 17, or $2^{16} + 1$, as is standard for many RSA implementations, then we set e' equal to e. We further discuss parameter settings in Sect. 5.

The prover's witness is the prime factorization of N. Recall that κ is a security parameter. The protocol will achieve statistical soundness error $2^{-\kappa}$.

Protocol $\mathcal{P}_{\mathsf{perm}\mathbb{Z}_N}$:

1. Both prover and verifier let

$$m_1 = \lceil \kappa / \log_2 \alpha \rceil \text{ and } m_2 = \left\lceil -\kappa / \log_2 \left(\frac{1}{\alpha} + \frac{1}{e'} \left(1 - \frac{1}{\alpha} \right) \right) \right\rceil . \quad (1)$$

   Notice that $m_2 \geq m_1$ since $e' > 1$.
2. The verifier chooses m_2 random values $\rho_i \in Z_N$ and sends them to Prover.
3. The Prover sends back

$$\sigma_i = (\rho_i)^{(eN)^{-1} \bmod \phi(N)} \pmod N$$

   for $i = 1 \ldots m_1$ (for convenience, we call this a "weird RSA signature") and

$$\sigma_i = (\rho_i)^{e^{-1} \bmod \phi(N)} \pmod N$$

   for $i = m_1 + 1 \ldots m_2$ (which is just a regular RSA signature).
4. The verifier accepts that (N, e) defines a permutation over all of $\mathbb{Z}_N$ if all of the following checks pass.
   (a) Check that $N > 0$ and N is not divisible by all the primes less than α. (Equivalently, one can let P be the product of all primes less than α (also known as $\alpha - 1$ primorial) and verify that $\gcd(N, P) = 1$).
   (b) Check that $e > 0$ and is e not divisible by all the primes less than e'. (In most implementations of RSA, e is a fixed prime, in which case the verifier can just check that $e = e'$).
   (c) Verify that $\rho_i = (\sigma_i)^{eN} \pmod N$ for $i = 1 \ldots m_1$.
   (d) Verify that $\rho_i = (\sigma_i)^e \pmod N$ for $i = m_1 + 1 \ldots m_2$.

Note that for many natural choices of parameters (e, κ, α), we have $m_1 = m_2$, and so step 4d disappears.

Theorem 4. *$\mathcal{P}_{\mathsf{perm}\mathbb{Z}_N}$ is a 2-message public-coin proof with perfect completeness, perfect honest-verifier zero-knowledge, and statistical soundness error $2^{-\kappa}$ for the promise problem*

$$\begin{aligned} \mathrm{L}_{\mathsf{yes}} &= \mathrm{L}_{\mathsf{paillier}\text{-}N} \cap \mathrm{L}_{\mathsf{perm}\mathbb{Z}_N^*} \cap \mathrm{L}_\alpha \cap \mathrm{L}_{e'} \\ \mathrm{L}_{\mathsf{no}} &= \overline{\mathrm{L}_{\mathsf{perm}\mathbb{Z}_N}} \cup \overline{\mathrm{L}_\alpha} \cup \overline{\mathrm{L}_{e'}} \end{aligned}$$

Proof. It is a standard fact (per Lemmas 3 and 4) that raising to a power f is a permutation of $\mathbb{Z}_N$ whenever $N \in \mathrm{L}_{\mathsf{paillier}\text{-}N}$ and $\gcd(f, \phi(N)) = 1$, and that the inverse of this permutation is raising to the power $(f^{-1} \bmod \phi(N))$. This fact, when we set $f = eN$ for $i = 1 \dots m_1$ and $f = N$ for $i = m_1 + 1, \dots, m_2$, gives perfect completeness. It also gives a perfect HVZK simulator who simply chooses σ_i and computes $\rho_i = (\sigma_i)^{eN} \bmod N$ for $i = 1 \dots m_1$ and $\rho_i = (\sigma_i)^e$ for $i = m_1 + 1, \dots, m_2$ (recall that, by definition, the simulator needs to work only for $(N, e) \in \mathrm{L}_{\mathsf{yes}}$).

To show soundness, suppose $(N, e) \in \mathrm{L}_{\mathsf{no}}$. If $x \in \overline{\mathrm{L}_\alpha} \cup \overline{\mathrm{L}_{e'}}$, the verifier will reject in steps 4a or 4b, and soundness holds. Therefore, assume $(N, e) \in \mathrm{L}_\alpha \cap \mathrm{L}_{e'}$. This means $(N, e) \notin \mathrm{L}_{\mathsf{perm}\mathbb{Z}_N}$.

Suppose $(N, e) \notin \mathrm{L}_{\mathsf{square}\text{-}\mathsf{free}}$. Since the smallest prime divisor of N is at least α, by applying Lemma 6, we know at most $1/\alpha$ fraction of $\mathbb{Z}_N$ will have an Nth root. By choosing m_1 elements of $\mathbb{Z}_N$ and verifying that they have Nth roots, we ensure that the chances that the prover passes Step 4c with N that is not square-free are at most $(1/\alpha)^{m_1} \leq 2^{-\kappa}$.

Now suppose $(N, e) \in \mathrm{L}_{\mathsf{square}\text{-}\mathsf{free}}$ but $(N, e) \notin \mathrm{L}_{\mathsf{perm}\mathbb{Z}_N}$. Since N is square free, the smallest prime divisor of N is at least α, and the smallest prime divisor of e is at least e', we can apply Lemma 7 to conclude that at most $1/\alpha + (1 - 1/\alpha)/e'$ fraction of $\mathbb{Z}_N$ have an eth root. By choosing m_2 elements of $\mathbb{Z}_N$ and verifying that they have eth roots, we ensure that the chances that the prover passes Steps 4c and 4d are at most

$$\left(\frac{1}{\alpha} + \frac{1}{e'}\left(1 - \frac{1}{\alpha}\right)'\right)^{m_2} \leq 2^{-\kappa}.$$

□

A Possible Optimization. Instead of choosing the ρ_i values from $\mathbb{Z}_N$, the prover could choose ρ_i values from $\mathbb{Z}_N^*$ (this requires m_2 gcd computations), and set a potentially lower $m_2 = \max(\lceil \kappa / \log_2 e' \rceil, m_1)$. The proofs of completeness and zero-knowledge proofs remain the same (because if (N, e) define a permutation over $\mathbb{Z}_N$, they also define a permutation when restricted to $\mathbb{Z}_N^*$). The proof of soundness changes in the last paragraph. Observe that, since $(\mathrm{L}_{\mathsf{square}\text{-}\mathsf{free}} \cap \mathrm{L}_{\mathsf{perm}\mathbb{Z}_N^*}) \subset \mathrm{L}_{\mathsf{perm}\mathbb{Z}_N}$, if $(N, e) \in \mathrm{L}_{\mathsf{square}\text{-}\mathsf{free}}$ but $(N, e) \notin \mathrm{L}_{\mathsf{perm}\mathbb{Z}_N}$, then $(N, e) \notin \mathsf{L}_{\mathsf{perm}\mathbb{Z}_N^*}$. Thus, per Lemma 8, the chances that the prover passes steps 4c and 4d are at most

$$\left(\frac{1}{e'}\right)^{m_2} \leq 2^{-\kappa}.$$

For example, for $\kappa = 128$ and $e = \alpha = 65537$, this optimization reduces the value of m_2 from 9 to 8. This reduction in m_2 is at the expense of $\gcd(\rho_i, N)$ computations, and so it may or may not improve overall performance, depending on the implementation and the parameter values. We emphasize, however, that the lower m_2 value will not give security $2^{-\kappa}$ without the gcd computations on the part of the verifier, so implementers of this optimization should ensure the verifier rejects if $\gcd(\rho_i, N) \neq 1$ for some i.

3.4 HVZK Proof for a Product of Two Primes

In this section, we consider the language

$$\mathsf{L}_{\mathsf{ppp}} = \{N > 0 \mid N \text{is odd and has exactly two distinct prime divisors}\}.$$

Note that the more interesting language is

$$\mathsf{L}_{\mathsf{pp}} = \{N > 0 \mid N \text{is odd and is a product of two distinct primes }\} = (\mathsf{L}_{\mathsf{ppp}} \cap \mathsf{L}_{\mathsf{square\text{-}free}}) \supset (\mathsf{L}_{\mathsf{ppp}} \cap \mathsf{L}_{\mathsf{pailler}\text{-}N}),$$

because it rules out prime powers as factors of N.

We obtain a two-round public-coin HVZK proof for the promise problem $\mathsf{L}_{\mathsf{yes}} = \mathsf{L}_{\mathsf{pp}}$ and $\mathsf{L}_{\mathsf{no}} = \overline{\mathsf{L}_{\mathsf{ppp}}}$ (note that only N not in $\mathsf{L}_{\mathsf{yes}} \cup \mathsf{L}_{\mathsf{no}}$ are those that have exactly two distinct odd prime divisors and are not square-free). We can obtain an HVZK proof for L_{pp} (with a similar gap for the case $p|q-1$) by combing the protocol in this section with the protocol for and $\mathsf{L}_{\mathsf{square\text{-}free}}$, similar to [AP18]. The combination can be space-saving, similar to Protocol $\mathcal{P}_{\mathsf{perm}\mathbb{Z}_N}$ in Sect. 3.3.

Let J_N denote the subset of $\mathbb{Z}_N^*$ with Jacobi symbol 1. Let QR_N denote the subset of J_N that consists of quadratic residues in $\mathbb{Z}_N^*$. The following is an HVZK protocol for for the promise problem ($\mathsf{L}_{\mathsf{yes}} = \mathsf{L}_{\mathsf{pp}}$, $\mathsf{L}_{\mathsf{no}} = \overline{\mathsf{L}_{\mathsf{ppp}}}$). Let κ be the statistical security parameter.

Protocol $\mathcal{P}_{\mathsf{ppp}}$

1. Both the Prover and the Verifier let $m = \lceil \kappa \cdot 32 \cdot \ln 2 \rceil$.
2. The Verifier chooses m random values $\rho_i \in J_N$ and sends them to Prover.
3. For every $\rho_i \in QR_N$, the Prover sends back $\sigma_i \in Z_N^*$ such that $\sigma_i^2 \bmod N = \rho_i$. Of the four square roots, the Prover chooses one at random. For other ρ_i, the prover sends back 0.
4. Verifier first checks that N is a positive odd integer and is not a prime or a prime power (see [Ber98, BLP07] and references therein). If these checks pass, then the Verifier accepts if the number of nonzero responses is at least $3m/8$, and for every nonzero σ_i, it holds that $\rho_i = (\sigma_i)^2 \bmod N$.

Note that our design choice to have the verifier pick values in J_N rather than in all of Z_N^* results in improved efficiency by a factor of four as compared to the hash-then-solve protocol presented in [AP18]. This is because when the verifier chooses elements in J_N, at least 1/2 of them have square roots for $N \in \mathsf{L}_{\mathsf{pp}}$, vs. 1/4 for $N \notin \mathsf{L}_{\mathsf{ppp}}$. In contrast, when the verifier chooses elements in all of Z_N^*, the fractions change to 1/4 and 1/8, respectively. But the number of repetitions m required to distinguish 1/4 from 1/8 is four times greater than the number of repetitions required to distinguish 1/2 from 1/4, for any fixed confidence level $2^{-\kappa}$ (this follows from bounds on the tail of the binomial distribution; see the proof of Theorem 5).

Theorem 5. *$\mathcal{P}_{\mathsf{ppp}}$ is a 2-message public-coin protocol for the promise problem ($\mathsf{L}_{\mathsf{yes}} = \mathsf{L}_{\mathsf{pp}}$, $\mathsf{L}_{\mathsf{no}} = \overline{\mathsf{L}_{\mathsf{ppp}}}$) with statistical completeness error $2^{-\kappa}$, computational honest-verifier zero-knowledge, and statistical soundness error $2^{-\kappa}$.*

Proof. In order to show completeness, we need to show that the honest prover will be able to carry out Step 3, and the verifier's checks in Step 4 will pass. Since the prover knows the factorization of $N = pq$, it can efficiently check if $\rho_i \in QR_N$ by determining if it is a quadratic residue module each prime divisor p and q of N.

Then, given that $\rho_i \in QR_N$, it is easy for the prover to compute σ_i such that $\sigma_i^2 \bmod N = \rho_i$. To do so, the prover computes $\beta_i = \rho_i \bmod p$ and $\gamma_i = \rho_i \bmod q$. Then the prover finds solutions $\pm b$ to $\sigma_i^2 \bmod p = \beta$, and $\pm c$ to $\sigma_i^2 \bmod q = \gamma$, using any of the available algorithms for finding square roots modulo primes. Finally, the prover uses the Chinese Remainder Theorem to obtain four solutions (corresponding to pairs $(b, c), (-b, c), (b, -c), (-b, -c)$) to $\sigma_i^2 \bmod N = \rho_i$. Thus, the prover can indeed carry out Step 3.

Let us now discuss why the verifier's checks in Step 4 will pass with probability close to 1. As discussed above if $\rho_i \in QR_N$ the prover can always send back valid σ_i's. So in order to achieve completeness, we need to make sure that among the ρ_i's sent from the Verifier to the Prover in Step 2, at least $3m/8$ of them are in QR_N. Since $N \in \mathsf{L}_{\mathsf{pp}}$, $|J_N| = \phi(N)/2$ while $|QR_N| = \phi(N)/4$ (it is in this step that we use the fact that that $N \in \mathsf{L}_{\mathsf{pp}}$ and not just in $\mathsf{L}_{\mathsf{ppp}}$; because $|J_N|$ when is N is a product of two prime powers can be more than twice $|QR_N|$ if one or both the powers is even).

By applying the classic Hoeffding bound [Hoe63, Theorem 2] for $m = \lceil \kappa \cdot 32 \cdot \ln 2 \rceil$, we see that Pr[the number of ρ_i's $\in QR_N < 3m/8$] $< e^{-2m(1/2-3/8)^2} = 2^{-2m/(64 \ln 2)} \leq 2^{-\kappa}$. Thus we conclude that our protocol has statistical completeness with error probability at most $2^{-\kappa}$.

To show soundness, suppose that $N \notin \mathsf{L}_{\mathsf{ppp}}$, i.e., N is even, a prime, a prime power, or has at least three prime divisors. If N is even, a prime, or a prime power, the verifier will reject. If N has at least three prime divisors, then at most 1/4 of the elements of J_N have square roots. But the prover can cheat only if $3m/8$ of the ρ_i values have square roots. Thus, probability of cheating is Pr[the number of squares is $\geq 3m/8$] $\leq e^{-2m(3/8-1/4)^2} \leq 2^{-\kappa}$ by the Hoeffding bound.

Finally, we argue that our protocol is computational honest-verifier zero-knowledge. We first recall the QR assumption [GM84].

Assumption 6 (QR assumption). *For any $N = pq$, a randomly chosen $\rho \in J_N$, and any PPT algorithm A,*

$$\Pr[\sigma = QR(\rho) \mid N = pq,\ \rho \leftarrow J_N,\ A(\rho, N) \rightarrow \sigma \in \{\pm 1\}] \leq 1/2 + negl(\kappa).$$

The HVZK simulator (which, by definition, needs to work only when $N \in \mathsf{L}_{\mathsf{pp}}$) will pick random values σ_i and square them getting ρ_i. For each number, it will flip a coin and, depending on the coin's output the simulator will either output (σ_i, ρ_i) or $(0, \rho_i')$ for a random $\rho_i' \in J_N$. Because of the QR assumption (the distributions of J_N and QR_N are computationally indistinguishable) the view of the simulator is computationally indistinguishable from that of an honest verifier interacting with a prover. □

3.5 HVZK Proof for a Blum Integer

In this section we consider the language $\mathsf{L}_{\mathsf{blum-powers}} = \{N > 0 \mid N = p^a q^b$ for primes $p \equiv q \equiv 3 \pmod 4\}$. Note, similar to Sect. 3.4, that the more interesting language is the language of Blum integers $\mathsf{L}_{\mathsf{blum}} = \mathsf{L}_{\mathsf{square\text{-}free}} \cap \mathsf{L}_{\mathsf{blum-powers}}$.

In this section we obtain a two-round public-coin HVZK protocol for the promise problem $(\mathsf{L}_{\mathsf{yes}} = \mathsf{L}_{\mathsf{blum}}, \mathsf{L}_{\mathsf{no}} = \overline{\mathsf{L}_{\mathsf{blum-powers}}})$. We can obtain a protocol for $\mathsf{L}_{\mathsf{blum}}$ (with a similar gap for the case $p \mid q-1$ as in Sect. 3.2) by combing the proofs for $\mathsf{L}_{\mathsf{blum-powers}}$ and $\mathsf{L}_{\mathsf{square\text{-}free}}$. Remarks at the beginning of Sect. 3.4 apply here, as well.

The protocol for $\mathsf{L}_{\mathsf{blum-powers}}$ is very similar to the protocol for $\mathsf{L}_{\mathsf{ppp}}$ but instead of considering square roots, we now consider 4th roots. Note that if N is a Blum integer then among the four roots of $\rho_i \in QR_N$, one and only one is a quadratic residue.

Protocol $\mathcal{P}_{\mathsf{blum-powers}}$
Same as protocol $\mathcal{P}_{\mathsf{ppp}}$ described in Sect. 3.4 but in step 3 the prover computes 4th roots instead and in step 4 the verifier checks 4th roots.

1. Both the Prover and the Verifier let $m = \lceil \kappa \cdot 32 \cdot \ln 2 \rceil$.
2. The Verifier chooses m random values $\rho_i \in J_N$ and sends them to Prover.
3. For every $\rho_i \in QR_N$, the Prover sends back $\sigma_i \in Z_N^*$ such that $\sigma_i^4 \bmod N = \rho_i$, choosing one at random from among four possibilities. For other ρ_i, the prover sends back 0.
4. Verifier first checks that N is a positive odd integer and is not a prime or a prime power (see [Ber98, BLP07] and references therein). The Verifier accepts if the number of nonzero responses is at least $3m/8$, and for every nonzero σ_i, it holds that $\rho_i = (\sigma_i)^4 \bmod N$.

Theorem 7. *$\mathcal{P}_{\mathsf{blum-powers}}$ is a 2-message public-coin protocol with statistical completeness error $2^{-\kappa}$, perfect honest-verifier zero-knowledge, and statistical soundness error $2^{-\kappa}$, for the promise problem* $(\mathsf{L}_{\mathsf{yes}} = \mathsf{L}_{\mathsf{blum}}, \mathsf{L}_{\mathsf{no}} = \overline{\mathsf{L}_{\mathsf{blum-powers}}})$.

Proof. Similar to the proof of Theorem 5, we get statistical completeness with error $2^{-\kappa}$. The prover knowing the factorization of N can efficiently compute the 4th roots for $N \in \mathrm{L}_{\mathsf{yes}}$, and completeness relies on receiving enough ρ_i's $\in QR_N$. Also we get the same statistical soundness error $2^{-\kappa}$.

Finally, $\mathcal{P}_{\mathsf{blum-powers}}$ achieves perfect honest-verifier zero-knowledge since -1 is always a Jacobi symbol 1 non-square. Then we can construct a simulator that, after computing ρ_i by raising a random σ_i to the fourth power, flips a coin and sends either $(0, -\rho_i)$ or (σ_i, ρ_i). □

4 Making Our Protocols Noninteractive via Fiat-Shamir

We use the Fiat-Shamir paradigm [FS86] to convert each of the 2-message public-coin HVZK interactive protocols presented above into a non-interactive zero-knowledge (NIZK) protocol. The transformation is very simple, because the first message in every protocol we present always consists of the verifier sending some challenges $\rho_1, \ldots, \rho_m$ to the prover. The challenges are uniformly distributed in some space with easy membership testing (such as $\mathbb{Z}_N$ or $\mathbb{Z}_N^*$, for example).

Thus, to make our protocols noninteractive, Prover samples ρ_i by herself using the random oracle. To make sure values ρ_i are in the correct space, such as $\mathbb{Z}_N$ or $\mathbb{Z}_N^*$, the prover performs rejection sampling for each ρ_i using a counter, trying multiple random-oracle outputs until obtaining the first one that lands in the desired space. Thus, each ρ_i is obtained by computing the output of the random oracle over the concatenation of (1) the protocol input—e.g., the RSA public key (N, e); (2) a salt given as a system parameter; (3) the index i; and (4) the counter value. If the result is in the correct space, the prover uses this ρ_i; if not, she increments the counter and tries again.

Thus, the protocol input and the salt determine the set of $\rho_i \in \mathbb{Z}_N$. The verifier can therefore compute ρ_i on his own, by following the same procedure as the prover, and subsequently perform verification. Note that the verifier, just like the prover, will need to perform rejection sampling.

The noninteractive proof then is simply the message that the prover sends to the verifier in the interactive protocol.

The security of this transformation is standard; we provide some formal details in Appendix B.

5 Specification, Implementation and Performance for NIZK of Permutations over $\mathbb{Z}_N$

Specification. Here we provide a more precise specification the protocol of Sect. 3.3 made non-interactive using the Fiat-Shamir paradigm as described in Sect. 4. The goal of this specification is to make the protocol precise enough for implementation and compatibility. The full specification is available in Appendix C. It assumes e is a fixed prime and thus sets $e' = e$. It takes in

α and the salt as system parameters. The random oracle used to deterministically select the ρ_i values is a "full-domain hash" [BR93] instantiated with the industry-standard MGF1 Mask Generation Function as defined in [MKJR16, Sec. B.2.1]. We use the industry-standard I2OSP and OS2IP to convert between octet strings and integers [MKJR16, Sec. 4.1] and the industry-standard RSASP to perform an RSA secret key operations [MKJR16, Sec. 5.2.1], and RSAVP for RSA public-key operations [MKJR16, Sec. 5.2.2].

Implementation. An open-source implementation of our specification in C#, based on the bouncycastle cryptographic library [bou], is publicly available [cod]. We hope that our implementation will become a part of bouncycastle.

Integration with TumbleBit. Our implementation has already been integrated into the open-source reference implementation of TumbleBit, which is currently being developed for production use [Ntu, Str17]. TumbleBit [HAB+17] is a unidirectional Bitcoin payment hub that allows parties to make fast, anonymous, off-blockchain payments through an untrusted intermediary called the Tumbler. The security of the TumbleBit protocol rests on the assumption that the Tumbler's RSA public key (N, e) defines a permutation over $\mathbb{Z}_N$. In the absence of this assumption, the Tumbler can steal bitcoin from payers.[2] Thus, in addition to publishing (N, e), a Tumbler publishes our NIZK proof that (N, e) defines a permutation, which is verified, during a setup phase, by any payer or payee who wants to participate in the protocol with this Tumbler. Integration with TumbleBit was easy. No modification to the existing TumbleBit protocol or codebase were required; instead, our NIZK was simply added to TumbleBit's setup phase.

Parameters and Performance for TumbleBit. When used with TumbleBit, our NIZK has parameters $\kappa = 128$, the RSA key length is $|N| = 2048$, the public RSA exponent is $e = e' = 65537$, and the salt is the SHA256 hash of the Genesis block of the Bitcoin blockchain.

The performance of our NIZK largely depends on our choice of the parameter α. A shorter α means that the verifier has to spend less time trying to divide N by primes less than α, but also increases m_1 and m_2, the number of RSA values

[2] Specifically, in TumbleBit, the Tumbler provides the payee Bob with a value z called a "puzzle," and a proof that its solution will transfer some of Tumbler's money to Bob. This solution is a value ϵ such that $z = \epsilon^e \bmod N$. The protocol crucially relies on uniqueness of ϵ, because the proof that the solution will unlock money applies to only one of the solutions of z. When Alice wants to pay Bob, she learns the solution to the puzzle in exchange for paying money to the Tumbler, and then gives that solution to Bob as payment. If RSA is not a permutation, then a malicious Tumbler can provide the payee Bob with a puzzle z that has two valid solutions $\epsilon_1 \neq \epsilon_2$, where $z = (\epsilon_1)^e = (\epsilon_2)^e \mod N$, and a proof that ϵ_1 transfers money. Then, to steal money, the Tumbler gives payer Alice the solution ϵ_2 in exchange for her money, which does not permit Bob to obtain the Tumbler's money and complete the transaction.

in the NIZK. The relationship between α and m_1, m_2 is determined by Eq. (1). Specifically for the TumbleBit parameters, we show this relationship in Fig. 1. To evaluate the performance of our NIZK, we choose the smallest value of α that corresponds a given pair of (m_1, m_2) values, and benchmark proving and verifying times for our NIZK for the RSA key length $|N| = 2048$ bits in Table 1 on a single-core of an Intel Xeon processor. We can see from the table that choosing $\alpha = 319567$ (so that $m_1 = 7$ and $m_2 = 9$) gives optimal performance, though performance for $\alpha = 65537$ is roughly similar and the optimal choice is likely implementation-dependent.

For the optimal choice of α, proving takes about 237 ms (a small fraction of the key generation cost, which is 2022 ms) and verifying takes about 713 ms. For comparison, verification of our NIZK is about 8 times faster than the folklore solution discussed in Sect. 1, which requires the verifier to spend 5588 ms to perform the Rabin-Miller primality test on a 2048-bit RSA exponent, and also slows down every public-key operation by a factor of about 60 because e is 2048 bits long (instead of $e = 65537$, which is 17 bits long). We should note that even though our solution is much faster than the folklore one, and adds only 12% to the prover's normal RSA key generation cost, it is still relatively expensive for the verifier: for comparison, the public key operation (encryption or signature verification) with $e = 65537$ takes only about 1.4ms.

From Table 1 we also see that verifying is generally slower than proving (until α gets so big that divisibility testing takes too long for the verifier). This follows because proving involves m_1 modular exponentiations (using RSASP), which can be done separately modulo p and modulo q for $N = pq$ (with the exponent reduced modulo $p-1$ and $q-1$), and then combined using the Chinese Remainder Theorem (CRT). Meanwhile, the verifier does not know p and q, and so cannot use (CRT); moreover, the exponent used for modular exponentiations (using RSAVP) is slightly longer than $\phi(n)$, but the verifier does not know $\phi(N)$ and so cannot reduce it. Thus, exponentiations performed by the verifier are slower than those performed by the prover.

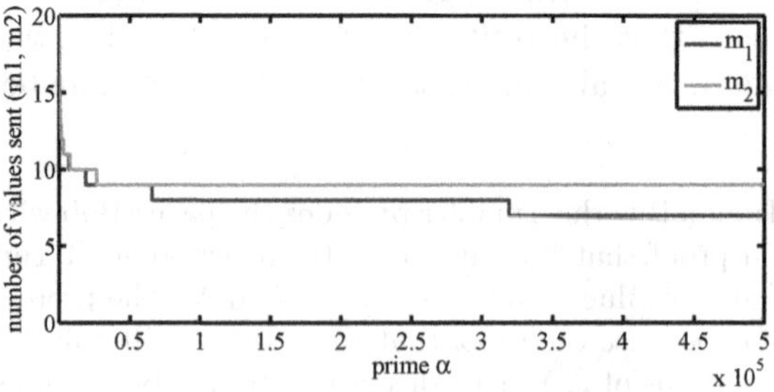

Fig. 1. Values of m_1 and m_2 versus the choice of parameter α for our NIZK, when $\kappa = 128$ and $e = e' = 65537$.

Table 1. Proving and verifying times for our C# implementation as observed on an Azure DS1 v2 virtual machine running Windows Server 2016 Datacenter (single-core 2.4 GHz Intel Xeon E5-2673 v3 Haswell processor, 3.5GiB RAM). Time is given in ms. Public exponent is $e = e' = 65537$ and security parameter is $\kappa = 128$.

Parameters			Permutation Proof	
α	m_1	m_2	Prove	Verify
41	24	24	632	2326
89	20	20	518	1925
191	17	17	443	1612
937	13	13	334	1216
1667	12	12	311	1127
3187	11	12	308	1042
3347	11	11	281	1025
7151	10	11	284	943
8009	10	10	256	948
19121	9	10	254	853
26981	9	9	233	854
65537	8	9	230	768
319567	7	9	237	713
2642257	6	9	234	956
50859013	5	9	230	6756

Acknowledgements. The authors thank Ethan Heilman, Alessandra Scafuro and Yehuda Lindell for useful discussions. This research was supported, in part, by US NSF grants 1717067, 1350733, and 1422965.

A Number-Theoretic Lemmas

We present number-theoretic lemmas that are useful for proving security of our protocols. Some of them are standard and are presented here only to make the presentation self-contained.

Let $\mathbb{Z}_N = \{0, 1, ..., N-1\}$ for any positive integer N and $\mathbb{Z}_N^*$ be the multiplicative group modulo N, i.e., the set of values in $\mathbb{Z}_N$ that are relatively prime to N, or else $\{x \in \mathbb{Z}_N | \gcd(x, N) = 1\}$. We use notation $p|N$ to denote that "p divides N".

Euler's phi or totient function (see, e.g., [Sho09, Section 2.6] for the relevant background) is defined for all positive integers N as:

$$\phi(N) = |\mathbb{Z}_N^*|.$$

If $N = pq$ where p, q are two distinct primes it holds that $\phi(N) = (p-1)(q-1)$. More generally, if the prime factorization of N is $N = p_1^{\alpha_1} \times \cdots \times p_k^{\alpha_k}$, then $\phi(N) = (p_1^{\alpha_1 - 1} \times \cdots \times p_k^{\alpha_k - 1}) \times ((p_1 - 1) \times \cdots \times (p_k - 1))$, with $\phi(1) = 1$ [Sho09, Theorem 2.11]. The following theorem is standard [Sho09, Theorem 2.13]:

Lemma 1 (Euler's theorem). *Let N be a positive integer and $a \in \mathbb{Z}_N^*$. Then $a^{\phi(N)} \bmod N = 1$.*

Given positive integers N and e, consider the map $x \mapsto x^e \bmod N$. We will first consider this map as restricted to $\mathbb{Z}_N^*$. The following lemma is standard.

Lemma 2. *The map* $x \mapsto x^e \bmod N$ *is a permutation of* $\mathbb{Z}_N^*$ *if and only if* $\gcd(e, \phi(N)) = 1$. *If the map is a permutation of* $\mathbb{Z}_N^*$*, then its inverse is the map* $x \mapsto x^d \bmod N$ *for* $d = e^{-1} \bmod \phi(N)$ *(which exists by [Sho09, Theorem 2.5] because* $\gcd(e, \phi(N)) = 1$*).*

Proof. Suppose $\gcd(e, \phi(N)) = 1$. Then let $d = e^{-1} \bmod \phi(N)$. Thus, $de = k\phi(N) + 1$ for some integer k. For every $x \in \mathbb{Z}_N^*$, $(x^e)^d \bmod N = x^{ed} \bmod N = (x^{\phi(N)})^k \cdot x \bmod N = 1^k \cdot x = x$, where the second-to-last equality follows from Lemma 1.

Now suppose $\gcd(e, \phi(N)) = g \neq 1$. Let p be a prime divisor of g. Then $p \mid \phi(N)$, and therefore $\mathbb{Z}_N^*$ contains an element $x \neq 1$ such that $x^p \bmod N = 1$ [Sho09, Theorem 6.42]. Therefore, $x^e \bmod N = (x^p)^{e/p} \bmod N = 1^{e/p} = 1$, and thus the map is not a permutation. □

A number N is *square free* if it can be written as $N = p_1 p_2 \dots p_k$ for *distinct* prime numbers p_i. (N is not square free if it is divisible by p^2, where p is some prime.)

Lemma 3. *For a positive integer* N*, if* $\gcd(N, \phi(N)) = 1$*, then* N *is square-free.*

Proof. Indeed, suppose $p^2 \mid N$ for some prime p. Then $p \mid \phi(N)$, so $\gcd(N, \phi(N)) \geq p > 1$. □

We now extend one direction of Lemma 2 to all of $\mathbb{Z}_N$ for the case of square-free N.

Lemma 4. *If for some positive integers* N *and* f*,* N *is square-free and* $\gcd(f, \phi(N)) = 1$*, then the map* $x \mapsto x^f \bmod N$ *is a permutation on* $\mathbb{Z}_N$*. Its inverse is computed as follows: for* $g = f^{-1} \bmod \phi(N)$ *(which exists by [Sho09, Theorem 2.5]) and for all* $x \in \mathbb{Z}_N$*,* $x^{gf} \bmod N = x$*.*

Proof. Let $N = p_1 p_2 \dots p_k$ for distinct prime numbers p_i. By the Chinese Remainder Theorem (CRT) [Sho09, Theorem 2.8], the ring $\mathbb{Z}_N$ is isomorphic to the product of rings $Z_{p_1} \times \cdots \times \mathbb{Z}_{p_k}$. It therefore suffices to show that $x^{ef} \bmod p_i = x$ for each i. Indeed, $fg = t\phi(N) + 1$ for some integer t, and therefore $x^{fg} = (x^{p_i - 1})^s \cdot x$ for some integer s, and the result follows by Fermat's little theorem [Sho09, Theorem 2.14] when $x \bmod p_i \neq 0$, and trivially when $x \bmod p_i = 0$. □

To extend the other direction of Lemma 2 to all of $\mathbb{Z}_N$ is a little more complicated.

Lemma 5. *If for some positive integers* N *and* f*, the map* $x \mapsto x^f \bmod N$ *is a permutation on* $\mathbb{Z}_N$*, then* $x \mapsto x^f \bmod N$ *is a permutation on* $\mathbb{Z}_N^*$ *(and thus* $\gcd(f, N) = 1$ *by Lemma 2) and either:*

- N *is square-free, or*
- $f = 1$

Proof. The first part of the lemma follows from the fact that when raised to the power f modulo N, elements of Z_N^* stay within Z_N^* (because if $\gcd(x, N) = 1$, then $\gcd(x^f \bmod N, N) = 1$). The second part of the lemma is proven as follows. Suppose N is not square-free and $f > 1$. Then let $p^2 \mid N$ for some prime p. The set $\{x \in \mathbb{Z}_N : x \text{ is divisible by } p\}$ contains N/p elements. The image of this set is contained in $\{x \in \mathbb{Z}_N : x \text{ is divisible by } p^2\}$, which contains only N/p^2 elements. Thus, the map is not injective.

The following lemma shows that one can validate if an integer N is square-free by checking if random values in $\mathbb{Z}_N$ have Nth roots. This lemma generalizes the result of Gennaro, Micciancio, and Rabin [GMR98, Section 3.1], which worked over $\mathbb{Z}_N^*$ and thus required a gcd computation every time a random value was selected.

Lemma 6. *Let N be a positive integer and p be a prime such that p^2 divides N (i.e., N is not square free). Then, the fraction of elements of $\mathbb{Z}_N^*$ that have an Nth root modulo N is at most $1/p$, and the fraction of elements of $\mathbb{Z}_N$ that have an Nth root modulo N is also at most $1/p$.*

Proof. Suppose x has an Nth root modulo N. Then there is a value r such that $r^N \equiv x \pmod N$. Hence, N divides $r^N - x$, which means p^2 divides $r^N - x$ (since p^2 divides N), and therefore r is the Nth root of x modulo p^2. Thus, in order to have an Nth root modulo N, x must have an Nth root modulo p^2. Since a uniformly random element x of $\mathbb{Z}_N$ is also uniform modulo p^2, and a uniformly random element x of $\mathbb{Z}_N^*$ is also uniform in $Z_{p^2}^*$ when reduced modulo p^2, it suffices to consider what fractions of $Z_{p^2}^*$ and of Z_{p^2} have Nth roots.

By Lemma 8 below, the number of elements of $Z_{p^2}^*$ that have Nth roots is at most $\phi(p^2)/e'$, where e' is the largest prime divisor of $\gcd(N, \phi(p^2)) = \gcd(N, p(p-1))$. Since $p|N$, we have $e' = p$. Thus, the number of elements of $Z_{p^2}^*$ that have Nth roots is at most $\phi(p^2)/p = p - 1$. This shows the first half of the conclusion.

If $x \in Z_{p^2} - Z_{p^2}^*$, then $p|x$. If x has an Nth root r modulo p^2, then $p^2|(r^N - x)$, hence $p|(r^N - x)$, hence $p|r^N$ (because $p|x$ and $p|(r^N - x)$), hence $p|r$ (because p is prime), hence $p^2|r^2$, hence $p^2|r^N$ (because $N > 1$), and hence $p^2|x$ (because $p^2|(r^N - x)$ and $p^2|r^N$). We therefore have that $x \in Z_{p^2}$ and $p^2|x$, which means that $x = 0$.

Thus, the total number of elements of Z_{p^2} that have an Nth root is at most $p - 1$ elements from $Z_{p^2}^*$ and one element from $Z_{p^2} - Z_{p^2}^*$ (namely, the element $x = 0$), for a total of at most p elements from Z_{p^2}. Thus, at most a $p/|Z_{p^2}| = 1/p$ fraction of elements of Z_{p^2} have Nth roots. It follows that at most a $1/p$ fraction of elements of $\mathbb{Z}_N$ has Nth roots. □

The following lemma shows that if we know that N is square free (which we can test using Lemma 6), then we can check whether raising to the power e is a permutation of $\mathbb{Z}_N$, by checking if random values in $\mathbb{Z}_N$ have eth roots.

Lemma 7. *Suppose $N > 0$ is a square-free integer so that $N = p_1 p_2 \dots p_k$ for distinct prime numbers p_i, and $e > 0$ is an integer. If raising to the power e modulo N is not a permutation over $\mathbb{Z}_N$, then the fraction of elements of $\mathbb{Z}_N$ that have a root of degree e is at most*

$$\frac{1}{p} + \frac{1}{e'}\left(1 - \frac{1}{p}\right),$$

where e' is the smallest prime divisor of e and p is the smallest prime divisor of N (these are well-defined, because if $N = 1$ or $e = 1$, then raising to the eth power is a permutation over $\mathbb{Z}_N$).

Proof. By Chinese Remainder Theorem (CRT) [Sho09, Theorem 2.8], the ring $\mathbb{Z}_N$ is isomorphic to the product of rings $Z_{p_1} \times \cdots \times Z_{p_k}$. Note that if raising to the power e modulo N is not a permutation over $\mathbb{Z}_N$, then there exist $x \not\equiv y \pmod{N}$ such that $x^e \equiv y^e \pmod{N}$. Let i be such that $x \not\equiv y \pmod{p_i}$ (it must exist by CRT); then raising to the power e modulo p_i is not a permutation of Z_{p_i}, because $x^e \equiv y^e \pmod{p_i}$ (by CRT).

Since a uniformly random element x of $\mathbb{Z}_N$ is uniform modulo p_i, it suffices to consider what fraction of Z_{p_i} has eth roots. By Lemma 8 below, the number of elements of $Z^*_{p_i}$ that have eth roots is at most $\phi(Z^*_{p_i})/e' = (p_i - 1)/e'$. The only element in $Z_{p_i} - Z^*_{p_i}$ is the element 0. So, in total, at most $(p_i - 1)/e' + 1$ elements of Z_{p_i} have eth roots. Since $p_i \geq p$,

$$\frac{(p_i - 1)/e' + 1}{p_i} = \frac{1}{e'} + \frac{1}{p_i}\left(1 - \frac{1}{e'}\right) \leq \frac{1}{e'} + \frac{1}{p}\left(1 - \frac{1}{e'}\right) = \frac{1}{p} + \frac{1}{e'}\left(1 - \frac{1}{p}\right).$$

□

The proofs of two lemmas above relied on the lemma below.

Lemma 8. *For any positive integers N and e, if raising to the power e modulo N is not a permutation over $\mathbb{Z}^*_N$, then $\gcd(e, \phi(N)) > 1$ and the number of elements of $\mathbb{Z}^*_N$ that have a root of degree e is at most $\phi(N)/e'$, where e' is the largest prime divisor of $\gcd(e, \phi(N))$.*

Proof. Suppose there exist x and y in $\mathbb{Z}^*_N$ such that $x^e \equiv y^e \pmod{N}$ but $x \not\equiv y \pmod{N}$. Then $x/y \not\equiv 1 \pmod{N}$ but $(x/y)^e \equiv 1 \pmod{N}$. Therefore, the multiplicative order of (x/y) is greater than 1 and divides e [Sho09, Theorem 2.12] and $\phi(N)$ [Sho09, Theorem 2.13], which implies that $\gcd(e, \phi(N)) > 1$. Let e' be the largest prime divisor of $\gcd(e, \phi(N))$.

Because e' is a prime that divides $\phi(N)$, $\mathbb{Z}^*_N$ contains an element z of order e' [Sho09, Theorem 6.42]. Therefore, the homomorphism that takes each element of $\mathbb{Z}^*_N$ to the power e has kernel of size at least e' (because this kernel contains distinct values $z, z^2, \dots, z^{e'}$ which are all eth roots of 1 because e' divides e). The image of this homomorphism contains exactly the elements that have roots of degree e, and the size of this image is equal to $\phi(N)$ divided by the size of the kernel [Sho09, Theorem 6.23], i.e., at most $\phi(N)/e'$. □

B Background on the Fiat-Shamir transform

Any efficient, interactive constant-round, public-coin, honest-verifier zero knowledge (HVZK) proof system can be converted into a noninteractive ZK argument[3] (NIZK) through the so called Fiat-Shamir (FS) transformation [FS86]. Applying FS allows us to replace the verifier V by instead calling a hash function on input the current transcript. The security of the resulting scheme holds in the random oracle [BR93] (RO), where a hash function H is evaluated through calls to an oracle that acts as a random function. The main idea in the security proof is that the simulator for HVZK can "program" the RO (i.e., the simulator decides the answer to each specific query). This allows the simulator to convert the entire transcript of a public-coin HVZK proof into a single message that is indistinguishable from the message computed by an honest NIZK prover. We first recall the definition of NIZKs in the RO and then state the Fiat-Shamir transformation theorem (definitions slightly modified from [FKMV12]).

Let S be a simulator that operates in two modes: $(h_i, st) \leftarrow \mathrm{S}(1, st, q_i)$ which on input a random oracle query q_i it responds with h_i (usually by lazy sampling), and $(\pi, st) \leftarrow \mathrm{S}(2, st, x)$ which simulates simulates the actual proof. (Note that calls to $\mathrm{S}(1, \cdots)$ and $\mathrm{S}(2, \cdots)$ share the common state st that is updated after each operation).

Definition 2 (NIZK). *Let* $(\mathrm{S}_1, \mathrm{S}_2)$ *be oracles such that* $S_1(q_i)$ *returns the first output of* $(h_i, st) \leftarrow \mathrm{S}(1, st, q_i)$ *and* $S_2(x, w)$ *returns the first output of* $(\pi, st) \leftarrow \mathrm{S}(2, st, x)$ *if* $(x, w) \in RL$.

A protocol $\langle \mathrm{P}^H, \mathrm{V}^H \rangle$ *is said to be a NIZK proof for language* L *in the random oracle model, if there exists a PPT simulator* S *such that for all PPT distinguishers* D *we have*

$$\mathtt{view}D^{H(\cdot),\mathrm{P}^H(\cdot,\cdot)} \approx \mathtt{view}D^{\mathrm{S}_1(\cdot),\mathrm{S}_2^H(\cdot,\cdot)}.$$

We now state and prove the following theorem for the Fiat-Shamir transformation [FKMV12]:

Theorem 8 (Fiat-Shamir NIZK). *Let* κ *be a security parameter. Consider a non-trivial constant round, public-coin, honest-verifier zero-knowledge (HVZK) interactive proof system* $\langle \mathrm{P}, \mathrm{V} \rangle$ *for a language* L. *Let* $H()$ *be a function with range equal to the space of the verifier's coins. In the random oracle model the proof system* $\langle \mathrm{P}^H, \mathrm{V}^H \rangle$*, derived from* $\langle \mathrm{P}, \mathrm{V} \rangle$ *by applying the Fiat-Shamir transform, is a noninteractive ZK argument.*

Proof. (sketch) All we need to show is that there exists a simulator S as required in Definition 2. This can be done by invoking the HVZK simulator associated with the underlying interactive proof system.

---

[3] As opposed to a proof system where soundness needs to hold unconditionally, in an argument system it is sufficient that soundness holds with respect to a computationally bounded adversary P^*.

We design S to work as follows:

- To answer to a query q to S_1, $S(1, st, q)$ lazily samples a lookup table kept in state st. It checks whether an answer for q was already defined. If this is the case, it returns the previously assigned value; otherwise it returns a fresh random value h and stores the pair (q, h) in the table.
- To answer to a query x to S_2, $S(2, st, x)$ calls the HVZK simulator of $\langle P, V\rangle$ on input x to obtain a proof π. Then, it updates the look up table by storing x, π. If the look up table happens to be already defined on this input, S returns failure and aborts.

Given that the protocol is non-trivial, the probability of failure in each of the queries to S_2 is negligible. □

C Detailed Specification for the NIZK of Permutations over $\mathbb{Z}_n$

The following specification is for the NIZK of Permutations over $\mathbb{Z}_n$, as described in Sect. 5. This specification assumes that the RSA exponent e is prime.

C.1 System Parameters

The system parameters are the RSA modulus length `len`, the security parameter κ (where by default $\kappa = 128$), a small prime α (about 16 bits long or less), and a publicly-known octet string `salt`.

C.2 Proving

System Parameters:

1. `salt` (an octet string),
2. α (a prime number)
3. κ (the security parameter, use 128 by default)
4. e, the fixed prime RSA exponent
5. `len`, the RSA key length.

Auxiliary Function: getRho, defined in Sect. C.4.

Input: Distinct equal-length primes p and q greater than α such that the RSA modulus is $N = pq$ is of length `len`, and e does not divide $(p-1)(q-1)$.

Output: $(N, e), \{\sigma_1, ..., \sigma_{m_2}\}$.

Algorithm:

1. Set m_1 and m_2 as in Eq. 1, Sect. 3.3, with $e' = e$.
2. Set $N = pq$.
3. Obtain the RSA secret key K as specified by [MKJR16, Sec. 3.2]:

$$K = (p, q, d_{NP}, d_{NQ}, q_{Inv})$$

4. Compute the "weird RSA" secret key corresponding to public key (N, eN) (with exponent eN and modulus N) in the [MKJR16, Sec. 3.2] as

$$K' = (p, q, d_{NP}, d_{NQ}, q_{Inv})$$

where p, q, q_{Inv} are the same as in the normal RSA secret key K and

$$d_{NP} = (eN)^{-1} \bmod (p-1) \qquad d_{NP} = (eN)^{-1} \bmod (q-1) \tag{2}$$

5. For integer $i = 1 \ldots m_2$
   (a) Sample ρ_i, a random element of Z_N, as

$$\rho_i = \text{getRho}((N, e), \texttt{salt}, i, \texttt{len}, m_2)$$

   (b) If $i \leq m_1$, let

$$\sigma_i = \text{RSASP1}(K', \rho_i)$$

   where RSASP1 is the RSA signature primitive of [MKJR16, Sec. 5.2.1]. In other words, σ is the RSA decryption of ρ_i using the "weird RSA" secret key K'.
   (It follows that σ_i is (eN)th root of ρ_i.)
   (c) Else let

$$\sigma_i = \text{RSASP1}(K, \rho_i)$$

   where RSASP1 is the RSA signature primitive of [MKJR16, Sec. 5.2.1]. In other words, σ is the RSA decryption of ρ_i using the regular RSA secret key K.
   (It follows that σ_i is eth root of ρ_i.)
6. Output $(N, e), \{\sigma_1, \ldots, \sigma_{m_2}\}$.

C.3 Verifying

System Parameters:

1. `salt` (an octet string),
2. α (a prime number)
3. κ (the security parameter, use 128 by default)
4. e, the fixed prime RSA exponent
5. `len`, the RSA key length

Auxiliary Function: getRho, defined in Sect. C.4.

Input: RSA public key (N, e) and $\{\sigma_1, ..., \sigma_{m_2}\}$.

Output: VALID or INVALID

Algorithm:

1. Check that N is an integer and $N \geq 2^{\texttt{len}-1}$ and $N < 2^{\texttt{len}}$. If not, output INVALID and stop.
2. Check that e is prime. If not, output INVALID and stop.
3. Compute m_1 and m_2 per Eq. (1), Sect. 3.3, with $e' = e$.
4. Check that there are exactly m_2 values $\{\sigma_1, ..., \sigma_{m_2}\}$ in the input. If not, output INVALID and stop.
5. Generate the vector Primes($\alpha-1$), which includes all primes up to and including $\alpha-1$. (This can be efficiently implemented using the Sieve of Eratosthenes when α is small.)
   For each $p \in$ Primes($\alpha - 1$):
   - Check that N is not divisible by p. If not, output INVALID and stop. (Alternatively, let `primorial` be the product of all values in Primes($\alpha-1$). `primorial` should be computed once and should be a system parameter. Check that gcd(`primorial`, N) = 1.)
6. For integer $i = 1 \ldots m_2$
   (a) Sample ρ_i, a random element of Z_N, as

   $$\rho_i = \text{getRho}((N, e), \texttt{salt}, i, \texttt{len}, m_2)$$

   (b) If $i \leq m_1$, check that

   $$\rho_i = \text{RSAVP1}((N, eN), \sigma_i)$$

   where RSAVP1 is the RSA verification primitive of [MKJR16, Sec. 5.2.2]. In other words, check that ρ_i is the RSA encryption of σ_i using the "weird RSA" public key (N, eN). If not, output INVALID and stop.
   (Thus, check that $\rho_i = \sigma_i^{eN} \bmod N$).
   (c) Else check that

   $$\rho_i = \text{RSAVP1}(PK, \sigma_i)$$

   In other words, check that the ρ_i is the RSA encryption of σ_i using the RSA public key (N, e). If not, output INVALID and stop.
   (Thus, check that $\rho_i = \sigma_i^e \bmod N$).
7. Output VALID.

C.4 Auxiliary function: getRho

This function is for rejection sampling of a pseudorandom element $\rho_i \in Z_N$. It is "deterministic," always producing the same output for a given input.

Input:

1. RSA public key (N, e).
2. `salt` (an octet string)
3. Index integer i.
4. Length of RSA modulus `len`
5. Value m_2, with $i \leq m_2$.

Output: ρ_i

Algorithm:

1. Let
$$|m_2| = \lceil \tfrac{1}{8}(\log_2(m_2 + 1)) \rceil$$
be the length of m_2 in octets. (Note: This is an octet length, not a bit length!)
2. Let $j = 1$.
3. While true:
   (a) Let PK be the ASN.1 octet string encoding of the RSA public key (N, e) as specified in [MKJR16, Appendix A].
   (b) Let EI = I2OSP$(i, |m_2|)$ be the $|m_2|$-octet long string encoding of the integer i. (The I2OSP primitive is specified in [MKJR16, Sec. 4.2].)
   (c) Let EJ = I2OSP$(j, |j|)$ be the $|j|$-octet long string encoding of the integer j, where $|j| = \lceil \frac{1}{8} \log_2(j + 1) \rceil$.
   (d) Let $s = PK||\texttt{salt}||EI||EJ$ be the concatenation of these octet strings.
   (e) Let ER = MGF1-SHA256$(s, \texttt{len})$ where H_1 is the MGF1 Mask Generation Function based on the SHA-256 hash function as defined in [MKJR16, Sec. B.2.1], outputting values that are `len` bits long.
   (f) Let ρ_i = OS2IP(ER) be an integer.
   (That is, convert ER to an `len` bit integer ρ_i using the OS2IP primitive specified in [MKJR16, Sec. 4.1].)
   (g) If $\rho_i \geq N$, then let $j = j + 1$ and continue; Else, break.
   (Note: This step tests if $\rho_i \in Z_N$.)
4. Output integer ρ_i.

References

[AP18] Auerbach, B., Poettering, B.: Hashing solutions instead of generating problems: on the interactive certification of RSA moduli. In: Abdalla, M., Dahab, R. (eds.) PKC 2018. LNCS, vol. 10770, pp. 403–430. Springer, Cham (2018). https://doi.org/10.1007/978-3-319-76581-5_14

[Ber98] Bernstein, D.J.: Detecting perfect powers in essentially linear time. Math. Comput. 67, 1253–1283 (1998)

[BFGN17] Benhamouda, F., Ferradi, H., Géraud, R., Naccache, D.: Non-interactive provably secure attestations for arbitrary RSA prime generation algorithms. In: Foley, S.N., Gollmann, D., Snekkenes, E. (eds.) ESORICS 2017. LNCS, vol. 10492, pp. 206–223. Springer, Cham (2017). https://doi.org/10.1007/978-3-319-66402-6_13

[BFL89] Boyar, J., Friedl, K., Lund, C.: Practical zero-knowledge proofs: giving hints and using deficiencies. In: Quisquater, J.-J., Vandewalle, J. (eds.) EUROCRYPT 1989. LNCS, vol. 434, pp. 155–172. Springer, Heidelberg (1990). https://doi.org/10.1007/3-540-46885-4_18

[BLP07] Bernstein, D.J., Lenstra, H.W., Pila, J.: Detecting perfect powers by factoring into coprimes. Math. Comput. **76**(257), 385–388 (2007)

[bou] bouncycastle c# api. https://www.bouncycastle.org/csharp/index.html

[BR93] Bellare, M., Rogaway, P.: Random oracles are practical: a paradigm for designing efficient protocols. In: Proceedings of the 1st ACM Conference on Computer and Communications Security, pp. 62–73. ACM (1993)

[BY96] Bellare, M., Yung, M.: Certifying permutations: noninteractive zero-knowledge based on any trapdoor permutation. J. Cryptol. 9(3), 149–166 (1996). https://cseweb.ucsd.edu/~mihir/papers/cct.html

[CM99] Camenisch, J., Michels, M.: Proving in zero-knowledge that a number is the product of two safe primes. In: Stern, J. (ed.) EUROCRYPT 1999. LNCS, vol. 1592, pp. 107–122. Springer, Heidelberg (1999). https://doi.org/10.1007/3-540-48910-X_8

[CMS99] Cachin, C., Micali, S., Stadler, M.: Computationally private information retrieval with polylogarithmic communication. In: Stern, J. (ed.) EUROCRYPT 1999. LNCS, vol. 1592, pp. 402–414. Springer, Heidelberg (1999). https://doi.org/10.1007/3-540-48910-X_28

[cod] Tumblebit setup implementation. https://github.com/osagga/TumbleBitSetup

[CPP07] Catalano, D., Pointcheval, D. and Pornin, T. Trapdoor hard-to-invert group isomorphisms and their application to password-based authentication. J. Cryptol. **20**(1), 115–149, 2007. http://www.di.ens.fr/~pointche/Documents/Papers/2006_joc.pdf

[DJ01] Damgård, I., Jurik, M.: A Generalisation, a simplification and some applications of Paillier's probabilistic public-key system. In: Kim, K. (ed.) PKC 2001. LNCS, vol. 1992, pp. 119–136. Springer, Heidelberg (2001). https://doi.org/10.1007/3-540-44586-2_9

[FKMV12] Faust, S., Kohlweiss, M., Marson, G.A., Venturi, D.: On the non-malleability of the Fiat-Shamir transform. In: Galbraith, S., Nandi, M. (eds.) INDOCRYPT 2012. LNCS, vol. 7668, pp. 60–79. Springer, Heidelberg (2012). https://doi.org/10.1007/978-3-642-34931-7_5

[FS86] Fiat, A., Shamir, A.: How to prove yourself: practical solutions to identification and signature problems. In: Odlyzko, A.M. (ed.) CRYPTO 1986. LNCS, vol. 263, pp. 186–194. Springer, Heidelberg (1987). https://doi.org/10.1007/3-540-47721-7_12

[GM84] Goldwasser, S., Micali, S.: Probabilistic encryption. J. Comput. Syst. Sci. **28**(2), 270–299 (1984)

[GMR98] Gennaro, R., Micciancio, D., Rabin, T.: An efficient non-interactive statistical zero-knowledge proof system for quasi-safe prime products. In: Gong, L., Reiter, M.K. (eds.) CCS 19, Proceedings of the 5th ACM Conference on Computer and Communications Security, San Francisco, CA, USA, 3–5 November 1998, pp. 67–72. ACM (1998). http://eprint.iacr.org/1998/008

[HAB+17] Heilman, E., Alshenibr, L., Baldimtsi, F., Scafuro, A. and Goldberg, S.: Tumblebit: an untrusted bitcoin-compatible anonymous payment hub. In: 24th Annual Network and Distributed System Security Symposium, NDSS. The Internet Society (2017). https://eprint.iacr.org/2016/575.pdf

[HMRT12] Hazay, C., Mikkelsen, G.L., Rabin, T., Toft, T.: Efficient RSA key generation and threshold paillier in the two-party setting. In: Dunkelman, O. (ed.) CT-RSA 2012. LNCS, vol. 7178, pp. 313–331. Springer, Heidelberg (2012). https://doi.org/10.1007/978-3-642-27954-6_20

[Hoe63] Hoeffding, W.: Probability inequalities for sums of bounded random variables. J. Am. Stat. Assoc. **58**(301), 13–30 (1963)

[KKM12] Kakvi, S.A., Kiltz, E., May, A.: Certifying RSA. In: Wang, X., Sako, K. (eds.) ASIACRYPT 2012. LNCS, vol. 7658, pp. 404–414. Springer, Heidelberg (2012). https://doi.org/10.1007/978-3-642-34961-4_25

[Lin17] Lindell, Y.: Fast secure two-party ECDSA signing. In: Katz, J., Shacham, H. (eds.) CRYPTO 2017. LNCS, vol. 10402, pp. 613–644. Springer, Cham (2017). https://doi.org/10.1007/978-3-319-63715-0_21

[LMRS04] Lysyanskaya, A., Micali, S., Reyzin, L., Shacham, H.: Sequential aggregate signatures from trapdoor permutations. In: Cachin, C., Camenisch, J.L. (eds.) EUROCRYPT 2004. LNCS, vol. 3027, pp. 74–90. Springer, Heidelberg (2004). https://doi.org/10.1007/978-3-540-24676-3_5

[MKJR16] Moriarty, K., Kaliski, B., Jonsson, J., Rusch, A.: RFC 8017: PKCS #1: RSA Cryptography Specifications Version 2.2. Internet Engineering Task Force (IETF) (2016). https://tools.ietf.org/html/rfc8017

[MPS00] MacKenzie, P., Patel, S., Swaminathan, R.: Password-authenticated key exchange based on RSA. In: Okamoto, T. (ed.) ASIACRYPT 2000. LNCS, vol. 1976, pp. 599–613. Springer, Heidelberg (2000). https://doi.org/10.1007/3-540-44448-3_46

[MRV99] Micali, S., Rabin, M.O., Vadhan, S.P.: Verifiable random functions. In: 40th Annual Symposium on Foundations of Computer Science, FOCS 1999, 17–18 October 1999, New York, NY, USA, pp. 120–130. IEEE Computer Society (1999)

[Ntu] Tumblebit implementation in.net core. https://github.com/NTumbleBit/NTumbleBit/

[Sho09] Shoup, V.: A Computational Introduction to Number Theory and Algebra, 2nd edn. Cambridge University Press (2009). http://www.shoup.net/ntb/ntb-v2.pdf

[Str17] Stratis Blockchain: Bitcoin privacy is a breeze: tumblebit successfully integrated into breeze, August 2017. https://stratisplatform.com/2017/08/10/bitcoin-privacy-tumblebit-integrated-into-breeze/

[WCZ03] Wong, D.S., Chan, A.H., Zhu, F.: More efficient password authenticated key exchange based on RSA. In: Johansson, T., Maitra, S. (eds.) INDOCRYPT 2003. LNCS, vol. 2904, pp. 375–387. Springer, Heidelberg (2003). https://doi.org/10.1007/978-3-540-24582-7_28

Shorter Pairing-Based Arguments Under Standard Assumptions

Alonso González[1(✉)] and Carla Ràfols[2]

[1] ENS de Lyon, Laboratoire LIP (U. Lyon, CNRS, ENSL, INRIA, UCBL), Lyon, France
alonso.gonzalez@ens-lyon.fr
[2] Universitat Pompeu Fabra and Cybercat, Barcelona, Spain
carla.rafols@upf.edu

Abstract. This paper constructs efficient non-interactive arguments for correct evaluation of arithmetic and boolean circuits with proof size $O(d)$ group elements, where d is the multiplicative depth of the circuit, under falsifiable assumptions. This is achieved by combining techniques from SNARKs and QA-NIZK arguments of membership in linear spaces. The first construction is very efficient (the proof size is $\approx 4d$ group elements and the verification cost is $\approx 4d$ pairings and $O(n + n' + d)$ exponentiations, where n is the size of the input and n' of the output) but one type of attack can only be ruled out assuming the knowledge soundness of QA-NIZK arguments of membership in linear spaces. We give an alternative construction which replaces this assumption with a decisional assumption in bilinear groups at the cost of approximately doubling the proof size. The construction for boolean circuits can be made zero-knowledge with Groth-Sahai proofs, resulting in a NIZK argument for circuit satisfiability based on falsifiable assumptions in bilinear groups of proof size $O(n + d)$.

Our main technical tool is what we call an "argument of knowledge transfer". Given a commitment C_1 and an opening x, such an argument allows to prove that some other commitment C_2 opens to $f(x)$, for some function f, even if C_2 is not extractable. We construct very short, constant-size, pairing-based arguments of knowledge transfer with constant-time verification for any linear function and also for Hadamard products. These allow to transfer the knowledge of the input to lower levels of the circuit.

1 Introduction

This paper deals with the problem of constructing non-interactive publicly verifiable arguments of knowledge under falsifiable assumptions to prove that a circuit ϕ is correctly evaluated in two different settings.

A. González—This author was supported in part by the French ANR ALAMBIC project (ANR-16-CE39-0006).
C. Ràfols—The research leading to this article was supported by a Marie Curie "UPF Fellows" Postdoctoral Grant and by Project RTI2018-102112-B-I00 (AEI/FEDER, UE).

S. D. Galbraith and S. Moriai (Eds.): ASIACRYPT 2019, LNCS 11923, pp. 728–757, 2019.
https://doi.org/10.1007/978-3-030-34618-8_25

In one such possible setting, all of the input of the circuit ϕ is known. In this case, the argument does not need to be zero-knowledge and can leak partial information. This is the typical situation in verifiable computation in which a resource-limited device delegates a costly computation to a more powerful machine.

Another important setting requires the input and output to be partially or totally hidden and the argument to be zero-knowledge. This is interesting from a theoretical perspective as CircuitSat is usually taken to be the standard NP complete problem. On the practical side, often the best way to prove a large, complicated statement in zero-knowledge is to encode it as a circuit and prove that it is satisfiable. Further, CircuitSat is considered a sort of benchmark to evaluate the efficiency of zero-knowledge proofs.

Succinct Non-Interactive Arguments of Knowledge or SNARKs in bilinear groups have been a phenomenal success in both of these scenarios [1,8,15,16,28]. These arguments are succinct, more specifically, they are constant size, that is, not dependent on the circuit size, and extremely efficient also concretely (3 group elements in the best constructions [16]). They are also very fast to verify, which is a very interesting feature in practice, as in many scenarios verification is performed many times. However, these constructions still suffer from some problems, like long trusted parameters, heavy computation for the prover and reliance on non-falsifiable computational assumptions. Further, it is a well-known fact that the latter is unavoidable for succinct arguments in the non-interactive setting [11].

Non-falsifiable assumptions offer great efficiency at the price of less understood security guarantees. The problem is that it is not possible to efficiently check if the adversary effectively breaks the assumption, which results in non-explicit security reductions [32] which inherently do not allow to choose concrete security parameters meaningfully. Therefore, it is interesting to construct arguments with properties similar to SNARKs (short proof size, fast verification) for correct circuit evaluation that avoid falsifiable assumptions.

When the input of the circuit is public, SNARKs can be used to prove that the circuit is correctly evaluated while avoiding falsifiable assumptions. Indeed, since it is possible to check if a prover breaks soundness (as the input is public), the tautological assumption "the scheme is sound" is already falsifiable. For the case where at least some part of the input is secret, the same trivial solution can be used if the prover additionally commits to the input with some commitment which is extractable under falsifiable assumptions.[1] However, these trivial solutions require circuit dependent assumptions.

The goal of this paper is to design efficient constructions both in terms of proof size and verification complexity from milder (falsifiable, circuit independent) assumptions.

[1] Essentially the only such commitment known is bit to bit encryption, e.g. Groth-Sahai commitments to bits.

1.1 Our Results

We construct an argument for proving that an arithmetic circuit $\phi : \mathbb{Z}_p^n \rightarrow \mathbb{Z}_p^{n'}$ is correctly evaluated. We give two instantiations, the first one with proof size $(3d+2)\mathbb{G}_1 + (d+2)\mathbb{G}_2$ and where verification requires $4d+6$ parings and $O(n+n'+d)$ exponentiations, for d the depth of the circuit. We give a less efficient scheme where both proof size and verification cost are approximately the double of the first construction, more concretely, the proof size is $(6d+3)\mathbb{G}_1+(2d+3)\mathbb{G}_2$ and the verification requires $8d+9$ pairings.

For the first construction, we need to rely on the knowledge soundness of QA-NIZK arguments of membership in linear spaces, which has only been proven in the generic group model [5]. The second argument is fully based on falsifiable assumptions. The first one is an assumption that falls into the Matrix Decisional Diffie-Hellman assumption framework of Escala et al. [4] extended in asymmetric groups, where the challenge matrix is given in both groups. The size of the matrix depends on q, for q being the maximum number of multiplicative gates with the same multiplicative depth in the circuit. The second assumption is also a q-type assumption and similar to the q-SFrac Assumption of [12].

For boolean circuits, the argument can be made zero-knowledge with $O((n - n_{pub}) + d)$ proof size, where n_{pub} is the public input size.

1.2 Our Techniques

Circuit Satisfiability can be represented as a set of quadratic and linear equations. It would seem that it suffices to find aggregated proofs of satisfiability of these equations to get sublinear proofs in the number of wires circuit wires. For instance, a natural strategy would be to commit to wires with shrinking commitments and use any constant-size QA-NIZK argument of membership in linear spaces (e.g. [25]) to give an aggregated proof that the affine constraints hold and use "aggregated" variants of GS Proofs [18] such as [2,14] for the quadratic constraints.

The reason why this approach fails is that when using shrinking commitments it is unclear what are the guarantees provided by QA-NIZK arguments since they are not proofs of knowledge (w.r.t. general PPT adversaries and not generic ones). Similarly, the arguments for quadratic equations are commit-and-prove schemes which require binding commitments to the solution of the equation.

Knowledge Transfer Arguments. Our solution is to divide the set of constraints into d sets of quadratic and affine constraints, one per multiplicative level of the circuit. Namely, if $\phi : \mathbb{Z}_p^n \rightarrow \mathbb{Z}_p^{n'}$ is an arithmetic circuit of depth d, we express correct evaluation at level i as the following system:

- (quadratic constraints) $c_{ij} = a_{ij}b_{ij}$ for $j = 1, \ldots, n_i$.
- (affine constraints) a_{ij}, b_{ij} are affine combinations of output wires of previous levels,

that is a_{ij}, b_{ij}, c_{ij} represent, respectively, the left, right and output of the jth gate at level i. Our technical innovation is to eliminate the need for binding commitments to the wires at all levels of the circuit by "transferring" knowledge of the input to lower levels.

More specifically, given adversarially chosen shrinking commitments L_i (resp. R_i, O_i) to all the left (resp. right, output) wires at level i, we first give a constant-size argument with constant-time verification which proves:

If $(\boldsymbol{a}_i, \boldsymbol{b}_i, L_i, R_i, O_i)$ is such that L_i, R_i open to $\boldsymbol{a}_i, \boldsymbol{b}_i$ then O_i opens to $\boldsymbol{c}_i = \boldsymbol{a}_i \circ \boldsymbol{b}_i$.

We think of this building block as a "quadratic knowledge transfer argument", as it shows that if an adversary knows an opening for left and right wires, it also knows an opening of the output wires at the next level. This property is formalized as a promise problem because the verifier of the argument never checks that L_i, R_i open to $\boldsymbol{a}_i, \boldsymbol{b}_i$ (otherwise the verification of the argument would be linear in the witness). Using a quadratic arithmetic program encoding [8] of the quadratic constraints we prove soundness under a certain q-assumption.

With this building block, the problem of constructing the argument is reduced to arguing that left and right wires are correctly assigned, i.e. proving that affine constraints are satisfied. We build a "linear knowledge transfer" argument with constant proof size and verification time showing that:

Given an opening of the commitments to the output wires $O_1, \ldots, O_i$ which is consistent with $L_1, \ldots, L_i$ and $R_1, \ldots, R_i$ then it is also consistent with L_{i+1} and R_{i+1}.

Correct evaluation of the circuit can be easily proven by combining these two building blocks. Since the input of the circuit is public and the shrinking commitments we use are deterministic, a consistent assignment $O_1, L_1, R_1, \ldots, O_d, L_d, R_d$ of the circuit wires is known by the reduction in the proof of soundness. A successful soundness adversary must output another assignment which disagrees with it starting from some level i. If the adversary outputs as part of its proof $L_1, \ldots, L_i, R_1, \ldots, R_i, O_1, \ldots, O_{i-1}, O_i^*$, with $O_i^* \neq O_i$, the reduction knows openings of L_i, R_i and it can break the soundness of the quadratic knowledge transfer argument. On the other hand, if it sends $L_1, \ldots, L_i^*, R_1, \ldots, R_i^*, O_1, \ldots, O_{i-1}$, where either $L_i^* \neq L_i$ or $R_i^* \neq R_i$, then it knows valid openings of O_j until level $i-1$ and it can break the soundness of the "linear knowledge transfer" argument.

To construct the linear knowledge transfer argument, we use QA-NIZK arguments of membership in linear spaces [14,20,21,25,27]. Although soundness of these arguments can be proven under standard assumptions, it turns out that traditional soundness is not what we need in this setting. Indeed, to see this, suppose we want to prove that two shrinking, deterministic commitments open to the same value. Let $\mathbf{M}, \mathbf{N}$ be the commitment keys. If $C_1 = \mathbf{M}\boldsymbol{w}$ and $C_2 = \mathbf{N}\boldsymbol{w}$ are commitments to the same value, obviously

$$\begin{pmatrix} C_1 \\ C_2 \end{pmatrix} \in \mathbf{Im} \begin{pmatrix} \mathbf{M} \\ \mathbf{N} \end{pmatrix}. \tag{1}$$

Let π a QA-NIZK proof of membership in linear spaces for (1). In our linear knowledge transfer argument, π should convince the verifier that:

"If $C_1 = \mathbf{M}\boldsymbol{w}$ for some known $\boldsymbol{w}$, and π verifies, then $C_2 = \mathbf{N}\boldsymbol{w}$."

The problem is that for any $\boldsymbol{w}'$ such that $C_1 = \mathbf{M}\boldsymbol{w} = \mathbf{M}\boldsymbol{w}'$, an adversary can set $C_2 = \mathbf{N}\boldsymbol{w}'$ and compute π honestly with $\boldsymbol{w}'$. In other words, the adversary can "switch witnesses" without breaking the soundness of the QA-NIZK argument. So standard soundness does not help to argue that the left and right wires are consistently evaluated with lower levels of the circuit.

On the other hand, the "witness switching attack" is easy to rule out, as it requires the attacker to know two openings for C_1, but this breaks the binding property of the first commitment. However, because the commitment is shrinking we do not know how to extract $\boldsymbol{w}'$ to get a reduction to the binding property unless we use the knowledge soundness property of the QA-NIZK Argument as proven (in the generic group model) in [5].

Soundness of the Linear Argument Under Standard Assumptions. One of our main technical contributions is to show that such witness switching attacks are not possible under a certain decisional assumption in bilinear groups. To get back to our example, our first observation is that, using the linear properties of the QA-NIZK arguments of membership in linear spaces, a break of the knowledge transfer property can be turned into a proof of membership $\pi^\dagger$ for a vector of the form $\left(\begin{smallmatrix}0\\C\end{smallmatrix}\right)$, where $C = C_2 - \mathbf{N}\boldsymbol{w} \neq 0$.

The crs of the QA-NIZK argument system is of the form $\mathbf{A}, \mathbf{B} = \mathbf{M}^\top\mathbf{K}_1 + \mathbf{N}^\top\mathbf{K}_2, \mathbf{KA}$, for some matrix $\mathbf{A}$ and a random matrices $\mathbf{K}_1, \mathbf{K}_2$. A proof for (C_1, C_2) must be of the form $C_1^\top\mathbf{K}_1 + C_2^\top\mathbf{K}_2$ (unless one solves some computationally hard problem). Intuitively, is not easy to construct $\pi^\dagger$ since it must be of the form $\pi^\dagger = C^\top\mathbf{K}_2$ and hence an adversary must somehow find an element in the kernel of $\mathbf{M}$ (which is in general a hard problem, otherwise the commitment is not binding) in order to eliminate any dependence on $\mathbf{K}_1$ in $\mathbf{B}$. However, in the security proof it is not clear how to extract such element in the kernel of $\mathbf{M}$, which is of the same size of $\boldsymbol{w}$, only from C and $\pi^\dagger$, which are of constant size. To bypass this problem, we assume that a stronger decisional assumption related to $\mathbf{M}$ holds, namely that it is hard to decide membership in the image of $\mathbf{M}^\top$ (a type of Matrix Diffie-Hellman assumption [4]). Specifically, we assume that $\mathbf{M}^\top\mathbf{K}_1$ is pseudo-random and, using this decisional assumption, we can jump to game where $\mathbf{K}_2$ is information theoretically hidden and then there is an exponentially low probability of computing $\pi^\dagger = C^\top\mathbf{K}_2$. To do this, we need to find a way around the problem that there is still some information about $\mathbf{K}_1$ which is leaked trough the crs of QA-NIZK arguments of [25] as $\mathbf{KA} = \left(\begin{smallmatrix}\mathbf{K}_1\mathbf{A}\\\mathbf{K}_2\mathbf{A}\end{smallmatrix}\right)$, where $\mathbf{A}$ is either a $(k+1) \times k$ matrix for general linear spaces or a $k \times k$ matrix when the linear spaces are generated by witness samplable distributions. To solve this, we use the fact that, information theoretically, part of $\mathbf{K}_1$ is never leaked through $\mathbf{KA}$ when $\mathbf{A}$ is a $(k+1) \times k$ matrix. We leave it as an open question to achieve a similar result when $\mathbf{A}$ is a $k \times k$ to exploit witness samplability.

Zero-Knowledge. In all our subarguments the verification equations are pairing product equations, so they can be made zero-knowledge with Groth-Sahai proofs [18]. However, our proof uses in a fundamental way that the input of the verification is public. Therefore, this only works when the commitment to the input is extractable. The resulting scheme is not practical as this is only possible with bit-by-bit commitments to the input. However, it can be easily extended to boolean circuits with a proof size of $O(n - n_{pub} + n' + d)$ group elements (where n_{pub} is the size of the public input), which is an interesting improvement over state-of-the-art, as all constructions in the crs model under falsifiable assumptions are linear in the circuit size (see [17] and concrete improvements thereof, mainly [14]).

1.3 Previous Work

CRS NIZK for NP from Falsifiable Assumptions. Groth, Ostrovsky, and Sahai [17] constructed a NIZK proof system for boolean CircuitSat only from standard assumptions. Both the the size of the proof (in group elements) and the verifier's complexity (in group operations) depend asymptotically on the circuit size. The construction can be extended to arithmetic circuits using [18]. Several concrete improvements in the proof size can be done with recent results in the QA-NIZK setting [14,20,21,25,27] but we are not aware of any asymptotic improvements.

A trivial approach to reduce the proof size is to encrypt the witness using fully homomorphic encryption [9] and let the verifier evaluate the circuit homomorphically. Building on this idea, and using hybrid fully homomorphic encryption, Gentry et al. [10] constructed a proof of size $n + \mathsf{poly}(\lambda)$. While this shows that it is theoretically possible to build proofs of size independent of the circuit size under standard assumptions, they need to give NIZK proofs for correct key generation of FHE keys and correct evaluations of the FHE encryption algorithm and decryption algorithms.[2] These NIZK proofs, in general, need to represent the statements as boolean circuits and therefore they are of lower practical interest. Furthermore, note that the verifier needs to homomorphically evaluate the circuit using the FHE scheme, so its runtime is proportional to the circuit size.

A very recent result constructs proofs of size proportional to the circuit size plus an additive overhead in the security parameter (as opposed to multiplicative as in our work) in pairing based groups [24]. For NC1, one of the constructions is of size n (independent of the circuit size) plus an additive overhead in the security parameter. Although the verifier's runtime is proportional to the circuit size, it may be possible to preprocess the circuit dependent part and add it to the crs so that the verifier's runtime is only proportional to the size of the input. On the downside, the size of the crs is $O(n^3)$ as well as the underlying security assumption which is a q-assumption with q of size $O(n^3)$. Furthermore, the

[2] Note that using the celebrated recent results of Peikert and Shiehian [35] this scheme can be based solely on the LWE assumption.

additive overhead might be large as it hides a NIZK proof (computed with [17]) for the correct decryption of a ciphertex. Such a NIZK proof requires representing the decryption algorithm as a boolean circuit and to commit to each circuit wire.

Verifiable Computation. Kalai et al. [22], based on [13] and the sum-check protocol of Lund et al. [29], constructed the first publicly verifiable non-interactive delegation scheme for boolean circuits from a simple constant size assumption in bilinear groups. Their crs is circuit dependent but it can be universal using a crs for the universal circuit.[3] The verifier's runtime is $O((n+d)\mathsf{polylog}(s))$, and the communication complexity is $O(d\cdot\mathsf{polylog}(\mathsf{s}))$ group elements, where s is the size of the circuit, and in most other parameters it is far from being efficient (crs size, prover complexity).

As explained in [22] there's a vast literature on verifiable computation (apart from the already mentioned) which can be roughly classified into a) designated verifier schemes [7,23], b) schemes under very strong assumptions: "knowledge of exponent" type (e.g. [8,34]), generic or algebraic group model (e.g. [16,30]), assumptions related to obfuscation, or homomorphic encryption [33] or c) interactive arguments [13]. Note that all these constructions are incomparable to ours as long as they either rely on arguably stronger assumptions (b) or are in a different model (a and c).

2 Preliminaries

Given some distribution $\mathcal{D}$ we denote by $x \leftarrow \mathcal{D}$ the process of sampling x according to $\mathcal{D}$. For a finite set S, $x \leftarrow S$ denotes an element sampled from the uniform distribution over S.

Bilinear Groups. Let $\mathcal{G}$ be some probabilistic polynomial time algorithm which on input 1^λ, where λ is the security parameter, returns the *group key* which is the description of an asymmetric bilinear group $gk = (p, \mathbb{G}_1, \mathbb{G}_2, \mathbb{G}_T, e, \mathcal{P}_1, \mathcal{P}_2)$, where $\mathbb{G}_1, \mathbb{G}_2$ and $\mathbb{G}_T$ are groups of prime order p, the elements $\mathcal{P}_1, \mathcal{P}_2$ are generators of $\mathbb{G}_1, \mathbb{G}_2$ respectively, $e : \mathbb{G}_1 \times \mathbb{G}_2 \to \mathbb{G}_T$ is an efficiently computable, non-degenerate bilinear map, and there is no efficiently computable isomorphism between $\mathbb{G}_1$ and $\mathbb{G}_2$.

Elements in $\mathbb{G}_\gamma$, are denoted implicitly as $[a]_\gamma = a\mathcal{P}_\gamma$, where $\gamma \in \{1, 2, T\}$ and $\mathcal{P}_T = e(\mathcal{P}_1, \mathcal{P}_2)$. With this notation, $e([a]_1, [b]_2) = [ab]_T$. Vectors and matrices are denoted in boldface. Given a matrix $\mathbf{T} = (t_{i,j})$, $[\mathbf{T}]_\gamma$ is the natural embedding of $\mathbf{T}$ in $\mathbb{G}_\gamma$, that is, the matrix whose (i, j)th entry is $t_{i,j}\mathcal{P}_\gamma$. We use the notation (a, b) to refer to a elements of $\mathbb{G}_1$ and b elements of $\mathbb{G}_2$.

$\mathbf{I}_n$ refers to the identity matrix in $\mathbb{Z}_p^{n\times n}$, $\mathbf{0}_{m\times n}$ to the all-zero matrix in $\mathbb{Z}_p^{m\times n}$ (simply $\mathbf{I}$ and $\mathbf{0}$, respectively, if n and m are clear from the context).

[3] There's the technicality that a verifier running in time sub-linear in the circuit size can not even read the circuit, which is part of the input of the universal circuit. For this reason, they restricted the circuits to be log space uniform boolean cicuits.

Lagrangian Pedersen Commitments. Given an arbitrary set $\mathcal{R} = \{r_1, \ldots, r_m\} \subset \mathbb{Z}_p$, we define the jth Lagrange interpolation polynomial as:

$$\lambda_j(X) = \prod_{\ell \neq j} \frac{(X - r_\ell)}{(r_j - r_\ell)}.$$

It is a well known fact that given a set of values x_j, $j = 1, \ldots, m$, $P(X) = \sum_{j=1}^{m} x_j \lambda_j(X)$ is the unique polynomial of degree at most $m-1$ such that $P(r_j) = x_j$. The Lagrangian Pedersen commitment in $\mathbb{G}_\gamma$ for some $\gamma \in \{1, 2\}$ to a vector $x \in \mathbb{Z}_p^m$ is defined as

$$\mathsf{Com}_{ck}(\boldsymbol{x}) = \sum_{i=1}^{m} x_j [\lambda_j(s)]_\gamma = [P(s)]_\gamma,$$

where the commitment key is $ck = ([\lambda_1(s)]_\gamma, \ldots, [\lambda_m(s)]_\gamma)$, for $s \leftarrow \mathbb{Z}_p$. It is computationally binding under the m-DLog assumption.

We also consider vectors of Lagrangian Pedersen commitments defined as $[P(\boldsymbol{s})]_\gamma = \sum_{i=1}^{m} x_i [\lambda_i(\boldsymbol{s})]_\gamma \in \mathbb{G}_\gamma^{k_s}$, where $\boldsymbol{s} \in \mathbb{Z}_p^{k_s}$ for some $k_s \in \mathbb{N}$ and $\lambda_i(\boldsymbol{s})$ is just $(\lambda_i(s_1), \ldots, \lambda_i(s_{k_s}))^\top$.

2.1 Cryptographic Assumptions

Definition 1. *Let $k \in \mathbb{N}$. We call $\mathcal{D}_{\ell,k}$ (resp. $\mathcal{D}_k$) a matrix distribution if it outputs in PPT time, with overwhelming probability matrices in $\mathbb{Z}_p^{\ell \times k}$ (resp. in $\mathbb{Z}_p^{(k+1)\times k}$). For a matrix distribution $\mathcal{D}_k$, we denote as $\overline{\mathcal{D}}_k$ the distribution of the first k rows of the matrices sampled according to $\mathcal{D}_k$.*

Assumption 1. *Let $\mathcal{D}_{\ell,k}$ be a matrix distribution and $gk \leftarrow \mathcal{G}(1^\lambda)$. For all non-uniform PPT adversaries $\mathcal{A}$ and relative to $gk \leftarrow \mathcal{G}(1^\lambda)$, $\mathbf{A} \leftarrow \mathcal{D}_{\ell,k}, \boldsymbol{w} \leftarrow \mathbb{Z}_p^k, [\boldsymbol{z}]_\gamma \leftarrow \mathbb{G}_\gamma^\ell$ and the coin tosses of adversary $\mathcal{A}$,*

1. *the Matrix Decisional Diffie-Hellman Assumption in $\mathbb{G}_\gamma$ ($\mathcal{D}_k$-MDDH_γ) holds if*

$$|\Pr[\mathcal{A}(gk, [\mathbf{A}]_\gamma, [\mathbf{A}\boldsymbol{w}]_\gamma) = 1] - \Pr[\mathcal{A}(gk, [\mathbf{A}]_\gamma, [\boldsymbol{z}]_\gamma) = 1]| \leq \mathsf{negl}(\lambda),$$

2. *the Split Matrix Decisional Diffie-Hellman Assumption in $\mathbb{G}_\gamma$ ($\mathcal{D}_k$-SMDDH_γ) holds if*

$$|\Pr[\mathcal{A}(gk, [\mathbf{A}]_1, [\mathbf{A}]_2, [\mathbf{A}\boldsymbol{w}]_\gamma) = 1] - \Pr[\mathcal{A}(gk, [\mathbf{A}]_1, [\mathbf{A}]_2, [\boldsymbol{z}]_\gamma) = 1]| \leq \mathsf{negl}(\lambda).$$

Two examples of interesting distributions are the following:

$$\mathcal{L}_k : \mathbf{A} = \begin{pmatrix} s_1 & 0 & \ldots & 0 \\ 0 & s_2 & \ldots & 0 \\ \vdots & \vdots & \ddots & \vdots \\ 0 & 0 & \ldots & s_k \\ 1 & 1 & \ldots & 1 \end{pmatrix} \qquad \mathcal{LG}_{\mathcal{R},k} : \mathbf{A} = \begin{pmatrix} \lambda_1^{\mathcal{R}}(s_1) & \lambda_1^{\mathcal{R}}(s_2) & \ldots & \lambda_1^{\mathcal{R}}(s_k) \\ \lambda_2^{\mathcal{R}}(s_1) & \lambda_2^{\mathcal{R}}(s_2) & \ldots & \lambda_2^{\mathcal{R}}(s_k) \\ \vdots & \vdots & \ddots & \vdots \\ \lambda_\ell^{\mathcal{R}}(s_1) & \lambda_\ell^{\mathcal{R}}(s_2) & \ldots & \lambda_\ell^{\mathcal{R}}(s_k) \end{pmatrix},$$

where $s_i \leftarrow \mathbb{Z}_p$ and $\mathcal{R} = \{r_1, \ldots, r_N\} \subset \mathbb{Z}_p$. The assumption associated to the first distribution is the k-Lin family. The assumption associated to the second one is new to this paper and is the $(\mathcal{R}, k)$-Lagrangian Assumption. In our construction, we will use the $\mathcal{LG}_{\mathcal{R},2}$-$\mathsf{SMDDH}_1$ assumption (for N the maximum number of gates of the same multiplicative depth). In the full version of this work we argue about the generic hardness of the $\mathcal{LG}_{\mathcal{R},2}$-$\mathsf{MDDH}_\gamma$ assumption in symmetric bilinear groups, which implies the generic hardness of $\mathcal{LG}_{\mathcal{R},2}$-$\mathsf{SMDDH}_1$ in asymmetric bilinear groups.

We note that for all interesting distributions $\mathcal{D}_k$, we can assume that the $\mathcal{D}_k$-MDDH Assumption is generically hard in k-linear groups and in particular, that every $k \times k$ minor is invertible with overwhelming probability.

The Kernel Diffie-Hellman Assumption [31] says one cannot find a non-zero vector in one of the groups which is in the co-kernel of $\mathbf{A}$. We also use a generalization in bilinear groups which says one cannot find a pair of vectors in $\mathbb{G}_1^{k+1} \times \mathbb{G}_2^{k+1}$ such that the difference of the vector of their discrete logarithms is in the co-kernel of $\mathbf{A}$.

Assumption 2. *Let $\mathcal{D}_{\ell,k}$ be a matrix distribution. For all non-uniform PPT adversaries $\mathcal{A}$ and relative to $gk \leftarrow \mathcal{G}(1^\lambda)$, $\mathbf{A} \leftarrow \mathcal{D}_{\ell,k}, \boldsymbol{w} \leftarrow \mathbb{Z}_p^k, [\boldsymbol{z}]_\gamma \leftarrow \mathbb{G}_\gamma^\ell$ and the coin tosses of adversary $\mathcal{A}$,*

1. the Find-Rep Assumption holds if

$$\Pr\left[\boldsymbol{r} \leftarrow \mathcal{A}(gk, [\mathbf{A}]_1, [\mathbf{A}]_2) : \boldsymbol{r}^T \mathbf{A} = \mathbf{0}\right] = \mathsf{negl}(\lambda),$$

2. the Kernel Matrix Diffie-Hellman Assumption holds in $\mathbb{G}_\gamma$ [31] if

$$\Pr\left[[\boldsymbol{r}]_{3-\gamma} \leftarrow \mathcal{A}(gk, [\mathbf{A}]_\gamma) : \boldsymbol{r}^\top \mathbf{A} = 0\right] = \mathsf{negl}(\lambda),$$

3. the Split Kernel Matrix Diffie-Hellman Assumption [14] holds if

$$\Pr\left[[\boldsymbol{r}]_1, [\boldsymbol{s}]_2 \leftarrow \mathcal{A}(gk, [\mathbf{A}]_1, [\mathbf{A}]_2) : \boldsymbol{r} \neq \boldsymbol{s} \wedge \boldsymbol{r}^\top \mathbf{A} = \boldsymbol{s}^\top \mathbf{A}\right] = \mathsf{negl}(\lambda).$$

The Find-Rep Assumption for the $\mathcal{LG}_{\mathcal{R},\ell,k}$ MDH Assumption is equivalent to solving k instances of the q-Dlog Assumption in both groups, in which the adversary receives q powers of s_i, $i = 1, \ldots, k$ in both groups and computes $s_i \in \mathbb{Z}_p$. This follows from the observation that if $\boldsymbol{r}$ is a solution of the Find-Rep problem, it can be associated to a polynomial which is 0 in s_i for all $i = 1, \ldots, k$ and its factorization allows to compute s_i.

We note that the Split Decisional and Split Kernel MDH Assumptions are generically hard in asymmetric bilinear groups for all distributions for which the non split variant is hard in symmetric bilinear groups whenever $k \geq 2$.

Finally, we introduce an assumption which is similar to the q-SFrac Assumption considered in [12], but in the source group.

Assumption 3 ($\mathcal{R}$-RSDH Assumption). *Let $\mathcal{R}$ be an arbitrary set of integers of cardinal q. The $\mathcal{R}$-Rational Strong Diffie-Hellman Assumption holds in $\mathbb{G}_1$ if the following probability is negligible in λ:*

$$\Pr\left[\begin{array}{c} e([z]_1, [1]_2) = e([w]_1, [t(s)]_2) \\ z \neq 0 \end{array} \middle| \begin{array}{c} gk \leftarrow \mathcal{G}(1^\lambda); \\ ([z]_1, [w]_1) \leftarrow \mathcal{A}\left(gk, \mathcal{R}, \left\{[s^i]_{1,2}\right\}_{i=1}^{q-1}, [s^q]_2\right) \end{array}\right],$$

where $t(s) = \prod_{r \in \mathcal{R}}(s - r)$, *and the probability is taken over* $gk \leftarrow \mathcal{G}(1^\lambda)$, $s \leftarrow \mathbb{Z}_p$ *and the coin tosses of adversary* $\mathcal{A}$.

It is important to note that it is possible to check if an adversary has succeeded in breaking the assumption, since the value $[t(s)]_2$ can be constructed as a linear combination of $\{[s^i]_2\}_{i=1}^q$ given $\mathcal{R}$.

The intuition why the assumption is generically hard is as follows. Since $[z]_1, [w]_1$ are given in the group $\mathbb{G}_1$, the adversary must construct them as a linear combinations of all elements it has received in $\mathbb{G}_1$, which are $([1]_1, [s]_1, \ldots, [s^{q-1}]_1)$. On the other hand, the adversary can only win if $z/t(s) = w$, but the adversary can only find a non-trivial solution generically if z is constructed as a (non-zero) multiple of $t(X) = \prod_{r \in \mathcal{R}}(X - r)$ evaluated at s. But this is not possible because in $\mathbb{G}_1$ it only receives powers of s of degree at most $q - 1$ and $t(X)$ is of degree q.

3 Arithmetic Circuits

Arithmetic circuits are acyclic directed graphs where the edges are called wires and the vertices are called gates. Gates with in-degree 0 are labeled by variables X_i, $i = 1, \ldots, n$ or with a constant field element, the rest of the gates are either labeled with $\times$ and are referred to as multiplication gates or with $+$ and are called addition gates. In this work we consider only fan-in 2 multiplication gates and the circuit is defined over a field $\mathbb{Z}_p$, where p is the order of some cryptographically useful bilinear group. Each circuit computes a function $\phi : \mathbb{Z}_p^n \to \mathbb{Z}_p^{n'}$.

Let $\mathcal{G}$ be the set of multiplicative gates of the circuit excluding multiplication-by-constant gates. We denote by m the cardinal of this set. For simplicity and without loss of generality, we may assume all outputs of the circuit to be the output of some multiplication gate.

For our construction of Sect. 5, we partition the set $\mathcal{G}$ of multiplicative gates of the circuit into different levels. More precisely, we define $\{\mathcal{G}_i\}_{i=1}^{d'}$, where $\mathcal{G}_i$, for $i = 1, \ldots, d'$, is the set of gates $G \in \mathcal{G}$ such that the maximum of gates in $\mathcal{G}$ evaluated in any path from the input of the circuit to an input of G is $i - 1$. The minimal such d' for which the partition exists is the multiplicative depth of the circuit, which we always denote by d. Further, we define $\mathcal{G}_0$ to be the set of n_0 variable inputs. If $G \in \mathcal{G}_i$, we say that G has multiplicative depth i. Let n_i be the cardinal of $\mathcal{G}_i$. With this notation, a circuit computes a function $\phi : \mathbb{Z}_p^{n_0} \to \mathbb{Z}_p^{n_d}$, i.e. $n = n_0$, $n' = n_d$ and the number of multiplication gates is $\sum_{i=1}^d n_i$.

We now consider an encoding of circuit satisfiability where the variables are divided according to their multiplicative depth. For each gate in $\mathcal{G}_i$, $i \in \{1, \ldots, d\}$ the circuit is correctly evaluated if the output of the gate is the product of two multivariate polynomials of degree 1 where the variables are outputs of gates of less multiplicative depth, that is, the output of gates in $\mathcal{G}_j$, for some j, $0 \leq j \leq i - 1$.

Lemma 1. *Let* $\phi : \mathbb{Z}_p^{n_0} \to \mathbb{Z}_p^{n_d}$, *be a circuit of multiplicative depth* d *and with* m *gates. For* $i \in \{1, \ldots, d\}$, *define* n_i *as the number multiplication gates at level* i. *There exist*

(a) variables C_{ij}, $i = 0, \dots, d$, $j = 1, \dots, n_i$,
(b) variables A_{ij}, B_{ij}, $i = 1, \dots, d$, $j = 1, \dots, n_i$,
(c) constants $f_{ij}, g_{ij}, f_{ijk\ell}, g_{ijk\ell} \in \mathbb{Z}_p$, $i = 1, \dots, d$, $k = 0, \dots, i-1$, $j = 1, \dots, n_i$, $\ell = 1, \dots, n_k$

such that, for every $(x_1, \dots, x_{n_0}) \in \mathbb{Z}_p^{n_0}$, if we set $C_{0j} = x_j$, for all $j = 1, \dots, n_0$, then $\phi(x_1, \dots, x_{n_0}) = (y_1, \dots, y_{n_d})$ and for each $i \in \{1, \dots, d\}$, A_{ij}, B_{ij}, C_{ij} are evaluated respectively to the left, the right and the output wires of the jth gate at level i, if and only if the following equations are satisfied:

1. *(Quadratic Constraints). For each $i = 1, \dots, d$, if $j = 1, \dots, n_i$: $C_{ij} = A_{ij}B_{ij}$.*
2. *(Affine Constraints) $A_{ij} = f_{ij} + \sum_{k=0}^{i-1} \sum_{\ell=1}^{n_k} f_{ijk\ell} C_{k\ell}$ and $B_{ij} = g_{ij} + \sum_{k=0}^{i-1} \sum_{\ell=1}^{n_k} g_{ijk\ell} C_{k\ell}$.*
3. *(Correct Output) $C_{dj} = y_j$, $j = 1, \dots, n_d$.*

Given an arithmetic circuit $\phi : \mathbb{Z}_p^{n_0} \to \mathbb{Z}_p^{n_d}$, we can define the witness for correct evaluation of $\phi(\boldsymbol{x}) = \boldsymbol{y}$ as a tuple $(\boldsymbol{a}, \boldsymbol{b}, \boldsymbol{c})$, where $\boldsymbol{a} = (\boldsymbol{a}_1, \dots, \boldsymbol{a}_d)$, $\boldsymbol{b} = (\boldsymbol{b}_1, \dots, \boldsymbol{b}_d)$, $\boldsymbol{c} = (\boldsymbol{c}_0, \dots, \boldsymbol{c}_d)$, $\boldsymbol{s}_i = (s_{i1}, \dots, s_{in_i})$ for any $s \in \{a, b, c\}$. The tuple is an an assignment to A_{ij}, B_{ij} and C_{ij} which satisfies the equations described in Lemma 1.

Using standard techniques due to [8], quadratic constraints can be written as a polynomial divisibility problem.

Lemma 2. *(QAP for the Hadamard Product) Let $(\boldsymbol{a}_i, \boldsymbol{b}_i, \boldsymbol{c}_i) \in (\mathbb{Z}_p^{n_i})^3$, $n_i \in \mathbb{N}$. Let $\mathcal{R} = \{r_1, \dots, r_N\} \subset \mathbb{Z}_p$ be a set of elements of $\mathbb{Z}_p$ for some $N \geq n_i$ and let $\lambda_i(X) = \prod_{j \neq i} \dfrac{X - r_j}{r_i - r_j}$. Define*

$$p_i(X) = \left(\sum_{j=1}^{n_i} a_{ij} \lambda_j(X) \right) \left(\sum_{j=1}^{n_i} b_{ij} \lambda_j(X) \right) - \left(\sum_{j=1}^{n_i} c_{ij} \lambda_j(X) \right).$$

Then, $\boldsymbol{c}_i = \boldsymbol{a}_i \circ \boldsymbol{b}_i$ if and only if $p_i(X) = h_i(X)t(X)$, where $t(X) = \prod_{r \in \mathcal{R}} (X - r)$ and $h_i(X) \in \mathbb{Z}_p[X]$ is a polynomial of degree at most $N - 2$.

Proof. By definition, $p_i(r_j) = a_{ij}b_{ij} - c_{ij}$, so $p_i(X)$ is divisible by $t(X)$ if and only if $a_{ij}b_{ij} - c_{ij} = 0$ for all $j = 1, \dots, n_i$.

On the other hand, for each i, affine constraints can be written also as polynomial relations. That is, for any set $\mathcal{R} = \{r_1, \dots, r_N\}$ such that $N \geq n_i$, there exist families of polynomials $\mathcal{V} = \{v_i, v_{ik\ell}\}$, $\mathcal{W} = \{w_i, w_{ik\ell}\}$ of degree $N - 1$ such that $(\boldsymbol{a}, \boldsymbol{b}, \boldsymbol{c})$ is a valid witness if and only if $\sum_{j=1}^{n_i} a_{ij}\lambda_j(X) = v_i(X) + \sum_{k=0}^{i-1} \sum_{\ell=1}^{n_k} c_{kl} v_{ik\ell}(X)$ and $\sum_{j=1}^{n_i} b_{ij}\lambda_j(X) = w_i(X) + \sum_{k=0}^{i-1} \sum_{\ell=1}^{n_k} c_{kl} w_{ik\ell}(X)$. It suffices to define $v_i(X) = \sum_{j=1}^{n_i} f_{ij}\lambda_j(X)$, $v_{ik\ell}(X) = \sum_{j=1}^{n_i} f_{ijk\ell}\lambda_j(X)$, $w_i(X) = \sum_{j=1}^{n_i} g_{ij}\lambda_j(X)$, $w_{ik\ell}(X) = \sum_{j=1}^{n_i} g_{ijk\ell}\lambda_j(X)$. The proof follows by evaluating the equations in the points $r_j \in \mathcal{R}$.

4 Arguments of Knowledge Transfer

In this section we construct what we informally name "knowledge transfer argument" for both linear and quadratic equations. The name captures the idea that these arguments ensure that if a valid opening is known for some committed value, then an opening is also known for another commitment and this second opening is a certain quadratic or linear function of the original opening.

Formally, the prover needs to prove membership in a language $\mathcal{L}$ of the form $(\boldsymbol{w}, C, D)$, where $\boldsymbol{w}$ is the opening of a shrinking commitment C. The statement is that "if C opens to $\boldsymbol{w}$, then D opens to $F(\boldsymbol{w})$". Since typically there is an exponential number of possible openings of C, the language would not make sense without $\boldsymbol{w}$, i.e. the statement "there exists an opening $\boldsymbol{w}$ of C such that $\boldsymbol{D}$ opens to $F(\boldsymbol{w})$" would most probably be always true.

Deciding membership in $\mathcal{L}$ can be done efficiently with a number of operations which is proportional to the size of the statement. Our verifier, however, does not use $\boldsymbol{w}$ for verification (i.e. it never checks that $\boldsymbol{w}$ is a valid opening of C) and does only a constant number of public key operations (ignoring the need to read $\boldsymbol{w}$ as part of the statement). When using these subarguments in the full argument for correct circuit evaluation, the verifier never reads $\boldsymbol{w}$ but $\boldsymbol{w}$ is uniquely determined by the context.

This is formalized as a promise problem defined by a language of good instances $\mathcal{L}_{YES}$ and of bad instances $\mathcal{L}_{NO}$. Completeness guarantees that proofs are accepted for all instances of $\mathcal{L}_{YES}$, while soundness guarantees that no argument will be accepted for instances of $\mathcal{L}_{NO}$. The promise is that "$\boldsymbol{w}$ is an opening of C" and nothing is claimed when $x \notin (\mathcal{L}_{YES} \cup \mathcal{L}_{NO})$ (i.e. when the promise does not hold). A formal definition of QA arguments for promise problems can be found in the full version of this work.

4.1 Argument for Hadamard Products

Let $m \in \mathbb{N}$. We give an argument for the promise problem defined by languages $\mathcal{L}_{YES}^{\mathsf{quad}}, \mathcal{L}_{NO}^{\mathsf{quad}}$, which are parameterized by $m \in \mathbb{N}$ and a Lagrangian Pedersen commitment key $ck = ([\boldsymbol{\Lambda}]_1, [\boldsymbol{\Lambda}]_2)$ and are defined as

$$\mathcal{L}_{YES}^{\mathsf{quad}} = \left\{ \begin{array}{l} (\boldsymbol{a}, \boldsymbol{b}, [L]_1, [R]_2, [O]_1) : \boldsymbol{c} = \boldsymbol{a} \circ \boldsymbol{b} \\ \text{and } [L]_1 = [\boldsymbol{\Lambda}]_1 \boldsymbol{a}, [R]_2 = [\boldsymbol{\Lambda}]_2 \boldsymbol{b}, [O]_1 = [\boldsymbol{\Lambda}]_1 \boldsymbol{c} \end{array} \right\},$$

$$\mathcal{L}_{NO}^{\mathsf{quad}} = \left\{ \begin{array}{l} (\boldsymbol{a}, \boldsymbol{b}, [L]_1, [R]_2, [O]_1) : \boldsymbol{c} = \boldsymbol{a} \circ \boldsymbol{b}, \\ [L]_1 = [\boldsymbol{\Lambda}]_1 \boldsymbol{a} \text{ and } [R]_2 = [\boldsymbol{\Lambda}]_2 \boldsymbol{b}, \\ \text{but } [O]_1 \neq [\boldsymbol{\Lambda}]_1 \boldsymbol{c} \end{array} \right\}.$$

Perfect Completeness. The argument described in Fig. 1 has perfect completeness as the values $[L]_1, [O]_1$ can be computed from $\{[\lambda_i(s)]_1 \ldots, [\lambda_m(s)]_1\}$, and $[R]_2$ from $\{[\lambda_i(s)]_2 \ldots, [\lambda_m(s)]_2\}$. Further, by definition, the polynomial $\ell(X)r(X) - o(X)$ takes the value $a_i b_i - c_i = 0$ at point $r_i \in \mathcal{R}$. Therefore,

$\underline{\mathsf{K}(gk,\mathcal{R})}$:
Sample $s \leftarrow \mathbb{Z}_p^*$;
Output crs $=$
$\Big(gk, \{[\lambda_1(s)]_\gamma, \ldots, [\lambda_m(s)]_\gamma\}_{\gamma\in\{1,2\}},$
$\{[s^i]_1\}_{i\in\{1,\ldots,m-2\}}, [t(s)]_2\Big)$.

$\underline{\mathsf{V}(\text{crs}, \boldsymbol{a}, \boldsymbol{b}, [L]_1, [R]_2, [O]_1, [H]_1)}$:
Check if:
$e([L]_1,[R]_2) - e([O]_1,[1]_2) = e([H]_1,[t(s)]_2)$;
output 1 in this case and 0 otherwise.

$\underline{\mathsf{P}(\text{crs}, \boldsymbol{a}, \boldsymbol{b})}$:
$\ell(X) = \sum_{i=1}^m a_i\lambda_i(X)$;
$r(X) = \sum_{i=1}^m b_i\lambda_i(X)$;
$o(X) = \sum_{i=1}^m c_i\lambda_i(X)$;
$h(X) = (\ell(X)r(X) - o(X))/t(X)$;
$[L]_1 = [\ell(s)]_1$; $[R]_2 = [r(s)]_2$;
$[O]_1 = [o(s)]_1$; $[H]_1 = [h(s)]_1$;
Output $[H]_1$.

Fig. 1. Our argument for Hadamard products. $\lambda_i(X)$ is the ith Lagrange polynomial associated to $\mathcal{R}$, a set of $\mathbb{Z}_p$ of cardinal m, $t(X)$ is the polynomial which has as roots all the elements of $\mathcal{R}$. Both $\boldsymbol{a}$ and $\boldsymbol{b}$ are m-dimensional vectors in $\mathbb{Z}_p$.

$\ell(X)r(X) - o(X)$ is divisible by $t(X)$, so $h(X)$ is well defined. Further, the degree of H is at most $m-2$ (since $\ell(X)r(X)$ has degree $2m-2$ and $t(X)$ has degree m) and thus $[H]_1$ can be computed from $\{[s]_1, \ldots, [s^{m-2}]_1\}$.

Computational Soundness. We argue that if $\mathcal{A}$ produces an accepting proof for $(\boldsymbol{a}, \boldsymbol{b}, \boldsymbol{c}, [L]_1, [R]_2, [O]_1) \in \mathcal{L}_{NO}^{\mathsf{quad}}$ then we can construct an adversary $\mathcal{B}$ against the $(\mathcal{R}, \boldsymbol{m})$-Rational Strong Diffie-Hellman Assumption. Given a challenge gk, $\{[s^i]_1\}_{i=1}^{m-1}, \{[s^i]_2\}_{i=1}^{m}$, adversary $\mathcal{B}$ can simulate the common reference string perfectly because $\lambda_i(X)$ is a polynomial whose coefficients in $\mathbb{Z}_p$ depend only on $\mathcal{R}$ of degree at most $m-1$. Therefore, $[\lambda_i(s)]_1, [\lambda_i(s)]_2$ can be computed from $\{s^i\}_{i=1}^{m-1}$ in both the source groups. On the other hand, $t(X)$ is a polynomial with coefficients in $\mathbb{Z}_p$ which depend only on $\mathcal{R}$ of degree at most m. So $[t(s)]_2$ can be computed in $\mathbb{G}_2$ given $\{[s^i]_2\}_{i=1}^m$.

Adversary $\mathcal{A}$ outputs $(\boldsymbol{a}, \boldsymbol{b}, \boldsymbol{c}, [L]_1, [R]_2, [O^\dagger]_1, [H^\dagger]_1)$ which is accepted by the verifier and $(\boldsymbol{a}, \boldsymbol{b}, \boldsymbol{c}, [L]_1, [R]_2, [O^\dagger]_1) \in \mathcal{L}_{NO}^{\mathsf{quad}}$, which in particular means that, for $L = \ell(s)$, $R = r(s)$, the equation

$$e([L]_1,[R]_2) - e([O^\dagger]_1,[1]_2) = e([H^\dagger]_1,[t(s)]_2) \quad (2)$$

holds but $O^\dagger \neq O(s)$.

Since adversary $\mathcal{B}$ received $\boldsymbol{a}, \boldsymbol{b}$ as part of $\mathcal{A}$'s output, it can run the honest prover algorithm and obtain O, H which satisfy that

$$e([L]_1,[R]_2) - e([O]_1,[1]_2) = e([H]_1,[t(s)]_2) \quad (3)$$

and $O = O(s)$.

Subtracting Eqs. (2) and (3), we get $e([O^\dagger - O]_1,[1]_2) = e([H^\dagger - H]_1,[t(s)]_2)$. Therefore, $([O^\dagger - O]_1, [H^\dagger - H]_1)$ is a solution to the $(\mathcal{R}, \boldsymbol{m})$-Rational Strong Diffie-Hellman Assumption.

We note that the verification algorithm never uses $(\boldsymbol{a}, \boldsymbol{b})$ which are part of the statement. When using the scheme as a building block, we omit $(\boldsymbol{a}, \boldsymbol{b})$ from the input of the verifier of the quadratic relations.

4.2 Argument for Linear Languages

Let gk be a bilinear group of order p and $\ell_1, \ell_2, n \in \mathbb{N}$ and $[\mathbf{M}]_1 \in \mathbb{G}_1^{\ell_1 \times n}$, $[\mathbf{N}]_1 \in \mathbb{G}_1^{\ell_2 \times n}$ be some matrices sampled from some distributions $\mathcal{M}, \mathcal{N}$. We give two different arguments for the promise problem defined by languages $\mathcal{L}_{YES}^{\mathsf{lin}}, \mathcal{L}_{NO}^{\mathsf{lin}}$, which are parameterized by $gk, [\mathbf{M}]_1, [\mathbf{N}]_1$ and are defined as:

$$\mathcal{L}_{YES}^{\mathsf{lin}} = \{(\boldsymbol{w}, [\boldsymbol{u}]_1, [\boldsymbol{v}]_1) : [\boldsymbol{u}]_1 = [\mathbf{M}]_1 \boldsymbol{w},\ [\boldsymbol{v}]_1 = [\mathbf{N}]_1 \boldsymbol{w}\}$$
$$\mathcal{L}_{NO}^{\mathsf{lin}} = \{(\boldsymbol{w}, [\boldsymbol{u}]_1, [\boldsymbol{v}]_1) : [\boldsymbol{u}]_1 = [\mathbf{M}]_1 \boldsymbol{w},\ [\boldsymbol{v}]_1 \neq [\mathbf{N}]_1 \boldsymbol{w}\}.$$

The arguments are simply the QA-NIZK Arguments of membership in linear spaces for general and witness samplable distributions as presented by Kiltz and Wee [25] (which generalize previous constructions [21,26]). Both arguments are very similar and can be easily written in a unified way. The idea is to use the arguments to prove that there exists a witness $\boldsymbol{w}$ such that $\begin{pmatrix} \boldsymbol{u} \\ \boldsymbol{v} \end{pmatrix} = \begin{pmatrix} \mathbf{M} \\ \mathbf{N} \end{pmatrix} \boldsymbol{w}$. Intuitively, assuming that it is hard to find non-trivial $(\boldsymbol{w}, \boldsymbol{w}')$ such that $[\boldsymbol{u}]_1 = [\mathbf{M}]_1 \boldsymbol{w} = [\mathbf{M}]_1 \boldsymbol{w}'$, this would prove that $[\boldsymbol{v}]_1 = [\mathbf{N}]_1 \boldsymbol{w}$. However, finding a security proof is not simple.

For witness samplable distributions, we only know a proof in the generic group model. The proof is a trivial consequence of the knowledge soundness property of QA-NIZK arguments which has already been used in previous works [5]. It has a proof size of k group elements when instantiated for the k-Lin Assumption.

Our main technical contribution is to prove soundness for the promise problem for general distributions (not necessarily witness samplable) assuming the hardness of the decisional problem for the distribution associated to matrix $\mathbf{M}$ (the $\mathcal{M}^\top$-MDDH Assumption). It has a proof size of $k+1$ group elements when instantiated for the k-Lin Assumption.

In Fig. (2) we describe the QA-NIZK argument of membership in linear spaces for witness samplable and general distributions (the only difference between these two cases is the definition of $\tilde{\mathcal{D}}_k$), as presented in [25]. The difference with the original presentation in [25] is that we separate the key $\mathbf{K}$ in blocks $\mathbf{K}_1, \mathbf{K}_2$ associated to $\mathbf{M}, \mathbf{N}$, which will be convenient for the proof. Perfect completeness, perfect zero-knowledge and computational soundness under any $\mathcal{D}_k$-KerMDH Assumption is proven [25].

Soundness of $\mathsf{Lin}_{\tilde{\mathcal{D}}_k}$, w.r.t. the language $\mathcal{L}_{NO}^{\mathsf{lin}}$, is a direct consequence of Lemma 3.

Lemma 3. *For any adversary $\mathcal{A}$ and for any $\mathbf{N} \in \mathbb{Z}_p^{\ell_2 \times n}$, let*

$$\epsilon_{\mathcal{A}} = \Pr\left[\begin{array}{c} \boldsymbol{v} \neq 0 \\ \boldsymbol{\pi} = \boldsymbol{v}^\top \mathbf{K}_2 \end{array} \middle| \begin{array}{l} \mathbf{M} \leftarrow \mathcal{M}; \mathbf{N} \leftarrow \mathcal{N}; \\ \text{crs} \leftarrow \mathsf{K}(gk, [\mathbf{M}]_1, [\mathbf{N}]_1); \\ ([\boldsymbol{v}]_1, [\boldsymbol{\pi}]_1) \leftarrow \mathcal{A}(\text{crs}, [\mathbf{M}]_1, [\mathbf{N}]_1) \end{array}\right].$$

$\underline{\mathsf{K}(gk, [\mathbf{M}]_1, [\mathbf{N}]_1)}$: // $\mathbf{M} \in \mathbb{Z}_p^{\ell_1 \times n}, \mathbf{N} \in \mathbb{Z}_p^{\ell_2 \times n}$
$\mathbf{K}_1 \leftarrow \mathbb{Z}_p^{\ell_1 \times \overline{k}}$; $\mathbf{K}_2 \leftarrow \mathbb{Z}_p^{\ell_2 \times \overline{k}}$;
$\mathbf{K} = \begin{pmatrix} \mathbf{K}_1 \\ \mathbf{K}_2 \end{pmatrix}$;
Sample $\mathbf{A} \leftarrow \widetilde{\mathcal{D}}_k$;
$[\mathbf{B}]_1 = [\mathbf{M}^\top \mathbf{K}_1 + \mathbf{N}^\top \mathbf{K}_2]_1$;
$\mathbf{C}_1 = \mathbf{K}_1 \mathbf{A}$; $\mathbf{C}_2 = \mathbf{K}_2 \mathbf{A}$; $\mathbf{C} = \mathbf{K}\mathbf{A}$
return crs $= (gk, [\mathbf{B}]_1, [\mathbf{A}]_2, [\mathbf{C}]_2)$.

$\underline{\mathsf{P}(\text{crs}, [\boldsymbol{u}]_1, [\boldsymbol{v}]_1, \boldsymbol{w})}$:
return $[\boldsymbol{\pi}]_1 = \boldsymbol{w}^\top [\mathbf{B}]_1$;

$\underline{\mathsf{V}(\text{crs}, [\boldsymbol{u}]_1, [\boldsymbol{v}]_1, [\boldsymbol{\pi}]_1)}$:
Check if:
$e([\boldsymbol{\pi}]_1, [\mathbf{A}]_2) =$
$e([\boldsymbol{u}^\top]_1, [\mathbf{C}_1]_2) + e([\boldsymbol{v}^\top]_1, [\mathbf{C}_2]_2)$

Fig. 2. The $\mathsf{Lin}_{\widetilde{\mathcal{D}}_k}$ argument for proving membership in linear spaces. The matrix $\mathbf{A}$ is either sampled from a distribution $\widetilde{\mathcal{D}}_k = \overline{\mathcal{D}}_k$ or from a distribution $\widetilde{\mathcal{D}}_k = \mathcal{D}_k$, such that the $\mathcal{D}_k$-KerMDH assumption holds. In the latter case $\overline{k} = k + 1$ while in the former case $\overline{k} = k$.

1. *When $\widetilde{\mathcal{D}}_k = \overline{\mathcal{D}}_k$ and $\mathcal{M}$ is witness samplable, if $\mathcal{A}$ is generic there exists a PPT adversary $\mathcal{B}$ such that $\epsilon_{\mathcal{A}} \leq \mathsf{Adv}_{\mathcal{M}\text{-}\mathsf{FindRep}}(\mathcal{B}) + \mathsf{negl}(\lambda)$.*
2. *When $\widetilde{\mathcal{D}}_k = \mathcal{D}_k$, there exists a PPT adversary $\mathcal{B}$ such that $\epsilon_{\mathcal{A}} \leq \mathsf{Adv}_{\mathcal{M}^\top\text{-}\mathsf{MDDH}}(\mathcal{B}) + 1/p$,*

where $\mathcal{M}^\top$ is the distribution which results from sampling matrices from $\mathcal{M}$ and transposing them.

Proof. (Lemma 3.1.) The proof is a direct consequence of the fact that scheme from Fig. 2 is an argument of knowledge in the generic group model, as proven by Fauzi et al. [5, Theorem 2]. Indeed, if this is the case there exists an extractor which given $\mathcal{A}$ outputs a witness $\boldsymbol{w}^*$ such that $\binom{\boldsymbol{0}}{\boldsymbol{v}} = \binom{\mathbf{M}}{\mathbf{N}} \boldsymbol{w}^*$. Since $\boldsymbol{v} \neq 0$, then $\boldsymbol{w}^* \neq 0$ and $\boldsymbol{w}^* \in \mathbb{Z}_p^n$ is a non-trivial element in the kernel of $\mathbf{M}$, breaking the $\mathcal{M}$-$\mathsf{FindRep}$ assumption[4].

Proof. (Lemma 3.2). The proof follows from the indistinguishability of the following games

Game_0: This game runs the adversary as in Lemma 3.
Game_1: This game is exactly as Game_0 but the crs is computed using algorithm K^*, as defined in Fig. 3, and the winning condition is

$$\boldsymbol{v} \neq 0 \text{ and } \boldsymbol{\pi} = (\boldsymbol{v}^\top (\mathbf{C}_2 - \mathbf{K}_{2,2}\underline{\mathbf{A}})\overline{\mathbf{A}}^{-1}, \boldsymbol{v}^\top \mathbf{K}_{2,2}),$$

where $\underline{\mathbf{A}}$ is the last row of $\mathbf{A}$ and $\overline{\mathbf{A}}$ is the first $k \times k$ block of $\mathbf{A}$.
Game_2: This game is exactly as Game_1 but $\boldsymbol{z} \leftarrow \mathbb{Z}_p^n$.

We now prove some Lemmas which show that the games are indistinguishable. Lemmas 4 and 5 show that the adversary has essentially the same advantage

4 For the distribution $\mathcal{M}^\top$ used in Sect. 5 this assumption is equivalent to the m-DLog assumption.

$\underline{\mathsf{K}^*(gk, [\mathbf{M}]_1, [\mathbf{N}]_1):}$ // $\mathbf{M} \in \mathbb{Z}_p^{\ell_1 \times n}, \mathbf{N} \in \mathbb{Z}_p^{\ell_2 \times n}$
Sample $\mathbf{A} \leftarrow \mathcal{D}_k$;
$\mathbf{C}_1 \leftarrow \mathbb{Z}_p^{\ell_1 \times k}$; $\mathbf{C}_2 \leftarrow \mathbb{Z}_p^{\ell_2 \times k}$; $\mathbf{C} = \begin{pmatrix} \mathbf{C}_1 \\ \mathbf{C}_2 \end{pmatrix}$; $\mathbf{K}_{1,2} \leftarrow \mathbb{Z}_p^{\ell_1}$; $\mathbf{K}_{2,2} \leftarrow \mathbb{Z}_p^{\ell_2}$;
$\mathbf{K}_{2,1} = (\mathbf{C}_2 - \mathbf{K}_{2,2}\underline{\mathbf{A}})\overline{\mathbf{A}}^{-1} \in \mathbb{Z}_p^{\ell_2 \times k}$; $[\boldsymbol{z}]_1 = [\mathbf{M}^\top]_1 \mathbf{K}_{1,2}$;
$[\mathbf{B}]_1 = ([\mathbf{M}^\top \mathbf{C}_1 \overline{\mathbf{A}}^{-1} - \boldsymbol{z}\underline{\mathbf{A}}\overline{\mathbf{A}}^{-1} + \mathbf{N}^\top \mathbf{K}_{2,1}]_1, [\boldsymbol{z}]_1 + [\mathbf{N}^\top]_1 \mathbf{K}_{2,2})$;
return crs $= (gk, [\mathbf{B}]_1, [\mathbf{A}]_2, [\mathbf{C}]_2)$.

Fig. 3. The modified crs generation algorithm used in Lemma 3.

of winning in any game. Lemma 6 says that the adversary has negligible probability of winning in Game_2. Lemma 3.2 follows from the composition of Lemmas 4, 5 and 6.

Lemma 4. *For any (unbounded) algorithm $\mathcal{A}$ we have* $\Pr[\mathsf{Game}_1(\mathcal{A}) = 1] = \Pr[\mathsf{Game}_0(\mathcal{A}) = 1]$.

Proof. If we define $\mathbf{K}_{1,1} = (\mathbf{C}_1 - \mathbf{K}_{1,2}\underline{\mathbf{A}})\overline{\mathbf{A}}^{-1}$ and $\mathbf{K} = \begin{pmatrix} \mathbf{K}_1 \\ \mathbf{K}_2 \end{pmatrix} = \begin{pmatrix} \mathbf{K}_{1,1} & \mathbf{K}_{1,2} \\ \mathbf{K}_{2,1} & \mathbf{K}_{2,2} \end{pmatrix}$, we observe that the output of K^* is well formed and the winning condition is the same as in the previous game, since

$$\begin{aligned}[\mathbf{B}]_1 &= ([\mathbf{M}^\top \mathbf{C}_1 \overline{\mathbf{A}}^{-1} - \boldsymbol{z}\underline{\mathbf{A}}\overline{\mathbf{A}}^{-1} + \mathbf{N}^\top \mathbf{K}_{2,1}]_1, [\boldsymbol{z}]_1 + [\mathbf{N}^\top]_1 \mathbf{K}_{2,2}) \\ &= ([\mathbf{M}^\top \mathbf{K}_{1,1} + \mathbf{N}^\top \mathbf{K}_{2,1}]_1, [\mathbf{M}^\top \mathbf{K}_{1,2} + \mathbf{N}^\top \mathbf{K}_{2,2}]_1) = [\mathbf{M}^\top \mathbf{K}_1 + \mathbf{N}^\top \mathbf{K}_2]_1, \qquad \text{and}\end{aligned}$$

$$\mathbf{K}\mathbf{A} = \begin{pmatrix} (\mathbf{C}_1 - \mathbf{K}_{1,2}\underline{\mathbf{A}})\overline{\mathbf{A}}^{-1} & \mathbf{K}_{1,2} \\ (\mathbf{C}_2 - \mathbf{K}_{2,2}\underline{\mathbf{A}})\overline{\mathbf{A}}^{-1} & \mathbf{K}_{2,2} \end{pmatrix} \begin{pmatrix} \overline{\mathbf{A}} \\ \underline{\mathbf{A}} \end{pmatrix} = \begin{pmatrix} \mathbf{C}_1 - \mathbf{K}_{1,2}\underline{\mathbf{A}} + \mathbf{K}_{1,2}\underline{\mathbf{A}} \\ \mathbf{C}_2 - \mathbf{K}_{2,2}\underline{\mathbf{A}} + \mathbf{K}_{2,2}\underline{\mathbf{A}} \end{pmatrix} = \mathbf{C},$$

and by definition $\boldsymbol{\pi} = (\boldsymbol{v}^\top(\mathbf{C}_2 - \mathbf{K}_{2,2}\underline{\mathbf{A}})\overline{\mathbf{A}}^{-1}, \boldsymbol{v}^\top \mathbf{K}_{2,2}) = (\boldsymbol{v}^\top \mathbf{K}_{2,1}, \boldsymbol{v}^\top \mathbf{K}_{2,2}) = \boldsymbol{v}^\top \mathbf{K}_2$.

Therefore we just need to argue that the distribution of $\mathbf{K}$ is the same in both games. But this is an immediate consequence of the fact that for every value of $(\mathbf{C}, \mathbf{K}_{1,1}, \mathbf{K}_{2,1})$ there exists a unique value of $(\mathbf{K}_{1,2}, \mathbf{K}_{2,2})$ which is compatible with $\mathbf{C} = \mathbf{K}\mathbf{A}$. Indeed, $\mathbf{C} = \mathbf{K}\mathbf{A} \iff \mathbf{C}_i = \mathbf{K}_{i,1}\overline{\mathbf{A}} + \mathbf{K}_{i,2}\underline{\mathbf{A}},\ i = 1,2 \iff (\mathbf{C}_i - \mathbf{K}_{i,2}\underline{\mathbf{A}})\overline{\mathbf{A}}^{-1} = \mathbf{K}_{i,1},\ i = 1,2$.

Lemma 5. *For any PPT algorithm $\mathcal{A}$ there exists a PPT algorithm $\mathcal{B}$ such that* $|\Pr[\mathsf{Game}_1(\mathcal{A}) = 1] - \Pr[\mathsf{Game}_0(\mathcal{A}) = 1]| \leq \mathsf{Adv}_{\mathcal{M}^\top\text{-}\mathsf{MDDH}}(\mathcal{B})$.

Proof. We construct an adversary $\mathcal{B}$ that receives the challenge $([\mathbf{M}^\top]_1, [\boldsymbol{z}^*]_1)$, where $\boldsymbol{z}^*$ is either $\mathbf{M}^\top \boldsymbol{r}$, $\boldsymbol{r} \leftarrow \mathbb{Z}_p^{\ell_1}$, or $\boldsymbol{z}^* \leftarrow \mathbb{Z}_p^n$. $\mathcal{B}$ computes the crs running $\mathsf{K}^*(gk, [\mathbf{M}]_1, [\mathbf{N}]_1)$ but replaces $[\boldsymbol{z}]_1$ with $[\boldsymbol{z}^*]_1$, and then runs $\mathcal{A}$ as in game Game_1. It follows that $\Pr[\mathcal{B}([\mathbf{M}^\top]_1, [\boldsymbol{z}^*]_1) = 1 | \boldsymbol{z}^* = \mathbf{M}^\top \boldsymbol{r}] = \Pr[\mathsf{Game}_1(\mathcal{A}) = 1]$ and $\Pr[\mathcal{B}([\mathbf{M}^*]_1, [\boldsymbol{z}^*]_1) = 1 | \boldsymbol{z}^* \leftarrow \mathbb{Z}_p^n] = \Pr[\mathsf{Game}_2(\mathcal{A}) = 1]$ and the lemma follows.

Lemma 6. *For any (unbounded) algorithm $\mathcal{A}$,* $\Pr[\mathsf{Game}_2(\mathcal{A}) = 1] \leq 1/p$.

Proof. We will show that, conditioned on $\mathbf{A}, \mathbf{C}, \mathbf{B}, \mathbf{M}, \mathbf{N}$, the matrix $\mathbf{K}_{2,2}$ is uniformly distributed. Since it holds that $\mathbf{BA} = (\mathbf{M}^\top, \mathbf{N}^\top)\mathbf{C}$, we get that the first k columns of $\mathbf{B}$, namely $\mathbf{B}_1$, are completely determined by $\mathbf{B}_2$, the last column of $\mathbf{B}$. Indeed

$$(\mathbf{B}_1, \mathbf{B}_2)\mathbf{A} = (\mathbf{M}^\top, \mathbf{N}^\top)\mathbf{C} \iff \mathbf{B}_1 = ((\mathbf{M}^\top, \mathbf{N}^\top)\mathbf{C} - \mathbf{B}_2\underline{\mathbf{A}})\overline{\mathbf{A}}^{-1}.$$

Hence, conditioning in $\mathbf{A}, \mathbf{C}, \mathbf{B}_2, \mathbf{M}, \mathbf{N}$ doesn't alter the probability. We have that $\mathbf{B}_2 = \boldsymbol{z} + \mathbf{N}^\top \mathbf{K}_{2,2}$, which consists of n equations on $n + \ell_2$ variables. It follows that there are ℓ_2 free variables. Then $\mathbf{K}_{2,2}$ is uniformly distributed and hence completely hidden to the adversary.

Note that

$$\boldsymbol{\pi} = \boldsymbol{v}^\top \mathbf{K}_2 \Longrightarrow \boldsymbol{\pi}_2 = \boldsymbol{v}^\top \mathbf{K}_{2,2},$$

where $\boldsymbol{\pi}_2$ is the last element of $\boldsymbol{\pi}$. Given that $\boldsymbol{v} \neq 0$, the last equation only holds with probability $1/p$ and so $\mathcal{A}$'s probability of winning.

The knowledge transfer property is a direct consequence of Lemma 3.

Theorem 1. *For any adversary $\mathcal{A}$ against the soundness of* Lin *with respect to* $\mathcal{L}_{NO}^{\mathsf{lin}}$*, it holds that:*

1. *When $\widetilde{\mathcal{D}}_k = \overline{\mathcal{D}}_k$, $\mathcal{M}$ is witness samplable, if $\mathcal{A}$ is generic then there exists a PPT adversary $\mathcal{B}$ such that* $\epsilon_{\mathcal{A}} \leq \mathsf{Adv}_{\mathcal{M}\text{-}\mathsf{FindRep}}(\mathcal{B}) + \mathsf{negl}(\lambda)$.
2. *When $\widetilde{\mathcal{D}}_k = \mathcal{D}_k$, there exist adversaries $\mathcal{B}_1$ and $\mathcal{B}_2$ such that*

$$\mathsf{Adv}_{\mathsf{Lin}}(\mathcal{A}) \leq \mathsf{Adv}_{\mathcal{D}_k\text{-}\mathsf{KerMDH}}(\mathcal{B}_1) + \mathsf{Adv}_{\mathcal{M}^\top\text{-}\mathsf{MDDH}}(\mathcal{B}_2) + 1/p.$$

Proof. Both for the witness samplable and the general case, given an adversary that produces a valid proof for a statement in $\mathcal{L}_{NO}^{\mathsf{lin}}$, successful attacks can be divided in two categories.

Type I: In this attack $[\boldsymbol{\pi}]_1 \neq [\boldsymbol{u}^\top]_1 \mathbf{K}_1 + [\boldsymbol{v}^\top]_1 \mathbf{K}_2$.
Type II: In this type of attack $[\boldsymbol{\pi}]_1 = [\boldsymbol{u}^\top]_1 \mathbf{K}_1 + [\boldsymbol{v}^\top]_1 \mathbf{K}_2$.

Type I attacks are not possible when $\overline{k} = k$, because proofs are unique, i.e. there is only one value of $\boldsymbol{\pi}$ which can satisfy the verification equation. Type I attacks are computationally infeasible when $\overline{k} = k + 1$, as they can be used to construct an adversary $\mathcal{B}_1$ against the $\mathcal{D}_k$-KerMDH assumption.[5] Adversary $\mathcal{B}_1$ receives a challenge $[\mathbf{A}]_2$ and then runs the soundness experiment for $\mathcal{A}$. When $\mathcal{A}$ outputs $([\boldsymbol{u}]_1, [\boldsymbol{v}]_1, [\boldsymbol{\pi}]_1)$, $\mathcal{B}_1$ outputs $[\boldsymbol{\pi}^\dagger]_1 = [\boldsymbol{\pi}]_1 - [\boldsymbol{u}^\top]_1 \mathbf{K}_1 - [\boldsymbol{v}^\top]_1 \mathbf{K}_2 \neq 0$. Since $[\boldsymbol{\pi}]_1$ is accepted by the verifier we get that $e([\boldsymbol{\pi}]_1, [\mathbf{A}]_2) = e([\boldsymbol{u}^\top]_1, [\mathbf{C}_1]_2) + e([\boldsymbol{v}^\top]_1, [\mathbf{C}_2]_2)$ and then $\boldsymbol{\pi}^\dagger \mathbf{A} = \boldsymbol{\pi}\mathbf{A} - \boldsymbol{u}^\top \mathbf{K}_1 \mathbf{A} - \boldsymbol{v}^\top \mathbf{K}_2 \mathbf{A} = \boldsymbol{\pi}\mathbf{A} - \boldsymbol{u}^\top \mathbf{C}_1 -$

[5] This part of the proof follows essentially the same lines of the first constant-size QA-NIZK arguments for linear spaces of Libert et al. [26] which were later simplified and generalized by Kiltz and Wee [25].

$\boldsymbol{v}^\top \mathbf{C}_2 = 0$. We conclude that the success probability of a type I attack is bounded by $\mathsf{Adv}_{\mathcal{D}_k\text{-}\mathsf{KerMDH}}(\mathcal{B}_1)$.

For type II attacks, for both types of distributions, since $[\boldsymbol{\pi}]_1 = [\boldsymbol{u}^\top]_1 \mathbf{K}_1 + [\boldsymbol{v}^\top]_1 \mathbf{K}_2$ is a valid proof for $\begin{pmatrix} [\boldsymbol{u}]_1 \\ [\boldsymbol{v}]_1 \end{pmatrix}$, then, by linearity of the verification equation, $\boldsymbol{\pi}^\dagger = \boldsymbol{\pi} - \boldsymbol{w}^\top \mathbf{B}$ is a valid proof for $\begin{pmatrix} 0 \\ [\boldsymbol{v}^\dagger]_1 \end{pmatrix} = \begin{pmatrix} [\boldsymbol{u}]_1 - [\mathbf{M}]_1 \boldsymbol{w} \\ [\boldsymbol{v}]_1 - [\mathbf{N}]_1 \boldsymbol{w} \end{pmatrix}$. Since $\boldsymbol{v} \neq \mathbf{N}\boldsymbol{w}$, we conclude that an attacker of type II can be turned into an attacker $\mathcal{B}_2$ for Lemma 3.

4.3 Extension to SMDDH Assumptions

In Sect. 5 the crs includes $\mathbf{M}$ in both groups, i.e. $[\mathbf{M}]_1, [\mathbf{M}]_2$. This implies that we need to prove Lemma 3 even when the adversary is given $[\mathbf{M}]_1, [\mathbf{M}]_2$. But this is not a problem, since we can build an adversary for Lemma 5 against the $\mathcal{M}^\top$-$\mathsf{SMDDH}_{\mathbb{G}_1}$ assumption. Similarly, we can prove that Theorem 1 holds, even when the adversary is given $[\mathbf{M}]_1, [\mathbf{M}]_2$, assuming the hardness of the $\mathcal{M}^\top$-SMDDH assumption.

4.4 Extension to Bilateral Linear Spaces

In Sect. 5 we need a QA-NIZK argument for bilateral linear spaces [14], which are linear spaces split between $\mathbb{G}_1$ and $\mathbb{G}_2$. In [14], a QA-NIZK argument for such languages is given, which is very close to the argument of membership in (unilateral) linear spaces of [25]. In Fig. (4) we describe the QA-NIZK argument of [14] adapted to matrices with 3 blocks. The proof of the knowledge transfer property is essentially the same as in the unilateral case and can be found in the full version of this work.

5 A New Argument for Correct Arithmetic Circuit Evaluation

In this section we describe our construction for proving correct evaluation of an arithmetic circuit. It makes use of two subarguments: a quadratic and a linear "knowledge transfer" subarguments. The reason why we use the term "knowledge transfer" is because these arguments will ensure that, if the prover knows a witness for the circuit evaluation up to level i which is also a valid opening up to level i of a set of shrinking commitments to the corresponding wires, it also knows a valid opening to the commitments of the wires at level $i+1$.

Since the input of the circuit is public, the idea is that these arguments allow to "transfer" the knowledge of the witness for correct evaluation (a consistent assignment to all wires) to lower levels of the circuit. Any adversary against soundness needs to break the "chain" of consistent evaluations at some point and thus, break the soundness of one of the two subarguments. This technique allows us to avoid using binding commitments to the wires at each level, while

$\underline{\mathsf{K}(gk, [\mathbf{M}]_1, [\mathbf{N}]_1, [\mathbf{P}]_2)}$:
// $\mathbf{M} \in \mathbb{Z}_p^{\ell_1 \times n}, \mathbf{N} \in \mathbb{Z}_p^{\ell_2 \times n}, \mathbf{P} \in \mathbb{Z}_p^{\ell_3 \times n}$
$\mathbf{K}_1 \leftarrow \mathbb{Z}_p^{\ell_1 \times \overline{k}}$; $\mathbf{K}_2 \leftarrow \mathbb{Z}_p^{\ell_2 \times \overline{k}}$; $\mathbf{K}_3 \leftarrow \mathbb{Z}_p^{\ell_3 \times \overline{k}}$
$\mathbf{K}^\top = (\mathbf{K}_1^\top, \mathbf{K}_2^\top, \mathbf{K}_3^\top)$;
Sample $\mathbf{A} \leftarrow \widetilde{\mathcal{D}}_k$; $\mathbf{\Gamma} \leftarrow \mathbb{Z}_p^{n \times \overline{k}}$
$[\mathbf{B}]_1 = [\mathbf{M}^\top \mathbf{K}_1 + \mathbf{N}^\top \mathbf{K}_2 + \mathbf{\Gamma}]_1$;
$[\mathbf{D}]_2 = [\mathbf{P}^\top \mathbf{K}_3 - \mathbf{\Gamma}]$;
$\mathbf{C}_1 = \mathbf{K}_1 \mathbf{A}$; $\mathbf{C}_2 = \mathbf{K}_2 \mathbf{A}$;
$\mathbf{C}_3 = \mathbf{K}_3 \mathbf{A}$: $\mathbf{C} = \mathbf{K}\mathbf{A}$
return crs $= (gk, [\mathbf{B}]_1, [\mathbf{D}]_2, [\mathbf{A}]_{1,2}, [\mathbf{C}_1]_2, [\mathbf{C}_2]_2, [\mathbf{C}_3]_1)$.

$\underline{\mathsf{P}(\text{crs}, [\boldsymbol{u}]_1, [\boldsymbol{v}_1]_1, [\boldsymbol{v}_2]_2, \boldsymbol{w})}$:
$\boldsymbol{\rho} \leftarrow \mathbb{Z}_p^{\overline{k}}$;
$[\boldsymbol{\pi}]_1 = \boldsymbol{w}^\top [\mathbf{B}]_1 + [\boldsymbol{\rho}]_1$;
$[\boldsymbol{\theta}]_1 = \boldsymbol{w}^\top [\mathbf{D}]_2 - [\boldsymbol{\rho}]_2$;
return $([\boldsymbol{\pi}]_1, [\boldsymbol{\theta}]_2)$.

$\underline{\mathsf{V}(\text{crs}, [\boldsymbol{u}]_1, [\boldsymbol{v}_1]_1, [\boldsymbol{v}_2]_2, [\boldsymbol{\pi}]_1, [\boldsymbol{\theta}]_2)}$:
Check if:
$e([\boldsymbol{\pi}]_1, [\mathbf{A}]_2) - e([\boldsymbol{u}^\top]_1, [\mathbf{C}_1]_2) - e([\boldsymbol{v}_1^\top]_1, [\mathbf{C}_2]_2) = e([\boldsymbol{\theta}]_2, [\mathbf{A}]_1) - e([\boldsymbol{v}_2^\top]_2, [\mathbf{C}_3]_1)$

Fig. 4. The $\mathsf{BLin}_{\widetilde{\mathcal{D}}_k}$ argument for proving membership in bilateral linear spaces. The matrix $\mathbf{A}$ is either sampled from a distribution $\widetilde{\mathcal{D}}_k = \overline{\mathcal{D}}_k$ or from a distribution $\widetilde{\mathcal{D}}_k = \mathcal{D}_k$, such that the $\mathcal{D}_k$-SKerMDH assumption holds. In the latter case $\overline{k} = k + 1$ while in former case $\overline{k} = k$. Since the $\mathcal{D}_1$-SKerMDH is false [14] for any $\mathcal{D}_1$, it should hold that $k \geq 2$.

still being able to define what it means to break soundness. Intuitively, the difficulty we have to circumvent is to reason about whether the openings of shrinking commitments satisfy a certain equation without assuming that the adversary is generic, as there are many possible such openings.

The reason why we use two arguments is natural given characterization of circuits given in Sect. 3. The variables A_{ij} (resp. B_{ij}, C_{ij}) describe correct assignments to the j-th left (resp. right, output) wire at level i. We use the quadratic knowledge transfer property to ensure that a certain value $\boldsymbol{O}_i$ is a valid (deterministic, not hiding) commitment to all the outputs at level i if $\boldsymbol{L}_{i-1}$ and $\boldsymbol{R}_{i-1}$ are valid commitments (i.e. consistent with the input) to all the right and left wires at the previous level. On the other hand, we encode the affine constraints as membership in linear spaces and use the linear knowledge transfer argument to ensure that $\boldsymbol{L}_i, \boldsymbol{R}_i$ are valid commitments to all left and right wires at level i if $\boldsymbol{O}_j$ for $j = 1, \ldots, i-1$ are valid commitments to the previous levels.

Throughout this section, R_ϕ represents a relation $R_\phi = \{(gk, \boldsymbol{x}, \boldsymbol{y}) : \phi(\boldsymbol{x}) = \boldsymbol{y}\}$ where gk is an asymmetric bilinear group of order p and $\phi : \mathbb{Z}_p^{n_0} \to \mathbb{Z}_p^{n_d}$ as described in Sect. 3 and $N = \max_{i=1,\ldots,d} n_i$ is the maximum number of multiplicative gates of same multiplicative depth. The construction is parameterized by a value k_s, following the dicussion in Sect. 4.2 on the security properties of the linear knowledge transfer argument.

This section is organized as follows: we first show how to encode affine constraints as membership in linear spaces, then we present the description of our argument in terms of the two subarguments and give the (sketched) proof of security, and finally we discuss its efficiency.

5.1 Encoding Affine Constraints as Membership in Linear Spaces

We translate the affine constraints described in the circuit encoding of Sect. 3 as membership of $([\boldsymbol{O}]_1, [\boldsymbol{L}]_1, [\boldsymbol{R}]_2)$ in a linear subspace of $\mathbb{G}_1^{n+(2d-1)k_s} \times \mathbb{G}_2^{dk_s}$.

We write in matrix form the expression of $(\boldsymbol{x}, [\boldsymbol{O}]_1, [\boldsymbol{L}]_1, [\boldsymbol{R}]_2)$ in terms of the internal wires of the circuit, following Sect. 3. The commitments to the output values $[\boldsymbol{O}]_1$ should satisfy that $[\boldsymbol{O}_i]_1 = [\boldsymbol{\Lambda}_i]_1 \boldsymbol{c}_i$, where $\boldsymbol{\Lambda}_i = (\lambda_1(\boldsymbol{s}), \ldots, \lambda_{n_i}(\boldsymbol{s}))$ and $\lambda_j(X)$ is the jth Lagrangian polynomial for some $\mathcal{R} = \{r_1, \ldots, r_N\} \subset \mathbb{Z}_p$ and the input $\boldsymbol{x} = \boldsymbol{c}_0$ is public. These constraints can be expressed in matrix form in Eq. (4):

$$\begin{pmatrix} \boldsymbol{x} \\ \boldsymbol{O}_1 \\ \boldsymbol{O}_2 \\ \boldsymbol{O}_3 \\ \vdots \\ \boldsymbol{O}_{d-1} \end{pmatrix} = \begin{pmatrix} \mathbf{I} & \mathbf{0} & \mathbf{0} & \mathbf{0} & \ldots & \mathbf{0} \\ \mathbf{0} & \boldsymbol{\Lambda}_1 & \mathbf{0} & \mathbf{0} & \ldots & \mathbf{0} \\ \mathbf{0} & \mathbf{0} & \boldsymbol{\Lambda}_2 & \mathbf{0} & \ldots & \mathbf{0} \\ \mathbf{0} & \mathbf{0} & \mathbf{0} & \boldsymbol{\Lambda}_3 & & \mathbf{0} \\ \vdots & \vdots & \vdots & & \ddots & \\ \mathbf{0} & \mathbf{0} & \mathbf{0} & \mathbf{0} & \ldots & \boldsymbol{\Lambda}_{d-1} \end{pmatrix} \begin{pmatrix} \boldsymbol{c}_0 \\ \boldsymbol{c}_1 \\ \boldsymbol{c}_2 \\ \boldsymbol{c}_3 \\ \vdots \\ \boldsymbol{c}_{d-1} \end{pmatrix} \qquad (4)$$

We denote the matrix on the right hand side of (4) as $\mathbf{M}$, so this equation reads $\begin{pmatrix} \boldsymbol{x} \\ \boldsymbol{O} \end{pmatrix} = \mathbf{M}\boldsymbol{c}$. On the other hand, the constraints satisfied by the left wires in terms of the output wires of previous levels can be written in matrix form as shown in Eq. (5):

$$\begin{pmatrix} \boldsymbol{L}_1 \\ \boldsymbol{L}_2 \\ \boldsymbol{L}_3 \\ \vdots \\ \boldsymbol{L}_d \end{pmatrix} = \begin{pmatrix} \mathbf{F}_{1,0} & \mathbf{0} & \mathbf{0} & \ldots & \mathbf{0} \\ \mathbf{F}_{2,0} & \mathbf{F}_{2,1} & \mathbf{0} & \ldots & \mathbf{0} \\ \mathbf{F}_{3,0} & \mathbf{F}_{3,1} & \mathbf{F}_{3,2} & \ldots & \mathbf{0} \\ \vdots & \vdots & \vdots & \ddots & \vdots \\ \mathbf{F}_{d,0} & \mathbf{F}_{d,1} & \mathbf{F}_{d,2} & \ldots & \mathbf{F}_{d,d-1} \end{pmatrix} \begin{pmatrix} \boldsymbol{c}_0 \\ \boldsymbol{c}_1 \\ \boldsymbol{c}_2 \\ \vdots \\ \boldsymbol{c}_{d-1} \end{pmatrix} + \begin{pmatrix} \hat{\boldsymbol{L}}_1 \\ \hat{\boldsymbol{L}}_2 \\ \hat{\boldsymbol{L}}_3 \\ \vdots \\ \hat{\boldsymbol{L}}_d \end{pmatrix}, \qquad (5)$$

that is, for each i, $\boldsymbol{L}_i = \sum_{k=0}^{i-1} \mathbf{F}_{i,k} \boldsymbol{c}_k + \hat{\boldsymbol{L}}_i$, where

$$\begin{aligned} \mathbf{F}_{i,k} &= \left(\sum_{j=1}^{n_k} f_{ijk1}\lambda_j(\boldsymbol{s}), \sum_{j=1}^{n_k} f_{ijk2}\lambda_j(\boldsymbol{s}), \ldots \sum_{j=1}^{n_k} f_{ijkn_k}\lambda_j(\boldsymbol{s})\right) \\ &= \left(v_{ik1}(\boldsymbol{s}), v_{ik2}(\boldsymbol{s}), \ldots v_{ikn_k}(\boldsymbol{s})\right) \end{aligned} \qquad (6)$$

and $\hat{\boldsymbol{L}}_i = \sum_{j=1}^{n_i} f_{ij}\lambda_j(\boldsymbol{s}) = v_i(\boldsymbol{s})$, for the constants which are defined in Lemma 1. We denote the matrix on the right hand side of Eq. (5) as $\mathbf{N}$, so this equation reads $\boldsymbol{L} = \mathbf{N}\boldsymbol{c} + \hat{\boldsymbol{L}}$. The constraints satisfied by the right wires in terms of the output wires of previous levels can be written in a similar form as shown in Eq. (7):

$$\begin{pmatrix} \boldsymbol{R}_1 \\ \boldsymbol{R}_2 \\ \boldsymbol{R}_3 \\ \vdots \\ \boldsymbol{R}_d \end{pmatrix} = \begin{pmatrix} \mathbf{G}_{1,0} & \mathbf{0} & \mathbf{0} & \ldots & \mathbf{0} \\ \mathbf{G}_{2,0} & \mathbf{G}_{2,1} & \mathbf{0} & \ldots & \mathbf{0} \\ \mathbf{G}_{3,0} & \mathbf{G}_{3,1} & \mathbf{G}_{3,2} & \ldots & \mathbf{0} \\ \vdots & \vdots & \vdots & \ddots & \vdots \\ \mathbf{G}_{d,0} & \mathbf{G}_{d,1} & \mathbf{G}_{d,2} & \ldots & \mathbf{G}_{d,d-1} \end{pmatrix} \begin{pmatrix} \boldsymbol{c}_0 \\ \boldsymbol{c}_1 \\ \boldsymbol{c}_2 \\ \vdots \\ \boldsymbol{c}_{d-1} \end{pmatrix} + \begin{pmatrix} \hat{\boldsymbol{R}}_1 \\ \hat{\boldsymbol{R}}_2 \\ \hat{\boldsymbol{R}}_3 \\ \vdots \\ \hat{\boldsymbol{R}}_d \end{pmatrix}, \qquad (7)$$

that is, for each i, $\boldsymbol{R}_i = \sum_{k=0}^{i-1} \mathbf{G}_{i,k}\boldsymbol{c}_k + \hat{\boldsymbol{R}}_i$, where

$$\begin{aligned}\mathbf{G}_{i,k} &= \left(\sum_{j=1}^{n_k} g_{ijk1}\lambda_j(\boldsymbol{s}), \sum_{j=1}^{n_k} g_{ijk2}\lambda_j(\boldsymbol{s}), \ldots \sum_{j=1}^{n_k} g_{ijkn_k}\lambda_j(\boldsymbol{s})\right) \\ &= \left(w_{ik1}(\boldsymbol{s}), w_{ik2}(\boldsymbol{s}), \ldots w_{ikn_k}(\boldsymbol{s})\right), \qquad (8)\end{aligned}$$

and $\hat{\boldsymbol{R}}_i = \sum_{j=1}^{n_i} g_{ij}\lambda_j(\boldsymbol{s}) = w_i(\boldsymbol{s})$. We denote the matrix on the right hand side of Eq. (7) as $\mathbf{P}$, so this equation reads $\boldsymbol{R} = \mathbf{P}\boldsymbol{z} + \hat{\boldsymbol{R}}$.

With the notation defined, satisfaction of the affine constraints can be written as $\begin{pmatrix} [\boldsymbol{O}']_1 \\ [\boldsymbol{L}]_1 - [\hat{\boldsymbol{L}}]_1 \\ [\boldsymbol{R}]_2 - [\hat{\boldsymbol{R}}]_2 \end{pmatrix} \in \mathbf{Im}\begin{pmatrix} [\mathbf{M}]_1 \\ [\mathbf{N}]_1 \\ [\mathbf{P}]_2 \end{pmatrix}$, where $[\boldsymbol{O}']_1 = \begin{pmatrix} [\boldsymbol{x}]_1 \\ [\boldsymbol{O}]_1 \end{pmatrix}$. That is, the linear constraints are satisfied if a certain vector is in a subspace generated by some matrix which depends on the circuit.

5.2 New Argument

In this section we describe our construction for proving correct evaluation of an arithmetic circuit.

$\mathsf{Setup}(R_\phi)$: Pick $\boldsymbol{s} \leftarrow \mathbb{Z}_p^{k_s}$. Generate $\mathrm{crs}_\phi = (crs_{\phi,1}, \ldots, \mathrm{crs}_{\phi,k_s})$, where $\mathrm{crs}_{\phi,i} \leftarrow \mathsf{Quad.K}(gk, \{[s_i^j]_1\}_{j=1}^{N-1}, \{[s_i^j]_2\}_{j=1}^{N})$ is the crs for the quadratic knowledge transfer argument defined in Fig. 1. Express affine constraints (Eqs. (4), (5), and (7)) which define circuit satisfiability as membership in the image of $([\mathbf{M}^\top]_1, [\mathbf{N}^\top]_1, [\mathbf{P}^\top]_2)^\top$ as explained in Sect. 5.1. Generate a crs for the bilateral linear knowledge transfer argument defined in Fig. 4 for $([\mathbf{M}^\top]_1, [\mathbf{N}^\top]_1, [\mathbf{P}^\top]_2)^\top$.

$\mathsf{Prove}(\mathrm{crs}, (\boldsymbol{x}, \boldsymbol{y}, \boldsymbol{a}, \boldsymbol{b}, \boldsymbol{c}) \in \mathcal{R}_\phi)$: Given the input $\boldsymbol{x}$, the output $\boldsymbol{y}$, and $(\boldsymbol{a}, \boldsymbol{b}, \boldsymbol{c})$ a valid assignment to left, right and output wires as described in Lemma 1, the prover proceeds as follows:

1. For each $i \in \{1, \ldots, d\}$, commit to $\boldsymbol{a}_i, \boldsymbol{c}_i$ in $\mathbb{G}_1^{k_s}$ and to $\boldsymbol{b}_i$ in $\mathbb{G}_2^{k_s}$ as: $[\boldsymbol{L}_i]_1 = \sum_{j=1}^{n_i} a_{ij}[\lambda_j(\boldsymbol{s})]_1 = [\boldsymbol{\Lambda}_i]_1\boldsymbol{a}_i$, $[\boldsymbol{R}_i]_2 = \sum_{j=1}^{n_i} b_{i,j}[\lambda_j(\boldsymbol{s})]_2 = [\boldsymbol{\Lambda}_i]_2\boldsymbol{b}_i$, $[\boldsymbol{O}_i]_1 = \sum_{j=1}^{n_i} c_{ij}[\lambda_j(\boldsymbol{s})]_1 = [\boldsymbol{\Lambda}_i]_1\boldsymbol{c}_i$.
2. (Quadratic Constraints) For each $i \in \{1, \ldots, d\}$, and each $j \in \{1, \ldots, k_s\}$, compute a proof $\Pi_{i,j}^{\mathsf{quad}}$ that the vector $\boldsymbol{a}_i \circ \boldsymbol{b}_i$, which is the componentwise product of the openings of $[L_{ij}]_1, [R_{ij}]_2$, is an opening of $[O_{ij}]_1$.
3. (Linear Constraints) Compute a proof Π^{lin} that $[\boldsymbol{L}_i]_1$ and $[\boldsymbol{R}_i]_2$ are commitments to the correct evaluation of all the left and right wires at level i, for all $i \in \{1, \ldots, d\}$, that is, that they satisfy the affine linear constraints which relate them to the outputs of gates at levels $j = 0, \ldots, i-1$.
4. Output $(\mathcal{C} = ([\boldsymbol{L}]_1, [\boldsymbol{R}]_2, [\boldsymbol{O}]_1), \Pi^{\mathsf{quad}}, \Pi^{\mathsf{lin}})$ as the proof, where $\Pi^{\mathsf{quad}} = \{\Pi_{i,j}^{\mathsf{quad}} : i = 1, \ldots, d,\ j = 1, \ldots, k_s\}$.

$\mathsf{Verify}(\mathrm{crs}, (\boldsymbol{x}, \boldsymbol{y}), (\mathcal{C}, \Pi^{\mathsf{quad}}, \Pi^{\mathsf{lin}}))$: Output 1 if the following two checks are successful and 0 otherwise:

1. Verify $\Pi^{\mathsf{quad}}, \Pi^{\mathsf{lin}}$.
2. Check that $[\boldsymbol{O}_d]_1 = \sum_{j=1}^{n_d} [\lambda_j(\boldsymbol{s})]_1 y_j$.

Security. Perfect completeness is obvious, because if $(\boldsymbol{x}, \boldsymbol{y}, \boldsymbol{a}, \boldsymbol{b}, \boldsymbol{c})$ is a valid witness for satisfiability, then it satisfies both linear and quadratic constraints because of the characterization of Sect. 3 and the definition of $\mathbf{M}, \mathbf{N}, \mathbf{P}$ presented in Sect. 5.1.

Let $\mathcal{A}$ be an adversary against the soundness of the scheme. We construct an adversary $\mathcal{B}_1$ against the quadratic knowledge transfer argument, $\mathcal{B}_{2,0}, \ldots, \mathcal{B}_{2,d-1}$ against the linear knowledge transfer argument.

Adversary $\mathcal{B}_1$ receives the common reference string of the quadratic subargument, which includes $(gk, \{[s^i]_1\}_{i=1}^{N-1}, \{[s^i]_2\}_{i=1}^{N})$ and samples $\alpha_j \leftarrow \mathbb{Z}_p^*$, $j = 2, \ldots, k_s$. It defines $s = s_1$, $s_j = \alpha_j s_j$ and computes the crs of the quadratic argument for s_j, $j = 1, \ldots, k_s$ from the received values. It then creates the common reference string of the full argument in the natural way, by defining the matrices $\mathbf{M}, \mathbf{N}, \mathbf{P}$ from the crs of the quadratic subargument and sampling the rest of the secret key. When it receives an accepting proof $(\mathcal{C} = ([\boldsymbol{L}]_1, [\boldsymbol{R}]_2, [\boldsymbol{O}]_1), \Pi^{\mathsf{quad}}, \Pi^{\mathsf{lin}})$ from adversary $\mathcal{A}$ for some statement $(\boldsymbol{x}, \boldsymbol{y})$, adversary $\mathcal{B}_1$ computes the full witness for correct evaluation $(\boldsymbol{a}, \boldsymbol{b}, \boldsymbol{c})$ from $\boldsymbol{x}$. The adversary searches for indexes i, j such that $[L_{ij}]_1$ and $[R_{ij}]_2$ are commitments to $\boldsymbol{a}_i$ and $\boldsymbol{b}_i$ but $[O_{ij}]_1$ is not a valid commitment to $\boldsymbol{a}_i \circ \boldsymbol{b}_i$, and it aborts if these indexes do not exist. From α_j, adversary $\mathcal{A}$ computes $\boldsymbol{\mu} = (\mu_1, \ldots, \mu_{n_i}) \in \mathbb{Z}_p^{n_i}$ such that $\lambda_\ell(s_j) = \mu_\ell \lambda_\ell(s)$ and $\nu \in \mathbb{Z}_p$ such that $\nu t(s_j) = t(s)$. It returns $(\boldsymbol{a}_i \circ \boldsymbol{\mu}, \boldsymbol{b}_i \circ \boldsymbol{\mu}, [L_{ij}]_1, [R_{ij}]_2, [O_{ij}]_1)$, as an instance of $\mathcal{L}_{NO}^{\mathsf{quad}}$ together with an accepting proof $[\nu H_{ij}]_1$.

Adversary $\mathcal{B}_{2,i}$, $i = 0, \ldots, d-1$ receives a common reference string of the linear subargument for the language associated to the first $i+1$ (resp. $i+2, i+2$) blocks of rows and the first $\sum_{j=0}^{i} n_i$ columns of $\mathbf{M}$ (resp. $\mathbf{N}, \mathbf{P}$). That is, $\mathbf{M}_i, \mathbf{N}_i$ are defined as:

$$\mathbf{M}_i = \begin{pmatrix} \mathbf{I} & & \mathbf{0} \\ & \mathbf{\Lambda}_1 & \\ & & \ddots & \\ \mathbf{0} & & & \mathbf{\Lambda}_i \end{pmatrix}, \qquad \mathbf{N}_i = \begin{pmatrix} \mathbf{F}_{1,0} & \mathbf{0} & \ldots & \mathbf{0} \\ \mathbf{F}_{2,0} & \mathbf{F}_{2,1} & \ldots & \mathbf{0} \\ \vdots & \vdots & \ddots & \vdots \\ \mathbf{F}_{i+1,0} & \mathbf{F}_{i+1,1} & \ldots & \mathbf{F}_{i+1,i} \end{pmatrix},$$

and $\mathbf{P}_i$ is defined similarly. Using the linear properties of the crs, $\mathcal{B}_{2,i}$ computes the common reference string of the full argument.[6] When it receives an accepting proof $(\mathcal{C} = ([\boldsymbol{L}]_1, [\boldsymbol{R}]_2, [\boldsymbol{O}]_1)\}_{i=1}^{d}, \Pi^{\mathsf{quad}}, \Pi^{\mathsf{lin}})$ from adversary $\mathcal{A}$ for some statement $(\boldsymbol{x}, \boldsymbol{y})$, adversary $\mathcal{B}_{2,i}$ computes the full witness $(\boldsymbol{a}, \boldsymbol{b}, \boldsymbol{c})$. It then checks if $[\boldsymbol{O}_1]_1, \ldots, [\boldsymbol{O}_i]_1$ are commitments to $\boldsymbol{c}_1, \ldots, \boldsymbol{c}_i$ but either $[\boldsymbol{L}_{i+1}]_1$ or $[\boldsymbol{R}_{i+1}]_2$ are not valid commitments to $\boldsymbol{a}_i$ or $\boldsymbol{b}_i$. If this is not the case, it aborts. Else it outputs $(\boldsymbol{c}_1, \ldots, \boldsymbol{c}_i, [\boldsymbol{O}_1]_1, \ldots, [\boldsymbol{O}_i], [\boldsymbol{L}_1]_1 - [\hat{\boldsymbol{L}}_1], [\boldsymbol{L}_{i+1}]_1 - [\hat{\boldsymbol{L}}_{i+1}], [\boldsymbol{R}_1]_2 - [\hat{\boldsymbol{R}}_1]_2, \ldots, [\boldsymbol{R}_{i+1}]_2 - [\hat{\boldsymbol{R}}_{i+1}]_2)$ together with its corresponding proof, which adversary $\mathcal{B}_{2,i}$ can compute from the proof given by adversary $\mathcal{A}$ and the secret values it sampled to extend the crs of the subargument to the full crs (this is possible using the linearity of the proof, full details are in the full version of this work.

For every successful adversary $\mathcal{A}$ at least one of the adversaries $\mathcal{B}_1, \mathcal{B}_{2,0}, \ldots, \mathcal{B}_{2,d-1}$ does not abort. This is because if the statement is false there must be

[6] We can assume w.l.o.g. that the crs for the linear knowledge transfer associated to $\mathbf{M}_i, \mathbf{N}_i, \mathbf{P}_i$ includes $\{[s_j]_{1,2}\}_{j=1}^{N-1}, [s_j^N]$, as this does not compromise security.

some point in the "chain" where either $[\boldsymbol{L}_i]_1, [\boldsymbol{R}_i]_2$ are honestly computed but $[\boldsymbol{O}_i]_1$ is not, or $[\boldsymbol{O}_i]_1$ is honestly computed but $[\boldsymbol{L}_{i+1}]$ or $[\boldsymbol{R}_{i+1}]$ is not.

The linear knowledge transfer argument at level i is based on the $\mathcal{L}_2$-$\mathsf{SKerMDH}$ and the $\mathcal{M}_i^\top$-$\mathsf{SMDDH}_{\mathbb{G}_1}$ assumptions. The latter reduces to the $\mathcal{LR}_{\mathcal{R},k_s}$-$\mathsf{SMDDH}_{\mathbb{G}_1}$ and the SXDH assumptions as proven in the full version of this work. Based on this proof, we can state the following Theorem.

Theorem 2. *Let $(gk, \phi : \mathbb{Z}_p^{n_0} \to \mathbb{Z}_p^{n_d}, \mathcal{R})$ be a bilinear group of order p, an arithmetic circuit and a set of $\mathbb{Z}_p$ of cardinal $N = \max_{i=1,\ldots,d} n_i$. For any adversary $\mathcal{A}$ against the soundness of the argument defined above there exist adversaries $\mathcal{B}_1, \mathcal{B}_2, \mathcal{B}_3, \mathcal{B}_4$ such that:*

$$\mathsf{Adv}_{\mathsf{snd}}(\mathcal{A}) \leq \mathsf{Adv}_{\mathcal{R}\text{-}\mathsf{RSDH}}(\mathcal{B}_1) + d\mathsf{Adv}_{\mathcal{L}_2\text{-}\mathsf{SKerMDH}}(\mathcal{B}_2) + dk_s\mathsf{Adv}_{\mathcal{LG}_{\mathcal{R},k_s}\text{-}\mathsf{SMDDH}_{\mathbb{G}_1}}(\mathcal{B}_3) + d\min(N - k_s, d)\log k_s\mathsf{Adv}_{\mathsf{SXDH}}(\mathcal{B}_4) + \frac{d(1+k_s)}{p}.$$

Note that the most efficient, secure choice is $k_s = 2$ and then the largest security loss factor is $d\min(N - k_s, d) \leq d \cdot N$, which is at most the number of multiplicative gates in the circuit.

5.3 Efficiency

In the most efficient instantiation, the proof size is $(3d+2, d+2)$ group elements and naive verification requires to compute $3d$ pairings for the quadratic relations and $2(n_0 + 3d + 4)$ for the linear part, and n_d exponentiations in $\mathbb{G}_1$ for the output. Using the "bilinear batching" techniques of Herold et al. [19] the number of pairings can be reduced to $n_0 + 3d + 4$ for the linear part. Since the input is known in $\mathbb{Z}_p$, n_0 pairings in this part can be replaced by n_0 exponentiations in $\mathbb{G}_T$. Finally, using standard batching techniques [6], the number of pairings for the quadratic part can be reduced to $d+2$. As a result the total number of pairings required for verification is $4d+6$, plus n_0 exponentiations in $\mathbb{G}_T$ and $O(n_0 + d + n_d)$ exponentiations in the source group.

In the instantiation which is secure under standard assumptions, the proof size is $(6d+3, 2d+3)$ group elements and naive verification requires to compute $6d$ pairings for the quadratic relations and $2(n_0 + 6d + 6)$ for the linear part, and using the same batching techniques the number of pairings required for verification is $8d + 9$.

5.4 Adding Zero-Knowledge

In this section we argue how to add zero-knowledge to the argument for correct arithmetic circuit evaluation of Sect. 5.2. The same discussion applies for the argument for boolean circuit satisfiability discussed in Sect. 6.1 for boolean circuits.

We have to distinguish two different situations. In the first one the input is public, and we can easily modify our proof so that it reveals nothing about the internal evaluation steps. When the input or part of the input must be secret, which is the most useful case, the circuit input cannot be part of the verifier's input, at least not in the clear. A natural idea is to let the prover commit to it. The problem is that our "knowledge transfer" idea requires the reduction in the soundness proof to know this secret input, which means that the commitment to the input must be extractable so that we can efficiently extend it to a vector of correct evaluations $(\boldsymbol{a}, \boldsymbol{b}, \boldsymbol{c})$. Even in a QA-NIZK setting where we can efficiently open the commitments, they are only F-extractable [3] (under falsifiable assumptions), which means that we can only extract in the source groups but not in $\mathbb{Z}_p$. This leaves us only with a couple of solutions, all of them unsatisfactory.

One of them is to commit to inputs bitwise and prove that this is done correctly. This is not acceptable in terms of concrete efficiency for arithmetic circuits, but it is a practical approach for boolean circuits.

The second one is to use a commitment to the input which is extractable under knowledge assumptions. Of course, then our construction is no longer secure under falsifiable assumptions, but it is interesting that it indicates a trade-off in SNARK constructions: longer proof size and verification costs ($\Theta(d)$ group elements/pairings, respectively) but weaker assumptions (only the input needs to be extracted and not the full witness).

In any case, we leave for future work to explore the possibilities of this or other mixed approaches (like using ROM based constructions for extracting the input). We now give the technical details on how to add zero-knowledge to our argument for correct circuit evaluation, distinguishing the two aforementioned situations.

Adding Zero-Knowledge to Correct Evaluation of Middle Wires. This step is straightforward. The argument is changed so that $[\boldsymbol{L}]_1, [\boldsymbol{R}]_2, [\boldsymbol{O}]_1$ are not given in the clear, but instead the prover gives GS commitments [18] to each of its components. For the quadratic argument, it gives a GS Proof that the verification equation is satisfied, that is, for each i it proves in zk that the pairing product equation:

$$e([L_i]_1, [R_i]_2) - e([O_i]_1, [1]_2) = e([H_i]_1, [T]_2)$$

is satisfied, where $[L_i]_1, [R_i]_2, [O_i]_1, [H_i]_1$ are hidden committed values.

For the linear argument, it suffices to give a GS proof of satisfiability of the verification equation in Fig. 4. In its most efficient instantiation, the verification equation in Fig. 4 consists of 2 pairing product equations and hence the GS proof consists of 8 elements of each group. An alternative, more efficient approach (which requires only $(2, 2)$ group elements) for the linear argument proves that the vectors of committed elements are in a certain linear (bilateral) space. The idea is quite simple but details are a bit cumbersome, so we explain it in the full version of this work.

Hiding the Input and Output. Finally, we discuss how to use our results in a scenario where not only the middle wires should be hidden but also the input and the output. In this case the prover should commit to the input and the output with perfectly binding commitments $(\boldsymbol{c}_x, \boldsymbol{d}_y)$.

The commitment to the input should be extractable. For instance, $\boldsymbol{c}_x$ can be just the concatenation of GS commitments to the inputs provided the prover submits also a proof of knowledge of their opening (giving additional bitwise commitments and a proof that $\boldsymbol{c}_x$ is of the right form, or a proof with knowledge assumptions or in the ROM). In any case, we require $\boldsymbol{c}_x$ to be algebraic, that is, it should be possible to write it as $\boldsymbol{c}_x = [\mathbf{E}]_1\boldsymbol{x} + [\mathbf{V}]_1\boldsymbol{r}$, where $\boldsymbol{r}$ is the vector of randomness and matrices $\mathbf{E}$, $\mathbf{V}$ are described in the commitment key (we can also allow $\boldsymbol{c}_x$ to have components in both $\mathbb{G}_1, \mathbb{G}_2$, in which case $\mathbf{E}$ and $\mathbf{V}$ will be split). The only difference with the case where the commitment is public is that in the first n_0 rows of $\mathbf{M}$ the identity matrix should be replaced by $\mathbf{E}$ and an additional column of the form $\left(\mathbf{V}, \mathbf{0}\right)^\top$ should be added.

The prover should also give a GS proof that $\boldsymbol{d}_y$ opens to the same value as $[\boldsymbol{O}_d]_1$.

6 Boolean Circuits

We extend our results to any boolean circuit $\phi : \{0,1\}^{n_0} \to \{0,1\}^{n_d}$. The gates of ϕ are assumed to have fan-in two but otherwise they can be of any type (excluding non-interesting or trivial gate types). The construction relies on the characterization of these gates as quadratic functions of the inputs. We list below the 10 gate types allowed for the circuit ϕ, along with its expression as a quadratic function. The list of gates is taken from [1], which observe that the last remaining 6 gate types depend mostly on one input and are not used often.

$\mathsf{AND}(a,b,c)$: $ab = c$	$\mathsf{NAND}(a,b,c)$: $1 - ab = c$
$\mathsf{OR}(a,b,c)$: $1-(1-a)(1-b) = c$	$\mathsf{NOR}(a,b,c)$: $(1-a)(1-b) = c$
$\mathsf{XOR}(a,b,c)$: $b(1-a) + a(1-b) = c$	$\mathsf{XNOR}(a,b,c)$: $1 - a(1-b) - b(1-a) = c$
$\mathsf{G}_1(a,b,c) = (c = \overline{a} \wedge b)$: $(1-a)b = c$	$\mathsf{G}_2(a,b,c) = (c = \overline{\overline{a} \wedge b})$: $1-(1-a)b = c$
$\mathsf{G}_3(a,b,c) = (c = a \wedge \overline{b})$: $a(1-b) = c$	$\mathsf{G}_4(a,b,c) = (c = \overline{a \wedge \overline{b}})$: $1 - a(1-b) = c$.

From this characterization we slice the circuit into several quadratic and affine constraints similar to the arithmetic case. As before, we partition the set of gates $\mathcal{G}$ of a given circuit ϕ into different subsets $\mathcal{G}_i$ according to the depth, n_i is cardinal of the gates at level i and we assume that gates at each level are ordered in some way and they are denoted as $G_{i1}, \ldots, G_{in_i}$.

For each level i, we define variables C_{ij}, $j = 1, \ldots, n_i$ which will encode the output of gate j at level i. The gate G_{ij} will be correctly evaluated if $C_{ij} = G_{ij}(A_{ij}, B_{ij})$, where $A_{ij} = C_{k_L \ell_L}$ and $B_{ij} = C_{k_R \ell_R}$ for some indexes $0 \leq k_L, k_R < i$, $1 \leq \ell_L \leq n_{k_L}$ and $1 \leq \ell_R \leq n_{k_R}$, which depend on i, j and which are specified by the circuit description. That is, the left wire of G_{ij} should be the output of the ℓ_Lth gate at level k_L and the right wire the output of the ℓ_Rth gate at level k_R.

Lemma 7. *Let $\phi : \{0,1\}^{n_0} \to \{0,1\}^{n_d}$, be a circuit of multiplicative depth d with n_i gates at level i. There exist*

(a) variables C_{ij}, $i = 0, \ldots, d$, $j = 1, \ldots, n_i$,
(b) variables A_{ij}, B_{ij}, $i = 1, \ldots, d$, $j = 1, \ldots, n_i$,
(c) constants $f_{ijk\ell}, g_{ijk\ell} \in \{0,1\}$, $i = 1, \ldots, d$, $k = 0, \ldots, i-1$, $j = 1, \ldots, n_i$, $\ell = 1, \ldots, n_k$,
(d) constants $\beta_{ij}, \gamma_{ij}, \epsilon_{ij}, \delta_{ij} \in \mathbb{Z}_p$, $i = 1, \ldots, d$, $j = 1, \ldots, n_i$, which depend on the type of gate G_{ij},

such that, for every $(x_1, \ldots, x_{n_0}) \in \{0,1\}^{n_0}$, if we set $C_{0,j} = x_j$, for all $j = 1, \ldots, n_0$, then $\phi(\boldsymbol{x}) = \boldsymbol{y}$ and A_{ij}, C_{ij} are evaluated to the left and output of the jth gate at level i, if and only if the following equations are satisfied:

1. (Quadratic constraints) For each $i = 1, \ldots, d$, for all $j = 1, \ldots, n_i$,

$$C_{ij} = A_{ij}B_{ij} + A_{ij}\beta_{ij} + B_{ij}\gamma_{ij} + \epsilon_{ij}, \tag{9}$$

2. (Affine constraints) $A_{ij} = \sum_{k=0}^{i-1}\sum_{\ell=1}^{n_k} f_{ijk\ell}C_{k\ell}$ and $B_{ij} = \sum_{k=0}^{i-1}\sum_{\ell=1}^{n_k} g_{ijk\ell}C_{k\ell}$.
3. (Correct Output) For all $j = 1, \ldots, n_d$, $C_{dj} = y_j$.

Proof. For the $(i,j)th$ circuit gate, a description of the circuit ϕ specifies the gate type and indexes $(k_{i,j,L}, \ell_{i,j,L})$ which indicate the left and right wire. Therefore, from the quadratic expression of boolean gates for boolean circuit satisfiability, correct evaluation of G_{ij} is expressed as:

$$C_{ij} = C_{k_{i,j,L},\ell_{i,j,L}}C_{k_{i,j,R},\ell_{i,j,R}}\alpha_{ij} + C_{k_{i,j,L},\ell_{i,j,L}}\beta_{ij} + C_{k_{i,j,R},\ell_{i,j,R}}\hat{\gamma}_{ij} + \epsilon_{ij},$$

for some $\alpha_{ij}, \beta_{ij}, \hat{\gamma}_{ij}, \epsilon_{ij} \in \mathbb{Z}$ which depend on the gate type. This can be rewritten as an equation over $\mathbb{Z}_p$ as:

$$C_{ij} = C_{k_{i,j,L},\ell_{i,j,L}}(C_{k_{i,j,R},\ell_{i,j,R}}\alpha_{ij}) + C_{k_{i,j,L},\ell_{i,j,L}}\beta_{ij} + (C_{k_{i,j,R},\ell_{i,j,R}}\alpha_{ij})(\alpha_{ij}^{-1}\hat{\gamma}_{ij}) + \epsilon_{ij}. \tag{10}$$

For any (i,j) we define the constant $f_{ijk\ell}$ and $g_{ijk\ell}$ to be 0 everywhere except for $f_{ijk_{i,j,L}\ell_{i,j,L}} = 1$ and $g_{ijk_{i,j,R}\ell_{i,j,R}} = \alpha_{ij}$. Therefore, if $A_{ij} = \sum_{k=0}^{i-1}\sum_{\ell=1}^{n_k} f_{ijk\ell}C_{k\ell} = C_{k_{i,j,L},\ell_{i,j,L}}$ and $B_{ij} = \sum_{k=0}^{i-1}\sum_{\ell=1}^{n_k} g_{ijk\ell}C_{k\ell} = C_{k_{i,j,R},\ell_{i,j,R}}$ and Eq. (10) which expresses correct evaluation of gate (i,j) can be rewritten as:

$$C_{ij} = A_{ij}B_{ij} + A_{ij}\beta_{ij} + B_{ij}\gamma_{ij} + \epsilon_{ij}, \tag{11}$$

where $\gamma_{ij} = \alpha_{ij}^{-1}\hat{\gamma}_{ij}$.

Obviously, this implies that if $c_{0,j} = x_j$, and the linear constraints are satisfied, then the rest of the output wires are also consistent with x_j and we conclude that $c_{n_d,j}$ is the output corresponding to this input. Therefore, if $c_{n_d,j} = y_j$, we can conclude that $\phi(\boldsymbol{x}) = \boldsymbol{y}$.

To achieve succinct ness, quadratic equations which encode correct gate evaluation are represented as a divisibility relation with the usual polynomial aggregation technique.

Lemma 8. *Let $\mathcal{R} \subset \mathbb{Z}_p$ be a set of cardinal N and let $\lambda_j(X)$ be the associated Lagrangian polynomials and $t(X)$ the polynomial whose roots are the elements of $\mathcal{R}$. Let $\phi : \{0,1\}^{n_0} \rightarrow \{0,1\}^{n_d}$, be any circuit such that $N = \max_{i=1,\ldots,d} n_i$. There exist some unique polynomials $u_{L,i}(X), u_{R,i}(X), u_{0,i}(X)$ of degree at most $N-1$ which are efficiently computable from the circuit description and such that for any tuple $(\boldsymbol{a}_i, \boldsymbol{b}_i, \boldsymbol{c}_i) \in (\{0,1\}^{n_i})^3$, if*

$$\ell_i(X) = \sum_{j=1}^{n_i} a_j \lambda_j(X), \qquad r_i(X) = \sum_{j=1}^{n_i} b_j \lambda_j(X), \qquad o_i(X) = \sum_{j=1}^{n_i} c_j \lambda_j(X),$$

it holds that $\boldsymbol{a}_i, vecc_i$ are consistent assignments to the left and output values of gates at level i if and only if $t(X)$ divides $p_i(X)$, where

$$p_i(X) = \ell_i(X) r_i(X) + \ell_i(X) u_{L,i}(X) + r(X) u_{R,i}(X) + u_{0,i}(X) - o_i(X).$$

Proof. The proof is a direct consequence of Lemma 7. Indeed, it suffices to define $u_{L,i}(X), u_{R,i}(X), u_{0,i}(X)$ to take the values $u_{L,i}(r_j) = \beta_{ij}$, $u_{R,i}(r_j) = \gamma_{ij}$ and $u_{0,i}(r_j) = \epsilon_{ij}$ for $j = 1, \ldots, n_i$ and 0 for $j = n_i + 1, \ldots, N$. Therefore, $p_i(r_j) = a_{ij} b_{ij} + a_{ij} \beta_{ij} + b_{ij} \gamma_{ij} + \epsilon_{ij} - c_{ij}$. This proves that if Eq. (11) is satisfied then $p_i(X)$ is divisible by $t(X)$, since it is 0 in all of its roots. Finally, the polynomials $u_{L,i}(X), u_{R,i}(X), u_{0,i}(X)$ can be efficiently computed from the circuit description, as they depend only on N and the type of each gate.

6.1 A New Argument for Correct Boolean Circuit Evaluation

From Lemma 7, we can design an argument for boolean circuit satisfiability based on falsifiable assumptions, similar as in Sect. 5. The argument is based on a quadratic and a linear "knowledge transfer" subarguments. The value $[R_i]_2$ is now defined as $[\boldsymbol{R}_i]_2 = \sum_{j=1}^{n_i} \alpha_{ij} b_{ij} \lambda_j(\boldsymbol{s})$. The linear argument is identical to the arithmetic case.

For the quadratic argument, now the prover needs to show (aggregating the proof at each level i for $j = 1, \ldots, n_i$) that the quadratic equations $C_{ij} = A_{ij} B_{ij} + A_{ij} \beta_{ij} + B_{ij} \gamma_{ij} + \epsilon_{ij}$ are satisfied, whereas before the equations were $C_{ij} = A_{ij} B_{ij}$. However, the security proof is almost identical to the arithmetic case.

Indeed, the verification equation of the quadratic argument is adapted to the new equation type, i.e. For each level $i = 1, \ldots, d$, and each $j = 1, \ldots, k_s$ given commitments $[L_{ij}]_1, [R_{ij}]_2, [O_{ij}]_1$, and some value $[H_{ij}]_1$ the quadratic argument checks if

$$e([L_{ij}]_1, [R_{ij}]_2) + e([L_{ij}]_1, [u_{L,i}(s_j)]_2) + e([u_{R,i}(s_j)]_1, [R_{ij}]_2) + e([u_{0,i}(s_j)]_1, [1]_2) \\ - e([O_{ij}]_1, [1]_2) = e([H_{ij}]_1, [T]_2),$$

where $u_{L,i}(X), u_{R,i}(X), u_{0,i}(X)$ are the polynomials associated to the gate constants at level i. To prove soundness, given an opening of $[L_{ij}]_1$ and $[R_{ij}]_2$ which is not consistent with $[O_{ij}]$, it suffices to compute $[O'_{ij}]_1, [H'_{ij}]_1$ consistent with these openings and subtract the two verification equations to find a solution to the $\mathcal{R}$-Rational Strong Diffie-Hellman Assumption.

Zero-Knowledge. The argument can be made zero-knowledge for the middle wires by proving with the GS proof system that the argument for correct circuit evaluation is satisfied, as discussed in Sect. 5.4 for the arithmetic case. The input can also be hidden provided it is encrypted with an extractable commitment. In the boolean case this can be done in a relatively efficient way, for example under the DDH Assyumption with GS commitments. The cost of giving the committed secret inputs and a proof that they open to $\{0, 1\}$ using the GS proof system is $(6(n_0 - n_{pub}), 6(n_0 - n_{pub}))$ group elements. It can be reduced to $(2(n_0 - n_{pub}) + 10, 10)$ group elements under standard assumptions using the results of González and Ràfols [14], but at the price of having a crs quadratic in n_0 and to $(2n_0 + 4, 6)$ with a linear crs under a non-standard (falsifiable) $(n_0 - n_{pub})$-assumption similar to the q-Target Strong Diffie-Hellman Assumption using the results of Daza et al. [2].

References

1. Danezis, G., Fournet, C., Groth, J., Kohlweiss, M.: Square span programs with applications to succinct NIZK arguments. In: Sarkar, P., Iwata, T. (eds.) ASIACRYPT 2014, Part I. LNCS, vol. 8873, pp. 532–550. Springer, Heidelberg (2014). https://doi.org/10.1007/978-3-662-45611-8_28
2. Daza, V., González, A., Pindado, Z., Ràfols, C., Silva, J.: Shorter quadratic QA-NIZK proofs. In: Lin, D., Sako, K. (eds.) PKC 2019, Part I. LNCS, vol. 11442, pp. 314–343. Springer, Cham (2019). https://doi.org/10.1007/978-3-030-17253-4_11
3. Escala, A., Groth, J.: Fine-tuning Groth-Sahai proofs. In: Krawczyk, H. (ed.) PKC 2014. LNCS, vol. 8383, pp. 630–649. Springer, Heidelberg (2014). https://doi.org/10.1007/978-3-642-54631-0_36
4. Escala, A., Herold, G., Kiltz, E., Ràfols, C., Villar, J.L.: An algebraic framework for Diffie-Hellman assumptions. J. Cryptol. **30**(1), 242–288 (2017)
5. Fauzi, P., Lipmaa, H., Siim, J., Zając, M.: An efficient pairing-based shuffle argument. In: Takagi, T., Peyrin, T. (eds.) ASIACRYPT 2017, Part II. LNCS, vol. 10625, pp. 97–127. Springer, Cham (2017). https://doi.org/10.1007/978-3-319-70697-9_4
6. Ferrara, A.L., Green, M., Hohenberger, S., Pedersen, M.Ø.: Practical short signature batch verification. In: Fischlin, M. (ed.) CT-RSA 2009. LNCS, vol. 5473, pp. 309–324. Springer, Heidelberg (2009). https://doi.org/10.1007/978-3-642-00862-7_21
7. Gennaro, R., Gentry, C., Parno, B.: Non-interactive verifiable computing: outsourcing computation to untrusted workers. In: Rabin, T. (ed.) CRYPTO 2010. LNCS, vol. 6223, pp. 465–482. Springer, Heidelberg (2010). https://doi.org/10.1007/978-3-642-14623-7_25
8. Gennaro, R., Gentry, C., Parno, B., Raykova, M.: Quadratic span programs and succinct NIZKs without PCPs. In: Johansson, T., Nguyen, P.Q. (eds.) EUROCRYPT 2013. LNCS, vol. 7881, pp. 626–645. Springer, Heidelberg (2013). https://doi.org/10.1007/978-3-642-38348-9_37
9. Gentry, C.: Fully homomorphic encryption using ideal lattices. In: Mitzenmacher, M. (ed.) 41st ACM STOC, pp. 169–178. ACM Press, May/June (2009)
10. Gentry, C., Groth, J., Ishai, Y., Peikert, C., Sahai, A., Smith, A.D.: Using fully homomorphic hybrid encryption to minimize non-interative zero-knowledge proofs. J. Cryptol. **28**(4), 820–843 (2015)

11. Gentry, C., Wichs, D.: Separating succinct non-interactive arguments from all falsifiable assumptions. In: Fortnow, L., Vadhan, S.P. (eds.) 43rd ACM STOC, pp. 99–108. ACM Press, June 2011
12. Ghadafi, E., Groth, J.: Towards a classification of non-interactive computational assumptions in cyclic groups. In: Takagi, T., Peyrin, T. (eds.) ASIACRYPT 2017, Part II. LNCS, vol. 10625, pp. 66–96. Springer, Cham (2017). https://doi.org/10.1007/978-3-319-70697-9_3
13. Goldwasser, S., Kalai, Y.T., Rothblum, G.N.: Delegating computation: interactive proofs for muggles. In: Ladner, R.E., Dwork, C. (eds.) 40th ACM STOC, pp. 113–122. ACM Press, May 2008
14. González, A., Hevia, A., Ràfols, C.: QA-NIZK arguments in asymmetric groups: new tools and new constructions. In: Iwata, T., Cheon, J.H. (eds.) ASIACRYPT 2015, Part I. LNCS, vol. 9452, pp. 605–629. Springer, Heidelberg (2015). https://doi.org/10.1007/978-3-662-48797-6_25
15. Groth, J.: Short pairing-based non-interactive zero-knowledge arguments. In: Abe, M. (ed.) ASIACRYPT 2010. LNCS, vol. 6477, pp. 321–340. Springer, Heidelberg (2010). https://doi.org/10.1007/978-3-642-17373-8_19
16. Groth, J.: On the size of pairing-based non-interactive arguments. In: Fischlin, M., Coron, J.-S. (eds.) EUROCRYPT 2016, Part II. LNCS, vol. 9666, pp. 305–326. Springer, Heidelberg (2016). https://doi.org/10.1007/978-3-662-49896-5_11
17. Groth, J., Ostrovsky, R., Sahai, A.: New techniques for noninteractive zero-knowledge. J. ACM **59**(3), 11 (2012)
18. Groth, J., Sahai, A.: Efficient noninteractive proof systems for bilinear groups. SIAM J. Comput. **41**(5), 1193–1232 (2012)
19. Herold, G., Hoffmann, M., Klooß, M., Ràfols, C., Rupp, A.: New techniques for structural batch verification in bilinear groups with applications to groth-sahai proofs. In: Thuraisingham, B.M., Evans, D., Malkin, T., Xu, D. (eds.) ACM CCS 2017, pp. 1547–1564. ACM Press, October/November (2017)
20. Jutla, C.S., Roy, A.: Shorter quasi-adaptive NIZK proofs for linear subspaces. In: Sako, K., Sarkar, P. (eds.) ASIACRYPT 2013, Part I. LNCS, vol. 8269, pp. 1–20. Springer, Heidelberg (2013). https://doi.org/10.1007/978-3-642-42033-7_1
21. Jutla, C.S., Roy, A.: Switching lemma for bilinear tests and constant-size NIZK proofs for linear subspaces. In: Garay, J.A., Gennaro, R. (eds.) CRYPTO 2014, Part II. LNCS, vol. 8617, pp. 295–312. Springer, Heidelberg (2014). https://doi.org/10.1007/978-3-662-44381-1_17
22. Kalai, Y., Paneth, O., Yang, L.: On publicly verifiable delegation from standard assumptions. Cryptology ePrint Archive, Report 2018/776 (2018). https://eprint.iacr.org/2018/776
23. Kalai, Y.T., Raz, R., Rothblum, R.D.: How to delegate computations: the power of no-signaling proofs. In: Shmoys, D.B. (ed.) 46th ACM STOC, pp. 485–494. ACM Press, May/June (2014)
24. Katsumata, S., Nishimaki, R., Yamada, S., Yamakawa, T.: Exploring constructions of compact NIZKs from various assumptions. In: Boldyreva, A., Micciancio, D. (eds.) CRYPTO 2019. Part III, volume 11694 of LNCS, pp. 639–669. Springer, Heidelberg (2019). https://doi.org/10.1007/978-3-030-26954-8_21
25. Kiltz, E., Wee, H.: Quasi-adaptive NIZK for linear subspaces revisited. In: Oswald, E., Fischlin, M. (eds.) EUROCRYPT 2015, Part II. LNCS, vol. 9057, pp. 101–128. Springer, Heidelberg (2015). https://doi.org/10.1007/978-3-662-46803-6_4

26. Libert, B., Peters, T., Joye, M., Yung, M.: Non-malleability from malleability: simulation-sound quasi-adaptive NIZK proofs and CCA2-secure encryption from homomorphic signatures. In: Nguyen, P.Q., Oswald, E. (eds.) EUROCRYPT 2014. LNCS, vol. 8441, pp. 514–532. Springer, Heidelberg (2014). https://doi.org/10.1007/978-3-642-55220-5_29
27. Libert, B., Peters, T., Yung, M.: Short group signatures via structure-preserving signatures: standard model security from simple assumptions. In: Gennaro, R., Robshaw, M. (eds.) CRYPTO 2015, Part II. LNCS, vol. 9216, pp. 296–316. Springer, Heidelberg (2015). https://doi.org/10.1007/978-3-662-48000-7_15
28. Lipmaa, H.: Progression-free sets and sublinear pairing-based non-interactive zero-knowledge arguments. In: Cramer, R. (ed.) TCC 2012. LNCS, vol. 7194, pp. 169–189. Springer, Heidelberg (2012). https://doi.org/10.1007/978-3-642-28914-9_10
29. Lund, C., Fortnow, L., Karloff, H.J., Nisan, N.: Algebraic methods for interactive proof systems. In: 31st FOCS, pp. 2–10. IEEE Computer Society Press, October 1990
30. Maller, M., Kohlweiss, M., Bowe, S., Meiklejohn, S.: Sonic: zero-knowledge snarks from linear-size universal and updateable structured reference string. Cryptology ePrint Archive, Report 2019/099 (2019). http://eprint.iacr.org/2019/099
31. Morillo, P., Ràfols, C., Villar, J.L.: The kernel matrix diffie-hellman assumption. In: Cheon, J.H., Takagi, T. (eds.) ASIACRYPT 2016, Part I. LNCS, vol. 10031, pp. 729–758. Springer, Heidelberg (2016). https://doi.org/10.1007/978-3-662-53887-6_27
32. Naor, M.: On cryptographic assumptions and challenges. In: Boneh, D. (ed.) CRYPTO 2003. LNCS, vol. 2729, pp. 96–109. Springer, Heidelberg (2003). https://doi.org/10.1007/978-3-540-45146-4_6
33. Paneth, O., Rothblum, G.N.: On Zero-testable homomorphic encryption and publicly verifiable non-interactive arguments. In: Kalai, Y., Reyzin, L. (eds.) TCC 2017, Part II. LNCS, vol. 10678, pp. 283–315. Springer, Cham (2017). https://doi.org/10.1007/978-3-319-70503-3_9
34. Parno, B., Howell, J., Gentry, C., Raykova, M.: Pinocchio: nearly practical verifiable computation. In: 2013 IEEE Symposium on Security and Privacy, pp. 238–252. IEEE Computer Society Press, May 2013
35. Peikert, C., Shiehian, S.: Noninteractive zero knowledge for NP from (plain) learning with errors. Cryptology ePrint Archive, Report 2019/158 (2019). https://eprint.iacr.org/2019/158

Author Index

Printed in the United States
By Bookmasters